Encyclopedia of
Plant Physiology

New Series Volume 19

Editors

A. Pirson, Göttingen

M. H. Zimmermann, Harvard

Photosynthesis III

Photosynthetic Membranes and Light Harvesting Systems

Edited by

L. A. Staehelin and C. J. Arntzen

Contributors

J. M. Anderson B. Andersson C. J. Arntzen R. Bachofen J. Barber J. Barrett R. E. Blankenship P. Böger J. Breton T. M. Bricker R. J. Cogdell J. Deisenhofer R. A. Dilley B. A. Diner T. J. Donohue P. L. Dutton R. C. Fuller E. Gantt N. E. Geacintov J. A. Guikema W. Haehnel G. Hauska J. E. Hearst A. J. Hoff D. Holten A. R. Holzwarth A. Joliot P. Joliot S. Kaplan K. Keegstra R. S. Knox D. J. Kyle C. Lütz P. Mathis R. E. McCarty B. A. Melandri H. Michel D. J. Murphy C. M. Nalin J. R. Norris I. Ohad D. R. Ort H. B. Pakrasi W. W. Parson R. C. Prince A. Rawyler I. J. Ryrie G. Sandmann K. Sauer H. Scheer P. Sétif L. A. Sherman P. A. Siegenthaler D. J. Simpson S. G. Sprague L. A. Staehelin H. Strotmann J. P. Thornber D. M. Tiede G. Van Brakel A. R. Varga G. Venturoli J. Whitmarsh V. Wiemken F.-A. Wollman C. F. Yocum H. Zuber

Springer-Verlag Berlin Heidelberg GmbH

Professor L. ANDREW STAEHELIN
Department of Molecular, Cellular,
and Developmental Biology
University of Colorado
Campus Box 347
Boulder, CO 80309
USA

Professor CHARLES J. ARNTZEN
E.I. du Pont de Nemours & Co.
Central Research and Development Department
Experimental Station
Bldg. 402, Room 3226
Wilmington, DE 19898
USA

With 163 Figures

ISBN 978-3-642-70938-8 ISBN 978-3-642-70936-4 (eBook)
DOI 10.1007/978-3-642-70936-4

Library of Congress Cataloging-in-Publication Data. (Revised for vol. 3) Main entry under title: Photosynthesis. (Encyclopedia of plant physiology; new ser., v. 5–6, 19) Includes bibliographies and indexes. Contents: 1. Photosynthetic electron transport and photophosphorylation, edited by A. Trebst and M. Avron. 2. Photosynthetic carbon metabolism and related processes, edited by M. Gibbs and E. Latzko. 3. Photosynthetic membranes and light harvesting systems, edited by L.A. Staehelin and C.J. Arntzen. 1. Photosynthesis – Collected works. I. Trebst, Achim. II. Alberte, R.S. III. Series. QK711.2.E5 vol. 5–6 581.1s [581.1′3342] 76-30357 [QK882] ISBN 978-3-642-70938-8

Originally published by Springer-Verlag Berlin Heidelberg New York Tokyo 1986
Softcover reprint of the hardcover 1st edition 1986

Preface

The *Encyclopedia of Plant Physiology* series has turned several times to the topic of photosynthesis. In the original series, two volumes edited by A. PIRSON and published in 1960 provided a broad overview of the entire field. Although the New Series has devoted three volumes to the same topic, the overall breadth of the coverage has had to be restricted to allow for greater in-depth treatment of three major areas of modern photosynthesis research: I. Photosynthetic Electron Transport and Photophosphorylation (Volume 5 edited by A. TREBST and M. AVRON, and published in 1977); II. Photosynthetic Carbon Metabolism and Related Processes (Volume 6 edited by M. GIBBS and E. LATZKO, and published in 1979); and III. Photosynthetic Membranes and Light-Harvesting Systems (this volume).

As we approached the organization of the current volume, we chose a set of topics for coverage that would complement the earlier volumes, as well as provide updates of areas that have seen major advances in recent years. In addition, we wanted to emphasize the following changes in the study of photosynthetic systems which have become increasingly important since 1977: the trend toward increased integration of biochemical and biophysical approaches to study photosynthetic membranes and light-harvesting systems, and a renewed appreciation of the structural parameters of membrane organization.

Due to the increased complexity of the field, we also decided to try a new format for our volume to better serve the following two purposes. First, we believe a review volume on photosynthetic membranes should serve as a reference source for nonspecialists interested in obtaining an overview of both oxygenic and anoxygenic photosynthesis. This need has been answered by the inclusion of five introductory chapters which summarize the main broad topic areas of the volume. We also recognize that a review volume should provide insight to the "state of the art" in specific research areas which have seen major recent advances. To this end, Chapters 6 through 11 have been organized such that each consists of a number of minireviews related to a common theme. All of the 43 minireviews are authored by highly regarded specialists, and focus on recent research highlights and interpretations of significant new findings. Great emphasis has been placed on the integration of the materials covered in the introductory chapters and in the minireviews. Extensive cross-referencing has been used to allow easy transitions by the reader from a general to a specialized coverage of a topic. Similarly, all minireviews contain references to the appropriate introductory chapters, as well as to other minireviews.

With students in mind, the authors of the introductory chapters have stressed integrative and comparative aspects of their topics. This type of approach is becoming more and more relevant in photosynthesis research, thanks to the convergence of information coming from structural, biophysical, and biochemical studies. Indeed, it is truly exciting to witness the progress being made toward the goal of a molecular understanding of the diverse biophysical and biochemical reactions associated with photosynthetic membranes and light-harvesting systems.

The specialized chapters in this volume begin with the topic of light harvesting by photosynthetic membranes. The minireviews of Chapter 6 summarize biochemical and structural studies of light-harvesting assemblages, with emphasis on the light-harvesting components of bacteria and algae, since higher plant chloroplast components are extensively reviewed in Chapter 3. Whereas Chapter 6 emphasizes the biochemical diversity in light-harvesting systems, Chapter 7 consists of minireviews which discuss unifying concepts governing light-harvesting events. All authors in this section are concerned with photon absorption and structural parameters of the pigment bed that determine the efficiency of excitation energy transfer to reaction centers.

The most fundamental result of photosynthetic light reactions is the conversion of excitation energy, derived from absorbed light, into stable chemical form. This occurs in the reaction center (RC). In the last 5 years there have been major advancements in the understanding of these processes – especially by those groups who have focused their work on photosynthetic bacteria. The identity of the cofactors (chlorophyll, pheophytin, quinones, etc.) involved in the initial charge separations and the events involved in charge stabilization are now highly defined. The minireviews of Chapter 8 present various aspects of this rapidly moving field, ranging from energetic considerations of the RC to discussions of similarities and differences among the different types of protein which comprise different reaction centers.

Chapter 9 makes a transition from the highly defined bacterial reaction centers into the less well understood photosystems I and II of green plants. The inclusion of several minireviews devoted to the complexity of reactions in photosystem II, for example, reflects the wide diversity in studies of a system capable of extracting electrons from water and catalyzing a stable charge separation that results in reduction of the plastoquinone pool. These studies extend from detailed understanding of the primary reactions to physiological adaptation of the process to light and chemical (herbicide) stresses.

The reaction centers of prokaryotic and eukaryotic photosynthetic membranes produce high energy electrons which are utilized in electron transport reactions. The energy released in these reactions is coupled to ATP synthesis. The membrane components and processes involved in the energy-coupling reactions are the topic of Chapter 10. Minireviews in this section are strongly biochemical in emphasis, with special reference to the structural organization of membranes and the protein complexes which mediate proton translocation and ATP biosynthesis. Specific enzymes involved in electron transport and inhibitors which affect them are reviewed in light of information obtained since 1977.

The last chapter of this volume (Chap. 11) deals with the use of integrative approaches to study processes associated with control of photosynthetic membrane assembly and maintenance. Chapter 11 includes subject material ranging from comparative structural analysis of photosynthetic membranes (developmental diversity) to the use of physical analysis of membranes or simulated membrane systems to characterize functional components. The minireviews of this chapter will be of increasing value as the field of membrane biosynthesis and assembly matures to use more information about physical and biochemical features of the photosynthetic membranes. We can anticipate the advent of use of genetic engineering tools to manipulate photosynthetic membranes, and the rapid expansion of knowledge in this area.

In summary, this volume is a selection of both overview chapters and numerous topical speciality reviews. It should be useful as a reference source and as a teaching aid for individuals interested in the rapidly expanding field of photosynthetic membranes.

Boulder and Wilmington, Spring 1986

L.A. Staehelin
C.J. Arntzen

Contents

4 Energy Transduction in Oxygenic Photosynthesis: an Overview of Structure and Mechanism

D.R. Ort (With 7 Figures)

8.2 Redox Center and Secondary Protein Structure Organization in Bacterial Reaction Centers and the Cytochrome b/c_1 Complex

D.M. TIEDE (With 2 Figures)

8.3 Energy Trapping in Photosynthesis as Probed by the Magnetic Properties of Reaction Centers

J.R. NORRIS and G. VAN BRAKEL (With 7 Figures)

8.4 X-Ray Diffraction Studies on a Crystalline Bacterial Photosynthetic Reaction Center: A Progress Report and Conclusions on the Structure of Photosystem II Reaction Centers

H. MICHEL and J. DEISENHOFER (With 3 Figures)

10.2 Mobile Electron Carries in Thylakoids

J. Whitmarsh (With 2 Figures)

10.3 Cytochrome b_6/f Complex – Quinol Interactions

P. Joliot and A. Joliot

10.4 Light-Driven Electron and Proton Transfer in the Cytochrome bc_1 Complex

R.C. Prince (With 2 Figures)

10.5 Plastocyanin

W. Haehnel (With 1 Figure)

List of Contributors

JAN M. ANDERSON
CSIRO
Division of Plant Industry
P.O. Box 1600
Canberra City
ACT 2601/Australia

B. ANDERSSON
Department of Biochemistry
University of Lund
P.O. Box 124
S-22100 Lund/Sweden

C.J. ARNTZEN
E.I. du Pont de Nemours & Co.
Central Research and Development
Department
Experimental Station
Bldg. 402, Room 3226
Wilmington, DE 19898/USA

R. BACHOFEN
Institut für Pflanzenbiologie
der Universität
Zollikerstraße 107
CH-8008 Zürich/Switzerland

J. BARBER
Department of Pure and
Applied Biology
Imperial College
Prince Consort Road
London SW7 2BB/United Kingdom

J. BARRETT
Rockefeller University
1230 York Avenue
New York, NY 10021/USA

R.E. BLANKENSHIP
Department of Chemistry
Arizona State University
Tempe, AZ 85287/USA

P. BÖGER
Lehrstuhl für Physiology
und Biochemie der Pflanzen
Universität Konstanz
Postfach 5560
D-7750 Konstanz/FRG

J. BRETON
Service de Biophysique
Département de Biologie
C.E.N. de Saclay
F-91191 Gif-sur-Yvette Cedex/France

T.M. BRICKER
Department of Chemistry
The University of Southern Mississippi
Hattiesburg, MS 39406-5043/USA

R.J. COGDELL
Department of Botany
University of Glasgow
Glasgow G12 8QQ/United Kingdom

J. DEISENHOFER
Max-Planck-Institut für Biochemie
Abt. Strukturforschung II
Am Klopferspitz
D-8033 Martinsried/FRG

R.A. DILLEY
Department of Biological Sciences
Purdue University
West Lafayette, IN 47907/USA

B.A. DINER
Institut de Biologie
Physico-Chimique
13, rue Pierre et Marie Curie
F-75005 Paris, France

T.J. DONOHUE
Department of Bacteriology
University of Wisconsin
1550 Linden Drive
Madison, WI 53706/USA

P.L. DUTTON
607 Goddard Laboratories
Department of Biochemistry
and Biophysics
37th & Hamilton Walk
Philadelphia, PA 19104/USA

R.C. FULLER
Department of Biochemistry
University of Massachusetts
Amherst, MA 01003/USA

E. GANTT
Smithsonian Environmental
Research Center
12441 Parklawn Drive
Rockville, MD 20852/USA

N.E. GEACINTOV
Chemistry Department
New York University
New York, NY 10003/USA

J.A. GUIKEMA
Division of Biology
Kansas State University
Ackert Hall
Manhattan, KS 66506/USA

W. HAEHNEL
Westf. Wilhelms-Universität
Münster
Lehrstuhl für Biochemie
der Pflanzen
Hindenburgplatz 55
D-4400 Münster/FRG

G. HAUSKA
Institut für Botanik
Universität Regensburg
Universitätsstr. 31
D-8400 Regensburg/FRG

J.E. HEARST
Department of Chemistry
430 Latimer Hall
University of California
Berkeley, CA 94720/USA

A.J. HOFF
Department of Biophysics
Huygens Laboratory
of the State University
NL-2300 RA Leiden/The Netherlands

D. HOLTEN
Department of Chemistry
Box 1134
Washington University
St. Louis, MO 63130/USA

A.R. HOLZWARTH
Max-Planck-Institut
für Strahlenchemie
Stiftstraße 34–36
D-4330 Mühlheim/Ruhr/FRG

A. JOLIOT
Institut de Biologie
Physico-Chimique
13, rue Pierre et Marie Curie
F-75005 Paris/France

P. JOLIOT
Institut de Biologie
Physico-Chimique
13, rue Pierre et Marie Curie
F-75005 Paris/France

S. KAPLAN
Department of Microbiology
University of Illinois
at Urbana-Champaign
407 South Goodwin Avenue
Urbana, IL 61801/USA

K. KEEGSTRA
Department of Botany
University of Wisconsin
Madison, WI 53706/USA

R.S. KNOX
Department of Physics
and Astronomy
University of Rochester
Rochester, NY 14627/USA

D.J. KYLE
Martek Corporation
9115 Guilford Rd.
Columbia, MD 21046/USA

C. LÜTZ
Botanisches Institut
der Universität Köln
1. Lehrstuhl
Gyrhofstraße 15
D-5000 Köln 41

P. MATHIS
Service de Biophysique
Département de Biologiè
C.E.N. de Saclay
F-91191 Gif-sur-Yvette Cedex/
France

R.E. McCarty
Section of Biochemistry
Molecular and Cell Biology
Cornell University
Ithaca, NY 14853/USA

B.A. Melandri
University of Bologna
Institute of Botany
Via Irnerio 42
I-40126 Bologna/Italy

H. Michel
Max-Planck-Institut
für Biochemie
Abteilung Membranbiochemie
Am Klopferspitz
D-8033 Martinsried/FRG

D.J. Murphy
Department of Botany
University of Durham
Durham, DH1 3LE/United Kingdom

C.M. Nalin
Department of Cell Biology
Roche Institute of
Molecular Biology
Roche Research Center
Nutley, NJ 07110/USA

J.R. Norris
Chemistry Division
Argonne National Laboratory
Argonne, IL 60439/USA

I. Ohad
Department of
Biological Chemistry
Institute of Life Sciences
The Hebrew University
of Jerusalem
91904 Jerusalem, Israel

D.R. Ort
Department of Plant Biology
289 Morrill Hall
University of Illinois
505 S. Goodwin Ave.
Urbana, IL 61801/USA

H.B. Pakrasi
Central Research and
Development Department
E.I. du Pont de
Nemours & Co.
Wilmington, DE 19898/USA

W.W. Parson
Department of Biochemistry SJ-70
University of Washington
Seattle, WA 98195/USA

R.C. Prince
Exxon Research
and Engineering Co.
Annandale, NJ 08801/USA

A. Rawyler
Laboratoire de
Physiologie végétale
Université de Neuchâtel
Ch. de Chantemerle 20
CH-2000 Neuchâtel/Switzerland

I.J. Ryrie
Department of Applied
Mathematics
Research School of
Physical Sciences
Australian National University
Canberra 2601/Australia

G. Sandmann
Lehrstuhl für Physiologie
und Biochemie der Pflanzen
Universität Konstanz
Postfach 5560
D-7750 Konstanz/FRG

K. Sauer
Chemistry Department
University of California
Berkeley, CA 94720/USA

H. Scheer
Botanisches Institut
der Universität
Menzinger Straße 38
D-8000 München 19/FRG

P. Sétif
Service de Biophysique
Département de Biologie
C.E.N. de Saclay
F-91191 Gif-sur-Yvette Cedex/
France

L.A. Sherman
Division of
Biological Sciences
University of Missouri-Columbia
Columbia, MO 65211/USA

P.A. SIEGENTHALER
Laboratoire de Physiologie végétale
Université de Neuchâtel
Ch. de Chantemerle 20
CH-2000 Neuchâtel/Switzerland

D.J. SIMPSON
Department of Physiology
Carlsberg Laboratory
Gl. Carlsbergvej 10
DK-2500 Copenhagen Valby/Denmark

S.G. SPRAGUE
West Virginia University
School of Medicine
Department of Biochemistry
Morgantown, WV 26506/USA

L.A. STAEHELIN
Department of Molecular Cellular and Developmental Biology
University of Colorado
Campus Box 347
Boulder, CO 80309/USA

H. STROTMANN
Botanisches Institut der Universität Düsseldorf
Universitätsstraße 1
D-4000 Düsseldorf/FRG

J.P. THORNBER
Department of Biology
University of California
405 Hilgard Ave.
Los Angeles, CA 90024
USA

D.M. TIEDE
Chemistry Division
Argonne National Laboratory
Argonne, IL 60439/USA

G. VAN BRAKEL
Biokemi
Odense Universitet
Campusvej 55
DK-5230 Odense M/Denmark

A.R. VARGA
University of Illinois at Urbana-Champaign
Department of Microbiology
Urbana, IL 61801/USA

G. VENTUROLI
University of Bologna
Institute of Botany
Via Irnerio 42
I-40126 Bologna/Italy

J. WHITMARSH
Department of Plant Biology
289 Morrill Hall
University of Illinois
505 S. Goodwin Ave.
Urbana, IL 61801/USA

V. WIEMKEN
Department of Botany
University
Hebelstraße 1
CH-4056 Basel/Switzerland

F.-A. WOLLMAN
Service de Photosynthése
Institut de Biologie Physico-Chimique
13, rue Pierre et Marie Curie
F-75005 Paris/France

C.F. YOCUM
Division of Biological Sciences
The University of Michigan
Ann Arbor, MI 48109-1048/USA

H. ZUBER
Institut für Molekularbiologie und Biophysik
ETH-Hönggerberg – HPM
CH-8093 Zürich/Switzerland

1 Chloroplast Structure and Supramolecular Organization of Photosynthetic Membranes

L.A. STAEHELIN

1 Introduction

All chemical and physical properties of cells have a structural basis. Thus, a complete understanding of photosynthetic membrane function requires a detailed knowledge of the structural organization of such membranes. During the last 15 years significant progress has been made in the comprehensive analysis of the photochemical reactions catalyzed by chlorophyll, in the identification of the molecules associated with these processes, and in the elucidation of the three-dimensional arrangement of these molecules in photosynthetic membranes. As all of these fields have progressed, a noticeable convergence of ideas has occurred, blurring the artificial boundaries between different areas of research, and bringing us closer to a true understanding of how light energy is converted into chemical energy forms. At the same time, interest in "nontraditional" photosynthetic organisms has begun to increase, partly out of general curiosity, and partly because such organisms may possess unique features that allow for better probing of basic photosynthetic mechanisms. Thus, two of the main goals of this review are to provide an overview of the structural diversity of photosynthetic membrane systems, and to highlight how architectural parameters help define functional parameters of photosynthetic membranes. The first section will deal with the general morphology of chloroplasts and of photosynthetic membranes of higher plants, algae and oxygen-evolving photosynthetic bacteria; the second, with the supramolecular organization of photosynthetic membranes in selected systems, and the third, with the structural responses of photosynthetic membranes to environmental stresses and with the relationship between dynamic changes in membrane organization and the regulation of functional activities.

2 Chloroplast Morphology

2.1 Light Microscopic Observations

Most higher plant chloroplasts have the shape of a circular or elongated lens and a maximum diameter of 3–10 μm (MÖBIUS 1920). A typical spinach leaf cell contains between 20 and 60 chloroplasts (BUTTERFASS 1979), each of which has an average volume of 34 μm^3 and dimensions of 5.5×2.1 µm (MENKE 1962). Algal chloroplasts are more diverse in structure: depending on the organism, they may be lens-shaped, fan-shaped, spiral-shaped, branched-reticulate-shaped, and cup-shaped. These and many intermediate forms have been described and exploited for taxonomic purposes (BISALPUTRA 1974; HORI and UEDA 1975; KIRK and TILNEY-BASSETT 1978). Numerous algae such as *Chlorella*, *Porphyridium*, and *Micrasterias* possess only one chloroplast per cell; at the other end of the scale is the giant alga *Acetabularia mediterranea*, whose cells have been estimated to contain up to 1.2×10^7 chloroplasts (SHEPPART 1965). The chloroplast boundary is formed by an envelope composed of two continuous membranes. The interior is filled with an aqueous matrix, the stroma, within which the internal chloroplast membranes are embedded. MENKE (1962) introduced the term "thylakoid" for these flattened chloroplast membrane sacs (Fig. 1).

2.2 The Chloroplast Envelope

2.2.1 Higher Plants

2.2.1.1 Morphology and Isolation

Studies with isolated, intact chloroplasts of vascular plants have demonstrated that the inner envelope membrane regulates the transport of metabolites into and out of the chloroplast, whereas the outer envelope membrane is highly permeable to many low molecular weight substances (HELDT and SAUER 1971; MURAKAMI et al. 1975; DOUCE and JOYARD 1979).

In thin section electron micrographs, the two envelope membranes appear as 6-µm layers separated by a space of between 10 and 20 µm (Fig. 1; GUNNING and STEER 1975). When examined by freeze-fracture electron microscopy (Fig. 2), the inner membrane displays many more particles than the outer membrane (SPREY and LAETSCH 1976; SIMPSON 1978b; CLINE et al. 1985), consistent with the greater number of transport and enzymatic activities associated with the inner membrane (DOUCE and JOYARD 1979; KEEGSTRA, Chap. 11.10, this Vol.). Biochemical studies of envelope membranes have been hampered by the fact that it is very difficult to physically separate the inner from the outer membranes (DOUCE and JOYARD 1979). The problems encountered in separating inner from outer envelope membranes have been traced to the presence of particle linkers between the two membranes that can be visualized both in thin sections (CARDE et al. 1982) and in freeze-fracture replicas (CLINE et al. 1985). Quantitative analysis of the particles seen on the fracture faces of purified envelope membrane fractions suggests that during mechanical separation of the two membranes, the linker molecules may cause some of the outer envelope membrane proteins to partition with the inner envelope membrane and vice versa (CLINE et al. 1985). The fractionation technique developed by CLINE et al. (1981) involves rupturing intact, isolated chloroplasts in a hypertonic medium by means of a freeze-thaw cycle. Under optimal conditions, a highly enriched

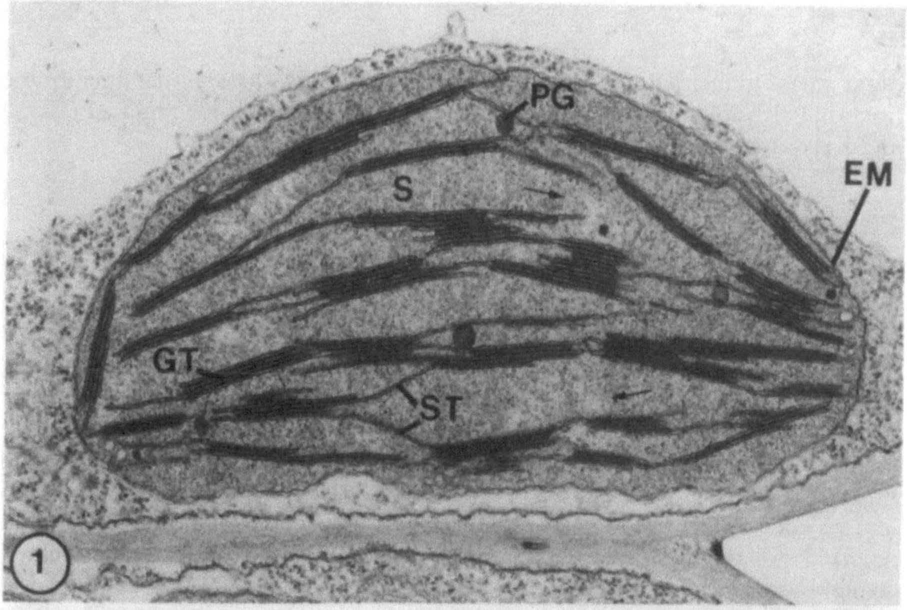

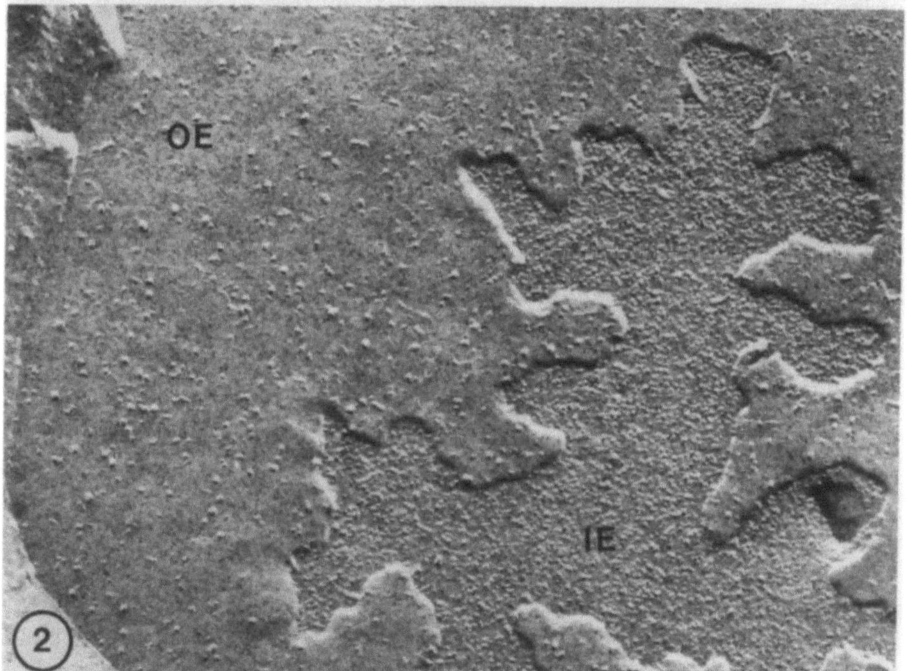

Fig. 1. Thin section electron micrograph of a tobacco chloroplast. Two envelope membranes (*EM*) delineate the chloroplast stroma (*S*), within which stacked grana thylakoids (*GT*) and unstacked stroma thylakoids (*ST*) can be recognized. Plastoglobuli (*PG*) and DNA containing regions (*arrows*) are also seen. The relationship between the different types of thylakoids is shown more clearly in Fig. 16. ×20,000

Fig. 2. Freeze-fractured chloroplast envelope membranes. The outer envelope (*OE*) membrane contains many fewer intramembrane particles than the inner (*IE*) one, consistent with the inner membrane having a higher protein to lipid ratio. ×80,000

outer membrane fraction and a reasonably pure (85–90%) inner membrane fraction (with some bound outer membrane fragments) can be obtained (Cline and Keegstra 1983; Cline et al. 1985).

2.2.1.2 Infoldings of the Inner Envelope Membrane

The inner envelope membrane of mature chloroplasts of C_3 plants is typically fairly smooth and is largely devoid of distinct infoldings. In contrast, conspicuous and sometimes quite long infoldings are seen in proplastids, during the greening of etioplasts, and during chloroplast development (reviewed by Thomson 1974; Douce and Joyard 1979). These observations suggested to early workers that new thylakoids may form by "budding" of the inner envelope membrane (Menke 1962). Although this idea is still often mentioned in discussions of research papers and in reviews (Wellburn 1982), it should be noted that no conclusive experimental evidence in support of this hypothesis has been reported. All recent studies addressing the relationship between inner envelope membrane and thylakoids agree on the fact that the structure and composition of the two membrane systems is distinctly different (Douce and Joyard 1979; Cline et al. 1981, 1985). Indeed, the cytoplasmically synthesized chloroplast proteins are even transferred to thylakoid membranes by different means (leader peptides) (Chua and Schmidt 1979; Keegstra 1984). In addition, were there connections between the intermembrane space of the envelopes and the thylakoid lumen, the ability of chloroplasts to generate considerable pH gradients across their thylakoid membranes (see Dilley, Chap. 10.7, this Vol.; Melandri and Venturali, Chap. 10.6, this Vol.) would be impaired. A more likely interpretation of the inner envelope membrane infoldings would be that they provide more space for accommodating membrane-bound enzymes involved in producing products needed for cell growth and development, e.g., membrane lipids (Douce and Joyard 1979; Keegstra, Chap. 11.10, this Vol.). Membrane expansion-induced infoldings may also develop in response to the transient accumulation of lipids in the inner membrane or by the transfer of cytoplasmically synthesized proteins through the inner membrane to the chloroplast stroma and the thylakoids. Finally, it should be noted that some of the smaller infoldings seen in thin sections could be caused by shrinkage of the plastids during processing for electron microscopy, and thus may be specimen preparation artifacts.

In contrast to the C_3 chloroplasts just discussed, those possessing a C_4 dicarboxylic acid pathway for carbon fixation exhibit very extensive tubular and vesicular elaborations of the inner membrane, termed peripheral reticulum (reviewed by Laetsch 1974). Peripheral reticulum membranes are observed in the periphery of the stroma of all mesophyll and bundle-sheath chloroplasts of C_4 plants, although they tend to be more developed in the former. The continuity of the peripheral reticulum and the inner envelope membranes has been proven by means of serial thin sections and by freeze-fracture electron microscopy (Sprey and Laetsch 1978). Both membrane systems contain the same complement of intramembrane particles, consistent with the notion that they are functionally equivalent structures. Although it is a common morphological marker for C_4 plants (Laetsch 1974), some C_3 plants have also been found to produce chloroplasts with a well-developed peripheral reticulum, but only in certain cell types, e.g., guard cells (PALLAS and MOLLENHAUER 1972). On the other hand, when C_4 plant cells, such as those of *Portulaca oleracea* (KENNEDY et al. 1977), are grown in tissue culture, the peripheral reticulum of most chloroplasts becomes much less conspicuous or even disappears. These and other observations suggest that the most likely function of the peripheral reticulum is to provide a greatly expanded inner envelope membrane system capable of accommodating the increased numbers of transport molecules and enzymes needed or the rapid exchange of metabolites between mesophyll and bundle-sheath chloroplasts and their surrounding cytoplasms.

2.2.2 Algae

The chloroplast envelopes of algae (see Table 1 for listing of main groups of algae) exhibit several structural features not found in those of higher plants. In common with higher plants, all algal chloroplasts are delineated by a basic envelope consisting of two parallel membranes, and in the Chlorophyceae (e.g., *Chlamydomonas*, *Chlorella*) and Rhodophyceae (e.g., *Porphyridium*, *Bangia*) only these two basic envelope membranes are seen. In some species, however, one or two additional membranes separate the chloroplast from the cytoplasm (see Fig. 3 for summary diagram; BISALPUTRA 1974; COOMBS and GREENWOOD 1976; GIBBS 1981). Thus, one additional membrane is found around dinoflagellate and euglenoid chloroplasts (GIBBS 1962, 1981; LEFORT-TRAN 1981), whereas two additional "extra-chloroplastic" membranes encompass the chloroplasts of Cryptophyceae, Haptophyceae, Chrysophyceae, Bacillariophyceae, Xanthophyceae, and certain Phaeophyceae (BISALPUTRA 1974; COOMBS and GREEN-

Table 1. Systematic organization of the main groups of eukaryotic algae

Division	Class	Examples
Chlorophyta	Chlorophyceae (green algae)	*Chlamydomonas*, *Chlorella*, *Scenedesmus*, *Ulva*, *Chara*, *Codium*, *Caulerpa*
	Prasinophyceae (green algae)	*Micromonas*, *Platymonas*
	Euglenophyceae (euglenoids)	*Euglena*, *Phacus*
Rhodophyta	Rhodophyceae (red algae)	*Porphyridium*, *Rhodella*, *Griffithsia*, *Spermothamnion*
Chromophyta	Cryptophyceae (cryptomonads)	*Cryptomonas*, *Chroomonas*, *Rhodomonas*
	Dinophyceae (dinoflagellates)	*Gonyaulax*, *Glenodinium*, *Peridinium*, *Prorocentrum*
	Chrysophyceae (golden-brown algae)	*Ochromonas*, *Synura*
	Xanthophyceae (yellow-green algae)	*Bumillariopsis*, *Tribonema*, *Vaucheria*
	Bacillariophyceae (diatoms)	*Phaeodactylum*, *Skeletonema*
	Phaeophyceae (brown algae)	*Acrocarpia*, *Fucus*, *Laminaria*
	Haptophyceae	*Chrysochromulina*
	Eustigmatophyceae	*Polydriella*

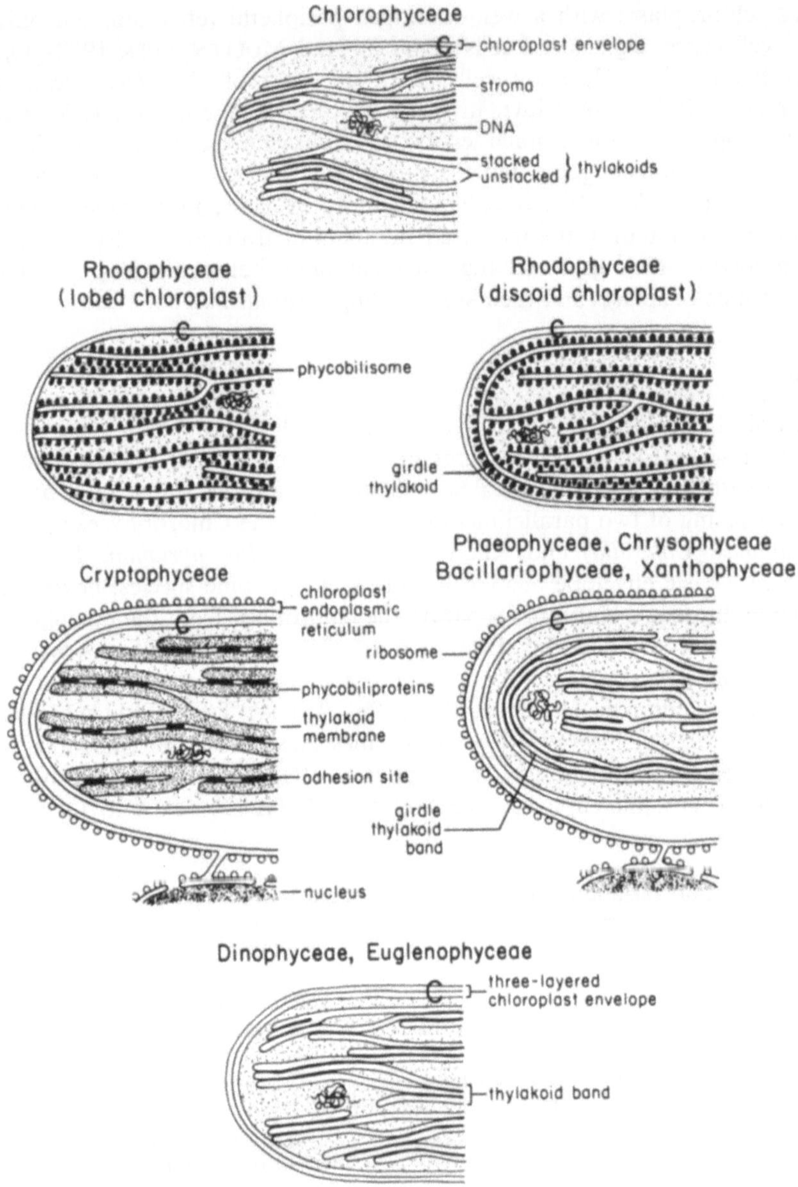

Fig. 3. Diagrams illustrating the general organization of thylakoids in the chloroplasts of different kinds of algae. Note that, with the exception of the red algae whose thylakoids possess surface-bound phycobilisomes, all algae have evolved light-harvesting systems that enable the thylakoids to become differentiated into stacked or adhering, and unstacked or nonadhering membrane regions. The functional significance of membrane stacking is discussed in Section 3.2.6 of this Chapter

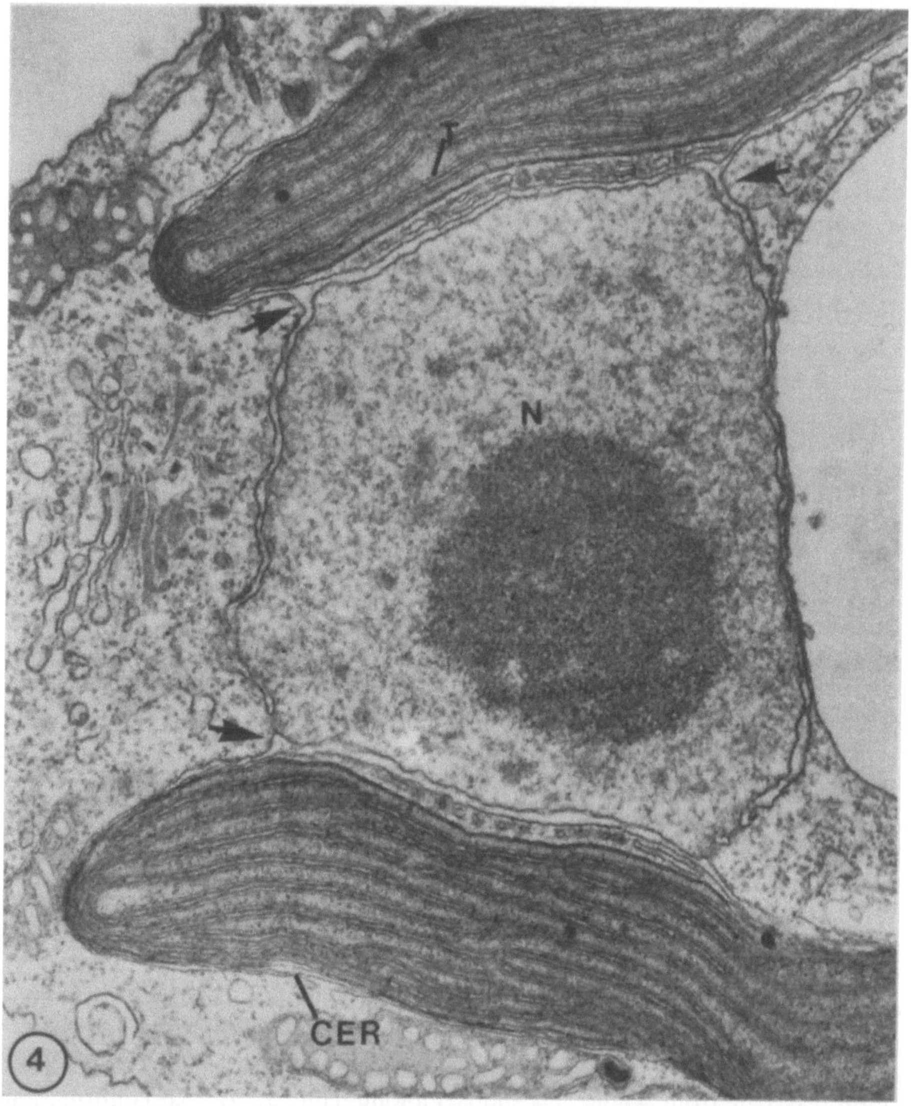

Fig. 4. Thin section electron micrograph showing a portion of the single-cell, chrysophycean (yellow green) alga *Ochromonas danica*. Of particular interest in the present context are the grouping of the thylakoids (*T*) into bands of three, the presence of a girdle-thylakoid band just inside the envelope membranes, and the continuity (*arrows*) of the chloroplast endoplasmic reticulum (*CER*) with the nuclear envelope. Nucleus (*N*) These features are all illustrated schematically in Fig. 3. × 35,000. (GIBBS 1970)

WOOD 1976; GIBBS 1981). Numerous electron microscopical studies (reviewed by GIBBS 1981) have shown that the latter extra-chloroplastic membrane system represents a special kind of endoplasmic reticulum, hence "chloroplast endoplasmic reticulum" (BOUCK 1965). It has ribosomes attached to its outer surface (Figs. 3, 4), and the outer membrane is continuous with the outer membrane

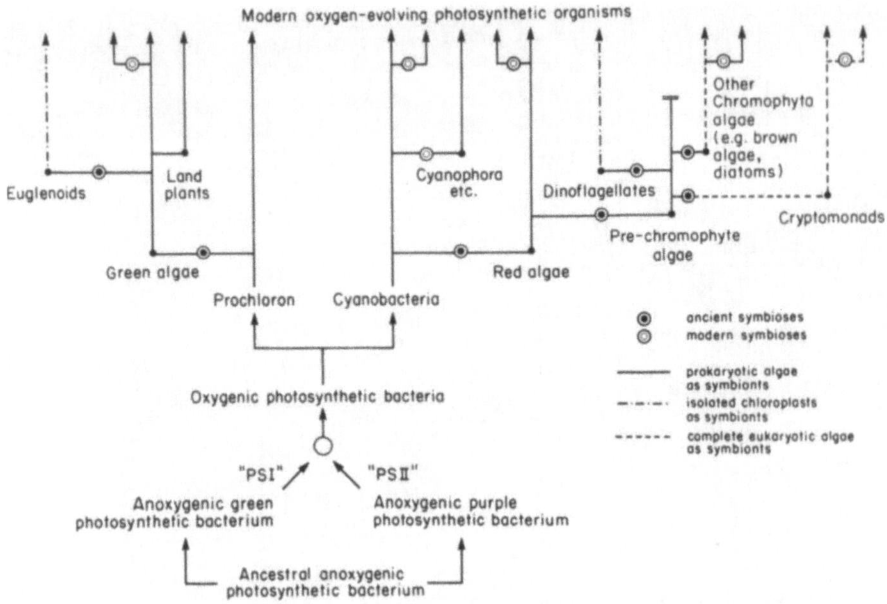

Fig. 5. Evolution of oxygen-evolving photosynthetic organisms, of chloroplasts and of modern photosynthetic endosymbionts. Based partly on hypotheses and diagrams of WHATLEY and WHATLEY (1981), GIBBS (1981), VAN VALEN (1982), AMESZ (1983), HEARST (Chap. 8.5, this Vol.) and DUTTON (Chap. 5, this Vol.)

of the nuclear envelope. Precursor forms of mastigonemes (flagellar surface appendages) often accumulate in the chloroplast endoplasmic reticulum before they are transported to the cell surface. Despite such observations, no convincing role related to chloroplast function has yet been attributed to the single and double extrachloroplastic membrane systems. Instead, WHATLEY et al. (1979) and GIBBS (1981) have postulated that they represent evolutionary relics. Their theories suggest that chloroplasts with more than two surrounding membranes evolved from the acquisition and partial degradation of eukaryotic rather than prokaryotic algal endosymbionts; those with four surrounding membranes derived from whole algal cells, and those with three membranes either from isolated chloroplasts or from whole cell endosymbionts (Fig. 5). A nice confirmation

Fig. 6. Portion of an ultrarapidly frozen and freeze-fractured *Chlamydomonas* cell showing patches of large particles (*arrows*) within the outer chloroplast envelope membrane. These particles appear to connect to the inner envelope membrane (not shown). Similar structures have been visualized in other green algae such as *Scenedesmus* and *Chlorella*. (STAEHELIN unpublished results). ×40,000

Fig. 7. Portion of an ultrarapidly frozen and freeze-fractured *Porphyridium cruentum* (red alga) cell. Rows of inner chloroplast envelope membrane fragments with characteristic particles (*arrows*) are seen attached to the outer envelope (*OE*) membrane. These rows coincide with lines of attachment of the margins of thylakoids to the inner envelope membrane (see Figs. 4, 17). Particles associated with the thylakoid attachment sites are aligned in both envelope membranes. ×40,000

6

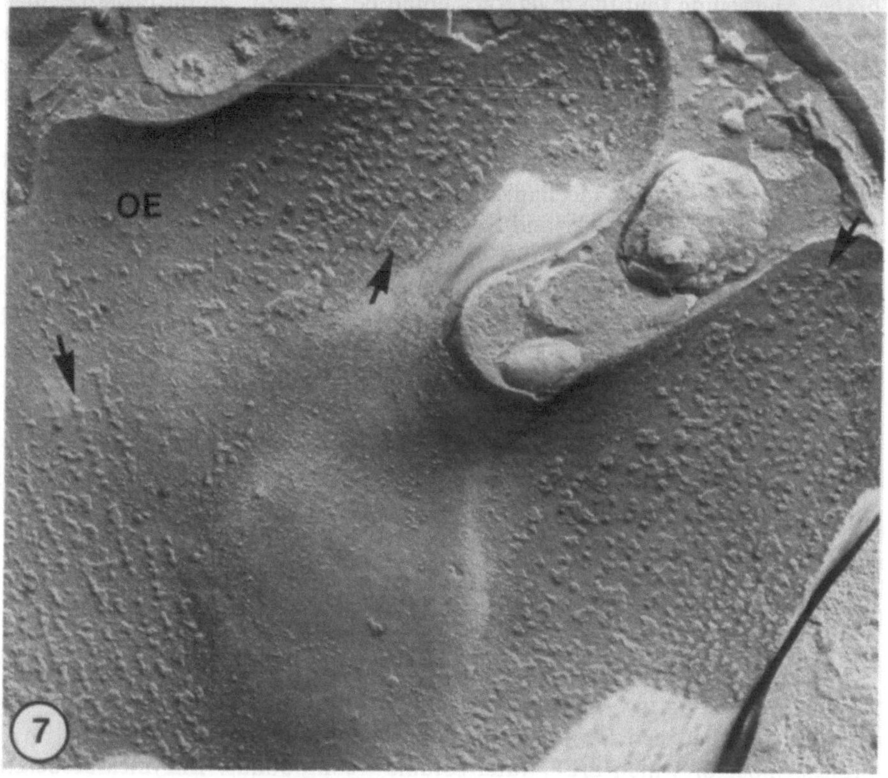
OE
7

of the whole eukaryotic cell endosymbiont theory can be found in the recent report of WILCOX and WEDEMAYER (1985) on a dinoflagellate with cyanophytic chloroplasts.

A limited number of freeze-fracture studies of the envelope and extra-chloroplastic membranes of algal chloroplasts have confirmed that the two membranes adjacent to the chloroplast stroma are structurally equivalent to the envelope membranes of vascular plants. In *Chlorella* (STAEHELIN 1966), *Euglena* (MILLER and STAEHELIN 1973; LEFORT-TRAN 1981), *Codium* (BRANDLEY 1981), *Chlamydomonas* (MELKONIAN and ROBENEK 1980; STAEHELIN unpublished results), and *Porphyridium* (STAEHELIN unpublished results) the inner envelope membrane contains between five and ten times more intramembrane particles (2500–300 μm^{-2}) than the outer envelope membrane, as is the case for higher plant chloroplasts. LEFORT-TRAN (1981) has reported an intermediate particle density (about 2000 μm^{-2}) for the extra-chloroplastic membrane of *Euglena*.

Two additional and unique structural features of the chloroplast envelope membranes of algae have been recently revealed in cells preserved by ultra-rapid freezing techniques (STAEHELIN unpublished results). In *Chlamydomonas* and *Scenedesmus*, characteristic patches of elongated particles (Fig. 6) are seen that seem to bridge the two envelope membranes. Ultrarapidly frozen *Porphyridium* cells reveal aligned rows of evenly spaced and distinctive particles in both the outer and inner envelope membranes (Fig. 7). These rows coincide with lines of contact between the margins of the thylakoids and the inner envelope membranes, suggesting that the thylakoids could be "riveted" to the envelope membranes to stabilize their spatial organization in the chloroplast. Whether these differentiated envelope membrane regions participate in the transfer of preproteins or other molecules from the cytoplasm to the thylakoid membranes has yet to be determined.

The sensitivity of the outer envelope membrane of isolated, intact chloroplasts of higher plants to mechanical and chemical perturbations has been recognized for some time (HELDT and SAUER 1971; DOUCE and JOYARD 1979; PRICE and REARDON 1982; CLINE et al. 1985). Thus, it is rather remarkable that intact chloroplasts isolated from *Caulerpa sedoides* are not only able to survive for several weeks in vitro, but can also divide repeatedly (GILES and SARAFIS 1972). Even more astonishing is the ability of isolated chloroplasts from siphonaceous algae (*Codium*, *Caulerpa*) to survive passage through the gut and phagocytosis by intestinal epithelial cells of marine slugs, thereby enabling such organisms to maintain chloroplasts as symbionts (TRENCH and OHLHORST 1976; BRANDLEY 1981). The biochemical basis of this stability would seem an intriguing topic of inquiry.

2.3 The Chloroplast Stroma

2.3.1 Matrix Components

The chloroplast matrix, known as stroma, encompasses all structures and molecules located between the inner envelope membrane and the thylakoid membranes, most of which are "water-soluble" compounds. Enzymes associated with the Calvin cycle, particularly ribulose bisphosphate carboxylase/oxygenase, appear to contribute the bulk of the proteinaceous components, but many other enzymes are present as well (ROBINSON and WALKER 1981; J.W. ANDERSON 1981). Thus, the stroma contains multiple copies of chloroplast DNA (about 30 per chloroplast in many types of higher plant cells), 70S ribosomes, mRNA's and all the other elements needed for protein synthesis (ELLIS 1976). Starch grains, usually ellipsoidal and up to 2 μm in length, and small, round, osmiophilic plastoglobuli are other common stromal components, while pyrenoids and eyespots are found only in the chloroplasts of certain algae (COOMBS and GREENWOOD 1976). Phytoferritin particles with a diameter of about 10 nm may also be discerned, particularly in chloroplasts of senescing leaves (BARTON 1970).

2.3.2 Plastoglobuli

Plastoglobuli increase in number in dark-grown plants and during leaf senescence, but decrease during membrane synthesis in the light (LICHTENTHALER and SPREY 1966; LICHTENTHALER 1968; BARR and ARNTZEN 1969). Chemical analysis of isolated and purified plastoglobuli has shown them to be composed primarily of lipophilic quinones and carotenoids such as plastoquinone, α-tocopherolquinone, α-tocopherylquinone and vitamin K_1. These observations may suggest that plastoglobuli serve as extralamellar pools of certain types of membrane lipids that accumulate during the degradation of old membranes, perhaps to be utilized during the formation of new ones.

2.3.3 Pyrenoids

A pyrenoid is a differentiated region of the chloroplast stroma, which occurs in the evolutionarily more primitive representatives of all algal divisions (BROWN 1967; KIRK and TILNEY-BASSETT 1978). The diameter of pyrenoids varies from 1–5 μm, and in thin sections they appear as uniformly dense or occasionally crystalline aggregates of finely granular material. In freeze-etch replicas of ultra-rapidly frozen *Porphyridium* cells, the pyrenoid is seen to consist of tightly packed granules with a diameter of about 10 nm (Fig. 8). Based on their location, two types of pyrenoid can be distinguished: those embedded within the center of chloroplasts, and which are characteristic of pyrenoid-containing Chlorophyta and Rhodophyta cells, and those that form bulbous protrusions and are typical of the Phaeophyta, Pyrophyta, and Cryptophyta (BISALPUTRA 1974; COOMBS and GREENWOOD 1976; KIRK and TILNEY-BASSETT 1978). Many pyrenoids are bordered and/or traversed by small numbers of thylakoids (Fig. 8), which appear to have a normal complement of intramembrane particles (STAEHELIN unpublished results), but others seem to have no direct contact with thylakoids at all. Biochemical studies of isolated and purified pyrenoids of *Eremosphaera viridis* (HOLDSWORTH 1971), *Micromonas squamata* (SALISBURY and FLOYD 1978) and *Bryopsis maxima* (SATOH et al. 1984) have revealed that their major constituent (85 to 90%) is ribulose bisphosphate carboxylase; the remaining 10 to 15% is composed of approximately ten other polypeptides. Using X-ray microanalysis, SALISBURY and FLOYD (1978) also detected copper, a metal

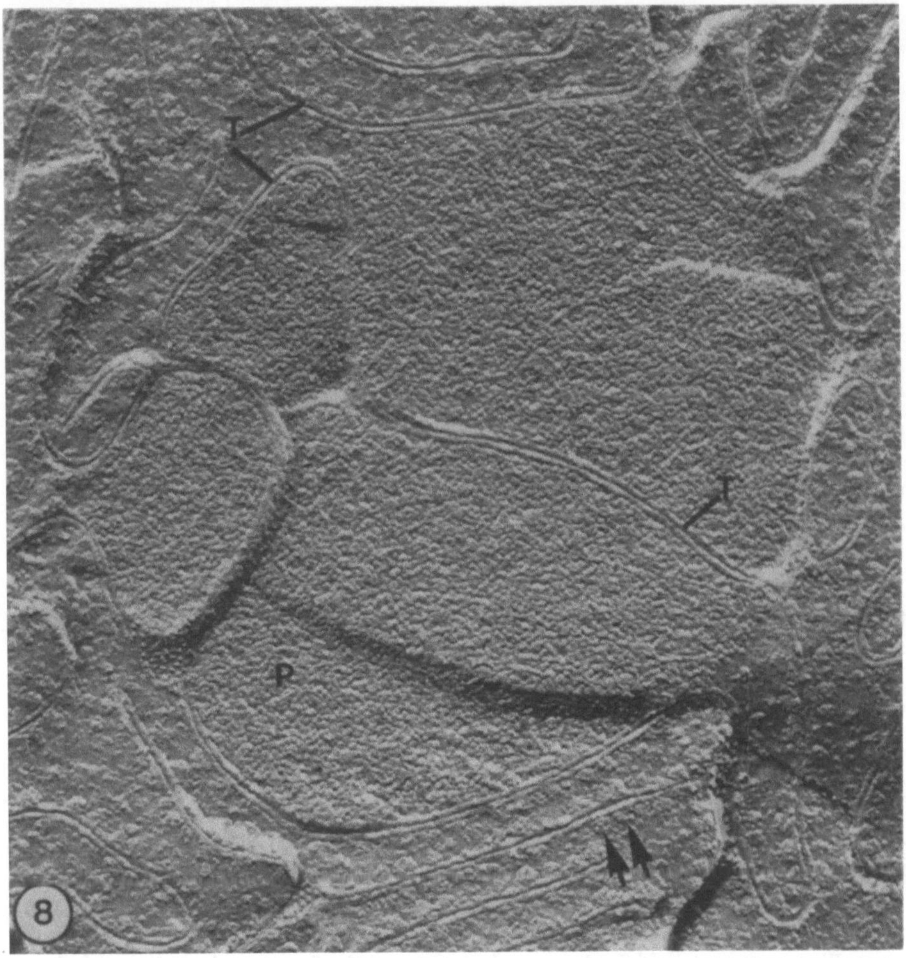

Fig. 8. Micrograph of a freeze-etched pyrenoid (*P*) of *Porphyridium cruentum* (red alga). Note the uniform size of the granules (about 10 nm in diameter) that seem to make up the bulk of the pyrenoid body. Thylakoids (*T*) cover most of the pyrenoid surface as well as transect the pyrenoid body. The granular substructure of the pyrenoid body contrasts with the relative smooth appearance of the stroma material between the phycobilisome (*arrows*) covered thylakoids. It is proposed that pyrenoids may serve both in the fixation of CO_2 and as protein reserve bodies. ×65,000

reported to be associated with noncrystalline preparations of ribulose bisphosphate carboxylase (WISHNICK et al. 1970).

The functional significance of pyrenoids is still an enigma. Due to their close spatial relationship with starch granules, a role in starch synthesis is usually assumed. However, pyrenoids are also found in algae that store water-soluble carbohydrates (Chrysophyceae, Bacillariophyceae, Haptophyceae, Phaeophyceae, Xanthophyceae), or whose starch is stored external to the chloroplast and not in the proximity of the pyrenoid. They are also absent from chloroplasts of many algae capable of accumulating stromal starch, and are present only during certain stages of the life cycles of others (EVANS 1968). Finally, *Euglena* cells made experimentally plastid-free, and thereby pyrenoid-free, are still able to synthesize and store paramylon, a starch analog. Based on these findings, one has to question the extent to which pyrenoids are involved in the synthesis of reserve carbohydrates. Based on their high protein content and occasional crystalline substructure, they may function both in CO_2 fixation and as protein reserves.

In the context of pyrenoids, it is intriguing to note that when chloroplasts of *Avena sativa* lose water during wilting, ribulose bisphosphate carboxylase molecules can aggregate in the stroma to form whorl-like masses of tightly packed fibrils, termed stroma centers (GUNNING 1965). SPREY (1977), on the other hand, has shown that by storing spinach leaves at 4 °C in the dark, crystalline aggregates of the same molecules can be made to accumulate within the lumen of thylakoids. How these molecules pass from the stroma across the thylakoid membrane and into the lumen is unknown.

2.3.4 Eyespots

All flagellated algae contain an eyespot capable of sensing the direction of a light source (KIRK and TILNEY-BASSET 1978; FOSTER and SMYTH 1980). In most of these organisms the red-colored eyespot is located within the peripheral stroma of the chloroplast where the envelope directly underlies the plasma membrane (Fig. 9). The Euglenophyta possess an eyespot in the cytoplasm (DODGE 1969). The most conspicuous elements of chloroplast-associated eyespots are the two-dimensional arrays of pigmented globules (on the average 100 nm in diameter). These globules are arranged in alternating layers with thylakoid membranes and granular matrix material (Fig. 9), with one or two such composite layers being most common, although up to nine have been observed in some algae (KIRK and TILNEY-BASSETT 1978). β-carotene is the major component of the eyespot granules in *Chlamydomonas* (OHAD et al. 1969). An in-depth treatment of the functional aspects of eyespots can be found in the excellent review of FOSTER and SMYTH (1980). As pointed out by these authors, all algal antennae are designed to see primarily in the forward direction, i.e., their field of view is limited to light beams coming from the front of the antenna. This is achieved by exploiting the principle of quarter wave interference reflectors. Such reflectors contain precisely spaced alternating thin layers of high and low refractive index (Fig. 9), which, upon being struck by light, can produce intensity maxima, one approximately a quarter wavelength out from the eyespot and one at the inner surface of each of the pigmented layers. Since the plasma membrane and the two envelope membranes are all positioned within this maxima region in *Chlamydomonas*, any one of these membranes could contain the actual photoreceptors. MELKONIAN and ROBENEK (1979, 1980) have examined the membranes associated with eyespots in *Tetraselmis* and *Chlamydomonas* by means of freeze-fracture electron microscopy. Their micrographs reveal that both the plasma membrane and the outer envelope membrane of the chloroplast are differentiated in the eyespot-region, both containing higher densities of intra-

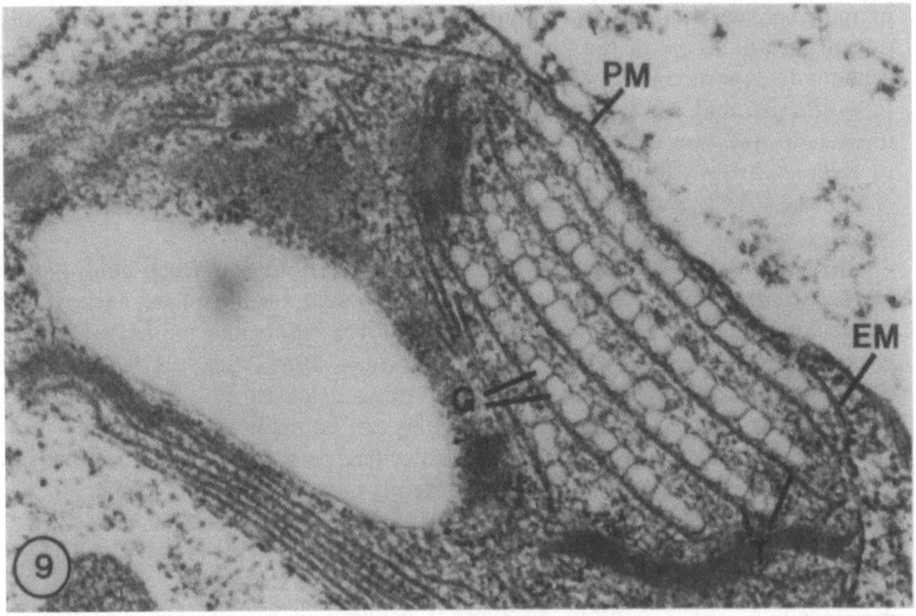

Fig. 9. Eyespot of *Chlamydomonas*. The eyespot is located within the periphery of the chloroplast where the chloroplast envelope membranes are attached to the inner surface of the plasma membrane. Four layers of pigment (β-carotene) granules (*G*) are seen; each layer is attached on its inner surface to a thylakoid (*T*) and is covered on its outer surface by a fibrillar mesh of defined width. The spacing of the layers indicates that they serve as quarter-wave reflectors, which amplify the light signal in the vicinity of the plasma membrane (*PM*) and the associated envelope membranes (*EM*), where the photoreceptor molecules appear to be located. ×20,000

membrane particles than the surrounding areas. From a functional point of view, the most likely location for the signal receptors would be the plasma membrane, since this would allow for an action potential-like mechanism to operate in signal transduction to the base of the flagellae. Indeed, spectral studies suggest that the receptors may contain rhodopsin-like molecules (FOSTER and SMYTH 1980).

2.4 Arrangement of Thylakoids

The diversity in morphology of photosynthetic membranes of oxygen-evolving plants, algae and bacteria (Figs. 3, 10) is rather confusing to most nonsystematists. To a significant extent this diversity appears to parallel differences in photosynthetic accessory pigment composition (Fig. 11), i.e., the pigments solely involved in the harvesting of light energy (THORNBER, Chap. 3; GANTT, Chap. 6.3; ANDERSON and BARRETT, Chap. 6.4, all this Vol.). Accessory light-harvesting pigments account for up to 99% of the pigments of photosynthetic membranes, and thereby define not only the color of a photosynthetic organism, but also

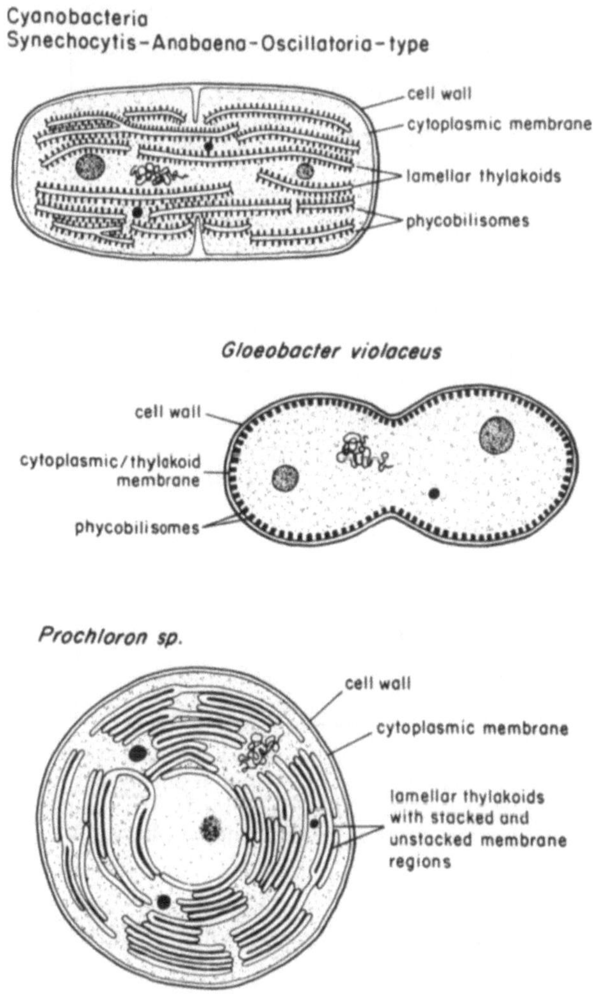

Fig. 10. Diagram illustrating the general organization of photosynthetic membranes found in oxygen-evolving prokaryotes. In most cyanobacteria the phycobilisome-studded thylakoids traverse the cytoplasm. The unusual cyanobacterium *Gloeobacter* lacks internal thylakoids; its photosystems are located instead in the cytoplasmic membrane to which the phycobilisomes are attached. The chlorophyll b-containing *Prochloron* thylakoids are differentiated into stacked and unstacked membrane regions

the segment of the spectrum that a given organism can photosynthetically exploit (see SAUER, Chap. 2, this Vol.). Their effects on membrane structure relate to the fact that the pigments are all lipidic in nature, and that the proteins associated with accessory pigments contribute about 50% of total membrane protein. The diagrams shown in Figs. 3, 5, 10, and 11 summarize in a simplified manner how the different thylakoid morphologies relate to evolutionary position and pigment composition of photosynthetic organisms.

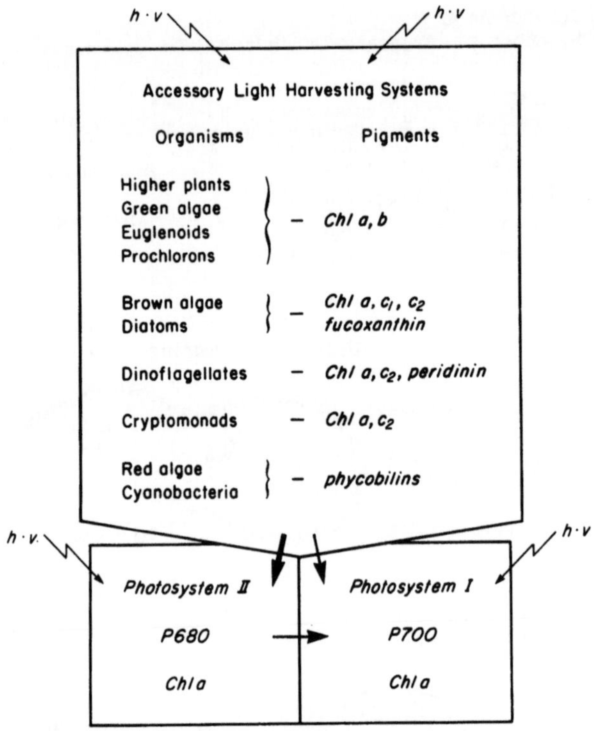

Fig. 11. Diagram showing the light-harvesting systems of oxygen-evolving photosynthetic organisms and the flow of excitation energy between these systems. P-680 and P-700 designate the reaction center chlorophylls of photosystems II and I, respectively. The major pigment types of the different light-harvesting systems are indicated in italics

2.4.1 Thylakoids of Oxygen-Evolving Prokaryotes

2.4.1.1 Cyanobacteria

Two major types of photosynthetic, oxygen-evolving prokaryote are known: cyanobacteria, also called blue-green algae, and prochlorons. Cyanobacterial thylakoids are lamellar in nature and traverse the cytoplasm (Figs. 10, 12). The exact organization of these thylakoids depends on the species as well as the physiological state of the cells (ECHLIN 1970; WOLK 1973). In some species the membranes are arranged into roughly concentric layers, in others they are more highly curved and exhibit a contorted appearance. A rather unique arrangement has been found in the unicellular cyanobacterium *Agmenellum quadruplicatum* (NIERZWICKI-BAUER et al. 1983). In this organism, individual thylakoids are maintained in a flattened and stretched configuration by means of longitudinal rows of linkers that attach the branched margins of the thylakoids to the inner surface of the cytoplasmic membrane analogous to the thylakoid/chloroplast envelope interactions seen in *Porphyridium*. Changes in the ultrastructure of cyanobacterial cells have been reported following nitrogen starvation (DE VASCONCELOS and FAY 1974; STEVENS et al. 1981a), carbon starvation

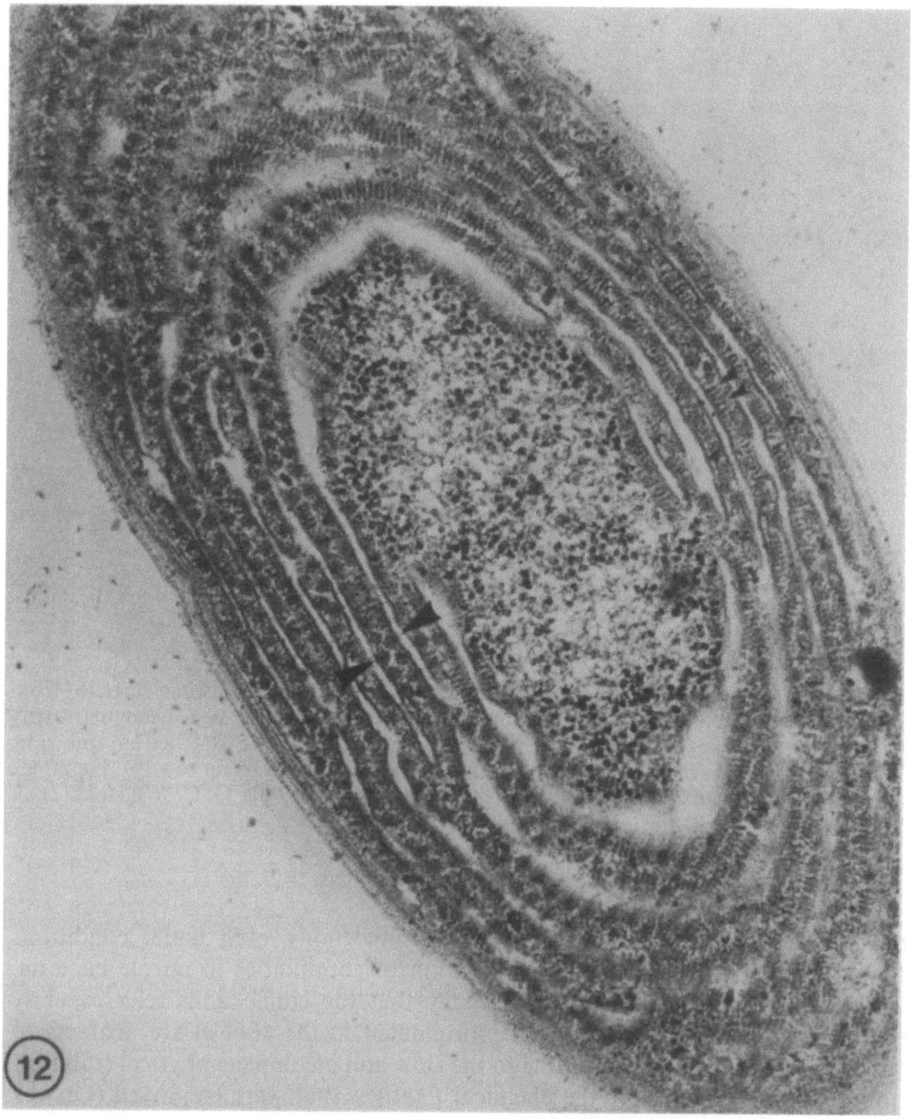

Fig. 12. Thin section electron micrograph of the cyanobacterium *Synechococcus lividus*. The thylakoids are covered with tightly packed hemidiscoidal phycobilisomes, some of which are seen in side view (*small arrowheads*) and others face-on (*large arrowheads*). × 60,000. (EDWARDS and GANTT 1971)

(L.S. MILLER and HOLT 1977), phosphorous starvation (STEVENS et al. 1981 b), and iron starvation (HARDIE et al. 1983).

Many reviewers (e.g., BISALPUTRA 1974) have claimed that the thylakoids of cyanobacteria can be physically continuous with the cytoplasmic membrane, as is the case for nonoxygenic purple bacteria (see SPRAGUE and VARGA, Chap. 11.1, this Vol.). However, our freeze-fracture micrographs of ultrarapidly frozen

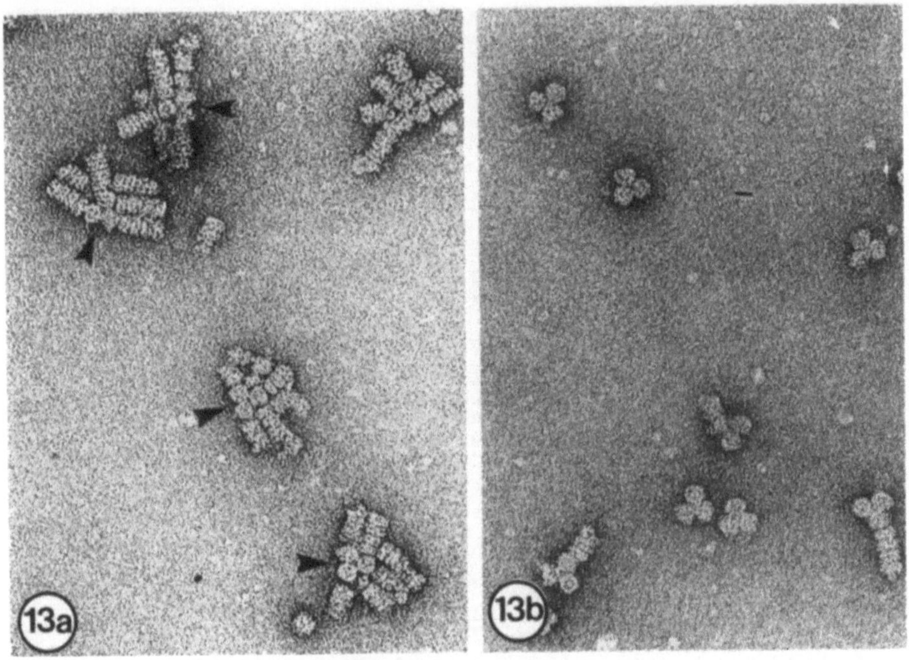

Fig. 13. Negatively stained, isolated phycobilisomes (**a**) and phycobilisome cores (**b**) from wild-type *Synechocystis* 6701 (cyanobacterium). The core consists of a triangular array of circular complexes to which rods of stacked disc elements are attached on one side. The noncovered side of the core (*arrowheads*) attaches through a linker to a PS II complex (see GANTT, Chap. 6.3, this Vol.). Both micrographs ×225,000. (R.C. WILLIAMS et al. 1980)

cyanobacteria have failed to reveal such connections, even under conditions that consistently lead to the demonstration of continuities in purple bacteria. Based on these findings, it is most likely that the connections seen by early electron microscopists (most reports originated in the 1960's) are artifacts of specimen preparation, attributable to the slow and incomplete physical stabilization of the membranes by the chemical fixatives that were employed (GILKEY and STAEHELIN 1986).

Phycobilisomes, large assemblies of intricately organized phycobiliprotein complexes (Fig. 13a, b), serve as the main light-harvesting antennae for photosystem II of cyanobacteria (see Sect. 3.3.2 and GANTT, Chap. 6.3, this Vol.). These hemidiscoidal structures are about 60 nm wide, 30 nm high, and 10 nm thick (GLAZER 1983), and attach to the cytoplasmic surfaces of the thylakoids. The density of phycobilisomes is regulated by light quality and intensity; low light conditions lead to the formation of more phycobilisomes and to their aggregation into rows. Certain cyanobacteria also possess the ability to change the composition of their phycobilisomes in response to changes in light quality (complementary chromatic adaptation; R.C. WILLIAMS et al. 1980; TANDEAU DE MARSAC 1983).

Figure 10 depicts in diagrammatic form the fine structural organization of the atypical cyanobacterium *Gloeobacter violaceus*. This cyanobacterium lacks thylakoids. Its elongated phycobilisomes, consisting of bundles of six parallel rod-shaped elements (each 12–14 nm in diameter and 50–70 nm thick), appear in thin sections to be attached to the inner surface of the cytoplasmic membrane (GUGLIELMI et al. 1981). Apparently, the elements of the photosynthetic electron transport chain are confined to this membrane. An analogous situation is found in the purple bacterium *Rhodospirillum tenue* (WAKIM et al. 1978; SPRAGUE and VARGA, Chap. 11.1, this Vol.).

2.4.1.2 Prochlorons

The discovery of the chlorophyll b-containing, photosynthetic prokaryote *Prochloron* by LEWIN (1976, 1977) has had a major impact on theories of evolution of photosynthetic organisms. All known members of the genus *Prochloron* are obligate symbionts of colonial ascidians (sea squirts) of the family Didemnidae, which inhabit reefs in the Pacific ocean and the Persian Gulf. Like cyanobacteria, the prochlorons carry out photosynthesis with two photosystems and evolve oxygen. However, instead of using phycobiliproteins as major light-harvesting antennae, they utilize a chlorophyll a/b light-harvesting complex (chl a/b LHC, also known as LHC II, see THORNBER, Chap. 2, this Vol.) that is spectrally and electrophoretically very similar to the chl a/b LHC of green algae and higher plants (WITHERS et al. 1978). For this reason STANIER and COHEN-BAZIRE (1977) have proposed that a *Prochloron*-like organism gave rise to the chloroplasts of green algae and higher plants, and the cyanobacteria to those of red algae (Fig. 5).

Cross-sectional views of the spherical *Prochloron* cells reveal a roughly concentric organization of the overlapping and branched thylakoids, and the presence of typical stacked and unstacked membrane regions (Fig. 10). In green algae and higher plants the chl a/b LHC has been shown to promote the type of membrane adhesion found between grana thylakoids (McDONNEL and STAEHELIN 1980; BARBER, Chap. 11.5, this Vol.). Thus, the discovery of appressed thylakoid membrane regions in *Prochloron* sp. (WHATLEY 1977; GIDDINGS et al. 1980; G. COX and DWARTE 1981) has provided yet another link between this prokaryote and higher plant chloroplasts. *Prochloron* cells derived from *Diplosoma virens* display a membrane-delineated, vacuole-like structure in their center (WHATLEY 1977; GIDDINGS et al. 1980). Whether this "vacuole" arises from specimen preparation/fixation-induced swelling of thylakoids, or whether it represents a unique membrane-bounded compartment, needs further clarification. No such compartment has been observed in a *Prochloron* species derived from *Didemnum molle* (G. COX and DWARTE 1981).

2.4.2 Higher Plant Thylakoids

2.4.2.1 Three-Dimensional Organization of Grana and Stroma Thylakoids

In higher plants, two types of thylakoid are distinguished, grana and stroma thylakoids. The name "granum" (plural "grana") was coined by early light microscopists to describe minute particles seen within chloroplasts. These grana were subsequently equated with cylindrical, closely packed stacks of thylakoids that can be resolved in the electron microscope. The membranes connecting the grana stacks became known as stroma lamellae or stroma thylakoids due to their association with the chloroplast stroma (MENKE 1962). Later studies revealed that all grana and stroma thylakoids within a chloroplast form a physically continuous membrane system, which is organized into a highly complex, three-dimensional network enclosing one single anastomosing chamber, usually referred to as the thylakoid lumen. The precise spatial relationship between grana and stroma thylakoids was established by means of reconstructions of serially sectioned chloroplasts (HESLOP-HARRISON 1963; WEHRMEYER 1964a, b; PAOLILLO 1970; THOMSON 1974). Three-dimensional models derived from such studies depict stacks of grana thylakoids being intersected at an angle by parallel, evenly spaced stroma thylakoids that spiral up and around the stacks. At each intersection between a granal and stromal thylakoid, a narrow neck-like membrane region can be seen to interconnect the two types of thylakoid. Many of these features can be verified in the freeze-fracture micrographs shown in Figs. 14, 15. In these pictures, virtually all the structural features previously revealed by tedious reconstruction work can be confirmed at one glance, thanks to a series of fortuitous fracture planes that pass both along and across the membranes. Thus the angle between the stroma and grana thylakoids, the neck-like connections between the two membrane systems, the number of spiralling stroma thylakoids around a given granum, and the even spacing between the parallel stroma thylakoids can be clearly discerned.

2.4.2.2 Normal Development of Grana and Stroma Thylakoids

The development of the intricate three-dimensional relationship between grana and stroma thylakoids has been elucidated by reconstructing from serial sections the morphology of *Lolium multiflorum* plastids from four developmental stages (MUSTARDY and BRANGEON 1978). The stages were obtained by sampling at 2-cm intervals along the length of third leaves from the base (stage 1, youngest

Fig. 14. Freeze-fracture electron micrograph of a single granum and associated stroma thylakoids from a pea chloroplast. Note the angle between the grana membranes (*GT*) and the stroma thylakoid (*ST*) on the right, which allows the stroma thylakoid to spiral up and around the granum. Where the stroma thylakoid crosses a grana thylakoid, the two membrane systems are continuous (*arrowheads*). On the left of the granum, five additional, evenly spaced stroma thylakoids (*ST*) are seen. Each one of these makes an angle with the grana thylakoids and is connected to the grana membranes as shown on the right. See also Fig. 15. ×120,000

Fig. 15. Freeze-fractured pea thylakoids illustrating the relationship between grana and stroma thylakoids as seen in a face-on view. The *circularly arranged arrowheads* point to crossover/attachment sites between two grana (*G*) stacks and surrounding stroma thylakoids (*ST*; see also Fig. 14). The angled stroma thylakoids resemble the blades of a conventional windmill, with the central grana membrane corresponding to the hub. The differences in freeze-fracture morphology between stacked (grana) and unstacked (stroma) membrane regions are seen more clearly in Fig. 21. ×65,000

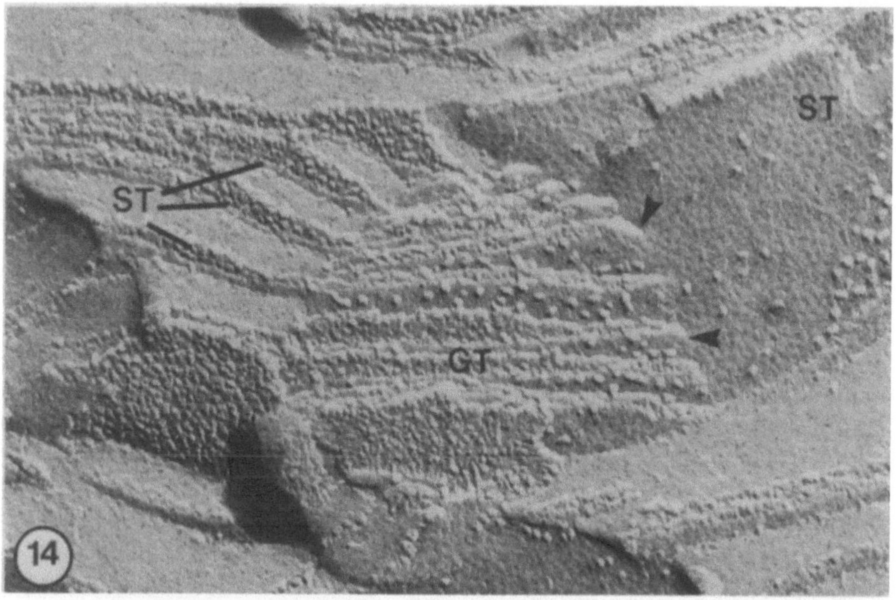
ST
ST
GT
14

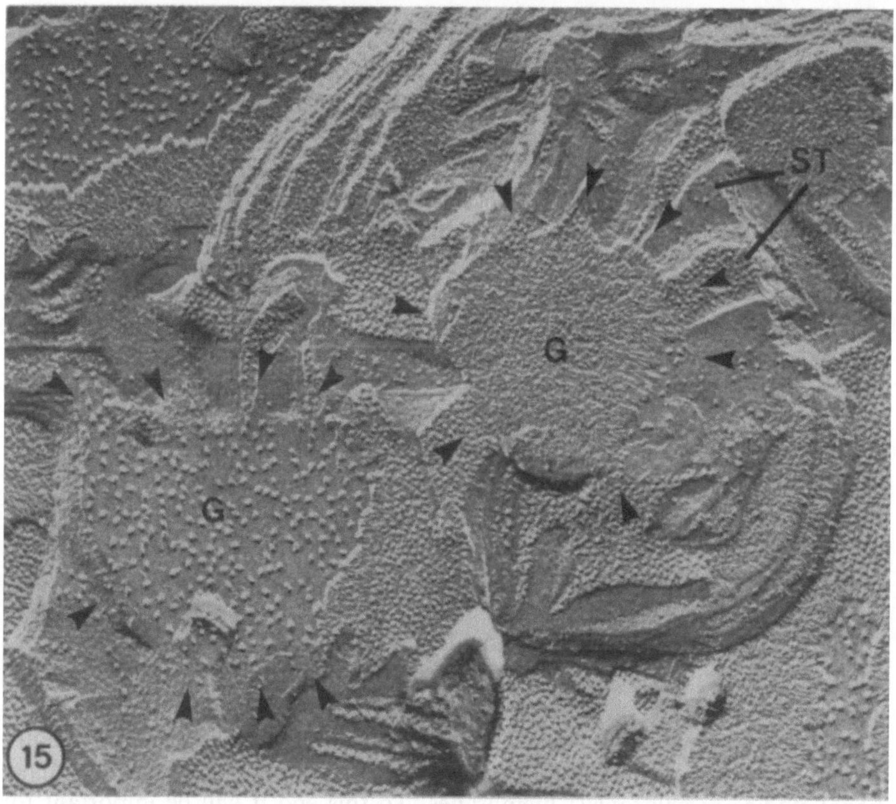
ST
G
G
15

form) to the tip (stage 4, mature form) of 3-week-old plants. Typical for the youngest plastids are single, highly perforated thylakoids. During stage 2, grana are initiated by membrane growth at specific sites in the margins of the perforations. Each tongue-like outgrowth initially expands into a typical disc-shaped grana thylakoid that overlaps the initiating one and becomes attached to it. Once this has happened, a second growth site is initiated within the margin of the same perforation adjacent to the first one. As this second tongue-like outgrowth expands over the first grana thylakoid, the initiating stroma thylakoid appears to be pulled up and around the margin of the forming grana stack. Continuation of this process leads to tilting of the initiating stroma thylakoid with respect to the plane of the grana thylakoids and thus to its spiral-like arrangement around the grana stack. The final assembly steps related to the multiplication of the stroma membranes around the margins of each granum are still somewhat obscure. MUSTARDY and BRANGEON (1978) suggest two possibilities: (1) branching of the so-called primary helix, which results in a secondary helix parallel to the parent one, or (2) splitting of parent lamellae and their subsequent fusion as secondary helices onto adjacent grana stacks. In all instances the pitch of the original spiralling stroma thylakoid is maintained.

2.4.2.3 Thylakoid Membrane Nomenclature

To describe the morphological features of thylakoid membranes of higher plants, WEIER et al. (1965, 1966) invented a nomenclature involving the use of terms such as fret, partition, loculus, and margin. Despite valiant attempts by reviewers and some textbook authors to explain and thereby popularize the Weier nomenclature, very few researchers have actually adopted it, largely because the names are difficult to remember, and because they can be replaced by more descriptive terms. Continued use of this nomenclature is therefore of questionable value.

In response to this unsatisfactory situation, a common usage nomenclature has begun to emerge, which also takes into consideration some of the new functional findings that have emerged during the last 15 years. Both from a structural and a functional point of view the most important spatial differentiation of higher plant and many algal thylakoids relates to their organization into stacked (or appressed) and unstacked (or nonappressed) membrane regions (Fig. 16). The "stroma thylakoids" are by definition always unstacked, their external surfaces exposed to the stroma. The definition of "grana thylakoids" is complicated by historical connotations. A granum, by definition, is a set of closely appressed thylakoids, and thus contains both stacked and unstacked membrane regions (Figs. 1, 16). Since, by all well-defined structural (STAEHELIN et al. 1977) and biochemical/functional (J.W. ANDERSON 1981) criteria, the unstacked membrane regions of a granum (marginal and top/bottom membranes) are equivalent to stroma thylakoids, the critical parameter of a granum is the stacked membrane regions. This fact has not always been appreciated by photosynthesis researchers in the past, but will need to be emphasized in future functional studies of chloroplast membranes. To this end it will be important for researchers to clearly define whether they are attributing a given function to a granum as a whole, or more specifically to the stacked membrane regions

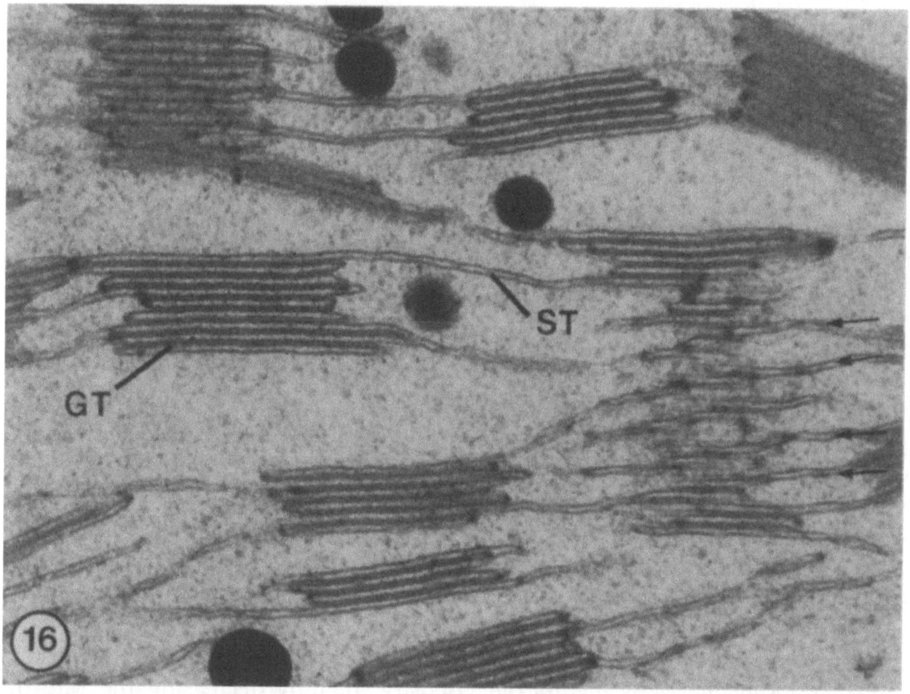

Fig. 16. Thin section through a portion of a spinach chloroplast illustrating interconnected stacked (grana; *GT*) and unstacked (stroma; *ST*) thylakoid membranes. Note the flattened, sac-like structure of the individual thylakoids. A tangentially sectioned granum is seen on the right and illustrates the angle between the grana and stroma membranes and the even spacing of the latter (*arrows*). × 80,000

of a granal stack. In this review, the terms "stacked" or "appressed" will be used alone or in conjunction with the word "grana", and "unstacked" or "nonappressed" alone or in conjunction with the word "stroma", to specify the type of thylakoid membrane to which reference is being made.

2.4.2.4 Naturally and Experimentally Induced Variations in the Ratio of Stacked to Unstacked Membrane Regions

The ratio of stacked to unstacked membrane regions in higher plant chloroplasts is variable. Within limits, there is an inverse relationship between the amount of thylakoid stacking and the intensity of growth light (J.M. ANDERSON 1982b). Plants grown under low light conditions invariably produce grana that have a larger diameter and contain more stacked membranes than those grown under high light conditions. For instance, in *Zea mays* high light mesophyll chloroplasts contain 61% stacked thylakoids and an average granum diameter of 0.35 μm; for low light chloroplasts the values are 73% and 0.43 μm (LICHTENTHALER et al. 1981). Interestingly, several types of herbicides such as fluometuron (WERGIN and POTTER 1975), methabenthiazuron (FEDTKE et al. 1977), bentazon

(MEIER et al. 1980) and atrazine (MATOO et al. 1984) have been shown to induce the formation of large, low-light-type grana in plants grown under medium-to-high-light conditions. Since most PS II reaction centers are located in stacked membrane regions and most PS I centers in unstacked membrane regions (J.W. ANDERSON 1981; see also Sect. 3 of this Chap.), and since all of these herbicides appear to interfere with electron transport between PS II and PS I (see SANDMANN and BÖGER, Chap. 10.10, this Vol.), the herbicide-induced increase in stacked regions appears to reflect an attempt by the plants to compensate for the reduction of electrons arriving at PS I. They do this by proportionately increasing the membrane areas containing PS II units and their associated antennae.

Another kind of chemical capable of increasing the ratio of stacked to unstacked membrane regions in chloroplasts is the substituted pyridazinone compound SAN-9785 (BASF 13-338; LEECH and WALTON 1983). Besides affecting carotenoid synthesis, this compound reduces the amount of polyunsaturated fatty acids, particularly linolenic acid (18:3), in leaves and chloroplasts when applied to young growing plants (ST JOHN 1976; KAHN et al. 1979; MORELAND 1980). Again the most simple explanation of the SAN-9785 effect would seem to be that more stacked membrane regions rich in PS II units and associated antennae are formed in an attempt to compensate for a decrease in the rate of electrons arriving at PS I. In this case, the loss of double bonds in membrane lipids probably increases membrane viscosity, thereby decreasing the rate of diffusion and consequently the electron transfer rate of the mobile electron carrier plastoquinone in the membrane (see also ORT, Chap. 4; WHITMARSH, Chap. 10.2; MURPHY, Chap. 11.11; SIEGENTHALER and RAWYLER, Chap. 11.9, all this Vol. for more details on mobile electron carriers and on membrane lipids).

Genuine shade plants of tropical rain forests, however, possess chloroplasts with an even higher ratio of stacked to unstacked membrane regions than those of high-light plants grown under low-light conditions. Each chloroplast of *Alocasia macrorrhiza*, for instance, contains four to five times more chlorophyll than a normal spinach chloroplast, and possesses grana stacks consisting of over 100 lamellae (GOODCHILD et al. 1972, J.M. ANDERSON et al. 1973). The photochemical components of these huge grana stacks have to compensate for the greater loss of PS II light versus PS I light in the leaf canopy (BJÖRKMAN and LUDLOW 1972).

Rather unusual "all-granal" chloroplasts have been reported for internal tissues of apple fruits (PHAN 1984) and for cultured cells of *Streptanthus tortuosus* containing virus-like particles (SJOLUND and SMITH 1974). In such chloroplasts virtually all thylakoids are organized into one or two huge grana (up to 25 μm in diameter) that can occupy as much as 60% of the chloroplast volume. In functional terms, however, the two types of chloroplast are quite different: the *Streptanthus* chloroplasts lack PS II activity, whereas the apple chloroplasts exhibit normal photosynthetic activity. Indeed, when the unusual internal tissue chloroplasts of apple fruits are compared with normal apple leaf chloroplasts in terms of O_2 emitted h^{-1} mg^{-1} chlorophyll, the former are over five times as active as the latter.

Another example of higher plant chloroplasts with consistently modified thylakoid morphology can be found in the bundle-sheath cells of advanced C_4 plants (panicoid grasses) such as maize, sorghum, and sugar cane (LAETSCH 1974). The most conspicuous characteristic of C_4 plants is the arrangement of the chlorophyll-containing tissue in concentric layers around the vascular tissue (Kranz-type anatomy). The outer mesophyll cells surround an inner layer of bundle-sheath cells. When compared with the typical chloroplasts of mesophyll cells, the bundle-sheath chloroplasts are seen to have few, if any, grana stacks.

It should be noted, however, that not all plants having the C_4 pathway for CO_2 fixation exhibit this dimorphism in thylakoid architecture. Among the dicotyledonous C_4 plants, *Euphorbia maculata* possesses the typical variation in structure of mesophyll and bundle-sheath chloroplasts; in *Amaranthus* and *Atriplex*, on the other hand, grana are present in both types of chloroplast (LAETSCH 1968, 1974). SCHMID et al. (1976) have demonstrated that the thylakoids of the chloroplasts in the labellum of the orchid *Aceras anthropophorum* are also devoid of grana. Biochemical studies revealed these chloroplasts to be able to perform cyclic photophosphorylation but unable to evolve oxygen or to carry out photosystem II-dependent photophosphorylation reactions.

2.4.3 Green Algal Thylakoids

The thylakoids of green algae are arranged in much the same manner as those of higher plants (Fig. 3). Typical for algae, the thylakoids tend to be long and extend from one end of the chloroplast to the other. Adjacent thylakoids (between two and six) form stacked membrane regions that, at the supramolecular level, are virtually indistinguishable from the stacked, grana membranes of higher plants (STAEHELIN et al. 1977; KAPLAN and ARNTZEN 1982). The main difference seems to lie in the variability of diameter and the regularity of arrangement of the stacked membrane regions. In *Chlamydomonas* there appears to be only a limited amount of regular arrangement; the thylakoids come together and move apart in an irregular manner. The thylakoids of *Nitella* (Characeae), on the other hand, often display an organization similar to higher plants with aligned stacks of thylakoids forming grana that are interconnected through typical, single stroma thylakoids. The fact that all green algae and higher plants possess stacked membrane regions, but that the arrangement of these stacked regions within chloroplasts is quite variable, suggests that thylakoid membrane stacking is of considerable functional importance, the specific arrangement of stacked regions much less so. A discussion of the functional significance of membrane stacking is presented in Section 4.2 of this chapter and in Chapter 11.5, this Volume of BARBER.

2.4.4 Euglenoid Thylakoids

The thylakoids of euglenoids display stacked and unstacked membrane regions similar to those of the more primitive types of green algae. The main difference is that euglenoid thylakoids are frequently grouped into bands of three, with single thylakoids crossing over from one band to another (Fig. 3).

2.4.5 Red Algal Thylakoids

Red algal thylakoids exhibit the simplest organization pattern of eukaryotic chloroplasts (Figs. 3, 4, 5). Their thylakoids, which resemble those of cyanobacteria, are single, mostly parallel and lack appressed membrane regions. The observed branching patterns are consistent with the notion that all thylakoids within a given chloroplast are interconnected. Two types of red algal chloroplast are distinguished: (1) those found in unicellular organisms such as *Porphyridium*, which possess a lobed chloroplast with thylakoids whose margins abut against, and in some instances even appear to be physically attached to the inner envelope membrane (Fig. 17, GANTT and CONTI 1965); and (2) the discoidal chloroplasts (Fig. 18) typical of more evolutionarily advanced, filamentous red algae, in which parallel internal thylakoids are surrounded by a girdle thylakoid concentric with the envelope (COOMBS and GREENWOOD 1976). The frequency with which connections are seen between the girdle and the inner thylakoids suggests that new internal thylakoids may arise as outgrowths of the girdle lamella.

Attached to the stromal surface of all of these thylakoids are phycobilisomes, the large, water-soluble, phycobilin-containing light-harvesting complexes typical of cyanobacteria, red algae and *Cyanophora* (GANTT 1981 and Chap. 6.3, this Vol., GLAZER 1983). The phycobilisomes of red algae come in two forms: hemidiscoidal or fan-shaped (e.g., *Rhodella*; MÖRSCHEL et al. 1977) with approximate dimensions of 55 nm × 35 nm × 10 nm, and hemispherical (e.g., *Porphyridium*; GANTT, chap. 6.3, this Vol.) with approximate dimensions of 48 nm × 32 nm × 32 nm. The contributions of GANTT (chap. 6.3, this Vol.); ZUBER (Chap. 6.1, this Vol.), and SCHEER (Chap. 7.6, this Vol.) discuss in greater detail the structure, biochemistry, and energy transfer properties of phycobilisomes. Suffice it to say here that the design of these light-harvesting antennae provides for a very efficient transfer of excitation energy to reaction centers located within the thylakoid membranes.

Within limits, the density of phycobilisomes associated with red algal thylakoids is inversely proportional to the intensity of growth light, similar to the situation in the cyanobacteria. In *Griffithsia pacifica*, for instance, the number of phycobilisomes drops from 67 μm^{-2} to 42 μm^{-2} when algae are transferred from 50 ft-c to 300 ft-c growth light conditions (WAALAND et al. 1974). Under low light conditions both types of phycobilisome, but particularly the hemidiscoidal ones, display a tendency to become organized into rows.

Fig. 17. Thin section through a *Porphyridium* (red alga) cell demonstrating the lobed configuration of the single chloroplast. Note how the margins of groups of parallel, phycobilisome-covered thylakoids (*T*) abut against the inner surface of the envelope membrane (*arrowheads*; see also Figs. 3 and 7). Rows of phycobilisomes (*PB*) are seen where thylakoids have been tangentially sectioned in the center of the micrograph. ×25,000

Fig. 18. Thin section through a chloroplast of the red alga *Spermothamnion tuneri*. A single girdle thylakoid (*GDT*) underlies the envelope membranes (*EM*). Phycobilisomes are seen attached to the thylakoids (*T*), and a branch point between an internal thylakoid and the girdle thylakoid is indicated by an *arrow*. See Fig. 3. ×50,000

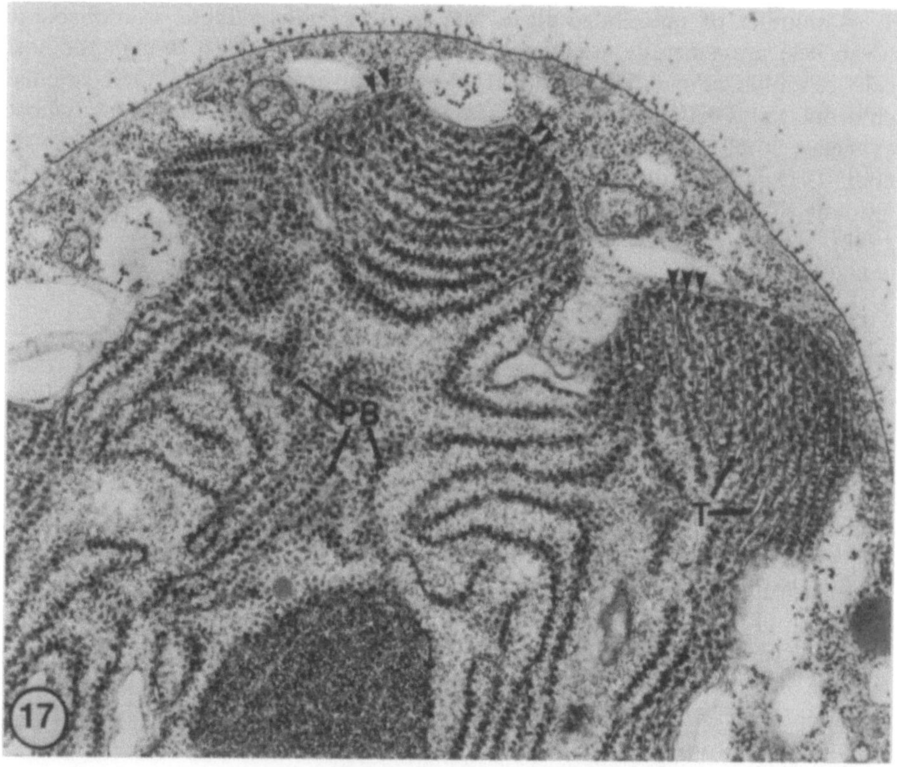
PB
T
17

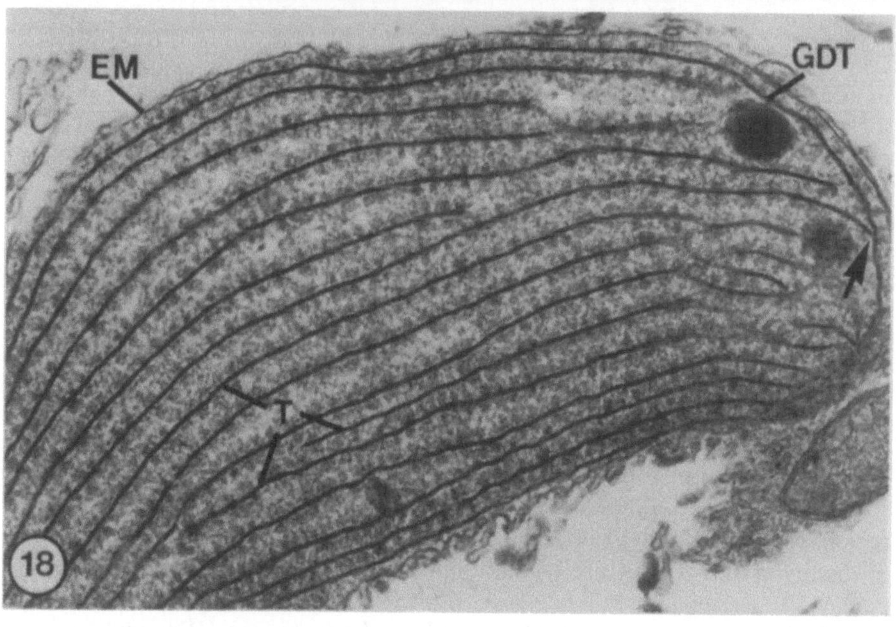
EM
GDT
T
18

A number of unicellular algae that possess endosymbiotic cyanobacteria (cyanelles) are generally assigned to the Rhodophyceae, even though the host cells resemble other kinds of eukaryotic phototroph devoid of their original chloroplasts. Two typical examples are *Glaucocystis* and *Cyanophora*, whose cyanelles closely resemble cyanobacteria in terms of thylakoid structure (TRENCH et al. 1978; GIDDINGS et al. 1983) but contain only about one tenth of the amount of DNA typical for free-living cyanobacteria (JAYNES and VERNON 1982).

2.4.6 Chromophytan Thylakoids

Although the least known of the algal divisions, the Chromophyta are more diverse, both in types of alga and morphology, than either the Rhodophyta or the Chlorophyta. The following major classes of algae have been assigned to the Chromophyta: Cryptophyceae, Dinophyceae, Chrysophyceae, Bacillariophyceae, Phaeophyceae, Xanthophyceae, and Haptophyceae (Table 1). In addition to having c-type chlorophylls (see ANDERSON and BARRETT, Chap. 6.4, this Vol.), all of these algae possess chloroplasts that are segregated from the cytoplasm by two additional membranes (only one in the Dinophyceae) known as the chloroplast endoplasmic reticulum (Figs. 3, 4). The following sections provide a very brief introduction to the characteristic arrangements of thylakoids in some of these algae. A more detailed treatment of the morphology of Chromophyta chloroplasts can be found in the reviews of DODGE (1973), BISALPUTRA (1974), and COOMBS and GREENWOOD (1976).

2.4.6.1 Cryptophyceae (Cryptomonads)

In addition to the aforementioned groups, phycobiliproteins also serve as accessory light-harvesting systems for the cryptophycean algae. However, besides differing chemically from those found in red algae, the phycobiliproteins of cryptomonads do not form phycobilisomes (GLAZER 1980). Instead they are found within the thylakoid lumen (DODGE 1973), where they add significantly to the overall thickness of a given thylakoid (Fig. 19). In thin sections, the densely staining material filling the lumen exhibits a fine-granular texture; only MÖRSCHEL and WEHRMEYER (1979) have found evidence for small supramolecular assemblages of these phycobiliproteins attached to the lumenal surface of these membranes (Fig. 19). Most frequently, the thickened thylakoids are arranged in pairs (Fig. 3), the association of which often appears looser than that between the stacked membranes of green algae, as evidenced in thin sections by the separation of adjacent thylakoids by a gap of 4–8 nm. The recent demonstration of small "adhesive patches" within the larger appressed membrane regions (DWARTE and VESK 1983) may explain how this "loose" association of the thylakoids is structurally maintained.

2.4.6.2 Phaeophyceae (Brown Algae) and Bacillariophyceae (Diatoms)

Thylakoids in these algae tend to be organized into bands of three with extensive regions of membrane adhesion (Fig. 3). A girdle band of thylakoids has also

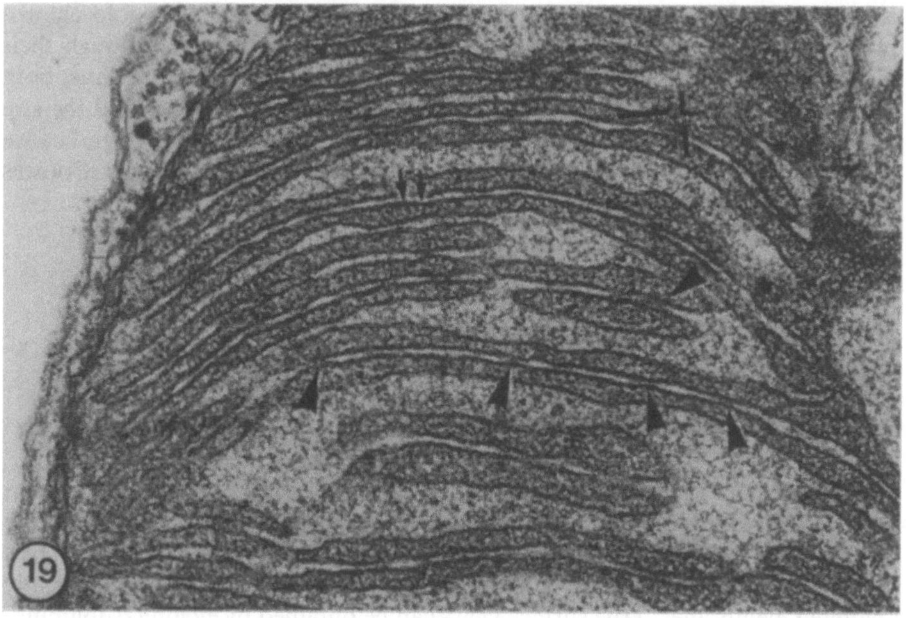

Fig. 19. Thin section through part of a chloroplast of *Hemiselmis rufescens*, a cryptophycean alga. The lumen (*L*) of the inflated thylakoids is filled with light-harvesting phycobiliproteins. Some of these pigment-proteins seem to be organized in the form of short rods (*arrows*) perpendicular to the plane of the membrane. Patches of denser material are evident where adjacent thylakoids adhere to each other (*arrowheads*; see also Fig. 3) ×90,000. (MÖRSCHEL and WEHRMEYER 1979)

been observed in many species, but this band does not seem to completely enclose the internal thylakoids.

2.4.6.3 Chrysophyceae (Golden-Brown Algae) and Xanthophyceae (Yellow-Green Algae)

The general arrangement of the chrysophytan thylakoids resembles the organization found in brown algae and diatoms (Fig. 3). Groups of three thylakoids that adhere to each other over long distances form typical thylakoid bands that are interconnected by single thylakoids that switch from one to another band. A girdle band, usually comprised of three appressed thylakoids, is also present.

2.4.6.4 Dinophyceae (Dinoflagellates)

Although the dinoflagellate thylakoids form distinctive bands, the overall disposition of these bands in the chloroplast stroma tends to be more variable than in the above-listed chromophytes. Furthermore, the average diameter of the tightly appressed membrane regions is reduced, and no girdle thylakoid is present (Fig. 3). The similarity between dinoflagellate and euglenoid chloroplasts in thin-section electron micrographs is quite striking, even though their pigment composition (e.g., chlorophyll c versus b) is very different. Thus, both types of chloroplast are surrounded by one additional membrane, and the size and disposition of the thylakoid bands is similar. The evolutionary significance of these and other similarities has been the subject of extensive debates (COOMBS and GREENWOOD 1976; WHATLEY and WHATLEY 1981; GIBBS 1981).

3 Supramolecular Organization of Thylakoids

3.1 Freeze-Fracture (-Etch) Electron Microscopy

One of the most important methods for studying the supramolecular organization of thylakoid membranes is the freeze-fracture (-etch) technique. The main advantage of this technique is that it can provide high resolution replica images of both surfaces and internal fracture faces of rapidly frozen and chemically unaltered membranes. True surface views can be obtained by etching (subliming) away the ice surrounding isolated, frozen membranes. Internal fracture faces are produced by mechanical splitting of the frozen bilayer membranes (Fig. 22). This type of fracturing occurs because the hydrophobic interactions that stabilize biological membranes are temperature-sensitive, and at -100 °C and below are reduced to such an extent that hydrophobic interfaces represent planes of weakness in frozen samples. As illustrated in Fig. 22, the two complementary membrane fracture faces are designated EF or E-face and PF or P-face. EF refers to the fracture face of the *e*xoplasmic membrane leaflet, PF to the fracture face of the *p*rotoplasmic membrane leaflet. The corresponding membrane *s*urfaces are labeled ES and PS. Since grana and stroma thylakoids give rise to different types of E- and P-faces, subscripts are used to designate whether a given fracture face or surface area corresponds to a *s*tacked (s) or *u*nstacked (u) membrane region (e.g., PFs, EFu etc.).

The particles seen on the fracture faces of thylakoid membranes are morphological equivalents of functional complexes of integral membrane proteins; those seen on the etched surfaces correspond to complexes of peripheral membrane proteins or to surface-protruding elements of integral protein complexes. To obtain a complete structural picture of a given membrane, the particles on complementary faces have to be summed and correlated with the surface features. When this is done in a careful, quantitative manner, freeze-fracture (-etch) electron microscopy can provide unique insights into the spatial organization and size of functional subunits in biological membranes. What follows is a brief summary of what has been learned from correlative freeze-fracture (-etch) and biochemical studies of thylakoid membranes of different types of photosynthetic organism. The discussion progresses from the best-researched higher plant to the least-known algal systems.

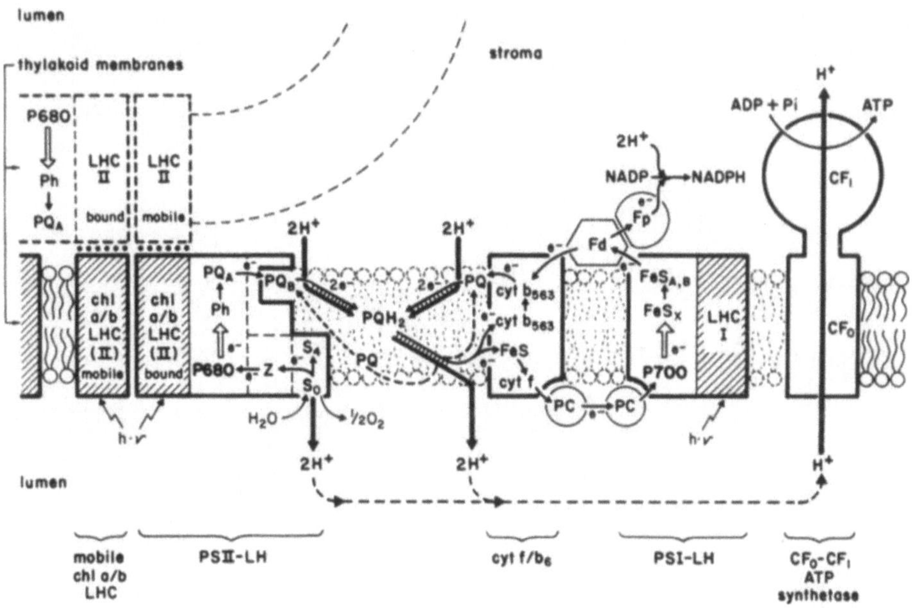

Fig. 20. Organization of chloroplast membrane components participating in the electron coupling reactions of photosynthetic electron transport. *Solid, thin arrows* electron transfer reactions; *open arrows* chemical transitions; *thick arrows* proton movements; *dashed arrows* recycling routes for PQ. Three structurally distinct protein complexes participate in the linear electron transport pathway from water to NADP: a PS II complex, which is linked by a lipid soluble pool of plastoquinone molecules to a cytochrome f/b_6 complex, and a PS I complex, which receives electrons from the cytochrome complex via the water-soluble protein, plastocyanin. Also indicated is a quinone electron transport cycle, or Q cycle, which causes the translocation of one additional proton for every two electrons passing through the linear electron transport chain. Protons deposited within the thylakoid lumen exit through the proton-translocating CF_0–CF_1 ATP synthetase complex. Light-harvesting pigment–protein complexes (LHC) serve both photosystems; these preferentially associate with either PS I or PS II and are designated LHC I or LHC II, respectively. One population of the LHC II is mobile and can serve either PS I or PS II by traveling laterally between the stroma lamellae (enriched in PS I centers) and grana stacks (enriched in PS II). Abbreviations: *chl a/b LHC* chlorophyll a/b light-harvesting complex; *S* and *Z* water-splitting and O_2-evolving enzymes that donate electrons to P-680; P-680, reaction center of PS II; *Ph* bound pheophytin a (primary electron acceptor of PS II); PQ_A and PQ_B special bound plastoquinone molecules (PQ_B is associated with the herbicide-binding protein and can exchange with pool PQ); *PQ* and PQH_2 plastoquinone and reduced plastoquinone; *FeS* Rieske iron sulfur protein; *cyt f* and b_{563} cytochromes f and b_{563} (b_6); *PC* plastocyanin; *P700* reaction center of photosystem I; FeS_x special FeS center that serves as the primary acceptor of photosystem I; $FeS_{A,B}$ two special FeS centers; *Fd* ferredoxin; *Fp* flavoprotein (ferredoxin-NADP reductase); *NADP* nicotinamide adenine dinucleotide phosphate; CF_0 and CF_1 basepiece and headpiece of ATP synthetase; *ADP* and *ATP* adenosine di- and triphosphate; $h \cdot \nu$ light energy. (Staehelin and Arntzen 1983)

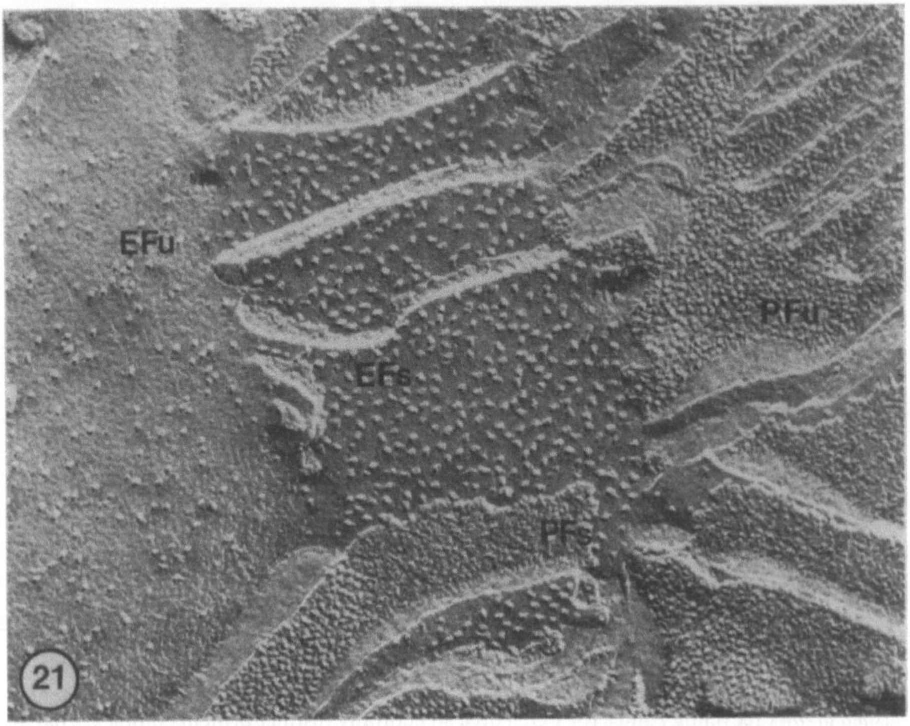

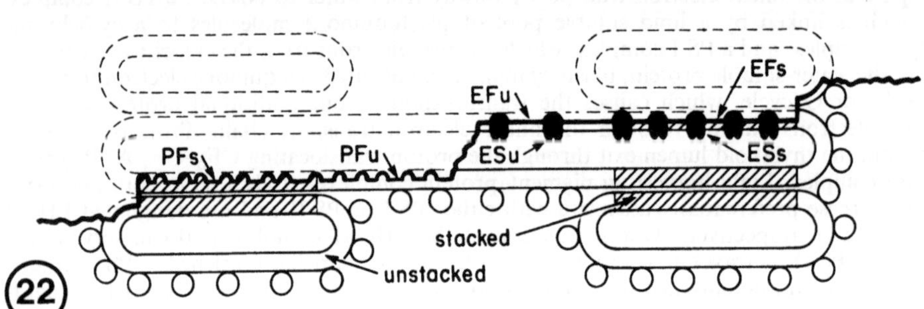

Fig. 21. Freeze-fractured isolated pea thylakoids. The fracture faces *EFs* and *PFs* belong to stacked regions, the faces *EFu* and *PFu* to unstacked ones. Note the physical continuity of the stacked and unstacked membrane regions, and the characteristic set of particles associated with each type of fracture face. See Fig. 22 for an explanation of the fracture faces and Fig. 23 for particle-size histograms. × 75,000

Fig. 22. Diagram illustrating how the faces *EFs*, *EFu*, *PFs*, and *PFu* seen in Fig. 21 arise during freeze-fracturing of thylakoid membranes

3.2 Higher Plant and Green Algal Thylakoids

Thylakoids contain all the functional elements that are needed for trapping and for transducing light energy into chemical energy forms (see SAUER Chap. 2, THORNBER Chap. 3, and ORT Chap. 4 all this Vol.) These include light-harvesting

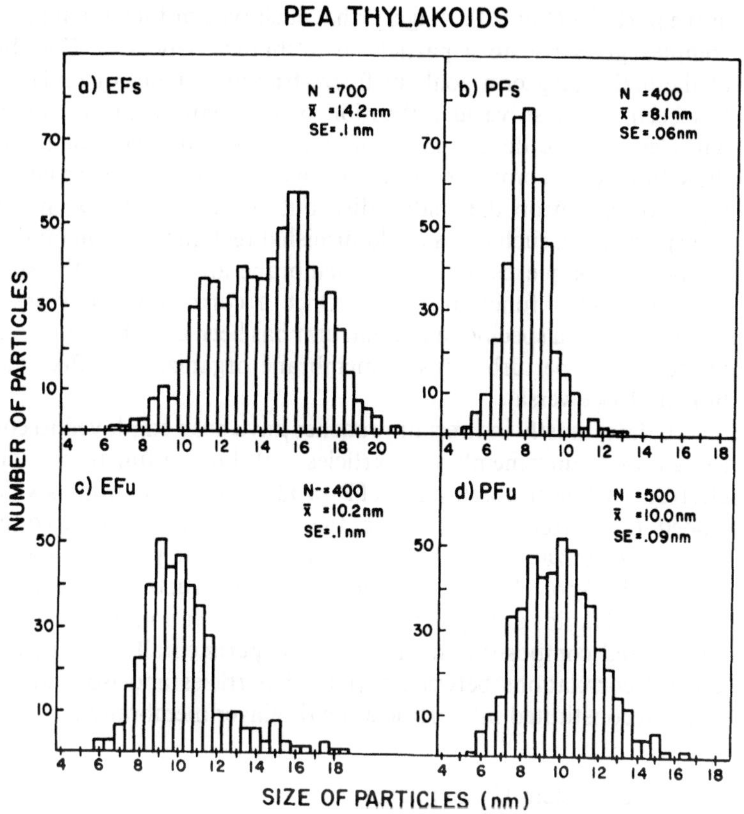

Fig. 23a–d. Histograms of particle sizes found on the four types of membrane fracture face of pea thylakoids

antennae, two types of reaction center, diverse electron carriers, a system for synthesizing ATP, and enzymes for extracting electrons from water. Biochemical and structural studies have revealed that these components are packaged into at least five major types of discrete intramembrane protein complexes, which are illustrated schematically in Fig. 20. Compositional and functional aspects of the different complexes are dealt with in greater detail in the overview chapters of Thornber (Chap. 3) and Ort (Chap. 4, both this Vol.) as well as in the majority of minireviews in this book. The main focus of the remaining sections of this review is on the three-dimensional organization of these complexes in thylakoid membranes, and on the functional implications of their nonrandom organization (Table 2).

A typical micrograph of freeze-fractured pea thylakoid membranes is shown in Fig. 21. Careful examination reveals that each type of fracture face (EFs, EFu, PFs, PFu) exhibits a distinct population of intramembrane particles, consistent with the current view that membrane stacking produces both a structural and functional differentiation of thylakoid membranes. The reproducibility of the images, as judged by the density and the distribution in sizes of the intramembrane particles (Fig. 23), suggests that each type of functional membrane protein complex gives rise to a particle of a characteristic size. For this reason, one of the main long-term goals of freeze-fracture studies of thylakoid membranes has been the positive identification of different categories of intramembrane particles and the development of supramolecular maps of the distribution of these functional complexes in the plane of the membrane. In addition to providing information on the spatial distribution of the components of the electron transport chain, such maps could furnish direct information on the exact number of specific complexes in a given membrane area and on the spacings between these and other functional units. Once a correlation between a specific type of particle and a specific functional unit has been established, it becomes possible to study how perturbations in membrane organization affect membrane function, and vice versa.

To date, mostly indirect approaches have been employed to correlate specific categories of intramembrane particles with known functional complexes of the electron transport chain and ATP production in chloroplasts. These have included (a) correlations between the grana/stroma thylakoid partitioning of clearly recognizable categories of intramembrane particles and/or biochemical activities; (b) correlations between changes in structure and changes in composition of thylakoids during development, (c) correlations between altered structure and altered composition of mutant or experimentally manipulated thylakoids, and (d) correlations between thylakoid particles and isolated functional complexes reconstituted into artificial lipid bilayer membranes.

3.2.1 Photosystem II Complexes

The sole type of intramembrane protein complex that has been unambiguously identified on freeze-fractured thylakoid membranes is the PS II complex. Its identification was made relatively easy by the fact that in higher plants and green algae only PS II complexes give rise to intramembrane particles that fracture with the E-face. The distribution of EF particles between grana (EFs) and stroma (EFu) thylakoid regions of spinach chloroplasts (approx. 85% to 15%, STAEHELIN 1976) closely matches the partitioning of PS II activities in biochemical fractionation experiments (Table 2; ANDERSSON and J.M. ANDERSON 1980; J.M. ANDERSON and MELIS 1983). Although these transmembrane EF particles vary in size from 10 to 18 nm, they can be shown to be related to each other due to the characteristic substructure of their protruding elements on the luminal (ES) membrane surface (Fig. 24; STAEHELIN 1976; SIMPSON 1978 and Chap. 11.6, this Vol.). As seen in Fig. 24, each protruding ES particle consists of two-lobed projections that can sometimes be further resolved into a total of four to six subunits.

Further support for the equivalency of EF particles and PS II units has come from studies of bundle-sheath chloroplasts (K.R. MILLER et al. 1977), a number of PS II mutants (SIMPSON et al. 1977 and Chap. 11.6, this Vol.; SJOLUND and SMITH 1974; K.R. MILLER and CUSHMAN 1979; OLIVE et al. 1979; WOLLMAN et al. 1981; POPOV et al. 1981; LETO et al. 1982; LACAMBRA et al. 1984), and of manganese-deficient chloroplasts (SIMPSON and ROBINSON 1984).

Table 2. Spatial distribution of chloroplast membrane components

Component	Stacked (grana) membranes	Unstacked (stroma) membranes	Reference
PS II	85	15	1, 2
PS I	<15	<85	1
Cyt b_6/f	50	50	3, 4, 5
Chl a/b LHC	More (70–90) (phosphorylation-dependent)	Less (10–30) (phosphorylation-dependent)	1, 6
ATPsynthase	0	100	7

(1) ANDERSSON and J.M. ANDERSON (1980); (2) STAEHELIN et al. (1977); (3) R.P. COX and ANDERSSON (1981); (4) J.M. ANDERSON (1982a); (5) ALLRED and STAEHELIN (1985); (6) KYLE et al. (1983); (7) K.R. MILLER and STAEHELIN (1976)

In all of these systems a decrease of PS II polypeptides and/or activities is matched by a decrease or total loss of EF particles. For instance, under conditions where the *Chlamydomonas* mutant F34SU3 exhibited a 50% decrease in PS II centers, a 50% decrease in EF particles was recorded, thereby suggesting that each particle contained one PS II reaction center (WOLLMAN et al. 1981). The fact that in most if not all of these PS II-deficient mutants a single mutation has led to the near total loss of a large protein complex is consistent with the hypothesis that when a mutation prevents a key polypeptide from being inserted into the thylakoid, other polypeptides of the same functional unit cannot be assembled, or if they assemble, the resulting nonfunctional complex tends to be recognized as such and is rapidly degraded. In this context, the substantial loss of EF particles (75%) and thus PS II reaction centers in thylakoids of spinach plants grown under Mn-deficient conditions (0.02 µM Mn vs. 10 µM Mn for control plants; SIMPSON and ROBINSON 1984) is most intriguing, since Mn is required for the activity of the water-splitting enzyme system (see YOCUM, Chap. 9.2, this Vol.), whereas exogenously added Mn^{2+} appears to be less critical for the general synthesis of thylakoid membrane polypeptides. It appears therefore that Mn is also required for the assembly of a stable PS II complex.

The variability in size of thylakoid EF particles has been shown to arise from PS II core complexes being surrounded by a variable number of bound chl a/b LHC units (ARMOND et al. 1977; KALOSAKAS et al. 1984). Thylakoids of intermittent light pea and bean plants that exhibit PS II activity but contain very little chl a/b LHC only exhibit 8 nm EF particles; upon transfer to continuous growth light conditions, these EF particles increase in size in a stepwise fashion (approx. 10.5 nm, 13 nm, 16 nm) as new chl a/b LhC units are inserted into the membranes and become associated with the 8 nm PS II cores. An increase in size of EF particles from 8 nm to 12 nm has also been observed during the greening of dark-grown and of intermittent light *Euglena* thylakoids (OPHIR and BEN-SHAUL 1974; DUBERTRET and LEFORT-TRAN 1981). The chl a/b LHC units that contribute to the size increases of EF particles are referred to as "bound" chl a/b LHC, in contrast to the "mobile" chl a/b LHC units (KYLE

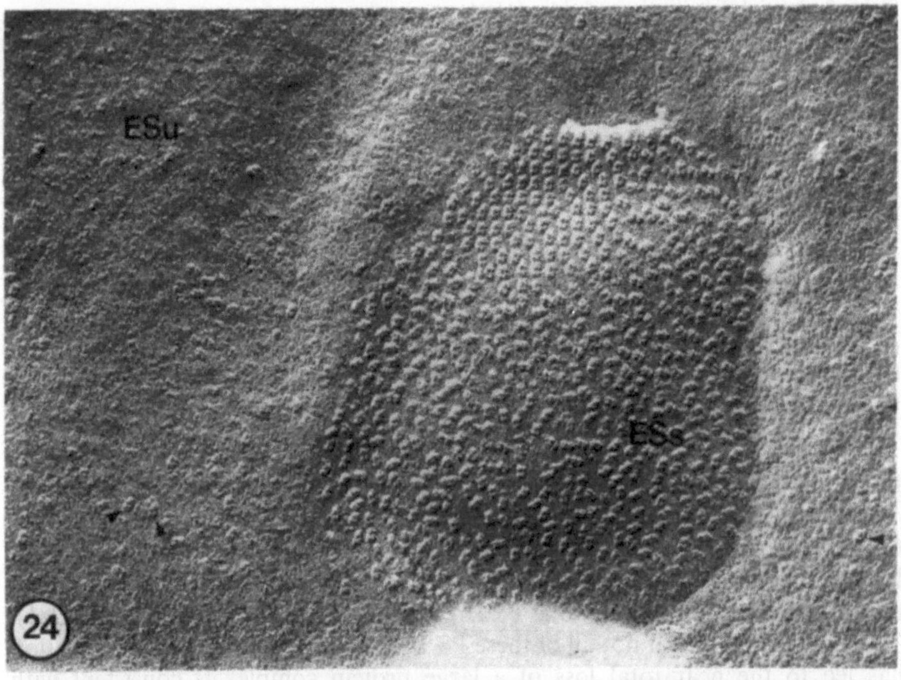
ESu
ESs
24

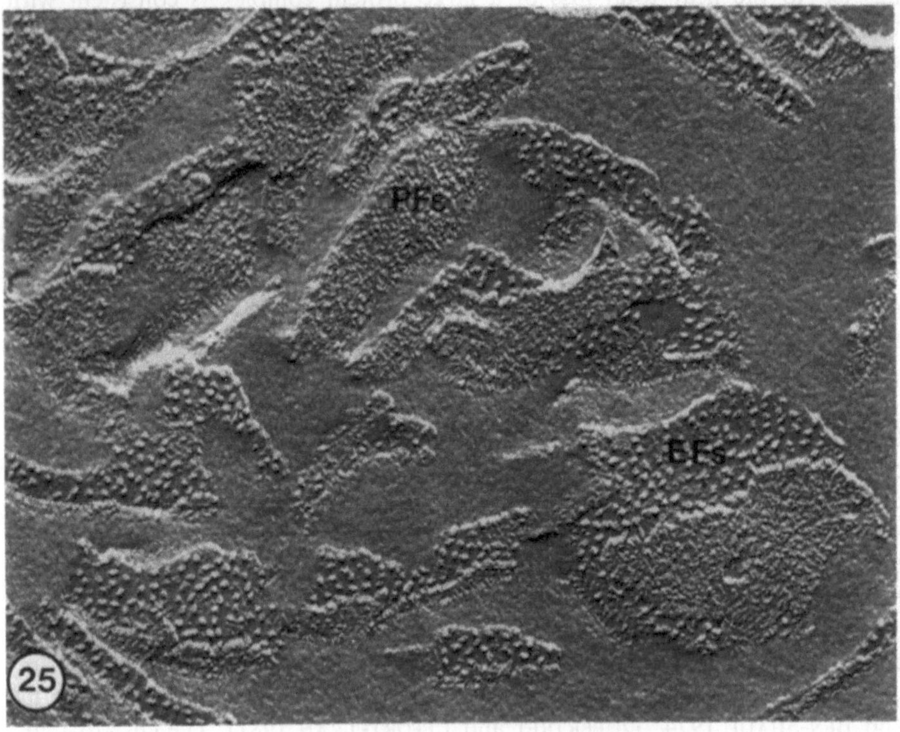
PFs
EFs
25

et al. 1983; STAEHELIN and ARNTZEN 1983) that are discussed in greater detail in Section 3.2.5. Chloroplasts of higher plants and of green algae that either lack or are deficient in chl b tend to have EF particles of reduced size. This relationship has been confirmed in studies of a chl b-less barley mutant (K.R. MILLER et al. 1976), chl b-deficient *Chlamydomonas* (GOODENOUGH and STAEHELIN 1971) and soybean (KECK et al. 1970) mutants, as well as for light-grown *Euglena* cells (K.R. MILLER and STAEHELIN 1973), which contain relatively low amounts of chl a/b LHC (THORNBER 1975). During low temperature acclimation of spinach, the EFs particles become more uniform in size (approx. 14 nm), leading to the formation of typical crystalline arrays of such particles (GARBER and STEPONKUS 1976; STAEHELIN et al. 1976).

EFu particles of thylakoids of higher plants and green algae are both smaller (approx. 10.5 nm average diameter) and more uniform in size than the related EFs particles, whose diameter ranges from 10 to 18 nm. This observation, together with results of unstacking–restacking studies, led to the postulate that stroma PS II units had a smaller antenna size and may differ from grana PS II units in other respects (STAEHELIN et al. 1977). This prediction has been confirmed by several investigators. A much higher light intensity is needed to saturate PS II activities in isolated stroma membranes than in grana membranes (ARMOND and ARNTZEN 1977). Under Mn-deficient growth conditions the loss of EFs particles is proportionally much greater than the loss of EFu particles (SIMPSON, Chap. 11.6, this Vol.). Analysis of the biphasic fluorescence kinetics of PS II by MELIS and HOMAN (1976) has revealed a fast and nonexponential α-component (PS IIα) and a slower β-component (PS IIβ). The two types of PS II reaction center complex differ both in the effective light-harvesting antenna size and in the apparent midpoint redox potential of their primary electron acceptor Q (MELIS and DUYSENS 1979; HORTON and CROZE 1979). Since Mg^{2+} ions affect the organization of PS IIα but not PS IIβ in the membrane, it was postulated that PS IIα could be localized in stacked grana membranes and PS IIβ in stroma thylakoids (MELIS and HOMAN 1978). Direct support for this hypothesis has come from the analysis of fractionated spinach thylakoids, which revealed

Fig. 24. Luminal surface of a spinach thylakoid exposed by deep-etching. The stacked membrane region (*ESs*) in the center of the micrograph can be clearly distinguished from the surrounding unstacked areas (*ESu*) due to the high density of large, protruding particles, each of which is divided into two lobes made up of subunits. The ESu areas contain relatively few such particles (*arrowheads*). Each large particle corresponds to a PS II complex, the protruding elements of which are structural equivalents of water-splitting enzymes. ×115,000

Fig. 25. Freeze-fracture electron micrograph of a typical oxygen-evolving PS II preparation. This sample was prepared according to the method of KUWABARA and MURATA (1982), but samples obtained using different procedures appear virtually identical. All oxygen-evolving PS II preparations of higher plants published to date consist of highly enriched, appressed grana membrane fragments. The *EFs* and *PFs* fracture faces resemble those of control membranes (Fig. 21). ×80,000

appressed grana membranes to be highly enriched in PS IIα and nonappressed stroma membranes in PS IIβ (J.M. ANDERSON and MELIS 1983). These findings are consistent with the idea that the larger EFs particles that contain more bound chl a/b LHC are the structural equivalents of PS IIα units, and the smaller EFu particles of PS IIβ units.

Attempts at isolating fully functional PS II complexes from higher plants and green algae in a form equivalent to EF particles have met with mixed success. Published isolation procedures fall into two categories: those that yield samples capable of evolving oxygen (see THORNBER, Chap. 3, this Vol.; YOCUM, Chap. 9.2, this Vol.; ANDERSSON and AKERLUND 1978; BERTHOLD et al. 1981; YAMAMOTO et al. 1982; KUWABARA and MURATA 1982), and those that have lost oxygen-evolving capability, PS II core-type complexes (MULLET and ARNTZEN 1981; CAMM and GREEN 1983; SATOH et al. 1983; WESTHOFF et al. 1983). Structural studies of the oxygen-evolving PS II preparations have revealed that all are composed of highly purified, paired, and appressed grana membrane fragments (Fig. 25; ANDERSSON and AKERLUND 1978; DUNAHAY et al. 1984). Thus, what have been referred to as oxygen-evolving PS II "particles" by some biochemists are not structures equivalent to EF particles of thylakoid membranes; they are membrane fragments highly enriched in EFs and PFs particles. This finding explains why such samples contain disproportionally large amounts of chl a/b LHC polypeptides as well as some other non-PS II polypeptides (DUNAHAY et al. 1984). These oxygen-evolving PS II membrane fragments have been used more recently as starting material for isolating the actual PS II complexes that correspond to EF particles. However, as indicated above, all published procedures for further purifying PS II complexes from PS II-enriched membrane fragments of higher plants lead to preparations of PS II cores containing the PS II reaction center, but incapable of evolving oxygen. Such PS II complexes reveal four or five polypeptides on denaturing SDS gels (SATOH et al. 1983; WESTHOFF et al. 1983).

Recently, SPRAGUE et al. (1985) have reconstituted purified PS II core complexes isolated according to the procedure of CAMM and GREEN (1983) into phospholipid and digalactolipid liposomes. One of their goals was to determine the size of the reconstituted PS II complexes in freeze-fracture replicas. In uncharged digalactolipid liposomes the PS II cores gave rise to 7.5 nm particles, similar in size to the EF particles (PS II core complexes) of intermittent light plastids which lack chl a/b LHC (ARMOND et al. 1977). In contrast, virtually no PS II complexes became incorporated into the phospholipid liposomes; instead they appeared to stick to the surface of the bilayers leading to highly convoluted networks of tubular membranes nearly devoid of intramembrane structures. It was concluded that proper insertion of PS II complexes into bilayer membranes requires a low phospholipid environment, as found in chloroplast membranes. More detailed discussions of the physicochemical properties of thylakoid lipids, their distribution between grana and stroma membrane regions, the lipid requirements of different protein complexes, and the results of reconstitution experiments can be found in the contributions of MURPHY (Chap. 11.11), SIEGENTHALER and RAWYLER (Chap. 11.9) and RYRIE (Chap. 11.7), all this Volume.

3.2.2 Photosystem I Complexes

As discussed in the preceding section, only PS II complexes can give rise to EF particles in higher plant and green algal thylakoids. It follows therefore that PS I complexes must give rise to a subfraction of PF particles, the other main subfractions being cytochrome b_6/f particles, CF_0–CF_1 particles and chl a/b LHC particles. Correlative evidence linking certain classes of PF particles with PS I complexes has been obtained from studies with PS I mutants and from studies of reconstituted PS I complexes. A consistent feature of freeze-fractured thylakoid membranes from a variety of PS I mutants is the loss of PFu particles in the 10–13 nm size range (Fig. 23). This has been shown to be the case for the maize mutant 1481 (MILLER 1980), the barley mutants viridis n^{34} and viridis zb^{63}, both of which are deficient in the P-700 reaction center protein (see SIMPSON, Chap. 11.6, this Vol.), and the *Chlamydomonas* mutants F14 and AC40, which are deficient in the PS I reaction center/core antenna (CPI) and the peripheral antenna (CPO) complex, respectively (OLIVE et al. 1983).

As discussed in greater detail in Chapter 9.7, this Volume, by WOLLMAN, a number of methods have been published for isolating and purifying PS I complexes. Of these, the PS I complexes prepared from pea thylakoids by the method of MULLET et al. (1980) appear to be among the most intact, since they retain the light-harvesting antenna chlorophylls, exhibit a chlorophyll to P-700 ratio of about 110, and contain at least 13 polypeptides. When such PS I-110 complexes are reconstituted into artificial phospholipid liposomes, particles with an average diameter of 10.6 nm are seen in freeze-fracture replicas (MULLET et al. 1980). Although this size falls into the size range of potential PS I particles on freeze-fractured thylakoids as defined by the mutant studies (10–13 nm), it is obviously close to the lower end of that range, suggesting that the in vivo PS I complexes may be slightly larger than the isolated ones. In this context it is interesting to note that a recently reported PS I complex isolated from octyl glucoside/SDS treated thylakoids not only contained four more polypeptides (in the 20 to 25 kDa range) than PS I-110 samples, but, when reconstituted into liposomes, also gave rise to freeze-fracture particles

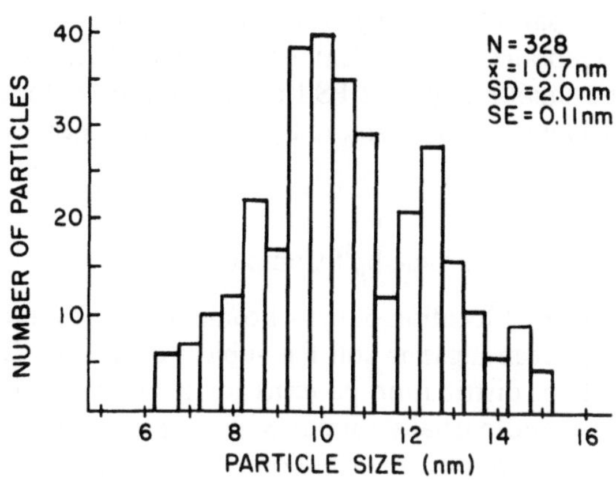

Fig. 26. Histogram of particle sizes of freeze-fractured, isolated PS I complexes reconstituted into liposomes. (DUNAHAY and STAEHELIN 1985)

falling into two size classes, 10 to 11 nm and 12 to 13 nm (Fig. 26; Dunahay and Staehelin 1985). The functional significance of these two size classes is still unclear at the present time, however, one has to entertain the possibility that PS I complexes of variable size (with variable amounts of bound antenna complexes) may exist in thylakoid membranes.

Studies of intermittent light plastids of pea have shown that smaller than normal but functionally active PS I core complexes can be assembled and maintained in thylakoid membranes under appropriate conditions (Armond et al. 1977). Thylakoids of intermittent light plastids exhibit normal PS I activity but lack chl b-containing light-harvesting antenna complexes (Mullet et al. 1980). Since the more peripheral light-harvesting complex of PS I known as LHC I, also contains some chl b (see Thornber, Chap. 3; Wollman, Chap. 9.7, both this Vol.), the PS I complexes of intermittent light plastids are also reduced in size. Thylakoids of such plastids possess mostly 6 to 9 nm PF particles and virtually no PF particles with diameters larger than 10 nm. When synthesis of chl b-containing proteins is induced by transfer to continuous light, a new class of 10 to 13 nm PFu particles is formed (Armond et al. 1977). These results suggest that the 10 to 13 nm PFu particles contain a 7 to 8 nm PS I core surrounded by chl b-containing antenna complexes. The number of antenna complexes bound to a PS I core unit could be variable, thereby giving rise to PS I particles of different size, as noted for PS II complexes.

Functional PS I complexes with a diameter of less than 10 nm have also been reported for the AC40 mutant of *Chlamydomonas*, which has been shown to lack the CPO antenna system of PS I (Olive et al. 1983). The CPO antenna system of *Chlamydomonas* (Wollman and Bennoun 1982; Ish-Shalom and Ohad 1983) appears to be equivalent to the LHC I system of higher plants (see Thornber, Chap. 3, this Vol.), which has also been termed CPI^{LH} on green gels (Dunahay and Staehelin 1985). Based on these findings, one may ask whether PS I core complexes deficient or completely lacking in peripheral antenna complexes exist in normal thylakoids. The potential size range of such complexes, 7 to 10 nm in diameter, makes identification of such particles on freeze-fractured thylakoid membranes extremely difficult, since the cytochrome b_6/f complex, the CF_0–CF_1 complex and the chl a/b LHC all produce PF particles in the same size range (see later sections of this chapter). Smaller PS I particles lacking peripheral antennae seem to be able to penetrate stacked membrane regions (Staehelin unpublished results) and may be responsible for the small amount of PS I activity (approx. 15%) associated with preparations of highly purified stacked grana membranes (Andersson and J.M. Anderson 1980).

3.2.3 Cytochrome b_6/f Complex

The exact location of cytochrome b_6/f in thylakoid membranes of higher plants and green algae is still the subject of debate. During the past few years three distinct distribution patterns for cytochrome b_6/f complexes have been postulated and partially supported by experimental evidence. The first theory postulates a uniform distribution between grana and stroma regions, the second,

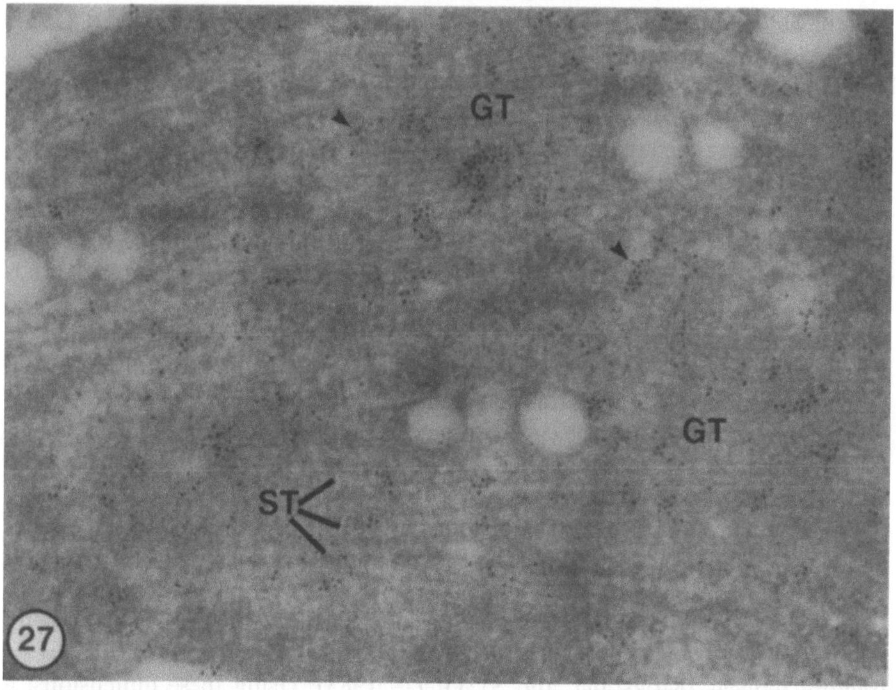

Fig. 27. Localization of the cytochrome b_6/f complex in thylakoid membranes of an intact spinach chloroplast by means of anti-cytochrome b_6/f antibodies coupled to ferritin. The labeling pattern indicates that cytochrome b_6/f complexes are located both in grana (*GT*) and stroma (*ST*) thylakoids. Some label is also seen at the interface between grana and stroma membranes (*arrowheads*). Quantitative analysis indicates close to a 1:1 ratio of labeling between the two regions. ×90,000. (ALLRED and STAEHELIN 1985)

exclusion from stacked grana thylakoids, and the third, a clustering of the complexes around the margins of the grana. Support for the first hypothesis has come from two membrane fractionation studies (R.P. COX and ANDERSSON 1981; J.M. ANDERSON 1982a) in which inside-out (grana) and right-side-out (stroma) membrane fractions were obtained by the phase-partitioning technique. The distribution of cytochromes b_6 and f between these two fractions was consistent with a uniform distribution between grana and stroma thylakoids. In contrast, HENRY and MOLLER (1981) using the same phase-partitioning method found nearly all cytochrome b_6/f in their stroma membrane fraction. An equivalent result was reported by WHITFORD et al. (1983), whose highly enriched oxygen-evolving PS II preparations (stacked grana membranes), contained very low amounts of the b_6 and f cytochromes. Based on an analysis of the kinetics of the electron transport chains of mesophyll and bundle-sheath chloroplasts of maize, GHIRARDI and MELIS (1983) postulated that the cytochrome b_6/f complexes are concentrated in the stroma membrane regions immediately adjacent to the stacked grana membranes. BARBER (1984) proposed a similar distribution, citing the differences in surface properties of appressed and nonappressed membrane regions as well as possible steric hindrance problems, as reasons for doubting the uniform distribution hypothesis of R.P. COX and ANDERSSON (1981) and J.M. ANDERSON (1982a). Neither GHIRARDI and MELIS (19837 nor BARBER (1984) have addressed the question as to how cytochrome b_6/f complexes could be held at the interface between grana and stroma thylakoids.

To overcome the limitations of the above-listed studies, ALLRED and STAEHELIN (1985) have recently examined the distribution of cytochrome b_6/f complexes in glutaraldehyde-fixed intact chloroplasts, using immunocytochemical labeling of thin sections. Their antibody labeling patterns show cytochrome b_6/f complexes to be present both in stacked grana and unstacked stroma thylakoids (Fig. 27), with a slightly greater density of label in the grana regions. Based on a 3:2 ratio of grana to stroma membrane areas, it is possible to calculate that approximately 2/3 of the cytochrome b_6/f complexes are present in stacked grana membranes. Although this last study has provided unambiguous information on the distribution of the cytochrome complexes, it does not indicate whether all complexes are equally active (see WHITMARSH, Chap. 10.2, this Vol. for a more detailed discussion).

The cytochrome b_6/f complex of spinach consists of four major polypeptides and one or more small hydrophobic polypeptides below 6 kDa (see HAUSKA, Chap. 10.1, this Vol.). Following reconstitution of this complex into phospholipid and galactolipid liposomes, intramembrane particles with an average diameter of 8.3 nm and an overall height of between 11 and 12 nm are seen in freeze-fracture replicas (MÖRSCHEL and STAEHELIN 1983). Using these dimensions, a molecular weight of approximately 280,000 was calculated for the reconstituted cytochrome b_6/f complex, consistent with a dimeric configuration. The 8.3 nm size of the cytochrome b_6/f complex is well within the range of both PFs and PFu particles of thylakoid membranes (Fig. 23).

3.2.4 CF_0–CF_1 ATP Synthase

The coupling factor complex CF_1 was the first protein complex of thylakoid membranes to be positively identified and localized. This peripheral segment of the larger CF_0–CF_1 complex was initially shown to be accessible to antibodies in isolated thylakoid preparations (MCCARTY and RACKER 1966). Subsequent electron microscopic studies correlated the CF_1 complex with knoblike structures that protruded from the stromal surface of the thylakoid membranes. The diameter of the protruding CF_1 knob is about 10 nm in negatively stained membrane samples (HOWELL and MOUDRIANAKIS 1967) and about 14 nm in freeze-etch preparations (K.R. MILLER and STAEHELIN 1976). A repeat of about 13 nm has been measured in negatively stained, monolayer arrays of purified CF_1 particles with each individual complex exhibiting a sixfold rotational symmetry. This hexagonal symmetry has been interpreted as being consistent with an $\alpha_3\beta_3\gamma\delta\varepsilon$ stoichiometry of subunits (AKEY et al. 1983; MCCARTY and NALIN, chap. 10.8, this Vol.).

The CF_1 units of intact, stacked thylakoids are located exclusively in nonappressed stroma membrane regions, including the grana margins (K.R. MILLER and STAEHELIN 1976; WELLBURN 1977). Spinach thylakoids possess about 700 CF_1 particles per square micron of stroma membrane. Upon in vitro unstack-

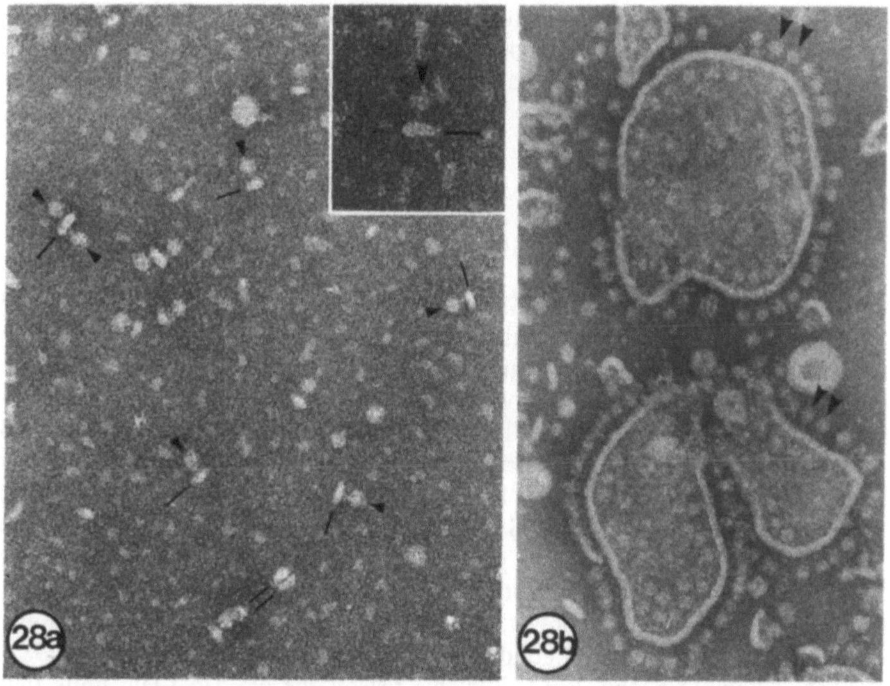

Fig. 28. Negatively stained isolated (**a**) and reconstituted (**b**) CF_0/CF_1 ATP synthase complexes. CF_1 "head" units marked with *arrowheads*, CF_0 basepieces marked with *lines*. Note the tendency of the hydrophobic basepieces to adhere to each other. ×200,000; insert ×300,000. (Micrographs courtesy of E. Mörschel, Marburg)

ing of thylakoids in low salt buffers, the CF_1 particles become randomly distributed over all membrane surfaces. Addition of >2 mM $MgCl_2$ leads to restacking of the membranes and to a complete resegregation of the CF_1 particles into the reformed stroma membrane regions (K.R. Miller and Staehelin 1976).

Very little is known about the hydrophobic CF_0 portion of the CF_0–CF_1 complex, although several proteins associated with the CF_0 basepiece have been identified (Pick and Racker 1979; Nelson et al. 1980, see also Chap. 10.8, this Vol. by McCarty and Nalin for a more detailed discussion of compositional aspects of the CF_0–CF_1 complex). A high resolution image of negatively stained, isolated CF_0–CF_1 complexes is shown in Fig. 28. The CF_1 headpiece, the neck region, and the CF_0 baseplate can be clearly discerned. In such preparations the CF_0 baseplates often seem to retain a membrane fragment-like configuration and to adhere to each other (Mullet et al. 1981 b). CF_0–CF_1 complexes have been reconstituted into phospholipid (Mullet et al. 1981 c) and digalactolipid (Mörschel and Staehelin 1983) liposomes, and shown to produce particles with a diameter of 9.5 nm in both types of samples. Since a large proportion of PFu particles of thylakoid membranes fall into this size range (Fig. 23) it has not been possible to positively identify the in vivo CF_0 complexes to date. One intriguing feature of the reconstituted complexes is that the apparent size of the CF_0 basepieces is considerably larger than would be expected from X-ray scattering data (PICK and RACKER 1979). This finding has led to the suggestion that the CF_0 complex may contain significant amounts of bound lipid (MÖRSCHEL and STAEHELIN 1983).

3.2.5 Chlorophyll a/b Light-Harvesting Complex

3.2.5.1 "Mobile" Versus "Bound" Forms of Chl a/b LHC

As already discussed in the section on PS II complexes, the size of EFs and EFu particles of thylakoid membranes appears to be related to the amount of chl a/b LHC that is tightly bound to the 8 nm PS II core particles (ARMOND et al. 1977; DUBERTRET and LEFORT-TRAN 1981). More recent studies have revealed an additional pool of chl a/b LHC units that are *not* tightly bound to the PS II complexes that give rise to EF particles (SIMPSON 1979; STAEHELIN and ARNTZEN 1979; OLIVE et al. 1981), even though they can transfer trapped light energy very efficiently to PS II reaction centers (STAEHELIN and ARNTZEN 1983). These separate units give rise to approximately 8 nm PF particles, the majority of which are located in stacked membrane regions, but whose distribution between grana and stroma thylakoids can change in a reversible manner in response to different light conditions (see Section 4,2 and KYLE et al. 1983; BARBER, Chap. 11.5, this Vol.). For this reason the chl a/b LHC units that produce 8 nm PF particles have been termed "mobile" chl a/b LHC to distinguish them from the PS II-"bound" chl a/b LHC units that are responsible for the size changes of EF particles mentioned above (STAEHELIN and ARNTZEN 1983). Freeze-fracture electron micrographs indicate that for each EFs particle (PS II core plus bound chl a/b LhC) there are about three mobile chl a/b LHC units (8 nm PF particles).

Because the idea of "mobile" and "bound" forms of chl a/b LHC originated only very recently (STAEHELIN and ARNTZEN 1983) and has not yet been uniformly accepted, most of the extensive literature on the chl a/b LHC (see THORNBER, Chap. 3, this Vol.) does not distinguish between the two forms. However, since the mobile chl a/b LHC appears to be present in thylakoid membranes in much larger quantities than the bound chl a/b LHC, and also forms more stable complexes, most literature reports on the chemical and physical properties of chl a/b LHC probably relate to the mobile form. For instance, the chlorophyll proteins associated with the mobile chl a/b LHC appear to give rise to the well-known CPII (CP27, $LHCP^3$, Chl a/b-P2 etc.) band and its oligomeric form CPII* ($LHCP^1$, Chl a/b-Ps**, CP64 etc.) on green gels of detergent-solubilized thylakoids (see THORNBER, Chap. 3; RYRIE, Chap. 11.7, both this Vol.). In contrast, the green gel-band-equivalent(s) of the bound chl a/b LHC has not yet been established. It is likely that CP29 (Chl a/b-P1), a relatively minor chlorophyll-protein complex, which runs just behind CP II (CP27) on mildly denaturing gels (see Fig. 3b–VI in THORNBER, Chap. 3, this Vol.; CAMM and GREEN 1980, 1983; MACHOLD and MEISTER 1979), is a component of the bound PS II light-harvesting complex. However, some CP II or CP II* complexes may also be components of large EF particles.

3.2.5.2 *Chl a/b LHC as Membrane Adhesion Factor*

Several lines of evidence have revealed that the mobile chl a/b LHC serves as the principle adhesion factor of thylakoid membranes, which explains its preferential association with stacked grana thylakoids (ANDERSSON and J.M. ANDERSON 1980). Plants grown under intermittent illumination develop chloroplasts with functional PS I and PS II units, but essentially devoid of both chl a/b LHC and grana stacks. When such plants are transferred to continuous light conditions, chl a/b LHC polypeptides (25–29 kDa) are synthesized and inserted into the thylakoid membranes, and grana formation is observed concomitantly (ARGYROUDI-AKOYUNOGLOU et al. 1971; ARMOND et al. 1977). Chloroplast mutants that lack one or several polypeptides of the chl a/b LHC have all been found to exhibit modified stacking behavior and to have lower densities of PFs particles (K.R. MILLER et al. 1976; ARNTZEN 1978; BURKE et al. 1979; RYRIE 1983; MICHEL et al. 1983; SIMPSON, Chap. 11.6, this Vol.). For instance in a chl b-less barley mutant in which one of the major chl a/b LHC polypeptides is absent, maintenance of grana stacks during thylakoid isolation requires larger amounts of divalent cations in the isolation buffer than in the case of wild-type thylakoids (BURKE et al. 1979; SIMPSON 1979; RYRIE 1983). On the other hand, stacked membrane regions essentially devoid of EFs particles but containing large numbers of 8 nm PFs particles have been observed in mutants that contain chl a/b LHC, but which demonstrate greatly reduced PS II activity (K.R. MILLER and CUSHMAN 1979; SIMPSON, Chap. 11.6, this Vol.). Similarly, shade-tolerant plant species of tropical rain forests contain dramatically increased amounts of chl a/b LHC and a much higher PFs to EFs particle ratio than sun-tolerant plant species such as spinach (GOODCHILD et al. 1972). Antibodies prepared against purified chl a/b LHC and applied to inside-out and right-side-out thylakoid membrane fractions indicate that the chl a/b LHC is exposed on both membrane surfaces, i.e., it forms a transmembrane complex (ANDERSSON et al. 1982).

3.2.5.3 *Reconstitution and Structural Characterization*

In recent years, several laboratories have reconstituted purified chl a/b LHC into liposomes (MC DONNEL and STAEHELIN 1980; MULLET and ARNTZEN 1980; RYRIE et al. 1980; SPRAGUE et al. 1985; RYRIE, Chap. 11.7, this Vol.). The reconstituted complexes give rise to freeze-fracture particles with diameters of approximately 8 nm, and when physiological concentrations of mono- and divalent cations are added to the chl a/b LHC liposomes, their behavior mimics cation-induced stacking of normal thylakoids (Fig. 29; see RYRIE, Chap. 11.7 and BARBER, Chap. 11.5, both this Vol., for a more detailed treatment of this topic). Cation-induced stacking of both thylakoids and chl a/b LHC liposomes can be largely abolished by limited trypsinization, which removes a small, highly basic fragment from the N-terminal end of the major chl a/b LHC polypeptide (JENNINGS et al. 1978; STEINBACK et al. 1979; CARTER and STAEHELIN 1980; MULLET et al. 1981b; RYRIE and FUAD 1982). Reconstitution of purified chl a/b LHC into intermittent light thylakoids devoid of this light-harvesting complex, restores the ability of such membranes to form stacks (DAY et al. 1984).

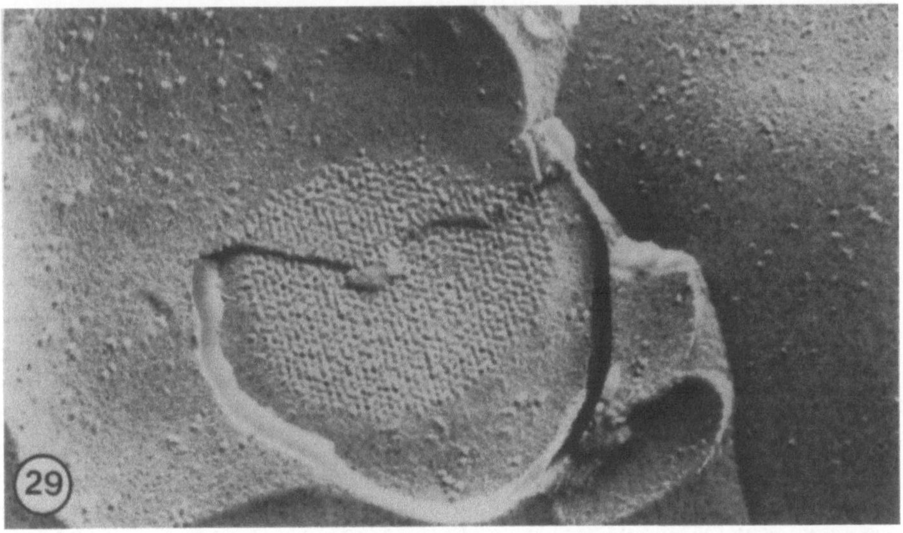

Fig. 29. Freeze-fractured liposomes reconstituted with mobile chl a/b LHC in the presence of 100 mM NaCl. Note the aggregation of the 8-nm particles into a hexagonal array in the area of membrane contact, consistent with the chl a/b LHC particles serving an adhesive function. Each particle corresponds to one trimeric chl a/b LHC unit, as shown in Fig. 30. ×100,000. (MCDONNEL and STAEHELIN 1980)

Most chl a/b LHC isolation methods make use of the cation-dependent adhesion properties of this complex to precipitate it out of solution and thereby separate it from contaminants. During this precipitation the chl a/b LHC units aggregate not only vertically but also laterally into semi-crystalline sheets that can be separated from each other by removal of the cations (MCDONNEL and STAEHELIN 1980). The tendency of chl a/b LHC particles to form two-dimensional crystalline arrays in vitro (Fig. 29) has recently been exploited by means of Fourier transform image analysis to learn more about the three-dimensional structure of the complex (LI and HOLLINGSHEAD 1982; KÜHLBRANDT et al. 1983; KÜHLBRANDT 1984; LI 1985). These crystalline sheets have a hexagonal lattice with a 12.5 nm unit cell dimension and belong to the P-321 space group. They are composed of interdigitating units which alternate in their orientation relative to the plane of the sheet. Three-dimensional maps of the arrays show that the chl a/b LHC is a highly asymmetric transmembrane protein composed of three structurally equivalent monomers (Fig. 30). The large surface area exposed on one side of the trimeric complex suggests its possible involvement in membrane–membrane interactions. Similar crystalline arrays of 8 nm particles have been observed in phospholipid-enriched thylakoid membranes (SIEGEL et al. 1981). This crystallization appears to be related to the charged nature of the zwitterionic phospholipid molecules used for the enrichment experiments (SPRAGUE et al. 1985); phospholipids are only a minor component of normal thylakoids (see ORT, Chap. 4; MURPHY, Chap. 11.11, both this Vol.).

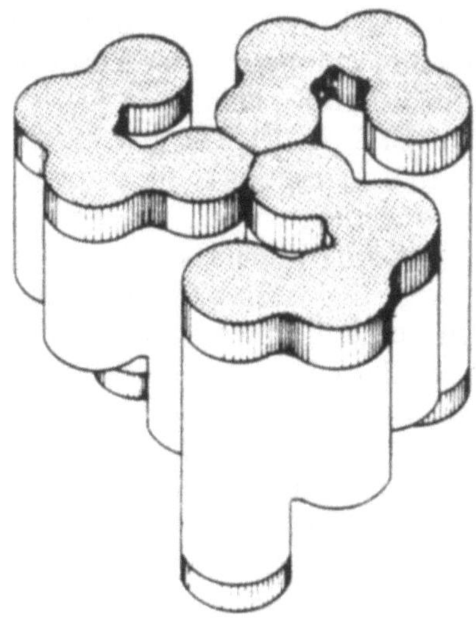

Fig. 30. Schematic drawing of one (mobile?) chl a/b LHC unit based on a three-dimensional map generated from Fourier transform analysis of crystalline arrays of purified pea complexes (see Fig. 29). The *shaded regions* show the part of the complex that would protrude from a hypothetical 4.5 nm lipid bilayer. The *top flat region* probably constitutes the surface that promotes membrane adhesion. (KÜHLBRANDT 1984)

3.2.6 Why Do Chloroplasts Need Grana?

The spatial differentiation of chloroplast membranes into stacked grana and unstacked stroma regions has intrigued photosynthesis researchers for several decades. The formation of grana stacks clearly relates to the functional specialization of the membranes. Chloroplast membranes serve one purpose: to trap radiant energy and produce stable high-energy chemical intermediates. It is clear that grana are not required to carry out this energy conversion, since partially developed membranes having no chl a/b LHC or grana are perfectly competent in terms of electron transport and photophosphorylation. Analyses of membrane development and the appearance of grana lead to the conclusion that the stacked membranes are correlated with increased efficiency of light trapping and utilization, particularly under limiting light conditions (ARMOND et al. 1976). This phenomenon has been explained by STAEHELIN and ARNTZEN (1983) as follows: Due to the fact that the chlorophylls associated with the chl a/b LHC absorb at shorter wavelengths than the reaction center chlorophylls of PS II and PS I, the chl a/b LHC is able to transfer excitation energy very efficiently to both types of reaction centers. This transfer is by means of resonance energy coupling of the dipole oscillations of the chromophores (SAUER, Chap. 2; KNOX, Chap. 7.1; HOLZWARTH, Chap. 7.3, all this Vol.). However, because resonance energy transfer reactions are statistical events that depend on a number of parameters, such as absorption band overlap and distance between and orientation among the chlorophyll molecules, several other types of excitation energy transfers also occur in chloroplast membranes such as chl a/b LHC → chl a/b LHC and PS II → PS I transfers (BUTLER 1977). The question that arises in this context is, "What prevents all the energy trapped by the different antennae systems from ending up in PS I?" This is a real problem, because the reaction center of PS I has the most red-shifted chlorophyll absorbing species and is also more efficient in light trapping than PS II, as evidenced by the lower room temperature fluorescence (W.P. WILLIAMS 1977). Evolution appears to have provided a "simple" structural solution to this problem in green plants (and many algae) by segregating most of PS II and chl a/b LHC into grana and most of PS I into stroma membrane regions. The spatial segregation of the two types of reaction center prevents excessive transfer of excitation energy from PS II to PS I. Nevertheless, this lateral segregation of reaction centers, each type competing for the available absorbed radiant energy, imposes certain limitations upon function. If the pigment bed was simply divided into grana versus stroma lamellae, the light-absorption properties of the chloroplasts would be constant. Because environmental light conditions are not constant, and because cellular demands for PS I-mediated noncyclic and cyclic electron flow frequently change, such an invariant pigment bed would be a disadvantage. By developing a mobile antenna system (the mobile chl a/b LHC) that can be variably partitioned between PS II-rich grana and PS I-rich stroma membranes, the major problem of regulating energy distribution between the two photosystems has been overcome and the efficiency of photosynthesis has been maximized.

3.2.7 Supramolecular Organization of Thylakoids – a Summary

In this section an attempt is made to summarize both in qualitative and quantitative terms current understanding of the distribution of functional units between stacked and unstacked regions of thylakoid membranes as determined by freeze-fracture (-etch) electron microscopical and by biochemical studies. The information presented in Fig. 31 and discussed below is based on data presented in this review, as well as on results reviewed by STAEHELIN et al. (1977), ARNTZEN (1978), J.M. ANDERSON (1982a, c), SIMPSON (1983) and ORT (Chap. 4, this Vol.).

All the major integral membrane protein complexes, PS I, PS II, cytochrome b_6/f, ATP synthase, and mobile chl a/b LHC, are transmembranous in nature, and each is partitioned between grana and stroma membrane regions in a unique manner (Table 2). Within limits the composition of thylakoid membranes can vary in response to changes in environmental parameters (see Sect. 4). These changes have been shown to involve alterations in the ratios of the different complexes, as well as alterations in composition and/or lateral distribution of chlorophyll-containing functional membrane units. It should also be emphasized that all membrane components appear to be in a dynamic equilibrium both in terms of composition and distribution, since individual complexes are held together by noncovalent bonds, and turnover of proteins and lipids is substantial, especially when a plant is exposed to light.

PS II. PS II complexes give rise to EFs and EFu particles. EFs particles correspond to PS IIα and EFu particles to PS IIβ units. Each functional PS II unit consists of a fully competent core complex (reaction center, water splitting apparatus, CP43 light-harvesting antennae, etc.) surrounded by variable amounts of *bound* chl a/b LHC. The differences in EF particle size can be attributed to the core complex (8 nm in diameter) being surrounded by different numbers of bound chl a/b LHC units. The EFs particles (PS IIα units) tend to be larger

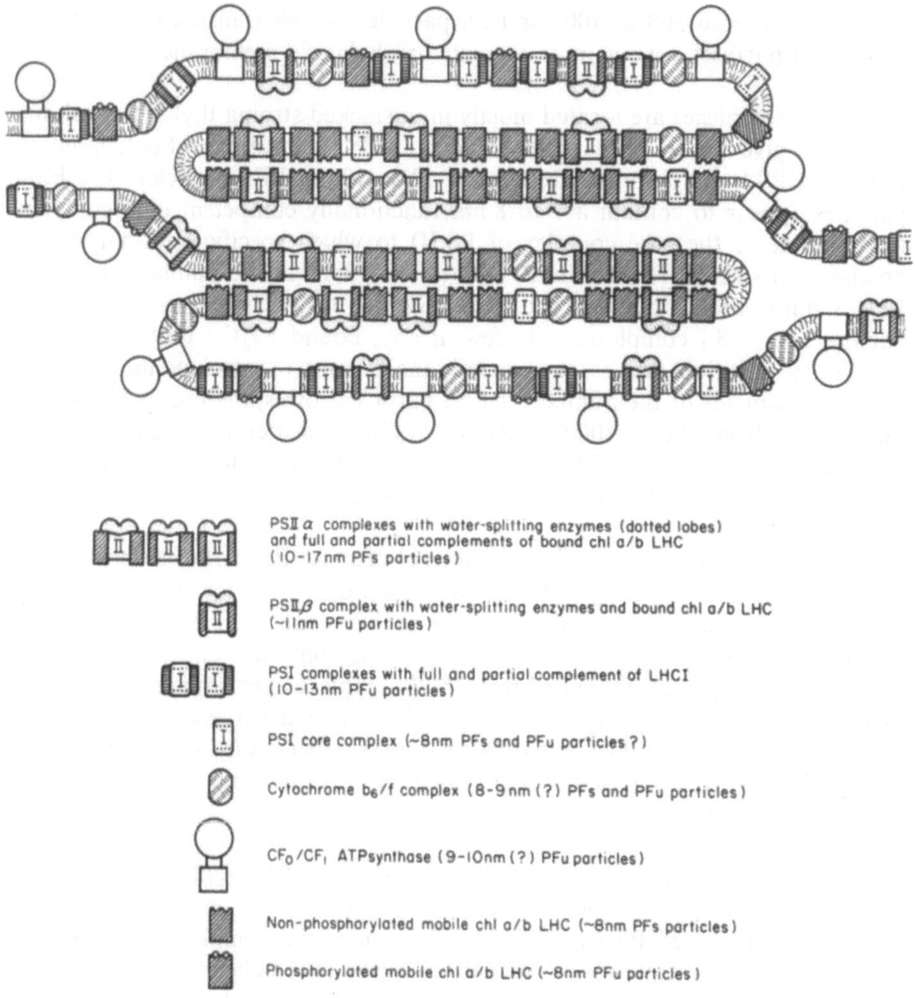

Fig. 31. Model of the supramolecular organization of chloroplast membranes of higher plants. Note the difference in composition of stacked, grana and unstacked, stroma membrane regions and the microheterogeneity of some of the complexes. The individual complexes are drawn approximately to scale, and an attempt has been made to reflect their normal ratios in thylakoid membranes. See Section 3.2.7 for a discussion of the special features of this model. Only a limited number of mobile chl a/b LHC particles are shown as being phosphorylated. Under State II conditions the number of phosphorylated particles would be significantly greater, under State I conditions there would be fewer

and more variable in size than the EFu particles (PS IIβ units). For unknown reasons, the latter appear to be unable to partition into stacked membrane regions and are less affected by Mn^{2+} deficiency. In intact thylakoids of normally grown plants (60% stacked grana membrane regions) 80 to 85% of the PS II units give rise to EFs particles and 15–20% to EFu particles. In absolute terms the density per square micrometer may vary from 1200 to 1700 for EFs particles and from 300 to 700 for EFu particles, which translates into 850 to 1300 PS II particles per square micrometer of thylakoid membrane.

PS I. PS I complexes are located mostly in unstacked stroma thylakoids (about 85%) and give rise to PF particles. Studies of PS I mutant thylakoids indicate that most PS I complexes correspond to 10 to 13 nm PFu particles. The PS I particles appear to contain a 7 to 8 nm functionally competent core complex (comparable to the core complex of PS II) to which specific light-harvesting antenna units (LHC I) are bound. Due to variations in the number of bound LHC I units, PS I complexes probably give rise to a whole spectrum of PF particle sizes. PS I complexes with few, if any, bound LHC I units appear to be able to partition into stacked membrane regions (STAEHELIN unpublished results). Estimates of the number of PFu particles corresponding to PS I complexes range from about 2100 to 3300 μm^{-2}. Grana-located PS I units may give rise to additional 250 to 400 PFs particles μm^{-2}. Averaged for the whole membrane the values would be 1000 to 1600 PS I particles per square micrometer.

Cytochrome b_6/f. Cytochrome b_6/f complexes produce PF particles, most likely in the 8 to 9 nm size range. Based on a roughly equal density distribution between stacked and unstacked membrane regions, and a PS II to cytochrome b_6/f ratio of about 1:1 (ORT, Chap. 4, this Vol.), one can estimate that there are between 850 and 1300 cytochrome b_6/f complexes per square micrometer in stacked membrane regions and an equivalent number per square micrometer in unstacked regions. If the complexes are dimers, as suggested by certain recent observations (ORT, Chap. 4, this Vol.), then the cytochrome b_6/f particle densities would be half the values shown above.

CF_0/CF_1 ATP Synthase. The CF_0 subunit of the CF_0/CF_1 ATP synthase complex produces 9–10 nm particles in reconstituted liposomes, and probably gives rise to similarly sized PFu particles in vivo, since the complex is exclusively located in stroma membranes. Calculations based on biochemical data suggest a CF_1 to PS II ratio of 1:2 (ORT, Chap. 4, this Vol.), which together with our PS II particle measurements would translate into 1000 to 1600 particles μm^{-2} of stroma membrane. Direct morphological measurements have produced a value of 700 CF_0/CF_1 complexes per square micrometer of unstacked stroma thylakoid.

Mobile chl a/b LHC. Mobile chl a/b LHC appears as an 8 nm PF particle on freeze-fractured thylakoid membranes. In its nonphosphorylated state it accumulates in stacked membrane regions, giving rise to PFs particles due to its double role as membrane adhesion factor and light-harvesting antenna. There may be as many as 3 mobile chl a/b LHC units per PS II (EFs) complex in stacked thylakoid regions. This translates into 3600 to 5100 PFs particles μm^{-2}. Under appropriate conditions (see Sect. 4.2.1, this Chap.) up to one fourth of these mobile chl a/b LHC particles can become phosphorylated and reversibly migrate to unstacked stroma regions. Under such conditions there will be one mobile chl a/b LHC unit per 1–2 stroma-located PS I complexes.

3.3 *Prochloron* Thylakoids

As already discussed in the section on the morphology of oxygen-evolving photosynthetic prokaryotes, *Prochloron* thylakoids have many features in common with those of higher plants and green algae. In particular, they utilize the same light-harvesting system, chl a/b LHC, and thin section images reveal the presence of stacked membrane regions (WHATLEY 1977; WITHERS et al. 1978). However, because *Prochloron* lives in association with ascidians up to 30 m below the sea surface where blue light, primarily absorbed by PS II, predominates, one might expect compensating modifications in thylakoid composition and organization. Changes have indeed been detected both at the chemical and structural level which result in increased delivery of excitation energy to PS I. In comparison with higher plants, *Prochloron* thylakoids have a much higher chl a/chl b ratio (4.4 to 6.9 versus 2.5 to 3.0) and contain only half the amount of chl a/b LHC (WITHERS et al. 1978). Furthermore, the membrane-bound kinase responsible for phosphorylating the mobile chl a/b LHC is active both in the light and in the dark (SCHUSTER et al. 1984), thereby maintaining a certain proportion of the mobile chl a/b LHC units constantly in a phosphorylated state. As discussed in Section 4.2 of this Chapter, in Chapter 11.5 (this Vol.) by BARBER, and in the review of STAEHELIN and ARNTZEN (1983), phosphorylated mobile chl a/b LHC units migrate to the unstacked, stroma membrane areas where most of their trapped energy can be delivered to PS I units.

Freeze-fracture micrographs of thylakoids of *Prochloron* sp. isolated from *Diplosoma virens* exhibit four types of fracture face that resemble those seen in higher plants (Fig. 32; GIDDINGS et al. 1980). The greatest similarities are observed on the E-faces: not only are the EFs particles more numerous than the EFu particles (3100 μm^{-2} versus 925 μm^{-2}), but particle size histograms demonstrate that the EFs particles fall into the same size categories (7.5, 10.5, 13, and 16 nm) as those of the green alga *Chlamydomonas* and of greening pea chloroplasts. This analogy suggests that *Prochloron* thylakoids also contain approximately 7.5 nm PS II core particles surrounded by variable amounts of bound chl a/b LHC.

However, in keeping with the reduced need for light-harvesting antennae for PS II reaction centers, there is a preponderance of smaller 7.5, 10.5, and 13 nm size classes of EF particles when compared to higher plant thylakoids. This suggests that the average PS II reaction center unit in *Prochloron* has fewer bound chl a/b LHC units. Furthermore, the very low PFs to EFs particle ratio in *Prochloron* (0.7 versus 2.3 in spinach) suggests that *Prochloron* may have few if any mobile chl a/b LHC units. This notion is further supported by the absence of mobile chl a/b LHC polypeptides in the 25 to 29 kDa range (SCHUSTER et al. 1984; ANDERSON Chap. 6.4, this Vol.). When all of these environmental adaptations are taken into consideration, the extent to which the organization of *Prochloron* thylakoids resembles those of higher plants is quite remarkable.

A freeze-fracture study of the thylakoids of a *Prochloron* sp. isolated from *Didenum molle* (G. COX and DWARTE 1981) has confirmed many of the findings of the GIDDINGS et al. (1980) study, but the authors also note some structural

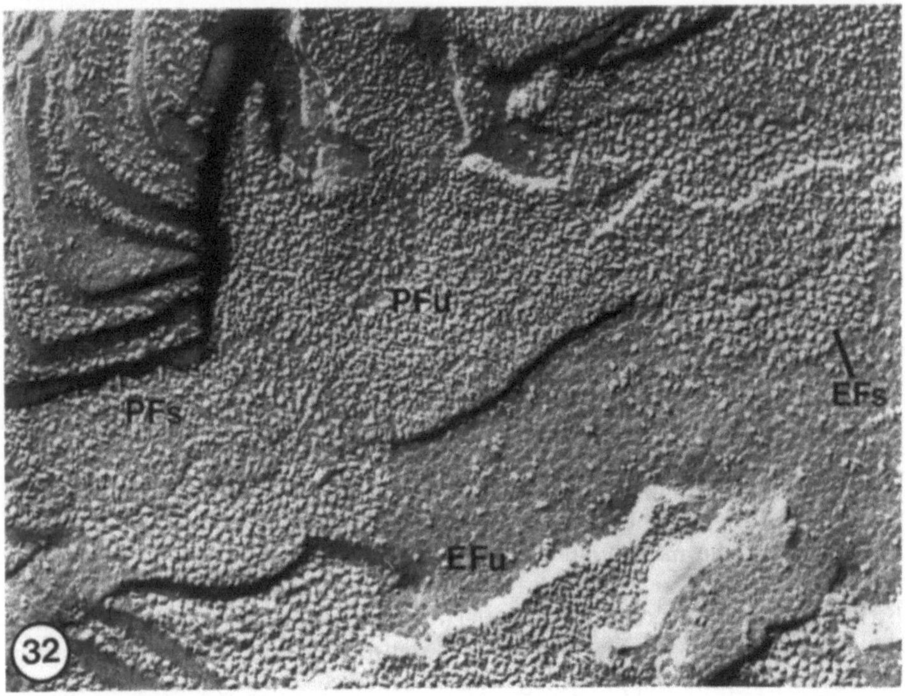

Fig. 32. Freeze-fractured thylakoid membranes of the chl a/b LHC-containing, oxygen-evolving prokaryote *Prochloron* (see also Fig. 10). The fracture face morphologies indicate that the thylakoids are differentiated in a manner similar to those of higher plants (cf. Figs. 21). See Fig. 22 for an interpretation of the fracture faces. ×60,000. (GIDDINGS et al. 1980)

differences (sizes and densities of particles, appearance of E-faces) between the thylakoids of the two *Prochloron* species. Careful examination of the micrographs from the latter study reveals several signs of less than optimal preservation of membrane structure, which makes interpretation of the pictures more difficult and less reliable. The differences in membrane preservation in the two studies probably relate to the fact that the G. COX and DWARTE (1981) *Prochloron* samples were stored for 1 or 2 days after collection before they were frozen, whereas the GIDDINGS et al. (1980) specimens were all frozen in less than 2 h following separation of the *Prochloron* cells from the didemnid host tissue. Thus, the significance of the reported differences is uncertain. Because *Prochloron* is the only chl b-containing prokaryote currently known, further studies of its photosynthetic apparatus would be highly desirable. Unfortunately, none of the attempts to grow *Prochloron* under laboratory conditions have met with much success, and experimental material still must be collected in tropical waters. A requirement for tryptophan in a cell maintenance (growth?) medium has been recently reported (PATTERSON and WITHERS 1982).

3.4 Thylakoid Membranes of Cyanobacteria, Red Algae and Cyanelles

3.4.1 Identification of Freeze-Fracture Particles

In common with higher plants and green algae, cyanobacteria, red algae, and cyanelles contain thylakoids with two photosystems, PS I and PS II. The two categories of organism differ in the types of pigment-proteins used as light-harvesting antennae, the former depending on chl b-containing complexes such as the chl a/b LHC, the latter on phycobiliproteins organized into phycobilisomes (Figs. 3, 10, 11; THORNBER, Chap. 3; GANTT, Chap. 6.3, both this Vol.). The physicochemical properties of these light-harvesting systems appear to be responsible for the large differences in structural organization of chl b and phycobilisome-containing thylakoids. As discussed above, the chl a/b LHC of higher plants and green algae is an integral membrane protein complex that also promotes membrane adhesion and thus the formation of grana stacks. In contrast, the phycobilisomes are large, water-soluble, peripheral membrane protein complexes that protrude from the membrane surface and physically prevent adjacent thylakoid membranes from approaching each other too closely (Figs. 3, 12, 17, 33; GANTT, Chap. 6.3, this Vol.). For this reason, thylakoids of phycobilisome-containing organisms lack stacked regions and exhibit only two types of fracture face, EF and PF. These appear similar to those of unstacked stroma chloroplast membranes of chl b-containing organisms in that the density of PF particles is much higher than that of EF particles (Fig. 34). For example, *Spermothamnion* thylakoids possess an EF particle density of $960 \pm 340\ \mu m^{-2}$ and a PF particle density of $3570 \pm 730\ \mu m^{-2}$ (STAEHELIN et al. 1978). However, the most important similarity relates to the fact that in both groups of organism PS II complexes give rise to EF particles, and PS I complexes to PF particles. This is not unexpected since, as discussed in greater detail in Chapter 11.4, this volume by BRICKER et al., the PS I and PS II complexes of cyanobacteria resemble those of higher plants both in terms of composition and general organization, even though their size is smaller due to the absence of bound chl b-containing light-harvesting complexes.

Three independent lines of evidence support the hypothesis that 10 nm EF particles correspond to PS II complexes in phycobilisome-carrying thylakoids; (a) they are absent from the thylakoids of cyanobacterial heterocysts that lack PS II activity (GIDDINGS and STAEHELIN 1979), (b) an increase in the PS II:PS I ratio from 0.6 in wild type to 1.2 in the phycobilisome-deficient III-C mutant of *Cyanidium caldarium* is matched by a doubling of the density of 10 nm EF particles (WOLLMAN 1979), and (c) rows of phycobilisomes can be shown to directly overlie rows of EF particles (see below). In those red algae (*Spermothamnion* and *Cyanidium*) and cyanelles (*Cyanophora*) for which particle size histograms of freeze-fractured thylakoids have been prepared, all EF particles seem to belong to one major size class with a modal distribution of about 10 nm (STAEHELIN et al. 1978; WOLLMAN 1979; GIDDINGS et al. 1983). On the other hand, those found on freeze-fractured thylakoids of the vegetative cells of the cyanobacterium *Anabaena variabilis* fall into two size classes centered around 7.5 nm and 10 nm (GIDDINGS and STAEHELIN 1979). Heterocyst thylakoids de-

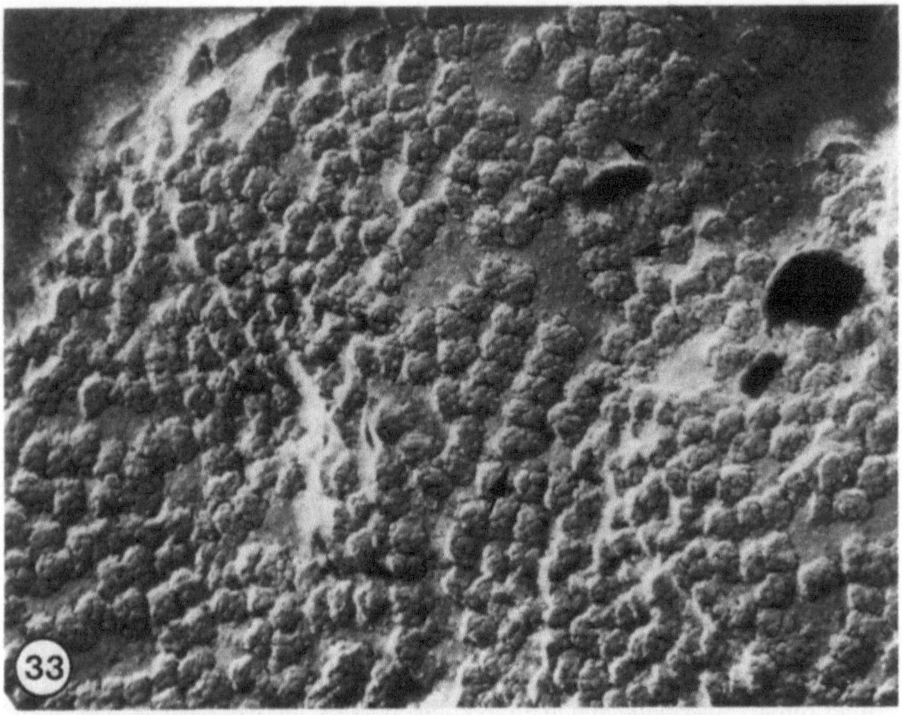
33

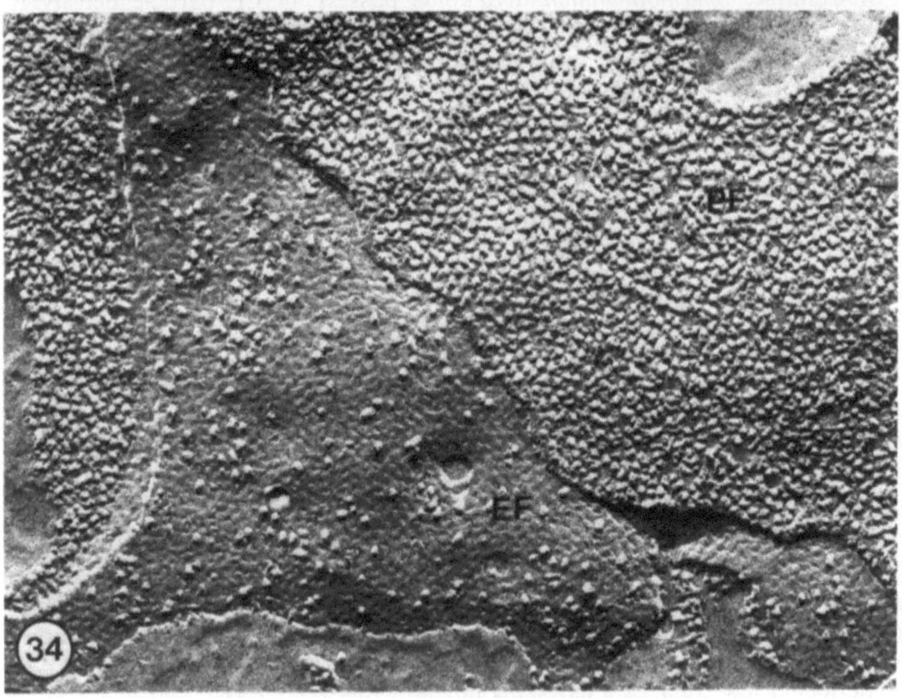
PF
EF
34

void of PS II activity lack the 10 nm EF particle size category, consistent with the idea of PS II complexes also giving rise to 10 nm EF particles on freeze-fractured cyanobacterial thylakoids. They do, however, contain two other categories of EF particles, one with a modal size of 7.5 nm (also found on the vegetative cell thylakoids), the other with a modal size of 5.5 nm. Since all cyanobacterial thylakoids have to serve both photosynthetic and respiratory functions (BINDER 1982), in contrast to red algae whose respiratory activities are segregated into mitochondria, the 7.5 nm EF particle category of cyanobacterial thylakoids may correspond to a membrane complex associated with the respiratory chain. The smaller 5.5 nm EF particles of heterocyst thylakoids could be related to the nitrogenase system.

The situation in *Gloeobacter violaceus* is somewhat more complex (GUGLIELMI et al. 1981). As summarized in Fig. 10, this cyanobacterium lacks thylakoids and has its phycobilisomes attached to the inner surface of the plasma membrane. It is therefore reasonable to assume that the components of the photosynthetic electron transport chain are intermixed with other enzymatic and transport complexes of the plasma membrane, making identification of specific particle categories even more difficult. Nevertheless, the EF particle-size histogram for the plasma membrane of *Gloeobacter* exhibits a major peak at 10.5 mm, consistent with the finding of PS II complexes giving rise to 10 mm EF particles in the thylakoids of other cyanobacteria and red algae.

Although methods have been developed to isolate not only PS II, but also PS I, cytochrome b_6/f, and CF_0/CF_1 complexes from cyanobacterial and red algal thylakoids (see BRICKER et al., Chap. 11.4, this Vol.), virtually no information is available on the actual size and distribution of the latter three complexes in freeze-fractured thylakoid membranes. Their compositional similarities with those isolated from higher plant thylakoids would tend to suggest that they give rise to PF particles of similar size, but this is about all that can be said at the present time. The PS I to PS II ratio in both cyanobacterial and red algal thylakoids is approximately 2.3 (KURSAR and ALBERTE 1983), versus about 1 for most in higher plants (see discussion by ORT, Chap. 4, this Vol.). Whether the greater activity of cyclic electron transport in cyanobacteria (GIMMLER 1977) could be partly responsible for the higher PS I to PS II ratio requires further study. Alternatively, more PS I units might be needed to ensure an equal distribution of excitations between the two photosystems, and to compensate for the fact that most of the energy trapped by the phycobilisomes is transferred to PS II reaction centers (see below).

Fig. 33. Surface view of an isolated *Porphyridium* (red alga) thylakoid covered with phycobilisomes (*arrows*) as revealed by freeze-etch electron microscopy (reverse contrast print). These phycobilisomes are of the hemispherical type (see also Figs. 17, 18) in contrast to the thinner hemidiscoidal ones illustrated in Figs. 12 and 13. The "ends" of the disc-containing rod elements (GANTT, Chap. 6.3, this Vol.) appear as round subunits on the individual phycobilisomes. ×120,000

Fig. 34. Freeze-fractured isolated thylakoids of the red alga *Spermathamnion turneri*. As in all thylakoids of oxygen-evolving organisms, the *PF* particles are three to four times as numerous as the *EF* particles. ×85,000

3.4.2 Relationship Between Phycobilisomes and EF Particles (PS II Complexes)

Since approximately 90% of the light energy absorbed by phycobilisomes is transferred initially to PS II reaction centers (LEY and BUTLER 1976; TEL-OR and MALKIN 1977; see also SCHEER, Chap. 7.6, this Vol.), it is reasonable to assume that phycobilisomes are closely associated with EF particles, the purported PS II units. In general this postulated relationship has been experimentally confirmed, although not every EF particle seems to be intimately associated with a phycobilisome. Numerous studies have demonstrated that the EF particles of phycobilisome-carrying thylakoids can be either randomly distributed or organized into more or less parallel rows with a spacing of approximately 50 nm (Fig. 35; NEUSHUL 1971; LEFORT-TRAN et al. 1973; LICHTLÉ and THOMAS 1976; STAEHELIN et al. 1978; GIDDINGS and STAEHELIN 1979; GOLECKI 1979; WOLLMAN 1979; G. COX et al. 1981; MÖRSCHEL and MÜHLETHALER 1983; GIDDINGS et al. 1983). As documented most clearly by LICHTLÉ and THOMAS (1976), whenever thin section electron micrographs reveal rows of phycobilisomes on thylakoid surfaces, freeze-fracture micrographs of equivalent samples reveal corresponding rows of EF particles, the spacing being the same between the phycobilisome and the EF particle rows. A more direct proof for the physical associa-

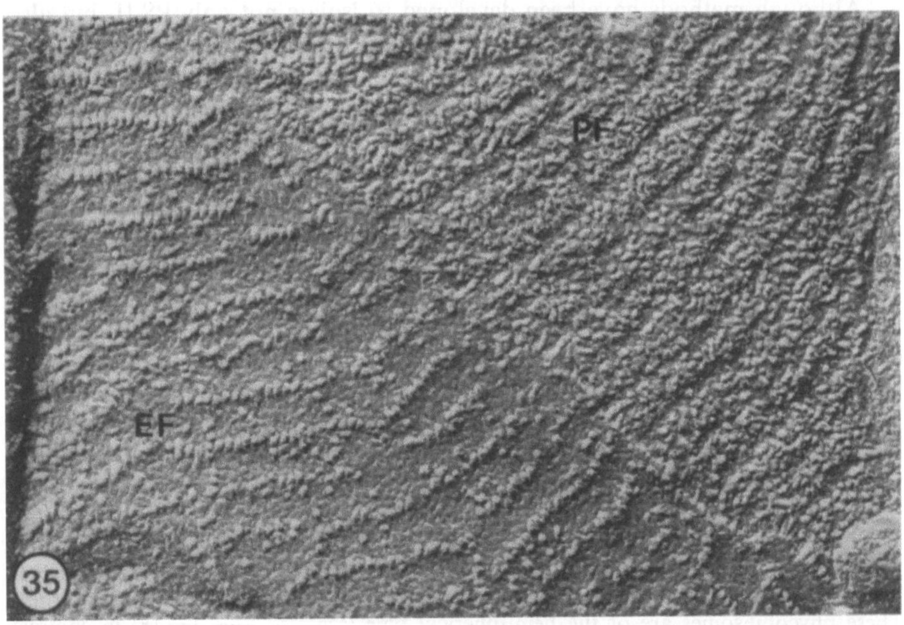

Fig. 35. Freeze-fractured cyanelle thylakoids of *Cyanophora paradoxa*. The *EF* particles correspond to PS II complexes, which become organized into rows when their associated fan-shaped phycobilisomes aggregate laterally into rows. The individual *EF* particles located between the rows are probably not directly linked to a phycobilisome. Bands of densely packed particles separated by particle-depleted grooves are seen on the P-face (*PF*). The grooves arise from the tearing away of the EF particle rows during the fracturing process. ×125,000

tion between EF particles and phycobilisomes has come from studies of deep-etched cyanelle thylakoids of *Cyanophora paradoxa* from which some of the phycobilisomes have been removed (GIDDINGS et al. 1983). In such samples it can be shown that the transmembrane EF particles give rise to dimeric protrusions on both the stromal (PS) and the lumenal (ES) membrane surfaces. Where only some of the phycobilisomes of a row have been washed away, it is possible to demonstrate that the phycobilisomes are indeed centered over the PS II particles that give rise to EF particles. It was further found that the periodicity of the hemidiscoidal phycobilisome rows (10 nm) matched the periodicity of the EF particle rows, suggesting a one-for-one association between phycobilisomes and EF particles. MÖRSCHEL and MÜHLETHALER (1983) have reported a 12 nm periodicity for rowed EF particles and for the overlying phycobilisomes of the thermophilic cyanobacterium *Mastigocladus laminosus*.

The extent to which EF particles and phycobilisomes are organized into rows seems to be related to the light conditions during growth. Most cells grown under relatively high light conditions and harvested at low cell densities display few if any EF particle rows, and the phycobilisomes appear more or less randomly organized. However, under reduced growth light conditions, as encountered in dense algal cultures, more phycobilisomes are added to the thylakoid surfaces to compensate for the loss of light energy. The resulting close packing of the phycobilisomes leads to the formation of phycobilisome and, concomitantly, of EF particle rows. In this context, it is interesting to note that the concentrically organized thylakoids of *Cyanophora* cyanelles exhibit a gradient of supramolecular organization from the outer to the inner ones, the "high light" outermost thylakoids displaying mostly randomly organized phycobilisomes and EF particles, and the shaded inner ones predominantly phycobilisomes and EF particles in rows (GIDDINGS et al. 1983). Both hemidiscoidal-type phycobilisomes, as found in cyanobacteria and cyanelles, and hemispherical-type phycobilisomes, as found in the red alga *Porphyridium* (see GANTT, Chap. 6.3, this Vol.), can become organized into rows. It should be noted, however, that the literature dealing with the morphology of thylakoids of cyanobacteria and red algae tends to overemphasize the extent to which phycobilisome and EF particle rows are present. This is presumably due to the fact that many morphologists are unaware of the extent and consequences of shading in dense cultures from which their samples are usually taken, and because the arrayed structures give rise to clearer, more easily interpretable and more striking images.

Although the transfer of excitation energy from phycobilisomes to PS II reaction centers is now well documented (LEY and BUTLER 1976; TEL-OR and MALKIN 1977; SCHEER, Chap. 7.6, this Vol.), few researchers have addressed the question whether all PS II complexes receive excitation energy from phycobilisomes. Two lines of evidence suggest that this is not always the case. DINER (1979) has determined that the phycobilisomes of *Cyanidium caldarium* can transfer excitation energy directly to just over half of the PS II reaction centers. Quantitative morphological studies of thylakoids of the red alga *Griffithsia pacifica* grown under different light conditions have led to similar conclusions. Cells grown under 50 ft.-c. light possess about 270 phycobilisomes μm^{-2} of thylakoid surface; those grown under 300 ft.-c. light about 165 phycobilisomes μm^{-2} (WAALAND et al. 1974). Freeze-fracture micrographs reveal that at 50 ft.-c. the thylakoids contain 670 EF particles μm^{-2} and at 300 ft.-c. 770 EF particles μm^{-2} (STAEHELIN et al. 1978). If all the EF particles were structural equivalents of PS II units, then the low light chloroplasts would have on the average one phycobilisome for every three PS II reaction centers, and the high light chloroplasts one phycobilisome for every four PS II complexes. A 1:2 phycobilisome to EF particle ratio has been reported for the red alga *Antithamnion glanduliferum* (LICHTLÉ and THOMAS 1976). Based on biochemical measurements, a 1:4 phycobilisome to PS II ratio has been calculated for the red alga *Neogardhiella* and close to a 1:2 ratio for the cyanobacterium *Anacystis* (KURSAR and ALBERTE 1983). Since both the measurements of phycobilisome to EF particle and phycobilisome to PS II reaction center ratios yield similar values, it seems highly likely that each EF particle is equivalent to a PS II complex.

It is unclear at the present time why many red algae and cyanobacteria produce significantly more PS II reaction centers than can be served directly by phycobilisomes. In this context, it is interesting to consider the fact that the organization of the phycobilisomes and the EF particles (PS II complexes) into tightly packed rows not only allows for a high packing density of the phycobilisomes, but potentially also for high efficiency excitation energy transfer between adjacent PS II units. Thus, any energy trapped by a given phycobilisome that cannot be utilized by the PS II unit attached to that phycobilisome could be easily transferred to and utilized by an adjacent PS II particle. However, freeze-fracture micrographs reveal that significant numbers of single EF particles are also located between the EF particle rows, thereby suggesting that at any one time a substantial fraction of PS II centers will not be driven by energy trapped by phycobilisomes.

3.5 Thylakoid Membranes of Chromophyta

The diverse group of algae known as Chromophyta (Table 1) make use of c-type chlorophylls (chl c, c_1, c_2) and fucoxanthin or peridinin in their light-harvesting antennae (Fig. 11; see ANDERSON, Chap. 6.4; THORNBER, Chap. 3, both this Vol.). The literature on the freeze-fracture morphology of Chromophyta thylakoids is still miniscule; only one or two studies per subgroup of algae have been published to date. Nevertheless, based on these few papers one can already state that in their basic supramolecular organization Chromophyta thylakoids resemble those of other oxygen-evolving photosynthetic organisms. The many similarities seem to reflect the conservation of structural and compositional features of PS I and PS II, and of the cytochrome b_6/f and CF_0/CF_1 complexes (see THORNBER, Chap. 3; ANDERSON, Chap. 6.4; DINER, Chap. 9.1, 10.8; WOLLMAN, Chap. 9.7; HAUSKA, Chap. 10.1; McCARTY and NALIN, Chap. BRICKER et al., Chap. 11.4, all this Vol.). The effect of the different types of pigment protein on the organization of Chromophyta thylakoids may best be summarized as producing "variations of a basic (structural) theme". Thin section electron micrographs reveal that the thylakoids of most Chromophyta are associated into fairly tight bands of three, with single thylakoids interconnecting adjacent bands (Figs. 3, 4). This type of association would seem to lead to a fairly inflexible ratio of 2:1 for appressed to nonappressed membrane regions, an organizational pattern that could conceivably interfere with the adaptation to different light environments (the amount of stacking in higher plants varies in response to both light quantity and light quality). Some of the recent freeze-fracture studies, however, suggest that "true" membrane stacking may be limited to only a variable fraction of the appressed membrane regions.

3.5.1 Cryptophyceae

By sequestering their biliproteins into the thylakoid lumen rather than organizing them into surface-exposed phycobilisomes (as in cyanobacteria and red algae), the Cryptomonads have gained the ability to differentiate their thylakoids into appressed and nonappressed membrane regions. Freeze-fracture micrographs of such thylakoids allow for a clear distinction of four types of fracture face, which, in analogy to those of green algae, can be referred to as EFs, EFu, PFs and PFu (Dwarte and Vesk 1983; Spear-Bernstein and Miller 1985). However, one important difference is that the EFs/PFs areas are limited to specially differentiated regions of already appressed membranes. As illustrated in Figs. 3 and 19 the appressed membranes seem to be held together by patches of more densely staining material. The typical EFs/PFs areas (Fig. 36) appear to coincide with these actual regions of membrane adhesion. Dwarte and Vesk (1983) have proposed that the fracture faces of the nonadhesive but appressed membrane regions be referred to as EFs_1 and PFs_1, but since the size and density of the particles of these fracture faces is very similar to those of the EFu and PFu faces, Spear-Bernstein and Miller (1985) have proposed that no distinction should be made between the two types of face. Further studies are obviously needed to prove beyond doubt that the two types of nonstacked thylakoid regions have an identical composition. For reasons of simplicity, the EFs_1/PFs_1 faces of Dwarte and Vesk (1983) will be considered equivalent to EFu/PFu faces in this review.

In *Chroomonas* and *Cryptomonas* (Dwarte and Vesk 1983), as well as in *Rhodomonas* (Spear-Bernstein and Miller 1985), the EFs and EFu regions of freeze-fractured thylakoids can be distinguished based on the density but not the size of the particles (Fig. 36). EFs particle densities range from 600 to 1000 μm^{-2}, EFu particle densities from 100 to 350 μm^{-2}. The average size for both EFs and EFu particles is reported to be 15 nm for *Chroomonas* and *Cryptomonas*, and 13 nm for *Rhodomonas*. Whether the size differences for the various organisms reflect real differences in composition or simply variability of measuring techniques is unclear at the present time. A systematic difference related to how particle sizes are measured is suggested by the fact that the average size for PF particles reported by Spear-Bernstein and Miller (1985) for *Rhodomonas lens* (PFs 9 nm, PFu 10 nm) is also 2 to 3 nm smaller than the values reported by Dwarte and Vesk (1983) for *Chroomonas* sp. and *Cryptomonas* sp. (PFs 11 nm, PFu 13 nm). The densities for PFs particles range from 2500 to 3500 μm^{-2}, and for PFu particles from 2100 to 4200 μm^{-2}. The many

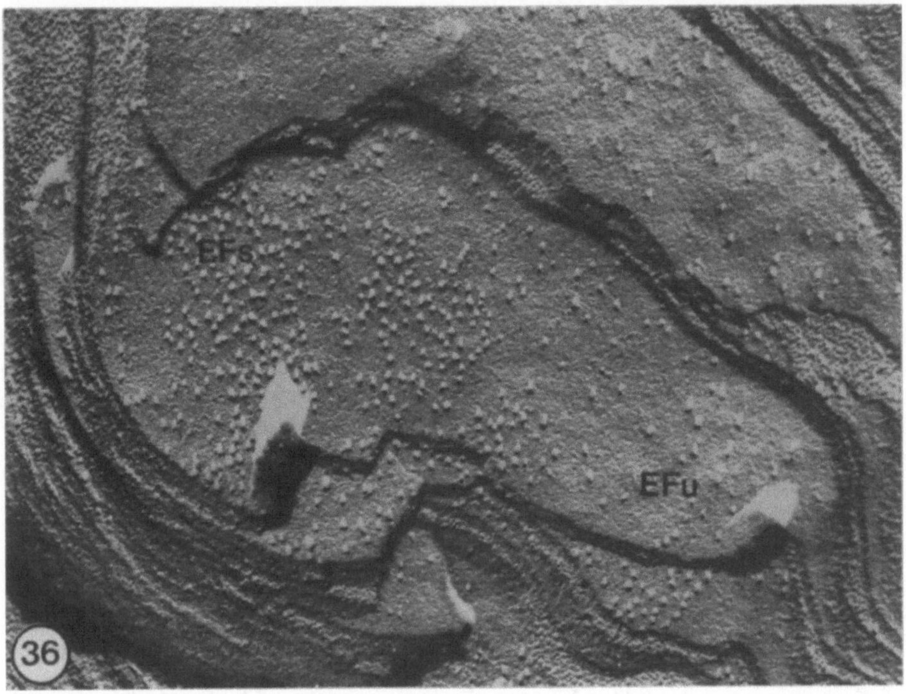

Fig. 36. Freeze-fractured thylakoid membranes of *Rhodomonas lens*, a cryptomonad alga. The membrane fracture face seen in the center is clearly differentiated into *EFs* and *EFu* regions. The patches of *EFs* particles coincide with membrane adhesion sites (cf. Figs. 3 and 19). ×65,000 (Micrograph courtesy of L. SPEAR-BERNSTEIN, Providence, RI)

similarities in the freeze-fracture morphology of the thylakoids of green algae and cryptomonads lends support to the notion that in these latter algae, PS II complexes also give rise to EF particles and PS I complexes to PF particles. Furthermore, the general orientation of the major integral membrane protein complexes seems to be unaffected by the "reverse" sidedness (luminal location) of the biliprotein light-harvesting antennae.

3.5.2 Phaeophyceae and Dinophyceae

Freeze-fractured thylakoids of the brown algae *Fucus serratus* (BERKALOFF et al. 1983) and *Sargassum* sp. (DWARTE and VESK 1982) reveal features similar to those seen in cryptomonads and diatoms (Fig. 37). EFs particles appear to be three to four times more numerous than EFu particles, but the average size of particles in each area is the same in both organisms, about 8.5 nm in *Fucus* and 10 nm in *Sargassum*. In neither system were the authors able to obtain replicas of sufficient quality to clearly distinguish PFs from PFu faces. For *Fucus* the average PF particle size is 8 nm, and the average PF particle density 3900 μm^{-2}.

Two reports on the freeze-fracture morphology of dinoflagellate thylakoids have been published to date. Unfortunately, neither study provides a complete

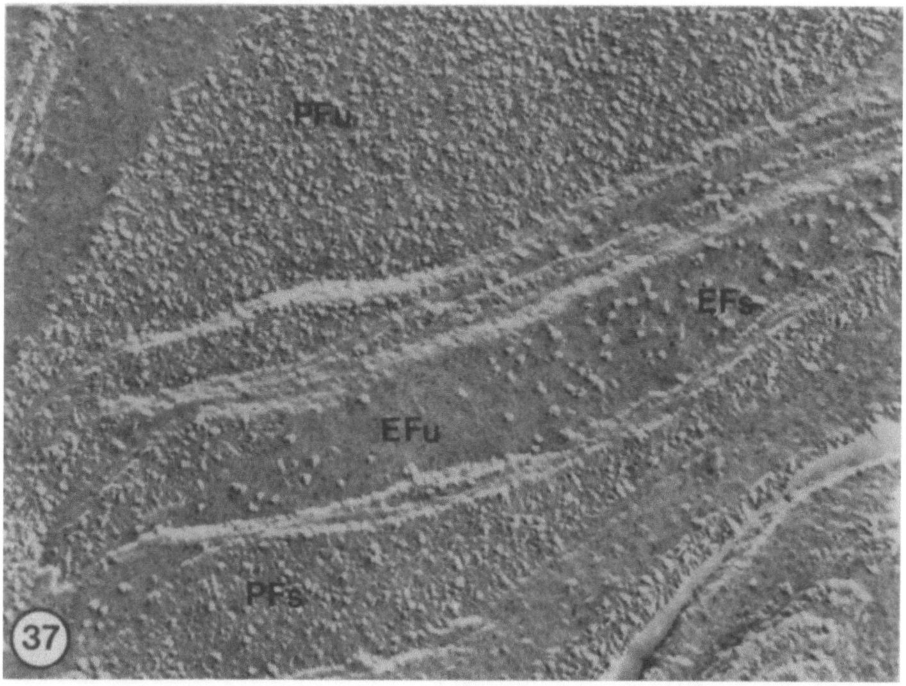

Fig. 37. Freeze-fractured thylakoids of the diatom *Skeletonema costatum*. Four types of fracture face, *EFs*, *EFu*, *PFs*, *PFu*, can be distinguished as in higher plants, consistent with the fact that brown algal and diatom thylakoids are differentiated into stacked and unstacked membrane regions (see Fig. 3). The individual fracture faces, however, have a somewhat different particle composition than those of higher plants (cf. Fig. 21). ×100,000

analysis of all four fracture faces. In the case of *Gonyaulax polyedra* (SWEENEY 1981) only EFs and PF (PFs + PFu) histograms are shown and for *Prorocentrum micans* (DWARTE and VESK 1982) only average EFs and EFu particle sizes are given. In common with other Chromophyta, the EFs and EFu particles seem to be of similar size, about 10 nm in both organisms, even though the density of EFs particles is two to five times larger. The average PF particle also has a diameter of approximately 10 nm. No changes in particle sizes were noted during the circadian cycle of *Gonyaulax* cells grown under constant light.

4 Experimental Modification of Thylakoid Membrane Structure

4.1 Structural Changes Associated with in Vitro Unstacking and Restacking of Thylakoids

As discussed in greater detail by BARBER in Chapter 11.5, this Volume, thylakoid membranes carry a net negative surface charge. Maintenance or formation of

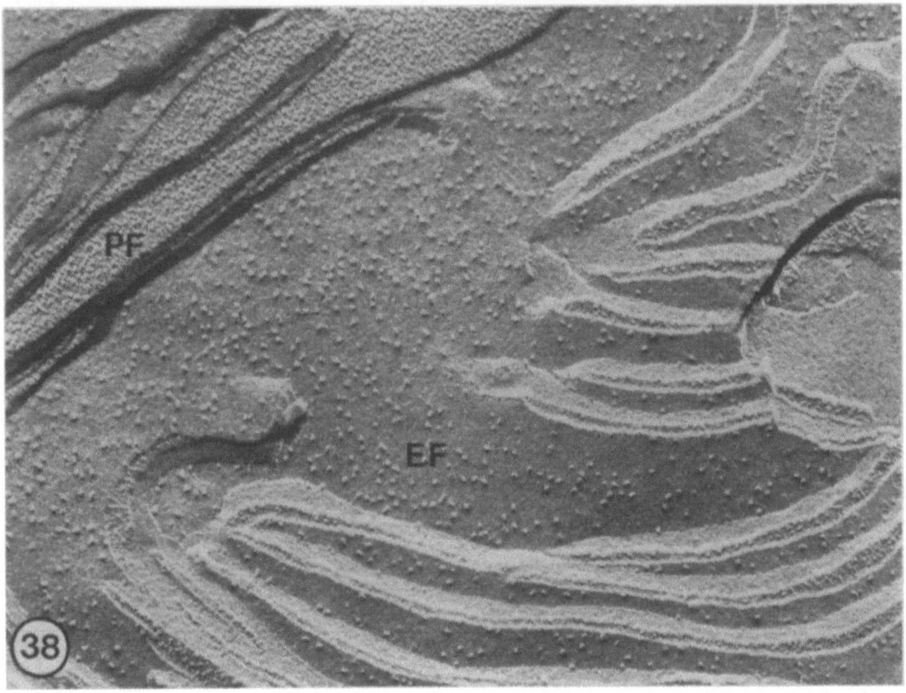

Fig. 38. Freeze-fracture micrograph of pea thylakoids experimentally unstacked for 30 min in a low salt medium (0.3 m sucrose, 50 mM Tricine). Note the sheet-like nature of the unfolded but still interconnected thylakoids, and the random distribution of all *EF* and *PF* particles. Addition of >3 mM $MgCl_2$ or >150 mM NaCl to such membranes leads to the reformation of stacked membrane regions and complete resegregation of the particles. ×60,000

stacked membrane regions depends upon domination by attractive van der Waals forces over electrostatic repulsion forces. Stacking occurs in the presence of cations that screen the electrostatic surface charges and thereby enable the membranes to approach each other more readily. Conversely, removal of these counterions leads to an increase in repulsive interactions between the negative charges on adjacent membranes and to a decrease or total loss of appressed membrane regions. Experimental unstacking of isolated thylakoids by transfer to zwitterionic buffers of low ionic strength (approximately 50 mM) was first demonstrated by IZAWA and GOOD (1966). Return of such thylakoids to >3 mM divalent or >100 mM monovalent cations leads to the reformation of membrane stacks.

Freeze-fracture studies have revealed that as membranes unstack in a low salt medium the lateral segregation of specific categories of intramembrane particles into grana and stroma membrane regions is lost due to intermixing and randomization of the two sets of particles (Fig. 38; OJAKIAN and SATIR 1974; STAEHELIN 1976). Complete resegregation of the particles occurs, however, when the thylakoids are resuspended in a high salt buffer (STAEHELIN 1976; STAEHELIN and ARNTZEN 1979).

Some confusion has arisen over the extent to which the structural organization of experimentally unstacked and restacked thylakoids resembles that of control thylakoids. Differences in the apparent precision with which unstacked thylakoids can be returned to a stacked configuration with a near normal appearance can be traced to the conditions used for unstacking. Careful analysis of the "unstacking" process has resolved the following steps (STAEHELIN et al. 1977): (1) Rapid separation of appressed membrane regions with no major changes in membrane conformation. (When restacked, such thylakoids are virtually indistinguishable from control thylakoids). (2) Increase in diameter of the tubular connections between grana and stroma thylakoids which allows for intermixing of the membrane protein complexes of these regions to occur. (Restacking of such thylakoids leads to the formation of stacked regions that still resemble grana, but whose stacked membranes are of variable size and are less precisely aligned than in normal grana stacks). (3) Complete unfolding and reorganization of the highly branched thylakoids into large membrane sheets as shown in Fig. 38. (Addition of cations leads to the formation of stacked membrane regions that tend to be large and variable in size. The membrane stacks bear little resemblance to control grana, but appear similar to those seen in more primitive green algae). At the supramolecular level, however, irrespective of the extent of initial unstacking-unfolding, when thylakoids are allowed to restack in the presence of >3 mM $MgCl_2$ for more than 30 min, virtually complete normal resegregation of all membrane protein complexes into stacked and unstacked membrane regions is observed (STAEHELIN 1976).

Different claims have been made as to how fast membrane restacking and resegregation of membrane particles can occur. Thus, STAEHELIN (1976) has reported that both at 4 °C and 20 °C, and following the addition of 2 mM $MgCl_2$ to experimentally unstacked thylakoids, membrane adhesion occurs fairly rapidly (less than 10 min), while complete segregation of the particles requires at least 30 min. BRIANTAIS et al. (1983), on the other hand, have found that in their samples (5 mM $MgCl_2$, 20 °C) restacking and segregation of the photosystems is complete within 3 min. This discrepancy probably relates to the fact that because the salt concentration used by STAEHELIN (1976) for restacking was suboptimal, the "sorting forces" responsible for resegregation of the protein complexes were weaker. The stacking-mediated segregation of the photosystems and other thylakoid membrane complexes can be explained by the surface electrical charge theory developed by BARBER and coworkers (BARBER 1980 and Chap. 11.5, this Vol.), which postulates that during stacking, protein complexes carrying a higher surface charge will be excluded from the stacked membrane regions. In this context, the PS I–LHC I, CF_0/CF_1, and phosphorylated mobile chl a/b LHC particles would belong to the high charge type, and the PS II-bound chl a/b LHC and nonphosphorylated mobile chl a/b LHC particles would carry a low net negative charge. The even partitioning of the cytochrome b_6/f complex between grana and stroma membranes suggests that its surface properties may fall somewhere between those of the two photosystems or that other partitioning effectors take precedent over surface charge effects.

4.2 State I/State II Transitions – Functional Regulation Mediated by Structural Rearrangements

When algae and plants are exposed sequentially to light of different spectral properties, their light-harvesting pigments undergo dynamic adaptations in response to the altered illumination, which in turn leads to an improved balance in energy distribution between PS I and PS II (BONAVENTURA and MYERS 1969; MURATA 1969; VERNOTTE et al. 1975; CHOW et al. 1981). These changes are usually referred to as "State I–State II" transitions because they reflect an alteration in the state of the photosynthetic pigment bed in response to excess light trapping by PS I or PS II, respectively. State I is defined as the condition that develops when a photosynthetic organism is exposed to light preferentially absorbed by PS I. Over a time scale of several minutes, an adaptation occurs that leads to more of the absorbed light energy being directed to PS II. The situation is reversed when PS II receives a disproportionate amount of the trapped energy (State II). It should be emphasized that these state changes are only important when light is limiting to the overall photosynthetic process. This is not uncommon under natural conditions, because plant growth patterns result in the shading not only of surrounding species but also of the lower leaves of individual plants (BJÖRKMAN 1981). In *Camellia japonica* a significant light gradient has even been reported to exist across individual leaves, leading to a functional adjustment of the thylakoids (of the type just discussed) of pallisade and spongy mesophyll chloroplasts (TERASHIMA and IOMEE 1984).

The physical basis for this regulation has recently been shown to involve the reversible lateral movement of mobile chl a/b LHC particles between PS II-rich stacked membrane regions and PS I-rich unstacked ones (Fig. 39; STAEHELIN et al. 1982; KYLE et al. 1983; BARBER, Chap. 11.5, this Vol.). The regulatory mechanism involves phosphorylation/dephosphorylation of the mobile chl a/b LHC units, which, as discussed above, are the principle adhesion factor of thylakoid membranes (BENNETT 1983; STAEHELIN and ARNTZEN 1983; CANAANI et al. 1984). Phosphorylation occurs on one or two threonine residues of a surface-exposed segment of the chl a/b LHC proteins. This trypsin-sensitive segment is rich in lysine/arginine residues and has been found to be necessary to elicit normal grana stacking, possibly via charge interactions with the adjacent membrane. In their nonphosphorylated state the chl a/b LHC particles promote membrane adhesion and therefore accumulate in the stacked membrane regions (Fig. 39). The addition of one or two phosphate groups to the cationic surface segment of the chl a/b LHC molecules increases their negative surface charge such that their ability to mediate membrane adhesion is impaired because of electrostatic repulsion. As a result, the chl a/b LHC particles are released from the stacked membrane regions and diffuse into the adjacent unstacked ones, which are highly enriched for PS I reaction centers. Since excitation energy transfer drops off by the 10^6 power of the distance, this shifting back and forth of light-harvesting antenna complexes enables plants to regulate the amount of light energy delivered to either PS I or PS II reaction centers. Quantitative analysis of freeze-fracture electron micrographs of unphosphorylated, phosphorylated and dephosphorylated thylakoids indicates that about 20 to

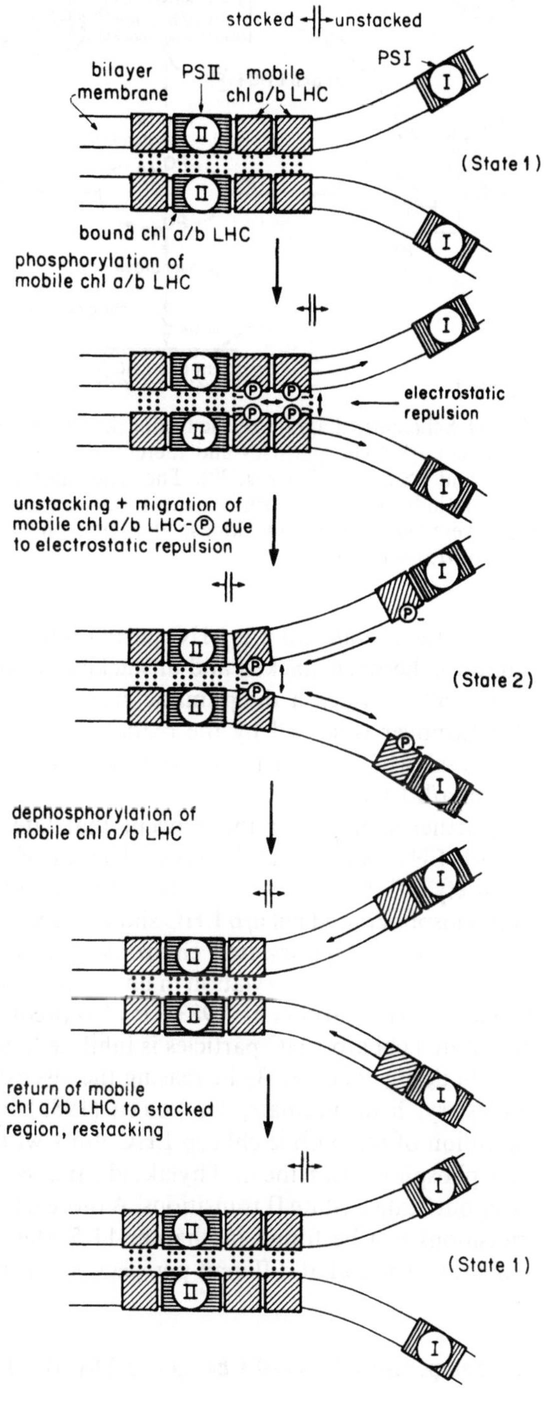

Fig. 39. Schematic diagrams illustrating how the reversible phosphorylation of the mobile chl a/b LHC particles affects membrane stacking and the distribution of these light-harvesting antennae between PS II-rich stacked membranes and PS I-rich unstacked membranes. This reversible movement of the mobile chl a/b LHC particles enables chloroplasts to regulate how much energy is funneled to PS II and PS I. (After STAEHELIN and ARNTZEN 1983)

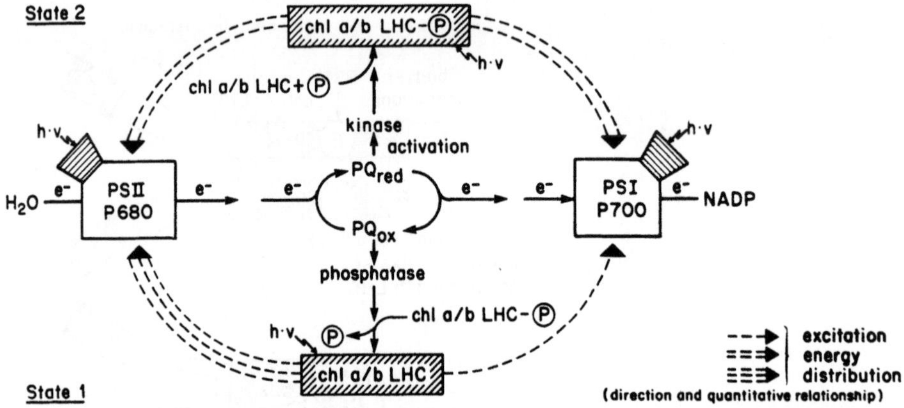

Fig. 40. Schematic diagram of the feedback mechanism that controls the phosphorylation of the chl a/b LHC particles and thereby the distribution of excitation energy between PS II and PS I (see also Fig. 39). The redox potential of the plastoquinone (*PQ*) pool serves as indicator of the turnover rate to the two photosystems. Accumulation of reduced PQ reversibly activates the kinase that phosphorylates the chl a/b LHC. (STAEHELIN and ARNTZEN 1983)

25% of the mobile chl a/b LHC particles particles participate in this reversible migration between stacked and unstacked membrane regions (KYLE et al. 1983). Figure 40 provides a diagramatic summary of how the rate of PS II versus PS I turnover is sensed by the oxidation-reduction state of the plastoquinone pool, and which, in turn, controls the phosphorylation/dephosphorylation of the chl a/b LHC.

Further support for the concept that lateral migration of phosphorylated chl a/b LHC is essential for state changes of the pigment bed has come from a variety of studies, two of which are mentioned here. The model predicts that phosphorylated chl a/b LHC should accumulate in stroma thylakoids. Fractionation of phosphorylated chloroplast membranes has revealed a ten times higher relative phosphorylation in unstacked, stroma membranes than in stacked grana ones (LARSSON et al. 1983). It also predicts that if migration of the phosphorylated chl a/b LHC particles is inhibited, no regulation of excitation energy distribution will occur. By increasing the viscosity of thylakoid membranes with cholesteryl hemisuccinate, HAWORTH (1983) has been able to prevent lateral migration of the mobile chl a/b LHC units without affecting the protein kinase that phosphorylates them. Thylakoids treated in this manner were unable to carry out State I–State II transitions. A more detailed treatment of State I–State II transitions can be found in chapter 11.5, this Volume by BARBER, which also provides a somewhat different perspective on this interesting topic.

4.3 Temperature-Induced Changes in Thylakoid Membrane Organization

4.3.1 Effects of Heat Stress

At high temperatures plants suffer a loss of photosynthetic activity that has its origin in both biochemical and biophysical processes (BERRY and BJÖRKMAN

1980; STIDHAM et al. 1982). Particularly affected are photophosphorylation, PS II-mediated electron transport rates, as well as certain soluble enzymes involved in CO_2 fixation. In addition, chlorophyll fluorescence-emission studies have revealed parallel losses in the efficiency of excitation energy transfer between pigment–protein complexes in the membranes (SCHREIBER and ARMOND 1978). These latter changes have been interpreted to result from a functional dissociation of the bulk light-harvesting chlorophylls from the reaction center complexes.

Freeze-fracture studies of heat-stressed chloroplasts have revealed that the functional dissociation of bulk antenna chlorophylls from reaction center chlorophylls is matched by changes in thylakoid structure. The temperature at which damage to the photosynthetic apparatus is detected is characteristic of a given plant. Plants that normally grow in cool environments tend to have a lower threshold temperature than plants growing in warmer environments. Some plants possess the ability to acclimate photosynthetically in response to changes in the prevailing growth temperature. *Nerium oleander* is a plant capable of both acclimating and tolerating relatively high growth temperatures of up to about 45 °C (ARMOND et al. 1980). The broad bean, *Vicia faba*, on the other hand, is a more temperate plant and is adversely affected by temperatures above 35 °C (GOUNARIS et al. 1984). Despite these differences, the chloroplast membranes of both types of plant exhibit the same structural responses to heat stress: (1) a decrease in the amount of stacked membrane regions coupled with structural alterations of such regions; (2) a decrease in size of EFs particles; and (3) the formation of phase-separated aggregates of non-bilayer-forming lipids (Fig. 41, ARMOND et al. 1980; GOUNARIS et al. 1984). The first two changes can be explained by a physical dissociation of bound chl a/b LHC units from PS II core complexes, and the subsequent exclusion of these subunits from the remaining stacked regions. The freeze-fracture appearance of the modified stacked regions suggests that they may be composed exclusively of mobile chl a/b LHC particles. It is not yet clear which specific parameters are responsible for the formation of aggregates of the nonbilayer-forming lipids. Most likely, the physical dissociation of the PS II complexes and/or the heat-induced partial denaturation of membrane proteins reduces the affinity between specific proteins and specific lipids (see SIEGENTHALER and RAWYLER, Chap. 11.9 and MURPHY, Chap. 11.11, both this Vol.), thereby lifting the constraints on the self-assembly of such lipids into three-dimensional arrays. In this context, it is of interest to note that pronase treatment of thylakoid membranes in the presence of Mg^{2+} also leads to the phase separation and aggregation of nonbilayer-forming lipids (CARTER and STAEHELIN 1980).

4.3.2 Chilling Effects

4.3.2.1 Higher Plants

Although there is a general consensus that chilling injury in plants is related to gel-to-liquid crystal phase transitions of membrane lipids, it is not clear which membrane system(s) is the primary locus of the injury. QUINN and W.P. WILLIAMS (1983) have critically reviewed the evidence in support of the hypothe-

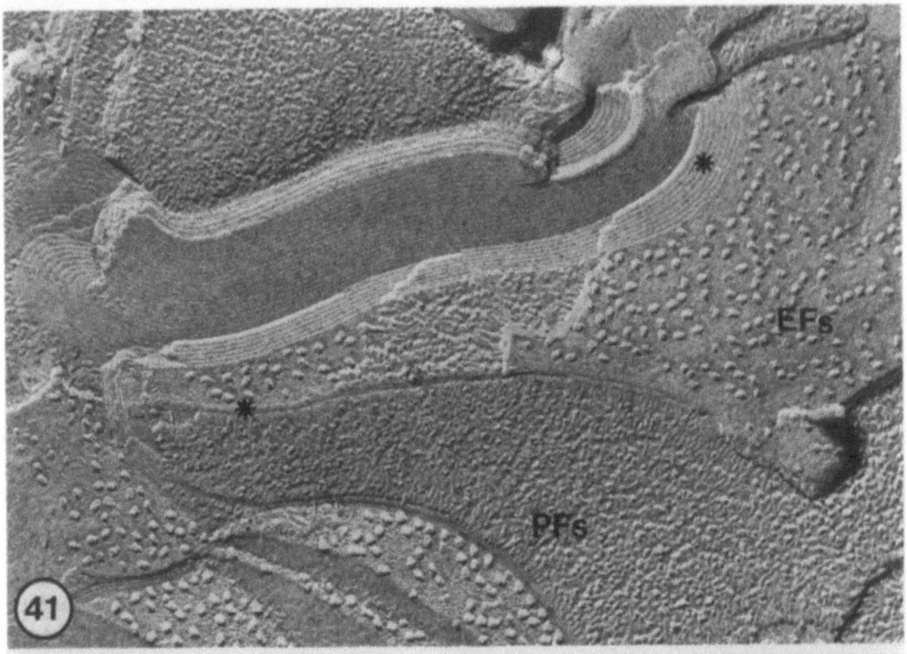

Fig. 41. Freeze-fractured pea thylakoids containing phase-separated aggregates (*) of non-bilayer-forming lipids (probably monogalactosyldiglycerides). The formation of such aggregates occurs when thylakoids have been subjected to a chemical (low pH, high salt, protease) or physical (high temperature, mechanical) stress. These isolated membranes were suspended for 2 h in 10 mM $MgCl_2$. Thylakoid fracture faces (EF_s, PF_s). ×95,000

Fig. 42. Freeze-fracture micrograph of an *Anacystis nidulans* (cyanobacterium) cell grown at 38 °C and cooled to 10 °C before being processed for electron microscopy. The chilling has induced a phase separation of membrane lipids, which manifests itself in the formation of smooth membrane patches devoid of particles (*). Note that such patches are seen both on the cytoplasmic membrane (*CM*) and the thylakoids (*T*). ×75,000

sis that a phase transition at about 10°–12 °C in chloroplast and mitochondrial membranes could be responsible for chilling injury in chilling-sensitive plants. They contend that because some chilling-resistant plants exhibit the same thylakoid membrane phase transition (measured as discontinuities of Arrhenius plots of Hill activity in chloroplasts) as chilling-sensitive plants, this phase transition is unlikely to represent a primary locus of injury. Furthermore, no evidence for chilling-induced, large-scale phase transitions of thylakoid membrane lipids has been obtained in freeze-fracture studies of higher plant chloroplasts. This is not unexpected, considering the high degree of unsaturation of the fatty acids associated with the mono- and digalactolipids, the two major types of chloroplast lipid (MURPHY, Chap. 11.11, this Vol.). Indeed, of the higher plant thylakoid lipids only the sulfolipids contain an appreciable amount of fully saturated fatty acid residues and exhibit a phase transition temperature above 0 °C.

4.3.2.2 Thermophilic Cyanobacteria

In contrast to higher plants, the membrane lipids of chilling-sensitive, thermophilic cyanobacteria are much more saturated and exhibit pronounced lipid phase transitions well above the freezing point. The onset of irreversible cell injury always seems to coincide with the onset of a phase transition (QUINN and W.P. WILLIAMS 1983; MURATA et al. 1984). The exact temperatures at which phase transitions are observed depends both on the organism and the growth temperature. In both *Anacystis nidulans* and *Anabaena variabilis* the thylakoid membranes enter their phase separation at a higher temperature than the plasma membrane. In *A. nidulans* grown at 38 °C the transition temperature for thylakoids is 23 °C, for the plasma membrane 13 °C (WADA et al. 1984). Interestingly, irreversible inactivation of photosynthesis only begins to appear at the temperature at which the lipids in the plasma membrane undergo a phase separation. This is also the temperature at which electrolytes begin to leak from the cells. Thus, chilling-induced inactivation of photosynthesis in cyanobacteria seems to be caused by the loss of solutes from the cytoplasm and not from the phase separation of thylakoid lipids, which is reversible (MURATA et al. 1984).

The structural changes associated with chilling-induced phase separation of membrane lipids in cyanobacteria have been investigated by a number of research groups (ARMOND and STAEHELIN 1979; BRAND et al. 1979; FURTADO et al. 1979; ONO and MURATA 1982). In all of these studies, both plasma membranes and thylakoid membranes cooled to below their phase transition temperature developed bilayer membrane patches devoid of intramembrane particles (Fig. 42). A quantitative analysis of the particles in the areas surrounding the patches performed by ARMOND and STAEHELIN (1979) indicated that the displacement of the integral membrane protein complexes occurs in both lateral and vertical directions. Returning the cells to their normal growth temperature (38 °C) resulted in the restoration of normal membrane morphology, suggesting that the proteins were not lost from the membrane. ONO and MURATA (1982) have also examined a cyanobacterium, *A. variabilis*, that is not sensitive to chilling. Their freeze-fracture micrographs revealed randomly organized thylakoid and plasma membrane particles both at the growth temperature as well as at chilling temperatures as low as 0 °C. By inference, this provides further circumstantial support for the hypothesis that chilling injury in cyanobacteria is caused by a phase separation of membrane lipids.

4.4 Prothylakoids and Prolamellar Bodies

Plastids that are formed in leaf cells of plants grown in the dark are known as "etioplasts". The prefix "etio" refers to the fact that these plastids are characteristic of etiolated (nonpigmented or whitened) leaves. They represent an arrested stage in the normal development of proplastids into chloroplasts. Exposure to white light leads to their transformation into chloroplasts. Etioplasts can be recognized by the presence of a three-dimensional tubular lattice known as the prolamellar body and associated prothylakoids (Fig. 43). The prothylakoid membranes are physically continuous with the wall elements of the prolamellar body. In nature, etioplasts are found only sparingly, due to the limited time that plants remain covered by earth after germination. When seen, their prolamellar bodies tend to be small and inconspicuous. The "classical" textbook etioplast with a large, semicrystalline prolamellar body (Fig. 43), is largly an "artifact" of laboratory growth conditions; optimal formation of prolamellar bodies requires about eight days of growth in the dark (LÜTZ,

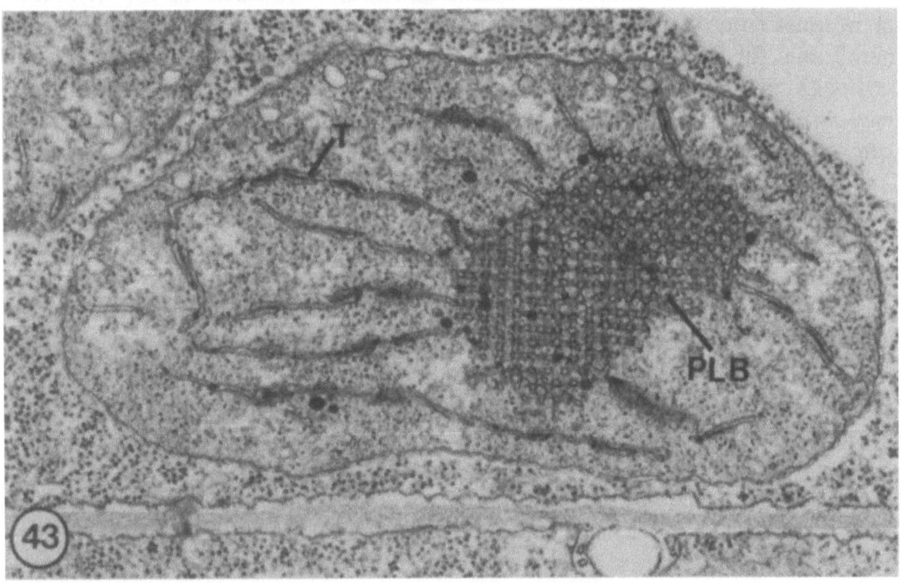

Fig. 43. Thin sectioned plastid from an intermittent light (7 days dark followed by 2 days of 2 min light/2 h dark treatments) pea plant. The plastid contains a large, well-organized prolamellar body (*PLB*), typical of seedlings kept for prolonged periods of time in the dark. Thylakoids (*T*) are attached to the prolamellar body margins. cf Fig. 44. ×40,000

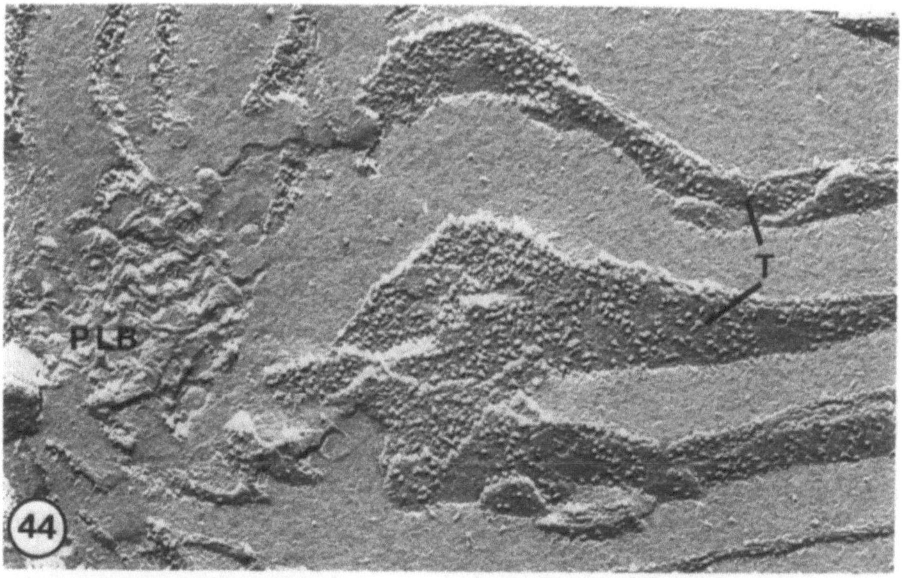

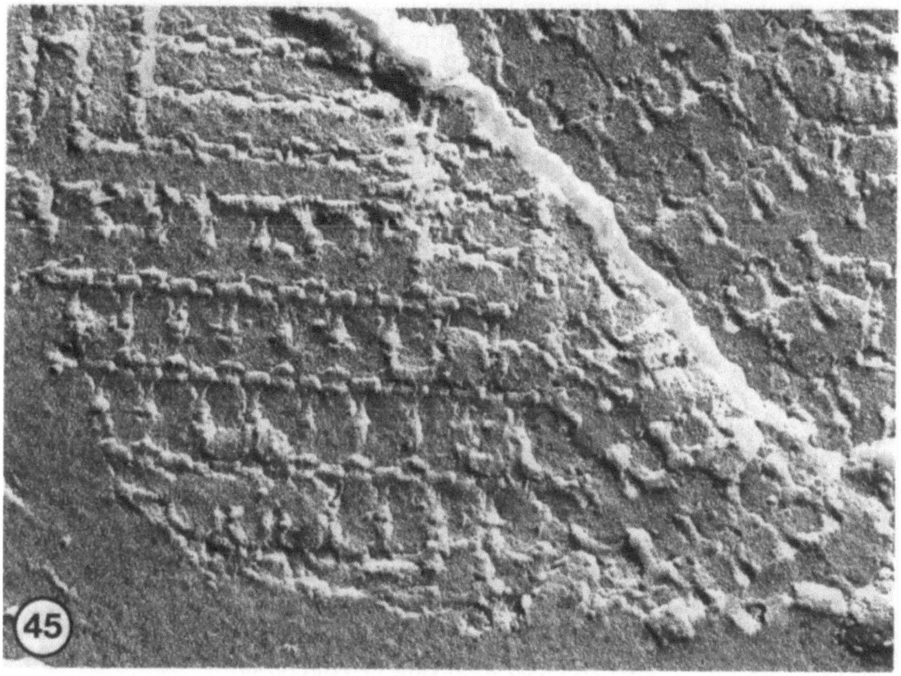

Fig. 44. Prolamellar body (*PLB*) with attached thylakoids (*T*) from an intermittent light plastid. The difference in composition of the two types of structures is striking. ×35,000

Fig. 45. Freeze-fractured, isolated prolamellar body from an etioplast. Note the absence of particles from the smooth tubular elements of the prolamellar body lattice. ×130,000

Chap. 11.8, this Vol.; KIRK and TILNEY-BASSETT 1978; WELLBURN 1982). Prolamellar bodies can only develop in cells undergoing differentiation and in which chloroplasts are actively forming new thylakoids.

As discussed in greater detail by LÜTZ in Chapter 11.8, this Volume, prolamellar bodies are composed mostly of lipids, mono-(MGDG) and di-(DGDG) galactolipids being the major lipid classes; phosphatidylglycerol is the main phospholipid. Contrary to some recent reports, no saponins are found in carefully prepared prolamellar body fractions. The high degree of curvature associated with the tubular lattice configuration of prolamellar body "membranes" is generally attrubuted to their high MGDG to DGDG ratio, since the cone shape of the MGDG molecules (see MURPHY, Chap. 11.11, this Vol.) favors rounded or tubular lipid assemblies over planar bilayer structures. "Ribosome-like" particles that are often seen in the stroma species of thin sectioned prolamellar bodies have been implicated as lattice stabilizers, even though their in vitro removal does not affect short-term stability of isolated prolamellar bodies (WELLBURN 1977).

A freeze-fracture electron micrograph of a prolamellar body from a pea etioplast is shown in Fig. 44. The most notable feature is the absence of intramembrane particles from the wall elements of the three-dimensional tubular lattice, confirming the largely lipidic nature of prolamellar bodies. In contrast, the attached prothylakoids contain large numbers of intramembrane particles (Fig. 45; SIMPSON 1978a). Some earlier studies suggested that certain enzymes such as protochlorophyllide reductase and thus the site of synthesis for chlorophyll precursors may be preferentially located in the prolamellar bodies. More recent work, however, has discounted this hypothesis by showing that most of this enzyme and its substrate protochlorophyllide are primarily located in the prothylakoids (RYBERG 1983). Similarly, CF_0/CF_1 ATP synthase complexes in etioplasts have also been found exclusively on prothylakoid membranes (WELLBURN 1977). Based on these studies, we must conclude that the primary function of a prolamellar body is to serve as a storage structure for thylakoid lipids, particularly MGDG and DGDG. Its crystalline substructure appears to exclude most, if not all, of the enzyme complexes associated with the physically continuous prothylakoid membranes. An excellent treatment of the bioenergetic, chemical, and ultrastructural changes associated with the conversion of etioplasts to chloroplasts can be found in the review of WELLBURN (1982).

Acknowledgements. Thanks are due to Drs. D. ALLRED, and T. DUNAHAY for reviewing the manuscript, to M. DEWIT for technical assistance, to J. LOGAN for preparing the drawings, and to Drs. D. ALLRED, E. GANTT, S. GIBBS, T. GIDDINGS, K. MILLER, E. MÖRSCHEL and R. WILLIAMS for supplying electron micrographs. This work was supported in part by grants NIH-GM 22912 and NSF-PCM 8118627.

References

Akey CW, Crepeau RH, Dunn SD, McCarty RE, Edelsten SJ (1983) Electron microscopy and single molecule averaging of subunit-deficient F_1-ATPase from *E. coli* and spinach chloroplasts. EMBO J 2:1409–1415

Allred DR, Staehelin LA (1985) Intra-thylakoid distribution of cytochrome b_6/f. Plant Physiol 78:199–202
Amesz J (1983) Photosynthesis: Photosystems in green plants and green bacteria. Prog Bot 45:89–105
Anderson JM (1982a) Distribution of the cytochromes of spinach chloroplasts between the appressed membranes of grana stacks and stroma-exposed thylakoid regions. FEBS Lett 138:62–66
Anderson JM (1982b) The significance of grana stacking in chlorophyll b-containing chloroplasts. Photochem Photobiophys 3:225–241
Anderson JM (1982c) The role of chlorophyll-protein complexes in the function and structure of chloroplast thylakoids. Mol Cell Biochem 46:161–172
Anderson JM, Melis A (1983) Localization of different photosystems in separate regions of chloroplast membranes. Proc Natl Acad Sci USA 80:745–749
Anderson JM, Goodchild DJ, Boardman NK (1973) Composition of the photosystems and chloroplast structure in extreme shade plants. Biochim Biophys Acta 325:573–585
Anderson JW (1981) Light-energy-dependent processes other than CO_2 assimilation. In: Hatch MD, Boardman NK (eds) The biochemistry of plants: photosynthesis. Academic Press, London New York, pp 473–500
Andersson B, Akerlund HE (1978) Inside-out membrane vesicles isolated from spinach thylakoids. Biochim Biophys Acta 503:462–472
Andersson B, Anderson JM (1980) Lateral heterogeneity in the distribution of the chlorophyll–protein complexes of the thylakoid membranes of spinach chloroplasts. Biochim Biophys Acta 593:427–440
Andersson B, Anderson JM, Ryrie IJ (1982) Transbilayer organization of the chlorophyll-proteins of spinach thylakoids. Eur J Biochem 123:465–472
Argyroudi-Akoyunoglou JH, Feleki, Akoyunoglou G (1971) Formation of two chlorophyll–protein complexes during greening of etiolated bean leaves. Biochem Biophys Res Commun 45:606–614
Armond PA, Arntzen CJ (1977) Localization and characterization of photosystem II in grana and stroma lamellae. Plant Physiol 59:398–404
Armond PA, Staehelin LA (1979) Lateral and vertical displacement of integral membrane proteins during lipid phase transition in *Anacystis nidulans*. Proc Natl Acad Sci USA 76:1901–1905
Armond PA, Arntzen CJ, Briantais JM, Vernotte C (1976) Differentiation of chloroplast lamellae. I. Light-harvesting efficiency and grana development. Arch Biochem Biophys 175:400–418
Armond PA, Staehelin LA, Arntzen CJ (1977) Spatial relationship of photosystem I, photosystem II, and the light-harvesting complex in chloroplast membranes. J Cell Biol 73:400–418
Armond PA, Björkman O, Staehelin LA (1980) Dissociation of supramolecular complexes in chloroplast membranes. A manifestation of heat damage to the photosynthetic apparatus. Biochim Biophys Acta 601:433–442
Arntzen CJ (1978) Dynamic structural features of chloroplast lamellae. Curr Top Bioenerg 8:111–160
Barber J (1980) An explanation for the relationship between salt-induced thylakoid stacking and the chlorophyll fluorescence changes associated with changes in spillover of energy from photosystem II to photosystem I. FEBS Lett 118:1–10
Barber J (1984) Lateral heterogeneity of proteins and lipids in the thylakoid membrane and its implications for electron transport. In: Sybesma C (ed) Advances in photosynthesis research, Vol III. Nijhoff, Junk, The Hague, pp 2:91–98
Barr R, Arntzen CJ (1969) The occurrence of α-tocophenylquinone in higher plants and its relation to senescence. Plant Physiol 44:591–598
Barton R (1970) The production and behavior of phytoferritin particles during senescence of *Phaseolus* leaves. Planta 94:73–77
Bennett J (1983) Regulation of photosynthesis by reversible phosphorylation of the light-harvesting chlorophyll a/b protein. Biochem J 212:1–13
Berkaloff C, Duval JC, Hauswirth N, Rosseau B (1983) Freeze-fracture study of thylakoids of *Fucus serratus*. J Phycol 19:96–100

Berry J, Björkman O (1980) Photosynthetic response and adaptation to temperature in higher plants. Annu Rev Plant Physiol 31:491–543

Berthold DA, Babcock GT, Yocum CF (1981) A highly resolved, oxygen-evolving photosystem II preparation from spinach thylakoid membranes. FEBS Lett 134:231–234

Binder A (1982) Respiration and photosynthesis in energy-transducing membranes of cyanobacteria. J Bioenerg Biomembr 14:271–288

Bisalputra T (1974) Plastids. In: Stewart WDP (ed) Algal physiology and biochemistry. Univ Calif, Berkeley, Los Angeles, pp 124–160

Björkman O (1981) Responses to different quantum flux densities. In: Lange OK, Nobel PA, Osmond CB, Ziegler H (eds) Encyclopedia of plant physiology, New Ser, vol 12A. Physiological plant ecology. Springer, Berlin Heidelberg New York, pp 57–107

Björkman O, Ludlow M (1972) Characterization of the light climate on the floor of a Queensland rain forest. Carnegie Inst Washington Yearb 71:85–94

Bonaventura C, Myers J (1969) Fluorescence and oxygen evolution from *Chlorella pyrenoidosa*. Biochim Biophys Acta 189:366–383

Bouck GB (1965) Fine structure and organelle associations in brown algae. J Cell Biol 26:523–537

Brand JJ, Kirkanski SJ, Ranurez-Mitchell R (1979) Chill-induced morphological alterations in *Anacystis nidulans* as a function of growth temperature. Planta 145:63–68

Brandley BK (1981) Ultrastructure of the envelope of *Codium australicum* (Silva) chloroplasts in the alga and after acquisition by *Elysia maoria* (Powell). New Phytol 89:679–686

Briantais JM, Vernotte C, Lavorel J, Olive J, Wollman FA (1983) Kinetics of cation-induced changes of photosystem II fluorescence and of lateral distribution of the two photosystems in the thylakoid membranes of pea chloroplasts. Biochim Biophys Acta 766:1–8

Brown RM (1967) The pyrenoid: its structure distribution and function. J Phycol (Suppl) 3:5–7

Burke JJ, Steinback KE, Arntzen CJ (1979) Analysis of the light-harvesting pigment-protein complex of wild-type and chlorophyll-b-less mutant barley. Plant Physiol 63:237–243

Butler WL (1977) Chlorophyll fluorescence: a probe for electron transfer and energy transfer. In: Trebst A, Avron M (eds) Photosynthesis I. Photosynthetic electron transfer and photophosphorylation. Encyclopedia of plant physiology, vol V. Springer, Berlin Heidelberg New York, pp 149–167

Butterfass T (1979) Patterns of chloroplast reproduction. Cell biology monographs, vol VI. Springer, Berlin Heidelberg New York

Camm EL, Green BR (1980) Fractionation of thylakoid membranes with the nonionic detergent octyl-β-D-glucopyranoside. Resolution of chlorophyll–protein complex II into two chlorophyll-protein complexes. Plant Physiol 66:428–432

Camm EL, Green BR (1983) Relationship between the two minor chlorophyll a – protein complexes and the photosystem II reaction center. Biochim Biophys Acta 724:291–293

Canaani O, Barber J, Malkin S (1984) Evidence for phosphorylation and dephosphorylation regulate the distribution of excitation energy between the two photosystems of photosynthesis in vivo: Photoacoustic and fluorimetric study of an intact leaf. Proc Natl Acad Sci USA 81:1614–1618

Carde JP, Joyard J, Douce R (1982) Electron microscopic studies of envelope membranes from spinach plastids. Biol Cell 44:315–324

Carter DP, Staehelin LA (1980) Proteolysis of chloroplast thylakoid membranes. II. Evidence for the involvement of the light-harvesting chlorophyll a/b protein complex in thylakoid stacking and for effects of magnesium ions on photosystem II-light-harvesting complex aggregates in the absence of membrane stacking. Arch Biochem Biophys 200:374–386

Chow WS, Telfer A, Chapman DJ, Barber J (1981) State 1–State 2 transition in leaves and its association with ATP-induced chlorophyll fluorescence quenching. Biochim Biophys Acta 638:60–68

Chua NH, Schmidt GW (1979) Transport of proteins into mitochondria and chloroplasts. J Cell Biol 81:461–483

Cline K, Keegstra K (1983) Galactosyl-transferases involved in galactolipid biosynthesis are located in the outer membrane of pea chloroplast envelopes. Plant Physiol 71:366–372

Cline K, Andrews J, Mersey B, Newcomb EH, Keegstra K (1981) Separation and characterization of inner and outer envelope membranes of pea chloroplasts. Proc Natl Acad Sci USA 78:3593–3599

Cline K, Keegstra K, Staehelin LA (1985) Freeze-fracture electron microscopic analysis of ultrarapidly frozen envelope membranes on intact chloroplasts and after purification. Protoplasma 125:111–123

Coombs J, Greenwood AD (1976) Compartmentation of the photosynthetic apparatus. In: Barber J (ed) The intact chloroplast. Elsevier, Amsterdam New York Oxford, pp 1–51

Cox RP, Andersson B (1981) Lateral and traverse organization of cytochromes in the chloroplast thylakoid membrane. Biochem Biophys Res Commun 103:1336–1342

Cox G, Dwarte DM (1981) Freeze-etch ultrastructure of a *Prochloron* species – the symbiont of *Didemnum molle*. New Phytol 88:427–438

Cox G, Benson D, Dwarte DW (1981) Ultrastructure of a cave-wall cyanophyte-*Gloeocapsa* NS4. Arch Mikrobiol 130:165–174

Day DA, Ryrie IJ, Fuad N (1984) Investigations of the role of the main light-harvesting chlorophyll-protein complex in thylakoid membranes. Reconstitution of depleted membranes from intermittent-light-grown plants with the isolated complex. J Cell Biol 97:163–172

Diner BA (1979) Energy transfer from the phycobilisomes to photosystem II reaction centers in wild-type *Cyanidium caldarium*. Plant Physiol 63:30–34

Dodge JD (1969) A review of the fine structure of algal eyespots. Br Phycol J 4:199–210

Dodge JD (1973) The fine structure of algal cells. Academic Press, London New York

Douce R, Joyard J (1979) Structure and function of the plastid envelope. Adv Bot Res 7:1–116

Dubertret G, Lefort-Tran M (1981) Functional and structural organization of chlorophyll in the developing photosynthetic membranes of *Euglena gracilis* Z. IV. Light-harvesting properties of system II photosynthetic units and thylakoid ultrastructure during greening under intermittent light. Biochim Biophys Acta 634:52–69

Dunahay TG, Staehelin LA (1985) Isolation of photosystem I complexes from octylglucoside/SDS solubilized spinach thylakoids: Characterization and reconstitution into liposomes. Plant Physiol 78:606–613

Dunahay TG, Staehelin LA, Seibert M, Ogilvie PD, Berg SP (1984) Structural, biochemical and biophysical characterization of four oxygen-evolving photosystem II preparations from spinach. Biochim Biophys Acta 764:179–193

Dwarte DM, Vesk M (1982) Freeze-fracture thylakoid ultrastructure of representative numbers of "chlorophyll c" algae. Micron 13:325–326

Dwarte D, Vesk M (1983) A freeze-fracture study of cryptomonad thylakoids. Protoplasma 117:130–141

Echlin P (1970) The photosynthetic apparatus in prokaryotes and eukaryotes. In: Charles HP, Knight BCJG (eds) Organization and control of prokaryotic and eukaryotic cells. Cambridge Univ Press, Cambridge, pp 221–248

Edwards MR, Gantt E (1971) Phycobilisomes of the thermophilic blue-green alga *Synechococcus lividus*. J Cell Biol 70:896–900

Ellis RJ (1976) Protein and nucleic acid synthesis by chloroplasts. In: Barber J (ed) The intact chloroplast. Elsevier, Amsterdam New York Oxford, pp 335–364

Evans LV (1968) Chloroplast morphology and fine structure in British fucoids. New Phytol 67:173–178

Fedtke C, Deichgräber G, Schnepf E (1977) Herbicide-induced changes in wheat chloroplast ultrastructure and chlorophyll a/b ratio. Biochem Physiol Pflanz 171:307–312

Foster KW, Smyth RD (1980) Light antennae in phototactic algae. Microb Rev 44:572–630

Furtado D, Williams WP, Brain APR, Quinn PJ (1979) Phase separations in membranes of *Anacystis nidulans* grown at different temperatures. Biochim Biophys Acta 555:352–357

Gantt E (1981) Phycobilisomes. Annu Rev Plant Physiol 32:327–347
Gantt E, Conti SF (1965) The ultrastructure of *Porphyridium cruentum*. J Cell Biol 26:365–381
Garber MP, Steponkus PL (1976) Alterations in chloroplast thylakoids during cold acclimation. Plant Physiol 57:681–686
Ghirardi ML, Melis A (1983) Localization of photosynthetic electron transport components in mesophyll and bundle-sheath chloroplasts of *Zea mays*. Arch Biochem Biophys 224:19–28
Gibbs SP (1962) Nuclear envelope-chloroplast relationships in algae. J Cell Biol 14:433–444
Gibbs SP (1970) The comparative ultrastructure of the algal chloroplast. Ann NY Acad Sci 175:454–473
Gibbs SP (1981) The chloroplast endoplasmic reticulum: Structure, function, and evolutionary significance. Int Rev Cytol 72:49–99
Giddings TH, Staehelin LA (1979) Changes in thylakoid structure associated with differentiation of heterocysts in the cyanobacterium *Anabaena cylindrica*. Biochim Biophys Acta 546:373–382
Giddings TH, Withers NW, Staehelin LA (1980) Changes in thylakoid structure of stacked and unstacked regions of the photosynthetic membranes of *Prochloron* sp., a prokaryote. Proc Natl Acad Sci USA 77:352–356
Giddings TH, Wasman C, Staehelin LA (1983) Structure of the thylakoids and envelope membranes of the cyanelles of *Cyanophora paradoxa*. Plant Physiol 71:409–419
Giles GL, Sarafis V (1972) Chloroplast survival and division in vitro. Nature New Biol 236:56–58
Gilkey JC, Staehelin LA (1986) Critical evaluation of ultrarapid freezing techniques for the preservation of cellular ultrastructure. J Electron Microsc Tech 3 in press
Gimmler H (1977) Photophosphorylation in vivo. In: Trebst A, Avron M (eds) Photosynthesis I. Encyclopedia of plant physiology, New Ser, vol V. Springer, Berlin Heidelberg New York, pp 448–472
Glazer AN (1980) Structure and evolution of photosynthetic accessory pigment systems with special reference to phycobiliproteins. In: Sigman D, Brazier MAB (eds) The evolution of protein structure and function. Academic Press, London New York, pp 221–244
Glazer AN (1983) Comparative biochemistry of photosynthetic light-harvesting systems. Annu Rev Biochem 52:125–157
Golecki J (1979) Ultrastructure of cell wall and thylakoid membranes of the thermophilic cyanobacterium *Anacystis nidulans*. Arch Biochem Biophys 220:155–166
Goodchild DJ, Björkman O, Pyliotis NA (1972) Chloroplast ultrastructure, leaf anatomy, and content of chlorophyll and soluble protein in rainforest species. Carnegie Inst Washington Yearb 71:102–107
Goodenough UW, Staehelin LA (1971) Structural differentiation of stacked and unstacked chloroplast membranes. Freeze-etch electron microscopy of wild-type and mutant strains of *Chlamydomonas*. J Cell Biol 48:594–619
Gounaris K, Brain ARR, Quinn PJ, Williams WP (1984) Structural reorganization of chloroplast thylakoid membranes in response to heat stress. Biochim Biophys Acta 766:198–208
Guglielmi G, Cohen-Bazire G, Bryant DA (1981) The structure of *Gloeobacter violaceus* and its phycobilisomes. Arch Microbiol 129:181–189
Gunning BES (1965) The fine structure of chloroplast stroma following aldehyde-osmium tetroxide fixation. J Cell Biol 24:79–93
Gunning BES, Steer MW (1975) Ultrastructure and the biology of plant cells. Arnold, London
Hardie LP, Balkwill DL, Stevens SE (1983) Effects of iron starvation on the ultrastructure of the cyanobacterium *Agmenellum quadruplicatum*. Appl Environ Microbiol 45:1007–1017
Haworth P (1983) Protein phosphorylation-induced State I–State II transitions are dependent on thylakoid membrane viscosity. Arch Biochem Biophys 226:145–154

Haworth P, Watson JL, Arntzen CJ (1983) The detection, isolation and characterization of a light-harvesting complex which is specifically associated with photosystem I. Biochim Biophys Acta 724:151–158
Heldt HW, Sauer F (1971) The inner membrane of the chloroplast envelope as the site of specific metabolite transport. Biochim Biophys Acta 234:83–91
Henry LE, Moller BL (1981) Polypeptide composition of an oxygen evolving photosystem II vesicle from spinach chloroplasts. Carlsberg Res Commun 46:227–242
Heslop-Harrison J (1963) Structure and morphogenesis of lamellar systems in grana-containing chloroplasts. I. Membrane structure and lamellar architecture. Planta 60:243–260
Holdsworth RH (1971) The isolation and partial characterization of the pyrenoid protein of *Eresmosphaera viridis*. J Cell Biol 51:499–512
Hori T, Ueda R (1975) The fine structure of algal chloroplasts and algal phylogeny. In: Tokida J, Hirose H (eds) Advance of phycology in Japan. Fischer, Jena, pp 11–42
Horton P, Croze E (1979) Characterizations of two quenchers of chlorophyll fluorescence with different and midpoint oxidation-reduction potentials in chloroplasts. Biochim Biophys Acta 545:188–201
Howell SH, Moudrianakis EN (1967) Hill reaction site in chloroplast membranes: non-participation of the quantasome particle in photoreduction. J Mol Biol 27:323–333
Ish-Shalom D, Ohad I (1983) Organization of chlorophyll–protein complexes of photosystem I in *Chlamydomonas reinhardii*. Biochim Biophys Acta 722:498–507
Izawa S, Good NE (1966) Effect of salts and electron transport on the conformation of isolated chloroplasts. II. Electron microscopy. Plant Physiol 41:544–552
Jaynes JM, Vernon LP (1982) The cyanelle of *Cyanophora paradoxa*: almost a cyanobacterial chloroplast. TIBS 7:22–24
Jennings RC, Forti G, Gerola D, Garlashi FM (1978) Studies on cation-induced thylakoid membrane stacking, fluorescence yield, and photochemical efficiency. Plant Physiol 62:879–884
Kahn MU, Lem NW, Chandorkar KR, Williams JP (1979) Effects of substituted pyridazinones (SAN 6706, SAN 9774, SAN 9785) on galactolipids and their associated fatty acids in the leaves of *Vicia faba* and *Hordeum vulgare*. Plant Physiol 64:300–305
Kalosakas K, Akoyunoglou G, Ohad I (1984) Relationship between the ultrastructure of freeze-fractured thylakoids and the composition of the thylakoid membrane during development of *Phaseolus vulgaris*. In: Sybesma C (ed) Advances in photosynthesis research, vol IV. Nijhoff Junk, The Hague, pp 6:645–648
Kaplan S, Arntzen CJ (1982) Photosynthetic membrane structure and function. In: Govindjee (ed) Photosynthesis: Comparative aspects of bacteria and green plants. Academic Press, London New York, pp 65–153
Keck RW, Dilley RA, Allen CF, Biggs S (1970) Chloroplast composition and structure differences in a soybean mutant. Plant Physiol 46:692–698
Keegstra K (1984) The role of the envelope membranes in chloroplast biogenesis. In: Randall DD, Blevins DG, Larson RL, Rapp BJ (eds) Current topics in plant biochemistry and physiology, vol III. Univ Missouri, Columbia, pp 90–99
Kennedy RA, Barnes JE, Laetsch WM (1977) Photosynthesis in C_4 plant tissue cultures. Significance of Kranz anatomy to C_4 acid metabolism in C_4 plants. Plant Physiol 59:600–603
Kirk JTO, Tilney-Bassett RAE (1978) The plastids. Elsevier/North-Holland Biomed Press, Amsterdam New York
Kühlbrandt W (1984) Three-dimensional structure of the light-harvesting chlorophyll a/b–protein complex. Nature 307:478–480
Kühlbrandt W, Thaler T, Wehrli E (1983) The structure of membrane crystals of the light-harvesting chlorophyll a/b–protein complex. J Cell Biol 96:1414–1424
Kursar TA, Alberte RS (1983) Photosynthetic unit organization in a red alga. Plant Physiol 72:409–414
Kuwabara T, Murata N (1982) An improved purification method and a further character-

ization of the 33 kd protein of spinach chloroplasts. Biochim Biophys Acta 680:210–215

Kyle DJ, Staehelin LA, Arntzen CJ (1983) Lateral mobility of the light harvesting complex in chloroplast membranes controls excitation energy distribution in higher plants. Arch Biochem Biophys 222:527–541

Lacambra M, Larsen U, Olive J, Bennoun P, Wollman FA (1984) Characterization of the thylakoid membranes of wild type and mutants of *Chlorella sorokiniana*. Photobiochem Photobiophys submitted

Laetsch WM (1968) Chloroplast specialization in dicotyledons possessing the C_4-dicarboxylic acid pathway of photosynthetic CO_2 fixation. Am J Bot 55:875–883

Laetsch WM (1974) The C_4 syndrome: a structural analysis. Annu Rev Plant Physiol 25:27–52

Larsson U, Jergil B, Andersson B (1983) Changes in the lateral distribution of the light-harvesting chlorophyll a/b–protein complex induced by its phosphorylation. Eur J Biochem 136:25–29

Leech RM, Walton CA (1983) Modification of fatty acid composition during chloroplast ontogeny and the effects on thylakoid appression and primary photochemistry. In: Thompson WW, Mudd JB, Gibbs M (eds) Biosynthesis and function of plant lipids. Am Soc Plant Physiol, Rockville, pp 56–80

Lefort-Tran M (1981) The triple-layered organization of the *Euglena* chloroplast envelope. Ber Dtsch Bot Ges 94:463–476

Lefort-Tran M, Cohen-Bazire G, Pouphile M (1973) Les membranes photosynthétiques des algues à biliproteins observées après cryodécapage. J Ultrastruct Res 44:199–209

Leto KL, Keresztes A, Arntzen CJ (1982) Nuclear involvement in the appearance of a chloroplast-encoded 32,000 dalton thylakoid membrane polypeptide integral to the photosystem II complex. Plant Physiol 69:1450–1458

Lewin RA (1976) Prochlorophyta as a proposed new division of algae. Nature 261:697–698

Lewin RA (1977) *Prochloron*, type genus of the Prochlorophyta. Phycologia 16:217–222

Ley AC, Butler WC (1976) Efficiency of energy transfer from photosystem II to photosystem I in *Porphyridium cruentum*. Proc Natl Acad Sci USA 73:3957–3960

Li J (1985) Light-harvesting chlorophyll a/b-protein: Three-dimensional structure of a reconstituted membrane lattice in negative stain. Proc Natl Acad Sci USA 82:386–390

Li J, Hollingshead C (1982) Formation of crystalline arrays of chlorophyll a/b-light harvesting protein by membrane reconstitution. Biophys J 37:363–370

Lichtenthaler HK (1968) Plastoglobuli and the fine structure of plastids. Endeavour 27:144–149

Lichtenthaler HK, Sprey B (1966) Über die osmiophilen globulären Lipideinschlüsse der Chloroplasten. Z Naturforsch 21b:690–697

Lichtenthaler HK, Bushman C, Doll M, Fietz HJ, Bach T, Kozel U, Meier D, Rahmsdorf U (1981) Photosynthetic activity, chloroplast ultrastructure, and leaf characteristics of high-light and low-light plants and of sun and shade leaves. Photosynth Res 2:115–141

Lichtlé C, Thomas JC (1976) Étude ultrastructurale des thylakoides des algues à phycobiliproteines, comparaison des résultats obtenus par fixation classique et cryodécapage. Phycologia 15:393–404

Machold O, Meister A (1979) Resolution of the light-harvesting chlorophyll a/b-protein of *Vicia faba* chloroplasts into two different chlorophyll-protein complexes. Biochim Biophys Acta 546:472–480

Matoo AK, St John JB, Wergin WP (1984) Adaptive reorganization of protein and lipid components in chloroplasts membranes as associated with herbicide binding. J Cell Biochem 24:163–175

McCarty RE, Racker E (1966) Effect of a coupling factor and its antiserum on photophosphorylation and hydrogen ion transport. Brookhaven Symp Biol 19:202–214

McDonnel A, Staehelin LA (1980) Adhesion between liposomes mediated by the chlorophyll a/b light-harvesting complex isolated from chloroplast membranes. J Cell Biol 84:40–56

Meier D, Lichtenthaler HK, Burkhard G (1980) Change of chloroplast ultrastructure in radish seedlings and the influence of the photosystem II-herbicide bentazon. Z Naturforsch 35c: 656–664

Melis A, Duysens LNM (1979) Biphasic energy conversion kinetics and absorbance difference spectra of photosystem II of chloroplasts. Evidence for two different photosystem II reaction centers. Photochem Photobiol 29: 373–382

Melis A, Homan PH (1976) Heterogeity of the photochemical centers in system II of chloroplasts. Photochem Photobiol 23: 343–350

Melis A, Homan PH (1978) A selective effect of Mg^{2+} on the photochemistry at one type of reaction center in photosystem II of chloroplasts. Arch Biochem Biophys 190: 523–530

Melkonian M, Robenek H (1979) The eyespot of the flagellate *Tetraselmis cordiformis* Stein (chlorophyceae): Structural specialization of the outer chloroplast membrane and its possible significance in phototaxis of green algae. Protoplasma 100: 183–197

Melkonian M, Robenek H (1980) Eyespot membranes of *Chlamydomonas reinhardii*: A freeze-fracture study. J Ultrastruct Res 72: 90–102

Menke W (1962) Structure and chemistry of plastids. Annu Rev Plant Physiol 13: 27–44

Michel H, Tellenbach M, Borschetti A (1983) A chlorophyll b-less mutant of *Chlamydomonas reinhardii* lacking in the light-harvesting chlorophyll a/b-protein complex but not its apoproteins. Biochim Biophys Acta 725: 417–424

Miller KR (1980) A chloroplast membrane lacking photosystem I. Changes in unstacked membrane regions. Biochim Biophys Acta 592: 143–152

Miller KR, Cushman RA (1979) A chloroplast membrane lacking photosystem II. Thylakoid stacking in the absence of the photosystem II particle. Biochim Biophys Acta 546: 481–497

Miller KR, Staehelin LA (1973) Fine structure of the chloroplast membranes of *Euglena gracilis* as revealed by freeze-cleaving and deep-etching. Protoplasma 77: 55–78

Miller KR, Staehelin LA (1976) Analysis of the thylakoid outer surface: Coupling factor is limited to unstacked membrane regions. J Cell Biol 68: 30–47

Miller KR, Miller GJ, McIntyre KR (1976) The light-harvesting chlorophyll protein complex of photosystem II. J Cell Biol 71: 624–638

Miller KR, Miller GJ, McIntyre KR (1977) Organization of the photosynthetic membrane in maize mesophyll and bundle-sheath chloroplasts. Biochim Biophys Acta 459: 145–156

Miller LS, Holt SC (1977) Effect of carbon dioxide on pigment and membrane content in *Synechococcus lividus*. Arch Mikrobiol 115: 185–198

Möbius M (1920) Über die Grösse der Chloroplasten. Ber Dtsch Bot Ges 38: 224–232

Mörschel E, Mühlethaler (1983) On the linkage of exoplasmic freeze-fracture particles to phycobilisomes. Planta 158: 451–457

Mörschel E, Staehelin LA (1983) Reconstitution of cytochrome f/b_6 and CF_0/CF_1 ATP synthetase complexes into phospholipid and galactolipid liposomes. J Cell Biol 97: 301–310

Mörschel E, Wehrmeyer W (1979) Elektronenmikroskopische Feinstrukturanalyse von nativen Biliproteidaggregaten und deren räumliche Anordnung. Ber Dtsch Bot Ges 92: 393–402

Mörschel E, Koller KP, Wehrmeyer W, Schneider H (1977) Biliprotein assembly in the disc-shaped phycobilisomes of *Rhodella violacea*. I. Electron microscopy of phycobilisomes in situ and analysis of their architecture after isolation and negative staining. Cytobiology 16: 118–129

Moreland DE (1980) Mechanisms of action of herbicides. Annu Rev Plant Physiol 31: 597–638

Mullet JE, Arntzen CJ (1980) Stimulation of grana stacking in a model membrane system. Mediation by a purified light-harvesting pigment-protein complex from chloroplasts. Biochim Biophys Acta 589: 100–117

Mullet JE, Arntzen CJ (1981) Identification of a 35–32 kilodalton polypeptide as a herbicide acceptor protein in photosystem II. Biochim Biophys Acta 635: 236–248

Mullet JE, Burke JJ, Arntzen CJ (1980) Chlorophyll-proteins of photosystem I. Plant Physiol 65:814–822

Mullet JE, Baldwin TO, Arntzen CJ (1981a) A mechanism for chloroplast thylakoid adhesion mediated by the chl a/b light-harvesting complex. In: Akoyunoglou G (ed) Photosynthesis III. Structure and molecular organization of the photosynthetic apparatus. Balaban Int Sci Serv, Philadelphia, pp 557–582

Mullet JE, Pick U, Arntzen CJ (1981b) Structural analysis of the isolated chloroplast coupling factor and the N-N′dicyclohexylcarbodiimide binding proteolipid. Biochim Biophys Acta 642:149–157

Murakami S, Torres-Pereira J, Packer L (1975) Structure of the chloroplast membrane-relation to energy coupling and ion transport. In: Govindjee (ed) Bioenergetics of photosynthesis. Academic Press, London New York, pp 555–618

Murata N (1969) Control of excitation transfer in photosynthesis. I. Light-induced change of chlorophyll a fluorescence in *Porphyridium cruentum*. Biochim Biophys Acta 172:242–251

Murata N, Wada H, Hirasawa R (1984) Reversible and irreversible inactivation of photosynthesis in relation to the lipid phrases of membranes in the blue-green algae (cyanobacteria) *Anacystis nidulans* and *Anabaena variabilis*. Plant Cell Physiol 25:1027–1032

Mustardy LA, Brangeon J (1978) Three-dimensional chloroplast intrastructure: Developmental aspects. In: Akoyunoglou G, Argyroudi-Akoyunoglou JH (eds) Chloroplast development. Elsevier/North-Holland Biomed Press, Amsterdam New York, pp 489–494

Nelson N, Nelson H, Schatz G (1980) Biosynthesis and assembly of the proton-translocating adenosine triphosphate complex from chloroplasts. Proc Natl Acad Sci USA 77:1361–1364

Neuschul M (1971) Uniformity of thylakoid structure in a red, a brown and two blue-green algae. J Ultrastruct Res 37:532–543

Nierzwicki-Bauer SA, Balkwill DL, Stevens SE (1983) Three-dimensional ultrastructure of a unicellular cyanobacterium. J Cell Biol 97:713–722

Ohad I, Goldberg I, Broza R, Schuldiner S, Gan-Zvi E (1969) Changes in lipid and pigment composition and photosynthetic activity during formation of chloroplast lamellae in a mutant of *Chlamydomonas reinhardtii* y-1. In: Metzner H (ed) Progress in photosynthesis research, vol I. Int Un Biol Sci, Tuebingen, pp 284–295

Ojakian GK, Satir P (1974) Particle movements in chloroplast membranes: quantitative measurements of membrane fluidity by the freeze-fracture technique. Proc Natl Acad Sci USA 21:2052–2056

Olive J, Wollman FA, Bennoun P, Recouvreur M (1979) Ultrastructure-function relationship in *Chlamydomonas reinhardtii* thylakoids by means of a comparison between the wild type and the F_{34} mutant which lacks the photosystem II reaction center. Mol Biol Rep 5:139–143

Olive J, Wollman FA, Bennoun P, Recouvreur M (1981) Ultrastructure of thylakoid membranes in *C. reinhardtii*. Evidence for variations in the partition coefficient of the light-harvesting complex-containing particles upon membrane fracture. Arch Biochem Biophys 208:456–467

Olive J, Wollman FA, Bennoun P, Recouvreur M (1983) Localization of the core and peripheral antennae of photosystem I in the thylakoid membranes of *Chlamydomonas reinhardtii*. Biol Cell 48:81–84

Ono TA, Murata N (1982) Chilling susceptibility of the blue-green alga *Anacystis nidulans*. III. Lipid phase cytoplasmic membrane. Plant Physiol 69:125–129

Ophir I, Ben-Shaul Y (1974) Structural organization of developing chloroplasts in *Euglena*. Protoplasma 80:109–127

Pallas EJ, Mollenhauer HH (1972) Physiological implications of *Vicia faba* and *Nicotiana tabacum* guard-cell ultrastructure. Am J Bot 59:504–514

Paolillo DJ (1970) The three-dimensional arrangement of intergranal lamellae in chloroplasts. J Cell Sci 6:243–255

Patterson GML, Withers N (1982) Laboratory cultivation of *Prochloron*, a tryptophan auxotroph. Science 217:1034–1035

Phan CT (1984) All-granal chloroplasts of apple fruit. In: Sybesma C (ed) Advances in photosynthesis research, vol III. Nijhoff/Junk, The Hague, pp 1:63–66
Pick U, Racker E (1979) Purification and reconstitution of the N-N′dicyclohexyl-carbodiimide-sensitive ATPase complex from spinach chloroplasts. J Biol Chem 25:2793–2799
Popov VI, Matorin DN, Gostinsky SV, Tageeva SV, Allakhverdov BL (1981) Ultrastructural organization of chloroplast membranes in mutants of *Pisum sativum* L. with impaired activity in the photosystems. Planta 151:512–524
Price CA, Reardon EM (1982) Isolation of chloroplasts for protein synthesis from spinach and *Euglena gracilis* by centrifugation in silica sols. In: Edelman M, Hallick RB, Chua NH (eds) Methods in chloroplast molecular biology. Elsevier/North-Holland Biomed Press, Amsterdam New York, pp 189–209
Quinn PJ, Williams WP (1983) The structural role of lipids in photosynthetic membranes. Biochim Biophys Acta 737:223–266
Robinson SP, Walker DA (1981) Photosynthetic carbon reduction cycle. In: Hatch MD, Boardman NK (eds) The biochemistry of plants: photosynthesis. Academic Press, London New York, pp 193–236
Ryberg M (1983) The localization of magnesium-protoporphyrin and proto-chlorophyllide in separated prolamellar bodies and prothylakoids of wheat treated with 8-hydroxyquinoline and δ-aminolaevulinic acid. Physiol Plant 59:617–622
Ryrie IJ (1983) Immunological evidence for apoproteins of the light-harvesting chlorophyll–protein complex in chlorophyll b-less mutant of barley. Eur J Biochem 131:149–155
Ryrie IJ, Fuad N (1982) Membrane adhesion in reconstituted proteoliposomes containing the light-harvesting chlorophyll a/b–protein complex: The role of charged surface groups. Arch Biochem Biophys 214:475–488
Ryrie IJ, Anderson JM, Goodchild DJ (1980) The role of the light harvesting chlorophyll a/b–protein complex in chloroplast membrane stacking. Cation-induced aggregation of reconstituted proteoliposomes. Eur J Biochem 107:345–354
Salisbury JL, Floyd GL (1978) Molecular, enzymatic and ultrastructural characterization of the pyrenoid of the scaly green monad *Micromonas squamata*. J Phycol 14:362–368
Satoh K, Nakatani HY, Steinback KE, Watson J, Arntzen CJ (1983) Polypeptide composition of a photosystem II core complex: Presence of a herbicide-binding protein. Biochim Biophys Acta 724:142–150
Satoh H, Okada M, Nakayama K, Miyaji K (1984) Purification and further characterization of pyrenoid proteins and ribulose-1,5-bisphosphate carboxylase-oxygenase from the green alga *Bryopsis maxima*. Plant Cell Physiol 25:1205–1214
Schmid GH, Jankowicz M, Menke W (1976) Cyclic photophosphorylation and chloroplast structure in the labellum of the orchid *Aceras anthropophorum*. J Microsc Biol Cell 26:25–28
Schreiber U, Armond PA (1978) Heat-induced changes of chlorophyll fluorescence in isolated chloroplasts and related heat damage at the pigment level. Biochim Biophys Acta 502:138–151
Schuster G, Owens GC, Cohen Y, Ohad I (1984) Light independent phosphorylation of the chlorophyll a/b-protein complex in thylakoids of the prokaryote *Prochloron*. In: Sybesma C (ed) Advances in photosynthesis research, vol III. Nijhoff/Junk, The Hague, pp 4:283–286
Sheppard DC (1965) Chloroplast multiplication and growth in the unicellular alga *Acetabularia mediterranea*. Exp Cell Res 37:93–110
Siegel CO, Jordan AE, Miller KR (1981) Addition of lipid to the photosynthetic membrane: Effects of membrane structure and energy transfer. J Cell Biol 91:113–125
Simpson DJ (1978a) Freeze-fracture studies on barley plastid membranes. I. Wild-type etioplast. Carlsberg Res Commun 43:145–170
Simpson DJ (1978b) Freeze-fracture studies on barley plastid membranes. II. Wild-type chloroplast. Carlsberg Res Commun 43:365–389
Simpson DJ (1979) Freeze-fracture studies on barley plastid membranes. II. Location of the light-harvesting chlorophyll protein. Carlsberg Res Commun 44:305–336

Simpson DJ (1983) Freeze-fracture studies on barley plastid membranes. VI. Location of the P700-chlorophyll a-protein. Eur J Cell Biol 31:305–314
Simpson DJ, Robinson SP (1984) Freeze-fracture ultrastructure of thylakoid membranes in chloroplasts from manganese-deficient plants. Plant Physiol 74:735–741
Simpson DJ, Hoyer-Hansen G, Chua NH, Wettstein D von (1977) The use of gene mutants in barley to correlate thylakoid polypeptide composition with the structure of the photosynthetic membrane. In: Hall DO, Coombs J, Goodwin TW (eds) Photosynthesis 1977. Proc 4th Int Congr Photosynth. Biochem Soc, London, pp 537–548
Sjolund RD, Smith DD (1974) Freeze-fracture studies of photosynthetically deficient "supergranal" chloroplasts in tissue cultures containing virus-like particles. J Cell Biol 60:285–292
Spear-Bernstein L, Miller KR (1985) Are the photosynthetic membranes of crystophyte algae inside-out? Protoplasma 129:1–9
Sprague SG, Camm EL, Green BR, Staehelin LA (1985) Reconstitution of light-harvesting complexes and photosystem II cores into galactolipid and phospholipid liposomes. J Cell Biol 100:552–557
Sprey B (1977) Lamellae-bound inclusions in isolated spinach chloroplasts. I. Ultrastructure and isolation. Z Pflanzenphysiol 83:159–179
Sprey B, Laetsch WM (1976) Chloroplast envelopes of *Spinacia oleracea* L. III. Freeze-fracturing of chloroplast envelopes. Z Pflanzenphysiol 78:360–371
Sprey B, Laetsch WM (1978) Structural studies of the peripheral reticulum in C_4 plant chloroplasts of *Portulaca oleracea* L. Z Pflanzenphysiol 87:37–53
Staehelin LA (1966) Die Ultrastruktur der Zellwand und des Chloroplasten von *Chlorella*. Z Zellforsch 74:325–350
Staehelin LA (1976) Reversible particle movements associated with unstacking and restacking of chloroplast membranes in vitro. J Cell Biol 71:136–158
Staehelin LA, Arntzen CJ (1979) Effects of ions and gravity forces on the supramolecular organization and excitation energy distribution in chloroplast membranes. Ciba Found Symp 61:147–175
Staehelin LA, Arntzen CJ (1983) Regulation of chloroplast membrane function: Protein phosphorylation changes the spatial organization of membrane components. J Cell Biol 97:1327–1337
Staehelin LA, Armond PA, Miller KR (1977) Chloroplast membrane organization at the supramolecular level and its functional implications. Brookhaven Symp Biol 28:278–315
Staehelin LA, Giddings TH, Badami P, Krzymowski WW (1978) A comparison of the supramolecular architecture of photosynthetic membranes of blue-green, red and green algae and of higher plants. In: Deamer DW (ed) Light transducing membranes. Academic Press, London New York, pp 335–355
Staehelin LA, Kyle DJ, Arntzen CJ (1982) Spillover is mediated by reversible migration of LHCP between grana and stroma thylakoids. Plant Physiol 69:69
Stanier RY, Cohen-Bazire G (1977) Phototrophic prokaryotes: the cyanobacteria. Annu Rev Microbiol 31:225–274
Steinback KE, Burke JJ, Arntzen CJ (1979) Evidence for the role of surface-exposed segments of the light-harvesting complex in cation-mediated control of chloroplast structure and function. Arch Biochem Biophys 195:546–557
Stevens SE, Balkwill DL, Paone DAM (1981a) The effects of nitrogen limitation on the ultrastructure of the cyanobacterium *Agmenellum quadruplicatum*. Arch Mikrobiol 130:204–212
Stevens SE, Paone DAM, Balkwill DL (1981b) Accumulation of cyanophycin granules as a result of phosphate limitation in *Agmenellum quadruplicatum*. Plant Physiol 67:716–719
Stidham MA, Uribe EG, Williams GJ (1982) Temperature dependence of photosynthesis in *Agropyron smithii* Rydb. II. Contribution from electron transport and photophosphorylation. Plant Physiol 69:929–934
St John JB (1976) Manipulation of the galactolipid fatty acid composition with substituted pyridazinones. Plant Physiol 57:257–259

Sweeney BM (1981) Freeze-fractured chloroplast membranes of *Gonyaulax polyedra* (Pyrrophyta). J Phycol 17:95–101
Tandeau de Marsac N (1983) Bull Inst Pasteur Paris 81:201–254
Tel-Or E, Malkin S (1977) The photochemical and fluorescence properties of whole cells, spheroplasts and spheroplast particles from the blue-green alga *Phormidium luridum*. Biochim Biophys Acta 459:157–174
Terashima I, Iomee Y (1984) Comparative photosynthetic properties of pallisade tissue chloroplasts and sponge tissue chloroplasts of *Camellia japonica* L: Functional adjustment of the photosynthetic apparatus to light environment within a leaf. Plant Cell Physiol 25:555–563
Thomson WW (1974) Ultrastructure of mature chloroplasts. In: Robards AW (ed) Dynamic aspects of plant ultrastructure. McGraw-Hill, London New York, pp 138–177
Thornber JP (1975) Chlorophyll-proteins: light-harvesting and reaction center components of plants. Annu Rev Plant Physiol 26:127–158
Trench RK, Ohlhorst S (1976) The stability of chloroplasts of siphonaceous algae in symbiosis with marine slugs. New Phytol 76:99–109
Trench RK, Pool RR, Logan M, Engelland A (1978) Aspects of the relation between *Cyanophora paradoxa* (Korshikoff) and its endosymbiotic cyanelles *Cyanocyta korschikoffiana* (Hall and Claus). I. Growth, ultrastructure, photosynthesis and the obligate nature of the association. Proc R Soc London Ser B 202:445–462
Valen LM van (1982) Phylogenies in molecular evolution: *Prochloron*. Nature 298:493–494
Vasconcelos L de, Fay P (1974) Nitrogen metabolism and ultrastructure in *Anabaena cylindrica*. I. The effect of nitrogen starvation. Arch Microbiol 96:271–279
Vernotte C, Briantais JM, Armond P, Arntzen CJ (1975) Pre-illumination effects on chloroplast structure and photochemical activity. Plant Sci Lett 4:116–123
Waaland JR, Waaland S, Bates G (1974) Chloroplast structure and pigment composition in the red alga *Griffithsia pacifica*: Regulation by light intensity. J Phycol 10:193–199
Wada H, Hirasawa R, Omada T, Murata N (1984) The lipid phase of thylakoid and cytoplasmic membranes from the blue-green algae (cyanobacteria) *Anacystis nidulans* and *Anabaena variabilis*. Plant Cell Physiol 25:907–911
Wakim B, Golecki JR, Oelze J (1978) The unusual mode of altering the cellular membrane content by *Rhodospirillum tenue*. FEMS Mikrobiol Lett 4:199–201
Wehrmeyer W (1964a) Zur Klärung der strukturellen Variabilität der Chloroplastengrana des Spinats in Profil und Aufsicht. Planta 62:272–293
Wehrmeyer W (1964b) Über Membranbildungsprozesse in Chloroplasten. II. Zur Entstehung der Grana durch Membranüberschiebung. Planta 63:13–30
Weier TE, Engelbrecht HP, Harrison A, Risley EB (1965) Subunits in the membranes of chloroplasts of *Phaseolus vulgaris*, *Pisum sativum*, and *Aspidistra* sp. J Ultrastruct Res 13:92–111
Weier TE, Bisalputra T, Harrison A (1966) Subunits in chloroplast membranes of *Scenedesmus quadricauda*. J Ultrastruct Res 15:38–56
Wellburn AR (1977) Distribution of chloroplast coupling factor (CF_1) particles on plastid membranes during development. Planta 135:191–198
Wellburn AR (1982) Bioenergetic and ultrastructural changes associated with chloroplast development. Int Rev Cytol 80:133–191
Wergin WR, Potter JR (1975) The effects of fluometuron on the ultrastructural development, chlorophyll accumulation and photosynthetic competence in developing velvetleaf seedlings. Pestic Biochem Physiol 5:265–279
Westhoff P, Alt J, Herrman RG (1983) Localization of the genes for the two chlorophyll a-conjugated polypeptides (mol. wt. 51 and 44 kd) of the photosystem II reaction center on the spinach plastic chromosome. EMBO J 2:2229–2237
Whatley JM (1977) The fine structure of *Prochloron*. New Phytol 79:309–313
Whatley JM, Whatley FR (1981) Chloroplast evolution. New Phytol 87:233–247
Whatley JM, John P, Whatley FR (1979) From extracellular to intracellular: the establishment of mitochondria and chloroplasts. Proc R Soc London Ser B 204:165–187
Whitford D, Gounaris K, Barber J (1983) Studies on cytochrome b-559 of the higher

plant thylakoid membranes. In: Sybesma C (ed) Advances in photosynthesis research, vol I. Nijhoff/Junk, The Hague, pp 4.497–4.504
Wilcox LW, Wedemayer GJ (1985) Dinoflagellate with blue-green chloroplasts derived from an endosymbiotic eukaryote. Science 227:192–194
Williams RC, Gingrich JC, Glazer AN (1980) Cyanobacterial phycobilisomes. Particles from *Synechocystis* 6701 and two pigment mutants. J Cell Biol 85:558–566
Williams WP (1977) The two photosystems and their interaction. In: Barber J (ed) Primary process of photosynthesis. Elsevier, Amsterdam New York Oxford, pp 99–144
Wishnick M, Lane MD, Scrutton MC (1970) The interaction of metal ions with ribulose 1,5-diphosphate carboxylase from spinach. J Biol Chem 245:4939–4947
Withers NW, Alberte RS, Lewin RA, Thornber JP, Brilton G, Goodwin TW (1978) Carotenoid, chlorophyll-protein composition and photosynthetic unit size of *Prochloron* sp., a prokaryotic green alga. Proc Natl Acad Sci USA 75:2301–2305
Wolk CP (1973) Physiology and cytological chemistry of blue-green algae. Bacteriol Rev 37:32–101
Wollman FA (1979) Ultrastructural comparison of *Cyanidium caldarium* wild type and III-C mutant lacking phycobilisomes. Plant Physiol 63:375–381
Wollman FA, Bennoun P (1982) A new chlorophyll-protein complex related to photosystem I in *Chlamydomonas reinhardtii*. Biochim Biophys Acta 680:352–360
Wollman FA, Olive J, Bennoun P, Recouvreur M (1981) Organization of the photosystem II centers and their associated antennae in the thylakoid membranes: a comparative ultrastructural, biochemical and biophysical study of *Chlamydomonas* wild type and mutant lacking in photosystem II reaction centers. J Cell Biol 87:728–735
Yamamoto Y, Ueda T, Shinkai H, Nishimura M (1982) Preparation of O_2-evolving photosystem II-subchloroplasts from spinach. Biochim Biophys Acta 679:347–350

2 Photosynthetic Light Reactions – Physical Aspects

K. SAUER

1 Introduction

Of the sunlight incident on the surface of the earth, about 2% is captured and stored by photosynthetic organisms. This volume describes the many facets of how that process is accomplished. In this chapter we consider the "physical part". This begins with the absorption of incident radiation by the photosynthetic pigments. The first response by the plant is mediated by electrons, which change orbitals as a consequence of photon absorption to produce an excited state of a molecule or of a collection of molecules. This excitation then migrates among antenna pigment molecules until it reaches a photosynthetic reaction center, where separation of the electron from a positively charged hole occurs.

The molecules that accomplish these events are organized in the photosynthetic membranes in a highly specific fashion to achieve the light energy conversion with excellent efficiency (SAUER 1975; STAEHELIN and ARNTZEN 1983; STAEHELIN, Chap. 1, this Vol.). An essential aspect of the organization is the orientation and arrangement of the reaction center electron donor and acceptor components in such a way that electrons are transferred from one surface of the membrane to the other (Fig. 1). This vectorial electron pump produces a transmembrane electric field that drives the migration of protons and other small ions across the membrane (THORNE and DUNIEC 1983). The electric field and ion gradient provide a direct source of chemical potential that is converted into the formation of high energy compounds such as ATP. In addition, the vectorial character provides for the reduction of terminal electron acceptors on one side of the membrane and the oxidation of electron donors on the other. The reduced and oxidized compounds, together with the ATP, then provide the chief biochemical intermediates that drive the metabolic biochemistry involved in carbon fixation, protein and lipid synthesis, and ultimately all life processes as we know them.

We shall begin our survey of photosynthesis by taking a closer look at the earliest events and at the milieu in which they occur. Two themes will permeate this analysis: the sequence in which the events occur, and the relation of the mechanism of the process to the molecular composition and arrangement of the photosynthetic membranes. Many aspects of these two themes will be presented in detail in subsequent chapters. Our task here is to provide an overview that will describe the connections and interrelations of the individual topics and to provide the reader with a guide to their role in the larger scheme.

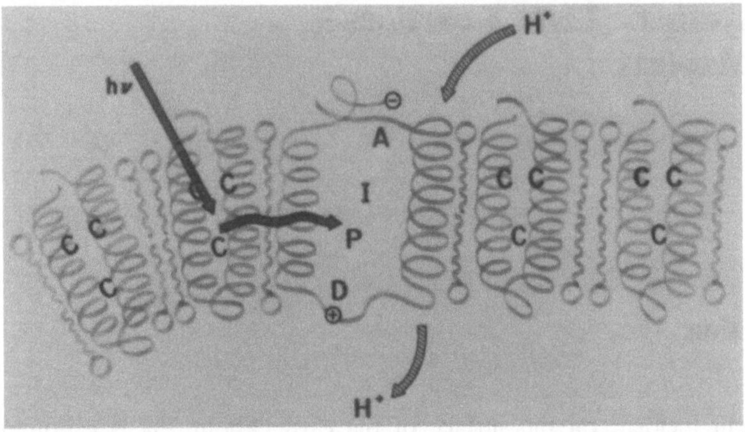

Fig. 1. Schematic diagram of the organization of a typical photosynthetic membrane. Light is absorbed by pigment molecules such as chlorophyll (*C*) in the light-harvesting antenna, and excitation is transferred to the reaction center. Excitation transfer to the primary electron donor (*P*) results in electron transfer through an intermediate (*I*) to an electron acceptor (*A*) on one surface of the membrane. Electron donors (*D*) on the opposite surface restore P to the reduced state. Because of the transmembrane orientation of the reaction center components, an electric field is set up across the membrane. Proton translocation also occurs because of the vectorial arrangement of electron/proton transfer components

2 Radiation – Sunlight

Sunlight incident on the surface of the earth occurs in the wavelength region between 300 nm and 1150 nm. At longer wavelengths water vapor and carbon dioxide in the atmosphere severely attenuate the light impinging from the sun, and at shorter wavelengths atmospheric absorption by ozone and other molecules is essentially 100% (ARCHER 1975). In the intervening window the intensity is maximal at around 600 nm[1], decreasing steadily to both longer and shorter wavelengths. Higher plant photosynthesis is accomplished by the absorption of light at wavelengths between 700 and about 350 nm, where approximately 56% of the surface sunlight intensity occurs. Wavelengths in the UV shorter than 350 nm are known to be potentially harmful to plants and to other biological organisms. Fortunately, this region of the solar spectrum contains only 4% of the solar irradiance. Other photosynthetic organisms, such as purple bacteria, are capable of utilizing photons with wavelengths as long as 1050 nm (THORNBER et al. 1978; THORNBER, Chap. 3; COGDELL, Chap. 6.2, both this Vol.). Thus, all but the extreme ends of the solar spectrum is useful for photosynthesis.

1 The wavelength of maximum solar power irradiance depends on whether the power density spectrum is expressed per unit wavelength or per unit frequency. The value stated is for plot of solar irradiance in W $cm^{-2}(\ln \lambda)^{-1}$ or W $cm^{-2}(\ln \omega)^{-1}$. When a logarithmic scale is used for wavelength or frequency, there is no difference in the wavelength at which the maximum occurs (CASTI and OTVOS 1986).

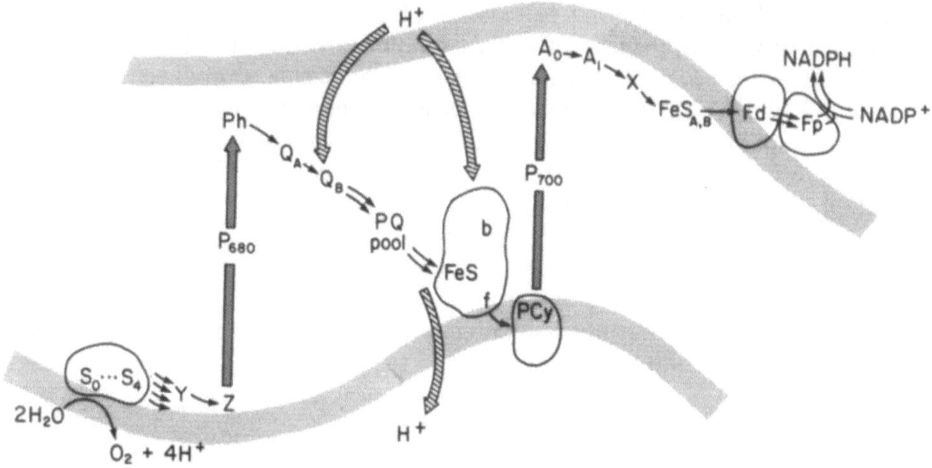

Fig. 2. In chloroplasts from higher plants and in other oxygen-evolving organisms, energy conversion is thought to occur by two light reactions acting in series (ORT, Chap. 4, this Vol.). Photosystem I (SETIF and MATHIS, Chap. 9.6, this Vol.) uses a primary donor (*P-700*) that transfers an electron to a low reduction potential acceptor (A_0) that quickly transfers it through a series of electron carriers, including several iron sulfur proteins (*X*, $FeS_{A,B}$, *Fd*) and a flavoprotein (*Fp*) to $NADP^+$ on the outside of the thylakoid membrane. Photosystem II (DINER, Chap. 9.1, this Vol.) has a primary donor (*P-680*) that receives electrons from a high potential quinone species (*Z*), an intermediate complex (*Y*), a Mn-containing water-splitting complex ($S_0 \ldots S_4$) and ultimately from water itself, which is oxidized and O_2 and H^+ are released at the inside thylakoid surface (YOCUM, Chap. 9.2; ANDERSSON, Chap. 9.1, both this Vol.). Electrons donated from PS II through pheophytin (*Ph*) to bound quinones Q_A and Q_B are accumulated by a plastoquinone (*PQ*) pool (ARNTZEN, Chap. 9.4). They are available to reduce a Rieske iron sulfur (*FeS*) center associated in a complex with cytochrome f and b_6, which in turn reduces soluble plastocyanin (*PCy*) that re-reduces P-700$^+$. Associated with this intermediate electron transfer is the translocation of H^+ from the outside to the inside of the thylakoid membrane (JOLIOT and JOLIOT, Chap. 10.3, this Vol.). Several components (Q_B, *Fp*) are known to accumulate electrons pairwise, and the water-splitting complex ($S_0 \ldots S_4$) accumulates four oxidizing equivalents

It is useful to examine the photon energy associated with this wavelength region from 1050 nm (114 kJ/einstein) to 350 nm (342 kJ/einstein). These are significantly smaller than the energy required to break a typical chemical bond. To break the O–H bond in water, for example, requires about 465 kJ mol^{-1}. Thus, the primary consequence of the absorption of solar radiation is *not* the direct breakage of chemical bonds. Radiation which is capable of such destructive action occurs farther in the UV spectral region and, fortunately, is screened by ozone, oxygen and other atmospheric components.

Because chemical rearrangements are the ultimate goal of photosynthesis, the limited photon energies in the visible solar spectrum lead to a requirement for the accumulation of energy from several successive photon-stimulated events before the chemistry takes place. In higher plant photosynthesis this accumulation occurs at the level of electron transfer (ORT, Chap. 4, this Vol.). The photosynthetic reaction centers transfer electrons one at a time, and these are then collected and transferred pairwise in quinone reduction and in $NADP^+$ reduction (Fig. 2). In the case of water oxidation, the transfer of four accumulated electrons is accomplished each time an O_2 molecule is produced (YOCUM, Chap. 9.2, this Vol.). In this manner sufficient energy is available from four einsteins of photons at 680 nm (4 × 175 kJ) or shorter wavelengths to split two moles of water to produce O_2 and a reduced product like H_2 (570 kJ) by the reaction (GREENBAUM et al. 1983)

$$2\,H_2O \rightarrow 2\,H_2 + O_2,$$

or to carry out the photosynthetic fixation of one mole of CO_2 into carbohydrate (485 kJ) by the reaction

$$CO_2 + H_2O \rightarrow (CH_2O) + O_2.$$

In the next sections we shall explore what is known about the manner in which this conversion of photon energy into stored chemical energy occurs.

3 The Excitation of Pigment Molecules

When photons of visible or ultraviolet light are absorbed by a molecule, this results in a redistribution of its electrons. The oscillating electromagnetic field associated with the radiation interacts with the charged electrons. Normally these electrons are constrained to occupy molecular orbitals having the lowest available energy, subject to the quantum mechanical restriction known as the Pauli Exclusion Principle. The consequence for most molecules is that the electrons are paired and there are no more than two electrons per orbital. Upon absorption of a photon by the molecular "chromophore" or color center, one of these electrons is promoted to an orbital that was not occupied in the ground state (Fig. 3). It is very likely that this process occurs without changing the spin quantum number of the electron; thus, an electron spin-paired (singlet) excited state is produced, assuming that the ground or initial state was a normal spin-paired molecule. The energy of the photon is thus converted into the increase in orbital electronic energy of the excited state of the molecule (BIRKS 1970).

The probability that photon absorption will occur is a quantum mechanical property of a molecule; one measure of this probability as a function of energy (or wavelength) is the absorptivity or extinction coefficient spectrum. A more graphical way to express this property is in terms of an absorption cross-section (BIRKS 1970). This is a measure of the target size that the molecule represents for the absorption of radiation. For example, the absorption cross-section of a chlorophyll a molecule at its absorption maximum in the red region of the spectrum is about 0.03 nm^2. This can be compared with the physical cross-section of a chlorophyll molecule, about 1.5 nm^2. The ratio of these two numbers

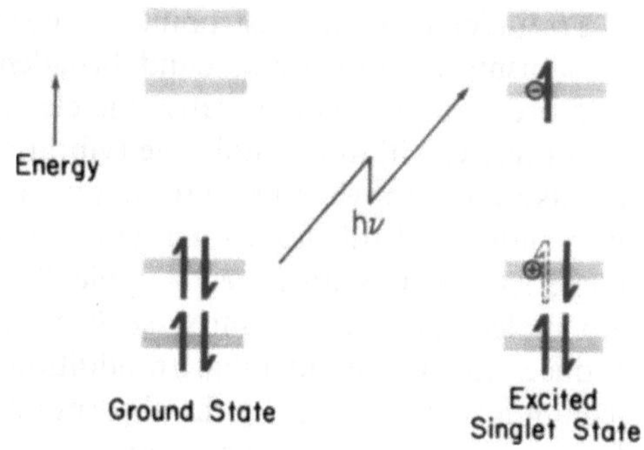

Fig. 3. Energy level diagram associated with the formation of a singlet excited state (molecular exciton). In the ground state all electrons are paired. The diagram shows the two highest-energy occupied molecular orbitals and the two lowest unfilled orbitals. Following the absorption of a photon by the molecule, one of the ground state electrons is transferred to a previously unoccupied orbital, leaving a positive hole in the lower orbital. If the electrons remain paired (spins opposed), then the excited state is called *singlet*

is a crude measure of the probability that a photon in the vicinity will be captured.

The size or extent of the molecular orbitals involved has important consequences for the characteristic energy of the radiation absorbed. Thus, a lone carbonyl group, such as the formyl substituent of chlorophyll b, has electron density localized in the immediate vicinity of the carbon and oxygen atoms. Even the most elementary quantum mechanical treatment of such a situation predicts that the electronic energy levels will be widely spaced, and indeed high energy photons in the UV region (280 nm) are required to excite a ground state electron of a carbonyl group to the lowest accessible unoccupied orbital. For a larger molecule such as chlorophyll, where the electrons are delocalized over the porphyrin ring, the energy levels are spaced closer together. Absorption of photons of lower energy can now occur, and the longest wavelength absorption is in the red (660 nm) (STRAIN and SVEC 1966). Other important factors, such as the solvent or matrix environment, nearby charged groups and interactions with other chromophores, govern the precise nature and structure of the experimental absorption spectra by exerting influence on the molecular orbitals (SEELY and JENSEN 1965; SHIPMAN et al. 1976).

For evolutionary survival it is advantageous for plants to develop ways of increasing the absorption cross-section of the pigments associated with a given reaction center and its associated biochemical apparatus, which is energetically expensive to produce. Thus, each reaction center is connected to an antenna of molecules (SEELY 1973; GLAZER 1983) that increases its photon capture cross-section in two principal ways. First, the cross-section for absorption is roughly additive, so there is an increase in proportion to the number of molecules in the antenna associated with each reaction center. Second, broadening the absorption bands increases the probability of capturing photons of energies (wavelengths) that are not absorbed efficiently by the isolated pigment molecules.

To understand this last point, we need to consider briefly some important mechanisms of absorption band broadening. Even in the isolated molecule, band broadening occurs because the electronic excitation is coupled to the motion of nuclei within the molecule (vibrations, rotations, etc.). For a large molecule like chlorophyll, there are many such possibilities, and the result is that many individual transitions merge together. This produces a continuous band of energies that is absorbed – typically 50 nm or more across. Interactions with molecules in the vicinity, such as the solvent or the membrane matrix, produces further broadening. In addition, electrical effects in the surrounding medium may cause shifts in the absorption maximum by influencing the energies of the ground and excited electronic states. Such effects are caused by the electric polarizability, the dipole moment and especially the placement of electrically charged groups in the surrounding medium (R.C. DAVIS et al. 1981; WARD et al. 1984; MAGGIORA and MAGGIORA 1984). This produces red or blue shifts in the absorption bands and, where several such environments exist in the same sample (membrane), it broadens the spectrum. Another important broadening mechanism will be considered next in connection with the phenomenon of excitation delocalization and transfer.

4 Excitation Transfer

The presence of the photosynthetic antenna pigment system increases the absorption cross-section associated with each reaction center by several hundredfold. At the same time it imposes the demand that efficient excitation transfer must occur from the molecules in the farthest reaches of the antenna. There are several mechanisms by which this can happen, including the occurrence of delocalized excitons, inductive resonance (Förster) transfer and perhaps others. These are discussed by KNOX (Chap. 7.1) HOLZWARTH (Chap. 7.3), GEACINTOV and BRETON (Chap. 7.4), SCHEER (Chap. 7.6), all this Volume; here we will present only an introduction to the subject.

An *exciton* may be considered to consist of an excited electron and its associated hole; the hole is the vacancy left in the ground state orbital when the electron is excited (BIRKS 1970). The simplest is the molecular exciton, where the electron and hole are located on the same molecule or chromophoric group. The molecular exciton is said to be *localized*, in the sense that it is confined to the region of the molecular orbitals involved. Molecular excitons may be in a singlet configuration, where the total electron spin angular momentum of the excited state is the same as it was in the ground state prior to absorption of the photon, i.e., all electrons are spin-paired; triplet excitons have two electrons unpaired (see HOFF, Chap. 7.2, this Vol.).

Where two or more identical or similar molecules are close to one another (<2 nm), interactions may arise that lead to the formation of a *delocalized exciton*. In this case the excited electron and its associated hole can range over the collective molecular orbitals of all of the molecules or chromophores in-

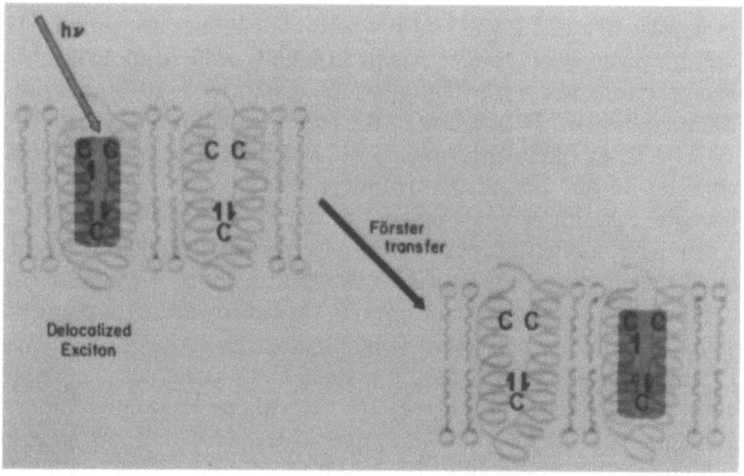

Fig. 4. A delocalized exciton occurs when a photon is absorbed by collection of interacting chromophores, such as a light-harvesting chlorophyll protein. The molecular orbitals are distributed over the set of interacting chromophores, rather than being localized on one molecule at a time. Excitation then migrates within the antenna by an inductive resonance (Förster transfer) mechanism. There may be 100 or more of these steps prior to reaching the reaction center

volved in the interaction (Fig. 4); however, the electron and hole remain correlated with one another, including the spin pairing if it is a singlet exciton. The delocalized exciton occurs in supramolecular orbitals that are different from those of the individual molecules when they are separated from one another. Not only are the exciton states delocalized over the interacting molecules, but the interactions lead to energy splittings that broaden the absorption band or, in some cases, cause it to split into several resolved components (SAUER 1975; SCHERZ and PARSON 1984). A particularly clear example is the case of the antenna bacteriochlorophyll protein complex from *Prosthecochlorus aestuarii*, whose structure has been partially determined by X-ray crystallography (MATTHEWS and FENNA 1980). The seven interacting BChl a molecules in each of the three subunits of this complex are at distances of the order of 1.4 nm from one another. The absorption spectrum, especially at low temperature, shows the resolved exciton absorption components that reflect the delocalization of excitation over the interacting array (PHILLIPSON and SAUER 1972).

In one sense delocalized excitons are not an example of excitation transfer, because the excitation is a collective property of the entire group of interacting chromophoric groups, even from the initial time of the photon absorption. Nevertheless, the delocalization constitutes an important part of the overall description of how excitation reaches the reaction centers, because there is ample evidence that delocalized excitons occur widely in the photosynthetic antennas among groups of 2, 3...7 or more chlorophyll and other pigment molecules. The interactions are greatest when the participating molecules (and their individual molecular orbital energies) are identical, but they will occur also between molecules like Chl a and Chl b where the energies are similar. The interactions are, however, short-range; except in highly ordered molecular crystals, delocalized excitons will not extend over the distances of 10–20 nm needed to connect the periphery of the antenna to the reaction center.

The most likely mechanism by which excitation hops from one pigment complex to another across distances greater than 2 nm is *inductive resonance transfer*, also known as Förster transfer (FÖRSTER 1965; KNOX Chap. 7.1, this Vol.). The interaction between an electronically excited pigment and its unexcited neighbors results in a downward transition of the initially excited group coupled with an upward transition in the nearby acceptor pigment by a through-space coulombic interaction (Fig. 4). Because induced moments are involved, the interaction falls off as the inverse sixth-power of the distance (R^{-6}). By contrast, the interaction energy for delocalized exciton coupling varies as R^{-3}. Förster transfer, despite the steeper dependence on distance, is nevertheless a longer-range interaction. For chlorophyll and related molecules it is known to be effective to distances of 8 to 10 nm (KNOX 1977). The range has not been determined for multi-chromophore pigment–protein complexes, such as those that constitute the photosynthetic light-harvesting antennas, nor is it known from experimental results where the boundary or transition between Förster transfer and exciton delocalization occurs. In fact, there are other mechanisms that have been proposed for excitation transfer at shorter distances. In one of these, where electrons and holes are simultaneously exchanged between an excited donor molecule (or group) and an unexcited acceptor molecule (DEXTER 1953), an exponential dependence on distance is expected. Good experimental evidence is lacking to distinguish which of these mechanisms is most relevant to the photosynthetic antenna. Much research still needs to be done in this important area.

An important feature to note about all of these excitation transfer mechanisms is that the negatively charged electron and the positively charged hole associated with it are in orbitals that occupy the same region of space in the molecule or interacting group of molecules. Separation of the positive from the negative charge into different regions does not occur until the excitation reaches the reaction center. That charge separation process is what we will consider next.

5 Electron Transfer – Charge Separation in the Reaction Center

The electron transfer or charge separation that occurs in photosynthetic reaction centers is like the more familiar redox reactions in which an electron is lost from an occupied orbital of the donor molecule (reductant) and transferred to an unoccupied orbital of the acceptor molecule (oxidant). The difference is that this occurs in the photosynthetic reaction center after it has become electronically excited, usually by excitation transfer from the antenna. Excited molecules are better electron donors than the same molecules in the ground state, because less energy is required to remove the electron from the higher energy orbital. Excited molecules can also be better electron acceptors, because of the hole available in a fairly low energy orbital. Which of these roles dominates in a particular situation then depends on what other electron donors or acceptors are in the vicinity. For this reason detailed knowledge of the structure and molecular arrangement of the components of the reaction center complex is essential for understanding how it works (M.S. DAVIS et al. 1979; PARSON and HOLTEN, Chap. 8.1; NORRIS and VAN BRAKEL, Chap. 8.3; MICHEL and DEISENHOFER, Chap. 8.4; DINER, Chap. 9.1; SETIF and MATHIS, Chap. 9.6, all this Vol.).

All known photosynthetic reaction centers contain chlorophyll or bacteriochlorophyll molecules, and in each case the evidence indicates that these molecules donate electrons upon excitation (COGDELL 1983). This means that there must be good electron acceptors in the vicinity. In those cases that are best documented, the initial acceptor appears to be either another chlorophyll (or bacteriochlorophyll) or the corresponding pheophytin. If both the electron donor and the acceptor have the same molecular structure, it will be difficult to know whether the prior excited electronic state occurred on one molecule or the other, or even on both together as a delocalized exciton. Nevertheless, as a result of the primary electron transfer step, the electron and the hole are now for the first time spatially separated from one another and localized on different molecular species. Simple coulombic considerations tell us that energy is required to perform this separation of positive and negative charge. The energy clearly comes ultimately from the previously absorbed photon, but a mechanism must be present to prevent the recombination of the electron and hole again onto the same molecule, which would dissipate the energy and preclude any possibility of storing it. Thus, the primary electron acceptor quickly (in less than 1 ns) transfers an electron to secondary acceptors, and the oxidized primary donor becomes reduced by secondary donors (PARSON and HOLTEN, NORRIS and VAN BRAKEL, Chap. 8.3, both this Vol.). This necessarily involves a more extensive assembly of components, all of which must be appropriately arranged in the reaction center complex. As the electron and hole get farther from one another, the coulombic interaction decreases (as R^{-1}) and the likelihood of a back reaction becomes less. (See PARSON and HOLTEN, Chap. 8.1, NORRIS and VAN BRAKEL, Chap. 8.3, both this Vol.).

6 The Reaction Center – Structure and Function

For the first time a picture is emerging of what a photosynthetic reaction center actually looks like. In a pioneering achievement MICHEL and coworkers have suceeded in crystallizing the reaction centers of the photosynthetic bacterium *Rhodopseudomonas viridis* (MICHEL 1982). Based on an analysis of the X-ray diffraction pattern of the crystals, a low resolution electron density map is

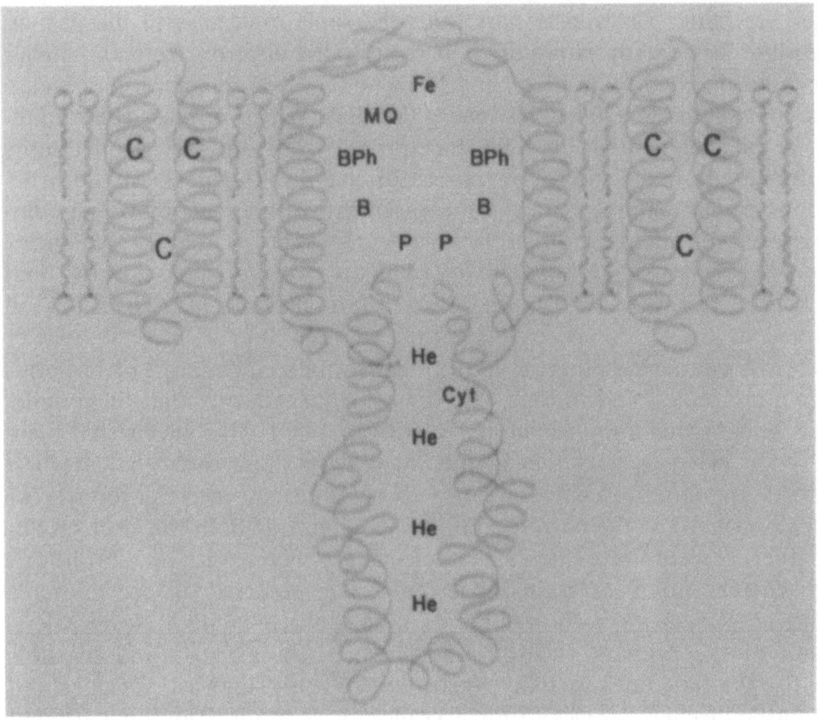

Fig. 5. Charge-separation in a photosynthetic reaction center occurs following the transfer of excitation from a nearby light-harvesting antenna complex (*C*) to the primary donor (*P*) of the reaction center. In the reaction center from the photosynthetic bacterium *Rps. viridis*, which is shown here pictorially, a pair of bacteriochlorophyll b (BChl b) molecules at the center (*P*) are at the site of this primary donation (MICHEL and DEISENHOFER, Chap. 8.4; NORRIS and VAN BRAKEL, Chap. 8.3, both this Vol.). Nearby are other BChl b (*B*) and bacteriophephytin b (*BPh*) molecules that serve as intermediate electron carriers in transferring the electron to a menaquinone (*MQ*) – iron complex located on one surface of the membrane. Electrons are subsequently extracted from cytochrome [four iron-containing hemes (*He*) roughly in a line] near the opposite membrane surface. The separation of charge thus produces a transmembrane electric field

now available (DEISENHOFER et al. 1984; MICHEL and DEISENHOFER, Chap. 8.4, this Vol.). In it one can identify a special pair of bacteriochlorophyll (BChl) molecules at the center that are thought to constitute the primary electron donor, P-960 in this organism (Fig. 5). Next to each of these and at short distances (1–2 nm) are another BChl and a bacteriopheophytin (BPh) molecule that are candidates for primary electron acceptors. Near one end of the complex and on a symmetry axis is the iron-quinone assembly that serves as a secondary electron acceptor, and opposite to it on the other side of the P-960 molecules is a chain of four heme groups of cytochromes that function as secondary electron donors. Here is a delightful example of the architects' dictum of form following function. If we adopt the likely hypothesis that the symmetry axis of the complex in the crystal is oriented perpendicular to the membrane when the reaction center is in situ, then we see that it is marvelously designed to transfer electrons from donors (cytochromes) at one surface of the membrane to acceptors (quinones) on the opposite surface. In this fashion the transmembrane electric field will arise, and the consequences of ion translocation will follow.

This first view of reaction center configuration structure is exhilarating, but it heightens other questions that need to be answered. Is this a representative reaction center, or will those of higher plants (PS I and PS II), cyanobacteria or green bacteria be different? Studies of optical properties in polarized light (PAILLOTIN et al. 1979; SHUVALOV and ASADOV 1979; GAST et al. 1983; ZINTH et al. 1983; BRETON, Chap. 7.5; TIEDE, Chap. 8.2; NORRIS and VAN BRAKEL, Chap. 8.3, all this Vol.) are suggestive that, in all reaction centers where a pheophytin serves as a primary electron acceptor, its porphyrin ring is oriented perpendicular to the membrane plane and perpendicular to the primary donor chlorophyll molecules. Other such comparisons lead one to expect that these similarities in optical and physical properties will be reflected at the structural level.

Another important question yet unanswered is how electrons move from one molecular component to another. The process is exceedingly rapid; the initial electron transfer step occurs in about 1 picosecond (1×10^{-12} s) (HOLTEN et al. 1980) and several following steps occur within a few nanoseconds (PARSON and HOLTEN, Chap. 8.1, this Vol.) The fact that some of these processes occur efficiently even at very low temperatures (≤ 4 K) has led to the proposal that electron tunneling plays an important role. Tunneling has generally been assumed to be possible only when the donor and acceptor are very close to one another (<1 nm), whereas electrons have to traverse 8–12 nm to span the membrane. Model studies, however, are now demonstrating that electrons can move easily across distances of 1.0 nm or more without any accompanying molecular motion (JORAN et al. 1984; WASIELEWSKI and NIEMCZYK 1984). Through an extension of such studies we may hope to gain a better understanding of the underlying physics.

7 Conclusion

We have been examining general aspects of the absorption of radiation by photosynthetic pigments, the transfer of the resulting electronic excitation among the antenna pigment molecules to the reaction center and then the initiation of the charge separation process in the reaction center. More detailed descriptions of these topics will be found in the following chapters. There are, of course, many other important aspects of photosynthetic energy conversion that will be discussed by the other authors in this volume. Of particular interest are the pairwise accumulation of electrons by quinone molecules and the coupling of this process to proton translocation (RICH 1984; ORT, Chap. 4; DUTTON, Chap. 5; JOLIOT and JOLIOT, Chap. 10.3; PRINCE, Chap. 10.4, all this Vol.), the production of strong oxidants capable of water oxidation and strong reductants that drive the reductive biochemistry associated with carbon fixation, the role of the transmembrane electric field in ion translocation and coupling of phosphorylation (THORNE and DUNIEC 1983; MELANDRI and VENTURALI, Chap. 10.6; DILLEY, Chap. 10.7, this Vol.), communication between cooperating PS I and PS II centers that appear to be located in different regions of the thylakoid membrane network in higher plants (BARBER, Chap. 11.5, this Vol.), and the nature of control processes that alter the organization and arrangement of the membrane components in response to changes in environmental factors (STAEHELIN and ARNTZEN 1983; STAEHELIN, Chap. 1, this Vol.). Despite the fact that final answers are not yet available for all of these questions, the readers of this volume will find that many exciting results have been obtained and that rapid progress is being made in expanding our understanding. Research has come a long way in the 8 years since appearance of Volume 5, this Series.

References

Archer MD (1975) Photochemical aspects of solar energy conversion. Photochemistry 6:739–764

Birks JB (1970) Photophysics of aromatic molecules. Wiley-Interscience, London

Casti TE, Otvos JW (1986) Solar irradiance: The effect of dimensionality on solar spectrum curves. J Chem Educ March 1986

Cogdell RJ (1983) Photosynthetic reaction centers. Annu Rev Plant Physiol 34:21–45

Davis MS, Forman A, Fajer J (1979) Ligated chlorophyll cation radicals. Their function in photosystem II of plant photosynthesis. Proc Natl Acad Sci USA 76:4170–4174

Davis RC, Ditson SL, Fentiman AF, Pearlstein RM (1981) Reversible wavelength shifts of chlorophyll induced by a point charge. J Am Chem Soc 103:6823–6826

Deisenhofer J, Epp O, Miki K, Huber R, Michel H (1984) X-ray structure analysis of a membrane protein complex: Electron density map at 3Å resolution and a model of the chromophores of the photosynthetic reaction center from *Rhodopseudomonas viridis*. J Mol Biol 180:385–398

Dexter DL (1953) A theory of sensitized luminescence in solids. J Chem Phys 21:836–850

Förster Th (1965) Delocalized excitation and excitation transfer. In: Sinanoglu O (ed) Modern quantum chemistry, part 3. Academic Press, London New York, pp 93–137

Gast P, Wasielewski MR, Schiffer M, Norris JR (1983) Orientation of the primary donor in single crystals of *Rhodopseudomonas viridis* reaction centers. Nature 305:451–452

Glazer AN (1983) Comparative biochemistry of photosynthetic light-harvesting systems. Annu Rev Biochem 52:125–157

Greenbaum E, Guillard RRL, Sunda WG (1983) Hydrogen and oxygen photoproduction by marine algae. Photochem Photobiol 37:649–655

Holten D, Hoganson C, Windsor MW, Schenck CC, Parson WW, Migus A, Fork RL, Shank CV (1980) Subpicosecond and picosecond studies of electron transfer intermediates in *Rhodopseudomonas sphaeroides* reaction centers. Biochim Biophys Acta 592:461–477

Joran AD, Leland BA, Geller GG, Hopfield JJ, Dervan PB (1984) Models for photochemical electron transfer at fixed distances. J Am Chem Soc 106:6090–6092

Knox RS (1977) Photosynthetic efficiency and exciton transfer and trapping. In: Barber J (ed) Topics in photosynthesis, vol II. Elsevier, Amsterdam New York Oxford, pp 55–97

Maggiora LL, Maggiora GM (1984) Protonated Schiff's base chlorophyll: A model for P700? Photochem Photobiol 39:847–849
Matthews BW, Fenna RE (1980) Structure of a green bacteriochlorophyll protein. Acc Chem Res 13:309–317
Michel H (1982) Three-dimensional crystals of a membrane protein complex. The photosynthetic reaction center from *Rhodopseudomonas viridis*. J Mol Biol 158:567–572
Paillotin G, Vermeglio A, Breton J (1979) Orientation of reaction center and antenna chromophores in the photosynthetic membrane of *Rhodopseudomas viridis*. Biochim Biophys Acta 545:249–264
Phillipson KD, Sauer K (1972) Exciton interaction in a bacteriochlorophyll-protein from *Chloropseudomonas ethylica*. Absorption and circular dichroism at 77 K. Biochemistry 11:1880–1885
Rich PR (1984) Electron and proton transfers through quinones and cytochrome bc complexes. Biochim Biophys Acta 768:53–79
Sauer K (1975) Primary events and the trapping of energy. In: Govindjee (ed) Bioenergetics of photosynthesis. Academic Press, London New York, pp 115–181
Scherz A, Parson WW (1984) Exciton interactions in dimers of bacteriochlorophyll and related molecules. Biochim Biophys Acta 766:666–678
Seely GR (1973) Effects of spectral variety and molecular orientation on energy trapping in the photosynthetic unit: A model calculation. J Theor Biol 40:173–187
Seely GR, Jensen RG (1965) Effect of solvent on the spectrum of chlorophyll. Spectrochim Acta 21:1835–1845
Shipman LL, Cotton TM, Norris JR, Katz JJ (1976) An analysis of the visible absorption spectrum of chlorophyll a monomer, dimer and oligomers in solution. J Am Chem Soc 98:8222–8230
Shuvalov VA, Asadov AA (1979) Arrangement and interaction of pigment molecules in reaction centers of *Rhodopseudomonas viridis*. Photodichroism and circular dichroism of reaction centers at 100 K. Biochim Biophys Acta 545:296–308
Staehelin LA, Arntzen CJ (1983) Regulation of chloroplast membrane function: Protein phosphorylation changes the spatial organization of membrane components. J Cell Biol 97:1327–1337
Strain HH, Svec WA (1966) Extraction, separation, estimation and isolation of the chlorophylls. In: Vernon LP, Seely GR (eds) The chlorophylls. Academic Press, London New York, pp 21–66
Thornber JP, Trosper TL, Strouse CE (1978) Bacteriochlorophyll in vivo: Relationship of spectral forms to specific membrane components. In: Clayton RK, Sistrom WR (eds) The photosynthetic bacteria. Plenum, New York London, pp 133–160
Thorne SW, Duniec JT (1983) The physical principles of energy transduction in chloroplast thylakoid membranes. Q Rev Biophys 16:197–278
Ward B, Chang CK, Young R (1984) Spectral shifts upon reversible modifications of CHO peripheral substituents in porphyrin, chlorin and bacteriochlorin. A phenomenological explanation for the red shift of protonated Schiff base. J Am Chem Soc 106:3943–3950
Wasielewski MR, Niemczyk MP (1984) Photoinduced electron transfer in meso-triphenyltriptycenylporphyrin-quinones. J Am Chem Soc 106:5043–5045
Zinth W, Kaiser W, Michel H (1983) Efficient photochemical activity and strong dichroism of single crystals of reaction centers from *Rhodopseudomonas viridis*. Biochim Biophys Acta 723:128–131

3 Biochemical Characterization and Structure of Pigment-Proteins of Photosynthetic Organism

J.P. THORNBER

1 Introduction

This article aims to give the reader a synopsis of how photosynthetic pigments are organized in plants and bacteria so that they perform efficiently. Ninety-five percent or more of the total pigment functions to harvest light energy and to transfer it to the remainder of the pigment which is located in a photochemical reaction center component where the primary photochemical event occurs (cf. Fig. 1). The combination of the antenna with those reaction center pigments to which they feed energy, is a photosystem. Energy transfer within a photosystem occurs with a very high efficiency as a result of the pigment molecules being appropriately oriented and spaced no closer than 10 Å or further apart than about 70 Å from each other (see Chap. 2 by SAUER, and the minireviews in Chap. 7, this Vol. for more detailed treatments of energy trapping). This organization is achieved by the noncovalent (in general) association of the pigment molecules with a variety of specific proteins. These pigment–protein complexes are embedded in the photosynthetic membrane or, at the very least, are tightly associated with the membrane surface (see STAEHELIN, Chap. 1, this Vol.).

The photosynthetic apparatus of eukaryotic plants and cyanobacteria (i.e., oxygen-evolving organisms) contains two different photosystems, each of which is a multicomponent pigment–protein complex. Other complexes in the membrane (see ORT, Chap. 3, this Vol.), which lack photosynthetic pigments (e.g., oxygen-evolving, cytochrome b_6/f and ATPase complexes), process the electrons and holes formed during the primary event in the photosystem to produce the ATP and NADPH required for carbon dioxide fixation and other metabolic processes. Unlike oxygen-evolving organisms, purple and green photosynthetic bacteria have only one photosystem, which uses light energy and photosynthetic pigments in a manner very similar to plants to produce the necessary ATP and, in these organisms, NADH (see DUTTON, Chap. 5, this Vol.).

It has long been recognized that the conversion of light energy into chemical energy during the photosynthetic process will be fully understood only when the molecular architecture describing the relationship of the chlorophyll and carotenoid molecules to each other and to other substances in the membrane is known. We are now approaching this goal after many years of unraveling a most complex situation. By using suitable surfactants and standard protein fractionation techniques, it has become possible to solubilize plant and bacterial photosynthetic membranes, and purify and characterize the individual pig-

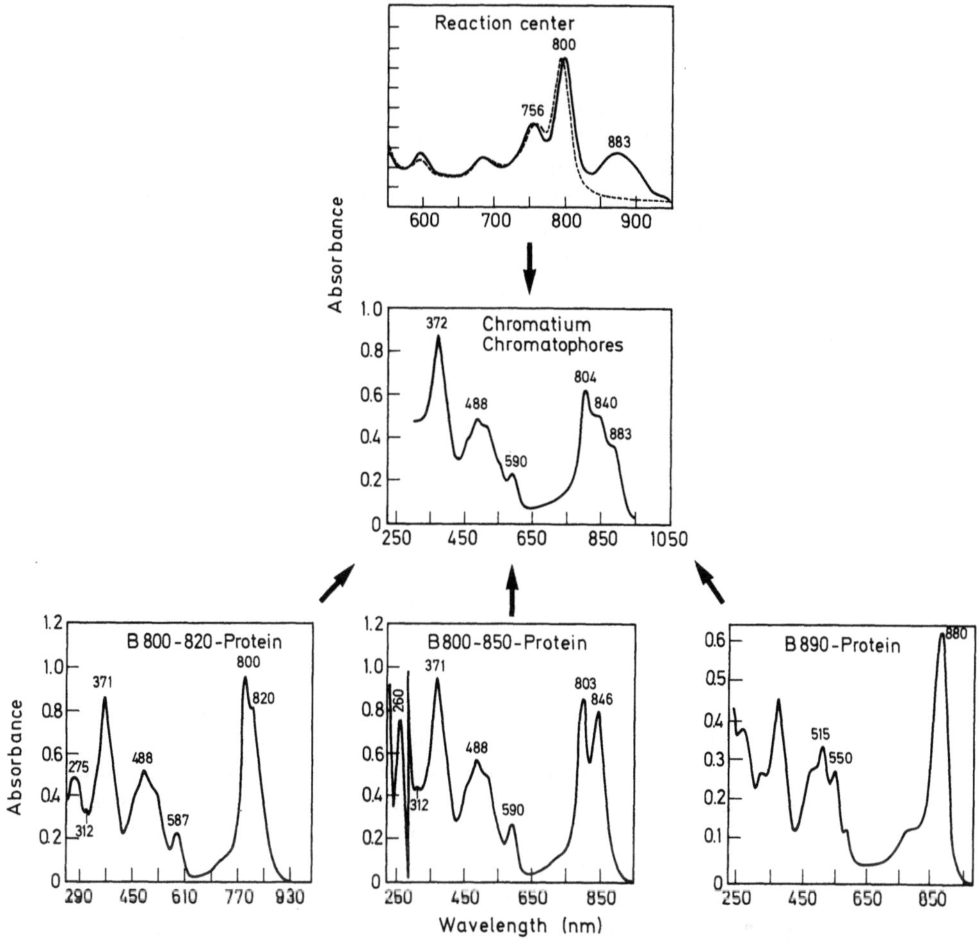

Fig. 1. Spectral analysis of the pigment-protein composition of a purple bacterium (*Chromatium vinosum*) (THORNBER 1970). Spectra of the three different antenna caroteno-bacteriochlorophyll-proteins (B-890-protein, B-800–850-protein and B-800–820-protein) of this organism are shown in the *bottom portion* of the figure, and that of the isolated photochemical reaction center of *Chromatium* is shown at the *top*. In the *center* is the spectrum of unfractionated photosynthetic membranes of this organism. The spectra of the four fractions sum to the spectrum of the whole membrane. A detailed analysis suggests that the ~883 nm band in chromatophores corresponds to the major absorbance band of the B-890-protein, the ~840 nm band to the 846 nm peak of the B-800–850-protein, and the 804 nm peak to the ~800 nm peaks in the B-800–850-protein and in the B-800–820 protein. The complex absorption band around 480 nm in intact membranes reflects primarily the absorbance contributed by carotenoid in the B-800–850- and B-800–820-proteins, but the more prominent shoulder at ~510 nm in the chromatophore spectrum is due to the 515 nm absorption resulting from the carotenoids in the B-890-protein. Since the photochemical reaction center constitutes only a small fraction of the total chromatophore pigment, its absorption bands are masked by the greater absorbance of the antenna, even for the 756 nm peak which occurs at a wavelength at which antenna absorbance is low. However, in *Rhodospirillum rubrum* in which the proportion of reaction center to antenna is greater and the B-800–850- and B-800–820-proteins are absent, the 756 and 800 nm bands of the reaction center can be seen in the chromatophore spectrum (PRINCE and THORNBER 1977).

Note the substantial wavelength separation between the different spectral forms of bacteriochlorophyll in this purple bacterium; the maximum separation between any two spectral forms of chlorophyll in green plants is 40 nm (see text)

mented complexes. It is the purpose of this article to describe the essence of each of these complexes and to direct the reader to other chapters and/or references that contain greater details on each pigment-protein.

In this chapter, I begin by describing the historical development of the subject because, in the case of plants, scientists who have been peripherally interested in the topic of chlorophyll-proteins have found it difficult to read the literature on the subject because of a bafflement over the terminology used to identify different chlorophyll-proteins. A chronological exposure of the subject will mitigate that confusion and provide a reader outside the area with the necessary background to read the earlier and the current literature on these complex proteins with deeper understanding and less puzzlement.

2 Development of the Subject

For much of this century two conflicting notions have prevailed about the chemical nature of the membrane constituent that is responsible for arranging the photosynthetic pigment molecules in the membrane to achieve the high efficiency of energy trapping and conversion: one was that the pigments were conjugated with protein(s) as occurs for other porphyrins, e.g., cytochromes and hemoglobin; the other was that the pigments were associated only with lipids, and that lipids, rather than protein molecules, provided the correct spacing and orientation required for efficient functioning of the pigments. Until the mid-1970's, each of these notions cycled in and out of favor, particularly in the case of the plant pigments. A student will be perhaps surprised to learn that the notion of chlorophyll being organized by proteins was quite readily accepted by researchers studying photosynthetic bacteria as early as the mid-1960's, while it took a further 10 years before the majority of those working on the oxygen-evolving organisms accepted that notion.

2.1 Historical Development of Our Understanding of Plant Chlorophyll-Proteins

SMITH and PICKELS (1941) obtained the first evidence that chlorophyll was complexed with protein in a component of the size usually expected for a protein. They observed that the chlorophyll in sodium dodecyl sulfate extracts of chloroplasts sedimented in the analytical ultracentrifuge together with a protein boundary of 2.6S. These authors observed no difference in the behaviors of chlorophyll a or b or the carotenoids upon centrifugation of the extract, which led them to suggest that the fundamental unit of the plant photosynthetic apparatus contained three chlorophyll a and one chlorophyll b molecules, carotenoid, and a protein. The reader should understand that there had been virtually no suggestion at this time of more than one photochemical reaction in plants.

The use of anionic detergents to solubilize chloroplast lamaellae has been continued ever since that time to investigate photosynthetic pigment organization. CHIBA (1960) obtained the first evidence for the multiplicity of pigmented complexes: he resolved a green and a yellow-green component upon free electrophoresis of sodium dodecyl sulfate extracts of plants. SIRONVAL et al. (1967) later repeated this observation by electrophoresing a similar detergent extract in agar gels. That chlorophyll was located in more than one molecular environment in the thylakoid membrane was compatible with the realization, in the early 1960's, that there are two different photosystems in plant photosynthesis. This notion of two different environments for chlorophyll was further strengthened in the mid-1960's by BOARDMAN and ANDERSON (1964) and BROWN and DURANTON (1964), both of whom observed that a chlorophyll a-enriched fraction was more readily extracted from chloroplast lamellae by detergents than was chlorophyll a- and b-containing material; the latter material was primarily associated with PS II, while the former was a representation of PS I. The advent of polyacrylamide gel electrophoresis in the mid-1960's permitted a finer resolution of proteins than that attained on agar, starch, or any other support. Its universal availability and the simplicity of its use resulted in a proliferation of data on the chlorophyll–protein composition of photosynthetic membranes. There are more details of this early phase in the development of the subject in THORNBER (1975).

OGAWA et al. (1966) and THORNBER et al. (1966) took advantage of polyacrylamide gel electrophoresis to provide further substantial evidence for the presence of two major chlorophyll-proteins in higher plant chloroplasts. OGAWA et al. (1966) used sodium dodecyl sulfate and THORNBER et al. (1966, 1967a) used sodium dodecyl benzenesulfonate, at various ratios with respect to the chlorophyll in higher plant chloroplast membranes, to obtain all of the chlorophyll and the membrane protein in aqueous solution. Electrophoresis of such detergent extracts on polyacrylamide gels resolved three chlorophyll-containing zones which were termed component or complex I, II, or III corresponding to their increasing electrophoretic mobility. Each pigmented zone could be extracted from the gel support. Complexes I and II were shown to be chlorophyll complexed with protein, while the fastest migrating band was concluded to be detergent-associated chlorophyll and carotenoid molecules (free pigment), that had probably been released from the two pigment–protein complexes. This most rapidly migrating colored zone contained no protein. Some 70% of the total chlorophyll in the leaf was contained in complexes I and II with about half the total chlorophyll being associated with complex II.

Complex I had a chlorophyll a/b ratio greater than 7, and in some preparations chlorophyll b was absent from this chlorophyll-protein. Complex II had a yellow-green color, contained most of the chlorophyll b of the starting extract, and had a chlorophyll a to b ratio between 1.1 and 1.8. The amino acid compositions of the two complexes showed that both proteins contained a high proportion of nonpolar amino acid residues, as expected for membrane proteins (THORNBER et al. 1967b) and, although the two amino acid compositions were similar, differences between their lys, his, pro, glu and ile contents were interpreted (THORNBER et al. 1966, 1967b) as showing that the apoproteins of the two complexes were not identical. The difference between the sedimentation coefficients of the two complexes (9S for complex I and 2–3S for complex II), and later the difference in their electrophoretic mobilities, were interpreted to indicate that complex I was larger than complex II. Since all the pigment sedimented with the protein boundary in both cases, the association of chlorophyll with protein in both complexes, first deduced by the co-electrophoresis of the pigment and protein, was substantiated. THORNBER et al. (1966, 1967a) and SIRONVAL et al. (1967) tested the function of these pigmented bands by observing which of the two chlorophyll-proteins was present in digitonin-prepared photosystem I and photosystem II particles (BOARDMAN and ANDERSON 1964). Both groups concluded that complexes I and II represented subunits of photosystem I and II respectively. No photochemical activities could be detected in either isolated chlorophyll-protein at that time. More details of the research during the period proceeding the polyacrylamide gel electrophoresis work can be found in THORNBER (1975).

There was no immediate acceptance by plant photosynthesis researchers that these studies proved that proteins organized the photosynthetic pigments. Rather, it was argued that the detergents had dispersed the membranes and dissolved the lipid-associated chlorophyll, which thereafter associated with the major proteins of the solubilized thylakoid membrane. The true function of these major proteins was believed to be the provision of structure for the membrane (CRIDDLE 1966). Thus, in the late 1960's, many researchers felt that the green zones observed on polyacrylamide gel electrophoresis were not representative of the organization of chlorophyll in vivo.

In equivalent research on photosynthetic bacteria, evidence was fast accruing and with wider acceptance that both antenna and reaction center pigments were associated with proteins and, furthermore, with different proteins. Thus, at the time the above work was being done, OLSON and ROMANO (1962) had obtained and studied a water-soluble bacteriochlorophyll a-protein from a green bacterium, and REED and CLAYTON (1968) later isolated a detergent-soluble photochemical reaction center from a purple bacterium. There was little doubt that protein was involved in the organization of both these components because (a) they retained the precise spectral characteristics known to be associated with the antenna and the reaction center components in situ (CLAYTON 1980), (b) protein denaturants destroyed such characteristics, and (c) the substantial shift in absorbance (30–100 nm for bacteriochlorophyll a, more for bacteriochlorophyll b) between bacteriochlorophyll in free solution and bacteriochlorophyll in situ, i.e., in the bacteriochlorophyll complexes, was best explained by proposing that it was due to bacteriochlorophyll interaction with protein rather than with other bacteriochlorophyll molecules; the latter would have produced smaller shifts than observed (THORNBER et al. 1978). The researchers studying the plant chlorophyll-proteins did not have this large spectral shift of chlorophyll to help them in an equivalent way; in many instances the shift between the in vivo and in vitro forms in plants is barely perceptible. This virtual lack of shift contributed greatly to the difficulty of obtaining acceptance of the reality of the fractionated chlorophyll-protein of plants, and in general, the very small shift has hindered elucidation of pigment organization within the plant complexes.

Further characterization of the two pigmented protein complexes of plants occurred during 1967–1974, and has been reviewed in detail by the author (THORNBER 1975). During this period, complex I and complex II were renamed the PS I and PS II chlorophyll-proteins or CP (chlorophyll-protein) I and CP II, respectively. While more detailed biochemical characterization of both was being pursued, a major effort was being devoted to further examining their function. In 1969 the author (THORNBER 1969) used an alternative purification procedure (ammonium sulfate/methanol treatment on Celite columns followed by hydroxylapatite chromatography) to obtain large amounts of a CP I-like component from sodium dodecyl sulfate extracts of a cyanobacterium. Subsequently this material was found (DIETRICH and THORNBER 1971) to contain a photochemically active P-700 (chlorophyll/P-700 $\simeq 40$). Hence, after SHIOZAWA et al. (1974) had shown that it was possible to obtain spectrally and functionally identical material from higher plants, the name of the CP I component was changed to the P-700-chlorophyll a-protein (SHIOZAWA et al. 1974). Concurrently, CP II was shown to be coded by nuclear DNA, and a chlorophyll b-less mutant of barley (chlorina f_2) was found that lacked this chlorophyll-protein (THORNBER and HIGHKIN 1974). These observations had various ramifications, one of which was that since the barley mutant was photochemically viable, CP II could not contain the photochemical reaction center of PS II. It was therefore proposed that CP II should be called the light-harvesting chlorophyll a/b-protein (THORNBER and HIGHKIN 1974; THORNBER 1975). More importantly, it appears to have been these observations, as well as similar studies on the same and on different mutants by others (ANDERSON and LEVINE 1974; GENGE et al. 1974 and CHUA et al. 1975), that mainly convinced plant photosynthesis researchers of the reality of the involvement of proteins in organizing the plant's photosynthetic pigments. Why the studies on chlorophyll-protein mutants rather than any other aspect of chlorophyll-protein research proved so convincing of their reality mystified this author; nevertheless the idea that proteins bring about the specific organization of the chlorophyll molecules has held firm since 1974.

Improvements in the polyacrylamide gel electrophoresis system (slab gels vs. tube gels) for the separation of polypeptides became available in the early 1970's. This allowed for the characterization of previously defined chlorophyll-proteins in terms of sizes and numbers of apoprotein(s). A polypeptide of about 68 kDa was identified as the P-700-chlorophyll-apoprotein by ANDERSON and LEVINE (1974), CHUA et al. (1975), MACHOLD (1975), BENGIS and NELSON (1975), and, similarly the light-harvesting chlorophyll a/b-protein was reported to have between one and three (depending on the plant species) apoproteins in the 25–30 kDa range.

In 1976 Brookhaven National Laboratory hosted the first symposium devoted entirely to research on chlorophyll-proteins and isolated photochemical reaction centers. The volume published in connection with the meeting gives an excellent picture of the state of knowledge at that time. Both the book and the meeting catalyzed a proliferation of information about chlorophyll-proteins.

There had been considerable debate during the previous decade about the origin of the chlorophyll in the free pigment zone following polyacrylamide gel electrophoresis of sodium dodecyl sulfate extracts of thylakoids. One notion was that some of a plant's chlorophyll may not be associated with protein in vivo, and hence would be located in this zone after fractionation, while another was that the chlorophyll was extracted by the detergent from the two known chlorophyll-proteins. At this meeting the author and colleagues proposed that some of this chlorophyll originated from the PS II reaction center complex, since there was no pigmented band on nondenaturing sodium dodecyl sulfate-polyacrylamide gels to equate to this component (THORNBER et al. 1977). Further, it was proposed that some of the chlorophyll in the free pigment zone possibly arose by dissociation of an antenna complex of photosystem I, an antenna system that complemented the P-700-chlorophyll a-protein, the latter being thought of as the heart of but not the whole of photosystem I. Interestingly, at this Brookhaven meeting, HAYDEN and HOPKINS reported the presence of an additional chlorophyll-protein which they later (1976) proposed was the photosystem II reaction center complex. This seminal observation was extended by DELEPELAIRE and CHUA (1981), who used low temperature (4 °C) polyacrylamide gel electrophoresis which requires a detergent, e.g., lithium dodecyl sulfate, that is soluble at 4 °C, rather than sodium dodecyl sulfate, which in the purest state is not soluble at low temperatures. DELEPELAIRE and CHUA (1981) observed two minor chlorophyll-protein zones (Fig. 3), termed by them CP III and CP IV, which, on the basis of their absence from some mutants lacking PS II activity, were proposed to be the PS II chlorophyll-proteins. These chlorophyll-proteins of approximately 50 kDa apparent size, each containing one apoprotein of 43–48 kDa, were also observed later by ANDERSON in a different gel system (ANDERSON et al. 1978) and by CAMM and GREEN (1980) and by others.

The first evidence for the existence of the putative photosystem I antenna chlorophyll-protein (see THORNBER and BARBER 1979; BOARDMAN et al. 1978) came from the work of MULLET et al. (1980). They fractionated a Triton X-100, a nonionic detergent, extract of thylakoid membranes by ultracentrifugation to obtain a PS I preparation, which was deduced to contain not only the expected PS I core complex having 65 chlorophylls and one P-700 (the P-700-chlorophyll a-protein), but also some chlorophyll a and b (about 45 molecules/P-700) associated with two polypeptides in the 20–25 kDa range (LHC I). Subsequent work by LAM et al. (1984a, b) and HAWORTH et al. (1983) has confirmed and extended this proposal (see later and also WOLLMAN Chap. 9.7, this Vol.).

As a result of the Brookhaven meeting in 1976, several groups put considerable effort into modifying the procedure for electrophoretic fractionation of detergent-solubilized thylakoids with an aim of retaining more of the total chlorophyll with protein. It was expected that more complexes would be observed, and that these complexes would be more representative of their in vivo state. Such improvements were achieved by altering the buffers, detergent concentrations, and temperatures of the electrophoresis, and were made simultaneously by ANDERSON et al. (1978), MARKWELL et al. (1978) and HENRIQUES and PARK (1978), and later by MACHOLD et al. (1979) (see Fig. 3). In general, the amount of chlorophyll in the free pigment zone was reduced from 30% to 12–18%, three light-harvesting chlorophyll a/b-protein bands of approx. 40, 60, and 80 kDa were observed, more of the total chlorophyll (about 20–25%) was retained in the PS I band(s), and PS II reaction center complexes were seen in some systems.

The greater than expected proliferation in the number of bands proved confusing to an interested observer not working in the area of chlorophyll-proteins. Such onlookers became even more disillusioned with the topic when it became apparent that the characteristics reported for each supposedly equivalent chlorophyll-protein differed, often only slightly however, between the different fractionation systems, even though the same plant species was used. The unsettling situation was reinforced by the absence of a generally accepted nomenclature system for the chlorophyll-proteins. Such a system could not be arrived at until each of the bands in each system had been better characterized. This occupied the period 1978–1982, during which time further slight variations applied to the existing fractionation systems resulted in yet more bands of uncertain relationship to the more fully characterized components. It became obvious that the different electrophoretic systems, and even other chromatographic or centrifugation systems, yielded slightly different representations of the same basic complexes. Much is to be gained by characterizing any one component of any one system, since each system has its advantages.

Most of these systems still gave some free pigment (~10%), nevertheless it was generally accepted that all of the chlorophyll, and probably all of the carotenoid, is associated with protein. In fact, MARKWELL et al. (1979), using a zwitterionic detergent, Deriphat 160, to replace sodium dodecyl sulfate in the gel buffers, observed that no free chlorophyll was generated upon electrophoresis of a sodium dodecyl sulfate extract of thylakoids.

At a meeting in Zurich in 1983 attended by many researchers working on plant chlorophyll-proteins, an attempt was made to simplify what was becoming a terminological nightmare for the chlorophyll-proteins. The participants felt confident in stating that there are four pigmented components in the photosynthetic apparatus of oxygen-evolving organisms:

CC I	previously named CP I or P-700-chlorophyll a-protein;
LHC I	a PS I antenna chlorophyll-protein;
CC II	previously CP III and IV, CPa or CP 47 and CP 43; and
LHC II	previously known as CP II, LHCP, or the light-harvesting chlorophyll a/b-protein.

Each of these components, with the possible exception of CC I, is composed of two or more individual and distinct chlorophyll-proteins. At the time the nomenclature was adopted it was felt that, while not perfect, it would serve until all the genes coding for the photosynthetic pigment–protein complexes had been identified and named, when a more universally acceptable nomenclature could be adopted.

2.2 Development of Our Knowledge of Pigment-Proteins of Photosynthetic Bacteria

As I indicated in the previous section, the unraveling of the nature of the bacterial chlorophyll-protein complexes was not as controversial as for the plant components. In the mid-1960's, characterization of a water-soluble bacteriochlorophyll a-protein from a green bacterium (OLSON et al. 1969) was in full swing. It was anticipated that elucidation of its structure would serve as a model for all other chlorophyll-protein complexes. This, however, has not turned out to be the case! During the same period, a group, headed by L.P. VERNON at the Kettering Research Laboratories in Ohio, was vigorously pursuing a study of pigment organization in the purple bacteria. Triton X-100 was used to solubilize chromatophore membranes of several purple bacteria, and the detergent extract fractionated by density gradient centrifugation. Two pigmented complexes were resolved from several bacteria, each of which had some, but not all, of the near infrared spectral forms of bacteriochlorophyll present in the membrane: One fraction was photochemically active, it contained the photochemical reaction center and had the longest-wavelength-absorbing (870–890 nm in the case of bacteriochlorophyll a-containing organisms) antenna pigments, and the other fraction contained the shorter wavelength-absorbing forms (800 and 850 nm for bacteriochlorophyll a-containing organisms) and had no activity other than that of harvesting light. Both fractions contained carotenoids. The basic analyses of these fractions in the variety of bacteria examined is summarized in THORNBER et al. (1978) (see also Fig. 1).

Towards the end of the 1960's REED and CLAYTON (1968) and GINGRAS and JOLCHINE (1969) presented photosynthesis researchers for the first time with isolated photochemical reaction centers. Both preparations came from carotenoidless mutants of *Rps. sphaeroides* (R-26) and *R. rubrum* (G-9), respectively. The following year THORNBER et al. (1969) showed that such mutants were not essential for isolating reaction centers, and the *Rps. viridis* reaction center became available. The availability of these and reaction centers from other purple bacteria presented the biophysicist and biochemist with ideal material to study the mechanism of the primary photochemical event. The advantage of such material for the biophysicist was primarily that isolated reaction centers were devoid of all pigments other than those involved in the primary event, while for the biochemist the preparations were devoid of all extraneous proteins (at least after the initial purification procedures had been improved). The biophysical data are described elsewhere in this volume (SAUER, Chap. 2; DUTTON, Chap. 5, and minireviews in Chap. 8, this Vol.). Later in this chapter our present knowledge of the biochemistry of these components is summarized, and other articles in this volume (ZUBER, Chap. 6.1; COGDELL, Chap. 6.2, this Vol.) and elsewhere (CLAYTON 1980; GINGRAS 1978; OKAMURA et al. 1982) will provide the reader with more details. The first reaction center preparations were separated from their associated antenna by treatment with the detergents, Triton X-100 or sodium dodecyl sulfate (*Rps. viridis*). Later, G. FEHER and R.K. CLAYTON introduced the field to a "new" detergent, lauryl dimethylamine oxide (LDAO) (see GINGRAS 1978), which proved more effective that other surfactants for dissociating membranes so that the reaction center is released as an essentially unaltered entity, from its association with most other membrane components. It is now reasonable to state that just about all purple photosynthetic bacteria, and even a gliding, green filamentous bacterium, *Chloroflexus aurantiacus* (BLANKENSHIP and FULLER, Chap. 8.6, this Vol.), have yielded or could yield their reaction centers to isolation. An overview of the membrane architecture of the different types of anoxygenic, photosynthetic bacteria can be found in Chapter 11.1 by SPRAGUE and VARGA, this Volume.

The excitement generated by the photochemical reaction center studies reduced the intensity of study of the antenna complexes in these bacteria. There was spasmodic progress in the first half of the 1970's: The author extended earlier work by the L.P. VERNON group by further fractionation and characterization of three pigmented caroteno-bacteriochlorophyll-protein complexes of *Chromatium vinosum* (Fig. 1): a RC-B-890-protein, a B-800–850-protein and a B-800–820-protein (B is used as an abbreviation for bulk chlorophyll, i.e., antenna chlorophyll) (THORNBER 1970). A study of the B-890-protein of *R. rubrum* was being made at this time by LOACH and his colleagues (LOACH et al. 1970), and CLAYTON and CLAYTON (1972) provided a detailed description of the B-800–850 complex in *Rps. sphaeroides*. In 1976, THORNBER et al. (1978) wrote a review of pigment organization in photosynthetic bacteria, in which the authors proposed that all *purple* photosynthetic bacteria contained a photochemical RC component closely associated with some 25–30 bacteriochlorophyll molecules in a B-890, B-870 or B-1015 complex. Several species of purple bacteria contained in addition a B-800–850 complex (e.g., *Rps. sphaeroides*, *Rps. capsulata*), while a few others (e.g., *Chromatium vinosum* and *Rps. acidophila*) also had a B-800–820 complex. The strategy for preparation of these components and the initial characterization of them is described in COGDELL and THORNBER (1979). It was mainly through the efforts of R.J. COGDELL's and of G. DREW's groups that a more complete biochemical description of these antenna bacteriochlorophyll-proteins was achieved between 1978–1982. This work is summarized later in this chapter and given in more detail by COGDELL in Chap. 6.2, this Volume. Recently, H. ZUBER's group in Switzerland has obtained the amino acid sequences of the B-800–850 and B-890-proteins of several purple bacteria. They have presented models of how these polypeptides might interact with bacteriochlorophyll and be arranged with respect to the membrane. The group has also proposed how the subunits of the complexes might interact to form the larger structure surrounding the reaction center (see ZUBER, Chap. 6.1, this vol.). Additional information on the supramolecular organization of the pigment-protein complexes of purple and green bacteria can be found in Chap. 11.1 of SPRAGUE and VARGA, this Volume.

3 The State of Knowledge of the Biochemical Organization of Photosynthetic Pigments in Plants and Bacteria

To summarize the current situation (Table 1) is a formidable task, and to do justice to all who have contributed to the topic in this short essay is difficult.

Table 1. The pigment-protein complexes of green plant thylakoid membranes

		Alternative names	Apparent size (kDa) of: Apo-protein(s)	Apparent size (kDa) of: Holo-complex[a]
PS I	CC I	CP I; PS I-40; P-700-chl a-protein; PS I reaction center	Two (or more) of 60–70	90–130
PS I	LHC I	–	22 and 23 (LHC Ia) 20 (LHC Ib)	?
PS II	CC II	CP III and IV; CPa; CP43 and CP47	43 and 47	~50
PS II		CP29	29	~30
PS II	LHC II	CP II; LHCP; LHC; Chl a/b-protein	Two (or more) of 25–28	~30

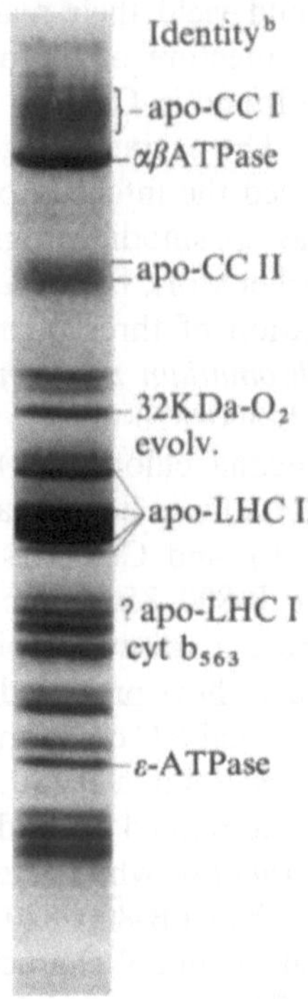

[a] Size of smallest pigmented form attained; i.e., not of higher oligomeric forms.
[b] Polyacrylamide gel electrophoresis of fully denatured *Lemna gibba* thylakoid polypeptides

As a consequence I have presented the subject matter from a personal viewpoint, and I have tried to give appropriate references so that the reader can obtain a more objective view of each topic than I may provide. I will first consider the organization of chlorophyll and carotenoid pigments in PS I (Sect. 3.1) and PS II (Sect. 3.2) of green plants, and second proceed to the organisms which do not contain chlorophyll b, but which do evolve oxygen during photosynthesis (Sects. 3.3 and 3.4). I conclude this section with a description of the arrays of pigments in the photosynthetic bacteria which do not evolve oxygen. As indicated above many workers in the field of chlorophyll-proteins now divide the chlorophyll-proteins in a photosystem into a core complex (CC) and a light-harvesting complex (LHC). For the present, this division seems adequate, but as more data accrue, it may be necessary to subdivide the two complexes. For example, in Chapter 1, this Volume, STAEHELIN draws a distinction between photosystem II-"bound" and "mobile" forms of LHC II (Chl a/b LHC) based on the fact that they are associated with different types of freeze-fracture particles.

3.1 Photosystem I of Green Plants

For photosystem I, the Core Component (CCI) is a chlorophyll-protein complex having an apparent size of approximately 110 kDa following electrophoresis on sodium dodecyl sulfate-polyacrylamide gels. It is sometimes referred to as CP 1 or the P-700-chlorophyll a-protein (THORNBER et al. 1977; see also Sect. 2.1). Associated with CCI is an antenna complex LHC I, a chlorophyll a and b-protein complex (chlorophyll a/b = 3.5–3.7) which should not be confused with the light-harvesting chlorophyll a/b-protein (LHC II) which is preferentially associated with PS II (cf. BENNETT 1983). Table 1 gives a synopsis of data on the higher plant chlorophyll–protein complexes.

3.1.1 Core Complex I (CC I)

This chlorophyll–protein complex has an electrophoretic mobility in nondenaturing sodium dodecyl sulfate-polyacrylamide gel electrophoresis equivalent to an apparent size of 90–130 kDa. CCI was originally thought to contain some 40 chlorophyll a molecules and 1–2 β-carotene molecules per P-700 which are associated with an apoprotein(s) in the 60–70 kDa region (THORNBER 1975; HILLER and GOODCHILD 1981; MALKIN 1982). On most electrophoretograms the region of the gel containing the apoprotein(s) is smeared, but on other occasions a doublet can be clearly seen (e.g., BAR-NUN et al. 1977; CHUA et al. 1975; DOS SANTOS and HALL 1982; VIERLING and ALBERTE 1983).

The subunit structure of a PS I reaction center preparation has been intensively studied by N. NELSON's group over the last decade (BENGIS and NELSON 1977; NECHUSHTAI and NELSON 1981; OKAMURA et al. 1982, see also WOLLMAN, Chap. 9.7, this Vol.). It was determined by them that there were two copies of the 70 kDa apoprotein per P-700, each of which was proposed to have 20 chlorophyll a molecules and one of which also contained P-700. A recent finding of FISH et al. (1984) and R. HERRMANN (personal communication) that the chloroplast genome contains two slightly different genes for the CCI apoprotein suggests, if both genes are expressed, that the two copies may not be identical. This would give a plausible explanation of how only one of two purportedly identical apoprotein molecules could contain P-700, and why a doublet is seen in the 60–70 kDa region, However, VIERLING and ALBERTE (1983) have reported that the two ~70 kDa apoproteins that they see in their CCI preparations are conformational variants of a single polypeptide species. Furthermore, they state that there are *four*, not two, copies of this 70 kDa polypeptide per P-700. This stoichiometry is supported by the recent results from A. GLAZER's group (LUNDELL et al. 1985) who also propose that the real size of the CC I holocomplex, albeit of a cyanobacterial CCI complex, is 450 kDa, whereas its apparent size on sodium dodecyl sulfate-polyacrylamide gel electrophoresis is 110–130 kDa. This discrepancy between the real and apparent sizes is understandable when it is realized that a tightly folded chlorophyll–protein complex will migrate through polyacrylamide gel on electrophoresis more rapidly than a fully denatured protein marker of the same molecular weight; however, a reader might perhaps be astounded by the extent of the difference between the real and the apparent sizes.

Most workers who study chlorophyll-proteins are now of the opinion that the number of chlorophyll a molecules associated with one P-700 in the CC I complex is greater than 40; the content of β-carotene is also greater than previously thought. Milder electrophoretic and/or density gradient centrifugation fractionation procedures (e.g., ANDERSON et al. 1978; LAM et al. 1984a, b; MULLET et al. 1980; LUNDELL et al. 1985) have given CC I or photosystem I (CC I + LHC I) preparations which have 80–130 chlorophyll a molecules per P-700 associated with the preparation. Frequently a photosystem I complex of a size greater than the CCI complex is observed by sodium dodecyl sulfate-polyacrylamide gel electrophoresis under the improved electrophoretic conditions (e.g., the CP Ia band of ANDERSON et al. 1978 and the CPI* of DUNAHAY and STAEHELIN 1985a). The relationship between these larger complexes and CC I is at present unclear.

While some groups have debated the precise number of chlorophyll molecules associated with CC I, they and others have been trying to remove as many of the non-P-700 chlorophyll a molecules from it as possible. In this way, photosynthesis researchers hope to produce and subsequently study CC I preparations more enriched for P-700 than have previously been available, so that the mechanism of the primary photochemical event in PS I can be delineated to the same extent as has been achieved for the purple bacteria (see SETIF and MATHIS, Chap. 9.6, this Vol.). One P-700-enriched CC I preparation (IKEGAMI and KE 1984), obtained in the form of a 160 kDa complex, contains only 10–15 molecules of chlorophyll a per P-700. It also has 12 atoms each of iron and sulfur (probably due to the presence of secondary electron carriers), but it possesses no carotenoid. Along this same line, there is a simple and rapid procedure which gives material having a chlorophyll to P-700 ratio of 20:1 (ALBERTE and THORNBER 1978). The fact that enrichment of P-700 cannot be made as great as occurs for P-870 or P-960 in the purple bacterial reaction center preparations (see Sect. 3.5.1) probably reflects a basic difference between pigment organization in their photochemically active sites: In purple bacteria all the ingredients required for performing the primary photochemical event are in an environment which contains nothing superfluous to its primary function. This environment, entrapped by specific proteins, is isolated as the reaction center preparation. This contrasts with the organization of PS I, where several antenna chlorophyll molecules apparently share the same protein environment as the pigments performing the primary event (the apoprotein is two to three times larger than the equivalent bacterial component). For this reason it has proved impossible, to extract from this chlorophyll-protein all of the antenna chlorophylls without affecting the P-700 environment. Some enrichment of P-700 with respect to total chlorophyll is achieved because a few antenna chlorophyll molecules can be extracted from the protein environment without destroying P-700.

Most preparations of PS I contain polypeptides in addition to the 70 kDa apoprotein(s) – N. NELSON's group initially noted five other subunits of 25, 20, 18, 16, and 8 kDa in their PS I reaction center preparations from higher plants (BENGIS and NELSON 1977; OKAMURA et al. 1982). These five subunits were termed subunits II–VI, respectively. Subunit VI was later found to be two subunits, 10 kDa (subunit VI) and 8 kDa (subunit VII). While the function of subunit I in binding chlorophyll a and P-700 is universally agreed upon, the functions of the other subunits remain to be unequivocally decided. The four smallest subunits are suggested to be Fe–S proteins, of which there are at least three in photosystem I (see SETIF and MATHIS chap. 9.6, this Vol.). The reason for believing this is that all Fe–S centers are lost upon removal of these subunits from the CC I complex (NELSON and NOTSANI 1977). However, the sodium dodecyl sulfate treatment used to remove these subunits, may well have made Fe–S centers in CC I undetectable rather than removed them, and hence the association of these centers with any specific subunit is still equivocal. The seven subunit preparation will photoreduce NADP when ferredoxin and ferredoxin-NADP reductase are added to it. Removal of subunit III prevents NADP reduction under these conditions (BENGIS and NELSON 1977). It was measured that there are two copies of subunit I and one copy each of subunits II–VII in the PS I reaction center preparation (OKAMURA et al. 1982). In equivalent preparations from two green algae (*Dunaliella* and *Chlamydomonas*) and one cyanobacterium (*Mastigocladus*), there are only four subunits (NECHUSHTAI and NELSON 1981; NECHUSHTAI et al. 1983). As in the case of the higher plant complex, the preparations are able to photoreduce NADP and contain three Fe–S centers (NECHUSHTAI et al. 1983). A PS I particle from another cyanobacterium (*Synechoccus sp.*) has been analyzed (LUNDELL et al. 1985) with slightly different results: Five subunits are present and the stoichiometry of subunits is I:II:III:IV:V=4:0.7:1:0.5:1.6 (cpd 2:1:1:1 for *Mastigocladus*).

NELSON's group has attempted to elucidate the organization of the subunits I–VII in the functional PS I reaction center preparation (NELSON and NOTSANI 1977). Upon heating their preparation to 100 °C for 5 min, they observed that subunits I, III, and V formed an aggregate lacking subunits II, IV, and VI. They interpreted this observation to indicate that subunits III and V are the nearest neighbors of subunit I (i.e., of the CC I apoprotein).

Further discussion of this chlorophyll-protein can be found in WOLLMAN, Chapter 9.7, this Volume.

3.1.2 Light-Harvesting Complex I (LHC I)

The light-harvesting complex I is a chlorophyll a/b-protein complex (chlorophyll a/b=3.5–3.7). Preliminary characterization of LHC I was reported by MULLET et al. (1980), who identified polypeptides of 22.5 and 24.5 kDa with a PS I peripheral antenna component. LHC I was proposed to be the site of origin of the 736 nm fluorescence of higher plant PS I observed at liquid nitrogen temperatures, while the CC I component contributed 722 and 690 nm fluorescence in the absence of LHC I (MULLET et al. 1980). Subsequent delineation of the complex came from HAWORTH et al. (1983) and LAM et al. (1984a, b) (see also ANDERSON, Chap. 6.4; WOLLMAN, Chap. 9.7, both this Vol.). All investigators in this area now agree that LHC I contains chlorophyll b in a greater proportion, with respect to chlorophyll a, than could be deduced in the seminal report. The discrepancy is apparently explained by the presence of water in the organic solvent solution of pigments used for fluorometric assay of chlorophylls a and b in the MULLET et al. (1980) work.

All the chlorophyll b in PS I preparations is proposed to be associated with LHC I (ANDERSON et al. 1983). No evidence has been presented for the presence of LHC II in PS I preparations. LHC II might, however, have been expected to occur because LHC II is proposed to migrate from photosystem II units in the grana stacks to photosystem I units in the stroma lamellae when it is phosphorylated during the State I–State II transition (see Sect. 3.2 below and STAEHELIN, Chap. 1; BARBER, Chap. 11.5, both this Vol.).

LHC I and LHC II, despite having similarly sized polypeptides and pigment compositions, show no cross-reactivity in immunological assays (LAM et al. 1984a). Recent work has shown that there are seemingly four polypeptides in the 20–26 kDa range associated with LHC I (HAWORTH et al. 1983; LAM et al. 1984a, b), and LHC I has been further resolved into two chlorophyll-proteins: LHC Ia from which arises a 680 nm fluorescence at 77 K and which contains two apoproteins of approximately 22 and 23 kDa, and LHC Ib which contains primarily one apoprotein of 20 kDa and emits 730 nm fluorescence (LAM et al. 1984a, b). The chlorophyll a/b ratio of both complexes is about 3.5.

A green algal chlorophyll-protein, termed CPO, shows several interesting equivalences to the LHC I component of higher plants: The green algal component has five small polypeptides between 19 and 27.5 kDa and a long wavelength fluorescence at 707 nm (ISH-SHALOM and OHAD 1983; WOLLMAN and BENNOUN 1982) (see WOLLMAN, Chap. 9.7; SIMPSON, Chap. 11.6, both this Vol.).

3.2 Photosystem II

3.2.1 CC II

One or two bands representing the pigment–protein(s) of CC II can be observed in many of the electrophoretic fractionation systems for chlorophyll-protein complexes (cf. Fig. 3); viz, complex 4 in the HAYDEN and HOPKINS (1976) system; CP47 and CP43 in the octyl glucoside system of CAMM and GREEN (1983); and CP III and CP IV in the lithium dodecyl sulfate procedure of DELEPELAIRE and CHUA (1981). GREEN and CAMM (1984) presented evidence that P-680, the photochemically active chlorophyll in photosystem II, is associated with the slightly larger of the two complexes. This conclusion was later confirmed by the work of NAKATANI et al. (1984) and YAMAGISHI and KATOH (1984). Alternative procedures, using chromatography and density gradient centrifugation, for purifying the CC II complex yield photochemically active material which contains other proteins (the herbicide-binding protein, another 32 kDa protein and cytochrome b-559) in addition to the two chlorophyll-proteins (e.g., SATOH et al. 1983; WESTHOFF et al. 1983).

About 40 chlorophyll a molecules and one P-680 are associated with CC II material (DINER and WOLLMANN 1980). Each chlorophyll-protein has an apoprotein of essentially the same size as the apparent size of the holocomplex, i.e., one of 40–45 kDa and the other slightly larger (about 47 kDa). The apoproteins apparently occur in a 1:1 stoichiometry. Chloroplast genes coding for these two proteins have been sequenced (MORRIS and HERRMANN 1984; HOLSCHUH et al. 1984). There are obvious similarities but certainly not complete homology between the two sequences. The correct sizes of the polypeptides deduced from their amino acid sequences are 44 and 51 kDa, i.e., they are slightly larger than observed on denaturing sodium dodecyl sulfate-polyacrylamide gels.

A considerable similarity exists between the CC II complex and the purple bacterial reaction center in their biochemical and photochemical properties. This correspondence is discussed in this volume by DUTTON (Chap.5), MICHEL and DEISENHOFER (Chapter 8.4) and by HEARST (Chapter 8.5). Further biochemical details of the CC II complex can be found in Chapter 9.1 of DINER and Chapter 9.5 of KYLE and ARNTZEN.

Very closely associated with CC II is a chlorophyll a/b-protein, known as CP29, which was first described by CAMM and GREEN (1980) (See Fig. 3). CP29 was termed such because of the size of its apoprotein, however, the holocomplex has about the same apparent size. The characteristics reported for CP29 show that it is not related to the major light-harvesting chlorophyll a/b-protein (LHC II): CP29 has a higher chlorophyll a/b ratio, and is more closely associated with the heart of photosystem II (i.e., CC II) than LHC II. Limited proteolysis experiments confirmed its difference (CAMM and GREEN 1983). Furthermore, the 29 kDa polypeptide does not cross-react immunologically with the LHC II polypeptides. It is also larger than the LHC II apoproteins, all of which have apparent sizes lying between 24 and 27 kDa. No data yet exist on the ratio of chlorophyll to protein in CP29, neither is much known about its carotenoid content or its amino acid composition. CP29, which functions only as a light-harvesting component, should perhaps not have been categorized under CC II but rather under LHC II; however, until more is known of its function it is perhaps safer to include it under CC II. In Chapter 1 of STAEHELIN in this Volume CP29 is postulated to be a major component of the PS II-"bound" form of the LHC II.

3.2.2 LHC II

Extensive recent reviews of this chlorophyll-protein have appeared (e.g., ARNTZEN 1978; THORNBER et al. 1979; HILLER and GOODCHILD 1981; BENNETT 1983; GLAZER 1983; ANDERSON 1984; MACHOLD 1984) and therefore I feel that I can be reasonably brief in my coverage of it, even though there is no chapter in this volume devoted solely to this much-investigated component. Furthermore, for the sake of brevity and clarity here, I shall refer to this chlorophyll-protein complex as LHC II to distinguish it from CP29 (see above).

The complex can be isolated by a variety of chromatographic, electrophoretic or gradient centrifugation procedures, each of which probably yields a slightly different form of the same basic unit (see, e.g., BENNETT et al. 1981). These slight differences probably explain some of the seemingly irreconcilable discrepancies between the data reported (e.g., pigment content) for this complex by various groups. LHC II constitutes about half of the chlorophyll and half of the protein in green plant thylakoids (THORNBER et al. 1967a). It functions not only as an antenna but it also contributes to the stacking of the thylakoids to form grana (ARNTZEN 1978; see also STAEHELIN, Chap. 1; BARBER, Chap. 11.5; RYRIE, Chap. 11.7, all this Vol.) and has been implicated in controlling the proportion of absorbed excitation energy directed to each of the two photosystems. Nuclear DNA contains a multigene family for the LHC II apoproteins (DUNSMUIR et al. 1983; TOBIN et al. 1984). Those genes that are expressed, are translated on cytoplasmic ribosomes as water-soluble precursor proteins and then processed to the mature size at some, as yet unidentified, stage in the uptake and assembly of the complex in the chloroplast (e.g., SCHMIDT et al. 1981; TOBIN 1981).

LHC II contains most of the chlorophyll b in the thylakoid. Preparations of LHC II generally have a chlorophyll a/b ratio of 1.0 to 1.4. A consensus composition for an LHC II unit is 4 chlorophyll a, 3 chlorophyll b and 1–2 xanthophyll molecules per polypeptide of 24–27 kDa (THORNBER et al. 1979; LICHTENTHALER et al. 1982); however, there is some controversy about the pigment to protein ratio. Some groups (e.g., BURKE et al. 1978) claim that there are about 13 chlorophyll molecules per polypeptide chain, while others favor a value of 6–7 (RYRIE et al. 1980; KÜHLBRANDT et al. 1983; see also THORNBER et al. 1979). In any case, the pigments are associated with the protein and not with each other (LUTZ et al. 1979), and three of the chlorophyll b molecules in a unit are close enough to interact excitonically in an area where they are surrounded by chlorophyll a molecules (KNOX and VAN METTER 1979). As indicated above, there are multiple apoproteins of sizes lying between 24 and 27 kDa, associated with LHC II; however, these apoproteins do not appear to occur in a simple stoichiometry, neither does it seem that there are the same number of slightly different-sized LHC II apoproteins in all green plants (see BENNETT 1983 for more details). The various apoproteins are of almost identical primary structure, as was initially shown by amino acid analysis, cross-reactivity with antibodies and partial fragmentation of the polypeptide chains with CNBr or proteases (CHUA and BLOMBERG 1979; HOOBER et al. 1980; SCHMIDT et al. 1981; BENNETT 1983), and later by sequence determination (e.g., DUNSMUIR et al. 1983). It is not yet known how many of the LHC II apoproteins associate with each other (and with chlorophylls a and b) to form the macro-unit (a 8-nm freeze-fractures particle) present in the membrane (see STAEHELIN, Chap. 1, this Vol.); however, a suggestion has been made that the complex may be trimeric in situ (KÜHLBRANDT et al. 1983). The specific function of each of the slightly different apoproteins in forming the macrostructure also remains to be determined. We might have to learn where each varient polypeptide is located in the multisubunit complex and how each interacts with lipids before the reason, if one exists, for the slight differences can be deduced.

Considerable effort has been made to discover more about the structure of the LHC II complex by forming the complex into crystalline arrays. So far, no three-dimensional crystals suitable for by X-ray crystallographic analysis have been reported; however, two-dimensional arrays of the complex reconstituted into artificial membranes have been made and examined in the electron microscopy (KÜHLBRANDT et al. 1983; KÜHLBRANDT 1984; LI and HOLLINGSHEAD 1982; LI 1985). Such work has substantiated a claim made in an earlier immunological study (ANDERSSON et al. 1982) that the polypeptide chain of the LHC II complex is exposed on both sides of the thylakoid membrane. Apparently the protein protrudes some 20 Å on one side of the membrane and some 7 Å on the other. Overall a unit of LHC II is 27 Å wide and 60–65 Å long and appears to occur as trimers. It has an alpha-helical content of 44% (NEBEDRYK et al. 1984). More will be said about the structure of this chlorophyll-protein in Section 4.

It is pertinent at this stage to say something about protein phosphorylation in plastids. Several reviews (e.g., BARBER 1982; BENNETT 1983; HAWORTH et al. 1983; HORTON 1983; STAEHELIN and ARNTZEN 1983) are available on this topic and the reader is also directed to Chapter 11.5 of BARBER. Many thylakoid proteins become phosphorylated by a protein kinase(s) when chloroplasts are exposed to light in the presence of ATP. These are mainly the LHC II apoproteins and other polypeptides associated with photosystem II (STEINBACK et al. 1982). It has been proposed that phosphorylation of the LHC II polypeptides is by a thylakoid-bound protein kinase, the activity of which is controlled by the redox state of the thylakoid plastoquinone pool (J.F. ALLEN et al. 1981). Much evidence has accrued relating such a phosphorylation to the State I–State II transition in plants (cf. BARBER, Chap. 11.5, this Vol.), a process by which the distribution of excitation energy to the two photosystems is regulated. Phosphorylation of LHC II was proposed to alter the organization of the pigment molecules so that excitation energy in LHC II is then fed to photosystem I rather than to photosystem II as it would in the nonphosphorylated condition (J.F. ALLEN et al. 1981; HAWORTH et al. 1982; BENNETT 1983; HORTON 1983). This notion has been amplified recently with a proposal (KYLE et al. 1983) that, in State I chloroplasts, phosphorylation of LHC II complex (or at least some part of it, see KYLE et al. (1984), results in phospho-LHC II units moving from the photosystem II-enriched grana to the photosystem I-enriched stroma lamellae (see ANDERSON 1982; BARBER 1982; STAEHELIN, Chapter 1; BARBER, Chap. 11.5; SIMPSON, Chap. 11.6, both this Vol. for discussions of distribution of chlorophyll-proteins between grana and stroma lamellae). This change is proposed to decrease the absorptive cross-section area of photosystem II and to increase that of photosystem I, driving the chloroplasts into a State II condition. A protein phosphatase operates to remove the phosphate from the phosphorylated LHC II polypeptide, which permits the return of the chlorophyll-protein to the grana and the chloroplast to State I. Much indirect evidence exists for this "mobile antenna" hypothesis (e.g., KYLE et al. 1983, 1984; LARSSON et al. 1983; STAEHELIN and ARNTZEN 1983), but surprisingly the direct association of LHC II with photosystem I, under State II conditions, has not been reported. It is reasonably well established that a threonine residue(s) located very close to the N-terminal residue is the site in the LHC II polypeptide chain that becomes phosphorylated under these conditions, and that probably all of the apoproteins of LHC II are phosphorylated at that site but not to the same extent (BENNETT et al. 1981).

Table 1 summarizes data on the plant chlorophyll-proteins.

3.3 Pigment-Protein Complexes of the Algae

The green (Chlorophyta) and the blue-green (Cyanophyta now Cyanobacteria) algae are the most investigated algal classes for their pigment-protein composition. Descriptions of such complexes in these algal groups are given either in Sections 3.1 and 3.2 (Chlorophyta) or in Section 3.4 (Cyanobacteria). What we are concerned with in this section are caroteno-chlorophyll–protein complexes in other algal classes. These membrane-associated components have been investigated by very few research groups, one of which (ANDERSON and BARRETT) gives a thorough documentation of the data available on them in Chapter 6.4, this Volume (see also review by GLAZER 1983). The reader may also want to consult Chapter 1 of STAEHELIN in which an overview of the different groups of algae is provided. It is unfortunate that there are not more studies made on the pigment-proteins in these lesser-known algal classes, because characterization of their major light-harvesting complex could greatly help in understanding the LHC I and II complexes of green plants, as well as in providing a particularly intriguing evolutionary picture. Furthermore, such studies on these algae, which have a greater involvement of carotenoids in the light-harvesting process than any other organism, could lead to a much better understanding of carotenoid organization and function in plants, about which we know very little at present (cf. COGDELL 1978). It was suggested at the 1976 Brookhaven meeting (THORNBER et al. 1977) that there probably existed a category of proteins, the light-harvesting chlorophyll-proteins, at least one member of which is present in each of the different classes of higher plants and algae. The pigment constitution of the member of this category in each major algal class was proposed. It was also suggested that while the pigment composition of the major light-harvesting complex might change between the different algal classes, that of the core complexes, CC I and CC II, has probably remained essentially invariant during evolution. Data obtained since then support this idea: a CC I component of essentially the same composition has so far been found in every algal group investigated, and the two chlorophyll a-proteins in the 40–50 kDa range associated with the heart of photosystem II (CC II) have been observed in a wide variety of algae (ANDERSON and BARRETT, Chap. 6.4, this Vol.).

With respect to the LHC's, those that have been studied have been obtained from only a few organisms representing a wide variety of algal groups: brown, diatom, dinoflagellate, cryptomad and red algae. Those organisms that contain chlorophyll c have been found to have one or more xanthophyll-chlorophyll a/c-proteins. The most intensively studied organism of these classes contains a fucoxanthin-chlorophyll a/c_2-protein which is associated with photosystem II, (i.e., with CC II), and a violaxanthin-chlorophyll a/c_1+c_2-protein which is probably a photosystem I component, i.e., LHC I for this organism. A large proportion of the fuxocanthin in the organism is associated with the photosystem II-associated complex (BARRETT and ANDERSON 1980). The relative proportions of the different pigments in supposedly homologous xanthophyll-chlorophyll a/c-proteins seem to vary quite substantially between the different algal classes (see ANDERSON and BARRETT Chap. 6.4, this Vol.). It appears likely that these complexes are each composed of different pigmented subunits of

13–18 kDa: ALBERTE et al. (1981) and BARRETT and THORNE (1981) have fractionated pigment-proteins by sodium dodecyl sulfate-polyacrylamide gel electrophoresis of this size from brown algae and diatoms (see also GUGLIEMELLI 1984), and, since their pigment composition is not identical with that of the larger LHC's present, they may represent slightly different, smaller pigment-protein building blocks for the larger complexes. Obviously, much remains to be solved about chlorophyll organization in these algal classes.

Dinoflagellates contain a particularly interesting xanthophyll and chlorophyll-containing LHC, a peridinin-chlorophyll a-protein. This water-soluble component has four carotenoid and one chlorophyll a molecule per protein unit. It has been proposed that the chlorophyll a molecule is surrounded by the four carotenoid molecules in the protein complex (see PREZELIN and HAXO 1976 and ANDERSON and BARRETT, Chap. 6.4, this Vol.). This is one of the few naturally occurring chlorophyll-protein complexes that is soluble in water. Others are a bacteriochlorophyll a-protein from the green bacteria (see Sect. 3.6) and some minor chlorophyll-protein that occur in a few higher plant species (see THORNBER 1975). While it is unusual for a chlorophyll-protein to be water-soluble, not all membrane-associated photosynthetic pigment-protein complexes are equally insoluble. The bilin pigments which constitute the phycobiliproteins in cryptomonads, red algae, and cyanobacteria are readily obtained in aqueous solution from these organisms without the assistance of detergents. These biliproteins constitute the major portion of the LHC in red and blue-green organisms (see GANTT, Chap. 6.3, this Vol.).

3.4 Pigment–Protein Complexes of the Cyanobacteria

The function, composition, and structure of the photosynthetic apparatus of the prokaryotic cyanobacteria more closely resemble those of higher plants than those of the other photosynthetic bacterial classes (see STAEHELIN, Chap. 1, this Vol.). Cyanobacteria are oxygen-evolving organisms having two photosystems of similar composition to those in eukaryotic plants. Structurally, they have a simple thylakoid membrane system lacking the stacked system of higher plants (see Figs. 10, 12 of STAEHELIN, Chap. 1 this Vol.). The most obvious biochemical difference from higher plants is in the nature of their major photosynthetic pigments: Cyanobacteria have bilin pigments whereas green plants have chlorophyll b; otherwise, both types contain chlorophyll a and carotenoids. Cyanobacteria are not alone among photosynthetic organisms in containing the bilin pigments. The bilin pigments occur covalently bound to proteins, as phycobiliproteins, in dinoflagellates, red algae and cyanobacteria (see Sect. 3.3, also GANTT, Chap. 6.3; ANDERSON and BARRETT, Chap. 6.4, both this Vol.). The phycobiliproteins are the most well-characterized biochemically of all the photosynthetic pigment-proteins (see ZUBER, Chap. 6.1; GANTT, Chap. 6.3; SCHEER, Chap. 7.6, all this Vol. and GLAZER (1983) for detailed reviews). They can be easily extracted from the thylakoid membranes with aqueous buffers, and fractionated and purified to homogeneity by standard protein purification techniques to give a product of constant composition on each occasion. This, unfortunately, is not the case for the hydrophobic chlorophyll-proteins. The phycobiliproteins are very closely associated with the CC II component in these algae. The reader is directed to the references just given for a full biochemical characterization of the phycobiliproteins.

Cyanobacterial membranes, washed free of phycobiliproteins, present a membrane system of simpler composition than that of higher plant thylakoids in which to investigate the organization of chlorophylls and carotenoids. The first chlorophyll-protein described in a cyanobacterium, was one which accounted for 30% of the total chlorophyll and which was chromatographically isolated from sodium dodecyl sulfate extracts (THORNBER 1969). It contained P-700 (chlorophyll a/P-700 = 40/1) and had an apparent size of 110 kDa on sodium dodecyl sulfate-polyacrylamide gels (DIETRICH and THORNBER 1971; THORNBER et al. 1977), i.e., it was a representation of CC I in these organisms but, unlike its higher plant counterpart, the photochemical activity of P-700 was readily detectable (cf. THORNBER et al. 1977). The amino acid composition and subunit structure of this chromatographically isolated material was described, and these and other data were used to derive a model that served for several years in the early 1970's as a postulate of chlorophyll organization in photosystem I in all organisms (see THORNBER et al. 1977). The analytical work was extended by REINMAN and THORNBER (1979) who used the improved higher plant electrophoretic fractionation systems available in 1978 (see Sect. 2 and Fig. 3), to examine chlorophyll-proteins in solubilized cyanobacterial membranes. These systems retained more of the total chlorophyll (approx. 90%) with their specific proteins after fractionation than had been achieved previously. It was observed that one cyanobacterium, *Phormidium luridum*, contained not only the 110 kDa P-700-chlorophyll a-protein but also a higher molecular weight complex(es) of 255 kDa also having photosystem I characteristics, as well as a PS II chlorophyll-protein of 58 kDa, equivalent in characteristics to that observed by HAYDEN and HOPKINS (1976) in higher plants. Furthermore, 80% or more of the solubilized chlorophyll was found to be associated with photochemically active PS I particles. Thus, the earlier chromatographic preparation (THORNBER 1969), representing 30% of the total chlorophyll and having a chlorophyll/P-700 = 40/1, either represented the isolation of the P-700-containing component rather than the entire PS I chlorophyll-protein complex, or a preferential loss of antenna chlorophyll occurred during isolation of the PS I chlorophyll–protein complex. Recent preparations (e.g., WILLIAMS et al. 1983) of the CC I complex from *Synechococcus* 6301, having chlorophyll/P-700 ratios of ~130/1 and a very similar subunit composition to earlier preparations having chlorophyll/P-700 ratios of 40, reveal that the latter explanation is the more likely one.

Research by others in the late 1970's and early 1980's, in particular by GUIKEMA and SHERMAN (see BRICKER et al., Chap. 11.4, this Vol.), has revealed an even more complex picture for the number of chlorophyll-protein bands present in cyanobacterial membranes as determined by nondenaturing sodium dodecyl sulfate-polyacrylamide gel electrophoresis: In addition to the PS II component (CC II), *five* PS I complexes of 90 kDa or greater were identified. Characterization of the polypeptide composition of each of these PS I chlorophyll-proteins, all of which contain the 60–70 kDa apoprotein of CC I seen in eukaryotic photosynthetic organisms, has been carefully done but no clear rationalization for their multiplicity has been forthcoming. However, there is a general agreement that there are fewer (about 3–4) nonpigmented polypeptides associated with cyanobacterial preparations of CC I than with the higher plant preparations which contain at least six (see NECHUSHTAI et al. 1983, for a direct comparison and BRICKER et al., Chapter 11.4 for a review of the subunit composition of the cyanobacterial preparations). Several of these nonpigmented proteins are thought to be Fe–S center-containing proteins that act as early electron acceptors in the PS I primary event.

In summary, PS I in cyanobacteria has a CC I complex which is very similar in composition to the higher plant CC I complex, except that a greater proportion of the prokaryotic complex can be readily observed in sodium dodecyl sulfate extracts as a considerably larger particle. No evidence has been presented for the presence of an LHC I component in cyanobacteria, presumably some of the phycobiliprotein plays this role. Photosystem II in cyanobacteria has a CC II complex of seemingly identical form to the equivalent complex in all other oxygen-evolving organisms, while the PS II antenna (LHC II) is largely constituted by the phycobiliproteins in these bacteria. The reader is referred to other articles in this Volume (STAEHELIN, Chap. 1; ZUBER, Chap. 6.1; GANTT, Chap. 6.3; SCHEER, Chap. 7.6; BRICKER et al., Chap. 11.4, all this Vol.) for a more detailed description of each of these components.

3.5 Purple Bacteria

3.5.1 Photochemical Reaction Centers

The photochemical reaction center, representing 1–2% of the total photosynthetic pigment in these bacteria, has been obtained completely free of antenna pigments and other extraneous proteins from many purple bacteria (FEHER and OKAMURA 1978; GINGRAS 1978; THORNBER et al. 1984a) and from one green filamentous bacterium, *Chloroflexus aurantiacus* (PIERSON and THORNBER 1983) which is thought to be the extant form of the organism that was the precursor of all photosynthetic organisms (STACKEBRANDT and WOESE 1981). The reasons for including a green bacterium's reaction center in the discussion in this section on purple bacteria will soon become apparent. (Further details about the reaction centers discussed here can be found in this Vol. in SAUER, Chap. 2; DUTTON, Chap. 5; PARSON and HOLTEN, Chap. 8.1; TIEDE, Chap. 8.2; NORRIS and VAN BRAKEL, Chap. 8.3; MICHEL and DEISENHOFER, Chap. 8.4; HEARST, Chap. 8.5; BLANKENSHIP and FULLER, Chap. 8.6; SPRAGUE and VARGA, Chap. 11.1).

The most thoroughly studied reaction center is that of *Rhodopseudomonas* (*Rps.*) *sphaeroides* (cf. CLAYTON 1980; OKAMURA et al. 1982). The simplest unit contains four molecules of bacteriochlorophyll a, two molecules of bacteriopheophytin a, one molecule of carotenoid, 1–2 molecule(s) of ubiquinone (UQ_{10}), one atom of iron, and one copy each of three polypeptides of molecular weights 28,000, 32,000 and 36,000 termed L, M, and H respectively (cf. SUTTON

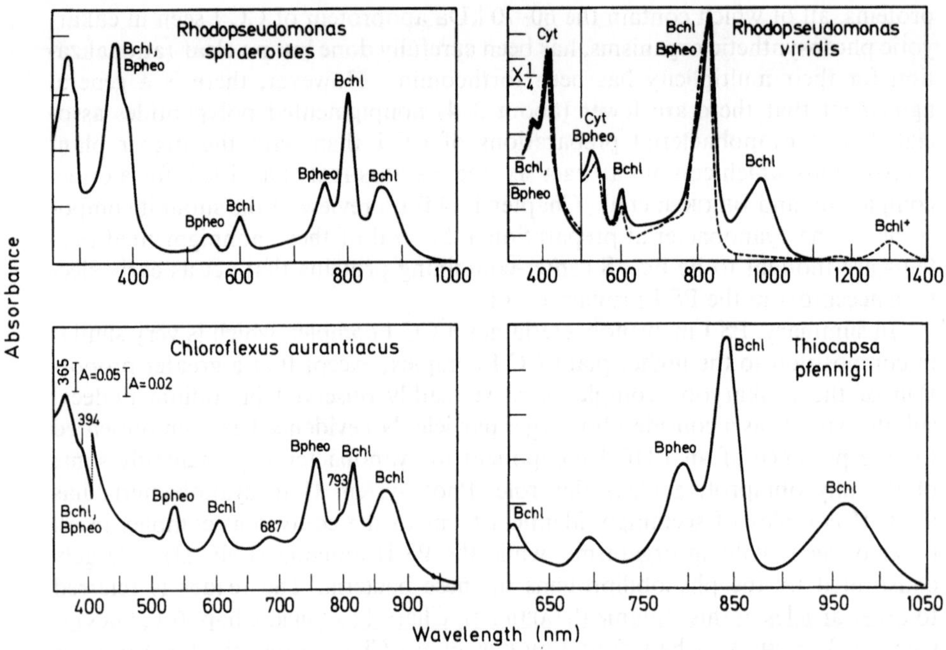

Fig. 2. Absorption spectra of some typical reaction center preparations from three purple and one green (*Chloroflexus aurantiacus*) bacteria. *Rhodopseudomonas sphaeroides* and *Chloroflexus aurantiacus* reaction centers contain bacteriochlorophyll a and bacteriopheophytin a, whereas the other two reaction centers contain the equivalent b forms, which absorb at longer wavelengths in vitro and in vivo. The assignment of the spectral forms to a reaction center pigment is shown in the figure. *Bchl* bacteriochlorophyll; *BPheo* bacteriopheophytin; *Cyt* cytochromes. The 690 nm peak in the two spectra on the lower line is due to a chlorin degradation product of the Bchl. The longest wavelength absorbance is that of the special pair of chlorophyll molecules forming the primary electron donor in the organism P-870 in Bchl a-containing organisms and P-960 in BChl b-containing organisms. The spectrum of the *Rhodopseudomonas viridis* reaction center is also shown (*dashed trace*) in its oxidized state, i.e., after it has donated an electron to an acceptor in the primary photochemical event: the peak at 1310 nm is due to the oxidized form of P-960 (oxidized spectra are not shown for the other reaction centers). Comparison of the relative heights of the three near-infrared peaks in the different reaction center preparations, which are due to the Qy absorption bands of the relevant pigments, reveals that the *Chloroflexus* component has a smaller absorbance for its central peak than the other reaction centers shown, primarily because it has a different pigment ratio than the others (see text). The absence of such a third peak in the *Rhodopseudomonas viridis* complex is due to its BPheo absorbing at relatively longer wavelengths than in the others, which results in the Bpheo peak merging with the central Bchl peak at 835 nm. The BPheo and BChl peaks at 540 and 600 nm are due to the absorbance of the Qx transitions of these pigments

et al. 1982). A simpler, but less stable, photochemically active unit can be obtained by treating the isolated reaction center with chaotropic agents which remove the H polypeptide but none of the pigment molecules. Thus, all the pigments in the reaction center are associated with the L or the M subunit or both. Recent crystallographic studies substantiate the last choice: All the components (bacteriochlorophyll, bacteriopheophytin and quinones) involved

Table 2. A comparison of the composition of some isolated photochemical reaction centers

Species	*Rps. sphaeroides*	*Rps. viridis*	*Chloroflexus aurantiacus*
Pigment content per P-870 or P-960 molecule	4 molecules BChl a 2 molecules BPheo a 2 molecules UQ_{10} 1 molecule carotenoid	4 molecules BChl b 2 molecules BPheo b 1 molecule UQ_{10} 1 molecule MQ_7 1 molecule carotenoid	3 molecules BChl a 3 molecules BPheo a No carotenoid
Subunit composition			
Number per P-870 or P-960 (and names)	3 (L, M and H)	4 (L, M, H and two c type cytochrome(s)[a]	2 (L and M?)
Apparent size of subunits from SDS-PAGE (from amino acid analyses)	H –28 (36) kDa M–24 (32) kDa L –21 (28) kDa	H –35 kDa M–28 kDa L –24 kDa Cytochrome – 38 kDa	 M–30 kDa L –28 kDa

[a] 4–5 hemes per P-960

in the charge separation in another purple bacterium (*Rps. viridis*) are positioned symmetrically about an axis which is oriented perpendicular to the plane of the membrane and is equidistant from the L and M subunits (see TIEDE, Chap. 8.2; NORRIS and VAN BRAKEL, Chap. 8.3; MICHEL and DEISENHOFER, Chap. 8.4, all this Vol.). The absorption spectrum of the *Rps. sphaeroides* reaction center and the assignment of chromophores to the absorption bands is depicted in Fig. 2.

By using comparative biochemical and biophysical approaches, it has been possible to assess those features of a reaction center's structure and function that are of general significance. Such data on several reaction centers are compared in Table 2 and Fig. 2. Many characteristics of the isolated reaction centers prove to be strikingly similar in the different purple bacteria and even in the filamentous green bacterial reaction center. In particular, the involvement of each of the bacteriochlorophyll, bacteriopheophytin and quinone molecules in the primary charge separation in purple bacteria seems to be identical (SAUER, Chap. 2; DUTTON, Chap. 5, both this Vol.), and even the mechanism in the green bacterial reaction center from *Chloroflexus aurantiacus* appears remarkably similar to that in the purple bacteria (see BLANKENSHIP and FULLER, Chap. 8.6, this Vol.). Secondly, the ratio of photosynthetic pigments to each other and to the primary donor seems constant in the purple bacterial components. All reaction centers isolated from the purple bacteria have four bacteriochlorophyll and two bacteriopheophytin per P-870 or P-960; however, the *Chloroflexus* reaction center has three bacteriochlorophyll and three bacteriopheophytin molecules, giving rise to the 805+865 and 760 nm bands (Fig. 2) respectively (PIERSON et al. 1983). Thirdly, all reaction centers isolated so far contain quinones and carotenoids – the former participate in charge separation and the latter primarily protect the reaction center pigments from photodestruction via the triplet state of bacteriochlorophyll (COGDELL 1978). The content, but not the type, of these biochemicals is very similar the various isolated reaction centers (Table 2). Lastly, the number, sizes, and primary structure of each of the L, M, and H subunits of a reaction center are very similar, often identical, in all purple bacteria (see HEARST, Chap. 8.5 this Vol. for comparison of primary structures). It is emphasized that the true sizes of a reaction center preparation's subunits are some 30% greater than those observed on sodium dodecyl sulfate-polyacrylamide gels (SUTTON et al. 1982). The spectra (Fig. 2) emphasize most of the differences that occur between any one bacterial reaction center and the others. Thus the presence of bacteriochlorophyll and bacteriopheophytin b rather than a in the *Rps. viridis* and *Thiocapsa pfennigii* complexes (SEFTOR and THORNBER 1984) results in these organisms having spectral forms of bacteriochlorophyll and bacteriopheophytin absorbing further into the near infra-red and being more separated spectrally than in their bacteriochlorphyll a counterparts (cf. Fig. 2). Note, however, that the two bacteriochlorophyll b-containing reaction centers differ spectrally from each other. The two bacteriochlorophyll b-RC's also contain c-type cytochromes (see Fig. 4.5), whereas the others that are shown do not. However, this is not a property of bacteriochlorophyll b-containing reaction centers alone, since isolated *Chromatium vinosum* reaction centers also contain cytochrome(s) (cf. LIN and THORNBER 1975). In any case, all reaction centers are associated with cytochromes in situ; it is the strength of the association which differs between the bacteria (cf. TIEDE, Chap. 8.2, this Vol.). The one major difference in the subunit composition of the reaction centers examined so far occurs between the *Chloroflexus* and the purple bacterial reaction centers: the former contains two polypeptides rather than the statutory three in the purple bacterial complex. The *Chloroflexus* polypeptides are probably equivalent to the L and M subunits of the purple bacterial reaction centers (PIERSON et al. 1983).

The mechanism of the primary photochemical event in bacteria has been mainly elucidated from studies on the *Rps. sphaeroides* and *Rps. viridis* reaction centers. Other articles in this volume (SAUER, Chap. 2; DUTTON, Chap. 5; PARSON and HOLTEN, Chap. 8.1; TIEDE, Chap. 8.2; NORRIS and VAN BRAKEL, Chap. 8.3) describe this mechanism in detail. In essence, it is generally believed that during the primary event, the excited state of P-870 or P-960, each of which is composed of a pair of bacteriochlorophyll molecules, transfers an electron to one of the two other bacteriochlorophyll molecules in the reaction center within a few picoseconds of excitation. During the next 200 picoseconds this electron migrates via a bacteriopheophytin molecule to a quinone. This chain of pigmented electron acceptors seems to be an essential feature of photochemical reaction center function in both bacteria and plants (see COGDELL 1983); rapid removal of the unpaired electron from the vicinity of the oxidized reaction center pigment minimizes wasteful charge recombination.

The ultimate aim of research on the photochemical reaction center is to obtain its three-dimensional structure, and its relative orientation in the membrane so that the photosynthetic energy conversion can be better understood. Until recently, progress in these aims came only from indirect experimental techniques (e.g., photoselection see HOFF, Chap. 7.2; TIEDE, Chap. 8.2; NORRIS and VAN BRAKEL, Chap. 8.3, all this Vol.) due to lack of crystals of reaction centers for X-ray analysis. However, crystals of the *Rps. viridis* and *Rps. sphaeroides* reaction centers have now been made (MICHEL 1982; ALLEN and FEHER 1984), and the detailed structure of the former reaction center is available (see MICHEL and DEISENHOFER, Chap. 8.4, this Vol.). When each of the spectral forms of the pigments and other biochemicals involved in the primary event can be identified in the structure, then we will have an understanding of the mechanism of the primary photochemical event in the detail researchers have been seeking for many years. This, however, will not be the end of the story, for knowledge of the structure in such detail is sure to raise a set of new and intriguing questions.

3.5.2 The Light-Harvesting Pigment-Protein Complexes

In Section 2.2 the purple bacterial antenna complexes were described as being of two distinct types: One is present in all purple bacteria, is intimately associated with the reaction center, and occurs in a fixed stoichiometry with respect to the reaction center. The B-890-protein of *Rhodospirillum* (*R.*) *rubrum* and the B-875-protein of *Rps. sphaeroides* are examples of this type (see COGDELL, Chap. 6.2; this Vol., and Table 3). The "B" terminology for naming the purple bacterial antenna complexes uses the near infra-red wavelength peak or peaks of the bacteriochlorophyll-proteins. In addition to the B-890-, B-875-type complexes, many species of purple bacteria contain a second spectrally and biochemically different class of bacteriochlorophyll-proteins of which the B-800–850-proteins of *Chromatium* (cf Fig. 1), *Rps. sphaeroides*, *Rps. capsulata* or *Rps. palustris* are examples. Since the amount of the B-800–850 complex, but not of the B-890 or B-875 complex, varies with respect to the reaction center concentration, it is not unreasonable to suppose that bacteria modulate the size of their light-gathering unit in response to environmental conditions, such as the incident light intensity, by changing their content of the B-800–850 complex (cf. THORNBER et al. 1978). Although this distinction between the two classes of antenna complex is still valid, more recent biochemical analyses of the antenna components of a wider range of purple bacteria reveal that each of these two classes has at least two subdivisions (Table 3). Two extensive reviews on these components are available in this Volume (ZUBER, Chap. 6.1; COGDELL, Chap. 6.2), and earlier reviews by CLAYTON and SISTROM (1978), and COGDELL and THORNBER (1980) might be consulted.

3.5.2.1 The B-800–850 Complex

This complex occurs in the membranes of *Rps. sphaeroides* as a multimer of a minimal unit consisting of three molecules of bacteriochlorophyll a, one mole-

Table 3. Comparison of the major antenna caroteno-chlorophyll-proteins in purple bacteria

Antenna Type	B-890-protein class		B-800–850-protein class		
	B-890-protein	B-875-protein	B-800–850-protein		B-800–820-protein
			Type I	Type II	
Examples of bacteria containing antenna type	*R. rubrum* *C. vinosum* *Rps. acidophila* ? *Rps. viridis*	*Rps. palustris* *Rps. capsulata* *Rps. sphaeroides* *Rps. gelatinosa*	*Rps. acidophila* 7750 and 7050 (high light grown)	*C. vinosum* *Rps. acidophila* 7050 (low light grown)	
BChl a: carotenoid	2:1	2:2	3:1	3:1	3:1
Number of polypeptides in isolated complex	2	2	2 or 3	2	2
Number of amino acid residues in polypeptides	52 and 54	52–58 and 47–48	54–60 and 52	~50–65	~50–65
Intensity of CD spectrum of long wave-length band	Strong	Weak	Strong	Strong	Strong
	All spectra indicate BChl dimer present in the band				

Note: The quantitative data are based on analysis of complexes isolated from only a few of the species given as examples

cule of carotenoid and one copy of each of two small (~6 kDa) polypeptides. No one has yet succeeded in isolating this minimum unit in a pigmented form; the smallest colored complex that has been isolated has a size greater than 100 kDa. Both apoproteins of the B-800–850 complex are as hydrophobic as the L and M apoproteins of the reaction center and are more hydrophobic than their plant counterparts (THORNBER et al. 1984a). Two bacteriochlorophyll a molecules in the complex contribute the 850 nm absorption band while the third bacteriochlorophyll molecule gives the 800 nm absorption. The carotenoid molecule and the 800 nm-absorbing bacteriochlorophyll molecule were found by COGDELL's group (WEBSTER et al. 1980; see also COGDELL and THORNBER 1980) to be associated with the smaller polypeptide (52 amino acid residues), while the other two bacteriochlorophyll molecules, interacting excitonically, were proposed to be conjugated with the larger apoprotein (54–60 amino acid residues). See ZUBER (Chap. 6.1, this Vol.) for a slightly different view on the association of pigments with the two chains). COGDELL in Chapter 6.2, this Volume describes the convention for naming the larger and smaller polypeptides.

The B-800–850 complex of *Rps. sphaeroides* (and some other organisms) is now (COGDELL et al. 1983; THORNBER et al. 1984a) referred to as a type I B-800–850 complex. This type is characterized by its 850 nm-absorption band being about one-and-a-half times as intense as the 800 nm band (cf. Fig. 1 with spectrum in COGDELL, Chap. 6.2, this Vol.). Other purple bacteria species contain type II B-800–850-complexes (see Table 3) in which the 850 nm absorption is of equal or lower intensity than the 800 nm band (see Fig. 1 and COGDELL, Chap. 6.2, this Vol.). Whenever the type II complex occurs in a bacterium there is also a B-800–820 antenna complex present (see Fig. 1). Their interrelationship is not yet clearly defined, but they appear to have very similar, perhaps identical, polypeptide compositions and pigment content.

3.5.2.2 *The B-890-Protein Complex* (Table 3)

The *R. rubrum* complex is the most fully characterized of this type (see COGDELL Chap. 6.2, this Vol.). Its minimum unit consists of two molecules of bacteriochlorophyll a and one molecule of carotenoid (usually spirilloxanthin) bound to two low molecular weight polypeptides. The amino acid sequence of both polypeptides is known; one has 52 and the other has 54 amino acid residues (see ZUBER, Chap. 6.1, this Vol.). The 890 nm peak gives an intense circular dichroism spectrum, typical of a bacteriochlorophyll dimer. Similar complexes have been characterized in *Chromatium* and *Rps. acidophila* (see COGDELL, Chap. 6.2, this Vol.).

The supposedly analogous B-875-protein from some other species (e.g., *Rps. sphaeroides*) (see spectrum in COGDELL, Chap. 6.2, this Vol.) differs from the B-890-complexes because not only is its maximum absorbance in the near infrared at shorter wavelengths, but it also contains two molecules of bacteriochlorophyll a and *two* molecules of carotenoid per pair of polypeptides; furthermore the 875 nm-band shows only a weak circular dichroism spectrum. Thus the B-890-protein and the B-875-protein complexes should be regarded as two different subclasses.

3.5.3 Topography of the Antenna and Reaction Center Complexes

ZUBER in Chapter 6.1, this Volume, gives the primary of all the B-800–850 or B-890 polypeptides sequenced to date. Each of the sequences of 50–60 amino acids exhibits three domains: A polar N-terminus located in the cytoplasm of the bacterium, a hydrophobic core of 20–23 residues which forms an alpha-helix which crosses the chromatophore membrane, and a polar C-terminus which protrudes into the periplasm (see BACHOFEN and WIEMKEN, Chap. 11.2, this Vol. for further details). In this manner the bacteriochlorophyll-protein molecules are prevented from flip-flopping in the membrane. It is proposed that each alpha-helix binds one, sometimes two, bacteriochlorophyll molecules via a histidine ligand (see ZUBER, Chap. 6.1, this Vol.). Each of the three subunits of the photochemical reaction center span the chromatophore membrane, and it is well established that the L and M subunits each cross the membrane five times.

3.6 Chlorophyll-Protein Complexes of the Green Bacteria

Green bacteria (Chlorobiaceae and the Chloroflexaceae), contain some bacteriochlorophyll a and carotenes but bacteriochlorophyll c, d, or e is their major photosynthetic pigment. The organization of each of these pigments has been reviewed by OLSON (1980), and GLAZER (1983) and for the Chloroflexaceae, in particular, by BLANKENSHIP and FULLER (Chap. 8.6, this Vol.). Much useful information for understanding the supramolecular organization of the photosynthetic apparatus of these organisms is also given in Chapter 11.1, this Volume by SPRAGUE and VARGA.

In essence all of the bacteriochlorophyll c, d, or e, much of the carotene and some of the bacteriochlorophyll a is contained in chlorosomes, which are organelle-like antenna structures that are attached to the cytoplasmic membrane where the rest of the photosynthetic pigments are located (see SPRAGUE and VARGA, Chap. 11.1, this Vol.). FEICK and FULLER (1984) have determined that of the four polypeptides associated with the chlorosome, one protein of 3.7 kDa is probably conjugated with the bacteriochlorophyll c. Furthermore, they have concluded that this small polypeptide occurs as a dimer which has 10–16 chlorophyll molecules associated with it. Such a pigment to protein ratio is so much greater than has been found in any other chlorophyll–protein complex, that it is very probable that a different method of organization occurs for the pigment molecules in this bacteriochlorophyll-protein than in most others (BLANKENSHIP and FULLER, Chap. 8.6, this Vol.). Perhaps the stacked oligomeric array suggested some years ago by CHOW et al. (1975) is used. In any case, it is now reasonably well established that bacteriochlorophyll c (and therefore probably d and e) are associated with protein, thus ending a long-standing thought that these photosynthetic pigments were unique in not being organized by proteins (cf. THORNBER et al. 1978). The bacteriochlorophyll a associated with the chlorosome of the Chlorobiaceae is housed in the well-described water-soluble bacteriochlorophyll a-protein, which forms a crystalline baseplate between the light-harvesting chlorosomes and the reaction centers in the cytoplasmic membrane (SPRAGUE and VARGA, Chap. 11.1, this Vol.). The crystal structure of this pigment protein has been published (MATTHEWS et al. 1979). The water-soluble bacteriochlorophyll a-protein of green bacteria has proved very important to the chlorophyll-protein field. It has served as a model for how chlorophyll-proteins are organized, and, for many years, its three-dimensional structure was used for calculations of chlorophyll–chlorophyll and chlorophyll–protein interactions. On the other hand, in the Chloroflexaceae, the equivalent bacteriochlorophyll a may be associated with a 5.8 kDa polypeptide to form a water-insoluble complex absorbing at 792 nm. Both this bacteriochlorophyll a-protein and the water-soluble one seem to function to funnel absorbed light energy from the chlorosome to a point in the cytoplasmic membrane where the reaction center and its most closely associated bacteriochlorophyll a-containing antenna are located (BLANKENSHIP and FULLER, Chap. 8.6; SPRAGUE and VARGA, Chap. 11.1, both this Vol.). In the case of *Chloroflexus aurantiacus*, it appears that the antenna most intimately associated with the reaction center has spectral properties almost identical to those of the purple bacterial antenna (i.e., a B-800–850 plus a B-875-like component may be present).

The very large ratio of antenna pigment to that in the reaction center (1000–2000:1) in green bacteria (FOWLER et al. 1971) has hindered progress in obtaining an understanding of the photosynthetic mechanism in green bacteria in equivalent detail to that obtained for the purple bacteria. Only recently has much progress been made in characterizing the biochemistry of the photochemical reaction center of these bacteria (see BLANKENSHIP and FULLER, Chap. 8.6, this Vol.). The first preparations (OLSON et al. 1976) contained bacteriochlorophyll a, carotenoids, several polypeptides and variable amounts of cytochromes b and c (see OLSON and THORNBER (1979) and OLSON (1980) for a summary of the earlier work on this component). Later, VASMEL et al. (1983) improved somewhat on the original preparations by purifying a core complex from *Prosthecochloris aestuarii* that contained only three polypeptides of 60, 43, and 32 kDa; unfortunately, this material had virtually no photochemical activity. Thereafter, HURT and HAUSKA (1984) greatly improved the situation by obtaining a photochemically active fraction from *Chlorobium limicola*, which also contained three polypeptides and the same pigments as the VASMEL et al., preparation. Most interestingly, the major polypeptide in the HURT and HAUSKA preparation is of 65 kDa, a size very close to that of the P-700-apoprotein, and, it is associated with bacteriochlorophyll. The second polypeptide of 24 kDa is a c-type cytochrome, while the third may be a Rieske Fe–S protein of 32 kDa. A ratio of 20 molecules of bacteriochlorophyll to one of P-840, the primary donor in these green bacteria, occurs in the isolated material. The *Chloroflexus* reaction center has characteristics very similar to those of the purple bacteria reaction center and is totally different from the reaction center in other green bacteria. However, the bulk of the *Chloroflexus* antenna is basically indistinguishable from the major antenna in any of the green bacteria (BLANKENSHIP and FULLER, Chap. 8.6, this Vol.). These relationships are probably reflective of the position of *Chloroflexus* in the evolutionary scheme of photosynthetic organisms (STACKEBRANDT and WOESE 1981).

4 Concluding Remarks

I have attempted to provide the reader with a summary of the available information on the biochemistry of both the reaction center (core) and the antenna (light-harvesting) complexes in all the various classes of photosynthetic organisms. Emphasis has been placed on those complexes that are integral membrane components, i.e., those that require the presence of a detergent to maintain their solubility upon isolation. I have presented a consensus view of the subject and only infrequently mentioned any controversy that may surround a component. This was done to keep the length of the article within reasonable bounds.

Controversy does exist, however, to varying degrees, for every one of the components mentioned, particularly with respect to their biochemical composition. This mainly arises because no standard procedure has yet proved universally acceptable for the purification of a particular complex. Consequently, the procedures adopted by one laboratory frequently differ from those adopted by others which result in products of slightly different compositions. There is then a tendency to rationalize a discrepancy as being due to one procedure yielding a purer product than the other, rather than to acknowledge that, for detergent-soluble proteins, alternative procedures can yield different, yet homogenous, forms of the same entity. In this connection it must be remembered that we are trained to accept that alternative ways of purifying *water-soluble* proteins must yield products of identical composition. The possibility of a much stronger association of an integral membrane component with its neighbors in the membrane milieu, compared to that between a water-soluble protein and its neighbors in the protoplasm, might explain, in part, why homogenous preparations of a pigment-protein obtained by two different methods can vary in composition. A further explanation is that the use of denaturing detergents in the purification can alter the extent of the association of integral and external biochemicals with the polypeptide chain. It is mainly for these reasons that investigation of photosynthetic pigment-protein complexes is a difficult task.

4.1 Isolation of the Pigment-Protein Complexes

Electrophoresis in polyacrylamide gels in the presence of a detergent is the most used fractionation system. The fine separation between components after electrophoresis (see Fig. 3) gives an investigator confidence in the purity of resolved bands, and this often outweighs the advantage of the higher yields obtainable from chromatographic/gradient centrifugation methods. Furthermore, the electrophoretic methods are easier to perform and require a minimal expense for equipment. The procedure is also more rapid, so that an investigator, starting with intact plant material in the morning, can have the fractionation completed by midday, compared to the other procedures which may take over 24-h to complete. Speed is often of the essence to prevent unwanted dissociation of the pigment-proteins after detergents have been added to the photosynthetic membranes.

In the electrophoretic procedures, detergents are added to the membrane generally at a certain ratio with respect to the chlorophyll content and, after centrifugation to remove any undissolved material, the detergent extract is immediately loaded into a well of a polyacrylamide gel underneath the reservoir buffer. It would be preferable to add the detergent at a specific ratio with respect to the lipid present, but estimating the lipid content cannot be done as rapidly or as simply as chlorophyll. Electrophoresis is usually performed at room temperature or at 4 °C. Lithium dodecyl sulfate is often used as the detergent for electrophoresis at the lower temperature because it is soluble at 4 °C, whereas sodium dodecyl sulfate is not. However, the heat generated in the gel during electrophoresis raises its temperature above the ambient 4 °C and consequently, for runs of less than a few hours, sodium dodecyl sulfate will remain in solution. When lower wattages are used for overnight electrophoresis, which generate less heat in the gel, lithium dodecyl sulfate is essential.

Some typical electrophoretic separations are shown in Fig. 3. Almost identical material was loaded onto each gel lane. Figure 3 clearly demonstrates that varying the fractionation procedure can yield homogenous forms of the same pigment-protein; however, it cannot be assumed that the structure and/or composition of a band in one system is identical to that of its representation in another system, as demonstrated by BENNETT et al. (1981) for LHC II. Nevertheless, such differences can be very useful in building a more complete description of any chlorophyll-protein.

Photosynthesis researchers have long sought the "perfect" detergent – one which completely solubilizes every protein component of the membrane without removing any of the pigment molecules or other prosthetic groups associated with them in situ. In addition, a search has been made for the "perfect" fractionation system – one which exhibits no free pigment, gives a clean separation between chlorophyll-protein bands and does not have either homogenous or heterogenous oligomeric forms of any complex. It is now apparent that these aims are too idealistic, and also that the "perfect" situation may not be recognizable until we know the answer to the whole chlorophyll-protein story. Sodium dodecyl sulfate seemed to come close to being the "perfect" detergent for some time: it generally solubilized all of the photosynthetic membrane of an organism and, as shown in Fig. 3, it does not generate much free pigment in the process; however, it does dissociate pigment from the complexes during polyacrylamide gel electrophoresis (Fig. 3) as well as subtly altering the arrangement of the chlorophyll molecules in some complexes (TAPIE et al. 1982). Triton X-100 does not have this last fault but, having no charged group in its structure, it cannot be used advantageously for electrophoretic fractionation of the membrane components. Octyl glucoside, when used in conjunction with lesser amounts of sodium dodecyl sulfate, looks the most promising detergent at present (cf. Fig. 3); however, it too can generate free pigment (TANG and SATOH 1984) and furthermore it does not completely solubilize the membranes.

In any case, the finding of a solubilization/fractionation system that does not generate free pigment is particularly important at this stage of unraveling the chlorophyll-protein story, because we need to characterize *every* chlorophyll-protein present in an organism. Such a system exists (Fig. 3) but the inferior resolution of bands by it can obscure any new complex migrating at approximately the same rate as a major chlorophyll-protein. To overcome this drawback researchers in the area use the better separations provides by those systems that generate 10% or more free pigment. However, a consequence of this choice can be to miss seeing a new pigment-protein (e.g., see DUNAHAY and STAEHELIN 1985b) because it is completely dissociated to free pigment and protein in that system. Thus fractionation systems can still be improved.

Free pigment seems to be generated by the fractionation system (particularly the electrophoretic ones) rather than by the membrane solubilization process. Some researchers think that the temperature of electrophoresis is the important factor in controlling free pigment generation, and it certainly is, above 50 °C. Others favor the view that the length of time of exposure of a chlorophyll-protein to a detergent during electrophoresis is most important (i.e., the shorter the time, the less free pigment), while still others believe that free pigment results from exposure of the complexes to changing ion concentrations in the noncontinuous buffer systems used in some electrophoretic separations. It is also thought that a perfect balance of detergent to protein must be maintained during fractionation, i.e., sufficient detergent must be in the buffer to keep the component soluble but not so much as to expose the complex to an excess of it. Within reasonable limits this explanation is unlikely to be correct since, in the author's experience, the majority of the free pigment is produced as the chlorophyll-proteins migrate from the solution in the well into the gel. The nonelectrophoretic fractionation systems may be the best way to decrease the amount of free pigment produced: several groups use sucrose gradient centrifugation to prepare chlorophyll-proteins, others use column chromatography while yet others (e.g., SATOH 1979; NECHUSHTAI and NELSON 1981) combine such procedures. In effect, however, the best way to prepare a chlorophyll-protein still eludes us because we have not had sufficient experience of these complex components.

Fig. 3i–vii. Photographs of seven different procedures for polyacrylamide gel electrophoresis of detergent-solubilized green plant thylakoid membranes. The dark bands are chlorophyll-containing components; the gels are *not* stained. **i** The electrophoretic system used by THORNBER et al. (1966, 1967a) and OGAWA et al. (1966) in the early studies on chlorophyll-proteins. *CP1* and *CP2* (and later *X*) were shown to be chlorophyll-protein complexes whereas *FP* (free pigment) is chlorophyll and carotenoid that is not complexed with protein. One-third of the total pigment occurs in the *FP* zone in this system. *CP1* is a representation of the photosystem I component, *CCI*, and *CP2* is a representation of LHC II (see text). **ii** Electrophoretic system (MARKWELL et al. 1978) in which more of the total chlorophyll is retained with protein after fractionation, i.e., F (*FP*) is reduced to about 13% of the total pigment applied. *A-1* is a photosystem I chlorophyll-protein complex (CC I + LHC I?), *AB-1*, *AB-2* and *AB-3* are representations of LHC II. CC II electrophoreses slightly more slowly than *AB-3* but is not separated from it. **iii** A system (MACHOLD et al. 1979) which fractionates a detergent extract of thylakoids into many zones but produces considerably more *FP* than **ii**. *Chl a-P1* is a form of CC I; *Chl a/b-P2* zones are LHC II fractions, *Chl a/b-P1* may be equivalent to CP29 (see text), *Chl a-P2* and *P3* are the two CC II chlorophyll-proteins (cf. text). **iv** The system of ANDERSON et al. (1978) showing two PS I components (*CP 1a* and *CP1*), three bands (*LHCP 1*, *LHCP 2*, and *LHCP 3*) purportedly trimer, dimer and monomer of LHC II respectively, and one CC II band (*CPa*). Free pigment is reduced (cf. **i**) to 10–13% of the total pigment. **v** The system of DELEPELAIRE and CHUA (1981) in which electrophoresis is performed at 4 °C and lithium dodecyl sulfate, which is soluble in water at 4 °C, replaces sodium dodecyl sulfate throughout the system. *CPI* and *CPII* are as under **i**, *CP III* and *IV* the two PS II chlorophyll a-proteins (CC II). **vi** The CAMM and GREEN (1980) octylglucoside/sodium dodecyl sulfate system as modified by DUNAHAY et al. (1984). Thylakoid solubilization is not 100% and there is some free pigment, but the multiplicity of well-resolved green zones makes this a powerful system for fractionation of thylakoid chlorophyll-proteins. **vii** Electrophoretic fractionation in a system (MARKWELL et al. 1979) in which the zwitterionic detergent (Deriphat 160) replaces sodium dodecyl sulfate in the gel and electrophoresis buffers but *not* in the thylakoid solubilization solution. *FP* is reduced to ≤3% and then is mainly carotenoid. Zones *N* and *O* are representations of photosystem I, *P* is a representation of LHC II; the nature of the *Q* and *R* bands is not known. An identical sodium dodecyl sulfate extract was loaded onto the gels photographed in **i**, **ii**, and **vii**, and therefore *FP* is mainly generated by the electrophoretic system and not by the preparation of the extract

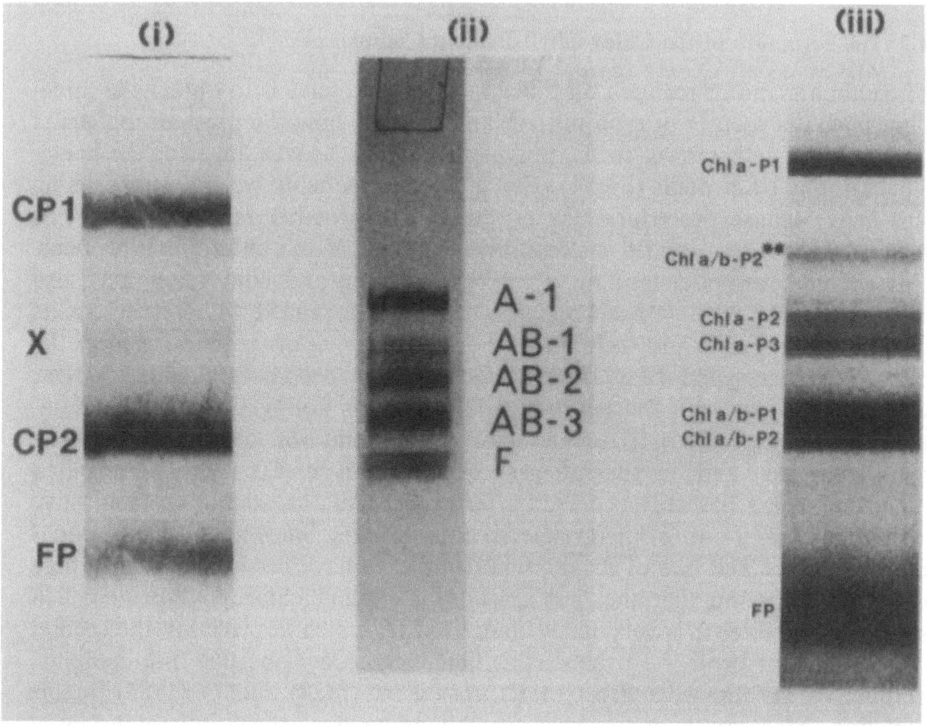
(i)
(ii)
(iii)
CP1
X
CP2
FP
A-1
AB-1
AB-2
AB-3
F
Chl a-P1
Chl a/b-P2**
Chl a-P2
Chl a-P3
Chl a/b-P1
Chl a/b-P2
FP

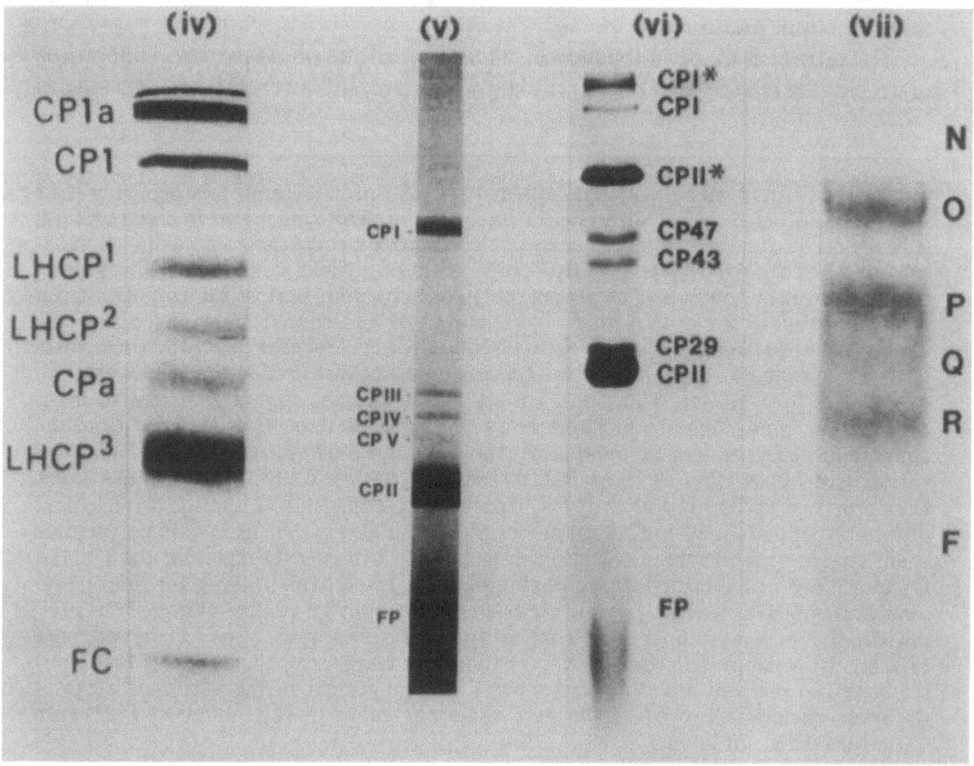
(iv)
(v)
(vi)
(vii)
CP1a
CP1
LHCP1
LHCP2
CPa
LHCP3
FC
CPI
CPIII
CPIV
CP V
CPII
FP
CPI*
CPI
CPII*
CP47
CP43
CP29
CPII
FP
N
O
P
Q
R
F

4.2 The Structure of the Chlorophyll-Protein Complexes

The ultimate aim of research on the chlorophyll-proteins is to obtain the three-dimensional structure of each protein and to learn how the pigment molecules are arranged with respect to the polypeptide chain, as was done for the bacteriochlorophyll a-protein (Fig. 4). For a few components we are approaching this goal because procedures for crystallization of other membrane proteins have recently been applied successfully to a few of the chlorophyll-proteins. For a complete understanding of energy trapping and conversion it is also necessary to know how each pigment-protein is arranged with respect to its nearest neighbors in the membrane. For the *Rps. viridis* reaction center, the phycobiliproteins and LHC II this aspect is also becoming clearer (GANTT, Chap. 6.3; MICHEL and DEISENHOFER, Chap. 8.4, this Vol.; GLAZER 1983; KÜHLBRANDT 1984; LI 1985). To achieve the present state of knowledge has been a difficult task, and, in general, a successful study of the more complicated plant complexes has only occurred after procedures for examining them have been first worked out for their bacterial counterparts. Since the purple bacterial reaction center and one of its antenna complexes have been crystallized, there is high expectation that the first crystals of a plant chlorophyll-protein will soon be available. It is very likely that LHC II, which is probably the second most abundant protein on the earth's land masses, will be this first complex, because of the intensive effort of KÜHLBRANDT (1984) and LI (1985) in this direction. Elucidation of its structure will be particularly useful to the large numbers of biochemists and molecular biologists, who are studying its composition, function, and biogenesis.

The advances made in the molecular biology of the photosynthetic apparatus in recent years (see the book edited by the present author, L.A. STAEHELIN

Fig. 4A–F. Current models of some antenna pigment-protein complexes depicting their structure and organization with respect to the membrane. *Upper two figures:* LHC II. **A** as deduced from electron microscopy of the complex incorporated into artificial membranes (KÜHLBRANDT 1984). LHC II is proposed to be a trimer composed of three structurally equivalent subunits. The shaded portions show that part of the complex which would protrude from a 45 Å thick lipid bilayer. **B** as deduced from analysis of the amino acid sequence (see text) by KARLIN-NEUMANN et al. (1985). Note the predictions of the three membrane-spanning alpha-helices and of the distribution of residues external to the 35 Å lipid bilayer. The model shows one of the nine subunits referred to in **A**. *Central two figures:* the antenna complexes of purple bacteria. **C** as deduced from the amino acid sequence and topographical studies of the B-890-protein of *Rhodospirillum rubrum* (see ZUBER Chap. 8.1 and BACHOFEN and WIEMKEN Chap. 11.2, both this Vol.). **D** as proposed by KRAMER et al. (1984) from spectroscopic studies and the arrangement of the polypeptide shown in **C**. Both models show a single alpha-helix in each polypeptide chain crossing the membrane. The *black* and *white arrows* in **D** represent the Qx and Qy transitions of the bacteriochlorophyll molecules, respectively; the zig-zag lines represent carotenoids. *Lower two figures:* the bacteriochlorophyll a-protein of a green bacterium (from X-ray crystallographic analysis by MATTHEWS et al. 1979). **E** arrangement of subunits in the protein seen from a position that emphasizes the extensive beta-sheet structure (no pigment molecules are shown). **F** arrangement of the porphyrin rings of the seven bacteriochlorophyll molecules inside one subunit. The amino acid sequence of this protein is not known

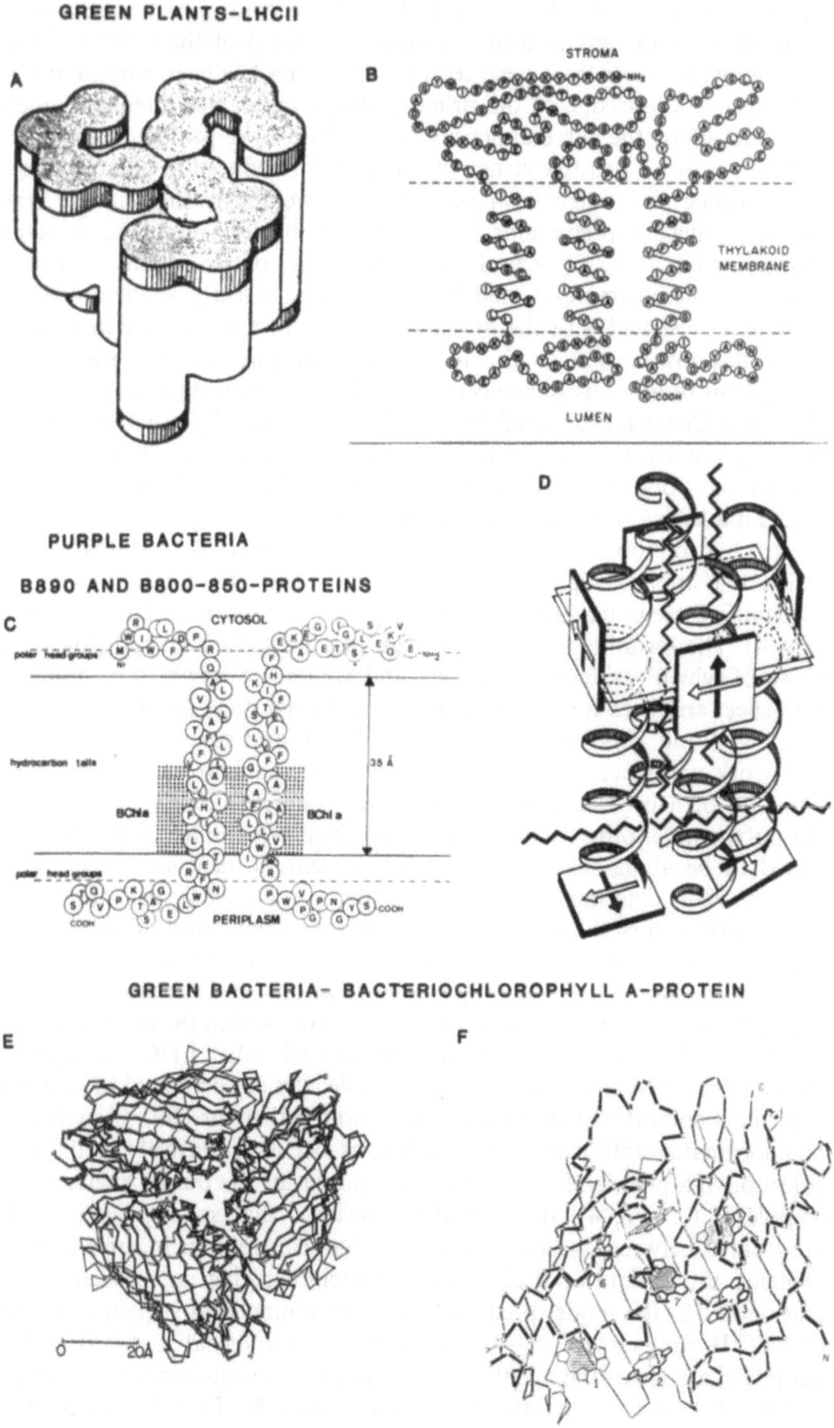
GREEN PLANTS-LHCII
A
B
STROMA
THYLAKOID MEMBRANE
LUMEN
PURPLE BACTERIA
B890 AND B800-850-PROTEINS
C
CYTOSOL
polar head groups
hydrocarbon tails
35 Å
BChla
BChl a
polar head groups
PERIPLASM
COOH
D
GREEN BACTERIA- BACTERIOCHLOROPHYLL A-PROTEIN
E
F
0 20Å

and R. HALLICK 1984) have greatly helped the chlorophyll-protein field. The publication of the amino acid sequences of several of the plant and bacterial chlorophyll-proteins from gene sequencing studies has been particularly beneficial, since a knowledge of the primary structure enables intelligent guesses to be made about the three-dimensional structure of a protein. The upper part of Fig. 4 shows a possible structure for LHC II, deduced in part from knowing the sequence of its structural gene. In the two illustrations in the center of Fig. 4, equivalent predictions for the purple bacterial antenna structure are shown; again, these are based on a knowledge of the amino acid sequence, which, for these complexes, had been determined directly on the protein.

When presented with the primary structure of a membrane-associated protein, it has now become routine to plot the hydrophobicity of a specific number of adjacent residues as this selected "window" is moved along the polypeptide chain (see CRAMER et al. 1985 for a review of such studies on several thylakoid-bound proteins). In this manner something can be learned about the organization of the polypeptide chain with respect to the lipid bilayer. Criteria have been established (EISENBERG et al. 1982a, 1982b; EISENBERG 1984) to identify in such hydrophobicity plots which domains of a protein probably form alpha-helices that cross the lipid bilayer. TOBIN et al. (1984) have used such approaches to identify three potential membrane-spanning regions in the LHC II apoprotein (Fig. 4). The resulting simple model can be embellished by applying consensus views of which residues are makers and which are breakers of alpha-helical, beta-sheet and beta-turn structures. This has been done for the LHC II protein (see Fig. 4 and KARLIN-NEUMANN et al. 1985), and some predictions made in that work will serve to show something of what might be deduced about the structure of a protein of known sequence. For example, very few regions of the polypeptide, other than the three membrane-spanning segments, were predicted to be alpha-helical, making the percent of such structures agree well with the value determined directly (NEBEDRYK et al. 1984); beta-turns predominate in that part of the sequence connecting the second and third helices; unlike the bacteriochlorophyll a-protein, there is very little beta-sheet structure predicted to occur in LHC II. Other considerations of EISENBERG et al. (1982b) enabled KARLIN-NEUMANN et al. (1985) to postulate that the membrane-crossing segments either interact with each other or with other LHC II apoproteins in the membrane. Most of the chlorophyll molecules are proposed to be associated with the membrane-spanning regions, although there are too many for each to be coordinated with a histidine residue, as occurs extensively in the bacteriochlorophyll a-protein (Fig. 4) or as is proposed for the purple bacterial antenna complexes (cf. Fig. 4). It is possible, however, that some pigment molecules are located between, rather than within, identical subunits in some chlorophyll-proteins. Other data on a chlorophyll-protein can then be applied to test and add details to the model predicted from the amino acid sequence. In the case of LHC II, for example, it is known from immunological studies that the polypeptide chain spans the membrane with the N-terminus exposed to the stroma, from electron microscope studies (KÜHLBRANDT 1984; LI 1985) that the polypeptide chain extends 20 Å on one side and about 10 Å on the other, and from linear dichroism studies that the alpha-helices are not absolutely perpendicular to the bilayer (Breton, Chap. 7.5, this Vol.); the last studies also provide the orientation of the porphyrin rings of the chlorophyll molecules. Combining all these biochemical, biophysical, and structural data can lead to a reasonably accurate picture of the three-dimensional structure of a chlorophyll-protein that should prove useful until it can be improved by X-ray crystallography of the complex. This supposes that crystals will be obtainable and that they will be suitable for X-ray analysis. Even if useful crystals do become available, it is probable that other techniques will have to be employed to determine the organization of the much larger light-harvesting arrays of chlorophyll-protein subunits that occur in situ.

One point that becomes increasingly obvious as a result of the recent structural studies and predictions of the folding of the chlorophyll-proteins is that more of their polypeptide chain is exposed outside the membrane bilayer than is typically depicted for such complexes which are generally shown almost completely buried in the bilayer. More than half of the LHC II residues lie outside the membrane milieu, with the majority of these being on the stroma side of the membrane. Similar investigations, as described above, for the CC I complex (Thornber unpublished) indicate that while the percent of the sequence located outside the membrane is only slightly greater than for LHC II, the portions extending from the membrane surface into the stroma and into the lumen may do so by some 40 Å. In general, the hydrophobicity of a protein as calculated from its amino acid composition is likely to be a good representation of how much of the protein is buried in the bilayer.

The pigment-proteins form a very interesting class of proteins, having an unusually large number of prosthetic groups (pigment molecules) associated with a single polypeptide molecule. This article has attempted to take the reader through the last two decades of research on the chlorophyll-proteins and to indicate the directions that the field is now taking. The immediate future for the field is very exciting. The colors of these components have long been esthetically pleasing to investigators working with them, but the satisfaction, soon to be realized, of knowing the three-dimensional structure of some complexes is even more appealing.

Acknowledgements. The author's research described in this article has largely been supported by grants from the National Science Foundation. I am grateful to my students (G. Peter, S. Nourizadeh and P. Chitnis), my postdoc (R. Nechushtai) and my colleague (Dr. E. Tobin) for helpful discussions, and to Camille Peterson and Mary Lawhead for help in preparing the manuscript.

References

Alberte RS, Thornber JP (1978) Rapid procedure for isolating photosystem I reaction centers in a highly enriched form. FEBS Lett 91:126–130

Alberte RS, Friedman AL, Gustafson DL, Rudnick MS, Lyman H (1981) Light harvesting systems of brown algae and diatoms. Isolation and characterization of chlorophyll a/c and chlorophyll a/fucoxanthin pigment-protein complexes. Biochim Biophys Acta 635:304–316

Allen JF, Bennett J, Steinback KE, Arntzen CJ (1981) Chloroplast protein phosphorylation couples plastoquinone redox state to distribution of excitation energy between photosystems. Nature 291:25–29

Allen JP, Feher G (1984) Crystallization of reaction center from *Rhodopseudomonas sphaeroides* – preliminary characterization. Proc Natl Acad Sci USA 81:4795–4799

Anderson JM (1982) The role of chlorophyll-protein complexes in the function and structure of chloroplast thylakoids. Mol Cell Biochem 46:161–172

Anderson JM (1984) Molecular organization of chloroplast thylakoid membranes. In: Sybesma C (ed) Advances in photosynthesis research, vol III. Nijhoff/Junk, The Hague, pp 1–10

Anderson JM, Levine RP (1974) Membrane polypeptides of some higher plant chloroplasts. Biochim Biophys Acta 333:378–387

Anderson JM, Waldron JC, Thorne SW (1978) Chlorophyll-protein complexes of spinach and barley thylakoids. Spectral characteristics of six complexes resolved by an improved electrophoretic procedure. FEBS Lett 92:227–233

Anderson JM, Brown JS, Lam E, Malkin R (1983) Chlorophyll b: an integral component of photosystem I of higher plant chloroplasts. Photochem Photobiol 38:205–210

Andersson B, Anderson JM, Ryrie IJ (1982) Transbilayer organization of the chlorophyll proteins of spinach thylakoids. Eur J Biochem 123:465–472

Arntzen CJ (1978) Dynamic structural features of chloroplast lamellae. Curr Top Bioenerg 8:111–160

Barber J (1982) Influence of surface charges on thylakoid structure and function. Annu Rev Plant Physiol 33:261–295

Bar-Nun S, Schantz R, Ohad I (1977) Appearance and composition of chlorophyll-protein complexes I and II during chloroplast membrane biogenesis in *Chlamydomonas reinhardtii* Y1. Biochim Biophys Acta 459:451–467

Barrett J, Anderson JM (1980) The P-700-chlorophyll *a*-protein complex and two major light-harvesting complexes of *Acrocarpia paniculata* and other brown seaweeds. Biochim Biophys Acta 590:309–323

Barrett J, Thorne SW (1981) Isolation of a F694-chlorophyll a-protein complex with low fluorescence yield and a chl c_2-protein and a fuxocanthin-protein from brown algae. In: Akoyunoglou G (ed) Photosynthesis, III. Structure and molecular organization of the photosynthetic membrane. Balaban, Philadelphia, pp 347–356

Bengis C, Nelson N (1975) Purification and properties of the photosystem I reaction center from chloroplasts. J Biol Chem 250:2783–2788

Bengis C, Nelson N (1977) Subunit structure of chloroplasts photosystem I reaction center. J Biol Chem 252:4564–4569

Bennett J (1983) Regulation of photosynthesis by reversible phosphorylation of the light-harvesting chlorophyll a/b protein. Biochem J 212:1–13

Bennett J, Markwell JP, Skrdla MP, Thornber JP (1981) Higher plant chlorophyll a/b-protein complexes: studies on the phosphorylated apoproteins. FEBS Lett 131:325–330

Boardman NK, Anderson JM (1964) Isolation from spinach chloroplasts of particles containing different proportions of chl a and b and their possible role in photosynthesis. Nature 293:166–167

Boardman NK, Anderson JM, Goodchild DJ (1978) Chlorophyll-protein complexes and structure of mature and developing chloroplasts. Curr Top Bioenerg 8:35–109

Brown JS, Duranton JG (1964) Partial separation of forms of chlorophyll a by SDS. Biochim Biophys Acta 79:209–211

Burke JJ, Ditto CL, Arntzen CJ (1978) Involvement of the light-harvesting complex in cation regulation of excitation energy distribution in chloroplasts. Arch Biochem Biophys 187:252–263

Camm EL, Green BR (1980) Fractionation of thylakoid membranes with the nonionic detergent octyl-β-D-glucopyranoside. Plant Physiol 66:428–432

Camm EL, Green BR (1983) Relationship between the two minor chlorophyll a-protein complexes and the photosystem II reaction center. Biochim Biophys Acta 724:291–293

Chiba Y (1960) Electrophoretic and sedimentation studies on chloroplast proteins solubilized by surface-active agents. Arch Biochem Biophys 90:294–303

Chow H, Serlin R, Strouse CE (1975) Crystalline molecular structure and absolute configuration of ethyl chlorophyllide a $2H_2O$: Model for the different spectral forms of chlorophyll a. J Am Chem Soc 97:7230–7236

Chua NH, Blomberg F (1979) Immunochemical studies of thylakoid membrane polypeptides from spinach and *Chlamydomonas reinhardtii*. Modified procedure for crossed immunoelectrophoresis of dodecyl sulfate protein complexes. J Biol Chem 254:215–223

Chua NH, Matlin K, Bennoun P (1975) A chlorophyll-protein complex lacking in photosystem I mutants of *Chlamydomonas reinhardtii*. J Cell Biol 67:361–377

Clayton RK (1980) Photosynthesis: Physical mechanisms and chemical patterns. IUPAB Biophys Ser. Cambridge Univ Press, Cambridge, pp 79–164

Clayton RK, Clayton BJ (1972) Relations between pigments and proteins in photosynthetic membranes of *Rhodopseudomonas sphaeroides*. Biochim Biophys Acta 283:492–504

Clayton RK, Sistrom WR (1978) The photosynthetic bacteria. In: Clayton RK, Sistrom WR (eds) Plenum, New York London

Cogdell RJ (1978) Carotenoids in photosynthesis. Philos Trans R Soc London, Ser B 284:569–579

Cogdell RJ (1983) Photosynthetic reaction centers. Annu Rev Plant Physiol 34:21–45

Cogdell RJ, Thornber JP (1979) The preparation and characterization of different types of light-harvesting pigment-protein complexes from some purple bacteria. Ciba Found Symp 61:61–79

Cogdell RJ, Thornber JP (1980) Light-harvesting pigment-protein complexes of purple photosynthetic bacteria. FEBS Lett 122:1–8

Cogdell RJ, Durant I, Valentine J, Lindsay JG, Schmidt K (1983) The isolation and partial characterisation of the light-harvesting pigment–protein complement of *Rhodopseudomonas acidophila*. Biochim Biophys Acta 722:427–435

Cramer WA, Widger WR, Herrmann RG, Trebst A (1985) Topography and function of thylakoid membrane proteins. Trends Biochem Sci 10:125–129

Criddle RS (1966) Protein and lipoprotein organization in the chloroplast. In: Goodwin TW (ed) Biochemistry of chloroplasts, vol I. Academic Press, London New York, pp 203–231

Delepelaire P, Chua NH (1981) Electrophoretic purification of chlorophyll a/b-protein complexes from *Chlamydomonas reinhardtii* and spinach and analysis of their polypeptide compositions. J Biol Chem 256:9300–9307

Dietrich WE, Thornber JP (1971) The P700 chlorophyll a-protein of a blue-green alga. Biochem Biophys Acta 245:482–493

Diner B, Wollman FA (1980) Isolation of highly active photosystem II particles from a mutant of *Chlamydomonas reinhardtii*. Eur J Biochem 110:521–526

Dos Santos CP, Hall DO (1982) Thylakoid polypeptides of light and dark aged chloroplasts. Plant Physiol 70:795–802

Dunahay TG, Staehelin LA (1985a) Isolation of photosystem I complexes from octyl glucoside/SDS solubilized spinach thylakoids: Characterization and reconstitution into liposomes. Plant Physiol in press

Dunahay TG, Staehelin LA (1985b) A new chlorophyll-protein from photosynthetic membranes. Plant Phys Suppl 77:37

Dunahay TG, Staehelin LA, Seibert M, Ogilvie PD, Berg SP (1984) Structural, biochemical and biophysical characterization of four oxygen-evolving photosystem II preparations from spinach. Biochim Biophys Acta 764:179–193

Dunsmuir P, Smith SM, Bedbrook J (1983) The major chlorophyll a/b binding protein of petunia is composed of several polypeptides encoded by a number of distinct nuclear genes. J Mol Appl Genet 2:285–300

Eisenberg D (1984) Three-dimensional structure of membrane and surface proteins. Annu Rev Biochem 53:595–623

Eisenberg D, Weiss RM, Terwilliger TC, Wilcox W (1982a) Hydrophobic moments and protein structure. Faraday Symp Chem Soc 17:109–120

Eisenberg D, Weiss RM, Terwilliger TC (1982b) The helical hydrophobic moment: a measure of the amphiphilicity of a helix. Nature 299:371–374

Feher G, Okamura MY (1978) Chemical composition and properties of reaction centers. In: Clayton RK, Sistrom WR (eds) The photosynthetic bacteria. Plenum, New York London, pp 349–386

Feick RG, Fuller RC (1984) Topography of the photosynthetic apparatus of *Chloroflexus aurantiacus*. Biochemistry 23:3693–3700

Fish LE, Kuck U, Bogorad L (1985) Two partially homogenous adjacent light-inducible maize chloroplast genes encoding polypeptides of the P700 chlorophyll α-protein complex of photosystem I. J Biol Chem 260:1413–1421

Fowler CE, Nugent NA, Fuller RC (1971) The isolation and characterization of a photochemically active complex from *Chloropseudomonas ethylica*. Proc Natl Acad Sci USA 68:2278–2282

Genge S, Pilger D, Hiller RG (1974) The relationship between chlorophyll b and pigment-protein complex II. Biochim Biophys Acta 347:22–30

Gingras G (1978) A comparative review of photochemical reaction center preparations from photosynthetic bacteria. In: Clayton RK, Sistrom WR (eds) The photosynthetic bacteria, Plenum, New York London, pp 119–131

Gingras G, Jolchine G (1969) Prog Photosynth Res 1:209–216

Glazer AN (1983) Comparative biochemistry of photosynthetic light-harvesting systems. Annu Rev Biochem 52:125–157

Green BR, Camm EL (1984) Evidence that CP47 (CPa-1) is the reaction center of photosystem. In: Sybesma C (ed) Advances in photosynthesis research, vol II. Nijhoff/Junk, The Hague, pp 95–98

Gugliemelli LA (1984) Isolation and characterization of pigment-protein particles from the light-harvesting complex of *Phaeodactylum tricornutum*. Biochim Biophys Acta 766:45–50

Haworth P, Kyle DJ, Horton P, Arntzen CJ (1982) Chloroplast membrane protein phosphorylation. Photochem Photobiol 36:743–748

Haworth P, Watson JL, Arntzen CJ (1983) The detection, isolation and characterization of a light-harvesting complex which is specifically associated with photosystem I. Biochim Biophys Acta 724:151–158

Hayden DB, Hopkins WG (1976) Membrane polypeptides and chlorophyll-protein complexes of maize mesophyll chloroplasts. Can J Bot 54:1684–1689

Henriques F, Park RB (1978) Polypeptide cross-linking in chloroplast membranes. Arch Biochem Biophys 189:44–50

Hiller RG, Goodchild DJ (1981) Thylakoid membrane and pigment organization. Biochem Plants 8:1–49

Holschuh K, Bottomley W, Whitfield PR (1984) Structure of the spinach chloroplast genes for the D2 and 44 kd reaction-centre proteins of photosystem II and for tRNA/ser (UGA). Nucleic Acids Res 12:8819–8834

Hoober JK, Millington RH, D'Angelo LP (1980) Structural similarities between the major polypeptides of the thylakoid membranes from *Chlamydomonas reinhardtii*. Arch Biochem Biophys 202:221–234

Horton P (1983) Control of chloroplast electron transport by phosphorylation of thylakoid proteins. FEBS Lett 152:47–52

Hurt EC, Hauska G (1984) Purification of membrane-bound cytochromes and a photoactive P-840 protein complex of the green sulphur bacterium *Chlorobium limicola* f. *thiosulfatophilum*. FEBS Lett 168:149–154

Ikegami I, Ke B (1984) A 160-kilodalton photosystem I reaction-center complex. Low-temperature absorption and EPR spectroscopy of the early electron acceptors. Biochim Biophys Acta 764:70–79

Ish-Shalom D, Ohad I (1983) Organization of chlorophyll-protein complexes of photosystem I in *Chlamydomonas reinhardtii*. Biochim Biophys Acta 722:498–507

Karlin-Neumann GA, Kohorn BD, Thornber JP, Tobin EM (1985) A chlorophyll a/b-protein encoded by a gene containing an intron with characteristics of a transposable element. J Mol Appl Genet 3:45–61

Knox RS, Van Metter RL (1979) Fluorescence of light-harvesting chlorophyll a/b protein complexes: implications for the photosynthetic unit. CIBA Found Symp 61:177–190

Kramer HJM, Van Grondelle R, Hunter CN, Westerhuis WHJ, Amesz J (1984) Pigment organization of the B-800–850 antenna complex of *Rhodopseudomonas sphaeroides*. Biochim Biophys Acta 765:156–165

Kühlbrandt W (1984) Three-dimensional structure of the light-harvesting chlorophyll a/b protein complex. Nature 307:478–480

Kühlbrandt W, Thaler T, Wehrli E (1983) The structure of membrane crystals of the light-harvesting chlorophyll a/b protein complex. J Cell Biol 96:1414–1424

Kyle DJ, Staehelin LA, Arntzen CJ (1983) Lateral mobility of the light-harvesting complex in chloroplast membranes controls excitation energy distribution in higher plants. Arch Biochem Biophys 222:527–541

Kyle DJ, Kuang T-Y, Watson JL, Arntzen CJ (1984) Movement of a sub-population of the light-harvesting complex (LHCII) from grana to stroma lamellae as a consequence of its phosphorylation. Biochim Biophys Acta 765:89–96

Lam E, Ortiz W, Mayfield S, Malkin R (1984a) Isolation and characterization of a light-harvesting chlorophyll a/b protein complex associated with photosystem I. Plant Physiol 74:650–655

Lam E, Ortiz W, Malkin R (1984b) Chlorophyll a/b proteins of photosystem I. FEBS Lett 168:10–14

Larsson UK, Jergil B, Andersson B (1983) Changes in the lateral distribution of the light-harvesting chlorophyll a/b protein complex induced by its phosphorylation. Eur J Biochem 136:25–29

Li J (1985) Light-harvesting chlorophyll a/b-protein: Three-dimensional structure of a reconstituted membrane lattice in negative stain. Proc Natl Acad Sci USA 82:386–390

Li J, Hollingshead C (1982) Formation of crystalline arrays of chlorophyll a/b light-harvesting protein by membrane reconstitution. Biophys J 37:363–370

Lichtenthaler HK, Prenzel U, Kuhn G (1982) Carotenoid composition of chlorophyll-carotenoid-proteins from radish chloroplasts. Z Naturforsch 37c:10–12

Lin L, Thornber JP (1975) Isolation and partial characterization of the photochemical reaction center of *Chromatium vinosum* (strain d). Photochem Photobiol 22:37–40

Loach PA, Sekura DL, Hadsell RM, Stemer A (1970) Quantitative dissolution of the membrane and preparation of the photoreceptor subunits from *Rhodospirillum rubrum*. Biochemistry 9:724–733

Lundell DJ, Glazer AN, Melis A, Malkin R (1985) Characterization of a cyanobacterial photosystem I complex. J Biol Chem 260:646–654

Lutz M, Brown JS, Remy R (1979) Resonance Raman spectroscopy of chlorophyll-protein complexes. CIBA Found Symp 61:105–120

Machold O (1975) On the molecular nature of chloroplast thylakoid membranes. Biochim Biophys Acta 382:494–505

Machold O (1984) Chlorophyll a/b-proteins in their relation to the light-harvesting complex. In: Sybesma C (ed) Advances in photosynthesis research, vol II. Nijhoff/Junk, The Hague, pp 107–114

Machold O, Simpson DJ, Lindberg Moller B (1979) Chlorophyll-proteins of thylakoids from wild-type and mutants of barley (*Hordeum vulgare* L.). Carlsberg Res Commun 44:235–254

Malkin R (1982) Photosystem I. Annu Rev Plant Physiol 33:455–479

Markwell JP, Reinman S, Thornber JP (1978) Chlorophyll-protein complexes from higher plants – a procedure for improved stability and fractionation. Arch Biochem Biophys 190:136–141

Markwell JP, Thornber JP, Boggs RT (1979) Higher plant chloroplasts: Evidence that all the chlorophyll exists as chlorophyll-protein complexes. Proc Natl Acad Sci USA 76:1233–1235

Matthews BW, Fenna RE, Bolognesi MC, Schmid MF, Olson JM (1979) Structure of a bacteriochlorophyll a-protein from the green photosynthetic bacterium *Prosthecochloris aestuarii*. J Mol Biol 131:259–285

Michel H (1982) 3-Dimensional crystals of a membrane protein complex. The photosynthetic reaction center from *Rhodopseudomonas viridis*. J Mol Biol 158:567–572

Morris J, Herrmann RG (1984) Nucleotide sequence of the gene for the P680 chlorophyll

a apoprotein of the photosystem II reaction center from spinach. Nucleic Acids Res 12:2837–2850
Mullet JE, Burke JJ, Arntzen CJ (1980) Chlorophyll proteins of photosystem I. Plant Physiol 65:814–822
Nakatani HY, Ke B, Dolan E, Arntzen CJ (1984) Identity of the photosystem II reaction center polypeptide. Biochim Biophys Acta 765:347–352
Nebedryk E, Andriaambinintsoa S, Breton J (1984) Transmembrane orientation of α-helices in the thylakoid membrane and in the light-harvesting complex. A polarized infrared spectroscopy study. Biochim Biophys Acta 765:380–387
Nechushtai R, Nelson N (1981) Purification properties and biogenesis of *Chlamydomonas reinhardii* photosystem I reaction center. J Biol Chem 256:11624–11628
Nechushtai R, Muster P, Binder A, Liveanu V, Nelson N (1983) Photosystem I reaction center from the thermophilic cyanobacterium *Mastigocladus laminosus*. Proc Natl Acad Sci USA 80:1179–1183
Nelson N, Notsani BE (1977) Function and organization of individual polypeptides in chloroplast photosystem I reaction center. In: Packer L, Papageorgiou GC, Trebst A (eds) Bioenergetics of membranes. Elsevier, Amsterdam New York Oxford, pp 233–244
Ogawa T, Obata F, Shibata K (1966) Two pigment-proteins in spinach chloroplasts. Biochim Biophys Acta 112:223–234
Okamura MY, Feher G, Nelson N (1982) Photosynthetic membrane structure and function. In: Govindjee (ed) Photosynthesis, vol I. Academic Press, London New York, pp 195–272
Olson JM (1980) Chlorophyll organization in green photosynthetic bacteria. Biochim Biophys Acta 594:33–51
Olson JM, Romano CA (1962) A new chlorophyll from green bacteria. Biochim Biophys Acta 59:726–728
Olson JM, Thornber JP (1979) Photosynthetic reaction centers. In: Capaldi R (ed) Membrane proteins in energy transduction. Dekker, New York, pp 279–340
Olson JM, Thornber JP, Koenig DF, Ledbetter MC, Olson RA, Jennings WH (1969) The bacteriochlorophyll-protein of green photosynthetic bacteria. Prog Photosynth Res 1:217–225
Olson JM, Giddings TH, Shaw EK (1976) An enriched reaction center preparation from green photosynthetic bacteria. Biochim Biophys Acta 449:197–208
Pierson BK, Thornber JP (1983) Isolation and spectral characterization of photochemical reaction centers from the thermophilic green bacterium *Chloroflexus aurantiacus* strain J-10-f1. Biochim Biophys Acta 764:148–159
Pierson BK, Thornber JP, Seftor REB (1983) Partial purification, subunit structure and thermal stability of the photochemical reaction center of the thermophilic green bacterium *Chloroflexus aurantiacus*. Biochim Biophys Acta 723:322–326
Prezelin RB, Haxo FT (1976) Purification and characterization of peridinin-chlorophyll a-proteins from the marine dinoflagellates, *Glenodinium* sp and *Gonyaulax polyedia*. Planta 128:133–141
Prince RC, Thornber JP (1977) A novel electron paramagnetic resonance signal associated with the 'primary' electron acceptor in isolated photochemical reaction centers of *Rhodospirillum rubrum*. FEBS Lett 81:233–237
Reed DW, Clayton RK (1968) Isolation of a reaction center fraction from *Rhodopseudomonas sphaeroides*. Biochim Biophys Acta 30:471–475
Reinman S, Thornber JP (1979) The electrophoretic isolation and partial characterization of three chlorophyll-protein complexes from blue-green algae. Biochim Biophys Acta 547:188–197
Ryrie IJ, Anderson JM, Goodchild DJ (1980) The role of light harvesting chlorophyll a/b protein complex in chloroplast membrane stacking. Cation-induced aggregation of reconstituted proteoliposomes. Eur J Biochem 107:345–354
Satoh K (1979) Properties of light-harvesting chlorophyll a/b-protein and photosystem I chlorophyll a-protein, purified from digitonin extracts of spinach chloroplasts by isoelectric focusing. Plant Cell Physiol 25:935–945

Satoh K, Nakatani HY, Steinback KE, Watson J, Arntzen CJ (1983) Polypeptide composition of a photosystem II core complex. Presence of a herbicide-binding protein. Biochim Biophys Acta 724:142–150

Schmidt G, Bartlett SG, Grossman AR, Cashmore AR, Chua N-H (1981) Biosynthetic pathways of two polypeptide subunits of the light-harvesting chlorophyll a/b protein complex. J Cell Biol 91:468–478

Seftor REB, Thornber JP (1984) The photochemical reaction center of the bacteriochlorophyll b-containing organism *Thiocapsa pfennigii*. Biochim Biophys Acta 764:148–159

Shiozawa JA, Alberte RS, Thornber JP (1974) The P700-chlorophyll a-protein. Isolation and some characteristics of the complex in higher plants. Arch Biochem Biophys 165:388–397

Sironval C, Clijsters H, Michel J-M, Bronchart R, Michel-Wolwertz M-R (1967) Sur la séparation de deux fractions à partir des membranes des chloroplastes (systemes I et II), sur leurs propriétés, sur l'organisation et le fonctionnement de ces membranes. In: Sironval (ed) Le chloroplaste. Masson, Paris, pp 99–123

Smith EL, Pickels EG (1941) The effect of detergents on the chlorophyll-protein compound of spinach as studied in the ultracentrifuge. J Gen Physiol 24:753–764

Stackebrandt E, Woese CR (1981) The evolution of prokaryotes. Soc Gen Microbiol Symp 32:1–31

Staehelin LA, Arntzen CJ (1983) Regulation of chloroplast membrane function. Protein phosphorylation changes the spatial organization of membrane components. J Cell Biol 97:1327–1337

Steinback KE, Bose S, Kyle DJ (1982) Phosphorylation of the light-harvesting chlorophyll-protein regulates excitation energy distribution between photosystem II and photosystem I. Arch Biochem Biophys 216:356–361

Sutton MR, Rosen D, Feher G, Steiner LA (1982) Amino-terminal sequences of the L, M and H subunits of reaction centers from the photosynthetic bacterium *Rhodopseudomonas sphaeroides* R-26. Biochemistry 21:3842–3849

Tang X-S, Satoh K (1984) Characterization of a 47-kilodalton chlorophyll-binding polypeptide (CP-47) isolated from a photosystem II core complex. Plant Cell Physiol 25:935–945

Tapie P, Haworth P, Hervo G, Breton J (1982) A quantitative comparison of the low temperature linear dichroism spectra of thylakoids and isolated pigment-protein complexes. Biochim Biophys Acta 682:339–344

Thornber JP (1969) Comparison of a chlorophyll a-protein complex isolated from a blue green alga with chlorophyll-protein complexes obtained from green bacteria and higher plants. Biochim Biophys Acta 172:230–241

Thornber JP (1970) Photochemical reactions of purple bacteria as revealed by studies of three spectrally different carotenobacteriochlorophyll-protein complexes isolated from *Chromatium*, strain D. Biochemistry 9:2688–2698

Thornber JP (1975) Chlorophyll proteins: Light-harvesting and reaction center components of plants. Annu Rev Plant Physiol 26:127–158

Thornber JP, Barber J (1979) Photosynthetic pigments and models for their organization in vivo. In: Barber J (ed) Photosynthesis in relation to model systems. Elsevier, Amsterdam New York Oxford, pp 27–70

Thornber JP, Highkin HR (1974) Composition of the photosynthetic apparatus of normal barley leaves and a mutant lacking chlorophyll b. Eur J Biochem 41:109–116

Thornber JP, Smith CA, Bailey JL (1966) Partial characterization of two chlorophyll-protein complexes isolated from spinach-beet chloroplasts. Biochem J 100:14–15

Thornber JP, Gregory RPF, Smith CA, Bailey JL (1967a) Studies on the nature of the chloroplast lamella. I. Preparation and some properties of two chlorophyll-protein complexes. Biochemistry 6:391–395

Thornber JP, Stewart JC, Hatton MWC, Bailey JL (1967b) Studies on the nature of chloroplast lamellae. II. Chemical composition and further physical properties of two chlorophyll-protein complexes. Biochemistry 6:2006–2014

Thornber JP, Olson JM, Williams DM, Clayton ML (1969) Isolation of the reaction center of *Rhodopseudomonas viridis*. Biochim Biophys Acta 172:351–354

Thornber JP, Alberte RS, Hunter FA, Shiozawa JA, Kan KS (1977) The organization of chlorophyll in the plant photosynthetic unit. Brookhaven Symp Biol 28:132–148

Thornber JP, Trosper TL, Strouse CE (1978) Bacteriochlorophyll in vivo: Relationship of spectral forms to specific membrane components. In: Clayton RK, Sistrom WR (eds) The photosynthetic bacteria. Plenum New York London, pp 133–160

Thornber JP, Markwell JP, Reinman S (1979) Plant chlorophyll-protein complexes: recent advances. Photochem Photobiol 29:1205–1216

Thornber JP, Cogdell RJ, Seftor REB, Pierson BK, Tobin EM (1984a) Comparative biochemistry of chlorophyll-protein complexes. In: Sybesma C (ed) Advances in photosynthesis research, vol II. Nijhoff/Junk, The Hague, pp 25–32

Thornber JP, Staehelin LA, Hallick RB (eds) (1984b) Biosynthesis of the photosynthetic apparatus. UCLA Symp, New Ser, vol 14. liss, New York

Tobin EM (1981) Photochrome-mediated regulation of messenger RNAs for the small subunit of ribulose 1,5-bisphosphate carboxylase and the light-harvesting chlorophyll a/b protein in *Lemna gibba*. Plant Mol Biol 1:35–51

Tobin EM, Wimpee CF, Silverthorne J, Stiekema WJ, Neumann GA, Thornber JP (1984) Phytochrome regulation of the expression of two nuclear-coded chloroplast proteins. In: Thornber JP, Staehelin LA, Hallick R (eds) Biosynthesis of the photosynthetic apparatus, vol 14. Liss, New York, pp 325–334

Vasmel H, Swarthoff T, Kramer HJM, Amesz J (1983) Isolation and properties of a pigment-protein complex associated with the reaction center of the green photosynthetic sulfur bacterium *Prosthecochloris aestuarii*. Biochim Biophys Acta 725:361–367

Vierling E, Alberte R (1983) P-700 chlorophyll-a protein. Purification, characterization and antibody preparation. Plant Physiol 72:625–633

Webster GD, Cogdell RJ, Lindsay JG (1980) Localization of the reaction-centre subunits in the intracytoplasmic membranes of *Rhodopseudomonas sphaeroides* and *Rhodopseudomonas capsulata*. Biochem Soc Trans 8:184–185

Westhoff P, Alt J, Hermann RG (1983) Localization of the genes for the two chlorophyll a-conjugated polypeptides (M.W. 51 and 44 kd) of the photosystem II reaction center on the spinach plastid chromosome. EMBO J 2:2229–2237

Williams RC, Glazer AN, Lundell DJ (1983) Cyanobacterial photosystem I: Morphology and aggregation behavior. Proc Natl Acad Sci USA 80:5923–5926

Wollman FA, Bennoun P (1982) A new chlorophyll-protein complex related to photosystem I in *Chlamydomonas reinhardti*. Biochim Biophys Acta 680:352–360

Yamagishi A, Katoh S (1984) A photoactive photosystem II reaction center lacking a chlorophyll-binding 40 kilodalton subunit from the thermophilic cyanobacterium *Synechococcus* sp. Biochim Biophys Acta 765:118–124

4 Energy Transduction in Oxygenic Photosynthesis: An Overview of Structure and Mechanism

D.R. ORT

1 Introduction

Our understanding of the construction and composition of the chlorophyll-containing membranes of photosynthetic organisms has advanced enormously in recent years. Nevertheless our understanding of the organization and orientation of the component parts of the membranes remains incomplete. Our understanding of the mechanisms involved in light capture, light energy conservation, electron transport, and ion movements continues to be limited by this inadequacy. Thus a remaining objective of photosynthesis research is the elucidation of the structure of the membrane in relation to the functions of its component parts. Recent advances on both parts are exciting, even though a complete understanding of how things are and how they function is still an unattained goal. This overview chapter will attempt to bring together current information about the organization of the membrane proteins and about the molecular processes carried on by the various proteins.

This book treats photosynthesis by oxygen-evolving organisms separately from photosynthesis from the non-oxygen-evolving bacteria (for the latter see Chap. 5 this Vol. by DUTTON). The separation of oxygenic and nonoxygenic photosynthesis is logical from many standpoints and much simplifies the discussion of each. However, it must be realized that the similarities far exceed the differences. Great insights have come from comparative studies of "bacterial" and "plant" photosynthesis (WRAIGHT 1982a). Parallels exist at many levels of structure and function, leaving water oxidation itself as the sole major improvisation of plants. For that matter, even the cyanobacteria, which are indisputably prokaryotic bacteria, carry on photosynthesis that is essentially the same as in higher plants, including oxygen production. The striking parallel between bacterial photosynthesis and plant photosynthesis points to the endosymbiotic origin of chloroplasts from invading prokaryotic algae (WHATLEY et al. 1979; STAEHELIN, Chap. 1, this Vol.). This endosymbiotic origin of chloroplasts is extensively suggested by morphological, biochemical, and genetic data. The relationship of modern-day cyanobacteria to the original invading symbiots is very strongly suggested in the case of the red algae (*Rhodophyta*) by analyses of ribosomal RNA (FOX et al. 1980). The phylogenetic relationship between bacteria and green plant chloroplasts is less well established, but this may only mean that the prokaryotic "ancestor" of green plant chloroplasts remains lost or unrecognized. A case has been made for members of the genus *Prochloron* as the progenitor of chloroplasts of green algae and higher plants (STANIER and COHEN-BAZIRE 1977; GIDDINGS et al. 1980). In any event, the concept of the symbiotic origin of the chloroplast allows us to appreciate the remarkable parallel between bacterial and plant photosynthesis.

For more than the past two decades, our view of oxygenic photosynthesis has been shaped by the familar "Z" scheme of electron transport (HILL and BENDALL 1960). The "Z" scheme has served as a convenient pedagogical device to introduce the notion of two photoacts operating in series and the idea of electrons moving energetically "down hill" from the strong reductants to the strong oxidants produced by each photosystem upon light absorption. A major inadequacy of the "Z" scheme has been the paucity of information concerning these "downhill" processes, particularly those processes transferring electrons between the two photosystems. Indeed, the simplified version of the "Z" scheme presented in textbooks may have encouraged students to think the transfer of electrons between the photosystems as being along a well-defined chain of intermediates when, in fact, electrons from many photosystem II reaction centers may be pooled as are electrons from sequential excitations of a single center. Figure 1 attempts to include in the "Z" scheme the concept that electrons are so pooled. This modified, more explicit scheme will be used to introduce the path of energy flow through oxygenic photosynthesis. Thereafter in this chapter, information about the mechanism of energy transduction and membrane organization will be developed to a point where even this modified scheme ceases to be a useful visualization aid.

The photosynthetic membrane of a plant performs a remarkable feat: it converts a portion of the energy available for incoming quanta of light into the phosphoryl group transfer potential of ATP and the redox potential of NADPH. In so doing the membranes are converting a very transient form of energy into forms which are stable over long periods of time, stable enough to be used much later by the energy-requiring biochemical processes of the plant cell, including the steps involving CO_2 reduction. The processes converting light energy into these stable products can, for purposes of this discussion, be separated into discrete energy transformations which can be illustrated by considering a single turnover of both photosystems occurring in unison. However, this is primarily a pedagogic device. Although both photosystems are necessary for photosynthesis, the two photoacts would rarely occur in unison because of the already mentioned pooling of electrons in intermediates between the photosystems.

The first step in the energy transformation is the absorption of light by any one of a large number of chlorophyll molecules or by associated "accessory" pigments. These arrays of pigment "antennae" must be large if light is to be captured efficiently as it must be if the light is limited. Thus the number of light-absorbing molecules vastly exceeds the number of reaction centers where the excitation is finally converted into a chemical reaction (see THORNBER, Chap. 3, this Vol.). The excitation of the pigment molecule originally absorbing the light is transferred from pigment molecule to pigment molecule over the pigment array until it is finally "trapped" by the specialized chlorophyll of the reaction centers (see SAUER, Chap. 2, this Vol.). The excited reaction center chlorophyll

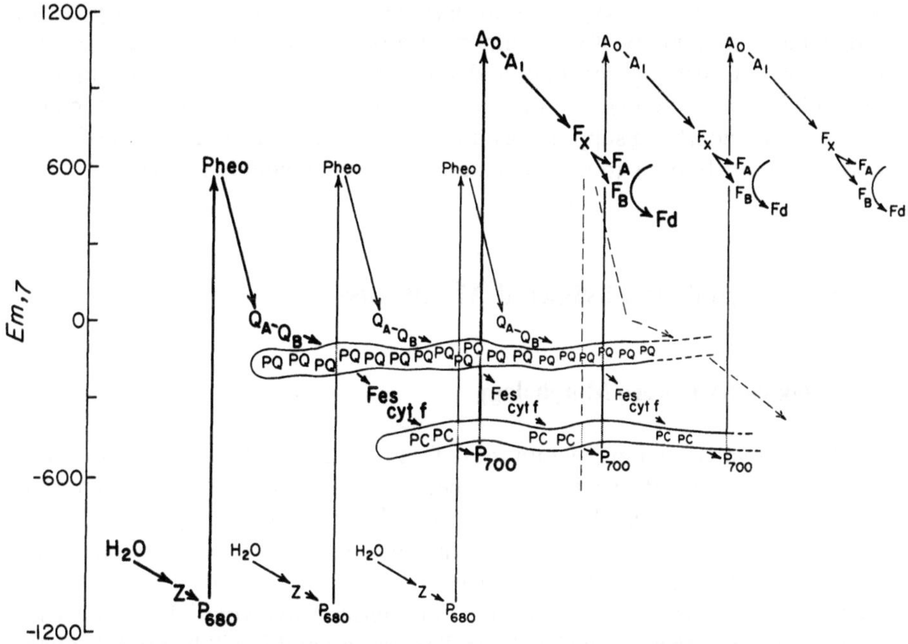

Fig. 1. A contemporary Z scheme depicting the sequence and oxidation-reduction midpoint potential ($E_{m,7}$) of the electron carriers in oxygenic photosynthesis. This figure differs from the familiar Z scheme format in order to convey the concept that photosynthetic electron transfer from H_2O to ferredoxin (*Fd*) is accomplished by three electron transport complexes operating in series interconnected by two pools of mobile carriers. Water is oxidized by the PS II complex and electrons are shuttled between the PS II complex and intersystem cytochrome b_6/f complex by a pool of the mobile lipophilic carrier plastoquinol (*PQ*). The cytochrome b_6/f complex reduces in turn a small pool of water soluble metalloproteins, plastocyanin (*PC*), that serves to convey reducing equivalents to the PS I complex. Thus, while there is a defined reaction sequence for photosynthetic electron transfer, the carriers do not function as a predetermined "chain"

then gives up an electron to an acceptor molecule whose chemical identity depends on which photosystem is considered. Although the photochemical reactions at the two photosystem reaction centers are different, both photosystems bring about transformations of the energy stored in the rearranged electrons and nuclei of the light-excited chlorophyll a, turning the excitation energy into a reduced electron acceptor molecule and an oxidized chlorophyll. This reaction takes place across the dielectric of the photosynthetic membrane. Thus part of the energy is stored in the form of a charge assymetry across the membrane. Reactions of the oxidized chlorophyll with hydrogen donors produce hydrogen ions without any loss of the charge assymetry, and thus the charge assymetry is converted to a transmembrane electrochemical potential of hydrogen ions. In a manner not yet understood, this hydrogen ion potential difference is used to drive the phosphoryl group transfer reaction of ATP synthesis.

This chapter emphasizes recent advances in our understanding of the above energy transformations. It seeks to be at once an overview of how the membrane-associated reactions of oxygenic photosynthesis occur, and a directory to the more detailed treatments of the component processes described in the appended mini-reviews. Clarifying concepts and the current state of our understanding are emphasized at the expense of tracing the origin of ideas and development of the crucial underlying supporting data. Consequently, the referencing of many important contributions to our understanding is unavoidably indirect.

2 The Chloroplast Thylakoid Membrane

2.1 Thylakoid Membrane Morphology

The chlorophyll-containing membranes of photosynthetic organisms responsible for the light-dependent reactions described above are conspicuous, flat lamellar vesicles, often occupying a large part of the volume of the chloroplasts of eukaryotic organisms. The outer surfaces of the lamellar membranes are often closely appressed to the outer surfaces of the other lamellae, although the extent of this "stacking" can be quite variable. The term "thylakoid" has been used widely to describe these specialized lamellar vesicles and it will be used throughout this discussion in this general sense without any further connotations of membrane morphology.

Thin section electron micrographs depicting the high incidence of membrane surface contacts are commonplace (for instance, see Chap. 1, this Vol. by STAEHELIN). Such surface contacts are observed both in the photosynthetic membranes of higher plants and in the membranes of algae. However, the three-dimensional interpretation of the thylakoid membrane organization is much less widely familiar even to those involved with research in photosynthesis. Owing to the enormous amount of information that has accumulated concerning the molecular characteristics of the thylakoid membrane and of the differential distribution of its components, it is becoming increasingly clear that features of chloroplast membrane superstructure must be incorporated into our thinking about the functioning of thylakoid membranes.

Information about the three-dimensional configuration of thylakoid membranes within the chloroplast has come from reconstructions based on electron micrographs taken of serial sections (PAOLILLO and FALK 1966; BRANGEON and MUSTÁRDY 1979), and of freeze etch replicas (e.g., see STAEHELIN, Chap. 1, Figs. 14 and 15, this Vol.) and, more recently, on scanning electron micrographs (MUSTÁRDY and JÁNOSSY 1979). There appears to be little disagreement regarding the basic three-dimensional appearance of the stacked regions of thylakoid membranes or about the interconnections of one stacked region with another. The central concepts, on which the current notions about thylakoid membrane superstructure are based, can be traced to work done nearly 20 years ago by PAOLILLO and FALK (1966). The simple appearance of an individual thin sectioned chloroplast provides little clue to the baroque three-dimensional architecture of stacked thylakoid membranes which is revealed by reconstructions based

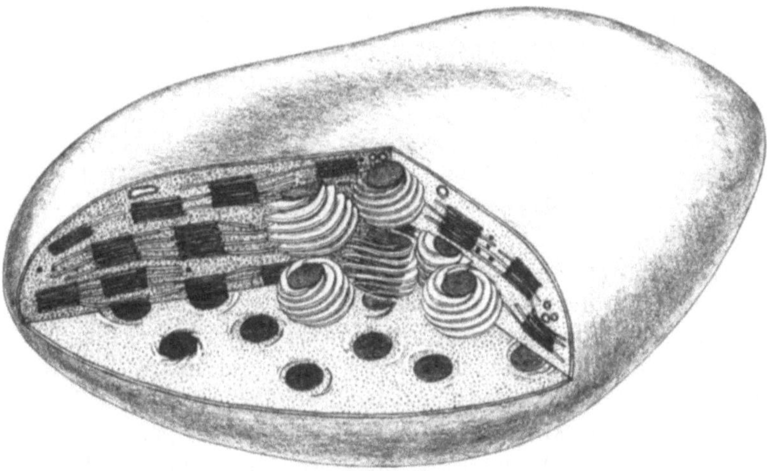

Fig. 2. The appearance of chloroplast thylakoid membrane architecture viewed in two and in three dimensions. This drawing of a sectioned chloroplast is intended to aid in visualization of the relationship of the appearance of thylakoid membranes as viewed in two dimensions with the appearance as viewed in three dimensions. The drawing is based on the familiar appearance of these membranes from thin section electron micrographs and on the less familiar appearance from scanning electron micrographs or from reconstructions based on serial thin sections. The drawing shows that the appressed membranes (i.e., grana stacks) seen in two dimensions arise from viewing in cross-section a cylindrical core of "discs" and that the helically arranged membrane sheets with their fret-like connections to the "discs" are the origin of the unappressed or stroma membranes observed in two dimensions

on serial sections. PAOLILLO and FALK (1966) proposed that multiple tilted membrane planes were arranged helically around the surface of a cylindrical core of stacked membranes, the so-called grana discs. These helically arranged membranes were envisioned as forming attachments with the individual "discs" of the grana stack. The lumen of the two membranes is thus continuous by virtue of these connections. The association is such that each "disc" is connected to a number of the helically arranged membranes and that each helically arranged membrane is connected to a number of different "discs". Neighboring grana stacks appear to be connected by narrow bridges between the helical membranes, thereby continuing the interconnection of lumen spaces. The drawing of a chloroplast depicted in Fig. 2 seeks to show the relationship of the appearance of thylakoid membranes in two dimensions with the appearance of stacked membranes in three dimensions. The drawings are based on the familiar appearance of these membranes in thin section (e.g. STAEHELIN, Chap. 1, this Vol.) and on the appearance from recently published scanning electron micrographs (MUSTÁRDY and JÁNOSSY 1979). The appearance of appressed membranes seen in two dimensions would arise from the cross-section of the cylindrical core of "discs", whereas the helically arranged membrane sheets and their fret-like connections to the "discs" would constitute the unappressed or stroma lamellae observed in two dimensions.

From a recent study of the development of the three-dimensional thylakoid membrane architecture during chloroplast ontogeny (BRANGEON and MUSTÁRDY 1979) came a model for the sequence of events leading to thylakoid membrane stacking. This model has several interesting implications for the interpretation of experimental results on membrane unstacking induced by cation depletion. During the initial stages of plastid differentiation the thylakoid membranes were observed to be simple flattened lamellar sheets and to be few in number. These sheets were seen, by three-dimensional reconstructions, to have multiple perforations. It was proposed that these perforations are the sites of grana initiation. The structural data were interpreted as indications that the periphery of each perforation is a point of membrane growth and that the development of the thylakoid membrane proceeds by an overlapping growth pattern above and below the perforated lamellar sheet. These "overlaps" are taken to be the origin of stacking and association of adjacent perforations by a twisting of the parent membrane forms the grana stack replete with a single helically arranged membrane sheet (i.e., the parent lamellar membrane). The single helix was observed to proliferate as development proceeded until each stack of discs was outfitted with five to ten helices each with multiple fusions to discs (i.e., the lumen of each helical membrane was seen as continuous with the lumens of numerous "discs" and conversely the lumen of each "disc" was seen as continuous with numerous helical membranes).

Two features of this proposal for grana stack formation stand out. First, the origin of membrane surface contact is envisioned as a growth process as opposed to an assembly that occurs subsequent to membrane growth. Secondly, the data are interpreted to indicate that the grana stack, once formed, is further decorated with multiple helically arranged membranes each with multiple fusion attachments to the "discs". Both of these features of the proposal suggest that the grana structure should be relatively stable and, once formed, largely irreversible. Yet it was reported by IZAWA and GOOD (1966) many years ago, and the observation has been confirmed countless times since, that the native membrane contacts disappear when isolated thylakoid membranes (envelope membranes absent) are suspended in low salt media. Viewed in two dimensions, the membranes take on the appearance of large flattened planes, sometimes loosely associated as pairs but with no structures resembling grana stacks. Freeze-etch electron microscopy reveals that randomization of fracture face particles accompanies the membrane unstacking (STAEHELIN et al. 1977). Thus it would seem that, in the absence of appreciable amounts of cations, those forces which constrained an extremely fluid membrane (see below) to an ordered and apparently static three-dimensional configuration are absent, and featureless planar sheets result. The readdition of salts induces interlamellar attractions which result in stacking of the lamellae. The stacking is often random but on occasion the salt restacked membranes appear, in two dimensions at least, remarkably reminiscent of the native state (IZAWA and GOOD 1966). Furthermore, segregation of fracture face particles between membrane appressed and unappressed regions is reestablished, suggesting that the very intricate thylakoid architecture of fully developed chloroplasts can be reversibly dissociated into planar membrane sheets and then reassociated to form native membrane contacts. However, careful analysis of cation depletion-induced unstacking by freeze-etch electron microscopy indicates that truly native reassociations are possibly only up to the time when the fret-like connections between the grana "discs" and the helically arranged membranes begin to broaden (STAEHELIN et al. 1977 and Chap. 1, this Vol.).

Associated with membrane stacking is an astonishing segregation of membrane components between regions of membrane appression and nonappression (STAEHELIN, Chap. 1, this Vol., KAPLAN and ARNTZEN 1982). Because of the necessity for certain of the physically separated functional complexes to communicate with one another via electron, proton or light energy transfer there is need to consider thylakoid membrane superstructure as a parameter in many of the photosynthetic processes discussed in this chapter and in the accompanying minireviews. However, at the outset it should be acknowledged that none of the fundamental processes of photosynthesis seems to have any obligate dependence on the stacking of the thylakoid membranes. Thylakoid membranes that are nongranal by virtue of genetic determination, or induced by low salt conditions, retain their integrity and are able to absorb light, carry out electron and proton transfer, synthesize ATP, and reduce NADP with seemingly no systematic alteration in either efficiency or rate. However, there are indications that the heterogeneous distribution of the pigment-protein complexes associated with stacking is a necessary feature of regulatory mechanisms that serve to balance the excitation energy between the two photosystems and thereby maintain efficient photosynthesis under low light conditions (STAEHELIN and ARNTZEN 1983).

Nevertheless, at present the significance of grana stacking and the advantage it confers to photosynthesis remain obscure. It may, as we all tend to assume, become clear as more is understood about the overall workings and regulation of the chloroplast. It is perhaps worthwhile to consider as an alternative the sentiment expressed in the statement "anything that is produced by evolution is bound to be a bit of a mess" (S. BRENNER in LEWIN 1984). The salient underlying thought is that evolution selects for survival, not for perfection, and consequently the survivor may carry some untidy vestiges as reminders of its evolutionary history.

2.2 Chemical Composition and Physical Properties of the Thylakoid Membrane

2.2.1 Composition

As with all other biological membranes[1], the thylakoid membrane features a family of diacyldiglycerides which associate to form a bilayer structure. The lipids are amphiphatic, that is, they contain a polar head group separated from the nonpolar acyl chains. A diverse group of more than 60 different polypeptides

1 An apparently unique exception to the bilayer organization of biological membranes is the plasma membrane of the thermo-acidphilic archebacterium *Thermoplasma*. The plasma membrane of this primitive organism appears to be a lipid monolayer formed by a $C_{40}H_{82}$ polyisoprenoid lipid (LANGWORTHY 1977).

are inserted into the bilayer, most of which are components of one of the four large amphipathic integral membrane protein complexes. In addition to the glycerol lipids of the bilayer, thylakoid membranes contain an abundance of chlorophyll in amounts equivalent to about 90 mM (calculated from the volume of the thylakoid membrane of 11.1 $l \cdot mol^{-1}$ chlorophyll [FLORES et al. 1983]). Unlike the glycerolipids, chlorophyll is very strongly associated with a particular group of membrane polypeptides (see THORNBER, Chap. 3, this Vol.).

2.2.1.1 Lipids

Owing to the apparent prokaryotic origin of chloroplasts (WHATLEY et al. 1979), perhaps it should not be surprising that the glycerol lipid composition of the thylakoid bilayer is very different from that of other plant membranes. The acyl lipid content of thylakoid membrane is about 35% w/w and 70% of this acyl lipid is contributed by two unusual and highly unsaturated glycolipids, monogalactosyldiacylglycerol (MGDG) and digalactasyldiacylglycerol (DGDG). While bearing no formal charge the head groups of these glycolipids have an overall polar character (Table 1). These two unusual acyl lipids are further distinguished by an exceptionally high content of polyunsaturated fatty acids (QUINN and WILLIAMS 1978). Thylakoid membranes are thought to contain only three other species of diacyl lipid. Two of these three species, phosphatidylglycerol (anionic) and phosphatylcholine (dipolar ionic), are commonplace, but the chloroplast sulfolipid, diacylsulfoquinovosylglycerol (anionic), is only very infrequently encountered elsewhere in nature. Table 1 depicts the structure of these five acyl lipid species and shows approximate figures for their relative abundance in thylakoid membranes. Chapters by MURPHY (11.11) and by SIEGENTHALER and RAWYLER (11.9) in this Volume describe the physical properties of thylakoid acyl lipids in more detail than given here, and the chapter by KEEGSTRA (11.10) offers an overview of the biosynthesis of chloroplast lipids.

The chloroplast thylakoid membrane glycolipids are unusual in the sense that they are not widely distributed among nonphotosynthetic species and are almost totally absent from the membranes on which the field of membrane biochemistry was founded (e.g., red cell plasma membranes, rod outer segment disc membranes, myelin sheath membranes, *E. coli* plasma membranes). The mono- and digalactosyldiacylglycerols comprise over 80% of the total diacyl lipids of the gliding thermophilic facultatively photosynthetic green bacterium *Chloroflexus aurantiacus* (KENYON 1978) and they are found in abundance in the photosynthetic membranes of certain cyanobacteria, but otherwise they occur rarely in nature, and then only in small amounts. However, from a different perspective, these glycolipids are not unusual, because by virtue of being the major diacyl lipids in thylakoid membranes they become the most abundant polar diacyl lipids of the biota (GOUNARIS and BARBER 1983).

There is currently considerable interest in specific functional roles of these lipids in photosynthetic membranes and in the functioning of photosynthesis (see also SIEGENTHALER and RAWYLER, Chap. 11.9; RYRIE, Chap. 11.7, both this Vol.). I am aware of three separate proposals for unique roles of particular

Table 1. The diacylglycerol lipids of thylakoid membranes

	Structure	Formal charge at neutral pH	% of total acyl lipid
Monogalactosyldiacylglycerol	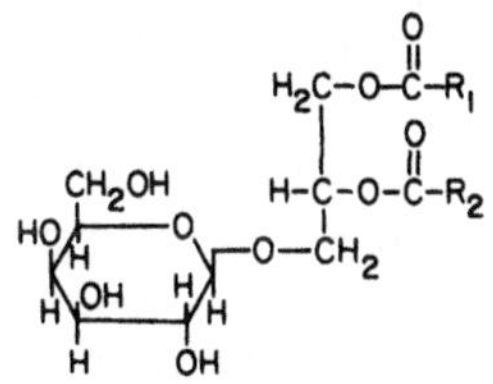	neutral	40
Digalactosyldiacyglycerol	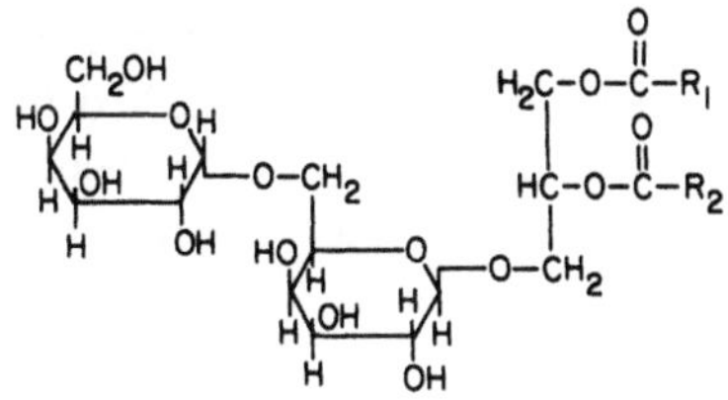	neutral	30
Diacylsulfoquinovosylglycerol	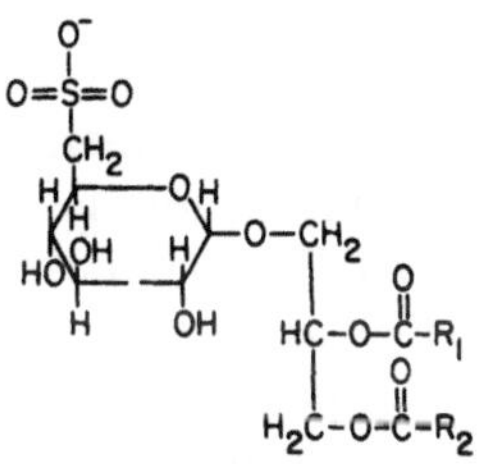	anionic	10
Phosphatidylglycerol	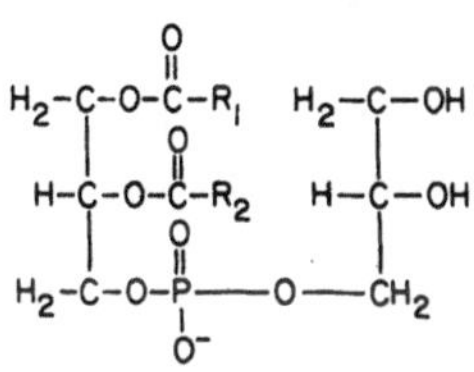	anionic	15
Phosphatidylcholine	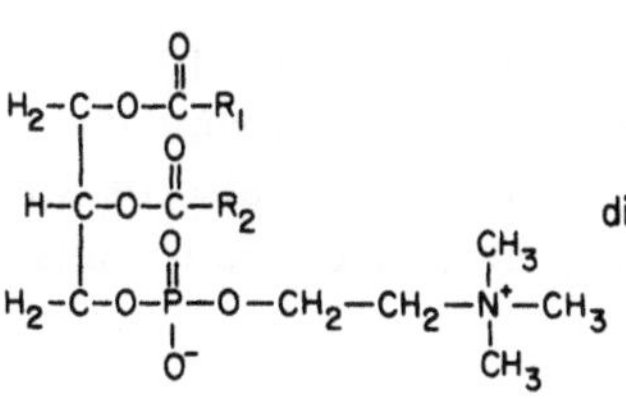	dipolar ionic	5

glycolipids and these proposals are not necessarily mutually exclusive of one another. Two of the suggestions require the specific association of these glycolipids with particular integral membrane polypeptide complexes. One thought is that specific lipid molecules are necessary to stabilize the insertion of protein complexes in the lipid bilayer (SPRAGUE et al. 1985 and DUBACQ and TRÉMOLIÈRES 1983). For instance, the cone-like molecular shape proposed for chloroplast MGDG may serve to properly seal the bilayer region surrounding globular integral membrane proteins (ISRAELACHVILI et al. 1980; MURPHY, Chap. 11.11, this Vol.). A second notion is that specific glycolipid-membrane protein interactions are necessary for the optimal enzymatic functioning of the protein (SIEGENTHALER and RAWYLER, Chap. 11.9, this Vol.). For instance, it has been reported that in reconstitution studies there is a requirement for MGDG for energy transfer among solubilized chlorophyll-protein complexes (SIEFERMANN-HARMS et al. 1982) as well as for the activation of the coupling factor complex (PICK et al. 1984). Membrane protein-acyl lipid interactions of the sort proposed above are generally visualized as requiring, although perhaps not absolutely, a stable lipid shell surrounding the membrane-embedded portion of the protein. There is ample evidence from well-studied systems to conclude that integral proteins do indeed perturb fluid lipid bilayer dynamics (e.g., CHAPMAN et al. 1979). Although these perturbations are short-range and do not extent beyond several lipid layers, it is nevertheless likely that no unperturbed lipid exists in a membrane that has a protein content as high as that of thylakoid membranes. However, the view that has evolved over the past decade of a long-lived shell of lipids bound to and surrounding integral proteins seems not to be supported by recent NMR evidence (e.g., BROWN et al. 1977; OLDFIELD et al. 1978; RICE et al. 1979). The NMR studies indicate that all lipids mix within 100 μs. It can *not* be concluded at this time that stable boundary lipid shells do not exist around certain proteins in some of the many membrane systems not yet investigated, for instance in thylakoid membranes. The weight of the evidence suggests, however, that the original EPR spin label studies and fluorescence polarization probe studies on which the notion of long-lived lipid boundary shells was based were misleading (PINK et al. 1984). Thus while stable lipid-protein associations cannot be precluded for thylakoid membranes, there seems, on the other hand, to be no strong precedent on which to base such notions.

The third proposal for a specific role of the thylakoid membrane glycolipids could also be interpreted as a requirement for a stable deployment of lipids in the bilayer. In line with the proposal (ISRAELACHVILI et al. 1980) for a cone-like molecular shape of chloroplast MGDG, it is thought that this lipid may stabilize the regions of acute membrane curvature that are very prevalent in the stacked regions of thylakoid membranes (MURPHY 1982; MURPHY and WOODROW 1983 and Chap. 11.11, this Vol.). If such deployments of MGDG do occur in native thylakoid membranes, these lipids must also be able to exist in planar bilayers in view of what has already been said about "unstacking" of thylakoid membranes by cation depletion.

2.2.1.2 Polypeptides

Much of this chapter will be devoted to the discussion of the four major functional polypeptide complexes of the thylakoid membrane. The photosystem I complex contains a reaction center and attendant antennae chlorophyll proteins, while the photosystem II complex contains, in addition, the polypeptides involved in water oxidation. Intersystem electron transfer is catalyzed by the cytochrome b_6/f complex and ATP formation by the coupling factor complex. The polypeptide components of these four complexes account for the bulk of the membrane protein both on a weight and on a molar basis. The structure and functioning of each of these complexes will be dealt with separately in sections that follow.

Knowledge of the abundance of the three electron transport complexes in the thylakoid membrane is essential to an understanding of the processes that work together to establish the mechanism and to determine the overall rate of the process. It is clear that there is no fixed stoichiometry of the complexes relative to one another or relative to total membrane chlorophyll that applies to all species of higher plants and algae. However, there is a remarkable constancy in the stoichiometry of the electron transport complexes within the single most well-studied species, *Spinacia oleracea* L. For spinach grown under a broad spectrum of high intensity light (i.e., $>600\ \mu E\ m^{-2}\ s^{-1}$ PAR) the ratio of the functional complexes is nearly 1:1:1 and there are about 600 chlorophyll molecules per complex (i.e., 1:1:1:600 chl) (WHITMARSH and ORT 1984; GRAAN and ORT 1984). It is essential to note that while a value of one active photosystem II center per 600 chlorophyll molecules in spinach chloroplasts is obtained from measurements of the ability of the chloroplasts to oxidize water (WHITMARSH and ORT 1984) or to reduce plastoquinone (GRAAN and ORT 1984), a value of one photosystem II center per 300 to 400 chlorophylls is obtained by estimation of light-induced reduction of the primary quinone acceptor (Q_A) (MELIS and ANDERSON 1983; WHITMARSH and ORT 1984) or by light-induced oxidation of the primary quinone electron donor (Z) (BABCOCK et al. 1983) of the photosystem II reaction center. While the discrepancy between these two sets of values remains to be resolved there is an important distinction in what is being measured that may point to an eventual resolution.[1] Measurements of the ability to oxidize water or to reduce plastoquinone require that the center be able to turn over repeatedly and at least one time per second. Measurements of light-induced generation of Q_A^- or Z^+ are essentially photostationary state measurements and assay the capability of the center to attain the state at least once. It may be that the former set of measurements gives a value for the active oxygen-evolving PS II centers and the latter measurements reveal, in addition, a population of PS II centers deficient in sustained O_2 evolution capacity. A population of low activity PS II centers might reflect PS II centers being repaired (e.g., recovery from photoinhibition – see KYLE and OHAD Chap. 9.5, this Vol.) or developmentally incomplete centers.

There are two independent reports (STROTMANN et al. 1973; FRASH et al. 1980) to indicate that there are approximately 850 chlorophyll molecules per coupling factor complex in spinach chloroplast membranes. These calculations were based on a coupling factor subunit stoichiometry of $\alpha_2\beta_2$, whereas recent evidence favors an $\alpha_3\beta_3$ stoichiometry (see Sect. 6.3.1 and MCCARTHY and NALIN, Chap. 10.8, this Vol.). Correcting for this change, a chlorophyll to coupling factor ratio of 1100 indicates only slightly more than 0.5 coupling factor per electron transport "unit".

1 The difference clearly does *not* arise from an inherent difference in the plant material since the estimation of photosystem II content by measurement of flash-induced water oxidation and by light-induced Q_A reduction showed a twofold difference within the same chloroplast membrane preparations (WHITMARSH and ORT 1984).

2.2.1.3 *Carbohydrates*

Carbohydrates are generally taken to be the third class of component found in biological membranes. Carbohydrates are never found free in biological membranes but always covalently attached to a membrane protein or lipid and always at the membrane surface, never in the hydrocarbon phase. Chloroplast thylakoid membranes have an abundance of carbohydrates by virtue of the high content of glycolipid. On the other hand, these membranes appear to be devoid of glycoproteins (KEEGSTRA and CLINE 1982).

2.2.2 Physical Properties

The negative surface charge of thylakoid membranes is a typical feature of biological membranes at physiological pH values. BARBER explains in his chapter (Chap. 11.5) devoted to this topic that the fixed charges on both surfaces of the thylakoid membrane are expected to give rise to elevated concentrations of electrically neutralizing ions (cations) immediately adjacent to the membrane's surface. For an idealized situation the electric potential producing this diffuse layer of counterions near to the membrane's surface declines exponentially with distance from the surface. The fact that thylakoid membrane surfaces are far from the ideal is nowhere more evident than in the measurement and in the prediction from electrostatic theory of the surface charge density. Experimentally obtained values for the surface charge density of the outer surface of thylakoid membranes range from about -0.005 to $-0.035\ \mathrm{C\,m^{-2}}$ at neutral pH (BARBER 1980).

Despite the complexities encountered in describing and quantitating the surface charge of the thylakoid membrane adequately, it is clear that the surface potentials and the nonuniform charge density have important consequences for certain functional aspects of photosynthesis, some of which will be dealt with separately later in this chapter, and some of which are mentioned by BARBER in Chapter 11.5, this Volume. Of particular interest in regard to thylakoid membrane structure is the involvement of surface charge in grana stacking. It is thought that the attraction of one thylakoid membrane surface for another is largely accounted for by van der Waal's forces between membrane proteins (BARBER et al. 1977). However, it does not seem probable that the electrostatic screening provided by the diffuse counterion layer would alone nullify the repulsive forces of the fixed charges (SCULLEY et al. 1980) and thereby allow the close surface to surface associations observed in grana stacks [$\sim$4 nm (NIR and PEASE 1973)]. In order to attain the low surface charge densities considered to be necessary for attractive van der Waal's forces to dominate, it has been suggested either that cations must interact with surface negative charges to form salts with the fixed charges (DUNIEC et al. 1981) or that there must be a net negative movement of high charge density particles away from stacked regions (BARBER 1980; BARBER, Chap. 11.5, this Vol.). Both of these alternative mechanisms retain as an essential feature the role of cations in screening surface charge.

The light-harvesting pigment protein complex of photosystem II (LHC II) has been implicated in numerous ways as being necessary for grana stack formation. Thus in accordance with the proposal of the forces governing stacking just mentioned, it must be LHC II which has the appropriate balance of repulsion arising from the fixed negative charge and attraction from van der Waal's forces so that electrostatic screening of fixed negative charges by cations is alone sufficient to allow the van der Waal's forces to dominate. Reconstitution studies show that phosphatidylcholine vesicles which do not associate even in the presence of cations can be made to do so by the incorporation of LHC II (MULLET and ARNTZEN 1980; STAEHELIN, Chap. 1, RYRIE, Chap. 11.7, both this Vol.). In addition, it also is now clear that the segment of membrane-exposed amino acids of LHC II required for cation-induced stacking can also undergo phosphorylation (BARBER 1982). Phosphorylation intensifies the negative surface charge density, exceeding the capacity of electrostatic screening, and stacking is thereby resisted. A recently proposed modification to this model for membrane adhesion is that the pigment proteins chiefly responsible for membrane adhesion are located in separate 8 nm particles not associated closely with the PS II center (STAEHELIN and DEWIT 1984).

The exceptionally high incidence of unsaturated fatty acids in the glycerol lipids of thylakoid membranes contributes significantly to the very fluid nature of these membranes over a relatively wide range of temperature. The abundant glycolipids of thylakoid membranes remain in a fluid state even at temperatures of several tens of degrees below zero. Somewhat surprisingly, reversible endothermic transitions, which almost certainly arise from glycerolipid phase transitions, are detectable by differential scanning calorimetry in thylakoid membranes of many plant species (LOW et al. 1984; RAISON and WRIGHT 1983). The size of these endothermic transitions indicates that generally only a small component (5 to 10%) of the total glycerolipid content is involved. The endotherm probably represents a coalescence of high melting point lipids into gel patches surrounded by liquid crystal state lipid, a condition often referred to as phase separation (see STAEHELIN, Chap. 1, this Vol.). The high melting point glycerolipid suggested as the origin of the above 0° phase separation is dipalmitoylphosphatidylglycerol. It is thought to induce the solidification of small amounts of otherwise low melting temperature lipids (RAISON and WRIGHT 1983).

There are factors other than the nature of the glycerolipids comprising the bilayer that are expected to influence the overall fluidity of the membrane. As pointed out previously, integral membrane proteins perturb the fluid lipid bilayer dynamics predicting a trend toward a less fluid membrane as the lipid to protein ratio declines. Indeed, the grana regions of thylakoids appear to be less fluid and have a lower lipid to protein ratio than nonappressed thylakoids (MURPHY and WOODROW 1983; GOUNARIS et al. 1983; MURPHY, Chap. 11.11, this Vol.). Opposing this trend, there are data to indicate that the long isoprenoid chain quinones characteristic of energy-conserving membranes have a fluidizing effect, caused by intercalation of the quinones with the fatty acids of the bilayer glycerolipids (FATO et al. 1984). In spinach thylakoid membranes the active plastoquinone pool is present at a concentration of about 1 mM [i.e., 1 plastoquinone 100 chl^{-1} (GRAAN and ORT 1984) and a thylakoid membrane volume of 11.1 l mol^{-1} chl (FLORES et al. 1983)]. The concentration of plastoquinone in the hydrocarbon phase of the membrane is likely to be two to three times this value.

The permeability of thylakoid membranes to inorganic ions is of significance due to the importance of ion movements in charge compensation for protein uptake as well as in phenomena such as the regulation of enzyme activities in the stroma. The conductance of the thylakoid membranes to larger molecules such as glucose or the various carbohydrate intermediates of the pentose phosphate cycle has important biochemical implications as well. Thylakoid membranes have undeservedly garnered the reputation for being unusually "leaky" to solutes in comparison to other biological membranes. In addition to situations in which "leakiness" was induced by isolation procedures or subsequent handling of isolated membranes (e.g., by freezing) the perception of "leakiness" was further fostered by a lack of recognition of the exceptionally high surface area to inner vesicle aqueous volume ratio and by the high conductance of thylakoids for a single physiological anion, Cl^-. The enclosed aqueous volume of isolated spinach thylakoid membranes suspended in 0.1 osmolar media is nearly equal to the volume of the thylakoid membrane itself, that is about 11 liters mol^{-1} chl (FLORES et al. 1983). In situ, judging from comparative electron micrographs (e.g., NIR and PEASE 1973), the aqueous enclosed volume may be only half the thylakoid membrane volume. With a surface area for one membrane surface in the neighborhood of 1.85×10^6 m^2 mol^{-1} chl (FLORES et al. 1983; THOMAS et al. 1956) the surface area to internal volume ratio is dramatically higher than other well-studied biological membranes (e.g., red cell plasma membranes, mitochondrial membranes or unilamellar lecithin vesicles) and, as a result, half-times for molecular species equilibrating across these diverse membrane types have little comparative value. Comparison of permeability coefficients reveals that thylakoid membrane conductance of most inorganic ions and organic metabolites is not unusually high. For instance, the thylakoid membrane permeability coefficient for sucrose was estimated to be 4.1×10^{-11} cm s^{-1} (FLORES et al. 1983) and for Na^+ 3×10^{-8} cm s^{-1} (BARBER 1972). The span of values from 2×10^{-10} to 2×10^{-11} cm s^{-1} for sucrose and from 5×10^{-6} to 5×10^{-10} cm s^{-1} for Na^+ have been measured for a range of other membrane types (JAIN and WAGNER 1980), showing that the permeability characteristics of the thylakoid membrane are in the range typical for biological membranes. However, the conductance of thylakoid membranes to Cl^- appears unusually high, at least in comparison with the conductance of other energy transducing membranes. The presence of a chloride transporter in thylakoid membranes has been suggested (VAMBUTAS and SCHECHTER 1983) but the supporting data are thus far not compelling.

2.3 Distribution of Thylakoid Membrane Components

2.3.1 Transmembrane Asymmetry

As far as I am aware, the transmembrane asymmetry of integral membrane proteins in native biological membranes is absolute. That is, the orientation of all of the molecules of a particular protein in a membrane is always the same. This uniformity of orientation is to be anticipated on the basis of the mechanisms which have been proposed for the insertion of polypeptides into membranes during their assembly. The same expectation naturally extends to thylakoid membranes and many experimental data support the anticipated asymmetry. This feature of biological membranes in general, and of thylakoid membranes in particular, provides the structural basis for the vectoral functioning of the photosynthetic membrane in light energy conversion.

The distribution of the various species of the diacyl lipids between the inner and outer leaflets of the thylakoid membrane bilayer is also asymmetric but in this case the asymmetry is not absolute. That is, each lipid species is present in both bilayer leaflets but in differing amounts. The alternative experimental approaches for obtaining quantitative data on transmembrane lipid asymmetry and the potential difficulties of each are discussed by SIEGENTHALER and RAWYLER (Chap. 11.9, this Vol.). These experimental difficulties may in part account for the somewhat different conclusions drawn by different laboratories regarding transmembrane distribution of diacyl lipids [see chapters in this volume by MURPHY (11.11) and by SIEGENTHALER and RAWYLER (11.9)]. Despite certain areas of disagreement, there appears to be consensus on several significant qualitative features of lipid distribution between the bilayer leaflets. The inner leaflet is thought to contain more than half of the total thylakoid diacyl lipid and to be enriched in MGDG and DSQ. Recall the proposal for the involvement of MGDG in stabilizing the inner radius of acute bends in the membrane which in itself would imply an enrichment of the inner leaflet with this lipid. The outer leaflet of the bilayer appears to contain more than half of the total PG, whereas the DGDG is relatively evenly distributed.

2.3.2 Lateral Asymmetry

A physical basis for the existence of lateral asymmetry of thylakoid membrane components is provided by the alternatives in the architecture of membrane regions that are appressed or are nonappressed. In fact, the need for the asymmetric distribution of certain membrane components in order to allow grana stack formation has already been alluded to in the earlier mention of the proposed roles for specific lipids in the sharply curved margins of the grana "discs" and in the need to exclude protein complexes of high surface charge density from appressed regions.

Grana stacks, in addition to having less lipid on a protein or chlorophyll basis, are reported to show pronounced lateral asymmetry of diacyl lipid composition as well (GOUNARIS et al. 1983). Despite consensus that granal membranes have a different diacyl lipid composition from nonappressed membranes, there seems to be a dependence of the results obtained on the method of membrane fractionation employed [see chapters in this Volume by MURPHY (11.1) and by SIEGENTHALER and RAWYLER (11.9)]. Nevertheless, there is agreement that grana stacks are depleted relative to the unfractionated thylakoids, in the two major thylakoid membrane diacyl lipids MGDG and DGDG.

There are numerous independent lines of evidence indicating that photosystem I and photosystem II complexes are segregated in different regions of the membrane (e.g., ANDERSON 1982; ANDERSSON and HAEHNEL 1982 and STAEHELIN, Chap. 1, this Vol.). This is in spite of the previously mentioned fact that the glycerolipid phase of the membrane has an unusually low viscosity. The majority of the photosystem II reaction centers with their attendant LHC II and water oxidation components are thought at present to be located in regions of membrane appression, their lateral mobility arrested by membrane surface to surface van der Waals interactions. The photosystem I reaction centers with their attendant antennae chlorophyll proteins are located predominantly in the portion of the membrane exposed to the stroma. This includes the outer margin of the grana stacks as well as the helically arranged membrane sheets surrounding the grana stacks. It has been suggested that the photosystem I complexes are excluded from the membrane-appressed regions by electrostatic repulsion arising from high surface charge density of the portion of the complex exposed from the outer membrane surface (BARBER 1982). Electron microscopy (MILLER and STAEHELIN 1976) studies favor a distribution of the coupling factor complex paralleling that of the photosystem I complex. The cytochrome b_6/f complex appears to be fairly uniformly distributed (COX and ANDERSSON 1981; PETERS et al. 1983; ALLRED and STAEHELIN 1985) and to retain a translational mobility which allows it to diffuse to and from appressed regions.

The lateral asymmetry of the membrane protein complexes, and most probably of the glycolipids as well, depends entirely on the existence of grana stacking. Naturally occurring, nongranal chloroplast thylakoids display only random distribution of the four membrane protein complexes (ARNTZEN 1978; SIMPSON, Chap. 11.6, this Vol.). Cation depletion-induced unstacking of granal chloroplasts also is followed by randomization of the complexes in the plane of the membrane (STAEHELIN et al. 1977; ARNTZEN 1978).

3 Photosystem II Complex

The identity of the photosystem II complex will, for the purposes of this chapter, be taken as the structural entity able to absorb light, oxidize water, reduce plastoquinone, and generate transmembrane asymmetry of electrical charge and chemical potential of hydrogen ions.

3.1 Composition and Structure

The photosystem II complex must contain the polypeptides for binding and orienting the light-harvesting pigments, the components for the initial light-driven electron transfer step between the primary electron donor and primary electron acceptor, the polypeptides involved in subsequent electron acceptor reactions, and the components of water oxidation and subsequent electron donor reactions. The photosystem II complex typically contains (e.g., from spinach) at least 11 different polypeptides (of uncertain stoichiometry), on the order of 200 chlorophyll a molecules, 100 chlorophyll b molecules, 50 carotenoid molecules, at least 3 plastoquinones (Q_A, Q_B, Z), 1 iron, 2 pheophytin a molecules, 1 or more cytochrome b-559 molecules, 4 manganese and an undetermined amount of chloride and perhaps calcium.

The polypeptides of the photosystem II complex that have so far been identified appear in Table 2. The reader is reminded that the tabulation is made at a time when research on photosystem II composition is intense and using numerous recent technical developments, and is therefore likely to need revision in the near future. Indeed, there is evidence for other polypeptides that are not listed (e.g., GREEN et al. 1982; LJUNGBERG et al. 1984) and at this stage it is not always possible to ascertain if different isolation procedures are yielding the same or similar but different polypeptides. The relative abundance of these polypeptides in the photosystem II complex is largely speculative and somewhat controversial at present. There is little basis for choosing one proposal in favor of others. Despite the immensity of that which remains to be understood about the composition of the PS II complex, the rate of progress in the past several years has only rarely been enjoyed in photosynthesis research.

The so-called core of the photosystem II complex is a set of five hydrophobic polypeptides which can be isolated from the other components of the thylakoid membrane after membrane disruption with one of several detergents. This complex contains about 50 chlorophyll a molecules which are associated with the 47 and 43 kDa polypeptides (CAMM and GREEN 1980). There is little doubt that this polypeptide doublet can perform the primary photochemical charge separation that defines the reaction center of photosystem II (DINER, Chap. 9.1; YOCUM, Chap. 9.2, both this Vol.). It may be, as some suggest (GREEN and CAMM 1984; NAKATANI et al. 1984), that the 43 kDa polypeptide functions only to bind chlorophyll a and transfer excitation energy, leaving the 47 kDa polypeptide alone as the true reaction center of photosystem II.

It is clear that the smallest polypeptide of the core complex (10 kDa) is the apoprotein of cytochrome b-559 (WIDGER et al. 1984a). Far from being clear, however, is the role of this enigmatic cytochrome. Despite its small size, cyt b-559 is envisioned as spanning the thylakoid membrane bilayer. The most recent notion is that the apoprotein of cytochrome b-559 may actually be an oligomer (possibly a tetramer) involving a 4.4 kDa polypeptide in addition to the 10 kDa polypeptide (WIDGER and CRAMER personal communication).

However compelling the underlying biochemical reasons may be for three photosystem II polypeptides of molecular mass about 33 kDa, there can be little doubt that some confusion is in store for future students of photosynthesis as a consequence. Two of the three are constituents of the core complex, one the apoprotein of the secondary quinone acceptor, Q_B (KYLE 1985), and the other involved in an uncertain way in water oxidation (METZ and SEIBERT 1984; GHANOTAKIS and YOCUM 1985; ANDERSSON, Chap. 9.3, YOCUM, Chap. 9.2, both this Vol.). The third polypeptide of this trio is a peripheral membrane protein, also with an as yet largely undefined role in water oxidation. The other two peripheral membrane proteins of the PS II complex, 23 kDa and 17 kDa, are

Table 2. The polypeptides of the four functional complexes of the thylakoid membrane

Complex		M_r[a] (kDa)	Association with functional components	Coding[b]
PS II Complex	Core	47 }	P-680, Pheo,	C
		43 }	Chl a, Z, Q_A	
		34	Mn(?)	C
		10	cyt b-559	C
		32	Q_B	C
	LHC II	28	Chl a,b	N
		26	Chl a,b	N
		24	Chl a,b	N
	Peripheral	33	Ca^{2+}(?), Cl^-(?)	N
		23	Ca^{2+}(?)	N
		17	–	N
PS I Complex	Core	70 (I)	P-700, A_0, Chl a	C
		25 (II)	–	N
		20 (III)	Plastocyanin	N
		18 (IV)	FeS	C(?)
		16 (V)	FeS	N
		8 (VI)	FeS	N
	LHC I	4 polypeptides from 19 to 24	Chl a+b	N
Cyt b_6/f complex		33, 34	Cyt f	C
		23	Cyt b_6	C
		20	Fe_2S_2	N
		17	–	C
Coupling Factor Complex	CF_1	59 (α)	–	C
		56 (β)	–	C
		37 (γ)	–	N
		18 (δ)	–	N
		13 (ε)	–	C
	CF_0	17 (I)	–	C
		11 (II)	–	N
		9 (III)	–	C
Ferredoxin-NADP oxidoreductase		33	FAD	N
Ferredoxin		11	FeS	N
Plastocyanin		10.5	Cu	N

[a] The molecular weights given are in most cases estimated from electrophoretic mobilities under conditions inducing dissociation of multimeric complexes. Due to differences in sample preparation and in electrophoresis conditions literature values range, sometimes significantly, from those presented here. In some cases it is not always possible to judge if different laboratories are reporting identical or similar but different polypeptides.

[b] The symbol C indicates that the gene coding for the polypeptide is within the chloroplast genome and the symbol N that the gene is within the nuclear genome

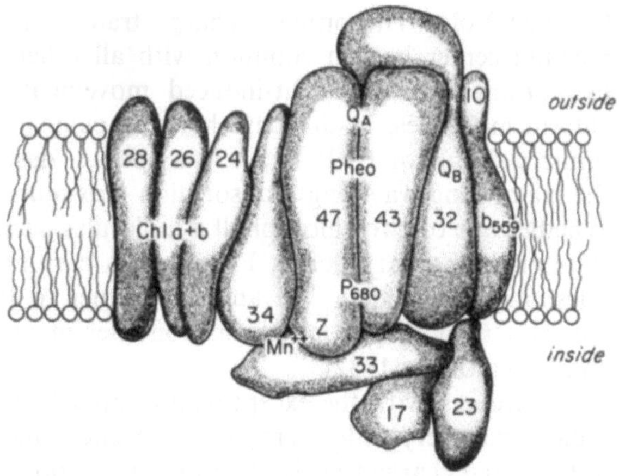

Fig. 3. A schematic sketch of the PS II complex depicting its array of different polypeptides and the proposed association of the redox active components. Only a single copy of each polypeptide is represented even though multiple copies of some are known to be present

required for water oxidation in the sense that water oxidation capability is impaired by their removal and each has reconstitutive properties under specific conditions (ANDERSSON, Chap. 9.3, this Vol., GOVINDJEE et al. 1985).

The polypeptides of the light-harvesting complex are believed to be associated one with another to form a native complex that binds chlorophyll a, chlorophyll b and carotenoids. The 26 kDa polypeptide is the most abundant of the three in most higher plant chloroplasts, but no consensus exists as to the stoichiometry of the three polypeptides. In fact there are indications that there is considerably more diversity in the molecular weight of these nuclearly coded pigment proteins than is indicated by the listing of three species in Table 2. Extensive diversity in LHC II structure is further supported by recent evidence of multiple LHC II genes (see THORNBER, Chap. 3, this Vol.). It is clear that the polypeptides of LHC II have significant structural homology. For instance, immunological cross-reactivity has been demonstrated for two major polypeptides of the analogous complex from *Chlamydomonas* (CHUA and BLOMBERG 1979). There are multiple copies of LHC II associated with each PS II center; 15 to 20 seems to be a typical value for plants such as spinach or peas grown at high light intensities.

The various different polypeptides of photosystem II are sketched as a complex in Fig. 3. This drawing is intended as an aid for the mind's eye of the PS II polypeptides and not as a structural model of the PS II complex. Those polypeptides which are thought to span the membrane are depicted as doing so. Only one copy of each polypeptide is represented and the components tentatively proposed as being associated with each polypeptide are indicated.

3.2 Primary Electron Transfer Events

One of the feats of the PS II centers is to mediate the one-electron charge transfer events of the reaction center, the four-electron oxidation of two water molecules and the two-electron reduction of plastoquinone (see DINER, Chap.

9.1, this Vol.). The primary charge transfer reaction of the photosystem II reaction center has, in common with all other sorts of reaction centers, the extraordinarily rapid light-induced movement of an electron between two chlorin molecules, in this case between chlorophyll a and pheophytin a. The primary donor of PS II is a Chl a species, designated P-680 to acknowledge the major long wavelength absorption maximum associated with its oxidation. Whether one or two chlorophyll a molecules participate in sharing the oxidized state of the primary donor, P^+-680, is a current and possibly partly semantic question (HOFF 1982). The anionic partner to P^+-680 in the primary radical pair is a reduced pheophytin a monomer most often designated I^- (KLIMOV and KRASNOVSKII 1982).

In keeping with the exceptionally rapid formation kinetics of the primary radical pair, very little energy of the excited chlorophyll donor, P*-680, is lost in the initial charge transfer event (VAN GORKOM 1985). However, stabilization of the photochemical charge separation in the time realm of biochemistry is accompanied by a large loss of free energy which is reflected in a comparision of the oxidation reduction midpoint potentials of I/I^- (about −600 mV) and its oxidant Q_A/Q_A^- (about −200 mV). Q_A is a plastoquinone molecule tightly bound to a reaction center polypeptide, most likely the 47 kDa polypeptide (DINER, Chap. 9.1, this Vol.). Q_A is reduced by I^- to create an anionic semiquinone. There are two uncommon features about the redox chemistry of this protein-bound reaction center plastoquinone: (1) it cannot be doubly reduced and (2) it does not, at least on the time scale relevant to electron transfer (i.e., ms), bind a proton when reduced. The Q_A molecule is in close enough proximity to the high spin nonheme ferrous iron atom of the reaction center to have electronic interactions (NUGENT et al. 1981) but the implications of these interactions are only poorly understood at the present time. The weight of current evidence suggests that the iron experiences no valence change in connection with quinone redox changes (PETROULEAS and DINER 1982).

The several hundred ps required for electron transfer between I^- and Q_A makes this reaction an order of magnitude slower than the primary charge transfer event. However, charge recombination between I^- and P^+-680 is remarkably at least 100 times slower still, which allows the stabilization of the charge separation on Q_A^- to occur with almost perfect quantum efficiency (DINER, Chap. 9.1, this Vol.).

3.3 The Acceptor-Side Reactions of Photosystem II Electron Transfer

Photosystem II complexes have two distinct plastoquinone acceptors that act in series. The first, Q_A, we have just discussed and identified its unusual one electron reduction characteristic. The second, Q_B, is reversibly associated (CROFTS and WRAIGHT 1983; ARNTZEN and PAKRASI, Chap. 9.4, this Vol.) with the 34 kDa polypeptide of the PS II core, a polypeptide which is notable as being the most rapidly turned-over protein of thylakoid membranes (KYLE and OHAD, Chap. 9.5, this Vol.). This secondary quinone acceptor acts to connect the single electron transfer events of the reaction center with the pool of free

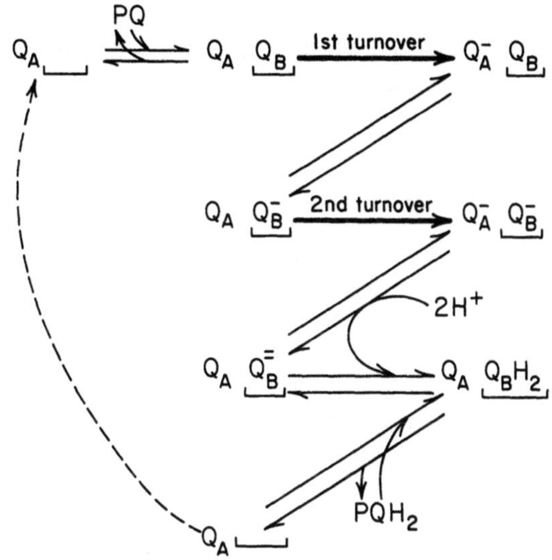

Fig. 4. The two-electron gate of photosystem II. Photosystem II centers must reconcile the two-electron reduction of plastoquinone with the one electron charge transfer events of the reaction center. This figure traces the sequence of events that occur during plastoquinone reduction. The operation of the two-electron gate is discussed in the text (Sect. 3.3.1)

plastoquinone in the membrane by operating as a two electron gate, allowing the passage of electrons only in pairs.

3.3.1 The Two-Electron Gate

The proposed operation of the two-electron gate of photosystem II is identical to that proposed for the purple photosynthetic bacteria (CROFTS and WRAIGHT 1983; DUTTON, Chap. 5, this Vol.) and is illustrated in Fig. 4. Subsequent to the first turnover of a reaction center, the electron is stabilized on the primary quinone acceptor Q_A and then, within a few tens of microseconds, is transferred to the secondary quinone acceptor, Q_B. This is an equilibrium reaction whose equilibrium constant is estimated to be about 20 (CROFTS and WRAIGHT 1983). Following a second turnover of the same PS II reaction center, there is a repetition of these same events, producing a doubly reduced Q_B. At some point in the overall process two protons are taken up from the external aqueous phase (i.e., stromal side) producing fully reduced and protonated plastoquinol. This mode of quinone reduction in which the intermediate reductive states are maintained bound to a protein ensures absolute asymmetry in proton uptake in keeping with the role of transmembrane proton asymmetries in energy coupling. The quinol is only loosely associated with the quinone-binding site and its dissociation from the PS II center is almost certainly the mode of reduction of the plastoquinone pool. The vacated quinone binding site on the 32 kDa polypeptide is then available to bind a plastoquinone from the membrane's pool.

It should be appreciated that it is the ability of the PS II acceptor complex to stabilize the semiquinone state of the secondary acceptor that allows quinone reduction to occur as two sequential one-electron steps and thereby interface the single electron events of the reaction center with the two-electron events of the plastoquinol pool. Indeed, the fact that plastoquinone reduction requires two sequential photoacts is the strongest kind of evidence that the quinone is protein-bound. If it were not, two semiquinones formed by different reaction centers on the first photoact would dismutate to form PQ and PQH_2.

3.3.2 Mode of Inhibition by Photosystem II Herbicides

The mechanism of action by herbicidal inhibitors of photosystem II electron transfer is elegantly accommodated within the mechanism of plastoquinone reduction summarized above: they prevent the oxidation of Q_A^-. It was independently and simultaneously proposed for the reaction center complex of purple photosynthetic bacteria (WRAIGHT 1982b) and for the photosystem II complex of plant thylakoids (VELTHUYS 1982) that the secondary quinone acceptor, Q_B, can be displaced from its binding site by the inhibitor molecule, I.

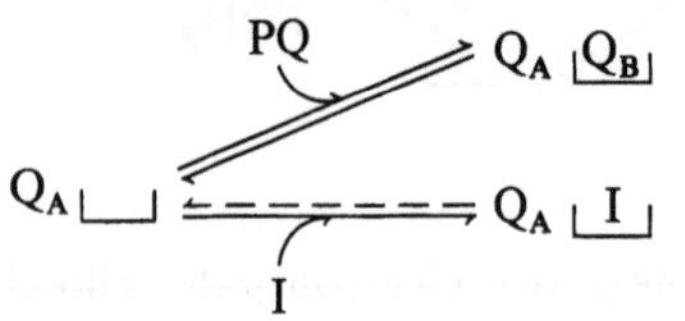

The action of the inhibitor appears to be competitive with plastoquinone with the rate of release of the inhibitor from the site many times slower than the rate of release for plastoquinone (e.g., VERMAAS et al. 1984a, b). The inhibitors, of which the triazines (e.g., atrazine) and ureas (e.g., diuron) are examples, cannot be reduced by Q_A^- and electron transfer beyond this point is thereby prevented.

As indicated earlier, plastoquinol (i.e., Q_BH_2) is only weakly bound to the quinone binding site of PS II, so this species is even more easily displaced by the inhibitor than is Q_B.

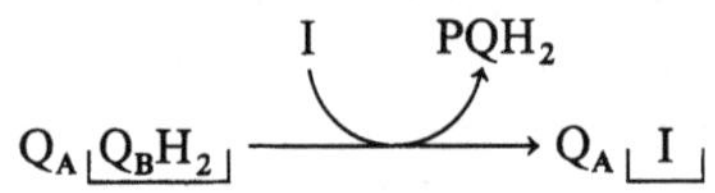

On the other hand, the semiquinone, Q_B^- is much more tightly bound than Q_B (CROFTS and WRAIGHT 1983). Owing to the equilibrium between Q_A and Q_B, an electron shared between the two molecules spends about 95% of the time on Q_B and the remainder of the time on Q_A. It appears that the inhibitor binds by displacing Q_B, trapping the electron on Q_A, thereby producing the illusion that the inhibitor has induced reverse electron flow from Q_B^- to Q_A (VELTHUYS and AMESZ 1974).

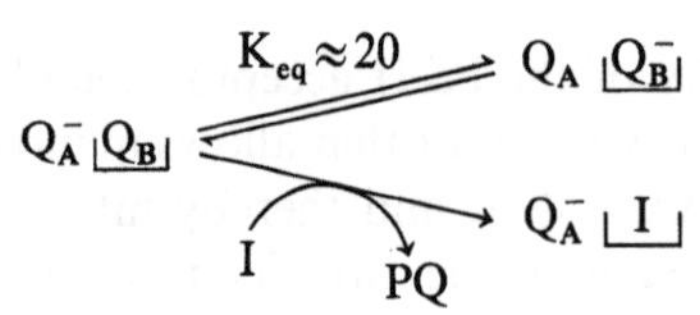

These notions about the mechanism of herbicide action have been quite successful in explaining resistance to triazine herbicides that has developed in more than 20 weed species on agricultural land treated annually with triazines. It has been demonstrated that a single amino acid substitution in the quinone binding region of the 32 kDa PS II protein (HIRSCHBERG et al. 1984) dramatically decreases the binding affinity of triazine herbicides (and of certain other herbicide families as well) to the PS II complex (PFISTER and ARNTZEN 1979; SANDMAN and BÖGER, Chap. 10.10, this Vol.). That is, these compounds no longer effectively compete with Q_B for the binding site, and inhibition of electron transfer is resisted. However, accompanying this amino acid substitution and resultant herbicide resistance is a very significant decrease in the equilibrium constant between Q_A^- and Q_B, which may be as much as 20-fold (ROBINSON 1985; VERMAAS et al. 1984c). If so, an electron shared between the quinone acceptors would spend half the time on Q_A and it may be predicted that at any instant in the light as many as one half of the photosystem II centers would be incapable of photochemistry (i.e., incapable of stabilizing the primary radical pair P^+-680 I^-). This is consistent with the observation that resistant weed biotypes have a 25 to 30% lower quantum yield than their susceptible counterparts (ORT et al. 1983).

3.4 Water Oxidation and Donor-Side Reactions

3.4.1 The Reduction of P^+-680

The reduction of P^+-680 is complex in the sense that the reduction kinetics are multiphasic with component phases ranging from a few tens of ns to the µs range (DINER, Chap. 9.1, this Vol.). The immediate donor of electrons to P^+-680 has been designated Z and is a one-electron component that mediates charge transfer between the reaction center and water oxidation. It remains uncertain as to whether there exist more than a single Z per reaction center and, if more than one, whether these operate in series or in parallel.

The weight of current evidence supports the interpretation that the light-induced EPR signal known as signal II_{vf} is caused by the formation of Z^+. Direct kinetic comparison of the rise time of signal II_{vf} with optical measurements of P^+-680 reduction are not possible due to the limited time resolution of EPR spectroscopy. However, an excellent kinetic correlation is obtained in chloroplasts in which the kinetics of P^+-680 reduction and signal II formation have been slowed by inhibition of water oxidation (BOSKA et al. 1983).

It has recently been proposed that Z^+ is a plastosemiquinone cation (PQH_2^+) produced by a one-electron oxidation without any accompanying deprotonation of Z, a specially bound plastoquinol (O'MALLEY et al. 1984; O'MALLEY and BABCOCK 1984). This proposal has the attractive feature that a semiquinone cation would have the extremely positive redox potential (perhaps 1.0 V for Z/Z^+) necessary if it is to be reduced by the water oxidation complex. The proposal that Z^+ is a plastosemiquinone cation is supported by in vitro model compound studies which simulate rather well the in vivo EPR spectrum of signal II (O'MALLEY et al. 1984; O'MALLEY and BABCOCK 1984). DINER reviews these and other data in his chapter on photosystem II (Chap. 9.1, this Vol.) and appropriately cautions that the chemical identity of Z/Z^+ should still be considered an open question in spite of the obvious attractiveness of the PQH_2/PQH_2^+ proposal.

3.4.2 Water Oxidation

As mentioned earlier, oxygenic photosynthesis seems to have little to call its own since each of the reactions and structures of the thylakoid membrane has at least a qualitatively identifiable ancestor in one or more of the nonoxygenic photosynthetic bacteria. Water oxidation is perhaps the only reaction unique to oxygenic photosynthesis, but represents a very significant exception and gives the process its name.

There are several major difficulties to be overcome in using water as a reductant: (1) the coupling of oxidation equivalents that are generated one at a time by the reaction center with the four electron oxidation of water to molecular oxygen; (2) prevention of highly reactive oxidation intermediates or the sheltering of such intermediates from unwanted reaction partners; (3) prevention of widely different free energy requirements for the successive oxidation steps ensuring that each of the four sequential steps will be within the energy available from a single photon.

The major facts which have been established about water oxidation are these: (1) four electrons are removed from two water molecules in four sequential one-electron steps, thus illustrating that each photosystem II center operates independently to oxidize water and to evolve molecular oxygen (JOLIOT and KOK 1975). The alternative that four PS II centers could cooperate in a concerted oxidation of water clearly does not occur; (2) molecular oxygen (i.e., diatomic oxygen) is produced following the removal of four electrons, but H^+ are released one each on the first and third electron removal with the remaining two released coincident with O_2 release (reviewed by GOVINDJEE et al. 1985). Thus the water oxidation complex never accumulates more than two net positive charges during the reaction sequence. These facts are the basis for the so-called "S-state" model of water oxidation (JOLIOT and KOK 1975). This familiar scheme is the basis for the interpretation of results on the mechanism of water oxidation and is presented yet again in Fig. 5.

The S-state is sequentially advanced as the photooxidized Z^+ removes an electron from the water-oxidizing enzyme complex (Fig. 5). States S_3 and S_2 are unstable and will relax in the dark within seconds to the S_1 state. The oxidation states S_0 and S_1 are stable with a typical distribution in dark-adapted membranes of 75% of PS II centers in S_1 and 25% in S_0 (KOK et al. 1970).

In spite of the success of the S-state model in accommodating and even predicting much about the behavior of the water oxidation system, only little is known with certainty about the molecular mechanism of water oxidation. An obligate role for Mn in water oxidation is well established. Numerous treatments (e.g., Tris treatment, NH_2OH treatment, heat treatment), which inhibit the water oxidation capacity of thylakoid membranes, have in common the

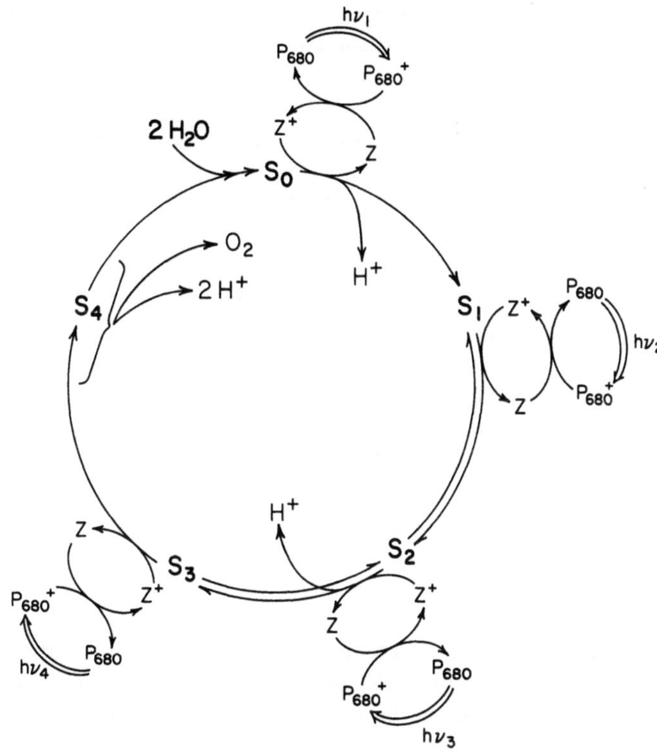

Fig. 5. The S-state model of water oxidation. The model shows that four electrons are removed from two water molecules as the S-state is sequentially advanced when photooxidized Z^+ removes one electron at a time from the water-oxidizing complex. This central feature of the model accommodates the experimental observation that each PS II center functions independently to oxidize water and evolve molecular oxygen. Molecular oxygen is produced following the removal of four electrons but hydrogen ions are released on each of the first and third electron removal with the remaining two released coincident with O_2 release

liberation of Mn bound to the PS II center (reviewed by YOCUM 1975 and Chap. 9.2, this Vol.). There are several classes or pools of Mn associated with the thylakoid membrane that are distinguished by the tenacity of binding, and this has led to a certain amount of uncertainty regarding the number of Mn atoms necessary for oxygen evolution. It is now clear that depletion of Mn^{2+} below 4 per reaction center at least partially impairs water oxidation (e.g., YOCUM et al. 1981).

The earliest interpretations for the role of Mn in water oxidation was as a charge accumulator, that is as the component which was the direct reductant for Z^+ and the direct oxidant of water. However, it is only very recently that direct physical evidence for redox changes in Mn associated with water oxidation has been obtained. The evidence comes from low temperature EPR studies (DISMUKES and SIDERER 1981; ZIMMERMANN and RUTHERFORD 1984) and EXAFS (Extended X-ray Absorption Fine Structure) investigations (KIRBY et al. 1981). This evidence is discussed by YOCUM (Chap. 9.2, this Vol.) and in recent reviews by RENGER and GOVINDJEE (1985), VAN GORKOM (1985) and GOVINDJEE et al. (1985). There remains considerable uncertainty as to both the nature of the manganese complex involved in the light-induced valence change and as to the oxidation states of Mn involved. An additional feature of these very exciting preliminary results on the role of Mn is the indication that electrons removed during the sequential advancement of the S states come not only from the Mn center but from its ligand groups as well. Most models favor water itself as an inner sphere ligand, implying that intermediate oxidation states of water are produced during the reaction sequence. This is a notion which is consistent with the H^+ release pattern.

The requirement for chloride ions in the reactions of water oxidation is clearly established and has been extensively investigated (HIND et al. 1969; KELLY and IZAWA 1978; CRITCHLEY 1983). In comparison to manganese, Cl^- is only loosely associated with the photosystem II complex and its role even more enigmatic. The requirement of chloride is in the range of 10 to 20 molecules per PS II center and certain other ions can fulfill the chloride requirement, but always with reduced efficiency (KELLY and IZAWA 1978). One plausible suggestion is that Cl^- is essential for stabilizing the manganese center of the water oxidation complex. The further suggestion that Cl^- may bridge adjacent manganese centers, thereby facilitating electron transfer, is an intriguing specific proposal for one possible mechanism for the stabilization (CRITCHLEY et al. 1982). This suggestion, however, is not without its critics and, of course, anions of some kind are obviously necessary to counter the increasing positive charge associated with Mn oxidation especially if a H^+ is not simultaneously released.

Ca^{2+} has recently been added to the list of inorganic ions required for efficient water oxidation (INOUE and ONO 1983). It is suggested that the functional Ca^{2+} is tightly associated with the three peripheral proteins of the PS II (GHANOTAKIS et al. 1984). Little can be said at this time about the specific role of Ca^{2+} in the mechanism of water oxidation.

4 Photosystem I Complex

A useful definition for discussion of the photosystem I complex of thylakoid membranes is that complex of polypeptides capable of the light-induced oxidation of plastocyanin (or Cyt c in some algae), reduction of ferredoxin and generation of a transmembrane asymmetry of electrical charge.

4.1 Composition and Structure

To accomplish their task, the polypeptides of the photosystem I complex must bind and orient the light-absorbing pigments, arrange the components of the reaction center involved with the initial electron transfer and charge generation events and appropriately position the molecules which act as donors and acceptors of electrons to and from the reaction center (SETIF and MATHIS, Chap. 9.6, this Vol.). The photosystem I complex contains at least 13 distinct polypeptides, about 200 chlorophyll molecules, an undetermined number of carotenoid molecules, three iron sulfur centers and perhaps one or more quinone molecules (WOLLMAN, Chap. 9.7, this Vol.).

The so-called core of the photosystem I complex from higher plants contains six distinct polypeptide subunits in contrast to only four subunits of green algae and cyanobacteria (OKAMURA et al. 1982). The subunits from higher plants are numbered I through VI in order of descending molecular mass (Table 2). The core complex is thought to contain a dimer of subunit I which constitutes the PS I reaction center and one copy of each of the remaining five polypeptides. There are strong indications that the two ~70 kDa polypeptides of the reaction center (subunit I) are in fact not identical, being either two truly different polypeptide species or minor structural variants of a single polypeptide (BAR-NUN et al. 1977; CHUA et al. 1975; THORNBER, Chap. 3, this Vol.). The detergent-induced loss of subunits IV, V, VI from the core complex correlates with the loss of EPR signals assigned to the three FeS electron acceptors of PS I (OKAMURA et al. 1982). Several lines of evidence demonstrate that subunit III facilitates the reduction of P^+-700 by plastocyanin, functioning to bind plastocyanin to the PS I complex (HAEHNEL et al. 1980). A functional role for subunit II in electron transfer has not been elucidated and indeed it may have no direct electron transport role.

Several of the seven or more remaining polypeptides of the PS I complex are involved in binding and orienting the light-harvesting chlorophyll and accessory pigments (THORNBER, Chap. 3, this Vol.). A particularly interesting recent development, aiding in the elucidation of the photosystem I pigment array, is the demonstration of polypeptides that bind chlorophyll b (HAWORTH et al. 1983; LAM et al. 1984a, b; ORTIZ et al. 1984). Formerly it had been thought that a distinguishing feature of the PS I complex was the absence of chlorophyll b-binding proteins. It is now apparent that three or four polypeptides in the molecular weight range of 19 to 24 kDa cooperate as complexes to bind both chlorophyll a and chlorophyll b. The chlorophyll a to b ratio of this isolated chlorophyll-binding complex is about 3.5 (HAWORTH et al. 1983) compared to a value of 10 estimated for the intact PS I complex (the situation is somewhat different for certain algae as noted by WOLLMAN in Chap. 9.7, this Vol.). It is clear that the chlorophyll a and b-binding proteins of PS I are distinct from those of PS II (LAM et al. 1984b). However, the pattern of synthesis of these proteins is exactly analogous to the pattern of synthesis of the PS II chlorophyll a and b-binding proteins, in that they are both synthesized on cytoplasmic ribosomes as larger precursor polypeptides, and both taken up and processed to the functional state by the chloroplast.

4.2 Primary Electron Transfer Events

As stated above, the PS I reaction center complex of higher plants consists of a doublet of 70 kDa polypeptides. The primary donor of the PS I reaction center is known as P-700 since the optical difference spectrum of the photo-oxidized primary donor shows a maximum bleaching near 700 nm. P^+-700, the oxidized state of the primary donor, is a weak oxidant in comparison to P^+-680 of PS II ($E_{m,7}$ about 480 mV for P^+-700 vs. about 1200 mV for P^+-680). The optical difference spectrum of P^+-700 identifies it as some species of a chlorophyll a cation, but lively discussion as to whether the primary donor is a chlorophyll monomer or dimer continues (WRAIGHT 1982a). Increasingly the interpretation appears to be converging on the conclusion that it is at once neither and both. The data and alternative arguments of this controversy are reviewed by SETIF and MATHIS in Chapter 9.6, this Volume and by RUTHERFORD and HEATHCOTE (1985) in a recent review published elsewhere.

Within 10 ps after the absorption of a photon by the antennae pigments of PS I, P*-700 loses an electron to an acceptor molecule bound to the reaction center complex. The orientation of the reaction center in the membranes ensures that the charge separation is perpendicular to the membrane surface, thereby producing a transmembrane electric potential difference. The chronology of the discovery of the various PS I electron acceptors to the actual sequence of the reactions in the chloroplast was almost inverse, since the instrumentation to detect these acceptors evolved with increasingly better time resolution. This has led to the curious and sometimes confusing reference to primary, "more" primary and "even more" primary acceptors. The first molecule known to be reduced by P*-700 is designated A_0. It is almost certainly a chlorin-type structure and based on both optical (SWARTHOFF et al. 1982) and ENDOR (FAJER et al. 1980) spectra is likely to be a chlorophyll molecule. A_0^- is an exceptionally reducing species relative to the first known electron acceptor of PS II. (The midpoint redox potential of A_0^- is estimated to the about −1100 mV compared to I^- of PS II measured to be about −600 mV.) Thus the charge separation span of PS I covers a much more negative range than the charge separation span of PS II.

The initial photochemical charge separation between P^+-700 and A_0^- is further stabilized by the transfer of the reducing equivalent from A_0^- to A_1, a process occurring on the time scale of tenths of ns. The chemical nature of A_1 is not known but is certainly *not* Chl a and probably not any other chlorin-type molecule (RUTHERFORD and HEATHCOTE 1985). There are features of the EPR hyperfine structure reminiscent of signal II (HEATHCOTE and EVANS 1980), raising the possibility that A_1^- may be yet another "special" semiquinone involved in photosynthetic electron transfer.

4.3 Electron Donors to Photosystem I

The immediate electron donors to photosystem I are small hydrophilic proteins that can be mobile within the lumen of thylakoid vesicles. In certain eukaroytic algae and many cyanobacteria the donor is a c-type cytochrome very similar to that found in bacteria and mitochondria. Plastocyanin functions as the immediate donor to PS I in higher plants and in most algae. In fact, certain species of cyanobacteria and eukaroytic algae contain both of these metalloproteins. The relative proportions are determined by the level of Cu and Fe available.

The crystallographic structure of plastocyanin has been determined (COLMAN et al. 1978; see also HAEHNEL, Chap. 10.5, this Vol.) and reveals two potential sites for electron transfer in and out of the metalloprotein, one site with hydrophobic character and the other with a rather high surface charge density. The requirement for cations for rapid P^+-700 reduction (HAEHNEL et al. 1980) implies a role for ionic screening, suggesting that the negatively charged region of plastocyanin may interact with a negatively charged region on subunit III of the PS I core complex.

4.4 The Acceptor-Side Components and Reactions of Photosystem I

Three distinct iron sulfur centers have been identified in PS I center preparations by low temperature EPR studies (MALKIN 1982; SETIF and MATHIS, Chap. 9.6, this Vol.). These have been designated F_A, F_B, and F_x, thought to be 4Fe−4S centers (EVANS 1982) and to be associated with subunits IV, V, and VI of the PS I core complex. These three iron sulfur centers appear to be present in amounts roughly stoichiometric with P-700 (RUTHERFORD and HEATHCOTE 1985). They are distinguished by exceptionally low redox midpoint potentials with measured values of: $F_x = -730$ mV, $F_B = -590$ mV and $F_A = -550$ mV (EVANS 1982). The uniform abundance and the descending midpoint potential of these electron acceptors might lead to the expectation that they operate in a linear series in the order listed. However, the revelant data are difficult to interpret and the mode of participation of these iron sulfur centers in photosystem I electron transfer is uncertain, but a simple linear sequence seems unlikely.

The earlier notion that F_B functions in a linear sequence prior to F_A has, in light of recent findings (reviewed by RUTHERFORD and HEATHCOTE 1985; SETIF and MATHIS, Chap. 9.6, this Vol.), given way to a model in which F_A and F_B are envisaged as operating in parallel, together constituting the terminal acceptor of the PS I center and thus jointly responsible for the reduction of soluble ferredoxin at the outer surface of the thylakoid membrane. In addition to the EPR signals associated with the redox reactions of these acceptors there is an optical change as well, known as P-430. A consistent interpretation of the origin of P-430 is that it arises from the combined contributions of both centers F_A and F_B (SETIF and MATHIS, Chap. 9.6, this Vol.).

The measured redox potential of F_x would logically place it as an acceptor prior to F_A and F_B in a sequential scheme. There is actually little direct evidence to support a role for F_x^- as the direct donor of electrons to F_A and F_B and there are in fact some indications that F_x may operate in parallel to the other two iron-sulfur centers (e.g., HIYAMA and FORK 1980). However, the most consistent interpretation of available data seems to place F_x in the electron transfer pathway between A_0 and the parallel iron sulfur acceptors F_A and F_B as depicted in Fig. 6.

Ferredoxin, soluble in the stromal phase of the chloroplast, couples electron transfer between the PS I center and NADP to form the NADPH necessary for CO_2 fixation. The reduction of NADP is catalyzed by ferredoxin-NADP

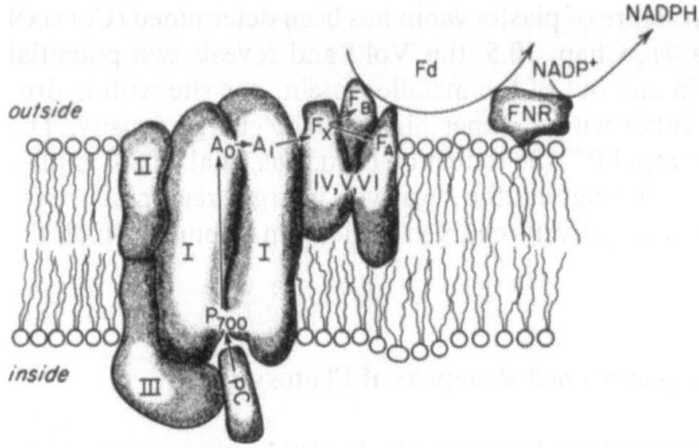

Fig. 6. Pathway for electron transfer through the photosystem I complex during oxidation of plastocyanin (*PC*) and the reduction of ferredoxin (*Fd*). The six different polypeptides depicted in the schematic drawing are viewed as forming the core of the PS I complex. A dimer of subunit I, the largest polypeptide of the core complex, is equivalent to the PS I reaction center since it is sufficient for the light-induced formation of the primary radical pair, P^{+}-700 A_0^-. The drawing also represents the reduction of $NADP^+$ by reduced ferredoxin via ferredoxin NADP oxidoreductase (*FNR*)

oxidoreductase (FNR) that is associated with the outer surface of nonappressed regions of the thylakoid membranes (Jennings et al. 1979). Dissociation of this flavoprotein protein from the membrane results in a substantial decrease in affinity of the enzyme for ferredoxin (Forti et al. 1983).

5 Intersystem Electron Transfer

The interaction between photosystem I and photosystem II centers occurs on two important levels: (1) the partitioning of light energy, and (2) the transfer of electrons. The phenomenon and mechanisms of the distribution of light energy absorbed by the various pigment protein complexes is among the subjects covered in detail in chapters in this volume by Sauer (Chap. 2) and by Thornber (Chap. 3). The discussion in this section will be concerned with the thermodynamically favored passage of electrons from carriers with low redox potentials to carriers with higher redox potentials, a process which serves to communicate reducing equivalents between the photosystem complexes. These complexes can be separated by relatively large average distances. These intersystem reactions are sometimes viewed as the dark electron transfer events of photosynthesis inasmuch as the photoacts have generated an oxidant and a reductant between which electron transfer requires no further light energy input. For instance, after the photosystems are turned over in unison by a brief flash of light, these reactions do indeed occur in the dark. The participants in the dark electron transfer reactions are plastoquinone, plastocyanin, and the cytochrome b_6/f complex, an integral membrane polypeptide complex containing an iron sulfur cluster in addition to the b- and c (cyt f)-type cytochromes.

5.1 Cytochrome b_6/f Complex

Photosynthetic and respiratory membranes from both prokaryotes and eukaryotes contain closely analogous structures that function to oxidize a low potential quinol and reduce a high potential metalloprotein (HAUSKA, Chap. 10.1, this Vol.). In the case of higher plant thylakoid membranes this complex oxidizes plastoquinol generated by PS II and reduces plastocyanin oxidized by PS I. The complex is most often called the cytochrome b_6/f complex but more properly termed plastoquinol: plastocyanin oxidoreductase.

5.1.1 Composition and Structure

The polypeptide composition of the quinol-metalloprotein oxidoreductase from photosynthetic membranes is substantially simpler than that found in mitochondrial membranes. The complex from mitochondria of various organisms typically has ten distinct polypeptides (HAUSKA, Chap. 10.1, this Vol.), whereas both the spinach thylakoid complex and the *Anabaena* complex have only four major subunits (Table 1).

The stoichiometry of the cytochrome b_6/f complex from spinach appears to be 1:1:1:1 for the 34+33, 23, 20, 17 kDa polypeptides. The 34/33 kDa polypeptide doublet appears to represent an artificial heterogeneity in the population of cytochrome f apoprotein since there seems to be only a single chloroplast gene coding for cyt f (ALT et al. 1983 and WILLEY et al. 1983). Considering the mechanism of interaction among the redox active components of the b_6/f complex, it is interesting that hydropathy plots suggest that the apoproteins of both cytochrome f and the Rieske iron-sulfur cluster (22 kDa polypeptide) contain only a single membrane-spanning α-helical region. Furthermore, it appears that the redox active group of each apoprotein (that is, the heme of cyt f and the Fe_2S_2 center of the Rieske iron-sulfur cluster) are attached to the polypeptides in a region exposed to the inner aqueous phase of the thylakoid vesicle (see Fig. 2 in HAUSKA Chap. 10.1, this Vol.). The two cytochrome b_6 hemes are bound by a single copy of the 23 kDa polypeptide. From the chloroplast gene sequence for this polypeptide a folding pattern based on hydropathy indices has been proposed. In this case the proposed structure shows five α-helices spanning the membrane with the two hemes positioned with their porphyrin rings parallel to the plane of the membrane and separated by a distance roughly equal to the hydrocarbon phase of the membrane (WIDGER et al. 1984b). Comparison of sequence homology of the combined amino acid sequences of the chloroplast cytochrome b_6 and the chloroplast 17 kDa polypeptide with the mitochondrial b cytochrome of yeast, human, mouse, and bovine show extensive homology between the 139-residue 17 kDa chloroplast polypeptide and the COOH-terminal end of the larger mitochondrial cyt b apoprotein (WIDGER et al. 1984b).

No positive identification of the cyt b_6/f complex in the thylakoid membrane has yet been made by freeze-fracture electron microscopy but incorporation of the isolated complex into liposomes produces 8.5 nm intramembrane particles. A molecular mass of 280 kDa was estimated from the size of these particles and interpreted to suggest a dimeric configuration of the cyt b_6/f complex (MÖRSCHEL and STAEHELIN 1983). However, in view of the subunit stoichiometry for the cyt b_6/f complex given by HAUSKA (Chap. 10.1, this Vol.) a molecular mass of 280 kDa indicates a trimeric configuration. Even so, a dimeric structure is supported by the recent finding that there are only half the number of binding sites for the quinol oxidase inhibitor DBMIB as there are molecules of cyt f (GRAAN and ORT 1985). A dimeric structure for the comparable mitochondrial oxidoreductase has also been proposed (SLATER 1981).

Plastoquinone is frequently co-isolates with the cyt b_6/f complex (see HAUSKA, Chap. 10.1, this Vol.). Approximately 0.6 plastoquinone A molecules are found to copurify with the complex (D.P. O'KEEFE personal communication). These data are also suggestive of a dimeric structure; however co-purification does not ensure specific binding or functional significance.

5.1.2 Mechanism of the Oxidation of Plastoquinol

The cytochrome b_6/f complex must reverse the accomplishment achieved by the two-electron gate of PS II in coupling the oxidation of the two electron-carrier plastoquinol with reduction of the monovalent carrier plastocyanin. This process is achieved by a group of one-electron redox components of the oxidoreductase. The pathway of electron flow through the b_6/f complex is still contentious, but models proposing a linear sequence for electron transfer through the b_6/f complex have recently declined in acceptance and given way to a variety of proposals that are variants on the so-called Q-cycle formulated originally by MITCHELL (1975).

In order to appreciate the principles of the Q-cycle and the potential advantages of such proposals in the interpretation of experimental data, it is useful to describe the simplest case, ignoring initially the anomalies and the embellishments to the model necessary to accommodate them. The result of the cycle of events in the b_6/f complex is to cause the net transfer of protons across the thylakoid membrane, the oxidation of quinol, and the reduction of plastocyanin.

In the Q-cycle there is envisioned a binding site for plastoquinol on the b_6/f complex near the inner surface of the thylakoid membrane, the Q_z site, and a binding site for plastoquinone near the outer surface, the Q_c site (Fig. 7). Under optimal conditions in the dark, the Q_z site is occupied by plastoquinol, the Q_c site by plastoquinone, the Rieske FeS center and cyt f are reduced and the b cytochromes are oxidized. The transfer of an oxidizing equivalent from photosystem I via plastocyanin results in the removal of one electron from the high potential carriers of the b_6/f complex. Cyt f and Rieske FeS appear to share the oxidizing equivalent (WHITMARSH et al. 1982) but for simplicity we can consider the FeS cluster to be fully oxidized and cyt f fully reduced. Q_zH_2 responds by reducing the Rieske iron sulfur center and the resulting

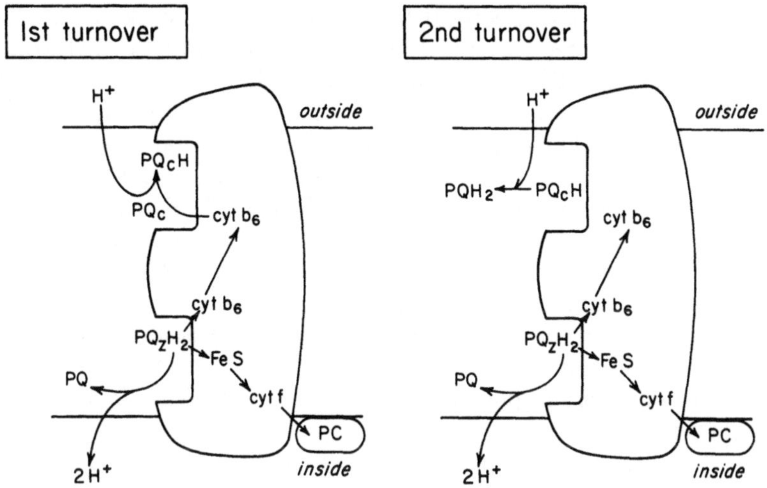

Fig. 7. The Q cycle model for the oxidation of plastoquinol by the cytochrome b_6/f complex. The Q cycle model proposes a binding site for plastoquinol on the b_6/f complex near the inner surface of the thylakoid membrane (Q_z site) and a binding site for plastoquinone near the outer surface (Q_c site). The net result of the first turnover of the cytochrome b_6/f complex is the oxidation of one molecule of quinol at the Q_z site with the attendant release of $2H^+$ to the inside of the thylakoid vesicle and the one electron reduction of quinone at the Q_c accompanied by H^+ uptake from the outer phase to form a semiquinone (PQ_cH). A second turnover of the same complex again fully oxidizes quinol to quinone at the Q_z site but the reaction at the Q_c site is the reduction of semiquinone to quinol ($PQ_cH \rightarrow PQH_2$). Thus overall, the Q cycle oxidizes two plastoquinols, reduces one plastoquinone and thereby translocates $4H^+$ for every $2e^-$ reaching P^+-700. The other salient feature of the Q cycle model is that it involves an electrically uncompensated movement of an electron across the membrane's dielectric during the cyt b-mediated electron transfer from Q_z to Q_c

Q_z semiquinone (Q_zH), a species expected to be substantially more reducing than Q_zH_2, is capable of reducing the nearby low potential cyt b_6. An electrically uncompensated electron is moved back across the membrane's dielectric as the low potential cyt b_6 reduces the high potential cyt b_6 and in so doing, creates a membrane potential. The last event in the sequence of reactions is the reduction of plastoquinone at the Q_c site by cyt b_6 and protonation with a hydrogen ion from the outer phase. Thus the net result of one turnover of the oxidoreductase is the complete oxidation of one molecule of quinol with the release to the inside of the vesicle of $2H^+$ and the reduction to the level of semiquinone of one plastoquinone molecule with the attendant uptake of an H^+ from the outer phase. A second turnover of the same complex results in an identical series of events, except that the reaction at the Q_c site is the reduction of semiquinone to quinol, causing it to dissociate from the site thereafter to be replaced by a plastoquinone from the membrane's pool. The reactions for two consecutive turnovers are represented schematically in Fig. 7.

The Q-cycle model for quinol oxidation differs from earlier linear models in two experimentally verifiable ways. (1) The linear model does not predict the generation of an electrical field, whereas the Q-cycle model does. (2) The Q-cycle achieves the movement of four protons across the thylakoid membrane for the net oxidation of one plastoquinol (i.e., 2 plastoquinols oxidized and 1 plastoquinone reduced), whereas a linear scheme only achieves the movement of two protons. Under particular sets of experimental conditions extra proton translocation beyond that expected for a linear scheme has been clearly demonstrated (VÉLTHUYS 1978; GRAAN and ORT 1983; RATHENOW and RUMBERG 1980). Demonstration of an electrogenic (i.e., electric field generating) reaction occurring between the photosystems was in fact the initial incentive for the proposal of a Q-cycle in chloroplasts (JOLIOT and DELSOME 1974). This latter feature is evident not only from the existence of a slow phase of the electrochromic band shift (JOLIOT and DELSOME 1974) but also from the slowing effect of a membrane potential on the turnover of the b_6/f complex (GRAAN and ORT 1983).

There are, however, many anomalies that are not possible to reconcile within the simple model described above, and many of these discordant observations are discussed in this volume by JOLIOT and JOLIOT (Chap. 10.3), by PRINCE (Chap. 10.4) and by RICH (1984).

The evidence is overwhelming that some form of a Q-cycle (or alternatively b-cycle, see Chap. 10.3, this Vol.) is operative in thylakoid membranes. Yet whether a Q-cycle always operates in steady-state electron flow is unresolved. It does not, for example, appear that extra proton translocation beyond that expected for a linear scheme occurs during steady-state electron transfer at high illumination levels, nor is there evidence for an intersystem electrogenic event under these conditions. Several investigators have supported the suggestion that the extra proton pumping associated with the electrogenic arm of the Q-cycle may be energetically feasible only when the transmembrane electrochemical potential is small (BOUGES-BOCQUET 1981, GRAAN and ORT 1983). The lack of extra proton pumping and of an intersystem electrogenic reaction during steady-state electron transfer could be accommodated in a mechanism where, at higher transmembrane potentials, a proton is carried outward with the electron destined for higher potential cytochrome b_6. The reaction would no longer be electrogenic and since no net (extra) proton translocation would occur, there would be little energy required. However, the data available at the present time can equally well justify alternative views.

5.1.3 Involvement of the Cytochrome b_6/f Complex in Cyclic Electron Flow Around Photosystem I

It is now clearly established that reduced ferredoxin produced by photosystem I has the alternate fate either for reducing NADP via ferredoxin-NADP oxidoreductase or of participating in cyclic electron transfer involving the cytochrome b_6/f complex.

Detection of the slow phase of the electrochromic band-shift during cyclic electron flow in intact chloroplasts (CROWTHER and HIND 1980) implies the pathway of electron flow through the b_6/f complex is identical whether the electrons originate from PS II or PS I via ferredoxin. The site at which the

PS I-reducing equivalents are returned via soluble ferredoxin to the cyt b_6/f complex is unknown, as is the mechanism of ferredoxin oxidation by the complex. Since the reduction of cyt b_6 by ferredoxin is prevented by DBMIB, an inhibitor of Q_zH_2 oxidation, it is likely that the interaction of ferredoxin with the b_6/f complex involves plastoquinol (CHAIN 1982). The identity of the ferredoxin: plastoquinone oxidoreductase could involve the Q_c site but this alone would not account for the inhibition by a Q_z inhibitor (i.e., DBMIB). Direct interaction with the Q_z site seems to be eliminated by virtue of its internal location which should make it inaccessible to external ferredoxin.

5.2 Mobile Electron Carriers

Although we do not yet understand the benefit of grana stacking and the attendant lateral heterogeneity of electron transport complexes, the stacking does appear to require that chloroplasts have a mechanism which allows reducing equivalents to communicate between distantly located complexes. Electron micrographs show the diameter of spinach grana stacks to be 400 to 500 nm. Assuming the extreme case of the uniform distribution of the b_6/f complex within the plane of the membrane and absolute heterogeneity of the photosystems with PS II centers confined to appressed membrane regions and PS I centers confined to unappressed regions, 100 nm seems to be a minimum average distance over which diffusion must occur (SELAK and WHITMARSH 1984). Of course the absolute heterogeneity of photosystems assumed in obtaining this value for the minimum average distance of diffusion is clearly not true, and this refinement of the model would tend to lessen the diffusion pathway. On the other hand, physical obstacles to diffusion such as membrane-protein complexes and narrow grana disc connections would effectively increase the length of the diffusion pathway. Sufficient information does not exist to evaluate these additional refinements quantitatively but their effects on the 100 nm estimate are antagonistic and would tend to cancel out.

Maximum rates of electron transfer between the photosystems indicates a half-time for turnover during steady-state electron transport of about 2 ms at room temperature. The minimum two-dimensional diffusion coefficient which would allow a component to diffuse 100 nm in this interval is about 10^{-8} cm^2 s^{-1} (SELAK and WHITMARSH 1984). In spinach thylakoid membranes there is very nearly an equal abundance of the three electron transport complexes (see Sect. 2.2.1.2 and Chap. 10.2, by WHITMARSH this Vol. for details). Even the b_6/f complex, the smallest integral membrane protein complex involved in electron transfer, would be expected to have a diffusion coefficient (HACKENBROCK 1981) one to two orders of magnitude too small to account for maximum observed rates of electron transfer between the photosystems. This leaves plastoquinone and plastocyanin as candidates for mobile carriers involved in lateral communication of reducing equivalents between the photosystems.

The communication of reducing equivalents between PS II centers and b_6/f complexes is almost certainly entirely by diffusion of plastoquinol, since hydrogen atom transfer from quinol to quinone within the membrane's quinone pool is not expected to occur at significant rates (CROFTS and WRAIGHT 1983). The lateral mobility of plastoquinone/quinol in the thylakoid membrane has not been measured directly and has often been assumed to have a diffusion coefficient similar to that determined for diacylglycerolipids, that is about 10^{-8} cm^2 s^{-1}. This value almost certainly overestimates the diffusivity of native quinones since the long rigid isoprenoid chain should retard the motion of quinones relative to diacylglycerolipids (WRAIGHT 1979). A diffusion coefficient in the range of 2 to 5×10^{-9} cm^2 s^{-1} was estimated for ubiquinone-10 at room temperature in mouse liver megamitochondria based on measurements of Fluorescence Recovery After Photobleaching using a fluorescent ubiquinone analog (GUPTE et al. 1984). Thus it is likely that plastoquinone in the thylakoid membrane has a diffusion coefficient substantially smaller than those characteristic of diacylglycerolipids and similar to that of ubiquinone in mitochondrial membranes. The somewhat low lateral mobility expected for plastoquinone is partially mitigated by a titer of about six readily photoreducible plastoquinone molecules for each PS II center and cytochrome b_6/f complex present in the membrane (i.e., 6:1:1 [GRAAN and ORT 1984]). The maximum flux due to diffusion of reducing equivalents between the PS II center and the b_6/f complex during steady-state electron transfer will depend upon the numbers of plastoquinol (from PS II to b_6/f) and plastoquinone (from b_6/f to PS II) molecules participating. Even so, the mobility of plastoquinone/quinol alone may not be adequate to carry out the long-distance transport of reducing equivalents necessary to account for the measured rates of steady-state electron transfer between the photosystems.

The small hydrophobic metalloprotein plastocyanin ($M_r = 10.5$ kDa, for higher plants) couples the redox activity of the b_6/f complex with the photochemical charge separation of the PS I center. In spinach thylakoid membrane vesicles there are two molecules of plastocyanin for each b_6/f complex and PS I center [i.e., 2:1:1 [GRAAN und ORT 1984)]. Following a period of some confusion, it has now become clear that plastocyanin is confined to the very narrow luminal space of the thylakoid membrane vesicle. The luminal space may be so narrow, in fact, that the distance separating the opposing membranes is very nearly the same as the largest dimension of the plastocyanin molecule, about 4 nm. From the very high surface area to enclosed volume ratio (FLORES et al. 1983) and the impeding effect that surface interactions have on water mobility, it is easy to understand why the viscosity of the internal aqueous phase is much higher than that of bulk water (BERG et al. 1979). Thus, the diffusion coefficient of plastocyanin in situ must be very much less than the value of about 10^{-6} cm^2 s^{-1} calculated from the Stokes-Einstein relationship for protein of this size in bulk water at room temperature. In fact, one extreme view is that plastocyanin is all but immobilized by further narrowing of the luminal space that occurs in response to illumination (HAEHNEL 1984). However, there are various sorts of experimental evidence to support the mobility of plastocyanin during electron transfer and there does not at this time appear to be any compelling reason to believe that the diffusion coefficient of PC at room temperature is not as great as 10^{-8} cm^2 s^{-1} (SELAK and WHITMARSH 1984). WHITMARSH, in Chapter

10.2, this Volume, devoted to mobile electron carriers in thylakoid membranes, develops more fully the concepts introduced here.

This discussion of the lateral mobilities of plastoquinone and plastocyanin might lead to the false impression that light-saturated photosynthetic electron transfer operates perilously near diffusion limitation. The fact that under conditions of maximum coupled electron transfer turnover the plastoquinone pool is nearly fully reduced and the plastocyanin pool nearly fully oxidized demonstrates that the lateral mobility of plastoquinone and plastocyanin is seldom if ever a significant contributing factor to limiting the rate of steady-state electron transfer. These data were long ago interpreted to show that plastoquinol oxidation is the dominant rate-determining step in electron transfer. Diffusion limitation would require that the offending carrier pool be roughly half oxidized and half reduced, since steady-state operation requires diffusion in both directions. Thus it seems inescapable that steady-state electron transfer is never diffusion-limited at room temperature and, based on the best estimates for the mobilities of plastoquinone and plastocyanin, there is no reason to expect it to be so.

6 Photophosphorylation

6.1 Mechanism and Energetic Considerations

The initially proposed interpretation of the coupling between photosynthetic electron transfer and the synthesis of ATP were patterned after the then seemingly analogous reactions in substrate level phosphorylations. Since that time the quite different "chemiosmotic" hypothesis has gained considerable experimental support and is now the main working hypothesis for the interpretation of experimental results related to photosynthetic and oxidative ATP synthesis (reviewed by ORT and MELANDRI 1982).

The energy-transducing membrane plays two essential roles in the chemiosmotic model for coupling. It serves both to orient the electron transfer complexes and ATP-synthesizing enyzme complex and to function as a permeability barrier between the inside and the outside of the enclosed vesicle. The protein complexes involved in the proton-coupled transfer reactions are envisioned as being oriented in the membrane so that hydrogen carriers (e.g., quinols) alternate with electron carriers (e.g., cytochromes). In this way a portion of the redox free energy can be captured when the oxidation of a hydrogen carrier by an electron carrier leads to the vectorial release of a proton to the inside of the thylakoid vesicle. This results in an increase in the hydrogen ion activity in the inside aqueous compartment. Similarly, the reduction of a hydrogen carrier by an electron carrier on the opposite side of the membrane causes an alkalization of that aqueous compartment. When accompanied by the charge-compensating movements of other ions across the thylakoid membrane, these proton-coupled redox reactions will result in a pH difference between the inner and outer aqueous phases. The coupling factor complex catalyzes reactions which tie the thermodynamically favorable movement of protons across the membrane via the coupling factor to the phosphorylation of ADP. This is a reversible reaction near equilibrium so the enzyme can either synthesize or hydrolyze ATP and in so doing either subtract from or add to the transmembrane pH difference.

A more complete description of the coupling process includes consideration of the role of the membrane potential in ATP synthesis. To the extent that proton movements are not charge-compensated by other ion movements, an electric potential will be created across the dielectric of the thylakoid membrane. In the case of inwardly directed proton movement resulting from electron transfer or ATP hydrolysis, the positive potential should contribute to the overall transmembrane free energy of the hydrogen ion in solution. Thus the total transmembrane free energy includes both a concentration term and an electrical term. The sum of these two terms constitutes the total protonmotive force. The protonmotive force (Δp) quantitatively expressed in the units of volts is as follow

$$\Delta p = \Delta\psi - 2.3\frac{RT}{F}\Delta pH.$$

When the energy coupling system has attained the state where the rates of ATP synthesis and hydrolysis are equal, this equilibrium will be such that the free energy of the proton-coupled redox reactions, the free energy of hydrolysis of ATP and the protonmotive force are equivalent. If the transport of n protons back across the membrane is coupled to the synthesis of one molecule of ATP, then in the units of kJ,

$$\Delta G_{ATP} = nF\,\Delta p,$$

where

$$\Delta G_{ATP} = \Delta G^{\circ}_{ATP} + RT\ln\frac{[ADP][P_i]}{[ATP]}.$$

The major principle of chemiosmotic coupling, that is the essential role of transmembrane proton activity differences, has been verified (Jagendorf 1975). In addition, there can be little doubt that the arrangement of redox components in photosynthetic membranes and the mechanism of their interaction allow for the direct transformation of redox energy into a transmembrane electrochemical potential difference of hydrogen ions. The equivalence, both kinetic and energetic, of the electrical and chemical components of the transmembrane proton activity potential in phosphorylating ADP seems well established. There is an enzyme complex present in photosynthetic and respiratory membranes that can transform the energy of the electrochemical potential of hydrogen ions into the phosphate group transfer potential of ATP. In recent years attention has turned to obtaining a more complete and detailed understanding of the molecular mechanism of the chemiosmotic energy coupling process in thylakoid membranes. In particular, understanding the relationship of the highly integrated individual events of the overall energy coupling process to the structure of the thylakoid membrane is among the highest research priorities. It must be admitted that an understanding of the mechanism of coupling of ATP formation with ion transport still eludes us.

6.2 "Phosphorylating Unit"

From the perspective of the relationship of function to structure, the process of energy transduction represents a significant departure from the photosynthetic processes so far discussed in this chapter. There is no single complex of polypeptides responsible for the transduction of redox energy into the group transfer potential of ATP. This discussion will consider the structural requirements of energy transduction as the "phosphorylating unit" of thylakoid membranes. While the actual identity of the "phosphorylating unit" in terms of numbers of components and surface area of the membrane involved remains uncertain, it can nevertheless be defined precisely as the minimum number of components which must act together to couple light-driven electron transfer to ATP formation.

It is no longer in question that a transmembrane electrochemical potential difference of H^+ is an obligate intermediate in the energy-coupling process between electron transfer and ATP formation. The asymmetry of proton activity across the thylakoid membrane is created by water oxidation activity of the PS II complex and the plastoquinol oxidase activity of the cytochrome b_6/f complex. Thus the minimum membership of the thylakoid membrane phosphorylating unit includes at least one of these electron transport complexes that generates the transmembrane proton asymmetry, a thylakoid membrane vesicle that resists proton leakage, and an ATP synthetase that can utilize the energy stored in the electrochemical potential to phosphorylate ADP.

6.2.1 Distribution and Stoichiometry of Components of the Phosphorylating Unit

Two features of thylakoid membrane architecture and composition eliminate the possibility of intimate physical association of the ATP synthetase (i.e., coupling factor complex) with each of the proton-producing redox reactions. The first of these two features is the lateral heterogeneity of the coupling factor complexes which are confined to nonappressed regions of the thylakoid membrane (MILLER and STAEHELIN 1976; ARNTZEN et al. 1969). This feature prevents close physical proximity of the coupling factor complex with PS II complexes and about one half of the cyt b_6/f complexes, since these are located in appressed regions of the membrane. The second feature is that there is present within the membrane only about a quarter as many coupling factors as there are proton-producing electron transport complexes. The published quantitations of the coupling factor content of spinach thylakoid membranes (STROTMANN et al. 1973; FRASH et al. 1980) require adjustment to correct for the previous underestimation of the M_r of the peripheral protein portion (i.e., CF_1) of the coupling factor complex (MORONEY et al. 1983; SÜSS and SCHMIDT 1982; McCARTY and NALIN, Chap. 10.8, this Vol.). The appropriate correction reveals that coupling factor is present in spinach thylakoid membranes in the amount of about 1 complex per 1100 chlorophyll molecules or about 1 CF_1:2 PS II centers:2 cyt b_6/f complexes.

Thus it seems quite evident that the coupling factor complex is remote from a large portion of the electron transport components that provide it with the energy for ATP synthesis. Analogous to the necessity for a diffusible carrier of reducing equivalents between distantly located redox complexes, remote coupling between the redox reactions and the ATP synthetase will require a diffusible intermediate as well.

6.2.2 Localized vs. Delocalized Coupling

The lateral separation in the membrane of the proteinaceous components participating in energy transduction is the harbinger of phosphorylating units that contain numerous electron transfer complexes that cooperate to energize ATP synthesis. Indeed, recent experiments demonstrate that at least 1000 proton-coupled redox components cooperate to produce a transmembrane chemical potential of accumulated protons sufficiently large to cause the net synthesis of ATP (HANGARTER and ORT 1985).

The extensive cooperation of electron transfer components in achieving the energetic requirement for the onset of ATP formation can be readily accommodated within the original proposal for entirely delocalized coupling. That is, the membrane's complement of proton-coupled redox components act together to acidify the internal aqueous phase of the thylakoid vesicle and the entire population of coupling factor complexes are equally accessible to accumulated hydrogen ions. Furthermore, delocalized coupling involves the rapid electrostatic and chemical equilibration of hydrogen ions produced by the proton-coupled redox reactions at the membrane surfaces into the surrounding bulk water phases. There seems little doubt that the current of hydrogen ions coupling the electron transfer reactions of the redox components to the ATP-synthesizing reactions of the coupling factor *can* be in equilibrium with the surrounding bulk water phases. For instance, experiments patterned after the classical work in Jagendorf's laboratory from the mid-1960's (reviewed by JAGENDORF 1975; ORT and MELANDRI 1982) demonstrated that light-driven electron transfer can indeed acidify the inner aqueous phase of thylakoid vesicles and the vesicles, once acid, can then be drawn on to produce ATP without further redox activity.

Despite the success of a delocalized coupling model in explaining many aspects of energy transduction by thylakoid membranes, delocalized coupling is not an entirely satisfying description of the normal energy translocation process carried out by thylakoid membranes. While it is clear that equilibration of protons from membrane interfacial regions into the surrounding bulk phases will eventually occur (as when the proton-coupled redox activity is deliberately separated in time from ATP synthesis as mentioned above), is is important to understand that this does not ensure that the acidification of the bulk inner phase is a necessary intermediate in phosphorylation. In other words, it does not ensure that the inner bulk phase plays an essential role in the communication of the sites of H^+ generation with those phosphorylating sites which use the H^+. There is experimental evidence from several different laboratories, indicating that the electrochemical potential of those protons involved in coupling need not be in equilibrium with the protons in the surrounding bulk water phase (ORT and MELANDRI 1982 see also MELANDRI and VENTURALI, Chap. 10.6; DILLEY, Chap. 10.7, both this Vol.). Perhaps the most intuitively obvious experimental demonstration of this point has come from an investigation of the effect that stabilizing the pH of internal aqueous phase of thylakoid vesicles has on the amount of proton-coupled electron transport necessary to achieve the minimum energetic state that is required for net formation of ATP (ORT et al. 1976; HORNER and MOUDRIANAKIS 1983; HANGARTER and ORT 1985). It was reasoned that if hydrogen ions produced by the redox reactions had to move through the inner aqueous bulk phase of the thylakoid vesicle, then the addition of alkaline pKa buffer to that phase must delay the formation of a ΔpH. In the absence of an electric potential, the delay in ΔpH formation should in turn delay the onset of ATP formation for a predictable number of additional turnovers of the proton-coupled redox reactions. Under conditions in which no transmembrane electric potential can form, the onset of ATP formation must await the titration of the endogenous alkaline buffering groups on the inner surface of the membrane. The size of this endogenous threshold buffering pool is about 60 mmol H^+ mol^{-1} chl in spinach thylakoid vesicles and about 15 turnovers of the electron transport components are required to titrate this pool (HANGARTER and ORT 1985). That is, in the absence of a membrane potential a minimum of 15 saturating single turnover flashes are necessary to initiate ATP formation. Introducing five times as many buffering groups than this into the inner aqueous phase caused no increase in the number of saturating flashes necessary for the onset of ATP formation. This observation indicates that the pathway of protons from sites of production at redox reactions to sites of utilization at coupling factor complexes does not include the whole of the inner aqueous phase.

To summarize, chloroplasts behave in a seemingly contradictory fashion; a large number (i.e., greater than 1000) of proton-coupled redox reactions cooperate to titrate the endogenous buffer on the inner surface of the membrane and thereby achieve the energetic threshold for ATP formation. However, the proton activity difference so formed seems not to be in equilibrium with the surrounding aqueous phases. Reconciliation may be found in the special properties of the water-membrane interface which promote local disequilibria (KELL 1979). There can be little question that water molecules adjacent to the membrane's surface are less randomly oriented than in the bulk. It has been suggested that this situation could create conditions in which the oriented water layer on the surface of the membrane would offer a kinetically preferred pathway for proton currents. However, this suggestion has its own failings in that it does not explain other experimental observations. Some of these failings are covered in chapters in this Volume by MELANDRI and VENTURALI (Chap. 10.6), by DILLEY (Chap. 10.7), and by ORT and MELANDRI (1982).

It is my predilection to believe that the protonmotive force that couples electron transfer to ATP formation in thylakoid membrane vesicles exhibits varying degrees of delocalization depending on the extent of energization. For instance, when energy input into the system is limited, as in low light or during the initial turnovers of electron transfer, it is essential to minimize dissipative losses including those losses which would result if protons were allowed to equilibrate with the surrounding bulk phases. Avoidance of such losses by directed diffusion might result from the generation of a strong "sink" for protons at a coupling factor complex that in turn results in a gradient in proton activity between the coupling factor and the nearest proton-generating redox complexes. Since there are no specialized proton conduits involved in this proposal for directed diffusion it should be viewed as a statistical process where an individual proton has a finite probability of equilibrating with surrounding bulk phase. Thus, acidification of the lumen of the thylakoid vesicle will gradually occur and with it a disappearance of the energy gap between any localized proton coupling currents and the bulk phase. Under conditions of high energy input the entire thylakoid vesicle may approach the extreme case of being isopotential for protons, but this extreme may never be fully attained.

6.3 Coupling Factor Complex

6.3.1 Structure

The coupling of the energy-requiring group transfer reaction of ADP phosphorylation with free energy available in the transmembrane electrochemical potential of accumulated protons is catalyzed by a multimeric protein complex known as the coupling factor (MCCARTY and CARMELI 1982). The coupling factor complex consists of a multisubunit peripheral membrane protein, CF_1, which is associated via ionic interactions with a multisubunit integral membrane protein, CF_0. CF_1 is present on the outer surface of nonappressed thylakoid membranes and contains the active sites for ATP synthesis. The CF_0 portion of the complex spans the membrane and acts to orient CF_1. It also acts as a proton conduit between the thylakoid vesicle lumen and the externally located CF_1.

The externally exposed portion of the coupling factor complex (i.e. CF_1) was the first thylakoid membrane protein complex to be identified and localized by electron microscopy. The CF_1 complex appears as a knob-like structure protruding from the stromal surface of nonappressed thylakoid membranes. The complex appears as a 13 to 14 nm structure in electron micrographs (STAEHELIN, Chap. 1, this Vol.).

Following a lengthy period in which the polypeptide subunit stoichiometry of the chloroplast enzyme was thought to be distinct from all other types of coupling factor complex (such as those from mitochondria or bacteria), it is now clear that the subunit stoichiometry of CF_1 closely mimics that of mitochondrial and bacterial F_1. The molecular weight of CF_1 is now agreed to be near 400,000 and the stoichiometry of its five polypeptide subunits of the $\alpha_3\beta_3\gamma\delta\varepsilon$. The individual molecular weights of the subunits of CF_1 are recorded in Table 1. The organization of these peptides into the functional complex is reviewed by McCarty and Nalin (Chap. 10.8, this Vol.).

The CF_0 portion of the coupling factor complex is embedded into the membrane and has yet to be identified in freeze-fracture replicas of thylakoid membranes. However, the CF_0 appears as a 9 to 10 nm intramembrane particle in freeze-fracture replicas of liposomes following incorporation of purified CF_1/CF_0 complexes (Mörschel and Staehelin 1983).

The subunit composition of CF_0 is not nearly so well defined as that of CF_1. The weight of available evidence suggests that three different polypeptides are involved in forming the complex (McCarty and Carmeli 1982; Nelson 1982). These have been designated I, II, and III and it appears that the smallest, subunit III, may be present in an abundance of six copies per CF_1. The estimated M_r values of these three highly hydrophobic polypeptides are compiled in Table 1.

6.3.2 Regulation of Catalytic Activity

The diurnal cycling of the photosynthetic energy substrate, light, presents chloroplasts with a situation not encountered by respiratory ATP-producing systems that have a more or less continuous supply of reduced carbon compounds. The activity of the proton ATPase of respiratory membranes is overt and the continuous maintenance of a protonmotive force is relied upon to maintain an equilibrium in the direction of net ATP synthesis. The situation is similar for many photosynthetic bacteria, since oxidative and photosynthetic redox reactions occur in the same membranes, sharing certain electron carriers and the coupling factor (Baccarini Melandri and Zannoni 1978). The consequence of overt activity in the chloroplast ATP synthetase/hydrolase enzyme would be complete draining of organelle ATP pools during dark periods when hydrolase activity cannot be held in check by the maintenance of a Δp. While the energetic consequences of this might be negligible in comparison to day-long energy storage, maintenance of optimal substrate pools for reinitiation of photosynthesis in the next light period probably depends critically on maintenance of ATP levels in the dark. Under normal circumstances chloroplasts contain about 1 to 1.5 mM total nucleotides and more than 25% remains in the form of ATP even after prolonged dark periods (Heber et al. 1982). The chloroplast coupling factor enzyme requires the maintenance of a Δp to remain in its active form (McCarty and Nalin, Chap. 10.8, this Vol.). Thus as the force needed to promote net ATP formation decays, so does the activity of the coupling factor enzyme. Of course, enzymes do not alter equilibrium constants and must catalyze reactions in both the forward and reverse directions, ensuring that the inactive state of the enzyme is inactive for both synthetase and hydrolase activity. This means that the inactivation of the catalytic function of the coupling factor by relaxation of the Δp must be reversed by the restoration of the Δp.

In addition to the transition of the enzyme complex between active and inactive states in response to membrane energization and the effect that thermodynamic factors have on the relative rates of the forward and reverse reactions, there are a number of other factors that conspire to regulate the activity of this complex enzyme. The sulfhydryl compound thioredoxin, which is present in the stroma and is reduced by photosystem I, has been shown to be involved in a secondary fashion in light activation of the coupling factor complex. The reduction of a disulfide bond on the γ subunit by light-generated reduced thioredoxin appears to stabilize that active state of the enzyme initially produced by membrane energization (NALIN and MCCARTY 1984).

There is an excellent correlation between the light-induced activation of the enzyme and the release of ADP from CF_1. The rebinding of ADP to CF_1 in the dark correlates closely with the relaxation of the enzyme to its inactive form. It is believed that the protein conformational change accompanying the transition of this enzyme between active and inactive states causes the release and subsequent rebinding of ADP (STROTMANN, Chap. 10.9, this Vol.). However, depletion of ADP alone does not induce activation.

There have been many data (see MCCARTY and NALIN, Chapter 10.8), particularly with isolated CF_1, implicating the involvement of the ε subunit in regulating the activity of the enzymes. For instance, activation of ATPase activity in isolated CF_1 by exposure to heat, alcohols, and certain detergents clearly involves either the removal or displacement of the ε subunit. Moreover, activation by ε subunit depletion can be reversed by reconsitution with purified ε subunit. While reorientation of the ε subunit within the complex may well be an element of the physiological regulation, its actual removal from the complex most certainly is not.

A complete cataloging of the allosteric regulators is well beyond the scope of the discussion here (see MCCARTY and CARMELI 1982). ATPase activity responds to anions including carboxylic acids, bicarbonate, and phosphate, as well as cations including Mg^{2+}, Mn^{2+} and Ca^{2+}. A great deal more must be understood about this enzyme and its mechanism before the interactions of these many factors will be understood.

A summary of chloroplast coupling factor regulation, admittedly oversimplified, includes a form of the enzyme that is inactive in the dark and thus is unable to deplete stromal ATP pools. Activation of the enzyme must preceed ATP synthesis. It is the consequence of Δp-induced conformational changes in CF_1, most probably accompanied by reorientation of the ε subunit and by release of bound ADP. Thioredoxin, reduced by PS I, stabilizes the active state enzyme by reduction of a disulfide bond on the γ subunit. Existence of the Δp that activated the enzyme exerts direct thermodynamic control over ATP hydrolysis; it acts to drive the reaction toward ATP synthesis. When illumination ceases, each of these processes eventually reverses but the dominant feature of the deactivation process may be a binding of ADP to CF_1 that occurs as the Δp dissipates.

6.3.3 Catalytic Mechanisms

The chemical reaction at the center of photophosphorylation is the transfer of a phosphoryl group from the donor molecule water to the acceptor molecule

ADP. As such, the bridge oxygen of the newly formed anhydride bond is that of the ADP molecule (SLATER 1953). Chemically, the hydrolysis of ATP is the reverse of this group transfer reaction, this time ATP being the group donor and water the group acceptor, in photosynthetic or oxidative ATP synthesis the enzymatic mechanism of coupling factors must utilize the negative free energy available in a transmembrane electrochemical potential of hydrogen ions to meet the positive free energy requirements of ADP phosphorylation. While the thermodynamic feasibility of coupling the energetically favorable efflux of protons from the thylakoid vesicle to the energetically unfavorable phosphorylation of ADP is clear, only the initial clues to the enzymatic mechanism of this reaction are known at present. The central issue is how the enzyme couples the chemical and electrical potential of the accumulated hydrogen ions with the making and breaking of covalent bonds.

There is at least one proposal (MITCHELL 1974) for the participation of the exiting protons directly in the chemistry of ADP phosphorylation. In this proposal, hydrogen ions leaving the membrane vesicle through the coupling factor complex protonate a bound phosphate to produce a positively charged phosphorous center at the active site of the enzyme. A subsequent nucleophilic attack of an ADP anion displaces $-O^{2-}$ from phosphate forming ATP. The unsuitability of $-O^{2-}$ as a leaving group leads to the further suggestion that two successive protonations of a phosphate ligand to Mg^{2+} produced $-OH_2$ as the ultimate leaving group. Modifications of this general proposal have been suggested (e.g., BOYER 1975). While this innovative proposal has numerous attractive features, there exists little experimental support for it.

An indirect involvement of the exiting protons in ATP synthesis is currently a much more popular alternative. Most notions about an indirect participation of protons in the mechanism of ATP formation have as a central feature energy-induced conformational changes in the coupling factor enzyme, changes that produce a dramatic decrease in the equilibrium constant for the phosphate group transfer reaction. The experimental data underlying the proposal for indirect proton involvement in the catalytic mechanism of ATP synthesis are composed of an exceptionally diverse and complex array of observations. Reviews in this Volume by STROTMANN (Chap. 10.9) and by MCCARTY and NALIN (Chap. 10.8) and elsewhere by MERCHANT and SELMAN (1985) and MCCARTY and MORONEY (1984) do an excellent job in deciphering and culling these observations to produce a coherent, although still incomplete, picture.

Currently the most useful hypothesis for the interpretation of experimental results pertaining to the enzymatic mechanism of coupling factor has evolved from initial proposals by BOYER (1977) and by SLATER (1977). They proposed that tightly bound ADP and P_i spontaneously form tightly bound ATP and that the energy of exiting protons is necessary to cause a conformational change in the enzyme that in turn causes the release of the tightly bound product. BOYER has continued to refine and develop the idea. One of the most salient embellishments accounts for experimentally observed cooperative interactions among catalytic sites of the enzyme complex. BOYER's model has come to be known alternatively as the "alternating site hypothesis" for the "energy-linked binding change mechanism". The model takes into account the $\alpha_3\beta_3$ subunit stoichiometry and views the enzyme as a symmetrical trimeric structure containing three cooperatively functioning catalytic sites (BOYER and KOHLBRENNER 1981). The binding constants at each site for substrates and products change in response to the energy-dependent conformational change of the enzyme as mentioned above. These changes are envisioned as occurring sequentially. The first step is the loose binding of ADP and P_i to a catalytic site which is followed in turn by conversion to tightly bound substrates and the formation of ATP by removal of H_2O. The final, third step in this catalytic "cycle" is the energy-dependent transition of tightly bound product to freely dissociable ATP. Completing its cycle, the site is now again in the appropriate state to bind ADP and P_i. The cooperativity of catalytic sites is accounted for in the proposal, in that a single energy-dependent conformational change is envisioned to alter the binding characteristics of all three sites simultaneously but in different ways. The unexpected feature of the proposal is that the anhydride bond formation proceeds without the input of energy, occurring instead as the result of a much lowered equilibrium constant of the reaction due to the environment at the active site.

7 Concluding Remarks

The aim of this chapter has been to deal with structural aspects of the constituents of the thylakoid membrane and at the same time with the mechanisms of the processes which, taken together, constitute oxygenic photosynthesis. Where possible, structure and mechanism have been presented together to emphasize their interdependence, but often the best which could be accomplished in light of the present inadequacies of information was a parallel but largely separate treatment.

I am encouraged to believe that we have entered a time of rapid growth in information regarding the molecular structure of the various functional components of photosynthetic membranes. This notion is most dramatically supported by the very recently reported X-ray crystallographic structure of the reaction center from the photosynthetic bacteria *R. viridis* (MICHEL and DEISENHOFER, Chap. 8.4, this Vol.). These molecular level structural data come at a time when they will contribute directly and immediately to the understanding of the mechanism of primary events of photosynthesis. Equally important, the structural information will play a crucial role in determining future directions of research into the primary photochemical and electron transfer processes. Owing to the parallelism between the PS II reaction center of plants and the reaction center of nonoxygenic photosynthetic bacteria, conceptual advances from the study of *R. viridis* reaction center will almost certainly contribute directly to the understanding of PS II.

X-ray diffraction analysis of crystallized membrane proteins is but one of several techniques making contributions to our understanding the molecular structure of photosynthetic membrane functional components. The enormous recent progress in isolating and sequencing genes coding for chloroplast thylakoid polypeptides has permitted estimation of their two-dimensional folding patterns based on hydropathy analysis. For instance, hydropathy plots of the individual polypeptide subunits of the cytochrome b_6/f complex produced information about the probable location of the 3 hemes and of the Fe_2S_2 center. This information is being taken into account in the construction of current schemes for electron transfer through the complex. Neutron scattering, electron scattering and electron microscopy, immunological techniques, resonance Raman spectroscopy, linear and circular dichroism, NMR, and EPR are all contributing to the accumulating bank of information about molecular structure.

It appears to me that studies concerning mechanisms and studies concerning the structure of photosynthetic membrane components are beginning to converge in a meaningful and synergistic fashion. As a consequence, the prospects for fundamental advances in our understanding of photosynthesis have never been more promising.

Acknowledgements. I am very grateful to Drs. N.E. Good, J.S. Boyer, R. Blankenship, C.F. Yocum and J. Whitmarsh for discussion, comments and critical reading of the chapter without which the number of errors would have been much greater.

References

Allred DR, Staehelin LA (1985) Intrathylakoidal distribution of cytochrome b_6/f. Plant Physiol, in press

Alt J, Westhoff P, Sears BB, Nelson N, Hurt E, Hauska G, Herrmann RG (1983) Genes and transcripts for the polypeptides of the cytochrome b_6/f complex from spinach thylakoid membranes. EMBO J 2:979–986

Anderson JM (1982) The significance of grana stacking in chlorophyll b-containing chloroplasts. Photobiochem Photobiophys 3:225–241

Andersson B, Haehnel W (1982) Location of photosystem I and photosystem II reaction centers in different thylakoid regions of stacked chloroplasts. FEBS Lett 146:13–17

Arntzen CJ (1978) Dynamic structural features of chloroplast lamellae. In: Vernon L, Sanadi R (eds) Current topics in bioenergetics, vol VIII. Academic Press, London New York, pp 111–160

Arntzen CJ, Dilley RA, Crane FL (1969) A comparison of chloroplast membrane surfaces visualized by freeze-etch and negative-staining techniques; and ultrastructural characterization of membrane fractions obtained from digitonin-treated spinach chloroplasts. J Cell Biol 43:16–31

Babcock GT, Ghanotakis DF, Ke B, Diner BA (1983) Electron donation to photosystem II in reaction center preparations. Biochim Biophys Acta 723:276–286

Baccarini Melandri A, Zannoni D (1978) Photosynthetic and respiratory electron flow in the dual function membrane of facultative photosynthetic bacteria. J Bioenerg Biomembr 10:109–138

Barber J (1972) Stimulation of millisecond delayed light emission by KCl and NaCl gradients as a means of investigating the ionic permeability properties of the thylakoid membranes. Biochim Biophys Acta 275:105–116

Barber J (1980) Membrane surface charges and potentials in relation to photosynthesis. Biochim Biophys Acta 594:253–308

Barber J (1982) Influence of surface charges on thylakoid structure and function. Annu Rev Plant Physiol 33:261–295

Barber J, Mills J, Love A (1977) Electrical diffuse layers and their influence on photosynthetic processes. FEBS Lett 74:174–181
Bar-Nun S, Schantz R, Ohad I (1977) Appearance and composition of chlorophyll-protein complexes I and II during chloroplast membrane biogenesis in *Chlamydomonas reinhardi* y-1. Biochim Biophys Acta 459:451–467
Berg SP, Lusczakoski DM, Morse II PD (1979) Spin label motion in the internal aqueous compartment of spinach thylakoids. Arch Biochem Biophys 194:138–148
Boska M, Sauer K, Buttner W, Babcock GT (1983) Similarity of EPR signal II_f rise and P_{680}^+ decay kinetics in Tris-washed chloroplast photosystem II preparations as a function of pH. Biochim Biophys Acta 722:327–330
Bouges-Bocquet B (1981) Factors regulating the slow electrogenic phase in green algae and higher plants. Biochim Biophys Acta 635:327–340
Boyer PD (1975) Energy transduction and proton translocation by adenosine triphosphatases. FEBS Lett 50:91–94
Boyer PD (1977) Coupling mechanisms in capture, transmission and use of energy. Annu Rev Biochem 46:957–966
Boyer PD, Kohlbrenner WE (1981) The present status of the binding-change mechanism and its relation to ATP formation by chloroplasts. In: Selman BR, Selman-Reimer S (eds) Energy coupling in photosynthesis. Elsevier/North-Holland Biomed Press, Amsterdam New York, pp 231–240
Brangeon J, Mustárdy LA (1979) The ontogenetic assembly of intra-chloroplastic lamellae viewed in 3-dimension. Biol Cell 36:71–80
Brown MF, Miljanich GP, Dratz EA (1977) Interpretation of 100- and 360-MHz proton magnetic resonance spectra of retinal rod outer segment disk membranes. Biochemistry 16:2640–2648
Camm EL, Green BR (1980) Fractionation of thylakoid membranes with the nonionic detergent octyl-β-D-glucopyranoside. Plant Physiol 66:428–432
Chain RK (1982) Evidence for a reductant-dependent oxidation of chloroplast cytochrome b-563. FEBS Lett 143:273–278
Chapman D, Gómez-Fernández JC, Goñi FM (1979) Intrinsic protein-lipid interactions physical and biochemical evidence. FEBS Lett 98:211–223
Chua N-H, Blomberg F (1979) Immunochemical studies of thylakoid membrane polypeptides from spinach and *Chlamydomonas reinhardii*: A modified procedure for crossed immunoelectrophoresis of dodecyl sulfate protein. J Biol Chem 254:215–223
Chua N-H, Matlin K, Bennoun P (1975) A chlorophyll-protein complex lacking in photosystem I mutants of *Chlamydomonas reinhardii*. J Cell Biol 67:361–377
Colman PM, Freeman HC, Guss JM, Murata M, Norris VA, Ramshaw JAM, Venkatappa MP (1978) X-ray crystal structure analysis of plastocyanin at 2.7 Å resolution. Nature 272:319–324
Cox RP, Andersson B (1981) Lateral and transverse organization of cytochromes in the chloroplast thylakoid membrane. Biochem Biophys Res Commun 103:1336–1342
Critchley C (1983) Further studies on the role of chloride in photosynthetic O_2 evolution in higher plants. Biochim Biophys Acta 724:1–5
Critchley C, Baianu IC, Govindjee, Gutowsky HS (1982) The role of chloride in O_2 evolution by thylakoids from salt-tolerant higher plants. Biochim Biophys Acta 682:436–445
Crofts AR, Wraight CA (1983) The electrochemical domain of photosynthesis. Biochim Biophys Acta 726:149–183
Crowther D, Hind G (1980) Partial characterization of cyclic electron transport in intact chloroplasts. Arch Biochem Biophys 204:568–577
Dismukes GC, Siderer Y (1981) Intermediates of a polynuclear manganese center involved in photosynthetic oxidation of water. Proc Natl Acad Sci USA 78:274–278
Dubacq JP, Trémolières A (1983) Occurrence and function of phosphatidylglycerol containing Δ3-trans-hexadecenoic acid in photosynthetic lamellae. Physiol Veg 21:293–312
Duniec JT, Israelachvili JN, Ninham BW, Pashley RM, Thorne SW (1981) An ion-exchange model for thylakoid stacking in chloroplasts. FEBS Lett 129:193–196

Evans MCW (1982) Iron-sulfur centers in photosynthetic electron transport. In: Spiro TG (ed) Iron-sulfur proteins, vol IV. Wiley, New York, pp 249–284
Fajer J, Davies MS, Forman A, Klimov VV, Dolan E, Ke B (1980) Primary electron acceptors in plant photosynthesis. J Am Chem Soc 102:7143–7145
Fato R, Bertoli E, Castelli GP, Lenaz G (1984) Fluidizing effect of endogenous ubiquinone in bovine heart mitochondria. FEBS Lett 172:6–10
Flores S, Graan T, Ort DR (1983) Measurement of the permeability of the chloroplast thylakoid membrane to amine buffers. Photobiochem Photobiophys 6:293–304
Forti G, Cappelletti A, Nobili RL, Garlaschi FM, Gerola PD, Jennings RC (1983) Interaction of ferredoxin and ferredoxin-NADP reductase with thylakoids. Arch Biochem Biophys 221:507–513
Fox GE, Stackebrandt E, Hespell RB, Gibson J, Maniloff J, Dyer TA, Wolfe RS, Balch WE, Tanner RS, Magnum LJ, Zablen LB, Blakemore R, Gupta R, Bonen L, Lewis BJ, Stahl DA, Leuhrsen KR, Chen KN, Woese CR (1980) The phylogeny of prokaryotes. Science 209:457–463
Frash W, Deluca CR, Kulzick MJ, Selman BR (1980) A quantitative estimation of chloroplast thylakoid-bound coupling factor 1 by rocket immunoelectrophoresis. FEBS Lett 122:125–128
Ghanotakis DF, Yocum CF (1985) Polypeptides of photosystem II and their role in oxygen evolution. Photosyn Res in press Vol 6
Ghanotakis DF, Topper JN, Babcock GT, Yocum CF (1984) Water-soluble 17 and 23 kDa polypeptides restore oxygen evolution activity by creating a high-affinity binding site for Ca^{2+} on the oxidizing side of photosystem II. FEBS Lett 170:169–173
Giddings Jr TH, Withers NW, Staehelin LA (1980) Supramolecular structure of stacked and unstacked regions of the photosynthetic membranes of *Prochloron sp.*, a prokaryote. Proc Natl Acad Sci USA 77:352–356
Gounaris K, Barber J (1983) Monogalactosyldiacylglycerol: The most abundant polar lipid in nature. Trends Biochem Sci 8:378–381
Gounaris K, Sundby C, Andersson B, Barber J (1983) Lateral heterogeneity of polar lipids in the thylakoid membranes of spinach chloroplasts. FEBS Lett 156:170–174
Govindjee, Kambara T, Coleman W (1985) Photosystem II and the oxygen-evolving complex. Photochem Photobiol 42:187–210
Graan T, Ort DR (1983) Initial events in the regulation of electron transfer in chloroplasts. The role of the membrane potential. J Biol Chem 258:2831–2836
Graan T, Ort DR (1984) Quantitation of the rapid electron donors to P_{700}, the functional plastoquinone pool, and the ratio of the photosystems in spinach chloroplasts. J Biol Chem 259:14003–14010
Graan T, Ort DR (1985) DBMIB binding sites indicate a dimeric configuration for the cytochrome b_6/f complex in spinach chloroplasts. Submitted
Green BR, Camm EL (1984) Evidence that CP 47 (CPa-1) is the reaction centre of photosystem II. In: Sybesma C (ed) Advances in photosynthesis research, vol II. Nijhoff/Junk, The Hague, pp 95–98
Green BR, Camm EL, van Houten J (1982) The chlorophyll-protein complexes of *Acetabularia*, a novel chlorophyll a/b complex which forms oligomers. Biochim Biophys Acta 681:248–255
Gupta S, Wu E-S, Hoechli L, Hoechli M, Jacobson K, Sowers AE, Hackenbrock CR (1984) Relationship between lateral diffusion, collision frequency, and electron transfer of mitochondrial inner membrane oxidation-reduction components. Proc Natl Acad Sci USA 81:2606–2610
Hackenbrock CR (1981) Lateral diffusion and electron transfer in the mitochondrial inner membrane. Trends Biochem Sci 6:151–154
Haehnel W (1984) On the lateral electron transport between the two light reactions in spinach chloroplasts. In: Sybesma C (ed) Advances in photosynthesis research, vol I. Nijhoff/Junk, The Hague, pp 545–548
Haehnel W, Hesse V, Pröpper A (1980) Electron transfer from plastocyanin to P_{700}. Function of a subunit of photosystem I reaction center. FEBS Lett 111:79–82
Hangarter R, Ort DR (1985) Evidence for extensive cooperation among electron transfer

complexes in achieving the energetic threshold for ATP synthesis. Eur J Biochem 149:503–510
Haworth P, Watson JL, Arntzen CJ (1983) The detection, isolation and characterization of a light-harvesting complex which is specifically associated with photosystem I. Biochim Biophys Acta 724:151–158
Heathcote P, Evans MCW (1980) Properties of the EPR spectrum of the intermediary electron acceptor (A_1) in several different photosystem I particle preparations. FEBS Lett 111:381–385
Heber U, Takahama U, Neimanis S, Shimizu-Takahama M (1982) Transport as the basis of the Kok effect. Levels of some photosynthetic intermediates and activation of light-regulated enzymes during photosynthesis of chloroplasts and green leaf protoplasts. Biochim Biophys Acta 679:287–299
Hill R, Bendall F (1960) Function of the two cytochrome components in chloroplasts: A working hypothesis. Nature 186:136–137
Hind G, Nakatani HY, Izawa S (1969) The role of Cl^- in photosynthesis: I. The Cl^- requirement of electron transport. Biochim Biophys Acta 172:277–289
Hirschberg J, Bleecker A, Kyle DJ, McIntosh L, Arntzen CJ (1984) The molecular basis of triazine-herbicide resistance in higher plant chloroplasts. Z Naturforsch 39c:412–420
Hiyama T, Fork D (1980) Kinetic identification of component X as P430: A primary electron acceptor of photosystem I. Arch Biochem Biophys 199:488–496
Hoff AJ (1982) Photooxidation of the reaction center chlorophylls and structural properties of photosynthetic reaction centers. In: Fong FK (ed) Light reaction path of photosynthesis. Springer, Berlin Heidelberg New York, pp 80–151
Horner RD, Moudrianakis EN (1983) The effect of permeant buffers on initial ATP synthesis by chloroplasts using rapid mix-quench techniques. J Biol Chem 258:11643–11647
Israelachvili JN, Marcelja S, Horn RG (1980) Physical principles of membrane organization. Q Rev Biophys 13:121–200
Izawa S, Good NE (1966) Effect of salts and electron transport on the conformation of chloroplasts. II Electron microscopy. Plant Physiol 41:544–552
Jagendorf AT (1975) Mechanisms of photophosphorylation. In: Govindjee (ed) Bioenergetics of photosynthesis. Academic Press, London New York, pp 413–492
Jain MK, Wagner RC (1980) Introduction to biological membranes. Wiley, New York
Jennings RC, Garlaschi FM, Gerola PD, Forti G (1979) Partition zone penetration by chymotrypsin, and the localization of the chloroplast flavoprotein and photosystem II. Biochim Biophys Acta 546:207–219
Joliot P, Delsome R (1974) Flash-induced 519 nm absorption change in green algae. Biochim Biophys Acta 357:267–284
Joliot P, Kok B (1975) Oxygen evolution in photosynthesis. In: Govindjee (ed) Bioenergetics of photosynthesis. Academic Press, London New York, pp 387–412
Kaplan S, Arntzen CJ (1982) Photosynthetic membrane structure and function. In: Govindjee (ed) Photosynthesis: Energy conservation by plants and bacteria, vol I. Academic Press, London New york, pp 65–151
Keegstra K, Cline K (1982) Evidence that envelope and thylakoid membranes from pea chloroplasts lack glycoproteins. Plant Physiol 70:232–237
Kell DB (1979) On the functional proton current pathway of electron transport phosphorylation. An electrodic view. Biochim Biophys Acta 549:55–99
Kelly P, Izawa S (1978) The role of chloride ion in PS II. I. Effects of chloride on PS II electron transport and on hydroxylamine inhibition. Biochim Biophys Acta 502:198–210
Kenyon CN (1978) Complex lipids and fatty acids of photosynthetic bacteria. In: Clayton RK, Sistrom WR (eds) The photosynthetic bacteria. Plenum, New York London, pp 281–313
Kirby JA, Goodin DB, Wydrzynski T, Robertson AS, Klein MP (1981) State of manganese in the photosynthetic apparatus. 2. X-ray absorption edge studies on manganese in photosynthetic membranes. J Am Chem Soc 103:5537–5542

Klimov VV, Krasnovskii AA (1982) Participation of pheophytin in the primary processes of electron transfer at the reaction centers of photosystem II. Biophysics (USSR) 27:186–198

Kok B, Forbush B, McGloin M (1970) Cooperation of charges in photosynthetic O_2 evolution. I. A linear four step mechanism. Photochem Photobiol 11:457–475

Kyle DJ (1975) The 32,000 dalton Q_B protein of photosystem II. Photochem Photobiol 41:107–116

Lam E, Ortiz W, Malkin R (1984a) Chlorophyll a/b proteins of photosystem I. FEBS Lett 168:10–14

Lam E, Ortiz W, Mayfield S, Malkin R (1984b) Isolation and characterization of a light-harvesting chlorophyll a/b protein complex associated with photosystem I. Plant Physiol 74:650–655

Langworthy TA (1977) Long-chain diglycerol tetraethers from *Thermoplasma acidophilum*. Biochim Biophys Acta 487:37–50

Lewin R (1984) Why is development so illogical? Science 224:1327–1329

Ljungberg U, Åkerlund H-E, Larsson C, Andersson B (1984) Identification of intrinsic polypeptides associated with the 23 and 33 kDa proteins of oxygen evolution. Biochim Biophys Acta 767:145–152

Low PS, Ort DR, Cramer WA, Whitmarsh J, Martin B (1984) Search for an endotherm in chloroplast lamellar membranes associated with chilling inhibition of photosynthesis. Arch Biochem Biophys 231:336–344

Malkin R (1982) Photosystem I. Annu Rev Plant Physiol 33:455–479

McCarty RE, Carmeli C (1982) Proton translocating ATPases of photosynthetic membranes. In: Govindjee (ed) Photosynthesis: Energy conversion by plants and bacteria, vol I. Academic Press, London New York, pp 647–695

McCarty RE, Moroney JV (1984) Functions of the subunits and regulation of chloroplast coupling factor 1. In: Martonosi A (ed) The enzymes of biological membranes, 2nd edn, Vol 4. Plenum Press, New York, pp 383–413

Melis A, Anderson JM (1983) Structural and functional organization of the photosystems in spinach chloroplasts. Antenna size, relative electron-transport capacity, and chlorophyll composition. Biochim Biophys Acta 724:473–484

Merchant S, Selman BR (1985) Photosynthetic ATPases: purification, properties, subunit isolation and function. Photosyn Res 6:3–31

Metz JG, Seibert M (1984) Presence in photosystem II core complexes of a 34 kilodalton polypeptide required for photolysis. Plant Physiol 76:829–832

Miller KR, Staehelin LA (1976) Analysis of the thylakoid outer surface: Coupling factor is limited to unstacked membrane regions. J Cell Biol 68:30–47

Mitchell P (1974) A chemiosmotic molecular mechanism for proton-translocating adenosine triphosphatases. FEBS Lett 43:189–194

Mitchell P (1975) The proton motive Q-cycle: A general formulation. FEBS Lett 59:137–139

Mörschel E, Staehelin LA (1983) Reconstitution of cytochrome f/b_6 and CF_0–CF_1 ATP synthetase complexes into phospholipid and galactolipid liposomes. J Cell Biol 97:301–310

Moroney JV, Lopresti L, McEwen BF, McCarty RE, Hammes GG (1983) The M_r of chloroplast coupling factor 1. FEBS Lett 158:58–62

Mullet JE, Arntzen CJ (1980) Simulation of grana stacking in a model membrane system. Mediation by a purified light-harvesting pigment-protein complex from chloroplasts. Biochim Biophys Acta 589:100–117

Murphy DJ (1982) The importance of non-planar bilayer regions in photosynthetic membranes and their stabilization by galactolipids. FEBS Lett 150:19–26

Murphy DJ, Woodrow IE (1983) Lateral heterogeneity in the distribution of thylakoid membrane lipid and protein components and its implications for the molecular organization of photosynthetic membranes. Biochim Biophys Acta 725:104–112

Mustárdy LA, Jánossy AGS (1979) Evidence of helical thylakoid arrangement by scanning electron microscopy. Plant Sci Lett 16:281–284

Nakatani HY, Arntzen CJ, Inoue Y (1984) Characterization of the photosystem II reac-

tion center polypeptide (CP 47). In: Sybesma C (ed) Advances in photosynthesis research, vol II. Nijhoff/Junk, The Hague, pp 99–102
Nalin CM, McCarty RE (1984) Role of a disulfide bond in the γ subunit in activation of the ATPase of chloroplast coupling factor 1. J Biol Chem 259:7275–7280
Nelson N (1982) Structure and function of the higher plant coupling factor. In: Barber J (ed) Electron transport and photophosphorylation. Elsevier, Amsterdam New York Oxford, pp 81–104
Nir I, Pease DC (1973) Chloroplast organization and the ultrastructural localization of photosystems I and II. J Ultrastruct Res 42:534–550
Nugent JHA, Diner BA, Evans MCW (1981) Direct detection of the electron acceptor of photosystem II: Evidence that Q is an iron-quinone complex. FEBS Lett 124:241–244
Okamura MY, Feher G, Nelson N (1982) Reaction centers. In: Govindjee (ed) Photosynthesis energy conservation by plants and bacteria, vol I. Academic Press, London New York, pp 195–272
Oldfield E, Gilmore R, Glaser M, Gutowski HS, Hschung JC, Kang S, King TE, Meadows M, Rice D (1978) Deuterium nuclear magnetic resonance investigation of the effects of proteins and polypeptides on hydrocarbon chain order in model membrane systems. Proc Natl Acad Sci USA 75:4657–4660
O'Malley PJ, Babcock GT (1984) EPR properties of immobilized quinone cation radicals and the molecular origin of signal II in spinach chloroplasts. Biochim Biophys Acta 765:370–379
O'Malley PG, Babcock GT, Prince RC (1984) The cationic plastoquinone radical of the chloroplast water splitting complex. Hyperfine splitting from a single methyl group determines the EPR spectral shape of signal II. Biochim Biophys Acta 766:283–288
Ono T-A, Inoue Y (1983) Requirement of divalent cations for the photoactivation of the latent water-oxidation system in intact chloroplasts from flashed leaves. Biochim Biophys Acta 723:191–201
Ort DR, Melandri BA (1982) Mechanism of ATP synthesis. In: Govindjee (ed) Photosynthesis: Energy conversion by plants and bacteria, vol I. Academic Press, London New York, pp 537–587
Ort DR, Dilley RA, Good NE (1976) Photophosphorylation as a function of illumination time II. Effects of permeant buffers. Biochim Biophys Acta 449:108–124
Ort DR, Ahrens WH, Martin B, Stoller EW (1983) Comparison of photosynthetic performance in triazine-resistant and susceptible biotypes of *Amaranthus hybridus*. Plant Physiol 72:925–930
Ortiz W, Lam E, Ghirardi M, Malkin R (1984) Antenna function of a chlorophyll a/b protein complex of photosystem I. Biochim Biophys Acta 766:505–509
Paolillo DJ, Falk RH (1966) The ultrastructure of grana in mesophyll plastids of *Zea mays*. Am J Bot 53:173–180
Peters FALJ, van Wielink JE, Sang HWWF, De Vries S, Kraayenhof FR (1983) Studies on well coupled photosystem I-enriched subchloroplast vesicles. Content and redox properties of electron transfer components. Biochim Biophys Acta 722:460–470
Petrouleas V, Diner BA (1982) Investigation of the iron components in photosystem II by Mössbauer spectroscopy. FEBS Lett 147:111–114
Pfister K, Arntzen CJ (1979) The mode of action of photosystem II-specific inhibitors in herbicide-resistant weed biotypes. Z Naturforsch 34c:996–1009
Pick U, Gounaris K, Admon A, Barber J (1984) Activation of the CF_0–CF_1 ATP synthase from spinach chloroplasts by chloroplast lipids. Biochim Biophys Acta 765:12–20
Pink DA, Chapman D, Laidlaw DJ, Wiedmer T (1984) Electron spin resonance and steady-state fluorescence polarization studies of lipid bilayers containing integral membrane proteins. Biochemistry 23:4051–4058
Quinn PJ, Williams WP (1978) Plant lipids and their role in membrane function, vol 34. Prog Biophys Mol Biol. Pergamon, London New York, pp 109–173
Raison JK, Wright LC (1983) Thermal phase transitions in the polar lipids of plant membranes. Their induction by disaturated phospholipids and their possible relation to chilling injury. Biochim Biophys Acta 731:69–78

Rathenow M, Rumberg B (1980) Stoichiometry of proton translocation during photosynthesis. Ber Bunsenges Phys Chem 84:1059–1062
Renger G, Govindjee (1985) The mechanism of photosynthetic water oxidation. Photosyn Res 6:33–55
Rice D, Meadows MD, Scheinman AO, Goñi FM, Gómez-Fernández JC, Moscarello MA, Chapman D, Oldfield E (1979) Protein-lipid interactions. A nuclear magnetic resonance study of Sarcoplasmic reticulum Ca^{2+}, Mg^{2+}-ATPase, lipophilin, and proteolipid apoprotein-lecithin systems and a comparison with the effects of cholesterol. Biochemistry 18:5893–5903
Rich PR (1984) Electron and proton transfers through quinones and cytochrome bc complexes. Biochim Biophys Acta 768:53–79
Robinson H (1985) Non-invasive measurements of photosystem II reactions in the field using flash fluorescence. In: Gensler WG (ed) Advanced agricultural instrumentation, design and usage. Nijhoff, The Hague, in press
Rutherford AW, Heathcote P (1985) Primary photochemistry in photosystem I, Photosynth Res 6:295–316
Sculley MJ, Duneic JT, Thorne SW, Chow WS, Boardman NK (1980) The stacking of chloroplast thylakoids. Quantitative analysis of the balance of forces between thylakoid membranes of chloroplasts and the role of divalent cations. Arch Biochem Biophys 201:339–346
Selak MA, Whitmarsh J (1984) Charge transfer from photosystem I to the cytochrome b/f complex: Diffusion and membrane lateral heterogeneity. Photochem Photobiol 39:485–489
Siefermann-Harms D, Ross JW, Kaneshiro KH, Yamamoto HY (1982) Reconstitution by monogalactosyldiacylglycerol of energy transfer from light-harvesting chlorophyll a/b-protein complex to the photosystems in Triton X-100 solubilized thylakoids. FEBS Lett 149:191–196
Slater EC (1953) Mechanism of phosphorylation in the respiratory chain. Nature 172:975–978
Slater EC (1977) Mechanism of oxidative phosphorylation. Annu Rev Biochem 46:1015–1026
Slater EC (1981) The cytochrome b paradox, the BAL-labile factor and the Q cycle. In: Skulachev VP, Hinkle PC (eds) Chemiosmotic proton circuits in biological membranes. Addison-Wesley, London, pp 69–104
Sprague SG, Camm EL, Green BR, Staehelin LA (1985) Reconstitution of light-harvesting complexes and photosystem II cores into galactolipid and phospholipid liposomes. J Cell Biol 100:552–557
Staehelin LA, Arntzen CJ (1983) Regulation of chloroplast membrane function: Protein phosphorylation changes the spatial organization of membrane components. J Cell Biol 97:1327–1337
Staehelin LA, Dewit M (1984) Correlation of structure and function of chloroplast membranes at the supramolecular level. J Cell Biochem 24:261–269
Staehelin LA, Armond PA, Miller KR (1977) Chloroplast membrane organization at the supramolecular level and its functional implications. Brookhaven Symp Biol 28:278–315
Stanier RY, Cohen-Bazire G (1977) Phototropic prokaryotes: the cyanobacteria. Annu Rev Microbiol 31:225–274
Strotmann H, Hesse H, Edelman K (1973) Quantitative determination of coupling factor CF_1 of chloroplasts. Biochim Biophys Acta 314:202–210
Süss K-H, Schmidt O (1982) Evidence for an $\alpha_3\beta_3\gamma\delta$I, II, ε, III_5 subunit stoichiometry of chloroplast ATP synthetase complex (CF_1–CF_0. FEBS Lett 144:213–218
Swarthoff T, Gast P, Amesz J, Buismann HP (1982) Photoaccumulation of reduced primary electron acceptors of photosystem I of photosynthesis. FEBS Lett 146:129–132
Thomas JB, Minnaert K, Elbers PF (1956) Chlorophyll concentrations in plastids of different groups of plants. Acta Bot Neerl 5:315–321
Vambutas V, Schechter S (1983) Chloride ion transport and its inhibition in thylakoid membranes. Arch Biochem Biophys 224:442–448

van Gorkom HJ (1985) Electron transfer in PS II. Photosynth Res 6:97–112
Velthuys BR (1978) A third site of proton translocation in green plant photosynthetic electron transport. Proc Natl Acad Sci USA 75:6031–6034
Velthuys BR (1982) The function of plastoquinone in electron transfer. In: Trumpower BL (ed) Function of quinones in energy conserving systems. Academic Press, London New York, pp 401–408
Velthuys BR, Amesz J (1974) Charge accumulation at the reducing side of system II of photosynthesis. Biochim Biophys Acta 333:85–94
Vermaas WFJ, Dohnt G, Renger G (1984a) Binding and release kinetics of inhibitors of Q_A^- oxidation in thylakoid membranes. Biochim Biophys Acta 765:74–83
Vermaas W, Renger G, Dohnt G (1984b) Exchange kinetics of herbicides and quinones at their common binding site near Q in photosystem II. In: Sybesma C (ed) Advances in photosynthesis research, vol IV. Nijhoff/Junk, The Hague, pp 13–16
Vermaas WFJ, Renger G, Dohnt G (1984c) The reduction of the oxygen-evolving system in chloroplasts by thylakoid components. Biochim Biophys Acta 764:194–202
Whatley JM, John P, Whatley FR (1979) From extracellular to intracellular: The establishment of mitochondria and chloroplasts. Proc R Soc London Ser B 204:165–187
Whitmarsh J, Ort DR (1984) Stoichiometries of electron transport complexes in spinach chloroplasts. Arch Biochem Biophys 231:378–389
Whitmarsh J, Bowyer JR, Crofts AR (1982) Modification of the apparent redox reaction between cytochrome f and the Rieske iron-sulfur protein. Biochim Biophys Acta 682:404–412
Widger WR, Cramer WA, Hermodson M, Meyer D, Gullifor M (1984a) Purification and partial amino acid sequence of chloroplast cytochrome b-559. J Biol Chem 259:3870–3876
Widger WR, Cramer WA, Herrmann RG, Trebst A (1984b) Sequence homology and structural similarity between cytochrome b of mitochondrial complex III and the chloroplast b_6/f complex: Position of the cytochrome b hemes in the membrane. Proc Natl Acad Sci USA 81:674–678
Willey DL, Huttly AK, Phillips AL, Gray JC (1983) Localization of the gene for cytochrome f in pea chloroplast DNA. Mol Gen Genet 189:85–89
Wraight CA (1979) The role of quinones in bacterial photosynthesis. Photochem Photobiol 30:767–776
Wraight CA (1982a) Current attitudes in photosynthesis research. In: Govindjee (ed) Photosynthesis: Energy conservation by plants and bacteria, vol I. Academic Press, London New York, pp 17–61
Wraight CA (1982b) The involvement of stable semiquinones in the two-electron gates of plant and bacterial photosynthesis. In: Trumpower BL (ed) Function of quinones in energy conserving systems. Academic Press, London New York, pp 181–197
Yocum CF (1985) Electron transfer on the oxidizing side of photosystem II. Components and mechanism. Photosynth Res Vol 6, in press
Yocum CF, Yerkes CT, Blankenship RE, Sharp RR, Babcock GT (1981) Stoichiometry, inhibitor sensitivy, and organization of manganese associated with photosynthetic oxygen evolution. Proc Natl Acad Sci USA 78:7507–7511
Zimmermann JL, Rutherford AW (1984) EPR studies of the oxygen evolving enzymne of photosystem II. Biochim Biophys Acta 767:160–167

5 Energy Transduction in Anoxygenic Photosynthesis

P.L. DUTTON

1 Introduction

THOMAS (1974) has supposed that a dispassionate extraterrestrial might perceive the Earth as populated by two symbiotic forms. The chloroplasts would be seen as cellular structures built from carbon compounds, ranging from the prokaryotic cyanobacteria at the simplest level through to those occupying the algae and green plants at the most elaborate. It would be clear that these species were equipped to harness solar energy to "fix" CO_2 by reduction to carbohydrate and thence to more elaborate molecules of the cell. Water would be identified as the source of electrons to reduce the CO_2, explaining the byproduct, dioxygen, that dominates the Earth's atmosphere. The partner to this symbiotic pair, the mitochondrion, would be recognized as respiratory bacteria in its simplest, prokaryotic form, through the likes of those found in yeast to those occupying animal and plant cells. The mitochondrion would be seen to utilize the chloroplast-derived, reduced carbon compound material, not only for its own biosynthetic needs but also, in collaboration with O_2, as a source of fuel to provide free energy to drive these biosynthetic reactions, reforming CO_2 and H_2O and completing the cycle. Thus while the mitochondrion is able to capitalize on the photosynthetic industry of the chloroplast, the chloroplast benefits from the CO_2 recycling activities of the mitochondrion. The plant-mitochondrial systems are outlined as a flow sheet in Fig. 1 A and B.

Important evolutionary precursors to the activities of chloroplasts and mitochondria are the predecessors of the present-day anoxygenic photosynthetic bacteria. Many of these phototrophic bacteria require conditions tending to the anaerobic and presumably, since the arrival of the oxygenic chloroplast and the "pollution" of the Earth with O_2, have retreated into aqueous niches that are low in or free of O_2 (PFENNIG 1978). Such conditions are necessary for the members of the purple sulfur (Chromatiaceae) and the green sulfur (Chlorobiaceae) families to find and utilize reduced sulfur compounds as a source of reducing equivalents with which to fix CO_2. There is, however, more choice for those species that can grow heterotrophically on reduced carbon compounds which provide both reducing equivalents and the carbon for biosynthetic purposes. Thus, members of the green nonsulfur (Chloroflexiaceae) can live phototrophically in aerated media while many of the purple nonsulfur (Rhodospirillaceae) bacteria have evolved to be facultative aerobes. Indeed, Rhodospirillaceae in the presence of oxygenated media may discard their phototrophic

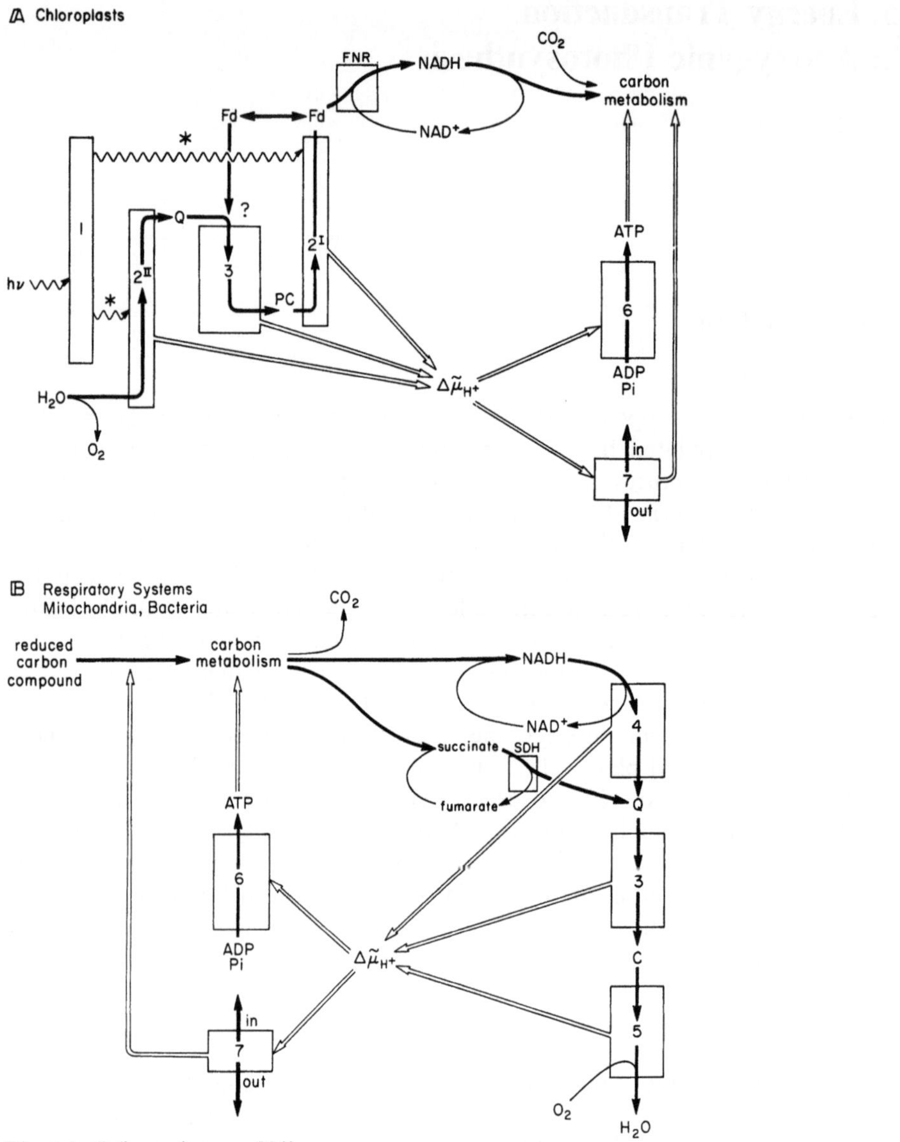

Fig. 1A–E (legend see p. 200)

way of life, secure under anaerobic conditions, in favor of one based on respiration, thence reflecting many of the features of their evolutionary child, the mitochondrion.

The extraterrestrial observer would be forgiven for missing the photosynthetic bacteria; indeed since they have little economic or nuisance value, most terrestrial observers are oblivious of them. In the laboratory, however, they have proven themselves to be of immense value especially in areas of common interest with the chloroplast and the mitochondrion. Many of the advances

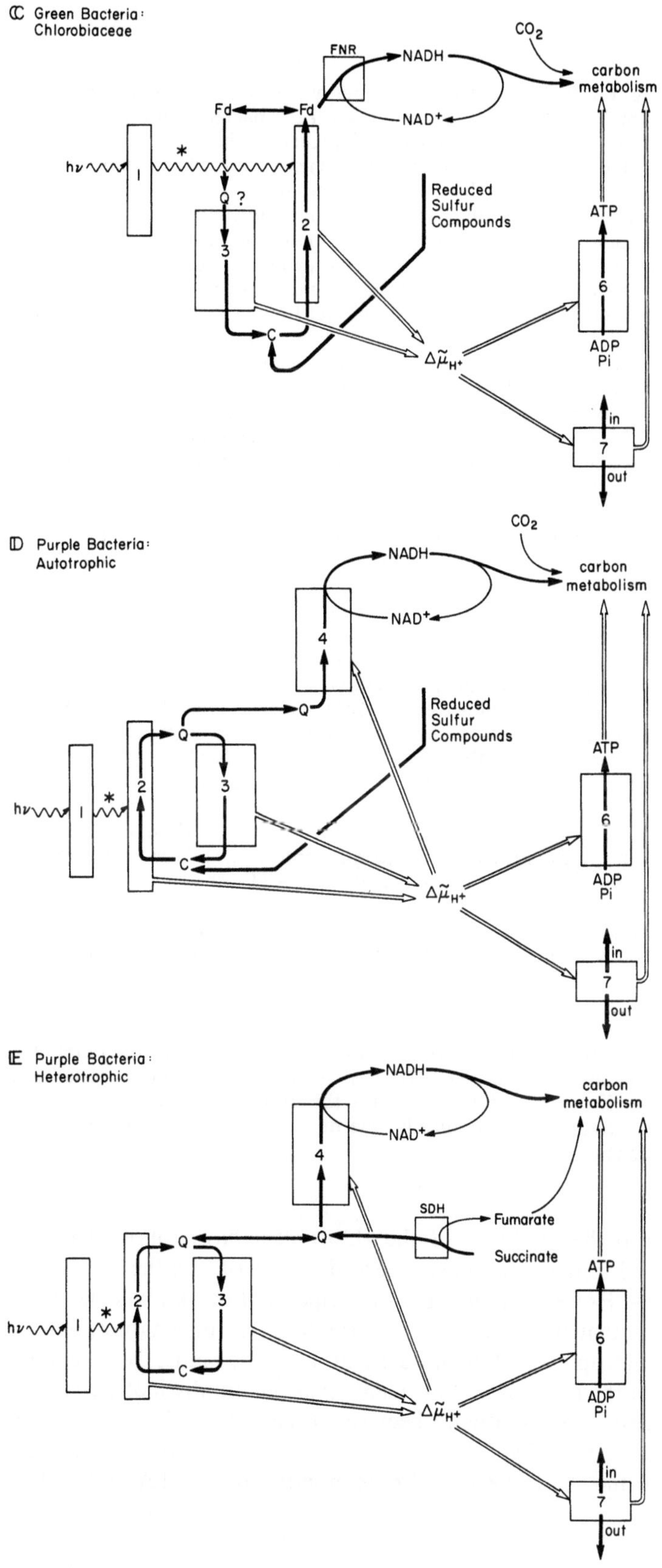
C Green Bacteria: Chlorobiaceae
FNR
NADH
NAD+
CO2
carbon metabolism
Fd
Fd
hν
*
1
Q ?
3
2
C
Reduced Sulfur Compounds
ATP
6
ADP Pi
Δμ̃H+
in
7
out
D Purple Bacteria: Autotrophic
NADH
NAD+
CO2
carbon metabolism
4
Q
Q
2
3
C
1
*
hν
Reduced Sulfur Compounds
Δμ̃H+
ATP
6
ADP Pi
in
7
out
E Purple Bacteria: Heterotrophic
NADH
NAD+
carbon metabolism
4
SDH
Fumarate
Succinate
Q
Q
2
3
C
1
*
hν
Δμ̃H+
ATP
6
ADP Pi
in
7
out

Fig. 1 A–E. Flowsheet of free energy and material in chloroplasts and respiratory (mitochondrial) systems and in photosynthetic bacteria. In this figure the *closed areas* represent classes of membrane-associated proteins where energy coupling occurs; that is, each is able to accept one form of free energy and deliver another. For simplicity, the *arrowheads* indicate the direction of physiological flow, but many of the parts of the systems can operate reversibly; only arrows of importance in the direction of photo- and oxidative phosphorylation are shown. *Wavy lines* represent light (*hν*) or singlet excited (*) transfer. *Thick lines* represent the flow of reducing equivalents or electrons; *thinner lines* associated with identified redox couples indicate the conversion of one state of a substrate component to another. *Open arrows* indicate net flow of free energy to and from the various energy coupling enzymes as $\Delta\mu_{H^+}$ or ATP or ion and solute gradients. The positioning of the components has no particular significance, although with respect to electron transfer the vertical scale most often indicates whether a reaction is favored (*downward*) or is not favored unless coupled to another free energy source. The *letters* between the areas indicate common kinds of components identified in connecting major redox proteins and substrates. *Q* a type of quinone; *c* a c-type cytochrome; *PC* plastoquinone. Area *1* represents the generic (i.e., no recognition of the variety, and heterogeneity) light-harvesting complex which absorbs light and delivers lowest level excited singlet energy to the reaction center. Area *2* represents the reaction center which accepts singlet excited state energy, and effects a charge separation which is translated into (a) transmembrane potential ($\Delta\psi$), a component of the $\Delta\mu_{H^+}$ and (b) a redox potential difference (ΔE_h) which initiates electron transfer into and out of the reaction center. Area *3* is a quinol-cytochrome c (plastocyanin in chloroplasts) oxidoreductase, generally called the bc_1 complex (or b_6f complex in chloroplasts). It accepts the reaction center generated ΔE_h in the form of a quinol (low E_h) and a cytochrome c (PC) (high E_h) and converts it into $\Delta\mu_{H^+}$ composed of a $\Delta\psi$ and ΔpH across the membrane. Area *4* represents the NADH-quinone oxidoreductase. In the respiratory system as shown in **B** it converts ΔE_h in the form of NADH (which it oxidizes to NAD^+) and quinone (which it reduces to quinol) into $\Delta\mu_{H^+}$. In **D** and **E** it is represented in the reverse direction as accepting $\Delta\mu_{H^+}$ and uses it to oxidize quinol to quinone and reduce NAD^+ to NADH for use in the cell. Area *5* represents cytochrome c oxidase which converts the ΔE_h provided by ferrocytochrome c and dioxygen into $\Delta\mu_{H^+}$. Area *6* represents the adenosine triphosphate-synthetase commonly called the ATP-ase which accepts $\Delta\mu_{H^+}$ and drives the phosphorylation of ADP to deliver ATP. Area *7* represents the transmembrane metabolite transporters which accept $\Delta\mu_{H^+}$ and catalyze the import and export of metabolites, substrates, and ions, leading to the formation of concentration gradients of these commodities.

A the chloroplast; **B** the mitochondrion. The succinate/fumarate couple is shown linked to the respiratory chain via succinate-dehydrogenase. There are other dehydrogenases that connect with the respiratory system at this point. **C** an arrangement that is commonly considered to exist in the green sulfur bacterial family Chlorobiaceae. **D** and **E** similar arrangements for the purple sulfur bacteria grown photoautotrophically and photoheterotrophically; the nonsulfur green Chloroflexus family may fit into these schemes

---

into the workings of the chloroplast have followed discoveries, descriptions, and insights that have emerged from work on the simpler photosynthetic bacteria. However, there are other aspects of our knowledge of photosynthetic bacteria that have received less attention. In this chapter, while comparisons to the chloroplast and the mitochondria will be made, focus will be directed principally on the features and problems that remain distinctly characteristic of the anoxygenic photosynthetic bacteria. Figure 1 C, D, and E presents three principal patterns of light-driven activities in photosynthetic bacteria and compares them to the chloroplasts and mitochondria in Fig. 1 A and B. In the figure the numbered areas represent different classes of energy-coupling enzymes that are common to chloroplasts, mitochondria, and photosynthetic bacterial families. These are introduced in the next sections.

2 The Photosynthetic Bacterial Membrane

The plasma membrane of photosynthetic bacteria can be seen in several different morphologies; these are discussed in depth in this Volume in Chapter 11.1 by SPRAGUE and VARGA and in Chapter 8.6 for the green bacteria by BLANKENSHIP and FULLER. Independently of the gross topography peculiar to the plasma membrane of each species, all are presumably closed structures with an inside containing the cellular cytoplasm and an outside facing the periplasmic space between the plasma membrane and the cell wall. The plasma membrane is not necessarily homogeneous and much is still to be learned both in general and with respect to specific cases. For example, in the facultative aerobic phototrophic bacteria, photosynthetic activity may be localized in a particular region of the membrane, which, while thought to be contiguous, is nevertheless segregated from other areas in which respiratory enzymes may dominate (see SPRAGUE and VARGA, Chap. 11.1, DONAHUE and KAPLAN, Chap. 11.3, this Vol.). Most studies on photosynthetic bacteria are done with fragments of the plasma membrane, which are obtained by breaking whole cells; hence there is always the potential problem of heterogeneity. Fortunately, however, several species have membranes with well-developed invaginations in which the photosynthetic activity resides, while regions of the membrane located closer to the cell wall are thought to accommodate the respiratory systems. Upon breakage, the invaginations pinch off and reseal to form small spherical particles called chromatophores, which range in diameter from 400–800 Å. The average thickness of the chromatophore membrane is in the 50–100 Å range, built about a phospholipid bilayer with its ~35 Å hydrocarbon core. These chromatophore membranes, because they are formed from the invaginations, are topologically inside-out with respect to the parent cell (see BACHOFEN and WIEMKEN, Chap. 11.2, this Vol.): what was disposed toward the cell cytoplasm is now an external surface; what was contained within the invagination on the periplasmic side is now in the interior of chromatophore. The choice to use these simpler membranes is obviously balanced against (1) the uncertainty that exists regarding areas designated "respiratory" and "photosynthetic" and (2) possible loss of soluble cofactors that interact with the membrane-bound constituents. Cofactors originally located on the cytoplasmic side are easily lost, while those on the periplasmic side occupying the invagination are less likely to be lost during breakage; the extent of the loss will obviously depend on the natural affinity of the cofactor for its binding site and the conditions of isolation.

As indicated, the acquisition of usefully homogenous chromatophores depends on species. However, it also depends on growth history, storage conditions, and mode of breakage. For example, *Rhodopseudomonas sphaeroides* grown photosynthetically at low to moderate light intensities will develop the highly invaginated membrane system, whereas with intense light these will not be so well developed. Under the former conditions, cells harvested in mid to late log phase and broken by extrusion through a French Press commonly yield a >90% population of sealed, inside-out chromatophores from which loss of soluble proteins originating in the periplasmic space has been usefully minimized; these membranes are fully able to support photophosphorylation and NADH production. In contrast, other species, such as *Rps. viridis*, in which the plasma membrane has a folded sheet arrangement in the cell, do not lend themselves to easy preparation of chromatophores that perform energy-linked functions for which a closed, intact membrane is required. It is important to note, however, that a similar situation is obtained in the case of *Rps. sphaeroides* if the cells are frozen prior to breakage; with this treatment, instead of a fairly homogenous population of sealed inside-out chromatophores, the great majority of the membranes can be in the form of sheet structures that do not support photophosphorylation.

It is now well acknowledged that a vital property of the plasma membrane, whether in the native cell or in the isolated chromatophore vesicle, is that it is closed and intrinsically relatively tight to the passage of ions and solutes. This goes beyond the membrane performing as a bag to maintain high reactant concentrations; it extends to the electrical performance of the membrane, which is the key function in photosynthesis and respiration (MITCHELL 1966, 1968).

The electrical capacitance ($C = \Delta Q/\Delta\psi$) of a membrane describes the number of charges (ΔQ) that have to be moved across a membrane to elicit a unit change of transmembrane potential ($\Delta\psi$). The capacitance of a pure phospholipid bilayer membrane is in the region of 0.5 $\mu F\ cm^{-2}$ (see JAIN 1972), while that of chromatophore membranes has been estimated to be about 1.1 $\mu F\ cm^{-2}$ (PACKHAM et al. 1978).[1] The resistance of a pure phospholipid membrane is remarkably high, about 10^{-9} ohm cm^{-2} (see JAIN 1972), while that of protein containing membranes like chromatophores is much less, nearer 10^{-6} ohm cm^{-2} (PACKHAM et al. 1978). These basic electrical characteristics prove sufficient to provide the chromatophore with the capability to develop voltages of 250 mV or more across the membrane with just two or three turnovers of the electron transfer systems, and to hold the voltage into the seconds time range, long enough to be utilized by physiological processes. For further discussions along these lines see JUNGE and JACKSON (1982).

3 Membrane-Associated Energy-Transducing Enzymes

Integrally associated with chloroplast and bacterial membranes are five main classes of energy-transducing enzymes that function to convert one form of free energy into another.

1 Recalculation of the data of PACKHAM et al. (1978), taking into account (a) that in their measurement they did not correct for the fact that only 30% of their reaction center $(BChl)_2^+$ was reduced by cytochrome c after the flash and (b) that this reaction contributes about 50% of the total transmembrane potential yields a membrane potential of 65 mV rather than 47 mV for a single turnover. This will yield a capacitance of near to 0.7 $\mu F\ cm^{-1}$ closer to that displayed by the pure phospholipid.

3.1 The Light-Harvesting System

The light-harvesting pigment protein network (area 1 in Fig. 1) serves to absorb light, and to convert it to the lowest-lying singlet excited state of (bacterio)-chlorophyll (*). The light-harvesting protein system also provides the framework for the mobilization of the singlet excited state. The system promotes efficient energy transfer from one chlorophyll molecule to another, so that before the excited state can decay it reaches and is captured by the second coupling device, the reaction center protein. The mechanistic and structural aspects of light absorption and energy transfer are well covered in this Volume in the overview Chapters 2 and 3 of SAUER and THORNBER, and further in Chapters 6 and 7, and will not be mentioned further in this chapter.

3.2 The Reaction Center

The reaction center protein (areas labeled 2 in Fig. 1) accepts singlet excited state energy from the light-harvesting system and converts it into electrochemical potential free energy of two recognizable forms: (a) an electric potential difference across the membrane (a membrane potential, $\Delta\psi$) and (b) a redox potential difference (ΔE_h) between an oxidant and a reductant. This dual energetic capability appears to be a property of all reaction centers whether in chloroplasts or bacteria. The secret of the duality of function lies in the fact that the light activates a charge separation involving a series of electron transfer reactions that in concert promote net vectorial electron transfer directed across the membrane. Contributions to this electron transfer come from several chlorin molecules arranged in the matrix of the reaction center protein (see TIEDE, Chap. 8.2 and MICHEL and DEISENHOFER, Chap. 8.4, this Vol.). The reaction centers of the purple sulfur Chromatiaceae (e.g., *Chromatium vinosum*) and the purple nonsulfur Rhodospirillaceae (e.g., *Rhodopseudomonas sphaeroides*, *Rps. viridis*) contain four bacteriochlorophylls (BChl) and two bacteriopheophytin (BPh) molecules (STRALEY et al. 1973). In the facultative anaerobic green bacteria Chloroflexaceae (e.g., *Chloroflexus aurantiacus*) the number is three BChl, and three BPh (PIERSON and THORNBER 1983; BLANKENSHIP et al. 1984). However, in the anaerobic green sulfur bacteria Chlorobiaceae (e.g., *Chlorobium limicola*, *Prosthecochloris aestuarii*) the BChl and BPh complement has not yet been quantitated.

The general view regarding these chlorins in bacteria is that two of the BChl are closely positioned in the protein such that they assume a dimeric character, $(BChl)_2$. This component, when speaking generally or when ill-defined in a particular photosynthetic reaction center, is commonly called P, followed by a number indicating the peak position of its lowest excited singlet band in the near infrared. $(BChl)_2$, or P, is the entity that is the locus of the singlet excited energy prior to photochemistry and is the key element of the transduction of light energy into electrochemical potential. In addition to $(BChl)_2$, one of the BPh has also been proven to be directly involved in early light-driven electron transfer in certain purple and green bacterial families. However, the BPh does not appear to act in an isolated fashion; it appears to be closely allied with one of the monomeric BChls, although precisely how is still not established; this (or these) vital components of early photochemistry are often collectively called I (see PARSON and HOLTEN, Chap. 8.1, NORRIS and VAN BRAKEL, Chap. 8.3 and BLANKENSHIP and FULLER, Chap. 8.6, this Vol.) The remaining chlorins (half of the content!) have not yet been functionally proven to be involved in electron transfer and are the objects of considerable speculation (see discussion of MICHEL and DEISENHOFER, Chap. 8.4, this Vol.).

Redox midpoint potentials[1] of the $(BChl)_2/(BChl)_2^{+\cdot}$ couple usually fall in the range 0.4 to 0.5 V range for purple bacteria while values for the green bacteria are somewhat lower (e.g., 0.25 V in *Chlorobium limicola* or *Prostheco-chloris aestuarii*; 0.36 V in *Chloroflexus aurantiacus*). The redox potentials of the BPh^-/BPh $(BChl^-/BChl)$ comprising I^-/I are not known with any certainty, but are in the region of or below -0.7 V.

The chlorin content of the chloroplast reaction centers of photosystem I or II is not known quantitatively, again due to the absence of well-defined isolated proteins. Nevertheless, spectrophotometric analysis suggests that the composition of the low potential end of the chloroplast photosystem II and the light-activated events are remarkably similar in character to those of the purple bacteria and green nonsulfur families. Less is known about the nature of the chlorophyll analog of the bacterial $(BChl)_2$ in photosystem II reaction centers which is designated P-680. However, it is clear that the oxidized photo-system II reaction center chlorophyll, $P\text{-}680^{+\cdot}$, is a much stronger oxidant than the $(BChl)_2^{+\cdot}$, displaying an E_{m_7} value estimated to be in the region of 1.1 V (see Chap. 9.1 by DINER and Chap. 9.2 by YOCUM, this Vol., for further details).

1 To describe oxidants and reductants in more quantitative terms we use a redox scale related to a standard. This standard is commonly the standard hydrogen electrode which has been given the value of 0 V. Thus the electrical difference between the redox couple in question, treated as a half-cell in electrical equilibrium with the hydrogen half-cell, is the redox potential (E_h) of the couple. $E_h = E_{m(x)} + RT/nF \ln [ox]/[red]$ where [ox] and [red] represent the concentrations of the oxidized and reduced forms of a redox couple related to the standard hydrogen half cell. R, T, and F represent the usual thermodynamic parameters and n is the number of electrons transferred per molecule during the oxidation and reduction reaction. $E_{m(x)}$ is that E_h value when [ox] = [red], under defined conditions of pH(x), is called the midpoint potential and is related to the standard potential of the half cell. Reductants have E_m values that are lower than oxidants; conditions where low E_h values prevail are reducing; high E_h values oxidizing. Electrons will tend to move downhill from redox components with low E_h values to compounds exhibiting higher E_h values. This is, in

$$A^{red} + B^{ox} = A^{ox} + B^{red},$$

$$E_h(B) - E_h(A) = E_m(B) - E_m(A) + (RT/nF)\ln[A^{ox}][B^{red}]/[A^{red}][B^{ox}],$$

and $\Delta G = -nF[E_h(B) - E_h(A)]$; also $\Delta G'_0 = -nF\,[E_{m_7}(B) - E_{m_7}(A)]$, where $\Delta G'_0$ is the standard under physiological, neutral conditions. It is worth remembering that the E_{m_7} of the H_2/H^+ couple is -0.42 V due to the -0.06 V/pH unit dependencies of this couple. Thus at pH 7, E_h values lower than this are needed to produce significant concentrations of H_2 from water; in contrast E_h values higher than $+0.82$ V are needed to yield significant concentrations of O_2 from water. These two couples provide a guide of the biological range in which aqueous life operates.

The reaction center of photosystem I (P-700) appears generally to be similar to that of the green sulfur bacteria, Chlorobiaceae, although P-700 has an E_{m_7} value of 0.5 V (see Chap. 9.6, this Vol. by SETIF and MATHIS for a full discussion), some 0.25 V higher than the analogous components of chlorobiaceae.

The reaction centers introduced above are sufficiently similar to permit the description of a "generic" reaction center.

a) Following light absorption the P becomes excited energetically defined by its lowest exited single state P*.

b) P*, is a powerful reducing agent and within 10 picoseconds (10^{-12} s) donates an electron to I. Once the electron has left P*, the molecule is no longer in its excited singlet state but it is now one electron short. As a result it becomes a powerful, positively charged oxidant, $P^{+\cdot}$. Conversely, I upon receiving an electron, becomes a powerful, negatively charged reductant, I^-.

c) Before useless return of an electron can occur between the above-discussed light-generated oxidant ($P^{+\cdot}$) and reductant (I^-), (this decay would take place in the 10^{-8} s half-time range), I^- transfers an electron to the first of a chain of associated oxidized secondary redox components located within the protein matrix at a point closer to one side of the membrane. The half-time of this reaction for bacterial and green plant reaction centers, where it has been measured, is in the range of $0.2–0.5 \times 10^{-9}$ s, much faster than the decay time. The fact that this transfer happens shows that these secondary redox molecules provide much more favorable sites (i.e., have higher E_m values) for the negatively charged electron; indeed, the difference in affinity (i.e., ΔE_m) between I^-/I and these secondary components must be more than enough to overcome the opposing coulombic forces involved in the physical separation of the negative charge from the positively charged oxidant, $P^{+\cdot}$.[1]

d) Once an electron has moved from I^- to secondary components, P^+ takes an electron from a reduced secondary redox component which is located at a point closer to the *other* side of the membrane; the half-time of this reaction varies greatly (10^{-8} s to 10^{-3} s) depending on the organism.

With these few steps the light-activated reaction center effects the separation of negative and positive charges within the protein directed across the profile of the membrane. The result is the formation of an electric potential across the protein which rapidly becomes delocalized over the membrane-aqueous surfaces, and is expressed as an electrical difference across the membrane from one aqueous phase to the other ($\Delta\psi$). In bacterial cells the reaction center-driven movement of electrons is vectorially directed away from the periplasmic space

1 It should be born in mind that E_m values are commonly obtained by redox potentiometry in the absence of their oppositely charged partners and are measured on a minutes time scale to achieve a level of redox equilibrium. The effective redox capabilities of the components on the sub-nanosecond time scale driven by the highly energetic singlet state may be different from that indicated by the E_m values; they may not achieve equilibrium and may be influenced by coupled events such as electric fields. This latter case adds a geometric factor in which the redox properties of a particular reductant or oxidant may be perceived differently, depending on the position of adjacent redox components with respect to the fields. This explains why energy levels obtained from the measurement of equilibrium between redox components on a functional basis will not necessarily gibe with those based on ΔE values.

toward the cytoplasm. Thus the internal aqueous cytoplasm becomes charged negative with respect to the external aqueous phase. In the case of the chromatophore, the polarity is opposite. The chloroplast is similar to the chromatophore where the external stroma becomes charged negative with respect to the internal, lumen space (see ORT Chap. 4, this Vol.). Once formed, the $\Delta\psi$ represents a form of free energy that is potentially available to drive *any* transmembrane reactions involving any charged species in the cell. These may involve charge changes that are directed across the membrane not only in any of the oxido-reductase enzymes, but also in ATPases and the substrate and ion translocases occupying the same membrane.

The value of $\Delta\psi$ generated immediately following a single turnover light activation of the reaction centers depends on several parameters. Some, such as membrane capacitance and reaction center density in the membrane, are external to the reaction center. Other parameters are intrinsic to the reaction center itself; these include the free-energy relationships in the redox reactions, the forward and decay of electron transfer kinetics and the spatial arrangements of the pigments and redox components associated with the reaction center (see TIEDE Chap. 8.2, this Vol.). A value of $\Delta\psi$ typical for the bacterium *Rps. sphaeroides* following a single turnover of all the reaction centers is in the region of 0.07 V (TAKAMIYA and DUTTON 1977).

As already indicated, all the light energy does not go into the charge separation. In addition to $\Delta\psi$, there is a large redox potential difference between the secondary, low E_m redox component(s) reduced by I^-, and the secondary, high E_m redox component oxidized by the $P^{+\cdot}$. The ΔE_h between the two redox components represents a considerable storage of potential free energy. The chemical character and the redox potentials of these secondary redox components vary, depending on the photosynthetic species; they represent the first biochemically stable products of the reaction centers and have profound influence on the physiology of the organism. This is discussed in the next section.

3.3 The Oxido-Reductases

Various oxidoreductases are, on a redox basis, able to connect with the reaction center-generated secondary reductants and oxidants and use them in a variety of ways. They can be recognized functionally as follows.

3.3.1 Direct $NAD(P)^+$ Reduction

Some reaction centers generate a reductant the redox potential of which is low enough and kinetically able to reduce $NADP^+$ or NAD^+ in a direct, exothermic sequence. The terms direct and exothermic imply a straightforward flow of electrons from the redox component reduced by the reaction center in a downhill sequence ending with NAD^+ or $NADP^+$ reduction.

The reaction center associated with photosystem I [area 2(I) in Fig. 1A] in the chloroplast is capable of generating NADPH directly. The I^- of this reaction center, linked via a series of redox components, mainly iron-sulfur clusters, finally reduces a soluble ferredoxin (E_{m_7} −0.42 V) located on the stromal side of the chloroplast (see ARNON 1977). The reduced ferredoxin in turn, via an FAD-containing ferredoxin-$NADP^+$ reductase (FNR), can reduce $NADP^+$ to NADPH (E_{m_7} −0.32 V) (see ORT Chap. 4, SETIF and MATHIS Chap. 9.6 and WOLLMAN Chap. 9.7, this Vol. for details).

The reaction center from the green photosynthetic bacterial family, Chlorobiaceae (area 2 Fig. 1C) similarly can reduce NAD^+ or $NADPH^+$ (the specificity is species-dependent) also via a ferredoxin and a ferredoxin-NAD^+ (or $NADP^+$) reductase enzyme located in the inner cytoplasmic side of the membrane (see OLSON et al. 1976; KNAFF and MALKIN 1976; SHIOI et al. 1976).

3.3.2 Direct Quinone Reduction

Other reaction centers do not produce a reductant that has a redox potential low enough and that is kinetically capable of reducing NAD^+ in a direct way. In these reaction centers the reducing equivalent on the light-generated I^- is, instead, funneled to reduce quinones (Q), which constitute a pool within the membrane surrounding the reaction center, to form quinols (QH_2).

Quinone pools (Q_P) exist in all photosynthetic and respiratory systems; plastoquinone in chloroplasts, ubiquinone in purple bacteria and mitochondria and menaquinone in green and some purple bacteria (reviewed by COLLINS and JONES 1981). In this regard the reaction centers of photosystem II [area 2(II) in Fig. 1A; see ORT Chap. 4, DINER Chap. 9.1, ARNTZEN Chap. 9.4], the reaction centers of the purple bacteria (area 2 in Fig. 1D, E) and of the green bacterial family Chloroflexaceae (not shown in Fig. 1; see BLANKENSHIP and FULLER Chap. 8.6, this Vol.) all transfer electrons to the Q pool. The process is catalyzed by a pair of quinones that occupy definite sites in the reaction center protein. The first quinone, Q_A, has an operating E_m for the Q_A/Q_A^- (n=1) couple in the various species that falls in the −0.13 (PS II) to −0.18 V (bacteria) range. Q_A accepts an electron from I^- in a 0.2–0.5 × 10^{-9} s half-time to form the Q_A^-. The Q_A^- donates an electron in the 10^{-4} s half-time range to the second quinone, Q_B, to form Q_B^-. The Q_B^- so formed is tightly associated with the site; it waits there for a second activation of the reaction center, and, when this occurs, receives a second electron from Q_A^- and two protons from the aqueous phase to form quinol (Q_BH_2). Although the details of the protonation states of Q_B are not understood, it does appear from the limited cases measured that the redox couples of Q_B at physiological pH values appear tens of millivolts higher than the associated Q_A^-/Q_A couple, thus making each of the electron transfers from Q_A to Q_B favorable. Once the Q_BH_2 is formed, it is only weakly associated with the Q_B site and this diffuses into the pool. The site then becomes available for occupancy by a quinone, so the process can be repeated. This neat piece of choreography, in which the quinone in the Q_B site acts as a "two-electron gate" between Q_A and the Q pool was first demonstrated by VERMEGLIO (1977) and WRAIGHT (1977). The two-electron gate is a mechanism that links the one electron per reaction center with the two-electron quinone pool. The requirement for the mechanism stems from the fact that the semiquinone in the pool is very unstable and cannot be formed via one electron transfer from Q_A to Q_B; that is, the semi-quinone/quinone E_m value is much lower in the pool than encountered in the protein bound species. However, the E_{m_7} of the QH_2/Q couple of the pool (i.e., the average E_m value for the two reactions) is much higher; for example the E_{m_7} (n=2) value for QH_2/Q in *Rp. sphaeroides* is +0.09 V, and values for other quinone pools in other species are not far away from this value. Thus the redox equilibrium between Q_A, Q_B and Q_P is well in the direction of electron transfer towards the quinone of pool. The biological reason for the *absence* of a stable semi-quinone in the pool may be that it renders the quinol/quinone couple unreactive thereby creating a stable pool of reducing equivalents in the membrane. As such, redox activity of the quinone pool will then be expected to be under the control of specific catalytic sites on the various oxidoreductases that share the pool.

A related device probably operates at the low potential end of the reaction centers of photosystem I and of the Chlorobiaceae. The light-generated reduced ferredoxin is not able to reduce the two-electron-NAD^+ or $NADP^+$ on a single electron basis because the E_m of the $NAD(P)^+$ and its "semiquinone" form is hundreds of millivolts lower than that of ferredoxin. However, the average for a two-electron reduction of NAD^+ or $NADP^+$ (E_{m_7} −0.32 V) is considerably higher than the −0.42 V value for the ferredoxin. The active component of the FNR that may serve as the two-electron gate between the one-electron ferredoxin and the two-electron nicotinamide adenine dinucleotides is most likely FAD. The reduced products of these secondary reactions, NADH or NADPH, represent a stable form of reducing equivalents analogous to the quinol.

3.3.3 The bc_1 Complexes

In contrast to the NADH and NADPH contained in the aqueous phase and available for a variety of biosynthetic processes, the immediate role for the quinol in chloroplasts, mitochondria, and bacteria appears to be as substrate for the quinol-cytochrome c (plastocyanin in chloroplasts) oxidoreductase which plays a vital and central role in respiration and photosynthesis (area 3 on all parts of Fig. 1). The enzyme is commonly called the bc_1 complex (b_6f complex in chloroplasts) and is discussed in general in this Volume in Chapter 10.1 (HAUSKA), for the chloroplast in Chapters 4 (ORT), 10.2 (WHITMARSH) and 10.3 (JOLIOT and JOLIOT) and for photosynthetic bacteria in Chapters 10.4 (PRINCE) and 8.2 (TIEDE). As the name implies, the enzyme catalyzes electron transfer from a quinol downhill to a ferricytochrome c (bacteria, mitochondria) or plastocyanin (chloroplasts).

As we shall discuss later, photosynthetic bacteria can have several forms of cytochrome c that serve as the secondary redox component to the light-generated reaction center, $P^{+\cdot}$. However, for the present discussion attention will be given only to the cytochromes c that link the reaction center with the bc_1 complex. These are water-soluble, and are located in the periplasmic space of the cell and commonly have E_{m_7} values in the 0.26–0.35 V range (that of plastocyanin is 0.35 V). Once oxidized by the reaction center, they diffuse along the membrane-aqueous interface (analogously to the quinol within the membrane) to the bc_1 complex (see WHITMARSH Chap. 10.2, this Vol. on diffusion of plastoquinone and plastocyanin in chloroplasts and of cytochrome c in mitochondria and in bacteria). As shown in Fig. 1D and E, the reaction center, Q, bc_1 complex and cytochrome c form a light-driven electron transfer cycle. This cyclic arrangement contrasts with the organization in chloroplasts where the b_6f complex serves as a "bridge" between photosystem II, which provides plastoquinol, and photosystem I, which provides oxidized plastocyanin, for the complex.

The ΔE_h (typically 0.3–0.35 V) available between QH_2/Q and ferro/ferricytochrome c (plastocyanin) for this process drives a sequence of reactions in the bc_1 (b_6f) complex that includes not only another transmembrane charge separation, adding to that generated by the reaction center, but also a mechanism that pumps protons across the membrane.

Considerable progress has been made in the recent past regarding the constitution of the bc_1-type complexes, although details on the mechanistic sides are only slowly emerging. Since there are several chapters on this topic in this Volume, I shall only briefly, in general form, mention how they link up with the quinol and cytochrome c and simply describe the currently most favored view of how they may operate. Cytochrome c (plastocyanin) makes contact at the high potential end of the complex by interacting with and oxidizing the protein bound cytochrome c_1(f); the cytochrome c_1 in turn oxidizes the Rieske FeS cluster which is important for quinone oxidation. The relationship of the quinone pool with the bc_1-complex is more complicated. The pool can interact on a redox basis with the protein at two apparently separate sites. One site, designated Q_z (or Q_o) serves to oxidize a quinol; the other site, designated Q_c (or Q_i), serves to reduce quinone. The Q_z site catalyzes the oxidation of the two-electron quinol to quinone in a remarkable concerted reaction in which one electron moves to the oxidized high potential Rieske FeS concomitant with the other moving to an oxidized low potential cytochrome b (e.g., in *Rps. sphaeroides* and mitochondria, cytochrome b-566; in chloroplasts, cytochrome b-563). This cytochrome b is thought in turn to reduce another cytochrome b, the redox properties of which are in the region of 100 mV higher (e.g., in *Rps. sphaeroides*, cytochrome b-560; in mitochondria, cytochrome b-562 and in chloroplasts another cytochrome b-563). The Q_c is considered to accept a single electron from the higher potential cytochrome b to form the $Q_c^{-}\cdot$ in a way analogous to that described for the Q_A reduction of Q_B to form $Q_B^{-}\cdot$. Like the Q_B it can be expected (although no evidence yet) that Q_c acts as a two-electron gate and waits stably in the site for another electron. This second electron comes from a second turnover of the bc_1 complex (also like the reaction center). The second turnover occurs when a second quinol is oxidized at the Q_z site followed by an electron transfer sequence like the first, through the two cytochromes b. Thus following this repetition, the $Q_c^{-}\cdot$ in currently viewed to accept another electron, incorporate protons and become a quinol, which is then considered to leave the site and enter the pool. The parallels of Q_B and Q_c are remarkable.

The overall process just described involves the oxidation of two quinols (at the Q_z site) and the reduction of one quinone (at the Q_c site); the requirement to go through this seemingly redundant repetition in the bc_1 complex is, as seems evident with the reaction center, another aspect of the interfacing of the two-electron quinones with the single-electron sequences within these redox proteins. Energetically, the important point is that the bc_1 complex affects the *net* oxidation of one quinol and reduction of two cytochromes c. Evidence is accumulating that supports the idea that the sequence involving electron transfer from Q_z through the b cytochromes to Q_c is directed vectorially across the membrane dielectric and generates $\Delta\psi$ (one charge per electron). Furthermore, the location of Q_B and Q_c nearer the cytoplasmic side of the membrane (Q reduction, proton binding) and Q_z nearer the periplasmic space (QH_2 oxidation, proton release) is considered to be the structural source of net vectorial proton translocation by the complex (see TIEDE, Chap. 8.2, this Vol.). These events in the bc_1 complex therefore change the $\Delta\psi$ value already formed by the reaction center and add a proton concentration gradient which can be seen as a ΔpH. Overall, per photon, each electron moved through the complete cycle elicits the movement of two charges and two protons across the membrane. Some people like to think about this from the standpoint of the oxidation of one quinol, and reduction of two cytochromes c; thus on this basis, two photons are required and the number of charges and protons moved are hence doubled. The overall result of the cooperation between reaction center and bc_1 complex is the generation of an electrochemical gradient of protons given by:

$$\Delta\mu_{H^+ L-R} = \Delta\psi_{L-R} + (RT/F)\ln [H^+_L]/[H^+_R], \\ \cong \Delta\psi_{L-R} - 0.06\,\Delta pH_{L-R}$$

where L and R represent opposite sides of the membrane.

The $\Delta\mu_{H^+}$ so formed is available for the cell to drive further membrane-linked energy-consuming reactions, as proposed originally by MITCHELL (1966, 1968). In general, experiments on the bc_1 complex from a variety of sources of all kinds support the above-described Q cycle model of MITCHELL (1975) with the added double-turnover amendment of the Q_c site introduced by Crofts (see CROFTS and WRAIGHT 1983). It should be said, however, that there is still room for changes in the way we view the function of the complex; more work in particular is needed with regard to the mechanism of proton translocation (see PRINCE et al. 1982; RICH 1984 and WIKSTROM and SARASTE 1984). Further, the actual values of $\Delta\psi$ and ΔpH have for years been a major bone of contention; indeed so has the issue or whether the $\Delta\mu_{H^+}$ is truly delocalized in the bulk aqueous phase on either side of the membrane, or whether there is a more local handling of the protons with respect to energy transmission to the other energy-requiring systems of the membrane (see MELANDRI and VENTUROLI Chap. 10.6, this Vol. and the review by WESTERHOFF et al. (1984a, b) for further discussion on this important matter).

3.3.4 NAD^+ Reduction Coupled to $\Delta\mu_{H^+}$

In bacterial species of the purple Chromatiaceae and Rhodospirillaceae and of the green Chloroflexaceae families, where quinol is the only biochemically available reduced product of the reaction center, the $\Delta\mu_{H^+}$-generating cyclic electron transfer system of the reaction center and bc_1 complex provides the free-energy requirement for the reduction of NAD^+. The other, equally important requirement (i.e., the catalytic vehicle) is provided by a second membrane redox enzyme that is energetically coupled to the $\Delta\mu_{H^+}$ generated by the cycle. This membrane redox enzyme is most probably one similar to the NADH-ubiquinone oxidoreductase shown in area 4 of Fig. 1B for mitochondria, and shown in Fig. 1D and E for photosynthetic bacteria. In the mitochondrion, as its name indicates, the enzyme is more commonly thought about in terms of a net electron transfer from NADH to ubiquinone. This is because in mitochondria and analogous respiratory systems NADH is obtained from reduced carbon compounds derived from imported metabolic sources. Furthermore, the ΔE_h (also 0.3–0.35 V available between NADH and ubiquinone) is converted into $\Delta\mu_{H^+}$ required by the cell. In the purple bacteria the physiological balance is reversed: $\Delta\mu_{H^+}$ is obtained via the light absorption and operation of the cycle, but NADH is required by the cell. CHANCE and HOLLUNGER (1961) first demonstrated in mitochondria that conditions could be arranged so that the NADH-ubiquinone oxidoreductase could be seen to operate in the reverse direction when provided with energy: Chance's "reversed electron flow". GEST (1963) was quick to recognize the value of this natural consequence of reversible systems with respect to the issue of NADH production in photosynthetic bacteria. Since his original suggestion there has been mounting evidence that NADH is produced in purple bacteria by a similar enzyme operating in reverse and driven by the light-driven activities of the reaction center and bc_1 complex as shown in Fig. 1D and E (see KNAFF 1978; MALKIN et al. 1981; DOI et al. 1982). Thus the overall process can be summarized, perhaps somewhat ideally, as follows:

a) The light-activated (two photons) reaction center reduces quinone to quinol and oxidizes two ferrocytochromes c to ferricytochrome c yielding $\Delta\psi$ and a ΔE_h in the 0.3 to 0.35 V range.

b) The bc_1 complex catalyzes the oxidation of quinol and reduction of two ferricytochromes c to regenerate the quinone and ferrocytochrome c, and to convert the ΔE_h into further $\Delta\psi$ and ΔpH.

c) As illumination continues, the free-energy associated with $\Delta\mu_{H^+}$ builds up across the membrane to a point approaching and opposing that associated with ΔE_h. As this progresses, the bc_1 complex begins to sense an energetic feedback from $\Delta\mu_{H^+}$, making the passage of reducing equivalents from quinol through the bc_1 complex to cytochrome c less feasible. However, the NADH-ubiquinone oxidoreductase also responds to the $\Delta\mu_{H^+}$ and, acting in a converse manner, the $\Delta\mu_{H^+}$ makes it more feasible energetically to catalyze the oxidation of quinol to quinone and to reduce NAD^+ to NADH. The stoichiometry of charges and photons moving back across the membrane per NAD^+ reduced has not been established in bacteria and it is not certain from studies with respiratory systems (for a recent study with mitochondria see FREEDMAN and LEMASTERS 1984).

d) The supply of electrons to replace those elevated to NAD^+ comes from oxidizable substrates as discussed in Section 3.3.6.

3.3.5 Direct Versus Indirect $NAD(P)^+$ Reduction

The majority of photosynthetic bacteria probably use the indirect "reverse electron flow" strategy for NAD^+ reduction described immediately above. This includes the large families of purple Rhodospirillaceae, and Chromatiaceae as well as most probably (no evidence) the green nonsulfur Chloroflexaceae. The temporal priority in these cases is clearly $\Delta\mu_{H^+}$ generation. As just described, the NADH ubiquinone oxidoreductase then taps into this source of free-energy along with other energy-requiring membrane systems such as solute transporters and ATPases. In comparison, at least in the bacterial world, the direct mode of NADH production as used by the Chlorobiaceae seems very much in the minority. Despite the capability of this family to effect direct NAD^+ reduction without requiring an intermediate $\Delta\mu_{H^+}$, the cells nevertheless require $\Delta\mu_{H^+}$ to drive ATP synthesis necessary for CO_2 reduction and other functions. In such bacteria $\Delta\mu_{H^+}$ is usually viewed (based on little experimental information) as being generated by the operation of a cyclic electron transfer system of the reaction centers and a bc_1 complex as shown in Fig. 1C.

The situation of direct NAD^+ reduction and $\Delta\mu_{H^+}$ requirement evident in the Chlorobiaceae seems to have its counterpart in the photosystem I reaction centers of chloroplasts. Thus, the chloroplast photosystem I reaction center, in addition to driving ferredoxin-mediated, direct $NADP^+$ reduction, provides the reduced ferredoxin for an array of functions (see ARNON 1977), one of which is the operation of a cycle about photosystem I in association with the b_6f complex (see SHAHAK et al. 1981; O'KEEFE 1983). The cycle in the chloroplast is commonly viewed as auxillary to the noncyclic flow from photosystem II through the b_6f complex to photosystem I, serving to boost $\Delta\mu_{H^+}$ to maintain ATP production and associated processes optimally matched with NADPH production.

The details of the connection between ferredoxin and the bc_1 or b_6f complexes of the Chlorobiaceae and chloroplasts are far from being understood. The chloroplast studies are incomplete, while in the case of Chlorobiaceae there is no solid evidence that such a cycle exists.

Another puzzling aspect regarding both chloroplast and Chlorobiaceae is the composition of the cycle. If the cycle in each case is geared to generating $\Delta\mu_{H^+}$, then why has it not taken fuller advantage of the ΔE_h that exists between ferredoxin and quinone? Insertion of an NAD(P)H-quinone oxido-reductase, akin to that found in mitochondria and the purple bacteria, operating in series with the quinol-cytochrome c (plastocyanin) oxidoreductase, would double the $\Delta\mu_{H^+}$ generation capacity of this cycle and would be equivalent to two thirds of the mitochondrial respiratory system!

3.3.6 The Supply of Electrons from Oxidizable Substrates

Having described the system of reaction centers, bc_1 complexes and NADH-ubiquinone oxidoreductases, and the net flows of electrons to nicotinamide adenine dinucleotides we can now deal with the role of the reaction center in the net import of electrons from reduced substrates.

Photosynthetic bacteria are perhaps best considered in the light of well-established patterns of mitochondria and chloroplasts. In mitochondria (and respiratory bacteria), after import of the reduced substrates across the membrane, contact with the respiratory chain is made on the inner, matrix (cytoplasm in bacteria) side of the plasma membrane. The supply enters the respiratory redox system mainly via NADH at the low potential end of the respiratory chain. Similarly well-known points of entry of electrons into the respiratory system of mitochondria and bacteria are provided by the several dehydrogenases that contact the same, inner side of the membrane but at the level of the quinone pool. Figure 1 B shows the common example of succinate oxidation to fumarate (E_{m_7} +0.03 V) via succinate dehydrogenase (SDH).

Anaerobic heterotrophic growth of photosynthetic bacteria on organic substrates such as succinate is possible by members of Chromatiaceae (e.g. *Chromatium vinosum*), the Rhodospirillaceae (e.g., *Rhodospirillum rubrum*, *Rhodopseudomas sphaeroides*) and Chloroflexaceae. As with the respiratory systems the substrate is imported into the cell, and via dehydrogenases makes delivery of reducing equivalents from the cytoplasmic side into the membrane Q pool as shown in Fig. 1 E in a manner similar to mitochondria (Fig. 1 B). The membrane-associated SDH of mitochondria contains, in addition to three different iron-sulfur centers and a flavin, a pair of protein-associated quinones called Q_S. The quinones of the pair are interesting; they both form relatively stable semiquinones (i.e., $Q^{-\cdot}-Q^{-\cdot}$) that are spin-coupled (see review by OHNISHI and SALERNO 1982). Membrane-associated SDH enzymes that appear similarly equipped have now been characterized in several photosynthetic bacterial species (see, for example, CARRITHERS et al. 1977; INGLEDEW and PRINCE 1977). However, apart from one demonstration (R.C. PRINCE unpublished) of its presence in *Rps. sphaeroides*, the Q_S pair has not been characterized. In general, details of the catalytic process of succinate oxidation by SDH are still not entirely resolved, although considerable progress has been made. Succinate is considered to reduce FAD on a 2H basis; the $FADH_2$ then interacts with the one-electron FeS clusters, and these make final contact with the Q pool through the action of the Q_S pair, perhaps in a way related to that discussed above for Q_A and Q_B. Thus, catalyzed by SDH, it is reasonable to consider that succinate supplies the quinone pool with reducing equivalents to maintain electrons in the cycle and to replace those driven to NAD^+. The net product is fumarate, which moves into the carbon metabolism of the cell (see GEST 1982).

Autotrophic growth of photosynthetic bacteria brings us closer to the situation in chloroplasts where reducing equivalents ear-marked for NADPH and CO_2 reduction are replenished by the action of the reaction center of photosystem II. Photosystem II has the capability to oxidize water to yield dioxygen (see ORT Chap. 4, DINER 9.1 and YOCUM 9.2). Photosynthetic bacteria of all species can, like chloroplasts, use CO_2 as a carbon source and grow autotrophically. However, instead of the vast reduced oxygen source (H_2O), bacteria utilize a variety of reduced sulfur compounds as a source of inorganic reducing equivalents. The parallels of H_2O and O_2 and H_2S and S were the basis of van Niel's unifying ideas on photosynthesis (VAN NIEL 1931, 1935) which still attract interest as further details of the two systems emerge (see GAFFRON 1963; WRAIGHT 1982). It is generally agreed (but see later) that the entry of reducing equivalents from reduced sulfur compounds into the light-driven electron transport system of photosynthetic bacteria occurs via a cytochrome c which interacts with the light-generated $(BChl)_2^{+\cdot}$ of the reaction center (Fig. 1 C and D). Thus reducing equivalents from these substrates are viewed as converging with the cyclic electron transfer system at the point of cytochrome c. Ignoring for now the cycle, this pattern of substrate-linked electron flow is clearly similar to that of the chloroplast photosystem II. Nevertheless, the important differences that exist between the two systems are readily apparent at the spectral and redox levels. Apart from differences in the peak positions of the absorbance and hence the energy of the light absorbed into the reaction center,[1] the E_{m_7} values of the $(BChl)_2/(BChl)_2^{+\cdot}$ couple in purple bacteria (in the 0.4 to 0.5 V range), fall far below that displayed by photosystem II (estimated 1.1 V) and well out of reach of even the average value for the four-electron H_2O/O_2 couple (E_{m_7} 0.82 V, n=4). Here lies the practical source of discrimination between the oxygenic chloroplasts and the anoxygenic photosynthetic bacteria. The chloroplast has equipped itself with a reaction center that absorbs light with some 0.4 V higher energy than that of the photosynthetic bacterial counterpart, and it is clear that the extra energy is tacked on to the oxidizing end to help achieve higher redox potentials. Furthermore, the chloroplast photosystem II has firmly associated with its reaction center complex a unique enzyme which comprises a highly choreographed "four-electron gate" designed to accumulate four oxidizing equivalents that are used to abstract four electrons from $2H_2O$ to yield O_2 (see YOCUM, Chap. 9.2, this Vol.). The basic strategies involved with this accumulation are the same as with the previously discussed two-electron gates in that (1) there is a thermodynamic incapability on the part of the reaction center P-680$^{+\cdot}$ to oxidize H_2O in separate, one-electron steps (for example, the E_{m_7} of $H_2O/OH^{\cdot}$, the first couple of the single electron oxidation step, is estimated at 2.35 V), (2) thermodynamic capability is achieved when the average of multiple electron steps is presented to H_2O, (E_{m_7} of $2H_2O/O_2$ is 0.82 V) and (3) the product, O_2, is stable in part due to the relatively low redox character of its first single electron reduction step (E_m of $O_2/O_2^{-\cdot}$ is -0.33 V) puts it out of reach of most reductants. The oxidizing potential of O_2 is realized fully only when it encounters an appropriately designed multi-electron donating enzyme such as the oxidases.

Thus in the absence of the ability to use reduced oxygen compounds, purple and green bacteria obtain reducing equivalents from the lower potential reduced sulfur compounds. Most common are sulfide and thiosulfate. There is, however, a daunting array of polymeric forms and oxyacids of sulfur that can serve as substrates or appear as stable or transient intermediates in the dissimilatory processes. In contrast to the membrane-associated, water-oxidizing enzyme attached to the photosystem II reaction center, which appears to maintain a tight control over the intermediates of water oxidation, the oxidation and reduction of sulfur compounds appears to be handled in a much freer way with intermediates appearing and disappearing in the cultural liquors during substrate dissimilation. The redox properties of the multiplicity of sulfur compounds vary in E_m values over a large range from $+0.25$ V to values that are well below

1 The actual values are 680 nm for the photosystem II, in the 850–890 nm range for BChl-a-containing purple bacteria, and approximately 960 nm for the BChl-b-containing *Rps. viridis*: these numbers translate into different energy levels of the lowest excited singlet state which governs the energy level of the light (approximately 1.8 eV for photosystem II; 1.2–1.4 eV for bacteria).

those of the NAD and NADP couples (see THAUER et al. 1977; TRÜPER and FISCHER 1982; FISCHER 1984).

The capability to use elemental sulfur itself as a substrate is used to distinguish the purple and green sulfur bacteria from the nonsulfur utilizing species. Thus while members of the Chlorobiaceae and Chromatiaceae can oxidize elemental sulfur, members of the Rhodospirillaceae or Chloroflexaceae cannot. The segregation based on elemental sulfur dissimilation (i.e., on the presence or absence of perhaps one enzyme?) seems arbitrary. Indeed, despite their predilection for photoheterotrophic growth, Rhodospirillaceae and Chloroflexaceae can use sulfide and/or thiosulfate as electron donors for autotrophic growth. However, as expected in these species, any sulfur produced in the process is deposited outside the cell and is discarded (see PFENNIG 1978; TRÜPER and PFENNIG 1978; and FISCHER 1984).

It is pertinent at this point to ask how the chloroplasts handle sulfur, since both reaction centers have the potential to oxidize reduced sulfur compounds. It is found that sulfide is an inhibitor of the water-splitting enzyme associated with the reaction center of photosystem II (just as it is with the oxygen-reducing mitochondrial type of cytochrome oxidase) and there has been no report of an alternative physiological route for electrons from reduced sulfur compounds into the photosystem II reaction center. On the other hand, the reaction center of photosystem I is functional in the presence of reduced sulfur compounds: work with the cyanobacterium *Oscillatoria limnetica* (COHEN et al. 1975) has shown that sulfide can be utilized as a source of electrons by the reaction center of photosystem I. However, like the Rhodospirillaceae and Chloroflexaceae, but unlike the more analogous Chlorobiaceae, it cannot oxidize it beyond sulfur which accumulates.

Some of the enzymes and redox groups involved in the oxidation of reduced sulfur compounds by photosynthetic bacteria are briefly introduced below. However, because of the varieties, exceptions, and elaborations in species, cytochromes, and sulfur compounds, as well as a great deal of uncertainty, the discussion is necessarily simplified. The discussion is aimed at providing a view of this counterpart to the reduced oxygen of plants that is often overlooked. For further details on the dissimilatory pathways and the enzymes and cofactors that may be involved in sulfur oxidation and reduction see a recent review by FISCHER (1984); further details of the redox chemistry of sulfur (and its involvement in sulfate-reducing bacteria) can be obtained from THAUER et al. (1977); more information on the cytochromes of photosynthetic bacteria can be obtained from the review of BARTSCH 1978).

a) Sulfide

Sulfide oxidation to sulfur (E_{m_7} of H_2S/S: -0.27 V, $n=2$)[1] has been proposed to be catalyzed by an FAD-containing diheme cytochrome c that is typified

1 The E_{m_7} values reported of the sulfur compounds are taken from the review by THAUER et al. (1977). They are reported there as E_0' values which were calculated from the $\Delta G_0'$ for each redox reaction determined in terms of coupled hydrogen oxidation/reduction. The primed constants usually imply measurements done at around neutral pH but not always at pH 7. Since most of the free energies and E_m values will be pH-dependent and since I have not been able to obtain source material, the indication of pH 7 must be taken with caution.

by the water-soluble flavocytochrome c-552 of *Chr. vinosum*. In *Chr. vinosum* the E_{m_7} of the hemes is close to 0 V while details of the FAD/$FADH_2$ couple and the stability of its semiquinone form are lacking. GRAY and KNAFF (1982) demonstrated the flavocytochromes' ability to act as a sulfide-cytochrome c (horse heart) oxidoreductase; they also showed that two horse heart cytochrome c molecules could form a functional complex with one flavocytochrome c-552. The electron transfer sequence proposed was that the FAD served to accept two reducing equivalents from sulfide to yield sulfur and $FADH_2$. The FAD here is presumably acting with the sulfide on a 2H basis while facilitating subsequent one electron transfers to each of the cytochrome c hemes presumably with the release of protons. The authors propose that in vivo the cytochrome c_2-like cytochrome c-551 (E_{m_7} 0.26 V; see VAN GRONDELLE et al. 1977) serves as the oxidant to the cytochromes c-552 hemes of the flavocytochrome and this mediates electron transfer to the *Chr. vinosum* reaction center-bound cytochrome c-555 (E_{m_7} 0.35 V) (see later) before transfer to the $(BChl)_2^{+\cdot}$ (E_{m_7} 0.48 V). The sequence appears to be a straightforward exothermic sequence that occurs in the periplasmic space of the cell.

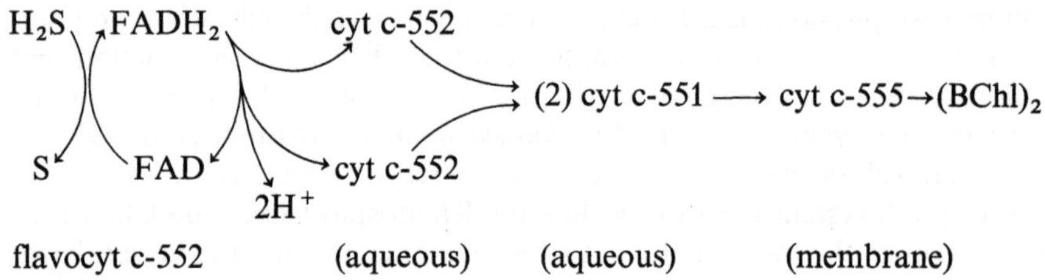

Similar (flavo) cytochromes c-552 and candidates for the role of cytochrome c-551 appear in other representative strains of Chromatiaceae and Chlorobiaceae.

Sulfide oxidation by Rhodospirillaceae (e.g., *Rsp. sulfidophila*) without involving sulfur production as a biproduct has been suggested to involve sulfite and thence to the highest oxidation state of sulfur, sulfate [E_{m_7} of $SO_3^=/SO_4^=$ is -0.54 V, (n=2)]. However, the details of the initial six-electron oxidation of sulfide to sulfite [E_{m_7} of $S^=/SO_3^=$ is -0.11 V (n=6)] are a mystery. One suggestion to explain the reaction involves a siroheme-containing enzyme. Such an enzyme is known in a variety of plants, fungi, and bacteria to catalyze the reverse reaction (i.e., sulfite reductase) reducing sulfite to sulfide. While a similar enzyme is induced in autotrophically grown *Chr. vinosum* (CO_2 and sulfide), it has not yet been found elsewhere. If this enzyme does indeed prove to be functional in photosynthetic bacteria and if it does operate in the opposite direction to oxidize sulfide, then there is the question of what will serve as the oxidant. However, in accepting that the enzyme may not be general to catalyze the reaction in photosynthetic bacteria, TRÜPER and FISCHER (1982) have suggested an alternative sequence for sulfide oxidation to sulfite. They propose that a flavocytochrome oxidizes the sulfide first to thiosulfate before further steps form sulfite, as discussed below.

b) Thiosulfate

Oxidants for the subsequent oxidation of thiosulfate in the enzyme thiosulfate: acceptor oxidoreductase yielding tetrathionate [E_{m_7} $S_2O_3^=/S_4O_6^=$ 0.024 V (n=2)] are as the name implies, again uncertain. What happens to the tetrathionite is also uncertain. In *Chr. vinosum*, several candidates for the immediate oxidant of thiosulfate have been presented including those mentioned for sulfide above. For example, VAN GRONDELLE (1977), provided evidence that suggested that cytochrome c-551 was on the oxidative pathway, although not necessarily the immediate oxidant. Indeed, several other oxidants have been proposed to mediate between enzyme and reaction center, including flavocytochrome c-552 from the "particulate" fraction, and the high potential iron-sulfur protein (HIPIP; E_{m_7} +0.35 V n=1) (FUKUMORI and YAMANAKA 1979). Whether these results represent a multiplicity of enzymes or lack of specificity cannot be decided from the experiments. Two other routes for thiosulfate dissimilation involve two "splitting enzymes" which yield sulfite as a product in each case: (1) Rhodanase (thiosulfate-cyanide sulfur transferase) which has been detected in several photosynthetic bacteria yields sulfite (and thiocyanate); and (2) a thiosulfate reductase catalyses an initial reduction of thiosulfate to sulfite and sulfide. [E_{m_7} for $S^=$; $SO_3^=/S_2O_3^=$ is −0.4 V (n=2)]; the electron donor, which presumably can deliver a redox potential in this region for this initial reductive investment is unknown. Once formed, sulfide and sulfite are then oxidized by other enzymes leading overall to the net oxidation of thiosulfate.

c) Sulfite

It is interesting to note that very few bacteria grow on sulfite, but it appears to be an important intermediate in all bacterial families. Sulfite can be oxidized by adenylsulfate (APS) and an ADP sulfurase. The APS reductase appears to have a similar FAD-diheme content as described for sulfide oxidation but this enzyme contains additional nonheme iron. In this case, in addition to reduction of the cytochrome c, adenosine monophosphate (AMP) is converted into APS [$SO_3^=$; AMP/APS, E_{m_7} −0.06 V (n=2)]. The reaction leads to the reduction of two cytochromes c and subsequent reactions are proposed to be in the same pattern as described earlier for sulfide oxidation leading to reaction center $(BChl)_2^{+\cdot}$ reduction. The APS is subsequently converted via ADP sulfurase to yield sulfate and adenosine diphosphate (ADP).

Sulfite can also be oxidized directly to sulfate via a sulfite-acceptor oxidoreductase-yielding sulfate (E_{m_7} for $SO_3^=/SO_4^=$ is −0.54 V n=2). As the name implies, the nature of the oxidant is again obscure; in this case cytochrome c′ has been suggested (see TRÜPER and FISCHER 1982). Another possibility for sulfite oxidation involves a combination with thiosulfate to form trithionite [E_{m_7} for $SO_3^=$; $S_2O_3^=/S_3O_6^=$ is +0.225 (n=2)] oxidant. However, while this redox reaction is proposed in nonphotosynthetic, sulfate-reducing bacteria, as far as I know it has not been demonstrated in phototrophes.

The foregoing discussion reveals a complex and uncertain situation. Reducing equivalents are offered in the cell from a variety of redox couples of widely disparate E_m values and in some cases involve dismutation and multiple reactants. The simplest set of oxidative pathways is present in species of the Rhodospirillaceae family which have been proposed to employ thiosulfate: acceptor oxidoreductase, the splitting enzymes, the "reverse" sulfite reductase (proposed for sulfide oxidation), and sulfite: acceptor oxidoreductase to finally yield sulfate. The Chromatiaceae and Chlorobiaceae are more elaborate; each contains in addition the APS reductase and the ADP sulfurase system for oxidation of sulfate to sulfate as well as enzymes that can handle sulfur. However, members of the same families display fairly disparate sets of reduced sulfur-oxidizing enzymes including some species that do not have any at all. The enzymology is still incomplete and, with only one or two promising exceptions, unambiguous information regarding substrate enzyme-oxidant/reductants in vivo is scant. Furthermore, the biochemical foundations seem very weak upon which are built the chains of redox components (such as cytochromes c, HIPIP etc.) that are proposed to guide reducing equivalents from sulfur-oxidizing enzymes to reaction center $(BChl)_2^{+\cdot}$. The source of substrate-linked electron flow shown in Fig. 1C and D reflects this widely held view that reducing equivalents enter the photosynthetic system at the level of cytochrome c immediate to the $(BChl)_2$ of reaction center. While this "noncyclic electron transfer" route cannot be dismissed, other perhaps more attractive possibilities might be considered too. For instance, more economical use of the reducing equivalents from the sulfur compounds could be made if entry into the system was made at the quinone pool level as established for succinate in Fig. 1E. This suggestion is possible on thermodynamic grounds, since all the redox couples discussed above (except the, unfounded, last one) have E_m values that are equal to or lower then that of the quinone pool. Thus it may be worth reviewing the admittedly scant information that we have regarding this alternative or additional possibility that redox equivalents enter at the level of the Q pool.

Most of the studies of noncyclic (substrate-linked) electron transfer have been done hand in hand with the measurement of $NAD(P)^+$ reduction. Using this rather distant assay with the purple photosynthetic bacteria there is no way of discriminating between the two alternatives. This is because in these species (Fig. 1D and E) NADH production requires $\Delta\mu_{H^+}$; thus both alternatives will be sensitive to uncouplers that dissipate $\Delta\mu_{H^+}$ generated by the cycle, and both will be sensitive to inhibitors of the bc_1 complex such as antimycin, myxothiazol or HOQNO that restrict $\Delta\mu_{H^+}$ formation. With Chlorobiaceae, however, the same criterion can be used to resolve the two alternatives: if the reduced sulfur compound donates electrons as shown in Fig. 1C, then NAD^+ reduction, apart from being insensitive to uncouplers (see KNAFF 1978), may be expected *not* to be sensitive to the usual inhibitors of the bc_1 complex. In contrast, if redox interaction is at the Q pool level, then, in order to reach the reaction center, it will first have to pass through the bc_1 complex and should be sensitive to inhibitors such as antimycin. The only work addressing this problem has been done with *Chlorobium limicola* by KNAFF and BUCHANON (1975), who followed NAD^+ reduction in the presence of sulfide as the reducing source. They provided evidence that the process was indeed inhibited by antimycin. However, at the time of the work the site of action of antimycin inhibition was identified as "closely associated with b-cytochromes"; hence b-type cytochrome was considered as a constituent of the noncyclic electron transfer pathway from sulfide to the reaction center. However, the mode of action of antimycin has since been clarified. The only point of interaction known for antimycin in mitochondria and bacteria is at the Q_c site of the bc_1 complex. It has been established in several systems that the existence of antimycin sensitivity correlates with the demonstrable presence of the Q_c semiquinone which is abolished on addition of the inhibitor. The abolishment probably results from the occupation of the site by antimycin to the exclusion of quinone, and the pronounced effects on the oxidation-reduction of cytochrome b associated with the bc_1 complex can be understood in these terms (see ROBERTSON et al. 1984a, b). Thus it is reasonable to consider that the "noncyclic electron transfer" pathway in *Chlorobium limicola* actually takes place by entry at the quinone level and subsequent passage through the energy-coupled bc_1 complex en route to the reaction center for direct light-driven delivery to $NAD(P)^+$. As such the functional organization shown in Fig. 1C for Chlorobiaceae may in fact look more like that of Fig. 1E.

A similar conclusion might be drawn regarding the Chromatiaceae. Information on the noncyclic and cyclic pathways emerged during the studies of VAN GRONDELLE et al. (1976, 1977) on *R. rubrum* and *Chr. vinosum*. The focus of the studies of these workers was on the action of the water-soluble cytochrome c (c-551 in *Chr. vinosum* as previously discussed) located in the periplasmic space. They concluded that the cytochrome c in each case diffused between reaction centers and what now is reasonable to consider is a bc_1 complex. They further suggested that the cytochrome c was a confluence of electrons from the light-driven cycle and a separate, noncyclic route from substrate, as shown in Fig. 1D. This work is often quoted (as earlier in this chapter) when discussing substrate oxidation. However, this issue was not the main point of the study and the experiments are far from complete. Furthermore, with more recent advances in the understanding of the intricate workings of the bc_1 complexes, especially with regard to the use of short activating light flashes, the sensitivity of observed kinetics of ambient conditions and the effects of inhibitors, there is ample room for a re-evaluation of this work in terms of the Q pool being the confluence and the "noncyclic pathway" being a passage through the bc_1 complex. If this proves to be true then, as is the case for succinate oxidation in Fig. 1E, the need to recognize a "noncyclic pathway" of substrate oxidation converging at cytochrome c, at least as a primary route for reducing equivalents into the system, becomes redundant. This could be true for all the photosynthetic bacteria.

The advantage of the proposed alternate arrangement would be that more free energy would be conserved from substrate oxidation. Instead of the large loss of free energy that will occur during oxidation of the substrate by the reaction center in the noncyclic pathway formalism (the ΔE_h is commonly between 0.5 and 1 V) the alternative view would diminish this "loss" to between zero and -0.5 V. It should be acknowledged that the traditional noncyclic mechanism is not devoid of a contribution to the energy of the cell; it will lead to the transfer of one charge across the membrane per substrate electron passing through the reaction center. However, the alternative proposal, which includes passage through the bc_1 complex, will provide an additional charge and $2H^+$ translocated per substrate electron entering the system.

This brings me to the final point regarding this poorly understood area. I am not aware of any studies that address either the possible *transport* of sulfur compounds across the cytoplasmic membrane, or deal with the consequences of their oxidation by dehydrogenases or oxidoreductases that are integral to the membrane. However, a more critical investigation of the membrane with alternative models in mind may reveal not only the "oxidants" that are currently lacking for several of the sulfur dissimilatory water soluble enzymes but also add a very interesting membrane energetic aspect to the dissimilatory processes of reduced sulfur compounds.

3.4 The Reaction Center – Cytochrome c Complex

During our previous discussions we have mentioned the nature of the electron donors associated with the reaction centers of chloroplast photosystem II (the hydrophobically associated water oxidizing system) and photosystem I (the water soluble plastocyanin). The proven, immediate secondary redox components to the reaction center $P^{+\cdot}$ or $(BChl)_2^{+\cdot}$ of photosynthetic bacteria appear uniformly to be c-type cytochromes. However, these also fall into two classes regarding the nature of their association with the reaction center proteins. Some, typified by the Rhodospirillaceaes family's *Rps. sphaeroides*, *Rps. capsulata*, and *R. rubrum*, interact directly with the reaction center and bc_1 complex and these have already been discussed as key parts of the light-activated cyclic system. In low ionic strength media these cytochromes display dissociation constants with the reaction centers and bc_1 complex in the micromolar range, but are easily removed from the membrane proteins by salt washing. Other species of bacteria possess hydrophobically linked cytochromes c which are not removed by salt washing. Neither are they necessarily dissociated from the reaction center even during the normal detergent treatments encountered during isolation and purification of the reaction centers. General information is limited, but it is clear that in the cases of *Chr. vinosum* and *Rps. viridis* (Thornber et al. 1969) the cytochromes remain firmly attached and functionally competent; in one other reported case, *Rps. gelatinosa* (Prince et al. 1978) the cytochromes can dissociate; a similar situation may exist for the cytochrome c of *Chloroflexus aurantiacus*, which is reported to display characteristics intermediate to the two extremes of cytochrome c-reaction center association (see Blankenship and Fuller, Chap. 8.6, this Vol.).

The hydrophobically linked cytochromes c represent a mystery. Functionally they do not appear to be an alternative to the water-soluble cytochrome c; for example, the soluble cytochrome c-551 in *Chr. vinosum* still appears to be required to diffuse between the bc_1 complex and the reaction center–cytochrome c complex (van Grondelle et al. 1977); a similar c_2-like cytochrome has been proven present in *Rps. viridis*. Furthermore, the hydrophobically linked cytochromes, when present, are evident in multiple forms based on both spectrometric and potentiometric criteria, and all appear to be oxidizable by the same reaction center $(BChl)_2^+$. Despite the mystery from the physiological standpoint, these heme proteins represent a fascinating and viable system for the basic exploration of biological electron transfer mechanisms per se. Certainly the three-dimensional structural information on *Rps. viridis* presented by MICHEL and DIESENHOFER (Chap. 8.4, this Vol.) will revitalize interest in these cytochromes regarding the redox and structurally related interactions between the hemes and $(BChl)_2$. Hopefully, attention will also be given to their physiological role.

3.4.1 General Observations

At present, information on these cytochromes in photosynthetic bacteria is far from complete. However, some common patterns can be deduced from the limited information that has been reported. Although a significant number of strains and species have been examined for cytochrome oxidation kinetics and spectra by KIHARA and associates (KIHARA and CHANCE 1969; KIHARA and DUTTON 1970; KIHARA and MCCRAY 1973), only four species have received quantitative analyses that necessarily blend a general perspective with a detailed view of the cytochrome oxidation-reduction that includes connected reactions and coupled events. The species are *Chr. vinosum* and *Rps. viridis* (see CASE et al. 1970; CASE and PARSON 1971; DUTTON et al. 1971; KENNEL and KAMEN 1971), *Rps. gelatinosa* (DUTTON 1971) and *Thiocapsa pfennigii* (PRINCE 1978). The results from work on these species are the basis of the following generalizations and suggestions that might be useful guidelines for further experimentation. See also DUTTON and PRINCE (1978) for further references.

a) The multiplicity of cytochromes is most commonly seen as high potential and low potential cytochromes. The high potential hemes display E_m values in the pH 7–8 range at 0.35 V for *Chr. vinosum*, *Rps. viridis* and *T. pfennigii* and 0.28 V for *Rps. gelatinosa*. The low potential hemes are considerably lower, in the 0.0 V region for *Chr. vinosum*, *Rps. viridis* and *T. pfennigii* and 0.13 V for *Rps. gelatinosa*. The spectra of the reduced forms of high potential cytochromes are at longer wavelengths than the low potential partners [*Chr. vinosum* 555, 553 nm; *Rps. viridis* 558, 553 nm; *T. pfennigii* 555, 550 nm; and *Rps. gelatinosa* 553, 548 nm (measured at 77 K)]. The oxidized forms of the high and low potential forms of cytochromes of *Chr. vinosum* are also distinct, as shown by EPR analysis (TIEDE et al. 1978).

b) In *Chr. vinosum*, *Rps. viridis*, and *T. pfennigii* it appears that there are two high and two low potential hemes per reaction center; in *Rps. gelatinosa* there may be one of each. It should be realized, however, that with the exception of *Rps. viridis* (see MICHEL and DEISENHOFER, Chap. 8.4, this Vol.) the ratios of hemes to reaction centers in the above species are not based on direct analysis and hence should not be taken as conclusive.

c) Kinetic analysis of the single-turnover flash-activated cytochrome c oxidation generally has led to the conclusion that both high potential hemes have an equal probability of being oxidized by the single oxidizing equivalent on the $(BChl)_2^{+\cdot}$ of the reaction center. The same conclusion is made for the low potential pair. It is also the case, however, that when all hemes are reduced prior to activation, the low potential hemes are the ones seen to undergo oxidation. The low potential hemes (where measured) display faster oxidation rates than the high potential forms, but it should be said that this alone cannot account for the apparently preferred oxidation of the low potential heme. In *Chr. vinosum* the high potential cytochrome c-555 is oxidized with a half-time of 2 μs while the low potential cytochrome c-553 is oxidized with a halftime of 0.8 μs, while in *Rps. gelatinosa* (at 77 K) the half-times are 60 μs and 5–10 μs respectively. The oxidation half-times for the high potential cytochrome c-553 high in *Rps. viridis* is 0.27 μs, while that of the low potential cytochrome c-553 remains to be established.

d) The high and low potential hemes also differ with respect to the temperature dependencies of their light-activated oxidation. The low potential cytochromes of species detailed above and many others tested by KIHARA display the remarkable property of undergoing light-induced oxidation down to cryogenic temperatures. The oxidation reaction is irreversible. The temperature dependency of the rate commonly shows a small activation energy at the higher temperatures (e.g., 3.5 kcal mol^{-1} for *Chr. vinosum*; 2.3 kcal mol^{-1} for *Rps. gelatinosa* which diminishes to zero at temperatures below 100 K. This pattern of behavior is the basis of several models which explain the event in terms of a single mechanism of thermally assisted electron tunneling (see HOPFIELD 1974; JORTNER 1980; and contrast with DEVAULT and CHANCE 1966).

The oxidation kinetics high potential cytochromes also shows a similar temperature dependency, but, as the temperature is lowered, there is an additional effect which diminishes the amplitude of the reaction. In *Chr. vinosum* for example, only 5% of the high potential cytochrome c-555 is measurable at 77 K; in *Rps. gelatinosa* at the same temperature about 50% is still competent to undergo oxidation (DUTTON 1971). It is worth noting that by "competent" we mean capable of oxidation in milliseconds or less following a short activation flash. Conversely, the incompetent fraction does not undergo oxidation at a significant rate even upon prolonged illumination at 77 K. Thus it appears that, in contrast to the low potential cytochrome c oxidation, a second kind of temperature-dependent parameter governs the transfer of electrons from the high potential cytochromes c to the light generated $(BChl)_2^{+\cdot}$. While the nature of this "freezing-out" effect is unknown, it seems clear that as the temperature is lowered the cytochrome can assume a configuration that renders it unable to transfer an electron to the $(BChl)_2^{+\cdot}$ (DUTTON 1971).

It is worth adding that in *Rps. gelatinosa*, which supports a significant population of high potential cytochrome c photooxidation at 77 K, the reaction is reversible; again, however, only a fraction of that oxidized is reversibly reduced. Thus in *Rps. gelatinosa* at 77 K, of the half of the high potential cytochrome c-553 that is competent to undergo oxidation about half is re-reduced in ~10 s in the dark; this reversible population can be repeatedly photo-oxidized. These observations suggest that, in contrast to the low potential cytochromes c, several clear configurational states may be resolvable and important in the governance of electron transfer between the hydrophobically linked high potential cytochromes c and the reaction center.

e) Indirect structural inferences regarding the location of the high and low potential cytochromes in the profile of the chromatophore membrane have been made from carotenoid bandshifts that accompany the cytochrome oxidation; the assumptions used by JACKSON and DUTTON (1973) are the common basis of such suggestions. Of the four species we are considering it is unfortunate that the only one which does not display measurable electrochromic shifts of the carotenoid complement, *Rps. viridis*, is the only one for which a useful structure exists; thus no correlation can be made to verify or modify the interpretations drawn from the bandshifts! In the other three species there are distinct and consistent differences in carotenoid bandshift responses that accompany the oxidation of high or low potential hemes. In all three cases, as found for example in *Rps. sphaeroides*, the $(BChl)_2$ to Q_A electron transfer induces a carotenoid bandshift. Furthermore, in all cases the bandshift is enhanced when the high potential cytochrome becomes oxidized, appropriate for the conclusions that the $(BChl)_2$ is positioned well into the low dielectric constant part of the membrane and that the high potential cytochrome c is placed closer to or on the outer side of the membrane. These two distinct electrochemical effects are the basis for the suggestion that electron transfer between the high potential cytochrome c to $(BChl)_2$ is electrogenic and partly but significantly contributes to the separation of charge across the membrane (see TIEDE, Chap. 8.2, this Vol.). Such a situation seems reasonably well established for the reaction center-cytochrome c system of *Rps. sphaeroides* (JACKSON and DUTTON 1973); confirming evidence for the conclusion has come from $\Delta\psi$-dependent shifts in the redox equilibria between cytochrome c_2 and $(BChl)_2$ (TAKAMIYA and DUTTON 1977), as well as from direct voltage measurements (PACKHAM et al. 1982; FEHER and OKAMURA 1984). However, when the low potential cytochrome c rather than the high potential cytochrome c is oxidized, a quite different extent of carotenoid bandshift is encountered. In *Rps. gelatinosa* and *T. pfennigii*, the amplitude of stably formed bandshift after the low potential cytochrome has become oxidized is diminished to a value which is about 30% of that observed after the high potential cytochrome is oxidized, and even less than that seen after the initial phase accompanying charge separation between $(BChl)_2$ and Q_A (DUTTON 1971; PRINCE 1978); in *Chr. vinosum* no bandshift was seen after cytochrome c-553 oxidation (CASE and PARSON 1973). While it must be acknowledged that the carotenoid bandshift is an empirically found membrane "voltmeter" and is not mechanistically wholly understood, and that positions of components placed in the profile of the membrane based on carotenoid band shift data have to be weighted with variation in dielectric properties and so on, the bandshifts have so far proven to be a surprisingly reliable guide to the detection of electrogenic events across the membrane. However, interpreting the bandshifts generated during the oxidation of the low potential cytochromes in the usual way, puts the low potential cytochromes for the *Rps. gelatinosa* and *T. pfennigii* in strata in the profile of the membrane between $(BChl)_2$ and Q_A; for *Chr. vinosum* the low potential of the cytochrome c-553 would be in a similar position to Q_A, a position that would be near to the side of the membrane opposite to that facing the periplasmic space where the high potential cytochrome c-555 is located (see DUTTON and PRINCE 1978; PRINCE 1978).

f) EPR studies on the *Chr. vinosum* hemes also yielded some unexpected conclusions. it was found that the hemes of the individual pairs of either the low and the high potential of cytochromes were not close enough to be magnetically coupled. However, at least one of the hemes of the cytochrome c-553 pair was close enough to interact with one of the hemes of the cytochromes c-555 pair; a distance (center to center) of 8 Å was estimated. It was also found that the heme planes of a pair were similarly oriented in the membrane but the planes of each pair of the cytochrome c-555 and c-553 hemes were orthogonal to each other; the heme plane for cytochrome c-555 was normal, while that of cytochrome c-553 was parallel to the plane of the membrane. Since none of the cytochromes interacted with any of the reaction center components, a distance of >23 Å was estimated for the distance from the $(BChl)_2$ to any of the hemes. TIEDE et al. (1978) placed the hemes around the reaction center to accommodate the distance estimates and to provide a simple parallel format for the kinetic equivalence of the hemes of each pair in their competition for the light-generated $(BChl)_2^{+\cdot}$. The parallel model appeared to most simply accommodate the remarkable, apparently independent (DUTTON 1971) photooxidation of either low potential heme shown in *Chr. vinosum* down to cryogenic temperatures. Similarly, the photooxidation of both low *and* (partially) high potential cytochrome of *Rps. gelatinosa* at 80 K also supported a parallel and physically independent arrangement of the cytochrome hemes with respect to the common oxidant $(BChl)_2$. However, for an explanation of the kinetically overriding dominance of the photooxidation of the low potential heme in these species (or those of *T. pfennigii* or *Rps. viridis*), the model required further elaboration; it was necessary to draw on phenomenon such as the redox-linked conformational changes reported by KENNEL and KAMEN (1971) in the polypeptide that accommodates the high and low potential hemes of *Chr. vinosum.*

3.4.2 The *Rps. viridis* Structure

The foregoing discussion is all well and good, but it contrasts sharply with the linear array of the four hemes of *Rps. viridis* as revealed by the X-ray diffraction derived 3-D structure of MICHEL and coworkers (see Chap. 8.4, this Vol., by MICHEL and DEISENHOFER). Although, obviously, *Rps. viridis* may be structurally different from *Chr. vinosum* and other species, an argument based on this premise will not help us in our current discussion; *Rps. viridis,* with its linear array, is able to perform virtually the same reactions as displayed by *Chr. vinosum.* Hence we are forced to examine the linear arrangement for alternative explanations. The main phenomena to be accommodated in a general model are (1) the apparent kinetic and thermodynamic equivalency of each pair of hemes, which includes the statistical nature of the redox potential-dependency of the amplitude of cytochrome oxidation following the first and second flashes (2) the apparent faster kinetics of the low potential cytochrome oxidation, (3) the kinetically overriding dominance of the low potential cytochromes and reduced prior to activation, (4) the "freezing out" of photooxidation displayed by the high potential cytochromes as the temperature is lowered, and (5) the success of photooxidation of the low potential cytochrome at temperatures as low as 4 K.

Some pointers to these issues are presented briefly below.

Apparent thermodynamics and kinetic equivalency in oxidation of the individual hemes of a pair that are linearly arranged to a single molecule of light-generated $(BChl)_2^{+\cdot}$ can be achieved in a number of ways. Ignoring the possibility that the hemes still, despite a linear spatial arrangement, react individually, as discussed above in Tiede's model, apparent equivalency can be obtained with only one of the hemes reacting with $(BChl)_2$ if the E_m values of the two hemes are the same and if the inter-heme electron transfer is considerably faster than the intrinsic oxidation rate of the heme proximal to the $(BChl)_2^{+\cdot}$. In *Chr. vinosum* where oxidation rates are close to $10^6\ s^{-1}$, a reasonable estimate would be $>10^7\ s^{-1}$; in *Rps. viridis*, where the oxidation rates are closer to $10^7\ s^{-1}$, the interheme rates have to be appropriately faster. These considerations can be extended to a chain containing four (or any number) of hemes, provided that the kinetic conditions are met. However, when there are electrochemically distinct pairs of hemes in a chain of four hemes there are further restrictions in the ordering of the hemes. Three arrangements are worth reviewing to illustrate these and allied possibilities, together with their problems.

a) The arrangement with the low and high potential cytochromes in the sequence: low, low, high, high, RC. This is an appealing arrangement considered by Michel and Deisenhofer (Chap. 8.4, this Vol.) based in part on the preliminary indication that a high potential heme may indeed be proximal to the reaction center. This is based on analysis of their structural data and is consistent with the heme proximal to the reaction center possessing the histidine and methionine residues as the fifth and sixth ligands to the iron that are typical of the high potential cytochromes c.

This arrangement, when only the two high potential cytochromes are reduced before activation, is simply equivalent to two hemes linearly arranged in the chain. Thus, providing inter-heme electron transfer is much faster than the intrinsic $(BChl)_2^{+\cdot}$-induced oxidation rate of the proximal cytochrome c-558, similar flash-induced oxidation kinetics of either of the two hemes will be obtained. Further, providing that the E_m values of the two hemes are the same, the observed statistical response seen in the amplitude of the flash-activated oxidation in first and second flashes will also be obtained. When, in addition to the high potential hemes, the low potential hemes are reduced before oxidation, the absence of an observed high potential cytochrome c oxidation (even as a transient) and the apparent overriding dominance of the oxidation of the low potential heme, can be accounted for if the low to high potential cytochrome electron transfer rate is again intrinsically much faster than the $(BChl)_2^{+\cdot}$-induced oxidation of the high potential heme. Furthermore, the requirements for the observed thermodynamic and kinetic equivalence of the oxidation of the low potential hemes can readily be satisfied in the same manner as mentioned above for the high potential counter parts.

This model, where the oxidation of the low potential hemes occurs in series following the oxidation of the high potential hemes, does not, without further consideration, account for the observed faster oxidation rate of the low potential hemes. However, the spatial organization of the model does provide an opportunity to explain the phenomenon that does not rest on conformational differences between the low (ox), low (ox), high (red), high (red), RC and the low (red), low (red), high (red), high (red) RC states. In such a model the rate differences could originate by an electrostatic interaction caused by redox-linked charge changes on the hemes (CASE and PARSON 1971). For instance, in the former state, with the low potential hemes oxidized, there are two extra positive charges in the array compared to the latter reduced state. Thus the rate of transfer from the high potential heme away from the low potential hemes toward the $(BChl)_2^{+\cdot}$ may be impeded. The presence of such interactions may have other consequences to the model, in particular with regard to the apparent E_m values of the hemes that come under the influence of each others' charges; these matters are mentioned later.

This linear model arranged in the sequence low, low, high, high, RC, however, promotes major changes in thinking when accommodating the low temperature behavior of cytochrome oxidation and raises several questions. For example, if at the photooxidation of the high (but not the low) potential cytochrome is frozen out at low temperatures, the chain of four hemes, linked efficiently on a redox basis as we have discussed, becomes defunct. It is necessary to propose a separate electron transfer strategy that will permit millisecond, temperature-insensitive electron transfer over the considerable ~40 Å distance from the nearer low potential heme edge to the $(BChl)_2$. In this regard it is conceivable that the high potential ferrocytochromes can serve to facilitate the reactions; such a possibility deserves further consideration.

b) An arrangement with the low and high potential cytochromes in the sequence: high, high, low, low, RC. This arrangement would, from a structural standpoint (i.e., distance), readily accommodate the low temperature oxidation of the low potential hemes and the freeze-out effect that could be viewed as occurring between the high and low potential hemes. However, a more serious impediment to this model exists in the thermodynamic barrier that the low potential hemes present to the passage of electrons from the distal pair of high potential hemes via the low potential hemes to the $(BChl)_2^{+\cdot}$. The approximate 0.35 V uphill free energy step (*Chr. vinosum*, *Rps. viridis* and *T. pfennigii*) from the high potential to low potential hemes, even with the overall *net* 0.13 V downhill free energy drop from the high potential hemes to the $(BChl)_2$, will dramatically slow the photooxidation of the high potential hemes. In *Chr. vinosum* for instance, with a room temperature rate of 10^6 s^{-1} for electron transfer from the proximal, low potential heme to the $(BChl)_2^{+}$, a rate in the region of only 1 s^{-1} would be encountered for the high potential cytochrome oxidation, irrespective of the inter-heme rates.[1] In this model the high potential hemes

1 In the redox equilibrium:

$$\text{high} \underset{k_{-1}}{\overset{k_1}{\rightleftharpoons}} \text{low} \overset{k_2}{\rightharpoonup} \text{RC}$$

the observed rate of oxidation k_{obs} is given by:

$$k_{obs} = k_1 k_2 / (k_{-1} + k_2)$$

will have to proceed independently if the observed rate is to be obtained, again, perhaps invoking the low potential ferricytochrome heme to facilitate transfer. However, with this qualification added, the model assumes the features of the model in which the hemes are arranged in parallel configuration about the RC.

c) An arrangement in which all the hemes individually possess identical electrochemistry but which, structured closely in linear array within the polypeptide, display substantial interheme charge interaction which leads to the observed splitting in redox titrations. However, such an extreme view on a purely electrostatic foundation to account for the redox properties is premature and may not be justified; for instance, simple calculations show that more than the observed two steps in the redox titration would be expected. Nevertheless, in a structure where the hemes are spaced center-to-center at approximately 15 Å, it would be worth investigating the basis of the presence or the reasons for the absence of charge interactions. For instance, the potential experienced 15 Å from one charge in a dielectric medium of average dielectric constant of 5 is ~0.2 V. The presence of four redox-linked charge changes (the E_m values of the hemes of *Chr. vinosum* have been shown to be pH-independent and sensitive to the ionic strength) yields greater effects as well as yielding a multiplicity of energy levels for the four hemes; further, the distribution of charges in the quartet will also be affected when the net charge on the heme is included (i.e., whether the heme is net positive or negative in the reduced form). For example, if all the hemes are positively charged when in the oxidized form, the dominant state after addition of two electrons will be that in which the central hemes are reduced. This will be expected because the remaining oxidized hemes with their extra positive charge should be repelled to the extreme positions of the linear array. Thus, in this model the central hemes will then be the high potential hemes. Alternatively, if all the hemes are neutral or already negative in the oxidized form, then the inverse will be the case. Further, mixed charge states can be considered. The distinct optical and EPR spectrometric characters of the high and low hemes, however, need to be accounted for in such models. While these characteristics of course may prove to arise simply from differences in the constitution of ligands to the heme or from the chemical environment, consideration of the possible electric field effects linked to oxidation-reduction and the differing basic charge environments in the region of each heme may be important. In a system for which a structure is known, an enquiry into these possibilities may be worth pursuing further than was previously warranted. See WARSHEL et al. (1984) and HONIG and HUBBELL (1983) for related discussions and references.

3.4.3 Further Notes on Conformations, States, and Reaction Rates

Finally, some additional comments that may be pertinent and illuminate the earlier discussions on frozen-in conformational states of the cytochrome-reaction center interactions:

a) Recently KLEINFELD et al. (1984) elegantly demonstrated that reaction centers may undergo structural changes during light-activated processes. Very significant changes were encountered after a flash in the rate of return of an electron from Q_A^{-} back to $(BChl)_2^{+\cdot}$ at cryogenic temperatures if the reaction centers were illuminated during the cooling process. Cooling in the dark yielded the well-known temperature-insensitive 20 ms half-time reaction rate; however, cooling in the light led to a chemical or structural alteration that slowed the reaction in a complicated fashion severalfold; the state induced during the cooling process was frozen in at 77 K for at least 2 days but appeared to relax relatively quickly above 90 K.

b) An examination of the ferricytochrome c-553 of *Chr. vinosum* also shows that the heme may report alterations in its environment (TIEDE et al. 1978). The EPR position of the characteristic g_z band of the EPR spectrum of ferricytochrome c-553 oxidized chemically before cooling to helium temperatures for analysis was shown to be different from that generated in the light at helium temperatures. Warming the sample from helium temperatures up to 100 K permitted the signal associated with the light-induced oxidized cytochrome to relax analogously to the previous example. After such a cycle of warming and recooling to helium temperature, the spectrum was the same as that of a sample oxidized chemically at room temperature.

c) Finally, a point observed by TIEDE et al. (1976) in their work with *Chr. vinosum* on the trapping of the I in the reduced state at low temperatures. They found that $I^{-\cdot}$ generation was straightforward at 200 K if Q_A was reduced prior to illumination; net generation of $I^{-\cdot}$ was successfully accomplished with concomitant oxidation of one heme of cytochrome c-553. However, if Q_A was oxidized prior to illumination so that one of the ferrocytochrome c-553 hemes was used to reduce Q_A before the trapping of $I^{-\cdot}$ could proceed using the other heme, the reaction involving the second heme did not proceed as expected. Only 20–30% of the anticipated cytochrome c-553 oxidation and I reduction occurred, indicating that the "equivalency" of the two hemes was not expressed.

3.5 Membrane Translocators

Several translocators associated with the plasma membrane of photosynthetic bacteria have been revealed. In general they are analogous to other bacterial systems and will only be briefly mentioned here. The movement of substrates and ions has been shown to be coupled to the light-generated $\Delta\mu_{H^+}$, which is low on the interior of the cell ($\Delta\psi$ negative, pH alkaline) relative to the outside; steady-state values in the region of 0.25 V are reported (see GROMET-ELHANAN 1977; MICHELS and KONINGS 1978; KNAFF and CARR 1979; DAVIDSON and KNAFF 1982). The transport systems clearly tap this source of free energy to direct the import and export of solutes into and out of the cell.

KNAFF and co-workers (for references see DAVIDSON and KNAFF 1982) have impressively demonstrated in *Chr. vinosum* the transport of amino acids, dicarboxylic acids, sugars, calcium, sodium, and potassium, and have related the observed transport to the regulation of ΔpH and $\Delta\psi$ values that prevail in the cells. Transport of amino acids has been shown in *Rps. sphaeroides* to be dependent on light- and respiratory-activated electron transfer (HELLINGWERF et al. 1975). Likewise, LOACH and colleagues have detailed the import of alanine and the growth substrate, malate, into cells of *R. rubrum* (ZEBROWER and LOACH 1981, 1982; LOACH et al. 1984). These latter workers have measured the efficiency of transport: the quantum yield for both L-alanine or L-malate was found to be 0.015 molecules transported per photon absorbed. This value at first glance is low, but it is not out of line with expectations if the overall cellular economy is taken into account. For instance, LOACH and coworkers (ZEBROWER and LOACH 1981, 1982) have revealed that transport is under the control of Mg^{2+} and that this may come into play in the natural habitat to prevent dissipation of gradients during the periods of light limitation. In addition to the control of activity, these studies have also raised the issue of local activities in the membrane that may be relevant to the questions regarding more localized handling of $\Delta\mu_{H^+}$ by the cell (see Chap. 10.6, this Vol., by MELANDRI and VENTUROLI).

3.6 The ATP Synthase

The adenosine triphosphate synthase (ATPase) is able to use the electrochemical gradient of protons generated by the reaction center and the bc_1 type complexes to drive the endothermic phosphorylation of adenosine diphosphate (ADP). The ATPases in mitochondria, chloroplasts, and photosynthetic bacteria are composed of two principle parts: (a) the F_0 unit which is an organization of subunits which in the membrane forms a proton-specific channel through the membrane, and (b) and F_1 unit which is an organization of subunits that sits over the aperture of the F_0 unit on the low electrochemical proton potential side of the membrane (see MCCARTY and NALIN, Chap. 10.8, this Vol., for a more detailed discussion of structure and composition). The F_1 units are seen as knobs that protrude from the inner side of the membrane into the cytoplasm of photosynthetic bacteria or into the matrix of mitochondria; as expected, they can be seen on the outside face of chromatophores, submitochondrial particles, and of chloroplasts (see STAEHELIN, Chap. 1, this Vol.). The F_1 unit is the catalytic site for the act of phosphorylation of ADP. The potential energy stored in the combined force of $\Delta\psi$ and ΔpH in terms of $\Delta\mu_{H^+}$ moves the positively charged H^+ back across the membrane through the F_0 unit to the F_1 unit. Admission through to the other low potential side is obligatorily coupled to the generation of ATP. PETTY and JACKSON (1979a) reported values of between 2 and 2.3 protons translocated per ATP synthesized following single turnover activation of chromatophores from several species of bacteria. However, the quantum yield was low, although considerably higher than for the solute translocators. The values reported were about 0.1 to 0.15 following single-turnover flash activation of dark adapted chromatophores; subsequent flashes appropriately spaced improved the yield severalfold (PETTY and JACKSON 1979a, b; and see also BALTSCHEFFSKY et al. 1982). Thus, as was indicated with the transporters, it seems clear that the relationship between the delocalized bulk phase $\Delta\mu_{H^+}$ generated by the cycle comprising the reaction center and the bc_1 complex and the function of the ATPase is not simple. The major issues rest on enzyme activity, energetic threshold of function, and whether there exists a more local protonic coupling between the ATPase and the components of the cycle. This is dealt with more fully by MELANDRI and VENTUROLI in Chapter 10.6, this Volume; as well as in the recent work of WESTERHOFF et al. (1984a, b).

Most of the enzymatic work with the F_0 and F_1 units of photosynthetic bacterial ATPase has been done on *R. rubrum* from which it has been isolated and purified. For further information on the proton ATPases in general the reader is referred to the reviews by CROSS (1981), MCCARTY and CARMELI (1982), and for recent progress with *R. rubrum* to the recent work of GROMET-ELHANAN and coworkers (see 1984a, b). See also STROTMANN, Chapter 10.9, this Volume, for a more detailed treatment of the enzymatic aspects of these reactions.

4 Concluding Remarks

In this chapter, I have attempted to provide an overview comparing and contrasting the molecular equipment and the physiological functions of anoxygenic photosynthetic bacteria, of oxygenic chloroplasts, and of oxygen-utilizing respiratory bacteria and mitochondria. Over the past 15 years or so the efforts of several laboratories have partly been directed toward describing the electrochemical and spectrometric properties of the redox centers in the proteins of these systems. With each new finding, whether it be a new component, a determined pK or E_m value, a line shape or a resolved kinetic sequence, we have enjoyed watching the islands of similarity grow into large areas of common ground. More recently protein sequences and secondary structure model building has added a vital complement to the game (see Chap. 8.5, this Vol., by HEARST). With sufficient commonality secure, the time seems to have arrived to start a critical review of the differences. I hope this chapter will serve, as it did for me in writing it, to reintroduce many of us to some of the parts that are unique to photosynthetic bacteria and to alert us to the realization that there is a great deal which appears in textbooks and our summarizing schemes that is far from settled, and that there is even more about which we are just plain ignorant.

Our view of the field of photosynthesis and respiration, in large measure due to MITCHELL's chemiosmotic hypothesis, has become modular, and hence naturally sectioned and simplified. This, in addition to the healthy progress in recent years in resolving and describing the function of the active components and the structural media of these related systems, has made the playing of molecular Tinker Toys a worthwhile pursuit. However, without a better understanding of metabolism and attendant bioenergetics and biochemistry of the oxidation of reduced substrates, in particular the sulfide compounds, we are in danger of missing some important pieces. This applies not only to mechanism, but also to the overall picture of evolution. In this latter regard (DICKERSON 1980) the reaction centers of photosynthetis and their electron donors and acceptors have assumed a central position in the general physiological and molecular view of evolution not only in the bacteria themselves and of chloroplasts (see Chap. 8.5, this Vol., by HEARST and a recent review by OKAMURA et al. 1982; YOUVAN and MARRS 1984), but also in other membrane proteins including cytochrome c oxidases (OKAMURA et al. 1982). Sequence homology in other membrane redox protein subunits, especially in the cytochrome c or cytochrome d (a chlorin) containing quinol oxidases, and perhaps the NADH-ubiquinol oxidoreductase and the bc_1 complex subunits, may also be worth looking at as their amino acid sequences are completed.

One of the most fascinating issues is the emergence of the oxygen-evolving system of chloroplasts. The answers regarding the evolutionary source of this vital enzyme may in part lie in the discovery of more species of photosynthetic bacteria. For example, the recently isolated bacterium *Heliobacterium chlorum* (GEST and FAVINGER 1983) contains a novel bacteriochlorophyll g. The chemical structure of this BChl, it has been noted, is only one simple chemical transformation away from chlorophyll a (GEST personal communication, and see also FULLER et al. 1985). The answers will also rest heavily on understanding the pigment-protein interactions that play a dominant role in governing spectrum and redox potential. However, it may also be worth looking closer at the bacterial reaction center itself for the source of the oxygen-evolving system. For instance, there are the monomeric BChls, the redox chemistry of which, at least as the $BChl/BChl^{+\cdot}$ couple, is not known. However, like those of the light-harvesting protein, we may expect it to be considerably higher than the $(BChl)_2/(BChl)_2^{+\cdot}$. Neither should we be content to agree that only the low potential end of the reaction center is similar to that of photosystem II regarding the Q_A (and Q_B) and the transition metal, Fe. The high potential end will also display a similarity if the recent evidence proves to be sound that the component identified functionally as $Z/Z^{+\cdot}$ is indeed a quinone (see Chaps. 9.1, this Vol., by DINER and 9.2 by YOCUM). If so, the oxygen-evolving enzyme will include a quinone interacting on the one side with a transition metal, Mn, and on the other perhaps via an intermediary component with P-680. Interesting in this regard (see OKAMURA et al. 1982) is the presence in bacterial reaction centers of a minor but significant natural amount of Mn as a replacement for the Fe associated with the Q_A and Q_B sites.

Another important similarity is shared between the Q_A site and the putative quinone of the $Z/Z^{+\cdot}$ site. Both species are shielded from contact with the bulk aqueous phase such that neither the protonation of Q_A^- (at equilibrium the pK of the $Q_AH/Q_A^{-\cdot}$ is ~9.8 in *Rps. sphaeroides*; see review by PRINCE and DUTTON 1978) nor the deprotonation of the putative $ZH_2^{+\cdot}$ state of the $Z/Z^{+\cdot}$ (in solution, $pK<0$) does not occur on a time scale of function; such contact would in each case diminish the respective reductive or oxidative potency of these light-generated semiquinones.

Despite the fact that the $Q^{-\cdot}$ and $QH_2^{+\cdot}$ are greatly separate in acid/base and redox performance, it seems not unreasonable to consider that minor mutations in the binding site of Q_A might lead to one that could accommodate the other. In this regard much will rest on the patterns of amino acid composition displayed by these and the other binding sites we have encountered during this chapter (Q_B, Q_S, Q_C, and Q_Z and Q_N) which confer on the occupant such distinctive redox properties. Of particular interest in this argument is the proposal by MICHEL and DEISENHOFER (Chap. 8.4, this Vol.) regarding the evolutionary development of the *Rps. viridis* reaction center and the emergence of the Q_A and Q_B sites. Their idea is that the bacteria reaction center evolved from a symmetric protein dimer which upon association yielded not only the $(BChl)_2$ but also the bifurcated pathways, each of which contain a monomeric BChl and BPh and a quinone. Mutations are proposed to change the arrangement yielding the distinctive characters of Q_A (shielded, tightly bound in the Q_A and $Q_A^{-\cdot}$ states) and Q_B (exchangable, key component of the two-electron gate, the operation of which includes the Q, $Q^{-\cdot}$ and QH_2 states). Q_B is proposed to have developed a redox connection to Q_A at the expense of its connection with its adjacent BPh- and BChl-containing pathways which them became, at least on a redox basis, dysfunctional. I wonder whether it would be any more elaborate from the evolutionary standpoint for the key elements of the chloroplast oxygenic system to have been assembled from the same basic pieces of the purple photosynthetic reaction center? Information on this matter should come from the subunit analysis of the oxygen-evolving enzyme. In this regard, the suggestion entertaining the identity of the $Z/Z^{+\cdot}$ component as a napthoquinone (see discussion by YOCUM in Chap. 9.2, this Vol.) rather than a plastoquinone is of some interest!

Acknowledgments. I am indebted in particular to DAVID KNAFF for many discussions, but also to HOWARD GEST, CLINT FULLER, BAZ JACKSON, DAN O'KEEFE and BOB BLANKENSHIP. The work was supported by grants from USPHS (GM 27309), NSF (PCM 82-09292) and DOE (DE-ACO2-80-ER 10590).

References

Arnon DI (1977) Photosynthesis 1950–1975 Changing concepts and perspectives. In: Trebst A, Avron M (eds) Photosynthesis I: Photosynthetic electron transport and phosphorylation. Encyclopedia of plant physiology, vol V. Springer, Berlin Heidelberg New York, pp 8–56

Baltscheffsky M, Baltscheffsky H, Boork J (1982) Evolutionary and mechanistic aspects of coupling and phosphorylation in photosynthetic bacteria. In: Barber J (ed) Electron transport and photophosphorylation. Elsevier/North-Holland Biomed Press, Amsterdam New York, pp 249–272

Bartsch RG (1978) Cytochromes in the photosynthetic bacteria. In: Clayton RK, Sistrom WR (eds) Plenum, New York London, pp 249–279

Blankenship RE, Mancino LJ, Feick R, Fuller RC, Machnicki J, Frank HA, Kirmaier C, Holton D (1984) Primary photochemistry and pigment composition of the reaction centers isolated from the green photosynthetic bacteria *Chloroflexus aurantiacus*. Proc Natl Acad Sci USA 79:6532–6536

Carruthers RP, Yoch DC, Arnon DI (1979) Isolation and characterization of bound iron-sulfur proteins from bacterial photosynthetic membranes. J Biol Chem 252:7561–7467

Case GD, Parson WW (1971) Themodynamics of the primary and secondary photochemical reactions in *Chr. vinosum*. Biochim Biophys Acta 253:187–202

Case GD, Parson WW (1973) Shifts of bacteriochlorophyll and carotenoid absorption bands linked to cytochrome c-555 photooxidation in *Chr. vinosum*. Biochim Biophys Acta 325:441–453
Case GD, Parson WW, Thornber JP (1970) Photooxidation of cytochromes in reaction center preparation from *Chr. vinosum* and *Rps. viridis*. Biochim Biophys Acta 223:12–128
Chance B, Hollunger G (1961) The interaction of energy and electron transfer reactions in mitochondria, vi. The efficiency of the reaction. J Biol Chem 236:1577–1585
Cohen Y, Padan E, Shilo M (1975) Facultative bacterial-like photosynthesis in the blue-green alga *Oscillatoria limnetica*. J Bacteriol 123:855–861
Collins MD, Jones D (1981) Distribution of isoprenoid quinone structural types in bacteria and their taxonomic implications. Bacteriol Rev 45:pp 316–354
Crofts AR, Wraight CA (1983) The electrochemical domain of photosynthesis. Biochim Biophys Acta 726:149–185
Cross RL (1981) The mechanism and regulation of ATP synthesis by F_1-ATPases. Annu Rev Biochem 50:681–714
Davidson VL, Knaff DB (1982) The electrochemical proton gradient in the photosynthetic purple sulfur bacterium *Chr. vinosum*. Photochem Photobiol 36:551–558
DeVault D, Chance B (1966) Studies of photosynthesis using a pulsed laser: Temperature depency of cytochrome oxidation rate in *Chr. vinosum* evidence for tuneling. Biophys J 6:825–847
Dickerson RE (1980) Cytochrome c and evolution of energy metabolism. Sci Am 242 (3):pp 136–153
Doi M, Takamiya K, Nishimura M (1982) Light and thiosulfate-dependent reduction of nicotinamide adenine nucleotides in whole cells of *Chromatium vinosum*. Plant Cell Physiol 21:1015–1022
Dutton PL (1971) Oxidation-reduction dependence of the interactions of cytochromes, bacteriochlorophyll and carotenoids at 77 K in chromatophores of *Chr. vinosum* and *Rps. gelatinosa*. Biochim Biophys Acta 226:63–80
Dutton PL, Prince RC (1978) Reaction center driven cytochrome interactions in electron and proton translocation and energy coupling. In: Clayton RK, Sistrom WR (eds) The photosynthetic bacteria. Plenum, New York London, pp 525–570
Dutton PL, Kihara T, McCray JA, Thornber JP (1971) Cytochrome c-553 and bacteriochlorophyll interaction at 77 K in chromatophores and a subchromatophore fraction preoperation in *Chr. vinosum*. Biochim Biophys Acta 226:81–87
Evans MCW (1969) Ferredoxin-NAD reductase and the reduction of NAD by *Chlorobium thiosulfatophilum*. In: Metzner H (ed) Progress in photosynthesis research, vol III. Laupp, Tübingen, pp 1474–1475
Evans MCW, Buchanan BB (1965) Photoreduction of ferredoxin and its use in CO_2 fixation by a subcellular fraction from a photosynthetic bacterium. Proc Natl Acad Sci USA 53:1420–1425
Feher CR, Okamura MY (1984) Structure and function of the reaction center from *Rps. sphaeroides*. In: Sybesma C (ed) Advances in photosynthesis research, vol II. Nijhoff/Junk, The Hague, pp 155–164
Fischer U (1984) Cytochromes and iron sulfur proteins in sulfur metabolism of phototrophic bacteria. In: Muller A, Krebs B (eds) Sulfur, its significance for chemistry, for the creo-, bio- and cosmosphere and technology. Studies in organic chemistry, vol V. Elsevier, Amsterdam New York Oxford, pp 383–407
Freedman JA, LeMasters JJ (1984) Thermodynamics of reverse transfer across site I: ATP/$2e^-$ is greater than one. Biochem Biophys Res Commun 125:8–13
Fukumori Y, Yamanaka T (1979) A high potential nonheme iron protein (HIPHP)-linked thiosulfate oxidizing enzyme from *Chr. vinosum*. Curr Microbiol 3:117–120
Fuller RC, Sprague SG, Gest H, Blankenship RE (1985) A unique photosynthetic reaction center from *Heliobacterium chlorum*. FEBS Lett 182:pp 345–349
Gaffron H (1963) Van Niel's theory: thirty years after. In: Gest H, San Pietro A, Vernon LP (eds) Bacterial photosynthesis. Antioch, Yellow Springs, OH, pp 3–14
Gest H (1963) Metabolic aspects of bacterial photosynthesis. In: Gest H, San Pietro A,

Vernon LP (eds) Bacterial photosyntheses. Antioch, Yellow Springs, OH, pp 129–150
Gest H (1982) In: Kaplan NO, Robinson A (eds) From cyclotrons to cytochromes. Academic Press, London New York, pp 305–321
Gest H (1983) Evolutionary roots of anoxygenic photosynthetic energy conversion. In: Ormerod JG (ed) The photosynthetic bacteria: Anaerobic life in the Light. Blackwell, London Oxford, pp 215–235
Gest H, Favinger JL (1983) Arch Microbiol 136:pp 11–16
Gray GO, Knaff DB (1982) The role of a cytochrome c-552 – cytochrome c complex in the oxidation of sulfide in *Chr. vinosum*. Biochim Biophys Acta 680:290–296
Gromet-Elhanan Z (1977) The electrochemical gradients and energy coupling in photosynthetic bacteria. Trends Biol Sci 2:274–277
Gromet-Elhanan Z, Khananshuili D (1984a) Chemical modification of essential amino-acid residues in the chromatophore F_1-ATPase and its isolated subunit. In: Sybesma C (ed) Advances in photosynthesis research, vol II. Nijhoff/Junk, The Hague, pp 595–598
Gromet-Elhanan Z, Khananshuili D (1984b) Characterization of two nucleotide binding sites on the isolated reconstitutively active subunit of the F_0–F_1 ATP synthase. Biochemistry 23:1022–1028
Grondelle R van, Duysens LNM, Wal van der JA, Wal van der HN (1977) Function and properties of a soluble c-type cytochrome c-551 in secondary photosynthetic electron transport in whole cells of *Chr. vinosum* as studied with flash spectroscopy. Biochim Biophys Acta 461:188–201
Grondelle R van, Duysens LNM, Wal van der HN (1976) Function of three cytochromes in photosynthesis of whole cells of *Rh. rubrum* as studied by flash spectroscopy: Evidence for two types of reaction centers. Biochim Biophys Acta 449:169–187
Hellingwerf K, Michels PAM, Dorpema J, Konings WN (1975) Transport of amino acids in membrane vesicles of *Rps. sphaeroides* by respiratory and cyclic flow. Eur J Biochem 55:397–406
Honig BH, Hubbell WL (1983) The stability of salt bridges in membrane proteins. Biophys J 41:203–210
Hopfield JJ (1974) Electron transfer between biological molecules by thermally assisted tunneling. Proc Natl Acad Sci USA 71:3640–3644
Ingledew WJ, Prince RC (1977) Thermodynamic Resolution of the iron-sulfur center of the succinic dehydrogenase in *Rps. sphaeroides*. Arch Biochem Biophys 178:303–307
Jackson JB, Dutton PL (1973) Kinetic and redox potentiometric resolution of the carotenoid band shifts in *Rps. sphaeroides* chromatophore. Biochem Biophys Acta 325:102–113
Jain MK (1972) The biomolecular lipid membrane. Van Nostrand-Reinhold, Princeton, NJ
Jortner J (1980) Dynamics of electron transfer in bacterial photosynthesis. Biochim Biophys Acta 594:pp 193–230
Junge W, Jackson JB (1982) The development of electrochemical potential gradients across photosynthetic membranes. In: Govindjee (ed) Photosynthesis: Energy conversion in plants and bacteria, vol I. Academic Press, London New York, pp 589–646
Kennel SJ, Kamen MD (1971) Iron-containing proteins in *Chr. vinosum* II: Purification properties of a cholate-solubilized cytochrome complex. Biochim Biophys Acta 253:153–156
Kihara T, Chance B 61969) Cytochrome photooxidation at liquid nitrogen temperatures in photosynthetic bacteria. Biochim Biophys Acta 189:116–124
Kihara T, Dutton PL (1970) Light-induced reactions of photosynthetic bacteria: Reaction in whole cells and cell free extracts at liquid nitrogen temperatures. Biochim Biophys Acta 205:196–204
Kihara T, McCray JA (1973) Water an cytochrome oxidation-reduction reactions. Biochim Biophys Acta 292:297–309

Kleinfeld D, Okamura MY, Feher G (1984) Electron transfer kinetics in photosynthetic reaction centers cooled to cryogenic temperatures in the charge-separated state: Evidence for light-induced structural changes. Biochemistry 23:pp 5780–5786

Knaff DB (1978) Reducing potentials in the pathway of NAD^+ reduction. In: Clayton RK, Sistrom WR (eds) The photosynthetic bacteria. Plenum, New York London, pp 629–640

Knaff DB, Buchanon BB (1975) Cytochrome b and photosynthetic sulfur bacteria. Biochim Biophys Acta 376:549–560

Knaff DB, Carr JW (1979) The energy linked carotenoid band shift in *Chr. vinosum*. Arch Biochem Biophys 193:379–384d

Knaff DB, Malkin R (1976) Iron sulfur proteins of the green photosynthetic bacterium *Chlorobium*. Biochim Biophys Acta 430:244–252

Loach PA, Parkes PS, Bustemante P (1984) Regulation of photosynthetic unit structure in *R. rubrum* whole cells. In: Sybesma C (ed) Advances in photosynthesis research. Nijhoff/Junk, The Hague, pp 189–197

Malkin R, Chain RK, Kraichoke S, Knaff DB (1981) Studies of the function of the membrane-bound iron-sulfur centers of the photosynthetic bacterium *Chromatium vinosum*. Biochim Biophys Acta 637:88–95

McCarty RE, Carmeli C (1982) Proton translocating ATPases of photosynthetic membranes. In: Govindjee (ed) Photosynthesis: Energy conversion by plants and bacteria, vol I. Academic Press, London New York, pp 647–695

Michels PAM, Konings WN (1978) The electrochemical proton gradient generated by light im membrane vesicles and chromatophores from *Rps. sphaeroides*. Eur J Biochem 85:147–155

Mitchell P (1966) Chemiosmotic coupling in oxidative and photosynthetic phosphorylation. Glynn Res, Bodmin, Cornwall

Mitchell P (1968) Chemiosmotic coupling and energy transduction. Glynn Res, Bodmin, Cornwall

Mitchell P (1975) Proton motive redox mechanism of the cytochrome bc_1 complex in the respiratory chain: protonmotive ubiquinone cycle. FEBS Lett 56:1–6

Morel A, Smith RC (1974) Relation between total quanta and total energy for aquatic photosynthesis. Limnol Oceanogr 19:591–600

Niel van CB (1931) On the morphology and physiology of the purple and green sulfur bacteria. Arch Microbiol 3:1–112

Niel van CB (1935) Photosynthesis of bacteria. Cold Spring Harbor Symp Quant Biol 3:138

Ohnishi T, Salerno JC (1982) In: Spiro TG (ed) Iron-sulfur proteins, vol 7. Wiley, New York, pp 285–327

Okamura MY, Feher G, Nelson N (1982) Reaction centers. In: Govindjee (ed) Photosynthesis: Energy conversion in plants and bacteria. Academic Press, London New York, pp 195–272

O'Keefe DP (1983) Sites of cytochrome b-563 reduction and the mode of action of DNP-INT and DBMIB in the chloroplast cytochrome b_{563}-f complex. FEBS Lett 162:349–354

Olson JM, Prince RC, Brune DC (1976) Reaction center complexes from green bacteria. Brookhaven Symp Biol 28:238–246

Packham NK, Berriman JA, Jackson JB (1978) The charging capacitance of the chromatophore membrane. FEBS Lett 89:205–210

Packham NK, Muller P, Dutton PL (1982) Photoelectric currents across planar bilayer membranes containing bacterial reaction center response under condition of single electron turnover. Biophys J 37:465–473

Petty KM, Jackson JB (1979a) Two protons are transferred per ATP synthesized after flush activation of chromatophores from photosynthetic bacteria. FEBS Lett 97:pp 367–372

Petty KM, Jackson JB (1979b) Correlation between ATP synthesis and the decay of the carotenoid bandshift after single turnover flash activation of chromatophores from *Rps. capslata*. Biochim Biophys Acta 547:pp 463–473

Petty KM, Jackson JB (1979c) Kinetic factors limiting ATP synthesis by chromatophores exposed to short flashes. Biochim Biophys Acta 547:474–483

Pfennig N (1978) General physiology and ecology of photosynthetic bacteria. In: Sistrom WR, Clayton RK (eds) The photosynthetic bacteria. Plenum, New York London, pp 3–18

Pierson BK, Thornber JP (1983) Isolation and spectral characterization of photochemical reaction centers from the thermophilic green bacterium *Chloroflexus aurantiacus*. Proc Natl Acad Sci USA 80:80–84

Prince RC (1978) The reaction center and associated cytochromes of *T. pfennigii*: Their thermodynamic and spectroscopic properties and their possible location within the photosynthetic membrane. Biochim Biophys Acta 501:195–201

Prince RC, Dutton PL (1978) Protonation and the reducing potential of the primary electron acceptor. In: Sistrom WR, Clayton RK (eds) The photosynthetic bacteria. Plenum, New York London, pp 439–453

Prince RC, Leigh JS, Dutton PL (1976) Thermodynamic properties of the reaction center of *Rps. viridis* in vivo.

Prince RC, Dutton PL, Clayton BJ, Clayton RK (1978) Properties of the reaction center of *Rps. gelatinosa* in situ and in a detergent solubilized form. Biochim Biophys Acta 502:pp 354–358

Prince RC, O'Keefe DP, Dutton PL (1982) The organization of the cyclic electron tranfer system in bacterial membranes: is this the hardware of a chemiosmotic system? In: Barber J (ed) Topics in photosynthesis vol 4. Elsevier, Amsterdam New York Oxford, pp 197–248

Rich P (1984) Electron proton transfers through quinones and cytochrome *bc* complexes. Biochim Biophys Acta 768:53–79

Robertson DE, Giangiacomo KM, DeVries J, Moser CC, Dutton PL (1984a) Two distinct quinone-modulated modes of antimycin sensitive cytochrome b reduction in the cytochrome bc_1 complex. FEBS Lett 178:343–350

Robertson DE, Prince RC, Bowyer JR, Matsuura K, Dutton PL, Ohnishi T (1984b) Thermodynamic properties of the semiquinone and its binding site in ubiquinonol cytochrome $c(c_2)$ oxidoreductase of respiratory and photosynthetic systems. J Biol Chem 259:1758–1763

Schmitt W, Schleifer C, Knobloch K (1981) The enzymatic system, Thiosulfate: Cytochrome *c* oxidoreductase from phototrophically grown *Chr. vinosum*. Arch Microbiol 130:334–338

Shahak Y, Crowther D, Hind G (1981) The involvement of ferredoxin-$NADP^+$ reductase in cyclic electron transport in chloroplasts. Biochim Biophys Acta 636:234–243

Shill DA, Wood PM (1984) A role for cytochrome c_2 in *Rps. viridis*. Biochim Biophys Acta 764:pp 1–7

Shioi Y, Takamiya K, Nishinura M (1976) Isolation and same properties of NAD reductase of the green photosynthetic bacterium *Prosthecochloris aesturii*. J Biochim (Tokyo) 79:361–371

Straley SC, Parson WW, Mauzerall DC, Clayton RK (1973) Pigment content and molar extinction coefficients of the photochemical reaction center from *Rps. sphaeroides*. Biochim Biophys Acta 305:597–609

Takamiya K, Dutton PL (1977) The influence of transmembrane potentials of the redox equilibrium between cytochrome c_2 and the reaction center in *Rps. sphaeroides* chromatophores. FEBS Lett 80:279–284

Thauer RK, Jungermann K, Decker K (1977) Energy conservation in chemotrophic anaerobic bacteria. Bacteriol Rev 41:100–180

Thekaekara MP, Drummond AJ (1971) Standard values for the solar constant and spectral components. Nature 229:6–9

Thomas L (1974) The lives of a cell: Notes of a biology watcher. Bantam Books, New York

Thornber JP (1970) Photochemical reactions of purple bacteria as revealed by three spectrally different caroteno-bacteriochlorophyll-protein complexes isolated from *Chr. vinosum*. Photochem Photobiol 172:351–354

Thornber JP, Olson JM, Williams DM, Clayton ML (1969) Isolation of the reaction center from *Rps. viridis*. Biochim Biophys Acta 447–467

Tiede DM, Prince RC, Dutton PL (1976) EPR and optical spectroscopic properties of the electron carrier intermediate between the reaction center bacteriochlorophylls and the primary acceptor in *Chr. vinosum*. Biochim Biophys Acta 449:447–467

Tiede DM, Leigh JS, Dutton PL (1978) Structural organization of the *Chr. vinosum* reaction center associated *c*-cytochromes. Biochim Biophys Acta 503:524–544

Trüper HG, Fischer U (1982) Anaerobic oxidation of sulfur compounds as electron donors for bacteria. Philos Trans R Soc London Ser B 298:529–542

Trüper HG, Pfennig N (1978) Taxonomy of rhodospirillales. In: Sistrom WR, Clayton RK (eds) The photosynthetic bacteria. Plenum, New York London, pp 19–27

Vermeglio A (1977) Secondary electron transfer reaction centers of *Rhodopseudomonas sphaeroides*: Out of phase peroidicity of two for the formation of ubisemiquinone and fully reduced quinone. Biochim Biophys Acta 459:516–524

Warshel A, Russel ST, Churg AK (1984) Macroscopic models for studies of electrostatic interactions in proteins: Limitations and applicability. Proc Natl Acad Sci USA 81:pp 4785–4789

Westerhoff HV, Melandri BA, Venturoli G, Azzone GF, Kell DB (1984a) Mosaic protonic coupling hypothesis for free-energy transaction. FEBS Lett 165:1–5

Westerhoff HV, Melandri BA, Venturoli G, Azzoni GF, Kell DB (1984b) A mininal hypothesis for membrane linked free energy transduction. The role of independent, small coupling units. Biochim Biophys Acta 768:257–292

Wikstrom M, Saraste M (1984) The mitochondrial respiratory chain in bioenergetics. In: Ernster L (ed) Bioenergetics. Elsevier, Amsterdam New York Oxford, pp 49–79

Wraight CA (1977) Electron acceptors of photosynthetic bacterial reaction centers: Direct observation of oscillatory behavior suggesting two closely equivalent quinones. Biochim Biophys Acta 459:525–531

Wraight CA (1982) Current attitudes in photosynthesis research. In: Govindjee (ed) Photosynthesis: Energy conversion by plants and bacteria. Academic Press, London New York, pp 17–61

Youvan DC, Marrs BL (1984) Molecular genetics and the light reactions of photosynthesis. Cell 39, pp 1–3

Zebrower M, Loach PA (1981) The role of Mg^{++} in the cooperative asociation of photoreceptor complexes in *R. rubrum*. in Energy coupling in photosynthesis. Selman, Selman-Peimer (eds) Energy coupling in photosynthesis. Elsevier/North-Holland, Biomed Press, Amsterdam New York, pp 333–340

Zebrower M, Loach PA (1982) Efficiency of light-driven metabolite transport in the photosynthetic bacterium *R. rubrum*. J Bacteriol 150:1322–1328

6 Comparative Biochemistry of Light-Harvesting Systems

6.1 Primary Structure and Function of the Light-Harvesting Polypeptides from Cyanobacteria, Red Algae, and Purple Photosynthetic Bacteria

H. Zuber

1 Introduction

Light energy absorbed during the primary processes of photosynthesis is transduced into electronic energy and, after charge separation in the reaction centers (RC), into the energy of the electrochemical protongradient (energy trapping, Clayton and Sistrom 1978). The majority of the pigment molecules of the photosynthesis apparatus serve as light harvesters. They are specifically bound and oriented within the light-harvesting, or antenna pigment-protein complexes. In these light-harvesting pigment-protein complexes, the pigment molecules are excited by photons. In a nonradiating-induced resonance transfer process leading up to the reaction centers, they transfer the excitation energy to other pigment molecules which are excited in turn. This fast process has an efficiency of almost 100%. The function and with it also the spectroscopic properties (absorption and fluorescence emission) of the light-harvesting pigment-protein complexes and pigments depends on the polypeptide components (see Sauer, Chap. 2; Scheer, Chap. 7.6, both this Vol.). These polypeptides determine: (a) the specific binding of the pigment molecules (type, number, position, distance and orientation) and thus also the specific interaction of the pigments for energy transfer; (b) the specific polypeptide environment of the pigment molecule (functional state, basis of the spectral shift); (c) the structural organization (architecture) of the pigment-protein complexes (intramolecular arrangement of polypeptides within the complex and intermolecular organization of the complexes on the basis of polypeptide-polypeptide (polypeptide-lipid) interactions).

For these reasons it is highly desirable to investigate the polypeptide structure of the antennae complexes before proceeding to the molecular interpretation of the physical processes of the capture and transfer of light energy. The primary

Fig. 1. Primary structures of the α- and β-polypeptide chains (subunits) of allophycocyanin (α-APC, β-APC), C-phycocyanin (α-CPC, β-CPC) or phycocyanin (α-PC, β-PC), and phycoerythrocyanin (α-PEC, β-PEC) from the cyanobacterium *Mastigocladus laminosus* and the red alga *Cyanidium caldarium*. Homologous amino acid residues between all phycobiliprotein polypeptide chains of the same organism are boxed

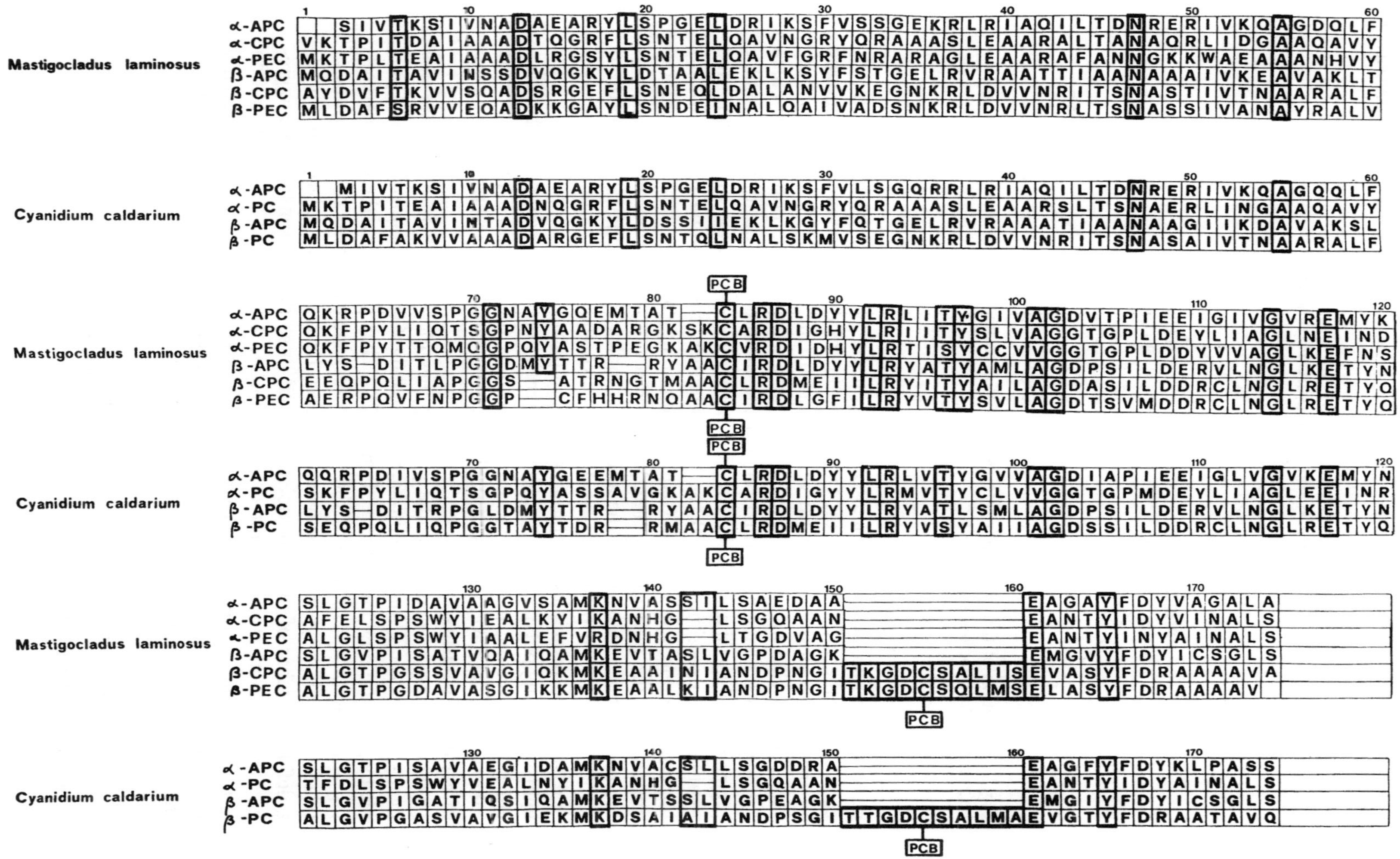

Mastigocladus laminosus
1 10 20 30 40 50 60
α-APC SIVTKSIVNADAEARYLSPGELDRIKSFVSSGEKRLRIAQILTDNRERIVKQAGDQLF
α-CPC VKTPITDAIAAADTQGRFLSNTELQAVNGRYQRAAASLEAARALTANAQRLIDGAAQAVY
α-PEC MKTPLTEAIAAADLRGSYLSNTELQAVFGRFNRARAGLEAARAFANNGKKWAEAAANHVY
β-APC MQDAITAVINSSDVQGKYLDTAALEKLKSYFSTGELRVRAATTIAANAAAIVKEAVAKLT
β-CPC AYDVFTKVVSQADSRGEFLSNEQLDALANVVKEGNKRLDVVNRITSNASTIVTNAARALF
β-PEC MLDAFSRVVEQADKKGAYLSNDEINALQAIVADSNKRLDVVNRLTSNASSIVANAYRALV
Cyanidium caldarium
1 10 20 30 40 50 60
α-APC MIVTKSIVNADAEARYLSPGELDRIKSFVLSGQRRLRIAQILTDNRERIVKQAGQQLF
α-PC MKTPITEAIAAADNQGRFLSNTELQAVNGRYQRAAASLEAARSLTSNAERLINGAAQAVY
β-APC MQDAITAVINTADVQGKYLDSSIIEKLKGYFQTGELRVRAAATIAANAAGIIKDAVAKSL
β-PC MLDAFAKVVAAADARGEFLSNTQLNALSKMVSEGNKRLDVVNRITSNASAIVTNAARALF
Mastigocladus laminosus
70 80 90 100 110 120
α-APC QKRPDVVSPGGNAYGQEMTAT—CLRDLDYYLRLITYGIVAGDVTPIEEIGIVGVREMYK
α-CPC QKFPYLIQTSGPNYAADARGKSKCARDIGHYLRIITYSLVAGGTGPLDEYLIAGLNEIND
α-PEC QKFPYTTQMQGPQYASTPEGKAKCVRDIDHYLRTISYCCVVGGTGPLDDYVVAGLKEFNS
β-APC LYS—DITLPGGDMYTTR—RYAACIRDLDYYLRYATYAMLAGDPSILDERVLNGLKETYN
β-CPC EEQPQLIAPGGS—ATRNGTMAACLRDMEIILRYITYAILAGDASILDDRCLNGLRETYQ
β-PEC AERPQVFNPGGP—CFHHRNQAACIRDLGFILRYVTYSVLAGDTSVMDDRCLNGLRETYQ
PCB
PCB
Cyanidium caldarium
70 80 90 100 110 120
α-APC QQRPDIVSPGGNAYGEEMTAT—CLRDLDYYLRLVTYGVVAGDIAPIEEIGLVGVKEMYN
α-PC SKFPYLIQTSGPQYASSAVGKAKCARDIGYYLRMVTYCLVVGGTGPMDEYLIAGLEEINR
β-APC LYS—DITRPGLDMYTTR—RYAACIRDLDYYLRYATLSMLAGDPSILDERVLNGLKETYN
β-PC SEQPQLIQPGGTAYTDR—RMAACLRDMEIILRYVSYAIIAGDSSILDDRCLNGLRETYQ
PCB
PCB
Mastigocladus laminosus
130 140 150 160 170
α-APC SLGTPIDAVAAGVSAMKNVASSILSAEDAA—EAGAYFDYVAGALA
α-CPC AFELSPSWYIEALKYIKANHG—LSGQAAN—EANTYIDYVINALS
α-PEC ALGLSPSWYIAALEFVRDNHG—LTGDVAG—EANTYINYAINALS
β-APC SLGVPISATVQAIQAMKEVTASLVGPDAGK—EMGVYFDYICSGLS
β-CPC ALGTPGSSVAVGIQKMKEAAINIANDPNGITKGDCSALISEVASYFDRAAAAVA
β-PEC ALGTPGDAVASGIKKMKEAALKIANDPNGITKGDCSQLMSELASYFDRAAAAV
PCB
Cyanidium caldarium
130 140 150 160 170
α-APC SLGTPISAVAEGIDAMKNVACSLLSGDDRA—EAGFYFDYKLPASS
α-PC TFDLSPSWYVEALNYIKANHG—LSGQAAN—EANTYIDYAINALS
β-APC SLGVPIGATIQSIQAMKEVTSSLVGPEAGK—EMGIYFDYICSGLS
β-PC ALGVPGASVAVGIEKMKDSAIAIANDPSGITTGDCSALMAEVGTYFDRAATAVQ
PCB

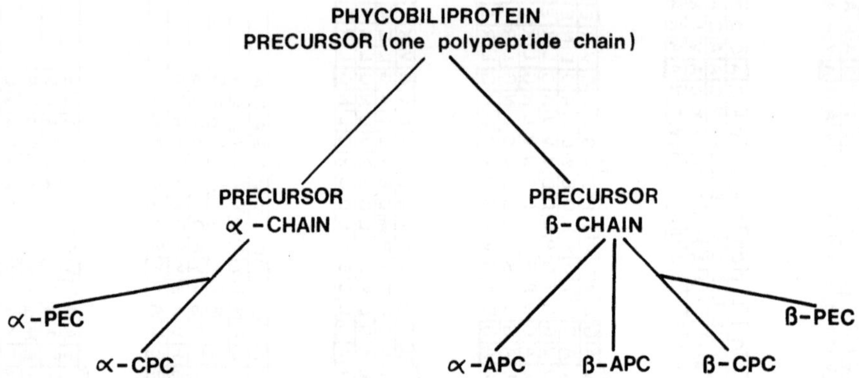

Fig. 2. Evolution of the phycobiliproteins (α- and β-polypeptide chains) from the cyanobacterium *Mastigocladus laminosus*

structure (amino acid sequence) of the light-harvesting polypeptides already contains all the information on the three-dimensional structure and function of these molecules. A comparative analysis of the primary structures should therefore provide an insight into (1) the possibilities of the polypeptide chain folding to the active conformation; (2) the number and position of possible pigment binding sites (which can also be studied by means of labeling and modification experiments); (3) the type, number, and position of functionally important amino acid residues (sequence homology analysis, modification, and mutation of the polypeptide chain); (4) the arrangement and organization of the polypeptide chain within the antenna complex (determination of the quarternary structure, topography: labeling and cross-linking experiments); (5) general structure principles of the antenna complexes; (6) the evolution of the light-harvesting polypeptides (protein differentiation to polypeptides with different light-harvesting properties) and thus also of the antenna complexes or, in other words, the development of the light-energy capturing and transfer systems of various organisms.

The primary structures of the polypeptides of several light-harvesting pigment-protein complexes of cyanobacteria, red algae (phycobiliproteins) and purple bacteria are now known (e.g., Figs. 1, 2)[1]. They show that the light-harvesting antennae (mw $< 10^6$ Da) of these micro-organisms consist of aggregates of relatively small polypeptides (mw $< 2.10^4$ Da). Each of these polypeptides binds a definite number of pigment molecules, and in so doing forms specific aggregates of pigment molecules for energy transfer. The aim of this contribution is to review hat is presently known about the structures and more particularly the primary structures of light-harvesting polypeptides and its implication for the structure and function of antenna complexes.

1 The one-letter code for the amino acid residues according to the IUPAC-IUB Commission on Biochemical Nomenclature (J Biol Chem (1968), 243:3557–3559; Eur J Biochem (1972), 27:201–207 is used in Figs. 1, 2 and 3.

2 The Polypeptides of the Extra-Membrane Light-Harvesting Pigment-Protein Complexes (Phycobiliproteins) of Cyanobacteria and Red Algae

2.1 Structural Organization of the Phycobiliproteins

Phycobilisomes are the main light-harvesting units of cyanobacteria and the chloroplasts of red algae (GANTT 1981). They absorb visible light within the range of 500–670 nm and represent 30–50% of the total light-harvesting capacity of these organisms (see GANTT, Chap. 6.3, this Vol.).

Phycobilisomes consist for the main part (85% of their mass) of the phycobiliproteins involved in energy transfer (light-harvesting pigment-protein complexes) and of a series of linker-polypeptides (MW about 10^4 Da–12×10^5 Da, GLAZER 1982). These latter polypeptides determine the arrangement and conformation of the phycobiliproteins in the phycobilisome, and the specific binding to the thylakoid membrane. In recent years a number of reviews have been published on the subject of phycobilisomes and phycobiliproteins (GANTT 1981; GLAZER 1981, 1982).

Phycobiliproteins are intensely colored, globular, water-soluble proteins. They owe their characteristic spectroscopic properties to open-chain tetrapyrrole prosthetic groups covalently bound by thio-ether bonds to cysteine residues (blue chromophore: phycocyanobilin (PCB), red chromophore: phycoerythrobilin (PEB) and in few cases phycourobilin (PUB). All the phycobiliproteins of cyanobacteria and red algae consist of two subunits (α,β) with molecular weights between 16×10^4 Da–22×10^3 Da. The monomer (α,β) basic structure (basic unit) has a tendency to aggregate to cyclic units (discs), mainly to trimers [$(\alpha\beta)_3$] and hexamers [$(\alpha\beta)_6$, the state of aggregation depends on the pH-value, ion and protein concentration]. The discs of the trimer and hexamer units (12 nm diameter, 3 or 6 nm thick, respectively) also asociate to form small stacks of limited length (3–4 discs). In vivo, these stacks, together with linker polypeptides, assemble to form the phycobilisomes. In the case of *Rhodella violacea* and *Mastigocladus laminosus* (NIES and WEHRMEYER 1981) the phycobilisomes are of the hemi-discoidal type and consist of six outer and three inner stacks. Three main types of phycobiliprotein can be distinguished on the basis of their spectral properties (main absorption maxima): (a) allophycocyanin (APC), 650–671 nm $(\alpha\beta)_3$, 1α-PCB, 1β-PCB; (b) phycocyanin (PC, 617–620 nm, $(\alpha\beta)_3$, $(\alpha\beta)_6$, C-PC: 1α-PCB, 2β-PCB; R-CP: 1α-PCB, 1β-PCB, 1β-PEB); (c) phycoerythrin (PE, 545–567 nm $(\alpha\beta)_6$, C-PE: 2α-PEB; 3β-PEB; B-PE: 2α-PEB, 4β-PEB, 2γ-PEB, 2γ-PUB; R-PE: 2α-PEB, 1β-PUB, 2β-PEB, 2γ-PEB, 2γ-PUB).

The stacks situated in the center of the phycobilisome consist of allophycocyanin; phycocyanin and phycoerythrin (phycoerythrocyanin) form the outer stacks. All cyanobacteria contain allophycocyanin and C-phycocyanin. Red algae contain allophycocyanin and R-, C-phycocyanin. Phycoerythrin is found in red algae (B-PE, R-PE) and in numerous cyanobacteria (C-PE, R-PE). Phycoerythrocyanin (PEC, 568–576 nm, 1α-PXB (unknown tetrapyrrole), 2β-PCB) has been found in cyanobacteria (*Mastigocladus laminosus*, *Anabaena* sp.).

The spectral properties of phycobiliproteins are determined by: (1) the type and number of the chromophores (pigments); (2) the type of the polypeptide chain (APC, PC, PE, PEC, α- and β-subunits), i.e., by the polypeptide environment of the chromophore; (3) the state of aggregation, i.e., by the conformational changes caused by the interaction of the polypeptide chains during the formation of the trimers and hexamers or by the interaction of the pigment molecules (see SCHEER, Chap. 7.6, this Vol.). In the course of the transition from monomeric (αβ) allophycocyanin (615 nm) to the trimeric $(\alpha\beta)_3$ form, the absorption maximum is red shifted (648 nm). Trimer and hexamer complexes with the linker polypeptides have specific absorption, fluorescence and CD spectra optimized for energy transfer. The open-chain tetrapyrroles, which can exhibit various configurations with different spectral properties (conformers), are structurally constrained to assume an extended configuration by interaction with the polypeptide chain. This increases absorption in the visible region of the spectrum.

2.2 Primary Structures of the Phycobiliproteins: Structural and Functional Implications

In the course of evolution of the phycobilisome antennae, the structures and thus also the primary structures of the phycobiliprotein units adapted to meet the requirements of (1) the specific polypeptide environment of the chromophore and (2) their particular aggregation properties. This is a process of protein differentiation by which each one of the α- and β-chains of the phycobiliproteins receives specific functional properties for energy transfer within the phycobilisome (ZUBER 1983).

The primary structure data now available on the three phycobiliproteins (APC, C-PC, PEC) of the phycobilisome from the cyanobacterium *Mastigocladus laminosus* (FRANK et al. 1978; SIDLER et al. 1981; FÜGLISTALLER et al. 1983), β-APC from *Anabaena variabilis* (DE LANGE et al. 1981), β-C-PC from *Synechococcus sp 6301* (FREIDENREICH et al. 1978) and APC and PC from the red alga *Cyanidium caldarium* (OFFNER et al. 1981; OFFNER and TROXLER 1983) provide a clear picture of this protein differentiation process, i.e., of the structure-function relation of phycobiliproteins and their α- and β-polypeptide chains (ZUBER 1983). All α- and β-chains are of approximately equal length (protein family with 160–172 amino acid residues, Fig. 1). The primary structures (amino acid sequences) of the α- and β-chains of the different phycobiliproteins from anyone organism (phycobilisome) differ greatly (low sequence homology, Fig. 1, Table 1). On the other hand, the sequence homology between the α- or β-polypeptide chains respectively of the same phycobiliprotein from different organisms (e.g., cyanobacterium *Mastigocladus laminosus* and red alga *Cyanidium caldarium*) is relatively high (APC ~80% PC, ~PE, PEC 70%). In the first instance, this shows protein differentiation to functionally distinct α- and β-polypeptides within the same phycobilisome. The slight species differences between the phycobiliproteins, even between prokaryotic cyanobacteria and eukaryotic red algae, however, suggests that the polypeptide chains and phycobiliproteins, once differ-

Table 1. Sequence homology between the α- and β-chains of C-phycocyanin (C-PC), phycoerythrocyanin (PEC) and allophycocyanin (APC) from the cyanobacterium *Mastigocladus laminosus*. Arrangement of the chains compared on the basis of sequence homologies

α-, β-Chains compared	Homology %	Position in the phycobilisome of the chains compared
β-CPC-β-PEC	67	External β-chains
α-CPC-α-PEC	62	External α-chains
α-APC-β-APC	38	**Central APC**
β-CPC-β-APC	39	External β-chains – central β-APC
β-PEC-β-APC	34	
β-CPC-α-APC	37	External β-chains – central α-APC
β-PEC-α-APC	36	
α-CPC-α-APC	30	External α-chains – central α-APC
α-PEC-α-APC	26	
α-CPC-β-CPC	27	**External CPC**
α-CPC-β-PEC	26	External α-, β-chains
α-PEC-β-CPC	26	External α-, β-chains
α-CPC-β-APC	24	External α-chains – central β-APC
α-PEC-β-APC	24	
α-PEC-β-PEC	21	**External** PEC

entiated in function, have been structurally and functionally conserved in an optimal state.

As shown by the comparison of the primary structures of the phycobiliproteins of *Mastigocladus laminosus* (Table 1), the main difference in the amino acid sequences is to be found between the α-chain of C-PC and PEC and the β-chain of C-PC and PEC or α-, β-APC, respectively. The structure difference probably reflects the special function of the β-chain of C-PC and PEC with "sensitizing" ("s") and fluorescing ("f") chromophores as compared to the α-chain with sensitizing ("s") chromophores ("energy transfer" between both chromophores (GLAZER 1981). Similarly, the β-chain of C-PC or PEC and the α-, β-APC seem to be related functionally. The structural/functional differentiation of the α- and β-chains regarding their energy transfer role within the monomers (αβ), trimers $(\alpha\beta)_3$ and hexamers $(\alpha\beta)_6$ appears superimposed on another criterium for protein differentiation: the position and function of the phycobiliproteins in the stacks of the phycobilisome. Similarities in the primary structure and three-dimensional structure (surface regions) should lead to a similarity in the aggregation behaviour during stack formation. The greatest sequence homology is to be found between the α- or β-chains of C-PC and PEC respectively, which both form the outer stacks (NIES and WEHRMEYER 1981) of the phycobilisome (62 and 67% respectively, Table 1). On the other hand, the lower sequence homology between the α- and β-chains of the "outer" phycobiliproteins and the α- and β-chains of APC suggests that the APC stack in the center of the phycobiliprotein also occupies a special position (function: focusing of the energy transfer to the reaction center).

Since the chromophore molecule (phycobilin, PCB, PEB, PUB) in APC, PC, PE, and PEC is covalently bound to the cysteine residues of the polypeptide chain, the primary structure analysis also reveals where the chromophore binding site is located in the polypeptide chain. As shown by both the total sequences (Fig. 1) and the partial sequences of bilipeptides (Glazer 1982), the main chromophore binding site is in a central position (position 84) in all α- and β-polypeptide chains [the central chromophore for energy transfer: PCB in APC, C-PC and β-chain of PEC, PXB in the α-chain of PEC, PEB (not shown) in PE (α,β)]. An additional chromophore binding site has been found in the β-chain of the "outer" phycobiliproteins (C-PC, PEC) close to the C-terminus (position 155). It is interesting that it is located within the region of a 10 amino acid insertion (relative to the α-chain and αβ-APC respectively). This chromophore binding site is probably related to the special function of the β-chain (carrier of the "s"chromophore). Functionally speaking, the important amino acid residues are those in the vicinity of the chromophore binding site at the central position 84 (Zuber 1983). The sequence homology in this region is relatively high: 46% (comparison of the three phycobiliproteins from *Mastigocladus laminosus*). Functionally important amino acid residues which might either be located in the immediate vicinity of the chromophore or interact with it (active site residues) are, for example, Arg 86, 93, Asp 87, Leu 92. A region of deletions/insertions of amino acid residues which also shows only low sequence homology lies at the N-terminal side of the chromophore binding site 84 (position 70–84). The structural differences between the individual polypeptide chains in this region should influence the spectral properties of the various phycobiliproteins.

On the basis of the primary data, one can estimate the possible number, size, and distribution of secondary structure elements (α-helices, β-structures) in the three-dimensional organization of the molecules (Zuber 1983). Here, the large number of long helices in APC, C-PC, and PEC, particularly in the N- and C-terminal region of the polypeptide chain, is noteworthy. The particular structural differences between central APC and the "outer" C-PC and PEC are also reflected in the secondary structure pattern.

Crystals of C-phycocyanin suitable for X-ray analysis have been prepared and the three-dimensional structure of this phycobiliprotein determined in this manner (3 Å resolution, Schirmer et al. 1985). The unit building the crystal structure is a trimer of (αβ) monomers. The dominant structural feature in the α- and β-subunits are 8 α-helices, arranged, interestingly, in a way similar to that of the helices in myoglobin and hemoglobin (phylogenetic relationship). The tetrapyrrole chromophores in position 84 of the α- and β-polypeptide chains are in close contact in the $(\alpha\beta)_3$ trimer (possible exciton interactions).

The sequence homology between the α- and β-chains also provides an insight into the phylogenetic relationship, i.e., the evolution of the α- and β-chains of phycobiliproteins and phycobilisomes (Zuber 1983). On the basis of the amino axid exchanges, one can make a hypothetical estimate as to the phylogenetic distances (PAM units) and thus establish a phylogenetic tree. The following evolutionary process can be advanced for the protein differentiation of the phycobiliproteins from *Mastigocladus laminosus* (Fig. 2):

a) From a one-chain precursor molecule, the α- and β-chains of a precursor phycobiliprotein developed by gene duplication.
b) From the β-chain precursor, the β-chain of phycocyanin and the α- and β-chains of APC developed by repeated gene duplication. The α-C-phycocyanin developed from the α-chain precursor.
c) At a later stage, the α- and β-PEC developed from the α- and β-C-PE. The phycobilisome from *Mastigocladus laminosus* evolved similarly to the phycobiliproteins, whereby the layers of the "outer" phycobiliproteins (C-PC, PEC) and the central APC were formed for heterogenous energy transfer from the outside to the center of the phycobilisome. This phycobilisome structure has been maintained in the eukaryotic chloroplasts of the red alga (high sequence homology of the phycobiliproteins, small species differences, c.f. above).

3 The Polypeptides of the Intra-Membrane Light-Harvesting Pigment-Protein Complexes of Purple Photosynthetic Bacteria

3.1 Types and Molecular Organization of the Antenna Complexes

It is generally assumed today that, within the antenna system of the purple photosynthetic bacteria inside the photosynthesic membrane (chromatophores), the light-harvesting BChl-protein complexes are specifically arranged in defined units around the reaction centers (see COGDELL, Chap. 6.2, this Vol.; CLAYTON and SISTROM 1978; COGDELL and THORNBER 1980; COGDELL and VALENTINE 1983). The light-harvesting complexes are hydrophobic intra-membrane entities and can only be isolated by solubilizing the membrane, e.g., with detergents. The kind, size, and intactness (spectrum of the native protein) of the complexes depends to a considerable extent on the conditions of solubilization (type and concentration of the detergent, pH value, etc.). The isolation and definition of the smallest antenna complex (basis unit) is difficult (SAUER and AUSTIN 1978). Purple photosynthetic bacteria (whole cells, chromatophores) exhibit characteristic near infrared (NIR) bands between 800–1000 nm, which are typical for the light-harvesting complexes. On the basis of the NIR absorption maxima (4 bands at 800 nm, 820 nm, 850 nm, 875/890 nm), purple bacteria (wild types) can be classified into three groups (CLAYTON and SISTROM 1978): Group I: *Rhodospirillum rubrum*, *Rhodopseudomonas viridis* (890 nm, 1015 nm); group II: *Rps. sphaeroides*, *Rps. capsulata*, *Rps. gelatinosa* (800 nm, 850 nm, 875 nm); group III: *Chromatium vinosum*, *Rps. acidophila* (800 nm, 820 nm, 850 nm, 875 nm). The various absorption maxima belong to two main types of light-harvesting pigment-protein complex, which are characteristic for the three groups of bacteria (most of the complexes have been isolated and characterized): complex type B-875–890 (B-1015 *Rps. viridis*) in the vicinity of the reaction center ("inner" complexes) and complex type B-800–850 and B-800–

820 around the B-875–890 complexes and the reaction centers respectively ("outer" complexes, variable in number). There appear to be sub-classes of both types of complexes (structure variations of the wild types), which partly also differ in their BChl a: carotenoid ratio (2:1, 2:2, 3:1) (COGDELL and VALENTINE 1983; THORNBER et al. 1983). These various complexes have been subjected to intensive examination from a spectral (absorption, fluorescence, CD) and a biochemical point of view (THORNBER, Chap. 3; COGDELL, Chap. 6.2, this Vol., CLAYTON and SISTROM 1978; COGDELL and THORNBER 1980; THORNBER et al. 1983) and have yielded the following results: B 890 complex of *R. rubrum*, *C. vinosum*, *Rps. acidophila*: 2 BChla, 1 carotenoid per 2 polypeptides, strong CD spectrum; B-875 complex of *Rps. sphaeroides*, *Rps. capsulata*, *Rps. gelatinosa*: 2 BChla, 2 carotenoids per 2 polypeptides, weak CD spectrum; B-800–850 complex of *Rps. sphaeroides*, *Rps. capsulata*, *Rps. gelatinosa*, *Rps. acidophila*, *C. vinosum*: 3 BChla, 1 carotenoid per 2 polypeptides, strong CD spectrum, and B-800–820 complex of *C. vinosum*, *Rps. acidophila*: 3 BChla, 1 carotenoid per 2 polypeptides. The strong CD spectra are typical of exciton-coupled BChla-dimers, the third BChla in the B-800–850 and B-800–820 complexes is the BChl monomer with an 800 nm absorption.

3.2 Primary Structures and the Structural and Functional Role of the Light-Harvesting Polypeptides

Similarly to the phycobiliproteins, the polypeptides of the light-harvesting BChl-protein complexes form a characteristic family of small (50–60 amino acid residues, in this instance) but hydrophobic polypeptides (Fig. 3). They can be extracted and isolated from the chromatophores or complexes with organic solvents. By this procedure the noncovalently bound BChl is split off. The primary structures of the light-harvesting polypeptides from *R. rubrum* (BRUNISHOLZ et al. 1981; GOGEL et al. 1983; BRUNISHOLZ et al. 1984a), *Rps. sphaeroides* (THEILER et al. 1984), *Rps. viridis* (BRUNISHOLZ et al. 1985), and part of the polypeptides from *Rps. capsulata* (TADROS et al. 1983; TADROS et al. 1984) and *Rps. gelatinosa* (BRUNISHOLZ et al. unpublished) have been determined (Fig. 3). It is apparent, as has already been shown by gel electrophoresis, that both types of complexes each contain two types of polypeptide (α, β as in phycobiliproteins, B-875–890-α, B-875–890-β, B-800–850-α, B-800–850-β). The α- and β-polypeptides exhibit only slight sequence homologies (7–13%), consistent with an early separation of the α- and β-chains (genes) in the course of evolution of their different functions). However, due to their higher sequence homology (13–28%), the α- or β-chains of B-870–890 and B-800–850, respectively, appear more closely related phylogenetically (later structural-functional protein differentiation of B-870–890 ("inner" complex) and B-800–850 ("outer" complex) in the light-harvesting RC system). As in the instance of the phycobiliproteins (phycobilisomes), the primary structure data also provide an indication as to the evolution of the intra-membrane light-harvesting system.

The α- and β-polypeptides of the individual complexes in the various purple bacteria have a relatively high sequence homology (e.g., B-870-α or B-870-β

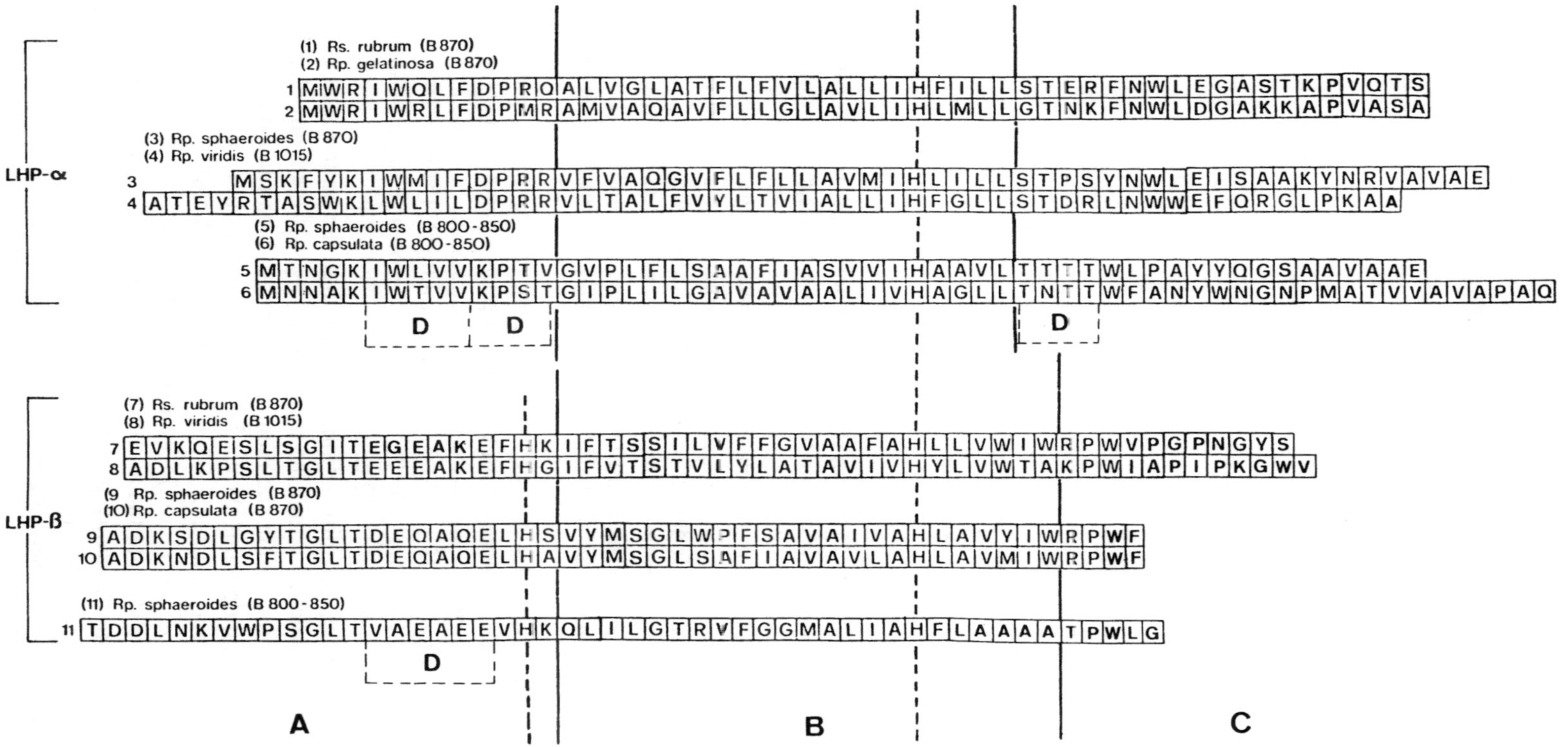

Fig. 3. Primary structures of the light-harvesting α- and β-polypeptides (LHP-α, LHP-β) of the B-875 (B-890) and B-800–850 complexes from *R. rubrum* (*1, 7*), *Rps. viridis* (*4, 8*), *Rps. sphaeroides* (*3, 5, 9, 11*), *Rps. capsulata* (*6, 10*) and *Rps. gelatinosa* (*2*). *A* polar-charged N-terminal domain; *B* hydrophobic central domain; *C* polar-charged C-terminal domain; *D* polar or charged cluster regions

Table 2. Sequence homology (phylogenetic relationship) between the B-870-α polypeptides or the B-870-β polypeptides of *R. rubrum, Rps. viridis, Rps. gelatinosa, Rps. capsulata, Rps. sphaeroides*

	Rps. viridis	*R. rubrum*	*Rps. gelatinosa*	*Rps. capsulata*	*Rps. sphaeroides*	
Rps. viridis		44%	35%	37%[a]	39%	
R. rubrum	40%		52%	48%[a]	42%	
Rps. gelatinosa	–	–		56%[a]	44%	B870-α
Rps. capsulata	32%	36%	–		78%	
Rps. sphaeroides	28%	34%	–	76%		
	B870-β					

[a] Sequence data of *Rps. capsulata* not shown in Fig. 2

respectively: 28–78%, Table 2). From the sequence homology (or inversely, from the sequence differences), one can deduce the relative phylogenetic relationship (or distance) between the light-harvesting complexes of the different purple bacteria (species differences): *Rps. viridis* – *R. rubrum* – *Rps. gelatinosa* – *Rps. capsulata* – *Rps. sphaeroides*. Certain antenna complexes were found to contain a third (γ) polypeptide, the function of which is still unknown.

As with the phycobiliproteins, the primary structure of the light-harvesting polypeptides provides an insight into the structure-functional organization of these polypeptides for energy transfer (ZUBER 1982; BRUNISHOLZ et al. 1981; BRUNISHOLZ et al. 1984a, b). Aligned to maximum sequence homology (also including the conserved His-residue, for example), the α- and β-polypeptide chains of the antenna complexes from the various purple bacteria show a typical organization of the primary structure into three domains: (a) a polar-charged N-terminal domain (α: 12–19 residues; β: 20–22 residues); (b) a central, hydrophobic domain (α: 21 residues, β: 23 residues; c) a polar-charged C-terminal domain (α: 18–25 residues, β: 4–12 residues).

This primary structure information indicates that the amino acid residues are distributed accordingly within the membrane (Fig. 4): the hydrophobic domain should be situated within the membrane (hydrocarbon tail region), while the C- and N-terminal domains should be located further out (polar head group region and membrane surface). The membrane thickness being 40–60 Å (30–40 Å hydrocarbon tail region), the hydrophobic domain of the α- and β-polypeptides could well fit into the membrane in the form of a transmembrane-α-helix. Data on IR spectroscopy indicate a preferred α-helix conformation of the polypeptides (*Rps. sphaeroides*) (THEILER and ZUBER 1984). In this instance, the polar and charged cluster regions of the N- and C-terminal sides of the

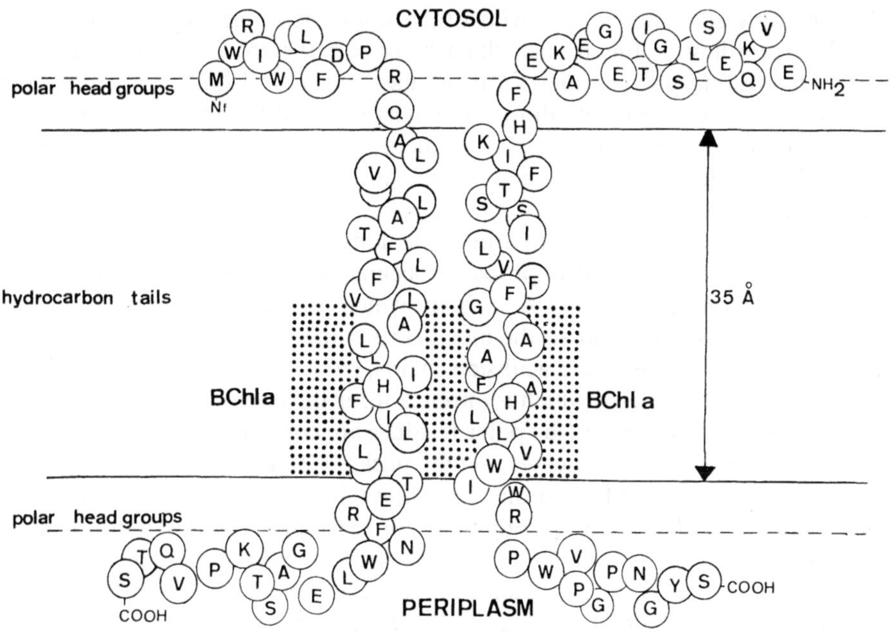

Fig. 4. Structure model of the transmembrane-orientated light-harvesting polypeptides from purple photosynthetic bacteria (amino acid residues of the α- and β-polypeptides from *R. rubrum*). An α-helical conformation of the hydrophobic domain within the hydrocarbon tail region is assumed. The transmembrane α-helix has a bipolar character: the cytoplasmic half carries some polar amino acid residues, the periplasmic side only nonpolar residues and the possible BChl a binding sites (exciton – coupled BChl pair) at the His residues (*hatched region*). The folding of the N- and C-terminal domain is hypothetical

hydrophobic domain would interact optimally with the polar head region. The transmembrane orientation of the α- and β-polypeptides was substantiated by the proteolytic degradation of these polypeptides in situ in the membrane (BRUNISHOLZ et al. 1984b; BACHOFEN and WIEMKEN, Chap. 11.2, this Vol.). In intact vesicles, with the cytoplasmic side of the membrane facing outward, parts of the N-terminal domains of the α- and β-polypeptides of *R. rubrum* were split off with proteolytic enzymes (proteinase K: α: 6 residues, β: 16 residues, chymotrypsin, trypsin, staph. proteinase: α: 3–7 residues). Accordingly, the N-terminal domain is exposed on the cytoplasmic surface of the membrane.

The α- and β-chains are found in a 1:1 ratio and should also function as a pair in the energy transfer system (basic structural unit). The conservative His-residues in the hydrophobic domain of the α- and β-polypeptides are probably BChl binding sites (penta-coordination with central Mg-atom). Thus the α-, β-polypeptide pair would supply the structural basis for two exciton-coupled BChl a-molecules (smallest elementary unit for energy transfer and spectral properties). In this instance, the conservative His-residue in the N-terminal domain of the β-chain could form a possible binding site for monomeric BChl. The sequence homology is particularly high in the vicinity of the possible His-

BChl-binding sites (active site, with other amino acid residues in contact with BChl). According to this model, the α-helix pair (heterodimers) of the α- and β-light-harvesting polypeptides form the basic structural unit of antenna complexes for: (a) the specific binding and arrangement of the BChl in the polypeptide-lipid environment; (b) the arrangement (aggregation) of the polypeptides in the BChl-protein complexes and in the entire light-harvesting system around the reaction centers.

In the α-helical rods of the hydrophobic domain of the α- and β-polypeptides the porphyrin ring of the BChl probably occupies a position, due to the configuration (orientation) of the His-side chain, which is roughly parallel to the helix axis and thus perpendicular to the membrane plane (THEILER and ZUBER 1984). The distribution of the amino acid side chains, particularly of the polar residues, on the surface of the helix, which is important for the interaction of the polypeptides in a lipid environment (the polar residues are concentrated on the cytoplasmic side of the α-helix) indicates a tight packing of the α-, β-chain helix pair (ZUBER 1982). This configuration also allows for a strong interaction of the two BChl molecules (exciton coupling). Furthermore, the distribution of the polar residues in the helices is consistent with an aggregation of the α-helix pairs into larger, probably cyclic units in the BChl protein complexes (B-870/890, B-800–850) (ZUBER 1982). Owing to the aggregation of antenna complexes, it may be assumed that the N- and C-terminal domains play an important role at the membrane surface (two-dimensional organization of the complexes around the reaction centers for heterogeneous energy transfer). Noncolored linker peptides could also be involved here (e.g., the "third" γ-polypeptides).

4 Concluding Remarks

The primary structure analysis of the light-harvesting polypeptides of cyanobacteria, red algae, and purple photosynthetic bacteria provides fundamental data on the structural organization of the light-harvesting system of these organisms. It also provides a basis for the three-dimensional structure analysis of the antenna complexes. Furthermore, a comparative analysis of the primary structures can reveal not only the functionally active structure elements but also the general structure principles of energy transfer. Indeed, the energy transfer in intramembrane and extramembrane light-harvesting systems seems to take place in structurally similar units (ZUBER 1982). The basic structural unit is a pair of nonidentical (α−β), polypeptides, which in turn carry a pair of pigment molecules. They are either already exciton-coupled (distances <20 Å) or the coherent excitons of the pigment molecules are formed by the specific aggregation of the basic structural units in the cyclic systems of the pigment-protein complexes. In both cases, however, the specific structures of the cyclic systems, which cause larger distances (>20 Å) between the pigment pairs (excitons) also suggest the probability of an inductive resonance transfer between the pigment pairs (BORISOV 1978; PEARLSTEIN 1982). The antenna system for heterogenous energy transfer is formed in the vicinity of the reaction centers by the specific aggregation of pigment protein complexes with different absorption maxima. Due to the different environments of the extra-membrane antenna systems in water (globular phycobiliproteins) and the intra-membrane antenna systems (integral pigment-proteins) it is obvious that functionally and structurally equivalent protein assemblies can take on very different forms. The primary structure data of the light-harvesting polypeptides of other antenna systems will in future illustrate both the scope of variability and the limits of this general structure principle for energy transfer in photosynthesis.

References

Borisov A Yu (1978) In: Clayton RK, Sistrom WR (eds) The photosynthetic bacteria. Plenum, New York London, pp 323–331

Brunisholz RA, Cuendet PA, Theiler R, Zuber H (1981) FEBS Lett 129:150–154

Brunisholz RA, Suter F, Zuber H (1984a) Hoppe Seyler's Z Physiol Chem 365:375–688

Brunisholz RA, Wiemken V, Suter F, Bachofen R, Zuber H (1984b) Hoppe Seyler's Z Physiol Chem 365:689–701

Brunisholz RA, Jay F, Suter F, Zuber H (1985) Biol Chem Hoppe-Seyler 366:87–98

Clayton RK, Sistrom WR (eds) (1978) The photosynthetic bacteria. Plenum, New York London

Cogdell RJ, Thornber JP (1980) FEBS Lett 122:1–8

Cogdell RJ, Valentine J (1983) Photochem Photobiol 38:769–772

Frank G, Sidler W, Widmer H, Zuber H (1978) Hoppe Seyler's Z Physiol Chem 359:1491–1507

Freidenreich P, Apell GS, Glazer AN (1978) J Biol Chem 253:212–219

Füglistaller P, Suter F, Zuber H (1983) Hoppe Seyler's Z Physiol Chem 364:691–712

Gantt E (1981) Annu Rev Plant Physiol 32:327–347

Glazer AN (1981) In: Hatch MD, Boardmann NK (eds) The biochemistry of plants, vol VIII. Academic Press, London New York, pp 51–96

Glazer AN (1982) Annu Rev Microbiol 36:173–198

Gogel GE, Parkes PS, Loach PA, Brunisholz RA, Zuber H (1983) Biochim Biophys Acta 746:32–39

Lange RJ de, Williams LC, Glazer AN (1981) J Biol Chem 256:9558–9566

Nies M, Wehrmeyer W (1981) Arch Microbiol 129:374–379

Offner GD, Troxler RF (1983) J Biol Chem 258:9931–9940

Offner GD, Brown-Mason AS, Ehrhardt MM, Troxler RF (1981) J Biol Chem 256:12167–12175

Pearlstein RM (1982) Photochem Photobiol 35:835–844

Sauer K, Austin LA (1978) Biochemistry 17:2011–2019

Schirmer T, Bode W, Huber R, Sidler W, Zuber H (1985) J Mol Biol 184:257–277

Sidler W, Gysi J, Isker E, Zuber H (1981) Hoppe Seyler's Z Physiol Chem 362:611–628

Tadros M, Suter F, Drews G, Zuber H (1983) Eur J Biochem 129:533–536

Tadros M, Suter F, Seydewitz HH, Witt I, Zuber H, Drews G (1984) Eur J Biochem 138:209–212

Theiler R, Zuber H (1984) Hoppe Seyler's Z Physiol Chem 365:721–729

Theiler R, Suter F, Wiemken V, Zuber H (1984) Hoppe Seyler's Z Physiol Chem 365:703–719

Thornber JP, Cogdell RJ, Seftor REB, Pierson BK (1983) In: Proc VIth Int Congr Photosynth, vol II. Brussels, p II.1.25

Troxler RF, Ehrhardt MM, Brown-Mason A, Offner GD (1981) J Biol Chem 256:12176–12184

Zuber H (1982) Abstr IVth Int Symp Photosynth Procaryotes, Bombannes, Fr, C 40

Zuber H (1983) In: Papageorgiou GC, Packer L (eds) Photosynthetic procaryotes: Cell differentiation and function. Elsevier, Amsterdam New York Oxford, pp 23–41

6.2 Light-Harvesting Complexes in the Purple Photosynthetic Bacteria

R.J. COGDELL

1 Introduction

Most purple bacterial antenna complexes can be conveniently grouped into one of two distinct classes. These two classes have been previously called light-harvesting complex I (LH I) and light-harvesting complex II (LH II). However, due to the lack of information in these titles, this nomenclature is now often replaced by B(Bulk)-890 or B-800–850 where the numbers refer to the appropriate bacteriochlorophyll long wavelength absorption maximum(a) (COGDELL and THORNBER 1979, 1980; THORNBER et al. 1978, 1984). So far the majority of species which have been studied in detail contain one type of antenna complex (LH I) which appears to be intimately associated with the reaction centre. This type of complex [e.g. the B-890 complex from *Rhodospirillum rubrum* (COGDELL et al. 1982; PICOREL et al. 1983) or the B-875 complex from *Rhodopseudomonas sphaeroides* (BROGLIE et al. 1980)] usually occurs in a fixed stoichiometry relative to the reaction centre, and is characterised by a single, strong bacteriochlorophyll (Qy) absorption maximum in the near infrared (SISTROM 1978; THORNBER, Chap. 3, this Vol.). Many species of purple bacteria also contain a second type of antenna complex (COGDELL and THORNBER 1979, 1980; THORNBER et al. 1978, 1984). This type of light-harvesting complex (LH II), of which the B-800–850 complexes from *Rps. sphaeroides* and *Rps. capsulata* are good examples (BROGLIE et al. 1980; CLAYTON and CLAYTON 1972; COGDELL and CROFTS 1978; FEICK and DREWS 1978; SAUER and AUSTIN 1978), is present in the photosynthetic membrane in variable amounts with respect to the reaction centres and is thought to be arranged more peripherally around them. This second group of complexes is characterised by having two strong bacteriochlorophyll absorption maxima in the near infrared. It has been shown that by varying the amount of this second type of light-harvesting complex the bacteria are able to modulate the size of their photosynthetic unit in response to various environmental conditions such as changes in light intensity (SISTROM 1978; SPRAGUE and VARGA, Chap. 11.1; DONOHUE and KAPLAN, Chap. 11.3, both this Vol.).

The preparation of an antenna complex begins with the isolation of the photosynthetic membrane. The membrane is then solubilised with a suitable, non-denaturing detergent and the antenna complexes are purified from the solubilised "soup", usually by conventional methods of protein purification (COGDELL and THORNBER 1979). These procedures are simplified by the fact that the antenna complexes are highly pigmented and are therefore readily followed by eye, during their isolation and purification.

The antenna pigments (bacteriochlorophyll and carotenoids) are noncovalently bound to their apoproteins. Monomeric bacteriochlorophyll a in 7:2 v/v acetone/methanol shows a single strong absorption band in the near infrared region at 772 nm (CLAYTON 1963). However when it is correctly bound to its apoprotein its near infrared absorption band is strongly red-shifted to between 800 and 900 nm (CLAYTON 1963). This red shift is even greater in bacteriochlorophyll b-containing organisms, where the free pigment absorbs at 790 nm and the antenna complex absorbs above 1000 nm (THORNBER et al. 1978). Any denaturation of the complex reverses this red-shift and so the absorption spectrum of an antenna complex is a sensitive indicator of its degree of intactness. It is of course important to verify that any preparation of an antenna complex is fully native before using it for further experimentation.

2 The B-890 Class of Light-Harvesting Complexes (LH I Complexes)

The B-890 complex from *R. rubrum* is the most fully characterised of this class (BRUNISHOLZ et al. 1981, 1984c; COGDELL et al. 1982; GOGEL et al. 1983; PICOREL et al. 1983; SAUER and AUSTIN 1978). The absorption spectrum is shown in Fig. 1. This complex exists in vivo as a multimer of a "minimal compositional unit", which consists of two molecules of bacteriochlorophyll a and one molecule of carotenoid (mainly spirilloxanthin) bound to two low molecular weight hydrophobic polypeptides (COGDELL et al. 1982; PICOREL et al. 1983). These apoproteins have been sequenced and shown to consist of 52 and 54 amino

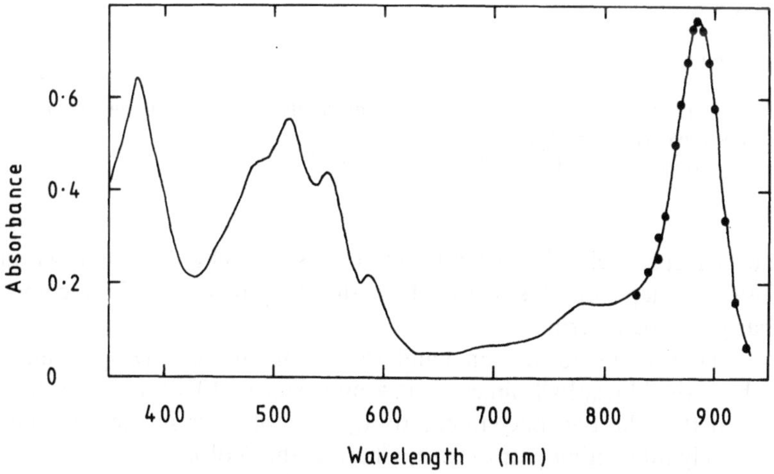

Fig. 1. The absorption spectrum of the B-890 complex from *R. rubrum* strain S1 (——) recorded in a Pye-Unicam SP8000; (-●-) recorded on a Pye-Unicam SP500. The complex was prepared as described in COGDELL et al. (1982) and resuspended in 20 mM Tris HCl pH 8.0, 0.1% LDAO

Table 1. A comparison of the major light-harvesting pigment-protein complexes in purple photosynthetic bacteria[a]

Antenna type	B-890 complex class		B-800–850 complex class		
	B-890-complex	B-875-complex	Type I	Type II	B-800–820-complex
Examples of bacteria containing antenna type	*R. rubrum* *Rps. viridis* *Chr. vinosum* *Rps. acidophila*	*Rps. sphaeroides* *Rps. capsulata* *Rps. palustris*	*Rps. sphaeroides* *Rps. capsulata* *Rps. gelatinosa* *Rps. acidophila* 7750	*Chr. vinosum* *Rps. acidophila* 7050	*Chr. vinosum* *Rps. acidophila* 7050
Bchla: carotenoid ratio	2:1	2:2	3:1	3:1	3:1
Nos. of polypeptide types in isolated complex	2 or 3	2	2 or 3	2	2
Intensity of CD spectrum of long wavelength band	Strong	Weak	Strong	Strong	Strong
Typical values of the millimolar extinction coefficient of the long wavelength Bchla absorption band[b]	*R. rubrum* B-890-complex 125 cm^{-1}	*Rps. sphaeroides* B-875-complex 126 cm^{-1}	*Rps. sphaeroides* B-800–850-complex 132 cm^{-1}	Not determined	Not determined

[a] The quantitative data are based on the analysis of only a few species and are only given as typical examples.
[b] The values are probably good to $\pm 10\ cm^{-1}$

acids respectively (BRUNISHOLZ et al. 1981, 1984a, b, c; GOGEL et al. 1983; ZUBER, Chap. 6.1, this Vol.). The state of aggregation of the complex in vivo has yet to be determined.

The 890 nm absorption band shows an intense CD spectrum, typical of a bacteriochlorophyll dimer (SAUER and AUSTIN 1978) and so it has been proposed that the two bacteriochlorophyll a molecules in the minimal unit form a closely interacting pair (SAUER, Chap. 2, this Vol.).

Some other species, however, such as *Rps. sphaeroides* and *Rps. capsulata*, contain a different type of "B-890" antenna complex (BOLT et al. 1981a, b; BROGLIE et al. 1980; BACHMANN et al. 1983; PETERS and DREWS 1983). The "minimal unit" of the B-875 complex from *Rps. sphaeroides* contains two molecules of bacteriochlorophyll a and two molecules of carotenoid (one extra carotenoid!) again bound to two low molecular weight hydrophobic polypeptides (BROGLIE et al. 1980; BACHMANN et al. 1983). These apoproteins have been sequenced and contain 58 and 48 amino acids (THEILER et al. 1984). Unlike the B-890 complex from *R. rubrum*, the B-875 complex from *Rps. sphaeroides* and *Rps. capsulata* shows only a weak CD spectrum in the 875 nm absorption band (BOLT et al. 1981a, b). This separation of the "B-890" group into two is therefore based upon differences in composition and spectral characteristics. These are summarised in Table 1.

3 The B-800–850 Complex Class of Light-Harvesting Complexes (LH II Complexes)

Just as with the B-890 complexes, the B-800–850 complexes are thought to exist within the photosynthetic membrane as aggregates of a "minimal compositional unit" (COGDELL and CROFTS 1978; FEICK and DREWS 1978; SAUER and AUSTIN 1978; SHIOZAWA et al. 1982). In *Rps. sphaeroides* this unit consists of three molecules of bacteriochlorophyll a and one molecule of carotenoid bound to two low molecular weight, hydrophobic polypeptides (COGDELL and CROFTS 1978; SAUER and AUSTIN 1978; COGDELL et al. 1980) (recently, however, it has been suggested that the carotenoid content of this complex may be higher (KRAMER et al. 1984)). The two apoproteins have been sequenced and contain 54 and 51 amino acids respectively (THEILER et al. 1984). So far the smallest pigmented oligomer that has been described has an apparent molecular weight of ~100,000 (COGDELL et al. 1980).

The absorption spectrum of the B-800–850 complex from *Rps. sphaeroides* is shown in Fig. 2. It has been proposed, from room temperature CD and 4th derivative analysis of the 800 and 850 nm absorption bands, that the 800 nm band represents monomeric bacteriochlorophyll and the 850 nm band dimeric bacteriochlorophyll. However, a recent low temperature CD analysis (KRAMER et al. 1984) has suggested that the situation is more complex, and that within

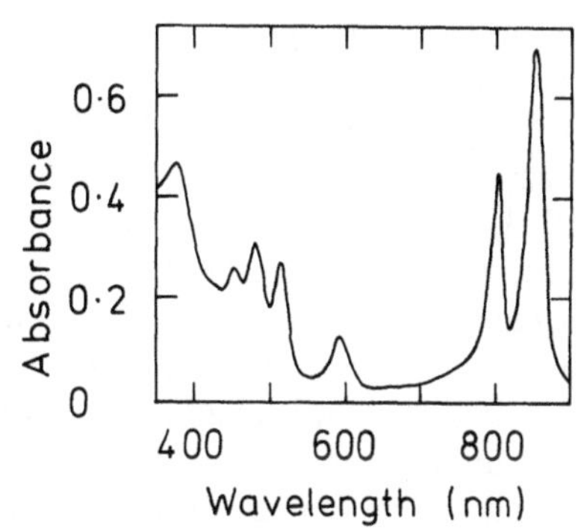

Fig. 2. The absorption spectrum of the type I B-800–850 complex from *Rps. sphaeroides* strain 2.4.1. The spectrum was recorded on a Pye-Unicam SP8–500. The complex was prepared as described in COGDELL and CROFTS (1978) and was resuspended in 20 mM Tris HCl. pH 8.0, 0.1% LDAO

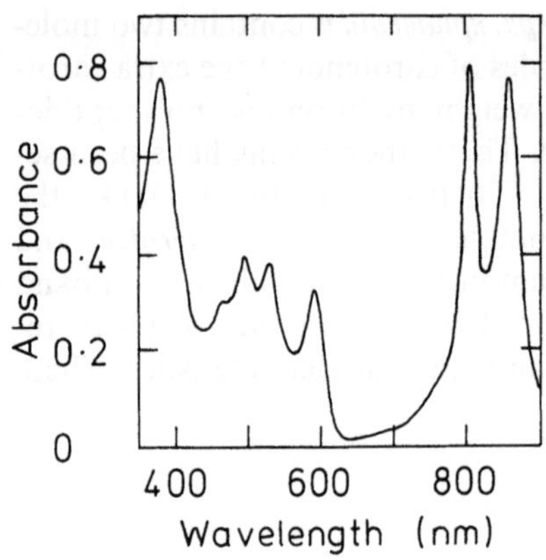

Fig. 3. The absorption spectrum of the type II B-800–850 complex from *Rps. acidophila* strain 7050. This spectrum was recorded in a Pye-Unicam SP8–500. The complex was prepared as described in COGDELL et al. (1983) and was resuspended in 20 mM Tris HCl pH 8.0, 0.1% LDAO

the multimeric structure the 800-nm-absorbing bacteriochlorophyll molecules may also interact together, rather than being strictly monomeric.

Interestingly, it is the carotenoid that is present within this type of antenna complex which is responsible for the well-known electrochromic carotenoid band shift (HOLMES et al. 1980; MATSURA et al. 1980; WEBSTER et al. 1980).

The analogous B-800–850 complex from *Rps. capsulata* has also been extensively studied (BOLT et al. 1981b; FEICK and DREWS 1978; SHIOZAWA et al. 1982). It is basically the same as the complex from *Rps. sphaeroides*, except that it also contains, within its "minimal unit", an additional, non-pigment-binding polypeptide. Similar colourless polypeptides have been described for the B-800–850 complex from *Rps. acidophila* 7750 and the B-1015 complex from *Rps. viridis* (COGDELL et al. 1983; BRUNISHOLZ et al. 1983).

The two B-800–850 complexes described above have been called type I B-800–850 complexes (COGDELL et al. 1983; THORNBER et al. 1984). Type I B-800–850 complexes are characterised by the 850 nm absorption band being ~1.5 times as intense as the 800 nm band (cf. Fig. 2). Some species, however, contain a B-800–850 complex in which the intensity of the 850 nm absorption band is equal to or lower than the 800 nm band (Fig. 3). This type of B-800–850 complex has been called a type II complex and is found in such species as *Chromatium vinosum* (THORNBER 1970) and *Rps. acidophila* 7050 (when this strain is grown at low light intensity; COGDELL et al. 1983). Unlike the type I complexes, the type II B-800–850 complexes show a large degree of spectral variability. The position of the "850 nm" absorption band can vary between 835 and 855 nm, apparently depending upon such factors as detergent concentration, pH, ionic strength etc. (COGDELL and THORNBER 1979; COGDELL et al. 1983). So far this sub-group of antenna complexes has received comparatively little study with perhaps the best characterised being the complex from *Rps. acidophila* 7050. In this species the type II B-800–850 complex has a "minimal unit" of three molecules of bacteriochlorophyll a, one molecule of carotenoid and two low molecular weight apoproteins (M_r 5–7000) (COGDELL et al. 1983). The apoproteins have yet to be sequenced and the size of the native aggregate is unknown. On the basis of composition the type I and type II B-800–850 complexes appear to be rather similar, it is their differing spectral characteristics

Fig. 4. The absorption spectrum of the B-800–820 complex from *Rps. acidophila* strain 7050. This spectrum was recorded in a Pye-Unicam SP8–500. The complex was prepared as described in COGDELL et al. (1983) and was resuspended in 20 mM Tris HCl pH 8.0, 0.1% LDAO

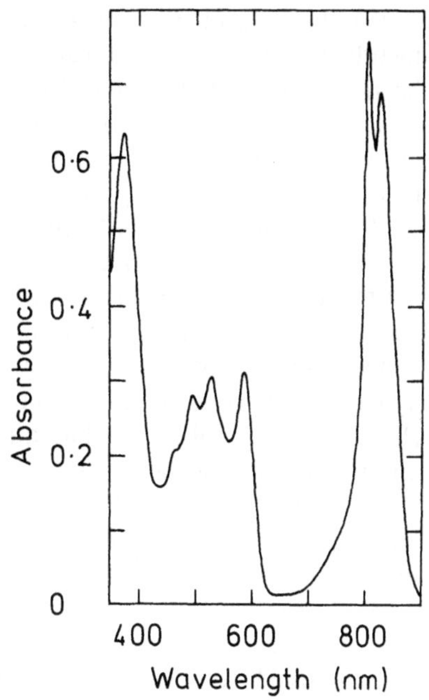

that distinguish them. Resonance Raman studies on a range of B-800–850 complexes support this division into two sub-groups (HAYASHI and TASUMI 1984; LUTZ and BRUNO 1983).

It has also been common observation that whenever a bacterium contains type II B-800–850 complexes that it also contains B-800–820 complexes (THORNBER et al. 1978) (Fig. 4). The status of this extra type of antenna complex remains to be clarified. It does, however, appear to be very closesly related to the type II B-800–850 complexes. In *Chr. vinosum* the B-800–820 complex and the type II B-800–850 complex have similar sizes and pigment content, while in *Rps. acidophila* 7050 they appear to have very similar apoproteins (COGDELL et al. 1983; THORNBER 1970).

4 General Remarks on the Structure of Bacterial Light-Harvesting Complexes

In all cases so far studied, the antenna complexes from the purple bacteria contain two types of apoproteins (ZUBER et al. 1983; ZUBER, Chap. 6.1, this Vol.; THORNBER et al. 1978; THORNBER, Chap. 3, this Vol.). A small number of pigment molecules (usually 3 or 4) are then non-covalently bound to these apoproteins forming a notional "minimal building block" from which the native, supramolecular aggregate is constructed.

A number of the light-harvesting apoproteins have now been sequenced (BRUNISHOLZ et al. 1983, 1981, 1984; GOGEL et al. 1983; TADROS et al. 1983; THEILER et al. 1984; ZUBER et al. 1983) and they fall into two structural classes called the α- and β-apoproteins. The α-apoproteins have a single conserved histidine while the β-apoproteins have a pair of conserved histidines. It has been suggested (ZUBER et al. 1983) that these conserved histidine residues could serve as fifth ligands for the magnesium atoms at the centre of the bacteriochlorophyll molecules. The "minimal unit" of each antenna complex appears to have one α- and one β-apoprotein (ZUBER, Chap. 6.1, this Vol.).

In general when a hydroplot is constructed from the amino acid sequence of an antenna apoprotein, three distinct structural domains are clearly defined (ZUBER et al. 1983; BRUNISHOLZ et al. 1984c). This tripartite character can be seen, for example, in the case of the B-890-α-apoprotein from *R. rubrum*. The B-890-α-apoprotein has a central core of about 23 amino acids which is very hydrophobic, while at either end the sequence is polar (ZUBER et al. 1983; ZUBER, chap. 6.1, this Vol.). This has quite naturally led to the suggestion that these apoproteins should lie across the photosynthetic membrane, with their central hydrophobic core imbedded in the membrane's non-polar interior. Also, by analogy with bacteriorhodopsin, it has been proposed that the trans-membrane segment of the apoprotein might be folded to form a single α-helix.

In *R. rubrum* these predictions have been verified experimentally. The presence of α-helices in proteins can be looked for, in the absence of structural data obtained from X-ray crystallography, by examining either the CD spectrum of the protein in the far UV (between 190 and 250 nm where peptide bonds absorb) or by using IR spectroscopy. Both these methods when applied to the antenna complexes indicate a high (70–85%) α-helical content (THEILER and ZUBER 1984, and COGDELL et al. unpublished observations). It is also possible using *R. rubrum* to obtain closed membrane vesicles that have both right-side-out (sphaeroplasts) and inside-out (chromatophores) polarity. This then allows both membrane faces to be interrogated for the presence of their antenna apoproteins (BACHOFEN and WIEMKEN, Chap. 11.2, this Vol.). Using mild proteolysis (BRUNISHOLZ et al. 1984a, b) and immunoprecipitation (COGDELL et al. unpublished observations), the antenna apoproteins have been clearly detected at both membrane surfaces. It appears that they lie across the photosynthetic membrane so that their N-termini are on the cytoplasmic side and their C-termini are on the periplasmic side.

This short chapter has briefly summarised current ideas on the structure and composition of antenna complexes from purple bacteria. More details on both their structure, function and intra-membrane organisation are presented in this volume by DUTTON (Chap. 5), THORNBER (Chap. 3) ZUBER (Chap. 6.1) BACHOFEN and WIEMKEN (Chap. 11.2), SPRAGUE and VARGA (Chap. 11.1), MICHEL (Chap. 8.4) and HEARST (Chap. 8.5). A discussion of the reaction centre and antenna complexes from green bacteria can be found in the contributions of BLANKENSHIP and FULLER (Chap. 8.6) and SPRAGUE and VARGA (Chap. 11.1).

References

Bachmann RC, Tadros MH, Oelze J, Takemoto JY (1983) Biochem Int 7:629–634
Bolt JD, Hunter CN, Niederman RA, Sauer K (1981a) Photochem Photobiol 34:653–656

Bolt JD, Sauer K, Shiozawa JA, Drews G (1981b) Biochim Biophys Acta 635:535–541
Broglie RM, Hunter CN, Delepelaire P, Niederman RA, Chua N, Clayton RK (1980) Proc Natl Acad Sci USA 77:87–91
Brunisholz R, Cuendet PA, Theiler R, Zuber H (1981) FEBS Lett 129:150–154
Brunisholz RA, Jay F, Zuber H (1983) In: Proceedings of the workshop on molecular structure and function of the light-harvesting pigment-protein complexes and photosynthetic reaction centres, Zurich, Switzerland, pp 54
Brunisholz R, Cogdell RJ, Valentine J, Zuber H (1984a) In: Abstr 16th Annu Meet USGEB, Zurich, Switzerland
Brunisholz RA, Wiemken V, Suter RF, Bachofen R, Zuber H (1984b) Hoppe-Seyler's Z Physiol Chem 365:689–701
Brunisholz RA, Suter F, Zuber H (1984c) Hoppe-Seyler's Z Physiol Chem 365:675–688
Clayton RK (1963) In: Gest H, San Pietro A, Vernon LP (eds) Bacterial photosynthesis. Antioch Press, Yellow Springs, Ohio, pp 495–500
Clayton RK, Clayton BJ (1972) Biochim Biophys Acta 283:492–504
Cogdell RJ, Crofts AR (1978) Biochim Biophys Acta 502:409–416
Cogdell RJ, Thornber JP (1979) In: Wolstenholme G, Fitzsimons DW (eds) Chlorophyll organisation and energy transfer in photosynthesis. Ciba Found Symp 61, New Ser. Exerpta Medica, Amsterdam Oxford New York, pp 61–79
Cogdell RJ, Thornber JP (1980) FEBS Lett 122:1–8
Cogdell RJ, Lindsay JG, Reid FP, Webster GD (1980) Biochim Biophys Acta 591:312–320
Cogdell RJ, Lindsay JG, Valentine J, Durant I (1982) FEBS Lett 150:151–154
Cogdell RJ, Durant I, Valentine J, Lindsay JG, Schmidt K (1983) Biochim Biophys Acta 722:427–435
Feick R, Drews G (1978) Biochim Biophys Acta 501:499–513
Gogel GE, Parkes PS, Loach PA, Brunisholz RA, Zuber H (1983) Biochim Biophys Acta 746:32–39
Hayashi H, Tasumi M (1984) In: Sybesma C (ed) Proc VIth Int Congr Photosynth Res. Nijhoff/Junk, The Hague, pp 211–214
Holmes NG, Hunter CN, Niederman RA, Crofts AR (1980) FEBS Lett 115:43–48
Kramer HJM, Grondelle R van, Hunter CJ, Westerhuis WHJ, Amesz J (1984) Biochim Biophys Acta 765:156–165
Lutz M, Bruno R (1983) In: Proceedings of the workshop on molecular structure and function of the light-harvesting pigment-protein complexes and photosynthetic reaction centres, Zurich, Switzerland, pp 62
Matsura K, Ishikawa T, Nishimura M (1980) Biochim Biophys Acta 590:339–344
Moskalenko AA, Erokin Yu E (1979) FEBS Lett 87:254–256
Peters J, Drews G (1983) FEMS Microbiol Lett 17:235–237
Picorel R, Belanger G, Gingras G (1983) Biochemistry 22:2491–2497
Sauer K, Austin LA (1978) Biochemistry 17:2011–2019
Shiozawa JA, Welte W, Hodapp N, Drews G (1982) Arch Biochem Biophys 213:473–485
Sistrom WR (1978) In: Clayton RK, Sistrom WR (eds) The photosynthetic bacteria. Plenum, New York London, pp 841–848
Tadros MH, Suter F, Drews G, Zuber H (1983) Eur J Biochem 129:533–536
Theiler R, Zuber H (1984) Hoppe-Seyler's Z Physiol Chem 365:721–729
Theiler R, Suter F, Wiemken V, Zuber H (1984) Hoppe-Seyler's Z Physiol Chem 365:703–719
Thornber JP (1970) Biochemistry 9:2688–2698
Thornber JP, Trosper TL, Strouse CE (1978) In: Clayton RK, Sistrom WR (eds) The photosynthetic bacteria. Plenum, New York London, pp 133–160
Thornber JP, Cogdell RJ, Pierson BK, Seftor REB (1984) J Cell Biochem 23:159–169
Webster GD, Cogdell RJ, Lindsay JG (1980) Biochim Biophys Acta 591:321–330
Zuber H, Brunisholz RA, Frank G, Fuglistaller P, Sidler W, Theiler R (1983) In: Proceedings of the workshop on molecular structure and function of the light-harvesting pigment-protein complexes and photosynthetic reaction centres, Zurich, Switzerland, pp 56–58

6.3 Phycobilisomes

E. GANTT

1 Introduction

The greatest variation in photosynthetic accessory pigments occurs in the algal groups (ANDERSON, Chap. 6.4; THORNBER, Chap. 3, both this Vol.). Accessory pigments, according to the generally accepted definition, are those pigments which absorb light to drive photosynthesis – in addition to chlorophyll a. When compared to green plants, the most radical departure in accessory pigments occurs in the cyanobacteria and red algae because they contain phycobiliproteins which are organized into unique antennae complexes, the phycobilisomes. This chapter provides a synopsis of phycobilisome characteristics with emphasis on their structural relationship to the photosystems, especially PS II.

Phycobilisomes are supramolecular complexes of phycobiliproteins and linker proteins. The phycobilisomes are attached to the stromal surface of thylakoids. Grana stacks and chlorophyll a/b complexes do not exist in these organisms. The phycobilisomes, however, can be regarded as functionally analogous to other light-harvesting complexes. Phycobilisomes have allophycocyanin on the inside proximal to the membrane followed by phycocyanin, and phycoerythrin, or phycoerythrocyanin when present, on the periphery. This structure enhances the energy transfer from pigments absorbing at shorter wavelength to longer wavelengths. The energy transfer properties of phycobiliproteins are discussed in this Volume in Chapter 7.6 by H. SCHEER, while their primary structure is presented in Chapter 6.1 by H. ZUBER. Involvement of phycobiliproteins in photosynthesis was clearly shown 100 years ago by elegant action spectra made by ENGELMANN (1884). By irradiating algal filaments with specific wavelengths he found action peaks of ca. 540–560 nm in red colored algae, and of ca. 620 nm in blue-colored algae. In the 1920's and 1930's, with the purification of phycoerythrins and phycocyanin (cf. RABINOWITCH 1951), it was possible to quantitatively assess the phycobiliprotein contribution to photosynthesis (DUYSENS 1952). The identity of allophycocyanin was established later (cf. HAXO et al. 1955) because it is generally less stable, often occurs in relatively lower concentration, and is spectrally easily masked by phycocyanin. Direct attachment of the phycobiliproteins to the thylakoids was anticipated from known high transfer efficiencies from phycobiliproteins to chlorophyll. With the identification of phycobilisomes and their isolation (cf. GANTT 1980), it was possible to show that they were directly attached to the photosynthetic lamellae, and are structurally and functionally efficient light-harvesting antennae.

Considerable progress has been made in the past years on phycobilisome structure, especially since the topic was last reviewed in this series by KROGMANN (1977). Recent reviews in the past few years have appeared from many laboratories (COHEN-BAZIRE and BRYANT 1982; GANTT 1980, 1981; GLAZER 1982, 1984; MACCOLL 1982, and TANDEAU DE MARSAC 1983), and provide references to up-dated progress. Published contributions have been extensive and are too numerous to include here, thus articles and reviews were selected with the intent of guiding the reader to the pertinent literature in this area. The structure and composition of phycobilisomes from various organisms is basically very similar, although there are differences in complexity. For illustrating the salient features of phycobilisomes, *Anacystis nidulans* (*Synechococcus* 6301) has been selected to represent the simple type, whereas *Porphyridium cruentum* (*P. purpureum*) serves as example of a more complex type. All have a central core and peripheral rods attached to the core. Both the core and rod components contain specific linker polypeptides (GLAZER 1984).

2 Stability of Phycobilisomes and Isolation

Although phycobilisomes do not have a delimiting membrane, they are stable in phosphate buffer of high molarity. Phycobilisome isolation is routinely carried out at room temperature by suspending cells in phosphate buffer (ca. 0.7 M, pH 7.0) and breaking them in a French pressure cell, sonication, or with a homogenizer. Phycobilisomes are then released by solubilizing the thylakoids with a nonionic detergent such as Triton X-100. Phycobilisomes remain in the supernate, following removal of large particles by centrifugation. Purified phycobilisomes can be obtained by centrifugation of this supernate on a sucrose-phosphate gradient (0.25–2.0 M sucrose) (GANTT 1980). Depending on the organism, the molarity of the phosphate and pH may be varied. Phycobilisomes have been isolated from dozens of species by modifications of this procedure. In fact, their isolation has been easier than their visualization by electron microscopy, especially in cyanobacteria. Low temperature (4°–10 °C) may lead to uncoupling of energy transfer and partial loss of allophycocyanin. Preparations of phycobilisomes remain stable for days, even weeks, if stored concentrated in phosphate buffer. An alternate isolation procedure, with higher molarity phosphate buffer (0.76–1.50 M) and polyethylene glycol (RIGBI et al. 1980) has the advantage of not requiring centrifugation. However, the presence of contaminating proteins and chlorophyll co-isolating with phycobilisomes can cause complications. Intact phycobilisomes have also been obtained without detergent treatment by treating *Griffithsia monilis* thylakoids with trypsin (HILLER et al. 1983). Phycobilisomes obtained within the first 10 min of trypsin treatment were indistinguishable from those prepared with Triton X-100.

The spectral composition of phycobilisomes isolated from the cyanobacterium *A. nidulans* and the red alga *P. cruentum* are seen in Fig. 1. The absorption spectra reflect the absence of phycoerythrin in *Anacystis*, and the paucity of

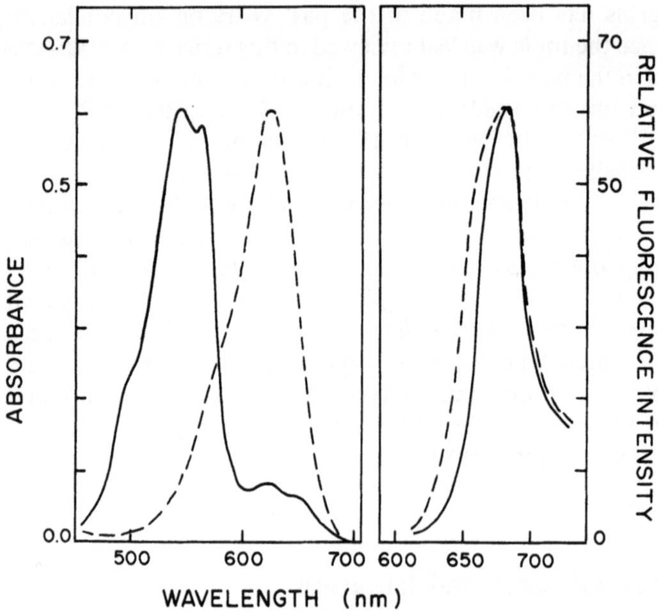

Fig. 1. The absorption spectra of phycobilisomes, at left, of *P. cruentum* phycobilisomes (—) with maxima of phycoerythrin ca. 545–565 nm, phycocyanin ca. 620 nm, and allophycocyanin 650 nm, and of *A. nidulans* (----) with phycocyanin and allophycocyanin (shoulder at ca. 650 nm) as the major pigments. The fluorescence emission at right, at ca. 675 nm in both species is characteristic of energetic tight coupling

phycocyanin and allophycocyanin in *Porphyridium*. It is significant that the fluorescence emission of both preparations occurs at ca. 675 nm. Such an emission is characteristic of most phycobilisomes regardless of the composition of the major phycobiliproteins, providing the phycobiliproteins are energetically well coupled. Absence of the 675–682 nm emission is usually attributable to the absence of, or deficiency in energetic coupling to the final long wavelength emitter(s).

3 Structure and Composition

The polypeptide composition (Fig. 2) of *Anacystis* is less complex than that of *Porphyridium*. In both, the low molecular weight chromophore-containing polypeptides comprise ca. 85% of the stainable polypeptides, while the uncolored linker polypeptides account for ca. 15%.

Determination of the phycobilisome size, shape, and distribution on the thylakoid surface requires cross-comparison from various approaches, especially electron microscopy. Several generalizations can be made from the few species thus examined. Phycobilisomes usually do not cover the entire thylakoid surface,

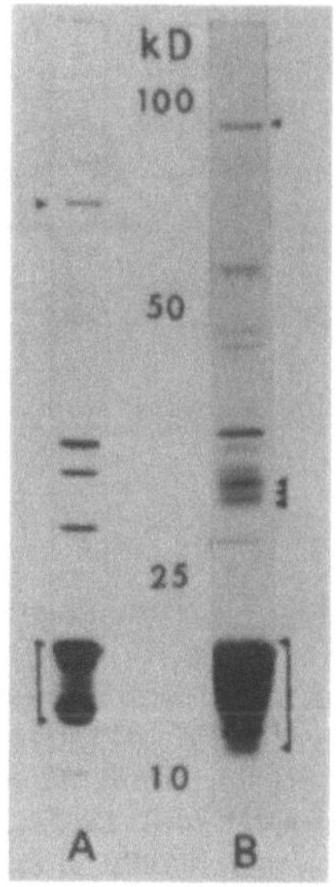

Fig. 2. SDS-PAGE gels (Coomassie stained) of *Anacystis* (**A**) and *P. cruentum* (**B**) reveal the grouping of phycobilin-containing polypeptides (α and β) in brackets at lower portion of the gels (15–21 kDa). The large molecular weight polypeptides (ca. 78 kDa *Anacystis* and 95 kDa *Porphyridium* with long wavelength emission) are on top. Between are the linker polypeptides, which are uncolored except for three red pigmented polypeptides (29–33 kDa) in *P. cruentum* (*arrowheads*)

but in cells grown at low light conditions they are more closely spaced than those in high light cells. Although phycobilisomes can be observed in rows, which may be parallel and closely spaced (COHEN-BAZIRE and BRYANT 1982; GANTT 1980), such ordered organization may be less common than generally assumed (STAEHELIN, Chap. 1, this Vol.).

In most cyanobacteria and in some red algae the phycobilisomes have a hemidiscoidal shape, i.e., they have the appearance of a fan (*A. nidulans*, *Nostoc sp.*, *Tolypothrix tenuis*). Several red algae (*P. cruentum* and *Griffithsia pacifica*) have phycobilisomes which are hemispherical in shape and are more massive. Cylindrically shaped phycobilisomes are rare and have only been reported for two cyanobacteria *Gloeobacter* (COHEN-BAZIRE and BRYANT 1982) and *Anabaena variablis* (ISONO and KATOH 1983). The hemidiscoidally shaped phycobilisomes are often closely spaced on the thylakoids. Phycobilisomes of *Anacystis* are ca. 60 nm wide, 30 nm high, and 10 nm thick. The thickness is equivalent to the diameter of a phycocyanin dodecamer.

Details of the phycobilisome structure are known for several cyanobacterial phycobilisomes, especially *Nostoc* sp. (ZILINSKAS 1982; ZILINSKAS and HOWELL 1983), *Synechocystis* (GLAZER 1982) and *Anabaena* (GLAZER 1982; ISONO and KATOH 1983). The structure of the *Anacystis* phycobilisome has been fully elucidated by GLAZER et al. (1983). Such a phycobilisome has about 370 chromophores. In this simple phycobilisome the central core consists of two disc-shaped allophycocyanin-containing complexes to which six rods are peripherally attached. Hemidiscoidal phycobilisomes more commonly have three disc-shaped complexes in the core (Fig. 3). The rods in *Anacystis* have been shown to contain phycocyanin and at least three uncolored linker polypeptides. The core has two uncolored polypeptides, two allophycocyanin types (660 nm and 675 nm emission), and a high molecular weight (75 kDa) blue polypeptide which has

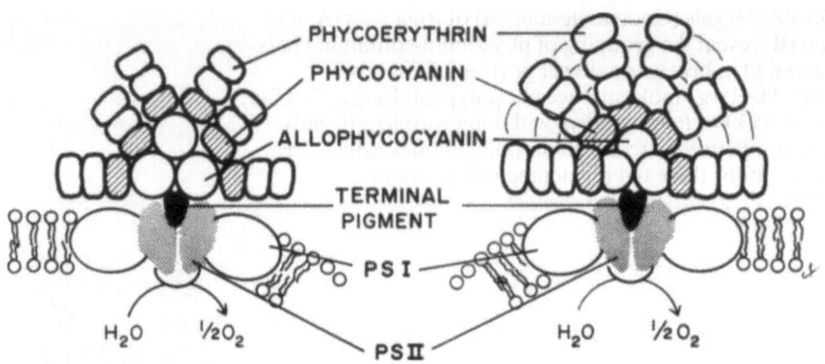

Fig. 3. Schematic arrangement of phycobiliproteins typical of many hemidiscoidal phycobilisomes (COHEN-BAZIRE and BRYANT 1982) on *left*, and on *right* a cut-away view of a more complex hemispherical phycobilisome (GANTT 1981). Rationale for presumed association of the large molecular weight terminal pigment near photosystem II antennae is from *Porphyridium* (REDLINGER and GANTT 1982; CLEMENT-METRAL and GANTT 1983; GANTT et al. 1985). A parallel section through the membrane with the terminal phycobilisome pigment in relation to photosystem I and II is shown in Fig. 4

been postulated to be involved in anchoring the phycobilisome to the thylakoid (GLAZER et al. 1983).

Certain cyanobacteria can chromatically adapt, i.e., they can synthesize phycoerythrin in green (or white) light, and phycocyanin in red light. Upon the addition of phycoerythrin the phycobilisome size may increase. However, in *Tolypothrix* or *Pseudoanabaena* 7409 (COHEN-BAZIRE and BRYANT 1982), red light induces synthesis of a second type of phycocyanin with the phycobilisome size remaining the same. A comprehensive review by TANDEAU DE MARSAC (1983) details the various degrees of chromatic adaptation. Chromatic adaptation in red algae has not yet been convincingly documented.

Phycoerythrin is generally the major phycobiliprotein in hemispherically shaped phycobilisomes. In *P. cruentum* the phycobilisome has dimensions of ca. 48 nm × 32 nm × 32 nm. Such a phycobilisome may contain between 65 and 75 phycobiliprotein dodecamers. The dodecamers are also arranged in rods which are attached to a core (Fig. 3). It can be estimated that at least 2700 chromophores are present in a phycobilisome of this type. The much larger chromophore number is not only attributable to an increased size, but also to the prevalence of phycoerythrin, which has more chromophores per apoprotein than does phycocyanin or allophycocyanin (SCHEER, Chap. 7.6, ZUBER, Chap. 6.1, both this Vol.). Phycoerythrin accounts for over 70% of the protein, whereas phycocyanin, allophycocyanin, and the linker polypeptides account for ca. 9%, 5%, and 15% of the protein, respectively, in *P. cruentum* phycobilisomes (GANTT 1981).

The nature and location of the long wavelength-emitting components (675–682 nm) has been of considerable interest. Two such components have been identified thus far (GLAZER et al. 1983). Of these the large molecular weight group has received the most attention. Large molecular weight polypeptides

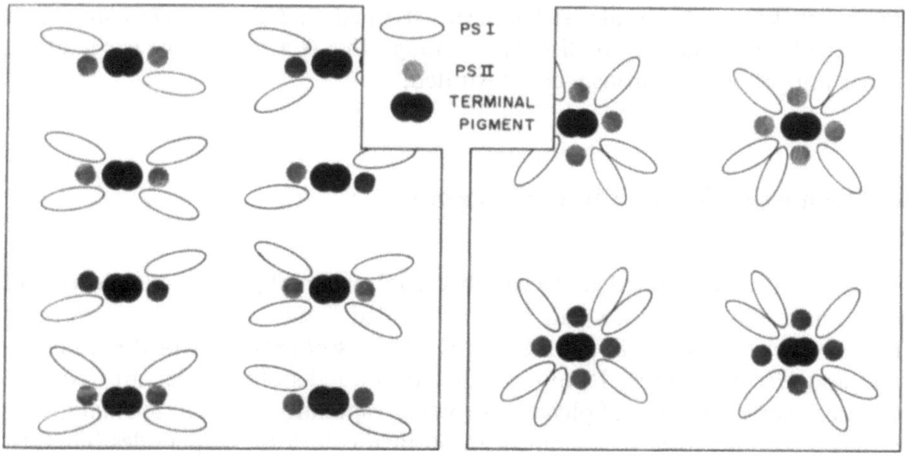

Fig. 4. Possible arrangement of the photosystems around the base of the phycobilisome. In cyanobacteria (such as *Anacystis*) on left, and in a red alga (*Porphyridium*) on right

in phycobilisomes have sizes ranging from ca. 75 kDa in *Anacystis* (GLAZER et al. 1983), to 95 kDa in *P. cruentum* (REDLINGER and GANTT 1982) and to 115 kDa in *Anabaena* (ISONO and KATOH 1983). All of these have a fluorescence emission at ca. 675–78 nm at room temperature, at liquid nitrogen temperature the emission is shifted to ca. 682–88 nm. They account for 1–2% of the Coomassie stainable protein. The 95 kDa polypeptide has been implicated as an effective energy linkage between phycobilisomes and thylakoids in *P. cruentum* (REDLINGER and GANTT 1982) and *Nostoc* (ZILINSKAS 1982). Antisera produced against the 94 kDa high molecular weight terminal polypeptide from *Nostoc* sp. has recently been found to be cross-reactive only with large molecular weight polypeptides in 12 cyanobacterial species and 4 red-algal species (GANTT et al. 1985). Although variable in size (75–115 kDa) they all have at least one immunologically reactive site, which appears to have been highly conserved over evolutionary development.

A 95 kDa polypeptide has been isolated from phycobilisomes and separately from phycobilisome-free thylakoids of *P. cruentum*. Although the components from the two sources had the same apparent molecular weight, the same protease digest patterns and were immunologically related, they were spectrally different (REDLINGER and GANTT 1982). The phycobilisome-derived polypeptide had a blue color and emitted at 675 nm, the one from the thylakoid was colorless. One possibility is that the chromophore was bleached or lost from the thylakoid-derived polypeptide. A protein which is partially embedded in the phycobilisome and partially in the thylakoids, as suggested in Fig. 3, would be a logical conduit for transfering energy as well as serving to anchor the phycobilisome to the thylakoid and assuring physical contact with photosystem II. However, results on phycobilisome release by trypsin treatment by HILLER et al. (1983) suggest that the 95 kDa polypeptides may not be directly involved in the attachment. They showed that the phycobilisomes released in the first 10 min had undegraded 95 kDa polypeptide. It has been assumed that the 95 kDa component is at the base of the phycobilisome. Although this is a reasonable expectation, it has not yet been rigorously demonstrated.

4 Reconstitution of Phycobilisomes

It has been demonstrated that cyanobacterial phycobiliprotein complexes and even entire phycobilisomes can be reconstituted from dissociated components (CANAANI and GANTT 1982; GLAZER et al. 1983; ISONO and KATOH 1983; ZILINSKAS and HOWELL 1983). It appears that uncolored polypeptides are required for the orderly assembly of phycobiliproteins, especially since Glazer's laboratory has reconstituted phycocyanin, with purified colorless polypeptides (GLAZER 1982; GLAZER et al. 1983). In vitro formation of phycobilisomes by combining core fractions with the rod fractions yielded energetically fully functional phycobilisomes with *Nostoc* sp. and *Fremyella diplosiphon* (CANAANI and GANTT 1982). Furthermore, intrageneric (or "hybrid") phycobilisomes have been obtained between these two species with up to 70% recovery. A colorless polypeptide (31 kDa) present in the rods of both these species, appears to be required for attachment of the rods to the cores. This is probably the same polypeptide in *Anacystis* that has been identified by GLAZER et al. (1983) as the 27 kDa rod terminator polypeptide. KIRILOVSKY et al. (1983) have even succeeded in reassociating phycobilisomes onto thylakoid membranes in *Fremyella*. The recombined phycobilisome-thylakoid complexes were capable of energy transfer to the thylakoid membranes. Their results suggest that the terminal high molecular weight pigment (95 kDa) had been retained on the thylakoid membrane.

5 Association with Photosystem II

It has long been known that phycobiliproteins transfer most of the energy initially to photosystem II. Thus, a close association of phycobilisomes with PS II was expected. The association of phycobilisomes with photosystem II has recently been established by isolation of photosystem II-phycobilisome particles (CLEMENT-METRAL and GANTT 1983). In *P. cruentum* such preparations routinely have oxygen evolution rates of up to 1500 μM O_2 mg^{-1} chl h^{-1} in a medium consisting of 0.5 M potassium phosphate, 0.26 m sodium citrate, 0.5 M sucrose, and 10 mM magnesium chloride. Photosystem I in these particles appears to be absent. Significantly, absorption by the phycobiliproteins, primarily by phycoerythrin, drives the photosystem II activity (CHERESKIN et al. 1984).

In red algae and cyanobacteria most of the chlorophyll is associated with PS I (KHANNA et al. 1983; KURSAR and ALBERTE 1983; LEY 1984). This is also reflected in the chlorophyll distribution in fractionated thylakoids of *Porphyridium* (REDLINGER and GANTT 1983). A major green PS I (CP I) complex contained 85% of the bound chlorophyll, and on denaturation resolved into the 68 kDa apoprotein band. Two PS II complexes, CP III and CP IV, collectively contained 15% of the bound chlorophyll. On denaturation CP III contained only a 52 kDa apoprotein, a possible reaction center II apoprotein, whereas CP IV contained primarily 40 kDa and 48 kDa apoproteins with occasional small amounts of 52 kDa.

Finally, the above results lead us to consider the possible structural arrangement of the phycobilisomes with PS I and PS II in the thylakoid. A diagramatic presentation in Fig. 4 of the possible arrangement of the photosystem around phycobilisomes, although a model yet to be tested, is supported by structural and physiological data. The model is significantly different from those proposed for higher plants in that PS I and PS II are spatially closely associated and are not sequestered into separate regions. The mechanism for regulation of excitation energy distribution appears to be different from grana-containing plants. According to BIGGINS et al. (1984) reversible protein phosphorylation does not appear to be involved in *Porphyridium* and *Anacystis*. If the photosystems are closely associated around the phycobilisomes, then small conformation changes at the base of the phycobilisome can cause a change in the distribution of excitation energy. The basis for such a model is detailed in BIGGINS et al. (1984).

An accurate stoichiometry of phycobilisomes: PS II reaction center: PS I reaction center in any one red alga or cyanobacterium is as yet premature. However, it appears that some red algae have more reaction center II: phycobilisome than cyanobacteria. The ratio of phycobilisomes to photosystem II have varied from 1–4, depending on the assay methods, organisms, and the conditions under which cells were grown (LEY 1984; GANTT 1980). In *Anacystis* the reaction center II to phycobilisome ratio has been determined as ca. 1.4 by KHANNA et al. (1983) and 1.7 by KURSAR and ALBERTE (1983). In red algae with large phycoerythrin-rich phycobilisomes the ratio was ca. 4 in *Neoghardiella bailyei* as well as in *P. cruentum* (KURSAR and ALBERTE 1983; LEY 1984). LEY (1984) further established that in *P. cruentum* all reaction centers II are coupled to phycobilisomes.

The surface topology of phycobilisomes on the stroma side of thylakoids has always suggested discrete and distinct functional domains within the thylakoids (cf. GANTT 1980). Results from freeze-fractured thylakoid membranes have also led to the suggestion that photosystem II particles are connected to the phycobilisomes. This assumption is inferred from (a) the similar spacing and occasional parallel alignment of 10 nm E-face particles in freeze-fractured membranes, and phycobilisomes in the stroma, (b) the reduction of 10 nm particles in thylakoids of heterocysts lacking photosystem II activity, and (c) the greater density of such particles in thylakoids with higher photosystem II activity (GANTT 1980; COHEN-BAZIRE and BRYANT 1982; STAEHELIN, Chap. 1, this Vol.). MÖRSCHEL and MÜHLETHALER (1983) in a freeze-fracture analysis of *Mastigocladus* have presented fractures which pass through the thylakoid membrane revealing putative PS II particles in line with parallel phycobilisome rows in fractured stroma regions. They concluded that each phycobilisome is associated with a 10 nm particle. GIDDINGS et al. (1983) came to similar conclusions in a study of *Cyanophora*, which has similar thin phycobilisomes. However, their micrographs show each 10 nm EF particle to have two protruding elements, thus suggesting that each phycobilisome is associated with an intramembrane particle with a binary substructure (two reaction centers?). If these freeze-fracture particles are photosystem components, then four 10 nm particles can be expected to be found in *P. cruentum* and *N. baileyeii.* It would be of great interest to establish the nature of the 10 nm particles by immunochemical labeling. The elucidation of the structural arrangement of phycobilisomes with respect to the photosystems is crucial to understanding excitation energy transfer which is generally much more efficient than in green plants.

Acknowledgments. Work cited from the author's laboratory was supported by the U.S. Department of Energy. Appreciation for the drawings is expressed to C.A. LIPSCHULTZ.

References

Biggins J, Campbell CL, Bruce D (1984) Biochim Biophys Acta 767:138–144
Canaani O, Gantt E (1982) Proc Natl Acad Sci USA 79:5277–5281
Chereskin B, Clement-Metral JD, Gantt E (1984) Plant Physiol 77:626–629
Clement-Metral JD, Gantt E (1983) FEBS Lett 156:185–188
Cohen-Bazire G, Bryant DA (1982) In: Carr NG, Whitton BA (eds) The biology of the cyanobacteria. Blackwell, Oxford London Edinburgh Melbourne, pp 143–190
Duysens LNM (1952) Ph D thesis, Utrecht
Engelmann TW (1884) Bot Z 42:81–93, 97–105
Gantt E (1980) Int Rev Cytol 66:45–80
Gantt E (1981) Annu Rev Plant Physiol 32:327–347
Gantt E, Lipschultz CA, Redlinger T (1985) In: Arntzen C, Bogorad L, Bonitz S, Steinback K (eds) Molecular biology of the photosynthetic apparatus. Cold Spring Harbor New York, 223–231
Giddings TH, Wasmann C, Staehelin LA (1983) Plant Physiol 71:409–419
Glazer AN (1982) Annu Rev Microbiol 36:173–198
Glazer AN (1984) Biochim Biophys Acta 768:29–51
Glazer AN, Lundell DJ, Yamanaka G, Williams RC (1983) Ann Microbiol Inst Pasteur 134B:159–180
Haxo F, O'hEocha C, Norris P (1955) Arch Biochim Biophys 54:162–173
Hiller RG, Post A, Stewart AC (1983) FEBS Lett 156:180–184
Isono T, Katoh T (1983) Plant Cell Physiol 24:357–368
Khanna R, Graham J-R, Myers J, Gantt E (1983) Arch Biochem Biophys 224:534–542
Kirilovsky D, Kessel M, Ohad I (1983) Biochim Biophys Acta 724:416–426
Krogmann DW (1977) In: Trebst A, Avron M (eds) Photosynthesis I: Encyclopedia of plant physiology, vol V. Springer, Berlin Heidelberg New York, pp 624–636
Kursar TA, Alberte RS (1983) Plant Physiol 72:409–414
Ley AC (1984) Plant Physiol 74:451–454
MacColl R (1982) Photochem Photobiol 35:899–904
Mörschel E, Mühlethaler K (1983) Planta 158:451–457
Rabinowitch EI (1951) Photosynthesis and related processes, vol II. Wiley, New York
Redlinger T, Gantt E (1982) Proc Natl Acad Sci USA 79:5542–5546
Redlinger T, Gantt E (1983) Plant Physiol 73:36–40
Rigbi M, Rosinski J, Siegelman HW, Sutherland JC (1980) Proc Natl Acad Sci USA 77:1961–1965
Tandeau de Marsac N (1983) Bull Inst Pasteur 81:201–254
Zilinskas BA (1982) Plant Physiol 70:1060–1065
Zilinskas BA, Howell DA (1983) Plant Physiol 71:379–387

6.4 Light-Harvesting Pigment-Protein Complexes of Algae

JAN M. ANDERSON and J. BARRETT

1 Introduction

The pigment-protein complexes which capture and convert solar energy in the aquatic environment are responsible for the considerable contribution of marine and freshwater algae to the total global photosynthetic biomass. Yet despite the great diversity of photosynthetic pigments in algae, their pigment-protein complexes, with the exception of the phycobiliproteins, have been little investigated. Nevertheless, recent investigations establish a basis for comparative studies of the functional, structural and evolutionary diversity of the algal light-harvesting pigment-proteins (see THORNBER, Chap. 3, this Vol. for an overview).

2 Light Environment

As most algae are submerged in water, their photosynthesis is influenced by the light-transmitting properties of water. Not only is the light intensity drastically diminished with increasing depth of water, but the spectral quality of the light is also greatly altered because of selective absorption and scattering by particles (KIRK 1983). With increasing depth of water there is rapid attenuation of far-red, red, and then blue light (Fig. 1). Further modifications to the light environment of algae are caused by the great fluctuations in daily or seasonal irradiance. Consequently, the light environment of algae is strikingly different to that of terrestrial plants which receive either full sunlight in open habitats, or in extreme shade only very limited irradiance enriched in far-red and green light but depleted in red and blue light.

3 Distribution of Light-Harvesting Pigments

All algae possess Chl a and β-carotene (rarely the α- or ε-isomer), and these pigments are essential for photosynthesis. Whereas land plants have only one pigment system, algae living in their light-limited, mainly blue-green environment, have evolved a great variety of light-harvesting pigment systems. These

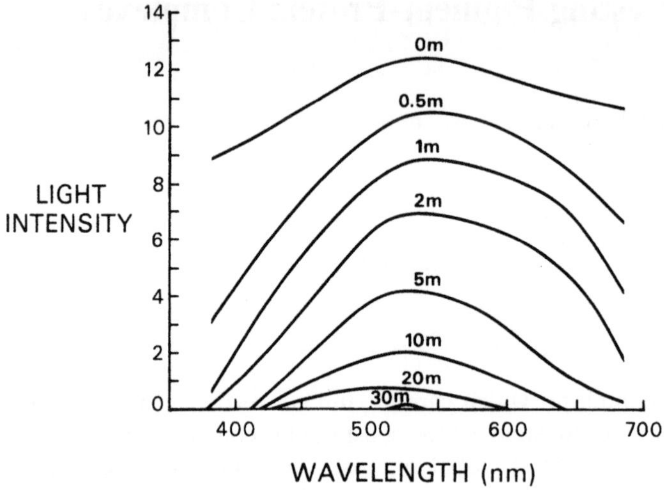

Fig. 1. Spectral composition of radiant energy at different depths in coastal waters off the east coast of Sweden. (LEVRING 1976)

various light-harvesting systems optimize light harvesting and enhance the survival of algae in specific ecological niches.

Systematically, the algae are grouped according to their photoaccessory pigments. The Chlorophyta (green algal line) which include the Euglenoids and the Chlorophyceae contain Chl b. The main xanthophyll in most classes is lutein but the Siphonales and some Prasinophyceae contain siphonaxanthin. The Chromophyta (brown algal line) possess Chl c_1 and c_2 and fucoxanthin or peridinin. Those algal classes containing fucoxanthin (brown algae, diatoms, golden-brown flagellates, chrysomonads and some chloromonads) contain Chl c_1 and Chl c_2, while the diatoms contain only Chl c_2 and peridinin. The Rhodophyta (red algae) and prokaryotic Cyanobacteria (blue-green algae) have phycobilins. The Cryptophyta is an unusual algal group which contains one single type of phycobiliprotein and also Chl c_2; in these algae α-carotene is more common than β-carotene.

4 The Structure and Spectra of Light-Harvesting Pigments

The chemical structures of the photosynthetic pigments (chlorophylls, carotenoids, and phycobilins) differ considerably, and these differences are manifested in the spectra of the pigments. The structures of Chl a and Chl b are well known, as is their characteristic high absorbance in the red (663 nm and 645 nm, respectively) and blue regions (430 nm and 445 nm, respectively). Chl c_1 and Chl c_2 (JEFFREY 1980) are not true chlorophylls, but magnesium complexes of porphyrins which have the isocyclic ring found in Chls a and b. Ring IV

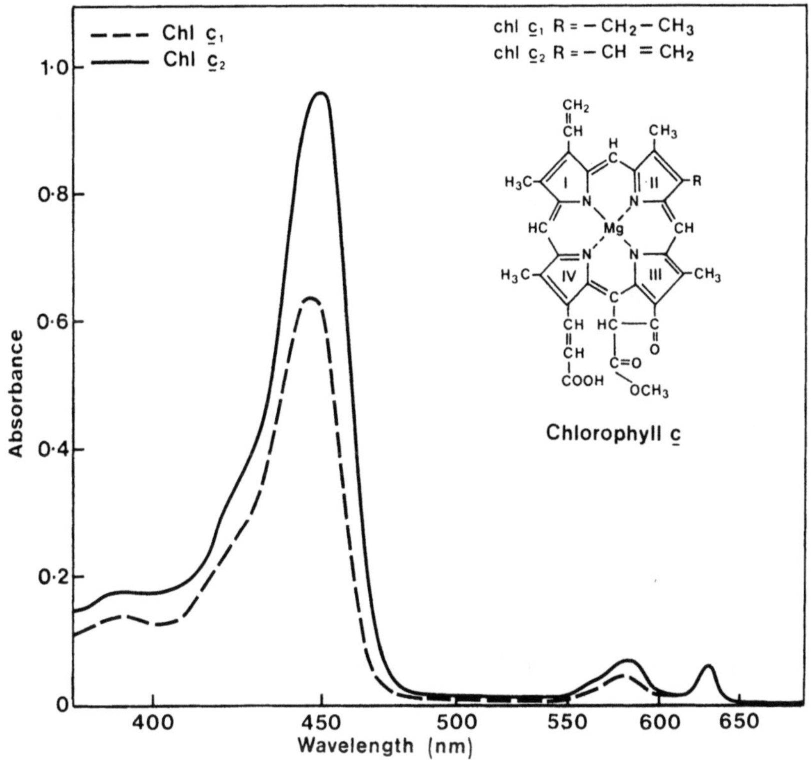

Fig. 2. Absorption spectra of Chl c_1 and Chl c_2 in ether. (JEFFREY 1980)

is not reduced (Fig. 2) and an unesterified acrylic side chain replaces the phytylated propionic side chain of Chl a and Chl b. Chl c_1 has a vinyl side chain at position 2 and an ethyl at position 4, while Chl c_2 has vinyls at both positions 2 and 4 of the macrocycle. Chl c_1 and Chl c_2 have very low red absorbance (628 nm) compared to Chl a and Chl b. Chl c_2 has relative greater absorbance at 448 nm than has Chl c_1 at 444 nm (Fig. 2, JEFFREY 1980).

Although over 60 carotenoids are found amongst the algae, each algal class contains only 4 to 5 main xanthophylls. Three polyoxygenated carotenoids, siphonoxanthin and the allenic xanthophylls, fucoxanthin and peridinin (Fig. 3) are of special importance for light harvesting in the sea. These xanthophylls when complexed to proteins have spectra which show large shifts to the red (60–90 nm), enabling these xanthophyll-protein complexes to capture the predominantly green light.

The phycobilins (open chain tetrapyrroles) are bound covalently to proteins through a thioether link. The phycobiliproteins, phycoerythrin, phycocyanin and allophycocyanin efficiently capture light in the 500–660 nm region (GANTT, Chap. 6.3, this Vol.).

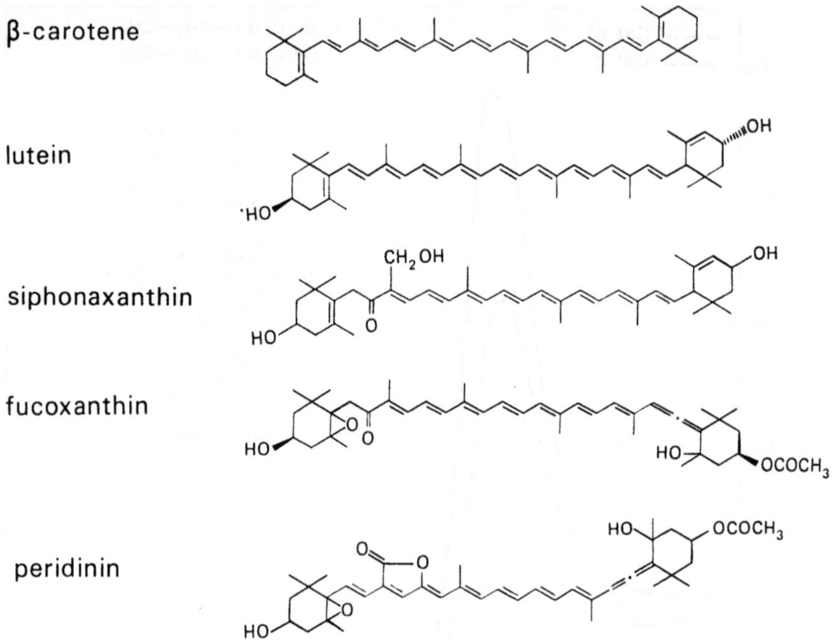

Fig. 3. Formulas of carotenoids of significance in light-harvesting in algae

5 Pigment-Protein Complexes

All of the chlorophylls, carotenes, and xanthophylls of algal thylakoid membranes are linked to a few specific proteins. Most of the algal pigment-proteins are embedded in the membranes as components of the supramolecular protein-complexes of PS II and PS I. Detergents are needed to release these intrinsic proteins from the membrane. An exception is the water-soluble peridinin-Chl a-protein of dinoflagellates. The major class of water-soluble proteins, the phycobiliprotins, are discussed by GANTT (Chap. 6.3, this Vol.).

More than 99% of the photosynthetic pigments act together as light-harvesting antennae for PS II and PS I (THORNBER, Chap. 3, this Vol.). The binding of pigments to proteins ensures efficient and very fast migration of excitation energy from the light-harvesting pigments to the reaction centers of PS II and PS I, P-680 and P-700 (SAUER, Chap. 2, this Vol.). The function of the protein is to maintain high concentrations of pigments in the correct orientations for maximal energy transfer without destructive energy losses due to quenching. The proteins define structural domains for various spectral species that facilitate energy migration down a favorable energy gradient to the longer wavelength pigments, P-680 and P-700.

There are two major classes of pigment-protein complexes: the inner antenna β-carotene-Chl a-proteins of PS II and PS I which are part of the functional core reaction center complexes of PS II and PS I, and the outer antenna pigment-proteins that form the light-harvesting complexes of PS II and PS I whose main function is to capture light. The molecular structure and spectral composition of the reaction center and inner core Chl a-proteins appear to be highly conserved amongst all algae and higher plants (THORNBER, Chap. 3, this Vol.). This accounts for the universal distribution of Chl a and β-carotene. In contrast, striking evolutionary changes have occurred in the outer antenna light-harvesting complexes (GLAZER 1983). This is seen in the diversity of pigments and their associated proteins, the variety of their structural organizations, as well as their location within or attached to the membrane. Indeed, the structural and physical properties of the different light-harvesting complexes themselves give rise to different thylakoid configurations and light-harvesting strategies in the various algal groups (STAEHELIN, Chap. 1, this Vol.).

6 Core Chl a-Proteins of PS I and PS II

β-Carotene-P-700-Chl a-Proteins of PS I. The intrinsic P-700-Chl a-protein(s) of PS I (CP I) have been found in all algal groups, although the extent of their biochemical, spectroscopic and structural characterization vary, and many PS I fractions include peripheral light-harvesting proteins (LARKUM and BARRETT 1983; REDLINGER and GANTT 1983). CP I contains P-700, some 40–60 Chl a molecules and 5–10 β-carotene molecules bound to a 68 kDa protein(s) which also has the primary Fe–S center of PS I. There is conflicting evidence regarding the number and molecular weight of the apoprotein, which is sometimes electrophoretically resolved as two distinct polypeptides of lower molecular weights. Since there are two chloroplast genes for this pigment-protein in spinach and maize (FISH et al. 1985), there may be two distinct proteins, only one of which contains P-700. The molar composition of CP I is not well established: Chl a/P-700 ratios of 40–65:1, Chl a/β-carotene ratios of 4:10 and Chl/protein ratios of 8:30 have been found (cf. LARKUM and BARRETT 1983). CP I has a red absorbance maximum at 678 nm. The fluorescence band of CP I at 77 K is at 680–690 nm. It is likely that the characteristic far-red low temperature fluorescence emission attributable to PS I (715–735 nm in algae) emanates from the light-harvesting complex of PS I (LHC I), rather than from CP I itself.

Chl a-Proteins of PS II. There is evidence that the two intrinsic Chl a-proteins of the core reaction center complex of PS II are distributed throughout all algal groups, although they have not all been isolated and characterized. P-680, phaeophytin, and the primary quinone acceptor Q_A of PS II, together with Chl a and β-carotene molecules are associated with a 47–51 kDa protein (CPa-1, CP 47). There is also another inner antenna Chl a-protein (CPa-2, CP 43) with an apoprotein of 41–43 kDa (e.g., *Chlamydomonas*, DELEPELAIRE and CHUA 1979; *Acetabularia*, GREEN et al. 1982; *Acrocarpia*, BARRETT and THORNE 1980; *Porphyridium*, REDLINGER and GANTT 1983; *Synechococcus* sp., YAMAGISHI and KATOH 1983). These two Chl a-proteins of *Chlamydomonas* are immunologically distinct (DELEPELAIRE and CHUA 1979). The molar composition of CPa-1 and CPa-2 are not yet known. These Chl a-proteins have red absorbance maxima at 673–675 nm. The characteristic low temperature fluorescence emission bands of algae and higher plants at 685 and 695 nm originate from PS II. F-695 emanates from the P-680-Chl a-protein, and F-685 from the other Chl a-protein (41–43 kDa apoprotein) (YAMAGISHI and KATOH 1983).

7 Light-Harvesting Complexes of the Chlorophyta

Higher plants, *Chlamydomonas reinhardtii*, *Codium fragile* sp. and possibly all other member of the Chlorophyta, possess two peripheral light-harvesting complexes: one is associated with PS II (LHC II) and the other with PS I (LHC I). While LHC II, comprising at least 50% of total pigment of most green algal thylakoids, was one of the first pigment-protein complexes to be isolated and characterized, the minor LHC I has only recently been isolated.

Lutein-Chl a/b-Proteins. LHC II has been isolated from many green algae using either Triton X-100 fragmentation followed by sucrose density gradient centrifugation in the absence of cations, or by mild SDS PAGE (cf. LARKUM and BARRETT 1983). This supracomplex has two major apoproteins of 27 and 25 kDa and a minor apoprotein of 29 kDa. It has a Chl a/Chl b ratio of 1.3–1.1 with lutein as the main xanthophyll. Another minor Chl a/b-protein (CP 29) with an apoprotein of 29 kDa and a higher Chl a/Chl b ratio has been isolated from *Acetabularia* (GREEN and CAMM 1982).

There appears to be a marked heterogeneity of Chl a/b-proteins of PS II since various electrophoretic systems result in the resolution of different LHC II complexes. This heterogeneity may be due in part to LHC II being encoded by multigene families. LHC II has absorption maxima at 672, 650, 470, and 438 nm. The fluorescence emission at 681 nm (77 K) has a very narrow bandwidth.

There is compelling evidence that LHC II (mobile chl a/b LHC; STAEHELIN, Chap. 1) is required for thylakoid stacking (KAPLAN and ARNTZEN 1982; BENNETT 1983). First, LHC II is located mainly in appressed thylakoids, and mutant chloroplasts lacking the main apoproteins of LHC II have no stacked membranes. Second, LHC II has an essential 2-kDa fragment exposed at the outer membrane surface, required for the cation-induced restacking of thylakoids. This stroma-exposed segment may be phosphorylated (BENNETT 1983).

LHC I has been isolated by SDS-PAGE from *Chlamydomonas* (ISH-SHALOM and OHAD 1983), and from higher plants by octylglucoside: sodium cholate detergent extraction of a Triton X-100 PS I complex (HAWORTH et al. 1983). Four polypeptides (25–20 kDa) appear to be associated with higher plant LHC I, although it is not yet known whether each protein binds Chl a and Chl b. The Chl a/Chl b ratio of ~3.6 (spinach and pea LHC I) is higher than that of LHC II; other molar compositional data are not yet known. LHC I has absorbance maxima at 672, 650, 475, and 438 nm. The fluorescence emission (77 K) is at 735 nm, the characteristic far-red fluorescence attributed to PS I for intact higher plant and some green algal thylakoids.

7.1 Siphonales

Siphonaxanthin-Chl a/b-Proteins. Some members of the Siphonales contain siphonaxanthin and/or its esterified form, siphonein, instead of lutein, as their main xanthophylls. Marine green algae also have a much higher Chl b content than most other Chlorophyceae (JEFFREY 1980). The presence of siphonaxanthin and Chl b in these marine green algal light-harvesting complexes provides enhanced absorbance in the 460–540 nm region compared to that found in lutein-containing light-harvesting complexes of most Chlorophyceae.

Siphonaxanthin-Chla/b-proteins of PS II have been isolated by mild SDS-PAGE from *Codium fragile* (ANDERSON 1983; BENSON and COBB 1983) and by Triton X-100/sucrose density gradient centrifugation in the absence of cations (ANDERSON 1985). The *Codium* PS II Chl a/b-protein complex has a low Chl a/Chl b ratio of 0.66 and contains both siphonaxanthin and siphonein (Fig. 4). These pigments are associated with four polypeptides of 35, 33, 29 and 27 kDa. The lower M_r proteins appear to be similar to those of green algal LHC II (ANDERSON 1985). The absorption spectrum (672, 650, 540, 510, 470, and 438 nm) shows enhanced Chl b content and 500–540 absorbance. The fluorescence emission spectrum has a maximum at 681 nm at 77 K and is similar to that of spinach LHC II despite the different pigment content.

Treatment of the *Codium* Triton X-100 PS I complex with octyl-glucoside: sodium cholate followed by sucrose gradient centrifugation yielded a PS I si-

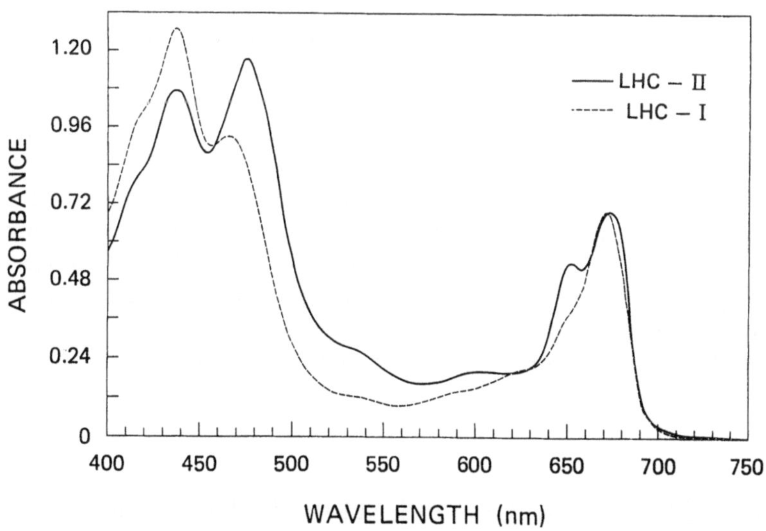

Fig. 4. Absorbance spectrum of the siphonaxanthin-Chl a/b-proteins of PS II (LHC II) and PS I (LHC I) from the marine green alga, *Codium*. (CHU and ANDERSON 1985)

phonaxanthin-Chl a/b-protein complex (Chu and Anderson 1985). The minor PS I Chl a/b-protein complex has a Chl a/Chl b ratio of 1.8 and only siphonaxanthin. The four apoproteins (25–20 kDa) of *Codium* LHC I have a similar migration in PAGE to those of spinach LHC I. The fluorescence emission of this PS I *Codium* complex is at 686 nm at 77 K.

Fluorescence excitation spectra of both *Codium* LHC II and LHC I clearly show that Chl b, siphonaxanthin, and siphonein in the isolated complexes transfer excitation energy with high efficiency to Chl a.

7.2 Prochlorophyta

Prochlorophyta (single genus, *Prochloron*), found in association with certain ascidians, is the sole group of prokaryotic oxygenic autotrophs that contain Chl b. This prokaryotic alga has no phycobiliproteins. Chl a/b-protein complexes have been isolated from a *Prochloron* sp. by SDS PAGE and SDS-sucrose gradient centrifugation (Hiller and Larkum 1985). The Chl a/b-protein complexes have higher Chl a/Chl b ratios ($\sim$2.3) than those of LHC II of *Chlamydomonas* and higher plants and may be analogous to the minor LHC II species, CP 29 of *Acetabularia* (Green and Camm 1982). The pigments of *Prochloron* LHC II are associated with a 34 kDa protein; significantly, there are no 25–29 kDa proteins analogous to those of higher plant and green algal LHC II. Schuster et al. (1983) demonstrated that the 34 kDa protein was phosphorylated in vitro in the dark or light by a membrane-bound kinase. The absorbance spectrum has a red maximum at 669 nm (Chl a) and Chl b (650 and 460 nm). The fluorescence emission maximum (77 K) is at 682 nm and has the usual narrow bandwidth observed for all the eukaryotic Chl a/b-proteins (Hiller and Larkum 1985).

Chl b was also associated with a *Prochloron* PS I complex isolated by Hiller and Larkum (1985). These prokaryotes then may also have a specific PS I Chl a/b-light-harvesting antenna as found in eukaryotes. It is likely that the *Prochloron* Chl a/b-proteins will also have a unique carotenoid composition, since zeaxanthin is the main xanthophyll present in *Prochloron* (Withers et al. 1978).

Clearly the molecular composition of the Chl a/b-proteins of prokaryotes and eukaryotes are diverse since they have not only different amounts of Chl b relative to Chl a and various carotenoids, but also the molecular weights of the apoproteins vary. Once the amino acid sequences of the apoproteins are deduced from their nucleotide sequences, the calculated textiary structures should provide information on the interactions of pigment and protein molecules.

8 Light-Harvesting Complexes of the Chromophyta

The principle feature of the light-harvesting complexes of the Chromophyta which include macro and micro-benthic algae, marine phytoplankton, and some

Table 1. Properties of Chl a/c-protein complexes of the Chromophyta and a Cryptomonad

Phyla	Species	Reference[a]	Pigment-protein	Molar ratios Chl a:Chl c: Carotenoid	Apoprotein (kDa)
Phaeophyta	*Acrocarpia paniculata*	1	Fucoxanthin-Chl a/c_2-protein	2:1	13–16
			Violaxanthin-Chl a/c_1/c_2-protein	3:1	–
		2	Fucoxanthin-protein		13–16
		2	Chl c-protein		13–16
	Several	3	Fucoxanthin-Chl a-protein		
			Chl a/c-protein	1:2	
Diatom	*Phaeodactylum tricornutum*	4	Fucoxanthin-Chl a/c-protein	2:1:4	15
			Fucoxanthin-Chl a/c-protein	5:1:4	15
		5	Fucoxanthin-Chl a/c-protein	2:1:5	17.5
Cryptophyta	*Chroomonas* sp.	6	Carotenoid-Chl a/c-protein	3:2	24, 20

[a] 1. Barrett and Anderson 1980, 2. Barrett and Thorne 1981, 3. Alberte et al. 1981, 4. Gugliemelli 1984, 5. Friedman and Alberte 1984, 6. Ingram and Hiller 1983.

freshwater algae, is the remarkable role of the allenic xanthophylls, fucoxanthin (all classes except dinoflagellates) and peridinin (dinoflagellates) in achieving maximal collection of the dominant green light of coastal and deeper waters. When fucoxanthin or peridinin are complexed to specific proteins, their light-harvesting capacity is extended some 80–100 nm to the red, as occurs with siphonaxanthin. These special carotenoids are a major evolutionary departure from the use of open or closed tetrapyrroles to harvest light for photosynthesis.

Gross difficulties in the isolation of the pigment-protein complexes from the brown seaweeds, diatoms, dinoflagellates and other phytoplanktons, all of which have special structural features designed to contend with the harsh marine environment, have delayed the successful application of isolation procedures used for green algae and higher plants. During the last decade, several studies have been made: some properties of the Chl c-containing complexes of Chromophyta are given in Table 1 as well as the Chl a/Chl c_2-protein complex of a cryptomonad for comparative purposes.

8.1 Phaeophyta

Two light-harvesting supramolecular complexes have been isolated from *Acrocarpia paniculata* by Triton X-100-sucrose gradient centrifugation (Barrett and Anderson 1980; Berkaloff et al. 1981; Duval et al. 1983). The major fucoxanthin-Chl a/c_2-protein complex isolated comprises about 60% of the brown algal pigments. A second light-harvesting complex, a violaxanthin-Chl a/c_1/c_2-protein complex is present in lesser amounts. The orange-brown fucoxanthin-Chl a/c_2-protein complex contains most of the fucoxanthin and has a Chl a/c_2 ratio of 2 (Table 1). Compared to that of chloroplasts, its absorbance spectrum shows

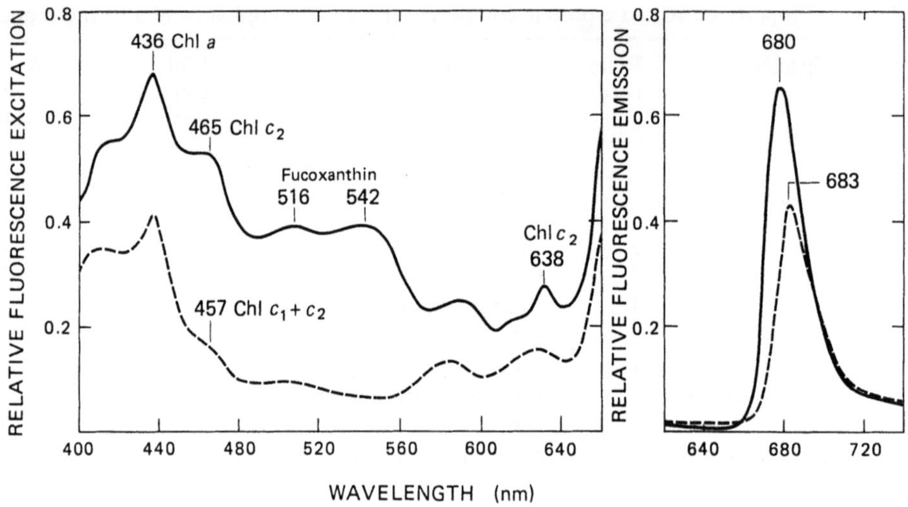

Fig. 5. Excitation and fluorescence emission spectra at 77 K of the brown algal *Acrocarpia* light-harvesting complexes: —— fucoxanthin-Chl a/c_2-protein; --- violaxanthin-Chl a/c_1/c_2-protein. (ANDERSON and BARRETT 1979)

enhanced absorbance at 465 and 434 nm due to Chl c_2 and 500–540 nm due to fucoxanthin. The green violaxanthin – Chl a/c_1/c_2-protein complex has a Chl a/Chl c_1 + Chl c_2 ratio of 3.0 and about equal amounts of Chl c_1 and Chl c_2 and violaxanthin (Table 1). Fluorescence excitation spectra of these light-harvesting complexes (Fig. 5) demonstrate that both Chl c_1, Chl c_2 and fucoxanthin transfer excitation energy to Chl a (ANDERSON and BARRETT 1979).

The fucoxanthin-Chl a/c_2-protein complex of various brown algae is clearly associated with PS II. The extraction of intact chloroplasts with digitonin followed by sucrose gradient centrifugation gave a brown supracomplex containing both the PS II Chl a-protein (fluorescence maximum at 77 K at 694 nm) and the fucoxanthin-Chl a/c_2-protein complex. The supramolecular PS II complex can be resolved into the two major subcomponents by treatment with cholate, followed by either sucrose density gradient centrifugation or by electrophoresis in agarose gels (BARRETT and THORNE 1980, 1981). The violaxanthin-Chl a/Chl c_1/c_2 complex appears to be associated with PS I, as this supramolecular complex remains with the thylakoid residue, as does the P-700–Chl a-protein complex, after removal of the PS II pigment system with digitonin or deoxycholate (BARRETT and THORNE 1980, 1981).

Using nondenaturing SDS PAGE, ALBERTE et al. (1981) isolated smaller pigment-proteins from brown algae and diatoms which were either fucoxanthin-Chl a-proteins or Chl a/c-proteins with 15 kDa apoproteins. BARRETT and THORNE (1980, 1981) have isolated single pigment-proteins: a fucoxanthin-protein and a Chl c_2-protein whose apoproteins range from 13–16 kDa. It is not known whether the supramolecular complex is made up of apoproteins (~15–18 kDa) which each bind Chl a, Chl c_2 and fucoxanthin molecules, each bind two pigments (Chl a and Chl c_2 and Chl a and fucoxanthin), or alternatively, each pigment is bound to individual specific apoproteins.

Recently CARON et al. (1985) have shown that the light-harvesting fractions isolated from digitonin-solubilized thylakoids of *Fucus serratus* contain the major part of the $C_{16:1}$-trans-phosphatidylglycerol, a lipid found mainly in the granal stacks of higher plants.

8.2 Diatoms

GUGLIEMELLI (1984) isolated two supramolecular light-harvesting complexes from the diatom *Phaeodactylum tricornutum* by an abrasive process and treatment of the thylakoids, so released, with sodium lauryl sarcosinate. A yellow fraction had relative molar proportions of Chl a: Chl c: fucoxanthin: other pigments of 3:1.3:6:2, and a M_r of 230 kDa (Sepharose chromatography). A green fraction had a M_r of 93 kDa, and lower amounts of Chl c and fucoxanthin relative to Chl a than found in the yellow fraction (Table 1). Both yellow and green fractions had apoproteins of about 15 kDa. The pigment-proteins of the yellow fraction contained about 23.5% pigment and had a M_r of 19.6 kDa, indicating 12 subunits per particle.

FRIEDMAN and ALBERTE (1984) isolated a major light-harvesting complex from *Phaedactylum tricornutum* by the use of zwitterionic detergent and sucrose density gradient centrifugation followed by DEAE-Sephacel ion exchange. This fucoxanthin-Chl a/c-protein had 2 Chl a:1 Chl c:5 fuxocanthin molecules per 17.5 kDa apoprotein, indicating 8 chromophores per apoprotein. Significantly this apoprotein had a greater cysteine content (1.5 mol%) compared with the *Chlamydomonas* LHC II apoprotein (0.6 mol%) and the water-soluble peridinin-Chl a-protein (0 mol%). The protein polarity values of the diatom and *Chlamydomonas* apoproteins were similar (~33–35%), consistent with their being intrinsic proteins. Monoclonal antibodies to the *Phaeodactylum* apoprotein showed that antigenically homologous species were present in other diatoms, suggesting this apoprotein may be the major diatom light-harvesting species (FRIEDMAN and ALBERTE 1984).

8.3 Chysophyceae and Xanthophyceae

The pigment-protein complexes of the yellow alga, *Synura petersenii* (Chrysophyceae) and the yellow-green alga *Tribonema aquale* (Xanthophyceae) have been separated by mild SDS PAGE following SDS/sodium deoxycholate solubilization (WIEDEMANN et al. 1983). In both cases, a PS I complex and PS II complex (CPa) were resolved as well as a band of higher mobility and 50% free pigment. In *Synura*, the fucoxanthin-Chl a/c-protein had a Chl a/Chl c absorbance ratio of 0.33 and marked absorbance due to fucoxanthin was evident, whereas the light-harvesting complex from *Tribonema* appeared to be less stable and little fucoxanthin absorbance was seen.

Some marine algae have 19'hexanoylfucoxanthin or other substituted fucoxanthins as their main xanthophyll. HAXO (1985) has shown that 19'-hexanoylfucoxanthin of the Coccolithophorid, *Emiliania huxleyi* (Prymnesiophyceae) markedly contributes to the photosynthetic action spectrum in the 500–550 nm region. Although none of the pigment-protein complexes containing fucoxanthin derivatives have yet been isolated, they are likely to have a significant role in light harvesting for some algae in the marine environment.

8.4 Cryptophyta

This algal group has intrinsically bound Chl c_2, as well as either phycocyanin or phycoerythrin located in the intrathylakoid lumen. Fluorescence studies have shown that Chl c_2 contributes to light harvesting in these algae (LICHTLÉ et al. 1980). A Chl a/c_2-protein fraction (with 40% of the total Chl a) has been isolated from a *Chroomonas* sp. by SDS PAGE or digitonin-sucrose gradient centrifugation; it also contained an unidentified carotenoid (Table 1) (INGRAM and HILLER 1983). The absorption spectrum has maxima at 672, 640, 588 (Chl c_2), 488 (carotenoid), 460 (Chl c_2) and 436 nm. Efficient energy transfer from Chl c_2 to Chl a occurred in this complex: the fluorescence emission peak is at 682 nm at 77 K.

8.5 Peridinin-Chl a-Proteins

Several water-soluble peridinin-Chl a-proteins have been isolated from dinoflagellates (HAXO et al. 1976; PREZELIN and HAXO 1976). These proteins contain either 4 peridinin and 1 Chl a molecules per 15.5 kDa or 32 kDa apoprotein, or 9 peridinin and 2 Chl a molecules pe 31.5 kDa apoprotein. They have high absorbance at 438 and 478 nm with a shoulder at 525 nm, and low peaks at 620 and 672 nm (Fig. 6). Peridinin transfers excitation energy to Chl a with

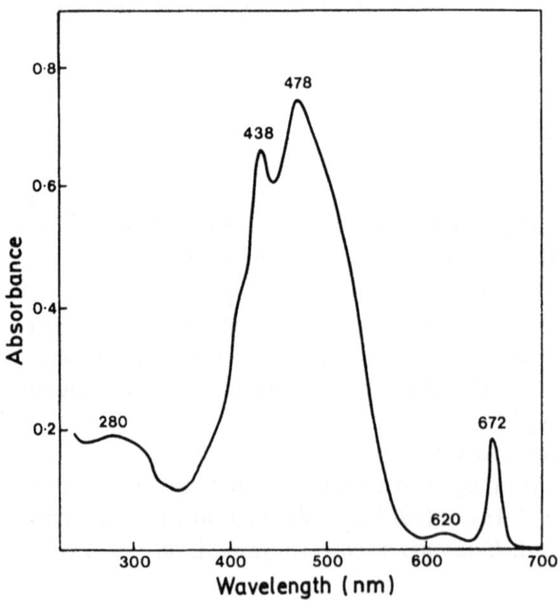

Fig. 6. Absorption spectrum of the peridinin-Chl a-protein from the dinoflagellate *Glenodinium*. (PREZELIN and HAXO 1976)

100% efficiency in the isolated peridinin-Chl a-proteins: the major fluorescence peak is at 672 nm, with minor 709 and 733 nm peaks. In a model for the mutual orientiation of the pigments, it was suggested that 4 peridinins surround a Chl a molecule, and two such units are aligned for the required energy transfer (SONG et al. 1976). It is not known whether the peridinin-Chl a-proteins are attached to the outer or inner thylakoid membrane surface, or to intrinsic Chl-proteins of PS II or PS I. The finding of two peridinin-Chl a-proteins by SDS PAGE suggests that some peridinin in certain dinoflagellates may be attached to intrinsic membrane proteins (PREZELIN and BOCZAR 1981).

8.6 Molecular Topology of Algal Pigment-Protein Complexes

The molecular topology of algal pigment-protein complexes is ill-defined, except for the phycobiliproteins. In spinach thylakoids, all of the Chl-proteins span the membrane and have segments exposed at the outer and inner thylakoid surfaces (ANDERSON 1984). Given the highly ordered arrangement of pigments needed for efficient energy transfer and the many pigments bound to each apoprotein, it may be essential that the algal proteins also span the membrane. Comparisons of the hydropathy index plots for the amino acid sequences deduced from the nucleotide sequences of plant Chl a-protein genes of PS II (MORRIS and HERRMANN 1984) and PS I (FISH et al. 1985) suggest these Chl a-proteins span the membrane many times as α-helices. Some of the α-helices contain histidines which may provide the ligands for Chl a molecules. The application of secondary and tertiary structure predictions based on the deduced amino acid sequences of algal pigment-proteins is an exciting area yet to be explored (see e.g. ZUBER, Chap. 6.1, this Vol.).

8.7 Structure of Algal Thylakoid Membranes

There is great variation in chloroplast structure amongst the different algal groups (STAEHELIN, Chap. 1, this Vol.; Figs. 3, 10): hence evolutionary diversification appears to have arisen at an early stage (COOMBS and GREENWOOD 1976; cf. LARKUM and BARRETT 1983).

1. Chlorophyta: The thylakoids lie parallel to one another with many areas of appressment (<20 Å) but the degree of membrane stacking is rather variable, both in the numbers of appressed membranes (2–6) and in the length of membrane appression. The uniformity of length of the grana and stroma thylakoids typical of higher plants is not found in the Chlorophyta. In the Prochlorophyta, the cells are largely filled with thylakoid membranes arranged closely together, with up to 20 thylakoids being appressed; a few single thylakoids are also present.

2. Chromophyta: The thylakoids are arranged basically as groups of three appressed membranes which extend along most of the length of the chloroplasts. These "triplets" are interconnected by single transverse thylakoids. This "triplet" thylakoid arrangement occurs in all classes of the Chromophyta.

3. Rhodophyta and Cyanobacteria: The thylakoid membranes in these algae, by contrast, are never appressed. The phycobilisomes are attached to the outer

surface of the thylakoids and this restricts the approach of neighbouring membranes (GANTT, Chap. 6.3, this Vol.). In the Cyanobacteria, the thylakoids are arranged variously throughout the cell, often aligned closely parallel to each other, but not appressed. Parallel alignment of thylakoids may also occur in Rhodophyta chloroplasts; in some cases the thylakoids are surrounded by girdle lamellae.

4. Cryptophyta: The thylakoids occur mainly in loosely associated pairs or sometimes triplets. Membrane appression is possible in this algal group, as the phycobiliproteins are not assembled into phycobilisomes.

9 Spatial Organization of PS II and PS I in Algal Thylakoid Membranes

In higher plants, there is a lateral heterogeneity in the distribution of the Chl-proteins between appressed and nonappressed membranes (ANDERSSON and ANDERSON 1980). Most PS I complexes are located in stromal thylakoids and most, but not all, PS II complexes are found in the appressed granal thylakoids. Biochemical and morphological studies have demonstrated that the stacked and unstacked membrane regions of higher plants and *Chlamydomonas reinhardtii* have different compositions and structural organization, reflecting a functional differentiation of these membranes (KAPLAN and ARNTZEN 1982; STAEHELIN, Chap. 1, this Vol.). It is not proven, however, that lateral heterogeneity in the distribution of the photosystems occurs in all Chlorophyta.

Little is known about compositional, functional, or structural differences in the appressed and nonappressed thylakoids of any of the classes of Chromophyta. Freeze-fracture studies of brown algal and cryptomonad thylakoids showed that the large freeze-fracture particles (EF's), interpreted as PS II complexes, occur mainly in stacked regions, possibly suggesting the occurrence of lateral heterogeneity in the Phaeophyta and Cryptophyta (DWARTE and VESK 1982, 1983).

In the Rhodophyta and Cyanobacteria there are only nonappressed membranes; their two freeze-fracture faces have a uniform distribution of particles, indicating that there is no compartmentation of the photosystems as in the Chlorophyta. Here PS II complexes with their associated phycobilisomes are able to interact with PS I complexes to allow substantial transfer of light energy from PS II to PS I (GANTT, Chap. 6.3, this Vol.).

10 Strategies of Light Harvesting in Algal Groups

The diversity in localization of the light-harvesting complexes in algae leads to different strategies for sharing light energy between the photosystems. The activity of the two photosystems under light-limiting conditions is controlled by two different mechanisms. One is a slow mechanism involving changes in the light-harvesting capacity of PS II and PS I that is effected by changes in the relative proportions of the components. These adaptations that are regulated by the nuclear and chloroplast genomes take hours or days to respond to changing conditions. In contrast, a fast mechanism, State I–State II changes (BONAVENTURA and MYERS 1969) occur in algae in seconds or minutes.

Long-Term Modulations in Light-Harvesting. Modulations in the relative proportions of chlorophyll-proteins will affect both light-harvesting and electron transport capacity. These may include:

1. total amounts of pigments associated with each photosystem;
2. relative amounts of light-harvesting complexes of PS II and PS I to inner core Chl a-proteins of PS II and PS I. This will determine the composition of PS II and PS I;
3. photosynthetic antenna units size of PS II and PS I;
4. amounts of P-680 relative to P-700.

Finally, there may be more or fewer membranes present in the algal cells.

In algae, as in higher plants, the proportions of reaction centers, light-harvesting antenna molecules, and electron carriers vary according to the irradiance which the algae receive during growth. In the Chlorophyta and related phyla, and possibly the Chromophyta, all four modulations listed above may occur. Under low light intensities, there is an increase in the relative proportions of light-harvesting Chl a/b-complexes relative to the core Chl a-proteins (cf. LARKUM and BARRETT 1983). This results in an increase in the spectral width of light-harvesting by each PS II unit. In the Rhodophyta and Cyanobacteria low light conditions result in more thylakoids and phycobilisomes per unit volume of chloroplast (cf. LARKUM and BARRETT 1983; RAPS et al. 1983). A significant feature of blue-green and red algae is that their P-680/P-700 ratios are normally <1, whereas they are often >1 in Chlorophyta. No P-680/P-700 ratios have yet been reported for Chl c-containing algae.

Short-Term Regulation of Excitation Energy Between the Photosystems. Algae have the ability to adapt to changes in spectral quality of ambient light by correcting imbalances in the rates of excitation of PS II and PS I within minutes or seconds. Recently, it has been proposed for higher plants (ALLEN et al. 1981; BENNETT 1983; BARBER, Chap. 11.5; STAEHELIN, Chap. 1, both this Vol.) that the State I–State II adaptation is due to lateral migration of mobile LHC II between appressed and stroma-exposed thylakoids depending on the redox state of the plastoquinone pool and the extent of phosphorylation of LHC II. Structural and biochemical evidence suggests that excess excitation of PS II relative to PS I causes the plastoquinone pool to become reduced, thereby activating a thylakoid kinase which will phosphorylate part of LHC II located at the edges of the grana stacks. The extra negative charge introduced causes some membrane destacking, and the phosphorylated LHC II would then be free to migrate in the nonappressed thylakoids, where it could interact with PS I complex and enhance excitation of PS I (KYLE et al. 1983).

Thylakoid proteins other than LHC II are phosphorylated (OWENS and OHAD 1982; BENNETT 1983). In *Chlamydomonas reinhardtii* many of the proteins of PS II complex are phosphorylated in vivo; OWENS and OHAD (1982) suggest that extensive phosphorylation of PS II complex may be required for its organization integrity. The extent of phosphorylation of the pigment-proteins of other algae is unknown.

The basis for the short-term control mechanism of excitation energy in most algae has yet to be established. Although the Chlorophyta do not have typical grana, they possess variable amounts of appressed and nonappressed thylakoids and LHC II; they may have the same short-term control mechanisms as those of higher plants. However, since the Chromophyta have no LHC II and apparently constant ratios of appressed to nonappressed thylakoids, different mechanism for the regulation of excitation energy will be required for this algal group. The Rhodophyta and Cyanobacteria have no LHC II and no stacked membranes, yet they are able to adapt very rapidly to changes in the spectral quality of ambient light. PS II complex is much closer to PS I complex to allow marked "spillover", and only a small conformational change would be required to change the distance between the intrinsic PS II and PS I complexes and the attached phycobilisomes (BIGGINS and BRUCE 1985).

11 Conclusions

Light capture is crucial for photosynthesis. The light environment of aquatic and particularly marine algae is complex due to striking variations in light intensity and spectral quality. In the course of evolution, unlike terrestial plants which evolved a single solution to light-harvesting, algae have produced a plurality of pigment-proteins and thylakoid structures. The general features of the main algal light-harvesting complexes have been characterized. Now the challenge is to establish interactions of the pigments with one another and their apoproteins and the topology of the supramolecular complexes.

References

Alberte RS, Friedman AL, Gustafson DL, Rudnick MS, Lyman H (1981) Biochim Biophys Acta 635:304–316
Allen JF, Bennett J, Steinback KE, Arntzen CJ (1981) Nature 291:21–25
Anderson JM (1983) Biochim Biophys Acta 724:370–380
Anderson JM (1984) In: Sybesma C (ed) Advances in photosynthesis research, vol III. Nijhoff/Junk, The Hague, pp 1–10
Anderson JM (1985) Biochim Biophys Acta 806:145–153
Anderson JM, Barrett J (1979) Ciba Found Symp 61:81–104
Andersson B, Anderson JM (1980) Biochim Biophys Acta 593:427–440
Barrett J, Anderson JM (1980) Biochim Biophys Acta 590:309–323
Barrett J, Thorne SW (1980) FEBS Lett 120:24–28
Barrett J, Thorne SW (1981) In: Akoyunoglou G (ed) Photosynthesis, vol III. Balaban, Philadelphia, pp 347–356

Bennett J (1983) Biochem J 212:1–13
Benson E, Cobb AH (1983) New Phytol 95:581–594
Berkaloff C, Duvel JC, Jupin H, Chrissovergis F, Caron L (1981) In: Akoyunoglou G (ed) Photosynthesis, vol III. Balaban, Philadelphia, pp 485–494
Biggins J, Bruce D (1985) Biochim Biophys Acta 806:230–236
Bonaventura C, Myers J (1969) Biochim Biophys Acta 189:366–383
Caron L, Dubacq JP, Berkaloff C, Jupin H (1985) Plant Cell Physiol 26:131–139
Chu ZX, Anderson JM (1985) Biochim Biophys Acta 806:154–160
Coombs J, Greenwood AD (1976) In: Barber J (ed) Compartmentation of the photosynthetic apparatus. The intact chloroplast. Elsevier/North-Holland Biomed Press, Amsterdam New York, pp 1–51
Delepelaire P, Chua NH (1979) Proc Natl Acad Sci USA 76:111–115
Duval JC, Jupin H, Berkaloff C (1983) Physiol Veg 21:1145–1157
Dwarte DM, Vesk M (1982) Micron 13:325–326
Dwarte D, Vesk M (1983) Protoplasma 117:130–141
Fish LE, Kück U, Bogorad L (1985) J Biol Chem 260:1413–1421
Friedman AL, Alberte RS (1984) Plant Physiol 76:483–489
Glazer AN (1983) Annu Rev Biochem 52:125–157
Green BR, Camm EL (1982) Biochim Biophys Acta 681:256–262
Green BR, Camm EL, Houten J van (1982) Biochim Biophys Acta 681:248–255
Gugliemelli LA (1984) Biochim Biophys Acta 766:45–50
Haworth P, Watson JL, Arntzen CJ (1983) Biochim Biophys Acta 724:151–158
Haxo FT (1985) J Phycol in press
Haxo FT, Kycia JH, Somers GF, Bennett A, Siegelman HW (1976) Plant Physiol 57:297–303
Hiller RG, Larkum AWD (1985) Biochim Biophys Acta 806:107–115
Ingram K, Hiller RG (1983) Biochim Biophys Acta 722:310–319
Ish-Shalom D, Ohad I (1983) Biochim Biophys Acta 722:498–507
Jeffrey SW (1980) In: Falkowski PG (ed) Primary productivity in the sea. Plenum Publ, New York, pp 33–58
Kaplan S, Arntzen CJ (1982) In: Govindjee (ed) Photosynthesis: Energy conversion by plants and bacteria, vol I. Academic Press, London New York, pp 65–151
Kirk JTO (1983) Light and photosynthesis in aquatic ecosystems. Cambridge Univ Press, Cambridge
Kyle DJ, Staehelin LA, Arntzen CJ (1983) Arch Biochem Biophys 222:527–541
Larkum AWD, Barrett J (1983) Adv Bot Res 10:1–219
Levring T (1976) In: Bainbridge R, Evans GC, Rockham O (eds) Light as an ecological factor. Blackwell, Oxford London Edinburgh Melbourne, pp 305–318
Lichtlé C, Jupin H, Duval JC (1980) Biochim Biophys Acta 591:104–112
Morris J, Herrmann RG (1984) Nucleic Acids Res 12:2837–2850
Owens GC, Ohad I (1982) J Cell Biol 93:712–718
Prezelin BB, Boczar BA (1981) In: Akoyunoglou G (ed) Photosynthesis III, vol. III. Balaban, Philadelphia, pp 417–426
Prezelin BB, Haxo FT (1976) Planta 128:133–141
Raps S, Wyman K, Siegelman HW, Falkowski PG (1983) Plant Physiol 72:829–832
Redlinger T, Gantt E (1983) Plant Physiol 73:36–40
Schiff JA (1981) Ann NY Acad Sci 361:166–192
Schuster G, Owens GC, Cohen Y, Ohad I (1983) In: Sybesma C (ed) Advances in photosynthesis research, vol III. Nijhoff/Junk, The Hague, pp 283–286
Song PS, Koka P, Prezelin BB, Haxo FT (1976) Biochemistry 15:4422–4427
Wiedemann I, Wilhelm C, Wild A (1983) Photosynth Res 4:317–329
Withers NW, Alberte RS, Lewin RA, Thornber JP, Britton G, Goodwin TW (1978) Proc Natl Acad Sci USA 75:2301–2305
Yamagishi A, Katoh S (1983) In: Inoue Y, Groffs AR, Govindjee, Murata N, Renger G, Satoh K (eds) The oxygen evolving system of photosynthesis. Academic Press, London New York, pp 39–48

7 Trapping Events in Light-Harvesting Assemblies

7.1 Theory and Modeling of Excitation Delocalization and Trapping

R.S. KNOX

1 Introduction

1.1 Excitation Delocalization and Trapping: Observations

Modern physics tells us that localization of anything at the atomic and molecular scale is unlikely, so let us address the meaning of "excitation delocalization." Consider the process of fluorescence quenching, in which a strongly emitting molecular species F in solution absorbs light and an increasing amount of species Q acts to decrease the fluorescence. When is the original excitation localized? Of course we can say it is "localized" on the entire population F and part of it is "delocalized" or "relocalized" to the population Q. What is generally meant, however, by those who model the system is that after absorption there is a certain probability that the excitation is localized at any given molecule of F, and that as time progresses, any excitation which was originally on a given molecule of F will be found, with a distribution of probabilities, elsewhere. "Elsewhere" could include other molecules of F as well as those of Q. As we will see in Section 2, there also exist more restricted usages of the concepts "delocalized" and "localized."

If the quenchers Q of the preceding discussion are present at concentrations equal to or greater than the fluorescers F, then it is overwhelmingly likely that "relocalization" will take place in one decisive transfer from an F to the nearest Q. A much more interesting case is that of very low concentrations of Q. Then, excitation originally at an F may wander over many F molecules before finding a Q. If this migration process is rapid compared with the fluorescence process, the action of the quencher is effectively enhanced, and the migration can be detected in a properly designed experiment. If there is no Q present, this delocalization can be detected through its effect on fluorescence polarization (see, e.g., KNOX 1975).

The migration of excitation throughout a molecular system is accomplished most quickly when the molecules are rather concentrated and regularly spaced. Thus, doped organic crystals have long served as controllable model systems to illustrate the phenomenon and test theoretical predictions. The most widely studied case is that of anthracene doped with tetracene (10^{-4} to 10^{-6} M). A review of early work from the 1930's through the middle 1960's is given by WOLF (1967). The central observation is the appearance, in crystals illuminated by sufficiently energetic photons, of the fluorescence of the dopant and the concomitant quenching of that of the host crystal. As the dopant concentration increases, these two fluorescences can be seen to increase and decrease, respectively.

While the above experiments are straightforward from a qualitative point of view, it is very difficult to sort out a simple quantitative picture. An obvious variable is the temperature, and indeed it changes the relative quantum yields of guest and host, but crystal defects strongly influence the results (e.g., "X-traps," WOLF 1967). These defects are present in even the purest and most carefully crystallized samples. Nonetheless, the following qualitative conclusion is undeniable: the host acts as a collector of excitation energy and this energy is delivered to the guests with high efficiency in many instances. One might think of the host molecules around the guests as forming a broad-band "antenna," like a radar dish. This terminology has become popular in the literature of photosynthesis, but one must keep in mind that the antenna is quite irreversible and would be a very poor transmitting device.

1.2 Historical Development of Theory and Modeling

Early approaches to the theory of the antenna effect were to make estimates of the rate constant k in the following simple kinetic equations involving the density of host and guest excitations (n_H and n_G, respectively):

$$\frac{dn_H}{dt} = I_H(t) - \beta_H n_H(t) - kn_H(t) \tag{1a}$$

$$\frac{dn_G}{dt} = I_G(t) - \beta_G n_G(t) + kn_H(t). \tag{1b}$$

$I_H(t)$ and $I_G(t)$ are the corresponding generation rates which, under direct optical excitation, are proportional to the light intensity, the absorption coefficients of host and guest, and their respective molecular concentrations. The rates β_H and β_G are the total radiative and nonradiative monomolecular decay rates of host and guest in the absence of transfer, and k is the overall rate of transfer from H to G, responsible for the antenna effect. In the steady state, with no direct excitations of the guests, $I_G(t) = 0$, these equations predict that the G:H fluorescence yield in the presence of transfer will be proportional to k/β_H. Since k is reasonably expected to be proportional to the relative density of guest molecules, this picture explains data such as those discussed in Section 1.1.

Nanosecond-region TRS (time-resolved spectroscopy) became available as a tool in the late 1960's and was applied to the anthracene-tetracene model system by POWELL and KEPLER (1969). An X-ray excitation pulse of 6 ns duration produced fluorescence whose time profiles, those of both host and guest, were seen explicitly by a fast detection system. These experiments opened up a new theoretical era. Previously, it had been sufficient to merely make estimates of k, either by fitting yields under more widely varying excitation conditions, or by theoretical estimates (see Sect. 3). Now, given the Powell-Kepler experiment, the time dependence was available and the full impliciation of Eqs. (1) could be tested. It was found immediately by POWELL and KEPLER that the equations were inadequate to explain their data, and that one needed somewhat anomalous theoretical forms of k, possibly including time dependence and parameters whose magnitudes were considered unusual by many researchers. A period of theoretical development followed; this has been reviewed briefly by the author (KNOX 1983, esp. part V). In recent times the Powell-Kepler experiments have been carried out with samples of greater purity, and with laser pulse excitations more readily characterized (see e.g., BRAUN et al. 1982). The time dependence is still difficult to model successfully, but steady-state data have been made to yield much new information on the excitation migration process (KENKRE and SCHMID 1983).

In the 1930's it was discovered that hundreds of chlorophyll molecules were involved in the creation of one set of photoproducts in the photosynthetic apparatus of green plants. The picture finally emerging by the mid-1960's was that of the "photosynthetic unit," a set of some 300 molecules of chlorophyll acting as the antenna, in the sense described above, transferring energy to traps called "reaction centers" (see THORNBER, Chap. 3, this Vol., for an overview). The latter generally do not fluoresce and some other attribute of the system after excitation must be observed – perhaps as direct as an excited-state reaction-center absorption change or as remote as the $[O_2]$ yield itself. Theoretical work on models of the photosynthetic unit proceeded rather independently of the work on organic crystals, generally basing itself on a discrete rather than a continuum model of the unit. Excitation hopped, explicitly, by random walk. The thesis of PEARLSTEIN (1966) was based on a combined viewpoint and remains today a primary reference. In the next section we examine the fundamentals on which it is based (see SAUER, Chap. 2, this Vol., for an overview of the physical aspects of light reactions).

2 Delocalization and Pairwise Transfer

2.1 States Produced by Light Absorption

If we are to study delocalization, there is some virtue in looking first at the localization which precedes it.

Consider an extremely dilute solution of identical molecules absorbing strongly at the wavelength of an incident beam. Chlorophyll a in a viscous solvent and a monochromatic beam at 665 nm provide an excellent example. The N molecules have random orientations with respect to the polarization of the beam. Modern quantum theory tells us that all molecules oriented similarly have an equal chance of being excited. The wave function of the system can be considered delocalized, with an amplitude of the order of $1/\sqrt{N}$ on each site. Those molecules with transition dipoles parallel to the beam polarization will have the largest excitation probabilities, and there will be many others that nearly always escape excitation.

The sufficiently dilute solution presents no localization-delocalization problem for two reasons: First, interactions between molecules are weak or nonexistent. Any excited molecule will decay before appreciable probability develops for a transition to another molecule. Thus the state of the system is delocalized in the sense that all parts are relatively equally subject to having been excited – and localized in the sense that the excitation, once on a given molecule (a fact we can learn only by sacrificing its identity to all others!) stays there during decay. Second, there is no correlation between the phases of the states of two excited molecules. This will be the case if the photon correlation length is small compared with the interchromophore distances. For purposes of this review we will consider the photon correlation length to be of the order of magnitude of the wavelength.

Consider now a somewhat less dilute solution, in which the molecules are generally closer together than a wavelength and indeed close enough to feel each others' influence weakly. In the chlorophyll example an appropriate concentration is $6 \times 10^{18}\,\mathrm{cm}^{-3}$, or 10 mM, the most probable nearest neighbor distance then being 3 nm. Both factors discussed immediately above are now altered. First, interactions between molecules to introduce an appreciable probability that an excitation on one molecule will be transferred to another initially unexcited molecule. This is the process first explained satisfactorily by FÖRSTER (1948) and it is the process in which delocalization is observed indirectly through a reduction in degree of fluorescence polarization compared with the infinitely dilute case. Excitations on subsets of molecules of given orientations (held rigid during the radiative life-time by thc viscosity of the solvent) delocalize to those with different orientations, and in particular populate those which were not likely to have absorbed light. Second, there may well be a local correlation between the phases of the states of excited molecules, and such a correlation can lead to rather serious corrections to the Förster depolarization theory (RAHMAN et al. 1979). However, these correlations have not as yet been called upon to explain any anomalies in thc data, and we tentatively conclude that they have a rapid decay brought about by vibrational and collisional fluctuations. Nonetheless, it seems clear that whenever the photon correlation length is greater than the distance between two molecules, the initial state is not necessarily localized in any sense.

Consider now the case of high concentration. If the system is random, near-neighbor pairs are usually most important and they can be discussed as an independent entity. The two factors we have been discussing now reinforce each other and transform the localization-delocalization question completely. The interaction between two nearby molecules is relatively large, as we will see shortly. When it is large, the proper excited states of a pair of molecules are delocalized and their energy level is split (an "exciton splitting"). De facto, the localized excited states do not lose their phase relationship because they must remain coupled as eigenstates. There is much unexplored territory between this case and the one discussed in the previous paragraph.

Another case of high concentration is the well-ordered crystal. Here, photon absorption may lead to a propagating wave packet of excitation which acts as a particle (see e.g., KNOX 1983). The extensive correlations required to maintain this packet will decay quickly under normal conditions (BAY and PEARLSTEIN 1963). An effective initial state resembling the "delocalized but localized" random solution situation is likely to develop quickly in, say, organic crystals at room temperature.

Before looking into the mechanisms which localize and delocalize states, let us remark that many other mechanisms of localization can exist. The most direct is the impurity or trap excited by light of appropriate wavelength. More complicated is the site selection inherent in an inhomogeneous molecular system subjected to monochromatic excitation. An exotic example of inhomogeneity is Anderson localization, in which disorder distributes excitation energies of an otherwise homogeneous large system in such a way that certain states are absolutely localized (DUKE 1979). This phenomenon has, however, no known application to photosynthesis.

2.2 Förster Theory of Pairwise Transfer

This section deals with the intermediate case discussed in Section 2.1, in which a relatively dilute but weakly interacting set of molecules is present. As we saw, it is reasonable here to think of an excitation deposited on one particular molecule, called the donor. If a second identical but unexcited molecule (the acceptor) is assumed to be nearby, and all others are neglected, we have established the initial conditions for "pairwise transfer." Although it had been recognized about 20 years earlier that the Coulomb interaction induced a transfer from the donor to the acceptor, FÖRSTER (1948) applied the irreversible formulation of quantum mechanics which correctly predicted the dependence of the process of donor-acceptor distance. The FÖRSTER theory has been reviewed numerous times and we cannot repeat the details (see FÖRSTER 1965; KNOX 1975). The following is a qualitative summary of a semiclassical version of the theory (KNOX 1975).

If the donor were alone in the medium, losing its excitation solely by emission of photons in a broad band of shape $F_D(\nu)$, it would decay at a rate proportional to $\int F_D(\nu)d\nu$. Here ν is the photon frequency. If now the acceptor were at a large distance away, it would be excited by the emitted radiation to an extent

$$k_{D\to A}^{(radiative)} = \int F_D(\nu)\frac{\sigma A(\nu)}{4\pi R^2}d\nu. \tag{2}$$

Here σ_A is the geometrical cross-section for photon absorption, proportional to the molar extinction coefficient, and it depends on the frequency at which the donor happened to emit. $4\pi R^2$ is the area of a sphere centered at the donor and having a radius equal to the donor-acceptor distance R. Since σ_A is of the order of nm^2 and this radiative process is an accurate picture only when $R \gtrsim 100$ nm, we are speaking of a very rare event. The most likely event is the escape of the photon through the sphere.

The processes leading to the radiation from the donor may be modelled as an electric dipole oscillating at the frequency ν. The farzone radiated power of a dipole depends on R^{-2}, this inverse square law being the physical basis for the form of Eq. (2). The difference for FÖRSTER transfer is that the acceptor is so close to the donor that it senses near-zone power flow, which electromagnetic theory tells us is $(\lambdabar/R)^4$ larger in magnitude. Here $\lambdabar$ is the reduced wavelength in a medium of refractive index n defined as $\lambdabar = c/2\pi n \nu$, about 100 nm. The physical processes causing the donor-acceptor transfer are otherwise exactly the same as in the far-zone case, so (2) is modified to

$$k_{D\to A}^{(\text{near-zone})} = \int F_D(\nu) \frac{\lambdabar^4 \sigma A(\nu)}{4\pi R^6} d\nu. \tag{3}$$

As shown earlier (KNOX 1975) this expression properly contains virtually all the essentials of Förster's famous results. Because we have used only order-of-magnitude arguments here, it lacks a geometrical factor depending on the orientations of donor and acceptor transition dipoles. But it has the R^{-6} dependence and the integral of the overlap between donor emission and acceptor absorption. The R^{-6} dependence is usually made explicit by rewriting (2) as

$$k_{D\to A}^{(\text{near-zone})} = \frac{1}{\tau}\left(\frac{\bar{R}_0}{R}\right)^6, \tag{4}$$

where $\bar{R}_0$ is a characteristic distance, computed from more precise expressions like Eq. (2), using independent experimental data, and having values typically 2.0 to 7.0 nm. The lifetime τ is usually taken as the excited state lifetime of the donor. When $R = \bar{R}_0$, the transfer rate competes equally with the monomolecular decay rate (by definition of $\bar{R}_0$). In contrast with the radiative case, the near-zone transfer probabilities are enormous. Seldom will a photon escape from the D-A system if $R < \bar{R}_0$.

The pairwise rate [Eq. (4)] has been tested extensively and has been found correct in great detail. The earliest tests were statistical and involved statistical averages in random solutions. FÖRSTER (1948) showed its general agreement with fluorescence polarization quenching studies. Note that the theory of Eqs. (2–4) does not depend on the molecules being identical. The acceptor may be a simple quencher. FÖRSTER (1949) showed the rates to be consistent with quenching of fluorescence. More recently STRYER (1978) has demonstrated the R^{-6} dependence explicitly by constructing D-A pairs of known and variable separation. A corresponding theory involving non-dipole-allowed transitions (DEXTER 1953) has been highly successful in explaining quenching effects in inorganic solid solutions.

2.3 Application: Annihilation of Excitons

The FÖRSTER theory predicts efficient energy transfer from an excited donor to any nearby acceptor having an absorption that overlaps the donor emission.

Transfer is particularly efficient if the acceptor has a slightly lower excitation energy than the donor, such that the normal Stokes shift of the donor fluorescence produces a very large spectral overlap. The acceptor's emission then overlaps the donor's absorption even less, and a one-way transfer is accomplished.

Suppose the acceptor is already excited before the donor loses its energy – that is, two excitations are present. The acceptor might be a molecule in a triplet state, which generally has a strong absorption band and will therefore act as a true acceptor. The donor will go to its ground state, as in the normal FÖRSTER process, and the acceptor will go to a yet higher state. From there it will ionize, return to an excited state, or drop to its ground state. The net result is loss of at least one excitation in a process which has become known as *exciton annihilation* (see GEACINTOV and BRETON, Chap. 7.4, this Vol.). Theoretical estimates for singlet-triplet annihilation rates were carried out by RAHMAN and KNOX (1973) for anthracene, rhodamine, and chlorophyll. When those molecules are in their lowest triplet state, they are surrounded by a "sphere of destruction" for singlets which ranges from 2 to 4 nm in radius. Similar calculations have been carried out recently by B. WITTMERSHAUS, D. GÜLEN, and the author (1986) for pairs of singlets. Although the FÖRSTER picture as applied to annihilation rates is only an approximate one, it appears to work well in quantifying an unusual example of energy transfer.

2.4 Beyond the Förster Theory: Coherence, Triplets, and the Monopole Approximation

Strong Coupling and Coherence. One frequently encounters a discussion of "strong coupling," wherein it is claimed that excitation energy may hop from donor to acceptor (a molecule identical to the donor, usually) at a rate proportional to R^{-3}. We take the view that this strong-coupling phenomenon is not a rate process, is very seldom observed at all in molecular pairs, and is of no clear interest to photosynthesis. Since such a great deal has been written about it in the photosynthesis literature we believe these few remarks are in order. A complete set of conditions for various coupling strengths was given by FÖRSTER (1965), and the viewpoint of this section is due to KENKRE and KNOX (1974).

The principal indication of strong coupling is a spectral splitting or shift of a monomer absorption peak. This, indeed, is proportional to R^{-3}. However, as we pointed out near the end of Section 2.1, the wave function of the system is *already* delocalized when one of the split states is produced. One needs no spatial energy transfer if the initial state is delocalized! The "rate" proportional to R^{-3} is the inverse of the time it would take for excitation to flow from the donor to the acceptor *if* it were localized to start with; but unless exceptional experimental measures are taken it will not be so localized. This elementary point has escaped notice in a remarkable number of papers.

Coherence, sometimes loosely defined as "delocalization," although it is much more, plays an important role in crystal exciton physics and has been characterized thoroughly in excited molecular pairs (BURLAND and ZEWALL

1979). Low temperature and narrow-band spectra are essential for such work. One measure of coherence has been applied to the case of transfer between chlorophylls at room temperature (KENKRE and KNOX 1974) with the conclusion that it vanishes in considerably less than 1 ps.

Triplets. Triplet excited states, which produce very weak molecular absorption bands and very long-lived phosphorescence emission, can transfer excitation energy by the same mechanisms as those discussed above for singlets. The only significant difference for our purposes is the size and R-dependence of the electronic interaction between donor and acceptor. At small R (<1.5 nm) the dominant mechanism is probably of the electron exchange type (DEXTER 1953), behaving as $e^{-R/a}$ where a is a molecular dimension. At larger R this effect dies out and the weak but long-range dipole-dipole terms behaving as R^{-6} survive. These terms will be present whenever the donor and acceptor have any optical importance. Short-range, virtually any triplet can propagate by the exchange mechanism.

The Monopole Approximation[1]. The derivation of the Förster transfer rate assumes that the electronic transitions occurring on the donor and acceptor involve point transition dipoles. This is often a poor assumption for large chromophores such as porphyrins and carotenoids. The electrons active in the optical transitions move around a π-electron system of at least 26 atoms in the chlorophyll chromophore. These atoms lie in a plane and extend over about 0.8 nm in a typical dimension. Therefore, chlorophyll pairs at $R<1.5$ nm generally will not satisfy the point-dipole assumption. The effect on the Förster theory is, however, mostly a matter of corrections to the order of magnitude and the detailed R dependence (CHANG 1977). We stress that unless its corrections drive the prediction into or out of the strong coupling limit, use of the monopole approximation does not affect the dynamical picture we have presented.

3 Excitation Flow to Traps

3.1 Principles: Adaptation of Pairwise Transfer Theory

One important aspect of FÖRSTER pairwise transfer theory remains to be discussed. The rate, Eq. (4), is a rate of flow of excitation probability onto the acceptor and therefore represents a rate of flow away from the donor. The quantum mechanical theory on which it is based (time-dependent perturbation theory) applies strictly only for short times, but it has been found possible to generalize as follows. If the probability of excitation of molecule i is called

1 The phrase "monopole approximation" arises from the fact that the distributed transition dipole moment can be described approximately in terms of the amounts (called "monopoles") associated with each of the C and N atoms over which it is distributed.

ρ_i then (in the absence of excitation sources, for simplicity) the probabilities are connected by the set of equations (one equation for each molecule i)

$$\frac{d\rho_i}{dt} = -\Sigma_j F_{ji}\rho_i + \Sigma_j F_{ij}\rho_j - \frac{1}{\tau_i}\rho_i, \tag{5a}$$

where F_{ij} is the rate of transfer from molecule j to molecule i as computed by perturbation theory. This is the Pauli Master Equation, applied long ago to the problem by FÖRSTER (1948). On the right side, the three terms represent, respectively, flow of probability away from i to all others; flow into i from all others; monomolecular decay. The rates F_{ij} are not necessarily symmetrical, but for identical molecules one generally does have $F_{ij} = F_{ji}$. A central point is that probabilities superimpose, as long as the linear equations hold true; thus excitation probability can flow onto a molecule from one or more neighbors. This holds for the case of a system containing a single excitation. The explicit introduction of multiple excitations requires rather severe approximations, as discussed later.

As shown explicitly by many authors (FÖRSTER 1948; PEARLSTEIN 1966) the many equations (5a) can be approximated by using a continuum description of the excitation density in which $\rho(r)$ is the probability per unit volume obeying

$$\frac{d\rho(\vec{r})}{dt} = -\frac{1}{\tau_0(\vec{r})}\rho(\vec{r}) + D\vec{\nabla}^2\rho(\vec{r}), \tag{5b}$$

where D is a diffusion constant, $\vec{\nabla}^2$ the Laplacian operator, and $\tau_0(\vec{r})$ a lifetime with possible local variations. The diffusion constant is related to the discrete case by $D = cFa^2$, where F is an average pairwise transfer rate, c an average coordination number, and a the nearest neighbor distance in the discrete molecular array. Typically $c = 6$, $a = 1$ nm, $F = 10^{11} s^{-1}$, $D \sim 6 \times 10^{-7}\, m^2\, s^{-1} = 0.0006\, cm^2\, s^{-1}$.

Replacing the discrete set of molecular sites by a continuum provides a very convenient mental picture, since the flow of excitation (while it lasts) is analogous to heat conduction. But there are quantitative problems. The trap must be simulated by an absorbing or partly reflecting internal boundary, forcing a particular symmetry on the problem if there is to be an easy solution of the equation.

The simplest case of a two-dimensional circular array with a circular trap at the center, with fast-pulse excitation, has been treated by many workers. The results are rather ambiguous to interpret (see KNOX 1983, pp. 230–234), especially in relation to the real system in which traps are not uniformly placed and in addition can compete with one another for the excitation.

We note a special case of an excitation trapping model in which there is no excitation diffusion and each antenna molecule transfers directly to traps that are randomly distributed. A fairly complicated average must be made, and the result is antenna fluorescence proportional to $\exp(-k\sqrt{t})$, where k is related to the parameters in F and to the trap concentration (FÖRSTER 1949, performed this average for an analogous case). Data from crystals and photosynthetic systems have frequently been fit to this "$\sqrt{t}$" theory, but it is likely that excitation diffusion is important and the situation is better handled by sums of exponentials (KNOX 1983). An exhaustive analysis of the case of random and periodic traps in regular crystals has been carried out by KENKRE and PARRIS (1983).

Equation (5b) can be integrated over the volume of the system in question to approach a connection with Eq. (1a). With $n_H^\circ \int \rho(\vec{r},t)d\vec{r} = n_H(t)$, the left side of (5b) and the first terms on the right are simple when τ_0 is constant. They are equal to the corresponding terms of Eq. (1a). The diffusion term can be handled mathematically by Green's theorem, assuming trapping surfaces. Unfortunately it does not come out proportional to $n_H(t)$, but rather proportional to the excitation density at the trapping surface. Nonetheless, it is implicitly assumed in all elementary treatments that there is an overall proportionality to n_H and that the last term is $-kn_H(t)$ of Eq. (1a). We have made this connection, incomplete as it is, to demonstrate the nature of the approximations necessary to proceed from the relatively exact Eqs. (5a) to the macroscopic form.

Excitation annihilation is in many ways analogous to trapping. We have seen earlier that transfer may take place from a donor to an *excited* acceptor – the latter is just another excitation on the host, say, and it acts as a "moving target." Such targets have density n_H and present a "trapping rate" γn_H, according to the simplest theory first used by KEPLER et al. (1963). Thus with $\beta \equiv \beta_H + k$, Eq. (1a) can be modified to

$$\frac{dn_H}{dt} = I_H(t) - \beta n_H - \gamma n_H{}^2.$$

Typical values of γ, where this equation has been applied to the chlorophyll systems of photosynthesis, are 10^{-9} to 10^{-8} $cm^3\ s^{-1}$ (BRETON and GEACINTOV 1980). At a momentary excitation density of 10^{18} cm^{-3}, therefore, the momentary annihilation rate is of the order of magnitude of the usual trapping and decay rate (10^{-10} s^{-1}) and is definitely observable. Despite its success in the interpretation of many time-dependent experiments, Eq. (6) has a very narrow range of validity according to more formal theory (KENKRE 1981). However, more elaborate versions of the phenomenological theory have been able to interpret many intensity-dependent yield effects (e.g., PAILLOTIN et al. 1979).

3.2 Applications to Photosynthesis

FRANCK and TELLER (1938) were the first to apply the exciton concept to the flow of excitation in the photosynthetic unit. As explained by ROBINSON (1967), their concept was correct but their model calculation mistakenly assumed a one-dimensional array of chromophores. With several hundred molecules, a random walk over such an array ending at a trap would consume about 10^3–10^4 steps, on average, by which time fluorescence would have occurred. PEARLSTEIN (1966, 1967) showed that many fewer steps ($\sim$0,1 N/*n*N rather than $\sim$0.1 N^2) were needed in a two-dimensional array which, anyway, was more appropriate to a membrane geometry.

There now seems to be no doubt that excitation transfer processes can accomplish delivery to the trap with adequate efficiency, and the ability of the trap to accept and retain excitation is often the rate-limiting step (KENKRE and SCHMID 1983). PEARLSTEIN (1982a) has capitalized on the fact that when the trap itself is the initial absorber of the light, it initiates a random walk on the lattice which has a very specific initial condition. His analysis shows that the basic parameter of the diffusion (a "first passage time") can be obtained from fluorescence data taken at two different excitation wavelengths. This recent theory is one which is directly testable and may therefore prove quite valuable.

We mention two particular model calculations, mainly to illustrate the range of thinking which has been done. SEELY (1973) computed the result of excitation flow over chromophore arrays which were highly specific as to transition dipole orientations and rates. One of his purposes was to show the ease with which energy could be redirected – by rotating one or two dipoles – in connection with the older ideas of spillover between photosystems. His calculation is well worth study by anyone interested in doing random walk modeling in photosynthesis.

SHIPMAN (1980) computed fluorescence time courses expected from a partially random distribution of chlorophylls. A minimum interchlorophyll distance of 1.2 nm was allowed, the distribution covered a cylindrical region of a membrane (4 nm radius about a trap at the center, 8 nm height), and the orientations of the transition moments were random. The density was chosen to be consistent with known chlorophyll-protein and membrane compositions. The resulting time courses, predicted for 6-ps excitation pulses, are quite reasonable insofar as they might apply to one homogeneous population of chlorophylls. This model seems a much better representation of the situation in the membrane than many totally random model calculations made in the earlier days of time-resolved spectroscopy. Further discussion of model calculations on lattices has been given by the author (KNOX 1977).

Problems of excitation transfer and delocalization are still being posed at the small-protein level. In the green bacterium *Prostecochloris aestuarii*, the smallest protein subunit (MATTHEWS et al. 1979) contains seven chromophores definitely strong-coupled and difficult to understand quantitatively despite its well-determined structure (PEARLSTEIN and HEMENGER 1978). In another well-characterized chlorophyllide-substituted hemoglobin, FÖRSTER-coupled transfer has been used to obtain transition moment orientations (MOOG et al. 1984). Finally, in a chlorophyll a/b protein whose structure is virtually unknown because it does not crystallize, both types of coupling appear to coexist: chlorophyll b is strong-coupled and chlorophyll a is FÖRSTER-coupled, at least according to current hypothesis (GÜLEN and KNOX 1984, and references therein).

4 Conclusions and Prospects

The general principles of excitation transfer and delocalization are now well understood. The meaning of the latter term is, however, not always kept precise.

Its practical meaning in the context of photosynthesis is that initially localized excitation can transfer rapidly over hundreds of chromophores and participate in very efficient collection processes ("antenna" effects). Its more formal counterpart, wherein "delocalization" implies some long-range wave function phase relationship or wave packet transport, is not known to be of comparable practical interest. However, this kind of delocalization is present in strongly-coupled chromophore groups, manifesting itself spectrally.

Together with structural analyses, which are largely limited to samples not in vivo, the understanding of excitation transfer processes will enable the determination of an increasingly detailed picture of the somewhat volatile molecular architecture of natural photosynthetic systems.

Acknowledgements. The author thanks C. HANZLIK and B. WITTMERSHAUS for helpful comments on this manuscript. The work was supported in part by NSF grant PCM-83-03004 and USDA grant 82-CRCR-1-1128.

References

Bay Z, Pearlstein RM (1963) Proc Natl Acad Sci USA 50:962–967
Braun A, Mayer U, Auweter H, Wolf HC, Schmid D (1982) Z Naturforsch 37a:1013–1023
Breton J, Geacintov NE (1980) Biochim Biophys Acta 594:1–32
Burland DM, Zewail AH (1979) Adv Chem Phys 40:369–484
Chang JC (1977) J Chem Phys 67:3901–3909
Dexter DL (1953) J Chem Phys 21:836–850
Duke CB (1979) Mol Cryst Liq Cryst 50:63–84
Förster Th (1948) Ann Phys Ser 62:55–57
Förster Th (1949) Z Naturforsch 4a:321–327
Förster Th (1965) In: Sinanoglu O (ed) Action of light and organic crystals, part III. Modern quantum chemistry, Academic Press, London New York, pp 93–137
Franck J, Teller E (1938) J Chem Phys 6:861–872
Gülen D, Knox RS (1984) Photochem Photobiophys 7:277–286
Gülen D, Wittmershaus B, Knox R (1986) Biophys J (in press)
Kenkre VM (1981) Z Phys B43:221–227
Kenkre VM, Knox RS (1974) Phys Rev Lett 33:803–806
Kenkre VM, Parris PE (1983) Phys Rev B27:3221–3234
Kenkre VM, Schmid D (1983) Chem Phys Lett 94:603–608
Kepler RG, Caris JC, Avakian P, Abramson E (1963) Phys Rev Lett 10:400–402
Knox RS (1975) In: Govindjee (ed) Bioenergetics of photosynthesis. Academic Press, London New York, pp 183–221
Knox RS (1977) In: Barber J (ed) Primary processes of photosynthesis, chap. 2. Elsevier/North-Holland Biomed Press, Amsterdam New York, pp 55–97
Knox RS (1983) In: DiBartolo B, Danko J (eds) Collective excitations in solids. Plenum, New York London, pp 183–245
Matthews BW, Fenna RE, Bolognesi MC, Schmid MF, Olson JM (1979) J Mol Biol 131:259–286
Moog RS, Kuki A, Fayer MD, Boxer SG (1984) Biochemistry 23:1564–1571
Paillotin G, Swenberg CE, Breton J, Geacintov NE (1979) Biophys J 25:513–534
Pearlstein RM (1966) Ph D Thesis, Univ Maryland
Pearlstein RM (1967) In: Olson JM (ed) Energy conversion by the photosynthetic apparatus. Brookhaven Symp Biol, no 19. Upton New York, pp 8–15
Pearlstein RM (1982a) Photochem Photobiol 35:835–844

Pearlstein RM (1982b) Govindjee (ed) Integrated approach to plant and bacterial photosynthesis, chap 4. Academic Press, London New York, pp 293–330
Pearlstein RM, Hemenger RP (1978) Proc Natl Acad Sci USA 75:4920–4924
Powell RC, Kepler RG (1969) Phy Rev Lett 22:636–639 (Erratum: same volume, p 1232)
Rahman TS, Knox RS (1973) Phys Status Solidi (b) 58:715–720
Rahman TS, Knox RS, Kenkre VM (1979) Chem Phys 44:197–211
Robinson GW (1967) In: Olson JM (ed) Energy converson by the photosynthetic apparatus. Brookhaven Symp Biol, no 19. Upton, New York, pp 16–48
Seely GR (1973) J Theor Bio 40:173–187
Shipman LL (1980) Photochem Photobiol 31:157–167
Stryer L (1978) Annu Rev Biochem 47:819–846
Wolf HC (1967) In: Bates DR, Estermann I (eds) Advances in atomic and molecular physics, vol III. Academic Press, London New York, pp 119–142

7.2 Excited State Kinetics of Chlorophyll Antenna Pigments

A.R. HOLZWARTH

1 Introduction

Considerable progress has been made during the last few years in the understanding of the organization and function of antenna pigments and reaction centers in higher plants and green algae. Investigations employing time-resolved spectroscopic methods and, in particular, fluorescence techniques have contributed significantly to this progress (see SAUER, Chap. 2, this Vol., for an overview). Chlorophyll (Chl) fluorescence, both stationary and time-resolved, provides a powerful intrinsic probe for energy transfer and charge separation kinetics of the photosynthetic apparatus. Fluorescence techniques are particularly useful because they can be applied in vivo on intact leaves and algal cells, as well as on isolated chloroplasts and thylakoid fragments.

The aim of the present report is to discuss the results relevant to an understanding of the first steps occurring in photosynthesis, which were obtained using time-resolved methods in the picosecond and nanosecond time range. In view of the limited space available, attention will be confined to the more recent developments. This intention is facilitated by the excellent reviews of the field by BRETON and GEACINTOV (1980) and KARUKSTIS and SAUER (1983a). The former also critically discussed the various artifacts that have confused the early discussions of picosecond results on photosynthetic systems. The reader is also referred to the review by BOSE (1982), and in this Volume KNOX, Chapter 7.1, HOFF, Chapter 7.2, GEANCINTOV and BRETON, Chapter 7.4, and NORRIS and VAN BRAKEL, Chapter 8.3.

2 Instrumentation

Although a detailed description of the various experimental methods is beyond the scope of the present report, it is important to realize that recent progress in this field is intimately related to improvements in measuring techniques. The methods presently available for recording fluorescence decays in the subnanosecond and picosecond time range are streak cameras (BRETON and GEACINTOV 1980), phase-shift techniques (MOYA and GARCIA 1983), and single-photon timing (SPT). The streak camera in combination with a single shot picosecond laser provided the highest time resolution until recently. However, this method

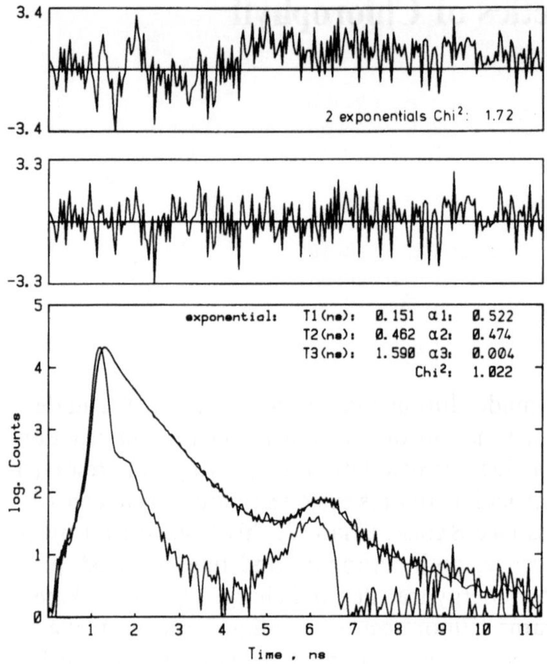

Fig. 1. Fluorescence decay of *Chlorella vulgaris* with open reaction centers (F_0) at λ_{em} = 682 nm on a semilogarithmic scale (*below*); λ_{exc} = 630 nm. The frames (*above*) show the weighted residuals for double and triple exponential model functions. Note the large differences in the statistical test ($\varkappa^2$-value) and the residuals for the two model functions. (HAEHNEL et al. 1983)

requires fairly high excitation intensities, generally in excess of 10^{13} photons cm^{-2}, which gives rise to artifacts such as singlet-singlet annihilation. The two other methods can operate with rather low intensities. During the last few years, developments in the SPT technique have made a considerably higher time resolution possible, and more attention has been paid to methods for judging the quality of the fits (HOLZWARTH et al. 1982; HAEHNEL et al. 1982; WENDLER and HOLZWARTH 1985). The narrowest SPT apparatus function using a conventional photomultiplier which has been obtained so far is approxim. 130 ps (WENDLER et al. 1984a; WENDLER and HOLZWARTH 1985) and channel-plate photomultipliers have been used which allow instrument response functions as low as about 50 ps (ZIMMERMAN et al. 1982; MURAO et al. 1982). With these instruments, fluorescence decays down to 10 ps can be measured. Examples of fluorescence decays obtained by the SPT technique are given in Figs. 1 and 2. When one is interested in recording the low intensity fluorescence decay of a photosynthetic antenna system free from distorting artifacts, the SPT technique is presently considered to be the best method (BRETON and GEACINTOV 1980). It is unsurpassed in sensitivity, signal-to-noise ratio and its ability to resolve complex multicomponent decays. The time resolution should now be sufficient to measure most energy transfer processes in photosynthesis. The synchronously pumped dye-laser, which is part of the picosecond SPT setup, provides the researcher with continuously tunable exciting light. This is required to selectively excite specific pigment pools. The SPT technique was introduced into photosynthesis research by the groups at Berkeley (SAUER and BREWINGTON 1978) and the Royal Institution in London (BEDDARD et al. 1979). The phase-

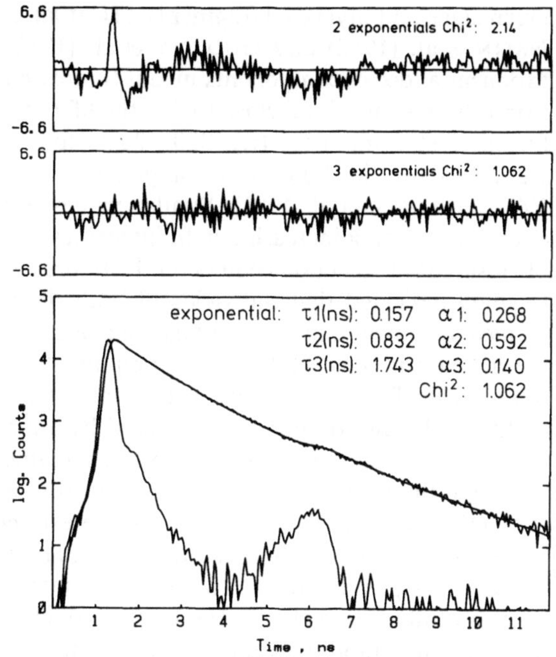

Fig. 2. Fluorescence decay of *Chlorella vulgaris* with partially closed reaction centers (F_0) at $\lambda_{em}=682$ nm on a semilogarithmic scale (*below*); $\lambda_{exc}=630$ nm. The frames (*above*) show the weighted residuals for double and triple exponential model functions. PS II centers were partially closed by 6 μM DCMU. (HAEHNEL et al. 1983)

shift technique is advantageous for measuring fluorescence lifetimes in short time intervals and to scan component-resolved spectra with intermediate to low excitation intensity (MOYA and GARCIA 1983). Single-shot techniques have merit when one is engaged in studying annihilation effects (see GEACINTOV and BRETON Chap. 7.4, this Vol.).

3 Room Temperature Fluorescence Kinetics

Although the time resolution was still limited to several hundred ps at the time, BEDDARD et al. (1979) were able to demonstrate by SPT that the fluorescence decay of in vivo Chl in chloroplasts does not follow a single exponential even at low excitation intensities where singlet singlet annihilation effects could be excluded. A two-component decay with lifetimes of approximately 400 ps and 1.4 ps was obtained. The relative amplitudes varied with the redox state of photosystem II. The use of the SPT technique in photosynthesis research has been propagated by mainly four groups, namely those at Berkeley (HAEHNEL et al. 1982; NAIRN et al. 1982; KARUKSTIS and SAUER 1983a–d; HAWORTH et al. 1983), Chicago (GULOTTY et al. 1982; LOTSHAW et al. 1982), San Diego (MAGDE et al. 1982; BERENS et al. 1984) and Mülheim (HAEHNEL et al. 1983; HOLZWARTH et al. 1984, 1985; WENDLER et al. 1984b). While the first three groups concentrated more on chloroplasts, the emphasis of the group at Mülheim was on green algae and thylakoid fragments. HAEHNEL et al. (1982), MAGDE et al. (1982), NAIRN et al. (1982) and GULOTTY et al. (1982) were the first to report three exponential decay components in Chl fluorescence from chloroplasts and algae. For a fluorescence level close to F_0, the lifetimes were in the range of 80–130 ps for the fast component (τ_f), 320–420 ps for the middle component (τ_m), and 1.2–2.3 ns for the long-lived component (τ_s) in chloroplasts (HAEHNEL et al. 1982; GULOTTY et al. 1982). Similar values were obtained in algae (HAEHNEL et al. 1982). Using increasing light intensities the latter authors studied the effect of closing PS II reaction centers on the fluorescence yield. Their detailed intensity study indicated that the yields of both the fast and middle components remained approximately constant as the closing of PS II centers progressed, while the increase in yield of the slow component was almost solely responsible for the variable fluorescence (F_{var}). In contrast to the yields, the lifetimes of the middle and slow components increased by approximately a factor of only two, while the lifetime of the fast component remained essentially constant.

These findings were basically confirmed in a study carried out by HAEHNEL et al. (1983) on green algae, where a three-exponential fit was generally required to describe the 682 nm fluorescence decay. However, in an attempt to approach as closely as possible to the two extreme states of PS II centers, i.e., fully upon (F_0) and fully closed (F_{max}), HAEHNEL et al. (1983) observed an opposite dependence of the amplitudes and yields of the slow and fast components on the state of PS II centers. At the F_0-level the yield of the slow component was found to be close to zero; it increased by a factor of 69 at the F_{max} level. In contrast, the yield of the fast component decreased by approximately an order of magnitude when going from open to closed centers.

For the first time now there appears to be agreement between at least four different laboratories on the general features of in vivo Chl fluorescence. This represents a major achievement compared to the situation a few years ago (BRETON and GEACINTOV 1980). There still remain some minor differences which might originate partly from differences in the biological materials under study (chloroplasts from various species and different green algae) and from the differences in time resolution of the SPT instruments used. While the gross features of the observed kinetics are the same, there seem to persist significant differences between the different laboratories in the interpretation of these findings, however. Both Sauer's group at Berkeley (HAEHNEL et al. 1982; NAIRN et al. 1982) and the group at Mülheim (HAEHNEL et al. 1983; HOLZWARTH et al. 1984) agree that at the F_0 level the major part of the short decay component arises from quenching of the Chl a antenna by the open PS II center. They also agree that the amount of long-lived fluorescence should reflect the percentage of closed centers. This component should be absent in fully open PS II units. Differences exist in the interpretation of the mechanism giving rise to the long-lived decay, however. SAUER and coworkers (HAEHNEL et al. 1982; NAIRN et al. 1982) adopted the Klimov mechanism, thereby attributing the long-lived component to a delayed fluorescence induced by charge recombination in centers with reduced secondary acceptor Q (KLIMOV et al. 1978). In view of the drastic decrease of the amplitude of the fast component upon closing PS II centers observed by HAEHNEL et al. (1983), these authors had doubts about the validity of this concept. This decrease has not been observed by the other groups (for a discussion related to the Klimov model see below). Again there was agreement between the two groups on the origin of the middle component, which was interpreted as to arise mainly from light-harvesting Chl a/b protein connected to PS II (LHC II). GULOTTY et al. (1982) gave a different interpretation by assigning the fast and middle components to PS II centers in different photochemical redox states. As has been pointed out by HAEHNEL et al. (1983), the assignment to LHC II would solve the apparent discrepancy between the known F_{max}/F_0 ratio of about 4 (BUTLER 1978) and the maximum yield of electron transport of approx. 0.95 (SUN and SAUER 1971). On the other hand, intermittent light-grown pea chloroplasts lacking the LHC II complex still showed the middle component in the fluorescence decay (HAEHNEL et al. 1982; KARUKSTIS and SAUER 1983b). A long fluorescence lifetime of the LHC antenna pigment would also not be consistent with the general idea that the LHC II should be tightly coupled. The time- and component-resolved emission and excitation spectra reported by WENDLER et al. (1984b) and HOLZWARTH et al. (1985) also gave direct evidence that the major part of the middle component is not related to LHC II. Thus the interpretation of the middle and long components is not settled presently.

It should be pointed out that any interpretation which leads to an assignment of one of the decay components to a single process might not be justified. First of all the resolution capability of the SPT technique is presently limited to three components in most cases. A first example of a four-exponential analysis has been reported recently (WENDLER et al. 1984b; HOLZWARTH et al. 1985). A full description of the complex kinetics of Chl antenna might require a considerably greater number of decay components. Thus, one or more of the three or even four components determined so far would still involve averaged descriptions of real decay components. Another reason lies in the fact that, depending on the underlying kinetic scheme, the measured fluorescence lifetimes will in general not be identical to the inverse energy transfer rate constants. To obtain more insight into the decay kinetics, variation in parameters that would influence the components differently are of great help. With this aim in mind, the three groups mentioned above moved in different directions. In a series of papers, SAUER and coworkers presented studies that relate the fluorescence kinetics of spinach chloroplasts to various parameters which cause structural changes. These parameters are, e.g., variation in Mg^{2+} concentration (NAIRN et al. 1982), redox potential (KARUKSTIS and SAUER 1983a, b), phosphorylation/dephosphorylation (HAWORTH et al. 1983) and etioplast to chloroplast differentiation (KARUKSTIS and SAUER 1983d). Only the most pertinent results can be summarized here: Addition of Mg^{2+} influenced the slow decay most strongly. The effect was opposite at the F_0- and the F_{max}-levels. At the F_0-level there was a decrease in τ_s while at the F_{max} level an increase was observed in both the lifetime and the yield. The other components were affected only slightly. NAIRN et al. (1982) concluded that addition of Mg^{2+} causes a decrease of spillover as well as an increase in the absorption cross section of PS II. They suggested that the transition from separate units to the interconnected state occurs in the range of 0.5–2 mM Mg^{2+}.

Similar changes in the slow decay component were observed upon change in redox potential (KARUKSTIS and SAUER 1983b). Interestingly, reductive titration increased the total yield by a factor of eight, i.e., τ_s was greater when Q was reduced chemically than when PS II centers were closed by increasing light intensity. A conclusive interpretation of these results was not possible, but the authors suggested that both Q_H and Q_L centers should act as electron acceptors in α-centers. Phosphorylation of LHC II (both at the F_0 and the F_{max} levels) again influenced mainly the slow component its yield being smaller in the phosphorylated state. An increase in the yield of the fast component, as shown by their data (HAWORTH et al. 1983), was not commented on by the authors. A shift from a sigmoidal to a hyperbolic intensity dependence of the slow phase yield was interpreted as being caused by a reorganization of the pigment bed due to the occurrence of a state transition. When the light intensity is increased, starting from a very low level, the lifetime of the slow phase first declines and then increases again. This has been demonstrated both by HAEHNEL et al. (1983) on green alga and by HAWORTH et al. (1983) on chloroplasts. HAWORTH et al. (1983) pointed out that this observation might indicate a heterogeneity in the fluorescence of the slow component.

In an extensive study of etioplast to chloroplast differentiation in pea, KARUKSTIS and SAUER (1983d) tried to associate the changes in the fluorescence yields of the kinetic components to known structural phenomena during chloroplast development. The initial fluorescence levels did not show cation-inducable changes, which was attributed to the lack of LHC II and grana. Upon transfer of leaves to continuous light LHC accumulates. In contrast to the slow decay, the differences of τ_m, the middle lifetime, at F_0 and F_{max} was found to be independent of cation concentration and, more interestingly, was present prior to grana formation. The authors attributed this phenomenon to the development of α and β centers and concluded that grana development was not a necessary requirement for this heterogeneity. The fast lifetime τ_f was found to be independent of this development. Their observations also indicated that the middle phase is more complex than arising just from LHC II. The assignment of τ_m to LHC II has been questioned recently (WENDLER et al. 1984b; HOLZWARTH et al. 1985).

In view of the known absorption differences between various Chl protein complexes, variation in the excitation wavelength and recording of time-resolved excitation and emission spectra should be a powerful method which would assist in the assignment of the different decay components. Studies of this kind were carried out be the group at Chicago (GULOTTY et al. 1982) using barley chloroplasts. Unfortunately these authors gave only relative amplitudes for the decay components as a function of excitation wavelength. Thus, a full spectral comparison with known Chl proteins is not possible. Nevertheless some trends were observed. At $\lambda_{em}=680$ the weight of the short-lived component was lower and the weight of the middle component higher than at $\lambda_{em} \geq 710$ nm. However, for excitation wavelengths greater than 680 nm, the amplitude of the short-lived component increased strongly. HOLZWARTH et al. (1984) reported a similar observation, together with a decrease in the lifetime of the short-lived component from approx 120 ps to 80 ps in *Chlorella*. This decay was interpreted as arising predominantly from PS I. The latter study was extended more recently by a four-component analysis of the time-resolved emission and excitation spectra of *Chlorella* (WENDLER et al. 1984b; HOLZWARTH et al. 1985). The spectral and kinetic contributions of PS I and PS II to the generally observed fast component from the three-exponential analysis of the fluorescence decays could thus be resolved for the first time. The PS I fluorescence with a lifetime of 80 ps showed a spectral emission maximum slightly below 700 nm. The contribution from PS II peaked in the range 682–685 nm and had a lifetime of approx. 180 ps in open centers. Direct information on PS I fluorescence at room temperature from intact cells was lacking so far.

4 Measurements of Thylakoid Fragments

Kinetic measurements on particles enriched in either PS II or PS I can, in principle, be of considerable value in helping assign the various decay components. There is, however, a basic problem involved in the realization of such an approach. Most isolation procedures require the use of detergents. This might lead to a distortion in the structural organization of the membrane proteins, eventually giving rise to a partial or complete energetic decoupling of Chl pigments (see THORNBER, Chap. 3, this Vol.). Obviously such phenomena will strongly influence the decay kinetics. Possibly a better approach is provided by the recent separation of thylakoids into an inside-out vesicle fraction (from grana) and a stroma-lamellae fraction highly enriched in PS I (ANDERSSON and ANDERSON 1980). Picosecond measurements on such systems have been carried out by HOLZWARTH et al. (1984). Inside-out vesicles showed behavior similar to intact chloroplasts (HAEHNEL et al. 1982) and intact alga (HAEHNEL et al. 1983) at the F_0 and F_{max} states. In contrast, stroma lamellae did not show any significant changes in fluorescence parameters upon addition of DCMU. These findings were in agreement with the expected heterogeneous distribution of photosystems in the two fragments. At the 680 nm emission wavelength, PS I showed essentially biexponential decay with lifetimes of 120–130 ps and 30–40 ps (largest amplitude component). The lifetime of 40 ps was also observed in PS I particles by SHUVALOV et al. (1979). Lifetimes in the range 60–130 ps were observed earlier also on various PS I particles (cf. BRETON and GEACINTOV 1980). Picosecond measurements of detergent-isolated P-700-enriched particles from spinach by absorption (KAMOGAWA et al. 1981) and fluorescence techniques (KAMOGAWA et al. 1983) were reported. The absorption measurement yielded a fast transient with a lifetime of $\simeq 25$ ps and a ns-component related to P-700 oxidation. The long-lived component was missing when P-700 was in the oxidized state prior to the measurement. Fluorescence decay was slower at 670–680 nm ($\simeq 70$ ps) and faster at 690–700 nm ($\simeq 10$ ps). The very short-lived decay was attributed to the high Chl/P-700 ratio in these particles (8–10 Chl/P-700). The wide range of lifetime values for these particles is probably related to the presence of various antenna pigments emitting in different wavelength regions (KYLE et al. 1983). This gives rise to a broad overall spectrum of PS I fluorescence, as is supported also by the time-resolved fluorescence spectra in *Chlorella* at room temperature (WENDLER et al. 1984b; HOLZWARTH et al. 1985).

5 Low Temperature Fluorescence Kinetics

At low temperature the fluorescence characteristics of thylakoids are radically different from those at room temperature. Emission attributed to PS I is now very strong ($\lambda_{em}=720$–735) (BUTLER and KITAJIMA 1975; CHO and GODVINDJEE 1970; RIJGERSBERG et al. 1979) and occurs in addition to the PS II fluorescence bands F-680 and F-695. MOYA and GARCIA (1983) have compiled most of the time-resolved data available up to 1982 for the low temperature Chl fluorescence. These data are rather diverse. One of the important open questions is whether there exists a rise time for the long wavelength fluorescence. MOYA and GARCIA (1983) have resolved the total fluorescence into up to seven components. Two of them (688 and 698 nm maxima) are attributed to PS II and can be correlated with room temperature data (see above). In the long wavelength region (>700 nm), two main bands were required to describe PS I fluorescence. The other bands are of unclarified origin. No rise times have been reported in this work, and it is not clear whether their equipment would have been adequate to recognize and resolve any such contribution. PELLEGRINO et al. (1983) have reported a relatively short rise time of $\simeq$20 ps for the 735 nm fluorescence at 90 K in spinach. This is in contrast to data reported by CAMPILLO et al. (1977) and more recent measurements by the Berkeley group (REISBERG et al. 1982). The latter authors used the SPT technique at very low excitation intensity and found a rise time of $\simeq$100 ps in spinach chloroplasts at 77 K for the long-wavelength emission. Their fluorescence lifetimes measured at 680 nm are in reasonable agreement, however, taking into account the single-exponential analysis by PELLEGRINO et al. (1983). It should be pointed out that a comparison of risetime data obtained by the different groups is difficult, since most authors working with streak cameras did not apply deconvolution to their data. No drastic differences between room and low temperature kinetics were observed for $\lambda_{em}=680$ nm (REISBERG et al. 1982). The long-lived component changed in yield by a factor of less than 2. This small change is surprising in view of the supposed charge-recombination origin of that component (KLIMOV et al. 1978; HAEHNEL et al. 1982). The seemingly contradictory data on fluorescence risetimes of the 730 nm fluorescence in chloroplasts might be rationalized by recent results of WITTMERSHAUS et al. (1985), who found biphasic rise kinetics.

6 Origin of the Long-Lived (Nanosecond) Decay

It has been suggested that charge separation between P-680 and the primary acceptor pheophytin can occur also in a closed PS II reaction center when the secondary acceptor Q is reduced (KLIMOV et al. 1978). The model implies that charge separation should occur at more or less the same rate as in open centers. The variable fluorescence yield should then arise from charge recombination repopulating partly the excited state of P-680 from the radical pair (P-680^{+}Ph^{-}). An activation energy of 0.04–0.08 eV has been determined for this process (Klimov et al. 1978) which should occur with a rate constant of $\approx$(2 ns)$^{-1}$ at room temperature. This is known in the literature as the Klimov model. The model has been widely accepted during the last few years. It was also suggested that upon charge recombination at low temperature the energy would become located on pheophytin. This should then give rise to the well-known fluorescence band F-695 (Breton 1982). The Klimov model for interpretation of the variable fluorescence has been questioned by Haehnel et al. (1983) on the basis of the decrease in the amplitude of the fast decay component τ_f upon closing PS II centers. This component should not disappear if the charge separation proceeds at the same rate as in open centers. Breton (1983) tried to find a way out of this dilemma by assigning the fast component of $\approx$130 ps lifetime (Haehnel et al. 1982, 1983) not to the primary charge separation but to the charge stabilization on Q^{-}. One might in the first place expect a reduced rate constant for charge separation by the mere presence of a negatively charged species (Q^{-}) close to the pheophytin anion. There exists, however, a more direct argument against the interpretation of the variable fluorescence as arising from charge recombination (Klimov et al. 1978; Breton 1983). It seems to have been overlooked that charge recombination with a temperature-dependent rate constant involving an activation energy of 0.04–0.08 eV (Klimov et al. 1978) or even up to 0.12 eV (Shuvalov et al. 1980) would have a negligible yield a low temperature. Using only the lower value of 0.04 eV for this activation energy, a decrease in rate constant of about 2 orders of magnitude would be obtained when going from room temperature to 77 K. Such a drastic decrease would not be consistent with the strong F-695 band under steady state conditions nor with the minor changes only in amplitude and yield of this component in the time-resolved experiment (Reisberg et al. 1982) at low temperature. Mathis and Schenk (1982) were also not able to confirm a temperature dependence of F_{var}. In a more recent study, even an increase in F_{var} at lower temperature was found (Mathis 1984). Furthermore, lifetimes for τ_s as large as 4 ns (Klimov et al. 1978; Shuvalov et al. 1980) have not been observed by other workers (Haehnel et al. 1982, 1983; Reisberg et al. 1982; Gulotty et al. 1982). It is therefore suggested that the slow component τ_s, responsible in large part for F_{var} at both room and low temperature, does not represent delayed fluorescence arising from charge recombination in closed PS II center but from direct population of the emitting states. It should be pointed out that this interpretation would still allow for charge separation to occur in a closed center, albeit at a much lower rate ($\approx$0.5 ns^{-1}) than in open centers. Also, the suggested pheophytin origin of the low temperature F-695 fluorescence (Breton 1982) would fit this model, provided that its excited state is generated by direct energy transfer from the Chl antenna. It would then play the role of a low temperature trap.

7 Conclusions and Outlook

The present understanding of in vivo Chl fluorescence kinetics involves two lifetimes for PS II centers. One applies to open centers and one to closed centers, with indication of a small degree of connectivity. Recent time-resolved experiments have also given evidence for a manifestation of heterogeneity in the fluorescence decays of PS II (KARUKSTIS and SAUER 1983b, c; WENDLER et al. 1984b; BERENS et al. 1984). Based on the data of HAEHNEL et al. (1983), a theoretical model has been proposed, which takes into account the α-β-heterogeneity of PS II reaction centers (BUTLER et al. 1983). In addition to the PS II components, a fast component from PS I with a lifetime of 80 ps has now been resolved (WENDLER et al. 1984b; HOLZWARTH et al. 1985). Measurements of Chl fluorescence kinetics, in particular using SPT techniques, during the last few years have extended our knowledge of the organization of the Chl antenna pigments to a significant extent. It seems that the necessary experimental tools are now available to insure rapid progress in this field in the future.

Acknowledgments. The author's work cited here has been supported in part by the Deutsche Forschungsgemeinschaft. I thank Drs. H. SCHEER and J. WENDLER, as well as Mr. H. GRÜN, for critically reading the manuscript.

References

Andersson B, Anderson JM (1980) Biochim Biophys Acta 593:427–440
Beddard GS, Fleming GR, Porter G, Searle GFW, Synowiec JA (1979) Biochim Biophys Acta 545:165–174
Berens SJ, Scheele J, Butler WL, Magde D (1984) In: Ultrafast phenomena. Springer, Berlin Heidelberg New York, pp 487–489
Bose S (1982) Photochem Photobiol 36:725–731
Breton J (1982) FEBS Lett 147:16–20
Breton J (1983) FEBS Lett 159:1–5
Breton J, Geacintov NE (1980) Biochim Biophys Acta 594:1–32
Butler WL (1978) Annu Rev Plant Physiol 29:345–378
Butler WL, Kitajima M (1975) Biochim Biophys Acta 396:72–85
Butler WL, Magde D, Berens SJ (1983) Proc Natl Acad Sci USA 80:7510–7514
Campillo AJ, Shapiro SL, Geacintov NE, Swenberg CE (1977) FEBS Lett 83:316–320
Cho F, Govindjee (1970) Biochim Biophys Acta 216:139–150
Gulotty RJ, Fleming GR, Alberte RS (1982) Biochim Biophys Acta 682:322–331
Haehnel W, Nairn JA, Reisberg P, Sauer K (1982) Biochim Biophys Acta 680:161–173
Haehnel W, Holzwarth AR, Wendler J (1983) Photochem Photobiol 37:435–443
Haworth P, Karukstis KK, Sauer K (1983) Biochim Biophys Acta 725:261–271
Holzwarth AR, Wendler J, Wehrmeyer W (1982) Photochem Photobiol 36:479–487
Holzwarth AR, Haehnel W, Wendler J, Suter GW, Ratajczak R (1984) In: Sybesma C (ed) Advances in photosynthetic research, vol I. Nijhoff/Junk, The Hague, pp 73–76
Holzwarth AR, Wendler J, Haehnel W (1985) Biochim Biophys Acta 807:155–167
Kamogawa K, Namiki A, Nakashima N, Yoshihara K, Ikegami I (1981) Photochem Photobiol 34:511–516
Kamogawa K, Morris JM, Takagi Y, Nakashima N, Yoshihara K, Ikegami I (1983) Photochem Photobiol 37:207–213
Karukstis KK, Sauer K (1983a) J Cell Biochem 23:131–158
Karukstis KK, Sauer K (1983b) Biochim Biophys Acta 722:364–371
Karukstis KK, Sauer K (1983c) Biochim Biophys Acta 725:246–253
Karukstis KK, Sauer K (1983d) Biochim Biophys Acta 725:384–393

Klimov VV, Allakhverdiev SI, Pashchenko VZ (1978) Dokl Akad Nauk SSSR 242:1204–1207
Klimov VV, Allakhverdiev SI, Demeter S, Krasnovskii AA (1979) Dokl Akad Nauk SSSR 249:227–230
Kyle DJ, Baker NR, Arntzen CJ (1983) Photobiochem Photobiophys 5:79–85
Lotshaw WT, Alberte RS, Fleming GR (1982) Biochim Biophys Acta 682:75–85
Magde D, Berens SJ, Butler WL (1982) In: Goldberg LS (ed) Picosecond lasers and applications. proc, SPIE, pp 80–86
Mathis P (1984) In: Sybesma C (ed) Advances in photosynthetic research. Nijhoff/Junk, The Hague, pp 155–158
Mathis P, Schenck C (1982) In: Solar energy research and development in photochemical photoelectrochemical and photobiological processes. Commission European Communities (ed.) Reidel, Dordrecht, pp 129–133
Moya I, Garcia R (1983) Biochim Biophys Acta 722:480–491
Murao T, Yamazaki I, Yoshihara K (1982) Appl Opt 21:2297–2298
Nairn JA, Haehnel W, Reisberg P, Sauer K (1982) Biochim Biophys Acta 682:420–429
Pellegrino F, Dagen A, Sekuler P, Alfano RR (1983) Photobiochem Photobiophys 6:15–23
Reisberg P, Nairn JA, Sauer K (1982) Photochem Photobiol 36:657–661
Rijgersberg CP, Amesz J, Thielen APGM, Swager JA (1979) Biochim Biophys Acta 545:473–482
Sauer K, Brewington GT (1978) In: Hall DO, Coombs J, Goodwin TW (eds) Proc 4th Int Congr Photosynth. Biochem Soc, London, pp 409–421
Shuvalov VA, Klevanik AV, Sharkov AV, Kryukov PG, Ke B (1979) FEBS Lett 107:313–316
Shuvalov VA, Klimov VV, Dolan E, Parson WW, Ke B (1980) FEBS Lett 118:279–282
Sun ASK, Sauer K (1971) Biochim Biophys Acta 234:399–414
Wendler J, Holzwarth AR (1985) to be published
Wendler J, Holzwarth AR, Wehrmeyer W (1984a) Biochim Biophys Acta 765:58–67
Wendler J, Haehnel W, Holzwarth AR (1984b) In: Ultrafast phenomena. Springer, Berlin Heidelberg New York
Wittmershaus BP, Nordlund TM, Knox WH, Knox RS, Geacintov NE, Breton J (1985) Biochim Biophys Acta 806:93–106
Zimmerman HE, Penn JH, Carpenter CW (1982) Proc Natl Acad Sci USA 79:2128–2132

7.3 Topology and Dimensions of Antenna Pigment Systems Probed by Exciton Annihilation

N.E. GEACINTOV and J. BRETON

1 Introduction

When photosynthetic systems are exposed to high intensity excitation, typically achieved with conventional pulsed lasers, the phenomenon of exciton annihilation manifests itself (BRETON and GEACINTOV 1980). Two types of bimolecular annihilation processes are recognized:

$$S_1 + S_1 \rightarrow S_0 + S_1 + \text{heat (or } S_0 + S_0) \quad (1)$$

$$S_1 + T_1 \rightarrow S_0 + T_1 + \text{heat}, \quad (2)$$

where S_0, S_1, and T_1 represent pigment molecules in the ground state, first excited singlet state, and triplet states respectively. In either case there is a loss of at least one singlet excitation. The first process, singlet-singlet exciton annihilation, is favored when a large density of excitons are created on time scales which are much shorter than the mean lifetime (τ) of singlet excitons (excitation pulse width $\Delta t \ll \tau$). The second process, singlet-triplet annihilation is favored when the rate of arrival of photons is much lower than the mean rate of decay of singlet excitons, i.e., $\Delta t \gg \tau$. Either one of these processes can be selected for study by choosing either single picosecond or microsecond duration laser pulses for excitation (GEACINTOV and BRETON 1982). In this review results on singlet-singlet annihilation phenomena are emphasized.

2 Principles of Annihilation Techniques

While the bimolecular annihilation of excitons observed upon high intensity laser pulse excitation is often considered to be an undesirable effect, these exciton quenching processes can provide interesting information about the topology of the different pigment systems, and the connectivity between them (see also KNOX, Chap. 7.1, this Vol., for a more theoretical treatment of this general topic). Because of the known heterogeneity of pigment systems (see THORNBER, Chap. 3, this Vol.), different types of singlet excitons may be present, each with their own decay time and giving rise to different emission spectra (MOYA et al. 1981). In addition, both free (mobile) and trapped excitons may be present in a given pigment system, each displaying different decay times and emission wavelength maxima, thus complicating the situation still further.

The general expression for the decay of singlet excitons in pigment bed i, following excitation by a very short pulse, can be written as follows, assuming a continuum model for simplicity (PAILLOTIN et al. 1979):

$$\frac{dn_i^s}{dt} = G - k_i\, n_i^s - \gamma_i(st) n_i^s\, n_i^t - \tfrac{1}{2}\, \gamma_i(ss) n_i^s\, n_i^s - \tfrac{1}{2} \gamma_i^1(ss) n_i^s\, n_i^s(trap), \tag{3}$$

where k_i represent all of the normal, low intensity decay rate constants in pigment bed i, the superscripts s and t denote singlet and triplet excitons respectively, while $\gamma_i(ss)$ and $\gamma_i(st)$ are the bimolecular singlet-singlet and singlet-triplet annihilation rate constants, and n_i^s and $n_i^s(trap)$ represent singlet exciton densities (free and trapped, respectively) in pigment bed i. G is the rate of arrival of singlet excitons from other pigment beds into pigment bed i.

The instantaneous fluorescence intensity is given by:

$$F(t) \propto \sum_i n_i^s + \sum_i n_i^s(trap). \tag{4}$$

The singlet exciton populations n_i must be summed over all fluorescence emitting pigment systems i and thus comprise all types of excitons, including mobile excitons in the antenna pigment systems, and excitons localized in smaller pigment pools, or trapped near the reaction centers.

3 Exciton Annihilation and Effects of Domain Size

We define a domain as a collection of pigment molecules to which excitons are confined. In principle, physical barriers may prevent excitons from moving from one domain to another. However, in photosynthetic systems it seems more likely that excitons are funneled from light-harvesting (LH) antenna molecules to smaller pigment pools near the reaction centers, where they tend to be confined because of lower probabilities of uphill energy transfer, particularly at low temperatures. At the other extreme, the size of large domains may be loosely defined as the number of molecules over which the excitons can range during their lifetimes. Experimentally, large and small domains can be distinguished from one another by the excitation intensity dependence of the fluorescence yields and decay profiles.

3.1 Small Domains

For a single pigment system the fluorescence yield Φ is predicted to follow the relationship

$$\Phi = (\Phi_0/Y)\,[1 - \exp(-Y)], \tag{5}$$

where Y is the number of photons absorbed per domain, and Φ_0 is the low intensity fluorescence yield. This equation for small domains (puddle model) can be derived on the basis of simple Poisson statistics (MAUZERALL 1976), or from a Master equation theory (PAILLOTIN et al. 1979). In these models all excitons created in a given domain, except the last one, are quenched according to Eq. (1). The last remaining exciton will decay with its normal, low intensity kinetics. Thus, while a severe quenching in the fluorescence yield may be observed, no changes in the fluorescence decay profiles are expected.

Indeed, NORDLUND and KNOX (1981) have observed that the lifetime of a monomeric form of a LH chlorophyll a/b protein containing ~6 chlorophyll molecules/particle is independent of the intensity of single picosecond laser pulses in the intensity range of 10^{14}–10^{18} photons cm^{-2} $pulse^{-1}$ (530 nm) corresponding to 10^{-3}–10 photon hits per particle. In the case of P-700-enriched photosystem I particles, the decay profiles are insensitive to the laser pulse excitation intensities, while the fluorescence yield decreases strongly with increasing pulse intensity (KAMOGAWA et al. 1983). In whole chloroplasts at low temperatures we have found also that the decay times of the PS I 735 nm fluorescence is relatively insensitive to pulse intensity (GEACINTOV et al. 1977; WITTMERSHAUS et al. 1985), while the yield decreases strongly with increasing laser pulse intensity (GEACINTOV et al. 1977). The C-705 fluorescence-emitting pigments in PS I (BUTLER 1978) appear to derive their energy mostly from other, larger LH antenna pigment systems (WITTMERSHAUS et al. 1985; CAMPILLO et al. 1977b), rather than by direct absorption. However, triplet-singlet quenching studies utilizing trains of picosecond laser excitation pulses indicate that there is a pigment pool of at least 30–40 molecules surrounding the C-705 molecules (GEACINTOV et al. 1978).

If the fluorescence yield in a given photosynthetic system obeys Eq. (5), the size of the domain can be estimated, since the value of $\Phi/\Phi_0 \simeq 0.63$ when there is one hit/domain (Y = 1). It follows that the larger the domain (or particle, if there is one domain per particle) the more the Φ(Y) curve will be displaced toward lower excitation intensities. Utilizing singlet-singlet annihilation, this approach was used by VAN GRONDELLE et al. (1983) to determine the size and structure of antenna complexes of photosynthetic bacteria, and their arrangement within the membranes. BRETON and GEACINTOV (1980) utilized the singlet-triplet exciton quenching technique to determine the sizes of monomeric LH chlorophyll a/b protein complexes (directly on electrophoretic gels), and LH aggregates suspended in solutions.

3.2 Large Domains

Explicit equations describing the excitation intensity dependence of the fluorescence yield in the presence of closed (PAILLOTIN et al. 1979) and open reaction centers (DEN HOLLANDER et al. 1983) have been derived. In the case of closed reaction centers within large domains (the lake model), neglecting the term G and for single pigment systems, integration of Eq. (3) gives:

$$\Phi = \Phi_0 (r/Y) \ln[1 + (Y/r)], \qquad (6)$$

where r is a parameter which is equal to the ratio of monomolecular/bimolecular (annihilation) decay rates.

In spite of the known heterogeneity of pigment systems, Eq. (6) provides an excellent fit to fluorescence yield vs. single picosecond laser pulse intensity curves obtained with spinach chloroplasts at various temperatures and in the presence of closed reaction centers. Domain sizes in excess of $\sim$600 chlorophyll molecules were deduced from the positions of the fluorescence yield curves [Eq. (6)] along the intensity (Y) axes (GEACINTOV et al. 1977; PAILLOTIN et al. 1979, 1983). This is surprising, particularly at low temperatures where different exciton populations with different emission lifetimes (MOYA et al. 1981), and thus different r-values, are known to exist. In fact, utilizing a series of different strains of the photosynthetic bacterium *Rps. sphaeroides*, each exhibiting different fluorescence lifetimes, CAMPILLO et al. (1977a) have shown that Eq. (6) is obeyed for these systems displaying the expected dependence of Φ on Y according to Eq. (6) for different exciton lifetimes, and thus values of r. For spinach chloroplasts at low temperatures, when Φ was measured at different wavelengths corresponding to PS II and PS I emissions (GEACINTOV et al. 1977), different fluorescence quenching curves were expected for these two photosystems since τ is different for the PS I (735 nm) and PS II (685 nm) emissions. The fact that Eq. (6) fits the experimental Φ(Y) curves with the same values of r for both photosystems (PAILLOTIN et al. 1979), even though the fluorescence decay times τ are known to be different, is presently not well understood.

Recently, singlet-singlet exciton annihilation phenomena were also utilized by BAKKER et al. (1983) to determine that the number of connected photosynthetic units (per domain) is 14–17 in *R. rubrum* and is 30 in *Rps. capsulata.*

4 Annihilation and Capture by Reaction Centers: Topological Effects

When the PS II reaction centers in dark-adapted chloroplasts are open, the fluorescence yield no longer obeys the simple relationship of Eq. (6), predicted for a lake model of the photosynthetic unit, in which the reaction center complexes are presumed to be uniformly distributed in a lake of LH antenna molecules (GEACINTOV et al. 1984). In such a model it is predicted that competition between singlet exciton-exciton annihilation and exciton capture by PS II reaction centers should lead to a decreased quantum efficiency of photochemical conversion, or closing of the reaction centers, at excitation intensities at which a decreased fluorescence quantum yield due to exciton-exciton annihilation is also observed. Indeed, this type of behavior is observed in photosynthetic bacteria (BAKKER et al. 1983). Furthermore, a lake model seems to be more appropriate for these systems since the exciton decay time is exponential in *R. rubrum* and increases uniformly as the reaction centers are progressively closed (FREIBERG et al. 1984). In contrast, the behavior of the fluorescence decay profiles in *Chlorella* and in spinach chloroplasts are much more complex under analogous conditions (HAEHNEL et al. 1982), and the efficiency of filling the reaction centers seems to be independent of exciton-exciton annihilation. A puddle model appears to be more appropriate than the lake model since the intensity dependence of q, the fraction of filled reaction centers, follows the simple expression for a puddle, or small domain model (MAUZERALL 1976; GEACINTOV et al. 1984):

$$q = 1 - \exp[-Y]. \tag{7}$$

The fluorescence begins to decline significantly due to exciton-exciton annihilation at excitation energies corresponding to only 0.1 hits per reaction center. Singlet-singlet annihilation does not seem to have any effect on the efficiency of the filling of the reaction centers, since the energy dependence of q follows Eq. (7) (GEACINTOV et al. 1984). However, at the highest intensities utilized in the work of DEPREZ et al. (1983), a reduction in q due to singlet-triplet annihilation is observed.

The lack of an effect of singlet-singlet annihilation on q may be rationalized in terms of the known heterogeneity of the fluorescence. According to Eq. (3), exciton annihilation cannot compete effectively with unimolecular exciton decay processes in pigment beds for which $k_i \gg \gamma_i(ss)n_i^s$ or $\gamma_i(ss)n_i^{s(trap)}$. We propose that the fast $\sim$100 ps fluorescence component observed in low excitation intensity fluorescence decay profile measurements (HOLZWARTH, Chapter 7.3) can be identified with transfer from the LH antenna pigment bed to PS II reaction centers and the surrounding PS II antenna pigment pools. We further assume that exciton annihilation in the large LH pigment domains is negligible before the excitons are trapped by these smaller pigment pools. Once the excitons are trapped in these smaller pools, exciton-exciton annihilation can occur and is subject to the quenching laws discussed for small domains in Section 3.1. This means that, no matter how many excitons reach a given PS II pigment pool, there will be one remaining exciton which can fill the reaction center.

The topology of photosynthetic membranes in this model resembles a large lake of LH pigments in which are embedded smaller pools of antenna pigments in proximity to the PS II reaction centers (see Fig. 31, STAEHELIN Chap. 1, this Vol.). If the fast fluorescence component in chloroplasts is less sensitive to bimolecular exciton annihilation effects than the longer fluorescence decay components (HOLZWARTH, Chap. 7.3, this Vol.), then this behavior should be reflected in the fluorescence decay profiles at different excitation intensities. Specifically, the fast component in the decay profiles should be less sensitive to the intensity of excitation, while a decrease in the amplitudes of the longer components due to annihilation should be observed with increasing pulse intensity. At this writing, detailed and accurate data on the intensity dependence of fluorescence decay profiles in green plants at room temperature are not available. However, the recent investigations of WITTMERSHAUS et al. (1985) on spinach chloroplasts at 77 K, are indeed consistent with these predictions. As the intensity of 530 nm, 30 ps excitation pulses is increased from 5×10^{13} to 2×10^{16} photons cm^{-2} $pulse^{-1}$, the amplitude of the intermediate relative to the fast phase is decreased. The decay kinetics are consistent with a theoretical model in which excitations are rapidly transferred from a large LH pigment pool to pools with a smaller optical absorption cross section, identified as the proximal PS II antenna pigments, in which exciton-exciton annihilations take place.

5 Mobile Excitons Versus Localized Excitons

Exciton annihilation techniques can discriminate between excitons which are mobile, and those which are trapped, or localized within small pigment pools. On the other hand, fluorescence decay profiles reflect *all* populations of excitons according to Eq. (4); the amplitudes extrapolated back to t=0, the time of the excitation by a delta-pulse, reflect the initial populations of excitons in different pigment sub-sets. Integrated fluorescence yields provide the least detailed information on the nature and heterogeneity of the fluorescence.

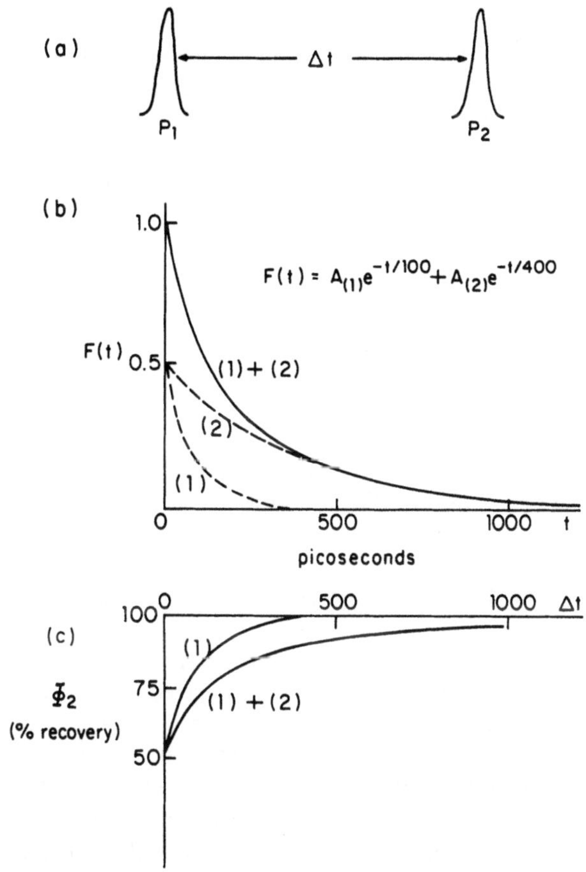

Fig. 1a–c. Schematic representation of double pulse experiments. **a** Two picosecond pulses, P_1 and P_2, are separated by a variable time interval Δt. **b** Decay of fluorescence generated by pulse P_1. The fluorescence is characterized by two components with equal amplitudes $A_{(1)}$, $A_{(2)}$, and lifetimes τ_1 and τ_2 of 100 and 400 ps respectively. *Dashed lines* decay of individual components; *solid line* observed fluorescence decay profile *F(t)*. **c** Φ_2 is the integrated fluorescence yield, due to pulse P_2, which is partially quenched by the excitons generated by pulse P_1 and which are still present at time Δt; (1) only exciton component (*1*) is active as a fluorescence quencher, while component (*2*) is due to trapped or localized excitons which are inactive as quenchers; (*1*)+(*2*) both exciton components are mobile and act as fluorescence quenchers. An initial quenching of Φ_2 of 50% has been assumed for purposes of illustration

If the initial density of excitations is sufficiently low, a condition which is usually satisfied with moderate excitation intensities of chloroplasts with 530 nm picosecond laser pulses, at least one of the excitons must be mobile if bimolecular exciton annihilation is to occur. This principle can be exploited to discriminate between those excitons [or fluorescence components, according to Eq. (4)] which are mobile and those which are localized, by utilizing a picosecond pulse pair excitation method (DOBEK et al. 1985).

In this method, the essential features of which are summarized in Fig. 1, the time delay Δt between two picosecond laser pulses P_1 and P_2 is varied, and the effect of the first pulse on the fluorescence yield Φ_2 of the second pulse is determined as a function of Δt. While the details of the experimental measurements are somewhat more complex than implied here, we have found that in dark-adapted spinach chloroplasts at room temperature (open reaction centers) the two pulses are decorrelated (no effect of P_1 on the fluorescence generated by P_2) as long as $\Delta t \gg 140 \pm 60$ ps. Accordingly, we have tentatively attributed the fast fluorescence decay phase of ~100 ps measured by others (HOLZWARTH, Chap. 7.3, this Vol.) to a mobile exciton phase, while the intermediate and slow phases are attributed to excitons localized in small pigment pools.

6 Conclusions

The energy transfer model which emerges from the presently available data can be summarized as follows:

1. Excitons generated in the LH antenna are rapidly transferred to smaller domains identified as the proximal PS II antenna pigment pools surrounding the PS II reaction centers. This is identified as a mobile exciton phase which may account for the ~100 ps component detected in low excitation intensity single photon counting measurements of fluorescence decay profiles. The similarity of this model with the one derived from freeze-fracture/biochemical studies (Fig. 31 of Chap. 1, STAEHELIN, this Vol.) is rather striking. Thus the "large LH pigment pool of PS II" in the model presented here appears to correspond to the pool of "mobile" chl a/b LHC units, and the "proximal antenna pigment pools surrounding the PS II reaction centers" to the PS II "bound" chl a/b LHC (together with the core antenna chlorophylls associated with CP 43.)

2. Excitations are also created within the PS II antenna; however, the initial number of excitons per pool is relatively low because of the smaller optical absorption cross-sections of these pools. Arrival of excitons from the larger LH pigment bed into these pools causes exciton-exciton annihilation within these proximal antenna pigment systems, and a rapid decrease of the density of excitons and the amplitude of the intermediate fluorescence component with increasing exitation intensity; the integrated fluorescence yield thus decreases as well.

3. The closing of the reaction centers in chloroplasts appears to be insensitive to singlet-singlet exciton annihilation. The capture of excitons by PS II reaction centers may be so efficient, once an exciton is localized within the smaller PS II antenna pool, that this process can compete effectively with annihilation. Similar considerations may apply to PS I reaction centers at low temperatures which, however, appear to have a lower connectivity to at least some of the LH antenna pigments (WITTMERSHAUS et al. 1985).

In bacterial photosynthetic systems, the energy transfer pathways appear to be different since exciton-exciton annihilation competes effectively with exciton capture by reaction centers. Experimental picosecond laser studies of the fluorescence yield suggest that the reaction centers are embedded in a lake of antenna pigments, and that excitons do not become localized in smaller pigment pools before annihilation occurs, as in chloroplasts.

Careful analysis of annihilation processes can provide information on the size of the pigment systems, on energy transfer pathways, and the topology of antenna pigment beds in photosynthetic systems.

Acknowledgements. This work was supported by the National Science Foundation, Grant PCM 83-08190, and in part by the Department of Energy, Contract AC02-76EV02386.

References

Bakker JGC, Grondelle R van, Hollander den WTF (1983) Biochim Biophys Acta 735:508–518

Breton J, Geacintov NE (1980) Biochim Biophys Acta 594:1–32

Butler WL (1978) Annu Rev Plant Physiol 29:345–378

Campillo AJ, Hyer RC, Monger TG, Parson WW, Shapiro SL (1977a) Proc Natl Acad Sci USA 74:1997–2001

Campillo AJ, Shapiro SL, Geacintov NE, Swenberg CE (1977b) FEBS Lett 83:316–320

Deprez J, Dobek A, Geacintov NE, Paillotin G, Breton J (1983) Biochim Biophys Acta 725:444–454

Dobek A, Deprez J, Geacintov NE, Paillotin G, Breton J (1985) Biochim Biophys Acta 806:81–92

Freiberg A, Godik VI, Timpmann K (1984) In: Sybesma C (ed) Advances in photosynthetic research, vol I. Nijhoff/Junk, The Hague, pp 45–48

Geacintov NE, Breton J (1982) In: Alfano AA (ed) Biological events probed by ultrafast laser spectroscopy. Academic Press, London New York, pp 157–191

Geacintov NE, Breton J, Swenberg CE, Paillotin G (1977) Photochem Photobiol 26:625–638

Geacintov NE, Swenberg CE, Campillo AJ, Hyer RC, Shapiro SL, Winn KR (1978) Biophys J 24:347–359

Geacintov NE, Paillotin G, Deprez J, Dobek A, Breton J (1984) In: Sybesma C (ed) Advances in photosynthetic research, vol I. Nijhoff/Junk, The Hague, pp 37–40

Grondelle R van, Neil Hunter C, Bakker JGC, Kramer HJM (1983) Biochim Biophys Acta 723:30–36

Haehnel W, Nairn JA, Reisberg P, Sauer K (1982) Biochim Biophys Acta 680:161–173

Hollander WTF den, Bakker JGC, Grondelle R van (1983) Biochim Biophys Acta 725:492–502

Kamogawa K, Morris JM, Takagi Y, Nakashima N, Yoshihara K, Ikegami I (1983) Photochem Photobiol 37:207–213
Mauzerall D (1976) J Phys Chem 80:2306–2309
Moya I, Mullet JE, Briantais JM, Garcia R (1981) In: Akoyunoglou G (ed) Photosynthesis I. Photophysical processes – membrane energization. Balaban, Philadelphia, pp 163–172
Nordlund TM, Knox WH (1981) Biophys J 36:193–201
Paillotin G, Swenberg CE, Breton J, Geacintov NE (1979) Biophys J 25:513–534
Paillotin G, Geacintov NE, Breton J (1983) Photochem Photobiol 37:475–478
Wittmershaus BP, Nordlund TM, Knox WH, Knox RS, Geacintov NE, Breton J (1985) Biochim Biophys Acta 806:93–106

7.4 Molecular Orientation of the Pigments and the Problem of Energy Trapping in Photosynthesis

J. BRETON

1 Introduction

In photosynthetic membranes the network of noncovalent bonding interactions which anchor the pigments to the protein backbone (LUTZ 1983) is responsible for the precise positioning of the pigments with respect to each other and to the protein moieties. These interactions, together with pigment-pigment interactions (within one complex and possibly also between adjacent complexes), determine the absorption and emission properties of the chromophores (see SAUER, Chap. 2, this Vol. for an overview). The amphiphilic nature of the Chl-protein complexes together with their transmembrane organization insure a well-defined orientation of these intrinsic membrane proteins with respect to the plane of the photosynthetic lamella (NABEDRYK and BRETON 1981; THORNBER, Chap. 3, this Vol.). In turn, this specific positioning of the protein scaffold has allowed a study of the precise orientations of the various antenna and reaction center pigments with respect to the membrane plane (BRETON and VERMEGLIO 1982).

In the early 1970's, linear dichroism (LD) measurements on oriented membranes of photosynthetic bacteria (MORITA and MIYASAKI 1971) and of green plants (BRETON and ROUX 1971; GEACINTOV et al. 1972) led to the first reports of a large orientation of the pigments in vivo. In the study by BRETON and ROUX it was already noticed that the long wavelength absorbing forms of Chl a were oriented closer to the membrane plane than the forms absorbing at shorter wavelengths (660–670 nm). It was later shown that the small dichroism observed around 660–670 nm was more indicative of an orientation of the corresponding chromophores at an angle close to the "magic angle" (35° away from the membrane plane) than of a random orientation. These studies also led to an estimation of $\sim$20–25° for the tilt angle of the long wavelength Q_Y transition moments of Chl a. A similar tilt has been described for the Q_Y transitions of the antenna of Bchl a- or Bchl b-containing organisms (BRETON 1974a; PAILLOTIN et al. 1979). By measuring the dichroism of the absorption changes linked to the photooxidation of the primary donors, it has been concluded that a similar tilt angle also characterized the long wavelength transitions of the traps of PS I (BRETON 1977) and of PS II (MATHIS et al. 1976; GANAGO et al. 1982) in green plants as well as of P-870 (VERMEGLIO and CLAYTON 1976) and of P-960 (PAILLOTIN et al. 1979) in photosynthetic bacteria. This similarity in the orientation of the long wavelength transitions of both the antenna and the traps led us to propose that such a geometrical arrangement of the pigments could play a role in the efficiency of the trapping process by increasing the localization of the exciton in the same plane as the one containing the traps (BRETON and GEACINTOV 1979). This hypothesis was also in line with the observation that the degree of mutual order increased toward the long wavelengths (GARAB and BRETON 1976; BECKER et al. 1976), thus suggesting that a rather parallel geometry could also exist between the antenna and the traps and contribute to the "focusing" of the exciton into the trap.

In the present chapter we will further examine this question in the light of recent experimental data on green plant material, and we intend to relate observations pertinent to both the orientation of the pigments in vivo and the energy migration and trapping processes. A discussion of the basic principles that underlie the measurement and the interpretations of polarized absorption and fluorescence spectra on oriented photosynthetic systems can be found in BRETON and VERMEGLIO (1982).

2 Orientation of the Absorption Dipoles in Vivo

The low temperature LD spectra of several Chl-protein complexes isolated from green plants have been analyzed and compared to the LD of isolated thylakoids (HAWORTH et al. 1982a, b; TAPIE et al. 1982, 1984a; TAPIE 1984). In the case of the light-harvesting complex (LHC), the dominant Chl-protein complex, the LD spectrum is identical to the difference between the LD spectrum of the wild-type and a mutant of barley which lacks LHC. Furthermore, the LD spectrum of native thylakoids can be equated to the sum of the LD spectra of LHC, PS I and PS II particles. These observations clearly indicate that the native orientations of the pigments can be preserved during the isolation and purification processes provided the appropriate detergent and procedure are selected for each complex. In Fig. 1 the LD and absorption spectra at 5 K are presented for LHC (Fig. 1a), PS I (Fig. 1b), PS II (Fig. 1c) and the thylakoid (Fig. 1d). The LD spectrum of the thylakoid is similar to the ones previously published at 77 K or 100 K except for an improved resolution, particularly noticeable in the long wavelength region.

For purified planar LHC aggregates, the main Chl a absorption band is split into two components at 675 and 670 nm with a prominent positive LD signal associated only with the 675 nm component. A very small shoulder under the 670 nm absorption band indicates an orientation of the corresponding chromophores at an angle slightly less than 35° from the plane. The main absorption band of Chl b peaking at 649 nm gives rise to a sharp negative LD signal indicative of an orientation of the corresponding chromophores rather perpendicular to the plane of the aggregate. A small absorption band at 661 nm is not correlated with a clear LD band, but rather seems to be associated with components at 656 nm (<0) and 664 nm (>0). Although part of the complexity of the LD spectrum in this region is probably caused by the overlap of vibrational components of the Chl a absorption, it seems also possible that excitonic

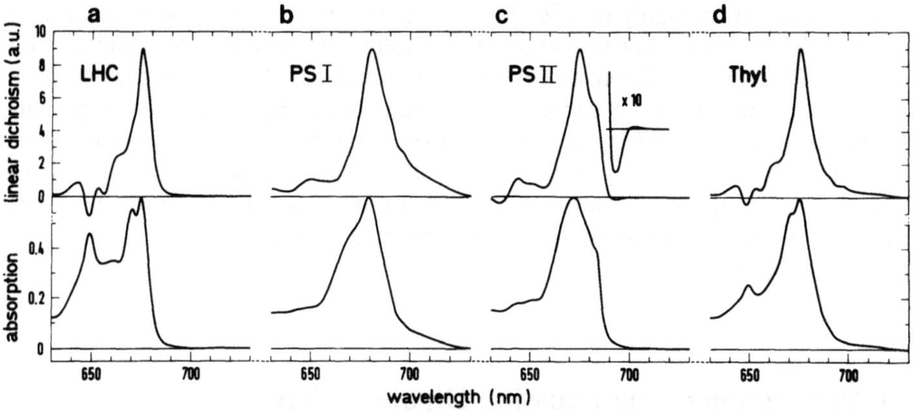

Fig. 1. Linear dichroism (*top*) and absorption (*bottom*) spectra taken at 5 K of **a** light-harvesting complex (*LHC*), **b** photosystem 1 (*PS I*), **c** photosystem 2 (*PS II*) and **d** thylakoids (*Thyl*) oriented by the polyacrylamide gel technique. The highly enriched PS II complex is from *Chlamydomonas* BF 4–14 and all the other samples are from pea

coupling between Chl b molecules contributes to the large number of bands in the LD spectrum.

For PS I particles prepared according to MULLET and ARNTZEN (1981), the LD peak at 680 nm corresponds to the main component seen in the absorption spectrum. A shoulder at 670 nm in the absorption spectrum gives no significant LD component, thus indicating an orientation close to the magic angle for this transition. The long wavelength side of the LD spectrum is complex, with several components, some of which are also observed in the absorption spectrum, notably the 696 nm and the 700–730 nm shoulders. The LD/A spectrum indicates that these two last sets of dipoles are oriented significantly closer to the main plane within the particle than the dipoles absorbing at 680 nm. It is probable that the 696 nm signal corresponds to the long wavelength transition of P-700. A small positive LD signal at 650 nm has been ascribed to Chl b included in LHC I, which exhibits an orientation very different from that observed in the main LHC.

For PS II particles, prepared according to various protocols, the low temperature LD spectra are all very similar (TAPIE et al. 1984a). The short wavelength absorption component is associated with a small positive LD signal, while the long wavelength wings of the absorption and LD spectra are rather complex. At 5 K the highly purified PS II particle from *Chlamydomonas* BF 4–14 (DINER and WOLLMAN 1980) kindly provided by Drs. DELEPELAIRE, DINER and WOLLMAN, shows a 683 nm shoulder in both the LD and absorption spectra. It is remarkable that this C-683 component is precisely located at the absorption maximum of P-680 under the same conditions (DEN BLANKEN et al. 1983). In addition a small shoulder at 680 nm in the absorption spectrum is associated with either no LD or more likely with a negative LD component which gives the 675–682 nm region of the LD spectrum its concave aspect. Furthermore a minute negative LD component at 692 nm has been detected in this (Fig. 1c) as well as in other highly purified PS II particles prepared according to different protocols. Negative LD components are clearly resolved at 680 and 692 nm in the LD/A spectra (TAPIE and BRETON unpublished). Thus, with the exception of the spectral species responsible for these small negative LD components in the spectra of the PS II particles, it appears that both in the intact thylakoids and in each of the three major isolated Chl-protein complexes, the short wavelength-absorbing Q_Y dipoles of Chl a are oriented nearly at the magic angle, while the long wavelength-absorbing dipoles are oriented closer to the long axis of the particles.

3 Orientation of the Emission Dipoles in Vivo

The low temperature emission of LHC is characterized by a main emission band at 680–681 nm with a shoulder at 698 nm, the relative amplitude of which is dependent upon the state of aggregation (MULLET and ARNTZEN 1981). By comparing on the same oriented sample the $A_{\parallel}/A_{\perp}$ and the $F_{\parallel}/F_{\perp}$ spectra, it is possible to determine which of the absorption component(s) is (are) involved in the emission. For LHC the polarized emission spectra (Fig. 2a) indicate that the 681 nm-emitting dipoles have the same orientation as the 675 nm component visible in the absorption and LD spectra (Fig. 1b). The $F_{\parallel}/F_{\perp}$ spectrum at 77 K is rather constant in the range 685–750 nm, but decreases sharply at shorter wavelength. A similar wavelength dependence is also observed at room temperature (BRETON, unpublished). These results indicate (1) that the dipoles responsible for the 670 nm absorption make a minor contribution to the total emission and (2) that the dipoles emitting at 698 nm and 680 nm both have the same orientation rather parallel to the plane of the LHC aggregate. The low temperature polarized emission from intact PS I particles is characterized by a single broad band peaking at 733 nm (Fig. 2b) and exhibiting the same anisotropy as the absorption dipoles responsible for both the 700–730 nm band and the 696 nm shoulder (TAPIE et al. 1984b). Although the absence of F-695 in various PS II preparations has been frequently reported in the past, there is compelling evidence that this emission band exists in native particles (with, however, some variability in the ratio F-685/F-695). The polarized fluorescence emission of PS II particles at 77 K is shown in Fig. 2c. F-685 appears rather parallel to the plane of the particle, while F-695 is clearly emitted from dipoles oriented perpendicular to that plane. Comparing $F_{\parallel}/F_{\perp}$ and $A_{\parallel}/A_{\perp}$ on the same oriented sample, this last emission can be correlated with the negative dip at 692 nm. Although we cannot exclude at this writing that F-685 is correlated with the set of dipoles absorbing at 675 nm, we rather favor the notion that this emission stems from the 683 nm absorption dipoles.

At room temperature the emission spectrum of chloroplasts exhibits a main band at 685 nm with a long wavelength shoulder. Working with oriented material it has been demonstrated that the fluorescence anisotropy $F_{\parallel}/F_{\perp}$ increases with wavelength within the main emission band (BRETON 1974b; GEACINTOV

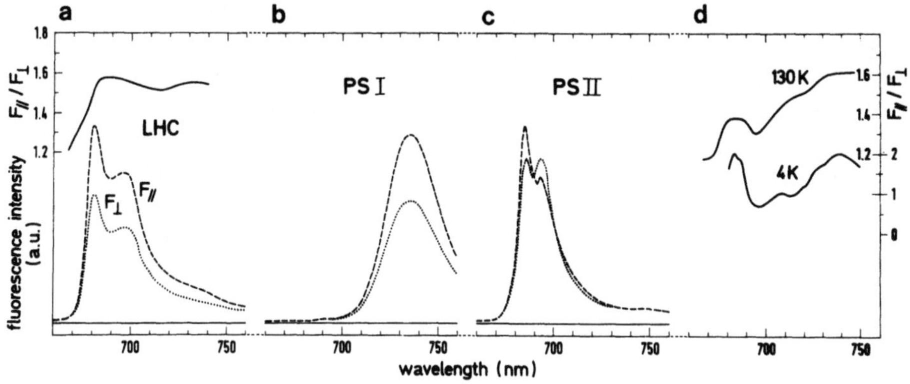

Fig. 2. Polarized fluorescence emission spectra taken at 77 K on the oriented samples of LHC **a**, PS I **b** and PS II **c** described in Fig. 1a, b and c respectively. The anisotropy ratio is shown in **a** (*top*) for LHC and in **d** for magnetically oriented spinach chloroplasts at 130 K. (After GARAB and BRETON 1976) and at 4 K. (After KRAMER and AMESZ 1982)

et al. 1974; BECKER et al. 1976). At 77 K the emission spectrum is characterized by three main bands peaking at 685 nm (F-685), 695 nm (F-695) and 735 nm (F-735) and a very small shoulder at about 680 nm (F-680). The first investigation of the low temperature polarized emission of oriented chloroplasts was carried out by GARAB and BRETON (1976) at a temperature of −140 °C, a temperature at which F-695 contribution is small. This study led to the conclusion that at least five emitting species (F-680, F-685, F-695, F-715, F-735) with different orientations were responsible for the emission spectrum. Noticing a pronounced dip at 695 nm in the $F_{\parallel}/F_{\perp}$ spectrum (Fig. 2d), they further concluded that the species responsible for this emission had either no anisotropy or was oriented preferentially perpendicular to the membrane plane. This observation was used to demonstrate that F-695 cannot arise from P-680, as proposed earlier (CHO and GOVINDJEE 1970), because an orientation at a small angle to the membrane plane had been observed for the Q_Y transition moment of P-680. More recently, similar experiments performed at lower temperatures have demonstrated that F-695 indeed originates from dipoles oriented preferentially perpendicular to the plane of the membrane (VASIN and VERKHOTUROV 1979; BRETON 1982; KRAMER and AMESZ 1982). This is especially clear in the work of KRAMER and AMESZ, because they achieved a high degree of orientation of the chloroplasts, and worked at various low temperatures, notably at 50 K, where F-695 is especially prominent. They essentially confirmed the conclusions drawn earlier by GARAB and BRETON regarding the orientation of the other main emission dipoles. Furthermore, they investigated the excitation spectra in polarized light as suggested by BRETON (1974b) for both PS II (F-695) and PS I (F-735) emissions, thereby gathering information on the orientation of the absorption dipoles feeding each of the two photosystems. For F-695 their difference polarized excitation spectrum at 5 K is almost identical to the LD spectrum of LHC (Fig. 1a), with the exception of a shoulder at 683 nm which is absent in the LD spectrum of LHC. In addition, their difference polarized excitation spectrum for F-735 is quite similar to our LD spectrum for intact PS I (Fig. 1b), although the contribution of the 700–730 nm component relative to the contribution of the species absorbing maximally at 680 nm is much larger (by a factor of about 3) in their case than in our LD spectra. This enhancement of the contribution of the long wavelength components (C-683 in PS II, C-700–730 in PS I) in the polarized excitation spectra compared to the LD spectra most probably reflects the decrease in the efficiency of the back-transfer of excitons from traps to the main antenna pigment bed at low temperature.

4 Assignement and Molecular Origin of the Fluorescence Bands in Vivo

While there is much evidence, including data from polarized spectroscopy, that F-735 originates from the C-700–730 pool of PS I pigments, the assignment of the other emission bands to specific subsets of pigments is more controversial. GULYAEV and TETEN'KIN (1980) attribute F-680 to C-670 of LHC, and KRAMER and AMESZ (1982) assign F-685 to C-675, whereas F-695 has been attributed to PS II traps (CHO and GOVINDJEE 1970), to C-683 of PS II (KRAMER and AMESZ 1982), to antenna close to the traps (BUTLER 1979), to LHC (MULLET and ARNTZEN 1981) or to the primary pheophytin (Ph) acceptor of PS II (BRETON 1982).

The identical anisotropy observed for C-675 and F-680 in LHC leads us to propose that F-680 in chloroplasts at 77 K is mostly emitted from C-675 of LHC with only a small contribution of C-670 which is responsible for the 670–680 rise in the $F_{\parallel}/F_{\perp}$ spectra (Fig. 2d). Although our data cannot exclude that C-675 of LHC in vivo emits specifically F-685, we note that in chloroplasts this component is very pronounced in the excitation spectra of both F-685 and F-695 (KRAMER et al. 1981), and accordingly we prefer to assign it to C-675 of LHC transferring its energy of the PS II emitters. The fluorescence anisotropy spectra of KRAMER and AMESZ (1982) show indications of the presence of at least two emission bands in the vicinity of 685 nm (Fig. 2d). This leads us to tentatively assign the shorter wavelength emission to C-675 of PS II and the longer (~687 nm) to the C-683 nm component seen in the 5 K spectra of PS II (Fig. 1c). The possibility that the 698 nm emission of isolated LHC can significantly contribute to F-695 nm in vivo can be excluded by the opposite orientation of these two emitters. Because we see no evidence in the 5 K LD spectra of thylakoids, LHC or PS II for a 683 nm absorption component with a negative LD, we provisionally dismiss the assignment of F-695 to this species (KRAMER and AMESZ 1982), and we prefer to ascribe this emission to a minor species responsible for both the small shoulder at 680 nm in the absorption spectra, and the negative dips at 692 nm and 680 nm in the LD spectra of

PS II (Fig. 1c). Noticing that the Q_Y transition of Ph is oriented perpendicular to the membrane both in plants (GANAGO et al. 1982) and in bacteria (VERMEGLIO and CLAYTON 1976; PAILLOTIN et al. 1979), and extending to low temperature the charge recombination model of SHUVALOV and KLIMOV (1976), we have proposed (BRETON 1982) that F-695 arises from direct radiative deactivation of Ph* populated by back-reaction in the radical pair P-680$^+$Ph$^-$. In view of the data presented here, we assign the 680 nm component seen in the absorption and the LD/A spectra of PS II to Ph molecule(s) with a perpendicular orientation. Furthermore we tentatively attribute the 692 nm dip in the LD spectrum of PS II (Fig. 1c) to a charge transfer band (P-680$^+$Ph$^-$)* slightly lower in energy than the state (P-680* Ph) and somewhat similar to the charge transfer bands recently described for bacterial reaction centers (MASLOV et al. 1983). In this scheme F-695 is emitted from the charge transfer state itself and keeps the same anisotropy as Ph. In addition, our hypothesis on the origin of F-695 has been extended (BRETON 1983) to propose that (1) a similar back-reaction mechanism can occur in open traps (P-680 PhQ) and (2) F-685 at low temperature could also arise from P-680* either directly or via a small pool of core antenna strongly coupled to P-680. On the basis of the polarization data presented here, we suggested that the few Chl a forming C-683 can play this role. With about 40 Chl a per PS II particle, we estimate that only 2 to 8 Chl a can contribute to C-683. It has been already noticed (GARAB and BRETON 1976; BRETON 1977) that the anisotropy of F-735 and P-700 are very similar. Furthermore, it has been shown that the Q_Y transition of P-700 and P-680 have the same orientation with respect to the plane of the membrane (MATHIS et al. 1976). The observation that at 5 K the anisotropy of F-685 and F-735 is identical (Fig. 2d) thus shows that the F-685-emitting dipoles at 5 K have the same orientation as P-680. This result indicates that F-685 may originate from P-680 either alone or via a few (≤ 6) antenna Chl a having the same orientation as P-680.

Thus with the exception of F-695 and of minor bands attributed to Ph in PS II, the low temperature polarized spectroscopy data point toward a general decrease of the tilt angle of the Q_Y dipole moments when going from short to long wavelength forms of Chl. Furthermore, these long wavelength forms in PS I (C-700–730) and in PS II (C-683) make the same angle ($\sim$20°–25°) with respect to the membrane plane as the Q_Y transitions of the photochemical traps. This property is not limited to green plants, as very similar observations have also been made in photosynthetic bacteria. We further note that the orientation perpendicular to the plane of the membrane of the Q_Y transition of the Ph molecules, primary acceptors in PS II and in photosynthetic bacteria, does not favor the direct energy transfer from the antenna onto the primary acceptor. These observations reinforce the hypothesis (BRETON and GEACINTOV 1979) that such a geometrical arrangement of the antenna and the trap must play a role in the efficiency of exciton trapping by the primary donor of the reaction center.

Acknowledgements. I thank my students P. TAPIE and Y. CHOQUET for their valuable contribution to the experimental work reported here.

References

Becker JF, Breton J, Geacintov NE, Trentacosti F (1976) Biochim Biophys Acta 440:531–544

Blanken HJ den, Hoff AJ, Jongenelis APJM, Diner BA (1983) FEBS Lett 157:21–27

Breton J (1974a) Biochem Biophys Res Commun 59:1011–1017

Breton J (1974b) In: Avron M (ed) Proc 3rd Int Congr Photosynth. Elsevier, Amsterdam New York Oxford, pp 229–234

Breton J (1977) Biochim Biophys Acta 459:66–75

Breton J (1982) FEBS Lett 147:16–20

Breton J (1983) FEBS Lett 159:1–5

Breton J, Geacintov NE (1979) Ciba Found Symp 61, New Ser. Excerpta Medica, Amsterdam, pp 217–236

Breton J, Roux E (1971) Biochem Biophys Res Commun 45:557–563

Breton J, Vermeglio A (1982) In: Govindjee (ed) Photosynthesis: Energy conversion by plants and bacteria, vol I. Academic Press, London New York, pp 153–194

Butler WL (1979) Ciba Found Symp 61, New Ser. Excerpta Medica, Amsterdam, pp 237–256

Cho F, Govindjee (1970) Biochim Biophys Acta 205:371–378

Diner BA, Wollman F-A (1980) Eur J Biochem 110:521–526

Ganago IB, Klinov VV, Ganago AO, Shuvalov VA, Erokhin YE (1982) FEBS Lett 140:127–130

Garab GI, Breton J (1976) Biochem Biophys Res Commun 71:1095–1102

Geacintov NE, Nostrand F van, Becker JF (1972) In: Forti G, Avron M, Melandri A (eds) Proc 2nd Int Congr Photosynth Res, vol I. Junk, The Hague, pp 283–290

Geacintov NE, Nostrand F van, Becker JF (1974) Biochim Biophys Acta 347:443–463

Gulyaev BA, Teten'kin VL (1980) Dokl Akad NauK SSSR (Engl Transl) 250:24–27

Haworth P, Tapie P, Arntzen CJ, Breton J (1982a) Biochim Biophys Acta 682:152–159

Haworth P, Tapie P, Arntzen CJ, Breton J (1982b) Biochim Biophys Acta 682:504–506

Kramer HJM, Amesz J (1982) Biochim Biophys Acta 682:201–207

Kramer HJM, Amesz J, Rijgersberg CP (1981) Biochim Biophys Acta 637:272–277

Lutz M (1983) In: Clark RJH, Hester RE (eds) Adv in infrared and Raman spectroscopies. Wiley, New York, pp 211–300

Maslov VG, Klevanik AV, Ismailov MA, Shuvalov VA (1983) Dokl Acad Nauk SSSR (Engl Transl.) 269:87–90

Mathis P, Breton J, Vermeglio A, Yates M (1976) FEBS Lett 63:171–173

Morita S, Miyasaki T (1971) Biochim Biophys Acta 245:151–159

Mullet JE, Arntzen CJ (1981) Biochim Biophys Acta 589:100–117

Nabedryk E, Breton J (1981) Biochim Biophys Acta 635:515–524

Paillotin G, Swenberg CE, Breton J, Geacintov NE (1979) Biophys J 25:513–534

Shuvalov VA, Klimov VV (1976) Biochim Biophys Acta 440:587–599

Tapie P (1984) Thèse Doct, Etat Univ Paris XI

Tapie P, Haworth P, Hervo G, Breton J (1982) Biochim Biophys Acta 682:339–344

Tapie P, Acker S, Arntzen CJ, Choquet Y, Delepelaire P, Diner B, Wollman F-A, Breton J (1984a) In: Sybesma C (ed) Proc 6th Int Photosynth Congr, vol I. Nijhoff/Junk, pp 693–696

Tapie P, Choquet Y, Breton J, Delepelaire P, Wollman F-A (1984b) Biochim Biophys Acta 767:57–69

Vasin YA, Verkhoturov VN (1979) Biofizika 24:260–263 (Transl Biophysics 24:269–273)

Vermeglio A, Clayton RK (1976) Biochim Biophys Acta 449:500–515

7.5 Excitation Transfer in Phycobiliproteins

H. SCHEER

1 Introduction

One of the major strategies in the competition of photosynthetic organisms for light is qualitative and quantitative adaptation in their light-harvesting apparatus (THORNBER, Chap. 3; STAEHELIN, Chap. 1; ANDERSON, Chap. 6.4, all this Vol.). This includes, in particular, the usage of pigments which harvest light efficiently in the spectral regions not covered by the chlorophylls a and b present in higher plants, e.g., the bacteriochlorophylls absorbing well beyond 700 nm (COGDELL, Chap. 6.2, this Vol.), and the phycobiliproteins and carotenoids absorbing in the "chlorophyll trough" between 470 and 630 nm (GANTT, Chap. 6.3, this Vol.). Among these, the phycobiliproteins are to date probably the best understood. They are not only the major light-harvesting pigments of cyanobacteria, red algae, and cryptophytes (see STAEHELIN, Chap. 1, this Vol., for a general characterization of these species), microorganisms which are responsible for a large fraction of the net photosynthetic production on earth, but they are also readily accessible and pleasing to the eye. Different aspects of phycobiliprotein research (GANTT 1979, 1981; GLAZER 1980, 1983; MACCOLL 1982; RÜDIGER 1979; SCHEER 1981, 1982; WEHRMEYER 1983; ZUBER 1978) and related topics (BRASLAVSKY et al. 1983; PRATT 1978, 1982; RÜDIGER 1980; RÜDIGER and SCHEER 1983) have been reviewed in the past 5 years. In this volume, GANTT (Chap. 6.3) describes the structure and function of phycobilisomes and ZUBER (Chap. 6.1) the primary structure of phycobiliproteins. This chapter provides a brief overview of the energy transfer in biliproteins. Due to space limitations, only few citations could be incorporated. They are a biased selection from a large body of work, and should be used as a source for further references.

2 Structure of Phycobiliproteins

Most phycobiliproteins contain one or both of the chromophores, phycocyanobilin (Fig. 1a) or phycoerythrobilin (Fig. 1b), but additional chromophores of yet unknown structure are present, for example, in red algae, phycoerythrins ("phycourobilin"), in phycoerythrocyanins, and in particular in biliproteins from cryptophytes. Both chromophores are linked to their apoproteins via one, and in some cases two (RAPOPORT and GLAZER 1984) stable thioether bond(s)

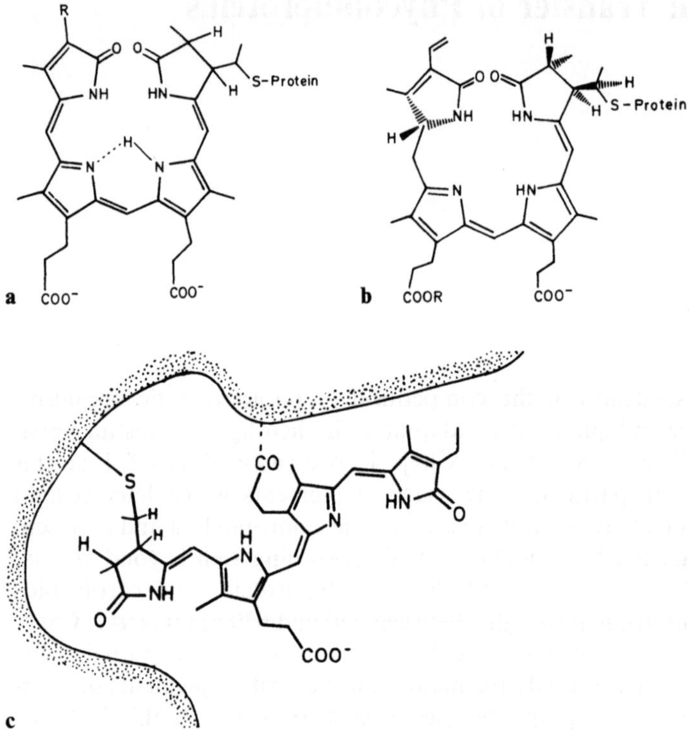

Fig. 1. Chromophores of phycobiliproteins. Molecular structures of **a** phycocyanobilin, the chromophore of PC and APC, **b** phycoerythrobilin, the main chromophore of phycoerythrins. A hypothetical conformation of the native phycocyanobilin in PC is shown in **c** (see SCHIRMER et al., 1985, for three dimensional structures of PC chromophores)

at rings A or D to cystein residues, as has been inferred from several independent lines of evidence (RÜDIGER 1979; LAGARIAS et al. 1979; ZUBER 1978; KÖST-REYES and KÖST 1979; FREIDENREICH et al. 1978), including a recent X-ray structure (SCHIRMER et al. 1985). Additional, more labile bonds have been postulated to exist based on conflicting results (see SCHEER 1982). Studies with a new, mild chromophore cleavage method have recently shown the absence of a second bond in two phycocyanins (KUFER et al. 1983).

The properties of the native biliprotein chromophores are profoundly influenced by noncovalent interactions with the protein (see SCHEER 1982). Both the intense visible absorption and fluorescence of the phycobiliproteins, which are crucial for their functions, are caused by these interactions, and are absent in the denatured pigments and in free chromophores (see below). The phycobiliproteins thus present a good example of "molecular ecology", e.g., the optimization of cofactors to specific functional needs by interactions with the apoproteins (SCHEER 1982).

All biliproteins are composed of two to three chromophore-bearing subunits. To date, the primary structures are known for more than ten different subunits, including the α- and β-subunits of all biliproteins from the cyanobacterium, *Mastigocladus laminosus* e.g. allophycocyanin (APC), phyocyanin (PC) and phycoerythrocyanin (PEC) (FÜGLISTALLER et al. 1983), and the red (?) alga, *Cyanidium caldarium* (APC and PC) (OFFNER and TROXLER 1983; see also ZUBER, Chap. 6.1, this Vol.). In the pigments from cyanobacteria and red alga, the α-subunit (≈17 kDa) bearing 1–2 chromophores and a slightly larger β-subunit carrying 1–4 chromophores, are always present in a 1:1 molar ratio. Additional chromophore-carrying γ subunits of generally much larger size are known in some APC's and in red algal PE's (phycoerythrin) in $(\alpha\beta)_n\,\gamma$ stoichiometry (n = 3 or 6) (see GLAZER 1983 and SCHEER 1982). The fundamental structural units, e.g., $\alpha\alpha'\beta$ (GANTT 1979; MÖRSCHEL and WEHRMEYER 1975) of the cryptophyte biliproteins have a different composition.

The chromophore composition is most simple in the cyanobacterial biliproteins, which carry only one type of chromophore on any given subunit. Subunits bearing two different chromophores are common in the more highly evolved eukaryotic red algae and cryptophytes. The chromophore-binding sequences are known for many phycobiliproteins (see above). Together with immunochemical studies and the complete peptide sequences cited above, an evolutionary family tree has been constructed which relates all phycobiliprotein polypeptides to a common ancestral gene (see ZUBER Chapter 6.1, GLAZER 1980; WEHRMEYER 1983).

Isolated biliproteins of cyanobacteria and red algae have a strong tendency to aggregation. Trimers (= heterohexamers) $(\alpha\beta)_3$ or hexamers (= heterododecamers) $(\alpha\beta)_6$ are generally observed (see MACOLL and BERNS 1981; SCHEER 1982). The former seem to be the largest aggregates in the absence of linkers. In vivo, they are organized within complex structures, the phycobilisomes (PB'somes) (GANTT and CONTI 1966). A single PB'some can bear up to 2700 chromophores of different types (GANTT Chap. 6.3, this Vol.) and represents the functional antenna unit of these organisms. In addition to the chromophore-bearing biliproteins, PB'somes contain several, generally colorless, "linker" peptides, which play a crucial role in their organization and their attachment to the membrane (see GLAZER 1983). Much of the basic PB'some structure with an APC core and outer rods containing PC and PE (see Fig. 2 in GANTT, Chap. 6.3, this Vol. for a diagram of the different types of PB'somes) has been obtained by combining high-resolution electron microscopy with biochemical methods (MÖRSCHEL et al. 1980; GLAZER et al. 1979; SIEGELMAN and KYCIA 1982; WANNER and KÖST 1980). PB'somes are located on the outer surface of the thylakoid membrane in a dense packing, but apparently without cooperation between different PB'somes.

Very little is known about the quaternary structure of the cryptophyte biliproteins. Cryptophytes do not have PB'somes; instead their biliproteins are located on the inner surface of the thylakoids and reveal virtually no substructure in the electron microscope (see Figs. 3 and 19 in STAEHELIN Chap. 1, this Vol.). This is paralleled by a decreased tendency of the isolated pigments to aggregate (see GANTT 1979; MACCOLL and GUARD-FRIAR 1983). Preliminary X-ray results have recently been published (MORISSET et al. 1984).

3 Photophysics and Photochemistry of the Chromophores

Denatured biliproteins, in which the chromophores are uncoupled from the protein, as well as free bile pigments of similar structures, have flexible chromophores which are predominantly present in solution in a cyclic-helical conformation (SCHEER 1982; BRASLAVSKY et al. 1983). These conformers have absorptions of moderate intensity in the visible range ($\varepsilon \approx 17{,}000\ M^{-1}\ cm^{-1}$ at neutral pH) (see BRANDLMEIER et al. 1981), low fluorescence yields ($\Phi \leq 10^{-3}$ see BRASLAVSKY et al. 1983) with corresponding fast excitation decays (SCHNEIDER, unpublished) and low phosphorescence intensities (LAND 1979). The deexcitation is governed by internal conversion, which is promoted by one or more of the following mechanisms: proton transfer (FALK and NEUFINGERL 1979; FRIEDRICH et al. 1981), vibrations, or rotations around single bonds (KUFER et al. 1983), and by photoreactions, e.g., to the labile 10 E isomers (BRASLAVSKY et al. 1983). This deactivation is only hindered if the above mechanisms are blocked, e.g., by a rigid fixation (KUFER et al. 1983), distortion of the chromophores (FALK and THIRRING 1980), or (photo)selection of more rigid and extended conformers (BRASLAVSKY et al. 1983). In these cases, the lifetimes of the excited states are increased to allow high yields of fluorescence, or photochemical reactions to products like the 4 E– or 15 E isomers which are stable at ambient temperatures.

In the biliproteins, such conformers are predominant. The chromophores are probably rigidly fixed in an extended and twisted conformation near the surface of the protein (SCHEER 1982). This rather general picture has been detailed recently by the current X-ray analysis of a PC (SCHIRMER et al. 1985). The energetically unfavorable conformations must be stabilized by noncovalent interactions with the polypeptide chains (SCHEER 1982). The absorption in the visible range is thus increased by a factor of ≈ 5, and fluorescence by several orders of magnitude to $\Phi \geq 0.5$ (see GRABOWSKI and GANTT 1978a). The phycobiliproteins are photochemically stable at ambient temperatures. It is, however, of interest to note that the photoreactive, light-sensing plant pigment phytochrome has a chromophore structure and protein interaction that resembles PC, and that partial denaturation renders phycobiliproteins to a certain extent photoreactive by partial uncoupling from the protein (BJÖRN 1979; BRASLAVSKY et al. 1983).

In spite of the large number of chromophores present in phycobiliproteins, there are generally only moderate interchromophore interactions. Strong coupling between chromophores has been established for the pigments in cryptophytes (HOLZWARTH et al. 1983; JUNG et al. 1980; KOBAYASHI et al. 1979), and may also be present in some of the cyanobacterial and red algal ones, but many of the isolated phycobiliproteins do not show good evidence of such coupling as judged from their CD spectra. However, aggregation is often paralleled by the rise of S-shaped CD signals which are indicative of strong coupling. Since the chromophores are probably located at or near the surface of the native biliproteins, this would mean that the nearest neighbors in situ may not be the chromophores on the same subunits or heterodimers, but rather on different ones [see STAEHELIN, Chap. 1, this Vol. for a discussion of phycobilisome packing in vivo and the recent x-ray results of SCHIRMER et al. (1985)].

4 The Energy Transfer Chain

The participation of phycobiliproteins in photosynthesis and, in particular, their association with photosystem II was originally recognized by action spectroscopy (EMERSON 1958). Energy transfer from biliproteins to chlorophylls has also been demonstrated in solution (FRACKOWIAK et al. 1979; KRASNOVSKII and EROKHINA 1969). The discovery of PB'somes (GANTT and CONTI 1966) marks the beginning of an understanding of phycobiliprotein function in cyanobacteria and red alga on a molecular basis. PB'somes collect light efficiently in the wavelength range between 480 and 630 nm, but their fluorescence is emitted almost exclusively (≥95%) by a small fraction of a minor constituent, e.g., APC (λ_{max}, absorption ≈665 nm, λ_{max}, emission ≈680 nm). If attached to the photosynthetic membrane, the energy is transferred further to the chlorophyll-containing reaction centers in the membrane, preferentially to PS II (GANTT 1981).

In the original PB'some model derived from electron microscopic and biochemical studies of the red alga, *Porphyridium cruentum*, a phycobilisome somewhat resembles an onion cut in half: a core, made up of APC, is surrounded by layers of PC and, further out, PE. Although this model has been considerably refined by applying improved techniques and working with organisms having less complex phycobilisomes (see above), the basic elements have remained unchanged: there is a morphological ordering from PE to PC to APC when going from the periphery to the center of the PB'some (see Fig. 2 in GANTT, Chap. 6.3, this Vol.).

This sequence corresponds perfectly to the decrease in excitation energy among the three pigments, which can thus form an energy transfer chain in which the fluorescence of any preceding member overlaps reasonably well with the absorption of the next one (Fig. 2). This model was refined by fluorescence polarization experiments first undertaken by TEALE and DALE (1970), which suggested part of an energy transfer chain to be present already within each individual biliprotein. The absorptions of PC's and PE's can be resolved into components of slightly different energies. These are assigned to individual chromophores of often the same molecular structure, but with slightly different absorption energies due to different interactions with the protein. The ones absorbing at shorter wavelengths act as sensitizers (s) to the ones absorbing and fluorescing at longer wavelengths (f chromophores), which provides an efficient fine-tuning of the energy transfer.

In cyanobacteria, each individual pigment represents only a rather short fragment of such an energy transfer chain. PC has, for example, one s chromophore on the β- and one f chromophore on both the α- and β-subunits. The full transfer chain is completed by the aggregation with APC and PE in the PB'some. The cryptophytes represent the other extreme. They do not have PB'somes, instead they have biliproteins which carry sufficiently different types of chromophores to establish an energy transfer over the full absorption spectrum of biliproteins. The red algae, finally, use a combination of both. They have PB'somes in which the major biliproteins already span a wide absorption range.

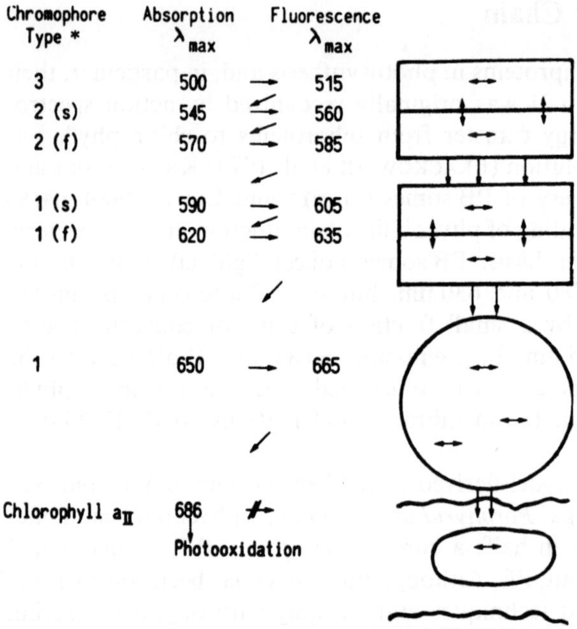

Fig. 2. Energy transfer scheme of a phycobilisome. *Right* a diagramatic representation of a PB'some with only one outer rod containing B-PE (*top double disk*) and C-PC (*lower double disk*). *Vertical arrows* correspond to longitudinal energy transfer, with the downward direction strongly favored due to a larger overlap integral; *horizontal arrows* to transversal energy transfer within the disks. *Left* approximate absorption and fluorescence maxima of the chromophores. *: 1 = phycocyanobilin, 2 = phycoerythrobilin, 3 = phycourobilin

The energy transfer chain is completed by the transfer to the chlorophylls within the photosynthetic membrane. Both the final donor in the PB'some and the acceptor in the membrane have not yet been firmly established. At present, the best candidates seem to be a large chromopeptide of the APC core (GINGRICH et al. 1983; REDLINGER and GANTT 1982; RUSCHKOWSKI and ZILINSKAS 1982) and a chlorophyll protein close to or even identical with P-680, respectively (CLEMENT-METRAL and GANTT 1983, see also GANTT, Chap. 6.3, this Vol.).

5 Dynamics of Energy Transfer

A large fraction of the energy transfer in phycobiliproteins proceeds probably via a FÖRSTER type process (see SAUER, Chap. 2; KNOX, Chap. 7.1, both this Vol.). This has been deduced from the efficiency of the energy transfer on one hand, and the low occurrence of strong couplings on the other. GRABOWSKI and GANTT (1978a, b) have critically investigated a series of biliproteins and calculated FÖRSTER's critical distances, which are mostly in the range of 40–60 Å, well beyond the diameter of the phycobiliprotein subunits (≈30 Å). The formalism of FÖRSTER (1949) has been developed for a randomly ordered and fluctuating system with high acceptor and low donor concentrations. Since probably none of these criteria applies to the phycobiliproteins (see above), a concise theoretical treatment of the process is not yet possible (BLUMEN and MANZ 1979; PEARLSTEIN 1982), but the critical radius is expected to be even larger based on still incomplete data on the chromophore orientations (see e.g., ZICKENDRAHT et al. 1980; SCHIRMER et al. 1985; GILLBRO et al. 1983). The second critical factor, i.e., the overlap integral between the donor fluorescence and the acceptor absorption, is also fine-tuned in the biliproteins (see above).

Recent applications of picosecond time-resolved techniques to the energy transfer process in biliproteins have principally verified this concept. The first detailed investigations by SEARLE et al. (1978) established in *Porphyridium cruentum* a sequential flow of excitation energy from PE via PC and APC to chlorophyll. The observed rates of excitation energy transfer could be fit by the $\exp(t^{-1/2})$ decay law characteristic for the classical FÖRSTER tansfer. Although the actual kinetics can be fit well with a concise set of rise and decay times for the PB'somes of this organism (WENDLER et al. 1984), the physical laws governing the decay have been questioned. The better signal-to-noise ratios of the current experiments still do not allow an unequivocal discrimination between multi-exponential $[\sum A \cdot \exp(kt)]$ and non-exponential decay laws, but generally favor the former. This is in particular supported by the observation of corresponding decay and rise time constants in pairs of donors and acceptors, respectively (WENDLER et al. 1984).

Qualitatively similar results have been obtained with entire PB'somes, as well as with isolated phycobiliproteins and their subunits from several cyanobacteria and red algae (HEFFERLE et al. 1984a, b; HOLZWARTH et al. 1982; KOBAYASHI et al. 1979; PELLEGRINO et al. 1981; SUTER et al. 1984; SWITALSKI and SAUER 1984; WONG et al. 1981; YAMAZAKI et al. 1984).

Photosynthetic antenna systems are optimized for high absorption cross-sections and cooperativity. They thus present a problem for the interpretation of time-resolved measurements, because $^1S-{}^1S$ annihilation can compete efficiently with energy transfer (see BRETON and GEACINTOV 1980; GEACINTOV and BRETON, Chapter 7.4). This process has been demonstrated in phycobiliproteins and investigated in detail by ALFANO's group (ALFANO 1982; PELLEGRINO et al. 1981; WONG et al. 1981). It can be distinguished from conventional excitation energy transfer ($^1S-{}^0S$) by its dependence on the intensity of the exciting light. More recent studies have thus focused on techniques which use only low photon fluxes, e.g., repetitive streak cameras (HEFFERLE et al. 1984a, b) or single-photon timing (HOLZWARTH et al. 1982; WENDLER et al. 1984; YAMAZAKI et al. 1984; HOLZWARTH, Chap. 7.3, this Vol.). An advantage of the former is the ready availability of polarized data, whereas the latter has an exceptionally high dynamic range and can thus detect even minor contributions.

Several criteria have been used to identify energy transfer in biliprotein antenna systems. The first is the analysis of time-resolved depolarization. Due to their rigid binding to proteins, rotational depolarization is generally thought to be absent on the subnanosecond time scale, and depolarization is thus believed to be due to energy transfer among chromophores (GILLBRO et al. 1983; HEFFERLE et al. 1983, 1984a, b). It should be noted, however, that segmental movements of the chromophores (HEFFERLE et al. 1984b) and nonparallel excitation and emission dipoles (SWITALSKI and SAUER 1984) have been implicated to rationalize the kinetics of some isolated biliproteins and their subunits. It is also not trivial to obtain energy transfer time from the raw data (HEFFERLE et al. 1984a). The second criterion is the correspondence of decay time constants in the donors with rise time constants in the acceptors, which are separated by wavelength selective excitation and emission (HOLZWARTH et al. 1982; WENDLER et al. 1984; YAMAZAKI et al. 1984).

The following conclusions can be drawn from the results obtained until now: the energy transfer from PE to the terminal APC chromophore proceeds at a time scale $\leq$100 ps (e.g., 60–70 ps in *Porphyridium cruentum*, WENDLER et al. 1984; YAMAZAKI et al. 1984). It follows the pigment sequence expected from their energetic and morphological ordering within the PB'somes. The energy transfer from PE to PC appears to be faster than that from PC to the APC core (SUTER et al. 1984). In isolated biliproteins, the transfer times increase with decreasing aggregation, e.g., from 70 ps in the trimer $(\alpha\beta)_3$ to $\approx$500 ps in monomeric $(\alpha\beta)_1$ PC from *Mastigocladus laminosus* (HEFFERLE et al. 1984b). The transfer times are also dependent on the species, the isolation procedure and temperature. An intriguing conclusion might be drawn from the comparison of isolated biliproteins with phycobilisomes, e.g., that the physiologically important longitudinal energy transfer toward the APC core is faster than the potentially wasteful transversal transfer within the platelets of the phycobilisome.

Currently, kinetic data are only available for a single cryptophytan pigment, i.e., PC-645 from *Chroomonas* sp. (HOLZWARTH et al. 1983; KOBAYASHI et al. 1979). The findings support again a somewhat different behavior of the cryptophytan pigments compared to the PB'some-forming biliproteins. Differences between the absorption and fluorescence excitation spectra have been discussed in terms of a heterogeneous population of f chromophores, albeit with the same decay kinetics, and the results support the presence of strongly coupled chromophores (JUNG et al. 1980).

6 Physiological Status

Photosynthetic organisms are dynamic systems with the capacity to react to environmental changes. These changes include the structure of the phycobilisomes, and their light-harvesting and energy-transfer capacities. The best investigated feature is the chromatic adaptation of many cyanobacteria, which involves changes in the composition and structure of the phycobilisomes in response to changes in the qualitity of light (BJÖRN 1979; GANTT 1981; GLAZER 1983). Other important factors are the nutritional and developmental status of the organisms (CSATORDAY 1978). Considerably less is known about the changes in the energy transfer kinetics in these systems. Using fluorescence spectroscopy CSATORDAY (1978) has followed the repigmentation of *Anacystis nidulans* cells after nitrogen starvation. This approach allows for the detection of particular biliproteins before they have been incorporated into the antenna system. HARNISCHFEGER and CODD (1978) have observed a decrease in biliprotein fluorescence after illumination of dark-adapted cells of the same organism, which was correlated with an increased chlorophyll fluorescence. Changes of the fluorescence yields (FORK et al. 1982) and kinetics (KARUKSTIS and SAUER 1984) upon treatment of algal cells with DCMU or desiccation, indicate the involvement of different chlorophyll containing antennas, and possibly also different PS II populations. This changed coupling between phycobilisomes and membrane-bound pigments probably contributes to the chromatic transients in cyanobacteria as well. These few examples have been chosen to illustrate that the physiological status of the cells (as well as the isolation procedures of PB'somes and their components) may produce significant changes in energy transfer reactions, which must be considered in the interpretation of the physico-chemical data.

Acknowledgements. The cited work of the author was supported by the Deutsche Forschungsgemeinschaft. I thank Drs. A.R. HOLZWARTH and S. SCHNEIDER for critically reading the manuscript.

Notes Added in Proof: The field has been advancing rapidly after finishing the manuscript for this mini review. Important aspects should be referenced.

The crystal structure has been determined for a second phycocyanin, viz. from *Agmenellum quadruplicatum*, which is hexameric instead of trimeric, but otherwise very similar to that of *M. laminosus* (T. SCHIRMER, W. BODE, M. SCHNEIDER, W. BODE, M. MILLER, M.L. HACKERT, J. Mol. Biol., in press). The details of chromophore structures and the linkage to the apoprotein have been determined by chemical methods, including the high-resolution NMR of bilipeptides (J.O. NAGY, J.E. BISHOP, A.V. KLOTZ, A.N. GLAZER, H. RAPOPORT, J. Biol. Chem. 260, 4864–4868 (1985)), milder degradation methods (W. KUFER, G. SCHMIDT, O. SCHMID, H. SCHEER, Z. Naturforsch., in press) and sequencing studies (W. SIDLER, B. KUMPF, H. ZUBER, W. RÜDIGER, Abstr. Vth. Int. Symp. on Photosynthetic Procaryotes (ed. H. ZUBER, p. 303, 1985, ETH Zürich). The results include the structure of the phycourobilin chromophore in R-PE and the characterization of singly and doubly bound chromophores, depending on the species and the chromophore attachment site.

An assignment of spectrally distinct chromophores to the different chromophore sites of PC from *M. laminosus* has been suggested on the basis of fluorescence polarization, circular dichroism and photochemical reactivities (M. MIMURO, P. FÜGLISTALLER, R. RÜMBELI, H. ZUBER, Abstr. Vth. Int. Symp. on Photosynthetic Procaryotes (ed. H. ZUBER, 1985, p. 276, ETH Zürich; W. JOHN, R. FISCHER, S. SIEBZEHNRÜBL, H. SCHEER, in Antennas and Reaction Centers of Photosynthetic Bacteria – Structure, Interactions and Dynamics (eds. M.E. MICHEL-BEYERLE, H. SCHEER, S. FISCHER, Springer, Berlin – Heidelberg – New York, in press).

Several new results have been presented on the excited state dynamics in biliproteins. Fast (7–10 ps) decay times of the short-wavelength fluorescence and a corresponding rise-term of the longer wavelength fluorescence has been demonstrated for PC 612 from the cryptophyte, *Hemiselmis virescens* (C.A. HANZLIK, L.E. HANCOCK, R.S. KNOX, D. GUARD-FRIAR, R. MACCOLL, J. Luminescence 34, 99–106 (1985). All groups have adopted the multi-exponential decay analysis, at least as a phenomenological approach, requiring up to 5 components depending on the system and on the signal-to-noise ratios of the data. The physical correctness of this approach has been supported by a global analysis of data sets obtained at different excitation and emission wavelengths with a single set of components (A.R. HOLZWARTH, in Antennas and Reaction Centers of Photosynthetic Bacteria – Structure, Interactions and Dynamics (eds. M.E. MICHEL-BEYERLE, H. SCHEER,

S. SCHNEIDER), Springer Verlag, Heidelberg, Berlin, New York, 1985, in press). This method allows at the same time the extraction of time-resolved spectral information. First data on the influence of linker peptides have shown only relatively small changes in the excited state dynamics of an R- and a C-PC (S. SWITALSKI, K. SAUER, pers. communication, J. WENDLER, W. JOHN, H. SCHEER, A.R. HOLZWARTH, unpublished results). The spectra of the individual components contributing to the emission have been studied in a chromatically adapting cyanobacterium, *Fremyella diplosiphon* (M. MIMURO, I. YAMAZAKI, T. YAMAZAKI, Y. FUJITA, Photochem. Photobiol. 41, 597–603 (1985). Particular attention has been paid to the relative importance of lateral and vertical energy transfer, with the latter being favored in green-light adapted cells containing PE, PC and APC. The uncoupling of phycobilisomes from the PSII-related chlorophylls has been induced by heat stress in *Anacystis nidulans* (P. MOHANTY, S. HOSHINA, D.C. FORK, Photochem. Photobiol. 41, 589–596 (1985). In an interesting application, tandem phycobiliprotein conjugates have been used as immunofluorescent probes with an unusually large Stoke's shift due to the energy transfer from PE to APC (A.N. GLAZER, J. STRYER, Biophys. J. 43, 383–386 (1983)).

References

Alfano RR (ed) (1982) Biological events probed by ultrafast laser spectroscopy. Academic Press, London New York (See in particular chaps 6: Breton J, Geacintov N; 17: Doukas et al.; 18: Shank and Greene)
Björn LO (1979) Q Rev Biophys 12:1–25
Blumen A, Manz J (1979) J Chem Phys 71:46–94
Brandlmeier T, Scheer H, Rüdiger W (1981) Z Naturforsch 36c:431–439
Braslavsky SE, Holzwarth AR, Schaffner K (1983) Angew Chem 22:670–689, Angew Chem Int Ed 20:656–674
Breton J, Geacintov NE (1980) Biochim Biophys Acta 594:1–32
Bryant DA (1982) J Gen Microbiol 128:835–844
Clement-Metral JD, Gantt E (1983) FEBS Lett 156:185–188
Csatorday K (1978) Biochim Biophys Acta 504:341–343
Emerson R (1958) Annu Rev Plant Physiol 9:1–24
Falk H, Neufingerl F (1979) Monatsh Chem 110:987–1001
Falk H, Thirring K (1980) Z Naturforsch 35b:376–380
Förster T (1949) Z Naturforsch 4A:321–327
Fork DC, Öquist G, Hoch GE (1982) Plant Sci Lett 24:249–254
Frackowiak D, Erokhina LG, Fiksinski K (1979) Photosynthetica 13:245–253
Freidenreich P, Apell GS, Glazer AN (1978) J Biol Chem 253:212–219
Friedrich J, Scheer H, Zickendraht-Wendelstadt B, Haarer D (1981) J Am Chem Soc 103:1030–1035
Füglistaller P, Suter F, Zuber H (1983) Hoppe-Seyler's Z Physiol Chem 364:691–712
Gantt E (1979) Biochem Physiol Protozoa 1:121–137
Gantt E (1981) Annu Rev Plant Physiol 32:327–347
Gantt E, Conti SF (1966) J Cell Biol 29:423–434
Gillbro T, Sundström A, Sundström V, Holzwarth AR (1983) FEBS Lett 162:64–68
Gingrich JC, Lundell DJ, Glazer AN (1983) J Cell Biochem 22:1–14
Glazer AN (1980) In: Sigman D, Brazier MAB (eds) The evolution of protein structure and function. Academic Press, London New York, pp 221–244
Glazer AN (1983) Annu Rev Biochem 52:125–157
Glazer AN, Williams RC, Yamanaka G, Schachman HK (1979) Proc Natl Acad Sci USA 76:6162–6166
Grabowski J, Gantt E (1978a) Photochem Photobiol 28:39–45
Grabowski J, Gantt E (1978b) Photochem Photobiol 28:47–54
Harnischfeger G, Codd GA (1978) Biochim Biophys Acta 502:507–513

Hefferle P, Nies M, Wehrmeyer W, Schneider S (1983) Photobiochem Photobiophys 5:325–334
Hefferle P, John W, Scheer H, Schneider S (1984a) Photochem Photobiol 39:221–232
Hefferle P, Geiselhart P, Mindl T, Schneider S, John W, Scheer H (1984b) Z Naturforsch 39c:606–616
Holzwarth AR, Wendler W, Wehrmeyer W (1982) Photochem Photobiol 36:479–487
Holzwarth AR, Wendler W, Wehrmeyer W (1983) Biochim Biophys Acta 724:388–395
Jung J, Song PS, Paxton RJ, Edelstein MS, Swanson R, Hazen Jr EE (1980) Biochemistry 19:24–32
Karukstis KK, Sauer K (1984) Biochim Biophys Acta 766:141–147
Kobayashi T, Degenkolb EO, Behrson R, Rentzepis PM, MacColl R, Berns DS (1979) Biochemistry 18:5073–5078
Köst-Reyes E, Köst HP (1979) Eur J Biochem 102:83–91
Krasnovskii AA, Erokhina LG (1969) Dokl Acad Nauk SSSR 186:957–960
Kufer W, Scheer H, Holzwarth AR (1983) Isr J Chem 23:233–240
Lagarias JC, Glazer AN, Rapoport H (1979) J Am Chem Soc 101:5030–5037
Land EJ (1979) Photochem Photobiol 29:483–487
MacColl R (1982) Photochem Photobiol 35:899–904
MacColl R, Berns DS (1981) Isr J Chem 21:296–300
MacColl R, Guard-Friar D (1983) Biochemistry 22:5568–5572
Mörschel E, Wehrmeyer W (1975) Arch Microbiol 105:153–158
Mörschel E, Koller K, Wehrmeyer W (1980) Arch Microbiol 125:43–51
Morisset W, Wehrmeyer W, Schirmer T, Bode W (1984) Arch Microbiol 140:202–205
Offner GD, Troxler RF (1983) J Biol Chem 258:9931–9940
Pearlstein RM (1982) Photochem Photobiol 35:835–844
Pellegrino F, Wong D, Alfano RR, Zilinskas B (1981) Photochem Photobiol 34:691–696
Pratt LH (1978) Photochem Photobiol 27:81–105
Pratt LH (1982) Annu Rev Plant 33:557–582
Rapoport H, Glazer AN (1984) In: Blauer G, Sund H (eds) Optical properties and structure of tetrapyrroles. De Gruyter, Berlin, pp 411–423
Redlinger T, Gantt E (1982) Proc Natl Acad Sci USA 79:5542–5546
Rüdiger W (1979) Ber Dtsch Bot Ges 92:414–426
Rüdiger W (1980) Struct Bond 40:101–140
Rüdiger W, Scheer H (1983) In: Shropshire W, Mohr H (eds) Photomorphogenesis. Encyclopedia of Plant Physiology Vol 16. Springer, Berlin Heidelberg New York, pp 119–151
Rusckowski M, Zilinskas BA (1982) Plant Physiol 70:1055–1059
Scheer H (1981) Angew Chem 93:230–250, Angew Chem Int Ed 20:241–261
Scheer H (1982) In: Fong FK (ed) Light reaction path of photosynthesis. Springer, Berlin Heidelberg New York, pp 7–45
Schirmer T, Bode W, Huber R, Sidler W, Zuber H (1985) J Mol Biol 184:257–277
Searle GFW, Barber J, Porter G, Tredwell CJ (1978) Biochim Biophys Acta 501:246–256
Siegelman HW, Kycia H (1982) Plant Physiol 70: 887–897
Suter GW, Mazzola P, Wendler J, Holzwarth AR (1984) Biochim Biophys Acta 766:269–276
Switalski SC, Sauer K (1984) Photochem Photobiol 40:423–427
Teale FWJ, Dale RE (1970) Biochem J 116:161–169
Wanner G, Köst HP (1980) Protoplasma 102:97–109
Wehrmeyer W (1983) In: Jensen U, Fairbrothers DE (eds) Proteins and nucleic acids in plant systematics. Springer, Berlin Heidelberg New York, pp 143–167
Wendler J, Holzwarth AR, Wehrmeyer W (1984) Biochim Biophys Acta 765:58–67
Wong D, Pellegrino F, Alfano RR, Zilinskas BA (1981) Photochem Photobiol 33:651–662
Yamazaki I, Mimuro M, Murao T, Yamazaki K, Yoshihara K, Fujita Y (1984) Photochem Photobiol 39:233–240
Zickendraht-Wendelstadt B, Friedrich J, Rüdiger W (1980) Photochem Photobiol 31:367–376
Zuber H (1978) Ber Dtsch Bot Ges 91:459–474

8 Physical Determinants of Charge Separation Processes

8.1 Primary Electron Transfer Reactions in Photosynthetic Bacteria: Energetics and Kinetics of Transient States

W.W. PARSON and D. HOLTEN

When bacteriochlorophyll absorbs a photon, it becomes an extremely active reductant (see SAUER, Chap. 2, this Vol. for an overview of the associated events). The effective midpoint potential (E_m) of the redox couple $BChl^*/BChl^+$ is more negative than that of $BChl/BChl^+$ by approximately $h\nu_0/e$, where BChl* and BChl are bacteriochlorophyll in the excited electronic state and ground state, $BChl^+$ is a π-cation radical, $h\nu_0$ is the o–o transition energy for excitation of BChl to BChl*, and e is the electronic charge (PARSON 1978). For the photochemically reactive BChl complex (P) in bacterial reaction centers (RC's), the E_m in the ground state is about +0.45 V and $h\nu_0$ of the lowest excited singlet state (P*) is about 1.38 eV. The E_m of P* is therefore approximately −0.9 V. This means that P* should be able to transfer an electron to another molecule of BChl or bacteriopheophytin (BPh), because the E_m values for $BChl^-/BChl$ and BPh^-/BPh are on the order of −0.85 V and −0.55 V (FAJER et al. 1975). However, the E_m values for the reductions of BChl and BPh have been measured only for the molecules in solution. The pigments in the RC could interact with charged amino acids, P^+, or the quinone (Q_A) that serves as a later electron acceptor.

Although BChl* or BPh* can react rapidly with electron acceptors in solution, the radical pairs that are formed tend to decay by back-reactions within a few picoseconds (HOLTEN et al. 1978a; HUPPERT et al. 1976; NETZEL et al. 1981). Bacterial RC's successfully avoid such back-reactions. In RC's isolated from *Rhodopseudomonas sphaeroides*, the radical pair $P^+Q_A^-$ is formed with a quantum yield of 102±4%, and lives for about 0.1 s even if electron transfer from Q_A^- to a secondary quinone is blocked (WRAIGHT and CLAYTON 1974). Part of the secret of this success appears to be that P^+ is far enough away from Q_A^- so that the orbital overlap between the two radicals is essentially nil. The movement of an electron from P* to Q_A occurs indirectly, through one or more intermediate electron acceptors that carry the electron quickly away from P^+.

The effectiveness of this principle was illustrated by MOORE et al. (1984), who studied photochemical charge separation in covalently linked complexes of a carotenoid (*Car*), porphyrin (*Por*) and quinone (*Q*). Excitation of *Por* initiated electron transfer to *Q*, generating Por^+Q^-. Without *Car*, Por^+Q^- decayed quickly; in its presence, electron transfer from *Car* to Por^+ left the complex in a relatively stable state, $Car^+\,Por\,Q^-$.

Indications that transient radical-pair states might precede the formation of $P^+Q_A^-$ in bacterial RC's came from studies in which photochemical electron transfer to Q_A was blocked by reducing Q_A chemically. Excitation with a short flash under these conditions generates a transient state (P^F) that has a lifetime of about 10 ns at 295 K (PARSON et al. 1975). The optical absorbance changes that occur suggest that P^F is a radical pair, P^+I^-, in which I^- involves both BChl and BPh (FAJER et al. 1975; SHUVALOV and PARSON 1981a). In RC's that contain BChl-a and BPh-a, such as those from *Rhodopseudomonas sphaeroides*, reduction of BPh is suggested by a bleaching of absorption bands at 545 and 760 nm. Absorbance decreases in bands at 600 and 800 nm suggest reduction of a BChl. RC's from *Rps. viridis*, which contain BChl-b and BPh-b, show analogous absorbance decreases at 545, 790, and 830 nm (HOLTEN et al. 1978b). In both types of RC, the formation of P^+I^- also causes the absorbance changes that are characteristic of P^+, and an additional absorbance increase in the 650-nm region. The 650-nm band is consistent with the formation of a π-anion radical of either BChl or BPh (FAJER et al. 1975).

The electron acceptor(s) can be trapped in the reduced state (I^-) by illuminating RC's continuously after the reduction of Q_A in the presence of a cytochrome (C) that can donate electrons to P^+ (SHUVALOV and KLIMOV 1976; TIEDE et al. 1976; OKAMURA et al. 1979). Reduction of P^+ by the cytochrome is an improbable event because the lifetime of P^+I^- is so short, but repeated cycling through the radical-pair state eventually can convert the reaction centers almost fully into the state $C^+ PI^- Q_A^-$.

The absorbance changes associated with the formation of state P^F (P^+I^-) with Q_A chemically reduced or removed depend on the temperature. With decreasing temperature, the absorbance changes suggesting reduction of BPh become more pronounced, relative to those suggesting reduction of BChl (SHUVALOV and PARSON 1981a). This suggests that P^+I^- is an equilibrium mixture of the two radical pairs $^1[P^+BChl^-]$ and $^1[P^+BPh^-]$. From the temperature dependence, $^1[P^+BChl^-]$ appears to lie about 0.03 eV above $^1[P^+BPh^-]$ in enthalpy.

Interpretations of the absorbance changes associated with state P^F are risky, because the exciton interactions among the four BChls and two BPhs of the reaction center are not well understood. Suppose that a BChl and a BPh are close together. Exciton interactions could result in a transfer of dipole strength from the BPh's absorption band near 760 nm to the BChl's band near 800 nm. Mixing with higher-energy (Q_x and Soret) transitions of both molecules could add further dipole strength to the 800-nm band. If the BPh undergoes reduction, the exciton interactions between the two molecules would vanish and the 800-nm band would be weakened. The bleaching in this band could be larger than the bleaching in the BPh's band at 760 nm. If the interactions of the two pigments change with temperature, or with time after oxidation or reduction of other components of the RC, the situation could be highly complex. In addition, if the charge-transfer state $^1[P^+BChl^-]$ is close to P* in energy, as suggested by observations discussed below, the two states will be strongly mixed in a quantum-mechanical sense. It therefore seems best to take an operational view of the transient states that are seen in the RC. Labels like "$^1[P^+BChl^-]$" are useful for simplicity when one attempts to sketch the pathway of electron flow, but may give an inaccurate picture of the molecular states that are involved.

One of the routes by which P^+I^- decays when electron transfer to Q_A is blocked is probably a direct back-reaction to the ground state, PI. Another route involves rephasing of the spins of the unpaired electrons on P^+ and I^-. Spin rephasing gives the radical-pair triplet character and allows it to collapse to a longer-lived triplet state (P^R). P^+I^- also can undergo a thermally activated back-reaction to P*. P* usually returns to P^+I^-, but it also can decay to the ground state nonradiatively or by fluorescing. The amplitude of this "delayed" fluorescence provides a measure of the free energy gap separating P* from P^F, which appears to be about 0.16 eV at 295 K (WOODBURY and PARSON 1984). The temperature dependence of the delayed fluorescence provides information on the enthalpy difference between P* and P^F, but the results on this point are difficult to interpret. The amplitude of the fluorescence increases with decreasing temperature between 270 and 200 K, and then decreases. The behavior at low temperatures suggests that the enthalpy gap between P* and P^F is less than 0.05 eV, which means that $^1[P^+BChl^-]$ is within about 0.02 eV of P* in enthalpy.

A much weaker fluorescence also can be detected during the lifetime of the triplet state P^R. From the temperature dependence of the fluorescence, P^R appears to lie about 0.4 eV below P* in enthalpy (SHUVALOV and PARSON 1981 a). Like P^F, P^R appears to consist of a mixture of at least two different states. These have been interpreted as being an excited triplet state of P (^3P) and a triplet radical-air state, $^3[P^+BChl^-]$ (SHUVALOV and PARSON 1981a). From the temperature dependence of absorbance changes associated with the two states, $^3[P^+BChl^-]$ lies about 0.03 eV above ^3P in enthalpy. However, these observations may be open to other interpretations, because the absorbance changes that accompany the formation of ^3P itself are temperature-dependent (SHUVALOV and PARSON 1981b; DEN BLANKEN and HOFF 1982). If the identification of $^3[P^+BChl^-]$ is correct, the enthalpy gap between the singlet and triplet radical pairs $^1[P^+BChl^-]$ and $^3[P^+BChl^-]$ is about 0.35 eV, which is almost as large as the gap between P* and ^3P. Because the singlet-triplet splitting probably depends largely on the exchange interaction between the two radicals, this means that the orbital overlap between P^+ and $BChl^-$ is extremely strong.

In *Rps. viridis*, the absorption band of P includes a long-wavelength shoulder that could reflect charge-transfer transitions to $^1[P^+BChl^-]$ (VERMEGLIO and PAILLOTIN 1982; SHUVALOV and KLEVANIK 1983; MASLOV et al. 1983). The shoulder lies about 0.02 eV below the main absorption band, in agreement with the estimate given above for the enthalpy gap between P* and P^F. The occurrence of direct transitions from the ground state to $^1[P^+BChl^-]$ also supports the conclusion that P^+ and $BChl^-$ are close together in the RC. The absorption shoulder and a corresponding feature of the fluorescence emission spectrum disappear if the electron acceptor complex is reduced to I^-.

The amount of orbital overlap of the BPh^- radical with P^+ has been probed by studying effects of magnetic fields on the quantum yield of the triplet state P^R. By interfering with the spin rephasing in P^+I^-, weak magnetic fields decrease the yield of P^R and slow the decay of P^+I^- (BLANKENSHIP et al. 1977; HOFF et al. 1977; SCHENCK et al. 1982). The spin rephasing probably occurs in the P^+BPh^- radical-pair, because the yield of P^R is greatest at low temperatures, when the population of P^+BPh^- exceeds that of P^+BChl^-. Magnetic fields disturb the rephasing by moving two of the three sublevels of $^3[P^+BPh^-]$ apart in energy so that they are no longer close enough to $^1[P^+BPh^-]$ to mix with it. At high magnetic fields (>10 kG) the field itself can drive rephasing and both decay rate of P^+I^- and the yield of P^R increase (BOXER et al. 1983; CHIDSEY et al. 1984). Analysis of the P^R yield as a function of the field strength indicates that $^1[P^+BPh^-]$ and $^3[P^+BPh^-]$ are separated by only about 10^{-7} eV (WERNER et al. 1978; MICHEL-BEYERLE et al. 1979; NORRIS et al. 1982; BOXER et al. 1983). The interaction between P^+ and BPh^- thus appears to be much weaker than that between P^+ and $BChl^-$. If the initial photochemical reaction is the creation of $^1[P^+BChl^-]$ from P*, and is followed by electron transfer from $BChl^-$ to BPh and then to Q_A, the progressive decrease in the overlap between P^+ and the anionic radical could contribute substantially to the efficiency of the photosynthetic apparatus.

The studies described thus far were performed largely with RC's in which electron transfer to Q_A was blocked. The triplet states that form under these conditions are almost certainly side products, rather than intermediates in the pathway that leads to $P^+Q_A{}^-$ in unblocked RC's. The spin rephasing that creates $^3[P^+BPh^-]$ is too slow to play a role in the normal electron transfer pathway. But what about the singlet radical pairs $^1[P^+BChl^-]$ and $^1[P^+BPh^-]$? To establish that these are true intermediates, one has to show that they are formed transiently with a high quantum yield in unblocked RC's.

When unblocked RC's are excited with a flash lasting on the order of 10 ps, the absorbance changes associated with state P^F occur during the flash (ROCKLEY et al. 1975; KAUFMANN et al. 1975; HOLTEN et al. 1978b; SHUVALOV et al. 1978; KIRMAIER et al. 1983a). Again, absorbance changes occur both in bands due to BChl and in bands due to BPh. The quantum yield appears to be high enough to support the conclusion that P^F is an intermediate in the pathway to $P^+Q_A{}^-$ (HOLTEN et al. 1980). The transient state decays with a lifetime of about 200 ps at 295 and 100 ps at 77 K (KIRMAIER et al. 1985). In the green bacterium *Chloroflexus aurantiacus*, the lifetime is 330 ps at 295 K and 220 ps at 77 K (KIRMAIER et al. 1983b and unpublished). Presumably, the decay of P^F reflects electron transfer from $BChl^-$ or PBh^- to Q_A. Because electron transfer to Q_A occurs at 77 K at least as rapidly as at 295 K, the immediate electron donor is probably the lower-energy radical-pair $^1[P^+BPh^-]$, rather than $^1[P^+BChl^-]$. The idea that Q_A interacts with BPh^- more directly than with $BChl^-$ agrees both with the observation that reduction of Q_A perturbs the absorption spectrum of the BPh more strongly than it does that of the BChl (SHUVALOV and ASADOV 1979; VERMEGLIO and CLAYTON 1977) and with the structural information on reaction center geometry presented in MICHEL and DEISENHOFER, Chap. 8.4, this Volume. The unusual temperature dependence of the kinetics may reflect the importance of nuclear overlap in determining the rate of electron transfer (KIRMAIER et al. 1985).

The kinetics of formation of $^1[P^+BPh^-]$ have been measured (HOLTEN et al. 1980; SHUVALOV and KLEVANIK 1983; KIRMAIER et al. 1983a). The 870-nm band of P is found to bleach essentially instantaneously when the RC is excited. The absorbance decreases at 545 and 760 nm, which probably signal the formation of BPh^-, occur after a distinct lag of about 5 ps at both 295 K and 110 K. Whether $^1[P^+BChl^-]$ is formed prior to the reduction of the BPh has been difficult to prove. The bleaching at 870 nm could reflect the formation of either P^+ or P^*, and attempts to detect the presence of $BChl^-$ have given conflicting results. Early reports (ROCKLEY et al. 1975; HOLTEN et al. 1978a) described a bleaching at 800 nm with a lifetime of about 30 ps, and SHUVALOV et al. (1978) suggested that this reflected the transient reduction of BChl. However, this relatively long-lived signal appears to be an artifact due to excitation of the RC's with more than one photon (AKHMANOV et al. 1980; HOLTEN et al. 1980). Measurements using weak excitation flashes at room temperature fail to show bleaching at 800 nm preceding the absorbance changes associated with BPh (BORISOV et al. 1983). Indications of such a bleaching have been obtained at 110 K (SHUVALOV and KLEVANIK 1983), but the observations involve relatively small differences between absorbance changes measured at different times after excitation. Such differences must be interpreted cautiously, because they include absorbance changes due to the decay of P^F to $P^+Q_A^-$. They could also include absorbance changes resulting from nuclear relaxations.

If $^1[P^+BChl^-]$ is an intermediate in the formation of $^1[P^+BPh^-]$, it might be possible to increase the lifetime of $BChl^-$ by reducing the BPh before the excitation. This has been attempted, with negative results (NETZEL et al. 1977; SCHENCK et al. 1981b; HOLTEN et al. 1978b). RC's were illuminated continuously in the presence of a cytochrome to reduce the electron acceptor complex to I^-, and then were excited with a 10-ps flash. There was no indication of additional photoreduction of BChl. (The flash did generate an unidentified transient state that had a lifetime of about 30 ps in *Rps. viridis* or 300 ps in *Rps. sphaeroides*. The transient state was first interpreted as P^*, but this interpretation proved incompatible with measurements of the fluorescence yield; SCHENCK et al. 1981b). These observations do not disprove that $^1[P^+BChl^-]$ preceeds $^1[P^+BPh^-]$ in the electron transfer sequence. If the BChl and BPh are close together, reduction of the BPh could prevent additional reduction of the BChl. As noted above, reduction of the acceptor complex to I^- alters the absorption and emission spectra of P in a manner that suggests the disruption of charge-transfer transitions.

In conclusion, there are many attractive features to the idea that the initial electron transfer steps in the RC's are:

$$P^* \xrightarrow{<5\ \text{ps}} {}^1[P^+BChl^-] \xrightarrow{\sim 5\ \text{ps}} {}^1[P^+BPh^-] \xrightarrow{200\ \text{ps}} P^+Q_A^-$$

However, the kinetic evidence for $^1[P^+BChl^-]$ in this scheme is still inconclusive.

References

Akhmanov SA, Borisov AY, Danielius RV, Gadonas RA, Kozlowski VS, Piskarskas AS, Razjivin AP, Shuvalov VA (1980) FEBS Lett 114:149–152

Borisov AY, Danielius RV, Kudzmauskas SD, Piskarskas AS, Razjivin AP, Sirutkaitis VA, Valkunas LL (1983) Photobiochem Photobiophys 6:33–38

Blanken HJ den, Hoff AJ (1982) Biochim Biophys Acta 681:365–374
Blankenship RE, Schaafsma TJ, Parson WW (1977) Biochim Biophys Acta 461:297–305
Boxer SG, Chidsey CED, Roelofs MG (1983) Annu Rev Phys Chem 34:389–417
Chidsey CED, Kirmaier C, Holten D, Boxer SG (1984) Biochim Biophys Acta 766:424–437
Fajer J, Brune DS, Davis MS, Forman A, Spaulding LD (1975) Proc Natl Acad Sci USA 12:4956–4960
Hoff AJ, Rademaker H, Grondelle R van, Duysens LMN (1977) Biochim Biophys Acta 460:547–554
Holten D, Windsor MW, Parson WW, Gouterman M (1978a) Photochem Photobiol 28:951–961
Holten D, Windsor MW, Parson WW, Thornber JP (1978b) Biochim Biophys Acta 501:112–126
Holten D, Hoganson C, Windsor MW, Schenck CC, Parson WW, Migus A, Fork RL, Shank CV (1980) Biochim Biophys Acta 592:461–477
Huppert D, Rentzepis PM, Tollin G (1976) Biochim Biophys Acta 440:356–364
Kaufmann KJ, Dutton PL, Netzel TL, Leigh JS, Rentzepis PM (1975) Science 188:1301–1304
Kirmaier C, Holten D, Parson WW (1983a) Biochim Biophys Acta 725:190–202
Kirmaier C, Holten D, Parson WW (1985) Biochim Biophys Acta 810:33–48
Kirmaier C, Holten D, Feick R, Blankenship RE (1983b) FEBS Lett 158:73–78
Maslov VG, Klevanik AV, Ismailov MA, Shuvalov VA (1983) Dokl Akad Nauk SSSR 269:1217–1221
Moore TA, Gust D, Mathis P, Mialocq J-C, Chachaty C, Bensasson RV, Land EJ, Doizi D, Liddell PA, Lehman WR, Nemeth GA, Moore AL (1984) Nature 307:630–632
Michel-Beyerle ME, Scheer EH, Seidlitz H, Tempus D, Haberkorn R (1979) FEBS Lett 100:9–12
Netzel TL, Rentzepis PM, Tiede DM, Prince RC, Dutton PL (1977) Biochim Biophys Acta 460:467–479
Netzel TL, Bergakamp MA, Chang C-K, Dalton J (1981) J Photochem 17:451–460
Norris JR, Bowman MK, Budil DE, Tang J, Wraight CA, Closs GL (1982) Proc Natl Acad Sci USA 79:5532–5536
Okamura MY, Issacson RA, Feher G (1979) Biochim Biophys Acta 546:394–417
Parson WW (1978) Photochem Photobiol 28:389–393
Parson WW, Clayton RK, Cogdell RJ (1975) Biochim Biophys Acta 387:265–278
Rockley MG, Windsor MW, Cogdell RJ, Parson WW (1975) Proc Natl Acad Sci USA 72:2251–2255
Schenck CC, Parson WW, Holten D, Windsor MW (1981b) Biochim Biophys Acta 635:383–392
Schenck CC, Blankenship RE, Parson WW (1982) Biochim Biophys Acta 680:44–59
Shuvalov VA, Asadov AA (1979) Biochim Biophys Acta 545:296–308
Shuvalov VA, Klevanik AV (1983) FEBS Lett 160:51–55
Shuvalov VA, Klimov VV (1976) Biochim Biophys Acta 440:587–599
Shuvalov VA, Parson WW (1981a) Proc Natl Acad Sci USA 78:957–961
Shuvalov VA, Parson WW (1981b) Biochim Biophys Acta 638:50–59
Shuvalov VA, Klevanik AV, Sharkov AV, Matveetz JA, Kryukov PG (1978) FEBS Lett 81:135–139
Tiede DM, Prince RC, Dutton PL (1976) Biochim Biophys Acta 449:447–469
Vermeglio A, Clayton RK (1977) Biochim Biophys Acta 461:159–165
Vermeglio A, Paillotin G (1982) Biochim Biophys Acta 681:32–40
Werner H-J, Schulten K, Weller A (1978) Biochim Biophys Acta 502:255–268
Woodbury NWT, Parson WW (1984) Biochim Biophys Acta 767:345–361
Wraight CA, Clayton RK (1974) Biochim Biophys Acta 333:246–260

8.2 Redox Center and Secondary Protein Structure Organization in Bacterial Reaction Centers and the Cytochrome b/c_1 Complex

D.M. TIEDE

1 Introduction

A great deal of information has been gained on the organization of redox centers in the bacterial reaction center (RC) and its protein secondary structures. One of the facts which is emerging is that on a ps time-scale, electron tunneling is likely to be occurring between redox groups which are fixed by the protein in a non-coplanar, and possibly perpendicular configuration. The most extensively studied RC's of *Rhodopseudomonas viridis* and *Rps. sphaeroides* show similar redox center organizations. Major differences have not yet been found with other species of purple bacteria. Space limitations permit only a small number of references to be cited for each of the structural determinations outlined in this review. The reader should also consult in this Volume SAUER (Chap. 2), DUTTON (Chap. 5), PARSON and HOLTEN (Chap. 8.1), NORRIS and VAN BRAKEL (Chap. 8.3), MICHEL and DEISENHOFER (Chap. 8.4).

Much less is known about the structure which supports the more complex proton and electron transporting activities carried out within the cytochrome b/c_1 complex (see also PRINCE, Chap. 10.4). Extensive analogies exist between the reaction mechanisms and inhibitor sensitivities of the mitochondrial and bacterial b/c_1 complexes. Information on the redox center organization for each is summarized here.

2 Redox Components: Definitions, Abbreviations

2.1 Redox Centers in the Reaction Center

The RC light-induced electron transfers have been reviewed by PARSON (1982) and HOFF (1982), as well as elsewhere in this volume DUTTON (Chap. 5; PARSON and HOLTEN, Chap. 8.1). The abbreviations used for the RC redox centers are as follows: two bacteriochlorophylls, BChl's, are closely associated to form a dimer, B_2, which acts as a single redox center, the primary electron donor. A monomeric BChl probably functions as the immediate electron acceptor for the B_2, and is labeled B. H represents the bacteriopheophytin (BPh), which

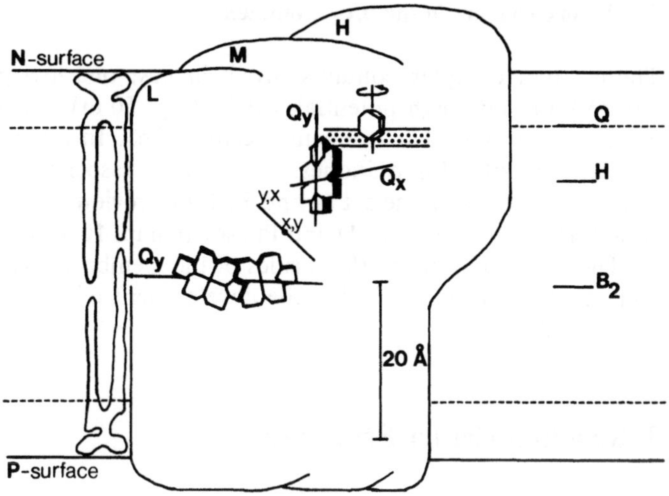

Fig. 1. Reaction center structure. The uncertainty in the Q_A orientation is indicated by the *arrow*. *Dotted region* marks the limits for the Fe^{2+} location. Two possible orientations for the acceptor B $Q_{x,y}$ transitions are labeled *x,y*

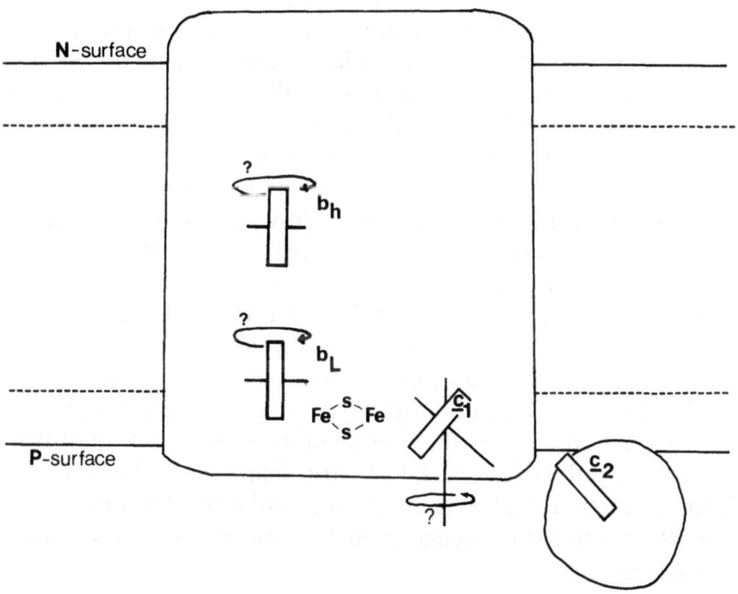

Fig. 2. Orientations of the redox centers in the b/c_1 complex. A distance scale is not drawn, since spacial relationships are not known. *Arrows* indicate some of the structural uncertainties

serves as the subsequent acceptor, and Q_A and Q_B are the first and second quinone acceptors. The RC also contains a ferrous iron, and monomeric BChl and BPh molecules which apparently do not undergo a redox change in the normal RC photochemistry.

2.2 Redox Centers in the b/c_1 Complex

The cyt b/c_1 complex contains two b hemes: one low potential cyt b_L (E_{m7} -80 mV) and one high potential, cyt b_h (E_{m7} 50 mV). There is a c-type heme, cyt c_1, and a binuclear iron-sulfur center, FeS. In addition, there are at least two binding sites for quinones. One, Q_c, is associated with the cyt b_h, and other, Q_z, reacts with the cyt b_L and FeS. For reviews, see CROFTS and WRAIGHT (1983), HAUSKA (Chap. 10.1) and PRINCE (Chap. 10.4), in this Volume.

The P- and N-refer to those sides of the membrane which become positively and negatively charged during the development of a membrane potential (see Figs. 1 and 2).

3 Reaction Center Structure

3.1 Protein Secondary Structure

Bacterial RC's are composed of three protein subunits, labeled L, M, H (OKAMURA et al. 1982). With RC's of *Rhodopseudomonas sphaeroides* R-26, the protein structures in the LMH complex (NABEDRYK et al. 1982) have been compared to those in the active LM complex (NABEDRYK et al. 1983), which contains all of the RC chromophores (4BChl, 2BPh, Fe^{2+}, Q). This UV circular dichroism analysis showed that both the LM and LMH contain a similar core of α-helix, representing $50 \pm 9\%$ and $48 \pm 4\%$ of the protein respectively, with the rest mainly composed of random protein structures. The polarized IR spectra on oriented membrane multilayers showed that the ensemble of the α-helix segments for both complexes are oriented in a transmembrane direction, making an angle on average of $\leq 30°$ to the membrane normal. This similarity in protein secondary structure between LMH and LM means that a similar organization must also be found in the H subunit (NABEDRYK et al. 1983). A similarity between LM and the pigmentless H subunit would show that protein structure at this level is not determined by prosthetic group content or photochemical activity, but possibly by the transmembrane organization of the subunits.

The overall 59 Å long, asymmetric shape of the RC shown in Fig. 1 comes from diffraction studies in oriented multilayers (PACHENCE et al. 1981) and is consistent with the molecular model of MICHEL and DEISENHOFER (Chapter 8.4) in most respects.

3.2 Redox Center Locations in the Membrane Profile

Fe Atom. The RC Fe^{2+} has been determined by resonance X-ray diffraction to lie within a narrow (2 Å) region at the edge of the hydrocarbon core of the membrane (BLASIE et al. 1983).

Q_A, Q_B. The positioning of the Q_A and Q_B near the aqueous interface (N-side) is determined by the fact that the quinones are situated in close proximity

(<10 Å) to the Fe^{2+} (see Sect. 3.3). This is further supported by the fact that the electron transfer between Q_A and Q_B is not electrogenic (PACKHAM et al. 1982) and the $Q_{A,B}$ reductions are associated with the uptake of a proton from the bulk aqueous phase (JACKSON and DUTTON 1973).

Donor, B_2. An electrical charging of the membrane capacitance occurs as a result of the $B_2^+Q^-$ charge separation. This potential is found to represent 50–60% of the total RC-generated response, with the remaining 40–50% arising with the cyt c_2 (in aqueous phase) to B_2 transfer (PACKHAM et al. 1982). Assuming a constant dielectric throughout the RC, these measurements place the B_2 0 to 3 Å away from the center of the membrane profile.

Acceptor H. TRISSL (1983) has provided a <1 ns time resolution of the electrical events associated with the RC charge transfer. The potential generated by the $B_2^+H^-$ transfer was found to be 2/3 of that generated by the $B_2^+Q^-$ separation. Again assuming a constant dielectric, this would suggest that the B_2H transmembrane separation is 2/3 of that for the B_2Q (TRISSL 1983).

3.3 Redox Center Separations

Fe−Q. The close proximity between Q_A, Q_B and the high spin Fe^{2+} is evidenced by a prominent magnetic coupling. BUTLER et al. (1984) have described the electronic structure for the Fe^{2+} and have simulated the Q^-Fe spectra as arising from an anisotropic antiferromagnetic coupling with $J_X = -0.13\ cm^{-1}$, $J_{Y,Z} = -0.58\ cm^{-1}$. The molecular structure of the QFe complex is still unknown, but EXFAS and X-ray absorption studies have shown that the Q lies outside the Fe first coordination sphere (OKAMURA et al. 1982). DISMUKES et al. (1984) have simulated a part of the QFe signal using an isotropic exchange $|J| = 0.12\ cm^{-1}$ and dipolar terms, suggesting a distance of 6.2–7.8 Å between Q and Fe. This simulation does not, however, take into account the complete set of experimental g-values described by BUTLER et al. (1984).

H−Q. The H and Q_A are close enough to experience a magnetic exchange interaction. In the ubiquinone-containing *Rps. sphaeroides* the couplings between H^- and Q^-Fe (OKAMURA et al. 1982) and between H^- and Q^- in iron extracted RC's (GAST and HOFF 1979) are measured to be about $1\text{–}4 \times 10^{-4}\ cm^{-1}$. In the menaquinone containing RC's of *Chromatium vinosum* and *Rps. viridis* the H^-Q^-Fe couplings are measured to be 50–100 times greater (TIEDE et al. 1976; PRINCE et al. 1977). Note that these exchange couplings are less than those between the Q Fe. The H^-Q^- distance dependent dipolar interactions may be on the order of 1–5 G (GAST et al. 1983), suggesting separations on the order of 15–25 Å.

B_2-Q_A. The B_2Q_A separation is large enough so that their EPR spectra do not show mutual perturbations (TIEDE et al. 1978). An upper limit for their magnetic interactions (<1 G) placed their minimum center-to-center separation at ≥23 Å.

B_2-B and B_2-H. The discoveries of reaction yield perturbations by magnetic fields and resonant microwave radiation have provided the means to assay interactions in the short-lived (<10 ns) $B_2^+B^-H$ and $B_2^+BH^-$ charge separated states (see PARSON 1982; NORRIS and VAN BRAKEL, Chap. 8.3, HOFF, Chap. 7.2, both this Vol.). This work indicates that the $B_2^+B^-$ are 1–2 orders of magnitude more closely associated than the $B_2^+H^-$. This conclusion was also drawn from an analysis of CD data (SHUVALOV and ASADOV 1979). Distances between these centers have been estimated from electron transfer rates and exchange interactions. These are somewhat uncertain, since the distance dependence of the orbital overlap is not well known, and will vary with orientation and the presence of intervening molecules (super exchange).

3.4 Redox Center Orientations with Respect to the Membrane Normal

Q_A. The similarity of the Q_A^-Fe EPR signal shapes in *Rps. sphaeroides*, *Rps. viridis* and *C. vinosum* (DUTTON et al. 1978) and angular dependence of the g 1.82 signal peak, with the maximum perpendicular to the membrane plane (TIEDE and DUTTON 1981; PRINCE, TIEDE, DUTTON, RUTHERFORD unpublished), suggests that the Q_AFe has a similar association and orientation in each case, independent of whether Q_A is a mena- or ubiquinone. Simulations by both BUTLER et al. (1984) and DISMUKES et al. (1984) indicate that the signal in the g 1.82 region will be composed of several transitions between mixed Q and Fe^{2+} spin states. The present orientation data do not permit a critical assessment of either of these simulations. Interpretation of the orientation data is further hindered by a lack of information on how molecular structures are related to the QFe magnetic properties.

However in a simpler system, with deuterated, Fe-extracted RC's, the Q_A^- g-anisotropy can be measured and identified with the quinone structure. Here the EPR dichroism showed that the plane of the Q_A ring is perpendicular to the plane of the membrane surface in RC's of *Rps. sphaeroides* (TIEDE and DUTTON 1981).

The structure of the Q_AFe in *Rhodospirillum rubrum* may be an exception to this general pattern (HALES and GUPTA 1979; DUTTON et al. 1978; GAST et al. 1983).

Acceptor H. The tetrapyrrole plane of H is oriented essentially perpendicular to the membrane surface. The Q_x (545 nm) transition lies parallel and the Q_y (800 nm *Rps. viridis*, 760 nm *Rps. sphaeroides*) perpendicular to the membrane surface. This is demonstrated most directly in the dichroism of trapped B_2H^- states in *Rps. viridis* (PAILLOTIN et al. 1979), but is also evident in spectra of oriented *Rps. sphaeroides* membranes (e.g., VERMEGLIO and CLAYTON 1976; VERMEGLIO et al. 1978). Attempts at indirectly resolving the orientation of the two BPh's based on absorption shifts in the $B_2^+Q^-$ state (RAFFERTY and CLAYTON 1979) have not been successful, since the sum of the predicted Q_y orientations severely contradict the measured values.

Acceptor B. The orientations of the acceptor B and the "voyeur" BChl are the least precisely determined of the RC chromophores. The sum of the B and BChl Q_Y (PAILLOTIN et al. 1979; VERMEGLIO and CLAYTON 1976) and Q_x (TIEDE and BRETON unpublished) transitions are found to lie <35° with respect to the membrane surface. Attempts to deconvolute absorption band-shifts in the $B_2^+Q_A^-$ and $B_2^+H^-Q^-$ states suggested that the two BChls are oriented differently. One, which shifts in $B_2H^-Q^-$, would have its Q_y lying in the membrane plane with the Q_x tilted out of plane; the other would have its Q_y tilted 50° from the membrane plane (PAILLOTIN et al. 1979), with the requirement that its Q_x lies in plane (TIEDE and BRETON unpublished). These possibilities are indicated in Fig. 1. A more definitive approach would be to measure directly the $Q_{x,y}$ dichroisms of the transient B^- states in oriented membranes.

Donor, B_2. A plane formed both by the optical $Q_{x,y}$ (VERMEGLIO and CLAYTON 1976) and the spin-polarized x,y magnetic axes is tilted 10°–20° from the membrane normal (HALES and GUPTA 1979; TIEDE and DUTTON 1981). This will likely be parallel to the BChl rings in a parallel plane B_2 model. Most recent evidence has supported a "staggered" parallel plane model (MAR and GINGRAS 1984; LUBITZ et al. 1984).

3.5 Angles Between Donor and Acceptor Molecular Planes

The angle between redox centers can be determined by making dichroism measurements with respect to two *independent* axes. We have done this here by combining information an orientation with respect to the membrane normal in oriented membranes with data on the orientation with respect to an optical axis in photoselection experiments. Both x and y directions must be measured to define the orientation of the BChl, BPh planes. So far, these measurements have been completed only for the B_2-H pair.

Angle Between B_2 and H Molecular Planes. Photoselection studies showed that the H Q_y (760 nm *sphaeroides*, 800 nm *viridis*) is oriented 90° away from the B_2 Q_y (860 nm *sphaeroides*, 960 nm *viridis*) (VERMEGLIO et al. 1978; KIRMAIER et al. 1983). This is an expression of the perpendicular orientation of B_2 and H planes with respect to the membrane. The angle between the B_2 and H planes is then determined by the angle between the H Q_x and B_2 Q_y. This angle has been measured in *Rps. sphaeroides* by photoselection-excitation within the BPh bands to be ~60° (VERMEGLIO et al. 1978). A similar angle has been found in *Rps. viridis* from the dichroism of the H $Q_{x,y}$ bands in the trapped B_2H^- state formed by photoselection at 960 nm (TIEDE et al. 1984).

These results allow us to conclude that although the B_2 and H planes are perpendicular to the membrane surface, they are also rotated at ~60° to each other. This means that the ps electron tunneling between B_2 and H must be occurring between donors and acceptors which are oriented in a non-coplanar fashion.

Another Electron-Tunneling Complex. The RC-associated, low potential cyt c-553 in *Chromatium vinosum* reduces the B_2^+ in a demonstrated low temperature, electron tunneling reaction. The center-to-center distance between the heme and the B_2^+ has been measured to be ≥ 25 Å (TIEDE et al. 1978). The cyt c-553 heme planes were also shown to be parallel to the membrane surface (TIEDE et al. 1978) while, as above, the B_2 plane is perpendicular to this surface. This gives another example where electron tunneling is occurring between nearly orthogonally oriented redox centers. The RC-associated high potential c cyt in *C. vinosum* were shown to be oriented perpendicular to the membrane surface, and located within ~8 Å of the low potential hemes (TIEDE et al. 1978).

4 Structure of the Cytochrome b/c_1 Complex

4.1 Location of Redox Centers in the Membrane Profile

FeS Cluster, cyt c_1, cyt b_L Hemes and Q_z. These components are placed near the aqueous interface on the cyt c_2 side of the membrane (P-side). This is taken from the fact that the electron transfer from Q_z to cyt c_2 is not electrogenic, suggesting that these components do not span the low dielectric profile of the membrane (BOWYER et al. 1980; GLASER and CROFTS 1984). This proximity to the interface has been substantiated by the measurement of magnetic interactions between these centers and paramagnetic cations added to the aqueous phase on the membrane P-side. In bacteria, the FeS cluster is estimated to the 14 Å away from this interface (PRINCE 1983). Similar observations have been made for the FeS, cyt c_1 (OHNISHI and SALERNO 1982), and cyt b_L (OHNISHI and VON JAGOW 1985) in mitochondria.

The cyt b_H Heme. The portion of the membrane potential contributed by the cyt b_L to cyt b_h electron transfer in bacterial chromatophores suggests that these components span 35–50% of the low dielectric membrane profile (GLASER and CROFTS 1984). This is supported by measurements in mitochondria of small magnetic interactions between the cyt b_h heme and paramagnetic cations in the aqueous phases, suggesting that the heme is buried in the middle of the membrane (OHNISHI and VON JAGOW 1985). This separation between cyt b_L and b_h hemes is in accord with proposals based upon the amino acid sequences, where the two heme-binding sites are predicted to lie 20 Å apart on opposite ends of transmembrane α-helix segments (WIDGER et al. 1984; HAUSKA, Chap. 10.1, this Vol.).

4.2 Redox Center Separation: cyt b_h − Q

Virtually no measurements have been made on the separation of redox centers in the bacterial b/c_1 complex. In the mitochondrial complex, an attenuation of the ferri cyt b_h EPR signal is associated with the presence of a semiquinone Q^- species, which led to the proposal that these two components are close enough to experience a magnetic exchange interaction (DE LA ROSA and PALMER 1983). This quinone could be bound at the Q_c site.

4.3 Orientation of Redox Centers

cyt b_h and b_L Hemes. Optical linear dichroism analysis indicates that the ferro cyt b_h and b_L heme planes are well-oriented at >55° to the membrane plane in both chromatophores (*Rps. sphaeroides*) and inner mitochondrial membranes (TIEDE et al. 1984). This agrees with EPR studies on oriented mitochondrial membranes which had shown that the ferri cyt b hemes are approximately perpendicular to the membrane surface (ERECIŃSKA et al. 1978). The addition of the inhibitor antimycin does not appreciably alter the orientation of the cyt b (TIEDE et al. 1984).

cyt c_1 Heme. In contrast to the optical dichroism from the approximately perpendicularly oriented b hemes, a dichroism is not identified for the c_1 heme in mitochondrial membranes, nor in chromatophores of *Rps. sphaeroides* (TIEDE et al. 1984). This would suggest either a random orientation, or a tilt near the "magic angle" of 55° between the heme plane and the membrane surface (i.e., 55° between the heme and membrane normals). Overlapping Bchl dichroism make the measurement difficult in chromatophores. These results disagree with earlier reports of a perpendicular orientation for the c_1 in mitochondria (ERECÍNSKA et al. 1978), and can be compared with an angle of 30° between cyt f plane and membrane surface in chloroplasts (CROWDER et al. 1982; BERGSTRÖM and VÄNNGÅRD 1982).

FeS Cluster. The Fe—Fe axis is oriented parallel to the membrane surface in bacterial chromatophores and chloroplasts (PRINCE 1983) and mitochondria (OHNISHI and SALERNO 1982). In chromatophores the inhibitors UHNQ and UHDBT cause a shift in the EPR absorption g-values, but do not alter the FeS orientation (PRINCE 1983).

There is no information on the orientation of these redox centers with respect to each other. The presence of transmembrane α-helix structures have been predicted from the protein sequence in the b protein of mitochondria (WIDGER et al. 1984), but direct measurements have not been made. These structural features are summarized in Fig. 2.

References

Bergström J, Vanngård T (1982) Biochim Biophys Acta 682:452–456
Blasie JK, Pachence JM, Tavormina A, Dutton PL, Stamatoff J, Eisenberger P, Brown G (1983) Biochim Biophys Acta 723:350–357
Bowyer JR, Dutton PL, Prince RC, Crofts AR (1980) Biochim Biophys Acta 592:445–460

Butler WF, Calvo R, Fredkin DR, Isaacson RA, Okamura MY, Feher G (1984) Biophys J 45:947–973
Crowder MS, Prince RC, Bearden A (1982) FEBS Lett 144:204–208
Crofts AR, Wraight CA (1983) The electrochemical domain of photosynthesis. Biochim Biophys Acta 726:149–185
Dismukes GC, Frank HA, Friesner R, Sauer K (1984) Biochim Biophys Acta 764: 253–271
Dutton PL, Prince RC, Tiede DM (1978) Photochem Photobiol 28:939–949
Erecínska M, Wilson D, Blasie JK (1978) Biochim Biophys Acta 501:63–71
Gast P, Hoff AJ (1979) Biochim Biophys Acta 548:520–535
Gast P, Groot A de, Hoff AJ (1983) Biochim Biophys Acta 723:52–58
Glaser EG, Crofts AR (1984) Biochim Biophys Acta 766:322–333
Hales BJ, Gupta AD (1979) Biochim Biophys Acta 548:276–286
Hoff AJ (1982) Mol Biol Biochem Biophys 35:80–151
Jackson JB, Dutton PL (1973) Biochim Biophys Acta 325:102–113
Kirmaier D, Holten D, Parson WW (1983) Biochim Biophys Acta 725:190–202
Lubitz W, Lendzian F, Scheer H, Gottstein J, Plato M, Möbius K (1984) Proc Natl Acad Sci USA 81:1401–1405
Mar T, Gingras G (1984) Biochim Biophys Acta 764:283–294
Nabedryk E, Tiede DM, Dutton PL, Breton J (1982) Biochim Biophys Acta 682:273–280
Nabedryk E, Tiede DM, Dutton PL, Breton J (1983) In: Sybesma C (ed) Proc 6th Int Congr Photosynth, vol III. Nijhoff/Junk, The Hague, pp 177–180
Ohnishi T, Jagow von (1985) Biophys J 47:241a
Ohnishi T, Salerno JC (1982) In: Spiro TG (ed) Iron-sulfur proteins, vol IV. Wiley, New York, pp 285–327
Okamura MY, Feher G, Nelson N (1982) In: Govindjee (ed) Photosynthesis, vol I. Academic Press, London New York, pp 195–272
Pachence JM, Dutton PL, Blasie JK (1981) Biochim Biophys Acta 635:267–283
Packham NK, Dutton PL, Mueller P (1982) Biophys J 37:465–473
Paillotin G, Vermeglio A, Breton J (1979) Biochim Biophys Acta 545:249–264
Parson WW (1982) Annu Rev Biophys Bioeng 11:57–80
Prince RC (1983) Biochim Biophys Acta 723:133–138
Prince RC, Tiede DM, Thornber JP, Dutton PL (1977) Biochim Biophys Acta 462:467–490
Rafferty CN, Clayton RK (1979) Biochim Biophys Acta 546:189–206
Rosa FF de la, Palmer G (1983) FEBS Lett 144:204–208
Shuvalov VA, Asadov AA (1979) Biochim Biophys Acta 545:296–308
Tiede DM, Dutton PL (1981) Biochim Biophys Acta 637:278–290
Tiede DM, Prince RC, Dutton PL (1976) Biochim Biophys Acta 449:447–467
Tiede DM, Leigh JS, Dutton PL (1978) Biochim Biophys Acta 503:524–544
Tiede DM, Rich P, Choquet Y, Breton J (1984) In: Proc 8th Int Biophys Congr, Bristol UK
Trissl HW (1983) Proc Natl Acad Sci USA 80:7173–7177
Vermeglio A, Clayton RK (1976) Biochim Biophys Acta 449:510–515
Vermeglio A, Breton J, Paillotin G, Cogdell R (1978) Biochim Biophys Acta 501:514–530
Widger WR, Cramer WA, Herrmann R, Trebst A (1984) Proc Natl Acad Sci USA 81:674–678

8.3 Energy Trapping in Photosynthesis as Probed by the Magnetic Properties of Reaction Centers

J.R. NORRIS and G. VAN BRAKEL

1 Introduction

Photosynthesis is a series of events and processes by which the electromagnetic energy of light is transformed into free chemical energy for synthesis of biological material. The primary processes of photosynthesis involve the absorption of light by an antenna system and the subsequent transfer of this energy to a chemical trap usually known as a reaction center (see SAUER, Chap. 2, this Vol. for an overview). Of relevance to this review is the use of magnetic resonance to reveal the structural and mechanistic features of photosynthesis that result in the harvesting of electromagnetic radiation.

2 Mechanistic Aspects of the Photosynthetic Process

2.1 Key Features

The properties of photosynthesis significant to this review are the following. (1) A pigment system results in the efficient capture of much of the visible and near infrared light spectrum. (2) This antenna transfers the trapped photons to a relatively small number of pigment molecules in which the energy resides in the form of excited singlet states (KNOX 1977, Chap. 7.1, this Vol.). (3) The chemical trapping of the energy begins when these pseudo-localized excited singlet states migrate to the reaction center complex. (4) The chemical trapping is a result of rapid formation of pairs of separated radicals. (5) The chemical trapping occurs with the transfer of an electron from a donor (bacteriochlorophyll in purple bacteria) to a nearby acceptor (bacteriopheophytin in purple bacteria; ROCKLEY et al. 1975; KAUFMANN et al. 1975; THURNAUER et al. 1975; PARSON et al. 1975; FAJER et al. 1975; SHUVALOV and PARSON 1981; BOWMAN et al. 1981; NORRIS et al. 1982; PARSON and HOLTEN, Chap. 8.1, this Vol.). (6) The entire process can exhibit a quantum yield near unity.

2.2 Primary Photochemistry

In bacterial systems, optical-absorption spectroscopy indicates that the first photo-induced radical pair charge separation is completed within approximately

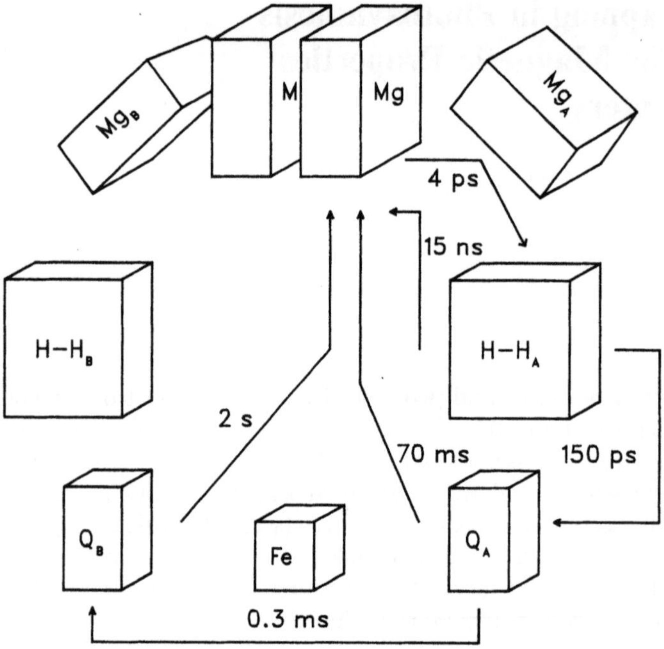

Fig. 1. Schematic approximation of the electron transfer components of the donor acceptor complex as found in bacterial reaction centers. The protein holding this assembly together is omitted. This drawing is based in part on X-ray structural information of *Rps. viridis* (DEISENHOFER et al. 1984; MICHEL and DEISENHOFER, Chap. 8.4, this Vol.) and on spectroscopic data on *Rps. sphaeroides*. *Boxes* represent aromatic ring systems associated with bacteriochlorophyll, (*M Mg*) and (*Mg*), and bacteriopheophytin, (*H*−*H*), and the quinones (*Q*) such as ubiquinone or menaquinone. One box does not represent a ring system, namely the Fe^2. The subscripts *A* and *B* represent two potential electron transfer pathways of which only the A pathway appears active

four picoseconds(ps) (ROCKLEY et al. 1975; KAUFMANN et al. 1975; SHUVALOV and PARSON 1981). In bacteria first charge-separated state will be referred to as the P^F radical pair. P^F is an excited state composed primarily of bacteriochlorophyll cation (M Mg) and bacteriopheophytin anion ($H-H_A$) located within the reaction center protein complex. Figure 1 depicts the donor-acceptor components of the reaction center. Shortly following the 4 ps charge separation, the electron jumps from the primary acceptor, bacteriopheophytin ($H-H_A$), to an additional acceptor molecule, a quinone-like moiety (Q_A), in ~150 ps. In other words, the observed natural lifetime of P^F is ~150 ps. However, if the secondary quinone acceptor Q_A is blocked, then the charge separation state P^F lives much longer, or ~15 ns as illustrated in Fig. 2. Picosecond studies on green plants will be ignored.

Because the initial charge separation act, as defined by the state P^F, is completed in a few ps, it might be assumed that donor and acceptor are in very close proximity. As a consequence of this assumption it would also be expected that charge separation annihilation would occur in considerably less time than

~15 nanoseconds(ns), as observed in quinone removed reaction centers. Thus, the close proximity as interpreted from the fast formation time of P^F in blocked reaction centers is inconsistent with an annihilation reaction as slow as the observed ~15 ns. As a result, a major question concerning the chemical trapping mechanism is how charge separation can occur within 4 ps in a system in which the reverse step can be prevented for as long as 15 ns. Such contradicting interpretations and observations on the reaction center chemistry may be resolved in terms of an additional, intermediate state that precedes P^F, as has been suggested previously (SHUVALOV and PARSON 1981; HABERKORN et al. 1979; SCHENCK et al. 1982).

Such an intermediate state is probably an additional chlorophyll acceptor that mediates electron transfer from the primary donor, bacteriochlorophyll, M Mg, to the primary acceptor, bacteriopheophytin, $H-H_A$, as depicted in Fig. 1. This possible intermediate state will be referred to as the bridging intermediate species, Mg_A. The exact arrangement of primary donor, bridging intermediate, and primary acceptor are crucial to the transfer of energy from antenna to trap, as well as the "permanent" trapping of this energy as chemical charge separation.

At this stage in the energy trapping scheme, the donor, the bridging intermediate and the acceptor are too close together to ensure "permanent" chemical trapping. Consequently, secondary and tertiary electron transfer is essential to complete the trapping process. These additional transfer steps increase the effective charge separation and thereby prevent the likelihood of back reaction. The undesirable back reactions correspond to charge separation annihilation, for example the electron of bacteriopheophytin anion returning to bacteriochlorophyll cation. Although the secondary chemistry is slower than the primary charge separation, it still proceeds with quantum yield near unity. The secondary state of the charge separation begins when the electron advances from the primary acceptor bacteriopheophytin to the secondary (Q_A) and tertiary (Q_B) acceptors that involve quinone-like molecules (Fig. 1). In an attempt to elucidate these aspects of photosynthesis we review primarily the oxidized primary donor, the reduced acceptor molecules, and the structural and mechanistic features that lead to energy trapping.

2.3 Consequences of Blocking Secondary-Tertiary Chemistry

In *Rhodopseudomonas sphaeroides* the quinones of the secondary (Q_A) and tertiary (Q_B) electron transfer chemistry can be removed chemically from the reaction center protein complex, blocking any further electron transport on the acceptor side of photosynthesis that would have occurred in intact systems. Or the quinones can be saturated with electrons before light excitation, again resulting in termination of secondary electron transfer. Instead of additional "forward" electron movement in these blocked systems charge separation is reversed. In other words, the electron on the primary acceptor back reacts by returning to the primary donor (Fig. 2). These blocked systems serve as "model systems" to test the current understanding of the primary chemical trapping mechanism.

This type of "blocked" reaction center is particularly useful in the study of the photosynthetic process by magnetic resonance techniques. In these depleted or blocked systems, a back-reaction occurs in 10 to 30 ns because the charge separation process cannot proceed beyond the excited state, P^F.

Even when quinone is removed to focus on the truly critical aspect of chemical trapping, the precise nature of the charge separation and the subsequent annihilation processes is complicated. For example, as depicted in Fig. 2 charge annihilation from P^F proceeds via two routes, one with singlet (all electrons paired) and the other with triplet electron spin multiplicity (Thurnauer et al. 1975) (two unpaired electrons).

The singlet annihilation directly restores the ground state of the donor, while the triplet channel proceeds mostly via the lowest excited triplet state of the donor, P^R, which has a lifetime of the order of 10 to 100 µs before ultimately restoring the singlet ground state of the system. The scheme as depicted in Fig. 2 and described here is too simple to explain all observations. Both back-reactions from $^3P^R$ to $^3P^F$, as well as additional states, perhaps states containing the bridging intermediate are probably involved (Shuvalov and Parson 1981; Boxer et al. 1985). The P^F annihilation is characterized by a lifetime as long as 25 ns. This slow rate is more consistent with electron transfer over relatively long distances, a conclusion in apparent contradiction to close proximity between donor and acceptor necessitated by the much faster (4 ps) forward charge separation. A particular aspect of this review is to describe the application of such magnetic techniques to probe the structural and mechanistic properties that provide the means to trap light energy through charge separation.

3 Components of the Reaction Center

3.1 Protein

Purified bacterial reaction centers have been obtained from a considerable number of species. They characteristically contain three polypeptide subunits, as revealed by SDS-polyacrylamide gel electrophoresis (SDS-PAGE). These subunits are labeled L, M, and H (light, medium, and heavy molecular weight) and occur in a stoichiometric ratio of 1:1:1 (Okamura et al. 1974; Dutton, Chap. 5; see also Hearst, Chap. 8.5; Bachofen and Wiemken, Chap. 11.2, all this Vol.).

The RC protein subunit structure of preparations from *Rp. sphaeroides* R-26, *Rhodospirillum rubrum* G-9 and *Chromatium vinosum* (see Sprague and Varga, Chap. 11.1, this Vol. for general characterization of cell types) appear to be remarkably similar (Feher and Okamura 1978), though slight variations occur. The apparent molecular weight determined by SDS-PAGE, for the three subunits is 21, 24, and 28 kDa, respectively (Okamura et al. 1974). Partial amino acid sequences of the L and H subunits have also been published (Sutton et al. 1982; Hearst, Chapter 8.5).

A recent, very promising development, largely determined by properties of the proteins, is the crystallization of purified RC from *Rps. viridis* MICHEL (1982), GAST et al. (1983), NORRIS and GAST (1985) and *Rps. sphaeroides* ALLEN and FEHER (1984), GAST and NORRIS (1984), NORRIS and GAST (1985). This opens up the possibility of detailed structural analysis of the reaction center and its protein environment by X-ray diffraction studies (see MICHEL and DEISENHOFER, Chapter 8.4, and MICHEL 1982; GAST et al. 1983; GAST and NORRIS 1984; NORRIS and GAST 1985; ALLEN and FEHER 1984; VAN WYK et al. 1984; ZINTH et al. 1983).

3.2 Primary Donor

When a quantum of light is absorbed, the bulk of the energy of the photon is funneled through the antenna system and arrives at the red-shifted donor of the reaction center. This red shift is partly responsible for the physical process of energy trapping. Within about 4 ps the primary donor, M Mg, (at least in bacteria) ejects an electron to the first acceptor, $H-H_A$, and thus chemical trapping of the light energy begins. Consequently, much research has been devoted to the nature and properties of this primary donor.

The most conspicuous property of the primary donor complex, P-870, has been the red-shifted wavelength of the absorption maximum, 860–870 nm in BChl a-containing bacteria (COGDELL, Chap. 6.2, THORNBER, Chap. 3, both this Vol.). The in vitro optical maximum of monomeric BChl a is near 760 nm. In BChl b-containing organisms the absorption maximum is 960 nm and correspondingly the primary donor has been designated P-960 (COGDELL, Chap. 6.2, THORNBER, Chap. 3, both this Vol.). These wavelengths are shifted considerably relative to the absorption maximum of isolated BChl a or b in liquid solution as well as to other components of the reaction center or antenna pigments. The traditional explanation for these shifts is a contribution of two effects: the primary donor consists of two Chl molecules folded over each other, and the resulting interaction between the two halves accounts for a part of the shift. The atomic structure determined by X-ray diffraction on single crystals of *Rps. viridis* shows that the two monomer chlorophylls interact via Mg-carbonyl bonds and macrocycle ring overlap (see MICHEL and DEISENHOFER, Chap. 8.4, this Vol.). The rest of the shift is likely caused by other molecules in the environment, most probably the proteins. Moreover, model systems of BChl dimers have red-shifted optical properties quite similar to the red-shifted primary donor SHERZ and PARSON (1984a, b).

Recently, there have been attempts to amplify these explanations for the red shift of the primary donor. These include measurements R.C. DAVIS et al. (1981) and calculations (ECCLES and HONIG 1983; PANCOSKA et al. 1983) on the effect of charges in the neighborhood of porphyrins. These charges may be present in vivo on amino acid residues of the surrounding protein. Also, the possibility that the red shift is produced by a chemical alteration of a chlorophyll molecule has been considered (PEARLSTEIN et al. 1982; HANSON 1984; HANSON et al. 1984).

From the initial X-ray studies on *Rps. viridis* clearly a dimer exists for the primary donor. The structure emerging from the X-ray work and the high resolution ENDOR (Electron Nuclear Double Resonance) studies indicates strong interactions between the two monomers of the dimer. One view is that of a super-molecule dimer as the donor for which exciton bands would be more difficult to calculate. Strong interactions have been included in the red-shift explanation of SHERZ and PARSON (1984a, b).

3.3 Acceptors

3.3.1 Bridging Intermediate and Primary Acceptor

The molecule that is currently regarded to be the first unambiguous recipient of the electron from the primary donor is one of the Bpheo, labeled $H-H_A$ in Fig. 1. However, an additional BChl, Mg_A of Fig. 1, is probably involved as the bridging species that "superexchanges" the electron from donor to the Bpheo labeled $H-H_A$ in Fig. 1. Certainly, it is our view that this bridging molecule is completely necessary to the chemical trapping process. Removal of this intermediate molecule without altering the location of the other donor or acceptor molecules should result in slow or negligible electron transfer. Without the fast 4 ps initial reaction, the chemical trapping cannot compete efficiently with physical de-trapping processes such as fluorescence etc. In addition, the structure emerging from X-ray diffraction studies, in our view, supports the necessity of a bridging molecule. Certainly, the recently successful model compounds that mimic photosynthetic charge separation work via a "bridging" moiety (WASIELEWSKI et al. 1985; WASIELEWSKI and NIEMCZYK 1984; MOORE et al. 1984). However, others may consider the in vivo participation of the BChl intermediate debatable.

3.3.2 Acceptor Complex Q_A-Fe-Q_B

The quinone-iron complex Q_A-Fe-Q_B of Fig. 1, sometimes called the primary acceptor, is actually the secondary-tertiary acceptor complex. Typical photochemistry of this complex under normal photosynthetic circumstances is given in Fig. 1 (KAUFMANN et al. 1975; BOWYER et al. 1979; INOUE et al. 1983; WRAIGHT 1979). The various quinones in purple photosynthetic bacteria have been discussed by PARSON (1978), HALSEY and PARSON (1974), PFENNIG (1978), BOWYER et al. (1979), BODMER et al. (1981), FEHER and OKAMURA (1978), PUCHEU et al. (1976), GAST et al. (1984) and WRAIGHT (1979). More details on the quinones can be found in the overview Chapter of DUTTON (this Vol.).

The function of the iron ion is still unclear, although many suggestions have been given (DUTTON et al. 1978; DISMUKES 1984). The magnetic interaction between Fe^2 and the reduced state of Q_A appears to be slightly more than with the tertiary acceptor, which some take as an indication that the iron is a little closer to the secondary acceptor, Q_A. The distance between the secondary acceptor and the iron is estimated to be $+7.0/-$ 0.8 Å from EPR data (DIS-

MUKES 1984). Also, the iron can be removed or replaced by Mn without interfering with the electron transfer from Bpheo to the secondary acceptor LOACH and HALL 1972; FEHER et al. 1972, 1974).

4 Probing of Chemical Trapping by Magnetic Resonance Techniques

4.1 Introduction

The application of magnetic resonance in photosynthesis is based on the fact that each electron transfer step involves products where an unpaired electron is located on a molecule for a considerable period of time. In this way, radicals are produced that are candidates for investigation by EPR and all related techniques. Usually, in this type of experiment, the normal flow of electrons is blocked somewhere in order to increase the yield and lifetime of a particular radical to facilitate detection such as the scheme of Fig. 2.

The normal process of photosynthesis involves energy trapping through charge separation, which implies that two radicals are being formed. Because of the nature of the interactions between the two electrons of two radicals in a radical pair, triplet states can be formed as a result of charge recombination. These radical pairs are suitable for study by optically detected magnetic reso-

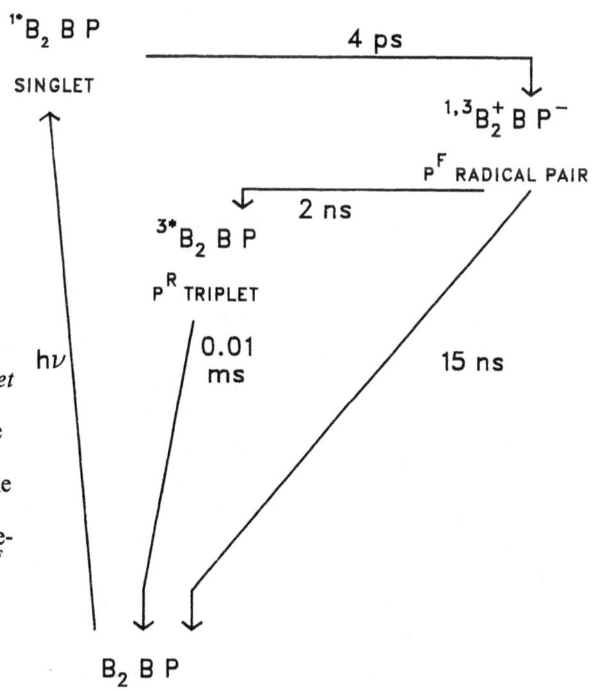

Fig. 2. The electron transfer scheme in Q_A removed or blocked reaction centers of *Rps. sphaeroides*. The *singlet* vs. *triplet* stages of the charge separation and subsequent annihilation are explicitly labeled. The effective lifetime of P^F is mostly due to the singlet channel, the reaction pathway labeled with a 15 ns lifetime. Note, however, that the P^F lifetime can extend to as long as 50 ns when the RC's are at low temperature and in the presence of magnetic fields

nance (ODMR) and any subsequent triplet states are also suitable candidates for both EPR and ODMR detection (CLARKE 1982; HOFF, Chap. 8.7, this Vol.).

Many techniques related and derived from EPR have been developed and applied to photosynthesis, often giving results not obtainable in any other way. ENDOR (Electron Nuclear Double Resonance) is perhaps most prominent among these. The results have been reviewed (NORRIS et al. 1978, 1979, but new work has been done since then LUBITZ et al. 1981; LENDZIAN et al. 1981; O'MALLEY and BABCOCK 1984a; LUBITZ et al. 1984, 1985). FDMR (Fluorescence Detected Magnetic Resonance) has been extensively used and reviewed (SCHAAFSMA 1982; HOFF 1982a). ADMR (Absorption Detected Magnetic Resonance) has been a new modification (DEN BLANKEN and HOFF 1982, 1983; DEN BLANKEN et al. 1983; HOFF, Chap. 8.7, this Vol.). Both FDMR and ADMR are extensions of ODMR. ESE (Electron Spin Echo) has found application, as reviewed in THURNAUER (1979). A new form of optically detected magnetic resonance (ODMR) known as reaction yield detected magnetic resonance (RYDMR) has been applied to the primary radical pair state of bacterial photosynthesis (BOWMAN et al. 1981; NORRIS et al. 1982; WASIELEWSKI et al. 1983a, b, 1984). The numerous results from all these techniques have given cause for ample theoretical contemplations and speculations. For a recent review see BOXER and ROELOFS (1983).

4.2 ENDOR

The nature of the primary donor has been extensively investigated by the ENDOR technique. A main question probed by ENDOR is whether the primary donor is a monomer or a dimer of Chl a in plants, or BChl in bacteria. Earlier ENDOR work in solid solutions of reaction centers has been reviewed (HOFF 1979, 1982b). In more recent studies of the group of Möbius (LUBITZ et al. 1981, 1984; LENDZIAN et al. 1981; MÖBIUS et al. 1982) the ENDOR technique has been extended to liquid solutions resulting in vastly improved spectral resolution. These workers interpret ENDOR in terms of a special pair of chlorophylls as the primary donor. BABCOCK and O'MALLEY have published low temperature ENDOR results on spinach chloroplasts (O'MALLEY and BABCOCK 1984a). Their work is notable because their interpretation leads to the prediction of a monomer structure of the primary donor in plants (O'MALLEY and BABCOCK 1984a), as well as in bacteria (O'MALLEY and BABCOCK 1984b).

Feher's group has measured liquid solution ENDOR of ^{14}N and ^{15}N monomer BChl a^+ and ^{15}N of RC of *Rps. sphaeroides* R-26 (LUBITZ et al. 1985). The N hyperfine coupling constants (hfcs) that are obtained can be compared to those measured with ESE (BOWMAN 1984) and they compare favorably.

The model that BABCOCK and O'MALLEY use to interpret their data is based on MO calculations published by PETKE and coworkers (PETKE 1980). These interpretations are similar to the proposals of M.S. DAVIS et al. (1979) for the primary donor of PS II, P-680. These authors assumed that the HOMO of Chl a^+ in P-680 is a mixed orbital caused by in vivo environmental factors. Based on this model, the sign of the hyperfine interaction can be predicted and then compared to experiment. Accordingly, the model advocated by BABCOCK and O'MALLEY predicts negative signs for the ^{15}N hfcs (LUBITZ et al. 1985). Positive signs have been determined by a general TRIPLE resonance experiment (LUBITZ et al. 1985). A personal communication with M. PLATO is cited to account for these positive signs (as well as for the observed magnitudes), based on advanced MO calculations. Thus the bulk of the evidence supports the special pair model for P-860.

4.3 EPR of Single Crystals

The full power of magnetic resonance for structural determinations can only be realized, however, when measurements are made on fully oriented samples, such as single crystals (GAST et al. 1983; GAST and NORRIS 1984; NORRIS and GAST 1985; ALLEN and FEHER 1984). The resonance positions of the triplet state of P^R can be determined precisely for single crystals (Fig. 3). For the crystals from *Rps. viridis* and *Rps. sphaeroides* the angles between the crystal

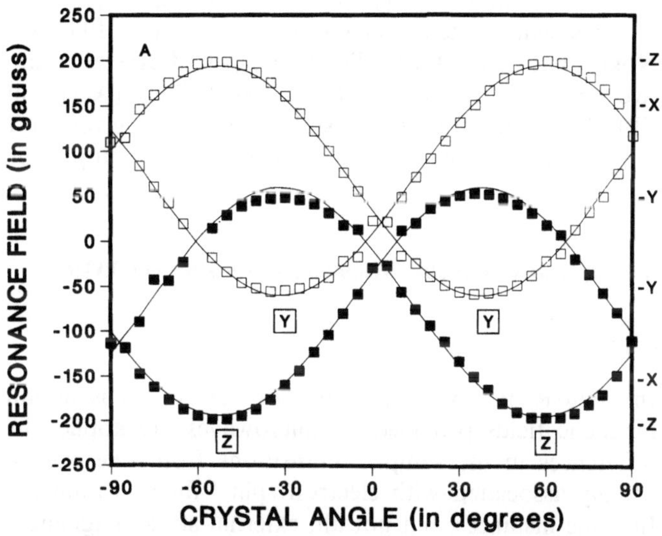

Fig. 3. EPR splittings for triplet P^R in single crystals of *Rps. sphaeroides* (GAST and NORRIS 1984). The vertical axis represents the distances between EPR resonance peaks as a function of the angle between crystal and magnetic field. Rotation is about the medium length growth axis of the crystal such that the applied, external magnetic field is perpendicular to the medium length crystal face axis. Note, even though there are four or eight triplet P^Rs per unit cell, at most angles four EPR peaks (the open and closed boxes) are observed, two resonances (TI0> to TI−1> and TI0> to TI+1>) for each angle, consistent with two magnetically inequivalent P^Rs per unit cell. *Solid lines* are calculated based on the angle that P^R makes with the unit cell axes. This precise angular dependence is necessary to locate P^R in the RC and is possible only in single crystals. The boxed Y and Z indicate crystal vs. field positions where the magnetic field is almost alligned with one of the three triplet axes in the crystal, i.e. the triplet y or z axis

Table 1. Triplet vs. cation EPR linewidths

Species	EPR Linewidth (Gauss)	
	Triplet	Cation
Rps. viridis	11.3	11.5
^{2}H *Rps. viridis*	6.8	4.6
Rps. sphaeroides R-26	10.3	9.6

axes and the triplet x, y, and z axes have been determined (GAST et al. 1983; GAST and NORRIS 1984; NORRIS and GAST 1985). The g-anisotropy of *Rps. sphaeroides* has also been published (ALLEN and FEHER 1984).

Interpretation of these splittings as a function of magnetic field and crystal angle accurately determines the angle of P^R and combined with structural data from X-ray should permit unambigous identification of P^R in molecular terms. EPR of the RC protein crystal from *Rps. viridis* revealed that the protein with a $P4_32_12$ symmetry (MICHEL 1982) has four magnetically nonequivalent sites, and has eight proteins per unit cell (GAST et al. 1983; GAST and NORRIS 1984; NORRIS and GAST 1985). The two types of crystal from *Rps. sphaeroides* have two and eight RC per unit cell, respectively (ALLEN and FEHER 1984). The EPR linewidths for two crystalline RC's are given in Table 1, together with the linewidths of the corresponding cation for comparison.

4.4 Optically Detected Magnetic Resonance (ODMR)

4.4.1 Introduction

In ODMR an optical property of the system is monitored while oscillating magnetic fields provided by microwaves are applied in resonance to energy differences in the sample. Transitions from one energy level to another are usually associated with electron spin "flips" within triplets or radical pairs. In some instances, but not all, external static magnetic fields are used to tune the size of the energy gaps in the sample to match the microwave frequency. If the microwaves alter the sample's optical properties by changing the population scheme of the energy levels, then ODMR is possible. Recently, optical absorption has been successfully employed to detect the radical pair state P^F and the triplet state P^R in the photosynthetic process. The radical pair state is believed to occur during every photochemical event whereas the triplet state is believed to be a by-product mainly of blocked photochemistry and consequently is not a main component of the photosynthetic process. However, the triplet state does provide a means to probe a variety of photosynthetic properties such as reaction products, energy trapping, fluorescence quenching, energy transfer, etc.

4.4.2 ODMR of the Photosynthetic Triplet State, P^R

The main purpose of studying the triplet state, P^R, is to learn about the mechanism of charge separation and the structural nature of the primary donor since the triplet state is believed to reside in the primary donor (see also HOFF, Chap. 8.7, this Vol.). This form of ODMR is performed at very low temperatures, ~2 K, and depends on the unequal population of the three triplet spin sublevels. Each spin sublevel has a different decay rate to the ground state. Since the microwaves can push molecules occupying on spin sublevel decaying at one rate to that of another spin sublevel decaying at a different rate, the lifetime of the triplet state changes with the application of resonant microwaves. Subsequent to microwave absorption by the system the amount of triplet state in the sample changes because the rate changes. The changing concentration of triplet is then detected by optical means.

Recently DEN BLANKEN and HOFF (HOFF, Chap. 8.7, this Vol.) have been using this effect they called ADMR (see p. 360) to obtain singlet-triplet (S−T) optical density difference spectra. DEN BLANKEN and HOFF find similar (S−T) absorption spectra for all systems they have observed. Their interpretation is

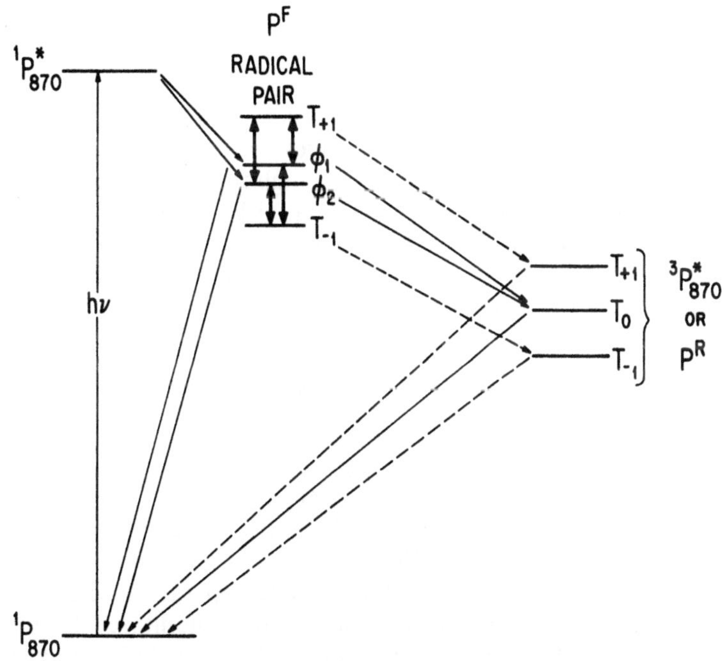

Fig. 4. RYDMR energy level diagram. *Single-headed arrows with solid lines* represent the decay pathway for P^F in the absence of microwave transitions represented by *double headed arrows*. The microwave transitions are effective only at certain magnetic fields when the external magnetic field splits the radical pair levels TI+> and TI−> to match the microwave energy giving rise to spectra such as in Fig. 6. The pathways represented by *dashed lines with single-headed arrows* are the new reactions that take place in the presence of microwaves. These new pathways can be detected by optically monitoring P^F or P^R, forming the basis of the RYDMR-ODMR

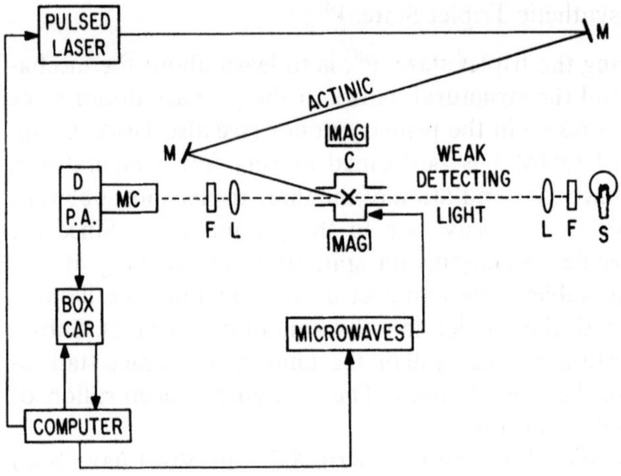

Fig. 5. Schematic representation of a RYDMR apparatus. *M* mirror; *MAG* external magnet; *F* light filter; *L* optical lens; *S* continuous optical source for measurement of optical density of sample X using the beam of light represented by a *dashed line*; *MC* monochromator; *D* optical photon detector; *P.A.* pre-amplifier. The state P^F is prepared by light from a pulsed laser while a microwave pulse is in the microwave cavity *C*

that the triplet state excitation resides on one of the monomer halves of the special pair donor for a time long enough for optical absorption to occur.

DEN BLANKEN and HOFF have data for both photosystem I and photosystem II primary donor in green plants. Again they interpret their (T−S) difference spectra in terms of dimeric species of chlorophyll. In this instance good agreement is found between a model system of Chl_2 and the in vivo systems.

4.4.3 Reaction Yield Detected Magnetic Resonance (RYDMR) of the Primary Radical Pair State, P^F

The experiment involves the application of 9 GHz microwave pulses to the P^F state in an external magnetic field close to the resonant field to the free electron as illustrated in Fig. 4. A simple representation of the experimental setup is shown in Fig. 5. While microwave pulses are applied to the reaction centers during the lifetime of P^F charge separation, the relative triplet yield as a function of external magnetic field is monitored by optical absorption spectroscopy. Typical RYDMR spectra are shown in Fig. 6. These room temperature microwave experiments, reaction yield detected magnetic resonance (RYDMR), have been described previously (BOWMAN et al. 1981; NORRIS et al. 1982; FRANKEVICH and PRISTUPA 1976). This form of RYDMR-ODMR (RUEDIN et al. 1972), which was recently applied to the study of photosynthesis, was originally developed by MOLIN and FRANKEVICH for the study of radical pairs outside the interest of photosynthesis (MOLIN et al. 1980; FRANKEVICH and PRISTUPA 1976; ANISIMOV et al. 1979a; ANISIMOV et al. 1979b).

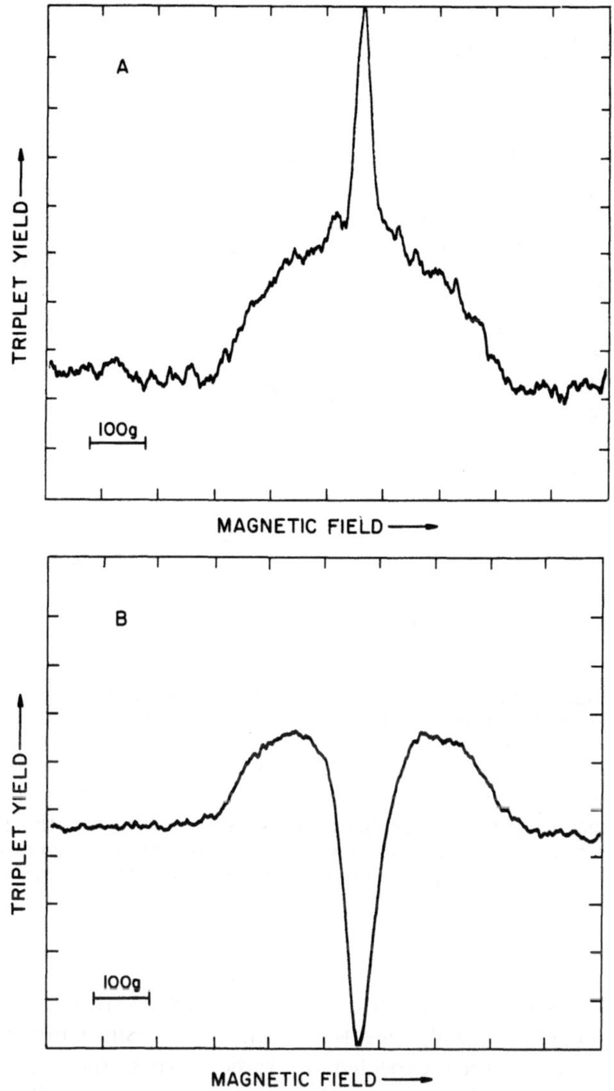

Fig. 6. Typical RYDMR spectra. *Vertical axis* uncalibrated optical density of P^R; *horizontal axis* external magnetic field. The microwave power is about 200 W in **A** whereas in **B** the microwave power is about 2000 W. (For a fuller explanation of these spectra see BOWMAN et al. 1981)

The RYDMR form of ODMR is applicable to photoproduced radical pairs such as associated with P^F. The interpretation of P^F as a radical pair was originally based on optical studies of P^F and also on standard EPR studies of the triplet state known as P^R. More detailed explanations can be found in references BOWMAN et al. (1981), NORRIS et al. (1982), DOKTOROV et al. (1982), LERSCH et al. (1982), TANG and NORRIS (1983), CHIDSEY et al. (1980), ROELOFS et al. (1982), BOXER et al. (1982a, b), CLOSS (1969), TANG and NORRIS (1982), HABER-

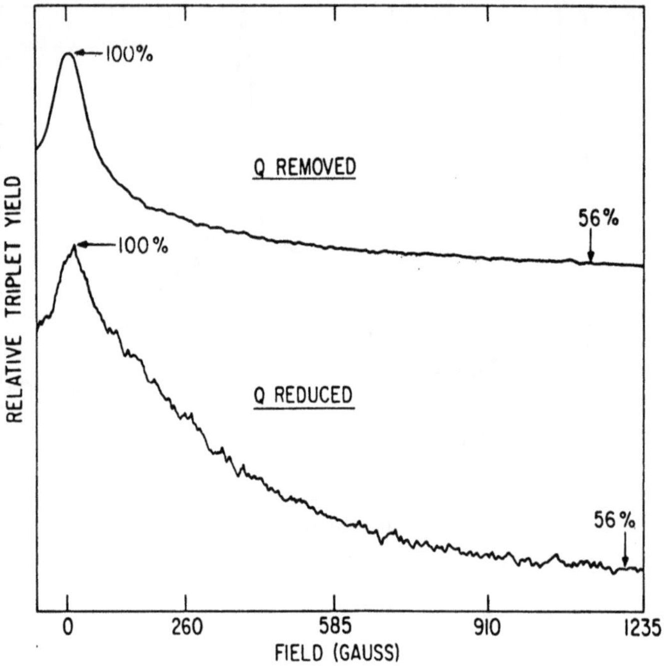

Fig. 7. Typical low field P^R triplet yield as a function of magnetic field. The change in triplet yield as a function of increased magnetic field has been reported in other studies (BOXER and ROELOFS 1983; CHIDSEY et al. 1980; OGRODNIK et al. 1982; BLANKENSHIP et al. 1977; HABERKORN and MICHEL-BEYERLE 1979; WERNER et al. 1978; BLANKENSHIP and PARSON 1979). At even higher magnetic fields the triplet yield increases with magnetic field. BOXER and co-workers have performed these high field experiments which are important because they are easier to interpret quantatively than the corresponding low field experiments. (BOXER and ROELOFS 1983; CHIDSEY et al. 1980)

KORN and MICHEL-BEYERLE (1979), MICHEL-BEYERLE et al. (1979, 1980), OGRODNIK et al. (1982), WERNER et al. (1978). Since microwaves affect the state P^F in the manner expected for radical pairs, the RYDMR experiment is direct proof that P^F is primarily a radical pair state. These experiments also prove that the radical pair scheme is responsible for the birth of triplets in the primary photochemical act when the secondary acceptors are blocked.

The observation of RYDMR-ODMR associated with the state P^F in *Rps. sphaeroides* has been interpreted in terms of an anisotropic magnetic dipole-dipole interaction between the spins of the triplet radical pair. The dipole-dipole interaction was invoked to explain the experimental observation that over 200 W of microwave power were required to observe these ODMR signals. Boxer and co-workers also found dipole-dipole interaction to be most likely involved in related experiments which probed triplet yield as a function of static external magnetic field. However, in other experiments this dipole-dipole interaction was not considered necessary for interpretation of triplet quantum yield changes as a function of magnetic field.

The ODMR experiments also were able to distinguish the lifetime of singlet radical pair from triplet radical pair. As expected from the theory of radical pairs, the lifetime of the P^F radical pair depends both on the size of the static external magnetic field and the size of the microwave magnetic field. Consequently, WASIELEWSKI et al. (1983b) demonstrated that the lifetime of P^F in blocked reaction centers could be shortened or lengthened by the application of resonant 9 GHz microwaves. The lifetime of singlet P^F is found to be about 20 ns whereas the triplet P^F has a lifetime of only about 2 ns. In addition, this ODMR work concluded that an isotropic exchange interaction of about 10 gauss accounted for the radical–radical interaction. The anisotropic dipolar interaction was found much larger, around 50 gauss.

The other type of experiment performed is the determination of the relative yield of the molecular triplet P^R, where the triplet occurs in the primary donor (D). The triplet yield is measured as a function of external magnetic field as illustrated in Fig. 7. The RYDMR apparatus of Fig. 5 is also suitable for this experiment, except that microwaves are not used. This molecular triplet of donor bacteriochlorophyll is formed by the self annihilation of the charge separation P^F. The external magnetic field has been varied from roughly zero gauss to 50000 gauss (BOXER and ROELOFS 1983; CHIDSEY et al. 1980; OGRODNIK et al. 1982; BLANKENSHIP et al. 1977; HABERKORN and MICHEL-BEYERLE 1979; WERNER et al. 1978; BLANKENSHIP and PARSON 1979).

5 Summary

Figure 1 depicts the presently known structural features necessary for charge separation. Since a dual pathway of charge separation does not seem to be in operation, the role of protein in the overall photochemical act must be unraveled. Progress on the protein aspect of photosynthesis is progressing rapidly due to the availability of single crystals of reaction center proteins MICHEL (1982), DEISENHOFER et al. (1984, MICHEL and DEISENHOFER, Chap. 8.4, this Vol.). Also the structure given in Fig. 1 is probably similar to the structure of the reaction centers in green plants (see DINER, Chap. 9.1, this Vol.). However, we already know certain structural features must be different. For example, since Chl a does not have a 2-acetyl group, any aggregation of chlorophylls in green plants must be different than that shown by X-ray for BChl b in *Rps. viridis* which involved Mg...2-acetyl interactions. Additional magnetic measurements requiring single crystals are also in progress, and promise to reveal more about the structure and mechanism of photosynthesis.

References

Allen JP, Feher G (1984) Proc Natl Acad Sci USA 81:4795–4799

Anisimov OA, Grigoryants VM, Molchanov VK, Molin YuN (1979a) Dokl Acad Nauk SSSR 248:380

Anisimov OA, Grigoryants VM, Molchanov VK, Molin YuN (1979b) Chem Phys Lett 66:265
Blanken HF den, Hoff AJ (1983) Biochim Biophys Acta 724:52–61
Blanken HJ den, Hoff AJ (1982) Biochim Biophys Acta 681:365–374
Blanken HJ den, Jongenelis APJM, Hoff AJ (1983) Biochim Biophys Acta 725:472–482
Blankenship RE, Parson WW (1979) Biophys J 25:205a
Blankenship RE, Schaafsma TJ, Parson WW (1977) Biochim Biophys Acta 461:297–305
Bodmer S, Snozzi M, Bachofen R (1981) In: Akoyunoglou G (ed) Photosynthesis II. Balaban, Philadelphia, pp 655–663
Bowman MK (1984) 8th DOE Solar Photochem Res Conf, Lake Geneva, Wisc
Bowman MK, Budil DE, Closs GL, Kostka AG, Wraight CA, Norris JR (1981) Proc Natl Acad Sci USA 78:3305–3307
Bowyer JR, Tierney GV, Crofts AR (1979) FEBS Lett 101:201–206
Boxer SG, Chidsey CED, Roelofs MG (1982a) J Am Chem Soc 104:2674–2675
Boxer SG, Chidsey CED, Roelofs MG (1982b) Proc Natl Acad Sci USA 79:4632–4636
Boxer SG, Chidsey CED, Roelofs MG (1983) Annu Rev Phys Chem 34:389–417
Chidsey CED, Roelofs MG, Boxer SG (1980) Chem Phys Lett 74:113–118
Clarke RH (ed) (1982) Triplet state ODMR spectroscopy. Techniques and applications to biophysical systems. Wiley, New York
Closs GL (1969) J Am Chem Soc 91:4552–4554
Davis MS, Forman A, Fajer J (1979) Proc Natl Acad Sci USA 76:4170–4176
Davis RC, Ditson SL, Fentiman AF, Pearlstein RM (1981) J Am Chem Soc 103: 6823–6826
Deisenhofer J, Epp O, Miki K, Huber R, Michel H (1984) J Mol Biol 180:385–398
Dismukes GC, Frank HA, Friesner R, Sauer K (1984) Biochim Biophys Acta 764:253–271
Doktorov AB, Anisimov OA, Burshtein AI, Molin YuN (1982) Chem Phys 71:1–8
Dutton PL, Prince RC, Tiede DM (1978) Photochem Photobiol 28:939–949
Eccles J, Honig B (1983) Proc Natl Acad Sci USA 80:4959–4962
Fajer J, Brune DC, Davis MS, Foreman A, Spaulding LD (1975) Proc Natl Acad Sci USA 72:4956–4960
Feher G, Okamura MY (1978) In: Clayton RK, Sistrom W (eds) The photosynthetic bacteria, chap 19. Plenum, New York London
Feher G, Okamura MY, McElroy JD (1972) Biochim Biophys Acta 267:222–226
Feher G, Isaacson RA, McElroy JD, Ackerson LC, Okamura MY (1974) Biochim Biophys Acta 368:135–139
Frankevich EL, Pristupa AI (1976) Pis'ma Zh Eksp Teor Fiz 24:304–308
Gast P, Norris JR (1984) FEBS Lett 177:277–280
Gast P, Wasielewski MR, Schiffer M, Norris JR (1983) Nature 305:451–452
Gast P, Michalski TJ, Hunt JE, Norris JR (1984) FEBS Lett in press
Haberkorn R, Michel-Beyerle ME (1979) Biophys J 26:489–498
Haberkorn R, Michel-Beyerle ME, Marcus RA (1979) Proc Natl Acad Sci USA 76:4185–4188
Halsey YD, Parson WW (1974) Biochim Biophys Acta 347:404–416
Hanson LK (1984) 8th DOE Solar Photochem Res Conf, Lake Geneva, Wisc, pp 5–7
Hanson LK, Chang CK, Ward B, Callahan PM, Babcock GT, Head JD (1984) J Am Chem Soc 106:3950–3958
Hoff AJ (1979) Phys Rep 54:75–200
Hoff AJ (1982a) In: Clarke RH (ed) Triplet state ODMR spectroscopy, chap 9. Wiley, New York, pp 367–425
Hoff AJ (1982) Biophys Struct Mech 8:107–150
Inoue Y, Crofts AR, Govindjee, Murata N, Renger G, Satoh K (eds) (1983) The oxygen evolving system of photosynthesis. Academic Press, London New York
Kaufmann K, Dutton PL, Netzel TL, Leigh JS, Rentzepis PM (1975) Science 188:1301–1304
Knox RS (1977) In: Barber J (ed) Topics in photosynthesis, chap 1, vol II. Elsevier, Amsterdam New York Oxford, pp 55–97

Lendzian F, Lubtiz W, Scheer H, Bubenzer C, Möbius K (1981) J Am Chem Soc 103:4635–4637
Lersch W, Ogrodnik A, Michel-Beyerle ME (1982) Z Naturforsch 37a:1454–1456
Loach PA, Hall RL (1972) Proc Natl Acad Sci USA 69:786–790
Lubitz W, Lendzian F, Möbius K (1981) Chem Phys Lett 81:235–241
Lubitz W, Lendzian F, Scheer H, Gottstein J, Plato M, Möbius K (1984) Proc Natl Acad Sci USA 81:1401–1405
Lubitz W, Isaacson RA, Abresch EC, Feher G (1985) Proc Natl Acad Sci USA in press
Michel H (1982) J Mol Biol 158:567–572
Michel-Beyerle ME, Scheer H, Siedlitz H, Tempus D, Haberkorn RE (1979) FEBS Lett 100:9–12
Michel-Beyerle ME, Scheer H, Seidlitz H, Tempus D (1980) FEBS Lett 110:129–132
Möbius K Fzöhling W, Lendzian F, Lubitz W, Plato M, Winscom CJ (1982) J Phys Chem 86:4491–4507
Molin YuN, Anisimov AO, Grigoryants VM, Molchanov VK, Salikhov KM (1980) J Phys Chem 84:1853
Moore TA, Gust D, Mathis P, Mialocq JC, Chachaty C, Benesasson RV, Land EJ, Doizi D, Liddel PA, Lehman WR, Nemeth GA, Moore AL (1984) Nature 307:630–632
Norris JR, Gast P (1985) J Photochem in press
Norris JR, Thurnauer MC, Bowman MK, Trifunac AD (1978) In: Dutton PL et al. (eds) Frontiers of biological energetics, vol I. Academic Press, London New York, pp 581–592
Norris JR, Scheer H, Katz JJ (1979) In: Dolphin D (ed) The porphyrins, chap 3, vol IV. Academic Press, London New York, pp 159–195
Norris JR, Bowman MK, Budil DE, Tang J, Wraight CA, Closs GL (1982) Proc Natl Acad Sci USA 79:5532–5536
Ogrodnik A, Kruger HW, Orthuber H, Haberkorn R, Michel-Beyerle ME, Scheer H (1982) Biophys J 39:91–99
Okamura MY, Steiner LA, Feher G (1974) Biochemistry 13:1394–1403
O'Malley PJ, Babcock GT (1984a) Proc Natl Acad Sci USA 81:1098–1101
O'Malley PJ, Babcock GT (1984b) In: Sybesma C (ed) Advances in photosynthesis research, vol I. Nijhoff/Junk, The Hague, pp 697–700
Pancoska P, Kapoun M, Skala L (1983) Photobiochem Photobiophys 6:39–46
Parson WW (1978) In: Clayton RK, Sistrom WR (eds) The photosynthetic bacteria, chap 25. Plenum, New York London, pp 455–470
Parson WW, Clayton RK, Cogdale RJ (1975) Biochim Biophys Acta 387:265–278
Pearlstein RM, Ditson SL, Davis RC, Fentiman AF (1982) Biophys J 37:112a
Petke JD, Maggiora GM, Shipman LL, Christoffersen RE (1980) Photochem Photobiol 31:243–257
Pfennig N (1978) In: Clayton RK, Sistrom WR (eds) The photosynthetic bacteria, chap 1. Plenum, New York London
Pucheu NL, Kerber NL, Garcia AF (1976) Arch Microbiol 109:301–305
Rockley MB, Windsor MW, Cogdell JJ, Parson WW (1975) Proc Natl Acad Sci USA 72:2251–2255
Roelofs MG, Chidsey CED, Boxer SG (1982) Chem Phys Lett 87:582–588
Ruedin Y, Schnegg PA, Jaccard C, Aegerter N (1972) Phys Status Solidi (b) 54:565
Schaafsma TJ (1982) Triplet state ODMR spectroscopy, chap 8. Wiley, New York, pp 291–365
Schenck CC, Blankenship RE, Parson WW (1982) Biochim Biophys Acta 680:44–59
Scherz A, Parson WW (1984a) Biochim Biophys Acta 776:653–665
Scherz A, Parson WW (1984b) Biochim Biophys Acta 776:666–678
Shuvalov VA, Parson WW (1981) Proc Natl Acad Sci USA 78:957–961
Sutton MR, Rosen D, Feher G, Steiner LA (1982) Biochemistry 21:3842–3849
Tang J, Norris JR (1982) Chem Phys Lett 92:136–140
Tang J, Norris JR (1983) Chem Phys Lett 94:77–80

Thurnauer MC (1979) Rev Chem Intermed 3:197–230
Thurnauer MC, Katz JJ, Norris JR (1975) Proc Natl Acad Sci USA 72:270–3274
Wasielewski MR, Niemczyk MP (1984) J Am Chem Soc 106:5043–5045
Wasielewski MR, Bock CH, Bowman MK, Norris JR (1983a) J Am Chem Soc 105:2903–2904
Wasielewski MR, Bock CH, Bowman MK, Norris JR (1983b) Nature 303:520–522
Wasielewski MR, Gast P, Norris JR (1984) In: Sybesma C (ed) Advances in photosynthesis research, vol I. Nijhoff/Junk, The Hague, pp 211–214
Wasielewski MR, Niemczyk MP, Svec WA, Pewitt EB (1985) J Am Chem Soc in press
Werner HJ, Schulten K, Weller A (1978) Biochim Biophys Acta 502:255–268
Wraight CA (1979) Photochem Photobiol 36:767–776
Wyk FGH van, Schaafsma TJ, Michel H (1984) In: Sybesma C (ed) Advances in photosynthesis research, vol II. Nijhoff/Junk, The Hague, pp 173–176
Zinth W, Kaiser W, Michel H (1983) Biochim Biophys Acta 723:128–131

8.4 X-Ray Diffraction Studies on a Crystalline Bacterial Photosynthetic Reaction Center: A Progress Report and Conclusions on the Structure of Photosystem II Reaction Centers

H. MICHEL and J. DEISENHOFER

1 Introduction

The earliest steps in the conversion of light energy into chemical free energy in photosynthesis involve the absorption of a photon by the light-harvesting antenna pigments, and transfer of the energy to the photosynthetic reaction center (RC). In the RC this energy (or energy derived from absorption of light by the reaction center itself) is used to transport an electron across the photosynthetic membrane (see SAUER, Chap. 2 and DUTTON, Chap. 5, this Vol. for an overview of the reactions).

In all known instances, the photosynthetic reaction centers are integral membrane protein complexes, whereas the light-harvesting antenna pigments can be contained either in integral membrane proteins or in water-soluble proteins which are attached to the photosynthetic membranes (see THORNBER, Chap. 3, this Vol.). Four types of reaction center are found in photosynthetic organisms: the photosystem I (PS I) and photosystem II (PS II) reaction centers of chloroplasts and cyanobacteria, the reaction centers from the purple photosynthetic bacteria and those from the green photosynthetic bacteria. There is some similarity between the reaction centers from PS I and those from the green photosynthetic bacteria, at least with respect to the first stable electron acceptor, which are iron-sulfur centers in both cases (OLSON and THORNBER 1979; BLANKENSHIP and FULLER, Chap. 8.6, this Vol.). There are also clear-cut similarities between the reaction centers of PS II and of the purple photosynthetic bacteria. In both types of reaction center the first stable electron acceptor is a quinone, Q_A, which is magnetically coupled to a ferrous iron atom, from where the electron is transferred to a secondary quinone, Q_B. In addition, both types of reaction center contain two pheophytin (OMATA et al. 1984) or bacteriopheophytin molecules (reviewed by FEHER and OKAMURA 1978; DUTTON, Chap. 5, this Vol.). In both instances it is possible to trap the electron on one of the pheophytin molecules (TIEDE et al. 1976; SHUVALOV and KLIMOV 1976; KLIMOV et al. 1977, 1980), indicating its role as intermediate electron carrier. An analogous spin-polarized triplet state of the primary donor P-680 in PS II reaction centers (RUTHERFORD et al. 1981) and P-870 (or P-960) in the reaction center of purple photosynthetic bacteria (DUTTON et al. 1971) strengthens the similarities. Finally, clear-cut sequence homologies exist between the Q_B (or herbicide binding or 32 kDa protein and the D-2 protein from PS II and the so-called L and M subunits of the reaction centers from purple photosynthetic bacteria (cf. ZURAWSKI et al. 1982; WILLIAMS et al. 1983; YOUVAN et al. 1984; RASMUSSEN et al. 1984; ROCHAIX et al. 1984; HEARST, Chap. 8.5, this Vol.).

The best-characterized reaction centers are those from purple photosynthetic bacteria (for reviews see FEHER and OKAMURA 1978; HOFF 1982; PARSON 1982; DUTTON, Chap. 5, PARSON and HOLTEN, Chap. 8.2, TIEDE, Chap. 8.2, all this Vol.). They consist of three protein subunits, which are called H (heavy), M (medium) and L (light). In several cases they contain an additional c-type cytochrome, as does the reaction center from the bacteriochlorophyll b containing purple photosynthetic bacterium *Rhodopseudomonas viridis*. The reaction center from *Rps. viridis* contains four bacteriochlorophyll b (BChl b) molecules, two bacteriopheophytin b (BPh b) molecules, one nonheme iron, one menaquinone as Q_A and one ubiquinone as Q_B (THORNBER et al. 1980; SHOPES and WRAIGHT 1983). Upon excitation, an electron moves within less than 20 picoseconds from the primary electron donor P-960 to one of the bacteriopheophytin b molecules and within 230 picoseconds to Q_A (HOLTEN et al. 1978; PARSON and HOLTEN, Chap. 8.1, this Vol.). The photooxidized primary donor is reduced by the cytochrome molecule in about 270 nanoseconds (HOLTEN et al. 1978). A monomeric bacteriochlorophyll b was postuled between the primary donor and the intermediate pheophytin acceptor (SHUVALOV and ASADOV 1979).

One of the secrets on which the high efficiency of biological light energy conversion is based is certainly the arrangement and structure of the pigments within the reaction center proteins. At present, the best method to obtain such structural information is X-ray crystallography. However, until recently X-ray structure analyses of membrane proteins were not possible, since three-dimensional crystals of X-ray quality were not available. By using a combination of small detergent and small amphiphilic molecules, which are probably bound to the hydrophobic protein surfaces via their alkane chains, it has become possible to crystallize isolated photosynthetic reaction centers from *Rps. viridis* and to obtain from such crystals high resolution X-ray diffraction data (MICHEL 1982). The reaction centers are photochemically active in the crystalline state (ZINTH et al. 1983; GAST et al. 1983; NORRIS and VAN BRAKEL, Chap. 8.3, this Vol.). Apart from this reaction center complex, only crystals of matrix porin from *Escherichia coli* have been obtained in a form suitable for high resolution structure analysis (GARAVITO et al. 1983).

The crystal structure of the reaction center of *Rhodopseudomonas viridis* has been determined at 3 Å resolution; a model of the arrangement of the heme groups, the bacteriochlorophyll and bacteriopheophytin pigments, and the iron quinone complex has been published (DEISENHOFER et al. 1984). In this communication we discuss the results of our analysis of the structure of the reaction center of *Rps. viridis,* and the implications of these results on the structure of PS II reaction centers as well as on the evolution of photosynthetic reaction centers.

2 Results and Discussion

2.1 The Electron Density Map and Its Interpretation

2.1.1 General Description

A stereo view of a section through a part of the electron density map corresponding to one RC complex, is shown in Fig. 1. The complex has an elongated shape with three distinguishable regions. The central part has the form of a cylinder with elliptical cross section (largest diameter ~70 Å, smallest diameter ~30 Å). The length of this cylinder is about 50 Å. A globular subunit is attached

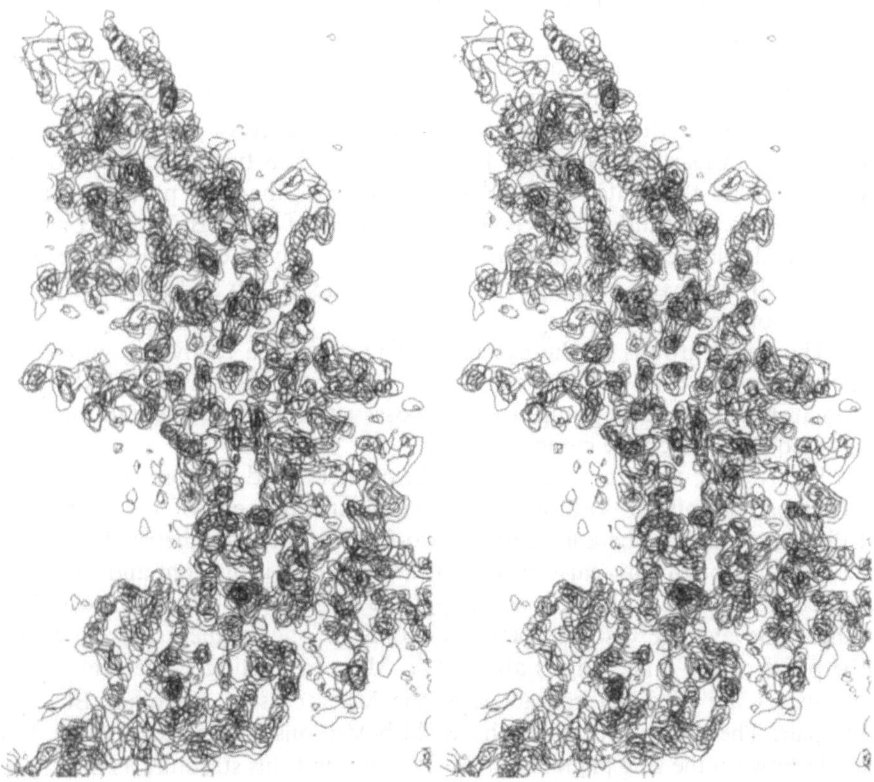

Fig. 1. Stereo drawing (scale 0.08 cm/Å) of 10 layers (distance between layers 1.2 Å) of the electron density distribution at 3 Å resolution. Contour lines start at 1.5 with increments of 1.8 (arbitrary units). The map was rotated so that the central local symmetry axis runs vertically in the middle layer. The section shows part of the central region with some chromophores; parts of the cytochrome (*top*), and of the H-subunit (*bottom*) can be seen

to each flat face of the cylinder. The centers of these peripheral subunits do not lie on the cylinder axis. The longest dimension of the whole RC is about 140 Å between the edges of the peripheral subunits. Within one of the globular subunits (the upper one in Fig. 1), the electron density map reveals four heme groups, indicating that this globular subunit is the cytochrome subunit. In the central region, electron densities corresponding to four BChl b and two BPh b, and a high electron density peak for the nonheme iron atom, are seen. Since these prosthetic groups are known to be associated with the L- and M-subunits, at least in *Rps. sphaeroides* (FEHER and OKAMURA 1978 and *Rhodospirillum rubrum* (WIEMKEN and BACHOFEN 1984; BACHOFEN and WIEMKEN, Chap. 11.2, this Vol.) we have to assume that these subunits form the central cylindrical part of the complex. The globular electron density attached to the central cylinder opposite the cytochrome subunit should therefore represents the H-subunit.

2.1.2 The Pigments and Their Arrangement

The electron density map allows clear distinction between BPh's and BChl's, since the electron densities attributed to the BPh's show a deep depression in the middle of the ring systems due to the absence of the magnesium ions (see Fig. 2 in DEISENHOFER et al. 1984). Electron density connections from protein chains to the center of each BChl b indicate that one protein side chain is bound as a ligand to the magnesium atom of each BChl b. Continuous electron density can be found for the complete phytyl chains of three BChl b and one BPh b, parts of the phytyl chains of one BChl b and one BPh b appear to be disordered.

The heme groups in the cytochrome part can be easily recognized from their shape and from the high electron density peaks associated with the central iron atoms. Each heme group seems to possess two sulfur bridges to the protein as is required for c-type cytochromes.

Figure 2 shows the arrangement of the pigments and the nonheme-iron atom. The most prominent feature in the spatial arrangement of the BChl b and BPh b molecules is the approximate two-fold rotation symmetry. The rotation axis runs through (or nearly through) the nonheme-iron atom and two clearly associated BChl b molecules, whose pyrrole rings I are stacked on top of each other. The average distance between atoms of one of these pyrrole rings to the plane of the other ring is about 3 Å. Most interestingly, the acetyl groups of rings I are in direct contact with the Mg atoms of the other BChl b in the pair. The distance between the BChl b-Mg ions is approximately 7 Å, the angle between the ring planes is ~15°. We interpret this structure as the "special pair", which was postulated on the basis of results of spectroscopic investigations (NORRIS et al. 1973, 1975; NORRIS, Chap. 8.3, this Vol.). The structure of the special pair excludes the possibility of a water molecule involved in the BChl b-BChl b contact as was postulated (for a review see NORRIS and KATZ 1978). To each BChl b of the special pair another BChl b is associated with Mg–Mg distances of ~13 Å and angles between the tetrapyrrole plains of about 70°. Each of these monomeric BChl b is in contact with a BPh b.

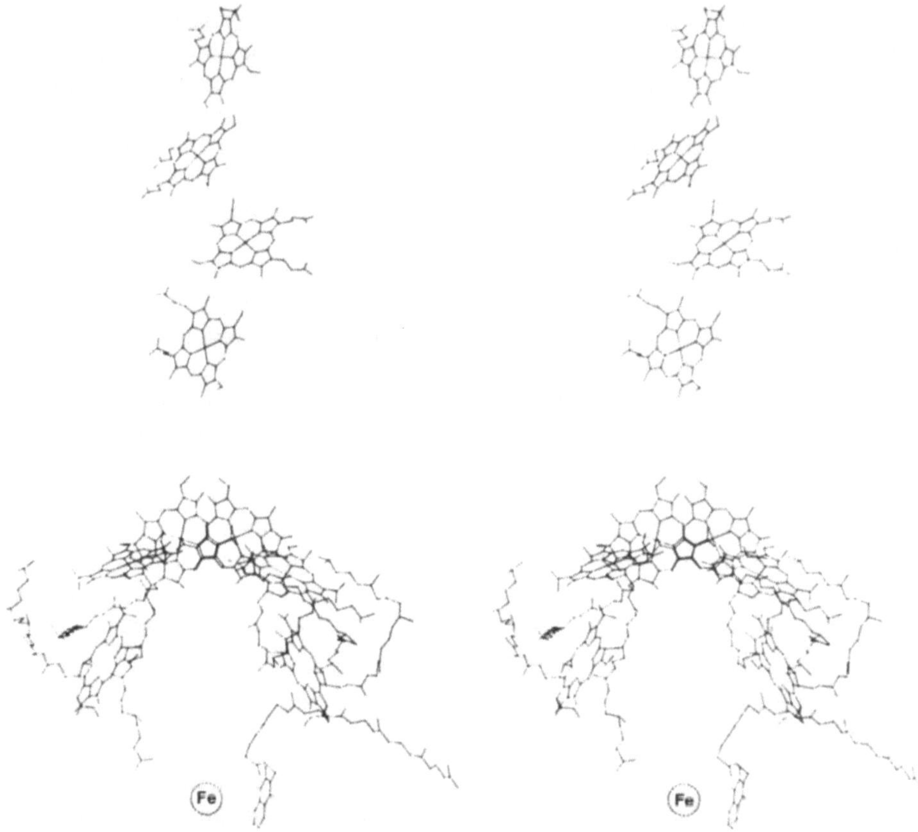

Fig. 2. Stereo drawing of the prosthetic groups of the RC showing 4 BChl b, 2 BPh b, 1 nonheme iron (*Fe*), 1 menaquinone, and 4 heme groups. The central local symmetry axis runs vertically in the picture plane. The plane of the membrane is assumed to be oriented perpendicular to the central local symmetry axis, i.e., horizontal and perpendicular to the picture plane. The BChl b, BPh b, quinone, and iron are located within the cylindrical central part of the RC complex

The distances between the ring centers are ~11 Å, the angles between the ring planes ~64°. The local twofold rotation axis relates the tetrapyrrole systems of these monomeric BChl b molecules, as well as the tetrapyrrole ring systems of the BPh b molecules. However, the phytyl side chains of the pigments do not obey the local twofold rotation symmetry.

From the nonheme-iron atom four strong electron density connections extend to four helices of the protein. At least in three cases the electron density is consistent with histidine side chains. Very close to one of these presumed histidines and the nonheme-iron atom (closest Fe-quinone distance ~7 Å), we find electron density, which we attribute to menaquinone. Menaquinone is the primary quinone, Q_A, in *Rps. viridis* (Shopes and Wraight 1983). Electron density for the quinone side chain extends to six of the isoprene units, it approaches one of the BPh b molecules to van der Waals distance. Very interestingly, in the symmetry-equivalent position of the quinone ring system, related by the twofold axis, an empty pocket in the protein is found. We consider this pocket as the likely binding site for the secondary quinone, Q_B, most of it being lost during isolation and crystallization of the RC. 1,10-O-phenanthroline, which is thought to be a competitive inhibitor for Q_B also in the bacterial system (STEIN et al. 1984), binds into this pocket (cooperation with O. EPP and R. HUBER unpublished).

The heme groups in the cytochrome subunit are positioned roughly along a line which is not parallel to the twofold rotation axis relating the BPh b and BChl b molecules. Only the heme closest to the special pair lies on this axis with distances of the heme-iron to both BChl b magnesium ions of ~21 Å.

The hemes can be grouped into two pairs, an inner pair that is close to the BChl molecules, and an outer pair, These pairs are also related by an approximate twofold rotation axis. The direction of this local heme symmetry axis differs by ~60° from that of the local symmetry axis through the special pair and the nonheme iron. The Fe–Fe distances within each heme pair are ~14 Å; the Fe–Fe distance between the neighboring hemes across the local twofold axis is ~16 Å. The positions of the heme groups strongly suggest that the photooxidized special pair can be reduced only via the closest heme.

2.1.3 The Protein Part

Most parts of the polypeptide chains of the four subunits could be traced with confidence and the limited sequence information which is available incorporated. The amino acid sequences known at present are 46, 30, and 28 residues at the N-termini of the L-chain, the M-chain and the cytochrome respectively (MICHEL et al. 1983), and about 80% of the sequence of the H-subunit, due to protein and gene sequencing (cooperation with K.A. WEYER and F. LOTTSPEICH unpublished). It is clear that the L- and M subunits, which form the central cylinder, consist of five long transmembrane helices each, which are also related by the twofold rotation axis. Several of the connections between these long presumed transmembrane helices on either side of the membrane are rather long and in part also helical. The main structural differences between the L- and M-subunits are seen at the amino- and carboxyterminal ends. Due to the relation by the twofold rotation axis running through the nonheme iron and the special pair, both subunits are equally involved in the formation of the special pair and the nonheme-iron quinone complex.

The cytochrome subunit appears to be attached only to the membrane surface (most likely on the periplasmic side; BACHOFEN and WIEMKEN, Chap. 11.2, this Vol.), and to not possess any intramembranous parts. Four of its five helices are related by the local twofold rotation axis running between the second and third heme group.

The H-subunit possesses one long transmembrane helix at its amino terminus, and a large cytoplasmic domain, whose center is not located directly on top of the central cylinder, but shifted parallel along the presumed membrane surface.

The number of helices found agrees with the number of helices predicted from amino acid sequences on the basis of hydrophobicity plots: one transmembrane helix had been predicted for the H subunit, and five transmembrane helices each for the L and M subunits of *Rhodopseudomonas capsulata* (YOUVAN et al. 1984), and for the M subunit of *Rps. sphaeroides* (WILLIAMS et al. 1984; HEARST, Chap. 8.5, this Vol.)

2.2 Functional Aspects

The arrangement of the BChl b and BPh b molecules suggests on initial examination that, starting from the special pair, the electron transfer can occur in two ways: at first to either of the monomeric BChl b molecules and then to the adjacent BPh b. However, there is good structural and spectroscopical evidence that only one way is used for electron transport (see also NORRIS and VAN BRAKEL, Chap. 8.3, this Vol.). Only one quinone is found in the electron density map, and its isoprenoid side chain approaches one of the BPh b molecules to van der Waals distance, so as to guide the electron from BPh to the quinone ring system. The absorption maxima of the two BPh b molecules are different, and it seems to be the BPh b molecule with its Q_y-transition moment at the longer wavelength that is involved in electron transfer (VERMEGLIO and PAILLOTIN 1982). Comparison of crystal spectra (ZINTH et al. 1983) and structure shows that this is the BPh b, which is approached by the isoprenoid side chain of the quinone (cooperation with W. ZINTH, W. KAISER and S. FISCHER). The electron is then transferred to the primary and then to the secondary quinone, which is exchanged for a new oxidized quinone after reception of two electrons and two protons.

The reduced ubiquinone equilibrates with the quinone pool in the membrane, and may then be reoxidized by the bc_1 complex under proton translocation and thus energy conservation. The existence of the bc_1-complex has still to be established in *Rps. viridis,* though antimycin-sensitive reactions have been found (JONES and SAUNDERS 1972). The electrons are then taken over by water-soluble cytochrome c_2 molecules in the periplasmic space. They can diffuse to the RC and reduce the high potential cytochrome hemes, which then reduce the photooxidized special pair (SHILL and WOOD 1984). Thus the cycle of light-driven electron flow is closed.

The lower potential cytochrome hemes must be the two outermost heme groups in order to establish a functional linear downhill electron flow to the photooxidized primary donor. These low potential hemes cannot be reduced by the cytochrome c_2. They are most likely involved in the reduction of low potential substrates, presumably reduced sulfur compounds. Thus electrons can be fed into the cycle of light-driven electron flow, or used for carbon dioxide fixation.

From the location of the four heme groups in the periplasmic space, it is unlikely that an energy-conserving site exists between the low and high potential cytochrome hemes, resulting in waste of some energy. It may be more

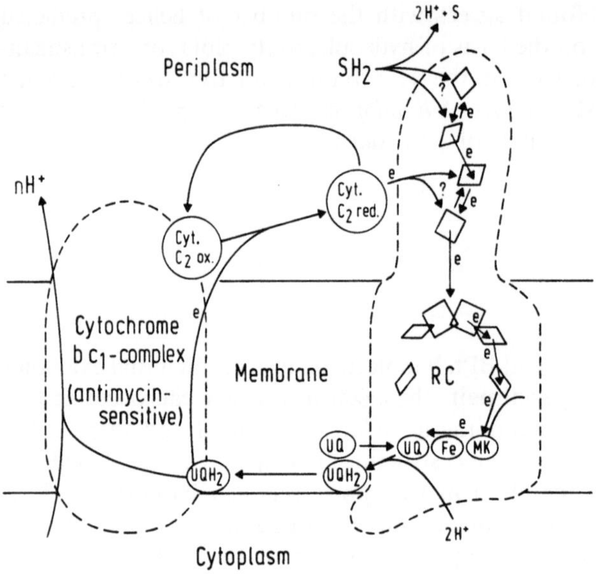

Fig. 3. Schematic drawing of cyclic and noncyclic photosynthetic electron transport in *Rhodopseudomonas viridis*

important to fill up the cellular electron pool rapidly than to use all available redox energy. Figure 3 illustrates the cyclic and noncyclic electron transport.

So far the function of the H-subunit is not precisely known. It has been suggested to be involved in the electron transfer from Q_A to Q_B in *Rps. sphaeroides* (DEBUS et al. 1981). Indeed the H-subunit approaches the presumed Q_B binding site very closely. From the structure it appears possible that the H-subunit interacts with the light-harvesting complexes from the cytoplasmic side so as to fix one of the light-harvesting complexes in a position to guarantee optimal energy transfer from the antenna to the RC. Alternatively, or in addition to this first function, the H subunits may serve as the target sites for the assembly of reaction centers (CHORY et al. 1984; SPRAGUE and BARGA, Chap. 11.1, DONOHUE and KAPLAN, Chap. 11.3, both this Vol.).

2.3 Evolutionary Aspects

The basic symmetric arrangement of the pigments and the L- and M-subunits, and the need for both protein subunits to generate the special pair and the electron-accepting quinone-iron complex, suggest the following hypothesis on the sequence of evolutionary events: the RC of the purple photosynthetic bacteria evolved from a symmetric protein dimer containing two BChl molecules forming a special pair, and possessing two equivalent ways to transfer electrons to the cytoplasmic side of the membrane. Both pathways consisted of one BChl molecule as first electron acceptor, then a BPh molecule as intermediate electron acceptor and a quinone molecule as first stable acceptor. Gene duplication and subsequent mutations could account for the origin of the asymmetric dimer. During evolution, the asymmetric dimer proved to be advantageous: one of the electron pathways lost its function in electron transport, a functional electron transfer between the two quinone molecules became possible. The quinone at the end of the still functional electron pathways remained firmly bound, and the other quinone became an exchangeable loosely bound quinone, Q_B. It became possible to exchange the Q_B molecule after accepting two electrons for an oxidized quinone from the quinone pool. Having Q_A as an additional electron storage device may have been one of the most important advantages when replacing a symmetric reaction center by an asymmetric one during evolution.

The BChl and BPh molecules no longer involved in electron transport may still be needed for the correct protein folding. They certainly can absorb light and transfer the energy to the special pair, and they may be involved in transferring energy of the absorbed light from the antenna pigments to the special pair.

2.4 Implications on the Structure of PS II Reaction Center

As mentioned in the introduction, considerable similarities exist between the RC of purple photosynthetic bacteria and the RC of PS II, namely a very similar electron-accepting quinone-iron complex, a similar spin-polarized triplet state of the primary donor, and a pheophytin molecule as intermediate electron acceptor. As discussed above, two symmetrically arranged subunits (L,M) cooperate to generate the primary donor and the electron acceptor in *Rps. viridis*. We therefore have to postulate the existence of two similar subunits in the RC of PS II cooperating in the formation of the primary donor and electron acceptor. Considerable sequence homologies have been found between the L-subunits of *Rps. sphaeroides* (WILLIAMS et al. 1983), the L- and M-subunits of *Rps. capsulata* (YOUVAN et al. 1984) and the so-called Q_B or herbicide-binding protein from chloroplasts (ZURAWSKI et al. 1982). The L-subunit from *Rps. sphaeroides* and the Q_B protein bind the herbicide azidoatrazine (DEVITRY and DINER 1984; PFISTER et al. 1981). It is therefore very likely that the L-subunit corresponds to the Q_B protein. The sequence homology suggests a very similar protein folding.

In this context it is interesting to note that a structure proposal for the Q_B protein has been made suggesting seven transmembrane helices (RAO et al. 1983). This is in contrast with the five transmembrane helices found in our structure analysis of the RC from *Rps. viridis*. A comparison of the proposed Q_B folding and the folding of the L or M subunit of our RC shows that two of the long connections between transmembrane helices (being partly helical) have been proposed to be transmembrane helices.

Considerable sequence homologies have also been found between the so-called D-2 protein from chloroplasts (RASMUSSEN et al. 1984; ROCHAIX et al. 1984) and the L and M subunits from purple photosynthetic bacteria. This D-2 protein we consider as the protein subunit from PS II RC corresponding to the M-subunit from the RC of purple bacteria, involved in the generation of the primary donor and the electron accepting quinone iron complex.

In summary we propose that the Q_B protein and the D-2 protein form a similar chlorophyll, pheophytin and quinone containing core in the RC of PS II as do the L and M subunits in the purple photosynthetic bacteria. This proposal is in contrast to the recent suggestions based on fluorescence and absorption measurements, assigning such a role to the so-called 47 kDa protein (NAKATANI et al. 1984; YAMAGISHI and KATOH 1984).

Acknowledgement. We thank the Deutsche Forschungsgemeinschaft (Sonderforschungsbereich 143) for financial support.

Note Added in Proof: Since the submission of the manuscript (Nov. 1984) the H-subunit (MICHEL et al., EMBO-J. 4, 1667–1672, 1985) and the L- and the M-subunits have been completely sequenced, the structure of the protein subunits has been published (DEISENHOFER et al., Nature 318, 618–624, 1985).

References

Chory J, Donohue TJ, Varga AR, Staehelin LA, Kaplan S (1984) J Bacteriol 159:540–554
Debus RJ, Okamura MY, Feher G (1981) Biophys J 33:19a
Deisenhofer J, Epp O, Miki K, Huber R, Michel H (1984) J Mol Biol 180:385–398
Dutton PL, Leigh JS, Seibert M (1971) Biochem Biophys Res Commun 40:406–413
Feher G, Okamura MY (1978) In: Clayton RK, Sistrom WR (eds) The photosynthetic bacteria. Plenum, New York London, pp 349–386
Garavito RM, Jenkins J, Jansonius JN, Karlsson R, Rosenbusch JP (1983) J Mol Biol 164:313–327
Gast P, Wasielewski MR, Schiffer M, Norris JR (1983) Nature 305:451–452
Hoff AJ (1982) In: Fong FK (ed) Molecular biology, vol 35. Biochemistry, and biophysics. Springer, Berlin Heidelberg New York, pp 80–151, 322–326
Holten D, Windsor MW, Parson WW, Thornber JP (1978) Biochim Biophys Acta 501:112–126
Jones OGT, Saunders VA (1972) Biochim Biophys Acta 275:427–436
Klimov VV, Klevanik AV, Shuvalov VA, Krasnovsky AA (1977) FEBS Lett 82:183–186
Klimov VV, Dolan E, Ke B (1980) FEBS Lett 112:97–100
Michel H (1982) J Mol Biol 158:567–572
Michel H, Deisenhofer J, Miki K, Weyer K, Lottspeich F (1983) In: Quagliariello E, Palmieri F (eds) Structure and function of membrane proteins. Elsevier, Amsterdam New York Oxford, pp 191–197
Nakatani HY, Ke B, Dolan E, Arntzen CJ (1984) Biochim Biophys Acta 765:347–352
Norris JR, Katz JJ (1978) In: Clayton RK, Sistrom WR (eds) The photosynthetic bacteria. Plenum, New York London, pp 397–418
Norris JR, Druyan ME, Katz JJ (1973) J Am Chem Soc 95:1680–1682
Norris JR, Scheer H, Katz JJ (1975) Ann NY Acad Sci 244:261–280
Olson JM, Thornber P (1979) In: Capaldi RA (ed) Membrane proteins in energy transduction. Dekker, New York, pp 279–340
Omata T, Murata M, Satoh K (1984) Biochim Biophys Acta 765:403–405
Parson WW (1982) Annu Rev Biophys Bioeng 11:57–80
Pfister K, Steinback KE, Gardner G, Arntzen CJ (1981) Proc Natl Acad Sci USA 78:981–985
Rao JKM, Hargrave PA, Argos P (1983) FEBS Lett 156:165–169

Rasmussen OF, Bookjans G, Stummann BM, Henningsen KW (1984) Plant Mol Biol 3:191–199
Rochaix JD, Pron M, Rahire M, Malnoe P (1984) Plant Mol Biol 3:363–370
Rutherford AW, Paterson DR, Mullet JF (1981) Biochim Biophys Acta 635:205–214
Shill DA, Wood PM (1984) Biochim Biophys Acta 764:1–7
Shopes RJ, Wraight CA (1983) Biophys J 41:40a
Shuvalov VA, Asadov AA (1979) Biochim Biophys Acta 545:296–308
Shuvalov VA, Klimov VV (1976) Biochim Biophys Acta 440:587–599
Stein RR, Castellvi AL, Bogacz JP, Wraight CA (1984) J Cell Biochem 24:243–259
Thornber JP, Cogdell RJ, Seftor REB, Webster GD (1980) Biochim Biophys Acta 593:60–75
Tiede DM, Prince RC, Reed GH, Dutton PL (1976) FEBS Lett 65:301–304
Vermeglio A, Paillotin G (1982) Biochim Biophys Acta 681:32–40
Vitry C de, Diner B (1984) FEBS Lett 167:327–331
Wiemken V, Bachofen R (1984) FEBS Lett 166:155–159
Williams JC, Steiner LA, Ogden RC, Simon MI, Feher G (1983) Proc Natl Acad Sci USA 80:6505–6509
Williams JC, Steiner LA, Feher G (1984) Proc Natl Acad Sci USA 81:7303–7307
Yamagishi A, Katoh S (1984) Biochim Biophys Acta 765:118–124
Youvan DC, Bylina EJ, Alberti M, Begusch H, Hearst JE (1984) Cell 37:949–957
Zinth W, Kaiser W, Michel H (1983) Biochim Biophys Acta 723:128–131
Zurawski G, Bohnert HJ, Whitfeld PR, Bottomley W (1982) Proc Natl Acad Sci USA 79:7699–7703

8.5 Primary Structure and Function of the Reaction Center Polypeptides of *Rhodopseudomonas capsulata* – the Structural and Functional Analogies with the Photosystem II Polypeptides of Plants

J.E. Hearst

Youvan et al. (1984) have recently reported the primary sequence of two fragments with lengths greater than 4000 basepair from *Rps. capsulata* which code for the light harvesting I polypeptides, B-870α and B-870β, the three reaction center polypeptides H, M, and L, and for some presumed regulatory functions relating to the synthesis of these components of the photosynthetic membrane (Zsebo and Hearst 1984). From the deduced amino acid sequences of the reaction center subunits it was concluded that the H subunit is a hydrophilic protein with a very hydrophobic amino terminus of 30 amino acids. The H subunit is a polypeptide 254 amino acids long with a molecular weight of 28,534 Da. The L and M subunits are both very hydrophobic proteins with 282 and 307 amino acids respectively. Their molecular weights calculate to be 31,565 and 34,440 Da, indicating that they have anomolous SDS PAGE electrophoretic mobilities because of their hydrophobic characters. Hydropathy plots suggest that the L and M subunits are transmembrane proteins which may cross the membrane five times.

Feher and Okamura (1978) have shown that the H subunit is not essential to the electron transfer activities of the reaction centers in *Rps. sphaeroides*. While the H, M, and L subunits always occur in equimolar quantities in reaction centers, the H subunit is 35 kilobasepairs away from the operon containing the LH B-870α, LH B870β, L subunit, and M subunit in *Rps. capsulata* (Youvan et al. 1983) and so the H subunit gene is clearly under separate transcriptional control from the other structural proteins in the reaction center. Chory et al. (1984) have suggested that the H subunit serves as the structural anchor in the membrane around which the reaction center forms and evidence is presented suggesting that the H subunit can be found in membrane fractions isolated even from aerobically grown cells.

We are therefore at a point where it is generally accepted that the only two polypeptides essential for reaction center activity are the L and M subunits in *Rps. sphaeroides*. These two subunits must contain binding sites for the strongly bound and weakly bound ubiquinones, Q_A and Q_B, for the Fe(II), and for the bacteriochlorophylls and the bacteriopheophytins. Of these interactions, I consider the sites for binding the quinones most crucial since they serve the roles of preserving or stabilizing very labile anion radicals which are one-electron intermediates in the quinone to hydroquinone reduction pathway. This model supposes that pigment binding functions are more variable and therefore less tightly conserved in evolution.

Hearst and Sauer (1983, 1984) have discovered that there is a highly conserved pattern of sequences of amino acids which is common to the L and

M subunits of *Rps. capsulata* and the Q_B-protein of chloroplast thylakoid membranes in spinach and tobacco. The conservation has survived an estimated three billion years of evolution (OLSON 1981). Such a striking conservation of amino acid sequence suggests that these portions of all three proteins, all of which are approximately 60% in from the amino terminus, are at the functional centers of these proteins. It is their hypothesis that these highly conserved sequences of amino acids are involved in quinone binding and function.

There is a large literature of physical measurements on thylakoid membranes from higher plants which establish the functional and compositional similarities between photosystem II reaction centers in these membranes and the reaction centers of *Rps. capsulata* and *Rps. sphaeroides.* A more detailed treatment of these topics can be found in the reviews of CROFTS and WRAIGHT (1983), SAUER (Chap. 2), THOMBER (Chap. 3), ORT (Chap. 4), DUTTON (Chap. 5), TIEDE (Chap. 8.2), PRUICE (Chapt. 10.4), MICHEL and DEISENHOFER (Chap. 8.4), and VERMAAS (1983). KLIMOV et al. (1977) reported the reduction of pheophytin in the primary light reaction of PS II by the chlorophyll complex P-680 (SHUVALOV et al. 1980). In a similar fashion after excitation with one photon, the purple bacteria transfer an electron from a bacteriochlorophyll dimer to bacteriopheophytin (URBAN and KLINGENBERG 1969). The semiquinone anion of Q_A was first reported in thylakoid membranes more than 10 years ago (WITT 1983, VAN GORKOM 1974). In green plants, Q_A and Q_B are known to be plastoquinones. In purple bacteria, Q_A, and Q_B are either ubiquinone or menaquinone. The semiquinone anions in purple bacteria have been examined by PULLES et al. (1976) and PRINCE and DUTTON (1978). KLIMOV et al. (1980) have reported an Fe(II) atom interacting with the quinones in the PS II reaction centers of chloroplasts. An analogous Fe(II) had been discovered earlier in purple bacteria (TIEDE et al. 1976 and OKAMURA et al. 1979).

While a very large literature establishes the physical similarities between purple bacteria reaction centers and photosystem II reaction centers of plants, algae, and cyanobacteria, efforts to isolate reaction centers with similar chemical compositions has led to confusion. BROGLIE et al. (1980) have demonstrated the successful isolation of purple bacterial reaction centers using lithium dodecyl sulfate treatment of photosynthetic membranes which have well-defined chemical composition. In particular, they always show an equimolar composition of the H, M, and L subunits. In higher cells, efforts to isolate reaction centers have resulted in far more complex protein gel electrophoresis patterns which suggest many more components. For example, WESTHOFF et al. (1983) have reported seven nonidentical polypeptides in the water-splitting photosystem II multisubunit structure which has been isolated from spinach. Three of these are termed "peripheral", are located in the thylakoid lumen, have molecular weights 34 kDa, 23 kDa, and 16 kDa, and appear to participate in water photolysis (AKERLUND et al. 1982, ANDERSSON, Chap. 9.3, this Vol.). The remaining four form a membrane spanning core that complexes all photosynthetic pigments of the reaction center. The proteins in this core are the 51 kDa and 44 kDa proteins which bind P-680 and the pheophytin which is the primary electron acceptor. The core also contains the 32 kDa herbicide binding protein which shows homology to the purple bacterial L and M reaction center subunits (HEARST and SAUER 1983, 1984) and it contains the 10 kDa cytochrome b-559.

MORRIS and HERRMANN (1984) have sequenced the 51 kDa protein and find no sequence corresponding to the hypothetical quinone binding site. The sequence of the 44 kDa chlorophyll a-complexed apoprotein of this core has been completed by MORRIS and HERRMANN, but is not yet available to me.

If the analogs to the L and M subunits of purple bacteria reaction centers exist in PS II reaction centers, then protein sequence homologies suggest that the 32 kDa herbicide binding protein, which binds Q_B, is an analog to the L subunit of the bacteria. The ARNTZEN laboratory has recently obtained azido atrazine binding to the L subunit of *Rps. sphaeroides* which also indicates this analogy to be correct (personal communication 1984). While the 44 kDa polypeptide of WESTHOFF et al. (1983) may be the other member of this pair, the difference in molecular weights suggests that this is not the case. On the physical side there is one additional piece of information which suggests a different solution to this apparent inconsistency. YAMAGISHI and KATOH (1984) have just developed a new isolation procedure for the PS II reaction center complex of the thermophilic cyanobacterium *Synechococcus* sp. and found that it resolved into two chlorophyll–protein complexes. The first, which they called CP2-b contained a chlorophyll binding 47 kDa polypeptide (presumably equivalent to the 51 kDa protein of higher plants), two polypeptides in the 28–31 kDa region, and a 9 kDa polypeptide which they attribute to cytochrome b-559. The second complex, which they called CP2-c, contained only a single 40 kDa chlorophyll binding protein (presumably equivalent to the 44 kDa protein of higher plants. CP2-b was able to promote the photoreduction of 2,6-dichlorophenolindophenol with diphenylcarbizide as the electron donor. CP2-c had only weak photoreduction activity in the same reaction leading the authors to conclude that CP2-b is the site of the PS II reaction center and that the 40 kDa polypeptide (the 44 kDa analog) is not essential for PS II electron transport. According to SATOH et al. (1983) isolated reaction center complexes from spinach contain the following components: two chlorophyll-binding proteins of molecular weight 47 kDa and 43 kDa, a 30 kDa herbicide binding protein, a 6 kDa apoprotein of cytochrome b-559 and a 34 kDa protein of unknown function. This agrees well with the conclusions of the HERRMAN laboratory except for the additional 34 kDa protein which is at the reaction center. [Attention should be drawn to the work of YOUVAN et al. (1984), in which the large underestimation of the molecular weights of these very hydrophobic reaction center proteins from SDS PAGE is clearly documented. One might expect similar low estimates in the works of YAMAGISHI and KATOH (1984), and of SATOH et al. (1983) which were mentioned above.]

The suggestion can be made here that all PS II reaction centers are likely to contain two similar protein subunits with molecular weights between 30 kDa and 40 kDa with roles analogous to those of the L and M subunits in the reaction centers of photosynthetic purple bacteria. Quinone binding is stressed here although common features of pigment binding might also prevail. The strongest evidence in favor of this position comes from the recent nucleic acid sequence data of ROCHAIX et al. (1984) and of RASMUSSEN et al. (1984). ROCHAIX et al. (1984) present the nucleic acid sequence of the psbD locus of the *Chlamydomonas reinhardii* chloroplast genome. The psbD gene codes for the D2 polypeptide which has been associated with photosystem II (CHUA and GILLHAM 1977) together with its partner, the D1 polypeptide, which is coded by the psbA gene and which corresponds to the 32 kDa herbicide binding protein or Q_B protein of higher plants. ROCHAIX et al. (1984) make the striking observation that while the overall homology between the Q_B protein and the D2 protein of *C. reinhardii* is only 27%, there are several domains where one can find homology ranging between 33 and 58%. Q_B contains 352 amino acids, while D1 contains 339 or possibly 350 amino acids. The putative quinone-binding region is highly conserved in both proteins, with Q_B corresponding best with the L subunit of *Rps. capsulata* and D2 corresponding best with the M subunit of *Rps. capsulata*.

RASMUSSEN et al. (1984) have sequenced the corresponding psbD gene from the higher plant chloroplast genome, *Pisum sativum* or pea. Their D2 protein is 353 amino acids long and shows an 84.7% homology to the D2 protein of *C. reinhardii*. Again the putative quinone-binding region is the most highly conserved region of the protein relative to the M subunit of *Rps. capsulata*.

Figure 1 shows an alignment of the amino acid sequences of the L subunit of *Rps. capsulata* (YOUVAN et al. 1984), the Q_B protein of spinach (ZURAWSKI et al. 1981) and the Q_B protein of *C. reinhardii* (ERICKSON et al. 1983). Overall, there is 27% homology between the L subunit and the Q_B protein of spinach, 26.2% homology between the L subunit and the Q_B protein of *C. reinhardii* and 92.3% homology between the two Q_B proteins. The most striking feature of this figure is the strong homology between all three proteins in the hypothetical quinone-binding region: N----PFHMLG----F------AMHG-LV-S.

Figure 2 shows an alignment of the amino acid sequences of the M subunit of *Rps. capsulata* (YOUVAN et al. 1984), the D2 protein of pea (RASMUSSEN et al. 1984) and the D2 protein of *C. reinhardii* (ROCHAIX et al. 1984). Overall, there is 20.8% homology between the M subunit and the D2 protein of pea, 19.5% homology between the M subunit and the D2 protein of *C. reinhardii* and an 84.7% homology between the two D2 proteins. The most striking feature of this figure is the strong homology between all three proteins in the hypothetical quinone binding region: F---H-N---NPFH----A----G-ALL-A-HGAT.

A comparison of these homologies suggests that in the putative strong site or M subunit site, the histidine-methionine and methionine-histidine pairs of the weak or L subunit site are not the only functionalities of importance (HEARST and SAUER 1984). Each strong binding site has only one of these two pairs of amino acid sequences, while the weak site in all cases has both pairs.

In conclusion, I am suggesting that the psbA and psbD loci of chloroplast genomes encode for the photosystem II analogs of the L and M reaction center subunits of *Rps. capsulata* and other photosynthetic purple bacteria. While the corresponding chloroplast proteins, D1 and D2, appear to have molecular weights of 32 to 34 kDa by SDS PAGE, they are very hydrophobic membrane bound proteins which may cross the membrane five or more times. Their true molecular weights will be between 38 and 40 kDa (ZURAWSKI et al. 1981 and RASMUSSEN et al. 1984). HEARST and SAUER (1983, 1984) have proposed that the most highly co-conserved sequences between the L subunit, the M subunit and the Q_B protein are involved in quinone binding. While WOLBER and STEIN-

```
MALLSFERKYRVPGGTLIGGSLFDFWVGPFYVGFFGVTTI      FFATLGFLLILWGAAMQGTWNP
MT AILERRESESLWGRFCNWI TSTENRLYIGWFGVLMIPTLLTATSVF IIAFIAAPPVDIDGIREPVSGSLLYGNNIISGAIIPTSA
MT AILERRENSSLWARFCEWI TSTENRLYIGWFGVIMIPCLLTATSVF IIAFIAAPPVDIDGIREPVSGSLLYGNNIITGAVIPTSN

QLISIFPPPVENGLNVAALDKGGLWQVITVCATGAFCSWALREVEICRKLGIGFHIPVAFSMAIFAYLTLVVIRPMMMGSWGYAFPYGIW
AIGLHFYPIWEAASVDEWLYNGGPYELIVLHFLLGVACYMGREWELSFRLGMRPWIAVAYSAPVAAATAVFLIYPIGQGSFSDGMPLGIS
AIGLHFYPIWEAASLDEWLYNGGPYQLIVCHFLLGVYCYMGREWELSFRLGMRPWIAVAYSAPVAAASAVFLVYPIGQGSFSDGMPLGIS

THLDWVSNTGYTYGNFHYNPFHMLGISLFFTTAWALAMHGALVLS       AANPVKGKTMRTPDHEDT       YFRDLMGYSVGTL
GTFNFMIVFQAEH NILMHPFHMLGVAGVFGGSLFSAMHGSLVTSSLIRETTENESANEGYRFGQEEETYNIVAAHGYFGRLIFQYASFN
GTFNFMIVFQAEH NILMHPFHMLGVAGVFGGSLFSAMHGSLVTSSLIRETTENESANEGYRFGQEEETYNIVAAHGYFGRLIFQYASFN

GIHRLGLLLALNAVFWSACCMLVSGTIYFDLWSDWWYWWV       NMPFWAD                       MAGGING
NSRSLHFFLAAWPVVGIWFTALGISTMAFNLNGFNFNQSVVDSQGRVINTWADIINRANLGMEVMHERNAHNFPLDLAAIEAPSTNG
NSRSLHFFLAAWPVIGIWFTALGLSTMAFNLNGFNFNQSVVDSQDRVLNTWADIINRANLGMEVMHERNAHNFPLDLASTNSSSNN
```

Amino Acid Sequence Alignments for:

L Subunit	*Rhodopseudomonas capsulata*	(282 amino acids)
Q_B Protein	*Spinacia oleracea*	(353 amino acids)
D1 Protein (psbA)	*Chlamydomonas reinhardii*	(352 amino acids)

Fig. 1. An amino acid sequence alignment for the L subunit reaction center polypeptide of *Rps. capsulata,* the herbicide binding thylakoid membrane protein, Q_B, of spinach and the D1 protein of *C. reinhardii* which is encoded by the psbA locus of the chloroplast genome. There are 76 matches and a 27.0% homology between L and Q_B, 74 matches and a 26.2% homology between L and D1, and 325 matches and a 92.3% homology between D1 and Q_B

```
MAEYQNFFNQVQVAGAPEMGLKEDVDTFERTPAGMFNILGWMGNAQIGPIYLGIAGTVSLAFGAAWFFTIGVWYWYQAGFDPFIFMRDLF
MTIALGKFTKDQNDLFDIMDDWLRRDRFVFVGWSGLLLFPCAYFAVGGWFTGTTFVTSWYTHGLASSYLEGCNFLTAAVSTP   ANSLAH
MTIAIGTYQEKRT  WFDDADDWLRQDRFVFVGWSGLLLFPCAYFALGGWLTGTTFVTSWYTHGLATSYLEGCNFLTAAVSTP   ANSMAH

FFSLEPPPAEYGLAIAPLKQGGVWQIASLFMAISVIAWWVRVYTRADQLGMGKHMAWAFLSAIWLWSVLGFWRPILMGSWSVAPPYGIFS
SLLLLWGPEAQGDLTRWCQLGGLWTFVALHGAFGLIGFMLRQFELARSVQLRPYNAIAFSGPIAVFVSVFLIYPLGQSGWFFAPSFGVAA
SLLFVWGPEAQGDFTRWCQLGGLWAFVALHGAFGLIGFMLRQFEIARSVNLRPYNAIAFSAPIAVFVSVFLIYPLGQSGWFFAPSFGVAA

HLDWTNQFSLDHGNLFYNPFHGLSIAALYGSALLFAMHGATILAVTRFGGERELEQIVDRGTASERAALFWRWTMGFNATMEG     IHR
IFRFILFFQGFH NWTLNPFHMMGVAGVLGAALLCAIHGATVENTLFEDGDGANTFRAFNPTQAE ETYSMVTANRFWSQIFGVAFSNKR
IFRFILFFQGFH NWTLNPFHMMGVAGVLGAALLCAIHGATVENTLFEDGDGANTFRAFNPTQAE ETYSMVTANRFWSQIFGVAFSNKR

WAIWMAVMVTLT        GGIGILLSGTVVDNWYVWAQVHGYAPVTP
WLHFFMLFVPVTGLWMSALGVVGLALNLRAYDFVSQEIRAAEDPEFETFYTKNILLNEGIRAWMATQDQPHENLIFPEEVLPRGNAL
WLHFFMLLVPVTGLWMSAIGVVGLALNLRAYDFVSQEIRAAEDPEFFFSIFIIPNHIINGSYFFNKSQKQIVYI
```

Amino Acid Sequence Alignments for :

M Subunit	*Rhodopseudomonas capsulata*	(307 amino acids)
D2 Protein (psbD)	*Pisum sativum*	(353 amino acids)
D2 Protein (psbD)	*Chlamydomonas reinhardii*	(352 amino acids)

Fig. 2. An amino acid sequence alignment for the M subunit reaction center polypeptide of *Rps. capsulata*, the D2 protein of pea, and the D2 protein of *C. reinhardii*. The two D2 proteins are encoded by the psbD loci of the chloroplast genomes of their respective organisms. There are 64 matches and 20.8% homology between the M and D2 pea, 60 matches and 19.5% homology between the M and D2 *Chlamydomonas*, and 287 matches and 84.7% homology between D2 *Chlamydomonas* and D2 pea

BACK (1984) have proven that the trypic peptide of the Q_B protein which contains the presumptive quinone binding site also contains the major site for the covalent linkage between azidoatrazine and the herbicide binding protein in thylakoid membranes, direct physical evidence relating to quinone binding remains elusive and requires further experimentation. In this spirit, it must be remembered that identification of the Q_B protein with the weak quinone-binding site, and the D2 protein with the strong binding site relates to a model and cannot yet be considered experimental fact.

Acknowledgements. This work was supported by the National Institutes of Health, Grant #GM30786 and by the Office of Energy Research, Office of Basic Energy Sciences, Biological Energy Research Division of the U.S. Department of Energy, Contract DE-AC03-76SF0098.

References

Akerlund HE, Jansson C, Andersson B (1982) Biochim Biophys Acta 681:1–10
Broglie RM, Hunter CN, Delepelaire P, Niederman RA, Chua N-H, Clayton RK (1980) Proc Natl Acad Sci USA 77:87–91
Chory J, Donohue TJ, Varga AR, Staehelin LA, Kalplan S (1984) J Bacteriol 159:540–554
Chua NH, Gillham NW (1977) J Cell Biol 74:441–452
Crofts AR, Wraight CA (1983) Biochim Biophys Acta 726:149–185
Erickson JM, Schneider M, Vallet JM, Dron M, Bennoun P, Rochaix JD (1983) In: Sybesma C (ed) Proc 6th Int Congr Photosynth, vol V. S. Nijhoff/Junk, The Hague, pp. 491–500
Feher G, Okamura MY (1978) In: Clayton RK, Sistrom WR (eds) The photosynthetic bacteria. Plenum, New York London, pp 349–386
Gorkom HJ van (1974) Biochim Biophys Acta 347:439–442
Hearst JE, Sauer K (1983) In: Sybesma C (ed) Proc 6th Int Congr Photosynth, vol II. S. Nijhoff/Junk, The Hague, pp 355–359
Hearst JE, Sauer K (1984) Z Naturforsch 39c:421–424
Klimov VV, Klevanik AV, Shuvalov VA, Krasnovsky A (1977) FEBS Lett 82:183–186
Klimov VV, Dolan E, Ke B (1980) FEBS Lett 112:97–100
Morris J, Herrmann RG (1984) Nucleic Acids Res 12:2837–2850
Okamura MY, Isaacson RA, Feher G (1979) Biochim Biophys Acta 546:394–417
Olson JM (1981) In: Fredrick JF (ed) Origins and evolution of eukaryotic intracellular organelles. NY Acad Sci, New York, pp 8–19
Prince RC, Dutton PL (1978) In: Clayton RK, Sistrom WR (eds) The photosynthetic bacteria. Plenum, New York London, pp 439–469
Pulles MPJ, Gorkom HJ van, Willemsen JG (1976) Biochim Biophys Acta 449:536–540
Rasmussen OF, Bookjans G, Stummann BM, Henningsen KW (1984) Plant Mol Biol 3:191–199
Rochaix J-D, Drom M, Rahire M, Malnoe P (1984) Plant Mol Biol 3:363–370
Satoh K, Nakatani HY, Steinback KE, Watson J, Arntzen CJ (1983) Biochim Biophys Acta 724:142–150
Shuvalov VA, Klimov VV, Dolan E, Parson WW, Ke B (1980) FEBS Lett 118:279–282
Tiede DM, Prince RC, Dutton PL (1976) Biochim Biophys Acta 449:447–467
Urban PF, Klingenberg M (1969) Eur J Biochem 9:519
Vermaas WFJ (1984) Thesis, Landbouwhogesch Wageningen
Westhoff P, Alt J, Hermann RG (1983) EMBO J 2:2229–2237
Witt K (1973) FEBS Lett 38:116–118

Wolber PK, Steinback KE (1984) Z Naturforsch 39c:425–429
Yamagishi A, Katoh S (1984) Biochim Biophys Acta 765:118–124
Youvan DC, Alberti M, Begusch H, Bylina EJ, Hearst JE (1983) Proc Natl Acad Sci USA 81:189–192
Youvan DC, Bylina EJ, Alberti M, Begusch H, Hearst JE (1984) Cell 37:949–957
Zsebo KM, Hearst JE (1984) Cell 37:937–947
Zurawski G, Bohnert HJ, Whitfield PR, Bottomley W (1981) Proc Natl Acad Sci USA 79:7699–7703

8.6 Membrane Topology and Photochemistry of the Green Photosynthetic Bacterium *Chloroflexus aurantiacus*

R.E. BLANKENSHIP and R.C. FULLER

1 Introduction

For the last half century, research on the photosynthetic bacteria has played a leading role in unraveling our understanding of the physical and chemical processes involved in photosynthesis. VAN NIEL's introduction of the comparative biochemical approach to the study of photosynthesis initiated an avalanche of research into the structure, biochemistry and biophysics of photosynthetic bacteria, concentrated primarily on the purple sulfur (Chromatiaceae) and nonsulfur (Rhodospirillaceae) bacteria. Research on the green photosynthetic bacteria lagged, primarily because of the difficulties associated with growing the organisms, and the rather fastidious obligate anaerobicity of the Chlorobiaceae. The discovery of the gliding thermophilic facultative green bacterium *Chloroflexus aurantiacus* (Chloroflexacae) (PIERSON and CASTENHOLZ 1974a, b) provided a new experimental tool with a wide flexibility of growth conditions (aerobic and photosynthetic) for study of not only the photochemistry of the green bacteria's unique antenna and reaction center systems, but also the development, topology, and biological control of their photosynthesis.

Both families of the green photosynthetic bacteria have their photochemical apparatus arranged in an unusual manner. The main light-harvesting antenna pigment BChl c, d or e is assembled within a nonmembranous structure, the chlorosome (OLSON 1980, STAEHELIN et al. 1978), formerly known as chlorobium vesicle (COHEN-BAZIRE et al. 1964, HOLT et al. 1966). Chlorosomes are closely associated with, but not derived from the bacterial cytoplasmic membrane (CM). The green bacteria also do not contain the specialized intracytoplasmic membranes (ICM) characteristic of the purple bacteria (see SPRAGUE and VARGA, Chap. 11.1; DONOHUE and KAPLAN, Chap. 11.3, both this Vol.).

Thus, the photosynthetic apparatus of *Chloroflexus* consists of two cytologically distinct structures, the chlorosome and cytoplasmic membrane. Methods have been developed for the purification of both chlorosomes and the CM (SCHMIDT 1980, SPRAGUE et al. 1981, FEICK et al. 1982). Characterization of the reaction center, which is located in the CM, has proceeded rapidly after development of a purification procedure (PIERSON and THORNBER 1983).

Chloroflexus has chlorosomes that are similar in many ways to those found in the Chlorobiaceae except for the smaller size of the internal rod elements and the different membrane attachment mechanism (SPRAGUE et al. 1981,

SPRAGUE and VARGA, Chap. 11.1, this Vol.). The reaction centers, however, appear to be very different from those of the Chlorobiaceae. The structure and function of the reaction centers, and the mechanism of carbon metabolism in *Chloroflexus* all appear to be very similar to those found in the purple nonsulfur bacteria, while the sulfur metabolism has not yet been thoroughly investigated (PIERSON and CASTENHOLZ 1974a, b, MADIGAN and BROCK 1975, SIREVÅG and CASTENHOLZ 1979, BRUCE et al. 1982, PIERSON and THORNBER 1983). Thus, this organism appears to occupy a unique position in that it carries attributes of widely separated types of bacteria. It is therefore perhaps not surprising that *Chloroflexus* appears to be an important organism with which to trace the evolution of photosynthetic energy conversion. Its 16 S RNA catalog indicates that it is only distantly related to any of the other photosynthetic bacteria, and represents the deepest division within the true bacteria (STACKEBRANDT and WOESE 1981).

2 The Chlorosome

The chlorosomes of *Chloroflexus* are structures 100 × 30 nm in size, attached to the cytoplasmic surface of the plasma membrane of the cell (Fig. 1 B). Freeze-fracture electron microscopy has revealed the chlorosomes to have varying thickness depending on the stage of development and to contain linear arrays of

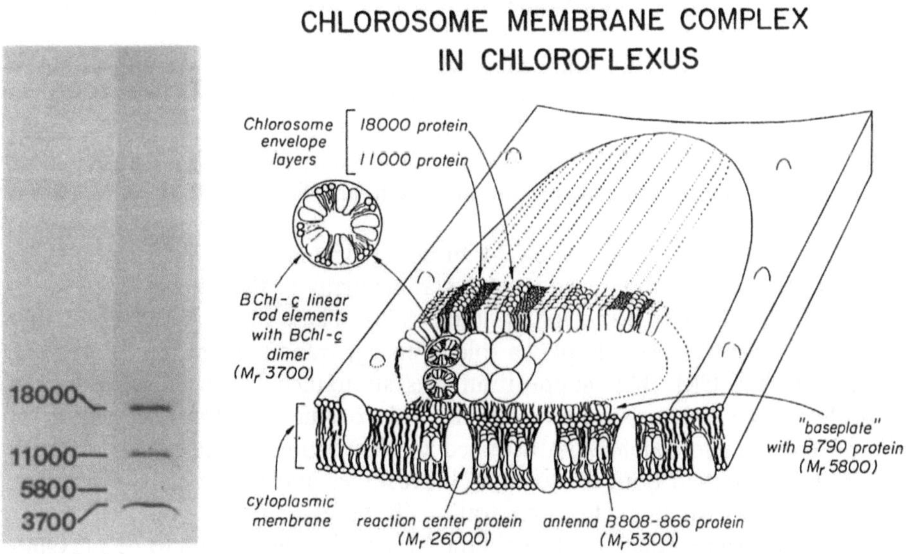

A B

Fig. 1. A SDS PAGE of pure chlorosomes. (T. REDLINGER and R.C. FULLER unpublished). **B** Topographical model of the chlorosome-cytoplasmic membrane complex. (FEICK and FULLER 1984)

rod-like structures thought to contain the antenna BChl c (STAEHELIN et al. 1978; SPRAGUE et al. 1981).

BChl c, the major antenna pigment in *Chloroflexus,* is entirely contained within the chlorosomes, along with carotenoids (chiefly β and γ) and a small amount of BChl a. BChl c is structurally and spectrally more similar to Chl a than BChl a, in that ring II is part of the conjugated π system (in BChl a this ring is reduced). Another notable feature of BChl c is the hydroxyethyl substituent at position 2 (GLOE and RISCH 1978). BChl c absorbs at 662 nm in most organic solvents, but its absorbance maximum is shifted to 740 nm in the chlorosome.

SDS PAGE analysis of purified chlorosomes has revealed one minor and three major polypeptides (Fig. 1 A). Studies involving proteolytic treatment, photolabeling with 3-azido-(2,7)-napthalene disulfonate, and chemical cross-linking with followed by two-dimensional PAGE have provided information on the molecular topography and the organization of proteins in the chlorosomes of *Chloroflexus* (FEICK and FULLER 1984). The results of these experiments are summarized in Fig. 1 B and reveal the following:

1. Of the three major proteins, two are located on the surface of the chlorosome. The first has an M_r of 11,000 and appears to be present as a trimer; the second has an M_r of 18,000 and appears as a cross-linked compound with the M_r 11,000 trimer. The two polypeptides are clearly closely associated on the surface of the chlorosome.

2. The third major polypeptide has an M_r of 3700 and appears to be associated with the major 740 nm absorbing BChl c. This polypeptide is present as a dimer and is associated with 10 to 16 BChl c's per dimer (FEICK and FULLER 1984).

3. A fourth minor chlorosome protein constituent has an M_r of 5800, is present as a trimer or pentamer after cross-linking, and is sequestered from protease digestion and photolabeling when the chlorosomes are still attached to the CM. This peptide is of extreme interest because it may be associated with a new spectral species of BChl a that absorbs at 790 nm in vivo (SCHMIDT 1980, SPRAGUE et al. 1981) and serves as a conduit for energy transfer from BChl c in the chlorosome to the antenna BChl a B-808–865 in the CM (BETTI et al. 1982).

The role of this chlorosome-CM bridging protein was first suggested by the appearance of an organized paracrystalline array at the chlorosome-attachment sites on the CM (STAEHELIN et al. 1978). These authors suggested that this "attachment" site could play a role in energy transfer from the chlorosome antenna to the CM. This supposition was strengthed by the discovery of a new BChl a (B-790) associated with the chlorosome and the observation of fluorescent emission at 802 nm ascribed to this pigment species. The 802 nm emission is observed both in isolated chlorosomes and whole cells, suggesting that B-790 is not an artifact of preparation (BETTI et al. 1982).

The chlorosome attachment site in the Chlorobiaceae appears to be a crystalline array of the water-soluble BChl a protein whose crystal structure has been determined (MATTHEWS and FENNA 1980). *Chloroflexus* appears to lack this protein entirely and utilize instead the B-790 pigment in the baseplate, and a B-808–865 antenna protein similar to that found in the purple bacteria as the acceptor of the excitation in the membrane.

Perhaps the most interesting aspect of the structure of the chlorosome is the small protein content relative to the large amount of BChl c. OLSON (1980) suggested on the basis of geometrical considerations that 12–14 BChl c molecules are associated with a 15 kDa peptide to form the linear rods observed in electron micrographs of chlorosomes in *Chlorobium limicola f. thiosulfatophilum*. FEICK and FULLER (1984) found that 10–16 BChl c molecules are associated with a dimer of M_r 3700 peptides, giving a weight ratio of pigment to protein of greater than one. This suggests that the organization of pigment in the chlorosome may be quite different from that found in the large number of chlorophyll proteins that have been isolated from a variety of organisms. In all these systems the mass of peptide is much larger than that of pigment and current evidence indicates relatively little direct (VAN DER WAALS) pigment-pigment contact (THORNBER, Chap. 3, this Vol.). Recent spectroscopic evidence strongly suggests, however, that the BChl c molecules in chlorosomes may form oligomeric arrays in which the hydroxyethyl group at position 2 coordinates to the Mg of another BChl c (SMITH et al. 1983). This unusual arrangement is entirely in agreement with all the structural and spectroscopic data now available. It thus appears that the mechanism of pigment organization in the chlorosome may be fundamentally different from that found in other antenna systems.

This antenna system in *Chloroflexus* is functionally analogous to the energy capture system found in the phycobilisome-thylakoid association in the cyanobacteria and red algae (GANTT, Chap. 6.3, this Vol.). These two types of photosynthetic organism have devised similar mechanisms for harvesting energy on one structure and transferring it to a reaction center on another. The end result is the same – energy capture and transfer – but it is carried out by two evolutionarily and chemically distinct structures.

3 Cytoplasmic Membrane

Methods have been described for purification of the cytoplasmic membranes away from the chlorosomes (SCHMIDT 1980, FEICK et al. 1982). The purest preparations (FEICK et al. 1982) contain essentially no 740 nm absorbing pigment, indicating that the contamination by chlorosomes is minimal. These preparations, which tend to form flat sheets rather than vesicles, have BChl a as their principal pigment, and exhibit near infrared absorption maxima at 808 and 866 nm primarily due to a low molecular weight ($M_r = 5300$) antenna pigment protein (FEICK and FULLER 1984). These purified membranes also contain respiratory enzymes and the reaction center. The absorption spectra of membranes from *Chloroflexus* and *Chlorobium* are quite different in that the spectrum of the former looks very similar to that of a typical purple bacterium, while the latter has a spectrum with maximum at 810 nm due to the presence of membrane-bound BChl a protein and has essentially no absorption beyond 840 nm (OLSON 1980, FEICK et al. 1982).

4 Reaction Center and Electron Transport Pathway

All available evidence indicates that the pathway of electron transport in *Chloroflexus* is in most respects very similar to that of purple photosynthetic bacteria. The Chlorobiaceae contain low-potential FeS proteins that function as early electron acceptors in a manner similar to that found in photosystem I of oxygen-evolving systems (AMESZ 1983, DUTTON, Chap. 5, MICHEL and DEISENHOFER, Chap. 8.4, this Vol.) Although whole cells of *Chloroflexus* contain FeS proteins (R. BLANKENSHIP and H. FRANK unpublished), there apparently are no dithionite-reducible membrane-bound FeS proteins. As discussed below, the early electron acceptors associated with the reaction center of *Chloroflexus* are a BChl/BPheo acceptor complex and a pair of menaquinones. This is very similar to the acceptors found in purple photosynthetic bacteria and PS II of O_2 evolving organisms (PARSON and HOLTEN, Chap. 8.1, MICHEL and DEISENHOFER, Chap. 8.4, both this Vol.). Structurally, the reaction center of *Chloroflexus* also appears similar to that of the purple bacteria, although there are some important differences between them (see below). The mechanism of electron transport that completes the photoreaction cycle in *Chloroflexus* is as yet very poorly understood.

BRUCE et al. (1982) carried out an extensive series of redox and kinetic experiments on isolated cytoplasmic membranes of *Chloroflexus*. They found that the photobleachable reaction center BChl (P-870) has a midpoint redox potential of $E_m = +340$ mV (vs. NHE). A membrane-bound photooxidizable c-type cytochrome of midpoint potential +260 mV serves as the electron donor to P-870$^+$. This cytochrome has been purified and found to contain two spectrally identical but thermodynamically distinct hemes with midpoint redox potentials of +265 mV and +140 mV (P. HUYNH and R. BLANKENSHIP unpublished). No b-type cytochromes have been detected in photosynthetically grown cells, although BRUCE et al. (1982) observed a shoulder at 562 nm in reduced-minus-oxidized difference spectra of nonpigmented aerobically grown cells. The nature of the respiratory electron transfer chain in *Chloroflexus* has not yet been studied in any detail.

An electron acceptor with $E_m = -50$ mV was detected by the failure to observe cytochrome c-554 photooxidation at low potentials (BRUCE et al. 1982), and was tentatively assigned to a quinone acceptor. Subsequent spectral studies with isolated reaction centers have established that this acceptor is a menaquinone (VASMEL and AMESZ 1983, BLANKENSHIP et al. 1983, 1984). Flash-induced binary oscillations of semiquinone absorption are observed when excess quinone is added to isolated reaction centers (R. BLANKENSHIP unpublished), indicative of a two-quinone acceptor system similar to that found in purple bacteria and the acceptor side of photosystem II. The electron transfer reaction between the two quinones is blocked by the inhibitor O-phenanthroline, but not by the herbicides atrazine and terbutryn, both of which block electron transfer at this site in purple bacteria and PS II (P. HANSEN and R. BLANKENSHIP unpublished).

The presence of electron acceptors prior to the quinones was established by using picosecond flash photolysis spectroscopy (KIRMAIER et al. 1983).

Bleaching of absorbance at 540, 755, and 805 nm occurs rapidly upon either 600 or 870 nm photon absorption. These absorbance changes relax with a $\tau =$ 310 ps to give a difference spectrum essentially identical to that observed on a millisecond time scale, where the system is in the state P-870$^+$ Q$^-$. The difference spectrum of the 310 ps component is very similar to that observed in *Rps. sphaeroides,* although the spectral resolution in *Chloroflexus* is somewhat better. The bleachings at 540 nm and 755 nm are indicative of BPheo involvement as an early electron acceptor. The additional bleaching at 805 nm suggests that BChl is also involved, and the pigments giving rise to these absorption changes are referred to as a BPheo/BChl acceptor complex, although the exact nature of their association remains to be determined (PARSON and HOLTEN, Chap. 8.1, this Vol.).

When normal photochemistry in *Chloroflexus* is blocked, a spin polarized ESR triplet is observed (BLANKENSHIP et al. 1984) that is diagnostic for a radical-pair mechanism of primary photochemistry (NORRIS, Chap. 8.3, this Vol.). The state of aggregation of P-870 is not yet certain. Triplet minus singlet absorption spectra have been interpreted as evidence in favor of a dimeric structure (DEN BLANKEN et al. 1983), although photodichroism experiments were more easily interpreted assuming P-870 as a monomeric pigment (KIRMAIER et al. 1984). The unusual pigment composition of *Chloroflexus* reaction centers (see below) allows for an interesting comparison of their properties with those of the more

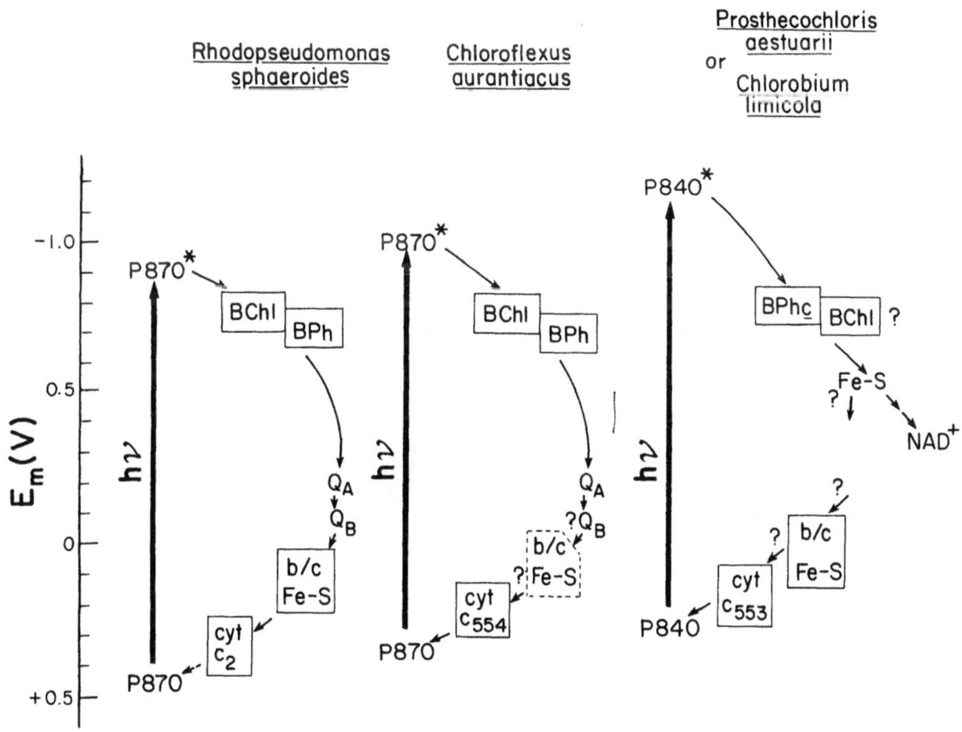

Fig. 2. Electron transfer pathways in *Rps. sphaeroides, Chloroflexus aurantiacus* and *Prosthecochloris aestuarii.* See text for details

widely studied purple bacteria. This characteristic may be important in assigning functions to those components of the reaction center whose role is as yet unknown.

A scheme summarizing the electron transfer pathways in *Chloroflexus, Rps. sphaeroides* and the green sulfur bacteria is shown in Fig. 2. The early reactions appear to be quite similar in the two organisms. The details of the secondary reactions subsequent to the quinones are as yet very poorly worked out; however the information presently available suggests that there may be some important differences in the reactions that serve to complete the cyclic electron transport chain. This region of the electron transport chain also appears to be somewhat different in those organisms (primarily purple sulfur bacteria) that have membrane-bound photooxidizable c-type cytochromes, for example the c-552–c-555 found in *Chromatium vinosum* and *Rps. viridis.*

5 Reaction Center Pigment and Protein Composition

The recent development of a reaction center isolation procedure (PIERSON and THORNBER 1983) has permitted a much more detailed analysis of the reaction center structure in *Chloroflexus*. The reaction center is similar in many ways to that found in many purple bacteria, with some important differences. The most apparent of these is the increased amount of BPheo relative to BChl.

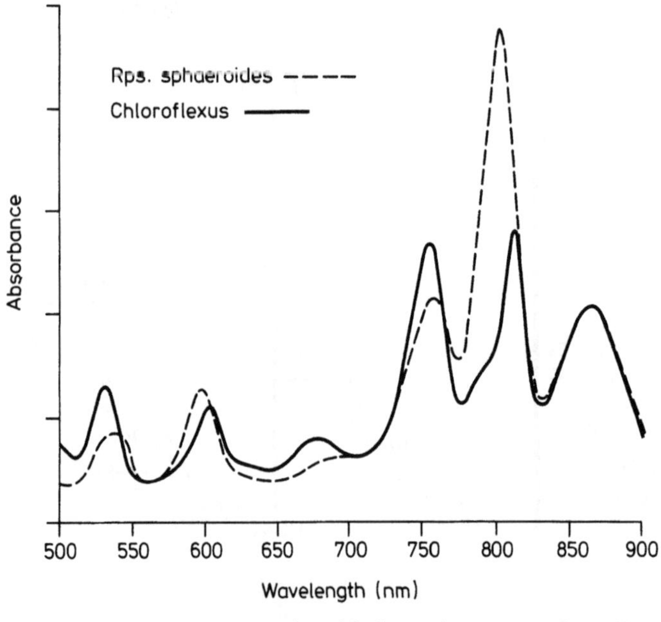

Fig. 3. Absorption spectra of purified reaction centers from *Rps. sphaeroides* and *Chloroflexus aurantiacus*. The spectra have been normalized at 865 nm, where the extinction coefficients for the two complexes are equal (see text)

In Fig. 3 the 535 nm and 755 nm peaks characteristic of BPheo are considerably larger relative to the 590, 800 (812), and 865 nm BChl peaks, when *Chloroflexus* and *Rps. sphaeroides* reaction centers are compared. This difference led PIERSON and THORNBER (1983) to propose that the pigment ratio in *Chloroflexus* reaction centers is BChl:BPheo=3:3 instead of BChl:BPheo=4:2 as found in purple bacterial reaction centers (STRALEY et al. 1973). Subsequent work (BLANKENSHIP et al. 1983, 1984) has generally confirmed the equimolar ratio of BChl and BPh (although see VASMEL et al. 1983). Recent experiments (L. MANCINO and R. BLANKENSHIP unpublished) in which the magnitude of the P-870$^+$ ESR signal induced by chemical oxidation was compared to the optical absorption changes have confirmed that the pigment composition is indeed 3 BChl and 3 BPheo, and have established a molar extinction coefficient of 184 mM^{-1} cm^{-1} at 812 nm and 135 mM^{-1} cm^{-1} at 864 nm.

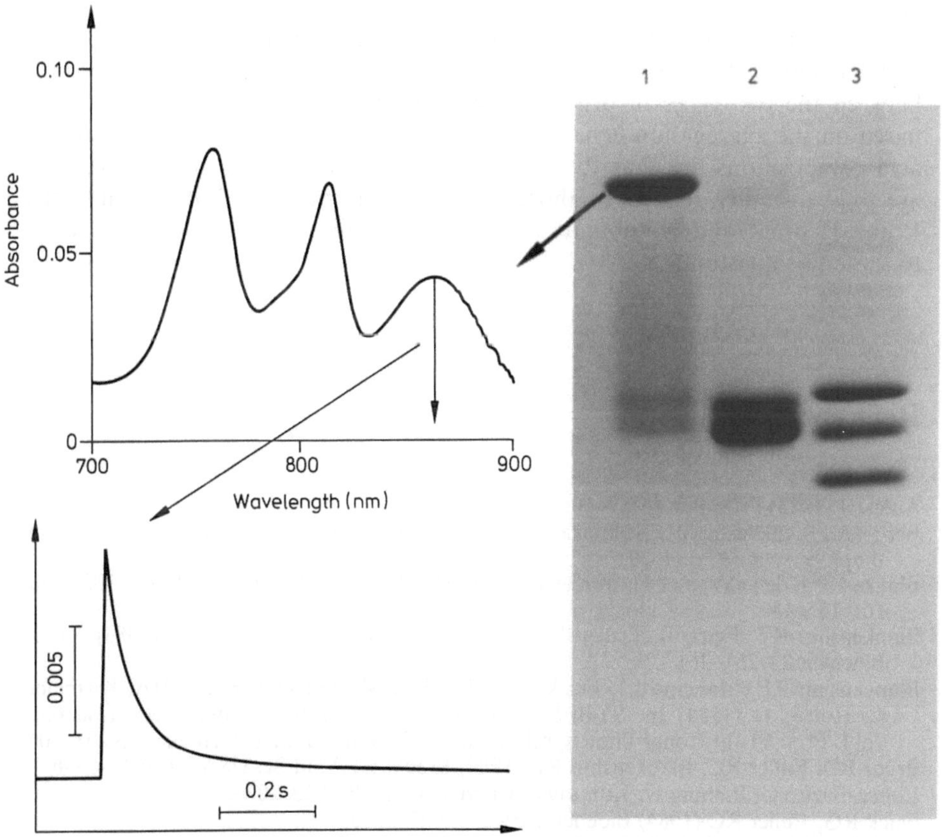

Fig. 4. SDS PAGE of reaction centers from *Chloroflexus* (lanes *1* and *2*) and *Rps. sphaeroides* (lane *3*). The sample in lanes *1* and *3* were treated with 3.3% SDS at room temperature; the sample in lane *2* was treated identically, except it was heated for 30 s at 85° C. The absorption spectrum and kinetic trace were taken on a slice from a gel that was not stained for protein. The kinetic trace represents the transmittance change at 865 nm induced by a 583 nm dye laser pulse. The 60 ms decay kinetics reflect the recombination between P-865$^+$ and Q^-. (L. MANCINO and R. BLANKENSHIP unpublished)

The reaction center isolated from the carotenoid-containing wild-type cell does not contain any bound carotenoid (PIERSON and THORNBER 1983), in contrast to the situation in purple bacteria, where reaction centers from wild-type cells invariably contain carotenoids. Similarly, the membrane-bound cytochrome c-554 is not part of the reaction center complex as isolated, again in contrast to those purple bacteria (e.g., *Chromatium vinosum, Rps. viridis*) that contain tightly bound c-type cytochromes.

The peptide composition of the *Chloroflexus* reaction center is also somewhat different from the LMH triad almost universally observed in purple bacteria. PIERSON et al. (1983) reported that the reaction center consists of two peptides of $M_r = 30$ k and 28 k. The intact complex migrates as $M_r = 52$ k band in an SDS gel. This is shown in Fig. 4, where the intact complex is compared to the heat-dissociated complex and the *Rps. sphaeroides* reaction center. The intact complex still retains its native absorption spectrum and even partial photoactivity.

The fact that the *Chloroflexus* reaction center contains only two peptides, and the purple bacteria usually contain three, makes structural comparison between the two types of organisms very interesting. Preliminary indications based on the aggregation behavior of the peptides in SDS gels and the amino acid compositions are that the two peptides in *Chloroflexus* reaction centers are only distantly related to those of the purple bacteria, but resemble the L and H or M and H subunits of *Rps. sphaeroides*. Further work is needed to settle this question.

References

Amesz J (1983) Prog Bot 45:89–105

Betti JA, Blankenship RE, Natarajan LV, Dickinson LC, Fuller RC (1982) Biochim Biophys Acta 680:194–201

Blanken HJ den, Vasmel H, Jongenelis APJM, Hoff AJ, Amesz J (1983) FEBS Lett 161:185–189

Blankenship RE, Feick R, Bruce BD, Kirmaier K, Holten D, Fuller RC (1983) J Cell Biochem 22:251–261

Blankenship RE, Mancino LJ, Feick R, Fuller RC, Machnicki J, Frank HA, Kirmaier C, Holten D (1984) In: SYBESMA C (ed) Advances in photosynthetic research, vol I. Proc VI Int Congr Photosynth, Brussels. Nijhoff/Junk, The Hague, pp 203–206

Bruce BD, Fuller RC, Blankenship RE (1982) Proc Natl Acad Sci USA 79:6532–6536

Cohen-Bazire G, Pfennig N, Kunisawa R (1964) J Cell Biol 22:207–225

Feick RG, Fuller RC (1984) Biochemistry 23:3693–3700

Feick RG, Fitzpatrick M, Fuller RC (1982) J Bacteriol 150:905–915

Gloe A, Risch N (1978) Arch Microbiol 118:153–156

Holt SC, Conti SF, Fuller RC (1966) J Bacteriol 91:311–322

Kirmaier C, Holten D, Feick R, Blankenship RE (1983) FEBS Lett 158:73–78

Kirmaier C, Holten D, Mancino LJ, Blankenship RE (1984) Biochim Biophys Acta 765:138–146

Madigan MT, Brock TD (1975) J Bacteriol 122:782–784

Matthews BW, Fenna RE (1980) Acc Chem Res 13:309–317

Olson JM (1980) Biochim Biophys Acta 594:33–51
Pierson BK, Castenholz RW (1974a) Arch Microbiol 100:5–24
Pierson BK, Castenholz RW (1974b) Arch Microbiol 100:283–305
Pierson BK, Thornber JP (1983) Proc Natl Acad Sci USA 80:80–84
Pierson BK, Thornber JP, Seftor REB (1983) Biochim Biophys Acta 723:322–326
Schmidt K (1980) Arch Microbiol 124:21–31
Sirevåg R, Castenholz R (1979) Arch Microbiol 120:151–153
Smith KM, Kehres LA, Fajer J (1983) J Am Chem Soc 105:1387–1389
Sprague SG, Staehelin LA, DiBartolomeis MJ, Fuller RC (1981) J Bacteriol 147:1021–1031
Stackebrandt E, Woese CR (1981) Soc Gen Microbiol Symp 32:1–31
Staehelin LA, Golecki JR, Fuller RC, Drews G (1978) Arch Microbiol 119:269–277
Straley SC, Parson WW, Mauzerall DC, Clayton RK (1973) Biochim Biophys Acta 305:597–609
Vasmel H, Amesz J (1983) Biochim Biophys Acta 724:118–122
Vasmel H, Meiburg RF, Kramer HJM, Vos LJ de, Amesz J (1983) Biochim Biophys Acta 724:333–339

8.7 Optically Detected Magnetic Resonance (ODMR) of Triplet States in Vivo

A.J. HOFF

1 Introduction

Triplet states are metastable states and as such were at one time thought to be involved in the primary act of photosynthesis, i.e., conversion of photon energy in separated, more or less stable charges. Although there is now abundant evidence that charge separation takes place from the exicted singlet state, the triplet state has kept an important place in the study of photosynthesis because (1) it is a versatile probe of local structure, and (2) it is a strong quencher of singlet excitations (RAHMAN and KNOX 1973), and as such serves as a probe of the energy transfer process. In this review, I will focus on the former property and discuss some of the newer results of magnetic resonance of the triplet state, especially those obtained with optical detection techniques.

2 Electron Spin Resonance of the Triplet State in Zero Magnetic Field

The triplet states considered here are excited states in which an electron is promoted from the higher occupied molecular orbital (HOMO) to the lowest unoccupied molecular orbital (LUMO) under spin reversal (intersystem crossing, ISC) by e.g. spin-orbit interaction. The results is two unpaired electrons (one in the HOMO, one in the LUMO) with parallel spin, i.e., the spin operator **S** equals unity and the multiplicity (i.e., the number of ways the spin vector $\hat{S}=\sqrt{S(S+1)}=\sqrt{2}$ in units of $\hbar/2$ can orient itself in a magnetic field) equal to $2S+1=3$. As the two unpaired electrons with parallel spins cannot occupy the same orbital (Pauli principle) their electrostatic repulsion is on the average less than in the first excited singlet state S_1 (where the two electrons have antiparallel spins), and the energy of the lowest triplet state (T_0) lies lower than that of S_1 by the so-called exchange energy J.

The two parallel spins of the triplet state represent two magnetic moments which interact magnetically via the dipole-dipole interaction. This causes a splitting of the triplet state into three triplet sublevels, which have eigenenergies equal to the principal values of the dipole-dipole interaction tensor $-X$, $-Y$ and $-Z$, with $X+Y+Z=0$ (Fig. 1). Commonly, two independent linear combi-

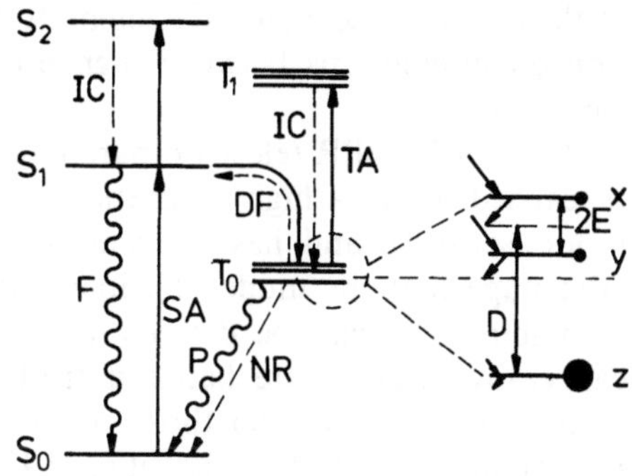

Fig. 1. Energy level diagram of the singlet and triplet manifold. S_0 singlet ground state; S_1 and S_2 singlet excited states; T_0 and T_1 first and second excited triplet states; *SA* and *TA* singlet and triplet absorption, respectively; *F* and *DF* fluorescence and delayed fluorescence, respectively; *P* phosphorescence; *NR* nonradiative transition; *IC* internal conversion. Enlarged T_0 levels: *X, Y* and *Z* eigenenergies of the dipole-dipole interaction; *D* and *E*, zero field splitting parameters. *Downward arrows* to the *right* populating probabilities; to the *left* decay rates; *filled circles* equilibrium populations

nations of the principal values are constructed, viz. $D=-\frac{3}{2}Z$ and $E=\frac{1}{2}(X-Y)$. The three triplet eigenstates $|x>$, $|y>$ and $|z>$ have populating and depopulating probabilities that depend on molecular symmetry and which in general are unequal. The relative equilibrium population n_i of the triplet sublevels under continous illumination is given by $n_i = \text{const.} \times p_i/k_i$ ($i=x, y, z$), where p is the populating probability and k the decay rate.

It can be shown (VAN DER WAALS and DE GROOT 1967) that, although in zero magnetic field there is no net magnetic dipole moment associated with any of the triplet eigenstates, there is a transition dipole moment present between any two of the eigenstates. Thus, an oscillating electromagnetic field of the proper frequency can induce transitons between the eigenstates, and one can carry out an EPR experiment in zero magnetic field. Generally, we will then obtain resonances at frequencies $(|D|+|E|)h^{-1}$, $(|D|-|E|)h^{-1}$ and $2Eh^{-1}$ (Fig. 1). The three transitions are polarized in directions that are mutually perpendicular: The $x \sim y$ transition moment is oriented along the z-axis of the dipole-dipole tensor (the z-spin axis), the $x \sim z$ transition moment along the y-spin axis and the $y \sim z$ transition moment along the x-axis.

The EPR transitions in zero magnetic field may be detected conventionally by detecting the absorption of the oscillating electromagnetic field, but much more sensitively by detecting one of the optical properties of the sample, i.e., the phosphorescence (phosphorence detected magnetic resonances, PDMR; SCHMIDT and VAN DER WAALS 1968), the fluorescence (fluorescence detected magnetic resonance, FDMR; VAN DORP et al. 1973), or the triplet-triplet ($T_n \leftarrow T_0$) and singlet-singlet ($S_n \leftarrow S_0$) absorbance (triplet absorption detected magnetic resonance, TADMR and absorption detected magnetic resonance, ADMR; CLARKE and HAYES 1972, CLARKE and CONNORS 1975, DEN BLANKEN and HOFF 1982). This is illustrated in Fig. 1. Optical detection has the additional advantage that the EPR resonance can be monitored as a function of wavelength. Especially in the case of absorbance-detected ODMR, this has given a wealth of information on interactions between pigments of the photosynthetic reaction centers (see below).

Obviously, EPR (electron paramagnetic resonance) of triplet states can also be carried out in a high magnetic field B_0 ($B_0 \gg |D|$). Although high field EPR is less sensitive and has a lower resolution than zero field EPR, it has the advantage that the mode of triplet formation (intersystem crossing or radical recombination) can be directly read from the spectral lineshape. Thus, one can easily discriminate between triplets formed in the antenna pigments by intersystem crossing and those of the primary donor that result from radical recombination (THURNAUER et al. 1975). In addition, high field EPR allows the determination of the orientation of the optical transition moment of the exciting light with respect to the principal axes of the dipole-dipole tensor by a magnetophotoselection experiment (KOTTIS and LEFEBVRE 1964). It has been shown recently, however, that such information can be obtained even more accurately by ADMR (DEN BLANKEN et al. 1984). It seems therefore, that for the determination of most triplet properties (|D| and |E| values, k_i's, spin axes orientation) ODMR offers more sensitivity and selectivity than high field EPR. In addition, it is a unique tool to measure triplet-minus-singlet absorbance difference spectra (DEN BLANKEN and HOFF 1982). In the next sections a number of recent applications of ODMR techniques in studies of the triplet states of photosynthetic materials will be summarized.

3 ODMR Techniques

Figure 2 displays a block diagram of an ODMR set-up. The experiments are conducted at low temperature (about 1.5 K) to inhibit spin-lattice relaxation.

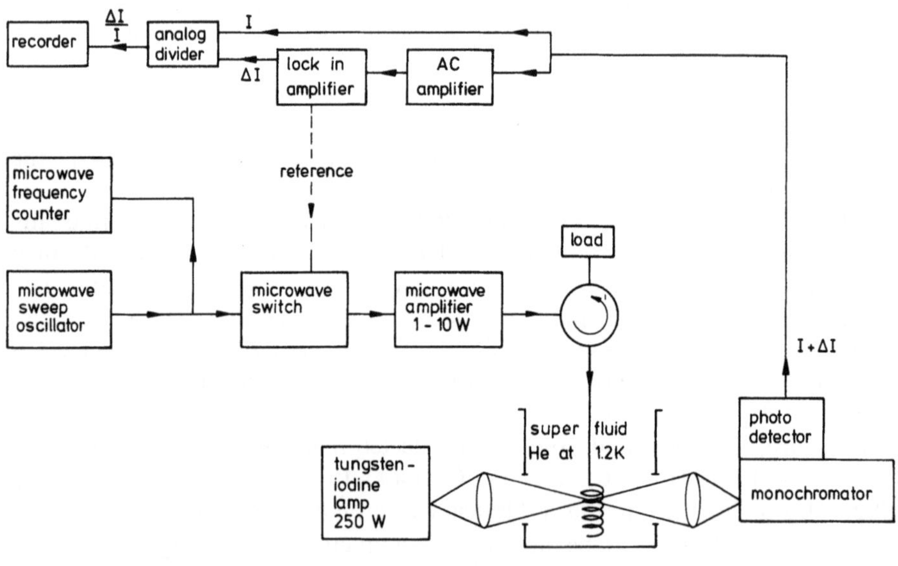

Fig. 2. Block diagram of the experimental ADMR set-up

The diagram shown is representative for an ADMR experiment, in which a lamp is used for both excitation and monitoring light. For FDMR a rectangular configuration for illumination and detection has some advantages when a four-window cryostat is available. Alternatively, the fluorescence may be detected as in the diagram shown using suitable cut-off filters, or using a light guide for top or bottom illumination and detection. For all pigments of interest here, the resonance frequencies of the oscillating magnetic field are in the lower microwave region (0.1–1 GHz). For most experiments microwave excitation is produced by means of a broad-band helix allowing easy optical access. Note that the microwaves are amplitude-modulated and the modulated optical signal is rectified by lock-in detection.

4 Zero Field Splitting Parameters and Decay Rates

The interest in the determination of |D| and |E| values lies in their relation to the spatial extent of the triplet wavefunction, viz. $D=\frac{3}{4}g^2\beta^2\left\langle\frac{r^2-3z^2}{r^5}\right\rangle$ and $E=\frac{3}{4}g^2\beta^2\left\langle\frac{x^2-y^2}{r^5}\right\rangle$, where x, y and z are the projections onto the three principal dipolar axes of the vector **r** that connects the two unpaired electrons, g is the electronic g-factor, β is the electronic Bohr magneton and $<>$ denotes an average over space. For planar aromatic molecules one expects D to be positive. As |D| and |E| are molecular characteristics, they may on one hand serve to identify a triplet state located on an isolated molecule, and on the other to obtain insight in delocalization of the triplet state on two or more pigments, on admixture of other (e.g., charge transfer) states, etc. Similar considerations hold for the k_i values, but here the connection with molecular structure is much less clear-cut.

For two strongly interacting molecules one may derive expressions that relate their configuration (angles between spin axes) to |D|, |E| and the k_i's (STERNLICHT and MCCONNELL 1961, HOCHSTRASSER and LIN 1968, CLARKE et al. 1977, BOWMAN and NORRIS 1978). These relations, however, are not unambiguous (HOFF 1982) and they do not take into account effects of ligands on the central magnesium (CLARKE et al. 1982, DEN BLANKEN and HOFF 1983a) and of small admixtures of charge transfer states (KOOYMAN and SCHAAFSMA 1980). Thus, although in principle the triplet parameters may offer detailed structural information, we are as yet not in a position to draw quantitative conclusions on geometry from the experimental values. They are useful, however, as a fingerprint of the triplet species under study, hence their values are listed in Table 1 for a number of triplets in photosynthetic reaction centers and antenna pigments, as well as for some pigments in vitro. Note that only two resonances are commonly observed, viz. at frequencies corresponding to the $|D|\pm|E|$ transitions. The 2 E transition is absent, presumably because of almost equal equilibrium population of the x and y triplet levels. It can be observed in a double resonance experiment, in which one of the $|D|\pm|E|$ transitions is continuously irradiated with resonant microwaves of fixed frequency (HOFF 1976).

Table 1 a. Representative zero field splitting parameters ($cm^{-1} \times 10^4$) of triplet states of chlorophylls in vivo and in vitro

Species	|D|	|E|	Reference
Rhodopseudomonas sphaeroides			
Strain R-26, cells	187.2±0.2	31.2±0.2	HOFF (1976)
Reaction centers	188.0±0.4	32.0±0.4	DEN BLANKEN et al. (1982)
Rhodospirillum rubrum			
Strain S1, cells	187.8±0.6	34.3±0.3	HOFF (1976)
Strain FRI, cells	187.9±0.6	34.3±0.3	HOFF and GORTER DE VRIES (1978)
Rhodopseudomonas capsulata			
Strain ATC 23782, cells	184.2±0.6	30.3±0.3	HOFF and GORTER DE VRIES (1978)
Chromatium vinosum			
Strain D, cells	177.4±0.6	33.7±0.3	HOFF and GORTER DE VRIES (1978)
Chloroflexus aurantiacus			
Reaction centers	197.7±0.7	47.3±0.7	DEN BLANKEN et al. (1983c)
Prosthecochloris aestuarii			
Reaction centers	208.3±0.7	36.7±0.7	VASMEL et al. (1984)
Rhodopseudomonas viridis			
Cells	156.2±0.7	37.8±0.7	DEN BLANKEN et al. (1983b)
Reaction centers	160.3±0.7	39.7±0.7	DEN BLANKEN et al. (1983b)
BChl a in methyltetra-hydrofuran	230.0±2.0	58.0±2.0	DEN BLANKEN and HOFF (1983a)
BChl b in methyltetra-hydrofuran	221.0±2.0	57.0±2.0	DEN BLANKEN and HOFF (1983a)
Photosystem I particles	281.7±0.7	38.3±0.7	DEN BLANKEN and HOFF (1983b)
Photosystem II particles	285.5±0.7	38.8±0.7	DEN BLANKEN et al. (1983c)
	288.8±0.7	42.2±0.7	DEN BLANKEN et al. (1983c)
Chl a in methyltetra-hydrofuran	281.0±6.0	39.0±3.0	KLEIBEUKER and SCHAAFSMA (1974)

5 Relative Orientation of Optical Transition Moments and Spin Axes

Quantitative data on the orientation of optical transition moments with respect to the spin axes are as yet only available from magnetophotoselection experiments. In addition, the orientation of the spin axes with respect to the normal

Table 1 b. Triplet sublevel decay rates in s^{-1}

Species	k_x	k_y	k_z	Reference
Rps. sphaeroides Strain R-26, cells	9,000±1,000	8,000±1,000	1,400±200	HOFF (1976)
R. rubrum Strain S1, cells	8,000±700	7,200±700	1,350±150	HOFF (1976)
C. aurantiacus	12,660±750	14,290±800	1,690±50	DEN BLANKEN et al. (1983c)
P. aestuarii	6,790±500	3,920±300	1,275±100	VASMEL et al. (1984)
Rps. viridis Cells	<16,000	<16,000	<2,600	DEN BLANKEN et al. (1983b)
Reaction centers	13,700±900	16,100±1,300	2,420±90	DEN BLANKEN et al. (1983b)
BChl a in MTHF	11,950±700	15,900±1,300	1,635±50	DEN BLANKEN and HOFF (1983a)
BChl b in MTHF	12,400±900	14,900±1,300	1,300±30	DEN BLANKEN and HOFF (1983a)
PS I particles	990±100	1,010±100	92±5	DEN BLANKEN and HOFF (1983b)
PS II particles	930±40	1,088±50	110±5	DEN BLANKEN et al. (1983a)

to the photosynthetic membrane has been determined with samples oriented by freezing in a high magnetic field, or by drying on a mylar support. The results of these experiments are collected in Table 2. Analogous experiments with linear dichroic ADMR spectroscopy were as yet only qualitatively analyzed (see Sect. 7). Note that the precise orientation of the spin axes in the chlorophyll molecule is unknown, only the orientation relative to the optical transition moments can be determined. For chlorophylls, these moments are commonly thought to be polarized in the plane of the molecule. For BChl a, the Q_y band (at 780 nm) and the Q_x band (at 600 nm) are thought to bisect the unsaturated and the saturated pyrole rings, respectively. For Chl a and Chl b the data are ambiguous (BAUMAN and WRÖBEL 1980). The chromophore in Zn-pyrochlorophyllide a substituted hemoglobin has its Q_y transiton dipole moment at $5° \pm 2°$ from the y-axis (through rings I and III), closer to ring V. It lies almost in the plane of the macrocycle (deviation not more than 5°) (MOOG et al. 1984).

In the interpretation of the magnetophotoselection data, the ordering of the x, y and z triplet sublevels of the primary donor triplet (^3P) is taken to be x, y, z to lower energy, corresponding to $D > O$ and $E < O$. That D of ^3P is positive (at least for *Rhodospirillum rubrum*) follows from elegant studies of THURNAUER (1979) of signal intensity as a function of temperature. The

Table 2. Orientation of spin axes[a] and optical transition moments

RC $BChl_2$ triplet	Angle with membrane normal[b] x	y	z	Reference
Rps. sphaeroides R-26	90°	10°–20°	70°–80°	TIEDE and DUTTON (1981)
Rps. sphaeroides wild type	90°	15°	75°	FRANK et al. (1984)
R. rubrum	90°	10°–20°	70°–80°	HALES and DAS GUPTA (1979)
Rps. viridis	50°	40°	85°	FRANK et al. (1979b)
Rps. palustris	45°	45°	75°	FRANK et al. (1979b)
RC carotenoid triplet				
Rps. sphaeroides wild type	40°	55°	80°	FRANK et al. (1984)

	Angle with transition moment[b, c] 860(990[d]) nm			546 nm			850 nm			830 nm			Reference
	x	y	z	x	y	z	x	y	z	x	y	z	
Rps. sphaeroides R-26	5°	90°	82°	65°	65°	35°							FRANK et al. (1979a)
				60°	60°	45°							BOXER and ROELOFS (1979)
Rps. viridis	60°	40°	65°				45°	60°	60°	60°	45°	60°	TROSPER et al. (1982)

[a] The ordering of the spin axes is chosen such that $\frac{1}{3}|D|+|E|=X, \frac{1}{3}|D|-|E|=Y$ and $|D|=-\frac{3}{2}Z$.

[b] To avoid an unwarranted impression of accuracy, angles are rounded off to the nearest decimal or quintal, i.e., the sum of the squares of the direction cosines need not sum to unity here.

[c] The angles were obtained from magnetophotoselection experiments by spectral simulation. In view of the fact that the peak positions of the lines were not always well-simulated (in a few cases the simulated positions are off by as much as 20%, taking into account the slight differences in scale of the published figures), the solution regions of the angles as given by the authors are not reproduced here.

[d] For *Rps. viridis*.

sign of E of monomeric Chl a,b and Pheo a,b is negative, that of BChl a,b and BPh a,b is positive (THURNAUER 1979), the sign of E in ^3P is unknown. (If P is a dimer and if the triplet state is delocalized, the sign and absolute value of E depends on the geometrical configuration of the constituent monomers).

6 Microwave-Induced Fluorescence (MIF) Spectra

Setting the microwave frequency on one of the resonances, one may scan the microwave-induced changes in fluorescence across the fluorescence spectrum (HOFF and GORTER DE VRIES 1981, BECK et al. 1981). In this way, only those fluorescence bands that respond to a particular triplet transition are detected. This helps in identifying triplet states in whole cells (BECK et al. 1981, KAISER et al. 1981), and it has been used to analyze the mechanism of energy transfer in the antenna system (HOFF and GORTER DE VRIES 1981).

Fluorescence of isolated bacterial reaction centers is weak, and FDMR experiments difficult to carry out. Using the MIF technique, DEN BLANKEN et al. (1982) were nevertheless able to discern two MIF bands, one at 905 and one at 935 nm (Fig. 3). The $|D|$ and $|E|$ values corresponding to the 905 nm band are those of the reaction center triplet, ^{3}P. The sign of the 905 nm band, however, is negative, which is unexpected if the fluorescence originates from P. Reversal of sign occurs when the fluorescing pigment is not P itself, but an associated pigment from which singlet excitation may travel to P by singlet energy transfer (HOFF and GORTER DE VRIES 1978). It is unlikely that this F-905 pigment is one of the reaction center B-800 pigments or one of the P-860 monomers whose absorbance appears to shift to 803 nm when a ^{3}P state is formed that is localized on one of the dimer constituents, because in that case the Stokes shift would amount to 100 nm, a highly unusual value for the chlorophylls. The 905 nm fluorescence may stem from a weak concentration of antenna bacteriochlorophylls (less than a few percent of P-860) that are strongly bound to the reaction center protein. Conversion of P-860 to the triplet state closes the reaction center and removes it as a strong quencher, which may strongly enhance the fluorescence of such an antenna pigment. Modulation of the concentration of ^{3}P by resonant microwaves may then very well result in a detectable MIF signal, even when the concentration of F-905 is well below that of P-860.

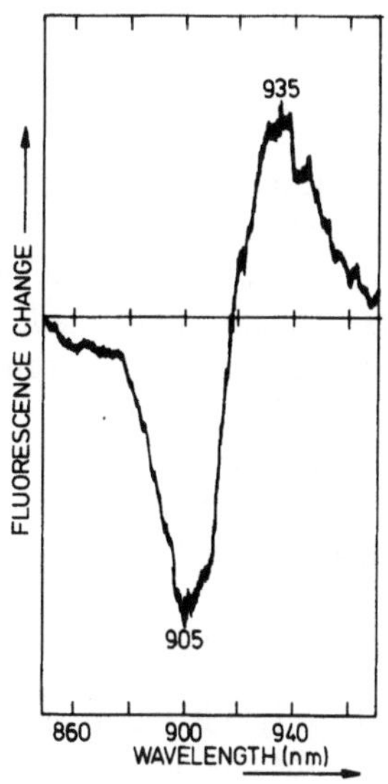

Fig. 3. Microwave-Induced Fluorescence spectrum of isolated reaction centers of *Rps. sphaeroides* R-26 at 1.2 K. The microwave frequency was set at 468 MHz (the $|D|-|E|$ transition), the spectral resolution was 1.3 nm

The sign of the 935 nm fluorescence indicates that it originates from P-860 proper, but the |D| and |E| values are not those of the bulk P-860 (as measured with ADMR or high field EPR). It was concluded that this fluorescence stems from a minority of reaction centers with a slightly different configuration, perhaps caused by the isolation procedure. Charge separation from these altered P-860 complexes may be somewhat less fast than in the intact, bulk P-860, giving rise to appreciable fluorescence that is modulated by the presence of ^3P. It was noted that the maximum of the positive MIF signal, when corrected for the presence of F-905, was close to the main reaction center fluorescence at 920 nm. (Note that this is blue-shifted by about 10 nm compared to the fluorescence of whole cells, measured at 4.2 K). Thus, it appears to be possible that the main fluorescence intensity of reaction centers is due to P-860 with a somewhat different structure than P-860 in the intact cell. Hence, caution should be exerted in analyzing experiments in which the fluorescence of isolated reaction centers is monitored as a probe of the primary photoprocesses.

7 Triplet-Minus-Singlet Absorbance Difference Spectra of Bacterial Reaction Centers

As mentioned above, absorbance-detected zero field resonance (ADMR) offers much better sensitivity than other detection schemes of zero field resonance for the detection of triplet states in photosynthetic reaction centers. That this was not immediately recognized (CLARKE and CONNORS 1975) was probably due to the fact that in the first experiments an inherently instable Xenon-arc lamp was used and that the method was applied to pigments of low solubility. It can be easily shown (DEN BLANKEN 1983) that for material of sufficient solubility to attain the optimal optical density of 0.87 cm^{-1}, the relative signal intensity in an ADMR experiment, $\left(\frac{\Delta I}{I}\right)_{\text{transmitted}}$, is comparable to the relative signal intensity of an FDMR experiment, $\left(\frac{\Delta I}{I}\right)_{\text{fluorescence}}$. However, the photon quantum noise in an absorbance measurement is much lower than in a fluorescence measurement, when a low fluorescence yield sample such as photosynthetic reaction centers is studied. In such a case the signal-to-noise ratio of an ADMR experiment can be made several orders of magnitude higher than that in FDMR. In addition to the accurate determination of the triplet parameters, this has allowed the accurate recording of triplet-minus-singlet absorbance difference (T – S) spectra.

The microwave-induced change in absorbance is to first order equal to (DEN BLANKEN and HOFF 1982):

$$[\Delta I(\lambda)/I(\lambda)]_{\text{transmitted}} = \alpha(\Delta S_0)[\varepsilon_{S_0}(\lambda) - \varepsilon_{T_0}(\lambda) - \varepsilon_{S,T}(\lambda)] \quad (1)$$

where $I(\lambda)$ is the intensity of the transmitted light as a function of wavelength, ΔS_0 the change in ground state population effected by the resonant microwaves,

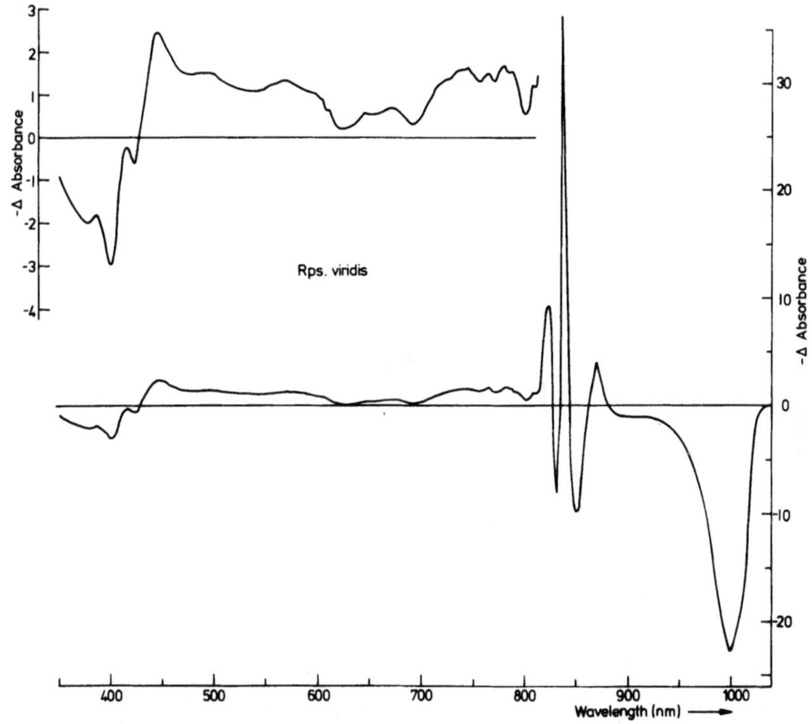

Fig. 4a. Triplet-minus-singlet absorbance difference spectrum at 1.2 K of reaction centers of *Rps. viridis* measured with absorbance detected zero field resonance (ADMR) of the triplet state. The microwave frequency was set at 356 MHz (the $|D|-|E|$ transition). Spectral resolution 6.4, 2.4, 1.6 and 2.4 nm in the 350–450, 450–790, 790–900, and 900–1050 nm region, respectively. Data points not indicated

$\varepsilon_{S_0}(\lambda)$, $\varepsilon_{T_0}(\lambda)$ and $\varepsilon_{S,T}(\lambda)$ the extinction coefficients of the singlet ground state, the first excited triplet state, and the changes in singlet ground state absorbance of neighboring pigments occasioned by the presence of a triplet state on the primary donor (or more generally a localized triplet state), respectively, and α a proportionality constant that depends on the pigment concentration, pathlength, etc. From Eq. (1), it follows that ADMR recorded as a function of the wavelength λ includes TADMR as a special case, and that by recording $\left(\frac{\Delta I(\lambda)}{I(\lambda)}\right)_{Tr}$ a triplet-minus-singlet spectrum is obtained that includes bleachings of the singlet ground state absorbance of a particular pigment (e.g., the primary donor) that result from the generation of the triplet state, the triplet-triplet absorbances within the triplet manifold, and the changes in absorbance of pigments that are electrostatically coupled to the triplet carrying pigment. In the bacterial reaction center, the latter may include band shifts of the acessory BChl's and the appearing band of monomeric BChl of the primary donor when its triplet state is located on one of the constituent BChl monomers and the exciton coupling between the monomers is weakened (see below).

The excellent sensitivity of ADMR has allowed the recording of low-temperature T – S spectra of in vivo material with unprecedented accuracy and resolu-

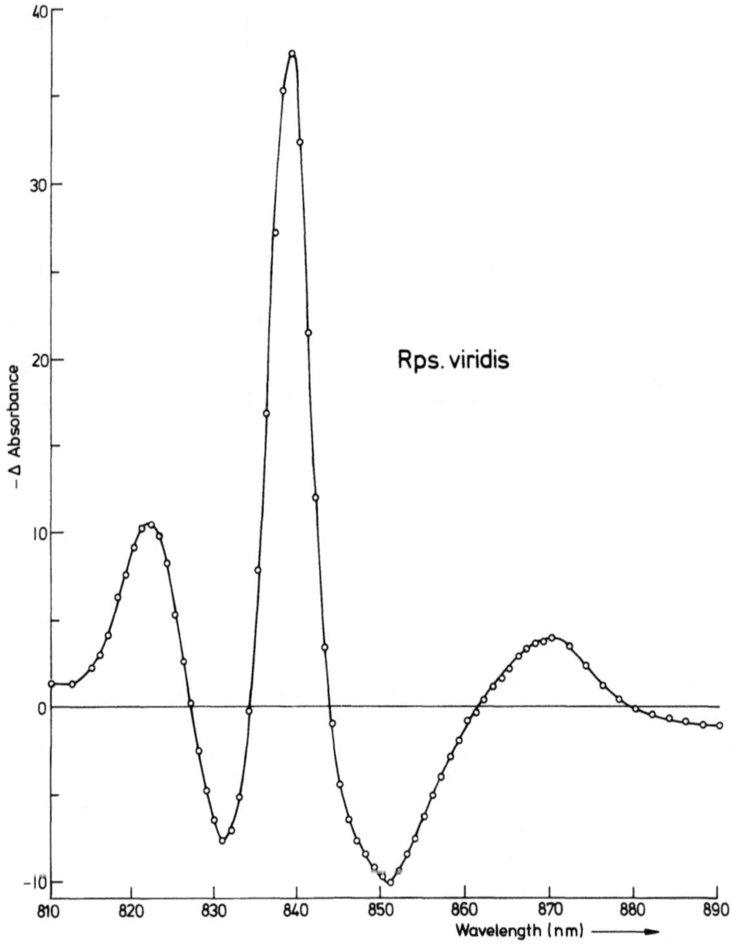

Fig. 4b. Enlarged portion of the 790–900 nm region of **a.** *Circles* are actual data points, the noise is less than one diameter. (DEN BLANKEN and HOFF 1982)

tion. In Fig. 4 a typical example is displayed. The spectrum is that of reaction centers of *Rps. viridis,* a BChl b-containing photosynthetic bacterium. It contains several noteworthy features. Firstly, the bleaching of the P-960 band (shifted to 990 nm at 1.5 K, the temperature of the measurement) is less broad than the bleaching observed by photooxidation of P-960. This is because the resonant microwaves select a particular slice out of the inhomogeneously broadened ADMR transition. It has been shown that there is a correspondence between the triplet parameters (D and E values) determining the exact resonance frequency and the optical transition wavelength (DEN BLANKEN and HOFF 1983c, DEN BLANKEN et al. 1983b). In all likelihood this correspondence exists because both the optical transition energy and the zero field splitting parameters are at least to a certain extent determined by the geometrical configuration of the

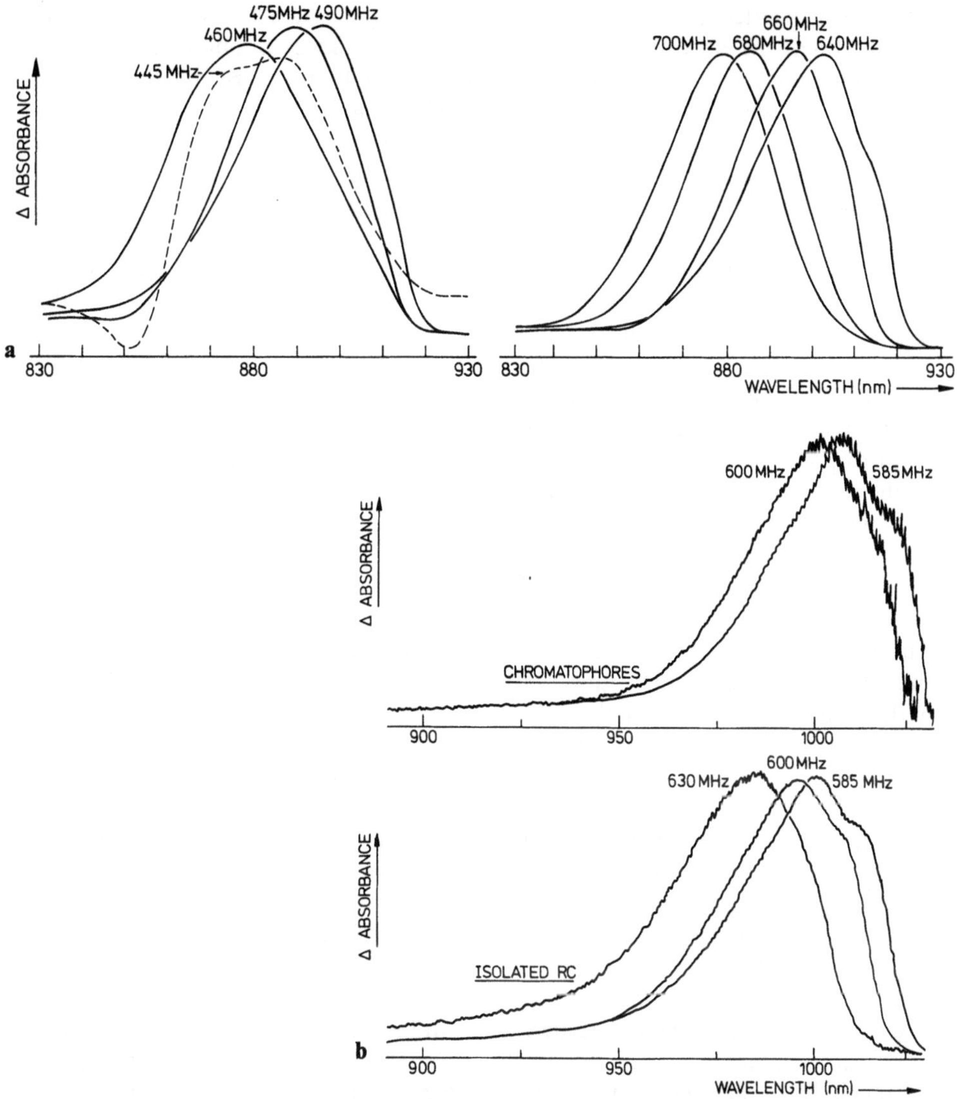

Fig. 5. **a** Long-wavelength band of the ADMR-monitored T−S spectrum of chromatophores of *Rps. sphaeroides* R-26 at 1.2 K. The resonant microwaves were set in the $|D| \pm |E|$ transitions at frequencies as indicated. The spectra are normalized, single scans; resolution 3.2 nm. The 445 MHz spectrum (*dashed line*) is the sum of T−S spectra of reaction center and antenna chlorophyll triplet states. (DEN BLANKEN and HOFF 1983c). **b** As in **a** for chromatophores and isolated reaction centers of *Rps. viridis*. (DEN BLANKEN et al. 1983b)

two BChl b monomers making up P-960. The effect is illustrated in more detail in Fig. 5 for *Rps. sphaeroides* R-26 and *Rps. viridis*. The higher resolution offered by the ADMR microwave selection makes visible that the bleaching of P-960 is composed of a more or less Gaussian band superimposed on a "pedestal", which is attributed to a so-called phonon wing, i.e., transitions shifted by the interaction of the electronic transitions and low-frequency phonons in the medium. Thus, the absorption band of P-960 (and of P-860 in BChl a-containing bacteria) is a superposition of many individual bands like that in Figs. 4 and 5. This partially resolves a long-standing discrepancy between the width of the BChl a Q_y transition in vitro (and that of the accessory BChl pigments in the reaction center) and the width of the long-wavelength absorption band of bacterial reaction centers.

Secondly, the long-wavelength bleaching of the bacterial primary donor shows for several BChl a- and b-containing species a distinct shoulder to the infrared (Fig. 5). This shoulder was earlier observed in low-temperature absorbance spectra of reaction centers of *Rps. viridis* and attributed to a charge transfer component (VERMEGLIO and PAILLOTIN 1982, MASLOV et al. 1983). Apparently, the shoulder is not confined to *Rps. viridis* and understanding its origin would seem to provide an important clue to the make-up of the primary donor, and to the mechanism of charge separation.

Thirdly, the ADMR-monitored T−S spectrum of Fig. 4 shows distinct features in the 830 nm region, where the accessory BChl b's absorb. The bands may be decomposed into two band shifts, a blue shift from 832 to 822 nm, a red shift from 852 to 870 nm and an appearing band (a "coloring") at 838 nm. The band shifts may be attributed to the two accessory BChl b pigments of the reaction center, which apparently have about equal interaction strength with P-960. This indicates that their distance to P-960 is about equal. The appearing band is attributed to a BChl b monomer belonging to P-960 whose absorbance is shifted from 990 nm to 838 nm (close to the in vitro absorbance of BChl b and in the region of the accessory BChl b pigments) because of the presence of a triplet state on its companion BChl b pigment within the P-960 dimer. It is well known that singlet-singlet electrostatic interaction between singlet excited pigments (the so-called exciton interaction) is much stronger than the interaction between a triplet excited pigment and its singlet neighbors because in the latter case the interaction is governed by the short-range exchange interaction. Thus, if the shift from the in vitro absorbance of BChl b at 795 nm in acetone to the 960 nm absorbance at room temperature of the primary donor in reaction centers is (partly) due to singlet-singlet exciton interaction, it should be much less when P-960 is excited to the triplet state. If this triplet state is localized (on an optical time scale) on one of the constituent monomers (SHUVALOV and PARSON 1981), then it follows that the remaining BChl b monomer should absorb not at 960 nm (990 nm at 1.5 K) but at a wavelength much closer to the monomeric in vitro absorbance.

The spectral region around 830 nm does not show any strong bleaching near 850 nm as observed in the oxidized-minus-reduced spectrum of P-960. This means that the earlier interpretation of this bleaching as arising from the blue exciton band of P-960 has to be abandoned (PAILLOTIN et al. 1979).

Features similar to those in Fig. 4 have been observed for BChl a-containing bacteria (be it with lesser resolution owing to the lesser spread in absorbances of BChl a compared to BChl b) (DEN BLANKEN and HOFF 1982, HOFF et al. 1985) and the general interpretation given above for *Rps. viridis* seems to hold equally well for the other species. This interpretation is strengthened by the analysis of linear dichroic T – S spectra, to which we will now turn.

8 Linear Dichroic Triplet-Minus-Singlet Absorbance Difference Spectra of Bacterial Reaction Centers

It was noted in Section 2 that the microwave transitions between any two $|u>$ and $|v>$ triplet sublevels in zero magnetic field are polarized along a direction parallel to $\mathbf{w}=\mathbf{u}\times\mathbf{v}$. Thus, in a polarized microwave field $\mathbf{B}_1$ the transition probability for an ADMR transition between $|u>$ and $|v>$ is proportional to $\mathbf{B}_1\cdot\mathbf{w}$. This means that in a randomly oriented sample, the microwaves then select a particular distribution of molecules. In other words, the distribution of triplet states in resonance is then not isotropic, and one may perform a linear dichroism experiment analogously to the more commonly employed photoselection technique (EL-SAYED et al. 1973, DEN BLANKEN et al. 1984). In principle, one may analyze such an experiment quantitatively, with similar formalism as in the photoselection technique (VERMEGLIO et al. 1978). Here we will confine ourselves to some qualitative features of LD-(T – S) spectra.

Figure 6 illustrates the T – S and LD-(T – S) spectra of *Rhodopseudomonas viridis, Rps. sphaeroides* R-26 and *Rhodospirillum rubrum.* One sees immediately that the long-wavelength bleaching exhibits a single polarization and cannot be composed of two, perpendicularly polarized, exciton bands. The intensity of the bleaching in the LD-(T – S) spectra has for low microwave power a ratio of 2:1 for the two ADMR transitions monitored (which are polarized perpendicularly), as expected for the long-wavelength band being polarized along one of the spin axes (Table 2). Note that the bleaching monitored at the $|D|+|E|$ transition is at a different wavelength than that monitored at the $|D|-|E|$ transition. This is a result of the microwave selection (Fig. 5).

The features in the 830 nm region in the LD-(T – S) spectra of *Rps. viridis* strongly support the earlier interpretation (Sect. 7) that they are composed of two band shifts and an appearing band. Band shifts will change intensity in the positive and negative lobe equally when going from one ADMR transition to the other, because both have the same direction of the transition moment. If bands of different origin are superimposed, they will change intensity independently when their transition moment is not parallel to that of the shifting band. This provides an easy check on the assignment of the various features. It is seen that in the LD-(T – S) spectrum at 597.4 MHz the appearing band is absent, clearly showing that it is oriented differently from the features to its left and right. These features are now unperturbed by the presence of the strong band at 838 nm, and it is clearly seen that both of them are due to a band shift.

The band shifts at both sides of the appearing band at 807 nm in the T – S spectrum of *R. rubrum* are partly obscured by this band. In the 465 MHz LD-(T – S) spectrum the appearing band is almost unperturbed by the band shifts, and has a width of 7 nm. Although this width is somewhat less than each

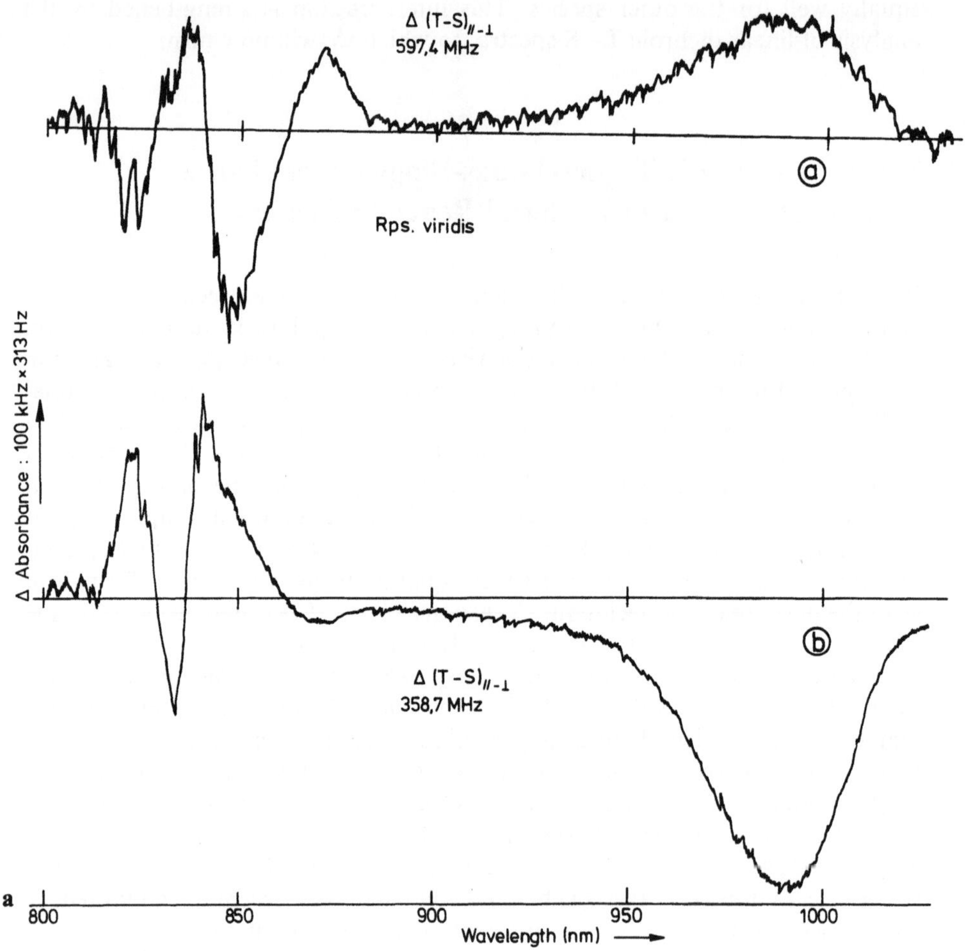

Fig. 6. Linear dichroic triplet-minus-singlet spectra of **a** *Rps. viridis*; **b** *Rps. sphaeroides* R-26 and **c** *R. rubrum* as monitored by ADMR at 1.2 K. Spectral resolution 2.4 nm. For **b** and **c** the T–S spectrum war recorded simultaneously with the LD-(T − S) spectra. The polarized microwaves at frequencies as indicated were amplitude-modulated at 335 Hz; the transmitted light intensity was modulated with a Morvue cell operating at 50 kHz (resulting in intensity modulations at 100 kHz). LD-(T − S) spectra were obtained by double lock-in detection, T–S spectra by lock-in detection at 335 Hz. (HOFF et al. 1985)

of the two components of the 812 nm band in the absorbance spectrum of *Rps. sphaeroides* R-26 (FEHER 1971), whose width is about 12 nm each, it is not inconsistent with a band of monomeric BChl a in an environment which allows little interaction with the matrix.

The interpretation of the strong appearing band at about 807 nm in the BChl a-containing bacteria and at 838 nm in *Rps. viridis* does not account

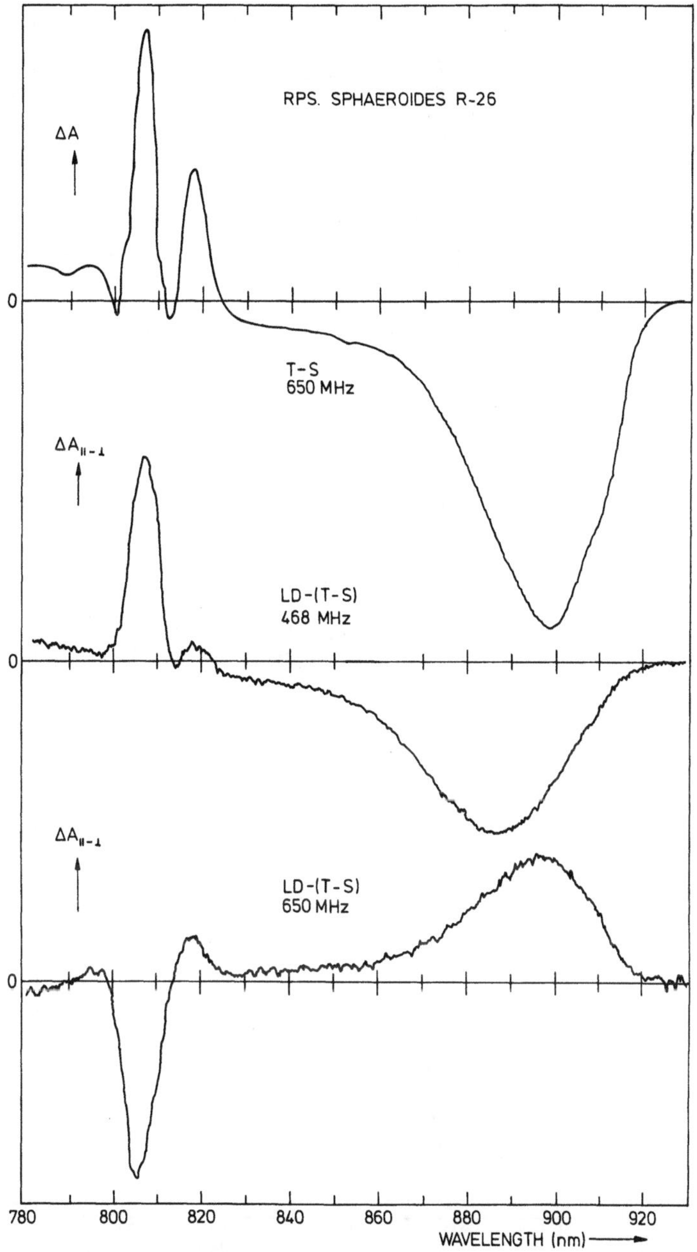

Fig. 6b

for the relatively low oscillator strength of this band. If there would be no intensity redistribution as a result of changes in electrostatic interactions induced by the triplet state, then one would expect that an appearing band of a BChl monomer that forms part of the dimeric P-860 (or P-960) complex would have

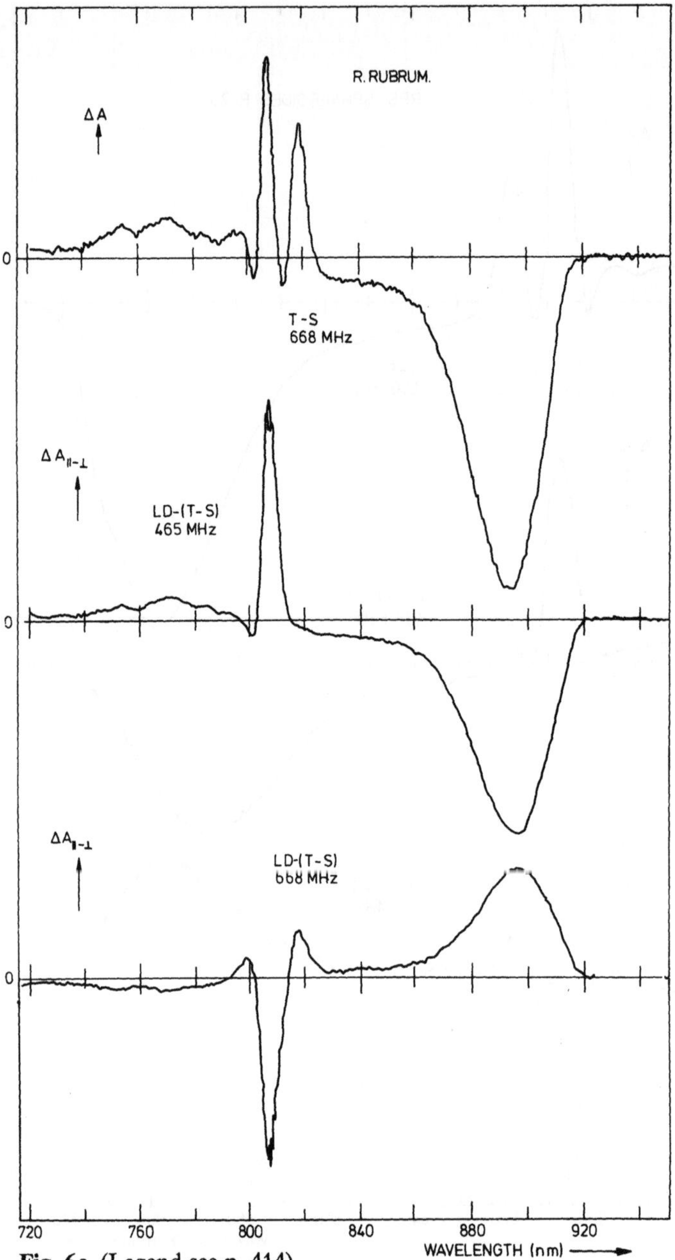

Fig. 6c. (Legend see p. 414)

about half the oscillator strength of P-860 (P-960). This is certainly not the case, and one has to invoke extensive redistribution of oscillator strength, or resort to a different explanation (in which case the simplicity of the model is sacrified) to account for the weakness of the appearing band.

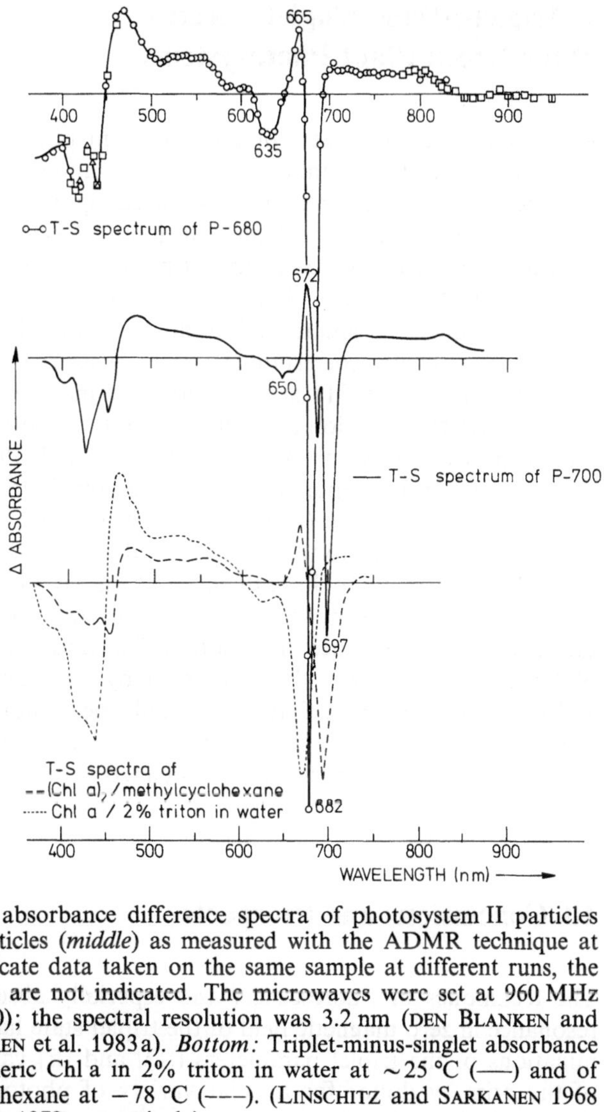

Fig. 7. Triplet-minus-singlet absorbance difference spectra of photosystem II particles (*top*) and photosystem I particles (*middle*) as measured with the ADMR technique at 1.2 K. *Different symbols* indicate data taken on the same sample at different runs, the data points for PSI particles are not indicated. The microwaves were set at 960 MHz (P-700) and 990 MHz (P-680); the spectral resolution was 3.2 nm (DEN BLANKEN and HOFF 1983b and DEN BLANKEN et al. 1983a). *Bottom:* Triplet-minus-singlet absorbance difference spectra of monomeric Chl a in 2% triton in water at ~25 °C (—) and of dimeric Chl a in methylcyclohexane at −78 °C (---). (LINSCHITZ and SARKANEN 1968 and PERIASAMY and LINSCHITZ 1979, respectively)

The foregoing shows that inspection only of LD-(T−S) spectra is already of considerable help in unraveling the T−S spectrum. Future quantitative analysis will yield precise information on the orientation of the various optical transition moments relative to the spin axes and will hopefully contribute to the determination of the structure of the primary donor and its accessory pigments.

9 Triplet-Minus-Singlet Spectra of the Green Plant Photosystems

At present reaction centers of the two plant photosystems cannot be isolated without appreciable "contamination" with antenna pigments. This makes the detection of photo-or microwave-induced optical changes considerably more difficults. Nevertheless, the ADMR technique has allowed the recording of accurate, high-resolution T−S spectra of both PS I and PS II particles (DEN BLANKEN and HOFF 1983b, DEN BLANKEN et al. 1983a). Figure 7 shows these spectra, together with T−S spectra of monomeric and dimeric Chl a in vitro. The PS I T−S spectrum corresponds to one measured with the optical flash technique (SETIF et al. 1981) and provides an independent check that the latter is due to triplet formation of P-700, and not to photoreduction of an intermediate acceptor as earlier proposed (SHUVALOV et al. 1979). The PS II T−S spectrum was not earlier reported. Both the in vivo T−S spectra correspond closely to the T−S spectrum of $(Chl\ a)_2$, but not to that of monomeric Chl a in vitro, especially with regard to the positive band at 672 nm (PS I) and 665 nm (PS II). Also, the features in the Soret region of the in vivo spectra are better reproduced in the $(Chl\ a)_2$ spectrum than in the Chl a spectrum. This correspondence was taken as strong evidence that both P-700 and P-680 are dimeric complexes of Chl a. See, however, the discussion by Setif and Mathis (Chap. 9.6, this Vol.) on other evidence that P-700 could be a monomer. Note that the PS I spectrum shows a band at 685 nm which can be interpreted as a band shift of a Chl a pigment that is closely associated with P-700.

10 Conclusions and Prospects

The techniques of fluorescence, and especially absorbance-detected magnetic resonance in zero magnetic field of the triplet state have allowed accurate determinations of the triplet parameters ($|D|$ and $|E|$ values, sublevel decay rates) of the primary donor for a great variety of photosynthetic reaction centers, thus producing a data base for future analysis of the primary donor. Absorbance detected zero field resonance has allowed the recording of accurate, high-resolution T−S and LD-(T−S) spectra that have offered insight into the interactions between the reaction center pigments and the "multiplicity" of the primary donor in plant photosystems.

ODMR of photosynthetic triplet states is a recent field, and rapid progress is being made. It is therefore difficult to enumerate future developments. A few areas in which important advances are likely to occur are, nonetheless, listed below.

Until now, the ODMR techniques have been applied at very low temperatures, and the temperature dependence of the various features in the T−S spectra could not be explored. The great enhancement in signal-to-noise ratio recently obtained for ADMR, however, will hopefully permit extending the application of ODMR to higher temperature regimes, thereby facilitating a comparison of T−S spectra obtained with ADMR and with flash spectroscopy at liquid nitrogen temperatures and above (SHUVALOV and PARSON 1981). Quantitative LD-(T−S) spectroscopy will allow the determination of pigment orientation in the reaction center and further help in the interpretation of T−S spectra. Employing absorbance detecting pulsed ODMR techniques such as the FDMR spin echo method developed by NISHI et al. (1978) may allow the study of dephasing mechanisms in P-860 and P-960, thus offering insight into their geometry. FDMR investigations of antenna complexes such as explored by SEARLE et al. (1981) may provide information an their coupling to the reaction center. Finally, a comparison of T−S spectra of various species of photosynthetic bacteria will enable researchers to probe variations in the design of reaction centers, thus allowing a classification based on reaction center structure, rather than on general morphology (see SPRAGUE and VARGA, Chap. 11.1, this Vol.)

Notes Added in Proof

Section 6. Recent fluorescence detected ODMR experiments on whole cells and isolated reaction centers of *Rps. viridis* bore out similar sign reversal of reaction center signals of the ODMR transition as a function of monitoring wavelength as shown in Fig. 3 for *Rps. sphaeroides* R-26 (ANGERHOFER, A., SCHUTZ, J.U. VON, WOLF, H.C. (1984) Z. Naturforsch. 39c, 1085–1090). However, the sign of the fluorescence detected ODMR signal of whole cells was positive, in agreement with DEN BLANKEN et al. (1983b). This poses a problem, as the interpretation of the authors of the sign reversal then predicts a negative sign of the ADMR signal of reaction centers, contrary to observation (DEN BLANKEN et al., 1983b). Note in this respect that the sign of ADMR-monitored T−S spectra does not reflect the sign of the ADMR-signal per se (DEN BLANKEN and HOFF, 1982; DEN BLANKEN et al., 1983b, 1984). The discrepancy between the signs of the antenna fluorescence and absorbance monitored ODMR signals in *Rps. viridis* was addressed by DEN BLANKEN et al. (1983b) and tentatively attributed to energy transfer from a vibrationally excited S_1-state of the antenna pigments to the reaction center.
Section 7. It has recently been shown by hole burning techniques (MEECH, S.R., HOFF, A.J., WIERSMA, D.A. (1985) Chem. Phys. Lett. 121, 287–292; BOXER, S.G., LOCKHART, D.J., MIDDENDORF, T.R. (1986) Chem. Phys. Lett., 123, 476–482) that P-860 is largely homogeneously broadened, the width of the hole conforming to the ODMR measured linewidth.
Section 8. A full quantitative analysis of the LD−(T−S) spectra of *Rps. viridis* and *Chloroflexus aurantiacus* has recently been carried out (LOUS, E.J., VASMEL, H., MEIBURG, R.F., HOFF, A.J., manuscript in preparation).
Section 9. High temperature T−S spectra have recently been obtained by a novel technique, Magneto-optical difference spectroscopy (MODS), in which changes in triplet concentration are effectuated by a small modulated magnetic field (HOFF, A.J., LOUS, E.J., MOEHL, K.W., DIJKMAN, J.A. (1985) Chem. Phys. Lett. 114, 39–43).

References

Bauman D, Wröbel D (1980) Biophys Chem 12:83–91
Beck J, Kaiser GH, Schütz JU von, Wolf HC (1981) Biochim Biophys Acta 634:165–173

Blanken HJ den (1983) Dissertation, Univ Leiden
Blanken HJ den, Hoff AJ (1982) Biochim Biophys Acta 681:365–374
Blanken HJ den, Hoff AJ (1983a) Chem Phys Lett 96:343–347
Blanken HJ den, Hoff AJ (1983b) Biochim Biophys Acta 724; 52–61
Blanken HJ den, Hoff AJ (1983c) Chem Phys Lett 98:255–262
Blanken HJ den, Zwet GP van der, Hoff AJ (1982) Biochim Biophys Acta 681:375–382
Blanken HJ den, Hoff AJ, Jongenelis APJM, Diner B (1983a) FEBS Lett 157:21–27
Blanken HJ den, Jongenelis APJM, Hoff AJ (1983b) Biochim Biophys Acta 725:472–482
Blanken HJ den, Vasmel H, Jongenelis APJM, Hoff AJ, Amesz J (1983c) FEBS Lett 161:185–189
Blanken HJ den, Meiburg RF, Hoff AJ (1984) Chem Phys Lett 105:336–342
Bowman MK, Norris JR (1978) Chem Phys Lett 54:45–48
Boxer SG, Roelofs MG (1979) Proc Natl Acad Sci USA 76:5636–5640
Clarke RH, Connors RE (1975) Chem Phys Lett 33:365–368
Clarke RH, Hayes JM (1972) J Chem Phys 57:569–570
Clarke RH, Connors RE, Frank HA, Hoch JC (1977) Chem Phys Lett 45:523–528
Clarke RH, Hotchandani S, Jagannathan SP, Leblanc RM (1982) Chem Phys Lett 89:37–40
El-Sayed MA, Gossett E, Leung M (1973) Chem Phys Lett 21:20–27
Dorp WG van, Schaafsma TJ, Soma M, Waals JH van der (1973) Chem Phys Lett 21:221–225
Feher G (1971) Photochem Photobiol 14:373–387
Frank HA, Bolt J, Friesner R, Sauer K (1979a) Biochim Biophy Acta 547:502–511
Frank HA, Friesner R, Nairn JA, Dismukes GC, Sauer K (1979b) Biochim Biophys Acta 547:484–501
Frank HA, Machnicki J, Toppo P (1984) Photochem Photobiol 39:429–432
Hales B, Das Gupta AD (1979) Biochim Biophys Acta 548:276–286
Hochstrasser RM, Lin T-S (1968) Chem Phys 49:4929–4945
Hoff AJ (1976) Biochim Biophys Acta 440:767–771
Hoff AJ (1982) In: Clacke RH (ed) Triplet state ODMR spectroscopy. Wiley-Interscience, New York, pp 367–425
Hoff AJ, Gorter de Vries H (1978) Biochim Biophys Acta 503:94–106
Hoff AJ, Gorter de Vries H (1981) Isr J Chem 21:277–282
Hoff AJ, Blanken HJ den, Vasmel H, Meiburg RF (1985) Biochim Biophys Acta 806:389–397
Kaiser GH, Beck J, Schütz JU von, Wolf HC (1981) Biochim Biophys Acta 634:153–164
Kleibeuker JF, Schaafsma TJ (1974) Chem Phys Lett 29:116–122
Kooyman RPH, Schaafsma TJ (1980) J Mol Struct 60:373–380
Kottis P, Lefebvre R (1964) Chem Phys 41:3660–3661
Linschitz H, Sarkanen K (1968) Am Chem Soc 80:4826–4832
Maslov VG, Klevanik AV, Ismailov MA, Shuvalov VA (1983) Dokl Akad Nauk SSSR 269:1217–1221
Moog RL, Kuki A, Fayer MD, Boxer SG (1984) Biochemistry 23:1564–1571
Nishi NN, Schmidt J, Hoff AJ, Waals van der JH (1978) Chem Phys Lett 56:205–207
Paillotin G, Vermeglio A, Breton J (1979) Biochim Biophys Acta 545:249–269
Periasamy N, Linschitz H (1979) J Am Chem Soc 101:1056–1057
Rahman TS, Knox RS (1973) Phys Status Solidi 58:715–720
Schmidt J, Waals JH van der (1968) Chem Phys Lett 2:640–642
Searle GFW, Koehorst RB, Schaafsma TJ, Lindberg Møller B, Wettstein D von (1981) Carlsberg Res Commun 46:183–194
Shuvalov VA, Parson WW (1981) Biochim Biophys Acta 638:50–59
Shuvalov VA, Dolan E, Ke B (1979) Proc Natl Acad Sci USA 76:770–773
Sternlicht H, McConnell HM (1961) J Chem Phys 35:1793–1800
Thurnauer MC (1979) Rev Chem Interm 3:197–230
Thurnauer MC, Katz JJ, Norris JR (1975) Proc Natl Acad Sci USA 72:3270–3274
Tiede DM, Dutton PL (1981) Biochim Biophys Acta 637:278–290

Trosper TL, Frank HA, Norris JR, Thurnauer MC (1982) Biochim Biophys Acta 679:44–50
Vasmel H, Blanken HJ den, Dijkman T, Meiburg RF, Amesz J, Hoff AJ (1984) Biochim Biophys Acta 767:200–208
Vermeglio A, Paillotin G (1982) Biochim Biophys Acta 681:32–40
Vermeglio A, Breton J, Paillotin G, Cogdell R (1978) Biochim Biophys Acta 501:514–530
Waals JM van der, Groot MS de (1967) In: Zahlan AB (ed) The triplet state. Cambridge Univ Press, Cambridge, pp 101–132

9 Photosystems I and II: Structure, Proteins, and Cofactors

9.1 The Reaction Center of Photosystem II

B.A. DINER

1 Introduction

Over the last few years it has become increasingly clear that considerable homology exists between the acceptor side components of photosystem II (PS II) and those of reaction centers from purple photosynthetic bacteria (PARSON and KE 1982, CROFTS and WRAIGHT 1983, DUTTON, Chap. 5, MICHEL and DEISENHOFER, Chap. 8.4, both this Vol.). While one might hardly have expected to see familiar faces among the high potential donors of PS II, recent evidence suggests that some of the same old dogs may have learned new tricks.

I will stress in this review what we have recently learned about electron transport from the secondary donor, Z, through to the first quinone acceptor, Q_A, and the participation under certain circumstances of alternate donors and acceptors. I will not discuss the secondary quinone acceptor, Q_B, but refer the reader of this Volume to Chapter 9.4 of ARNTZEN and PAKRASI and to the broader reviews of CROFTS and WRAIGHT (1983), VERMAAS and GOVINDJEE (1981) and ORT (Chap. 4) for more details on this topic.

2 Primary Reactants

2.1 Transmembrane Charge Separation

In the primary photoreaction of photosystem II (PS II), the primary electron donor, P-680, a chlorophyll a (Chl a) in its lowest excited singlet state, transfers an electron to a primary acceptor Ph, a pheophytin a (Ph a), and then to the first quinone acceptor, Q_A, a plastoquinone. The electron crosses the thylakoid membrane from P-680 near the inner surface to Q_A near the outer surface in an electrogenic reaction with a rate constant estimated at $>2{,}5 \times 10^9\ s^{-1}$ by ns spectroscopy (SHUVALOV et al. 1980) and $>5 \times 10^9\ s^{-1}$ by a macroscopic electrode technique (TRISSL et al. 1982 and personal communication), possibly limited by exciton transfer to the reaction center (see below). This rate, >10-fold more rapid than the fluorescence lifetime for Chl a in vivo, uncoupled to reaction centers, (2–3 ns, KARUKSTIS and SAUER 1983), is a necessary condition for the high quantum yield of the photoreduction of Q_A (~0,90, THIELEN and VAN-GORKOM 1981 b).

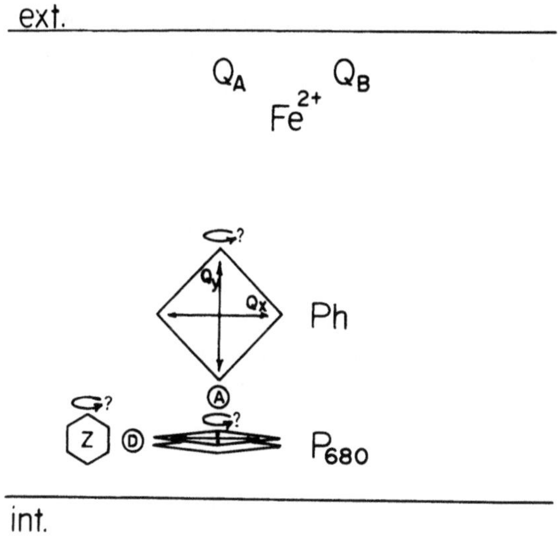

Fig. 1. Orientation and location of PS II reaction center components in the thylakoid membrane. The orientation of Z, P-680, and Ph are known with respect to the membrane plane (see also text) but not with respect to each other. *D* and *A* are tentative additional donors and acceptors, respectively

The individual rate constants for electron transfer from P-680 to Ph and from Ph^- to Q_A are not known. However, in bacterial reaction centers which contain similar components, P-870 (bacterio-Chl a), I (bacterio-Ph a) and Q_A (ubiquinone) these rate constants have been estimated to be $>10^{11}\ s^{-1}$ for P-870 $I \rightarrow P^+$-870 I^- and $\sim 5 \times 10^9\ s^{-1}$ for $I^- Q_A \rightarrow IQ_A^-$ (ROCKLEY et al. 1975, KAUFFMANN et al. 1975).

The major electrogenic step in PS II corresponds to the electron transfer from Ph^- to Q_A as determined by detection of the membrane polarization using macroscopic electrodes (TRISSL, personal communication) and by fluorescence measurements in the presence of an external electric field (MEIBURG et al. 1983). The latter have estimated this transfer as being responsable for 90% of the membrane potential contributed by PS II. Assuming homogeneous membrane permittivity, this estimation is consistent with the estimated 6–7 Å (by EPR) which separate an oxidized secondary donor to P-680 and Ph^- (RUTHERFORD and THURNAUER 1982) but contrasts with the estimated 8–9 Å (by EPR) which separate Ph^- and Q_A^- (KLIMOV and KRASNOVSKII 1981).

2.2 P-680 (also called Chl a_{II})

P-680 was identified as a photooxidizable Chl a based on optical spectroscopy (DÖRING et al. 1969, VAN GORKOM et al. 1975). There has been some controversy, however, as to its state of aggregation (monomer, dimer, or even trimer). Based on the linewidth (DAVIS et al. 1979) of the EPR spectrum of P-680$^+$ (g = 2,0027, ΔHpp = 8 G, KE et al.1982), the zero-field splitting of the spin-polarized P-680 triplet formed by P-680$^+$ Ph^- recombination (RUTHERFORD et al. 1981 b), the infrared spectrum of P-680$^+$ (MATHIS and SETIF 1981) and the triplet-minus-singlet spectrum of P-680 (by ADMR, DEN BLANKEN et al. 1983), there is general agreement that the plus charge and the triplet state are localized on a monomeric species. The triplet-minus-singlet spectrum of P-680 (DEN BLANKEN et al. 1983) and the $P\text{-}680^+ - P\text{-}680$ spectrum (VAN GORKOM et al. 1974) show an absorbance increase at 665 nm in favor of the simultaneous appearance of a ground-state monomeric Chl a and a monomeric Chl a triplet or cation, both derived from a ground-state dimer. A pair of Chl a molecules thus appear to interact strongly in their singlet states and weakly in the triplet state or as a cation (DEN BLANKEN et al. 1983, HOFF, Chap. 7.2, this Vol.).

Linear dichroism studies (MATHIS et al. 1976) on the orientation of P-680 have shown the Qy transition moment to be within 20° of the membrane plane. Orientation studies (RUTHERFORD 1985) of the spin-polarized triplet of P-680 favor a localization of the plane of at least one of the Chl a parallel to that of the membrane.

2.3 Pheophytin (Ph, also called I)

The role of pheophytin as primary electron acceptor was demonstrated through the observation of its photoreduction at redox potentials low enough to reduce Q_A (≤ -300 mV) (KLIMOV and KRASNOVSKII 1981). The midpoint potential (-610 mV), the optical, EPR ($g=2{,}0035$, $\Delta Hpp=12{,}5$ G, KLIMOV and KRASNOVSKII 1981) and ENDOR (FAJER et al. 1980) spectra of the reduced species are all consistent with a monomeric anion of reduced pheophytin a. OMATO et al. (1984) have reported two pheophytin molecules per reaction center in PS II particles, though, as in photosynthetic bacteria, only one of these appears to be photoactive (KLIMOV et al. 1980a).

Linear dichroism studies of the photoactive Ph in subchloroplast particles using the squeezed-gel technique (GANAGO et al. 1982) have shown the Qy and Qx transition moments to be oriented approximately perpendicular and parallel, respectively, with respect to the membrane plane. The plane of the molecule is thus normal to the membrane.

An acceptor even more primary than pheophytin has been proposed (RUTHERFORD 1981) based on formation of a radical ($g=2{,}00$, $\Delta Hpp=16$ G) following illumination at 200 K in PS II particles poised at $E_h = -645$ mV (Ph reduced).

2.4 Q_A (also called Q, X-320, Q_H, Q_1, A_1)

Q_A was first detected through the dependence of the fluorescence yield on its redox state, high (F_{max})$-Q_A$ reduced, low (F_O)$-Q_A$ oxidized (DUYSENS and SWEERS 1963). STIEHL and WITT (1968) were the first to detect absorbance changes (X-320) in the UV consistent with the formation of a plastosemiquinone on the acceptor side of PS II. VAN GORKOM (1974) showed convincingly, through comparison with UV in vitro spectra, that Q_A is indeed reduced to the plastosemiquinone anion with no protonation of the semiquinone occurring on the seconds time scale even at pH 4. The pH dependence of -60 mV/pH unit for the E_m of Q_A/Q_A^- (see below) requires that a proton be taken up (though not necessarily by the semiquinone itself) at least on the minutes time scale of the redox titrations. The spectrum of Q_A^- minus Q_A (VAN GORKOM 1974) shows some differences with that of the in vitro spectrum of $PQ^- - PQ$. Aside from red-shifting induced by detergent treatment, these differences arise from local electrochromic effects of Q_A^- on reaction center chlorophyll in the red and blue regions of the spectrum, and a definite band shift (C-550, isosbestic point at 545 nm) of pheophytin (VAN GORKOM 1974). For reasons that are unclear, the observation of the pheophytin band shift, upon reduction of Q_A, requires the presence of β-carotene (COX and BENDALL 1974).

Under certain conditions (detection <15 K, high microwave power) a doublet EPR signal is observed for the state $Ph^- Q_A^-$ (KLIMOV et al. 1980b). By analogy with bacterial reaction centers, this signal was attributed to an interaction of Ph^- with a semiquinone-iron complex. This complex was detected directly in PS II particles (NUGENT et al. 1981) showing a similar EPR signal ($g = 1.82$) to that observed in bacterial reaction centers. The g-value of this signal has recently been shown to shift from 1.82 to 1.90 on going from pH 6 to 9 in O_2-evolving PS II membranes (RUTHERFORD and ZIMMERMANN 1984), probably due to the deprotonation of a site on the protein near Q_A^- (see also Sect. 2.6). Extraction of the iron using $LiClO_4$ and o-phenanthroline gave rise to an EPR signal for Q_A^- with $g = 2{,}0044$ and $\Delta H_{pp} = 9$ G (KLIMOV et al. 1980) typical of a semiquinone. Detection of the iron directly, by Mössbauer spectroscopy of PS II particles (PETROULEAS and DINER 1982), indicated a high spin ($S = 2$) Fe^{2+} with distorted octahedral coordination in an environment similar to that of the iron in bacterial reaction centers. The quinone appears to be located outside the first coordination sphere of the Fe^{2+} in bacteria (DISMUKES et al. 1984) a situation which probably exists as well for PS II. The iron appears not to participate directly in electron transfer and its role is at present unknown.

2.5 Fluorescence

Following the discovery of pheophytin as primary electron acceptor in PS II and in bacterial reaction centers, SHUVALOV and KLIMOV (1976) proposed that most of the "variable fluorescence", that dependent on the redox state of Q_A, and thought to be emitted as a consequence of competition between fluorescence lifetime and energy trapping by the reaction center, was instead a delayed fluorescence emission occurring through charge recombination between P^+ and Ph^- and reformation of $^1P^*$-680. Such luminescence was already known for recombination of more stable secondary charge pairs occurring at greater than microsecond times. This proposal is consistent with such observations as the temperature (KLIMOV et al. 1978) and magnetic field (RADEMAKER et al. 1979) dependencies of the fluorescence yield, and the quenching of fluorescence to the F_O level associated with the reduction of Ph (KLIMOV and KRASNOVSKII 1981).

With the development of single photon counting techniques and picosecond spectroscopy, it became possible to test this model more directly. Consistent with the model, SHUVALOV et al. (1980) showed, in subchloroplast particles, that upon excitation with 20–30 ps pulses at $E_h = -450$ mV (Q_A reduced) the fluorescence yield (20 °C) decayed with a 4 ns lifetime, as did the optical signal of Ph^- following a 3 ns laser pulse.

HAEHNEL et al. (1982) found in spinach chloroplasts (20 °C) a somewhat faster fluorescence decay component of 1–2 ns which they attributed to $P\text{-}680^+$ Ph^- recombination and which was virtually absent at the F_O level (Q_A oxidized). The amplitude of this component was stimulated 5 to 20-fold at the F_{max} level (all Q_A reduced), approximately what one would predict for a quantum yield of 0.90–0.95 for the primary photoreaction of PS II. Experiments on a chlorophyll b-less barley mutant (SEARLE et al. 1979, KARUKSTIS and SAUER 1984) and on partially or fully greened pea chloroplasts (KARUKSTIS and SAUER 1983) showed that the observation of this component and its stimulation on going from F_O to F_{max} were considerably augmented by the presence of both light-harvesting chlorophyll a/b (LHC) and of Mg^{2+}. These results indicate that the observation and stimulation of the 1–2 ns component depends largely on certain structural aspects of the LHC antenna.

Two other decay components of ~100 and ~400 ps were also observed (HAEHNEL et al. 1982). Of the two, only the latter component shows a dependence on the redox state of Q_A, increasing ~twofold from F_O to F_{max} in the absence of LHC, but invariant in the presence of LHC plus Mg^{2+} (KARUKSTIS and SAUER 1984). The shortest decay component of ~100 ps has been attributed to the rates of energy transfer from the peripheral PS I and PS II antenna to their reaction centers (GULOTTY et al. 1982), a process independent of the redox state of Q_A. This assignment for the 100 ps component implies that the rate limiting step of ≤100–150 ps, measured for the reduction of Q_A by TRISSL (see above), is the energy transfer to the PS II reaction center.

If we accept the SHUVALOV-KLIMOV hypothesis for the origin of the variable fluorescence, then these observations suggest that the intrinsic $P\text{-}680^+ Ph^-$ recombination time is less than 2 ns, and that recombination probably reexcites the same or, by energy transfer, neighboring reaction centers producing a new charge separation. In this case the apparent $P\text{-}680^+ Ph^-$ lifetime at room temperature is determined by loss processes in the light-harvesting antenna, responsable for the ~400 ps and the 1–2 ns components. BRETON (1982) has proposed, however, that as the temperature is lowered, there is increasing localization of the exciton on pheophytin, responsable for the 695 nm fluorescence emission band.

Lifetimes ranging from 30 to 150 ns have also been attributed to the $P\text{-}680^+$ Ph^- radical pair (DUYSENS 1983, DINER and DELOSME 1983b). These may either reflect a heterogeneity in the $P\text{-}680^+Ph^-$ recombination with components >2 ns or a secondary electron transfer on the donor side which would stabilize Ph^-. If we accept that $P\text{-}680^+$ is a quencher of fluorescence (BUTLER et al. 1973), then the observation that fluorescence decay components of <2 ns increase in amplitude upon reduction of Q_A implies that most of $P\text{-}680^+$ has been reduced in <2 ns by Ph^- (recombination) and/or a secondary donor.

2.6 Redox Potentials

Measurements of fluorescence presumably arising from $P\text{-}680^+Ph^-$ recombination as a function of temperature indicated an energy difference of 0.04–0.08 eV between P-680$^+$Ph$^-$ and ^{1}P*-680 in chloroplasts (KLIMOV et al. 1978) suggesting an energy for P-680$^+$Ph$^-$ approximately 1.75 eV above the ground state of P-680 Ph.

Redox titration of the light-induced photoreduction of pheophytin (KLIMOV and KRASNOVSKII 1981) and of the spin-polarized triplet arising from P-680$^+$ Ph$^-$ recombination (RUTHERFORD et al. 1981a) both gave E_m of approximately −610 mV for Ph/Ph$^-$, similar to that observed in vitro. KLIMOV et al. (1978) assumed that this E_m was the same with Q_A oxidized or reduced and together with the 1.75 eV for P-680$^+$Ph$^-$ (see above) estimated an E_m of ~1.1 V for P-680$^+$/P-680. This potential is sufficiently positive to drive the oxidation of water even in the presence of light-driven acidification of the thylakoid interior ($E_{m\,O_2/H_2O} \simeq 0.9-1.0$ V).

Redox titration of Q_A have involved detection of the fluorescence yield (component Q_H, HORTON and CROZE 1979), C-550 and oxidation of cytochrome b-559 at 77 K (ERIXON and BUTLER 1971, KNAFF 1975), and the flash-induced electrochromic shift (MALKIN 1978, DINER and DELOSME 1983a). Most of these measurements agree on a pH-dependent midpoint potential (−60 mV/pH unit) with an E_{m7} of ~0 V for Q_A/Q_A^-. A value of −300 mV (THIELEN and VAN GORKOM 1981a) was obtained in the absence of mediators and may not have been at redox equilibrium.

As it appears that Q_A^- does not take up a proton before transferring an electron to secondary acceptor Q_B (FOWLER and KOK 1974), the true operating potential of Q_A/Q_A^- is likely to be that prevailing above the pK of Q_A^- (H^+) (−130 mV, KNAFF 1975) where the E_m is no longer pH-dependent. The lack of protonation may also explain why it is so difficult to place a second electron on Q_A.

MEIBURG et al. 1983 have argued, based on electric field-induced charge recombination between P-680$^+$$Q_A^-$, that the potential difference between PhQ_A^- and Ph$^-$$Q_A$ cannot be greater than 350 mV, considerably less than the ~600 mV indicated above. If these authors are right, then the measured E_m of Ph/Ph$^-$ may indeed be too negative (possibly because of coulombic interaction with Q_A^- in the redox titrations, see above).

3 Secondary Reactions, Donor Side

3.1 Reduction Kinetics of P-680$^+$

The kinetics of reduction of P-680$^+$ have been measured using fluorescence detection (SONNEVELD et al. 1979) (P-680$^+$ is a quencher of fluorescence, BUTLER et al. 1973a) as well as optical spectroscopy (VAN BEST and MATHIS 1978, ECKERT et al. 1984). VAN BEST and MATHIS (1978) demonstrated a half-time reduction of 30 ns on the first flash in dark-adapted chloroplasts and algae (i.e., in the state S_1 of the oxygen evolving complex, OEC). In the steady state 50–60% of the P-680$^+$ is reduced with biphasic kinetics of 30 and 400 ns followed by multiphasic kinetics in the microsecond range (ECKERT et al. 1984). The kinetics of reduction of P-680$^+$ depend on the oxidation state (S-states) of the OEC (DELOSME 1972, SONNEVELD et al. 1979, BOUGES-BOCQUET 1980). They are slowest upon excitation of state S_3 and fastest upon excitation of states S_0 and S_1. These different times have been suggested to arise from a pair of secondary donors to P-680, Z_1 and Z_2 operating in parallel in different S-states (JOLIOT and JOLIOT 1972, BOUGES-BOCQUET 1980).

Following treatment by >1 mM NH_2OH, 0.8 M Tris pH 8,0, or low pH, the reduction kinetics of P-680$^+$ are greatly slowed to microsecond times (see BOUGES-BOCQUET 1980). Following Tris treatment of spinach chloroplasts, the kinetics show a weak pH dependence ($t_{1/2}$=4.6 µs at pH 7, 14 µs at pH 5) (CONJEAUD and MATHIS 1980).

3.2 Signal II

EPR signal II (g=2.0044 and ΔH_{pp}=19 G) associated with PS II has been shown to be composed of several kinetic components in O_2-evolving chloroplasts. These are IIu, stable in the dark; IIs (stable for hours) formed in 1–2 s through oxidation by states S_2 and S_3 of the OEC (BABCOCK and SAUER 1973); IIvf appearing in ≤3 µs (instrument limited, BOSKA and SAUER 1984) and decaying with kinetics that are dependent on the S-states (BABCOCK et al. 1976). Following Tris treatment, signal IIvf disappears and is replaced by signal IIf appearing in µs times and decaying in hundreds of millisecond times in the absence of added reductants and at shorter times depending on the concentration and the nature of a wide variety of artificial electron donors (e.g., hydroquinone, NH_2OH, benzidine; BABCOCK and SAUER 1975). Recently, BOSKA et al. (1983) have shown that the rise time of EPR signal IIf matches the decay kinetics of P-680$^+$, detected optically by CONJEAUD and MATHIS (1980). These measurements indicated that the reduction of P-680$^+$ was coupled to the formation of signal IIf (oxidation of Z). DINER and DE VITRY (1984) and DEKKER et al. (1984) have recently reported the UV and visible spectra of $Z^+ - Z$. The former authors also observed in PS II particles that while the half rise times of Z^+ and the decay times of P-680$^+$ were similar and evolved in parallel as a function of pH, there was an apparent lag in the rise of Z^+ at pH 5 not observed in the decay of P-680$^+$. These measurements are suggestive of an intermediate donor between Z and P-680, also supported by ns kinetic measurements of P-680$^+$ reduction (BRETTEL et al. 1984). These authors observed biphasic kinetics ($t_{1/2}$ of 50 and 260 ns, 1:1) upon 3 ns flash excitation in states S_2 and S_3, the 2nd phase >10^3 times shorter than the decay time of Signal IIvf (BABCOCK et al. 1976).

3.3 Identity of Z (also called D_1, Y)

EPR signals IIu, IIs, and IIf all appear to have identical spectra. Signal IIvf is much more difficult to measure owing to its short lifetime. It is most apparent upon excitation of PS II in state S_3 of the OEC (lifetime of IIvf ~1 ms, BABCOCK et al. 1976). The rough spectrum, recorded point by point following flash excitation, closely resembles those obtained for the other signal II, suggesting that they are the same species (BLANKENSHIP et al. 1975) despite the >120 mV lower E_m of signal IIf as compared to IIvf (with state S_1 in OEC, YERKES et al. 1983) and differences in rise kinetics (BOSKA et al. 1983) and microwave power saturation (WARDEN et al. 1976). In support of this conclusion, the optical spectrum of Z^+-Z in the blue (corresponding to signal IIvf, LAVERGNE 1984), produced upon excitation of PS II in state S_1, is identical to that (corresponding to signal IIf) produced upon illumination of PS II particles (DINER and DE VITRY 1984) or Tris-washed chloroplasts (DEKKER et al. 1984).

Based on the kinetic behavior of donor side components at low and at ambient temperature JOLIOT and JOLIOT (1972) and BOUGES-BOCQUET (1980, review) suggested the presence of two parallel donors Z_1 and Z_2 to P-680$^+$, both connected to the OEC, Z_1 would operate for transitions S_0–S_1 and S_1–S_2 and Z_2 for S_2–S_3 and S_3–S_4-S_0. LAVERGNE (1984) has recently observed somewhat different spectra in the blue for Z^+ upon excitation of PS II in states S_1 and S_3 of the OEC. The blue spectra are, however, dominated by a local electrochromic shift of the reaction center pigments and the presence of a positive charge in the OEC for states S_2 and S_3 (itself responsable for an electrochromic shift, DEKKER et al. 1984) could alter the blue spectrum of Z^+-Z compared to that in states S_0 and S_1 where the charge is absent.

A number of authors have proposed, based on extraction and reconstitution studies (KOHL and WOOD 1969) and EPR data (g- value and hyperfine structure) that signal II arises from a plastosemiquinone anion (HALES and DAS GUPTA 1981) or plastochromanoxyl radical (KOHL and WOOD 1969). O'MALLEY et al. (1984) have recently proposed that signal II might arise from a plastosemiquinone cation (QH_2^+) formed by one-electron oxidation of the quinol but without deprotonation. A theoretical analysis and studies using oriented membranes led these authors to propose as well that the plane of the ring was perpendicular to that of the membrane.

The principal arguments in favor of QH_2^+ are its very positive redox potential, a necessary requirement for Z^+ (within 60 mV for signal IIvf of the E_m of P-680$^+$/P-680, (BOUGES-BOCQUET 1980) and $E_{m\,8.5}=760$ mV for IIs and 1.0 V for IIf (BOUSSAC and ETIENNE 1984) and satisfactory simulation of the in vivo EPR spectrum using model compounds (O'MALLEY et al. 1984). The local electrochromic shift observed in the blue for Z^+-Z (see above) would also suggest that Z does not itself lose a proton upon oxidation. A proton is probably released elsewhere (Bohr effect), but the evidence is contradictory. RENGER and VOELKER (1982) and FÖRSTER and JUNGE (1984) have reported proton release ($t_{1/2}=1$ ms and 100 µs, respectively) following oxidation of Z in Tris-washed chloroplasts. YERKES et al. (1983), however, observed that the equilibrium constant for P-680$^+$Z=P-680Z$^+$ is independent of pH from pH 6.5 to 8.5. A weak pH dependence (~0.3 decades/pH unit) for the rate of P-680$^+$ reduction has also been reported (CONJEAUD and MATHIS 1980) but upon correction for differences between surface and bulk pH these rates tend toward 1 decade/pH unit (CONJEAUD personal communication) consistent with proton release.

The chemical identity of signal II is still an open question, as neither the EPR nor the optical spectrum of Z^+-Z is sufficiently diagnostic. The original observation of KOHL and WOOD (1969) of extraction and reconstitution of a deuterated plastoquinone (PQ), narrowing EPR signal II, is the strongest evidence for attributing this signal to plastosemiquinone. This experiment merits repeating. OMATO et al. (1984) have reported 1.85 PQ-9 for 50 Chl (50 Chl/PS II center) in PS II particles, and so attribute one PQ-9 to Q_A and the other to Z. The quantitation of the number of signal II spins per center should be done to justify such a conclusion. Our own extraction and quantitation experiments (CARLES and DINER 1982, DE VITRY unpublished observations) on PS II particles with 1 signal II spin/center (BABCOCK et al. 1983) are more in favor of 1 PQ-9 per center arising from Q_A (Q_B absent). If Z is indeed a plastoquinol, then it is likely to be either tightly bound (difficult to extract) and thus not PQ-9.

3.4 Other Donors

3.4.1 Carotenoid

Under certain conditions (presence of lipophylic anions (VELTHUYS 1981, SCHENCK et al. 1982), low pH, or low temperature (SCHENCK et al. 1982) a carotenoid functions as electron donor to P-680$^+$ but on a parallel pathway to that occupied by Z. The oxidation of the carotenoid is characterized by a bleaching between 450 and 530 nm and an absorbance increase at 990 nm (SCHENCK et al. 1982). The flash yield for formation of the oxidized carotenoid (car$^+$) increases upon lowering the temperature below 200 K, becoming maximal (0.08 car$^+$/center-saturating flash, assuming an extinction coefficient of 1.75×10^5 M^{-1} cm^{-1} at 990 nm) and constant between 100 K and 15 K. The flash yield is stimulated at low and particularly at ambient temperatures (up to 0.8 car$^+$/center-saturating flash in spinach chloroplasts) by the addition of ANT2p or other lipophylic anions, many of which are so-called ADRY reagents, introduced by RENGER (1972) and which accelerate the decay times of states S_2 and S_3 of the OEC (YOCUM, Chap. 9.2, this Vol.). Two explanations have been proposed to explain the stimulation by the lipophylic anions: (1) that they stabilize the formation of a plus charge on the carotenoid in a hydrophobic environment (VELTHUYS 1981), and (2) that these are first oxidized by Z$^+$ and then in turn oxidize the carotenoid (GHANOTAKIS et al. 1982). The kinetics of formation of car$^+$ (SCHENCK et al. 1982) appear to be too rapid, however, compared to what one would predict from the second-order rate constants for Z$^+$ reduction by the ADRY reagents (GHANOTAKIS et al. 1982). Furthermore, stimulation by ADRY reagents of carotenoid oxidation at 77 K (SCHENCK et al. 1982) where ADRY$^+$ is not free to diffuse also argues against the second mechanism in experiments involving a small number of isolated flashes. This mechanism may well predominate, however, during extended illumination, which inactivates the rapid carotenoid photooxidation.

3.4.2 Cytochrome b-559

Another component, closely coupled to P-680, is cytochrome b-559. This species is photooxidized upon continuous illumination at 77 K (KNAFF and ARNON

1969) or at −55 °C in states S_2 and S_3, which cannot be further oxidized at this temperature (VERMEGLIO and MATHIS 1973). Cytochrome b-559 is also oxidized upon illumination at 25 °C in the presence of tetraphenylboron (VELTHUYS 1981), FCCP and TMPD (YERKES and CROFTS 1984), all of which probably oxidize the cytochrome following their oxidation by Z^+ or car^+. VELTHUYS (1981) has also observed an oxidation-induced reduction of oxidized cyt b-559 with reduction kinetics of the oxidized cytochrome close to those observed for reduction of car^+. The two-electron donor responsible for this mechanism was most likely the tetraphenylboron itself. In addition to this apparent kinetic link between the carotenoid and the cytochrome, both show a similar temperature dependence for their photooxidation SCHENCK et al. 1982, BUTLER et al. 1973). Furthermore COX and BENDALL (1974) have shown that low temperature oxidation of cyt b-559 could be restored to hexane-extracted lyophilized chloroplasts by the addition of β-carotene. These experiments, while not constituting a proof, are suggestive of the carotenoid and cyt b-559 associated with the same side pathway. The physiological role of neither species is known, though BUTLER and MATSUDA (1983) have suggested a proton-shuttling mechanism for cyt b-559 linked to the S states, similar to a model proposed by WIKSTRÖM et al. (1981) for cytochrome oxidase. The notion of high ($E_{m7} \simeq 380$ mV) and low ($E_{m7} \simeq 80$ mV) potential forms of this cytochrome may simply reflect a protonated or unprotonated state, respectively. The presence of lipid also appears to stabilize the high potential state (BUTLER and MATSUDA 1983).

4 Alternate Acceptors

4.1 Low Potential Acceptor, Q_L

The titration of the fluorescence yield indicated a low potential component (not associated with C-550) titrating with an $E_{m7.8}$ of ~ −250 mV and termed Q_L by HORTON and CROZE (1979). A high potential component (Q_H, $E_{m7.8}$ = −45 mV) observed in the same titration is certainly Q_A. A component with an E_{m8} = −235 mV was also observed (MALKIN 1978) with millisecond time resolution in a titration of the flash-induced carotenoid band shift (ΔA_{520}), an indicator of the transmembrane electric field. DINER and DELOSME (1983a) showed, however, that there was no −250 mV wave in titrations of the flash-induced ΔA_{520} if detected at 50 μs after the actinic flash. Q_L is thus not involved in a primary transmembrane electron transfer. A slow ΔA_{520} component was, however, detected on the millisecond time scale which titrated at −250 mV (pH 7.5) in agreement with the observation of MALKIN (1978). The slow ΔA_{520} was also shown to be activated by PS I and not by PS II. This slow component was suggested to be associated with the slow electrogenic phase of the cyt b_6/f complex and is consistent with recent findings of GIRVIN and CRAMER (1984) (see JOLIOT and JOLIOT, Chap. 10.3, this Vol.).

Eckert and Renger (1980) and Joliot and Joliot (1981a) have reported the detection of a second primary PS II electron acceptor (X_a and Q_2, respectively). According to Joliot and Joliot (1981c) at least some of these are present in the same centers as Q_A. Q_2 was shown by Joliot and Joliot (1981b) to be a quencher of fluorescence and did not give rise to a C-550 signal. Both groups showed that the acceptor was not a plastoquinone (Joliot and Joliot 1981b) and that it appeared not to participate in a trans-membrane electron transfer as shown for Q_L. In light of the above results it is likely that $X_a \equiv Q_2 \equiv Q_L$.

Horton (1981) has shown that Q_H (Q_A) and Q_L are present either in the same center or belong to different centers in the same pigment bed. In favor of a separate center model are the results of Rutherford and Mathis (1983), who have observed upon illumination at 200 K that the split EPR signal of Ph^- (spin coupled to Q_A^--Fe(II)) titrates as Q_H and Q_L with an E_{m7} of -50 and -275 mV. This observation also implies that both Q_H and Q_L closely interact with pheophytin. It is curious then that only Q_H (Q_A) and not Q_L give rise to a pheophytin band shift (C_{550}) upon reduction (Diner and Delosme 1983a).

Evans and Atkinson (1983) have reported a -430 mV wave in the titration of the light-induced spin polarized P-680 triplet arising from radical pair recombination, in PS II particles at pH 10. This wave was ascribed to the titration of a component U acting as electron acceptor to Ph. This component was, however, not observed by Rutherford et al. (1981a) in a similar titration in different PS II particles. As the E_m of Q_L shows a pH dependence of -60 mV/pH unit (Horton and Croze 1979), this midpoint potential could be consistent with Q_L (assuming a $pK \geq 10$).

4.2 Q_{400} (also called A_H, R and Q_2)

An electron acceptor (Q_{400}, $E_{m7} = 400$ mV) has been described (Ikegami and Katoh 1973) which is rapidly ($t_{1/2} \leq 5$ µs) reduced following illumination (Bowes et al. 1979), in a reaction which precedes the DCMU block. This acceptor was implicated in a high probability double-hit photoreaction for O_2 production using 20 µs actinic flashes (Velthuys and Kok 1978).

While one might speculate that this acceptor could be the iron of the quinone-iron acceptor complex oxidized to Fe^{3+}, Q_{400} has only been observed in the presence of $K_3Fe(CN)_6$. Ghanotakis et al. (1983) have suggested that Q_{400}^+ is $Fe(CN)_6^{-3}$ itself oxidizing Q_A^-. However, the rates of Q_A^- oxidation that they obtain, in steady-state measurements, are much slower than those observed by Bowes et al. (1979) following single-flash excitation. Such a model would also have to explain why the E_m of Q_{400} varies by -60 mV/pH unit and why the addition of DCMU before $Fe(CN)_6^{-3}$ prevents the observation of Q_{400}^+. If ferricyanide is nonetheless responsible, then it would have to occupy a high affinity site which would take up a proton upon reduction and would be blocked by prior addition of DCMU. The ultimate test of the identity of Q_{400}^+ would be, of course, the spectroscopic detection of its oxidation and reduction.

4.3 α- and β-Centers

All of the acceptors we have described, $Q_H(Q_A)$, Q_L and Q_{400} appear to be associated with what have come to be called α-centers. These are present in granal regions of the thylakoids and participate in energy transfer between PS II centers. They also show a PS II action spectrum implicating Chl b. β-centers have been described as being responsible for a slow phase in the fluorescence induction curve in the presence of DCMU. They are present in the stromal regions of the thylakoid membrane, do not show energy transfer between centers and appear to have a far-red spectrum more like that of PS I (THIELEN and VAN GORKOM 1981a). The primary acceptor of β-centers, Q_β, is apparently a quencher of fluorescence and has purportedly an $E_{m8.3}$ of 115 mV (THIELEN and VAN GORKOM 1981a). It has been reported to show an absorbance change characteristic of C-550 and of plastosemiquinone (MELIS and DUYSENS 1979) and not to be involved in a transmembrane electron transfer (THIELEN and VAN GORKOM 1981a). The attribution of both a C-550 signal and an $E_{m8.3}$ of 115 mV is, however, contradicted by titrations of C-550 which do not show a titration wave at 115 mV.

4.4 Still More Heterogeneity

There are also reports of PS II centers which do not show the two-electron gate behavior characteristic of the secondary acceptor, Q_B (non-B centers, LAVERGNE 1982, and β-centers, THEILEN and VAN GORKOM 1981a), and those in which the secondary acceptor stores an electron for ~1 s rather than the usual minutes lifetime observed for Q_B^- (BOUSSAC and ETIENNE 1982). There is at present no unified model which incorporates all the diverse properties described for the primary and secondary acceptors of PS II. It is likely, however, that the lateral heterogeneity in the thylakoid membranes (different electron transport components present in granal and stromal regions) is in part responsable for this diversity (see STAEHELIN, Chap. 1, this Vol. for discussion of structural equivalents of PS II_α and PS II_β centers).

5 Polypeptides

PS II particles devoid of LHC generally contain 40–50 Chl/reaction center (DINER and WOLLMAN 1980, SATOH et al. 1983b) included in two Chl-protein complexes, CP III (51 kDa) and CP IV (48 kDa) in the case of *Chlamydomonas* (DELEPELAIRE and CHUA 1979), CP 47 (47 kDa) and CP 43 (43 kDa) in higher plants (MACHOLD et al. 1979). NAKATANI (1983) has reported some photoreduction of Ph at low potential upon illumination of CP 47, purified on polyacrylamide gels in the presence of lithium dodecyl sulfate. This observation would place the primary charge separation $P\text{-}680\,Ph \rightarrow P\text{-}680^+\,Ph^-$ on this polypeptide alone, as similar experiments on CP 43 were negative. This observation was recently supported by the isolation of a photoactive PS II preparation with CP 47 as the only Chl-protein complex present (YAMAGISHI and KATOH 1984). DE VITRY et al. (1984) have also shown that while photoreduction of Ph in *Chlamydomonas* particles could be observed in the absence of CP IV, photoreduction of Q_A was only observed in the presence of both CP III and CP IV. YAMAGISHI and KATOH (1983) have recently reported the isolation of a dimer of CP 47 from *Synechococcus*. A similar oligomer of CP III has also been observed in *Chlamydomonas* (LDS-PAGE run at 4 °C, DINER and DELEPELAIRE unpublished) however, it is not clear if this association exists in vivo or is formed during isolation.

Cytochrome b-559 is associated with a polypeptide of 10 kDa (WIDGER et al. 1983, probably as a dimer). The localization of Z has not yet been definitively established; however, if stable light-induced pheophytin reduction requires the presence of Z, then the results of NAKATANI (1983) would argue for placing Z as well as Ph on CP 47.

References

Babcock GT, Sauer K (1973) Biochim Biophys Acta 325:483–503

Babcock GT, Sauer K (1975) Biochim Biophys Acta 396:48–62

Babcock GT, Blankenship RE, Sauer K (1976) FEBS Lett 61:286–289

Babcock GT, Ghanotakis DF, Ke B, Diner BA (1983) Biochim Biophys Acta 723:276–286

Best JA van, Mathis P (1978) Biochim Biophys Acta 505:178–188

Blanken HJ den, Hoff AJ, Jongenelis APJM, Diner BA (1983) FEBS Lett 157:21–27

Blankenship RE, Babcock GT, Warden JT, Sauer K (1975) FEBS Lett 51:287–293

Boska M, Sauer K (1984) Biochim Biophys Acta 765:84–87

Boska M, Sauer K, Buttner W, Babcock GT (1983) Biochim Biophys Acta 722:327–330

Bouges-Bocquet B (1980) Biochim Biophys Acta 594:85–103

Boussac A, Etienne AL (1982) Biochim Biophys Acta 682:281–288

Boussac A, Etienne AL (1984) Biochim Biophys Acta 766:576–581

Bowes JM, Crofts AR, Itoh S (1979) Biochim Biophys Acta 547:320–335

Breton J (1982) FEBS Lett 147:16–20

Brettel K, Schlodder E, Witt HT (1984) Biochim Biophys Acta 766:403–415

Butler WL, Matsuda H (1983) In: Inoue Y, Crofts AR, Govindjee, Murata N, Renger G, Satoh K (eds) The oxygen evolving system of photosynthesis. Academic Press, London New York, pp 113–122

Butler WL, Visser JWM, Simons HL (1973) Biochim Biophys Acta 292:140–151

Carles C, Diner BA (1982) Congr Printemps Soc Chim Biol Abstr No 14

Conjeaud H, Mathis P (1980) Biochim Biophys Acta 590:353–359

Cox RP, Bendall DS (1974) Biochim Biophys Acta 347:49–59

Crofts AR, Wraight CA (1983) Biochim Biophys Acta 726:149–183

Davis MS, Forman A, Fajer J (1979) Proc Natl Acad Sci USA 76:4170–4174

Dekker JP, Gorkom HJ van, Brok M, Ouwehand L (1984) Biochim Biophys Acta 764:1301–1309

Delepelaire P, Chua NH (1979) Proc Natl Acad Sci USA 76:111–115

Delosme R (1972) In: Forti G, Avron M, Melandri A (eds) Proc 2nd Int Congr Photosynth. Nijhoff/Junk, The Hague, pp 187–195

Diner BA, Delosme R (1983a) Biochim Biophys Acta 722:443–451

Diner BA, Delosme R (1983b) Biochim Biophys Acta 722:452–459
Diner BA, Vitry C de (1984) In: Sybesma C (ed) Proc 6th Int Congr Photosynth. Nijhoff/Junk, The Hague, pp I.4.407–I.4.411
Diner BA, Wollman FA (1980) Eur J Biochem 110:521–526
Dismukes GC, Frank HA, Friesner R, Sauer K (1984) Biochim Biophys Acta 764:253–271
Döring G, Renger G, Vater J, Witt HT (1969) Z Naturforsch 24b:1139–1143
Duysens LNM (1983) In: Inoue et al. (eds) The oxygen evolving system of photosynthesis. Academic Press, London New York, pp 3–13
Duysens LNM, Sweers HE (1963) In: Miyachi S (ed) Studies on microalgae and photosynthetic bateria. Univ Tokyo Press, Tokyo, pp 353–372
Eckert HJ, Renger G (1980) Photochem Photobiol 31:501–511
Eckert HJ, Renger G, Witt HT (1984) FEBS Lett 167:316–320
Erixon K, Butler WL (1971) Biochim Biophys Acta 234:381–389
Evans MCW, Atkinson YE (1983) In: Inoue Y et al. (eds) The oxygen evolving system of photosynthesis. Academic Press, London New York, pp 55–61
Fajer J, Davis MS, Forman A, Klimov VV, Dolan E, Ke B (1980) J Am Chem Soc 102:7143–7145
Förster V, Junge W (1984) In: Sybesma C (ed) Proc 6th Int Congr Photosynth. Nijhoff/Junk, The Hague, pp II.4.305–II.4.308
Fowler CF, Kok B (1974) Biochim Biophys Acta 357:299–307
Ganago IB, Klimov VV, Ganago AO, Shuvalov VA, Erokhin YE (1982) FEBS Lett 140:127–130
Ghanotakis D, Yerkes CT, Babcock GT (1982) Biochim Biophys Acta 682:21–31
Ghanotakis DF, Babcock GT, Yerkes CT (1983) Arch Biochem Biophys 225:248–255
Girvin ME, Cramer WA (1984) Biochim Biophys Acta 767:29–38
Gorkom HJ van (1974) Biochim Biophys Acta 347:439–442
Gorkom HJ van, Tamminga JJ, Haveman J, Linden IK van der (1974) Biochim Biophys Acta 347:417–438
Gorkom H van, Pulles MPJ, Wessels JSC (1975) Biochim Biophys Acta 408:331–339
Gulotty RJ, Fleming GR, Alberte RS (1982) Biochim Biophys Acta 682:322–331
Haehnel W, Nairn JA, Reisberg P, Sauer K (1982) Biochim Biophys Acta 680:161–173
Hales BJ, Das Gupta A (1981) Biochim Biophys Acta 637:303–311
Horton P (1981) Biochim Biophys Acta 637:152–158
Horton P, Croze E (1979) Biochim Biophys Acta 545:188–201
Ikegami I, Katoh S (1973) Plant Cell Physiol 14:829–836
Joliot P, Joliot A (1972) In: Forti G (ed) Proc 2nd Int Congr Photosynth. Elsevier, Amsterdam New York Oxford, pp 315–322
Joliot P, Joliot A (1981a) In: Akoyunoglou G (ed) Proc 5th Int Congr Photosynth. Balaban, Philadelphia, pp 885–899
Joliot P, Joliot A (1981b) FEBS Lett 134:155–158
Joliot P, Joliot A (1981c) Biochim Biophys Acta 638:132–140
Karukstis KK, Sauer K (1983) Biochim Biophys Acta 725:384–393
Karukstis KK, Sauer K (1984) Biochim Biophys Acta 766:148–155
Kauffmann KJ, Dutton PL, Netzel TL, Leigh JS, Rentzepis PM (1975) Science 188:1301–1304
Klimov VV, Krasnovskii AA (1981) Photosynthetica 15:592–609
Klimov VV, Allakhverdiev SI, Paschenko VZ (1978) Dokl Akad Nauk SSSR 242:1204–1207
Klimov VV, Dolan E, Ke B (1980a) FEBS Lett 112:97–100
Klimov VV, Dolan E, Shaw ER, Ke B (1980b) Proc Natl Acad Sci USA 77:7227–7231
Knaff DB (1975) FEBS Lett 60:331–335
Knaff D, Arnon D (1969) Proc Natl Acad Sci USA 63:956–962
Kohl DH, Wood PM (1969) Plant Physiol 44:1439–1445
Lavergne J (1982) Photobiochem Photobiophys 3:273–285
Lavergne J (1984) FEBS Lett 173:9–14
Machold O, Simpson DJ, Moller BL (1979) Carlsberg Res Comun 44:235–254

Malkin R (1978) FEBS Lett 87:329–333
Mathis P, Setif P (1981) Isr J Chem 21:316–320
Mathis P, Breton J, Vermeglio A, Yates M (1976) FEBS Lett 63:171–173
Meiburg RE, Gorkom HJ van, Dorssen RJ van (1983) Biochim Biophys Acta 724:352–358
Melis A, Duysens LNM (1979) Photochem Photobiol 29:373–382
Nakatani HY (1983) In: Inoue Y et al. (eds) The oxygen evolving system of photosynthesis. Academic Press, London New York, pp 49–54
Nugent JHA, Diner BA, Evans MCW (1981) FEBS Lett 124:241–244
O'Malley PJ, Babcock GT, Prince RC (1984) Biochim Biophys Acta 766:283–288
Omato T, Murata N, Satoh K (1984) Biochim Biophys Acta 765:403–405
Parson WW, Ke B (1982) In: Govindjee (ed) Photosynthesis: Energy conversion by plants and bacteria, vol I. Academic Press, New York London, pp 331–355
Petrouleas V, Diner BA (1982) FEBS Lett 147:111–114
Rademaker H, Hoff AJ, Duysens LNM (1979) Biochim Biophys Acta 546:248–255
Renger G (1972) Biochim Biophys Acta 256:428–439
Renger G, Voelker M (1982) FEBS Lett 149:203–207
Rockley MG, Windsor MW, Cogdell RJ, Parson WW (1975) Proc Natl Acad Sci USA 72:2251–2255
Rutherford AW (1981) Biochem Biophys Res Commun 102:1065–1070
Rutherford AW (1985) Biochim Biophys Acta 807:189–201
Rutherford AW, Mathis P (1983) FEBS Lett 154:328–334
Rutherford AW, Thurnauer MC (1982) Proc Natl Acad Sci USA 79:7283–7287
Rutherford AW, Zimmermann JL (1984) Biochim Biophys Acta 767:168–175
Rutherford AW, Mullet JE, Crofts AR (1981a) FEBS Lett 123:235–237
Rutherford AW, Paterson DR, Mullet JE (1981b) Biochim Biophys Acta 635:205–214
Satoh K, Koike H, Inoue Y (1983a) Photobiochem Photobiophys 6:267–277
Satoh K, Nakatani HY, Steinback KE, Watson J, Arntzen CJ (1983b) Biochim Biophys Acta 724:142–150
Schenck CC, Diner BA, Mathis P, Satoh K (1982) Biochim Biophys Acta 680:216–227
Searle GFW, Treadwell CJ, Barber J, Porter G (1979) Biochim Biophys Acta 545:496–507
Shuvalov VA, Klimov VV (1976) Biochim Biophys Acta 440:587–599
Shuvalov VA, Klimov VV, Dolan E, Parson WW, Ke B (1980) FEBS Lett 118:279–282
Sonneveld A, Rademaker H, Duysens LNM (1979) Biochim Biophys Acta 548:536–551
Stiehl HH, Witt HT (1968) Z Naturforsch 23b:220–224
Thielen APGM, Gorkom HJ van (1981a) FEBS Lett 129:205–209
Thielen APGM, Gorkom HJ van (1981b) Biochim Biophys Acta 637:439–446
Trissl HW, Kunze U, Junge W (1982) Biochim Biophys Acta 682:364–377
Velthuys BR (1981) FEBS Lett 126:272–276
Velthuys B, Kok B (1978) In: Hall DO (ed) Proc 4th Int Congr Photosynth. Biochem Soc, London, pp 397–407
Vermaas WFJ, Govindjee (1981) Photochem Photobiol 34:775–793
Vermeglio A, Mathis P (1973) Biochim Biophys Acta 314:57–65
Vitry C, de, Wollman FA, Delepelaire P (1984) Biochim Biophys Acta 767:415–422
Warden JT, Blankenship RE, Sauer K (1976) Biochim Biophys Acta 423:462–478
Widger WR, Cramer WA, Hermodson M, Gillifor M, Meyer D, Farchaus J, Liedtke B (1983) In: Inoue Y et al. (eds) The oxygen evolving system of photosynthesis. Academic Press, London New York, pp 123–133
Wikström M, Krak K, Saraste M (1981) Annu Rev Biochem 50:623–655
Yamagishi A, Katoh S (1983) In: Inoue et al. (eds) The oxygen evolving system of photosynthesis. Academic Press, London New York, pp 39–48
Yamagishi A, Katoh S (1984) Biochim Biophys Acta 765:118–124
Yerkes CT, Crofts AR (1984) In: Sybesma C (ed) Proc 6th Int Congr Photosynth. Nijhoff/Junk, The Hague, pp I.4.489–I.4.496
Yerkes CT, Babcock GT, Crofts AR (1983) FEBS Lett 158:359–363

9.2 Electron Transfer on the Oxidizing Side of Photosystem II: Components and Mechanism

C.F. YOCUM

1 Enzymatic Oxidation of H_2O by PS II

The mechanism for PS II-catalyzed oxidation of water was originally proposed to consist of four discrete S-states which acted as a charge accumulator (KOK et al. 1970):

$$S_0^0 \rightarrow S_1^{+1} \rightarrow S_2^{+2} \rightarrow S_3^{+3} \rightarrow S_4^{+4} + 2\,H_2O \rightarrow S_0 + 4\,H^+ + O_2$$

Proton release accompanies three S-state transitions (SAPHON and CROFTS 1977), however, and if these protons do not arise from membrane Bohr éffects, then it seems likely that they must be produced as a consequence of the sequential oxidation of water. The S-state model should therefore be viewed as a multi-step enzymatic reaction in which substrate is bound at an early stage in the mechanism and is oxidized sequentially to produce the final product. Known properties of the S-states, enumerated below, are not inconsistent with an enzymatic mechanism:

1. Each S-state center operates as an independent unit; intermediates of H_2O oxidation do not exchange off one enzyme and onto another;
2. S-state decay is a sequential reversal of the forward reaction ($S_3 \rightarrow S_2 \rightarrow S_1$, rather than $S_3 \rightarrow S_1$);
3. The active site of the S-state enzyme resides in a "cleft" which insulates reaction intermediates from bulky or charged oxidants and reductants; substrate analogs (NH_3) or analogs of proposed intermediates (NH_2OH or H_2O H_2O_2) can bind to the active site to interrupt the normal sequence of the reaction.

This review is limited to a brief analysis of the components of the S-state system, and to selected considerations relates to the activities of the oxygen-evolving complex. A broader view of PS II may be obtained by consulting the contributions in this volume of SAUER (Chap. 2), ORT (Chap. 4), DINER (Chap. 9.1), ANDERSSON (Chap. 9.3), ARNTZen (Chap. 9.4).

2 Components of the Oxygen-Evolving Complex

The known components of the oxygen-evolving complex include Mn, Ca, and Cl, along with several components detected by EPR or optical absorption spectroscopy. These species may be grouped in a reaction sequence as follows:

S_n ⟶	$Z/Z^{\cdot +}$ ⟶	P-680/P-680$^+$
Mn, Cl, Ca	ESR Signal II_f	EPR
M^+ or L^+, Y-320	ESR Signal II_{vf}	EPR
EPR multiline signal		
g=4.1 EPR signal		

Of these various components, Mn was the first to be implicated in O_2 evolution activity. Although the ion has for some time been known to be an essential constituent of the oxygen-evolving complex, some controversy has existed as to the size of the functional pool of Mn associated with the oxidizing side of PS II. The actual number required for a functional oxygen-evolving complex appears to be 4 Mn per PS II reaction center (YOCUM et al. 1981). Reports claiming lower stoichiometries (2 Mn, for example) derive from preparations with low activity or can be explained by reaction mechanisms which do not involve H_2O oxidation (VELTHUYS 1983)). It is now generally agreed that the 4-Mn cluster of the oxygen-evolving complex cannot be detected by EPR at room temperature or by NMR measurements of H_2O proton relaxation rates owing perhaps to the presence in S_1 and S_0 of higher oxidation states of Mn and of ligands other than water bound to the metals.

Several procedures which inhibit O_2 evolution such as Tris buffer in the light (CHENIAE and MARTIN 1978) or NH_2OH in the dark (SHARP and YOCUM (1981) convert the tightly bound Mn of PS II into a soluble or loosely bound form where it can be detected by magnetic resonance probes. Quantitation of Mn release after destruction of the oxygen-evolving complex reveals a heterogeneity in the 4-Mn cluster with respect to inhibitor sensitivity (Tris labilizes 2 out of 4 Mn, while NH_2OH labilizes 3 out of 4 atoms). A model accounting for these results requires that all 4 Mn of the PS II reaction center be associated with the oxygen-evolving complex (YOCUM et al. 1981). Another result of studies on amine-Mn interactions is based on the fact that at low concentrations amine-free bases can produce a reversible inhibition of activity (no functional Mn is released). VELTHUYS (1975) has shown that the mechanism of this inhibition involves binding of the amine to the S_2 and S_3 states; the molecular identity of the binding site has been identified as Mn by GHANOTAKIS et al. (1983). The most likely explanation for binding of amines to Mn, first advanced by VELTHUYS, is that these metals in the S_2 and S_3 states exist in oxidation states higher than Mn (II) and as such have associated ligand field stabilization energies which would enhance their amine binding capacity in comparison to the Mn (II) oxidation state.

Although Mn in the oxygen-evolving complex is not detected by EPR at room temperature, DISMUKES and SIDERER (1981) have discovered a complex multiline signal at cryogenic temperature whose properties are consistent with a Mn cluster in which higher oxidation states of the metal are present. This signal is under intensive investigation in several laboratories, and the characteristics of the signal have been reported to include the following:

- The signal amplitude (illumination at room temperature, EPR detection at low temperature) is maximal on the 1st and 5th flashes (DISMUKES and SIDERER 1981, ZIMMERMANN and RUTHERFORD 1984).
- The signal decays with a lifetime similar to that of the S_2 state (BRUDVIG et al. 1983).
- The signal can be generated in the presence of DCMU, where limited turnovers would optimize the concentration of S_2 (ANDREASSON et al. 1983).
- The signal is a broad species, spanning about 1500 G, and contains at least 19 lines.

DISMUKES and SIDERER (1981) have proposed that this low-temperature multiline signal originates from a binuclear Mn complex (Mn (III$<^{O}_{O}>$Mn (IV), although other formulations, for example a Mn (II)–Mn (III) dimer or a four-metal cluster containing both Mn (III) and Mn (IV) can also be used to model the multiline signal. What is of foremost importance is that an intermediate of water oxidation can be trapped and scrutinized by a powerful biophysical probe, and the preliminary results support the existence of higher oxidation states of Mn in the S_2 state. Mention should also be made here of other EPR signals detected in the $g=4$ region at cryogenic temperatures (CASEY and SAUER 1984, ZIMMERMANN and RUTHERFORD 1984) which appear at the present time to be candidates for the precursors to the S_2 and S_3 states, and the $g=2$ multiline signal. These $g=4$ signals may originate from Fe, but further experimentation is necessary before a rigorous assignment of their molecular origin can be made.

Along with Mn, Cl^- is a well-established cofactor of the oxygen-evolving complex, and its effects on activity have been extremely well characterized, as described below:

- Although Cl^- is the preferred anion, Br^- and NO_3 can be substituted; other anions (F^- or ^-OH, for example) are ineffective substitutes (KELLY and IZAWA 1978, CRITCHLEY 1983).
- The Cl^--depleted oxygen-evolving complex is more sensitive to NH_2OH inhibition (KELLY and IZAWA 1978) and to alkaline inactivation (THEG and HOMANN 1982).
- Transitions beyond S_2 are blocked by Cl^- depletion (ITOH et al. 1984).
- Inhibitory amines (NH_3, Tris) compete with Cl^- for a binding site on Mn (SANDUSKY and YOCUM 1984).

The points enumerated above suggest a common site on Mn in the S_2 state where NH_3, Cl^- and ^-OH compete for binding, along with certain other anions (F^-, for example) if they are present. CRITCHLEY et al. (1982) pointed out that the ionic radius of an effective anion (1.81 Å for Cl^- vs. 1.4 Å for F^- or ^-OH) is a critical factor which governs O_2 evolution activity. It has been proposed that Cl^- acts as a bridging ligand between Mn atoms in the oxygen-evolving complex and as such facilitates electron transfer. Although this proposal must be explored further, it explains why displacement of Cl^- from Mn by NH_3 (an ineffective bridging ligand) inhibits activity (SANDUSKY and YOCUM 1984)).

A third inorganic ion which is required for efficient functioning of the oxygen-evolving complex is Ca^{2+}, although much of the literature concerning the site of action of the ion is confused. INOUE and ONO (1983), however, identified binding sites for Ca^{2+} on the oxidizing side of PS II based on an examination of the conditions required for photoactivation of latent oxygen-evolving centers. Other divalent cations cannot substitute for Ca^{2+} and the functional pool is tightly bound to the oxidizing side of PS II, as evidenced by the failure of the ionophore A23187 to inhibit the photoactivated system. The site of Ca^{2+} binding on the oxidizing side of PS II has been identified by GHANOTAKIS et al. (1984) as a complex of the water-soluble polypeptides (33, 24, and 17 kDa) involved in O_2 evolution activity (see ANDERSSON, Chap. 9.3, this Vol.). Although the precise role for Ca^{2+} on the oxidizing side is not yet clear, it seems likely that the tightly bound ion is required for a structural role in maintaining the catalytic activity of the Mn cluster of the oxygen-evolving complex.

In addition to the inorganic ions described above, other species, detected by EPR or optical absorption spectroscopy are essential components of the oxygen-evolving complex. All O_2-evolving membranes examined to date contain a dark-stable free radical (signal II_s) detected by room temperature EPR. The molecular identity of this species and of $Z^{\cdot +}$ as well is thought to be a plastohydroquinone cation radical, $PQH_2{}^+$ (O'MALLEY and BABCOCK 1984, see also DINER, Chap. 9.1, this Vol.). Signal II_s does not possess kinetic properties consistent with a direct role in O_2 evolution activity, but a second pair of EPR-detectable signals (originally called II_{vf} and II_f) are intimately involved in electron transfer from the oxygen-evolving complex to P-680$^+$. Both signals are light-induced and decay in the dark with kinetics which correlate with the presence (II_{vf}, $t_{1/2} \leqq 1$ ms) or absence (II_f, $t_{1/2}$ up to 500 ms) of O_2 evolution activity (BABCOCK and SAUER 1975, WARDEN et al. 1976). Additional experiments by BOSKA et al. (1983) and BOSKA and SAUER (1984) have correlated the rise times of signal II_f and II_{vf} with reported decay times for P-680$^+$, and have independently correlated the P-680$^+$ decay time in Tris-treated samples by EPR. All of the observations cited above are consistent with the assignment of the Signal II_f or II_{vf} species to the function commonly denoted by $Z/Z^{\cdot +}$, i.e. the donor to P-680$^+$, as shown below:

$$\text{ZP-680}^+ \xrightarrow[t_{1/2} \leqq 15\,\mu s]{\text{fast}} Z^{\cdot +}\,\text{P-680} \begin{cases} \xrightarrow{O_2\ \text{evolution}} \text{Sn}Z^{\cdot +} \xrightarrow[(II_{vf})]{t_{1/2} \leqq 1\ \text{ms}} S_{n+1}\,Z \\ \xrightarrow{\text{Tris-washed}} S_{\text{inh.}}\,Z^{\cdot +} \xrightarrow[(II_f)]{t_{1/2} \leqq 500\ \text{ms}} S_{\text{inh.}}\,Z \end{cases}$$

where $S_{\text{inh.}}$ denotes the Tris-inhibited S-state system. Results from measurements of fluorescence decay kinetics also support this formulation (ITOH et al. 1984), and in the interests of nomenclatural clarity, the Signal II terminology should be dropped in favor of $Z/Z^{\cdot +}$ with an appropriate notation to describe decay kinetics.

Room temperature optical absorption spectroscopy has also been utilized to detect signals ascribed to the formation of $Z^{\cdot +}$ and to analyze the properties of signals which are proposed to originate from intermediates in the oxygen-evolving complex. DINER and DE VITRY (1984) and DEKKER et al. (1984) have described the optical characterization of $Z^{\cdot +}$ with somewhat differing results. According to DINER and DE VITRY (1984) the $Z^{\cdot +}$ optical spectrum contains maxima at about 260, 300, 350, and 395 nm. The best fit for this species is proposed to be a substituted naphthohydroquinone cation radical. DEKKER et al. report a spectrum for $Z^{\cdot +}$ with maxima at 260, 275, 300, and 390–450 nm and show that this spectrum is essentially independent of pH. In contrast to DINER and DE VITRY (1984), DEKKER et al. (1984) agree with O'MALLEY and BABCOCK (1984) in assigning a plastohydroquinone cation radical structure to $Z^{\cdot +}$. The present discrepancies in structure assignments may be due to the deconvolutions necessary to separate the $Z^{\cdot +}$ optical absorption signal from other optical changes (due to Q_A^- and perhaps Chl radicals), and to the fact that model quinone cation radicals in acid solution may not generate the same spectra as the $Z^{\cdot +}$ species which resides in a membrane-bound environment. What is important is that there is agreement from optical and EPR measurements that $Z^{\cdot +}$ is a quinone cation radical and does not release or bind protons during its oxidation/reduction cycle.

Optical absorption spectroscopy has also identified transients which are proposed to arise from the oxygen-evolving complex. DEKKER et al. (1984) report the presence of a signal they denote as M^+ which has a broad positive absorption band at 305 nm. This species appears on the 1st flash ($S_1 \rightarrow S_2$) of a 4-flash sequence, remains stable on the 2nd flash ($S_2 \rightarrow S_3$) and decays rapidly on the 3rd flash ($S_3 \rightarrow S_4 \rightarrow S_0$). This signal has also been reported by VELTHUYS (1981) who called it L^+, and who showed that it was abolished by NH_2OH. In later studies, VELTHUYS (1983) showed that in the presence of an alternate donor to $P\text{-}680^+/Z^{\cdot +}$, the 3rd flash decay of L was retarded, which suggests that L (M^+) is a signal from the oxygen-evolving complex, rather than from $Z^{\cdot +}$. DEKKER et al. (1984), whose optical characterization of M^+ in PS II membranes is probably more refined than Velthuys' studies on L^+ with intact thylakoids, ascribe the $M \rightarrow M^+$ transition to the formation of an Mn (III)–Mn (IV) dimer based on work with oxy-liganded Mn (III)–Mn (IV) gluconate complexes whose optical properties approximate those of M^+.

RENGER and WEISS (1983) have reported on an optical signal in trypsinized thylakoid membranes with a positive absorption band at 320 nm and a negative trough at 270 nm which they have termed Y. Like the $M \rightarrow M^+$ transition, Y-320 appears on the 1st flash and then, in contrast to M^+, appears on successive flashes but decays with more rapid kinetics on the 2nd, 3rd, and 4th flashes. The amplitudes of the residual absorption of flash induced Y-320 after 4.5 ms, when compared to the Y-320 amplitude on the 1st flash, produces a maximum on flashes 3 and 7, revealing the period-4 oscillation expected for a component of the oxygen-evolving complex. The spectrum of Y-320 is different from M^+

and it will be of interest to see whether Y-320 is retained in PS II membranes, with the same optical and kinetic properties as have been found in the trypsinized thylakoid system.

3 Factors Influencing the Sequential Oxidation of H_2O

In its unperturbed state, the components of the oxygen-evolving complex function to produce O_2 after the removal of 4 electrons from 2 molecules of water. This sequence is readily disrupted by amine binding or Cl^- depletion, but also by the addition of one or another of a select group of reagents which can reset the redox poise of the complex or affect the lifetimes of higher S-states. As will be seen, these reagents are all membrane-permeant and in certain cases bear a resemblance to either water or to possible intermediates of the H_2O oxidation mechanism. An example of the latter is H_2O_2 which has been shown to produce a 2-flash delay in the appearance of the O_2 "gush" (VELTHUYS and KOK (1978), although this abnormality is followed by the normal 4-flash oscillation in O_2 yield. This observation is attributed to the formation of a new state (S_{-1}) of the oxygen-evolving complex, and is also observed with low concentrations of NH_2OH. Since H_2O_2 is readily degraded by catalase, VELTHUYS and KOK suggested that their results were due to a "catalase-like" activity associated with the oxygen-evolving complex.

RADMER (1983) has summarized the results of extensive studies on the effects of NH_2OH and several of its derivatives on the oxygen-evolving complex. Increasing concentrations of NH_2OH eventually supercede H_2O as the substrate for the oxygen-evolving complex and a flash-induced evolution of N_2 is observed in place of the oscillatory O_2 pattern. At concentrations of NH_2OH where Mn extraction occurs the N_2 yields are the same on each flash. At lower NH_2OH concentrations, a N_2 "gush" is observed on the first flash followed by decreasing N_2 yields on subsequent flashes. At these NH_2OH concentrations ($\leqq \sim 40$ µM), O_2 yields increase as the N_2 yields decrease. A reasonable explanation for these results is that NH_2OH binds to S_1 and reduces it to S_0 with release of 1/2 N_2, after which a second NH_2OH binds to the newly formed S_0 state. Examination of a number of NH_2OH derivatives showed that the shape of these molecules was critical for binding to the oxidation site in PS II; this site was presumed to be the H_2O binding site based on the similarity between $\mathrm{H{-}\overset{\displaystyle H}{\overset{|}{N}}{-}\overset{\displaystyle H}{\overset{/}{O}}}$ and $\mathrm{H{-}\overset{\displaystyle H}{\overset{|}{O}}\,\overset{\displaystyle H}{\overset{|}{O}}{-}H}$. From these considerations the H_2O binding site was estimated to be a "cleft" with two dimensions of 4 Å and 2.5 Å. The third dimension (based on the O–N bond distance, presumably between two metals) was estimated to be 1.47 Å, although it would be unusual to find two metals closer together than about 2 Å.

Finally, a group of lipophilic anions (anilinothiophenes, certain uncouplers like FCCP and CCCP, all termed ADRY reagents by RENGER 1972) exert a complex set of effects on O_2 evolution activity. These reagents can affect the lifetime of the S-states in such a way as to lower the O_2 yield produced by a train of flashes; increased dark intervals between flashes (in the presence of a lipophilic anion) produced decreased O_2 yields when compared to a control sample. These data support a site of action for the "ADRY" compounds in the higher S-states, in which an unspecified mechanism discharges the higher S-state to a lower S-state. Other effects have been associated with the lipophilic anions (loss of Signal II_s and changes in the redox potential of cytochrome b-559), and it is now clear from studies on the accelerated rereduction of $Z^{\cdot +}$ in Tris-washed preparations by several of the lipophilic anions (GHANOTAKIS et al. 1982) that these compounds contain oxidation-reduction activity. To explain the efficient activity (at low concentrations) of these compounds in shortening the lifetimes of higher S-states and reducing $Z^{\cdot +}$, a model has been proposed in which the lipophilic anions are oxidized by the S_2 and S_3 states or by $Z^{\cdot +}$, and rereduced by the rapid oxidation of pigments (Chl, carotenoids), cytochrome b-559 or other, unspecified endogenous reductants. Although $Z^{\cdot +}$ is the target of the lipophilic anions in Tris-washed membranes, the binding site for these reagents in the oxygen-evolving complex has never been identified.

4 Summary and Conclusions

The oxygen-evolving complex of PS II is a unique enzyme containing Mn and Cl^- in an active site which is stabilized by Ca^{2+}. This active site is connected by the $Z/Z^{\cdot +}$ couple to PS II photochemistry and functions so as to catalyze the linear 4-step oxidation of H_2O to molecular O_2. Reversible inhibition of the active site is produced by the binding of amine-free bases, while certain redox-active compounds, like NH_2OH or H_2O_2 can bind to the active site to produce a disruption of the normal reaction sequence. Selected properties of the intermediate S-states of the oxygen-evolving complex are catalogued in Table 1. Inspection of the table shows that among all the S-states, S_2 is unique. In the S_2 state, both EPR and optical absorption spectroscopy detect signals which can be attributed to the presence of higher oxidation states of Mn, and Cl^- binds to this S-state in competition with amines. Lipophilic anions (ADRY reagents) and presumably NH_2OH destabilize the S_2 state. From these observations and the available data on properties of various synthetic complexes of Mn, it seems reasonable to conclude that at the level of S_2, H_2O oxidation reactions have occurred, and that this S-state contains a bound intermediate of the sequential H_2O oxidation reaction; the chemical nature of this intermediate and the Mn oxidation states present in S_2 are sufficient to assure that two subsequent one-electron oxidations will produce a molecule of O_2.

In the absence of more refined evidence for the chemical nature of the S-states, conclusions regarding the mechanism of H_2O oxidation by PS II must

Table 1. Properties of individual S-states

Lifetime	S_o Dark stable	S_1 Dark stable	S_2 $t_{1/2} \leq 90\,s$	S_3 $t_{1/2} \leq 90\,s$
Optical Signal	–	–	M^+	M^+
EPR Multiline Signal	–	–	+	–
Binding of Cl^-	?	?	+	?
Binding of NH_3	?	?	+	+
ADRY Reactive	–	–	+	+

remain speculative. A number of hypothetical models are available, however (see e.g. KUSUNOKI 1983), and a common property of models of the H_2O oxidation reaction is the presence at the proposed active site of two atoms of Mn. Most hypothetical reaction mechanisms consist of sequential 1-electron oxidations of the substrate, with resulting intermediates derived from the observed 1,0,1,2 pattern of proton release. The one-electron oxidation of H_2O deserves careful scrutiny in these models, since it presents a formidable thermodynamic barrier:

$$H_2O \rightarrow H^+ + e^- + {}^{\cdot}OH, \; E_0' = +1.8 \text{ V}$$

This redox potential is approximately 0.6–0.8 V more positive than the potentials that have been estimated for $Z/Z^{\cdot +}$ and P-680/P-680$^+$. Among the elements which must ultimately be explained by a complete model for the operation of the oxygen-evolving complex is the demonstrated requirement for four, rather than two atoms of Mn in the active site and the proton release pattern, which occurs in an acid environment on the interior of the thylakoid membrane. For the latter phenomenon, BUTLER (BUTLER and MATSUDA 1983) has formulated an ingenious scheme which involves cytochrome b-559. In this formulation, the cytochrome would participate in oxidation-reduction reactions in PS II and would also facilitate deprotonation of H_2O during its oxidation. The reduced low-potential form of 559 would be protonated during H_2O oxidation and cycle to the high potential form during oxidation, followed by deprotonation. Rereduction of the high potential form would generate the low potential form for reprotonation. In this formulation, b-559 protonation would occur at the oxidizing side of PS II, but reduction would occur at the reducing side of PS II. This hypothesis merits further exploration since b-559 can serve as both an electron donor to, and acceptor from PS II, and the cytochrome is always found closely associated with the "core-complex" of PS II.

The other point raised above, that of two hypothesized Mn atoms at the active site of the oxygen-evolving complex, as opposed to the four atoms found on the oxidizing side of PS II, requires further comment. Other enzymes (laccase and cytochrome oxidase, for example), which catalyze the four-electron reduction of O_2 to H_2O, are well characterized and contain four metal centers (four Cu or 2 Cu and 2 heme Fe). Extensive investigations into the redox mechanisms of these enzymes (summarized by MALMSTRÖM 1982) indicate that the metal centers may be divided into two clusters of two metals each. One cluster (for example Cu_B and heme a_3 of cytochrome oxidase) bind O_2 and reduce it with electrons supplied by the Cu_A-heme a cluster. It is not unreasonable to assume that a similar arrangement is functional in the oxygen-evolving complex of PS II:

$$(2\,H_2O)(Mn)_2 \xrightarrow{e^-} (Mn)_2 \xrightarrow{e^-} Z/Z^{\cdot +} \qquad \text{(PS II)}$$

$$O_2\text{-(heme } a_3/Cu_B) \xleftarrow{e^-} \text{(heme } a/Cu_A) \xleftarrow{e^-} \text{cyt c} \quad \text{(Cytochrome oxidase)}$$

There are differences between the two enzymes, the most obvious being the metal sites and the fact that the oxidase binds O_2 and releases H_2O, whereas the oxygen-evolving complex binds water and releases O_2. Both enzymes are incapable, in theory, of bypassing the thermodynamic barriers imposed by one-electron oxidation of water to $^{\cdot}OH$ (PS II) or the one-electron reduction of O_2 to $O_2^{\cdot -}$ (oxidase). In the oxidase system, the thermodynamic barrier is overcome by a single, two-electron reduction of the bound substrate, and this point might be a pertinent consideration in the formulation of future models for the mechanism of H_2O oxidation by PS II. Such models will only be precise when protein ligands to Mn have been identified, and biophysical probes have rigorously identified the chemical nature of the S-states. The present rate of progress in unraveling the structure and catalytic properties of the oxygen-evolving complex suggest that a fuller understanding of this important enzyme may not be far away.

References

Andreasson L-E, Hanson O, Vanngard T (1983) Chem Scr 21:71–74

Babcock GT, Sauer K (1975) Biochim Biophys Acta 376:315–328

Boska M, Sauer K (1984) Biochim Biophys Acta 765:84–87

Boska M, Sauer K, Buttner W, Babcock GT (1983) Biochim Biophys Acta 722:327–330

Brudvig GW, Casey JL, Sauer K (1983) Biochim Biophys Acta 723:366–371

Butler WL, Matsuda H (1983) In: Inoue Y, Crofts AR, Govindjee, Murata N, Renger G, Satoh K (eds) The oxygen evolving system of photosynthesis. Academic Press, London New York, pp 113–122

Casey JL, Sauer K (1984) Biochem J 45:217a

Cheniae GM, Martin IF (1978) Biochim Biophys Acta 502:321–344

Critchley C (1983) Biochim Biophys Acta 724:1–5

Critchley C, Baianu IC, Govindjee, Gutowsky HS (1982) Biochim Biophys Acta 682:436–445

Dekker JP, Gorkham JH van, Brok M, Ouwehand L (1984) Biochim Biophys Acta 764:301–309

Diner BA, Vitry C de (1984) In: Sybesma C (ed) Proc 6th Int Congr Photosynth. Nijhoff/Junk, The Hague, pp 407–411

Dismukes GC, Siderer Y (1981) Proc Natl Acad Sci USA 78:274–278

Ghanotakis DF, Yerkes CT, Babcock GT (1982) Biochim Biophys Acta 682:21–31

Ghanotakis DF, O'Malley PJ, Babcock GT, Yocum CF (1983) In: Inoue Y, Crofts AR, Govindjee, Murata N, Renger G, Satoh K (eds) The oxygen evolving system of photosynthesis. Academic Press, London New York, pp 91–101

Ghanotakis DF, Topper JN, Babcock GT, Yocum CT (1984) FEBS Lett 170:169–173
Inoue Y, Ono T-A (1983) Biochim Biophys Acta 723:191–201
Itoh S, Yerkes CT, Koike H, Robinson HH, Crofts AR (1984) Biochim Biophys Acta 766:612–622
Kelly P, Izawa S (1978) Biochim Biophys Acta 502:192–210
Kok B, Forbush B, McGloin M (1970) Photochem Photobiol 11:457–475
Kusunoki M (1983) In: Inoue Y, Crofts AR, Govindjee, Murata N, Renger G, Satoh K (eds) The oxygen evolving system of photosynthesis. Academic Press, London New York, pp 165–172
Malmström BG (1982) Annu Rev Biochem 51:21–59
O'Malley PJ, Babcock GT (1984) Biochim Biophys Acta 765:370–379
Radmer R (1983) In: Inoue Y, Crofts AR, Govindjee, Murata N, Renger G, Satoh K (eds) The oxygen evolving system of photosynthesis. Academic Press, London New York, pp 135–144
Renger G (1972) Biochim Biophys Acta 256:428–439
Renger G, Weiss W (1983) Biochim Biophys Acta 722:1–11
Sandusky PO, Yocum CF (1984) Biochim Biophys Acta 766:603–611
Saphon S, Crofts AR (1977) Z Naturforsch 32C:317–326
Sharp RR, Yocum CF (1981) Biochim Biophys Acta 635:90–104
Theg S, Homann P (1982) Biochim Biophys Acta 679:221–234
Velthuys BR (1975) Biochim Biophys Acta 396:392–401
Velthuys BR (1981) In: Akoyunoglou G (ed) Photosynthesis II. Electron transport and photophosphorylation. Balaban, Philadelphia, pp 75–85
Velthuys B (1983) In: Inoue Y, Crofts AR, Govindjee, Murata N, Renger G, Satoh K (eds) The oxygen evolving system of photosynthesis. Academic Press, London New York, pp 83–90
Velthuys BR, Kok B (1978) Biochim Biophys Acta 502:211–221
Warden JT, Blankenship RE, Sauer K (1976) Biochim Biophys Acta 423:462–478
Yocum CF, Yerkes CT, Blankenship RE, Sharp RR, Babcock GT (1981) Proc Natl Acad Sci USA 78:7507–7511
Zimmermann JL, Rutherford AW (1984) Biochim Biophys Acta 767:160–167

9.3 Proteins Participating in Photosynthetic Water Oxidation

B. ANDERSSON

1 Introduction

For many years photosystem II electron transport was described mainly in biophysical terms due to a limited knowledge of biochemical components. However, in recent years rapid progress has been made concerning the biochemistry of photosystem II. The focus of this chapter is on the identification of proteins associated with the water oxidation side of photosystem II. At present, approximately ten polypeptides are believed to play some role in photosynthetic water oxidation. Some of these polypeptides are electrostatically bound to the thylakoid membrane (extrinsic proteins), while others are anchored into the lipid bilayer by hydrophobic interactions (intrinsic proteins).

2 Extrinsic Proteins of Water Oxidation

2.1 The 33, 23, and 16 kDa Proteins. Identification and Connection with Water Oxidation

The original observations which led to the identification of the 33, 23, and 16 kDa proteins and suggested their association with oxygen evolution were reported around 1980. Two major experimental approaches led to the identification of these three proteins, manipulations of oxygen-evolving photosystem II preparations and mutational studies. For the former approach, the availibility of inside-out vesicles and so-called photosystem II particles was crucial. (It should be noted, however, that in this context the term photosystem II "particle" is misleading and should therefore be dropped, because all such preparations derived from higher plants consist of highly enriched, stacked grana membrane fragments (DUNAHAY et al. 1984). The most appropriate term would seem to be "photosystem II membrane preparations.") Both preparations have properties which make them highly suitable for compositional studies on the water oxidation machinery. These include exposure of the inner thylakoid surface, high rates of oxygen-evolving activity, and a relatively simple polypeptide composition due to a pronounced photosystem II enrichment.

A pioneering experiment was carried out by ÅKERLUND (1981), who subjected thylakoid vesicles of opposite sidedness to high concentrations of NaCl.

When subsequently assayed for oxygen evolution in a low salt medium, the right-side-out vesicles had retained their oxygen-evolving activity while the inside-out vesicles were inactivated. ÅKERLUND explained this differential inhibition by a salt-dependent, reversible release of extrinsically bound proteins from the inner thylakoid surface. In the case of inside-out vesicles, such proteins would be lost into the supernatant, while they would be trapped in the lumen of right-side-out vesicles, ready to rebind during low salt conditions. Indeed this model was shown to be valid, since readdition of the desalted supernatant to the washed inside-out vesicles under low salt conditions resulted in a partial restoration of the lost oxygen-evolving activity. This supernatant contained polypeptides of 23 and 16 kDa. In an earlier study, KUWABARA and MURATA had isolated a 33 kDa protein which was also suggested to function in photosystem II (MURATA et al. 1983). This conclusion received further support by the observation that the 33 kDa protein was released, together with the 23 and 16 kDa proteins, from inside-out vesicles and photosystem II membrane preparations by alkaline Tris (ÅKERLUND 1983, YAMAMOTO and NISHIMURA 1983).

Following the early approach of ÅKERLUND (1981) several procedures for releasing proteins from everted thylakoid membranes have now been developed (Table 1). All of these treatments lead to a complete or severe loss of oxygen evolving activity but the release pattern for the 33, 23, and 16 kDa proteins, and manganese varies. Washings based on high concentrations of NaCl release both the 23 and 16 kDa proteins but leave the 33 kDa protein and manganese in the membrane. With 2.5 M urea, the 33 kDa protein is preferentially released together with approximately half of the manganese. Alkaline Tris releases all three proteins together with most of the manganese. Recently, washing with high concentrations of $CaCl_2$ was introduced as a way to release the three proteins without affecting the manganese. As will be discussed in Section 2.4, alkaline Tris washings also release a 10 kDa polypeptide from everted thylakoid preparations.

For obvious reasons, correlations between inhibition and protein release can only give indirect evidence regarding the requirement for a certain protein in oxygen evolution. In contrast, protein-dependent restoration of a lost activity gives much more direct evidence for its involvement in that activity. Such reconstitution has so far been most convincingly demonstrated for the 23 kDa protein. Following its removal by salt washing, readdition of this protein under low salt conditions to depleted inside-out vesicles or photosystem II membranes restored most of the lost activity (ÅKERLUND 1983, LINDBERG-MØLLER and HØJ 1983, MURATA et al. 1983). In most studies, reconstitution has not been possible with the 16 kDa protein. The exception are cholate/NaCl treated photosystem II membranes where the 16 and 23 kDa proteins stimulated oxygen evolution in a cooperative manner (FUKUTAKA et al. 1983). For quite some time no stimulation of oxygen evolution could be obtained with the 33 kDa protein, probably because its release was accompanied by loss of manganese (Table 1). This problem has been overcome by the introduction of $CaCl_2$ and NaCl/urea washings. Readdition of the 33 kDa protein to $CaCl_2$-washed photosystem II membranes devoid of all three proteins, resulted in some 25% reconstitution of oxygen evolution (ONO and INOUE 1984). A small restoration was also found with

Table 1. Inhibition and reconstitution of water oxidation in everted thylakoid preparations in relation to the presence of the 33, 23, and 16 kDa proteins

Treatment of everted thylakoids	Oxygen evolution (%)	Protein content (%)			Mn (%)	Reconstitution		Reference
		33	23	16		%	with	
1 M Tris-HCl pH 9.3	0	0	0	0	15	0	–	Murata et al. (1983)
0.25 M NaCl pH 7.4	25	100	30	15	100	60	23 kDa	Andersson et al. (1984b)
1 M NaCl pH 6.5	40	100	0	0	100	100	23 kDa	Murata et al. (1983)
1 M NaCl/50 mM cholate	0	50	0	0	–	40	23+16 kDa	Fukutaka et al. (1983)
2.5 M urea pH 6.5	0	0	60	70	40	–	–	Murata et al. (1983)
1 M NaCl/2.5 M urea	0	0	0	0	50	5	33 kDa	Miyao and Murata (1983)
1 M CaCl pH 6.5	2	0	0	0	90	25	33 kDa	Ono and Inoue (1984)

The hundred percentage values correspond to untreated inside-out vesicles or photosystem II particles. The degree of reconstitution was related to the activity in completely untreated particles.

the 33 kDa protein after NaCl/urea treatment of photosystem II membranes (MIYAO and MURATA 1983).

The other main approach for identifying proteins associated with water oxidation involves characterization of the polypeptide composition of mutants with impaired oxygen evolution and decreased levels of manganese. In an early study, BENNOUN et al. (1981) reported on a *Chlamydomonas* mutant with impaired oxygen evolution which lacked polypeptides at approximately 21, 18, and 6 kDa. The extrinsic nature of these polypeptides was demonstrated by their release from photosystem II membranes by guanidine extraction. These 21 and 18 kDa polypeptides are likely to correspond to the 23 and 16 kDa proteins released from everted spinach thylakoids. More recently, CAMMARATA et al. (1984) found that the inhibited oxygen-evolving activity and low amount of manganese in two *Chlamydomonas* mutants correlated with the absence of a 23 kDa polypeptide. Mutational support for the importance of the 16 kDa protein in water oxidation has come from an *Oenothera* photosystem II mutant which specifically lacks this protein (WESTHOFF et al. 1984). A *Scenedesmus* mutant with impaired oxygen evolution and heavily reduced manganese content lacked a 34 kDa protein (BISHOP 1984). However, this protein appears to be distinct from the 33 kDa protein of MURATA (see Sect. 3).

2.2 Properties, Biogenesis and Organization in the Thylakoids

Each of the 33, 23, and 16 kDa proteins can now be isolated to purity from crude protein extracts or supernatants after appropriate washings of photosystem II membrane preparations (FUKUTAKA et al. 1983, LINDBERG-MØLLER and HØJ 1983, MURATA et al. 1983, YAMAMOTO and NISHIMURA 1983, JANSSON 1984). The separation usually depends on ion-exchange chromatography, but isoelectric focusing or chromatofocusing has also been applied. Gel filtration suggests that each of the three proteins exists in monomeric form after isolation. They all contain a large proportion of charged and hydrophilic amino acids, which is consistent with their high water solubility. Notably, the 33 kDa protein is rich in lysine, which makes it distinct from the herbicide-binding 32 kDa protein which lacks this amino acid. The isoelectric point of the 33, 23, and 16 kDa proteins is 5.1–5.2, 6.4–7.3 and 8.5–9.5 respectively. When isolated as described above, none of the proteins contains any metals and shows no absorption in the visible region of the spectrum indicative of prosthetic groups. However, results from ABRAMOWICZ and DISMUKES (1984) indicate that the 33 kDa protein contains two manganese when isolated under oxidizing conditions. The 33, 23, and 16 kDa proteins are present in equimolar amounts in the thylakoids. For each of the three proteins there is approximately 1 polypeptide per 200 chlorophylls (MURATA et al. 1983, ANDERSSON et al. 1984b). This value gives one or two copies of each protein for every P-680, depending on what reaction center content is assumed. As revealed by thylakoid fractionation experiments the 33, 23, and 16 kDa proteins are preferentially located in the appressed grana thylakoid region together with other photosystem II components (ANDERSSON et al. 1984b, DUNAHAY et al. 1984). The three proteins are all located on the

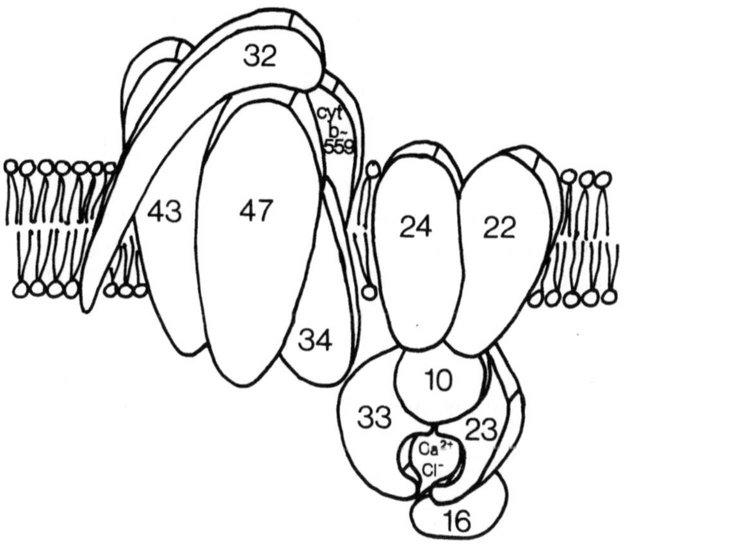

Fig. 1. Hypothetical arrangement of photosystem II proteins, which appear to be associated with the water oxidation complex

inner thylakoid surface (ÅKERLUND 1983, YAMAMOTO and NISHIMURA 1983, ANDERSSON et al. 1984b), since (1) the 33 and 16 kDa proteins were completely digested by trypsin in everted but not in normal thylakoids, (2) antibodies against each of the proteins agglutinated only inside-out vesicles, and (3) the proteins were released by salt or Tris washings from inside-out but not from right-side-out vesicles. The three proteins all appear extrinsic since they can be detached from the inner thylakoid surface by manipulations of the pH or ionic strength of the medium without any need for detergents. The release patterns (Table 1) suggest that the binding of the 23 and 16 kDa proteins involves only electrostatic forces, while the binding of the 33 kDa protein also includes hydrogen bonds and hydrophobic interactions. Sequential rebinding studies on washed inside-out vesicles (ANDERSSON et al. 1984b) and photosystem II membranes (MURATA et al. 1983) demonstrated that the 16 kDa protein is bound to the 23 kDa protein, which in turn has a binding site on the 33 kDa protein (Fig. 1). The 33 and 23 kDa proteins both have binding sites on the actual membrane. Nearest-neighbor analyses have localized these binding sites to three polypeptides of 24, 22, and 10 kDa (LJUNGBERG et al. 1984b).

Using in vitro translation and immunoprecipitation, it has been shown that the 33, 23, and 16 kDa proteins are all encoded by nuclear DNA (WESTHOFF et al. 1984). They are synthesized in the cytoplasm as precursors of 39, 33, and 26 kDa respectively and post-translationally imported through the three chloroplast membranes while being processed to their mature forms. The 33 and 23 kDa proteins but not the 16 kDa protein are present in etioplasts of barley; the latter is found only in trace amounts (RYRIE et al. 1984). Thus the synthesis of the three proteins seems to be independently regulated. Interestingly, the 16 kDa protein is formed concomitantly with the onset of oxygen evolution during greening.

2.3 Specific Functions in Water Oxidation

The 33 kDa protein has mostly been linked to manganese. Release of this protein from everted thylakoids is usually accompanied by loss of manganese, which is not the case if only the 23 and 16 kDa proteins are released (Table 1). However, a strict quantitative correlation between the removal of 33 kDa protein and of manganese was only recently obtained when MIYAO and MURATA (1984b) analyzed urea-treated photosystem II particles pre-washed with NaCl. They suggested that two out of four manganese in photosystem II are associated with the 33 kDa protein. Support for this conclusion has come from ABRAMOWICZ and DISMUKES (1984), who, by ensuring oxidizing conditions during the protein extraction, isolated a 33–34 kDa protein carrying two manganese while otherwise resembling the protein of MURATA. Whether this result means that the extrinsic 33 kDa protein is the mangano-protein that carries out the redox reactions of the S-clock in oxygen evolution remains to be elucidated. In fact, other recent results suggest that the role for the 33 kDa protein is structural rather than catalytic. Using $CaCl_2$ treatment of photosystem II membranes ONO and INOUE (1984) managed to release all of the 33 kDa protein while leaving essentially all manganese on the membrane. Moreover, in such particles some oxygen evolution could be restored simply by adding $CaCl_2$ (ONO and INOUE 1984), suggesting that oxygen evolution can proceed in the absence of the 33 kDa protein. It was concluded that the 33 kDa protein is necessary to maintain an optimal environment around the manganese site. A similar conclusion was reached by MIYAO and MURATA (1984a).

A specific role for the 23 kDa protein was inferred from experiments on mangrove thylakoids (ANDERSSON et al. 1984a). In mangrove thylakoids that had lost their 23 and 16 kDa proteins, the rate of oxygen evolution was drastically reduced at low chloride concentrations. The activity of the thylakoids lacking the two proteins could be restored by high concentrations of chloride. Readdition or pure spinach 23 kDa protein to inside-out mangrove vesicles, depleted of the 23 and 16 kDa proteins, lowered their chloride requirement for optimal oxygen evolution 50–100-fold to give maximal rates at only 3 mM NaCl. Based on these results it was concluded that one role of the 23 kDa protein is to increase the chloride affinity for the water oxidation site, which by itself appears to have a low affinity for chloride. It should be stressed that this function of the 23 kDa protein should not be restricted to the salt-tolerant mangroves, since high chloride concentrations were also able to replace the 23 kDa protein in spinach inside-out vesicles (ANDERSSON et al. 1984a) and photosystem II membrane preparations (NAKATANI 1984). More recently AKABORI et al. (1984) has shown that the 16 kDa protein is also involved in the binding of chloride. In reconstitution experiments involving spinach photosystem II membranes washed in cholate/NaCl, the 23 kDa protein alone could only maintain high rates of oxygen evolution at concentrations above 3 mM chloride. For optimal rates at submillimolar concentrations of chloride the 16 kDa protein had to be present in addition to the 23 kDa protein.

Another crucial finding was that the 23 and 16 kDa proteins can be replaced by high, nonphysiological concentrations of calcium (GHANOTHAKIS et al. 1984,

MIYAO and MURATA 1984a). In a very elegant study, GHANOTHAKIS et al. (1984) showed that the 23 and/or 16 kDa proteins create a high-affinity binding for calcium at the water oxidation site. Addition of calcium was essential for reconstitution with these proteins if the photosystem II particles and the protein extract had been dialyzed against EGTA. At present it is not clear whether the calcium effect is attributed to the 23 or the 16 kDa protein. A more detailed discussion of the salt requirements of the water-splitting apparatus can be found in this Volume in Chap. 9.2 of YOCUM.

In conclusion, most of the results considered above suggest that the 33, 23, and 16 kDa proteins are essential regulatory and or structural subunits of the water oxidation machinery of chloroplast membranes.

2.4 A 10 kDa Polypeptide

In addition to the 33, 23, and 16 kDa protein, there is also a 10 kDa polypeptide released from everted thylakoid preparations by alkaline Tris (LJUNGBERG et al. 1984a). This polypeptide, which is not cytochrome b-559, can be resolved only on SDS-PAGE systems with a high resolution in the low molecular weight region, and therefore escaped detection in the earlier washing studies (Table 1). The 10 kDa protein is released from inside-out vesicles but not from thylakoids of normal orientation. Moreover it is highly enriched in the grana partition region as is the case with photosystem II components. It is not released by high concentrations of $CaCl_2$ or NaCl. Immunoprecipitation nearest-neighbor analyses suggests the 10 kDa protein to be in close proximity to both the 23 and 33 kDa proteins (LJUNGBERG et al. 1984b). Thus, clearly a fourth extrinsic protein located at the inner thylakoid surface and with a possible role in oxygen evolution must be considered.

3 Intrinsic Proteins of Water Oxidation

Indirect evidence suggests that in addition to extrinsic proteins there are also intrinsic proteins associated with the water oxidation process. These include a 34 kDa polypeptide, the 47 kDa reaction center polypeptide of photosystem II and cytochrome b-559. In addition there are two intrinsic polypeptides (22 and 24 kDa) which are structurally associated with the 33 and 23 kDa proteins.

By studying chloroplast mutants BISHOP (1984) has been able to correlate the change in electrophoretic mobility of a 34 kDa polypeptide in *Scenedesmus* with loss of oxygen evolution, decrease in manganese, and loss of the high potential form of cytochrome b-559. Revertants of this mutant showed control rates of activity and normal values for manganese and high potential cytochrome b-559 H.P. Decrease in oxygen evolution and manganese have also been associated with a 34 kDa polypeptide in a maize mutant (BRICKER et al. 1983). This polypeptide has generally been considered identical to the 33 kDa protein of MURATA, but more recent considerations argue against this. The 34 kDa polypeptide seems to be of plastid origin (BISHOP 1984) and has been shown to be intrinsic (BRICKER et al. 1983) in contrast to the 33 kDa protein of MURATA, which is of nuclear origin and extrinsic. The "mutant 34 kDa polypeptide" probably correlates with the newly identified 34 kDa subunit of the photosystem II core complex, which has been postulated to play a role on the donor side of photosystem II (SATOH et al. 1983). Such a correlation has recently been supported by polypeptide analyzis of photosystem II core complexes isolated from wild-type and mutant chloroplast of *Scenedesmus* (METZ and SEIBERT 1984).

Based on the amino acid sequence of the photosystem II reaction center polypeptide (47 kDa), it has been predicted that this polypetide has a large hydrophilic loop exposed to the thylakoid lumen (MORRIS and HERRMANN 1984). It is tempting to speculate that this part of the polypeptide could be directly involved in water oxidation. Such an arrangement would provide the very close association between the primary reactions of photosystem II and water oxidation suggested from kinetic data. ONO and INOUE (1984) suggest that manganese is bound to the 47 kDa protein and that this binding is stabilized by the extrinsic 33 kDa protein.

Cytochrome b-559 has generally been assumed to play some role on the donor side of photosystem II (BUTLER and MATSUDA 1983). It is well known that treatments which lower the potential of cytochrome b-559 often lead to deactivation of oxygen evolution conversely, restoration of the high potential form coincide with increased rates of oxygen evolution. It has been suggested that the cytochrome could participate in a process that involves changes both in redox properties and in midpoint potential. During such a reaction, cytochrome b-559 would reduce a component on the acceptor side of photosystem II and serve as a proton translocator on the donor side (BUTLER and MATSUDA 1983). A *Scenedesmus* mutant with no spectral indication of cytochrome b-559 and lacking only a 10 kDa polypeptide showed loss of both oxygen evolution and manganese (BISHOP 1984 and personal communication). This observation provides support for some involvement of cytochrome b-559 in the process of water oxidation as well as in manganese coordination.

Immunoprecipitation studies of partly solubilized photosystem II particles have revealed that two intrinsic polypeptides of 24 and 22 kDa are nearest neighbors of the extrinsic 33 and 23 kDa proteins (LJUNGBERG et al. 1984b). The results suggest that these intrinsic polypeptides provide the binding sites at the inner thylakoid surface for the 33 and 23 kDa proteins. Whether this structural role for the 24 and 22 kDa proteins is complemented by some functional role remains to be seen.

4 Hypothetical Architecture of the Water Oxidation Complex

Based partly upon data presented in the previous sections, a possible arrangement of the intrinsic and extrinsic proteins associated with water oxidation is illustrated in Fig. 1. This multicomponent complex should contain catalytic, structural, and regulating subunits which all are essential for optimal in vivo water oxidation, although oxygen evolution in vitro may not require all of these subunits. Maybe some subunits are not essential for all oxygen-evolving organisms; i.e., the 23 and 16 kDa proteins appear to be missing in cyanobacteria (STEWART et al. 1985). It is possible that the photosystem II core complex contains the catalytic subunit of oxygen evolution, while the extrinsic 33, 23, 16, and possibly the 10 kDa protein may be part of an assessory or regulating subcomplex. The location of manganese is not shown in the diagram since, at present, there are too many contradictory reports. The 47, 34, 33, and 23 kDa polypeptides, and cytochrome b-559 have all been suggested to be directly or indirectly associated with the manganese. Perhaps there is no discrete manganoprotein; instead manganese may be associated with several polypeptides which all are required for its proper orientation at the catalytic site of water oxidation.

Future research should lead to a significant refinement of this model, particularly, with respect to the assignment of functional roles to the various subunits. Additional polypeptides may be discovered and together with other findings will hopefully create a more profound understanding of the calcium-chloride-manganese-polypeptide interactions that are of central importance to the water-splitting apparatus of oxygen-evolving photosynthetic organisms.

References

Abramowicz DA, Dismukes GC (1984) Biochim Biophys Acta 765:309–318

Akabori K, Imaoka A, Toyshima Y (1984) FEBS Lett 173:36–40

Åkerlund H-E (1981) In: Akoyunoglou G (ed) Photosynthesis, vol II. Balaban, Philadelphia, pp 465–472

Åkerlund H-E (1983) In: Inoue Y, Croft AR, Govindjee, Murata N, Renger G, Satoh K (eds) The oxygen evolving system of photosynthesis. Academic Press, London New York, pp 201–208

Andersson B, Critchley C, Ryrie IJ, Jansson C, Larsson C, Anderson JM (1984a) FEBS Lett 168:113–117

Andersson B, Larsson C, Jansson C, Ljungberg U, Åkerlund H-E (1984b) Biochim Biophys Acta 766:21–28

Bennoun P, Diner BA, Wollman F-A, Schmidt G, Chua NH (1981) In: Akoyunoglou G (ed) Photosynthesis, vol III. Balaban, Philadelphia, pp 839–849

Bishop NI (1984) In: Sybesma C (ed) Advances in photosynthesis research, vol I. Nijhof/Junk, The Hague, pp 321–328

Bricker TM, Metz JG, Miles D, Sherman LA (1983) Biochim Biophys Acta 724:447–455

Butler WL, Matsuda H (1983) In: Inoue Y, Crofts AR, Govindjee, Murata N, Renger G, Satoh K (eds) The oxygen evolving system of photosynthesis. Academic Press, London New York, pp 113–122

Cammarata K, Tamura N, Sayre R, Cheniae G (1984) In: Sybesma C (ed) Advances in photosynthesis research, vol I. Nijhof/Junk, The Hague, pp 311–320

Dunahay TG, Staehelin LA, Seibert M, Ogilvie PD, Berg SP (1984) Biochim Biophys Acta 764:179–193

Fukutaka E, Imaoka A, Kozo A, Toyoshima Y (1983) FEBS Lett 158:217–221

Ghanotakis DF, Topper JN, Babcock GT, Yocum CF (1984) FEBS Lett 170:169–173

Jansson C (1984) In: Sybesma C (ed) Advances in photosynthesis research, vol I. Nijhof/Junk, The Hague, pp 375–378
Lindberg-Møller B, Høj PB (1983) Carlsberg Res Commun 48:161–185
Ljungberg U, Åkerlund H-E, Andersson B (1984a) FEBS Lett 175:255–258
Ljungberg U, Åkerlund H-E, Larsson C, Andersson B (1984b) Biochim Biophys Acta 767:145–152
Metz JG, Seibert M (1984) Plant Physiol 76:829–832
Miyao M, Murata N (1983) FEBS Lett 164:375–378
Miyao M, Murata N (1984a) FEBS Lett 170:350–354
Miyao M, Murata N (1984b) Biochim Biophys Acta 765:253–257
Morris J, Herrmann RG (1984) Nucleic Acids Res 12:2837–2850
Murata N, Miyao M, Kuwabara T (1983) In: Inoue Y, Crofts AR, Govindjee, Murata N, Renger G, Satoh K (eds) The oxygen evolving system of photosynthesis. Academic Press, London New York, pp 213–222
Nakatani HY (1984) Biochem Biophys Res Commun 120:299–304
Ono T-A, Inoue Y (1984) FEBS Lett 168:281–286
Ryrie IJ, Young S, Andersson B (1984) FEBS Lett 177:269–273
Satoh K, Nakatani HY, Steinback KE, Watson J, Arntzen CJ (1983) Biochim Biophys Acta 724:142–150
Stewart AC, Ljungberg U, Åkerlund H-E, Andersson B (1985) Biochim Biophys Acta 808:353–362
Westhoff P, Jansson C, Klein-Hitpass L, Berzborn R, Larsson C, Bartlett S (1985) Plant Mol Biol 4:137–143
Yamamoto Y, Nishimura M (1983) In: Inoue Y, Crofts AR, Govindjee, Murata N, Renger G, Satoh K (eds) The oxygen evolving system of photosynthesis. Academic Press, London New York, pp 229–238

9.4 Photosystem II Reaction Center: Polypeptide Subunits and Functional Cofactors

C.J. ARNTZEN and H.B. PAKRASI

1 Introduction

The photosystem II (PS II) of oxygenic photosynthetic organisms is a lipoprotein complex that is located in the thylakoid membranes and is composed of several integral and peripheral membrane proteins. Associated with these proteins are several essential cofactors such as chlorophyll, pheophytin, quinones and nonheme iron molecules. The whole complex mediates electron transfer from water to the plastoquinone (PQ) pool of the noncyclic electron transport chain.

As already pointed out in this Volume by ORT (Chap. 4) and by DUTTON (Chap. 5), the functional properties of the reaction center (RC) of purple photosynthetic bacteria and PS II are very similar (see also Chap. 9.1; NORRIS and VAN BRATEL, Chap. 8.3; MICHEL and DEISENHOFER, Chap. 8.4; CROFTS and WRAIGHT 1983). Both utilize a special RC (Bacterio) chlorophyll (P-860 or P-680 for bacterial or PS II RC's, respectively), which donates an electron to a pheophytin (Ph) molecule. The subsequent electron transfers are to two bound quinones acting in series, designated Q_A and Q_B. The secondary acceptor Q_B is exchangeable with the bulk PQ pool. In PS II, the oxidized P-680$^+$ molecule receives electrons from the immediate donor species, Z, proposed to be a quinone molecule also (O'MALLEY, see also ORT, Chap. 4, this Vol., for an overview).

The chapters of ANDERSSON (9.3) and YOCUM (9.2) in this Volume discuss the components and mechanisms on the oxidizing side of PS II (from H_2O to the reaction center). The focus of this chapter is on the polypeptides and cofactors associated with the "core complex" of PS II. Operationally, this complex will be defined as the structural unit which mediates transfer of electrons from an exogenous donor (for the reaction center chlorophyll) to the PQ pool.

2 Polypeptides of the PS II Core Complex

Over the last several years, major progress has been made in obtaining detergent-derived submembrane preparations retaining photochemical activity for PS II and containing a small number of polypeptides (SATOH and BUTLER 1978, DINER and WOLLMAN 1980, SATOH 1983, SATOH et al. 1983, BRICKER et al. 1985, SPRAGUE et al. 1985). These core preparations contain as few as five polypep-

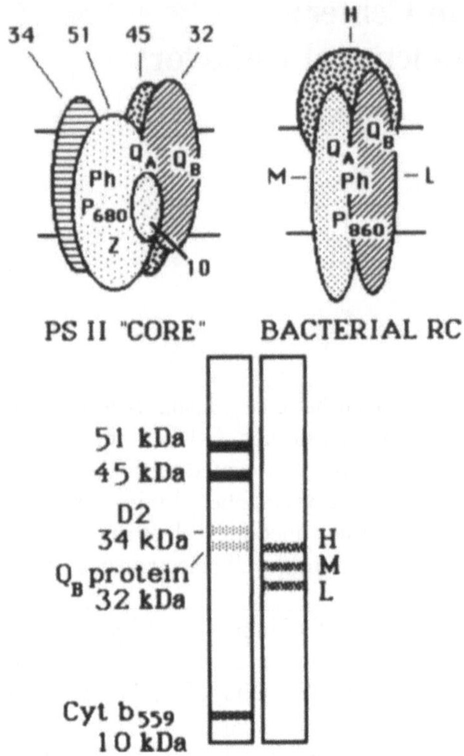

Fig. 1. Stylized drawings depicting the structural organization of the photosystem II core complex of oxygenic photosynthetic organisms and the reaction center (*RC*) of purple bacteria. When subjected to SDS-PAGE (shown diagrammatically in the *lower portion* of the figure), these complexes resolve into five and three polypeptide bands, respectively. The central question being addressed in this review is: Which PS II polypeptides determine the functional properties of the bound cofactors involved in the light-driven electron transfer? Since the bacterial RC performs similar functions and also contains very similar bound cofactors, it is compared for analogous binding sites and/or properties. Z is the immediate electron donor to the PS II reaction center chlorophyll, P-680. *Ph* is (bacterio-)pheophytin. Q_A and Q_B are bound quinones which act in series as electron acceptors from Ph^-. P-860 is the bacterial RC chlorophyll. *Z* is shown on the same polypeptide (51 kDa) as P-680 from the consideration of extremely fast electron transfer from Z to P-680. However, neither the chemical identity or exact location of Z is known. See text for more detail

tides. This is shown schematically in Fig. 1. [It should be noted that the apparent molecular weights indicated in the figure are only estimates. It is known that varying the conditions used in sodium (or lithium) dodecyl sulfate-polyacrylamide gel electrophoresis (SDS or LDS-PAGE) affects the relative mobilities of membrane-integral hydrophobic proteins (SATOH et al. 1983). In addition, there are species-specific variations of molecular weights for some proteins]. In this chapter we shall designate these polypeptides by their nominal relative M_r's of (1) 51 kDa, (2) 45 kDa, (3) 34 kDa, (4) 32 kDa and (5) 10 kDa. This kind of generalization is justified on the basis of the following observations:

1. The gene sequences of all five proteins are highly conserved between cyanobacteria, green algae, and higher plants (KYLE 1985, WILLIAMS, VERMAAS, PAKRASI and ARNTZEN unpublished observations) and

2. The 51 kDa, 45 kDa and the 10 kDa proteins are even immunologically related in such widely separated species as peas, *Chlamydomonas* (DELEPELAIRE and CHUA 1979) and cyanobacteria (PAKRASI, RIETHMAN and SHERMAN 1985).

In contrast to the difficulty encountered in the development of "clean" PS II core complex isolation protocols, the reaction centers of purple photosynthetic bacteria are isolated with greater ease. They have long been known to contain three polypeptides designated as heavy (H), medium (M) and light (L) proteins (OKAMURA et al. 1982). It is known that the H subunit can be removed from this complex without any significant loss of photochemical capability of the reaction center. Since the bacterial and PS II RC's perform similar reactions, they may contain polypeptides which have analogous functions. We shall address this concept in the following sections.

2.1 51 kDa Polypeptide

The highest apparent molecular weight 51 kDa polypeptide is one of the two dominant proteins of the PS II core complex. It binds chlorophyll a and is the apoprotein of the Chl-protein complex CP 47 (also designated as CPa-1, CAMM and GREEN 1983, NAKATANI et al. 1984). Current evidence favors the presence of the reaction center chlorophyll being associated with this polypeptide. The most convincing data in support of this theory have come from the work of NAKATANI et al. (1984), who demonstrated that this purified, pigmented protein exhibits a small light-induced functional capacity when isolated by non-denaturing LDS-PAGE. In particular, they measured reversible changes in absorption attributed to P-680 oxidation and Ph reduction. CP 47 is also characterized by a 695 nm fluorescence emission (77 K) believed to originate from the photoreduced Ph (BRETON 1982). No evidence was gained for Q_A reduction by or association of quinones with the purified 51 kDa protein. However, it cannot be ruled out that Q_A binding might be regulated by this protein in the intact complex.

YAMAGISHI and KATOH (1984) have isolated a chlorophyll-protein complex containing the 51 kDa, 34 kDa, 32 kDa and the 10 kDa polypeptides from a thermophilic cyanobacterium. This complex could mediate light-induced electron transport from diphenylcarbazide, an artificial donor to PS II RC, to dichlorophenolindophenol, an acceptor from Q_B. The reaction requires the presence of Z, the immediate donor of P-680. Hence, the 51 kDa polypeptide is present in preparations which have binding sites for P-680, Ph, the quinones Q_A and Q_B, and presumably Z.

The gene encoding the 51 kDa protein has been isolated and sequenced for spinach chloroplasts (MORRIS and HERRMANN 1984). The derived amino acid sequence of this protein leads to a molecular mass of 56.2 kDa. The protein sequence of this hydrophobic Chl-binding polypeptide indicates the presence of seven membrane spanning regions.

2.2 45 kDa Polypeptide

The other dominant protein of the PS II core complex is a 45 kDa polypeptide that also binds Chl a and is the apoprotein of the Chl-protein complex CP 43 (also called CPa-2; CAMM and GREEN 1983, NAKATANI et al. 1984). This protein displays a fluorescence emission band at 685 nm at 77 K and is thought to serve an antenna role in PS II RC (NAKATANI et al. 1984). According to YAMAGISHI and KATOH (1984) it can be removed from the PS II complex without a significant loss in photoactivity. However, DE VITRY et al. (1984) have shown that in PS II particles from *Chlamydomonas,* the presence of the 45 kDa polypeptide is essential for the photoreduction of Q_A. Charge separation between P-680 and Ph is possible with only the 51 kDa polypeptide present, whereas both the 51 kDa and the 45 kDa polypeptides are needed (in a 1:1 molar ratio) to have Q_A reduction. However, this may simply mean that the 45 kDa component helps in maintaining the structure of the PS II complex, thereby influencing the properties of the Q_A binding site, even if that site is on another protein.

The gene encoding the 45 kDa protein from spinach chloroplast has been characterized recently (HOLSCHUH et al. 1984, ALT et al. 1984). The derived amino acid sequence for the 45 kDa polypeptide indicates a protein of molecular weight of either 51.8 kDa or 44.8 kDa, with the ambiguity based upon uncertainty as to the true translational start site. The gene sequences for the 51 kDa and the 45 kDa polypeptides show structural similarity, which is not unexpected on the basis of their Chl-binding properties. However, the two proteins are immunologically unrelated (DELEPELAIRE and CHUA 1979).

In conclusion, the presently available evidence indicates that the 45 kDa polypeptide does not contain the PS II RC and serves primarily as an antenna for the RC.

2.3 Q_B (32 kDa) Protein

Two proteins in the 30–34 kDa region are found in the PS II core preparations. Both of them stain poorly with Coomassie blue. In earlier studies they were designated as D1 and D2 where D referred to diffuse bands (CHUA and GILLHAM 1977). A third protein in the same molecular weight range is also associated with PS II, but is not present in the core complex (KUWABARA and MURATA 1982). This latter protein stains well with Coomassie blue and is part of an extrinsic protein complex involved in the oxidation of water (See ANDERSSON, Chap. 9.3, this Vol.). The relative positions of D1 and D2 polypeptides are strongly affected by gel conditions; the apparent molecular weight of D1 can be shifted by 2 to 3 kDa simply by changing the urea concentration in the SDS-PAGE system (NAKATANI et al. 1983, KYLE et al. 1983). One of these polypeptides (D1) has been identified via labeling with photoaffinity herbicides which interfere with Q_B function (see below); it is therefore designated as the Q_B-protein. The other D2 protein plays an as yet unknown role in PS II function; this will be discussed in more detail in Section 2.4 below. It should be noted that we cannot rule out the existence of even another poorly staining polypeptide in this region of the gel (BRICKER et al. 1985).

The Q_B-protein is probably one of the most well-studied proteins in thylakoid membranes. In this section we shall tabulate some of the interesting features of this protein. For more detailed discussion, the reader is referred to recent reviews by EDELMAN and REISFELD (1978), HIRSCHBERG et al. (1984), KYLE (1985) and KYLE and OHAD (Chap. 9.5, this Vol.).

The 32 kDa protein of PS II core was initially identified as a very rapidly labeled protein (STEINBACK et al. 1981a). The rate of turnover of this protein has been shown to be light-regulated (EDELMAN and REISFELD 1978, ARNTZEN et al. 1984). At very high intensities of light, the rate of degradation of the protein becomes large compared to the rate of integration of new 32 kDa proteins into functional PS II complexes, resulting in a net loss of photosynthetic capacity of the organisms. This phenomenon is referred to as photoinhibition of photosynthesis (ARNTZEN et al. 1984). Concurrent with the loss of the 32 kDa protein during photoinhibition, there is a loss of Q_B function (OHAD et al. 1984) indicating that the 32 kDa protein takes part in the binding of the secondary acceptor Q_B of PS II reaction center.

The Q_B-protein in higher plants contains no lysine residues as compared to the D-2 protein that contains lysine (KYLE 1985). The Q_B-protein in cyanobacteria contains lysine, however (MULLIGAN et al. 1984, CURTIS and HASELKORN 1984), indicating that the absence of this amino acid in higher plant proteins is not a required functional feature. The protein has surface-exposed domains, as evidenced by protease sensitivity (STEINBACK et al. 1981b) of the protein in situ. Trypsin cleaves the Q_B-protein in at least two different sites (MARDER et al. 1984).

2.3.1 Quinone-Herbicide Interactions at the Q_B Site

A large number of commercial herbicides inhibit electron flow on the reducing side of photosystem II (see also SANDMANN and BÖGER, Chap. 10.10, for a more detailed discussion of chloroplast-acting herbicides). Compounds such as atrazine, metribuzin, diuron, bromacil, ioxynil, and dinoseb all belong to different chemical families but have a common mode of action: they block electron transfer from Q_A^- to Q_B (ARNTZEN et al. 1982). It is now well accepted (see ORT, Chap. 4, this Vol.) that herbicides such as atrazine and diuron, which have been well characterized, displace Q_B from its binding site (WRAIGHT 1981, VELTHUYS 1981, CROFTS and WRAIGHT 1983, VERMAAS et al. 1983).

2.3.2 Photoaffinity Tags of the Q_B Site

Herbicide binding to thylakoids or to PS II-enriched membrane fragments is reversible (VERMAAS et al. 1984). This fact made it impossible to relate the herbicide-binding sites to any specific polypeptide in early studies, since the herbicide was displaced during protein isolation. The problem was overcome by the introduction of functionally active photoaffinity herbicides (PFISTER et al. 1981, MULLET and ARNTZEN 1982, GARDNER 1981). ^{14}C-azido-atrazine was found to bind covalently to a single high-affinity site. By autoradiography, this site was shown to be on a 32 kDa polypeptide. Since atrazine competes for binding with Q_B, which exchanges with the soluble PQ pool in the membrane, the 32 kDa protein was designated the Q_B protein (HIRSCHBERG et al. 1984). More recently an azidotriazinone photoaffinity label (related in chemical structure to metribuzin) was also found to specifically label the same protein, confirming the fact that all the Q_B-antagonist herbicides act within a similar domain (OETTMEIER et al. 1984).

Experiments have also been conducted using an azido-quinone which functions as a Q_B analog. When covalently linked to the Q_B site by photoaffinity activation, this quinone reduced atrazine binding but not ioxynil binding affinity. It did not decrease the absolute number of binding sites for either herbicide (VERMAAS et al. 1983). It was concluded that the herbicides and quinones do not occupy identical binding sites within the Q_B-protein (see also OETTMEIER et al. 1984), although binding of one affects that of the other (possibly by some allosteric mechanisms).

2.3.3 Mutations Affecting Quinone and Herbicide Binding at the Q_B Site

In 1979 it was reported that naturally occurring weed biotypes had developed atrazine-resistance via a change in the properties of PS II in the "mutant" strain (ARNTZEN et al. 1979). The modified chloroplasts showed a loss of herbicide-binding sites (PFISTER et al. 1979) and a reduced rate of electron transfer from Q_A^- to Q_B (PFISTER and ARNTZEN 1979, BOWES et al. 1980). Both wild-type and atrazine-resistant weed species have the 32 kDa Q_B-protein, but only in the former was the protein tagged by ^{14}C-azido-atrazine (PFISTER et al. 1981).

The gene encoding the Q_B-protein (called the *psb*A gene) is chloroplast-encoded. It has been cloned and sequenced from several higher plant and algal species (reviewed in KYLE 1985). When derived amino acid sequences of the Q_B-protein of atrazine-resistant and -susceptible weed species were compared, the resistant plants were found to contain only a serine to glycine substitution in this protein of *Amaranthus* (HIRSCHBERG and MCINTOSH 1983) and *Solanum nigrum* (HIRSCHBERG et al. 1984 and GOLOUBINOFF et al. 1984). Herbicide-resistant mutants of *Chlamydomonas* have also been isolated and characterized (GALLOWAY and METS 1984). Analysis of the Q_B-protein in these mutants was achieved by sequencing the *psb*A gene from each isolate (ERICKSON et al. 1985). It was found that three different single amino acid changes in the Q_B-protein resulted in herbicide resistance. A serine to alanine substitution at amino acid 264 in the protein resulted in both diuron and atrazine resistance, as well as an apparent reduction in Q_A^- to Q_B electron transfer. [The serine to glycine conversion at the same site in the protein in the higher plant species (HIRSCHBERG et al. 1984) gave only atrazine resistance with reduced electron transfer rates.] In two other mutants, the electron transport properties were not detectably altered when either atrazine or diuron resistance was achieved (a phenylalanine to tyrosine conversion at amino acid residue number 255 gave mainly atrazine resistance, whereas a valine to isoleucine conversion at amino acid residue number 219 gave mainly diuron resistance).

2.4 The 34 kDa D2 Protein

As mentioned in Section 2.2, PS II core complexes contain a second polypeptide in the 30–34 kDa region (dependent on the gel system used) that is distinctly different from the Q_B-protein (SATOH et al. 1983, BRICKER et al. 1985). The exact function of this second protein is yet to be determined. DELEPELAIRE has recently described the presence of a 34 kDa protein (D2) in the PS II core complex of *Chlamydomonas* (DELEPELAIRE 1983); furthermore it was demonstrated to be present in variant forms designated D2.1 and D2.2. The presence of the phosphorylated D2.1 form was dependent on the redox state of the plastoquinone pool in the chloroplasts subjected to electrophoretic analysis (DELEPELAIRE 1984). This implies that the D2 protein is influenced (at least indirectly via kinase activation; see STAEHELIN and ARNTZEN 1983) by the flow of electrons through the PS II complex. ^{14}C-labeled azidoatrazine binds to the Q_B-protein of *Chlamydomonas* (OHAD et al. 1984) and not to the D2 protein. Hence, D2 is not functionally analogous to the Q_B-protein. Compared to the 32 kDa protein, it does not turn over rapidly. However, similar to the Q_B-protein, D2-protein has surface-exposed regions that are subject to proteolysis by trypsin (DELEPELAIRE 1984).

Recently a gene, *psb*D, from the chloroplast genome of *Chlamydomonas* was cloned and sequenced (ROCHAIX et al. 1984). This gene has been suggested to code for the D2-protein. The *psb*D genes from spinach and pea plastid genomes have also been characterized (RASMUSSEN et al. 1984, HOLSCHUH et al. 1984, ALT et al. 1984). To this date, the only indication of the possible involvement of the *psb*D gene product in PS II comes from the finding that in spinach plastids, the sequences of the *psb*D gene and the gene for the 45 kDa protein overlap each other. Moreover, these two genes are cotranscribed (ALT et al. 1984, HOLSCHUH et al. 1984), indicating a possible functional relationship between these gene products. The *psb*D gene sequence is very well conserved between *Chlamydomonas,* spinach, and pea. Additionally, it shares significant homology with *psb*A (the gene for Q_B-protein), a fact that has led to the hypothesis of a possible quinone-binding function of the *psb*D gene product (ROCHAIX et al. 1984). The most important piece of information missing from the literature is the one that directly links the *psb*D gene with the D2 polypeptide. Undoubtedly, the D2-protein is an important component of the PS II core complex. Since the sequence of the *psb*D gene is known, antibodies can now be raised against specific synthetic peptides corresponding to stretches of *psb*D sequence. These antibodies can then be used (a) to identify the *psb*D gene product (is it really the D2 protein?) and (b) to examine possible alteration in specific functions associated with D2 by the addition of the antibodies. Another approach will be to modify the *psb*D gene by recombinant DNA methods and examine the mutants. In the near future, we may expect to find considerable research activities toward the elucidation of the functional role of the D2 protein in PS II.

2.5 10 kDa Protein

The smallest polypeptide in PS II core preparations runs at 9–10 kDa. This protein has been shown to be the apoprotein of cytochrome b-559 (WIDGER et al. 1984, BRICKER et al. 1985). The role of this protein in PS II function is at present enigmatic. This cytochrome is found in two redox forms, termed high and low potential cyt b-559. Detailed discussions on the different hypotheses regarding the possible function of cyt b-559 may be found in two excellent review papers (CRAMER and WHITMARSH 1977, BENDALL 1982). Recently two new procedures have been described to isolate large amounts of cyt b-559 from higher plant chloroplasts (WIDGER et al. 1984, METZ et al. 1983). In addition, the gene for cyt b-559 has been isolated from spinach plastids (HERRMANN et al. 1984, WESTHOFF et al. 1985). Analysis of the cloned DNA indicates the presence of two neighboring genes encoding a 9.4 kDa and a 4.4 kDa protein, respectively. Two polypeptides of similar molecular weight (approximately 10 kDa and 6 kDa) have been observed in the cyt b-559 preparations mentioned above (WIDGER et al. 1984, METZ et al. 1983). On the basis of these observations it has been proposed that cyt b-559 is a heterodimer of the 9.4 kDa and 4.4 kDa polypeptides (BABCOCK et al. 1985).

3 Photosystem II: A Membrane–Protein Complex

As mentioned in Section 2, PS II has many functional characteristics that are very similar to those of the reaction center complex of purple photosynthetic bacteria (see also in this Vol. DUTTON, Chap. 5; HEARST, Chap. 8.5; MICHEL and DEISENHOFER, Chap. 8.4; CROFTS and WRAIGHT 1983). In this context, it is appropriate to discuss some of the recent findings about the better-characterized bacterial reaction center, as this may shed some new light on the general research area on PS II.

As indicated earlier, bacterial reaction centers can now be fractionated down to a complex that contains only the L and M subunits, and still maintains photochemical activity (OKAMURA et al. 1982). Research in two different directions has provided us with a detailed insight into the structure and function of this photoactive complex. First, the genes encoding the L and M subunits of *Rhodopseudomonas sphaeroides* and *Rps. capsulata* have been cloned. Their complete nucleotide and derived amino acid sequences have also been published (WILLIAMS et al. 1983, 1984, YOUVAN et al. 1984). There are close to 80% homologies among the amino acid sequences of the L subunits and the M subunits of the two species. The calculated sizes of the L and M subunits of *Rps. capsulata* are 31.4 kDa and 34.4 kDa, respectively. Genes for L and M subunits are closely linked in the bacterial genome.

A second approach to the study of the bacterial RC has been very successfully pursued by MICHEL and coworkers (Chap. 8.4, this Vol.; DEISENHOFER et al. 1984) by obtaining an X-ray diffraction map (at 3 Å resolution) of crystals of *Rps. viridis* reaction center. They have reported that the L and M subunits form the central cylindrical part of the complex, with each having five bilayer spanning regions. The cylinder contains high density regions corresponding to the four BChl and two bacterial pheophytins known to be in the RC, as well as a nonheme iron. Close to this iron is a region of electron density thought to correspond to the bound Q_A. Q_B was lost during RC crystal preparation.

Although the crystal structure and gene sequence analyses have not been done for the reaction center of the same organism, these types of joint approaches have enabled the examination of the association of different cofactors, e.g., Q_A, Q_B and nonheme iron to specific regions of L or M subunits.

Herbicides such as atrazine (or the closely related terbutryn) inhibit electron transport in bacterial reaction centers (STEIN et al. 1984). Their mode of action is to displace Q_B, in direct analogy to herbicide action in PS II. It was shown that herbicide-resistant mutants of *Rps. sphaeroides* can be isolated, and the resistance is correlated to a specific loss of herbicide binding with an apparently concomitant reduction in Q_A^- to Q_B electron transfer (again in direct analogy to PS II, BROWN et al. 1984). Use of the photoaffinity azidoatrazine label demonstrated that the L subunit is the herbicide receptor in the bacterial RC (DE VITRY and DINER 1984, BROWN et al. 1984). By the analysis of the DNA sequences of the genes for the L subunit from mutant and wild-type *Rps. sphaeroides,* GILBERT and ARNTZEN (unpublished results) have shown that the resistance is due to a single amino acid change (again in direct analogy to PS II). Clearly the L subunit of the bacterial RC functions analogously to the 32 kDa Q_B-protein of PS II. It is, therefore, not surprising that the spinach Q_B-protein and the bacterial L subunit share about 25% amino acid homology (YOUVAN et al. 1984, WILLIAMS et al. 1984). Additionally, the *psb*D gene (presumed to code for D2 protein; see Sect. 2.4 above) and the gene for the M subunit also share sequence homologies to almost the same level. Based on these homologies, it has been suggested that Q_B- and D2-proteins may form a complex similar to the LM complex and thus the reaction center of PS II may actually be present in this Q_B-D2 protein complex (see HEARST, Chap. 8.5; MICHEL and DEISENHOFER, Chap. 8.3, both this Vol.). Since the 30–34 kDa polypeptides of the PS II core complexes have never been observed to bind Chl molecules, we conclude that this speculation remains unlikely at present. Thus the roles of the Q_B- and D2-proteins in the binding of P-680, the PS II reaction center Chl molecule(s), are still at the level of a hypothesis.

It is important to remember that the correct quaternary structure of the entire membrane-integral core complex is extremely important in maintaining optimum functional properties of PS II. Changing or deleting any one of the five proteins of the core complex may have an effect on components attached to a different protein. As an example, RENGER et al. (1984) have shown that the treatment of pea thylakoids with a lysine-specific protease significantly reduces the binding capacity of atrazine, although the 32 kDa Q_B-protein (that binds atrazine) does not have any lysine residue and is thus not affected by treatment with this protease. Hence, action of the protease on a different protein can alter the functional characteristics of Q_B-binding. Similarly, in the bacterial reaction center, the presence of the H subunit has been found to be necessary for the proper electron transport rates through Q_A and Q_B (DEBUS et al. 1985), although the LM complex can undergo stable charge separation in the absence of the H subunit. In the light of these considerations, it is entirely possible that the actual binding sites of quinone type and other cofactors may be shared among or influenced by more than one polypeptide in the complex.

Notes Added in Proof: Recently DEISENHOFER et al. (Nature, 1985) have published a more detailed structural description of the reaction center of *R. viridis*. Interestingly, the amino acid residues in the L and M subunits of the bacterial reaction center that were found to bind RC BChl, Q_a and Q_b molecules are completely conserved in Q_B and D_2 proteins, respectively. This new information on the homology at the level of specific key amino acids offers stronger suggestion about the roles of Q_B and D_2 proteins in the reaction center of PS II.

References

Alt J, Morris J, Westhoff P, Herrmann RG (1984) Curr Genet 8:597–606

Arntzen CJ, Ditto CL, Brewer PE (1979) Proc Natl Acad Sci USA 76:278–282

Arntzen CJ, Pfister K, Steinback KE (1982) In: LeBaron HM, Gressel J (eds) Herbicide resistance in plants. Wiley, New York, pp 185–214

Arntzen CJ, Kyle DJ, Wettern W, Ohad I (1984) In: Hallick R, Staehelin LA, Thornber JP (eds) Biosynthesis of the photosynthetic apparatus: Molecular biology, development and regulation. Liss, New York, pp 313–324

Babcock GT, Widger WR, Cramer WA, Oertling WA, Metz JG (1985) Biochemistry 24:3638–3645

Bendall DS (1982) Biochim Biophys Acta 683:119–151

Bowes J, Crofts AR, Arntzen CJ (1980) Arch Biochem Biophys 200:303–308

Breton J (1982) FEBS Lett 147:16–20

Bricker TM, Pakrasi HB, Sherman LA (1985) Arch Biochem Biophys 237:170–176

Brown A, Gilbert C, Guy R, Arntzen CJ (1984) Proc Natl Acad Sci USA 81:6310–6314

Camm EL, Green BR (1983) J Cell Biochem 23:171–179

Chua N-H, Gillham NW (1977) J Cell Biol 74:441–452

Cramer WA, Whitmarsh J (1977) Ann Rev Plant Physiol 28:33–172

Crofts AR, Wraight C (1983) Biochim Biophys Acta 726:149–185

Curtis SE, Haselkorn R (1984) Plant Mol Biol 3:249–258

Debus RJ, Feher G, Okamura MY (1985) Biochemistry 24:2488–2500

Deisenhofer J, Epp O, Miki K, Huber R, Michel H (1984) J Mol Biol 180:385–398

Delepelaire P (1983) Photobiochem Photobiophys 6:279–291

Delepelaire P (1984) EMBO J 3:701–706

Delepelaire P, Chua N-H (1979) Proc Natl Acad Sci USA 76:111–115

Diner BA, Wollman FA (1980) Eur J Biochem 110:521–526

Edelman M, Reisfeld A (1978) In: Akoyunoglou G (ed): Chloroplast development. Elsevier/North-Holland. Biomed Press, Amsterdam New York, pp 641–652

Erickson JM, Rahire M, Mets L, Rochaix J-D (1985) Science 228:204–207

Galloway RE, Mets LJ (1984) Plant Physiol 74:469–474

Gardner G (1981) Science 211:937–940

Goloubinoff P, Edelman M, Hallick RB (1984) Nucleic Acids Res 12:9489–9496

Herrmann RG, Alt J, Schiller B, Widger WR, Cramer WA (1984) FEBS Lett 176:239–244

Hirschberg J, McIntosh L (1983) Science 222:1346–1349

Hirschberg J, Bleecker A, Kyle DJ, McIntosh L, Arntzen CJ (1984) Z Naturforsch 39c:412–420

Holschuh K, Bottomley W, Whitfield PR (1984) Nucleic Acids Res 12:8819–8834

Kuwabara T, Murata N (1982) Biochim Biophys Acta 680:210–215

Kyle DJ (1985) Photochem Photobiol 41:107–116

Kyle DJ, Ohad I, Guy R, Arntzen CJ (1983) In: Inoue Y, Crofts AR, Govindjee, Murata N, Renger G, Satoh K (eds) The oxygen evolving system of photosynthesis. Academic Press, London New York, pp 91–101

Marder JB, Goloubinoff P, Edelman M (1984) J Biol Chem 259:3900–3908

Metz JG, Ulmer G, Bricker TM, Miles D (1983) Biochim Biophys Acta 725:203–209
Morris J, Herrmann RG (1984) Nucleic Acid Res 12:2837–2850
Mullet JE, Arntzen CJ (1981) Biochim Biophys Acta 655:236–248
Mulligan B, Schultes N, Chen L, Bogorad L (1984) Proc Natl Acad Sci USA 81:2693–2697
Nakatani HY, Satoh K, Steinback KE, Arntzen CJ (1983) In: Miyamoto J (ed) IUPAC pesticide chemistry, human welfare and the environment. Pergamon, Oxford New York, pp 171–176
Nakatani HY, Ke B, Dolan E, Arntzen CJ (1984) Biocheim Biophys Acta 765:347–352
Oettmeier W, Masson K, Soll H-J, Draber W (1984) Biochim Biophys Acta 767:590–595
Ohad I, Kyle DJ, Arntzen CJ (1984) J Cell Biol 99:481–485
Okamura MY, Feher G, Nelson N (1982) In: Govindjee (ed) Photosynthesis, vol I. Academic Press, London New York, pp 195–272
O'Malley PJ, Babcock GT (1984) Biochim Biophys Acta 765:370–379
Pfister K, Arntzen CJ (1979) Z Naturforsch 34c:996–1009
Pfister K, Radosevich SR, Arntzen CJ (1979) Plant Physiol 64:995–999
Pfister K, Steinback KE, Gardner G, Arntzen CJ (1981) Proc Natl Acad Sci USA 78:981–985
Rasmussen OF, Bookjans G, Stummann BM, Hennigsen KW (1984) Plant Mol Biol 3:191–199
Renger G, Hagemann R, Vermaas WFJ (1984) Z Naturforsch 39c:362–367
Rochaix J-D, Dron M, Rahire M, Malnoe P (1984) Plant Mol Biol 3:363–370
Satoh K (1983) In: Inoue Y, Crofts AR, Govindjee, Murata N, Renger G, Satoh K (eds) The oxygen evolving system of photosynthesis. Academic Press, London New York, pp 27–38
Satoh K, Butler WL (1978) Plant Physiol 61:373–379
Satoh K, Nakatani HY, Steinback KE, Watson J, Arntzen CJ (1983) Biochim Biophys Acta 724:142–150
Sprague SG, Camm EL, Green BR, Staehelin LA (1985) J Cell Biol 100:552–557
Staehelin LA, Arntzen CJ (1983) J Cell Biol 97:1327–1337
Stein RR, Castellvi AL, Bogacz JP, Wraight CA (1984) J Cell Biochem 24:243–259
Steinback KE, McIntosh L, Bogorad L, Arntzen CJ (1981a) Proc Natl Acad Sci USA 78:7463–7467
Steinback KE, Pfister K, Arntzen CJ (1981b) Z Naturforsch 36:98–108
Velthuys BR (1981) FEBS Lett 126:277–281
Vermaas WFJ, Arntzen CJ, Gu L-Q, Yu C-A (1983) Biochim Biophys Acta 723:266–275
Vermaas WFJ, Dohnt G, Renger G (1984) Biochim Biophys Acta 765:74–83
Vitry C de, Diner B (1984) FEBS Lett 167:327–331
Vitry C de, Wollman FA, Delepelaire P (1984) Biochim Biophys Acta 787:415–422
Westhoff P, Alt J, Widger WR, Cramer WA, Herrmann RG (1985) Plant Mol Biol 4:103–110
Widger WR, Cramer WA, Hermodson M, Meyer D, Gullifor M (1984) J Biol Chem 259:3870–3876
Williams JC, Steiner LA, Ogden RC, Simon MI, Feher G (1983) Proc Natl Acad Sci USA 80:6505–6509
Williams JC, Steiner LA, Feher G, Simon MI (1984) Proc Natl Acad Sci USA 81:7303–7307
Wraight CA (1981) Isr J Chem 21:348–354
Yamagishi A, Katoh S (1984) Biochim Biophys Acta 765:118–124
Youvan DC, Bylina EJ, Alberti M, Begusch H, Hearst JE (1984) Cell 37:949–957

9.5 The Mechanism of Photoinhibition in Higher Plants and Green Algae

D.J. KYLE and I. OHAD

1 Introduction

The photosynthetic capacity of an organism can be severely reduced following exposure to light intensities in excess of that required to saturate photosynthesis. This phenomenon is referred to as photoinhibition, and precedes the loss of chlorophyll by photooxidation (POWLES 1984). The light intensity at which photoinhibition occurs is variable since it is dependent upon the light intensity required to saturate photosynthesis. Thus, in well-watered, sun-adapted species, higher light intensities (2–3 mE m^{-2} s^{-1}) may be required for photoinhibition (POWLES and CRITCHLEY 1980) as compared to generally low light intensities (0.1–1 mE m^{-2} s^{-1}) required for water-stressed (POWLES and BJÖRKMAN 1983) or shade-adapted species (POWLES and THORNE 1981). Therefore, it is apparent that although light is the ultimate cause of the photoinhibition, many other parameters may be involved in the expression of the light stress, such as the rates of CO_2 fixation and photosynthetic electron transport.

2 Phenomenology of Photoinhibition

2.1 Higher Plants

Changes in photosynthetic capacity of cells or tissues can be easily monitored using in vivo chlorophyll fluorescence as an intrinsic probe of the photosynthetic membrane. CRITCHLEY and SMILLIE (1981) used this technique to study light damage in intact cucumber leaves and concluded that the high light treatment led to an inactivation of photosystem II (PS II). Measurements of fluorescence yields from PS II and PS I at 77 K by POWLES and BJÖRKMAN (1982) in leaves of *Nerium oleander* confirmed that photoinhibition is accompanied by the inactivation of the primary photochemistry of PS II with little or no effect on PS I.

When attached leaves were exposed to photoinhibitory conditions, a severe decrease in both the saturated rate (POWLES and CRITCHLEY 1980) and quantum yield for CO_2 assimilation (POWLES et al. 1983) was observed. This loss of ability to fix CO_2 occurs concomitantly with a loss of electron transport capacity around PS II (CRITCHLEY 1981). Clearly, a loss in the ability to carry out electron transport as a consequence of photoinhibition will reduce the concentrations of NADPH required for CO_2 fixation. Reduction of the intercellular CO_2 concentration (which may occur upon stomatal closure), however, results in a more rapid and extensive photoinhibition (POWLES et al. 1979). Thus, under conditions in which CO_2 fixation is maximal, higher light intensities are required for the photoinhibition of PS II and ultimate reduction of the quantum yield for CO_2 fixation. When CO_2 fixation is already limited, however, the photosynthetic membrane becomes more sensitive to photodamage. In both cases, photoinhibition of PS II occurs, but the light intensity required for the effect varies, depending on the organism's ability to fix CO_2.

Photorespiration, as well as CO_2 fixation, regenerates the terminal acceptor of photosynthetic electron transport and OSMOND (1981) has speculated that photorespiration might be an important mechanism in preventing photoinhibition under conditions of high illumination when CO_2 is limiting. This concept is supported by the observation that increased O_2 partial pressures (at zero external CO_2) lessened the extent of photoinhibition as a consequence of the photorespiratory cycling of the terminal acceptor of electron transport (POWLES et al. 1979).

2.2 Algal Cells

The ease of regulating nutrient, CO_2, and O_2 partial pressures, and incident light intensities makes algal cell cultures ideal models for the study of photoinhibition. The uniformity of light exposure in a stirred algal suspension is a requirement for the study of primary sites of photodamage. These conditions cannot be easily studied in higher plant leaves, since the light attenuation across a leaf results in photoinhibition on the light-exposed surface but not on the shaded surface of the leaf (POWLES and BJÖRKMAN 1982). Light attenuation, however, also occurs in a water column and loss of photosynthetic capacity (photoinhibition) and fluorescence of phytoplankton populations was observed during midday in the upper water layers (higher light intensities) but not in the lower water layers in a lake under natural conditions (BELAY 1981).

A loss of photosynthesis in *Chlorella* in culture following high light treatment has been observed (MYERS and BURR 1940, KOK 1956). Using similar conditions, SETLIKOVA et al. (1984) have more recently shown a photoinhibition response using other algal species. Algal cultures offer the additional advantages of radioactive pulse labeling and the use of various inhibitors of electron transport and/or protein synthesis. Using these techniques, the primary site of photoinhibition damage in *Chlamydomonas* was shown to occur in PS II (KYLE et al. 1984). The specific site of photoinhibition in vivo has been further identified as the secondary acceptor of PS II (Q_B) by ARNTZEN et al. (1984). This conclusion was based on the following observations: (1) high light treatment inhibits electron transport from H_2O to diaminodurene (PS II function requiring Q_B), but not that from H_2O to silicomolybdate (PS II function not requiring Q_B) (KYLE et al. 1985, EDELMAN et al. 1984); (2) the number of quinone-binding sites per unit chlorophyll (Q_B-proteins), as assayed by herbicide-binding sites, is reduced following photoinhibition (KYLE et al. 1984); (3) the apoprotein binding the Q_B quinone (32 kDa protein), pulse-labeled prior to photoinhibitory treatment, is lost during photoinhibition (OHAD et al. 1984); (4) fluorescence kinetics measurements indicate that photoinhibited *Chlamydomonas* cells have a reduced rate of Q_A to PQ (mediated by Q_B) electron transport (KYLE et al. 1984). It appears, therefore, that the primary site of photoinhibition damage in algal cells is an inhibition of electron flow through Q_B. Extrapolation of this observation to higher plants may require caution, as less clear data are presently available in higher plants.

2.3 Photoinhibition in Chloroplasts

Photodamage occurs very rapidly and extensively in isolated chloroplasts or thylakoids. This damage includes both loss of electron flow as well as photooxidation of membrane pigments. YAMASHITA and BUTLER (1969) reported a photobleaching of chlorophyll using Tris-washed chloroplasts, which was prevented by the addition of a PS II donor during illumination, and concluded that the strong oxidant produced by PS II caused the damage. High light effects on photophosphorylation can be alleviated by including antioxidants (vitamin K; FORTI and JAGENDORF 1960) or cyclic PS I elicitors (phenazine methyl sulfate; AVRON 1960). JONES and KOK (1966) studied the action spectrum for photoinhibition of isolated thylakoids and concluded that the visable light damage is due to quanta absorbed by PS II. More recently, KRAUSE et al. (1978) showed a hypersensitivity to light by isolated chloroplasts in the absence of CO_2, which was alleviated by the addition of bicarbonate. Both KRAUSE et al. (1978) and CORNIC et al. (1982) emphasize the requirement of oxygen for photoinhibition, since the damage was reduced when illumination was carried out in 100% N_2. These data contrast with the results obtained with intact leaves where oxygen prevented photoinhibition (POWLES et al. 1979), but the discrepancy can be explained by a photorespiratory depletion of Calvin cycle intermediates from isolated chloroplasts upon illumination in the presence of O_2. When this undirectional transport of intermediates out of the chloroplasts was prevented by incubation in a medium containing excess quantities of these intermediates, photoinhibitory loss of CO_2-fixation was prevented (CORNIC et al. 1982).

3 Mechanism of Photoinhibition

FORTI and JAGENDORF (1960) have suggested that photodamage of chloroplast membranes may involve active oxygen species, since the damage can be minimized by the addition of antioxidants. However, one would expect damage by oxygen radicals to occur uniformly throughout the membrane unless the site of the generation of the radical was highly focused and contained. Experiments using *Chlamydomonas* (OHAD et al. 1984) or *Spirodela* (EDELMAN et al.

1984) have demonstrated a light intensity-dependent turnover (removal and replacement) of the 32 kDa apoprotein of the Q_B acceptor of PS II. Furthermore, the primary site of the light-induced loss of photosynthetic activity appears to be at the position of electron transport through this secondary quinone acceptor. We therefore conclude that light causes damage to the Q_B apoprotein, and when the rate of damage exceeds the rate of repair, loss of Q_B electron transport function is observed (photoinhibition).

How does light damage the Q_B protein? Light generates the substrate (electrons) for this protein and its enzymological activity basically involves three steps (see DINER, Chap. 9.1, and ARNTZEN and PAKRASI, Chap. 9.4 for further details): (1) the acceptance of an electron from the reduced primary quinone (Q_A^-) and transfer of that electron to an oxidized plastoquinone from the bulk pool; (2) once reduced, the quinone from the bulk pool is then stabilized (tightly bound), as the anion radical ($Q_B^{-\bullet}$) at a binding pocket of the protein; (3) a second electron (from Q_A^-) is then transferred to the bound quinone anion forming the doubly reduced anion ($Q_B^{=}$) which is subsequently protonated and released from the apoprotein to the bulk pool as the free quinol (PQH_2). ARNTZEN et al. (1984) have suggested that reactive species generated in step 2 ($Q_B^{-\bullet}$) or step 3 ($Q_B^{=}$) above, have a certain probability of interacting with a good leaving group in the binding pocket of the Q_B protein. We suggest that this leaving group may be a histidine or methionine in the putative quinone-binding site, as postulated by HIRSCHBERG et al. (1984). The idea is supported by the observation that when quinone anion formation is blocked by the addition of a competitive inhibitor (diuron or atrazine), the degree of photoinhibition is reduced (ARNTZEN et al. 1984). Although these data are suggestive that the quinone anion potentiates the damage, we cannot rule out the possibility that an oxygen radical, generated by the interaction of molecular oxygen with the quinone anion, is the damaging species. This question, however, can be addressed using purple photosynthetic bacteria which have a photochemical reaction center analogous to PS II and, although light-induced quinone radical formation occurs, these bacteria are requisite anaerobes during photosynthetic growth. Turnover of the bacterial reaction center polypeptides, including the one analogous to the Q_B protein of chloroplasts (L-subunit in *Rhodospirillum rubrum*) does not occur at any extraordinarily high rate during photoinhibition under anaerobiosis (KYLE and LOACH unpublished), suggesting that oxygen is required for the damage of the Q_B protein.

4 Recovery from Photoinhibition Damage

Once damaged, the Q_B apoprotein must be either repaired, or removed and replaced. OHAD et al. (1984) have shown that prelabeled Q_B apoprotein is physically removed from the membrane during photoinhibition of *Chlamydomonas* cells, and recovery of photoinhibited cells required chloroplast-directed protein synthesis. ARNTZEN et al. (1984) and SETLIKOVA et al. (1983) have shown that a much more rapid and extensive loss of Q_B activity occurs if chloramphenicol is included during the photoinhibition treatment, due to the inhibition of replacement of the damaged Q_B protein. These data suggest that the operative mechanism in *Chlamydomonas* involves removal and replacement using protein synthesized de novo (protein turnover) rather than a repair of the damaged protein. More recently KYLE et al. (1985b) have shown that recovery of photosynthetic activity in damaged cells is not only contingent on the ability to synthesize replacement Q_B protein, but also on the removal of the damaged protein.

The light-induced rapid turnover of the Q_B protein necessitates efficient transcription and translation of the gene coding for the Q_B protein (designated *psb*A). The *psb*A gene is present in only a single copy per chloroplast genome in all the higher plants thus far studied (reviewed by KYLE 1984). Although only a single gene is transcribed, the mRNA transcript levels are among the most abundant in higher plant chloroplasts and do not appear to be strongly light-regulated in green tissue (EDELMAN and REISFELD 1978). We feel, therefore, that the regulation of the appearance of newly synthesized protein in the membrane is not at the transcriptional level, but rather at the translational or post-translational level. Nevertheless, the high level of mRNA transcripts coding for the Q_B protein reflects a situation which is primed for rapid de novo protein synthesis when the signal (protein damage) is recognized (see ARNTZEN and PAKRASI, Chap. 9.4, this Vol.).

5 Synergism Between Light and Other Stresses

We have presented the concept that photoinhibition results when the rate of damage of the Q_B protein exceeds its rate of repair (via removal and replacement). Furthermore, we have suggested that the turnover of the Q_B-protein is a natural consequence of its enzymatic function which involves the stabilization of quinone radicals. CROFTS et al. (1984) have recently proposed that the protonation of the doubly reduced semiquinone anion ($Q_B^=$) and the release of the quinol (PQH_2) into the bulk pool is an equilibrium reaction:

$$Q_A\,Q_B^= \underset{2H^+}{\overset{2H^+}{\rightleftharpoons}} Q_A{:}PQH_2$$

Thus, as the concentration of the reduced quinol increases, one has an increased temporal stability of the quinone anions and therefore an increased probability of a damaging interaction. As a consequence, any limitation to the regeneration of the terminal acceptor of electron transport would result in a condition of light sensitivity. Indeed *Chlamydomonas* mutants deficient in CO_2 fixation are highly photosensitive (SPREITZER and METS 1981). The implication of this observation with respect to higher plant stress is obvious. Many environmental stresses induce stomatal closure, resulting in a CO_2 limitation of carbon fixation. POWLES and BJÖRKMAN (1983) have shown that drought stress renders *Nerium*

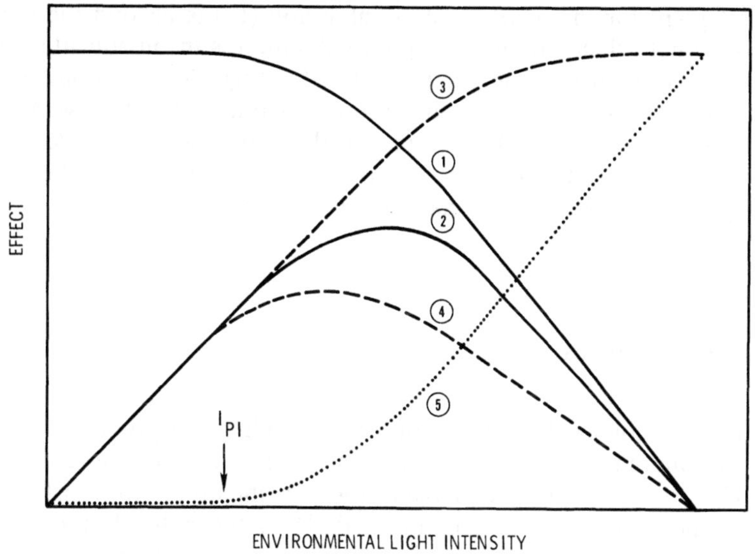

Fig. 1. A hypothetical representation of the effect of light on a photosynthesizing cell indicating the onset of photoinhibition (I_{PI}). The photosynthetic rates (*solid lines*) are indicated for the maximum potential (curve *1*) and actual rate at a given light intensity (curve *2*). Light saturation occurs as curve 2 approaches curve 1. The maximum photosynthetic potential is reduced at the intensity in which photoinhibition (*dotted line,* curve *5*) becomes significant. This light intensity (I_{PI}) is defined by the Q_B-protein turnover rate (*dashed lines*) as the point where the rate of damage (curve *3*) exceeds the rate of repair (curve *4*)

oleander hypersensitive to photoinhibition. A similar enhanced sensitivity to photoinhibition has been observed during cold stress (VAN HASSELT and VAN BERLO 1980). Thus, under ideal growth conditions, high light intensities (full sunlight) may be required to cause an accelerated damage or photoinhibition of the light reactions, but under most naturally variable conditions an environmental stress which reduces CO_2-fixing capacity will result in a hypersensitive condition to light, and damage will ensue.

These principals are outlined in diagrammatic form in Fig. 1. The Q_B protein damage will be a function of electron flow (therefore light intensity) up to a maximum rate at which point the PQ pool is fully reduced. The repair, however, involves several coordinated events including removal of the damaged protein, synthesis, insertion, and integration of new protein, and although it would be a function of the damage, should reach a maximum value at some point reflecting the rate-limiting step of these processes. As long as the rate of repair can keep pace with the rate of damage, no net loss of electron flow is observed (photoinhibition is minimal). As one approaches the maximum potential for photosynthesis (light-saturated rates), however, the PQ pool becomes proportionally more reduced and the rate of damage begins to exceed the rate of repair. At this point photoinhibition becomes significant and the electron flow capacity and the maximum potential for photosynthesis decreases. If this set point for the onset of photoinhibition (I_{PI}) is beyond the range of normal daily light fluctuations, then photoinhibition is inconsequential. However, any decrease in the maximum potential (limiting CO_2 fixation) or any decrease in the rate of repair (limiting protein synthesis due to low ATP/ADP ratios or low nitrogen availability) can shift the I_{PI} to much lower intensities; intensities at which the maximum potential for photosynthesis can be severely limited by photoinhibition under natural daylight conditions.

6 Conclusion

The physiological phenomenon of photoinhibition probably occurs by a common mechanism in higher plants, algae, isolated chloroplasts and thylakoids. The primary focus of damage is a loss of the ability to transfer electrons from the primary quinone acceptor of PS II to the bulk plastoquinone pool. The enzyme responsible for this transfer (Q_B-apoprotein) has a certain probability of being irreversibly damaged as a consequence of this enzymatic function. The rate of damage increases with the electron flow rate (light intensity) and residency time of the quinone anion apecies on the Q_B apoprotein (reduction of the plastoquinone pool). Repair of the damage requires de novo protein synthesis and when the rate of damage exceeds the rate of repair, a net loss in photosynthetic capability (photoinhibition) results. The set point of light intensity at which photoinhibition occurs (I_{PI}) is variable and affected by other environmental stresses.

References

Arntzen CJ, Kyle DJ, Wettern W, Ohad I (1984) In: Hallick R, Staehelin LA, Thornber JP (eds) Biosynthesis of the photosynthetic apparatus: Molecular biology, development and regulation. UCLA Symp Ser vol 14, pp 318–324
Avron M (1960) Biochim Biophys Acta 44:41–48
Belay A (1981) New Phytol 89:61–74
Cornic G, Woo KC, Osmond B (1982) Plant Physiol 70:1310–1315
Critchley C (1981) Plant Physiol 67:1161–1165
Critchley C, Smillie RM (1981) Aust J Plant Physiol 8:133–141
Crofts AR, Robinson HH, Snozzi M (1984) In: Sybesma C (ed) Advances in photosynthesis research, vol I. Nijhoff/Junk, The Hague, pp 461–468
Edelman M, Reisfeld A (1978) In: Akoyunoglou G (ed) Chloroplast development. Elsevier/North-Holland Biomed Press, Amsterdam New York, pp 641–652
Edelman M, Matoo AK, Marder JB (1984) In: Ellis RJ (ed) Chloroplast biogenesis. Cambridge Univ Press, Cambridge, in press
Forti G, Jagendorf AT (1960) Biochim Biophys Acta 44:34–40
Hasselt PR van, Berlo HAC van (1980) Physiol Plant 50:52–56
Hirschberg J, Bleecker A, Kyle DJ, McIntosh L, Arntzen CJ (1984) Z Naturforsch 39c:412–420

Jones LW, Kok B (1966) Plant Physiol 41:1037–1043
Kok B (1956) Biochim Biophys Acta 21:234–244
Krause GH, Kirk M, Heber V, Osmond CB (1978) Planta 142:229–233
Kyle DJ (1984) Photochem Photobiol 41:107–116
Kyle DJ, Ohad I, Arntzen CJ (1984a) Proc Natl Acad Sci USA 81:4070–4074
Kyle DJ, Ohad I, Arntzen CJ (1985) In: Key J, Kosuge T (eds) Cellular and molecular biology of plant stress. Liss, New York, pp 51–69
Myers J, Burr GO (1940) J Gen Physiol 24:45–67
Ohad I, Kyle DJ, Arntzen CJ (1984) J Cell Biol 99:481–485
Osmond B (1981) Biochim Biophys Acta 639:77–98
Powles SB (1984) Annu Rev Plant Physiol 35:15–44
Powles SB, Critchley C (1980) Plant Physiol 65:1181–1187
Powles SB, Björkman O (1982) Plant 156:97–107
Powles SB, Björkman O (1983) Carnegie Inst Washington Yearb 81:75–82
Powles SB, Thorne SW (1981) Planta 152:471–477
Powles SB, Osmond B, Thorne SW (1979) Plant Physiol 64:982–988
Powles SB, Berry JA, Björkman O (1983) Plant Cell Environ 6:117–123
Setlikova E, Masojidek J, Nedbal L, Setlik I (1984) Proc 6th Int Congr Photosynth, Brussels, pp 255–258
Spreitzer RJ, Mets L (1981) Plant Physiol 67:565–569
Yamashita T, Butler WL (1969) Plant Physiol 44:1342–1346

9.6 Photosystem I Reaction Center and Its Primary Electron Transfer Reactions

P. SÉTIF and P. MATHIS

1 Introduction

The photosystem I (PS I) reaction center is the site in the photosynthetic membrane of oxygen-evolving organisms, where a light-activated electron transfer permits the reduction of ferredoxin. The cycling of PS I is insured by electrons coming from photosystem II. In addition to this directional (noncyclic) electron transfer, PS I also realizes a cyclic electron transfer that pumps protons for the synthesis of ATP. The PS I reaction center is a highly integrated membrane protein complex that consists of hydrophobic polypeptides, antenna pigments, and electron carriers. Its size and composition are not exactly known (see WOLLMAN, Chap. 9.7, this Vol.). So-called PS I particles have been extracted from photosynthetic membranes by various detergents. They retain PS I photoactivity, but none can be safely described as the PS I reaction center. The primary electron donor P-700 is coupled to an antenna system from which it receives excitation energy, to a chain of low-potential electron acceptors, and to a secondary donor which permits its re-reduction (see ORT, Chap. 4, this Vol. for an overview).

Although many properties of the PS I center are not yet known, it is clear that its structural and functional organization is very similar in all oxygenic organisms. It shares also many properties with the reaction center of some chlorobacteria of the *Chlorobium* type (see BLANKENSHIP and FULLER, Chap. 8.6; SPRAGUE and VARGA, Chap. 11.1, both this Vol.). PS I can be excited with light of rather long wavelength, up to 730 nm, where the photon energy is 1.7 eV. The final stored energy, corresponding to an electron going from the secondary donor plastocyanin to ferredoxin, is around 0.80 eV. The energetic yield is thus 48%, not counting the membrane potential. The primary acceptors of PS I are capable of reducing H^+ into H_2 via a hydrogenase: besides being of basic interest for understanding PS I center function, this property is of great importance for researchers interested in utilizing photosynthetic systems for the production of artificial fuels. Due to the brevity of this review, we have not been able to provide a comprehensive coverage of the relevant literature. The reader interested in more details is referred to the reviews of KE (1973), MATHIS and PAILLOTIN (1981) and MALKIN (1982).

2 The Primary Donor P-700

The primary donor, P-700, was named by KOK, who discovered a light-induced bleaching at 700 nm, which appeared to reflect the oxidation of a chlorophyll-type species. During the 1970's a general consensus developed that P-700 consisted of two molecules of chlorophyll a. This hypothesis is now being seriously challenged by more recent findings.

2.1 Basic Properties of P-700

P-700 can be oxidized by light or by chemical oxidants in a 1-e^- reaction with a pH-independent E_m of +490 mV (versus NHE), which may vary with the state of the reaction center and the temperature (SETIF and MATHIS 1980). The absorption spectrum of P-700 is not known, but the difference spectrum ($P\text{-}700^+ - P\text{-}700$) has been measured (HIYAMA and KE 1972). It displays major bleachings around 700 and 435 nm and small positive bands around 800 and 450 nm; a small bleaching at 680 nm is of variable amplitude. At low temperature, narrow peaks develop, positive at 695 and negative at 685 nm (Fig. 1). Oxidation of P-700 is accompanied by a circular dichroism difference spectrum (PHILIPSON et al. 1972). $P\text{-}700^+$ is a radical which has been extensively studied by magnetic methods (HOFF 1979). The EPR spectrum is a broad and unstructured single gaussian line at $g = 2.0025$ with a width of 7.2 G (Fig. 1). The spin relaxation of $P\text{-}700^+$ is rather complex (ROSE and BEARDEN 1980). Proton ENDOR spectra of $P\text{-}700^+$ reveal hyperfine splittings (NORRIS et al. 1974) which have been assigned by comparison with the chlorophyll a cation in vitro (SCHEER et al. 1977, O'MALLEY and BABCOCK 1984).

The singlet excited state of P-700 is always efficiently quenched, and its fluorescence has not been observed unambiguously. The triplet state ^{3}P-700 can be efficiently populated due to charge recombination with the primary acceptor (FRANK et al. 1979, RUTHERFORD and MULLET 1981). At low temperature, this triplet has a characteristic EPR spectrum (Fig. 1), indicating that the T_0 level is preferentially populated. The difference absorption spectrum due to the formation of ^{3}P-700 has been measured by flash absorption spectroscopy and by absorption-detected magnetic resonance (SÉTIF et al. 1981, DEN BLANKEN and HOFF 1983, HOFF, Chap. 7.2., this Vol.). The ^{3}P-700 decay is temperature independent below 100 K ($t^1/_2 = 0.8$ ms) and is greatly accelerated at higher temperature ($t^1/_2 \simeq 6$ μs) (MATHIS et al. 1978).

2.2 Chemical Nature of P-700

The spectroscopic properties of P-700 and especially the absorption difference spectra, are consistent with the hypothesis that it is made of one or several molecules of chlorophyll a. A different type of chlorophyll named Chl-RCI

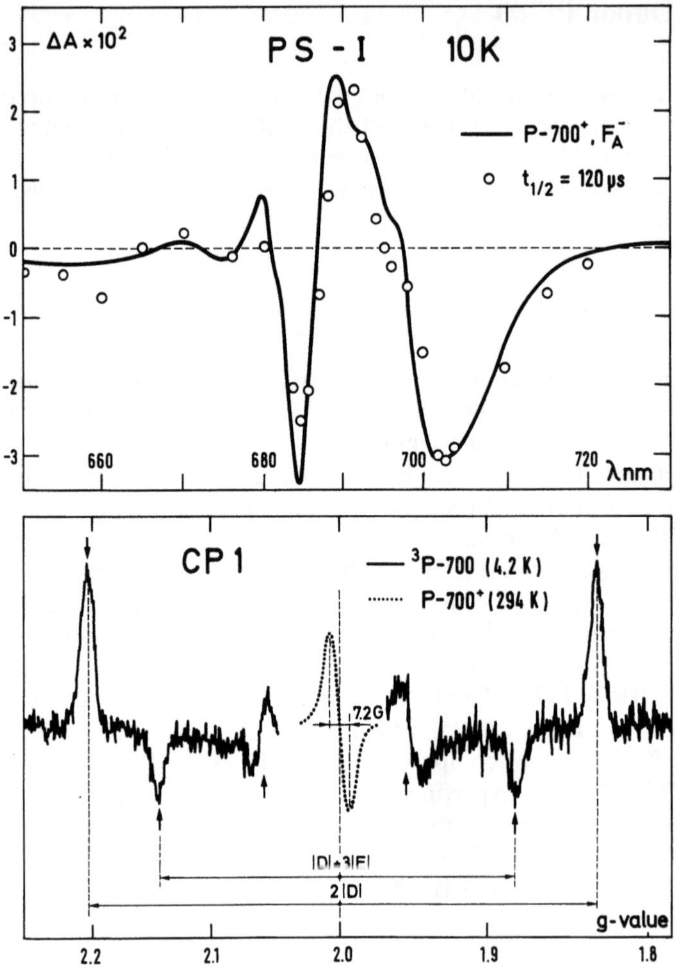

Fig. 1. *Top:* 10 K difference absorption spectrum of PS I particles prepared with digitonin and deoxycholate and subjected to continuous illumination (*continuous line:* formation of a stable state P-700$^+$, F_A^-) and to flash excitation (*circles:* formation of P-700$^+$, A_1^- decaying with $t^1/_2 = 120$ µs) (see Sétif et al. 1984). *Bottom:* EPR spectra of P-700$^+$ (*middle:* enlarged horizontal scale) and of ^{3}P-700, in CP 1 particles. (See Sétif et al. 1981)

has recently been isolated and found to be stoichiometric with the PS I reaction center (Dörnemann and Senger 1982). It appears to be a 13^2-hydroxy-20-chloro chlorophyll a derivative (Scheer et al. 1983). It has not been proved, but it is very probable that it constitutes a natural constituent of the PS I reaction center. Its relationship to P-700 is still unclear. Assuming provisionally that P-700 is made of Chl a, a major question is whether it contains one or several (say two) molecules of Chl a (monomer versus dimer hypothesis). This question is still unanswered, but it is clear that the controversy arises from the fact that we know much more, based on spectroscopy, about P-700$^+$ and ^{3}P-700 than about the neutral ground state P-700, and that the interactions between one Chl a and its surroundings should be different in these various states.

2.2.1 Two Chlorophyll Molecules

During the 1970's it was widely accepted that P-700 was a dimer of Chl a. This was supported by a comparison of the width (ΔH) of the EPR spectrum of P-700$^+$ ($\Delta H = 7.2$ G) and of Chl a$^+$ in vitro ($\Delta H = 9.3$ G). The linewidth is essentially due to hyperfine coupling and the narrowing, which is close to a factor of $\sqrt{2}$, could be interpreted by assuming that the unpaired electron is delocalized over two molecules (NORRIS et al. 1971). More recent results do not support this interpretation. At present the dimer hypothesis is supported mostly by absorption experiments. Application of the exciton theory to Chl predicts that if two molecules are closely spaced, the individual Q_Y transitions are replaced by a set of two new Q_Y transitions at respectively shorter and longer wavelengths, the intensity and energy of which are a function of the geometry of the pair. This model provides a good interpretation of the difference absorption and CD spectra between P-700 and P-700$^+$:

$$\text{P-700 (Chl}-\text{Chl)}_{697,683} \rightarrow \text{P-700}^+ (\text{Chl}^+, \text{Chl}_{686})$$

(PHILIPSON et al. 1972). SCHAFFERNICHT and JUNGE (1981) interpreted in a similar manner their linear dichroism data. Evidence for a dimer was also found in the difference spectrum between ^{3}P-700 and P-700 by DEN BLANKEN and HOFF (1983), who proposed that the high-energy transition in the dimer is at 650 nm, and the monomeric band of ^{3}P-700 at 672 nm (see also HOFF, Chap. 7.2, this Vol.). These wavelengths differ largely from the previous ones. This disagreement may reflect an intrinsic weakness of the dimer model, but it may also be due to the variable contribution of electrochromic absorption shifts.

In principle, these spectral data could give some insight into the geometry of the dimer. Various in vitro systems have been prepared, where two Chl molecules are held together by bifunctional ligands or by a covalent linkage (KATZ et al. 1978, BOXER 1983). They mimic some properties of P-700. It is worth pointing out that the electronic interaction which is invoked in the exciton theory does not imply any chemical link between the two Chl a, which may be held closely spaced ($\simeq 5$ Å) by the protein moiety.

2.2.2 One Chlorophyll Molecule

Recent magnetic resonance measurements have shown rather unambiguously that in P-700$^+$ the unpaired electron is localized on only one Chl for the time of measurement (10^{-7} s or less): the EPR line-shapes of P-700$^+$ and Chl a$^+$ are identical when evaluated by the 2nd-moment method after enrichment in ^{13}C (WASIELEWSKI et al. 1981b). The hyperfine coupling constants measured by proton ENDOR are also similar for P-700$^+$ and for immobilized Chl a$^+$ in vitro (O'MALLEY and BABCOCK 1984). These authors propose that P-700 includes only one Chl a, whose Q_Y transition is strongly shifted toward 700 nm by interaction with the protein environment. Explanation of the ENDOR data requires that ring V is strongly modified in P-700. Along these lines, WASIELEWSKI et al. (1981 a) proposed that P-700 consists of one Chl a with ring V in the enol form. In addition to the ESR and ENDOR data on P-700$^+$, this last hypothesis accounts for the midpoint oxidation potential which is high for Chl a (+0.86 V) and much lower for an enol model (+0.50 V), like P-700.

Despite all the data that have been accumulated on P-700, its chemical structure is still not defined. Probably all the present information can be understood if one assumes that the neutral ground state involves two (or more) electronically interacting Chl, and the unpaired electron is localized on one Chl in the state P-700$^+$. The zero-field splitting parameters D and E of ^{3}P-700 can probably not help in deciding between a monomeric or a dimeric structure (BOXER 1983, DEN BLANKEN and HOFF 1983).

3 Photosystem I Electron Acceptors

Since the simultaneous discovery of an iron-sulfur center A by EPR at low temperature and of P-430 by flash absorption spectroscopy at room temperature in 1971, a large amount of work has been devoted to the membrane-associated acceptors of PS I. Some recent reviews have summed up the current knowledge on PS I (MALKIN 1982, MATHIS and PAILLOTIN 1981) while others have dealt with more specific topics, i.e., the iron-sulfur centers (EVANS 1982), and the application of EPR and ENDOR to photosynthesis (HOFF 1979, 1982, 1984).

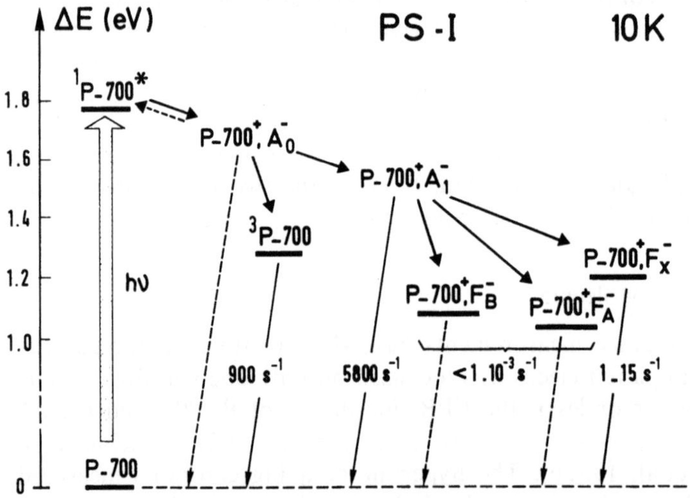

Fig. 2. States of the PS I reaction center, placed according to their approximative energy. *Continuous arrows* represent steps which effectively take place at 10 K (with respective rates). *Dotted arrows* represent steps which are hypothetical at 10 K

To date, low-temperature EPR studies have revealed the existence of five electron acceptors in most PS I particles: the iron-sulfur centers $A(F_A)$, $B(F_B)$ and $X(F_X)$ and the species A_0 and A_1. At room temperature, flash absorption spectroscopy has revealed at least three species: A_1, A_2, and P-430. Most recent efforts have been aimed at identifying and mapping these various low-potential electron carriers.

3.1 Terminal Acceptors: F_A and F_B

On the basis of many different observations, including chemical analysis of iron and labile sulfide, absorption, EPR and Mössbauer spectroscopies, these low-potential electron carriers $E_m(F_A) = -540$ mV; $E_m(F_B) = -590$ mV, in spinach; were identified as two interacting 4 Fe−4 S centers (EVANS 1982). From the comparison of the X- and Q-band EPR spectra of ($F_{A_1}^-$, F_B^-), AASA et al. (1981) derived an interaction energy of about 0.01 cm^{-1} between F_A^- and F_B^-. This is in agreement with the K-band spectrum obtained by HOFF (1979).

Illumination of PS I at low temperature places at most one electron in the pool (F_A, F_B). First studies made with spinach in aqueous media reported that F_A becomes preferentially reduced. Together with the redox potentials of F_A and F_B, this was taken as evidence that the electron migrates first to F_B and then from F_B^- to F_A. This linear scheme has been questioned recently, because in many other organisms (*Phormidium laminosum*, CAMMACK et al. 1979, *Dunaliella*, HOOTKINS et al. 1981, barley, NUGENT et al. 1981) and also in spinach in the presence of glycerol (EVANS and HEATHCOTE 1980), both centers become partly reduced and their relative E_m is changed. Moreover there seems to be no electron exchange between F_A and F_B below 215 K (CHAMOROVSKY and CAMMACK 1982a). These observations support a parallel functioning of F_A and F_B, in agreement with some chemical inactivation experiments (GOLBECK and WARDEN 1982), but in contrast with others (MALKIN 1984).

Changes in the conformation of F_A and F_B have been deduced from an analysis of photoreduction and recombination at low temperature (CHAMOROVSKY and CAMMACK 1982a) and of EPR orientations under different states of reduction (AASA et al. 1981).

There are good arguments in favor of the idea that both F_A and F_B constitute the species P-430, which is the terminal acceptor in PS I particles at physiological temperatures (however, see HIYAMA and FORK 1980). A single saturating flash can transfer one electron in P-430, but this is not true at low temperature: at 10 K in spinach PS I, the irreversible charge separation P-700 ... $F_A \longrightarrow$ P-700$^+$ ... F_A^- has a yield of at most 20% (SÉTIF et al. 1984, see also CROWDER and BEARDEN 1983) and takes place in about 65% of the reaction centers, while the other centers undergo only a charge separation which reverses with $t^1/_2 = 120$ μs (SÉTIF et al. 1984). Two recent studies put limits for the time of F_A reduction at 10–25 K: $t^1/_2 < 1$ ms (CROWDER and BEARDEN 1983) and $t^1/_2 >$ 0.4 ms (SÉTIF et al. 1984). At room temperature, the time of F_A or F_B reduction has been inferred from the decay of a special electron spin echo signal: $t^1/_2 \simeq$ 170 ns (THURNAUER and NORRIS 1980, THURNAUER et al. 1982).

3.2 F_X and A_2

When F_A and F_B are chemically prereduced, illumination at low temperature reduces another species named F_X, which is probably a special 4 Fe–4 S iron-sulfur protein. The E_m of F_X has been recently measured directly at −705 mV (CHAMOROVSKY and CAMMACK 1982b). F_X has often been considered to be an intermediate in the reduction of F_A or F_B, but two recent studies by flash EPR and absorption conclude that, at low temperature, F_X is not associated with electron flow between P-700 and F_A or F_B (CROWDER and BEARDEN 1983, SÉTIF et al. 1984).

When P-430 is reduced, in flash experiments at room temperature, a single flash reduces a species named A_2, which then recombines with P-700$^+$ with $t^1/_2 = 250$ µs; it was proposed that A_2 and X are the same species (SAUER et al. 1978). This is also supported by flash absorption, showing that the difference spectrum due to the reduction of A_2 resembles that of an iron-sulfur protein (SAUER et al. 1978, GOLBECK et al. 1978, KOIKE and KATOH 1982). Recently we have shown more specifically that the A_2 reaction at 294 K becomes observable when F_A and F_B are reduced and disappears when F_X is also reduced (SÉTIF et al. 1983). In contrast to the photoreduction at low temperature, that of A_2 at room temperature is highly efficient. Does this mean that $A_2(F_X)$ is on the main path of electron flow at physiological temperatures?

3.3 More Primary Electron Acceptors: A_0, A_1

Flash absorption experiments conducted when A_2 was pre-reduced (Triton particles at low potential) or inactivated (CP1 particles prepared with SDS) revealed the existence of a more primary acceptor named A_1 (SAUER et al. 1978, MATHIS et al. 1978). Under similar conditions, a triplet state was detected by EPR, during illumination at low temperature (FRANK et al. 1979, RUTHERFORD and MULLET 1981). Because of its peculiar polarization pattern (Fig. 1), previously encountered in bacterial reaction centers, it was proposed to be the triplet state of P-700, formed through charge recombination of the primary biradical. A detailed optical and EPR study on CP 1 particles permitted identification of ^{3}P-700 and determination of its decay rate at low temperature (SÉTIF et al. 1981, 1982, GAST et al. 1983). At 10 K, the yield of ^{3}P-700 formation was estimated to be about 0.6 from optical studies (SÉTIF et al. 1981) in contrast with a value of 0.05 obtained by EPR (GAST et al. 1983). This discrepancy has not yet been resolved. Detailed EPR studies on Triton PS I particles correlated the amount of light-induced ^{3}P-700 with the initial redox state of the primary acceptors (GAST et al. 1983, BONNERJEA and EVANS 1982). Based on these findings the existence of two early acceptors named A_0 ($g \simeq 2.002$, $\Delta H = 11.5$ G) and A_1 ($g \simeq 2.005$, $\Delta H \simeq 10.5$ G) was postulated. ^{3}P-700 is probably formed through charge recombination between P-700$^+$ and A_0^-.

At low temperature (5–30 K) a 120 µs charge recombination involving P-700$^+$ was observed under mildly reducing conditions (MATHIS and CONJEAUD 1979). When F_A and F_B are pre-reduced, this 120 µs phase is the dominant process over the slow (50–800 ms) relaxation ascribed to the state (P-700$^+$, X^-): 70–80% versus 10–15% (SÉTIF et al. 1984). When X is prereduced, the 120 μs phase is replaced by a 20 μs decay, with a similar difference spectrum in the red (Fig. 1) (SÉTIF and MATHIS unpublished). We ascribe the 120 μs phase to the decay of (P-700$^+$, A_1^-). The recombination lifetime of this biradical has not yet been determined at room temperature. The lifetime of the biradical (P-700$^+$, A_0^-) may vary from 200 ps (normal conditions) to the nanosecond range when forward electron transfer is inhibited (SHUVALOV et al. 1979, SONNEVELD et al. 1981). The chemical nature of A_0 and A_1 is still unknown. A_0 is often considered to be a Chl a. This is based on the low reduction potential of Chl a (−0.88 V, FUJITA et al. 1978) and on the g-value of A_0^-, similar to that of Chl a. These arguments are rather weak. Absorption experiments could be interpreted similarly, although maximum beachings were reported around 700 nm (SHUVALOV et al. 1979) or 670 nm (SWARTHOFF et al. 1982) and that a correlation with A_0 reduction has not been clearly shown. A_1 is certainly not a Chl a, as shown by the high value of the g factor (2.005) and by the absence of absorption bleaching in the red when A_1 is photoreduced (SÉTIF et al. 1984).

In time-resolved EPR experiments, spin polarization in PS I (CIDEP) was first interpreted in terms of interaction between P-700$^+$ and two reduced acceptors: F_X^- and a Chl a^- (refs in FURRER and THURNAUER 1983). These authors detected room temperature CIDEP transients at the X- and K-bands and showed the participation of two different radicals, probably P-700$^+$ and A_1^-, in a polarized spectrum which resembles low-temperature CIDEP spectra obtained with F_A and F_B reduced, and attributed to P-700$^+$ and A_1^- (GAST et al. 1983, MCCRACKEN and SAUER 1983) as suggested by MCINTOSH et al. (1979). The older data should be re-evaluated in the more recent context of primary partners (P-700, A_0, A_1) and of heterogeneity of secondary acceptors (F_X, F_B, F_A) (see HOFF 1984).

4 Structural Aspects

4.1 Electron Carriers and Polypeptides

Our description so far has been concerned with redox centers. These are held together by hydrophobic proteins that insure a favorable positioning for efficient primary charge separation and electron transfer. The treatment of photosynthetic membranes with SDS permits isolation of a complex (CP I) with about 50 Chl a, and whose activity is restricted to P-700 and A_0. It contains only two polypeptides of 60–70 kDa per one P-700 (BENGIS and NELSON 1977, LAGOUTTE et al. 1981). In some experiments the two polypeptides appear to be slightly different (VIERLING and ALBERTE 1983, WOLLMAN, Chap. 9.7, this Vol.). The site of insertion of P-700 and A_0 on the CP I protein is still unknown.

Most PS I particles prepared without SDS contain the bound iron-sulfur centers (F_A, F_B, F_X). They also contain small polypeptides of 8 to 25 kDa, some of which are of the usual size for iron-sulfur proteins (6–12 kDa; YASUNOBU and TANAKA 1973). Based on labeling experiments with ^{59}Fe or ^{35}S, iron sulfur centers have been proposed to be located on polypeptides of MW about 8 kDa (GUIKEMA and SHERMAN 1982, LAGOUTTE et al. 1984). This hypothesis is reinforced by the observation that, in PS I particles, all cysteine residues belong to the 8 kDa and 60–70 kDa proteins (LAGOUTTE et al. 1984). These assignments will be easier to make when a method becomes available for selectively dissociating the various iron-sulfur proteins.

4.2 Geometrical Aspects

By spectroscopy with polarized light on oriented membranes, it was found that the 700 nm absorption transition of P-700 lies nearly in the plane of the thylakoid membrane, while a 660 nm transition is tilted by 50° (BRETON and VERMEGLIO, 1982, GEACINTOV and BRETON, Chap. 7.3; BRETON, Chap. 7.4, this Vol.). Photoselection experiments led to a similar conclusion for the 700 nm band and also showed that some β-carotene molecules are very precisely oriented with respect to P-700 (SCHAFFERNICHT and JUNGE 1981). The secondary acceptors F_A, F_B, F_X are oriented with respect to the membrane plane, as shown by EPR of oriented thylakoids (DISMUKES and SAUER 1978, PRINCE et al. 1980, AASA et al. 1981, HOOTKINS and BEARDEN 1983). Orientation data imply that the studied redox centers are firmly positioned in the RC, and that the RC has a well-defined position with respect to the plane of the membrane.

Distances between redox centers have not been determined. Some numbers have been derived from rates of back-reaction, but they must be taken with caution.

4.3 Functional Accessibility

The mechanism of electron transfer in and out of the PS I center is still a matter of debate. $P\text{-}700^+$ is probably re-reduced directly by plastocyanin (HAEHNEL, Chap. 10.5, this Vol.). The requirement for a small polypeptide is still disputed (HAEHNEL et al. 1980, BHARDWAJ et al. 1982). Ferredoxin is generally thought to accept electrons from the terminal acceptor of PS I, but its mode of interaction with the terminal bound acceptor(s) has not been elucidated. In thylakoids, the sites for electron donation to P-700 and for ferredoxin reduction are located at the internal (lumenal) and external membrane surfaces, respectively.

5 Conclusion

Our understanding of the PS I RC is hampered by three major factors: (1) the large number of antenna pigments associated with one P-700, (2) the low redox potential of the acceptors, which adds uncertainty to equilibrium redox potentiometry measurements, and (3) the integration of the secondary acceptors into one complex that is difficult to dissociate. All of these properties, however, appear to have been well preserved in the evolution of oxygenic photosynthetic organisms; furthermore, no mutants have yet been discovered that would greatly help advance research on the PS I reaction center.

References

Aasa R, Bergström J, Vänngård T (1981) Biochim Biophys Acta 637:118–123
Bengis C, Nelson N (1977) J Biol Chem 252:4564–4569
Bhardwaj R, Burkey KO, Tripathy BC, Bross EL (1982) Plant Physiol 70:424–429
Blanken HJ den, Hoff AJ (1983) Biochim Biophys Acta 724:52–61
Bonnerjea J, Evans MCW (1982) FEBS Lett 148:313–316
Boxer SG (1983) Biochim Biophys Acta 726:265–292
Breton J, Vermeglio A (1982) In: Govindjee (ed) Photosynthesis: Energy conversion by plants and bacteria, vol I. Academic Press, London New York, pp 153–194
Cammack R, Ryan MD, Stewart AC (1979) FEBS Lett 107:422–426
Chamorovsky SK, Cammack R (1982a) Biochim Biophys Acta 679:146–155
Chamorovsky SK, Cammack R (1982b) Photobiochem Photobiophys 4:195–200
Crowder M, Bearden A (1983) Biochim Biophys Acta 722:23–35
Dismukes GC, Sauer K (1978) Biochim Biophys Acta 504:431–445
Dörnemann D, Senger H (1982) Photochem Photobiol 35:821–826
Evans MCW (1982) In: Spiro (ed) Iron-sulfur proteins, vol IV. Wiley, New York, pp 249–284
Evans MCW, Heathcote P (1980) Biochim Biophys Acta 590:89–96
Frank HA, McLean M, Sauer K (1979) Proc Natl Acad Sci USA 76:5124–5128
Fujita I, Davis MS, Fajer J (1978) J Am Chem Soc 100:6280–6282
Furrer R, Thurnauer MC (1983) FEBS Lett 153:399–403
Gast P, Swarthoff T, Ebskamp FCR, Hoff AJ (1983) Biochim Biophys Acta 722:163–175
Golbeck JH, Warden JT (1982) Biochim Biophys Acta 681:77–84
Golbeck JH, Velthuys BR, Kok B (1978) Biochim Biophys Acta 504:226–230
Guikema J, Sherman L (1982) Biochim Biophys Acta 681:440–450
Haehnel W, Hesse V, Pröpper A (1980) FEBS Lett 111:79–82
Hiyama T, Fork DC (1980) Arch Biochem Biophys 199:488–496
Hiyama T, Ke B (1972) Biochim Biophys Acta 267:160–171
Hoff AJ (1979) Phys Rep 54:75–200
Hoff AJ (1982) Biophys Struct Mech 8:107–150
Hoff AJ (1984) Q Rev Biophys 17:153–282
Hootkins R, Bearden A (1983) Biochim Biophys Acta 723:16–29
Hootkins R, Malkin R, Bearden A (1981) FEBS Lett 123:229–234
Katz JJ, Norris JR, Shipman LL, Thurnauer MC, Wasielewski MR (1978) Annu Rev Biophys Bioeng 7:393–434
Ke B (1973) Biochim Biophys Acta 301:1–33
Koike H, Katoh S (1982) Photochem Photobiol 35:527–531
Lagoutte B, Sétif P, Duranton J (1981) In: Akoyunoglou G (ed) Proc 5th Int Congr Photosynth, vol III. Balaban, Philadelphia, pp 237–243
Lagoutte B, Sétif P, Duranton J (1984) FEBS Lett 174:24–29
Malkin R (1982) Annu Rev Plant Physiol 33:455–479
Malkin R (1984) Biochim Biophys Acta 764:63–69
Mathis P, Conjeaud H (1979) Photochem Photobiol 29:833–837

Mathis P, Paillotin G (1981) In: Hatch MD, Boardman NK (eds) The biochemistry of plants, vol VIII. Academic Press, London New York, pp 97–161
Mathis P, Sauer K, Remy R (1978) FEBS Lett 88:275–278
McCracken JL, Sauer K (1983) Biochim Biophys Acta 724:83–93
McIntosh AR, Manikowski H, Bolton JR (1979) J Phys Chem 83:3309–3313
Norris JR, Uphaus RA, Crespi HL, Katz JJ (1971) Proc Natl Acad Sci USA 68:625–628
Norris JR, Scheer H, Druyan ME, Katz JJ (1974) Proc Natl Acad Sci USA 71:4897–4900
Nugent JHA, Møller BL, Evans MCW (1981) Biochim Biophys Acta 634:249–255
O'Malley PJ, Babcock GT (1984) Proc Natl Acad Sci USA 81:1098–1101
Philipson KD, Sato VL, Sauer K (1972) Biochemistry 11:4591–4595
Prince RC, Crowder MS, Bearden AJ (1980) Biochim Biophys Acta 592:323–337
Rose KA, Bearden A (1980) Biochim Biophys Acta 593:342–352
Rutherford AW, Mullet JE (1981) Biochim Biophys Acta 635:225–235
Sauer K, Mathis P, Acker S, Best JA van (1978) Biochim Biophys Acta 503:120–134
Schaffernicht H, Junge W (1981) Photochem Photobiol 34:223–232
Scheer H, Katz JJ, Norris JR (1977) J Am Chem Soc 99:1372–1381
Scheer H, Wieschoff H, Schaefer W, Cmiel E, Nitsche B, Schiebel HM, Schulten HR (1983) In: Sybesma C (ed) Proc 6th Int Congr Photosynth, Brussels, vol II. Nijhoff/Junk, The Hague, pp 81–84
Sétif P, Mathis P (1980) Arch Biochem Biophys 204:477–485
Sétif P, Hervo G, Mathis P (1981) Biochim Biophys Acta 638:257–267
Sétif P, Quaegebeur JP, Mathis P (1982) Biochim Biophys Acta 681:345–353
Sétif P, Mathis P, Lagoutte B, Duranton J (1983) In: Sybesma C (ed) Proc 6th Int Congr Photosynth, Brussels, vol I. Nijhoff/Junk, The Hague, pp 589–591
Sétif P, Mathis P, Vänngård T (1984) Biochim Biophys Acta 767:404–414
Shuvalov VA, Ke B, Dolan E (1979) FEBS Lett 100:5–8
Sonneveld A, Duysens LNM, Moerdijk A (1981) Biochim Biophys Acta 636:39–49
Swarthoff T, Gast P, Amesz J, Buisman HP (1982) FEBS Lett 146:129–132
Thurnauer MC, Norris JR (1980) Chem Phys Lett 76:557–561
Thurnauer MC, Rutherford AW, Norris JR (1982) Biochim Biophys Acta 682:332–338
Vierling E, Alberte RS (1983) Plant Physiol 72:625–633
Wasielewski MR, Norris JR, Crespi HL, Harper J (1981a) J Am Chem Soc 103:7664–7665
Wasielewski MR, Norris JR, Shipman LL, Lin CP, Svec WA (1981b) Proc Natl Acad Sci USA 78:2957–2961
Yasunobu KT, Tanaka M (1973) In: Lovenberg W (ed) Iron-sulfur proteins, vol II. Academic Press, London New York, pp 27–130

9.7 Photosystem I Proteins[1]

F.-A. WOLLMAN

1 Introduction

Photosystem I (PS I) was first defined by KOK (1956), who noted that both algae and plants exhibited a light-induced absorption change at 705 nm, which was correlated with the formation of strong reductants (reduced ferredoxins) and the concurrent oxidation of P-700, the primary donor of this photosystem. Several electron transport components are involved in this process, namely P-700, the early acceptors A1 (attributed to chlorophyll a; SHUVALOV et al. 1979, HEATHCOTE and EVANS 1980) and A2 (chemical nature undetermined), and two bound iron-sulfur centers A and B which are both of the 4 Fe–4 S type (CAMMACK and EVANS 1975). All of these compounds are organized within protein-containing complexes located in the thylakoid membranes. A detailed review of their functional properties can be found in Chapter 9.6, this Volume by SÉTIF and MATHIS.

Biochemical and genetic analysis of chloroplasts has revealed two types of protein complex in photosystem I: the photosystem I reaction center protein complex (RC I protein complex), and several additional proteins which give full competence to PS I in a mature chloroplast. Among the latter are proteins that elicit electron flow to and from the reaction center, including plastocyanin (HAEHNEL, Chap. 10.5, this Vol.), soluble ferredoxin, and ferredoxin NADP reductase, which will not be covered in the present review. In addition, recent advances in the field of PS I research have led to the characterization of a PS I peripheral antenna system, named either LHC I or CP 0 depending on the species and/or laboratories (MULLET et al. 1981, WOLLMAN and BENNOUN 1982), and which I will refer to as the LHC I protein complex.

As a preliminary comment, it should be noted that the reported number and molecular weights of the subunits of the RC I and LHC I protein complexes varies significantly. This is largely due to differences in the electrophoresis conditions used to separate the subunits. Thus the number and molecular weights of the subunits of the PS I complex should be taken as approximations only.

Other reviews which include a discussion of PS I complexes have been published recently (MALKIN 1982, ARNTZEN and KAPLAN 1983, OKAMURA et al. 1983, GLAZER 1983). The reader is also referred to the following articles in this Volume: STAEHELIN, Chap. 1; THORNBER, Chap. 3; ORT, Chap. 4; SÉTIF and MATHIS, Chap. 9.6; SIMPSON, Chap. 11.6; SIEGENTHALER and RAWYLER, Chap. 11.9).

1 This article is dedicated to the memory of Bernadette Bouges-Bocquet

2 The RC I Protein Complex

The subunits of the RC I protein complex are postulated to be involved in the formation of the site at which a stable charge separation occurs, and must therefore include the electron transport components ranging from P-700 to the iron sulfur centers A and B. In addition, they are essential for the functional stability of PS I in the thylakoid membrane. By comparing the polypeptide composition of purified PS I complexes from the green alga *Chlamydomonas reinhardtii* with the polypeptides lacking in PS I mutants. GIRARD et al. (1980) have been able to define the polypeptide subunits of the RC I complex. These include a large subunit of 66 kDa and several subunits of lower molecular weight, ranging from 4 to 20 kDa. A similar set of polypeptides has been shown to be absent from PS I mutants from barley (HILLER et al. 1980) and to comprise the PS I complexes isolated from higher plants (BENGIS and NELSON 1977, MULLET et al. 1980a, NECHUSHTAI et al. 1981, WESTHOFF et al. 1983, DUNAHAFY and STAEHELIN 1985) and cyanobacteria (TAKAHASHI et al. 1982, NECHUSHTAI et al. 1983, WILLIAMS et al. 1983).

The large subunit of the RC I complex is the apoprotein of the chlorophyll–protein complex I (CP I) which is readily purified from intact thylakoids through sodium-dodecylsulfate polyacrylamide gel electrophoresis (SDS-PAGE) (THORNBER et al. 1979). Its amino acid composition is remarkably similar in different species (THORNBER 1969, NELSON and NOTSANI 1977, SÉTIF et al. 1980, VIERLING and ALBERTE 1983a). P-700 and the acceptor A1 are localized within CP I (MATHIS et al. 1978) in a stoechiometric ratio of about one P-700 per 40 to 60 chlorophyll a molecules as opposed to less than 30 chlorophylls per CP I apoprotein (THORNBER et al. 1979, SÉTIF et al. 1980, VIERLING and ALBERTE 1983a). These differences in stoichiometric ratios may be accounted for by either partial denaturation of P-700 in the isolated CP I or by the existence of more than one copy of the apoprotein per P-700. This aspect has been further discussed in the context of the observation that the apoprotein of CP I does not migrate as a single band but rather as two bands, the separation of which depends on the electrophoretic conditions (CHUA et al. 1975, ANDERSON and LEVINE 1974). The question has then been raised as to whether this is due to proteolytic degradation or whether two subunits of different electrophoretic mobility exist in CP I. The two bands from either *C. reinhardtii* (NECHUSHTAI and NELSON 1981) or barley (VIERLING and ALBERTE 1983a) yield similar peptide maps upon proteolytic digestion or CNBr cleavage. However, no interconversion of one form to another was observed by VIERLING and ALBERTE (1983), which led them to conclude that two conformational variants of a single polypeptide exist in CP I. Some posttranslational modification of the apoprotein other than glycosylation (KEEGSTRA and CLINE 1982) or phosphorylation (BENNETT 1977), must be considered.

A stoichiometry of two copies of the large subunit per copy of other small subunits has been reported for several RC I complexes (BENGIS and NELSON 1977, NECHUSHTAI and NELSON 1981, DE VITRY et al. 1983). More than 95% of the large subunit can be converted to a 140 kDa polypeptide upon treatment of the RC I complex from Swiss chard with a chemical cross-linker (OKAMURA et al. 1983). These experiments suggest that the RC I protein complex contains a dimer of the large subunit. This view has been challenged by VIERLING and ALBERTE (1983), who favor a tetrameric organization of the large subunit in the RC I complex. However, the recent report by IKEGAMI and KE (1984a) of a RC I protein, that consists of five small subunits associated with the large P-700-containing subunit, and which exhibits an overall ratio of one P-700 per 160 kDa of protein, does not support the conclusion of VIERLING and ALBERTE.

The knowledge of the specific functions of the other RC I subunits is still rather poor. Involvment of subunit III. MW 20 kDa, of the RC I complex in electron donation of plastocyanin to P-700 has been described in Swiss chard and spinach (BENGIS and NELSON 1977, HAEHNEL et al. 1980). However, in the absence of the 20 kDa subunit a significant amount of electron transfer from plastocyanin to P-700 can still be observed (BHARDWAJ et al., TAKABE et al. 1983). The polypeptides that contain the iron-sulfur centers A and B have not yet been positively identified. Disappearance of the EPR signals corresponding to the centers was observed upon depletion of three or four subunits from the RC I protein complex in Swiss chard (BENGIS and NELSON 1977) and pea (RUTHERFORD and MULLET 1981), respectively. ^{59}Fe incorporation into thylakoid membrane polypeptides has also been used to localize putative apoproteins of the Fe–S centers. The corresponding PS I polypeptides were reported to be 18 kDa and 15 kDa in barley (MOLLER et al. 1981) but less than 10 kDa in *Anacystis* (GUIKEMA and SHERMAN 1982). Consistent with the latter identification is a report by LAGOUTTE et al. (1984), who show that an 8 kDa subunit of the RC I complex from spinach is highly enriched in cystein residues. A recent report on the selective action of diazonium benzene sulfonate (DABS) on the B centers (MALKIN 1984) opens the way to the identification of its apoprotein using radioactive DABS.

Chlorophyll a and β-carotene are associated with the RC I complex and are located in CP I. The ratio of chlorophylls to P-700 in RC I protein complex is the same as in the isolated CP I. This ratio can, however, be decreased below 40 upon further detergent treatment of the isolated complex (BENGIS and NELSON 1977, MULLET et al. 1980a) reaching about seven chlorophylls per P-700 upon solvent extraction (IKEGAMI and KATOH 1975). A β-carotene–Chl a interaction is responsible for the occurrence of far-red absorbing and emitting chlorophyll a components in the PS I reaction center (OQUIST et al. 1980). The fluorescence emission at 77 K of the RC I protein is characterized by a peak in the 717–722 nm region (MULLET et al. 1980a, TAPIE et al. 1984). Further pigment depletion reveals emission peaks at shorter wavelengths, 694–705 nm, the amplitude of which depends on the redox state of the reaction center (MULLET et al. 1980a, IKEGAMI and KE 1984b).

3 The LHC I Protein Complex

The existence of a Chl a/b protein complex, mainly serving as a PS II peripheral antenna in green algae and higher plants, was first reported by THORNBER et al. (1967). Throughout the 1970's a similar antenna system was postulated for

PS I (THORNBER et al. 1979, BOARDMAN and ANDERSON 1978), but only in 1980 did the first report on the specific involvment of several thylakoid polypeptides in the organization of a PS I peripheral antenna appear (MULLET et al. 1980a, b). Two lines of evidence supported this idea: a PS I complex isolated from pea thylakoids and containing 110 chlorophylls per P-700, the PS I 110 particle, can be converted into a RC I protein complex containing 65 ChL/P-700 by Triton extraction of several 22.5–24.5 kDa polypeptides: paralleling this loss of chlorophylls and polypeptides is a change in the 77 K emission peak from 730 nm to 717 nm. On the other hand, the same set of 21.5–24.5 kDa polypeptides is lacking in intermittent light-grown cucumber cotelydons and in a Chl b-less barley mutant which both show no 730 nm emission (at variance with the corresponding normal chloroplasts). Further studies undertaken in the same laboratory led to the purification of LHC I using PS I 110 complexes treated with Triton (MULLET et al. 1981) or a combination of Zwittergent 16 and dodecyl β-D-maltoside (HAWORTH et al. 1983). The resulting LHC I complex contains two polypeptide doublets in the 23 kDa and 21 kDa regions and a fifth component of lower molecular weight, about 10 kDa.

Independent evidence for the existence of a LHC I protein complex was reported in *C. reinhardtii* (WOLLMAN and BENNOUN 1982). It consists of five subunits – MW: 27.5, 27, 25, 23, 18 kDa – which display a 2:4:1:2:3 stoichiometry after isolation of the protein by SDS-PAGE (DE VITRY et al. 1983). The five LHC I subunits are lacking in the AC 40 mutant of *C. reinhardtii* (WOLLMAN and BENNOUN 1982) which has a PS I antenna of reduced size (OLIVE et al. 1983). According to ISH-SHALOM and OHAD (1983) the 27 kDa subunit is selectively retained in the Y-Ip mutant of *C. reinhardtii*. Thus, this subunit might constitute an independent chlorophyll binding protein which, upon association with the four other subunits of LHC I, provides a suitable organization for the efficient transfer of energy between the peripheral antenna (27 kDa subunit) and the reaction center in PS I. The number of antenna chlorophylls associated with the LHC I protein complex in a PS I unit is 50 and 80 in pea and *C. reinhardtii*, respectively (MULLET et al. 1980a, WOLLMAN and BENNOUN 1982). Chlorophyll b constitutes a significant part of the pigments associated with LHC I (ANDERSON et al. 1983). Reported Chl a/Chl b ratios vary from 3.7 to 6.3 (WOLLMAN and BENNOUN 1982, HAWORTH et al. 1983, METZ et al. 1984).

Whereas the RC I protein complex and its chlorophyll-binding subunit, CP I, seem ubiquitous among photosynthetic organisms evolving oxygen (THORNBER, Chap. 3, this Vol.), variations in spectral properties of LHC I from different organisms suggest that green algae and higher plants have evolved different peripheral antennae for PS I. For instance corresponding emission peaks at 77 K are at 730–735 nm for pea (HAWORTH et al. 1983) and at 707 nm for *C. reinhardtii* (WOLLMAN and BENNOUN 1982). In a comparative study of the orientation of the different holochromes in LHC I from the two sources, TAPIE et al. (1984) show that they display the same general organization except for a small number of pigment molecules, C705, that appears to be responsible for the 730 nm emission typical of LHC I from higher plants. Thus the folding of the protein subunits must be rather similar in both cases, and it is likely that sequence homologies between subunits of the two LHC I proteins will be observed.

Yet another difference relates to the fact that LHC I from *C. reinhardtii* thylakoids migrates as a chlorophyll-protein complex, CP 0, of lower mobility than CP I upon SDS-PAGE at 4 °C (WOLLMAN and BENNOUN 1982). This is not the case for higher plant LHC I. A CP Ia complex, which consists of CP I plus several other subunits (ANDERSON et al. 1978, REMY and HOARAU 1978) is, however, observed in both systems. CP Ia has the same polypeptide composition as the PS I 110 particles in higher plants (ANDERSON et al. 1983, KUANG et al. 1984) and thus corresponds to the association of LHC I and RC I protein complexes in the form of a PS I photosynthetic unit. A similar observation was made on *C. reinhardtii* by ISH-SHALOM and OHAD (1983), although the presence of the small subunits of the RC I protein in CP Ia was not reported by these authors.

Recent studies on higher plant thylakoids have finally revealed the presence of LHC I protein subunits in chlorophyll–protein complexes separated by electrophoresis (KUANG et al. 1984, LAM et al. 1984, METZ et al. 1984, REMY and AMBART-BRETEVILLE 1984). According to METZ et al. (1984) such a chlorophyll–protein complex has the same electrophoretic mobility as CP a_2 (CP 43; a major chlorophyll–protein complex in the RC II protein) when maize thylakoids are treated with lithium dodecylsulfate (LDS). Similarly, DUNAHAY and STAEHELIN (1985) have isolated a green band from spinach, CP I^{LH}, which contains the LHC I polypeptides and connigrates with CP43 in octylglucoside/SDS gels. On the other hand, KUANG et al. (1984) and REMY and AMBART-BRETTEVILLE (1984) show that LHC I complexes, separated from PS I complexes in pea thylakoids, migrate in the position of LHCP2 in the presence of SDS (LHCP2 in the presence of SDS (LHCP2 and LHCP1 are oligomeric forms of LHCP3, the main Chl a/b protein associated with PS II peripheral antenna (ANDERSON et al. 1978, ANDERSON 1980). None of these LHC I complexes isolated electrophoretically contains the five subunits of the LHC I protein as defined by HAWORTH et al. (1983). The reported numbers and molecular weights of the LHC I apoproteins differ widely, ranging from 4 subunits of 24–22.5 and 21–20.5 kDa (KUANG et al. 1984) to 3 subunits in the 25–26 kDa region (METZ et al. 1984, DUNAHAY and STAEHELIN 1985), or 2 subunits of 24 and 29 kDa (REMY and AMBART-BRETTEVILLE 1984) or 1 subunit of 20 kDa (LAM et al. 1984). Whereas most of the 77 K fluorescence of the LHC I complex isolated by KUANG et al. (1984) is emitted in the 725–735 nm region, an additional emission at 681 nm is observed by REMY and AMBART-BRETTEVILLE (1984), which is the only emission peak remaining in the LHC I complex isolated by LAM et al. (1984). This might be due to an increasing modification of the pigment–protein interaction in LHC I as the number of protein subunits retained in the LHC I complex decreases.

An additional similarity between LHC I complexes in higher plants and in green algae arises from the electrophoretic comigration of LHC I and LHCP2 recently described in pea thylakoids: a similar comigration had been already described in *C. reinhardtii* which is responsible for the apoprotein profile of the CP II band (DELEPELAIRE and CHUA 1981, WOLLMAN and BENNOUN 1982).

4 Architecture of the PS I Protein Complexes

Based on aggregation properties of the subunits of the RC I protein complex upon heating. NELSON and NOTSANI (1977) concluded that the plastocyanin receptor subunit (MW: 20 kDa) and the 16 kDa subunit, associated with the reducing side of the RC I protein, are the closest neighbors of the P-700-containing subunit. The latter subunit protrudes at the external surface of the thylakoid (BENGIS and NELSON 1975) and is the only integral membrane polypeptide among the seven subunits of the RC I protein in *C. reinhardtii* (DELEPELAIRE 1983). Freeze-fracture studies of PS I mutant thylakoids have shown that the RC I protein is located in stroma thylakoid PF particles with diameters greater than 105 Å both in higher plants (MILLER 1980, SIMPSON, Chap. 11.6) and in green algae (OLIVE et al. 1983, LACAMBRA et al. 1984). In the latter case the large size of the PFu particles was attributed to the association of the RC I and LHC I proteins in a same particle, each protein corresponding to particles 80 Å in diameter. These results are consistent with the previous report on PS I 110 from pea, which yield particles with an average diameter of 106 Å in freeze-fractured reconstituted liposomes (MULLET et al. 1980a, DUNAHAY and STAEHELIN 1985). A more detailed discussion of PS I complexes and freeze-fracture particles can be found in Chapter 1 of STAEHELIN.

A cyanobacterial RC I protein complex has been characterized by negative staining (WILLIAMS et al. 1983). It appears as a prolate ellipsoid, the dimensions of which, 18 × 8 nm, require further investigation to be reconciled with the freeze-fracture data. Detailed morphological studies of purified LHC I and RC I proteins are needed to obtain more conclusive information on the architecture of PS I complexes.

5 Site of Synthesis and Assembly of Subunits of the PS I Protein Complexes

The large subunits of the RC I protein complex is encoded in the chloroplast genome and translated on chloroplast ribosomes as substantiated by several studies using site-specific inhibitors (CHUA and GILLHAM 1977, NECHUSHTAI and NELSON 1981, DELEPELAIRE 1983) or protein synthesis in intact chloroplasts (GREEN 1980, ZIELINSKI and PRICE 1980). The structural gene of this subunit has been recently mapped on the chloroplast genome from spinach (WESTHOFF et al. 1983). The latter study demonstrates that it exists as a single copy and is located on the plastid chromosome close to the gene coding for the 44 kDa chlorophyll-binding polypeptide of PS II.

The LHC I subunits, and all but one of the small subunits of the RC I protein complex are translated on cytoplasmic ribosomes and encoded in the nuclear genome of *C. reinhardtii* (NECHUSHTAI and NELSON 1981, DELEPELAIRE 1983). This is at variance with a report on the RC I protein of *Spirodela*, in which at least two of the small subunits were found to be chloroplast translates

(NECHUSHTAI et al. 1981). It has been established that the RC I protein subunit II from spinach, MW: 22 kDa, which is immunologically related to the 25, 19 and 16 kDa subunits of the RC I protein from Swiss chard. *C. reinhardtii* and *M. laminosus* respectively, is synthesized on cytoplasmic ribosomes as a precursor of 26 kDa (WESTHOFF et al. 1983).

Insertion of the LHC I and RC I protein complexes into the thylakoid membrane occurs independently. This is shown by (1) the existence of *C. reinhardtii* mutants lacking either one of the two complexes but containing normal amounts of the other protein complex (WOLLMAN and BENNOUN 1982), (2) the insertion of the RC I protein complex but not of the LHC I protein complex in intermittent light-grown seedlings (MULLET et al. 1980b), and (3) the insertion of the LHC I protein complex but not of the RC I protein complex upon greening of the Y-1p mutant strain of *C. reinhardtii* in the presence of chloramphenicol which inhibits chloroplast translation (GERSHONI et al. 1982).

The formation of photosystem I units during thylakoid biogenesis has been studied by several groups (see for instance, BAR NUN et al. 1977, HILLER et al. 1978, MULLET et al. 1980b, GERSHONI et al. 1982). Light triggers the synthesis of the large RC I subunit and both P-700 activity and CP I formation can be observed within 1 h of the onset of illumination (VIERLING and ALBERTE 1983b) although some P-700 activity may occur in the absence of detectable CP I (GERSHONI et al. 1982). A sequential model of formation of photosystem I units has been proposed by the latter authors. Accumulation of CP I in thylakoid membranes is still observed for some time upon return to darkness (VIERLING and ALBERTE 1983b). As pointed out by these authors, this is consistent with the continuing synthesis of the RC I protein in intermittent light-grown seedlings (DAVIS et al. 1976), and might be explained by accumulation of chlorophyll during the time of illumination. The mode of assembly of the subunits in each PS I complex has still to be worked out. GIRARD et al. (1980) showed that at least 13 complementation groups scattered throughout the nuclear genome contain genes responsible for either the synthesis, regulation, transport, or assembly of the RC I subunits. These authors report that the same set of polypeptides is lacking in PS I mutants of either chloroplastic or nuclear origin, which supports the idea of a concerted insertion of the RC I subunits into the thylakoid membrane. This view seems conflicting with reports on the formation of CP I in isolated intact chloroplasts, or with the detection of unassembled RC I subunits at the thylakoid membrane level in the presence of site-specific inhibitors. However, the former case could be explained by preceding accumulation or recycling of cytoplasmic translation products in the plastid. On the other hand, detection of labeled polypeptides in the thylakoid membrane fraction in the presence of site-specific inhibitors by no means implies their accumulation to a normal level in the plastid membrane. It was suggested that there is no tight coupling in the expression of nuclear and chloroplast genes which code for different subunits of several functional chloroplast membrane complexes (ELLIS 1981). A post-translational regulation, involving rapid degradation of the unassembled RC I protein subunits must then be considered. Turnover analysis of unassembled polypeptides should be able to provide new information on the regulation of thylakoid protein insertion.

Acknowledgments. The author whishes to thank P. Delepelaire for stimulating discussions during the preparation of this review.

References

Anderson JM (1980) Biochim Biophys Acta 591:113–126
Anderson JM, Levine RP (1974) Biochim Biophys Acta 357:118–126
Anderson JM, Waldron JC, Thorne SW (1978) FEBS Lett 92:227–233
Anderson JM, Brown JS, Lam E, Malkin R (1983) Photochem Photobiol 98:205–210
Arntzen CJ, Kaplan R (1983) In: Govindjee (ed) Photosynthesis: Energy conversion by plants and bacteria. Academic Press, London New York, pp 65–151
Bar Nun S, Schantz R, Ohad I (1977) Biochim Biophys Acta 459:451–467
Bengis C, Nelson N (1975) J Biol Chem 2783–2788
Bengis C, Nelson N (1977) J Biol Chem 252:4564–4569
Bennet J (1977) Nature 269:344–346
Bhardwaj R, Burkey KO, Baishnab C, Tripathy BC, Gross EL (1982) Plant Physiol 70:424–429
Boardman NK, Anderson JM (1978) In: Akoyunoglou G, Argyroudi-Akoyunoglou JH (eds) Chloroplast development. Elsevier, Amsterdam New York Oxford, pp 1–14
Cammack R, Evans MCW (1975) Biochim Biophys Res Commun 67:544–549
Chua NH, Gillham NW (1977) J Cell Biol 74:441–452
Chua NH, Matlin K, Bennoun P (1975) J Cell Biol 67:361–377
Davis DJ, Armond PA, Gross E, Arntzen CJ (1976) Arch Biochem Biophys 175:64–70
Delepelaire P (1983) Photobiochem Photobiophys 6:279–291
Delepelaire P, Chua NH (1981) J Biol Chem 256:9300–9307
Dunahay TG, Staehelin LA (1985) Plant Physiol in press
Ellis RJ (1981) Annu Rev Plant Physiol 32:111–137
Gershoni JM, Schochat S, Malkin S, Ohad I (1982) Plant Physiol 70:637–644
Girard J, Chua NH, Bennoun P, Schmidt G, Delosme M (1980) Curr Genet 2:215–221
Glazer AN (1983) Annu Rev Plant Physiol 52:125–157
Green BR (1980) Biochim Biophys Acta 609:107–120
Guikema J, Sherman L (1982) Biochim Biophys Acta 681:440–450
Haehnel W, Propper A, Krause H (1980) FEBS Lett 111:79–82
Haworth P, Watson JL, Arntzen CJ (1983) Biochim Biophys Acta 724:151–158
Heathcote P, Evans MCW (1980) FEBS Lett 111:381–385
Hiller RG, Pilger TBG, Genge S (1978) In: Akoyunoglou G, Argyroudy-Akoyunoglou JH (eds) Chloroplast development. Elsevier, Amsterdam New York Oxford, pp 215–220
Hiller RG, Lindberg-Moller B, Hoyer-Hansen G (1980) Carlsberg Res Commun 45:315–318
Ikegami I, Katoh S (1975) Biochim Biophys Acta 376:588–592
Ikegami I, Ke B (1984a) Biochim Biophys Acta 764:70–79
Ikegami I, Ke B (1984b) Biochim Biophys Acta 764:80–85
Ish-Shalom D, Ohad I (1983) Biochim Biophys Acta 722:498–507
Keegstra K, Cline K (1982) Plant Physiol 70:232–237
Kok B (1956) Biochim Biophys Acta 22:399–401
Kuang TY, Argyroudy-Akoyunglou JH, Nakatani HY, Watson J, Arntzen CJ (1984) Arch Biochem Biophys 235:618–627
Lacambra M, Larsen U, Olive J, Bennoun P, Wollman FA (1984) Photobiochem Photobiophys 8:191–205
Lagoutte B, Sétif P, Duranton J (1984) FEBS Lett 174:24–29
Lam E, Oritz W, Mayfield M, Malkin R (1984) Plant Physiology 74:650–665
Malkin R (1982) Annu Rev Plant Physiol 33:455–479

Malkin R (1984) Biochim Biophys Acta 764:63–69
Mathis P, Sauer K, Remy R (1978) FEBS Lett 88:2575–2578
Metz JG, Krueger RW, Miles D (1984) Plant Physiol 75:238–241
Miller KR (1980) Biochim Biophys Acta 592:143–152
Moller BL, Hoyer Hansen G, Hiller RG (1981) In: Akoyunoglou G (ed) Photosynthesis: Structure and molecular organization of the photosynthetic membrane. Balaban, Philadelphia, pp 245–256
Mullet JE, Burke JJ, Arntzen CJ (1980a) Plant Physiol 65:814–822
Mullet JE, Burke JJ, Arntzen CJ (1980b) Plant Physiol 65:823–827
Mullet JE, Leto K, Arntzen CJ (1981) In: Akoyunoglou G (ed) Photosynthesis: Chloroplast development. Balaban, Philadelphia, pp 557–568
Nechusthai R, Nelson N (1981) J Cell Biol 256:11624–11628
Nechusthai R, Nelson N, Matoo A, Edelman K (1981) FEBS Lett 125:115–119
Nechusthai R, Muster P, Binder A, Liveanur V, Nelson N (1983) Proc Natl Acad Sci USA 80:1179–1183
Nelson N, Notsani BE (1977) In: Packer L, Papageorgiou GC, Trebst A (eds) Bioenergetics of membranes. Elsevier, Amsterdam New York Oxford, pp 233–244
Okamura MY, Feher G, Nelson N (1983) In: Govindjee (ed) Photosynthesis: energy conversion by plants and bacteria. Academic Press, London New York, pp 195–272
Olive J, Wollman FA, Bennoun P, Recouvreur M (1983) Biol Cell 48:81–84
Oquist G, Samuelsson G, Bishop NI (1980) Physiol Plant 50:63–70
Remy R, Ambard-Bretteville F (1984) FEBS Lett 170:174–180
Remy R, Hoarau J (1978) In: Akoyunoglou G, Argyroudi-Akoyunoglou JH (eds) Chloroplast development, Elsevier, Amsterdam New York Oxford, pp 235–240
Rutherford BAW, Mullet JE (1981) Biochim Biophys Acta 635:225–235
Sétif P, Acker S, Lagoutte B, Duranton J (1980) Photosynth Res 1:17–27
Shuvalov VA, Dolan E, Ke B (1979) Proc Natl Acad Sci USA 76:770–773
Simpson DJ (1982) Carlsberg Res Commun 47:215–225
Takabe T, Ishikawa H, Niwa S, Itoh S (1983) J Biochem (Tokyo) 94:1901–1911
Takahashi Y, Koike H, Katoh S (1982) Biochem Biophys 219:209–218
Tapie P, Choquet Y, Breton J, Delepelaire P, Wollman FA (1984) Biochim Biophys Acta 767:57–69
Thornber JP (1969) Biochim Biophys Acta 172:230–241
Thornber JP, Gregory RPF, Smith CA, Bailey JL (1967) Biochemistry 6:391–394
Thornber JP, Markwell JP, Reinman S (1979) Photochem Photobiol 29:1205–1216
Vierling E, Alberte RS (1983a) Plant Physiol 72:625–633
Vierling E, Alberte RS (1983b) J Cell Biol 97:1806–1814
Vitry C de, Wollman FA, Delepelaire P (1983) CR Acad Sci 297:277–280
Westhoff P, Alt J, Nelson N, Bottomley W, Bunemann H, Herrmann RG (1983) Plant Mol Biol 2:95–107
Williams RC, Glazer AN, Lundell DJ (1983) Proc Natl Acad Sci USA 80:5923–5926
Wollman FA, Bennoun P (1982) Biochim Biophys Acta 680:352–360
Zielinski RE, Price CA (1980) J Cell Biol 85:435–445

10 Energy Transfer and Energy-Coupling Processes

10.1 Composition and Structure of Cytochrome bc_1 and b_6f Complexes

G. HAUSKA

1 Introduction

Cytochrome bc complexes acting as quinol-cytochrome c (or plastocyanin) oxidoreductases are central constituents of many electron transport chains (ORT, Chap. 4; DUTTON, Chap. 5, this Vol.). Although mitochondrial respiration is the net reversal of chloroplast photosynthesis, they both oxidize quinol via a cytochrome bc complex. In chloroplasts the cytochrome b_6f complex functions between the two photosystems, and it reduces plastocyanin with plastoquinol. In mitochondria the cytochrome bc_1 complex reduces cytochrome c with ubiquinol. Both the b_6f type and the bc_1 type are also found in prokaryotes. Cyanobacteria contain a cytochrome b_6f complex for oxygenic photosynthesis, like chloroplasts (BRICKER et al., Chap. 11.4, this Vol.). A cytochrome bc_1 complex is found in various facultative aerobic bacteria, like *Paracoccus denitrificans* and the Rhodospirillaceae (PRINCE, Chap. 10.4, this Vol.). Cyanobacteria are also able to grow by respiration, and there is good evidence that photosynthetic and respiratory electron transport are linked by the cytochrome b_6f complex in the same membrane (BINDER 1982). Similarly, photosynthetic and respiratory electron transport in Rhodospirillaceae are connected by a cytochrome bc_1 complex (BACCARINI-MELANDRI and ZANNONI 1978). These "zwitter"-systems sharing the cytochrome bc complex are interesting with respect to the phylogenetic relationship of photosynthesis and respiration. A cytochrome bc complex might occur also in the phototropic green sulfur bacteria, like *Chlorobium* (HURT and HAUSKA 1984), which are considered to be among the earliest organisms to have evolved on earth.

Active cytochrome bc complexes of defined polypeptide composition have been isolated first from mitochondria, and more recently from chloroplasts of higher plants (HURT and HAUSKA 1981, 1982a; HAUSKA 1985), from the cyanobacterium *Anabaena variabilis* (KRINNER et al. 1982) and from *Rhodopseudomonas sphaeroides* (GABELLINI et al. 1982; YU et al. 1984). All these preparations catalyze quinol-cytochrome c (or plastocyanin) oxidoreduction with the inhibitor sensitivities of the parent membranes: the bc_1 types of mitochondria or *Rhodopseudomonas* are sensitive to antimycin and myxothiazol, while the b_6f types are not. All preparations show the phenomenon of "oxidant-induced reduction of cytochrome b", and, when incorporated into liposomes, function as electrogenic proton translocators (HURT et al. 1983).

The work on isolated bc complexes published up to 1982 has been reviewed in detail (HAUSKA et al. 1983). Therefore the present article will focus only on the essential aspects of composition and structure, including more recent findings up to mid 1984. The latter relate mostly to the elucidation and interpretation of amino acid sequences. Related contributions in this Volume are by TIEDE (Chap. 8.2) on the structure of the bacterial cytochrome bc_1 complex, by PRINCE (Chap. 10.4), by JOLIOT and JOLIOT (Chap. 10.3), and by WHITMARSH (Chap. 10.2) on mechanistic aspects of bc-complexes, and by RYRIE (Chap. 11.7) on the reconstitution into lipid membranes.

2 Composition

2.1 Redox Components

The isolated cytochrome bc complexes have the universal composition of 2 hemes cytochrome b, 1 heme cytochrome c_1 or f, and 1 high-potential Fe_2S_2 center named after Rieske. Lower content of heme b in a preparation from *Rhodopseudomas sphaeroides* R26 (YU et al. 1984) can be partially explained by loss during isolation, and partially by the use of extinction coefficients different from the values determined by GABELLINI and HAUSKA (1983a). In addition some, but not all of the bc preparations retain quinone, but this is not required for oxidoreductase activity (HURT and HAUSKA 1982b; VON JAGOW et al. 1984).

2.1.1 Cytochrome b

Two forms of cytochrome b, which differ spectroscopically as well as in redox potential, are contained in all the bc complexes (HAUSKA et al. 1983; PRINCE, Chap. 10.4, this Vol.). These two forms have long been known in mitochondria. When isolated under mild conditions, the bc_1 complex from beef heart mitochondria contains a high-potential form with an α-peak at 562 nm and an E_m7 of 90 mV, and a low-potential form with an α-peak at 566 plus a shoulder at 558 nm, and with an E_m7 of −30 mV. Two similar forms are known for chromatophores of Rhodospirillaceae and have also beed detected in the bc_1-complex isolated from *Rps. sphaeroides* GA (GABELLINI et al. 1982; GABELLINI and HAUSKA 1983a). The midpoint potentials are 50 and −90 mV, at pH 7. Harsher treatments abolish the differences of the two forms, especially during attempts to isolate cytochrome b.

Two spectrally distinct forms of cytochrome b_6 in the b_6f-complex have been detected only recently (HURT and HAUSKA 1983; CLARK and HIND 1983). In this case the form with the higher redox potential shows a split α-peak. The two forms have E_m7 values of about −50 and −150 mV, which are some 100 mV more negative than the ones for their bc_1-counterparts. They do not correspond to values reported for chloroplast membranes earlier (BÖHME and CRAMER 1973), and their physiological significance is under debate (HAUSKA et al. 1983). In this context it is noteworthy that the two forms of cytochrome b_6 did not change properties after incorporation into liposomes (HURT unpublished). In the cytochrome b_6f complex isolated from the cyanobacterium *Anabaena variabilis*, the two spectral forms of cytochrome b_6 are also present (HURT and HAUSKA 1983), having potentials more positive (about 10 and −100 mV at pH 7) than found for the chloroplast b_6f complex. Recently a cytochrome bc fraction with high purity was isolated from the green sulfur bacterium *Chlorobium limicola* f. *thiosulfatophilum* (HURT and HAUSKA 1984). The b-type component had the α-peak at 562 nm, and again showed redox heterogeneity. If two equally contributing components are assumed, E_m7 values of about 90 and −50 mV are obtained, similar to the mitochondrial values (HURT unpublished).

The two forms of cytochrome b have also been detected by EPR in the mitochondrial (CASE and LEIGH 1976) and in the chloroplast bc-complex. For the latter, it is not agreed whether both forms are represented by low spin signals around G=3.5 (BERGSTRÖM et al. 1983), or whether the high-potential form is presented by a high-spin signal around G=6, which earlier has been ascribed to denatured cytochrome b_6 (SALERNO et al. 1983).

In conclusion, heterogeneity of cytochrome b is a common feature of all the isolated bc complexes so far, probably reflecting two forms with different heme environment. This is confirmed by the elucidation of the primary structures of mitochondrial and plastidal cytochrome b, showing that these proteins contain two heme binding sites (WIDGER et al. 1984; SARASTE 1984; see below). In all cases studied, both E_m values of cytochrome b are pH-dependent in the physiological range, which might indicate a role of cytochrome b in proton translocation (HAUSKA et al. 1983).

2.1.2 Cytochrome c_1 or f

Redox titrations of cytochromes c_1 or f correspond to single n=1 components (HAUSKA et al. 1983; PRINCE, Chap. 10.4, this Vol.). The midpoint potentials are pH-independent in the physiological range. They are at 230 mV for cytochrome c_1 in the mitochondrial bc_1 complexes, at 290 mV for cytochrome c_1 in the bc_1 complex from *Rps. sphaeroides* GA, and at 340 mV for cytochromes f in the b_6f complex from chloroplasts and cyanobacteria. The α-peaks are at 553 nm for the cytochromes c_1 and algal cytochrome f, at 554 for chloroplast cytochrome f, and at 556 for cyanobacterial cytochrome f (HAUSKA et al. 1983). They are asymmetric and split into two at low temperature in all cases, except for cytochrome c_1 from *Rps. sphaeroides*. The c-type component of the bc fraction isolated from *Chlorobium* (α-peak 550.5 nm) has a midpoint potential of 220 mV (HURT and HAUSKA 1984).

2.1.3 The Rieske FeS Center

This high potential Fe_2S_2 protein was first discovered by RIESKE in the mitochondrial bc_1 complex, and is present in all isolated bc complexes which are active in quinol-cytochrome c oxidoreduction (HAUSKA et al. 1983; PRINCE, Chap. 10.4, this Vol.). It has a characteristic EPR spectrum and its detection in biological material can be taken to indicate the occurrence of a bc complex.

Its midpoint potential in the bc_1 complexes from mitochondria and from *Rps. sphaeroides*, and in the b_6f complex from chloroplasts is 280–290 mV. This becomes pH-dependent above pH 8. A Rieske FeS center with an E_m7 of 140 mV has been observed in *Chlorobium* (KNAFF and MALKIN 1976), but this is lost during the reported isolation of a bc fraction from this organism (HURT and HAUSKA 1984). A "Rieske-type" protein has been recently isolated from *Thermus thermophilus* (FEE et al. 1984) which contains only 2 cysteins per Fe_2S_2 cluster. ENDOR measurements suggest that one of the two irons is ligated to nitrogen, probably histidine. This difference from other Fe_2S_2-proteins such as the ferredoxins, i.e., having both irons bound by two cysteins, might explain the unusually high redox potential of the Rieske FeS proteins. The primary structure of the Rieske FeS protein from *Neurospora crassa* mitochondria shows a domain with 4 cysteins, however (HARNISCH et al. 1985). This domain is highly conserved in the Rieske FeS protein from *Rps. sphaeroides* (GABELLINI et al. 1984).

2.1.4 Quinone

It still remains an open question whether special quinone molecules play a crucial role during turnover – i.e., whether quinol is a component and not only a substrate of the bc-complexes. A stable semiquinone anion in membranes from mitochondrial and from *Rps. sphaeroides* (ROBERTSON et al. 1984) and in the isolated bc_1-complex from mitochondria (see HAUSKA et al. 1983) has been detected. But this might just reflect a relatively firm binding of the intermediary semiquinone form, as well as decreased pK values of transiently bound quinol (RICH 1984; HAUSKA et al. 1983), and need not indicate a special quinone, permanently separated from the quinone pool, as advocated by C.A. YU and L. YU (1981, 1982b). The occurrence of a stabilized semiquinone at a site termed Q_i or Q_c, as opposed to an instable semiquinone at Q_o or Q_z, is crucial to Mitchell's Q-cycle for explaining energy conservation via the bc-complexes (MITCHELL 1976). Stabilized semiquinones are also featured in alternative models (WIRKSTRÖM et al. 1981; MATSUURA et al. 1982; PAPA 1982). Possible mechanisms of energy conserving electron flow in bc complexes have been reviewed repeatedly (BOWYER and TRUMPOWER 1981; WIKSTRÖM et al. 1981; CROFTS and WRAIGHT 1983; HAUSKA et al. 1983; RICH 1984), and are also discussed in this Volume of by PRINCE (Chap. 10.4) and JOLIOT and JOLIOT (Chap. 10.3).

No "stable" plastosemiquinone could be detected in the isolated b_6f complex from chloroplasts (PRINCE and HAUSKA, unpublished), under the conditions for detection of the ubisemiquinone in bc_1 complexes (ROBERTSON et al. 1984). Therefore, it a plastosemiquinone operates at the Q_i site also during the reaction cycle of the b_6f complex, it is at least ten times less stabilized, compared to the ubisemiquinone in the bc_1 complexes, which is below the limit of detection by EPR. A less stable semiquinone would cope with the more negative E_m values for cytochrome b_6 relative to the mitochondrial and bacterial cytochrome b in the bc_1 complexes (see above). Alternatively, in contrast to the bc_1 complexes, the Q_i site might be operative in the b_6f complex under certain conditions only (RICH and HEATHCOTE 1984; HAUSKA et al. 1983; RICH 1984; JOLIOT and

JOLIOT, Chap. 10.3, this Vol.), which might not be met under the redox poised conditions of the EPR measurement.

A specialized quinone molecule operating in the b_6f complex is indicated by the finding that plastoquinone is retained almost stoichiometrically during isolation. Furthermore, a first chemical analysis revealed that not plastoquinone A, but rather a plastoquinone C-like form is predominantly retained (HURT and HAUSKA 1982b). However, plastoquinone A reconstituted activity in the fully extracted complex. It is possible, therefore, that the slightly more hydrophilic plastoquinone C is selected by chance during the isolation procedure. Plastoquinol-1 also reconstituted activity in fully extracted preparations of the b_6f-complex. Reoxidation of cytochrome b_6, after oxidant-induced reduction, remained slow, however (HURT and HAUSKA 1982b). It seems that plastoquinone 9 is required at the Q_i site for efficient reoxidation. Similarly, it was observed that duroquinol could not replace ubiquinol 10 at the Q_i site of the mitochondrial bc_1 complex (ZHU et al. 1982).

In summary, the bc complexes consist of two high-potential redox centers – the Rieske FeS-protein and cytochrome c_1 or f, and of two low-potential centers – the two hemes in cytochrome b. The quinone system is operating with an overall E_m7 value of about 100 mV (PRINCE, Chap. 10.4, this Vol.), probably at two sites, or stages, with different semiquinone stability. Its function at the two sites or stages is linked to cytochrome b reduction and reoxidation, and this is somehow coupled to electrogenic proton translocation.

2.2 Polypeptides

Figure 1 presents the polypeptide composition of the various bc complexes. The mitochondrial bc_1 complex (Fig. 1A) shows a more complicated pattern than the others. Ten subunits have been identified. Only seven are resolved on the gel of Fig. 1, but the 32 kDa polypeptide of cytochrome c_1 sits on top of the diffuse band of cytochrome b, and the smallest subunit can be separated into three bands in the presence of urea (MARRES and SLATER 1977; CAPALDI et al. 1977). The polypeptide at 27 kDa carries the Fe_2S_2 center (TRUMPOWER 1981). It is easily removed by Triton X-100 (LI et al. 1981; ENGEL et al. 1980) and can be rebound to the deficient complex reconstituting oxidoreductase activity and proton translocation in liposomes (ENGEL et al. 1983). Recently the isolation of each of the ten subunits from the bovine complex (VON JAGOW et al. 1984), and the reconstitution of the *Neurospora crassa* oxidoreductase from a cytochrome bc_1 subcomplex plus the two core proteins plus the Rieske FeS protein (LINKE and WEISS 1984) have been reported.

The cytochrome b_6f complexes from spinach chloroplasts and *Anabaena variabilis* consist of four major polypeptides – cytochrome f corresponding to the doublet at 34 and 31 kDa, cytochrome b_6 (23 kDa), the Rieske FeS-protein at 20 and 22 kDa, and a 17 kDa subunit (Fig. 1C, D). In addition small hydrophobic polypeptides below 6 kDa are present which can be seen after silver staining, and can be extracted from the complex by acidic acetone (HURT and HAUSKA 1982a). CLARK et al. (1984b) reported that the ferredoxin-$NADP^+$

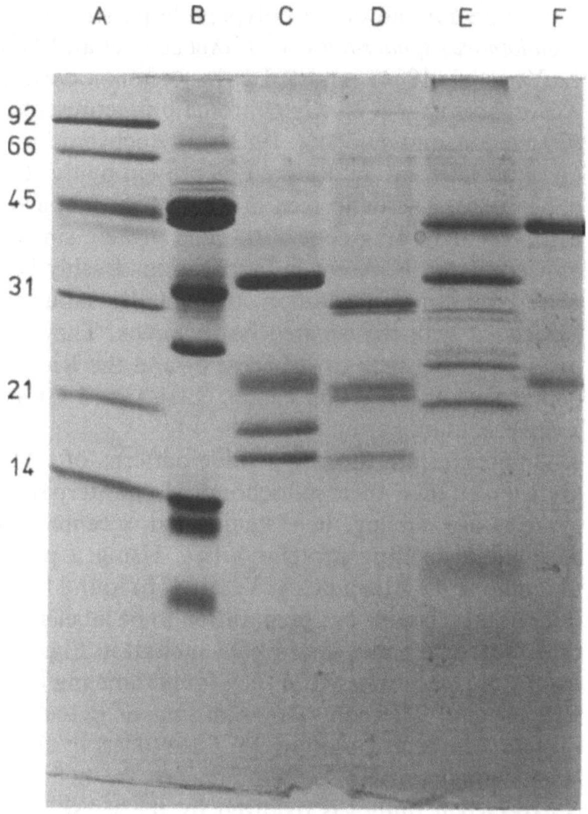

Fig. 1. Polypeptide patterns of bc complexes. SDS-PAGE on 15% gel after LAEMMLI (1970), stained with Coomassie blue. *A* standard; *B* bc_1 complex from beef heart mitochondria, 0.3 nmol c_1 (courtesy Drs. ENGEL and VON JAGOW, Munich); *C* b_6f complex from spinach, 0.3 nmol f (HURT and HAUSKA 1982a); *D* b_6f complex from *Anabaena variabilis*, 0.1 nmol f (KRINNER et al. 1982); *E* bc_1 complex from *Rhodopseudomonas sphaeroides* GA, 0.2 nmol c (GABELLINI et al. 1982); *F* bc fraction from *Chlorobium limicola* f. thiosulfatophilum, 0.1 nmol c. (HURT and HAUSKA 1984)

oxidoreductase (FNR), a 37 kDa protein, was a further subunit of the spinach b_6f-complex. An inactive preparation from pea chloroplasts, lacking the Rieske FeS protein, but containing FNR, was obtained by PHILLIPS and GRAY (1983). Indeed, if the chloroplast membranes are not extensively washed with NaBr before extraction, FNR remains associated with the b_6f complex to a large extent, even when our procedures are carefully followed (HURT and HAUSKA 1982a; HAUSKA 1985). However, the presence of FNR does not affect plastoquinol-plastocyanin oxidoreductase activity, and an active b_6f complex lacking FNR, but retaining the Rieske FeS center can be obtained also from pea and other higher plant plastids (HAUSKA 1985). Therefore, FNR is not a subunit of the b_6f complex with regard to the minimum composition for oxidoreductase activity. Its association with the b_6f complex is interesting, however, with regard to ferredoxin-mediated cyclic electron flow around photosystem I.

Figure 1E shows the polypeptide pattern of the bc_1 complex from *Rhodopseudomonas sphaeroides* GA (GABELLINI et al. 1982). L. YU and C.A. YU (1982a, L. YU et al. 1984) reported a very similar pattern for the bc_1 complex from *Rps. sphaeroides* R26. Cytochrome b is represented by the 40 kDa polypeptide (GABELLINI and HAUSKA 1983b), cytochrome c_1 by the 34 kDa polypeptide (WOOD 1980) and the Rieske FeS protein by the 25 kDa polypeptide (GABELLINI et al. 1984). The band seen at 21 kDa in the particular preparation of Fig. 1E is a proteolytic fragment of the FeS protein since it cross-reacts with a specific antiserum and is absent in preparations freshly isolated in the presence of protease inhibitors. In addition, two smaller subunits (10 and 8 kDa) are always associated with the isolated bc_1 complex. They appear as diffuse, weak bands in Fig. 1E. Between cytochrome c_1 and the Rieske FeS protein the three polypeptides of the reaction center (FEHER and OKAMURA 1978) are still seen as a slight contamination.

In conclusion, the polypeptide patterns of the photosynthetic bc complexes are simpler than their mitochondrial counterparts. In particular, the two core proteins are missing. In assigning redox centers to polypeptides, the question of quinone binding proteins arises. Using a photoaffinity label derivative of ubiquinone, L. YU and C.A. YU (1982b) found both the 38 and 17 kDa polypeptides of their bovine bc_1 preparation to be labeled, which corresponds to labeling of cytochrome b and the 14 kDa subunit in Fig. 1B. With their bc_1-preparation from *Rps. sphaeroides* R26 they found labeling of cytochrome b and the Rieske FeS protein. The equivalent labeling of cytochrome b_6 and the Rieske FeS protein has been observed by OETTMEIER et al. (1982) using a photoaffinity label derivative of plastoquinone for the b_6f complex. This labeling pattern is expected if quinol is oxidized by the Rieske FeS center at the Q_o site, and the resulting unstable semiquinone reduces cytochrome b, as postulated in Mitchell's Q cycle (see Sect. 2.1.4 above). it does not necessarily reflect tight binding of the quinone to the proteins, as assumed by C.A. YU and L. YU (1981).

The inactive bc fraction from *Chlorobium* contains only two polypeptides (Fig. 1F). The 40 kDa species represents cytochrome b and the 23 kDa species cytochrome c (HURT and HAUSKA 1984). Most recently, the isolation of an ubiquinol oxidase complex from *Paracoccus denitrificans* has been reported by TRUMPOWER (1984), which contains seven polypeptides – two for cytochrome oxidase (44 and 32 kDa; LUDWIG 1980), a novel membrane-bound cytochrome c-552 (22 kDa), cytochrome b (39 kDa), cytochrome c_1 (62 kDa; LUDWIG et al. 1983), the Rieske FeS protein (20 kDa) and an additional subunit of 57 kDa which might be a core protein. The "supercomplex" is particularly interesting, since it catalyzes the reaction of ubiquinol with oxygen in the absence of a soluble cytochrome c at very high rates.

3 Structure

The stoichiometry of the polypeptides is still debated, and so is the size of the functional unit (HAUSKA et al. 1983). However, NALECZ et al. (1984) demonstrated most recently that the dimeric form of the mitochondrial bc_1 complex can be reversibly dissociated at low salt without loss of activity.

3.1 Genes and Primary Structures

Of the ten polypeptides of the mitochondrial bc_1 complex only cytochrome b is synthesized within mitochondria. Its gene is split by introns in yeast mitochondria (NOBREGA and TZAGOLOFF 1980), but not in mitochondria from higher organisms (ANDERSON et al. 1981; SARASTE 1984; WIDGER et al. 1984). All the other polypeptides are synthesized on cytoplasmic ribosomes and are transported into mitochondria where they are processed to mature forms (TEINTZE et al. 1982). Of the four major polypeptides of the plant b_6f complex, on the other hand, only the Rieske FeS protein is synthesized on cytoplasmic ribosomes as a precursor which is transported into and processed in chloroplasts, while cytochromes f and b_6 and the 17 kDa subunit are synthesized in the chloroplast (ALT et al. 1983; PHILLIPS and GRAY 1984a). The genes for cytochrome b_6 and the 17 kDa subunit are adjacent, separated by 1000 base pairs, and are transcribed into a common mRNA (ALT et al. 1983). The gene for cytochrome f maps separately (ALT et al. 1983; WILLEY et al. 1983). It is translated into a precursor polypeptide 4 kDa larger than mature cytochrome f (ALT et al. 1983; WILLEY et al. 1984a, b; ALT and HERRMANN 1984), probably reflecting processing during insertion from the ribosomes in the stroma into the thylakoid membrane, with the major, heme-carrying part facing the lumen. Recently also the genes for cytochrome b, cytochrome c_1 and the Rieske FeS protein of the bc_1 complex of *Rps. sphaeroides* GA were studied and found to constitute an operon in the order: FeS-protein/cytochrome b/cytochrome c_1 (GABELLINI et al. 1984).

The genes of cytochrome b_6 and the 17 kDa polypeptide of the spinach b_6f complex have been sequenced (HEINEMEYER et al. 1984; PHILLIPS and GRAY 1984b). Comparison of these sequences and their hydropathy plots (KYTE and DOOLITTLE 1982) with the sequences of mitochondrial cytochrome b from six species, including fungi, higher plants, and mammals (WIDGER et al. 1984; SARASTE 1984) led to the following conclusions:

1. The primary structures of the mitochondrial cytochromes b are highly conserved. This is also the case for the partial sequence known for cytochrome b from *Rps. sphaeroides* (GABELLINI et al. 1984).
2. Cytochrome b_6 is highly homologous to a little more than the N-terminal half, and the 17 kDa subunit of the b_6f complex is homologous to part of the C-terminal half of mitochondrial cytochrome b. The molecular masses calculated from the sequences are 23.7 and 15.2 kDa for cytochrome b_6 and the other subunit, which is close to the M_r values of 23 and 17 kDa derived from SDS-PAGE (Fig. 1C). The molecular mass of mitochondrial cytochrome b is around 42 kDa for all the species, which is very different from the value of 30 kDa obtained on SDS-PAGE.
3. Hydropathy plots suggest nine transmembrane helices for mitochondrial cytochrome b, 5 for cytochrome b_6 and 3 for the 17 kDa subunit.

4. Four histidines are conserved throughout in the b_6 stretch of the sequences – two in the second transmembrane helix and two in the fifth helix, counting from the N-terminus. Thirteen amino acids separate the two histidines in all cases, except for helix V of cytochrome b_6, where an extra threonine is inserted. It is possible that this difference is responsible for the different redox potentials of cytochrome b_6 and mitochondrial b.
5. The two hemes are fixed at opposite membrane surfaces between helices II and V, by salt bridges from conserved Lys or Arg-residues to the carboxyethyl groups of the ring. They are perpendicular to the membrane plane.

The gene of cytochrome f has also been sequenced (WILLEY et al. 1984a, b; ALT and HERRMANN 1984). The sequences of cytochrome f and c_1 (WAKABAYASHI et al. 1982) show much less homology than the ones of the b-type cytochromes. Furthermore, the partial sequence just elucidated for cytochrome c_1 from *Rps. sphaeroides* shows little homology (GABELLINI personal communication). The heme binding site cys-X-Y-cys-his is conserved, however, and hydropathy patterns suggest a common arrangement in the membranes, with only one hydrophobic helix at the C-terminus, spanning the membrane and fixed by positive charges on either surface (Fig. 2). The sequence of the Rieske FeS protein so far is known for *Neurospora crassa* (HARNISCH et al. 1984), and partly for *Rps. sphaeroides* (GABELLINI et al. 1984). Again, only little homology is observed except for a stretch of about 50 residues, in the C-terminal region, which is conserved to 80%. This contains four cys, which probably bind the Fe_2S_2 cluster. The sequence shows one hydrophobic stretch suitable for spanning the membrane, close to the N-terminus (Fig. 2).

3.2 A Topographical Model

Studies on the topographical structure of the mitochondrial bc_1-complexes and the b_6f-complexes, including proteolysis, labeling and immunological techniques, are numerous and have been summarized in detail in our previous review (HAUSKA et al. 1983). Analysis of two-dimensional crystals by neutron diffraction and electron microscopy have led to a rather detailed model for the bc_1 complex from *Neurospora crassa* (KARLSON et al. 1983). However, the mitochondrial complex is complicated by the presence of the core proteins and other polypeptides (Fig. 1), and does not appear to represent the minimum structure for a H^+-translocating quinol-cytochrome c oxidoreductase, which might just comprise cytochrome c_1 or f, the Rieske FeS protein and cytochrome b (=cytochrome b_6 plus the 17 kDa subunit for b_6f complexes). A topographical minimum model has been proposed by us (HAUSKA et al. 1983). Figure 2 presents a refined version for the b_6f complex, accounting for the hydropathy analyses of the primary structures discussed above. In the upper section, the arrangement of the individual subunits in the membrane are depicted, the upper surface representing the stromal, the lower the lumenal surface of the thylakoids. These correspond to the matrix/cytoplasmic and to the intermembrane/periplasmic surfaces in mitochondria/bacteria. In the lower part of Fig. 2 the subunits are

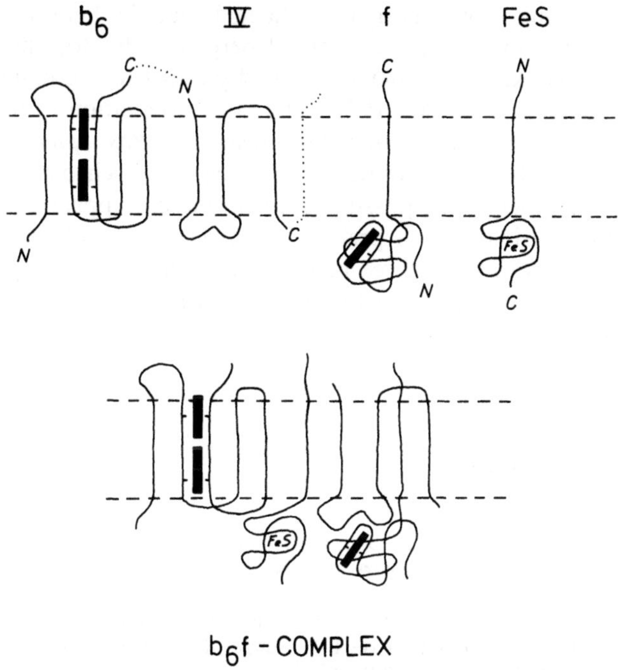

Fig. 2. Proposed topographical arrangement of the b_6f-polypeptides in the membrane

grouped together. It is interesting to note that cytochrome f or c_1, and the Rieske FeS protein are inserted in opposite directions, which is pertinent to the mechanism of insertion. The location of the N-terminus of cytochrome b_6 is not known yet and is arbitrary in Fig. 2. The polypeptide stretches of mitochondrial cytochrome b which are missing from cytochrome b_6 plus subunit-IV (17 kDa) of the chloroplast, when aligned for maximal homology, are shown by dots. As discussed above, helix II and helix V of cytochrome b_6 bind the two hemes with their planes perpendicular to the membrane plane. The upper bar represents the high potential form, the lower bar the low potential form of these hemes (see above). This arrangement is in accordance with a transmembrane electron pathway in cytochrome b, as has been suggested (MITCHELL 1976). An electrogenic step related to cytochrome b has been detected recently by measuring the slow phase of the carotenoid shift in membranes from *Rps. sphaeroides* (GLASER and CROFTS 1984, see also JOLIOT and JOLIOT, Chap. 10.3). The placement of the b hemes as in Fig. 2 is suggested by effects of electric potentials on the two forms of cytochrome b in mitochondria (MITCHELL 1976). Using paramagnetic shift agents and electron paramagnetic resonance (EPR) spectroscopy, it has been concluded, on the other hand, that in mitochondria both hemes are close to the lower surface in Fig. 2 (CASE and LEIGH 1976, see PRINCE, Chap. 10.4, this Vol.). A more recent refinement of these measurements can be interpreted to prove the model in Fig. 1, however (OHNISHI personal communication). LD- and EPR-spectroscopy of oriented bc_1 complexes from mitochondria again indicated that the hemes of cytochrome b and also of cytochrome c_1 are oriented perpendicular to, and that the Fe–Fe axis of the Rieske FeS center is oriented parallel to the plane of the membrane (ERECINSKA and WILSON 1979). The same orientation of the Fe–Fe has been found in the FeS centers of chloroplasts and *Rps. sphaeroides* (PRINCE et al. 1980; PRINCE 1983). In contrast, the heme of cytochrome f (as shown in Fig. 2) and of another heme, possibly cytochrome b_6, appear to be tilted (CROWDER et al. 1982). Finally, in a preliminary investigation of oriented b_6f complex by LD spectroscopy, all three hemes were found to be perpendicular to the plane of the membrane, as in the mitochondrial complex (HURT and BRETON unpublished).

References

Alt J, Herrmann RG (1984) Curr Genet 8:551–557

Alt J, Westhoff P, Sears BB, Nelson N, Hurt E, Hauska G, Herrmann RG (1983) EMBO J 2:979–986

Anderson S, Bankier AT, Barrell BG, Bruijn MHL De, Coulson AR, Drouin J, Eperon JL, Nierlich DP, Roe BA, Senger F, Schreier PH, Smith AJH, Staden R, Joung IG (1981) Nature 290:457–465

Baccarini-Melandri A, Zannoni D (1978) J Bioenerg Biomembr 10:109–138

Bergström J, Andreasson LA, Vänngard T (1983) FEBS Lett 164:71–74

Binder A (1982) J Bioenerg Biomembr 14:271–286

Böhme H, Cramer WA (1973) Biochim Biophys Acta 325:275–283

Bowyer JR, Trumpower BL (1981) In: Skulachev VP, Hinkle P (eds) Chemiosmotic proton circuits in biological membranes. Wesley, Reading, MA, pp 105–122

Capaldi RA, Bell RL, Branchek T (1977) Biochem Biophys Res Commun 74:425–433

Case GD, Leigh JS (1976) Biochem J 160:769–784

Clark RD, Hind G (1983) Proc Natl Acad Sci USA 80:6249–6253

Clark RD, Hawkesford MJ, Coughlan SJ, Bennett J, Hind G (1984b) FEBS Lett 174:137–142

Crofts AR, Wraight CA (1983) Biochim Biophys Acta 726:149–185

Crowder MS, Prince RC, Bearden A (1982) FEBS Lett 144:204–208

Engel WD, Schägger H, Jagow G von (1980) Biochim Biophys Acta 592:211–222

Engel WD, Michalski C, Jagow G von (1983) Eur J Biochem 132:395–402

Erecinska M, Wilson DF (1979) Arch Biochem Biophys 192:80–85

Fee JA, Findling KL, Yoshida T, Hille R, Tarr GE, Hearshen DO, Dunham WR, Day EP, Kent TA, Münck E (1984) J Biol Chem 259:124–133

Feher G. Okamura MY (1978) In: Clayton RK, Syström WR (eds) The photosynthetic bacteria. Plenum, New York London, pp 349–386

Gabellini N, Hauska G (1983a) FEBS Lett 153:146–149

Gabellini N, Hauska G (1983b) FEBS Lett 154:171–174

Gabellini N, Bowyer JR, Hurt E, Melandri BA, Hauska G (1982) Eur J Biochem 126:105–111

Gabellini N, Harnisch U, McCarthy JEG, Hauska G, Sebald W (1984) 3rd Eur Bioenerg Conf Short Rep 3:73–74

Glaser EG, Crofts AR (1984) Biochim Biophys Acta 766:322–332

Harnisch U, Weiss H, Sebald W (1985) Eur J Biochem 149:95–99

Hauska G (1985) Methods Enzymol Biomembranes part M, in press

Hauska G, Hurt E, Gabellini N, Lockau W (1983) Biochim Biophys Acta 726:97–133

Heinemeyer W, Alt J, Herrmann RG (1984) Current Genet 8:543–459

Hurt E, Hauska G (1981) Eur J Biochem 117:591–599
Hurt E, Hauska G (1982a) J Bioenerg Biomembr 14:405–424
Hurt E, Hauska G (1982b) Biochim Biophys Acta 682:466–473
Hurt EC, Hauska G (1983) FEBS Lett 153:413–419
Hurt EC, Hauska G (1984) FEBS Lett 168:149–154
Hurt EC, Gabellini N, Shahak Y, Lockau W, Hauska G (1983) Arch Biochem Biophys 225:879–885
Jagow G von, Engel WD, Schägger H (1984) 3rd Eur Bioenerg Conf Short Rep 3:173–174
Karlson B, Hovmöller S, Weiss H, Leonard K (1983) J Mol Biol 165:287–302
Knaff D, Malkin R (1976) Biochim Biophys Acta 430:244–252
Krinner M, Hauska G, Hurt E, Lockau W (1982) Biochim Biophys Acta 681:110–117
Kyte J, Doolittle R (1982) J Mol Biol 157:105–132
Li Y, Vries S de, Leonard K, Weiss H (1981) FEBS Lett 135:277–280
Linke P, Weiss H (1984) 3rd Eur Bioenerg Conf Short Rep 3:143–144
Ludwig B (1980) Biochim Biophys Acta 594:177–190
Ludwig B, Suda K, Cerletti N (1983) Eur J Biochem 137:597–602
Marres CAM, Slater EC (1977) Biochim Biophys Acta 462:531–548
Matsuura K, Packham NK, Tiede DM, Mueller P, Dutton PL (1982) In: Trumpower BL (ed) Functions of quinones in energy conserving systems. Academic Press, London New York, pp 277–284
Mitchell P (1976) J Theor Biol 62:327–367
Nalecz MJ, Bolli R, Azzi A (1984) 3rd Eur Bioenerg Conf Short Rep 3:157–158
Nobrega FG, Tzagoloff A (1980) J Biol Chem 255:9828–9837
Oettmeier W, Masson K, Soll HJ, Hurt E, Hauska G (1982) FEBS Lett 144:313–317
Papa S (1982) J Bioenerg Biomembr 14:69–86
Phillips AL, Gray JC (1983) Eur J Biochem 137:553–560
Phillips AL, Gray JC (1984a) Eur J Biochem 138:591–595
Phillips AL, Gray JC (1984b) Mol Gen Genet 194:477–484
Prince RC (1983) Biochim Biophys Acta 723:133–138
Prince RC, Crowder MS, Bearden AJ (1980) Biochim Biophys Acta 635:132–148
Rich PR (1984) Biochim Biophys Acta 768:53–79
Rich PR, Heathcote P (1984) In: Sybesma C (ed) Advances in photosynthesis research. Proc 6th Int Congr Photosynth, Brussels 1983, vol II. Nijhoff/Junk, The Hague, pp 363–366
Robertson DE, Prince RC, Bowyer JR, Matsuura K, Dutton PL, Ohnishi T (1984) J Biol Chem 259:1758–1763
Salerno JC, McGill JW, Gerstle GC (1983) FEBS Lett 162:257–261
Saraste M (1984) FEBS Lett 166:367–372
Teintze M, Slaughter M, Weiss H, Neupert W (1982) J Biol Chem 257:10364–10371
Trumpower BL (1981) Biochim Biophys Acta 639:129–155
Trumpower BL (1984) 3rd Eur Bioenerg Conf Short Rep 3:25–26
Wakabayashi S, Matsubara H, Kim CH, King TE (1982) J Biol Chem 257:9335–9344
Widger WR, Cramer WA, Herrmann RG, Trebst A (1984) Proc Natl Acad Aci USA 81:674–678
Wikström M, Krab K, Saraste M (1981) Annu Rev Biochem 50:623–655
Willey DL, Huttly AK, Phillips AL, Gray JC (1983) Mol Gen Genet 189:85–89
Willey DL, Auffret AD, Gray JC (1984a) Cell 36:555–562
Willey DL, Howe CJ, Auffret AD, Bowman CM, Dyer TA, Gray JC (1984b) Mol Gen Genet 194:416–422
Wood PM (1980) Biochem J 189:385–391
Yu CA, Yu L (1981) Biochim Biophys Acta 639:99–128
Yu L, Yu CA (1982a) Biochem Biophys Res Commun 108:1285–1292
Yu L, Yu CA (1982b) J Biol Chem 257:10215–10221
Yu L, Mei QC, Yu CA (1984) J Biol Chem 259:5752–5760
Zhu QS, Berden JA, Vries S de, Slater EC (1982) Biochim Biophys Acta 680:69–79

10.2 Mobile Electron Carriers in Thylakoids

J. WHITMARSH

1 Introduction

Electron transfer from H_2O to NADP involves three integral membrane protein complexes operating in series: photosystem II, cytochrome b/f, and photosystem I. Electrons are transferred between these large protein complexes by smaller molecules, with communication between photosystem II and the cytochrome b/f complex mediated by a pool of plastoquinone that is located in the hydrophobic region of the thylakoid membrane, while electron transfer from the cytochrome b/f complex to photosystem I is mediated by the peripheral protein plastocyanin, the oxidation and reduction of which appears to occur at the inner surface of the membrane. Electron transfer from the photosystem I complex to NADP requires ferredoxin, a soluble protein found in the aqueous stromal phase of the chloroplast, and ferredoxin-NADP oxidoreductase, a protein found on the outer membrane surface. (For an overview of photosynthetic electron transport see ORT, Chap. 4, this Vol. CROFTS and WRAIGHT 1983, and HAEHNEL 1984a). This scheme would seem to provide a clear picture of electron transport if it were not for the fact that the thylakoid membrane exhibits a stunning degree of lateral heterogeneity (STAEHELIN, Chap. 1, this Vol.). This lateral heterogeneity, which is exhibited both in the overall structure of the membrane and in the distribution of the reaction center complexes within the membrane, raises perplexing questions concerning the lateral mobility of the electron transport components. As an example, membrane fractionation studies have led to recent models of the thylakoid membrane in which the majority of the photosystem II complexes are spatially separated from photosystem I by distances that average 1000 Å or more (ANDERSON 1982a; ANDERSSON and HAEHNEL 1982; ANDERSON and MELIS 1983). In these models electrons must be transferred rapidly over the long distances between the two photosystems to account for linear electron transport. Calculations based upon these data indicate that the diffusion coefficients necessary to account for the observed rates of electron transfer in thylakoid membranes (SELAK and WHITMARSH 1984; MILLNER and BARBER 1984) would have to be as large or larger than those measured for the fastest carriers in other bioenergetic membranes. These considerations point toward diffusion as a limiting factor in electron transport, a conclusion that is at odds with data indicating that photosynthetic electron transport is not diffusion-limited. At present the extent of electron transfer between the two membrane regions is unknown. The experimental evidence is indirect and has led to a variety of proposals, ranging from the suggestion

(MILLNER and BARBER 1984) that plastoquinone may have a diffusion coefficient three orders of magnitude greater than that measured for ubiquinone in mitochondrial membranes (GUPTE et al. 1984), to the possibility that there is no rapid shuttling of molecules between the two membrane regions. In addition to requirements for lateral mobility that may be imposed by the spatial separation of the reaction centers, the suggestion that the reaction centers exhibit a variable stoichiometry requires a mobile carrier to achieve communication between components that do not exhibit a one-to-one correspondence. While considerations such as these have led to the belief that mobile electron carriers are required in photosynthetic electron transport, at least over short distances, the significance of lateral diffusion to the overall process of electron transport is a question that remains unanswered.

This review focuses on electron transport in higher plants and is intended first to delineate the requirements and constraints on the lateral mobility of electron transport components imposed by the membrane composition, membrane topography, and the kinetics of electron transport, and second, to examine the experimental work aimed at establishing which components are mobile and at determining their rates of migration. At the outset it is fair to say that there are no direct measurements of lateral diffusion coefficients of any of the electron transport components in thylakoid membranes; in fact there is disagreement in the field over which components are mobile and which are not. Nevertheless, some degree of diffusion seems to be required to account for a variety of data, and the problem finally comes down to which components are mobile, how fast do they move, and how far do they roam? To answer these questions we need a clearer view of the structure and composition of the membrane, better data as to location and stoichiometries of the electron transport components, as well as a more complete understanding of the electron transfer reactions.

2 Diffusion in Two Dimensions – Definitions

We can define diffusion as the movement of molecules due to random thermal motion. For molecules that are constrained to a membrane, our interest is limited to two-dimensional lateral motion in a fluid. For a single molecule this motion can be viewed as a rapid vibrational motion at a particular site in the membrane that results in many collisions with nearest neighbors before the molecule moves on to a new site. In this manner there are numerous collisions with nearest neighbors before new nearest neighbors are encountered. It is the random movement from site to site that results in diffusional mobility. If initially there is a heterogeneous distribution of molecules in the membrane, then diffusion will result in their randomization. The equations that describe the time course of the movement of the diffusing molecules can be derived from kinetic theory and are known as Fick's Laws of Diffusion (CRANK 1975). Fick's First Law, which expresses the fact that the net movement is from a region of higher concentration to a region of lower concentration, states that the rate of transfer of diffusing molecules through a unit area is proportional to the concentration gradient. In one dimension Fick's First Law is given by

$$J = -D\frac{\partial C}{\partial x}, \tag{1}$$

where J is the number of molecules diffusing per unit time through unit area, C(x) is the concentration of the diffusing molecules at position x, and D is the diffusion coefficient. This relationship serves to define the coefficient of diffusion D. For each molecular species in a particular solvent there is a characteristic diffusion coefficient. Assuming the diffusion coefficient is constant in time and in space, most diffusion problems are analyzed using Fick's Second Law, which follows from the First Law by applying the principle of conservation of mass. In one dimension Fick's Second Law is given by

$$\frac{\partial C}{\partial t} = D\frac{\partial^2 C}{\partial x^2}. \tag{2}$$

Because of this relationship analysis of diffusional mobility is often reduced to establishing the diffusion coefficient. For systems that obey Fick's Laws, and for geometries that allow solutions to the differential equation, knowing the diffusion coefficient is sufficient to determine the net movement of the diffusing molecules.

The simplest and most often-used relationship between the rate of movement of a diffusing molecule and the diffusion coefficient is that derived by EINSTEIN, which states that the average of the square of the displacement of a molecule is directly proportional to the diffusion coefficient multiplied by the diffusion time (EINSTEIN 1956; DANIELS and ALBERTY 1967). In two dimensions the equation[1] is

$$\langle r^2 \rangle = 4D\tau, \tag{3}$$

where r is the distance measured between the location of the molecule at time $t=0$ and the location at time $t=\tau$. This means that if a molecule with a diffusion coefficient D limited to two-dimensional motion starts at the origin ($r=0$) at time $t=0$, then after a time τ the average distance the molecule has moved is given by $(4D\tau)^{1/2}$ (more accurately, the root-mean-square of the distance) (Table 1). As we will see in Section 4, this relationship has been used to estimate the diffusion coefficient of various electron carriers in thylakoids. So far as thylakoid membranes are concerned, this approach has limited value because of the difficulty in determining the migration distance, and because the diffusion coefficient may vary significantly in different regions of the heterogeneous membrane.

1 In one dimension the equation is $\langle x^2 \rangle = 2D\tau$, and in three dimensions, $\langle R^2 \rangle = 6D\tau$ (EINSTEIN 1956).

Table 1. The average distance of diffusion from the origin in two dimensions using the relation $<r^2> = 4D\tau$ [Eq. (3)] is shown in Å

Diffusion coefficient $cm^2 s^{-1}$	Time				
	100 µs	300 µs	1 ms	3 ms	10 ms
1×10^{-10}	20	35	63	109	200
3×10^{-10}	35	60	109	190	346
1×10^{-9}	63	109	200	346	632
3×10^{-9}	109	190	346	600	1,095
1×10^{-8}	200	346	632	1,095	2,000
3×10^{-8}	346	600	1,095	1,897	3,464
1×10^{-7}	632	1,095	2,000	3,464	6,324
3×10^{-7}	1,095	1,897	3,464	6,000	10,954
1×10^{-6}	2,000	3,464	6,324	10,954	20,000

More direct measurements of diffusion coefficients in membranes are limited because of the small distances involved. One of the more successful techniques developed to deal with the problem of small membrane areas is fluorescence recovery after photobleaching (FRAP) (CHERRY 1979; EDIDIN 1981). The technique allows microscopic observation of the spatial motion of a fluorescent probe as a function of time. The membrane area observed is determined by a laser flash, and as a consequence is at least as wide as the wavelength of the laser beam, limiting this technique to membrane areas of a square micron or greater. To date, this technique has not been applied to photosynthetic membranes. This is due in part to the complexity of thylakoid morphology and because stacked membrane regions have a diameter of approximately 0.5 µm. The possibility of using giant chloroplasts or fused thylakoid membranes remains to be explored. (Other techniques used to determine diffusion coefficients are described in reviews by CHERRY (1979) and EDIDIN (1981). In Table 2, diffusion coefficients are shown for a number of membrane-associated molecules that transfer electrons. Although chloroplast components are not well represented, analogous electron-carrying molecules in mitochondria and in photosynthetic bacteria have been studied.

In the absence of kinetic data, diffusion coefficients can be estimated using the Stokes-Einstein equation (EINSTEIN 1956) that relates the diffusion coefficient to the radius of the diffusing molecule (a) and the viscosity (η) of the solvent.

$$D = \frac{kT}{6\pi a\eta} \tag{4}$$

where k is the Boltzmann constant and T the absolute temperature. This equation is applicable for a diffusing molecule that is larger than the solvent molecules and is suspended in a homogeneous three-dimensional fluid. For a large molecule in a two-dimensional fluid, for example a protein in a lipid membrane,

Table 2. Lateral diffusion coefficients for selected membrane associated electron transport components

Component	Membrane	D (cm^2/s)	Reference
UQ analog[a]	Megamitochondria	3.7×10^{-9}	GUPTE et al. (1984)
UQ analog[a]	Fused inner mitochondrial	2.3×10^{-9}	GUPTE et al. (1984)
Lipids	Model and biological	1×10^{-8}	HOUSLAY and STANLEY (1982)
Cytochrome c[b]	Fused inner mitochondrial	1.9×10^{-9}	GUPTE et al. (1984)
Cytochrome c	Phospholipid vesicle	$0.3–1 \times 10^{-9}$	OVERFIELD and WRAIGHT (1980)
Cytochrome c_2[c]	Phospholipid vesicle	$0.3–1 \times 10^{-8}$	OVERFIELD and WRAIGHT (1980)
Plastocyanin[b]	Phospholipid vesicle	5×10^{-8}	FRAGATA et al. (1984)
Cytochrome P-450[b]	Lipid multibilayers	2×10^{-8}	WU and YANG (1984)
Cytochrome b/c_1[d]	Fused inner mitochondrial	4.4×10^{-10}	GUPTE et al. (1984)
Cytochrome oxidase[e]	Fused inner mitochondrial	3.7×10^{-10}	GUPTE et al. (1984)
Cytochrome oxidase	Inner mitochondrial fused with liposomes	3×10^{-9}	KAWATO et al. (1982)
Cytochrome oxidase	Lipid vesicles	2×10^{-9}	KAWATO et al. (1981)

[a] Ubiquinone analog: Q_0C_{10} NBDHA (see GUPTE et al. [1984] for the structure).
[b] Labeled with a fluorescent probe.
[c] cytochrome c_2 from *Rps. sphaeroides*.
[d] Complex III.
[e] Complex IV.

SAFFMAN and DELBRÜCK (1975) have derived the following relationship between the diffusion coefficient and the membrane viscosity (η_m)

$$D = \frac{kT}{4\pi h \eta_m}\left(ln\frac{h\eta_m}{a\eta_w} - 0.5772\right). \tag{5}$$

In this derivation the protein is assumed to be a cylinder of radius a in a membrane of width h, where η_w is the viscosity of the aqueous phase. In this derivation the diffusion coefficient is independent of the amount of protein extending into the aqueous phase. The extent to which the SAFFMAN-DELBRÜCK equation is successful in predicting the diffusion coefficient is outside the scope of this review (see for example PETERS and CHERRY 1982; VAZ et al. 1984).

It is worth noting that the temperature dependence of the diffusion coefficient is due not only to its direct dependence on the absolute temperature, but also through the temperature dependence of the fluid viscosity. For example, in water the viscosity increases by a factor of 1.35 as the temperature changes from 20° to 10 °C. In olive oil the viscosity increases by a factor of 1.64 over the same temperature range.

Although here we will focus on diffusion as a mechanism for the transfer of reducing equivalents between different membrane redox components, the overall analysis must consider the chemical reaction as well. In a second-order bimolecular reaction, diffusion enables the reactants to come close enough together to interact. In this process, diffusion is limiting when the rate of the reactants coming together is slower than the chemical reaction that occurs once they are together. Such reactions are referred to as diffusion-limited (or diffusion-controlled) reactions. In contrast, bimolecular reactions that require a number of encounters of the reactants as nearest neighbors before a reaction occurs are said to be reaction-limited (WILKINSON 1980; HARDT 1979; AXELROD 1983; GUPTE et al. 1984).

So far we have ignored several factors that can prevent rapid diffusion (HOUSLAY and STANLEY 1982). In the next section we will discuss constraints on free diffusion which may prevent rapid lateral mobility of the redox components in thylakoid membranes.

3 Topology and Composition of the Thylakoid Membrane

The thylakoid membrane is composed of lipids forming a bilayer in which the major electron transport complexes, antenna pigment-proteins, plastoquinone, as well as other molecules, are tightly packed. The membrane forms a closed vesicle dividing the aqueous phase of the chloroplast into two separate regions, the stroma (outer) phase and the lumen (inner) phase. Since the reaction centers and the cytochrome b/f complex each span the lipid bilayer, with segments extending into the inner and the outer aqueous phase, their lateral motion is restricted to the plane of the membrane. Likewise, the hydrophobic nature of plastoquinone limits its motion to the hydrophobic core of the membrane (ULRICH et al. 1985). Only for plastocyanin, which is located in the inner aqueous space, is there the possibility of diffusion relatively independent of the membrane, although as discussed below, even for plastocyanin the membrane probably plays the dominant role in determining its lateral mobility.

The architecture of thylakoid membranes is complex in that some regions of the membrane are stacked (appressed) against one another, while others are unstacked (nonappressed), with one surface exposed to the chloroplast stroma (Fig. 1) (STAEHELIN, Chap. 1, this Vol., PAOLILLO 1970; BRANGEON and MUSTARDY 1979; BERZBORN et al. 1981). As discussed in the previous section, the factors that determine the lateral mobility of a given molecule are many. For movement within a homogeneous region the diffusion coefficient depends

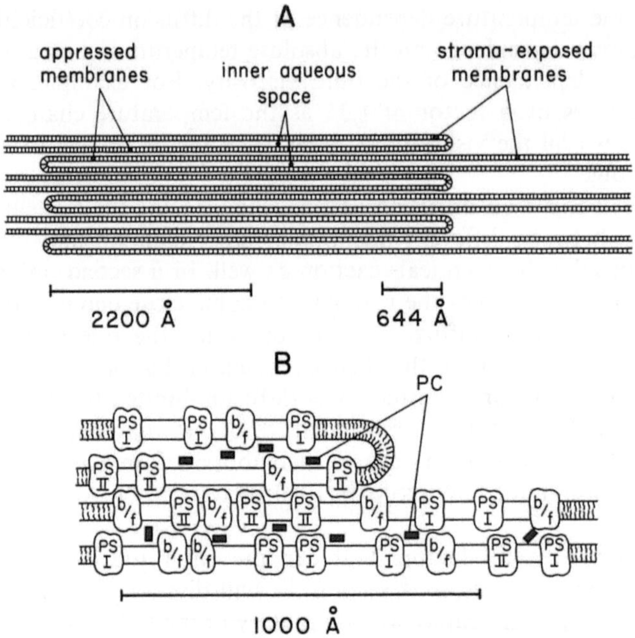

Fig. 1 A, B. Representation of the thylakoid membrane. **A** Cross-section of a granum stack extending into the stroma-exposed membranes. Note that the distance of 644 Å represents the distance that plastocyanin must penetrate the appressed membrane in order to cover 50% of the appressed membrane area. **B** Enlarged view of the edge of a granum extending into stroma-exposed membranes. The photosystem I and II complexes, the cytochrome b/f complex and plastocyanin are shown at approximately the concentration calculated in the text. Other protein complexes are omitted

on the local environment. For long-range interactions the structural and compositional heterogeneity of the membrane becomes a factor. Thus, the diffusion coefficient of a molecule may change from one membrane region to another, or there may be a restrictive barrier between different membrane regions.

The proportion of appressed compared to nonappressed membranes depends on several factors. For example, in extreme shade plants the number of membranes in a stack is greater than in a sun-adapted plant (ANDERSON et al. 1973). In plants grown at different light intensities the number of membranes in a stack is observed to increase as the intensity of the growth light decreases (BERZBORN et al. 1981). In peas the proportion of stacked membranes decreases in thylakoids exposed to light under conditions that result in the phosphorylation of membrane proteins (KYLE et al. 1983). This latter observation, which has to do with the distribution of excitation energy between the two photosystems, provides a demonstration of the fluidity of the thylakoid membrane, albeit on a time scale slow compared to electron transfer reactions. In plants grown under high light the thylakoid membrane is divided almost evenly between appressed and nonappressed regions. When the ratio is determined by measuring the linear length of appressed versus exposed membranes using electron micrographs, BERZBORN et al. (1981) found that the ratio of appressed/stroma-exposed membranes was 40/60 in spinach, while in peas KYLE et al. (1983) found that the ratio changed from 67/33 in nonphosphorylated membranes to 52/48 in membranes that had been phosphorylated in the light.

Viewed from above, the grana stacks are circular in shape, with diameters that generally fall between 4000 and 5000 Å (see Fig. 15 of STAEHELIN, Chap. 1, this Vol.). The thickness of a single bilayer is 40–60 Å. If the polar head groups are ignored and just the fatty acid chains counted, the hydrophobic center of the bilayer would be expected to be 30 to 40 Å thick. However, the width of the hydrophobic core is open to question. It may be that plastoquinone is situated in the center of the lipid bilayer, between the ends of the acyl chains (ALONSO et al. 1981; KATSIKAS and QUINN 1981), in which case the width of the membrane may be larger (JOHANNSSON et al. 1981).

For plastocyanin mobility the width of the inner aqueous space may be more important than the composition of the membrane. Thin-section electron micrographs of chloroplasts fixed in the dark indicate that the distance from one membrane to another across the inner aqueous space is 40 to 90 Å (e.g., MILLER and NOBEL 1982; FELLOWS and BOYER 1976). This distance decreases for chloroplasts fixed in the light. Since plastocyanin is an 10.5 kDa protein with minimum and maximum dimensions of 28 and 40 Å (COLMAN et al. 1978; HAEHNEL, Chap. 10.5, this Vol.), the narrow confines of the inner aqueous space, together with the intrusion of bits of integral membrane proteins into the space, may severely restrict lateral motion. Evidence for a restricted inner aqueous space is provided also by EPR measurements, indicating that the viscosity of the inner aqueous phase was severalfold higher than that of free water (BERG et al. 1979). However, an analogous protein, cytochrome c in mitochondria, seems to diffuse quite rapidly along the membrane surface attached to cardiolipin molecules (FROUD and RAGAN 1984). This raises the possibility of motion of a protein along one membrane surface without interference by the opposite membrane, even in a narrow space.

Our current knowledge of the membrane architecture makes it difficult to determine the accessibility of different membrane regions to one another. In electron micrographs continuity between the appressed and stroma-exposed membranes seems apparent. However, based on electron micrographs of serial sections of a chloroplast, a model of the thylakoid membrane has been developed where communication between stacked and unstacked regions is limited to narrow lanes, known as frets, that regularly emerge from the grana, extending into stroma membranes (see Figs. 14, 15 of STAEHELIN, Chap. 1, this Vol., PAOLILLO 1970; BRANGEON and MUSTARDY 1979; MUSTARDY and JANOSSY 1979). Based on the models shown by BRANGEON and MUSTARDY (1979), it appears for a molecule moving toward the outer diameter of a stacked region, about 50% of the time it would be in a membrane region that extended into the stroma-exposed membrane, while the remainder to the time the molecule would encounter a curved edge leading down to yet another appressed membrane (Fig. 2). This geometry may severely curtail interaction between the two regions, so much so that lateral mobility between the two phases may be dominated by physical inaccessibility rather than the long distances involved.

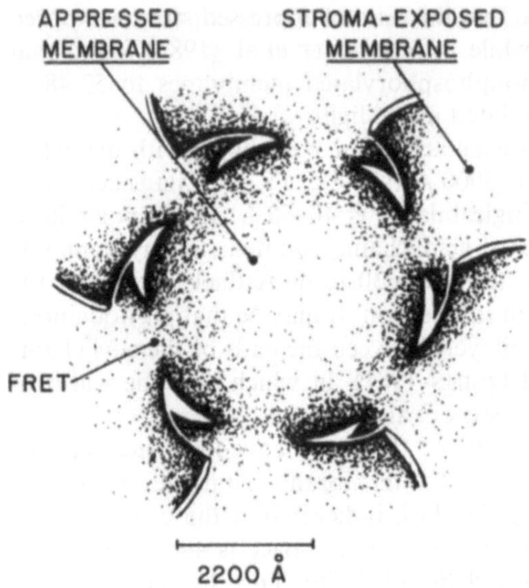

Fig. 2. A single appressed membrane viewed from above that attempts to show communication between appressed and stroma-exposed membranes

Another factor that is likely to reduce lateral motion within the membrane is the extremely high concentration of protein. On a weight-to-weight basis the thylakoid membrane is approximately 70% protein and 30% lipid (FORD et al. 1982; MURPHY and WOODROW 1983). Although the lipids are mainly unsaturated, allowing for a high degree of movement of their acyl chains, the membrane proteins are so crowded that rapid lateral motion may be impossible. This is true for both appressed and stroma-exposed membranes, even though the two membrane regions exhibit differences in their composition (FORD et al. 1982; MURPHY and WOODROW 1983; ARONSON et al. 1983; MURPHY, Chap. 11.11, this Vol.). In appressed membranes, assuming that the integral proteins are globular with a diameter of 100 Å, MURPHY and WOODROW (1983) have calculated that the maximum edge-to-edge distance is 11 Å. Such a dense protein matrix would cause a diffusing molecule to follow a tortuous path, a constraint that is especially important for plastoquinone. However, conclusions based on calculations such as these are necessarily equivocal. There is uncertainty in the organization of the proteins in the membrane; for example, to what extent do the integral proteins extend into the aqueous space, thereby altering the protein-to-lipid ratio in the hydrophobic core? In addition the effect that the protein matrix would have on diffusion is not easy to predict (SAXTON 1982; KELL 1984). For these reasons, the possibility of rapid longrange plastoquinone diffusion cannot be discounted.

Use of the equation derived by EINSTEIN [Eq. (3)] to determine diffusion coefficients from kinetic data requires an estimate of the average distance between reactants, which most often requires a knowledge of the lateral distribution of the components and their concentration in the thylakoid. Values reported in the literature for the amount of electron carriers cover a wide range and are the source of some controversy.[1] For the sake of discussion the values reported by WHITMARSH and ORT (1984) will be used here. In normally developed spinach the ratio of the photosystem II oxygen-evolving complexes to photosystem I to cytochrome b/f to chlorophyll is found to be 1:1:1:600, within 10%. Although the major protein complexes exhibit a one to one stoichiometry the smaller electron carriers are more plentiful. Several types of experiments reveal approximately six molecules of plastoquinone per photosystem II to undergo fast light-induced redox reactions (STIEHL and WITT 1969; MARSHO and KOK 1970; GRAAN and ORT 1984), while the amount of plastocyanin in the thylakoid is one to two molecules per photosystem I (HAEHNEL 1977a; PETERS et al. 1983; GRAAN and ORT 1984).)

In comparative studies of different membrane fractions, the photosystem II complexes are found predominantly in the appressed membrane regions, while stroma-exposed membranes are enriched in photosystem I complexes (SANE et al. 1970; ARNTZEN et al. 1972). In some studies it has been argued that PS I is totally excluded from the appressed membrane region, while 85% of PS II is in appressed membranes and the remaining 15% is in the stroma-exposed membranes (ANDERSON 1982a; ANDERSSON and HAEHNEL 1982; ANDERSON and MELIS 1983; STAEHELIN, Chap. 1, this Vol.). One qualification involved in these models is that the membrane fractionation procedure often yields a small fraction of the original thylakoid. For example, the amount of the appressed inside-out membrane fraction recovered typically contain less than 15% of the total chlorophyll. In general the tacit assumption is made that the recovered fraction is representative of all the appressed membranes. It may be that a third class of membrane must be considered, intermediate between the stroma-exposed and the appressed fraction that is recovered.

In contrast, several studies of the lateral distribution of the cytochrome b/f complex show it to be randomly distributed between appressed and nonappressed regions (COX and ANDERSSON 1981; ANDERSON 1982b; TAKANO et al. 1982; ANDERSON and MALKIN 1982; F.A.L.J. PETERS et al. 1983; MANSFIELD and BENDALL 1984; ALLRED and STAEHELIN 1985). An alternative interpretation of some of these data has led to a model in which the cytochrome b/f complex is concentrated in an annular region around the grana (BARBER 1983; GHIRARDI and MELIS 1983). This suggestion is based in part on the observation that in inside-out membrane fractions derived from appressed regions, no cytochrome b/f was detected (HENRY and MÖLLER 1981). The possibility that the cytochrome b/f complex was lost during the membrane fractionation procedure, or that the fractionation procedure selects for an aberrant fraction of the membrane is difficult to discount. It is significant that in other studies (cited above) the cytochrome b/f complex was detected in a similar inside-out, appressed membrane fraction. Furthermore, recent immunocytochemical labeling studies of ALLRED and STAEHELIN (1986) using intact chloroplasts provides direct evidence for location of the cytochrome b/f complex in both membrane regions. Nevertheless, the possibility that the cytochrome b/f complex occupies the interface be-

1 The amount of photosystem II in normally developed spinach is a controversial issue (see for example, MCCAULEY et al. 1984, and WHITMARSH and ORT 1984).

tween the two membrane regions is an interesting one. In this geometry plastoquinone could move about freely in the stacked regions and plastocyanin in the unstacked regions, while electrons could migrate from one membrane region to another, crossing the interface via the cytochrome b/f complex.

Plastoquinone appears to be present in both the appressed and the nonappressed membrane regions (JENNINGS et al. 1983). The lateral distribution of plastocyanin is difficult to determine, because it is readily lost in the medium during membrane disruption. PETERS et al. (1983) have data for photosystem I-enriched subchloroplast particles from which they argue that plastocyanin is exclusively, or at least preferentially located in the stroma-exposed regions. This interpretation is not supported by the earlier observation of ARNTZEN et al. (1972), in which they detected plastocyanin in the grana membranes.

The distance between the complexes can be calculated using the above information. Because of uncertainties in the distribution and stoichiometries of the components, the distances are likewise subject to considerable uncertainties. Nevertheless, such calculations are useful in establishing the minimum rates of diffusion necessary to achieve measured rates of electron transfer. Consider a thylakoid membrane that satisfies the following conditions: one photosystem II per 600 chlorophyll, 85% of the photosystem II complexes are in the appressed regions, 50% of the membrane area is appressed, and the membrane area per chlorophyll is 200 $Å^2$ (BERZBORN et al. 1981; GREGORY and BRADBEER 1973). Under these conditions the average membrane area per photosystem II in the appressed regions is 7×10^4 $Å^2$. This leads to a maximum center to center distance of nearest neighbor photosystem II complexes of 260 Å. For photosystem II complexes in the nonappressed regions the values are 4×10^5 $Å^2$/PS II and a maximum nearest neighbor distance of 630 Å. Under the assumption that the cytochrome b/f complex is evenly distributed the values are 1.2×10^5 $Å^2$/cyt b/f and a maximum nearest neighbor distance of 350 Å. In the appressed membrane regions the maximum edge-to-edge distance from a photosystem II complex to the nearest cytochrome b/f complex would be less than 200 Å.

The concentration of plastocyanin in the inner aqueous space can be estimated by comparing the amount of plastocyanin on a chlorophyll basis with the amount of water on a chlorophyll basis enclosed by the thylakoid membrane. Using a value of 1 plastocyanin/300 chlorophyll (see above) and 9 ml H_2O/ 1 mmol chlorophyll (FLORES et al. 1983), the concentration of plastocyanin is 370 μmol. It is noteworthy that the inner aqueous volume estimated from electron micrographs is approximately one-half of that indicated by the enclosed water, yielding a higher plastocyanin concentration.

It should be noted that these calculations are subject to uncertainties beyond those already mentioned. The assumption of evenly distributed components is unrealistic, for randomly distributed complexes the nearest neighbor estimates would be reduced by one-half (CHANDRASEKHAR 1943). Additionally, if the complexes formed dimers, as has been suggested for the cytochrome b/f complexes (STAEHELIN and DEWIT 1984; GRAAN and ORT personal communication), these distances would be significantly altered.

4 Lateral Diffusion Between Appressed and Stroma-Exposed Membrane Regions

In recent models of the thylakoid in which photosystem I is excluded from the appressed membrane regions, a mobile component diffusing from appressed to stroma-exposed between the two membrane regions is required to account for electron flow from photosystem II to photosystem I. The likely candidates for shuttling electrons are plastoquinone and plastocyanin. Based on the membrane geometry discussed in the previous section and the kinetics of the interaction of the different components we can estimate the minimum diffusion coefficient necessary to account for the data. In order to provide the most stringent conditions it is best to consider kinetics induced by short flashes rather than to extrapolate from steady state rate measurements. The reasons for this are twofold. First, flash-induced kinetic measurements provide direct measurements of turnover times; these rates are frequently faster than turnover rates calculated from steady-state data. Second, in cases where a large fraction of the component is observed turning over in a single turnover flash, problems associated with membrane heterogeneity may be revealed by biphasic kinetics. By way of example we will consider the transfer of an oxidizing equivalent from photosystem I to the cytochrome b/f complex, that is, diffusion by plastocyanin from the stroma-exposed membranes to cytochrome b/f complexes located in the appressed membranes. In thylakoid membranes P-700, plastocyanin, cytochrome f and the Rieske FeS center are in their reduced state in the dark. Upon illumination P-700 is photooxidized, leading to the oxidation of plastocyanin, which subsequently oxidizes the rapidly equilibrating donors cytochrome f and the Rieske FeS center (WHITMARSH et al. 1982). Using thylakoid membranes isolated from spinach, SELAK and WHITMARSH (1984) showed that in a 20-ms actinic flash all the cytochrome f present in the membrane is oxidized with a half-time of 1 ms or less, and that within 2 ms 75% of the cytochrome was oxidized. The overall reaction, including binding and debinding, electron transfer between the interacting components, and diffusion would have to occur within this time. The minimum diffusion coefficient for plastocyanin necessary to account for these data can be calculated using the equation derived by EINSTEIN [Eq. (3)] that relates the average square displacement ($<r^2>$) to the time of diffusion (τ). The problem is to determine the maximum time it takes plastocyanin to diffuse a known distance. If we accept that the cytochrome b/f complexes are evenly distributed between appressed and stroma-exposed membranes and that photosystem I is located only in the stroma-exposed membranes, and take into account that cytochrome f in the stroma is probably oxidized more rapidly than that in the grana, then 2 ms is the maximum half-time for the oxidation of at least 50% of the cytochrome f in the appressed membranes. Estimating the diffusion distance required for equilibration between the two membrane regions is more arbitrary. If a single grana stack is viewed from above as a circle with a diameter of 4000 Å, a mobile component penetrating from the circumference need only travel 0.3 of the radius (approximately 600 Å) to cover half of the grana area (see Fig. 1A). Since, in addition to penetrating into the appressed region, plastocyanin would have to diffuse from its site of oxidation in the stroma to the grana edge, we have chosen 1000 Å as the minimum average distance required. These values give a minimum diffusion coefficient of 1.2×10^{-8} cm^2 s^{-1}. The actual diffusion coefficient could be larger, since the estimated diffusion distance was the shortest that could possibly allow interaction between grana and stroma lamellae, and the interaction time was the maximum to allow interaction at the furthest extremes of the diffusion path. This value is reasonable for a small molecular weight peripheral protein (Table 2) and, under the above assumptions, is consistent with plastocyanin shuttling electrons between the two membrane regions.

Since the data suggesting the total exclusion of photosystem I from the appressed region are open to alternative interpretations, the possibility that some photosystem I complexes are present in the appressed membranes must be considered. In this case it could be argued that it is not necessary to invoke the diffusion of plastocyanin between different membrane regions to account for the data. For example, cytochrome f in the appressed membranes could be oxidized by a small amount of P-700 located in the appressed regions turning over several times, while P-700 in the stroma-exposed regions could oxidize the cytochrome f located in the stroma-exposed membranes. In this case the vast majority of photosystem I complexes would necessarily be involved solely in cyclic electron transport. A problem with this suggestion is that it does not appear to be compatible with experimental observations of quantum requirements for oxygen evolution that are very near the theoretical minimum assuming no cyclic electron transport. One method to circumvent the problem of multiple turnovers of P-700 is to use short actinic flashes that promote only a single turnover of each reaction center. Using this approach it is possible to observe the oxidation of 62% of the total cytochrome f present in the thylakoid membrane in a single flash with a half-time of 240 μs. Arguing along the same lines as developed above, a minimum diffusion coefficient for plastocyanin of 6×10^{-8} cm^2 s^{-1} was estimated (SELAK and WHITMARSH 1984).

Numerous workers have done similiar calculations to estimate diffusion coefficients for other electron transport components in thylakoid membranes, including plastoquinone (e.g., TAKANO et al. 1982; HAEHNEL 1982; MILLNER and BARBER 1984, see also HAEHNEL, Chap. 10.5, this Vol.). MILLNER and BARBER (1984) conclude that the lateral mobility of plastoquinone should be adequate to account for the observed rates of electron transfer from photosystem II located in appressed membranes to cytochrome b/f complexes located in stroma-exposed membranes.

At this point it is important to stress that using the equation derived by EINSTEIN [Eq. (3)] to estimate diffusion coefficients is indirect and that the results are often equivocal. As the example discussed above shows, the calculation is based on several assumptions, any one of which may be wrong. Indeed, in the existing data I am aware of, there is no direct, unequivocal demonstration that any electron transport component shuttles rapidly between the stroma-exposed and appressed membrane regions. Nevertheless, because the diffusion coefficients calculated above are the minimum values necessary to account for the kinetic data, they can be used to judge the likelihood that the cytochrome b/f complex (or photosystem II or I) acts as a rapidly diffusing carrier between the two membrane regions. Looking at the diffusion coefficients of other large membrane-bound protein complexes (Table 2), one would predict that the diffusion coefficients for photosystem II, photosystem I, and the cytochrome b/f complex would be an order of magnitude too small. In addition, the high density of protein in the membrane, together with possible interactions between one membrane surface and another, would seem to conspire to make the lateral diffusion of the large photosynthetic complexes very slow, even under the most optimistic assumptions, so that it is reasonable to discount diffusion of reaction centers or of cytochrome b/f complexes as significant factors in electron transfer between appressed and stroma-exposed membrane regions. However, because diffusion coefficients for these complexes have not been measured, we cannot rule out surprises.

5 Lateral Diffusion Within a Localized Region

Here the problem is simplified to the extent that we can ignore spatial constraints between different membrane regions and consider a homogeneous membrane. If we assume that the electron transport complexes in appressed membranes are randomly distributed, then the distance between reactants would be less than 200 Å. This short distance means that diffusion of the photosystem II or cytochrome b/f complex could be a significant factor in electron transport. Using a diffusion coefficient of 3×10^{-10} cm^2 s^{-1}, a reasonable value for a large integral complex (Table 2), then the average distance of diffusion in 3 ms would be 190 Å (Table 1). This simple calculation does not take into account the various factors discussed earlier that may severely curtail lateral mobility, and it must be stated that experimental evidence demonstrating rapid diffusion of photosystem II, photosystem I, or the cytochrome b/f complex, even over short distances, is lacking.

The most convincing example of a rapidly mobile electron carrier in thylakoids is provided by plastoquinone. The current view is that plastoquinone acts as a mobile pool between photosystem II reaction centers and cytochrome b/f complexes. This view is supported by a variety of experiments, including the observation that in the light six plastoquinone molecules are reduced for each photosystem II reaction center (STIEHL and WITT 1969). The early evidence for the mobility of plastoquinone was based on inhibition studies of noncyclic electron transport. The results could not be explained by independent electron transport chains, the data indicated that at some point electron sharing was occurring (ORT, Chap. 4, this Vol., DUYSENS 1971; MALKIN and MICHAELI 1971; WILLIAMS 1971; SIGGEL et al. 1972; OUITRAKUL and IZAWA 1973). For example, it was shown by SIGGEL et al. (1972) that inhibition of the majority of the photosystem II reaction centers did not block the light-induced reduction of photosystem I reaction centers, indicating that one photosystem II reaction center could transfer electrons to several photosystem I reaction centers. Further support for this view is provided by evidence that the interaction between the acceptor side of photosystem II and plastoquinone is bimolecular. The notion is that plastoquinone bound at the Q_B site in photosystem II readily exchanges with the mobile plastoquinone in the membrane (see in this Vol. ORT, Chap. 4, ARNTZEN and PAKRASI, Chap. 9.4, KYLE and OHAD, Chap. 9.5, CROFTS and WRAIGHT 1983).

While there is strong evidence that plastoquinone is mobile, the rate of its diffusion is difficult to establish. HAEHNEL (1976) addressed this problem by measuring the kinetics of the reduction of P-700 under conditions where the only electrons available to reduce P-700 had to come from photosystem II. Since P-700 is at the end of a sequence of electron transfer steps, there is a measurable lag between the actinic flash and the onset of P-700 reduction. In spinach thylakoids the lag was short, 2–3 ms. This lag represents the maximum time allowed for the migration of plastoquinone from photosystem II to the cytochrome b/f complex. Depending on the choice of distance between the two complexes, a minimum diffusion coefficient can be calculated, but, as was the case discussed above, establishing the distance of migration is the most uncertain element. For example, in a contrasting view, BOUGES-BOCQUET (1981) and JOLIOT and JOLIOT (1981a, b) have proposed that under some conditions there can be a rapid direct electron transfer from photosystem II to the cytochrome b/f complex that does not involve the mobile plastoquinone pool. [This suggestion is similar to HAEHNEL'S (1982) proposal for the interaction between the cytochrome b/f complex and photosystem I discussed below.]

To understand the behavior of plastocyanin in mediating electron transport between cytochrome f and P-700, particular attention must be paid to the experimental conditions. Under certain conditions the interaction between plastocyanin and P-700 is consistent with a bimolecular reaction, in which plastocyanin and, presumably to a lesser extent, the photosystem I complex are freely diffusing reactants. However, under more physiological conditions the reaction kinetics are complex, and can be fitted by disparate models, including one in which freely diffusing plastocyanin does not have a role in rapid electron transport.

Because plastocyanin has a small extinction coefficient and a broad oxidized minus reduced absorption spectrum, its interaction with P-700 is most often investigated by measuring the reduction of P-700$^+$ by plastocyanin following a light flash. Early measurements of P-700$^+$ reduction kinetics did not yield first-order reaction kinetics, and led to the proposal that the interaction was second-order (WOOD and BENDALL 1975; BOUGES-BOCQUET 1975; DELOSME et al. 1978). Because of its small size and location in the inner aqueous space, plastocyanin was suggested to be a mobile electron carrier enabling a pooling of electrons between the oxidation of plastoquinol and P-700.

This view was supported by HAEHNEL (1977b, 1982), who showed that the effect of DBMIB, an inhibitor of plastoquinone oxidation by the cytochrome b/f complex, on the reduction kinetics of P-700 could be most easily explained assuming a pooling of electrons by an electron transport component acting between the quinol oxidation site and P-700 (see below); plastocyanin was the obvious candidate. A problem with this argument is that DBMIB appears to be mobile during the reaction time, i.e., it moves from one cytochrome b/f complex to another within 1 ms (VELTHUYS 1982; JONES and WHITMARSH personal observation), so that the question of electron sharing between plastoquinol and P-700 remains unanswered.

Strong support for plastocyanin as a mobile carrier comes from studies in which photosystem I reaction centers or photosystem I-enriched subchloroplast particles provide the experimental material (HAEHNEL et al. 1980a; OLSON and COX 1982; TAKABE et al. 1983; OLSEN and PEDERSEN 1983). In these experiments the rate of P-700$^+$ reduction following a light flash was measured as a function of the concentration of plastocyanin. The reaction obeyed second-order kinetics. Plots of the inverse of the half-time of P-700 reduction versus the plastocyanin concentration revealed a linear relationship (WILKINSON 1980) yielding second order rate constants ranging from 1×10^5 to 1.5×10^8 M^{-1} s^{-1}. In addition to demonstrating that plastocyanin was capable of interacting rapidly with P-700 by diffusion, the importance of the membrane surface was studied by changing the ionic strength of the reaction medium. Because both plastocyanin and the lumen-exposed surface of the thylakoid membrane are negatively charged, the rate of P-700 reduction is strongly dependent on the ionic strength of the reaction medium (e.g., TAMURA et al. 1980; TAKABE et al. 1983; OLSEN and COX 1982).

While there is agreement that there are conditions under which plastocyanin acts as a freely mobile electron carrier, its behavior under more physiological conditions is a source of controversy. The general observation is that the reduction of P-700 by plastocyanin following a light flash is not first-order. However, since under physiological conditions the reactants are not easily altered (e.g., changing the concentration of plastocyanin), the data are fewer, and can be fitted to second-order reaction kinetics (WOOD and BENDALL 1975; DELOSME et al. 1978; OLSEN et al. 1980; OLSEN 1982; OLSEN and PEDERSEN 1983) or to two first-order reactions or a combination of a first-order and a second-order reaction (HAEHNEL Chap. 10.5, this Vol.; HAEHNEL et al. 1980a, b; HAEHNEL 1982). OLSEN and colleagues (OLSEN et al. 1980; OLSEN and COX 1982; OLSON and PEDERSEN 1983) have studied the behavior of plastocyanin at subzero temperatures in an organic/aqueous media. The reduction kinetics of P-700$^+$, although obviously biphasic, fit a second-order reaction quite well. Further support for a bimolecular reaction is provided by the observation that the rate of P-700$^+$ reduction by plastocyanin is slowed 2.0- to 2.4-fold as the viscosity of the inner aqueous phase is increased, indicating a diffusion-controlled reaction. This work provides convincing evidence for a mobile pool of plastocyanin in thylakoid membranes at subzero temperatures in a nonphysiological medium, the difficulty is in extrapolating these conclusions to the action of plastocyanin in thylakoids at room temperature under physiological conditions.

The problem with interpreting the room temperature kinetics of P-700$^+$ reduction is that they are not adequate to distinguish between two first-order and a second-order reaction. To overcome this problem HAEHNEL (1982, 1984b) has studied the redox kinetics of cytochrome f, P-700, and to some extent plastocyanin, as a function of partial inhibition of electron transport. This is a common way to distinguish electron transport between noninteracting chains and between linear reactions that can share electrons between components.

For example, consider the reduction of P-700 by electrons from plastoquinol. In the case of noninteracting electron transport chains inhibition of 50% of the cytochrome b/f complexes would result in a 50% decrease in the amount of P-700 reduced and no change in the rate of its reduction. In the case of a freely diffusing pool of plastocyanin, inhibition of 50% of the cytochrome b/f complexes would still allow the reduction of all of the P-700, but the rate would be slowed by two-fold. The experimental procedure consisted of measuring the reduction of P-700$^+$ by electrons donated by plastoquinol following a long flash of light as a function of plastocyanin inhibition (HAEHNEL 1982). In the control experiments using inhibited thylakoids the reduction kinetics of P-700$^+$ were monophasic. In experiments using thylakoids in which plastocyanin was impaired, the reduction of P-700$^+$ was obviously biphasic, indicating a fraction of P-700$^+$ that was insensitive to inhibition and a fraction that was sensitive to plastocyanin inhibition. Similarly, the oxidation kinetics of cytochrome f were monitored (HAEHNEL 1984b), and provided evidence that some cytochrome f was not accessible to the remaining active plastocyanin. These observations form the basis for a model of electron transport from plastoquinol to P-700 that is quite different from the ones discussed above. To account for these data, HAEHNEL proposed that plastocyanin forms a complex with both the photosystem I complex and the cytochrome b/f complex. In this model the question of whether photosystem I and the cytochrome b/f complex each combine with an individual molecule of plastocyanin, or whether two complexes combine with a single molecule of plastocyanin, appears to be left open. Because of the distribution of photosystem I, this model would seem to apply to the stroma-exposed membrane regions. Support for complexed plastocyanin as an activator of the cytochrome b/f complex is provided by the observation that the steady-state rate of electron transport is highly sensitive to plastocyanin inhibition (HAEHNEL 1982), which would not be the case if plastocyanin acted as a rapid mobile pool.

There are two aspects of these data that need to be examined with care. One is the question of whether inhibited plastocyanin may still bind to the cytochrome b/f complex and competitively inhibit electron transport (HAEHNEL 1982). The second is that the extent of cytochrome f oxidation observed by HAEHNEL in a long flash is half of that observed under similar conditions in other studies (WHITMARSH and CRAMER 1979; SELAK and WHITMARSH 1984).

6 Concluding Remarks

In a process as complex as photosynthesis, the factors that limit and regulate the rate of electron transport are many, and no single element exerts total control. Recently it has been suggested that diffusion of plastoquinone may be a significant rate-limiting factor. Based on this possibility, diffusion-controlled electron transport has been considered as a limiting factor in carbon reduction. While this review draws attention to the fact that the role of diffusion in electron transport is not well characterized, there are convincing arguments that diffusion is not rate-limiting. At the heart of these arguments is evidence that the oxidation of plastoquinol is the dominant rate-limiting reaction in electron transport. While there is no evidence that diffusion of either plastoquinone or plastocyanin is limiting the rate of electron transport, uncertainties with regard to the importance of lateral diffusion in photosynthesis would be removed if more direct techniques for measuring diffusion coefficients were applied to thylakoid membranes. A beginning in this direction is represented by the work of FRAGATA et al. (1984), in which the diffusion coefficient of plastocyanin was measured in lipid vesicles using the method of fluorescence recovery after photobleaching (Table 2).

Acknowledgements. I wish to thank Drs. J. CHEESEMAN, W.A. CRAMER, R.A. DILLEY, D.W. KROGMANN, D.R. ORT, and C. WRAIGHT for helpful discussions and their critical comments.

References

Allred DR, Staehelin LA (1986) Plant Physiol in press
Alonso A, Gomez-Fernandez JC, Aranda FJ, Belda FJF, Goni FM (1981) FEBS Lett 132:19–22
Anderson JM (1982a) Photobiochem Photobiophys 3:225–241
Anderson JM (1982b) FEBS Lett 138:62–66
Anderson JM, Malkin R (1982) FEBS Lett 148:293–296
Anderson JM, Melis A (1983) Proc Natl Acad USA 80:745–749
Anderson JM, Goodchild DJ, Boardman NK (1973) Biochim Biophys Acta 325:573–585
Andersson B, Haehnel W (1982) FEBS Lett 146:13–17
Arntzen CJ, Dilley RA, Peters GA, Shaw ER (1972) Biochim Biophys Acta 256:85–107
Aronson H, Waggoner C, More J, Bery S (1983) Biochim Biophys Acta 725:519–528
Axelrod D (1983) J Membr Biol 75:1–10
Barber J (1983) Plant Cell Environ 6:311–322
Berg SP, Lusczakoski DM, Morse PD (1979) Arch Biochem Biophys 194:138–148
Berzborn RJ, Müller D, Roos P, Andersson B (1981) In: Akoyunoglou G (ed) Photosynthesis III. Structure and molecular organisation of the photosynthetic apparatus. Balaban, Philadelphia, pp 107–120
Bouges-Bocquet B (1975) Biochim Biophys Acta 396:382–391
Bouges-Bocquet B (1981) In: Akoyunoglou G (ed) Proc 5th Int Congr Photosynth, vol II. Balaban, Philadelphia, pp 19–27
Brangeon J, Mustardy L (1979) Biol Cell 36:71–80
Chandrasekhar S (1943) Rev Mod Phys 15:1–89
Cherry RJ (1979) Biochim Biophys Acta 559:289–327
Colman PM, Freeman HC, Guss JM, Murata M, Norris VA (1978) Nature 272:319–324
Cox RP, Andersson B (1981) Biochem Biophys Res Commun 103:1336–1342
Crank J (1975) The mathematics of diffusion, chaps 1 and 10. Clarendon, Oxford
Crofts AR, Wraight CA (1983) Biochim Biophys Acta 726:149–185
Daniels F, Alberty RA (1967) Physical chemistry, chap 11. Wiley, New York
Delosme R, Zickler A, Joliot P (1978) Biochim Biophys Acta 504:165–174
Duysens LNM (1971) In: Forti G, Avron M, Melandri A (eds) Proc 2nd Int Congr Photosynth Res. Junk, The Hague, pp 19–25
Edidin M (1981) In: Finean M (eds) Membrane structure, chap 2. Elsevier/North-Holland Biomedical Press, Amsterdam New York, pp 39–82

Einstein A (1956) In: Fürth R (ed) Investigations on the theory of the Brownian movement. Dover, New York
Fellows RJ, Boyer JS (1976) Planta 132:229–239
Flores S, Graan T, Ort DR (1983) Photobiochem Photobiophys 6:293–304
Ford RC, Chapman DJ, Barber J, Pedersen JZ, Cox RP (1982) Biochim Biophys Acta 681:145–151
Fragata M, Ohnishi S, Adada K, Ito T, Takahashi M (1984) Biochemistry 23:4044–4051
Froud RJ, Ragan CI (1984) Biochem J 217:561–571
Ghirardi ML, Melis A (1983) Plant Physiol 74:993–998
Graan T, Ort DR (1984) J Biol Chem 259:14003–14010
Gregory P, Bradbeer JW (1973) Planta 109:317–326
Gupte S, Wu E-S, Hoechli M, Jacobson K, Sowers AE, Hackenbrock CR (1984) Proc Natl Acad Sci USA 81:2606–2610
Haehnel W (1976) Biochim Biophys Acta 440:506–521
Haehnel W (1977a) Biochim Biophys Acta 459:418–441
Haehnel W (1977b) In: Packer L, Papageorgiou GC, Trebst A (eds) Bioenergetics of membranes. Elsevier/North-Holland Biomedical Press, Amsterdam New York, pp 317–328
Haehnel W (1982) Biochim Biophys Acta 682:245–257
Haehnel W (1984a) Annu Rev Plant Physiol 35:659–693
Haehnel W (1984b) In: Sybesma C (ed) Advances in photosynthesis research. Nijhoff/Junk The Hague, pp 545–548
Haehnel W, Hesse V, Pröpper A (1980a) FEBS Lett 111:79–82
Haehnel W, Pröpper A, Krause H (1980b) Biochim Biophys Acta 593:384–399
Hardt SL (1979) Biophys Chem 10:239–243
Henry LEA, Möller BL (1981) Carlsberg Res Commun 46:227–242
Houslay MD, Stanley KK (1982) Dynamics of biological membranes. Wiley, New York, pp 85–90
Jennings RC, Garlaschi FM, Gerola PD (1983) Biochim Biophys Acta 722:144–149
Johannsson A, Keightley CA, Smith GA, Richards CD, Hesketh TR, Metcalfe JC (1981) J Biol Chem 256:1643–1650
Joliot P, Joliot A (1981a) FEBS Lett 134:155–158
Joliot P, Joliot A (1981b) In: Akoyunoglou G (ed) Proc 5th Int Congr Photosynth, vol III. Balaban, Philadelphia, pp 885–889
Katsikas H, Quinn PJ (1981) FEBS Lett 133:230–234
Kawato S, Sigel E, Carafoli E, Cherry RJ (1981) J Biol Chem 256:7518–7527
Kawato S, Lehner C, Müller M, Cherry RJ (1982) J Biol Chem 257:6470–6476
Kell DB (1984) TIBS 9:86–88
Kyle DJ, Staehelin LA, Arntzen CJ (1983) Arch Biochem Biophys 222:527–541
Malkin S, Michaeli G (1971) In: Forti G, Avron M, Melandri A (eds) Proc 2nd Int Congr Photosynth Res. Junk, The Hague, pp 149–167
Mansfield RW, Bendall DR (1984) Biochim Biophys Acta 766:62–69
Marsho TV, Kok B (1970) Biochim Biophys Acta 223:240
McCauley SW, Taylor SE, Dennenberg RJ, Melis A (1984) Biochim Biophys Acta 765:186–195
Miller MM, Nobel PS (1972) Plant Physiol 49:535–541
Millner PA, Barber J (1984) FEBS Lett 169:1–6
Murphy DJ, Woodrow IE (1983) Biochim Biophys Acta 725:104–112
Mustardy LA, Janossy AGS (1979) Plant Sci Lett 16:281–289
Olsen LF (1982) Biochim Biophys Acta 682:482–490
Olsen LF, Cox RP (1982) Biochim Biophys Acta 679:436–443
Olsen LF, Pedersen JZ (1983) Photobiochem Photobiophys 5:1–10
Olsen LF, Cox RP, Barber J (1980) FEBS Lett 122:13–16
Ouitrakul R, Izawa S (1973) Biochim Biophys Acta 305:105–118
Overfield RE, Wraight CA (1980) Biochemistry 19:3328–3334
Paolillo DJ Jr (1970) J Cell Sci 6:243–255
Peters R, Cherry RJ (1982) Proc Natl Acad Sci USA 79:4317–4321

Peters FALJ, Wielink JE van, Sang HWWF, Vries S de, Kraayenhof R (1983) Biochim Biophys Acta 722:460–470
Saffman PG, Delbrück M (1975) Proc Natl Acad Sci USA 72:3111–3113
Sane PV, Goodchild DJ, Park RB (1970) Biochim Biophys Acta 216:162–178
Saxton MJ (1982) Biophys J 39:165–173
Selak MA, Whitmarsh J (1984) Photochem Photobiol 39:485–489
Siggel U, Renger G, Stiehl HH, Rumberg B (1972) Biochim Biophys Acta 256:328–335
Staehelin LA, DeWit M (1984) J Cell Biochem 24:261–269
Stiehl HH, Witt HT (1969) Z Naturforsch 246:1588–1598
Takabe T, Ishikawa H, Niwa S, Itoh S (1983) J Biochem 94:1901–1911
Takano M, Takahashi M-A, Asada K (1982) Arch Biochem Biophys 218:369–375
Tamura N, Yamamoto T, Nishimura M (1980) Biochim Biophys Acta 592:536–545
Ulrich EL, Girvin ME, Cramer WA, Markley JL (1985) Biochemistry 24:2501–2508
Vaz WLC, Goodsaid-Zalduondo F, Jacobson K (1984) FEBS Lett 174:199–207
Velthuys BR (1982) In: Trumpower B (ed) Function of quinones in energy transducing systems. Academic Press, London New York, pp 401–408
Whitmarsh J, Cramer WA (1979) Biophys J 26:223–234
Whitmarsh J, Ort DR (1984) Arch Biochem Biophys 231:378–389
Whitmarsh J, Bowyer JR, Crofts AR (1982) Biochim Biophys Acta 682:404–412
Wilkinson F (1980) Chemical kinetics and reaction mechanisms, chap 6. Van Nostrand Reinhold, New York
Williams WP (1971) In: Forti G, Avron M, Melandri A (eds) Proc Int Congr Photosynth Res. Junk, The Hague, pp 745–752
Wood PM, Bendall DS (1975) Biochim Biophys Acta 387:115–128
Wu E-S, Yang CS (1984) Biochemistry 23:28–33

10.3 Cytochrome b_6/f Complex – Quinol Interactions[1]

P. JOLIOT and A. JOLIOT

1 Introduction

The photosynthetic electron transfer chain is associated with three transmembrane protein complexes, Photosystem II (PS II), cytochrome b_6/f complex and photosystem I (PS I). A plastoquinone (PQ) pool, which freely diffuses in the lipophilic domain of the membrane, establishes a link between PS II and the cytochrome b_6/f complex. Electron transfer between the cytochrome b_6/f complex and PS I probably involves the diffusion of plastocyanin in the internal aqueous phase of the thylakoid. Until 10 years ago, the generally accepted model of photosynthetic electron transfer was the classical Z-scheme in which the two electrons of plastoquinol are sequentially transferred to cytochrome f via the Fe–S protein (see ORT, Chap. 4, this Vol.). From the measurement of field-indicating absorption changes at 515 nm, it was deduced that electron acceptors and donors of both photosystems are located on the outer and the inner faces of the membrane, respectively (see review by WITT 1971). Based on the Z-scheme and on this structural model, which agrees with Mitchell's chemiosmotic theory (1966), one can predict that one proton is pumped through the membrane for each electron transferred from PS II to PS I. However, this model also predicts that the electron transfer process from PS II acceptors to PS I donors would be associated with a large dissipation of free energy, since the difference of redox potentials between the primary PS II quinone acceptor Q_A and the primary PS I donor P-700 is close to 400 mV. The overall equilibrium constant between Q_A and P-700 would be higher than 1×10^6 under uncoupled conditions or 1×10^3 in the presence of a transmembrane proton gradient of 3 pH units (180 mV). The apparent equilibrium constant measured in weak light, i.e., in conditions in which any kinetic limitation can be excluded, is actually lower than 10 (JOLIOT et al. 1968). Similarly, the apparent equilibrium constant between plastoquinone and P-700 is close to 1. The classical Z-scheme does not account for these data.

2 Slow Electrogenic Phase

Flash illumination of *Chlorella* induces a biphasic increase of the transmembrane potential which is measured by the field-indicating change at 515 nm (WITT and MORAW 1959; JOLIOT and DELOSME 1974). The fast phase (phase a, $t\frac{1}{2}$ <100 ns) is associated with the photochemical charge separation occurring in both PS I and PS II centers, while the slow phase (phase b $t\frac{1}{2} \sim 10$ ms) is only

1 This article is dedicated to the memory of Bernadette Bouges-Bocquet

dependent upon PS I activity. For instance, phase b is not modified when using mutants which lack PS II and is absent when using PS I-depleted mutants. The amplitude of phase b is equal to the fraction of phase a which is associated with PS I primary charge separation (JOLIOT and DELOSME 1974). BOUGES-BOCQUET (1977) established that phase b was kinetically correlated with the reduction of the secondary PS I donors cytochrome f and plastocyanin by a proton carrier R′ which is probably a plastoquinone, as shown by the inhibitory effect of 2,5-dibromo-3-methyl-6-isopropyl-p-benzoquinone (DBMIB). Later, VELTHUYS (1978) showed that the slow increasing phase could be observed in isolated chloroplasts when a fraction of the plastoquinone pool was reduced prior to a flash illumination. Under oxidizing conditions, VELTHUYS (1980) observed that the amplitude of phase b oscillates with a period-2 in synchrony with the formation of plastoquinol through the two-electron gate mechanism (BOUGES-BOCQUET 1973). The electrogenic process responsible for phase b is associated with the reoxidation of the plastoquinol formed either by PS II or through the reducing side associated with the chloroplast respiratory chain characterized in living cells by BENNOUN (1982). BOUGES-BOCQUET (1980) observed period-2 oscillations of opposite phase to those observed by VELTHUYS following treatment of Chlorella cells with 0.1 mM dithionite to reduce an appreciable fraction of the plastoquinone pool. No abnormal damping of these oscillations was observed, which suggests that plastoquinol formed by PS II rapidly reacts with PS I secondary donors before equilibrating with the plastoquinone pool. The amplitude of phase b oscillations reported by VELTHUYS and by BOUGES-BOCQUET is small compared to that of phase a (<20%). Upon illumination by a series of flashes 7 s apart, we recently observed (JOLIOT and JOLIOT 1985) large phase b oscillations in *Chlorella sorokiniana* briefly treated with 0.3 mM benzoquinone and then washed. Under these conditions, the ATPases and the chloroplast respiratory chain are blocked. The plastoquinone pool is fully oxidized and the PS I donors are fully reduced. In the presence of ferricyanide, which oxidizes the secondary PS II acceptor Q_B^- in Q_B, the amplitude of phase b decreases to close to zero upon the first flash. A large phase b is induced by the second flash and corresponds to 1.2 electrons transferred via the slow electrogenic process. Analysis of the spectral changes due to the PS I donors (P-700, plastocyanin, cytochrome f) indicates that during the 7 s interval between flashes, all the electrons present in plastoquinol are transferred to PS I donors. Phase b kinetics under these conditions are strikingly slow ($t_{\frac{1}{2}} \sim 40$–70 ms) when compared to the kinetics observed in quinone-untreated cells, or in chloroplasts in reducing conditions ($t_{\frac{1}{2}} \sim 4$–10 ms). Surprisingly enough, the kinetics are first-order and their half-times are independent of the flash number, i.e., of the amount of plastoquinol formed by PS II.

3 Proton Translocation Associated with the Intersystem Electron Transfer

FOWLER and KOK (1976) and FOWLER (1977) were the first to report a H^+/e^- ratio close to 2 for the intersystem electron transfer reaction, i.e., twice what is expected from the classical Z-scheme. This ratio of 2 was determined with a pH electrode by measuring the proton uptake on the outside of the thylakoid membrane in the absence of uncoupler, or the proton release in the internal phase in the presence of uncoupler. Using dye indicators, VELTHUYS (1978) confirmed that the H^+/e^- ratio was higher than 1, and moreover showed (VELTHUYS 1980) that during a flash series, the proton release at the inner face of the membrane approximately correlates in phase and in kinetics with the slow electrogenic phase. Nevertheless, some aspects of Fowler's experiments highlight certain problems. Thus, FOWLER observed small period-2 oscillations for the proton uptake in contrast to large ones seen for the proton release at the inner face of the membrane. Under the highly oxidizing conditions used by FOWLER (1 mM ferricyanide) the secondary PS II acceptor Q_B is in its oxidized form. Hence, no plastoquinol is formed after the first flash and consequently no intersystem electron transfer can occur. The uptake of 2 protons observed by FOWLER after the first flash is then unexpected. FOWLER also reported an increase of two more protons when ferricyanide is replaced by methylviologen while the expected value should be one. One cannot exclude a calibration error of factor 2, a possibility already raised by SAPHON and CROFTS (1977). Recents measurements by HOPE and MATTHEWS (1983) give an average H^+/e^- ratio of 1 for the proton uptake in the presence of ferricyanide. It can be concluded that a large uncertainty exists in the value of the H^+/e^- ratio associated with the intersystem electron transfer.

4 Reverse Electron Flow

AVRON and co-workers extensively studied the process of reverse electron flow observed when a large proton gradient is generated by ATP hydrolysis. SCHREIBER and AVRON (1979) reported a large reduction of the plastoquinone pool when ATP is added to chloroplasts, the ATPases of which were activated by dithiothreitol (DTT) and pre-illumination. Under these conditions, a large proton gradient is developed and cytochrome f becomes partially oxidized (RIENITS et al. 1974). Despite the presence of a strong reductant (DTT) and of a mediator such as phenazine methosulfate (PMS), no plastoquinone reduction is observed in the absence of a proton gradient, which means that this process is kinetically and not thermodynamically limited. In terms of the classical Z-scheme, the reduction of plastoquinone by PS I donors occurs exclusively at the inner face of the membrane, and should not be dependent upon a proton gradient. Reverse electron flow provides a new argument in favor of the existence of an additional electron or proton transmembrane process, coupled to the reactions between plastoquinol and PS I donors.

5 Cytochrome b_6/f Complex

The cytochrome b_6/f complex which is utilized in similar forms by most energy-converting membrane systems (HAUSKA, Chap. 10.1; PRINCE, Chap. 10.4; DUT-

TON, Chap. 5, this Vol.) has been isolated from chloroplasts by HURT and HAUSKA (1981). Cytochrome f is located close to the inner face of the membrane as no electrogenic phase is associated with its PS I-induced oxidation. In contrast, analysis of the sequence of the cytochrome b_6 protein has revealed that one cytochrome b_6 heme is probably located close to the outer face of the membrane and the second close to the inner face (SARASTE and WIKSTRÖM 1983; SARASTE 1984; WIDGER et al. 1984).

RUMBERG (1966) was the first to propose that the oxidation of plastoquinol could lead to a concerted reduction of both cytochromes f and b_6. This process can be related to the oxidant-induced reduction of mitochondrial cytochrome b (see review by WIKSTRÖM 1973). HURT and HAUSKA (1982) reported that cytochrome b_6 of the isolated cytochrome b_6/f complex can be reduced by a pulse of ferricyanide in the presence of plastoquinol, a process similar to that observed for mitochondrial cytochrome b/c complex.

VELTHUYS (1979) confirmed RUMBERG's interpretation by demonstrating a cytochrome b_6 reduction upon illumination by a pair of flashes of chloroplasts under oxidizing conditions. A generally accepted interpretation for the reduction of cytochrome b_6 proposed particularly by MITCHELL (1975) is shown in Scheme 1.

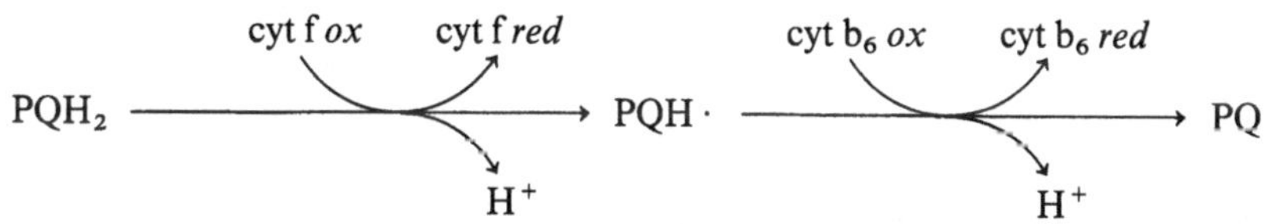

Scheme 1

In this model, the redox potential of the $PQH_2/PQH\cdot$ couple is supposed to be more positive than the potential of the couple $PQH\cdot/PQ$, the value of which must be sufficiently negative to promote cytochrome b_6 reduction (see Sect. 6).

Redox changes of cytochrome b_6 have been studied by CROWTHER and HIND (1980) in chloroplasts maintained under reducing conditions and subjected to short flashes of illumination. In such samples a well-developed slow electrogenic phase is observed. A fast reduction of about 0.3 molecules of cytochrome b_6 is achieved in less than 3 ms, while the reoxidation occurs in the 15 to 45 ms range, a time much longer than the rise time of the slow electrogenic phase (~5 ms). SELAK and WHITMARSH (1982) and JONES et al. (1984) observed that in the presence of 2-heptyl-4-hydroxyquinoline N-oxide (HQNO) or 2-n-nonyl-hydroxyquinoline N-oxide (NQNO) respectively, the flash-induced reduction of cytochrome b_6 was much larger than in the control and corresponded to 0.6 cytochrome b_6 molecules per complex. Moreover, these inhibitors strongly decreased the amplitude of phase b. The authors concluded that a major fraction of the slow electrogenic phase is associated with the cytochrome b_6 oxidation which would be inhibited by HQNO or NQNO.

JOLIOT and JOLIOT (1984a) confirmed and extended VELTHUYS' results (1979) by measuring in chloroplasts under oxidizing conditions (ferricyanide) the reduction of cytochrome b_6 and of the PS I donors, P-700, plastocyanin, and cytochrome f. A group of two flashes induced the formation of almost one plastoquinol molecule via the two-electron gate mechanism occurring on the PS II acceptor side. The oxidation of a part of this plastoquinol leads to a reduction ($t1/2 \sim 10$ ms) of both cytochrome b_6 and PS I donors in agreement with Scheme 1. A subsequent slow phase in the reduction of the PS I donors is then observed and is not associated with further reduction of cytochrome b_6. After a group of four flashes, a new plastoquinol is formed and a large and slow ($t\frac{1}{2} \sim 100$–200 ms) reduction of PS I donors is observed without significant changes in the redox state of cytochrome b_6. We do not know at present if this slow reduction of PS I donors involves cytochrome b_6 turnover or if all the electrons are sequentially transferred via the Fe–S-cytochrome f pathway.

LAVERGNE (1983) reported that cytochrome b_6 can be photoreduced under conditions where the classical oxidant-induced process can be excluded. The experiment is performed on a *C. sorokiniana* mutant lacking PS I, i.e., with PS I donors which remain always reduced. A large cytochrome b_6 reduction is observed upon illumination in continuous light. This reduction requires that a fraction of the plastoquinone pool is reduced and that a large membrane potential is developed across the membrane. LAVERGNE'S result shows that cytochrome b_6 is reduced by an electron carrier G which presents characteristic spectral changes in the blue region of the spectrum. The equilibrium constant between G and cytochrome b_6 depends upon the membrane potential, which shows that G is located closer to the outer face of the membrane than cytochrome b_6. It is very likely that G is interacting with plastoquinol at a site close to the outer face of the membrane distinct from the site where the oxidant-induced reduction of cytochrome b_6 occurs.

Relative location of PS II and cytochrome b_6/f complex: Most PS II centers are known to be located in appressed thylakoid membrane regions, while PS I is located mainly in the nonappressed regions (STAEHELIN, Chap. 1, this Vol.). The distribution of the cytochrome b_6/f complex between appressed and nonappressed membrane regions is a matter of controversy. When algae or chloroplasts are given two flashes under oxidizing conditions, i.e., when the only available plastoquinol is formed by PS II, no lag longer than 1 ms is observed in the reduction of cytochrome b_6 (JOLIOT and JOLIOT 1984a) or in the slow electrogenic phase (JOLIOT and JOLIOT 1985). CROFTS and WRAIGHT (1983) estimated the time of diffusion of plastoquinol from appressed to nonappressed regions to be 100 ms. One has then to conclude that a large fraction of the cytochrome b_6/f complex is close to PS II, i.e., located in the appressed region of the membrane. Recent immunocytochemical staining of intact chloroplasts have confirmed this distribution (ALLRED and STAEHELIN 1985). As already mentioned, the phase b oscillations observed by BOUGES-BOCQUET (1980) and the measurements of phase b turnover (BOUGES-BOCQUET 1981b) also suggest that plastoquinol is directly transferred from PS II to cytochrome b_6/f complex, without equilibrating with the plastoquinone pool.

6 Redox Potentials of the Carriers Involved in the Cytochrome b_6/f Complex and in the Slow Electrogenic Phase

The data in the literature which are obtained either by direct titration or by measurement of equilibrium constants between the different carriers lead to a confusing picture. An obvious technical limitation in the redox titration is the rate of equilibration between soluble mediators and electron carriers, most of which are included in a highly hydrophobic environment. One must consider with great caution all titrations performed under repetitive illumination, where true equilibrium conditions are not satisfied.

Cytochrome f. Titration of cytochrome f leads to an E_{m7} value close to 360 mV (see review by BENDALL 1982). Assuming a redox potential of 450 to 480 mV for P-700, one can predict a large equilibrium constant ($20<K<100$) between cytochrome f and P-700, which contracts with the low value ($K \sim 2$) measured by JOLIOT and JOLIOT (1984b) under mild reducing conditions. However, under oxidizing conditions, these authors measured a larger equilibrium constant ($K \sim 20$) during the course of reduction of PS I donors by plastoquinol molecules formed by PS II.

Fe–S Protein. The generally admitted E_{m7} value for the Fe–S center is +280 mV, i.e., 60 mV lower than the E_{m7} value for cytochrome f (see review by TRUMPOWER 1981). JOLIOT and JOLIOT (1984b) measured the oxidation of PS I donors using weak continuous light in the presence of 3-(3,4-dichlorophenyl)-1,1-dimethylurea (DCMU) and ascorbate and did not observe a lag in the cytochrome f oxidation kinetics. One must conclude that the redox potential of the Fe–S center is either lower than 280 mV (i.e., Fe–S fully oxidized in the presence of ascorbate) or is close to the redox potential of cytochrome f.

Cytochrome b_6. The value of the redox potential of cytochrome b_6 is also controversial (see review by RICH 1984). GIRVIN and CRAMER (1984) recently measured E_{m7} values of −30 mV for the titration in the dark and −75 mV for the photoreducible cytochrome b_6.

Slow Electrogenic Phase. The titrations of GIRVIN and CRAMER (1984) are the more reliable since they were performed with long intervals between flashes. These authors reported an E_{m7} of +100 mV with a slope corresponding to $n=2$ and a pH dependence of 60 mV per pH unit, values which are characteristic for a plastoquinol-plastoquinone couple. As previously reported by BOUGES-BOCQUET (1981a), the amplitude of the slow phase markedly decreases with the pH. The striking feature of the titrations of GIRVIN and CRAMER is that the amplitude and the half-time of phase b are independent of the redox potential between 50 mV and −200 mV. This agrees with previous titrations by DINER and DELOSME (1983), who also observed a PS I-dependent slow electrogenic phase in the ms range with titrates at a potential as low as −240 mV.

If we admit that both titrations and equilibrium constant determinations are reliable, we have to assume that the redox potentials of the different carriers can vary greatly depending upon the experimental conditions. Among possible interpretations are (1) structural changes of the protein with which the electron carriers are associated; (2) change in the electric charge of sites located in the vicinity of the electron carrier. The charged species could be a proton which can be fixed either directly on the electron carrier itself or on a site of the apoprotein sufficiently close to the electron carrier (Bohr effect). Alternatively, mutual electrostatic interactions could occur between several electrons carriers included in the same protein complex. Such a mechanism has been proposed by JOLIOT and JOLIOT (1984b) for interactions between cytochrome b_6 and cytochrome f.

7 Models

At the time we are writing this review, the most popular models for the intersystem electron transfer derive from the Q-cycle model of MITCHELL (1966). The cytochrome b_6/f complex is believed to include two sites able to fix plastoquinone which are located close to the outer (site C) and the inner (site Z) faces of the membrane (Scheme 2).

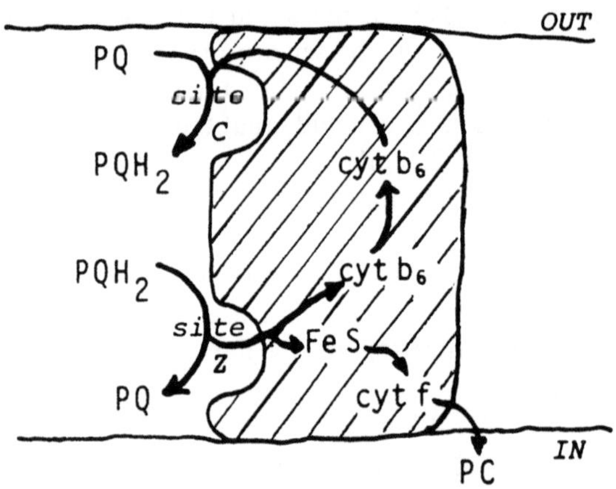

Scheme 2

Plastoquinol is first oxidized at site Z by a concerted reaction which leads to the reduction of cytochrome f and of the cytochrome b_6 located close to the inner face of the membrane. GLASER and CROFTS (1984) in chromatophores and JONES et al. (1984) in chloroplasts proposed that the electrogenic step is mainly associated with the electron transfer between the two b_6 cytochromes. The second cytochrome b_6 located close to the outer face is then reoxidized by a plastoquinone molecule fixed at the site C. Most of the arguments which favor the Q-cycle model concern reactions occurring at site Z (oxidant-induced reduction of cytochrome b_6). Reactions occurring at site C are inferred only indirectly.

The following data remain difficult to interpret in terms of the Q-cycle, at least without invoking additional hypotheses (see also discussion of WHITMARSH, Chap. 10.2, this Vol.): (1) GIRVIN and CRAMER (1984) observed that phase b is independent of the redox state of cytochrome b_6 prior to the flash excitation. Recently, JOLIOT and JOLIOT (1985) measured phase b and flash-induced electron transfer reactions in living algae after a long anaerobic incubation. Despite the fact that most cytochrome b_6 had been reduced, a first flash induced a phase b of normal amplitude and kinetics. Contrary to GIRVIN and CRAMER (1984), we observed a large photooxidation of cytochrome b_6 ($\sim$0.9 molecule per cytochrome b_6/f complex), the duration of which largely exceeded that of phase b. A second flash induced a transient reduction of cytochrome b_6 previously oxidized by the first flash. The flash-induced redox changes of cytochrome f and phase b both for amplitude and kinetics remained identical after the first or the second flash. The experiments performed under reducing conditions in isolated chloroplasts and living algae suggest that phase b is associated with the electron transfer from plastoquinol to PS I donors and does not depend upon cytochrome b_6 turnover. (2) SELAK and WHITMARSH (1982) reported that in the presence of HQNO, the flash-induced reduction of cytochrome b_6 precedes the reduction of cytochrome f. Similarly, JOLIOT and JOLIOT (1985) observed that in algae under oxidizing or reducing conditions, the initial rate of reduction of cytochrome b_6 is much faster than the reduction of cytochrome f. These results are in apparent contradiction with the process of concerted oxidation of plastoquinol depicted in Scheme 1. This contradiction can be solved if one assumes that the electron transfer from the Fe–S protein to cytochrome f is the time-limiting process. (3) In the Q-cycle model, the formation of a plastoquinol molecule at site C implies that two electrons are sequentially transferred at this site. A first possibility is that the cytochrome b_6/f complex acts as a storage device. In chloroplasts, under oxidizing conditions, one electron is actually stored on cytochrome b_6 after a group of two flashes (JOLIOT and JOLIOT 1984a). A different behavior is observed for quinone-treated algae, where the amplitude of phase b is proportional to the amount of plastoquinol formed by PS II after each flash of a series, and where no electron storage in the carriers of the intersystem electron transfer chain can be detected (see Sect. 2). If a Q-cycle type of mechanism occurs, some sort of cooperativity should be involved. Such cooperativity could occur at site C, assuming for instance that two cytochrome b_6/f complexes form a duplex (MITCHELL and MOYLE 1985). Another possibility is that two plastoquinol molecules react sequentially at site Z with the same cytochrome b_6/f complex. The main difficulty with any hypothesis which involves cooperativity is that it will be in contradiction with the first-order kinetics observed for the slow electrogenic phase, the half-time of which remains independent of the amount of available plastoquinol.

In a second class of model, the electrogenic phase is directly associated with a movement of protons via a transmembrane proton channel (see review by WIKSTRÖM et al. 1981). It has been reported that N,N'-dicyclohexylcarbodiimide (DCCD), which is known as an inhibitor of proton channels blocks both proton pumping and electron transfer in the cytochrome b_6/f complex (see review by RICH 1984). As the required concentration of DCCD is very high, it seems premature on this basis to conclude to the involvement of a proton channel in the cytochrome b_6/f complex. A possible of a proton pump coupled with a redox reaction is given in Scheme 3.

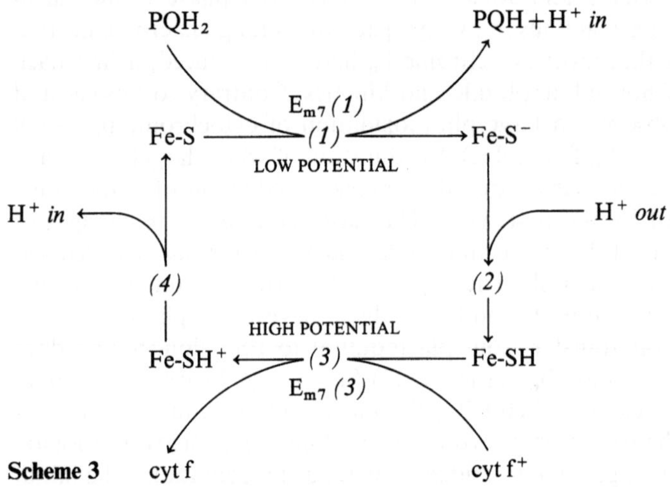

Scheme 3

The basic features of such a mechanism are the following: (1) a redox couple exists in protonated and deprotonated forms with corresponding high and low redox potentials respectively. Protons are fixed either directly on the electron carrier or on an apoprotein site close enough to modify the redox properties of the carrier by electrostatic interactions. (2) In the reduced states, proton exchange essentially occurs with the external aqueous phase while in the oxidized states, the exchange occurs with the internal aqueous phase. To satisfy these conditions, a mobile structural device is needed, the mechanism of which has yet to be clearly defined. Several models have been proposed which are based on the principle of proton pumping. Among them, the b-cycle (WIKSTRÖM and KRAB 1980; WIKSTRÖM et al. 1981) resembles the Q-cycle, except that a single quinone site is included in the cytochrome b_6/f complex. The reoxidation of cytochrome b_6 is then associated with the reduction of a semiquinone formed at this site. However, the b-cycle model cannot answer in a simple manner the question raised by the titrations of GIRVIN and CRAMER (1984). PAPA et al. (1982) proposed a model in which the Fe–S center plays a central role. Although the suggested sequence Fe–S→quinone→cytochrome f for the electron transfer seems rather unlikely to us, the authors present some arguments which suggest that the Fe–S center could act as a hydrogen carrier.

We do not think that any of the models discussed in the literature can account for all the available experimental data. Despite the criticism developed above, it seems difficult to reject the Q-cycle model which fits fairly well with the recently determined structures of the cytochrome b_6/f complex and which is supported by much spectroscopic evidence. We favor models in which the slow electrogenic phase could result both from proton-pumping processes and from a transmembrane electron transfer involving the 2 b_6-cytochromes. JOLIOT and JOLIOT (1985) present a model in which the Fe–S protein is coupled to a proton pump (Scheme 3). In this model, the electrogenic step (Reaction 2) is supposed to be the time-limiting process, which explains why the reduction of cytochrome f could be slower than the reduction of cytochrome b_6. Reactions (3), (4) and (1) should be completed in less than 300 μs, which corresponds to the delay in the time-course of the slow electrogenic phase. Under oxidizing conditions, the first electron is transferred from plastoquinol to oxidized Fe–S (Reaction 1) and the second to cytochrome b_6. It is not excluded that this second electron is finally transferred to the Fe–S center by a slow process which does not involve the Q-cycle, as already suggested by RICH (1984). Under reducing conditions, the Fe–S protein is first oxidized by transferring an electron to PS I donors [Reactions (3) and (4)]. Then, plastoquinol is oxidized by Reaction (1) as under oxidizing conditions. Cytochrome b_6, which would then be in its reduced state, transfers one electron to the plastosemiquinone which is re-reduced to plastoquinol as proposed in the b-cycle model (WIKSTRÖM and KRAB 1980).

BOUGES-BOCQUET (1981a) showed that the amplitude and kinetics of the slow electrogenic phase are highly dependent upon the values of the proton gradient and the membrane potential. CROWTHER and HIND (1980) also reported that in chloroplasts maintained under reducing conditions and illuminated by a series of flashes at low frequency (2 Hz), the addition of nigericin was required to observe the full extent of phase b. Then it is likely that the difference between $E_{m7}(1)$ and $E_{m7}(3)$ which limits the value of the electrochemical proton gradient against which this proton-pumping device could work is relatively low (probably <100 mV).

In this type of model, the yield of the electrogenic phase should depend upon the proportion of protons which are taken up from the outer medium, as postulated in Scheme 3, to those taken up from the internal phase. The stoichiometry between the number of protons pumped to the number of electrons transferred could have any value between 0 and 1. On the other hand, if both the Q-cycle and the proton-pumping processes could operate simultaneously, one can expect that the amplitude of phase b could eventually exceed that of phase a associated with PS I reaction (JOLIOT and JOLIOT 1985).

One must note that the involvement of the Fe–S center in a proton-pumping device is based only on indirect arguments. In a paper now in press, JOLIOT and JOLIOT propose a direct involvement of a plastoquinone associated with the cytochrome b_6/f complex in a proton-pumping mechanism, an hypothesis suggested by WIKSTRÖM (personal communication).

Acknowledgments. The authors are indebted to Dr. JERÔME LAVERGNE for his critical reading of the manuscript and numerous helpful comments.

References

Allred DR, Staehelin LA (1985) Plant Physiol 78:199–202
Bendall DS (1982) Biochim Biophys Acta 683:119–151
Bennoun P (1982) Proc Natl Acad Sci USA 79:4352–4356
Bouges-Bocquet B (1973) Biochim Biophys Acta 314:250–256
Bouges-Bocquet B (1977) Biochim Biophys Acta 462:371–379
Bouges-Bocquet B (1980) FEBS Lett 117:54–58
Bouges-Bocquet B (1981a) Biochim Biophys Acta 635:327–340
Bouges-Bocquet B (1981b) In: Akoyunoglou G (ed) Photosynthesis II. Electron transport and photophosphorylation. Balaban Int Sci Serv, Philadelphia, pp 19–27
Crofts AR, Wraight CA (1983) Biochim Biophys Acta 726:149–185
Crowther D, Hind G (1980) Arch Biochem Biophys 204:568–577
Diner BA, Delosme R (1983) Biochim Biophys Acta 722:443–451
Fowler CF (1977) Biochim Biophys Acta 459:351–363
Fowler CF, Kok B (1976) Biochim Biophys Acta 423:510–523
Girvin ME, Cramer WA (1984) Biochim Biophys Acta 767:29–38
Glaser EG, Crofts AR (1984) Biochim Biophys Acta 766:322–333
Hope AB, Matthews DB (1983) Aust J Plant Physiol 10:363–372
Hurt E, Hauska G (1981) Eur J Biochem 117:591–599
Hurt E, Hauska G (1982) Photobiochem Photobiophys 4:9–15
Joliot P, Delosme R (1974) Biochim Biophys Acta 357:267–284
Joliot P, Joliot A (1984a) Biochim Biophys Acta 765:210–218
Joliot P, Joliot A (1984b) Biochim Biophys Acta 765:219–226
Joliot P, Joliot A (1985) Biochim Biophys Acta 806:398–409
Joliot P, Joliot A, Kok B (1968) Biochim Biophys Acta 153:635–652
Jones RW, Selak MA, Whitmarsh J (1984) Biochem Soc Trans 12:879–880
Lavergne J (1983) Biochim Biophys Acta 725:25–33
Mitchell P (1966) Biol Rev 41:445–502
Mitchell P (1975) FEBS Lett 59:137–139
Mitchell P, Moyle J (1985) In: Lenaz G (ed) Coenzyme Q. Willey, New York, pp 145–163
Papa S, Guerrieri F, Lorusso M, Izzo G, Capuano F (1982) In: Trumpower BL (ed) Function of quinones in energy conserving systems. Academic Press, London New York, pp 527–539
Rich PR (1984) Biochim Biophys Acta 768:53–79
Rienits KG, Hardt H, Avron M (1974) Eur J Biochem 43:291–298
Rumberg B (1966) In: Thomas JB, Goedheer JC (eds) Currents in photosynthesis. Donker, Rotterdam, pp 375–382
Saphon S, Crofts AR (1977) Z Naturforsch 32:810–816
Saraste M (1984) FEBS Lett 166:367–372
Saraste M, Wikström M (1983) In: Quagliarello E, Palmieri F (eds) Structure and function of membrane protein. Elsevier, Amsterdam New York Oxford, pp 139–144
Schreiber U, Avron M (1979) Biochim Biophys Acta 546:436–447
Selak MA, Whitmarsh J (1982) FEBS Lett 150:286–292
Trumpower BL (1981) Biochim Biophys Acta 639:129–155
Velthuys BR (1978) Proc Natl Acad Sci USA 75:6031–6034
Velthuys BR (1979) Proc Natl Acad Sci USA 76:2765–2769
Velthuys BR (1980) FEBS Lett 115:167–170
Widger WR, Cramer WA, Herrmann RG, Trebst A (1984) Proc Natl Acad Sci USA 81:674–678
Wikström M (1973) Biochim Biophys Acta 301:155–193
Wikström M, Krab K (1980) In: Sanadi R (ed) Currents topics in bioenergetics, vol II. Academic Press, London New York, pp 51–101
Wikström M, Krab K, Saraste M (1981) Annu Rev Biochem 50:623–655
Witt HT (1971) Q Rev Biophys 4:365–477
Witt HT, Moraw R (1959) Mitt Z Phys Chem Neue Folge 20:254–282

10.4 Light-Driven Electron and Proton Transfer in the Cytochrome bc_1 Complex

R.C. PRINCE

1 Introduction

This chapter will mainly address the cytochrome bc_1 complexes of *Rhodopseudomonas sphaeroides* and *Rps. capsulata*, both members of the Rhodospirillaceae. As discussed elsewhere in this volume, the cytochrome bc_1 complex of these organisms seems very similar to the complex in mitochondria (both animal and plant) and the cytochrome b_6f complex of chloroplasts, and it is a widely held belief that insights gleaned on one system may be applicable to all (see HAUSKA, Chap. 10.1; TIEDE, Chap. 8.2, this Vol.).

In photosynthetic bacteria, the bc_1 complex participates in a cyclic electron pathway, receiving electrons, via ubiquinone, from the photochemical reaction center and returning them via cytochrome c_2 (see DUTTON, Chap. 5, this Vol.). Thus the system functions without the need of external oxidant or reductant, and can be studied under conditions of known ambient redox potential. Furthermore, although the photochemical reactions of the reaction center are very rapid (<ns), cytochrome c_2 donates electrons to the oxidized bacteriochlorophyll primary donor species of the reaction center in the tens of microseconds time range, so that a brief flash of light (<20 μs) effectively elicits only a single turnover of the reaction center, and the bc_1 complex is presented with only a single oxidizing and a single reducing equivalent per reaction center. These and other more familiar advantages peculiar to bacterial systems have allowed considerable progress in understanding electron flow in the bc_1 complex. Comprehensive reviews by CROFTS and WRAIGHT (1983) and PRINCE et al. (1982b) provide a survey of the older literature, so this chapter will address more recent findings and attempt to indicate the areas of current controversy.

2 Components of the Cytochrome bc_1 Complex

The bc_1 complex is a multicomponent system containing two b-type cytochromes, a cytochrome c_1, an iron-sulfur protein, and ubiquinone.

2.1 Cytochrome c_1

Despite the existence of membrane-bound c-type cytochromes in mitochondrial bc_1 complexes and the chloroplast b_6f complex, cytochrome c_1 was not identified in the bacterial complex until 1980 because of its spectral and thermodynamic similarity to the soluble cytochrome c_2 (WOOD 1980). The optical spectrum of its α-band is only some 2 nm to the red of that of cytochrome c_2, and there is still some controversy as to its precise midpoint potential. Estimates range from 265 to 350 mV near pH 7 (see WOOD 1980; CROFTS and WRAIGHT 1983; GABELLINI et al. 1982) and a detailed study, particularly as a function of pH, has not been published. As we shall see below, the precise E_m values of cytochromes c_1, c_2 and the Rieske iron-sulfur cluster are needed to reach a clear understanding of their function in the bc_1 complex.

2.2 Cytochromes b

There are two b-type cytochromes in the bc_1 complex, distinguished by both their spectral and thermodynamic properties. Cytochrome b_H, for high potential, has an E_{m7} of +50 mV, and a pK on its reduced form at pH 7.4 (PETTY and DUTTON 1976). Its α-band absorption maximum is relatively symmetrical with a maximum at 560 nm. The addition of the inhibitor antimycin shifts the absorption spectrum of cytochrome b_H about 2 nm to the red (VAN DEN BERG et al. 1979). Cytochrome b_L, for low potential, has an E_{m7} of approximately −90 mV, and requires one electron and one proton for its reduction in the pH range 6–8 (see PETTY and DUTTON 1976). It has not been studied over a wider pH range. Its α-band absorption is split, with maxima at 558 and 566 nm. The addition of the inhibitor myxothiazol is reported to shift the absorption maximum of cytochrome b_L about 1 nm to the blue in *Rps. sphaeroides* (MEINHARDT and CROFTS 1982) although it shifts b_L to the red in the mitochondrial complex (VON JAGOW and ENGEL 1981). It also seems to raise the E_{m7} of b_L some 50 mV (MEINHARDT and CROFTS 1982).

2.3 The Rieske Iron-Sulfur Protein

Named for its discoverer, the Rieske cluster is seen by EPR at low temperatures. It has G_x 1.81, G_y 1.90, G_z 2.03 in the reduced form when the ubiquinone pool is oxidized, but the G_x feature broadens markedly when the quinone pool becomes reduced (MATSUURA et al. 1983a). The Rieske cluster has an E_{m7} of +280 mV and exhibits a pK on its oxidized form at pH 8, so that at more alkaline pH its redox reaction involves an electron and a proton (PRINCE and DUTTON 1976). Various α-hydroxy hydrophobic quinones bind to the Rieske cluster and shift its EPR spectrum and raise its E_m. In doing so they eliminate the effect of reducing the quinone pool described above (see MATSUURA et al. 1983a). These inhibitors are displaced, in turn, by myxothiazol (VON JAGOW et al. 1984) or diphenylamine (PRINCE and DUTTON unpublished).

2.4 Quinones

Chromatophore membranes contain some 50 molecules of ubiquinone per bc_1 complex and their role remains controversial. En masse they exhibit an E_{m7}

of +90 mV for the Q/QH_2 couple, with no evidence for any significant amount of semiquinone even at the midpoint (TAKAMIYA and DUTTON 1979). Distinct roles for some of these quinones are becoming apparent; as discussed elsewhere in this Volume (DUTTON, Chap. 5; MICHEL and DEISENHOFER, Chap. 8.4), the reaction center has long-lived associations with quinones in the Q_A and Q_B binding sites (see ARNTZEN and PAKRASI, Chap. 9.4, this Vol.), but there is disagreement over the existence of tightly bound quinones in the bc_1 complex. In particular, there is disagreement as to whether the quinones in the complex can be regarded as permanent fixtures on the timescale of electron flow, exchanging only slowly with the pool, or whether exchange between the pool and the binding site is an obligatory part of the normal function of the complex.

The rate of electron transfer through the bc_1 complex is stimulated more than tenfold as the E_h is lowered from 180 mV to 110 mV at pH 7, and this has been variously interpreted as indicating the reduction of a "special" quinone, Q_z, in the complex, or the reduction of the first few quinones of the "pool" being sufficient to "prime the system" for optimum flow (see PRINCE et al. 1982b; CROFTS and WRAIGHT 1983). In either case the functional quinone has to bind to a binding site (the Q_z site) for its reactivity, so the debate concerns how long the quinone resides in this site, and whether, once oxidized, it is re-reduced in situ, or comes out of the site to be replaced by reduced quinone from the "pool".

Unfortunately, there has been no distinctive spectroscopic probe of the quinone at the Q_z site, and its thermodynamic properties have been assessed only indirectly from the rates of cytochrome c_1 and c_2 reduction, and the rate of the carotenoid bandshift associated with the reactions of the bc_1 complex. The inherent inaccuracies of such an approach have made it inpossible to convincingly define the residency time at the Q_z site. SNOZZI and CROFTS (1984) have done some elegant experiments manipulating the quinone to phospholipid ratio of the membrane, and have shown that the reactions of the complex slow down as the Q "pool" has a larger phospholipid "sea" to swim in. They conclude that the Q-pool is in very rapid equilibrium with the Q_z site. On the other hand, ROBERTSON et al. (1984) have studied an antimycin-sensitive semiquinone, Q_c^-, and this has thermodynamic properties very similar to those expected of a Q_z in only very slow equilibrium with the pool. Obviously the relationship between Q_c and Q_z needs to be clarified: are they two manifestation of the same quinone perhaps at different binding sites, or two distinct entities?

Another semiquinone has been detected by DEVRIES et al. (1981) in the mitochondrial bc_1 complex. This semiquinone is insensitive to antimycin, and is only seen under nonequilibrium conditions, which imply that it results from the one electron oxidation of Q_zH_2 by the Rieske iron-sulfur cluster. This has not yet been detected in bacterial systems.

3 Stoichiometries and Interactions in the bc_1 Complex

Membranes of cells grown photosynthetically contain about 0.7 bc_1 complexes per reaction center as assayed by the concentration of antimycin binding sites (VAN DEN BERG et al. 1979), the concentration of Q_z (PRINCE et al. 1978; BASH-

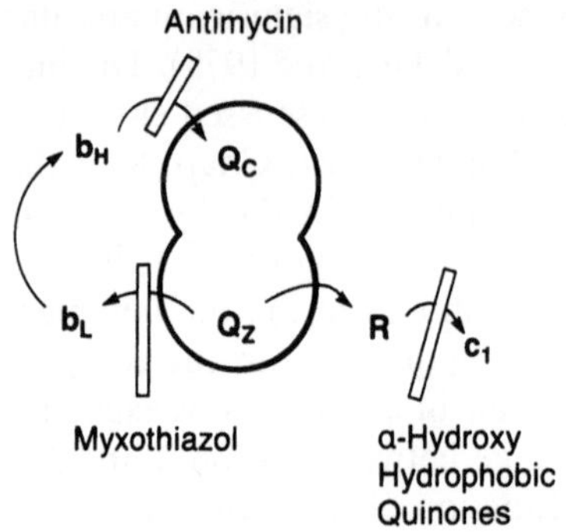

Fig. 1. The sites of action of some common inhibitors of the bc_1 complex. The components of the bc_1 complex are shown surrounding a central "quinone domain" with b_H b_L, c_1 and R signifying the cytochromes and the Rieske iron sulfur protein. The binding sites for Q_c and Q_z are both shown, although it is not clear whether they are occupied alternately by the same quinone, are both permanently occupied, or are in rapid equilibrium with the quinone pool. The pathway of electron flow, shown in more detail in Fig. 2, is indicated by *arrows*, with the effects of the inhibitors shown as *blocks*. As discussed in the text, these inhibitors may work by displacing quinone from its binding site(s) in the "quinone domain"

FORD et al. 1979) and the concentration of the Rieske iron sulfur cluster (PRINCE 1983), although this may increase to about 2:1 in cells grown on 2% oxygen, and much higher still in cells grown on saturating oxygen to repress bacteriochlorophyll synthesis (MARRS and DUTTON unpublished observation). In photosynthetically grown cells there is a substantial excess of b type cytochromes with similar thermodynamic properties to cytochrome b_H, and possibly cytochrome b_L, but no role has been forthcoming for these components (O'KEEFE and DUTTON 1981).

Many of the effects of inhibitors of electron flow can be rationalized as competing with quinone binding sites in the complex, as illustrated in Fig. 1. Thus antimycin displaces quinone from the Q_c binding site (ROBERTSON et al. 1984) and in doing so shifts the absorption spectrum of cytochrome b_H (VAN DEN BERG et al. 1979). Myxothiazol competes for the Q_z binding site on cytochrome b_L and in doing so shifts the absorption spectrum and E_m of cytochrome b_L (MEINHARDT and CROFTS 1982); it would thus be expected to eliminate the semi-quinone of DE VRIES et al. (1981). Hydrophobic α-hydroxy quinones (such as 2-hydroxy, 3-undecyl, 1,4-naphthoquinone (UHNQ) and 5-undecyl, 6-hydroxy, 4,7-dioxobenzothiazole (UHDBT) bind to the Q_z binding site on the Rieske center, and in so doing shift the EPR spectrum and the E_m of the Rieske center (BOWYER et al. 1980; MATSUURA et al. 1983a).

4 Pathways of Electron Flow Within the bc_1 Complex

In the last 10 years, schemes of a linear sequence of electron flow in the bc_1 complex have gradually been replaced by models based on the Q-cycle originally

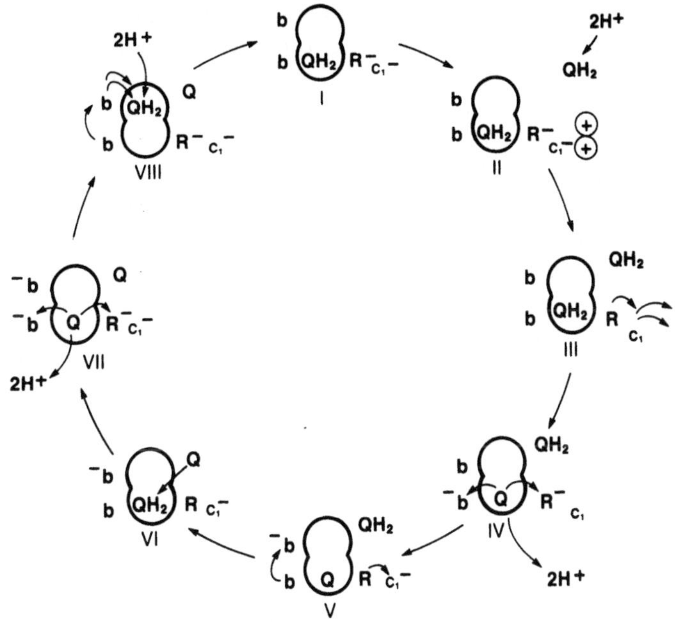

Fig. 2. Electron and proton movements in the bc_1 complex. The components of the complex are shown as in Fig. 1, with the reduced forms shown with a minus charge (e.g., R^-) or protons (e.g., QH_2). The subscripts on the b-cytochromes and the quinones are omitted for simplicity. The complex is presented with two oxidizing equivalents carried by cytochrome c_2 (+) and two reducing equivalents (on QH_2) per flash of light (*II*). Apart from proton uptake in *II* and *VIII*, the convention used is that the *arrows* indicate the reaction that has just occurred to produce the redox state shown. Under optimal conditions the cycle is complete in a few milliseconds

proposed by MITCHELL (see MITCHELL 1976; PRINCE et al. 1982b; CROFTS et al. 1983; JOLIOT and JOLIOT, Chap. 10.3, this Vol.). CROFTS et al. (1983) have rationalized many findings by considering a model based on a stoichiometry of two reaction centers per bc_1 complex, so that a bc_1 complex is presented with two oxidizing and two reducing equivalents per saturating flash of light.

It is perhaps simplest to consider the reactions under optimal conditions; when the Q_z site is occupied by a reduced quinone before the flash, and the Rieske cluster, cytochromes c_1 and c_2 are reduced, and cytochromes b_H and b_L are oxidized (Fig. 2I). The flash of light excites two reaction centers per bc_1 complex, so the net effect is to take two electrons from the Rieske-c_1-c_2 triumvirate of essentially equipotential donors. Of course the precise E_m values of these centers will define the distribution of the remaining electron between them, and this is still contentious, as discussed above. Nevertheless, for simplicity let us assume that the remaining electron is on the cytochrome c_2, and the Rieske cluster and cytochrome c_1 are fully oxidized (Fig. 2III). In that case Q_zH_2 is presented with four oxidized parties, cytochromes b_H, b_L, c_1 and the Rieske cluster. From inhibitor studies and kinetic analysis we can define the electron pathway; Q_zH_2 reduces the Rieske cluster, which in turn reduces cytochrome c_1. The semiquinone of Q_z so generated (see DEVRIES et al. 1981) reduces cytochrome b_L which in turn reduces cytochrome b_H (Fig. 2IV, V). At this juncture the model of CROFTS et al. (1983) predicts the reduction of Q_z, or its replacement, by a reduced quinone emanating from the reaction centers (Fig. 2VI). Whether this involves an electron from each reaction center, or two electrons from one of the two on alternating flashes (the so-called quinone gate) is discussed elsewhere in this Volume (DUTTON, Chap. 5). Whichever occurs, the reduced quinone at the Q_z site is presented with an oxidized Rieske cluster and an oxidized cytochrome b_L as before, and promptly reduces them both (Fig. 2VII). The two b cytochromes are then proposed to collaborate to reduce Q_z, via Q_c, (Fig. 2VIII) and the system returns to its starting state. In the process, the two oxidations of Q_zH_2 are thought to have deposited a total of four protons in the aqueous phase on the cytochrome c_2 side of the membrane, and to have been reprotonated on reduction from the other side of the membrane. Two of the latter protons were on the reduced quinone emanating from the reaction center (Fig. 2, II), two picked up on the reduction of the quinone by the b-cytochromes (Fig. 2, VIII).

The uptake of protons by the complex has been studied in some detail (PETTY et al. 1979). Many anomalies were found, particularly in that the rate of proton uptake did not seem to correlate with the rates of electron transfer through the complex. Furthermore the proton binding reactions displayed surprisingly low pK values. The stoichiometry of protons per electron was carefully determined, but unfortunately this was before the realization that the functional unit might be two reaction centers and one bc_1 complex, and the complex may pick up more protons than previously suspected, or shown in Fig. 2 (see also MATSUURA et al. 1983b). Almost nothing is known about the release of protons by the complex. Obviously the time is right for a thorough reevaluation of proton movements by the complex.

When the initial conditions are more oxidizing than shown in Fig. 2, so that Q_z is not prereduced, the kinetics of electron flow through the complex are much slower. CROFTS et al. (1983, and see also SNOZZI and CROFTS 1984) have detected a lag in the kinetics of reduction of the b cytochromes, which they interpret as the time required for the reductant generated in the reaction center to migrate to the Q_z site to "prime" the complex. Under these conditions there is only one "turnover" at the Q_z site (Fig. 2I–VI), and the cytochrome b_H stays reduced for many tens of milliseconds (O'KEEFE and DUTTON 1981). Nevertheless, the full reactions of the complex still seem to occur (MATSUURA et al. 1983b). On the other hand, the complex is severely inhibited if the starting potential is low enough to prereduce cytochrome b_H (O'KEEFE and DUTTON 1981). Under such conditions this cytochrome goes oxidized, and stays oxidized, and Q_A of the reaction center becomes reduced which inhibits useful photochemistry. Space precludes a more thorough discussion of these phenomena, but they are discussed at length in the original papers.

5 Electrogenic Events in the bc_1 Complex

As discussed briefly above, the net result of the cycle of events in the bc_1 complex is to move protons to the inside of the chromatophore from the external aqueous phase. Figure 2 shows the movement of four protons per complete turnover of the complex (per two electrons moving) although there is some evidence that there could be more (e.g., see discussion in MATSUURA et al. 1983b). MITCHELL's original chemiosmotic predictions (1966) suggested that these protons moved in an electroneutral fashion, accompanying electrons, and that the electrogenic events were transmembrane electron transfers. This is certainly the case in the reaction center, where the reduction of Q_A and oxidation of cytochrome c_2, on opposite sides of the membrane, results in the generation of a significant membrane potential (see JACKSON and DUTTON 1973). It has long been assumed that a similar transmembrane electron transfer, specifically the oxidation of cytochrome b_H, occurred in the bc_1 complex. Recently GLASER and CROFTS (1984) have shown that the reduction of cytochrome b_H involves an electrogenic step across 35–50% of the membrane and it seems likely that this is the movement of an electron. Whether the remaining electrogenic event is the further movement of an electron, or the movement of a proton in the other direction, is still unclear (see PRINCE et al. 1982b; CROFTS and WRAIGHT 1983 for further discussion).

6 Electron Flow in the bc_1 Complex of Other Photosynthetic Bacteria

Many other photosynthetic bacteria possess at least some of the components of the bc_1 complex (see DUTTON and PRINCE 1978), and at this stage it seems reasonable to expect that these complexes may be similar to those of *Rps. sphaeroides* and *Rps. capsulata*. This is particularly true for *Rhodospirillum rubrum*, where VAN DER WAL and VAN GRONDELLE (1983) have presented evidence for the function of a Q-cycle that could be rather like that shown in Fig. 2. The existence of bc_1 complexes in organisms such as *Rps. viridis* (see SHILL and WOOD 1984), *Chromatium vinosum* (see GAUL and KNAFF 1983) and *Chlorobium limicola* f. thiosulfatophilum (see HURT and HAUSKA 1984) has also been suggested recently.

7 Light-Induced Electron Flow in the Mitochondrial bc_1 Complex

It seems appropriate to mention here that the inherent advantages of studying light-induced electron flow in the bacterial bc_1 complex can be extended to mitochondrial and chloroplast systems. Such experiments, first developed by DUTTON and his colleagues (PACKHAM et al. 1980), involve mixing photochemical reaction centers isolated from *Rps. sphaeroides*, horse heart cytochrome c and the isolated mitochondrial bc_1 complex. A flash of light then drives electron flow in the isolated bc_1 complex, providing an alternative system to test models such as those of Fig. 2 (MATSUURA et al. 1981; PRINCE et al. 1982a; ZHU et al. 1983).

References

Bashford CL, Prince RC, Takamiya K, Dutton PL (1979) Biochim Biophys Acta 545:223–235
Berg WH van den, Prince RC, Bashford CL, Takamiya K, Bonner WD, Dutton PL (1979) J Biol Chem 254:8594–8604
Bowyer JR, Dutton PL, Prince RC, Crofts AR (1980) Biochim Biophys Acta 592:445–460
Crofts AR, Wraight CA (1983) Biochim Biophys Acta 726:149–185
Crofts AR, Meinhardt SW, Jones KR, Snozzi M (1983) Biochim Biophys Acta 723:202–218
DeVries S, Albracht SPJ, Berden JA, Slater EC (1981) J Biol Chem 256:11996–11998
Dutton PL, Prince RC (1978) In: Clayton RK, Sistrom WR (eds) The photosynthetic bacteria. Plenum, New York London, pp 525–570
Gabellini N, Bowyer JR, Hurt E, Melandri BA, Hauska GA (1982) Eur J Biochem 126:105–111
Gaul DF, Knaff DB (1983) FEBS Lett 162:69–75
Glaser E, Crofts AR (1984) Biochim Biophys Acta 166:322–333
Hurt EC, Hauska G (1984) FEBS Lett 168:149–154
Jackson JB, Dutton PL (1973) Biochim Biophys Acta 325:102–115
Jagow G von, Engel WD (1981) FEBS Lett 136:19–24
Jagow G von, Ljungdahl PO, Graf P, Ohnishi T, Trumpower BL (1984) J Biol Chem 259:6318–6326
Matsuura K, Packham N, Mueller P, Dutton PL (1981) FEBS Lett 131:17–22
Matsuura K, Bowyer JR, Ohnishi T, Dutton PL (1983a) J Biol Chem 258:1571–1579
Matsuura K, O'Keefe DP, Dutton PL (1983b) Biochim Biophys Acta 722:12–22
Meinhardt SW, Crofts AR (1982) FEBS Letts 149:217–222
Mitchell P (1966) Chemiosmotic coupling in oxidative and photosynthetic coupling. Glynn Research, Cornwall
Mitchell P (1976) J Theor Biol 62:327–267
O'Keefe DP, Dutton PL (1981) Biochim Biophys Acta 635:149–166
Packham NK, Tiede DM, Mueller P, Dutton PL (1980) Proc Natl Acad Sci USA 77:6339–6343
Petty KM, Dutton PL (1976) Arch Biochem Biophys 172:346–353
Petty KM, Jackson JB, Dutton PL (1979) Biochim Biophys Acta 546:17–42
Prince RC (1983) Biochim Biophys Acta 723:133–138
Prince RC, Dutton PL (1976) FEBS Lett 65:117–119
Prince RC, Bashford CL, Takamiya K, Berg WH van den, Dutton PL (1978) J Biol Chem 253:4137–4142
Prince RC, Matsuura K, Hurt E, Hauska G, Dutton PL (1982a) J Biol Chem 257:3379–3381
Prince RC, O'Keefe DP, Dutton PL (1982b) In: Barber J (ed) Electron transport and photophosphorylation. Elsevier, Amsterdam New York Oxford, pp 197–248
Robertson DE, Prince RC, Bowyer JR, Matsuura K, Dutton PL, Ohnishi T (1984) J Biol Chem 259:1758–1763
Shill DA, Wood PM (1984) Biochim Biophys Acta 764:1–7
Snozzi M, Crofts AR (1984) Biochem Biophys Acta 166:451–463
Takamiya K, Dutton PL (1979) Biochem Biophys Acta 546:1–16
Wal HN van der, Grondelle R van (1983) Biochim Biophys Acta 725:94–103
Wood PM (1980) Biochem J 189:385–391
Zhu QS, Wal HN van der, Grondelle R van, Berden JA (1983) Biochim Biophys Acta 725:121–130

10.5 Plastocyanin

W. HAEHNEL

1 Introduction

Plastocyanin is a copper protein with a molecular weight of 10,500 (KATOH 1960 and KATOH et al. 1962) which functions as electron carrier between the cytochrome b_6/f complex and the photosystem I (PS I) complex. This minireview will be limited to aspects of the structure of the cytoplasmically synthesized plastocyanin, the molecular mechanism of the extremely fast electron transfer reactions and the degree of the lateral mobility of plastocyanin which is essential for the understanding of the functional organization of the photosynthetic electron transport in higher plants. Previous reviews on plastocyanin have been published by KATOH (1977) and BOULTER et al. (1977) and should be consulted, for example, for the molecular properties. The discussion on the reaction site of plastocyanin in the electron transfer sequence (KATOH 1977) seems to be settled today. Several papers on different aspects have been published in a special symposium on plastocyanin (HALL et al. 1978). Isolation procedures have been discussed by ELLEFSON et al. (1980) and by YOCUM (1982). Reviews on broader fields contain substantial information on plastocyanin about amino acid sequences (RAMSHAW 1982), its exchange with cytochrome c-553 in algae (SANDMANN and BÖGER 1983), and its function in electron transport (HAEHNEL 1984a).

2 Synthesis

Plastocyanin is coded in the nucleus, synthesized on cytoplasmic ribosomes, and transported post-translationally into the chloroplast (GROSSMAN et al. 1982). Immunoprecipitation with antibodies raised against plastocyanin showed that it is synthesized after induction by light (SANDMANN and BÖGER 1981) as a precursor of $M_r = 21{,}000$ (HERRMANN et al. 1983) or 25,000 (GROSSMAN et al. 1982). If all the additional amino acids in this precursor are located at the amino terminus, this presequence would be the longest so far documented with $M_r = 11$–$15{,}000$. The presequence is probably involved in the transport through the two envelope membranes and the processing may be energy-dependent (NELSON and SCHATZ 1979). Unlike most other plastidial proteins of nuclear origin, the hydrophylic plastocyanin has to pass three membranes, both envelope membranes and the thylakoid membrane, to reach its functional site in the thylakoid lumen. Maturation of the primary translation product of plastocyanin requires the addition of Cu to the precursor polypeptide, but little is known about the insertion mechanism. The isolation of the cDNA clone of plastocyanin by HERRMANN et al. (1983) may help to answer these questions.

Plastocyanin was first discovered in the green alga *Chlorella ellipsoidea* by KATOH (1960) and has been found in chloroplasts of all higher plants (KATOH et al. 1961, PLESNIČAR and BENDALL 1970, RAMSHAW et al. 1973), in most but not all algae, in some cyanobacteria (blue-green algae; GORMAN and LEVINE 1966, SANDMANN et al. 1983), but not in anoxygenic photosynthetic bacteria. In some algae such as *Chlamydomonas* and *Scenedesmus* as well as in several *Chlorella* strains, plastocyanin is replaced by soluble cytochrome c-553 (formerly called cytochrome c-552) if the cells are grown in a Cu-deficient nutrient medium (SANDMANN et al. 1983). An equivalent function of cytochrome c-553 and plastocyanin (WOOD 1978) is supported by the efficient reduction of P-700 by cytochrome c-553 in vitro (HAUSKA et al. 1971) and in vivo (NANBA and KATOH 1983). Analysis of plastocyanin sequences from a large variety of species has indicated that plastocyanin is evolving at least twice as fast as cytochrome c (RAMSHAW 1982), which should result in an efficiently functioning molecule.

3 Structure

3.1 Refined Crystal Structure

The X-ray crystal structure of poplar plastocyanin (COLMAN et al. 1978) is now solved at a resolution of 0.16 nm (GUSS and FREEMAN 1983) which is precise enough to localize all atoms of the amino acid side chains and the hydrogen bonds. The polypeptide chain shown in Fig. 1 comprises eight strands and seven connecting loops. All strands have segments in β-conformation except strand 5, which is irregular and includes a short α-helix formed by residue 51 to 56. The intramolecular hydrogen bonds connect the strands into antiparallel and parallel β-sheets (CHOTHIA and LESK 1982, GUSS and FREEMAN 1983) in such a way that the molecule (approximate dimensions of $4.0 \times 3.2 \times 2.8$ nm) resembles more a β-sandwich, with one face on the left-hand and the other on the right-hand side of Fig. 1, than a flattened cylinder (MURATA et al. 1982).

All charged residues and the majority of the uncharged polar residues are exposed to the surrounding medium, while most nonpolar residues and aromatic ones are buried. Comparison of 67 plant plastocyanins shows 64 conserved or conservatively substituted residues in a total of 99. The sequence of plastocyanin from three algae is considerably less conserved (cf. GUSS and FREEMAN 1983). Invariant and highly conserved residues indicate structural features which may be essential for the function of the protein.

This is particularly true for the copper-binding site (FREEMAN et al. 1978). The ligands of the Cu(II) atom are N of His 37, S of Cys 84, N of His 87,

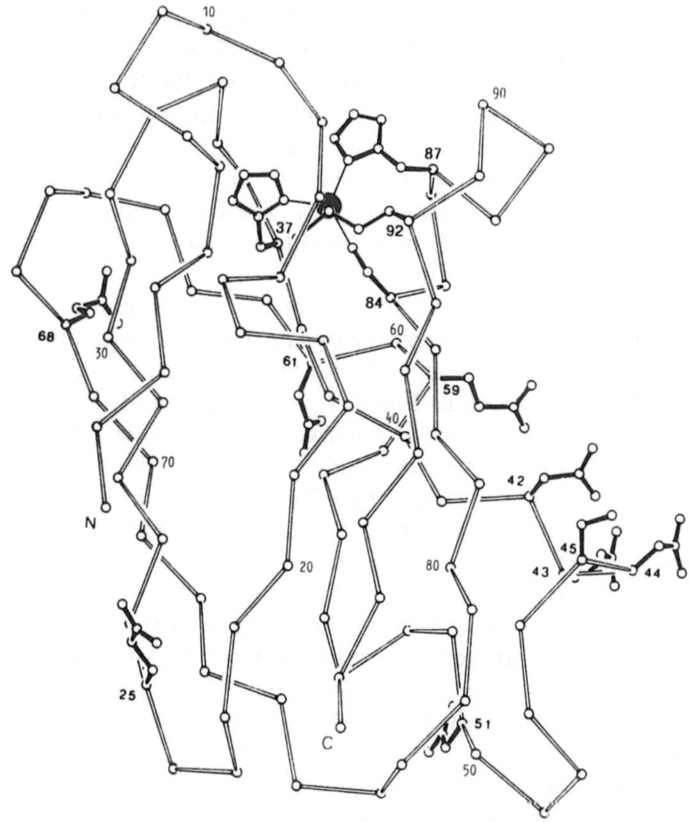

Fig. 1. Molecular structure of poplar plastocyanin (FREEMAN 1981). The polypeptide backbone is represented by the C_α atoms with every tenth numbered. The four Cu-binding side-chains and the conserved acidic side-chains of plant plastocyanins are shown. (Residue *45* is Ser in polar but Glu in other plant plastocyanins)

and S of Met 92 at a distance of 0.204, 0.213, 0.210, and 0.290 nm, respectively. The coordination is considerably distorted from a tetrahedral geometry with an exceptionally long bond between Cu(II) and S(Met 92) (GUSS and FREEMAN 1983). Studies of the Cu EXAFS have shown the same distance of the Cu ligands in solution as in the crystal structure (SCOTT et al. 1982). The special geometry of the Cu center causes the high redox potential of Cu(I)/Cu(II)-plastocyanin with $E_0' = 370$ mV (KATOH et al. 1962) as compared to Cu^+/Cu^{2+} with $E_0 = 170$ mV. The conformation of the Cu center is stabilized by a hydrogen bridge from the amide group of the conserved Asn 38 at strand 4 to S of Cys 84 at strand 7 (GUSS and FREEMAN 1983).

The Cu site is located in a pocket of three loops (Fig. 1) formed by largely conserved hydrophobic residues. In the middle of this flat hydrophobic patch, one edge of the imidazole ring of His 87 is exposed to the solvent. It is the only Cu-binding residue extending to the surface and therefore a likely site for an electron transfer. ^{1}H NMR measurements have shown a binding of the redox-inactive $Cr(CN)_3^{3-}$ near His 87 and indicate at least for the negative $Fe(CN)_6^{3-}$ an electron transfer via this residue (COOKSON et al. 1980). Another feature which could have a function in electron transfer is a sequence of four invariant aromatic residues extending into the interior of the molecule from the Cu ligand side chain of Met 92 to a site at the surface opposite to the Cu site (GUSS and FREEMAN 1983). The last residue of this sequence is Tyr 80, which is fixed by a hydrogen bond to residue 76 and has two conserved residues in its immediate environment.

The most extensively studied feature is the acidic patch of plant plastocyanins. Plastocyanin from spinach or *Populus nigra* has a ratio of (Asp + Glu) : Lys = 15 : 6. Arginine is absent. Only three lysines but nine acidic residues are conserved in plant plastocyanins. Figure 1 shows that six of these acidic residues (42–45, 59, 61) form an elongated negatively charged patch. In the middle of this patch is the conserved aromatic residue of Tyr 83 which is exposed to the solvent. ^{1}H NMR line-broadening of the Tyr 83 protons by binding of positively charged inorganic complexes (COOKSON et al. 1980) suggests that electron transfer may take place from this site (LAPPIN et al. 1979) to the Cu atom, with the negative patch as recognition site (COLMAN et al. 1978).

The structure of the reduced Cu(I)plastocyanin at pH values greater than 7 is not very different from that of Cu(II)plastocyanin (FREEMAN 1981). The small changes in the Cu-ligand bonds between 0 and 0.02 nm are consistent with a sufficiently low Franck-Condon reorganization barrier for a rapid outer-sphere electron transfer. However, although the structure of oxidized plastocyanin shows negligible changes at decreasing pH values, the coordination of the Cu(I) site is changed considerably by protonation of the His 87 imidazole ring and by a simultaneous increase of the Cu – N(His 87) distance from 0.23 to 0.33 nm (FREEMAN 1981). This is consistent with the redox-titration of KATOH et al. (1962), which has shown the protonation of reduced plastocyanin with a pK of 5.4. The resulting coordination geometry of the Cu center is trigonal-planar, which stabilizes Cu(I) (FREEMAN 1981). Therefore a reaction of the HCu(I)-plastocyanin should be slow not only with inorganic complexes (SEGAL and SYKES 1978) but also with its biological redox partners at the low pH inside the thylakoids. FREEMAN (1981) has proposed that this mechanism could control the electron transport rate in the light, by overriding the rate-limiting plastoquinol oxidation at the cytochrome b_6/f complex (SIGGEL 1974, RICH 1984).

3.2 Chemical Modification

Exchange of Cu against Hg by incubation with $HgCl_2$ (KIMIMURA and KATOH 1972) or release of Cu (BERG and KROGMANN 1975) by alkaline KCN treatment (IZAWA et al. 1973) is a specific way to inactivate plastocyanin in situ with negligible effect on other electron transport reactions. Substitution of Cu with Ag in the growth medium of *Scenedesmus* induces the synthesis of inactive silver plastocyanin (BOHNER et al. 1981).

Most chemical modifications of amino acid residues introduce changes at the acidic patch. FARVER and PECHT (1981) found a tight binding of Cr^{3+} to the acidic residues 42–45 after reduction of plastocyanin by Cr^{2+}. The negatively charged carboxyl groups can be converted to positively charged amino groups by covalent binding of ethylenediamine (BURKEY and GROSS 1981 b and 1982). Upon binding of up to four ethylenediamine residues, the redox potential of plastocyanin is increased by 26 mV (TAKABE et al. 1984). This modification changes the tertiary conformation of tyrosine and phenylalanine residues but not of the Cu center, as indicated by the absorption and circular dichroic spectra (DRAHEIM et al. 1985). A third mechanism for introducing positive charges at the acidic patch is by redox-inactive inorganic complexes as Cr(1,10-phenanthroline)$_3^{3+}$ (COOKSON et al. 1980).

The goal of these modifications to understand the molecular mechanism of electron transfer, has remained elusive, due to their different effects on the reaction rate of plastocyanin. Binding of Cr^{3+} decreases the rate of plastocyanin oxidation by PS I reaction center by a factor of 3 as compared to that of the native protein (FARVER et al. 1982). The conclusion drawn from this finding was that the negative patch is the binding site for the transfer of electrons from the Cu center via Tyr 83 to P-700. Consistent with the proposal of COLMAN et al. (1978), a second physiologically significant electron transfer site was assumed to be centered at the hydrophobic patch. An alternative interpretation has been offered by BEOKU-BETTS et al. (1983), who, based on a modest inhibition of the electron transfer from cytochrome f to plastocyanin after binding of one Cr^{3+} or a positive inorganic complex, have suggested that the negative patch provides also the binding site for a positive region on cytochrome f. In contrast to these results, modification by ethylenediamine increased the oxidation rate of plastocyanin by PS I particles, even in the presence of 5 mM Mg^{2+} ions (BURKEY and GROSS 1981 b), and decreased the rate of reduction by cytochrome f (TAKABE et al. 1984). This is not consistent with the suggestion that the negative patch is a specific recognition site. The functional significance of the acidic patch is also questioned by the sequence of algal plastocyanin, e.g., from *Anabaena variabilis* (AITKEN 1975) where only two residues (42 and 51) are conserved as acidic. In addition, a function of Tyr 83 in the electron transfer is not consistent with the effect of tetranitromethane modification of one of the three tyrosyl residues of spinach plastocyanin, most likely the only one exposed to the solvent (Tyr 83) (DAVIS and SAN PIETRO 1979). Neither the reaction rates nor the spectrum or the redox potential of the NO_2-plastocyanin showed changes as compared to control plastocyanin.

The rather small inhibition by binding of Cr^{3+} or of positively charged complexes to the acidic patch may be ascribed to changes in the overall dipole moment or in the local charge distribution, as shown for cytochrome c (KOPPENOL and MARGOLIASH 1982), the analogous protein of mitochondria (HAUSKA, Chap. 10.1, this Vol., HAUSKA et al. 1983). A modification of only one lysine of cytochrome c outside the interaction domain at the exposed heme edge decreased the reaction rate to the same extent as the mentioned modification of plastocyanin. The slow reaction rate between plastocyanin and the positively charged cytochrome c indicates the absence of a precise molecular fit for electron transfer (AUGUSTIN et al. 1983) in the 1:1 complex of the two proteins (GEREN et al. 1983). The molecular mechanism of electron transfer of plastocyanin needs to be examined in greater detail.

4 Reduction of P-700

4.1 Surface Charges

Due to the presence of immobilized surface charges, the surface potential of thylakoid membranes plays a significant role in electron transfer reactions (BARBER 1982, Chap. 11.5, this Vol.). The effect of monovalent (WOOD and BENDALL 1975) and of low concentrations of divalent cations (LOCKAU 1979, HAEHNEL et al. 1980a), and the effect of the isoelectric point of the protein (DAVIS et al. 1980) indicate that the electron transfer from isolated plastocyanin to P-700$^+$ is a function of the surface potential and the net charge of the electron donor (TAMURA et al. 1980 and 1981). Modeling of the data by means of the Gouy-Chapman diffuse double layer theory indicates a low surface charge density of between -0.85 and -1.08 μC cm^{-2} for the reduction site of P-700$^+$ in sonicated chloroplasts and in Triton particles of PS I (ITOH 1979, TAMURA et al. 1980 and 1981).

When the pH is dropped from 6 to 5, the reaction rate of plastocyanin with P-700$^+$ decreases in both digitonin treated chloroplasts (WOOD and BENDALL 1975) and in PS I particles (TAMURA et al. 1981). This decrease is not consistent with a decrease in repulsion between plastocyanin (pI=3 (SIEGELMAN et al. 1976), pI=3.94 for spinach (HAEHNEL unpublished) and the negatively charged vicinity of P-700$^+$. The formation of inactive HCu(I)-plastocyanin (FREEMAN 1981) could provide an explanation for this effect. However, the similarity of the pH profile of P-700$^+$ reduction by both ferrocyanide (ITOH 1979) and plastocyanin (TAMURA et al. 1981) suggests that the decrease in the rate reflects the pH dependence of the acceptor site of P-700. In addition, the reduction rate of P-700$^+$ in the PS I particles increased again by two orders of magnitude at pH values decreasing from 5 to 4 in the presence of low concentrations of monovalent cations. This result is consistent with the presence of opposite charges of plastocyanin and PS I in this pH range, and cannot be explained by an inactivation of reduced plastocyanin. Further studies are needed to understand the effect of the protonation of His 87 (FREEMAN 1981) on the reaction rate.

4.2 Kinetics

Different excitation conditions and different kinetic analyses make it difficult to compare the data on the oxidation of purified plastocyanin by PS I, which would be easy on the basis of the second-order rate constant k_2. In most studies the concentration of plastocyanin [Pc] exceeds that of P-700 considerably, and

k_2 can be estimated from the halftime $t^1/_2$ of the pseudo first-order reduction of P-700$^+$: $k_2 = \ln 2/(t^1/_2 \cdot [Pc])$. In the Michaelis-Menten approach, the pseudo first-order relation between the rate and the concentration of plastocyanin in the range of $[Pc] \ll K_m$ can be used to estimate k_2 from the maximal rate V_{max}, the apparent K_m value, and the concentration of oxidized P-700 [P-700$^+$], which equals that of total P-700 at saturating excitation $k_2 = V_{max}/(K_m \cdot [P\text{-}700^+])$. Depending on the units of V_{max}, the value may have to be multiplied by the chlorophyll concentration used in the experiments. In contrast to second-order kinetics with a reaction rate strictly proportional to the plastocyanin concentration (e.g, WOOD and BENDALL 1975, TAKABE et al. 1983), the Michaelis-Menten reaction kinetics show a substrate saturation (e.g., DAVIS et al. 1980, BURKEY and GROSS 1981 a), which should originate from a mechanism with a rate-limiting step after the formation of the complex. The discrepancy may be due to some rate limitation not related to electron transfer from plastocyanin to P-700$^+$ (cf. WOOD and BENDALL 1975). In some investigations, total plastocyanin may not have remained reduced during illumination, as pointed out by LIEN and SAN PIETRO (1979). However, even if this effect cannot be avoided, k_2 may be estimated as given above.

The values published for electron transfer from plastocyanin to P-700$^+$ at approximately optimal ion concentrations are indicative of three different situations for the transfer of electrons that are related to the degree of purification of PS I.

Digitonin particles of PS I: $0.8–1.5 \times 10^8\ M^{-1}\ s^{-1}$
(WOOD and BENDALL 1975, HAEHNEL et al. 1980a, OLSEN and COX 1982).

Sonicated chloroplasts and Triton-PS I particles: $1–2 \times 10^7\ M^{-1}\ s^{-1}$
(WOOD and BENDALL 1975, TAMURA et al. 1980 and 1981),

Purified PS I reaction center: $2–3 \times 10^5\ M^{-1}\ s^{-1}$
(DAVIS et al. 1980, BHARDWAJ et al. 1982, TAKABE et al. 1983).

The extremely high rate constant of $1.5 \times 10^8\ M^{-1}\ s^{-1}$ for the digitonin samples is close to that expected for a diffusion-controlled reaction, and indicates a situation which is probably close to that in vivo. Characteristic for such preparations is a maximal reaction rate at mono- and divalent cation concentrations as low as 30 and 3 mM, respectively, (COX and OLSEN 1982). Similar salt requirements have been found with Triton-PS I particles (TAMURA et al. 1981). The reason for the lower rate constant of $2 \times 10^7\ M^{-1}\ s^{-1}$ for the sonicated chloroplasts and Triton-PS I particles is not understood, but may be ascribed to a limited accessibility of the reaction site of P-700$^+$ in sonicated chloroplasts and a change in the local lipid or protein conformation by Triton X-100. BENGIS and NELSON (1977) have shown that removal of a subunit of the PS I reaction center, designated subunit III, was accompanied by a considerable decrease in the reduction rate of P-700$^+$ by plastocyanin.

4.3 Subunit III of PS I

The function of subunit III in the transfer of electrons to P-700$^+$ is still an enigma. One problem is related to its identification on gels of PS I reaction center preparations obtained in different laboratories from different plants

(THORNBER, Chap. 3; ORT, Chap. 4; WOLLMAN, Chap. 9.7 all this Vol.; BENGIS and NELSON 1977, NECHUSHTAI et al. 1981, MULLET et al. 1980). Subunit III ($M_r = 20{,}000$) changes position with subunit IV ($M_r = 18{,}000$; NECHUSHTAI et al. 1981) when electrophoresed on slab gels with higher resolution (MULLET et al. 1980) than tube gels (BENGIS and NELSON 1977). The P-700-chlorophyll a protein complex investigated by TAKABE et al. (1983) was assembled from only 5 subunits. It showed essentially the same characteristics of the P-700$^+$ reduction by plastocyanin as that isolated by LIEN and SAN PIETRO (1979) and DAVIS et al. (1980) and the PS I reaction center from *Chlamydomonas* with 4 subunits (NECHUSHTAI and NELSON 1981). The stimulation of the slow reaction rate in these preparations by an extreme concentration of 100 mM divalent cations but not by monovalent cations at a pH value greater than 6 resembles that of PS I reaction centers from higher plants after depletion of subunit III (NECHUSHTAI and NELSON 1981). BHARDWAJ et al. (1982) and TAKABE et al. (1983) found after removal of the other subunits from the reaction center protein of $M_r = 70{,}000$ (BENGIS and NELSON 1977) almost no change of the rate and type of interaction between plastocyanin and P-700. They concluded that the binding site of plastocyanin is at the PS I reaction center protein and that subunit III is not involved significantly in the electron transfer from plastocyanin to P-700$^+$. However, this type of interaction does not seem to reflect the interaction in situ with a rate three orders of magnitude faster. Most data would be consistent with a function of subunit III to provide the proper conformation for the fast electron transfer of plastocyanin (HAEHNEL et al. 1980a). But subunit III has not yet been purified, and its actual contribution has not been shown by reconstitution. There may be other structural features of PS I which provide the efficient recognition site for plastocyanin.

An unknown electron carrier between plastocyanin and P-700 has been suggested by BOUGES-BOCQUET and DELOSME (1978) and ascribed to subunit III. But HAEHNEL et al. (1980a) found that the reduction of P-700 in digitonin particles of PS I after a short flash is proportional to the concentration of reduced plastocyanin (up to 120 μM, HAEHNEL unpublished) which excludes the function of an unknown electron carrier and indicates the direct reduction of P-700$^+$ by plastocyanin.

5 Oxidation of Cytochrome f

Cytochrome f is a subunit of the integral cytochrome b_6/f complex (for details see ORT, Chap. 4; HAUSKA, Chap. 10.1, this Vol., BENDALL 1982, HAUSKA et al. 1983). The nucleotide sequence of the cytochrome f gene in pea (WILLEY et al. 1984) and in spinach (ALT and HERRMANN 1984) indicate a topology of cytochrome f with one hydrophobic span anchored in the membrane and 75% of the polypeptide chain including the heme-containing region extending into the intrathylakoid space. This is consistent with a location of plastocyanin inside the thylakoid as concluded by HAUSKA et al. (1971) and confirmed by the accessibility of P-700 in inside-out thylakoid vesicles (HAEHNEL et al. 1981).

The electron transfer from purified cytochrome f to plastocyanin is extremely fast with a constant of $3.6 \times 10^7\ M^{-1}\ s^{-1}$ for oligomeric cytochrome f from parsley (WOOD 1974) and $6 \times 10^7\ M^{-1}\ s^{-1}$ for monomeric cytochrome f from *Brassica rapa* (NIWA et al. 1980) and Japanese radish (TANAKA et al. 1981) at pH 7 and an ionic strength of 0.1 M. At decreasing concentrations of monovalent and divalent cations NIWA et al. (1980) and TAKABE et al. (1984) report an increase of the rate constant up to $1.8 \times 10^8\ M^{-1}\ s^{-1}$. These authors conclude that positive amino groups near the reaction site of the negatively charged cytochrome f control the reaction rate. However, the results of TANAKA et al. (1981) do not show any effect of the ionic strength or polycations on the reaction rate. The contradiction in these results is not understood.

6 Reactions in Situ

6.1 Organization of Electron Transport

The lateral heterogeneity in the distribution of the integral electron transport complexes in the thylakoid membrane of higher plants raises several questions pertaining to the long range electron transport from PS II in appressed to PS I in exposed membrane regions (STAEHELIN, Chap. 1, this Vol., ANDERSON 1981). For the cytochrome b_6/f complex the data are contradictory and indicate either a uniform distribution or a preferential location in the stroma lamellae region (see discussion in Chap. 1 of STAEHELIN, this Vol.). The function of this complex is closely related to its location and to a possible diffusion of plastocyanin and plastoquinone (HAEHNEL 1984a). (1) If plastoquinol does not diffuse fast enough with respect to the electron transfer times (GUPTE et al. 1984) plastocyanin would have to shuttle rapidly for linear electron transport between appressed and exposed membrane regions. Only cytochrome b_6/f complexes in the appressed regions could participate in linear electron transport. (2) If the diffusion of plastoquinol is fast (FATO et al. 1984) and that of plastocyanin slow, only cytochrome b_6/f in nonappressed regions can function in linear electron transport. (3) If both plastoquinol and plastocyanin diffuse rapidly, the location of any complex is not important for its participation in linear electron transport.

One problem in finding an answer to these questions is that the diffusion coefficient for both plastoquinone and plastocyanin has not been measured in situ. The discussion has had to rely on indirect evidence. However, plastocyanin may mediate, even at a limited diffusion rate, a trans-luminal electron transfer with the same probability as lateral electron transfer between neighboring complexes. Thus it could also connect cytochrome b_6/f complexes in appressed membranes with P-700 in opposite stroma-exposed end-membranes.

6.2 Diffusion Coefficient

Cytochrome f is oxidized in situ after a short flash with a halftime of 0.1 ms in *Chlorella* (BOUGES-BOCQUET 1977) and of 0.3 ms in spinach chloroplasts (HAEHNEL et al. 1980b). This is the total time required for electron transfer from plastocyanin to P-700$^+$, a subsequent diffusional step of plastocyanin from PS I to the cytochrome b_6/f complex, and the electron transfer from cytochrome f to oxidized plastocyanin. Using this oxidation time t of cytochrome f, the lower limit of the diffusion coefficient D for a two-dimensional random walk diffusion of plastocyanin can be calculated from the mean square displacement $\overline{x^2} = 4 \times D \times t$ (c.f. HAEHNEL 1984a, SELAK and WHITMARSH 1984). Assuming an average distance between cytochrome b_6/f complexes in stacked and PS I in unstacked membrane regions of 200 nm, a diffusion coefficient greater than 0.3–1.0 $\times 10^{-6}$ cm^2 s^{-1} is estimated. This is more than one order of magnitude greater than the diffusion coefficient of 3×10^{-8} cm^2 s^{-1} found by photobleaching of labeled plastocyanin in phospholipid bilayer membranes (FRAGATA et al. 1983). Even under free diffusion conditions the isotropic diffusion coefficient of plastocyanin would not be greater than 4–12 $\times 10^{-8}$ cm^2 s^{-1} as estimated from Stoke's equation for the viscosity in the thylakoid lumen, which is 10 and 30 times higher than in water in the dark and in the light, respectively (NESBITT and BERG 1982). Thus, a long-range electron transfer between appressed and nonappressed membrane regions does not seem to be possible within the oxidation time of cytochrome f.

6.3 Kinetics in Situ

The fast reduction of P-700$^+$ by its immediate electron donors has been analyzed by two exponential decays with halftimes of 20 and 200 µs (HAEHNEL et al. 1971). The two kinetic components have been associated with a reduction of P-700$^+$ by bound and more mobile plastocyanin (HAEHNEL et al. 1980b). Shielding of surface charges and a high osmolarity of the solution appeared to be necessary for the formation of the complex between plastocyanin and P-700 in situ. BOTTIN and MATHIS (1984) studied the reaction of soluble plastocyanin with digitonin particles of PS I and confirmed the formation of a complex. On the other hand, OLSEN and COX (1982) argued that the resolution may not be sufficient to discriminate second-order kinetics from two superimposed exponential decays. Second-order kinetics are evidence for freely mobile plastocyanin.

In an approach that avoids the ambiguity of the kinetic analysis, the effect of partial inhibition of plastocyanin on the absorbance changes of P-700, cytochrome f, and on the linear electron transport has been studied (HAEHNEL 1982 and 1984b). The reduction of P-700$^+$ after long flashes can provide detailed information on the location of a diffusible component relative to the inhibition site and the rate-limiting plastoquinol oxidation. Partial inhibition of plastocyanin gave rise to a biphasic reduction of P-700$^+$ with a new slow phase (HAEHNEL 1982) and a decrease in the amplitude of cytochrome f (HAEHNEL

1984b). These findings are evidence that neither total P-700 nor total cytochrome f is accessible to the residual active plastocyanin and are not consistent with a function of plastocyanin as a long-range shuttling electron carrier. A short-range mobility or a limited pool of plastocyanin molecules is indicated by the time course of cytochrome f oxidation, which shows second-order kinetics as a function of active plastocyanin (HAEHNEL 1984b).

GRAAN and ORT (1984) determined the amount of electron carriers functioning between plastoquinone and P-700 in a novel approach, and, consistent with the previous values of MALKIN and BEARDEN (1973) and HAEHNEL (1977), found two plastocyanin molecules per P-700. They could correlate the loss of plastocyanin by sonication with the decrease in the electron capacity of the electron carriers. This result indicates that after sonication the residual plastocyanin is sufficiently mobile to mediate the oxidation of total cytochrome f and the Rieske-FeS protein. Consistent with this result, measurements of linear electron transport at partial inhibition of plastocyanin indicate a long-range mobility of plastocyanin after EDTA treatment of the thylakoids at low ionic strength (HAEHNEL 1984b).

Consistent also with other data would be a control of the mobility of plastocyanin by the luminal distance between the thylakoid membranes. In intact chloroplasts it is 6–8 nm in the dark and only 2–4 nm in the light (MURAKAMI and PACKER 1970). Thus, the luminal distance in the light being not greater than the size of plastocyanin suggests that plastocyanin may be trapped between the luminal sides of the membrane in intact thylakoids. Other restrictions to the mobility of plastocyanin are possible from interactions with the PS I and the cytochrome b_6/f complex which both protrude into the lumen (SIMPSON 1982, MÖRSCHEL and STAEHELIN 1983).

References

Aitken A (1975) Biochem J 149:675–683

Alt J, Herrmann RG (1984) Curr Genet 8:551–557

Anderson JM (1981) FEBS Lett 124:1–10

Augustin MA, Chapman SK, Davies DM, Sykes AG, Speck SH, Margoliash E (1983) J Biol Chem 258:6405–6409

Barber J (1982) Annu Rev Plant Physiol 33:261–295

Bendall DS (1982) Biochim Biophys Acta 683:119–151

Bengis C, Nelson N (1977) J Biol Chem 252:4564–4569

Beoku-Betts D, Chapman SK, Knox CV, Sykes AG (1983) J Chem Soc Chem Commun 20:1150–1152

Berg SP, Krogmann DW (1975) J Biol Chem 250:8957–8962

Bhardwaj R, Burkey KO, Tripathy BC, Gross EL (1982) Plant Physiol 70:424–429

Bohner H, Sandmann G, Böger P (1981) Biochim Biophys Acta 636:65–69

Bottin H, Mathis P (1984) In: Sybesma C (ed) Advances in photosynthesis research, Vol I. Nijhoff/Junk, The Hague, pp 1.4.569–1.4.572

Bouges-Bocquet B (1977) Biochim Biophys Acta 462:362–370

Bouges-Bocquet B, Delosme R (1978) FEBS Lett 94:100–104

Boulter D, Haslett BG, Peacock D, Ramshaw JAM, Scawen MD (1977) In: Northcote DH (ed) International review of biochemistry, vol 13. Plant biochemistry II. Univ Park Press, Baltimore, pp 1–40

Burkey KO, Gross EL (1981a) Biochemistry 20:2961–2967
Burkey KO, Gross EL (1981b) Biochemistry 20:5495–5499
Burkey KO, Gross EL (1982) Biochemistry 21:5886–5890
Chothia C, Lesk AM (1982) J Mol Biol 160:309–323
Colman PM, Freeman HC, Guss JM, Murata M, Norris VA, Ramshaw JAM, Venkatappa MP (1978) Nature 272:319–324
Cookson DJ, Hayes MT, Wright PE (1980) Biochim Biophys Acta 591:162–176
Cox RP, Olsen LF (1982) In: Barber J (ed) Electron transport and photophosphorylation. Elsevier/North-Holland Biomed Press, Amsterdam New York, pp 49–79
Davis DJ, San Pietro A (1979) Anal Biochem 95:254–259
Davis DJ, Krogmann DW, San Pietro A (1980) Plant Physiol 65:697–702
Draheim JE, Anderson GP, Pan RL, Rellick LM, Duane JW, Gross EL (1985) Arch Biochem Biophys 237:110–117
Ellefson WL, Ulrich EA, Krogmann DW (1980) Methods Enzymol 69:223–229
Farver O, Pecht I (1981) Proc Natl Acad Sci USA 78:4190–4193
Farver O, Shahak Y, Pecht I (1982) Biochemistry 21:1885–1890
Fato R, Battino M, Castelli GP, Lenaz G (1984) FEBS Lett 179:238–242
Fragata M, Takahashi M, Asada K, Ito T, Ohnishi S (1983) In: Sybesma C (ed) Abstr 6th Int Congr Photosynth, vol I. Nijhoff/Junk, The Hague, p 289
Freeman HC (1981) In: Laurent JP (ed) (IUPAC) Coordination chemistry-21. Pergamon, Oxford New York, pp 29–51
Freeman HC, Norris VA, Ramshaw JAM, Wright PE (1978) FEBS Lett 86:131–135
Geren LM, Stonehuerner J, Davis DJ, Millett F (1983) Biochim Biophys Acta 724:62–68
Gorman DS, Levine RP (1966) Plant Physiol 41:1648–1656
Graan T, Ort DR (1984) J Biol Chem 259:14003–14010
Grossman AR, Bartlett SG, Schmidt GW, Mullet JE, Chua N-H (1982) J Biol Chem 257:1558–1563
Gupte S, Wu E-S, Hoechli L, Hoechli M, Jacobson K, Sowers A, Hackenbrock CR (1984) Proc Natl Acad Sci USA:2606–2610
Guss JM, Freeman HC (1983) J Mol Biol 169:521–563
Haehnel W (1977) Biochim Biophys Acta 459:418–441
Haehnel W (1982) Biochim Biophys Acta 682:245–257
Haehnel W (1984a) Annu Rev Plant Physiol 35:659–693
Haehnel W (1984b) In: Sybesma C (ed) Advances in photosynthesis research, vol I. Nijhoff/Junk, The Hague, pp 1.4.545–1.4.548
Haehnel W, Döring G, Witt HT (1971) Z Naturforsch 26b:1171–1174
Haehnel W, Hesse V, Pröpper A (1980a) FEBS Lett 111:79–82
Haehnel W, Pröpper A, Krause H (1980b) Biochim Biophys Acta 593:384–399
Haehnel W, Berzborn R, Andersson B (1981) Biochim Biophys Acta 637:389–399
Hall DO, Coombs J, Goodwin TW (eds) (1978) Proc 4th Int Congr Photosynth. Biochem Soc, London, pp 753–819
Hauska GA, McCarty RE, Berzborn RJ, Racker E (1971) J Biol Chem 246:3524–3531
Hauska G, Hurt E, Gabellini N, Lockau W (1983) Biochim Biophys Acta 726:97–133
Herrmann RG, Westhoff P, Alt J, Winter P, Tittgen J, Bisanz C, Sears BB, Nelson N, Hurt E, Hauska G, Viebrock A, Sebald W (1983) In: Ciferri O, Dure III L (eds) Structure and function of plant genomes. Plenum, New York London, pp 143–153
Itoh S (1979) Biochim Biophys Acta 548:579–595
Izawa S, Kraayenhof R, Ruuge EK, Devault D (1973) Biochim Biophys Acta 314:328–339
Katoh S (1960) Nature 186:533–534
Katoh S (1977) In: Trebst A, Avron M (eds) Encyclopedia of plant physiology, vol V. Springer, Berlin Heidelberg New York, pp 247–252
Katoh S, Suga I, Shiratori I, Takamiya A (1961) Arch Biochem Biophys 94:136–141
Katoh S, Shiratori I, Takamiya A (1962) J Biochem 51:32–40
Kimimura M, Katoh S (1972) Biochim Biophys Acta 283:279–292
Koppenol WH, Margoliash E (1982) J Biol Chem 257:4426–4437

Lappin AG, Segal MG, Weatherburn DC, Sykes AG (1979) J Am Chem Soc 101:2297–2301
Lien S, San Pietro AS (1979) Arch Biochem Biophys 194:128–137
Lockau W (1979) Eur J Biochem 94:365–373
Malkin R, Bearden AJ (1973) Biochim Biophys Acta 292:169–185
Mörschel E, Staehelin LA (1983) J Cell Biol 97:301–310
Mullet JE, Burke JJ, Arntzen CJ (1980) Plant Physiol 65:814–822
Murakami S, Packer L (1970) J Cell Biol 47:332–351
Murata M, Guss JM, Freeman HC (1982) Protein Nucleic Acid Enzyme (Tokyo) 27:579–593
Nanba M, Katoh S (1983) Biochim Biophys Acta 725:272–279
Nechushtai R, Nelson N (1981) J Bioenerg Biomembr 13:295–306
Nechushtai R, Nelson N, Mattoo AK, Edelman M (1981) FEBS Lett 125:115–119
Nelson N, Schatz G (1979) Proc Natl Acad Sci USA 76:4365–4369
Nesbitt DM, Berg SP (1982) Biochim Biophys Acta 679:169–174
Niwa S, Ishikawa H, Nikai S, Takabe T (1980) J Biochem (Tokyo) 88:1177–1183
Olsen LF, Cox RP (1982) Biochim Biophys Acta 679:436–443
Plesničar M, Bendall DS (1970) Biochim Biophys Acta 216:192–199
Ramshaw JAM (1982) In: Boulter D, Parthier B (eds) Encyclopedia of plant physiology, vol 14A. Springer, Berlin Heidelberg New York, pp 229–290
Ramshaw JAM, Brown RH, Scawen MD, Boulter D (1973) Biochim Biophys Acta 303:269–273
Rich PR (1984) Biochim Biophys Acta 768:53–79
Sandmann G, Böger P (1981) Photosynth Res 2:281–289
Sandmann G, Böger P (1983) In: Läuchli A, Bieleski RL (eds) Encyclopedia of plant physiology, New Ser, vol 15. Inorganic plant nutrition. Springer, Berlin Heidelberg New York, pp 563–596
Sandmann G, Reck H, Kessler E, Böger P (1983) Arch Microbiol 134:23–27
Scott RA, Hahn JE, Doniach S, Freeman HC, Hodgson KO (1982) J Am Chem Soc 104:5364–5369
Segal MG, Sykes AG (1978) J Am Soc 100:4585–4592
Selak MA, Whitmarsh J (1984) Photochem Photobiol 39:485–489
Siegelman MH, Rasched IR, Kunert K-J, Kroneck P, Böger P (1976) Eur J Biochem 64:131–140
Siggel U (1974) In: Avron M (ed) Proc 3rd Int Congr Photosynth. Elsevier, Amsterdam New York Oxford, pp 645–654
Simpson DJ (1982) Carlsberg Res Commun 47:215–225
Takabe T, Ishikawa H, Niwa S, Itoh S (1983) J Biochem (Tokyo) 94:1901–1911
Takabe T, Ishikawa H, Niwa S, Tanaka Y (1984) J Biochem (Tokyo) 96:385–393
Tamura N, Yamamoto Y, Nishimura M (1980) Biochim Biophys Acta 592:536–545
Tamura N, Itoh S, Yamamoto Y, Nishimura M (1981) Plant Cell Physiol 22:603–612
Tanaka K, Takahashi M, Asada K (1981) Plant Cell Physiol 22:33–39
Willey DL, Auffret AD, Gray JC (1984) Cell 36:555–562
Wood PM (1974) Biochim Biophys Acta 357:370–379
Wood PM (1978) Eur J Biochem 87:9–19
Wood PM, Bendall DS (1975) Biochim Biophys Acta 387:115–128
Yocum CF (1982) In: Edelman M, Hallick RB, Chua N-H (eds) Methods in chloroplast molecular biology. Elsevier/North-Holland Biomed Press, Amsterdam New York, pp 973–981

10.6 Local and Delocalized Interactions in Energy Coupling

B.A. MELANDRI and G. VENTUROLI

1 Introduction

This short review will discuss the mechanism of energy coupling in photosynthetic membranes. The attention has been focused on the most recent studies, keeping discussion of earlier results to a minimum. For this reason, and given the purpose of this minireview, a minimal reference list has been included. For a more extensive discussion the reader should refer to pertinent reviews (JAGENDORF 1975, WITT 1975, SCHLODDER et al. 1982, BALTSCHEFFSKY et al. 1982, ORT and MELANDRI 1982, ORT, Chap. 4; DILLEY, Chap. 10.7, this Vol.)

2 Steady-State Coupling in Chloroplasts

To a reasonable degree of certainty, the major features responsible for coupling between proton translocation and photophosphorylation have been demonstrated in chloroplasts. Light-induced electron flow catalyzes an uncoupler-sensitive H^+ uptake (NEUMANN and JAGENDORF 1964) resulting in a pH difference across the thylakoid membrane (ΔpH) as detected by 9-aminoacridine fluorescence quenching (SCHULDINER et al. 1972) and by the distribution of ^{14}C-methylamine cation (ROTTENBERG et al. 1972). Phosphorylation was found to decrease the steady-state extent of H^+ uptake and the magnitude of ΔpH (PICK et al. 1973, PORTIS and MCCARTY 1974) consistent with the concept of the involvement of a proton-translocating ATPase. Approximately third-order dependences of the rate of ATP synthesis on the internal H^+ concentration, $[H^+_{in}]$, were obtained when the membrane energization was modulated by light intensity or uncoupler addition (PORTIS and MCCARTY 1974). These sharp dependencies appeared to be consistent with the relatively small effect of photophosphorylation on the steady-state level of ΔpH. A threshold value of ΔpH, in the range 2.0–2.5 units, was observed (PICK et al. 1974), below which no significant phosphorylation could be detected. The existence of a critical value of ΔpH, which appeared to be somewhat dependent on the specific method employed to modulate the high energy state of the membrane, was also supported by studies in which pseudo-steady states of phosphorylation were induced by trains of single turnover xenon flashes fired at variable frequency (GRÄBER and WITT 1976). Taking advantage of the different relaxation times of ΔpH and of the transmembrane electrical potential difference, as measured by the electrochromic absorption changes of carotenoids, these two parameters could be varied independently, and their thermodynamic equivalence in relation to the phosphorylation reaction was, at least qualitatively, established.

The relationships between $[H^+_{in}]$ and the electron flow rate have been extensively investigated by PORTIS and MCCARTY (1976) under a variety of experimental conditions. The electron transfer rate was found to have a linear relationship to $[H^+_{in}]$, when this parameter was changed by light intensity and by inhibitors of electron flow in the absence of phosphorylating substrates. The proportionality constant relating $[H^+_{in}]$ to the electron flow rate appeared to be modulated by the permeability of the membrane to H^+, suggesting that the rate of electron flow was mainly controlled by the level of the H^+ gradient. These observations led to the formulation of a quantitative model for steady-state coupling in which a transmembrane ΔpH, determined by electron transfer-linked H^+ influx and H^+ efflux through the ATPase coupling factor and the membrane leaks, was assumed to be the driving force for phosphorylation. This model also provided an explanation for the observed decrease of the P/2e ratios with decreasing light intensity. A constant P/2e ratio, equal to 1.3, was obtained when the electron transfer rate measured under phosphorylating conditions was corrected for basal electron flow as determined in parallel experiments. In subsequent studies (MCCARTY and PORTIS 1976, DAVENPORT and MCCARTY 1981) this approach was extended to the effect of energy transfer inhibitors on the apparent P/2e ratio. The previously determined relationship between P/2e ratio and $[H^+_{in}]$, as predicted by the model, was confirmed under these conditions and the same value for the intrinsic P/2e ratio was obtained. Assuming an H^+/e ratio equal to 2 (JUNGE and AUSLÄNDER 1974), the calculated intrinsic P/2e implies an H^+/P ratio equal to 3, a value which seems to be consistent with the third-order dependence of photophosphorylation on $[H^+_{in}]$, if $[H^+_{in}]$ is assumed to be the limiting factor for phosphorylation.

This simple picture of protonic coupling, which identifies the driving force for phosphorylation with a bulk-to-bulk phase electrochemical potential difference for protons ($\Delta\tilde{\mu}_{H^+}$), is, however, now being questioned due to recent results of steady-state measurements. DE KOUCHKOVSKY and HARAUX (1981) examined the effects of hydrogen-deuterium exchange on energy coupling: in ^2H_2O, at equal measured ΔpH, the redox chain control was increased, and the observed P/2e ratio decreased (DE KOUCHKOVSKY et al. 1982). These observations have been interpreted as providing evidence for a heterogeneity of pH along the membrane surface. The postulated proton diffusion barriers give rise to local pH gradients that differ in magnitude at the points of H^+ influx (redox chain) and efflux (passive leaks and coupling factors). Indications of a close interaction between electron flow processes and ATP synthesis came from the finding that, for a given measured ΔpH, the system I redox chain was found to be more efficient in driving ATP synthesis than the system II chain (HARAUX et al. 1983). As also observed in bacterial chromatophores (see Sect. 3), a more pronounced variation in the rate of phosphorylation with $\Delta\tilde{\mu}_{H^+}$ is obtained when this parameter is depressed by reducing light intensity than by the addition of uncouplers (SIGALAT et al. 1983).

3 Steady-State Coupling in Bacterial Chromatophores

Many experimental results have accrued to indicate that in chromatophores from photosynthetic bacteria the steady-state coupling between electron flow and ATP synthesis involves a transmembrane electrical potential difference ($\Delta\psi$) as well as a transmembrane pH gradient (JACKSON et al. 1968, CASADIO et al. 1974). The maximal affinity of the phosphorylation reaction attainable by chromatophores in continuous light (ΔG_p) has been measured and compared with the level of the transmembrane electrochemical potential difference for protons. Essentially identical values for ΔG_p have been obtained in many laboratories (LEISER and GROMET-ELHANAN 1977, KELL et al. 1978, BACCARINI-MELANDRI et al. 1977). In contrast, when different methods are employed for measuring $\Delta\psi$ and ΔpH in steady state, greatly varying values for the transmembrane H^+ electrochemical potential difference are obtained, ranging from 400 to 100 mV. The determination of the stoichiometry of H^+/ATP from the $\Delta G_p/\Delta\tilde{\mu}_{H^+}$ ratio is further complicated by the fact that, according to nonequilibrium thermodynamics, this parameter yields an underestimation of the stoichiometric coefficient, which is dependent on the coupling efficiency of the energy-transducing system (CAPLAN 1971).

Utilizing the field-indicating carotenoid shift for $\Delta\psi$ measurements, and the fluorescence quenching of 9-aminoacridine for the evaluation of ΔpH, the relationship between the rate of ATP synthesis and $\Delta\tilde{\mu}_{H^+}$ has been studied under conditions, in which both the ionic permeability of the membrane and the rate of electron transfer were perturbed (BACCARINI-MELANDRI et al. 1977, CASADIO et al. 1978). In the presence of uncouplers, a roughly linear correlation between the rate of ATP synthesis and $\Delta\tilde{\mu}_{H^+}$ was found, and an apparent energy threshold comparable to that observed in chloroplasts was detected. A markedly different correlation emerged, however, when the rate of electron flow was impaired either by adding antimycin or by limiting the intensity of the actinic light: under these conditions the rate of photophosphorylation was in fact drastically reduced without a parallel decrease of $\Delta\tilde{\mu}_{H^+}$. The lack of an unequivocal relationship between the rate of ATP synthesis and $\Delta\tilde{\mu}_{H^+}$ as evidenced by these results, strongly suggests that redox reactions are more directly involved in phosphorylation than predicted by a minimal protonic coupling scheme, which assumes delocalized proton gradients as the relevant driving forces (see also DILLEY, Chap. 10.7, this Vol.)

In the framework of chemiosmosis, the existence of a delocalized protonic pool implies that addition of an inhibitor of the primary proton pump will have relatively less effect on the overall flux in the presence of an inhibitor of the secondary pump, and vice versa. These conclusions can be tested by studying the overall effect of a simultaneous inhibition of electron flow and ATP synthesis. The results of such dual inhibitor titrations in bacterial chromatophores (HITCHENS and KELL 1982a, b) conflicted with these expectations, and were taken as evidence that the high energy protons released by individual electron transport chains were not released into the entire coupling membrane vesicle, but were only available to particular H^+-ATP synthases. In line with this interpretation of the dual inhibitor titration experiments, uncoupling agents were found to be more potent in decreasing the rate of ATP synthesis in the presence than in the absence of energy transfer inhibitors (HITCHENS and KELL 1983a, b). COTTON and JACKSON (1983) failed, however, to observe any kinetic effect of F_0-inhibitors on the titration of ATP synthesis with uncouplers, when a different experimental protocol was employed in order to minimize time-dependent changes in the rate of photophosphorylation. The straightforward interpretation of this type of experiment is hampered by the complex response of the reaction rates to the extent of $\Delta \tilde{\mu}_{H^+}$; parallel evaluation of the protonic gradient would be advisable for these experiments.

4 Transient Coupling Phenomena in Chloroplasts

The perturbation of a steady-state condition of chloroplast membrane, e.g., by giving a flash of light or a diffusion potential pulse, is a method suitable for obtaining information on the storing capacity of the membrane and on the size of the precursor pool for ATP synthesis. This approach has been adopted by many laboratories, since the pioneering work of HIND and JAGENDORF (1963) demonstrated that ATP could be formed in the dark by pre-illuminated chloroplasts. These early data suggested a good correlation between the time course for the amount of ATP formed ($[X_E]$) and the decay of the protonic gradient, as monitored with a glass electrode; this correlation was also supported by the stimulation of $[X_E]$ by permeating internal buffers, with pK's of 5 to 6 (AVRON 1972), and by ionic diffusion potentials, positive inside (SCHULDINER et al. 1973). Subsequently, experiments on dark-to-light transitions were performed with the aim of evaluating the kinetics of formation of the protonic gradient and its correlation with the induction of ATP synthesis (ORT et al. 1976, ORT and DILLEY 1976). When the formation of $\Delta\psi$ was prevented by valinomycin and K^+, hydrophilic buffers did not delay photophosphorylation, even though they prevented the formation of ΔpH for many turnovers. Evidence of this sort was taken as a proof that bulk hydrated protons were not involved as precursors of ATP, but that, possibly, parallel protonic pathways could be operative in this function. When the formation of ΔpH and the initiation of photophosphorylation were time-resolved with trains of single turnover flashes, it was determined that as few as 7 nmol H^+ mg^{-1} Chl were necessary to fill the pool of ATP precursors in the absence of valinomycin, and 35 nmol H^+ mg^{-1} Chl in its presence (GRAAN et al. 1981). These figures seem to be at odds with the large endogenous buffering capacity of thylakoids, especially when large concentrations of hydrophylic buffer are equilibrated across the membrane. Evidence for a transient compartmentation of protons in a domain inaccessible to hydrophilic buffers has been recently obtained by HONG and JUNGE (1983) and THEG and JUNGE (1983). The relevance of these protons for energy coupling has been questioned, however, by these authors. A different interpretation suggests that the onset of photophosphorylation is not related to the extent of ΔpH on an energetic basis, but rather kinetically and in connection with inactive-active transitions of the H^+-ATPase (HARRIS and CROFTS 1978). This implies that the apparent threshold of ΔpH for photophosphorylation has no relation with the free energy change for ATP formation, but might be related to an energy-dependent conformational change of this enzyme. If this is true, the amount of protons involved in this function should be very small, and the phenomenon should be affected also by electrostatic components, since it is destabilized by valinomycin and K^+.

5 Transient Coupling Phenomena in Chromatophores

ATP synthesis by pre-illuminated chromatophores could be demonstrated in *Rhodospirillum rubrum* (LEISER and GROMET-ELHANAN 1975); the amount of ATP formed was dependent on the ambient pH, and presented an optimum around pH 8.0. Postillumination phosphorylation was also stimulated by the addition of permeant ionic species (K^+ and valinomycin or thiocyanate) during the light stage (GROMET-ELHANAN and LEISER 1975). It was suggested that the stimulation resulted from an increase of ΔpH due to the collaps of $\Delta\psi$; consistent with this conclusion, no ATP was formed for $\Delta pH \leq 2.1$ units, similarly to what observed under continuous illumination. Diffusion potentials, positive inside, exerted a stimulatory effect on ATP synthesis, even when the light-induced ΔpH was above the minimal threshold level, suggesting again an energetic equivalence between $\Delta\psi$ and ΔpH.

Single turnover resolution of the coupling kinetics was obtained by the use of luciferine-luciferase luminescence (MELANDRI et al. 1980). The amount of ATP formed per turnover of the electron transfer chain was found to be a single-valued function of $\Delta\psi$, both when the frequency of flashing was changed, or when ATP was synthetized during the postillumination period. ATP was already formed after only two turnovers (corresponding to no more than 200 H^+ per chromatophore, or 4 H^+ per photosynthetic unit). This response was, however, observed only if $\Delta\psi$ was allowed to form rapidly; when valinomycin and K^+ were added, ATP synthesis lagged for 20 to 25 turnovers, until a sufficiently high ΔpH was reached. These observations were consistent with continuous light experiments (using 100–200-ms-long flashes) and direct determinations of $\Delta\tilde{\mu}_{H^+}$ (MELANDRI et al. 1978). The induction time was also found to be longer at high concentration ratios of ATP/ADP. This suggests that the threshold observed had an energetic meaning and was not related to activation-deactivation cycles of the ATP synthetase.

The synthesis of ATP in flashing light revealed, however, some peculiar properties not consistent with a chemiosmotic mechanism: no ATP was formed in postillumination after about 200 ms in the dark, even in the presence of a large membrane potential. However, ATP was synthesized under these conditions following one single isolated turnover flash, demonstrating that the lack of synthesis was not due to energy limitation (under several experimental conditions a single turnover of the photosynthetic chain could not produce a $\Delta\tilde{\mu}_{H^+}$ large enough to drive ATP synthesis per se). Moreover, the ATP yield per flash decreased following addition of antimycin A or of DCCD, with a response that was linear to the extent of inhibition of electron transfer or of the ATPase, but independent of the frequency of flashing, i.e., of the average rate of electron transfer (VENTUROLI and MELANDRI 1982). These observations are difficult to reconcile with a conventional chemiosmotic model formed by mutually independent and reversible energy transducers. It appears, on the contrary, as if energy transduction occurs in a rigidly coupled fashion, through rather direct interactions between multienzyme complexes. These conclusions were also reached for *R. rubrum* chromatophores, in which the ATP yield per flash was linearly correlated with the amount of cytochrome c present in the preparation, and therefore with the rate of electron transfer, rather than with the extent of $\Delta\tilde{\mu}_{H^+}$ (DEL VALLE-TASCON et al. 1978).

6 ATP Synthesis Driven by Artificial Ion Gradients

A direct demonstration of ATP formation at the expense of a pH gradient came from "acid base transition" experiments performed in chloroplasts (JAGENDORF and URIBE 1966) and subsequently in reconstituted systems including the purified ATPase of the thermophilic bacterium PS3 (SONE et al. 1977) and the chloroplast CF_0-F_1 incorporated into asolectin liposomes (GRÄBER et al. 1984a).

Kinetic analysis of phosphorylation driven by acid base transitions has been attempted utilizing rapid quenched-flow techniques (GRÄBER et al. 1984b). This experimental approach allowed the study of the rates of ATP synthesis at short reaction times (20 200 ms) where a definite imposed ΔpH as well as the concentrations of the substrates remained constant. Under these conditions, rates of ATP synthesis comparable to those measured in the light were observed. Considerable rates of phosphorylation could be observed only at an imposed pH exceeding 2.5 units, a value which again compares favorably with the threshold level detected in light-induced ATP synthesis. The use of rapid quench-flow techniques has also permitted the investigation of ATP synthesis under conditions in which transmembrane $\Delta\psi$'s generated by K^+-valinomycin diffusion potentials were superimposed on the transmembrane ΔpH's induced by acid base transitions. When the magnitudes of the diffusion potentials were evaluated according to the Goldman equation, using literature data of the permeability coefficients, a single dependence of the rate of ATP synthesis on artificially generated $\Delta\tilde{\mu}_{H^+}$ was found, regardless of the relative contribution of ΔpH or $\Delta\psi$. These studies show quite unambiguously that a transmembrane electrochemical potential difference for protons can efficiently catalyze proton transport coupled ATP synthesis in chloroplasts. In rapid acid base transition experiments performed with reconstituted CF_0F_1 liposomes, rates of phosphorylation of only up to a few percent of the maximal rates attainable in chloroplasts could be obtained (GRÄBER et al. 1983a).

Phosphorylation induced by a pulsed external electric field across a chloroplast suspension placed between two macroscopic electrodes has been demonstrated by WITT's group (WITT et al. 1976). The amount of ATP generated was found to be related in a linear manner to the number of pulses, to the field intensity and to the pulse duration, yielding apparent rates comparable to those of photophosphorylation. A detailed analysis of the electrochromic absorption changes of the membrane pigments induced by the external electric field under conditions of different osmolarity and ionic strength of the suspension, provided information on the transmembrane field generation (SCHLODDER and WITT 1980). It was concluded that the external field induced a transmembrane potential difference by accumulation of ions from the bulk phases at both sides of the membrane, with a rise and decay time in the μs domain. Since the membrane polarization depends on the electrodynamic properties of the suspension, as well as on the geometric parameters of the vesicles, the evaluation of the induced transmembrane potential difference is somewhat uncertain. At an external electric field strength of 10^3 V cm^{-1}, an average $\Delta\psi$ around 200 mV has been calculated, in agreement with the estimated values of the protonmotive force during light-induced phosphorylation.

Only one report on ATP synthesis driven by ion diffusion potentials in a bacterial system has appeared. In *R. rubrum* chromatophores, the addition of internal buffers (succinate or pyridine) was required to observe minimal ATP formation (about 4–5 nmol ATP mg^{-1} Bchl), even when the artificially imposed ΔpH exceeded three units, i.e., was substantially larger than the minimal energetic threshold (LEISER and GROMET-ELHANAN 1974). A large synergistic stimulation was, however, obtained when a K^+ and valinomycin pulse was superimposed to the acid-base transition (up to 40–50 nmol ATP mg^{-1} Bchl). In this case the ATP yield was larger, the larger the ΔpH. Data like these demonstrate that ATP can be formed with energy contained in an ionic gradient. The amount of ATP formed has, however, hardly any relation to the free energy change of the gradient, but relates instead to the energy storage capacity of the vesicles. The amount of ATP formed in an acid base transition in the presence of 41 mM pyridine (pK = 5.2, close to the inner pH of chromatophores) was 4.6 nmol mg^{-1} Bchl. This quantity is much smaller than the amount of buffer present in the inner aqueous lumen (15 μl mg^{-1} Bchl, as evaluated with the nitroxide-labeled probe tempone in *Rhodopseudomonas sphaeroides* chromatophores). If pyridine buffered the inner protons entirely at pH = 5.2, a yield of at least 150 nmol mg^{-1} Bchl should be expected for a stoichiometry of 2 H^+ per ATP. The yield of ATP, on the contrary, never exceeded a few nmol mg^{-1} Bchl (or 50 nmol mg^{-1} Bch, if a K^+ diffusion potential was superimposed). It is doubtful, therefore, that the inner aqueous protons are involved in driving ATP synthesis, even when buffered by relatively hydrophobic species like pyridine. This conclusion is somewhat reminiscent of that drawn for chloroplasts subjected to a dark-to-light transition in the presence of permanent hydrophylic buffers (ORT and DILLEY 1976).

7 Interpretation of the Phenomena of Localized Coupling

Abundant evidence is available, both in chloroplasts an chromatophores, that the ATP synthase is a regulated enzyme, which can be present in an active or inactive state (for reviews see SHAVIT 1980, MCCARTY and CARMELI 1982, ORT, Chap. 4, this Vol.). Light has been shown to induce an activation of the ATPase activity; in chloroplasts this active state can be stabilized for several minutes by thiol-reducing compounds. Several ligands interfere with the active-inactive transition: phosphate stabilizes the active state in chloroplasts, and stimulates the V_{max} of the ATPase in chromatophores; ATP inhibits light activation in the bacterial system, and accelerates its decay in chloroplasts. In all systems the activation process seems to involve marked conformational changes. This is evidenced by an enhanced susceptibility of the ATPase to inhibition by N-ethyl-maleimide, by an increased ability to exchange tritium with water (RYRIE and JAGENDORF 1971), by variations in the rotational mobility of a conjugated eosin label, and by an increased accessibility of oxygen to a site capable of quenching its triplet excited state (WAGNER and JUNGE 1980).

In principle the inactive-active transition of the ATP synthase can be considered to be the limiting step for ATP synthesis. Models have been proposed in which the rate of phosphorylation is determined by the average number of active enzymes. In the model proposed by GRÄBER et al. (1984b), the activation is proposed to be induced by the binding of H^+_{in} and OH^-_{out}, or, alternatively, by the translocation of H^+ from the inner to the outer compartment of chloroplasts. In both cases, the fraction of active enzymes will depend in a complex exponential way on $\Delta\tilde{\mu}_{H^+}$. Alternatively, MILLS and MITCHELL (1984) have proposed that activation involves protonation of the enzyme, with different pK's on both sides of the membrane. This implies that activation of CF_0-CF_1 involves the establishment of a dual pH optimum on both sides of the membrane, and is not simply dependent on a transmembrane $\Delta\tilde{\mu}_{H^+}$. Whatever the model, the effector inducing the activation is always postulated to be a bulk and delocalized proton; these models, seem therefore inadequate to account for the phenomenon of localized interactions between electron transfer and ATP synthesis discussed above. The direct involvement of thiol reduction in the activation process, and the recent demonstration that in intact chloroplasts activation can be brought about by the thioredoxin system, suggests that, particularly in chromatophores, the photosynthetic generation of reducing agents might induce an activation process linked to electron transfer in a local intramembrane pathway.

An alternative explanation suggests that the localized interaction could be due to the existence of intramembrane proton currents, directly coupling single or a small number of electron transfer chains with ATP synthases. These ideas have been taken into consideration in the formulation of several recent models (PROCHASKA and DILLEY 1978, HARAUX and DE KOUCHKOVSKY 1983, WESTERHOFF et al. 1984a, b). A common feature of these proposals is that the energy-transducing complexes are not randomly distributed across the membrane lipids, but rather laterally ordered to form coupling units, including a reduced number of both types of transducer. These units are considered to be isolated from the aqueous bulk phases on both sides of the membrane by diffusion barriers high enough to permanently limit the equilibration of intramembrane proton domains with the bulk phases. An analog simulation of these protonic circuits, utilizing exclusively ohmic components, has been able to reproduce, under steady-state conditions, most of the localized effects associated with energy transduction (Westerhoff et al. 1984b). These networks need to be further elaborated by introducing capacitative and non-ohmic components to make them more representative of the behavior of real membranes. According to these models, the functional organization of photosynthetic membranes in terms of protonic currents, is much more complex than that postulated by Mitchell in his original membrane-coupling hypothesis. This complexity may form the basis of energy-transduction mechanisms involving localized protonic interactions.

Acknowledgments. We wish to acknowledge the kind collaboration of our colleagues Peter Gräber and Wolfgang Junge, who made preprints of their work available to us. This work was partially supported by Consiglio Nazionale delle Ricerche, grant no. 82.02028.04

References

Avron M (1972) In: Forti G, Avron M, Melandri BA (eds) Proc 2nd Int Congr Photosynth. Nijhoff/Junk, The Hague, pp 861–871
Baccarini-Melandri A, Casadio R, Melandri BA (1977) Eur J Biochem 78:389–402
Baltscheffsky M, Baltscheffsky H, Boork J (1982) In: Barber J (ed) Electron transport and photophosphorylation. Elsevier/North-Holland Press, Amsterdam New York, pp 249–272
Caplan S (1971) Curr Top Bioenerg 4:1–65
Casadio R, Baccarini-Melandri A, Zannoni D, Melandri BA (1974) FEBS Lett 49:203–207
Casadio R, Baccarini-Melandri A, Melandri BA (1978) FEBS Lett 87:323–328
Cotton NPJ, Jackson JB (1983) FEBS Lett 161:93–99
Davenport JW, McCarty RE (1981) In: Selman BR, Selman-Reiner S (eds) Energy coupling in photosynthesis. Elsevier/North-Holland Biomed Press, Amsterdam New York, pp 141–148
Graan T, Flores S, Ort DR (1981) In: Selman BR, Selman-Reiner S (eds) Energy coupling in photosynthesis. Elsevier/North-Holland Biomed Press, Amsterdam New York, pp 25–34
Gräber P, Witt HT (1976) Biochim Biophys Acta 423:141–163
Gräber P, Rogner M, Samoray D, Hauska G (1984a) Proc IV Int Congr Photosynth Vol II, (GC Sybesma ed). Martinus Nihoff/Dr W Junk Publishes, The Hague, pp 427–430
Gräber P, Junesh U, Schatz GH (1984b) Ber Bunsen Ges Phys Chem 88:599–608
Gromet-Elhanan Z, Leister M (1975) J Biol Chem 250:90–93
Haraux F, Kouchkovsky Y De (1983) Physiol Veg 21:563–576
Haraux F, Sigalat C, Moreau A, Kouchkovsky Y De (1983) FEBS Lett 155:248–252
Harris DA, Crofts AR (1978) Biochim Biophys Acta 502:87–102
Hind G, Jagendorf AT (1963) Proc Natl Acad Sci USA 49:715–722
Hitchens GD, Kell DB (1982a) Biochem J 206:351–357
Hitchens GD, Kell DB (1982b) Biosci Rep 2:743–749
Hitchens GD, Kell DB (1983a) Biochem J 212:25–30
Hitchens GD, Kell DB (1983b) Biochim Biophys Acta 723:308–316
Hong YQ, Junge W (1983) Biochim Biophys Acta 722:197–208
Jackson JB, Crofts AR, Stedingk L von (1968) Eur J Biochem 6:41–54

Jagendorf AT (1975) In: Govinjee (ed) Bioenergetics of photosynthesis. Academic Press, London New York, pp 414–492
Jagendorf AT, Uribe E (1966) Proc Natl Acad Sci USA 55:170–177
Junge W, Ausländer W (1974) Biochim Biophys Acta 333:59–70
Kell DB, Ferguson SJ, John P (1978) Biochim Biophys Acta 502:111–126
Kouchkovsky Y de, Haraux F (1981) Biochem Biophys Res Commun 99:205–212
Kouchkovsky Y de, Haraux F, Sigalat C (1982) FEBS Lett 139:245–249
Leiser M, Gromet-Elhanan Z (1974) FEBS Lett 43:267–270
Leiser M, Gromet-Elhanan Z (1975) J Biol Chem 250:84–89
Leiser M, Gromet-Elhanan Z (1977) Arch Biochem Biophys 178:79–88
McCarty RE, Carmeli C (1982) In: Govindjee (ed) Energy conversion by plants and bacteria. Academic Press, London New York, pp 647–695
McCarty RE, Portis AR (1976) Biochemistry 15:5110–5114
Melandri BA, Santis A de, Venturoli G, Baccarini-Melandri A (1978) FEBS Lett 95:130–134
Melandri BA, Venturoli G, Santis A de, Baccarini-Melandri A (1980) Biochim Biophys Acta 592:38–52
Mills JD, Mitchell P (1984) Biochim Biophys Acta 764:93–104
Neumann J, Jagendorf AT (1964) Arch Biochem Biophys 107:109–119
Ort DR, Dilley RA (1976) Biochim Biophys Acta 449:95–107
Ort DR, Melandri BA (1982) In: Govindjee (ed) Energy conversion by plants and bacteria. Academic Press, London New York, pp 537–587
Ort DR, Dilley RA, Good NE (1976) Biochim Biophys Acta 449:108–124
Pick U, Rottenberg H, Avron M (1973) FEBS Lett 32:91–94
Pick U, Rottenberg H, Avron M (1974) In: Avron M (ed) Proc 3rd Int Cong Photosynth. Elsevier, Amsterdam New York Oxford, pp 967–974
Portis AR, McCarty RE (1974) J Biol Chem 249:6250–6254
Portis AR, McCarty RE (1976)J Biol Chem 251:1610–1617
Prochaska LJ, Dilley RA (1978) Arch Biochem Biophys 187:61–71
Rottenberg H, Grünwald T, Avron M (1972) Eur J Biochem 25:54–63
Ryrie IJ, Jagendorf AT (1971) J Biol Chem 246:3771–3779
Schlodder E, Witt HT (1980) FEBS Lett 112:105–113
Schlodder E, Gräber P, Witt HT (1982) In: Barber J (ed) Electron transport and photophosphorylation. Elsevier/North-Holland Biomed Press, Amsterdam New York, pp 105–175
Schuldiner S, Rottenberg H, Avron M (1972) Eur J Biochem 25:64–70
Schuldiner S, Rottenberg H, Avron M (1973) Eur J Biochem 39:455–462
Shavit N (1980) Annu Rev Biochem 49:111–138
Sigalat C, Kouchkovsky Y de, Moreau A, Haraux F (1983) In: Sybesma C (ed) Abstr 6th Int Congr Photosynth Brussels. Nijhoff/Junk, The Hague, pp 211–216
Sone N, Yoshida M, Hirata H, Kagawa Y (1977) J Biol Chem 252:2956–2996
Theg SM, Junge W (1983) Biochim Biophys Acta 723:294–307
Valle-Tascon S del, Grondelle R van, Duysens LNM (1978) Biochim Biophys Acta 504:26–39
Venturoli G, Melandri BA (1982) Biochim Biophys Acta 680:8–16
Wagner R, Junge W (1980) FEBS Lett 114:327–333
Westerhoff HV, Melandri BA, Veturoli G, Azzone GF, Kell DB (1984a) FEBS Lett 165:1–5
Westerhoff HV, Melandri BA, Venturoli G, Azzone GF, Kell DB (1984b) Biochim Biophys Acta 768:257–292
Witt HT (1975) In: Govinjee (ed) Bioenergetics of photosynthesis. Academic Press, London New York, pp 493–554
Witt HT, Schlodder E, Gräber P (1976) FEBS Lett 69:272–276

10.7 Proton-Membrane Interactions in Chloroplast Thylakoids

R.A. DILLEY

1 Introduction

In Chapter 10.6 of this Volume, MELANDRI and VENTUROLI discuss the role of the protonmotive force in energy coupling in thylakoids and bacteria. In this review it will be taken as a given that ATP formation is driven by the protonmotive force. The issue I will discuss is the interaction of protons with the membrane material such as protein acid base dissociation groups. Although the issue is not well understood, it appears that protons interact with membrane components in localized domains, sequestered from rapid equilibrium with the bulk aqueous phases (see ORT, Chap. 4, this Vol.). Some of the background data supporting this hypothesis will be presented. The techniques used to measure sequestered proton pools will be discussed, as well as which proteins participate, and how such proton pools may be related to proton movement in energy coupling.

2 Proton Uptake and Endogenous Buffering

Chloroplast thylakoid electron transfer results in proton release, in water oxidation, and in proton uptake associated with the reduction and oxidation of plastoquinone (PQ), yielding a measured $H^+/e-$ ratio of close to 2. A "Q cycle" operating in the PQ-cyt b-563-cyt f region may also contribute to the proton uptake (see JOLIOT and JOLIOT, Chap. 10.3; ORT, Chap. 4, this Vol.). In the steady state at saturating light intensity, the extent of H^+ uptake can be near 1 H^+ per chlorophyll at pH_{out} ~6, and near 0.2–0.3 H^+ per chlorophyll at pH_{out} ~8. The internal lumen pH drops to ca. 4.7 to 5.0. Given a lumenal aqueous volume of, say, 10 μl μmol^{-1} chl, this means that the free H^+ concentration is about 10^{-4} μmol H^+ μmol^{-1} chl, or that 99.95% of the accumulated protons are bound to buffering groups. WALZ et al. (1974) showed that the internal phase is heavily buffered by groups (probably carboxyl) having pK's near 5, with a much weaker buffering action in the pH range of 6 to 9. The location of the buffering groups is not understood. The simplest view is that the buffering groups all lie at the aqueous phase-membrane interface, thus having no barrier for proton diffusion up to and away from the groups. However, one of the conclusions we have come to from work in this laboratory (and supported by recent results from other workers) is that some of the higher pK_a buffering groups must be located in sequestered protein domains rather than being exposed directly to the lumenal space (discussed below).

In the 1960's questions were asked about the possible role of the membrane buffering groups in energy coupling (DILLEY 1971), but no clear answers were apparent, except to say that the existence of massive buffering power near pH 5 provides a reservoir for proton accumulation. Such a reservoir could be useful in "storing" accumulated protons at a sufficiently acidic pH to provide a somewhat steady supply of protons for driving ATP formation in the face of sudden, short fluctuations in light intensity (as occurs in natural conditions due to leaf movement, shading, etc.). No evidence has been offered in support of the notion that protonation-driven conformational changes of non-CF_1 membrane proteins are in any way involved in the primary energy storage-exchange reactions of ATP formation. Thus, interest in any mechanistic implications of proton buffering by membrane proteins waned until very recently.

There is now some interest in the possibility that protons may move through, along or in membranes via hydrogen bonding chains (NAGLE and TRISTRAM-NAGLE 1983), but it has not yet been possible to measure such phenomena. The methods used to detect proton–membrane interactions will be discussed next, followed by a brief review of the evidence thus far obtained that suggests the occurrence of sequestered, intramembrane proton-buffering domains.

3 Chemical Probes for Detecting Membrane Proton-Binding Groups

Proton binding to proteins can be studied using straightforward pH titration methods, yielding information about pK_a's. Titrations of native compared to denatured proteins can indicate the presence of buried acid base groups (CLELAND 1977). Chemical modification, and temperature dependence and solvent perturbation of the pK_a can help identify the acid base group(s) involved (CLELAND 1977). Energy-transducing organelles can create large proton gradients through active H^+ pumping into membrane-bounded compartments. The pH of the enclosed aqueous space can easily be measured by weak acid or weak base distribution methods. Determining where the acid base buffering groups are located is not so easy. Some membrane proteins may have charged groups deeply embedded in the membrane rather than only sticking into the aqueous phases, as for example in bacteriorhodopsin. Very few membrane proteins have been studied as thoroughly as bacteriorhodopsin, so not much is known about the distribution of ionizable groups in membranes. Sidedness studies with chemical modification reagents can be useful (LASZLO et al. 1982) for detecting buried groups, but the results are difficult to interpret. We have used chemical modification probes and direct pH measurements of proton exchange after adding protonphores, in combination with techniques for controlling the pH of the inner and outer bulk aqueous phases, to deduce information about putative sequestered, intramembrane proton-buffering domains.

4 Sequestered Proton-Buffering Domains

Thylakoid membranes show a metastable buffered proton pool that is: (1) sequestered within the membrane phase (not the inner aqueous phase), (2) easily released by proton leakiness inducers, i.e., uncouplers or heating, and (3) readily restored by proton-releasing electron transfer reactions or ATPase proton pumping. What is referred to here is not the major pK_a ~5 buffering array, rather the (quantitatively) minor buffering array having pK_a values near 8.

The metastable thylakoid proton pool can be experimentally detected by a direct pH measurement such as shown in Fig. 1 of BAKER et al. (1982). At an external pH of 8.3 or higher, in the dark, the addition of an uncoupler releases H^+ to the outside (BAKER et al. 1981, 1982). HOMANN's group at Tallahassee has independently published similar results and has devoted a considerable amount of effort to the characterization of the phenomenon (JOHNSON et al. 1983, and references therein). They showed that the amount of uncoupler-releasable, metastable protons can be considerably greater than the 20–40 nmol H^+ mg^{-1} chl routinely found by DILLEY's group. The magnitude of the proton release after uncoupler addition was greater when the thylakoids were suspended in a low osmotic potential medium.

Proton accumulation by the electron transfer reactions or by means of ATPase proton pumping (BAKER et al. 1982) were equally effective in reprotonating the buffering pool after a brief heat transition dumped the proton pool. A better way to show the reversible dumping and re-filling of the metastable buffering pool is to dump the pool into pH >8.3 media by adding a low concentration of desaspidin or m-chlorocarbonylcyanide phenylhydrazone (Cl-CCP) to thylakoids followed by bovine serum albumin (BSA) which effectively rebinds those two uncouplers (DILLEY and SCHREIBER 1984). After BSA addition, a brief light treatment restores the metastable proton pool.

Obviously, from the data showing proton effux in the dark due to addition of uncoupler, nothing about sequestering of the proton pool is suggested, i.e., the protons released by the uncoupler or heat could have come from the lumen buffering groups. Other experiments (LASZLO et al. 1984a) show that the lumen is not the source of the protons observed in experiments such as shown in Fig. 1 of BAKER et al. (1982). That was tested using a "pH-clamp" type of experiment which allowed the lumen pH to be poised above the pK_a of the dissociating amine groups (in the dark). For that experiment, thylakoids were incubated in pH 8.8 buffer for 50 min, in the dark, prior to determining the ΔpH by the ^{14}C-methylamine distribution technique. Aliquots were taken to determine the protonation state of the buffering groups in the absence and the presence of uncouplers. The lumen pH equilibrated close to the external pH (above the apparent pK_a of the buffering groups) with or without uncoupler addition, yet the metastable proton pool was maintained, in the minus uncoupler case, in the protonated state. The data can only be explained by postulating that the amine groups in question are contained in a buried domain that does not readily equilibrate with either bulk aqueous phase unless an uncoupler is added.

Acetic anhydride reacts avidly with the $-NH_2$ but not the $-NH_3^+$ form of amines and we have used this property to correlate the uncoupler-induced loss of buffered protons with the extra acetylation of thylakoid amine groups using ^3H-acetic anhydride (BAKER et al. 1981, 1982, LASZLO et al. 1984a). The thylakoids we prepare have approximately 30 nmol mg^{-1} chl^{-1} of NH_3^+ functions which can be deprotonated and acetylated, the protons coming from intramembrane, sequestered buffering groups. The direct pH measurement of proton efflux from the metastable pool gave about 24 nmol H^+ mg^{-1} chl released, thus there is a good correlation between the appearance of NH_2 groups and H^+ efflux upon addition of an uncoupler.

Acetylation of the additional amine groups "exposed" by uncoupler or heat treatment leads to inhibition of PS II water oxidation (BAKER et al. 1981). This provides a convenient, rapid assay for measuring the protonation state of the metastable buffering pool. The reason for the inhibition is not entirely understood, but it is known that three proteins (the 17, 22, and 23 kDa polypeptides) closely associated with water oxidation are rich in lysine residues and show heavy acetylation increases concomitant with the inhibition (LASZLO et al. 1984b). The acetyl group on certain amines apparently perturbs the proper function of those groups in the water oxidation activity. Just as proton leakiness in the dark at pH >8 induces the occurrence of additional acetylation of $-NH_2$ groups and inhibition of water oxidation due to the additional acetylation, proton pumping by PS I or PS II (BAKER 1982, BAKER et al. 1981) or by the ATPase in the dark (BAKER et al. 1982), protects against the inhibitory effect of acetic anhydride and maintains the acetylation level at the lower levels. The amine groups involved are obviously in rapid equilibrium with the protons released by the pumps. Possibly the PS II-associated amines are involved with the Cl^- requirement for PS II, as suggested by THEG and HOMANN (1982). Chloride loss and subsequent PS II inhibition, as well as the "light protection" against Cl^- loss, etc. occurs under conditions identical to the proton depletion and acetylation events, or the protection against those events by proton accumulation.

How good is the evidence that the amine buffering groups discussed above are sequestered within membrane domains rather than being directly exposed to the inner bulk aqueous phase? The pH clamp experiment mentioned above suggests that, as does the PS II Cl^--requirement study of THEG and HOMANN (1982). Another confirmatory line of evidence, but using very different techniques, is the work of JUNGE and colleagues with the neutral red pH-indicating dye. This dye reports pH changes within the thylakoid lumen, although it binds closely to the inner surface of the membrane. HONG and JUNGE (1983) found that frozen and thawed thylakoids showed quite a different response of the neutral red signal to the permeant buffer phosphate than did freshly prepared membranes. The results with the latter suggested that the protons from water oxidation were not buffered by the addition of P_i to the inner aqueous phase, whereas frozen thylakoids gave a neutral red response clearly showing that added P_i buffered out some of the dye protonation absorbance change. Their interpretation was that in "good" membranes, i.e., unperturbed by harsh treatment, protons move in restricted domains separated from the inner aqueous phase. Freeze-thaw or aging treatments can alter the way protons move, allowing them rapid access to the inner bulk phases.

Theg and Junge (1983), using a different protocol, came to a similar conclusion concerning the existence of a sequestered buffering pool, specifically associated with photosystem II. They also measured the rapid absorption change due to neutral red pH indicator dye, but they used fresh thylakoids which were treated so as to cause depletion of the metastable proton pool. When they depleted the pool with an uncoupler, the first seven or eight flashes of actinic light failed to produce the usual rapid, PS II-dependent, neutral red absorption change. One explanation for their findings is that the depleted buffering array absorbed the protons produced by the first seven or eight flashes and thus the events that give the absorption change did not occur. After filling the buffering pool, the subsequent proton release did cause the rapid absorption change. A similar explanation can be invoked to understand our experiments on the influence of proton depletion treatments on the length of the onset lag in ATP formation, as discussed in Section 5.

About seven to ten membrane proteins contribute to the pool of anhydride-reactive $-NH_2$ groups (Laszlo et al. 1984b). It is of more than passing interest that a major contribution to the sequestered amine groups comes from three proteins closely associated with the PS II water-oxidizing system (17, 22, and 33 kDa polypeptides see above ref. and Anderson, Chap. 9.3; Yocum, Chap. 9.2, this Vol.), and the 8 kDa CF_0 proton channel-forming protein (Tandy et al. 1982). This is consistent with (but does not prove) the hypothesis that the target amine groups are part of a sequestered pathway for proton movement from one of the proton sources, water oxidation, to the CF_0–CF_1 sink for protons involved in energy coupling.

5 Sequestered Proton Pools and Energy Coupling

Another correlation consistent with the above hypothesis is found in the study of Dilley and Schreiber (1984), who measured the lag in the onset of ATP formation as affected by the state of protonation or deprotonation of the sequestered amine buffering pool. The lag was observed to increase from near 20 single turnover flashes to about 30 flashes when the sequestered protons were released from the membrane prior to the activating flash train. The effect was reversed, i.e., the lag returned to about 20 flashes when proton-depleted membranes were given a short light treatment (followed by 8 min of dark) prior to adding ADP, P_i and starting the flash regime. It was suggested that protons needed to attain the threshold protonmotive force have access to (or pass through) the sequestered buffering pool on their way to the CF_0–CF_1 complex. When the buffering pool is filled from the beginning, the phosphorylation onset lag is less by an amount of flashes such as would have been required to protonate those buffering groups. The observed lag increase of near ten flashes is equivalent to about

25–35 nmol H^+ μmol^{-1} chl, given that a single flash yields about 3 nmol H^+ μmol^{-1} chl. This value, near 30 nmol H^+ μmol^{-1} chl, is close to the amount of protons we observe effluxing from the thylakoids in the H^+ depletion treatment. Of course, those results could be obtained whether the buffering groups were located in an intramembrane sequestered domain or in the lumen space. But the interpretation seems reasonable when one takes into account the results of the pH clamp experiment on the location of the buffering array, the work of ORT et al. (1976) showing the lack of influence of lumen phase exogenous buffers on the lag in onset of ATP formation, and the results of THEG and JUNGE (1983) discussed above.

More work must be done to critically assess the relationship of the sequestered buffering groups to proton movement into the lumen and to the CF_0–CF_1 complex, but the system is one that is amenable to such experimentation.

References

Baker GM, Bhatnagar D, Dilley RA (1981) Biochemistry 20:2307–2315
Baker GM, Bhatnagar D, Dilley RA (1982) J Bioenerg Biomembr 14:249–264
Cleland WW (1977) In: Meister A (ed) Advances in enzymology, vol 45. Wiley New York, pp 273–387
Dilley RA (1971) Curr Top Bioenerg 4:237–271
Dilley RA, Schreiber U (1984) J Bioenerg Biomembr 16:173–193
Hong YQ, Junge W (1983) Biochim Biophys Acta 722:197–208
Johnson JD, Pfister VR, Homann PH (1983) Biochem Biophys Acta 723:256–265
Laszlo JA, Millner PA, Dilley RA (1982) Arch Biochem Biophys 215:571–581
Laszlo JA, Baker GM, Dilley RA (1984a) J Bioenerg Biomembr 16:37–51
Laszlo JA, Baker GM, Dilley RA (1984b) Biochim Biophys Acta 764:160–169
Nagle JF, Tristram-Nagle S (1983) J Membr Biol 74:1–14
Ort DR, Dilley RA, Good NE (1976) Biochim Biophys Acta 449:108–124
Tandy NE, Dilley RA, Hermodson MA, Bhatnagar D (1982) J Biol Chem 257:4301–4307
Theg SM, Homann PH (1982) Biochim Biophys Acta 679:221–234
Theg SM, Junge W (1983) Biochim Biophys Acta 723:294–307
Walz D, Goldstein L, Avron M (1974) Eur J Biochem 47:403–407

10.8 Structure, Mechanism, and Regulation of the Chloroplast H^+-ATPase (CF_1-CF_0)

R.E. McCarty and C.M. Nalin

1 Introduction

ATP synthesis in chloroplasts is catalyzed by a proton-translocating enzyme that consists of two parts: coupling factor 1 (CF_1) and an aggregate of hydrophobic proteins, termed CF_0. CF_1 contains the active sites of the H^+-ATPase (ATP synthase), whereas CF_0 functions to bind CF_1 to the membrane and as a proton channel. CF_1 is present on the outer surface of thylakoid membranes exposed to the stroma. Its extrinsic nature is confirmed by the ease of its removal from the membrane, its accessibility to antisera, and by its appearance in the electron microscope. CF_0, in contrast, can be removed from thylakoid membranes only when the membranes are destroyed.

Relatively pure (>95%) CF_1 may be conveniently obtained in large amounts from an inexpensive source. Moreover, stable thylakoids that are devoid of CF_1 and that catalyze good rates of ATP synthesis when reconstituted with CF_1 have been prepared (Nelson and Eytan 1979). The availability of these preparations and the ease of energization of the thylakoid membrane simplify the analysis of proton linked ATP synthesis and hydrolysis.

The coverage of CF_1-CF_0 in this review is necessarily limited and will emphasize recent information about the enzyme's structure and regulation. Several reviews cover CF_1-CF_0 in more depth (Nelson 1982, McCarty and Carmeli 1982, McCarty and Moroney 1985). The reader may also want to consult Chapter 10.6 of Melandri and Venturoli, and Chapter 10.7 of Dilley in this Volume to learn more about proton-membrane interactions, and Chapter 10.9 of Strotmann that discusses nucleotide binding sites of the ATPase complex.

Although CF_1 is similar in structure and function to F_1s from other coupling membranes (see Senior and Wise 1983), some properties of the chloroplast enzyme suggest that it differs in key ways from other F_1s. Several of these apparent differences have been reconciled in recent years. The re-evaluation of the molecular weight of CF_1 (400,000) indicates that the subunit stoichiometry of CF_1 closely resembles that of mitochondrial and bacterial F_1. In addition to its structural similarities, CF_1 catalyzes oxygen exchange reactions during turnover that parallel those of mitochondrial F_1 and that are modulated by substrate, indicating the possibility that CF_1 contains multiple interacting catalytic sites. The catalytic site of CF_1 is likely to resemble that of the mitochondrial enzyme, since efrapeptin, a peptide antibiotic that interacts with the catalytic site of F_1, also interacts with CF_1 (NALIN unpublished).

2 Structure of CF_1-CF_0

2.1 Subunit Stoichiometry

Until recently an $\alpha_2\beta_2\gamma\delta\varepsilon_2$ subunit composition for CF_1 was favored (NELSON 1982), whereas the evidence for bacterial and mitochondrial F_1's clearly pointed to an $\alpha_3\beta_3\gamma\delta\varepsilon$ composition (SENIOR and WISE 1983). However, the revised molecular weight value for CF_1 (MORONEY et al. 1983), structure determination by electron microscopy (AKEY et al. 1983, TIEDGE et al. 1983), and incorporation of radioactivity into CF_1 subunits in plants grown in a $^{14}CO_2$ atmosphere (SÜSS and SCHMIDT 1982) suggest that the most tenable stoichiometry is $\alpha_3\beta_3\gamma\delta\varepsilon$. Although SÜSS and SCHMIDT (1982) found 1 δ/CF_1, as many as 3 δ subunits/CF_1 have been detected (BERZBORN and ROOS 1984). Based on an analysis of tryptophane fluorescence of CF_1, BÉLIVEAU et al. (1982) calculated that CF_1 contains 2 ε subunits. Since very minor contamination of CF_1 could significantly raise the apparent tryptophane content of CF_1 preparations, 1 ε/CF_1 (SÜSS and SCHMIDT 1982) may be a more reasonable value. Also, the binding of one ε/CF_1 is sufficient to inhibit totally the ATPase activity of ε-free CF_1 (M.L. RICHTER unpublished).

Although CF_0 has not yet been purified and its subunit composition is not well defined, reconstitutively active CF_1-CF_0 contains three additional proteins, labeled I–III in order of decreasing molecular weight (NELSON and HAUSKA 1979, PICK and RACKER 1979). Subunit III, the DCCD-binding protein, has been purified. The amount of protein recovered, DCCD titrations, and radiolabeling indicate that subunit III is present in four to six copies per CF_1 (NELSON 1982). Subunits I and II were found in 0.3 and 0.7 to 0.9 copies/CF_1, respectively (SÜSS and SCHMIDT 1982). The suggestion by SÜSS and MANTEUFFEL (1983) that subunits I and II are proteolytic fragments of chlorophyll–protein complexes requires further documentation.

2.2 Organization of CF_1-CF_0

Image analysis of electron micrographs shows CF_1 to have six electron-dense structural units arranged in a pseudo-hexagonal array with a stain-excluding central core (AKEY et al. 1983). The size of a single molecule determined by this method (129 Å) is in close agreement with that of *E. coli* F_1 determined by neutron scattering (SATRE and ZACCAI 1979).

Using fluorescently labeled antibodies, nucleotides, and maleimides (CERIONE et al. 1983, SNYDER and HAMMES 1985), distances between various locations within CF_1, and from these sites to the membrane, have been determined by fluorescence energy transfer. A useful working model for the three-dimensional structure of the complex is emerging (Fig. 1).

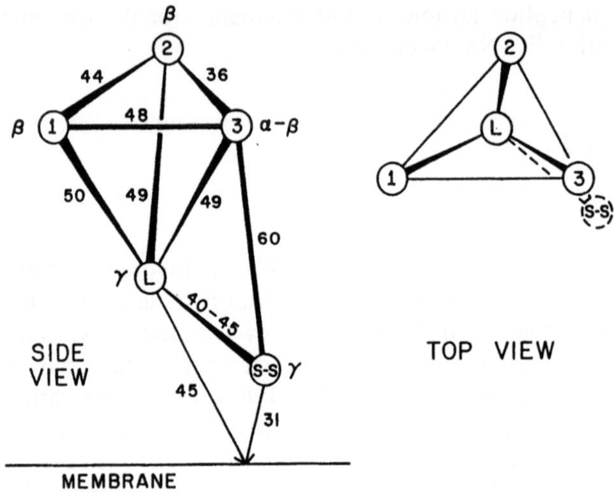

Fig. 1. A model for CF_1 structure based on fluorescence energy transfer. *1, 2, 3* nucleotide binding sites. *L* sulfhydryl group,; *SS* disulfide in the γ subunit (Nalin and McCarty 1984). The accessible sulfydryl group in the γ subunit is not shown (Snyder and Hammes 1984)

2.3 Roles of the Subunits

Two distinct nucleotide binding sites on CF_1 were shown by reaction with photoaffinity nucleotide analogs to be present in the β subunit. Another site may be at the α–β interface (see also Strotmann, Chap. 10.9, this Vol.). Since one of the sites may be catalytic, these results, combined with those from inhibitor studies (Deters et al. 1975) and the remarkable degree of conservation of the primary structure of the β subunit, suggest that the catalytic site(s) is present on the β subunits. The α subunits may play a role in catalysis, regulation or in the binding of CF_1 to CF_0.

The γ subunit is required for ATP synthesis and hydrolysis by membrane-bound CF_1-CF_0, but not in intact form for the ATPase activity of soluble CF_1 (Deters et al. 1975). The γ subunit plays an important role in regulation and may be involved in proton translocation (McCarty and Moroney 1985).

Although the δ subunit of bacterial F_1s is required for rebinding F_1 to the membrane, it is not an absolute requirement for the specific reassociation of CF_1 with thylakoids (Patrie and McCarty 1984). The δ subunit is required for maximal rates of photophosphorylation and to block fully the proton channel through CF_0.

The ε subunit of CF_1 is also not essential for rebinding the enzyme to thylakoids (Patrie and McCarty 1984, Richter et al. 1984). In this regard CF_1 differs from F_1 from *E. coli* (Senior and Wise 1983). The ε subunit of CF_1, however, resembles bacterial ε in its ability to inhibit ATPase activity. The ε subunit is also required for ATP synthesis and for CF_1 to block the F_0 proton channel (Richter et al. 1984).

3 Mechanism and Regulation

3.1 Substrates

The substrates for CF_1 are metal-nucleotide complexes and P_i. The metal ion may be essential for positioning the β-phosphoryl group of ADP for nucleophilic attack on P_i (FRASCH and SELMAN 1982, HOCHMAN and CARMELI 1981). Although Mg^{2+} is the preferred cation for ATP hydrolysis and synthesis by membrane-bound CF_1, the order of effectiveness of metal in supporting ATP hydrolysis by the soluble enzyme is $Mn^{2+} > Mg^{2+} > Ca^{2+} > Co^{2+}$. However, because free metal ions inhibit ATPase activity with the same order of effectiveness, activity in the presence of Ca^{2+} is usually higher than that with Mg^{2+}. Certain detergents (PICK and BASSILIAN 1982), organic solvents (SAKURAI et al. 1981), HCO_3^- and certain dicarboxylic acids relieve metal ion inhibition by an unknown mechanism(s). Since excess Mg^{2+} has little effect on the ATPase activity of the membrane-bound enzyme, solubilization of the enzyme induces this effect.

3.2 H^+/ATP Stoichiometry

A minimum of 3 H^+ are required to drive ATP synthesis when a ΔpH across thylakoid membranes is 3.5 and the phosphorylation potential ($\Delta G'_{ATP}$) is 14.5 kcal mol^{-1} (AVRON 1978). The maximum phosphorylation efficiency (ATP/2e) for electron flow from water to a photosystem I acceptor under oxidizing conditions is 1.33. Since the maximum ATP/2e is the ratio of the H^+/2e for electron transport and the H^+/ATP for CF_1-CF_0 and H^+/2e is 4, the H^+/ATP is likely to be 3 (DAVENPORT and MCCARTY 1984).

3.3 Regulation of Activity

The rate of ATP hydrolysis by soluble or membrane-bound CF_1 is very low. A bewildering array of treatments (reviewed in MCCARTY and MORONEY 1985) enhance ATPase activity.

Removal or displacement of the ε subunit appears to be the key to activation by heat, alcohols, and certain detergents. After heating, ε becomes susceptible to removal from CF_1 by either $(NH_4)_2SO_4$ fractionation (HOLOWKA and HAMMES 1977) or by a chromatographic procedure (PATRIE and MCCARTY 1984). Ethanol treatment of CF_1 bound to DEAE-cellulose removes the ε subunit (RICHTER et al. 1984), as does HPLC of CF_1 on an ion exchange column in the presence of ethylene glycol (FINEL et al. 1984). All ε-deficient CF_1 preparations are active ATPases that are sensitive to purified ε subunit. Octylglucoside treatment also labilizes CF_1–ε interactions (PATRIE and MCCARTY 1984).

Dithiothreitol activation of the ATPase activity of CF_1 has been correlated to reduction of a disulfide bond in the γ subunit (ARANA and VALLEJOS 1982,

Nalin and McCarty 1984). In contrast to previous results (Arana and Vallejos 1982), we found that CF_1 contains only one disulfide bond. Since reduction does not markedly affect the affinity of ε for CF_1, and since reductive activation and that due to ε displacement are additive, the ε subunit is not directly involved in reductive activation (Nalin and McCarty 1984, Richter and Patrie unpublished).

Low levels of proteases activate CF_1 ATPase activity without digestion of the ε subunit (Moroney and McCarty 1982). Protease activation of the oxidized enzyme correlated most closely to cleavage of the α subunits.

There are several indications that the ε and γ subunits interact. The γ subunit undergoes conformational changes upon energization of thylakoids (McCarty and Moroney 1985) and upon reduction of its disulfide bond. Since the dithiols formed by reduction of the disulfide bond are >60 Å away from the nucleotide sites (Snyder and Hammes 1985), changes in the structure of the γ subunit are felt over long distances. Subtle changes in α–γ or γ–ε interactions may influence activity.

Reduction of the disulfide bond in the γ subunit (Ketcham et al. 1984) is also likely to be involved in the physiologically significant reductive activation of CF_1 (Mills and Mitchell 1982, Vallejos et al. 1983). Loss or displacement of the ε subunit is not likely to occur in the activation of membrane-bound CF_1, since this process would uncouple ATP synthesis.

The mechanism(s) of the inhibition by ADP of the ATPase activity of membrane-bound CF_1 (see Strotmann, Chap. 10.9, this Vol.) and that of inhibition of the ATPase activity of reduced CF_1 that occurs when the enzyme rebinds to the membrane remain to be elucidated.

3.4 Cooperativity During Catalysis

Cooperative interactions between nucleotide binding sites have been shown to be an essential feature of H^+-ATPases (Cross 1981). CF_1 also exhibits cooperativity, based on measurement of substrate modulation of oxygen exchange reactions that occur during ATP synthesis and hydrolysis. The extent of oxygen exchange indicates the number of times reversible bond formation occurs at a catalytic site prior to product release. Therefore, if substrate binding at one site is required for product release at an interacting site, as the concentration of substrate is reduced, additional reversals will lead to greater extent of exchange. During ATP synthesis, decreasing the ADP and P_i concentration increases the extent of exchange per ATP formed (Hackney et al. 1979). Similarly, decreasing the concentration of ATP during hydrolysis in the presence of Mg^{2+} increases the number of water oxygens exchanged into product P_i. No exchange occurs during the hydrolysis of Ca^{2+}-ATP indicating that product release is fast relative to the rate of reversal of the bond cleavage step in the presence of Ca^{2+} (Kohlbrenner and Boyer 1983).

Although these results are consistent with the binding-change mechanism involving multiple catalytic sites, cooperativity between a single catalytic site and a regulatory site (Bruist and Hammes 1982) has been proposed. Although only a single nucleotide binding site in CF_1 capable of turnover at rates consistent with catalysis has so far been detected, additional nucleotide sites may be found.

4 Genetic Analysis of CF_1-CF_0

To date, only a small portion of the chloroplast genome has been sequenced and the gene products identified. For CF_1, the α, β, and ε subunits are encoded by chloroplast DNA, based on in vitro translation of chloroplast mRNA selectively hybridized to specific fragments of chloroplast DNA. As with other chloroplast gene products, the three subunits are not encoded as a single operon, but are distributed on chloroplast DNA (WESTHOFF et al. 1981). The genes that code for the β (atp B) and ε (atp E) subunits are cotranscribed as a single mRNA (ZURAWSKI et al. 1982). The order of transcription of these genes (atp B before atp E) is the same as for the genes that code for these subunits in *E. coli*. The gene that encodes the α subunit is located 40 kilobases removed from the β and ε genes. The tobacco gene for the α subunit has been sequenced (DENO et al. 1983). Subunits I and III of CF_0 have been identified as chloroplast gene products by in vitro translation (NELSON et al. 1980), and although the genes have been mapped, sequences are not, as yet, available. The remaining subunits of CF_1-CF_0 are synthesized on cytoplasmic ribosomes and transported into the chloroplast. By analogy to other organellar proteins, these subunits are likely to be synthesized as larger precursors and subsequently transported into the chloroplast and processed to the mature form. Genes for precursor forms of the γ and δ subunits have been identified (WATANABE and PRICE 1982).

5 Future Directions

The inability to prepare the α, β and γ subunits of CF_1 in reconstitutively active form has been an impediment to the analysis of the functions of these subunits. It is relatively easy to cause CF_1 to fall apart, but difficult to prevent the formation of nonspecific aggregates. CF_0 components I and II should be purified and their roles in the complex clarified. The rates of ATP synthesis by vesicles reconstituted with purified CF_1-CF_0 are low in comparison with the turnover of the enzyme in thylakoids. Improvements in the reconstitution procedures and in the isolation of CF_1-CF_0 may overcome this relatively serious problem.

Mechanistically, many questions remain to be answered. Identification of the catalytic sites, cooperativity, and proton involvement in catalysis are areas that will see continued attention. In addition, biosynthesis and assembly of the complex in vivo may lead the way to a better understanding of the organization of the entire photosynthetic apparatus. In this review, we have focused on advances made since the last edition of this series. By the next edition, some of these questions will likely be answered.

References

Akey CW, Crepeau RH, Dunn SD, McCarty RE, Edelstein SJ (1983) EMBO 2:1409–1415

Arana JL, Vallejos RH (1982) J Biol Chem 257:1125–1127

Avron M (1978) FEBS Lett 96:225–232

Béliveau R, Moroney JV, McCarty RE (1982) Arch Biochem Biophys 214:668–674

Berzborn RJ, Roos P (1984) In: Sybesma C (ed) Proc 6th Int Congr Photosynth. Nijhoff/Junk, The Hague, vol II, pp 491–494

Bruist MF, Hammes GG (1981) Biochemistry 20:6298–6305

Bruist MF, Hammes GG (1982) Biochemistry 21:3370–3377

Cerione RA, McCarty RE, Hammes GG (1983) Biochemistry 22:769–776

Cross RL (1981) Annu Rev Biochem 50:681–714

Davenport JW, McCarty RE (1984) Biochim Biophys Acta 766:363–374

Deno H, Sinozaki K, Sugiura M (1983) Nucleic Acids Res 11:2185–2191

Deters DW, Racker E, Nelson N, Nelson H (1975) J Biol Chem 250:1041–1047

Finel M, Rubinstein M, Pick U (1984) FEBS Lett 166:85–89

Frasch WD, Selman BR (1982) Biochemistry 21:3636–3643

Hackney DD, Rosen G, Boyer PD (1979) Proc Natl Acad Sci USA 76:3646–3650

Hochman Y, Carmeli C (1981) Biochemistry 20:6287–6292

Holowka DA, Hammes GG (1977) Biochemistry 16:5538–5545

Ketcham SR, Davenport JW, Warncke K, McCarty RE (1984) J Biol Chem 259:7286–7293

Kohlbrenner WE, Boyer PD (1983) J Biol Chem 258:10881–10886

McCarty RE, Carmeli C (1982) In: Govindjee (ed) Photosynthesis: Energy conversion by plants and bacteria. Academic Press, New York, London, vol I, pp 647–695

McCarty RE, Moroney JV (1985) In: Martonosi A (ed) The enzymes of biological membranes, 2nd edn. Plenum Press, New York, vol 4, pp 383–414

Mills JD, Mitchell P (1982) Biochim Biophys Acta 679:75–83

Moroney JV, McCarty RE (1982) J Biol Chem 257:5910–5914

Moroney JV, Lopresti L, McEwen BF, McCarty RE, Hammes GG (1983) FEBS Lett 158:58–62

Nalin CM, McCarty RE (1984) J Biol Chem 259:7275–7280

Nelson N (1982) In: Barber J (ed) Electron transport and photophosphorylation. Elsevier, Amsterdam New York Oxford, pp 81–104

Nelson N, Eytan E (1979) In: Mukohata Y, Packer L (eds) Cation flux across biomembranes. Academic Press, London New York, pp 409–415

Nelson N, Hauska G (1979) In: Lee CF, Schatz G, Ernster L (eds) Membrane bioenergetics. Addison-Wesley, New York, pp 189–202

Nelson N, Nelson H, Schatz G (1980) Proc Natl Acad Sci USA 77:1361–1364

Patrie WJ, McCarty RE (1984) J Biol Chem 259:11121–11128

Pick U, Bassilian S (1982) Biochemistry 24:6144–6152

Pick U, Racker E (1979) J Biol Chem 254:2793–2799

Richter ML, Patrie WJ, McCarty RE (1984) J Biol Chem 259:7371–7373

Sakurai H, Shinohara K, Hisabori T, Shinohara K (1981) J Biochem (Tokyo) 90:95–102

Satre M, Zaccai G (1979) FEBS Lett 102:244–248
Senior AE, Wise JG (1983) J Membr Biol 73:105–124
Snyder B, Hammes GG (1984) Biochemistry 23:5787–5797
Süss K-H, Manteuffel R (1983) FEBS Lett 153:134–140
Süss K-H, Schmidt O (1982) FEBS Lett 144:213–218
Tiedge H, Schäfer G, Mayer F (1983) Eur J Biochem 132:37–45
Vallejos RH, Arana JL, Ravizzini RA (1983) J Biol Chem 258:7317–7321
Watanabe A, Price CA (1982) Proc Natl Acad Sci USA 79:6304–6308
Westhoff P, Nelson N, Bünemann H, Herrmann RG (1981) Curr Genet 4:109–120
Zurawski G, Bottomley W, Whitfeld PR (1982) Proc Natl Acad Sci USA 79:6260–6264

10.9 Evaluation of Results on Nucleotide-Binding Sites of the ATPase Complex[1]

H. STROTMANN

1 Introduction

The chloroplast ATP synthase complex is composed of two main constituents, the membrane-integral CF_0 and the extrinsic CF_1, which is located on the outer surface of the thylakoid membrane. While CF_0 enables proton conduction through the membrane, CF_1 is the catalyst of reversible hydrolysis of ATP, and therefore contains binding sites for the substrates phosphate, ADP, and ATP. In addition, CF_1–nucleotide interactions play a role in the regulation of the enzyme. In the last decade effort have been made to discriminate and characterize the binding sites. Information has been obtained from studies of both membrane-bound and isolated CF_1. Due to the brevity of this review, the author has had to limit himself to a discussion of those papers that appeared most relevant at the time of preparation of this manuscript in 1984.

2 Definitions

CF_1 contains multiple nucleotide-binding sites which are usually characterized by affinities and exchange velocities. Sites which exhibit a K_d of 10^{-7} M or less are called tight sites and distinguished from loose sites, which have a higher K_d. Sites which bind and release nucleotides at a rate compatible with the overall rate of the catalytic process may be catalytic sites; they are distinguished from noncatalytic sites, which show an exchange velocity below the catalytic turnover rate. Needless to say, proper characterization of a catalytic site requires that its catalytic activity must be proven.

These definitions of binding sites are, however, rather meaningless if we look at the membrane-bound ATPase, because affinities change with the energy state of the thylakoids. For instance, a site containing tightly bound ADP or ATP in de-energized chloroplasts becomes a loose site (increase of K_d by a factor of 100 or more) upon membrane energization. Since this reaction is related with activation of the ATPase (see MELANDRI and VENTEROLI, Chap. 10.6; DILLEY, Chap. 10.7; MCCARTY and NALIN, Chap. 10.8, this Vol.),

[1] Dedicated to my friend and teacher Prof. August Ried, on the occasion of his 60th birthday

the respective site is often called a regulatory site and discriminated from the catalytic site. These denominations suggest a permanent spatial and functional diversity of the two sites. This has not yet been proven by clear experimental evidence. Theoretically, the catalytic or regulatory function could be fulfilled by the same site in two different conformational states.

3 The Number of Nucleotide-Binding Sites

Nucleotide binding has been studied extensively with isolated CF_1. The results are difficult to compare because of the different enzyme preparations used, the techniques employed in binding studies, and differences in other aspects of experimental protocols. Most recent reports on this subject suggest that a total of three nucleotides can be bound per CF_1 (CARLIER and HAMMES 1979, BRUIST and HAMMES 1981, GIRAULT et al. 1982). Isolated CF_1 contains up to 1 ADP per enzyme molecule, which does not dissociate during isolation (SHOSHAN et al. 1978, CARLIER and HAMMES 1979, GIRAULT et al. 1982). In contrast, the isolated CF_1-CF_0 complex has been reported to contain 1 ADP and 1 ATP (CERIONE and HAMMES 1981). The ADP can be slowly exchanged with medium ADP or ATP, in the latter case yielding bound ADP and free P_i (BRUIST and HAMMES 1981). The exchange requires nucleotide binding to another site and is essentially a one-turnover process (CERIONE and HAMMES 1981). Since the rate of exchange for the heat-activated enzyme is 1000 times lower than the rate of ATP hydrolysis, it is assumed that the exchange site is not identical with the catalytic site. GIRAULT et al. (1982) suggest that the nondissociable and poorly exchangeable nucleotide may be equivalent with the tightly bound nucleotide found in membrane-associated CF_1, which is exchanged only upon membrane energization (see below). In addition, two exchangeable ADP- or ATP-binding sites have been identified by GIRAULT et al. (1982). The authors propose that these two sites, which might cooperate in an alternating mode, form the catalytic part of the enzyme.

Two dissociable sites that could be catalytic sites, have also been detected by HAMMES's group. The K_d values were determined as 2–4 μM (BRUIST and HAMMES 1981). In their experiments however, only one exchangeable site was present when a nonexchangeable ATP-specific site was occupied by ATP in the presence of Mg^{2+}. Positive cooperativity between the putative catalytic sites was deduced from studies with the nucleotide analogs 2′,3′-O-(1,4,6-trinitrophenyl)-ADP or -ATP (TNP-ADP, TNP-ATP). At low substrate concentrations, substrate binding was enhanced by these analogs (BRUIST and HAMMES 1982).

Illuminated thylakoids were found to bind up to 1.7 nmol labeled adenine nucleotides mg^{-1} chlorophyll under steady-state conditions when supplied with (^{14}C)ADP and P_i (BICKEL-SANDKÖTTER 1983). This is equivalent to 1.3 mol nucleotide mol^{-1} CF_1. A fraction of the bound nucleotides can be chased by the addition of excess unlabeled ADP in the presence of FCCP, while another fraction is not exchanged (BICKEL-SANDKÖTTER and STROTMANN 1981). Several lines of evidence suggest that the exchangeable nucleotides participate in catalysis of ATP formation, and that the tightly bound nucleotides reside on enzyme molecules that are catalytically inactive (BICKEL-SANDKÖTTER 1983). At low ADP concentration (5 μM) the steady-state fraction of exchangeable (=loosely bound) nucleotides increases at the expense of nonexchangeable (=tightly bound) nucleotides when the light intensity is increased and vice versa. Under such conditions it seems that there is one nucleotide bound per enzyme molecule in either of the two forms. Under steady-state phosphorylation conditions, up to 1 ATP per CF_1 was found to be bound by the hexokinase/glucose accessibility assay (ROSEN et al. 1979).

Upon de-energization of the membranes, the level of tightly bound nucleotides increases in a biphasic manner (STROTMANN et al. 1979b). In part, this reaction is due to conversion of loosely bound to tightly bound nucleotides (SCHUMANN 1984). Ultimately, the de-energized membranes contain 1 tightly bound nucleotide per CF_1, which cannot be removed by washing or exchanged by externally added nucleotides. It consists either of ADP or ATP (VINKLER et al. 1978, STROTMANN 1984). The nucleotides can only be remobilized by new membrane energization.

Determination of the number of binding sites by nucleotide-binding studies depends on the correct value of the molecular weight of CF_1 and the CF_1 to chlorophyll ratio, respectively. In the studies communicated above, a molecular weight of 325,000 (FARRON 1970) was used, which has also been taken as the bases for calculating the CF_1 : chlorophyll ratio (1:860, STROTMANN et al. 1973). However, the recently re-determined molecular weight is about 400,000 (MERCHANT et al. 1983, MORONEY et al. 1983). Accordingly, the stoichiometry of bound nucleotides per enzyme may have been underestimated by 25%, both in isolated and in membrane-bound CF_1. In the studies with thylakoid membranes variations of the CF_1 : chlorophyll ratio, due to the origin of the plant material (BERZBORN et al. 1981) have also to be considered. A number of three nucleotide binding sites or a multiple of three seems to be compatible with the now accepted subunit stoichiometry of $\alpha_3\beta_3\delta\varepsilon$ (YOSHIDA et al. 1979, SÜSS and SCHMIDT 1982, MERCHANT et al. 1983), assuming that the three copies of the large subunits are functionally equivalent.

For assessment of the total number of binding sites, nucleotide binding is, however, an inadequate instrument, particularly if, for example, sites or subunits communicate in a cooperative manner. Several results suggest that indeed site cooperativity is involved in nucleotide interactions with the membrane-bound enzyme. Based on water–oxygen exchange experiments, HACKNEY et al. (1979) concluded that ATP which was formed on a catalytic site is expelled if ADP and P_i are bound at a second site. Furthermore, light-dependent release of tightly bound nucleotides was shown to be strongly accelerated if medium ADP or ADP+P_i were bound to another site (STROTMANN 1984, SCHUMANN 1984). These and other results may explain why in the thylakoid-associated ATPase only one or two sites are occupied by a nucleotide at the same time.

In summary, we may conclude that three nucleotide binding sites, as defined by nucleotide-binding studies with isolated CF_1, is the minimum, but not necessarily the total number of sites. In this context it should be mentioned that up to six nucleotide binding sites have been detected in comparable ATPases from mitochondria and bacteria.

4 Localization of Nucleotide-Binding Sites

Subunit localization of nucleotide-binding sites has been studied by employing several different covalently binding photoaffinity labels. In experiments with isolated CF_1, BRUIST and HAMMES (1981) found that the ATP analog 3′-O-(3-[N-(4-azido-2-nitrophenyl)amino]-propionyl)-ATP (arylazido-ATP) in the presence of Mg^{2+} was incorporated into the β-subunit, whereas the corresponding ADP derivative was detected in both α and β subunits. WAGENVOORD et al. (1981), using 8-azido-ADP, also recovered the label in both subunits. The photoreactive benzophenone derivative 3′-O-(4-benzoyl)benzoyl-ADP (BzADP) was found to be bound in β, when the reaction was carried out with the membrane-bound enzyme (BAR-ZVI et al. 1984). Using BzATP, KAMBOURIS and HAMMES (1985) found that all three nucleotide binding sites of isolated CF_1 could be labeled with this analog. The label was primarily detected in the β subunit; some label, however, was also present in α. The authors concluded that the three sites may be in a similar location in β near the interface between the α and β subunits. CZARNECKI et al. (1982) reported that 2-azido-ADP introduced into the "tight" site of membrane-bound ATPase, was exclusively detected in the β-subunit. This compound is probably the most useful photolabel for two reasons: in contrast to 8-azido-ADP, the 2-azido-ADP exhibits a conformation (preferred anti-conformation) which is well recognized by the enzyme and accepted as a substrate in photophosphorylation. In contrast to the ribose derivatives arylazido-ADP and BzADP, the photoreactive group is situated close to one of the areas of the nucleotide which is involved in binding to the enzyme (SCHLIMME et al. 1979). Hence, with 2-azido-ADP there may be the best chance to link the label to a region of the polypeptide chain which actually forms the nucleotide binding site.

Tightly bound 2-azido-ATP formed during photophosphorylation was also recovered in the β-subunit and additionally detected in the same peptide pieces after proteolytic digestion of the subunit (ABBOTT et al. 1984), demonstrating that the tight sites for ADP and ATP are identical in structure.

5 Catalytic Mechanism

ATP formation includes cleavage of one oxygen atom from inorganic phosphate so that the bridge oxygen of the newly formed β–γ phosphate bond is contributed by the ADP molecule (AVRON and SHARON 1960). Correspondingly, after ATP hydrolysis water-oxygen is found in the hydrolysis product P_i. The stereochemistry of the ATPase reaction has been investigated with isolated F_1-ATPases from mitochondria (WEBB et al. 1980) and the thermophilic bacterium PS 3 (SENTER et al. 1983) by using adenosine 5′-(O-3-thiotriphosphate) (ATPγS) which contained S, ^{16}O and ^{18}O as substituents on γ-phosphorus. F_1-catalyzed hydrolysis of this ATP analog in $(^{17}O)H_2O$ yielded ADP and chiral thiophosphate, containing ^{16}O, ^{17}O, and ^{18}O. The configuration of thiophosphate showed that hydrolysis had proceeded with substituent inversion, indicating an "in-line" phosphoryl transfer. The results also suggest that no phosphorylated covalent intermediate is involved in reversible hydrolysis of ATP by F_1-ATPase, since substituent inversion is a characteristic feature of enzyme-catalyzed one-step phosphoryl transfer reactions, while two-step transfers proceed with double inversion, i.e., substituent retention.

Following the rule of microreversibility of enzymatic reactions, we have to accept that the synthesis of ATP proceeds in the same way as its hydrolysis in the opposite direction. A hypothesis for the molecular mechanisms of ATP formation which fulfills the observed stereochemistry was proposed 10 years ago by MITCHELL (1974).

The enzymatic reaction requires a precise orientation of the substrate molecules within the catalytic site in order to facilitate the necessary transition state. From studies with chiral thiophosphate analogs of ADP and ATP, we know that the phosphate chain is stereospecifically oriented, partly through a specific Mg-complex structure and partly via ionic bonds to positively charged groups in the catalytic site (STROTMANN et al. 1979a). The basic amino acids arginine and lysine seem to be involved as suggested by chemical modification of the ATPase with butanedione and phenylglyoxal (SCHMID et al. 1977, VIALE et al. 1982) or pyridoxal phosphate, respectively (SUGIYAMA and MUKOHATA 1979).

The most intriguing problem related to the function of the ATPase is how energy from the transmembrane electrochemical proton gradient can be used to force endergonic net ATP synthesis. The current view is that energy is not invested in the chemical reaction itself, but primarily directed to the protein in order to facilitate substrate binding and release of the product ATP, which is spontaneously formed at the catalytic site. This idea was developed and strongly supported by sophisticated experiments by P. BOYER and his colleagues.

In a recent version (BOYER and KOHLBRENNER 1981) of the so-called energy-linked binding change mechanism, a trimeric structure of the ATPase complex containing three, in principle identical, catalytic sites is assumed. The sites change their binding properties for the substrates and products in a sequential mode by conformational changes (Fig. 1) which are brought about by energy uptake from the electrochemical proton gradient. In the first step ADP and P_i are bound by the active site which is in a low affinity state. In a second step the loosely bound substrates are transferred to tightly bound substrates and converted to the products H_2O and tightly bound ATP. The third reaction is the transition of tightly bound to loosely bound ATP, which in this state is able to dissociate from the site. The site, having lost its ATP, is reconverted to a site which again is able to pick up ADP and P_i. At substrate saturation,

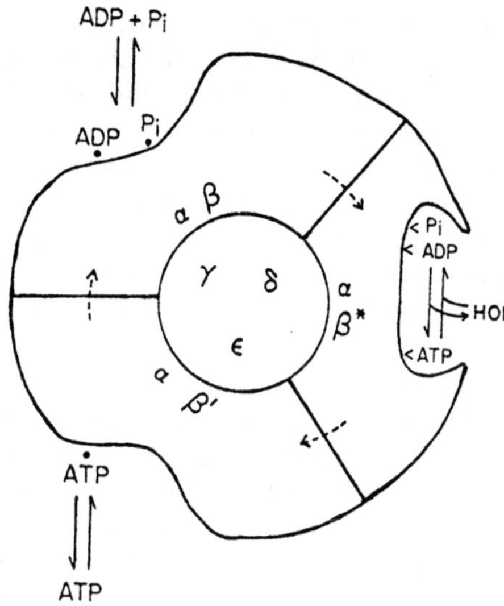

Fig. 1. A three-site binding-change mechanism for ATP synthase (BOYER and KOHLBRENNER 1981). See text, for details

the three sites are believed to work in a strictly cooperative manner, i.e., they are in three different states at the same time, as depicted in Fig. 1.

The hypothesis implies that reversible ATP formation on the enzyme is a reaction which proceeds with an equilibrium constant close to 1. Actually in isolated mitochondrial F_1 performing single site catalysis, the equilibrium constant $E \cdot ADP \cdot P_i / E \cdot ATP$ was determined to be 0.5 (GRUBMEYER et al. 1982). The idea is furthermore supported by the results (a) that intermediate phosphate-water-oxygen exchange which monitors reversible ATP hydrolysis on the enzyme is not inhibited by uncouplers in submitochondrial particles (ROSING et al. 1977), and (b) that tightly bound ATP can be formed from tightly bound ADP in a one-turnover reaction on isolated CF_1 and membrane-bound ATPase without energy investment at high P_i concentration and low pH (FELDMAN and SIGMAN 1982, 1983) and on mitochondrial F_1 in the presence of dimethylsulfoxide (SAKAMOTO and TONOMURA 1983). Attempts to measure in membrane-bound ATPase conformational changes involved in the catalytic cycle, e.g., by covalently linked fluorescent probes (WAGNER and JUNGE 1980), are still in an initial stage. Changes in affinities of the catalytic site toward ADP and ATP have been deduced from steady-state kinetics of ATP synthesis and hydrolysis. In energized chloroplasts, the apparent affinity for ADP is at least tenfold higher than the affinity for ATP; this difference is, however, abolished when the chloroplasts are converted to a de-energized state (FRANEK and STROTMANN 1981).

6 Roles of Tightly Bound Nucleotides

The notation tight binding means fast association and slow dissociation of a nucleotide molecule from its binding site. At the extreme, binding may be an essentially irreversible process, e.g., when pre-illuminated chloroplasts are supplied with ADP after membrane de-energization (STROTMANN and BICKEL-SANDKÖTTER 1977). The bound ADP is neither removed by extensive washing nor exchanged by externally added ADP or ATP. One might imagine that in a tightly bound state the nucleotide molecule is captured by closure of the site due to a conformational change of the protein. It appears that this is a spontaneous reaction, while remobilization of the tightly bound nucleotide requires energization of the chloroplast (MAGNUSSON and MCCARTY 1976, STROTMANN and BICKEL-SANDKÖTTER 1977).

The functional role of tightly bound nucleotides is a matter of controversy. In the energy-linked binding change hypothesis, transitorily tightly bound ATP and ADP are involved as intermediates in the catalytic cycle. Recently SMITH et al. (1983) reported that tightly bound ATP on the membrane-bound enzyme undergoes slow but continuous oxygen exchange with water in the surrounding medium, indicating that the site possesses some residual catalytic activity. The formation of tightly bound ^{32}P-ATP from ^{32}P-P_i and ADP in a catalytically competent manner was found to accompany photophosphorylation even in the presence of an ADP-regenerating hexokinase system (ROSEN et al. 1979). Accordingly, the tightly bound ATP could not have originated from the medium, but might have been formed at the same site before. The kinetic competence of the system has been questioned by AFLALO and SHAVIT (1982), who were unable to chase the ^{32}P-ATP rapidly enough using unlabeled phosphate or arsenate. Before discussing this issue any further, another role of tightly bound nucleotides must be considered.

It is well known that in de-energized chloroplasts the ATPase is usually a latent, i.e., catalytically inactive enzyme. The catalytic activity is unmasked by energization of the membranes, hydrolysis requires, in addition, exposure to thiol compounds like DTT (MCCARTY and RACKER 1968). In the intact chloroplast, the role of DTT is probably fulfilled by reduced thioredoxin (MILLS et al. 1980). Membrane energization, but not necessarily thiol modulation, is also a prerequisite for switching into the ATP synthesizing mode of operation. In de-energized membranes, the catalytically inactive ATPase contains tightly bound ADP and ATP. These nucleotides are released or exchanged (depending on the absence or presence of medium nucleotides) when the thylakoids are energized (HARRIS and SLATER 1975, MAGNUSSON and MCCARTY 1976, STROTMANN and BICKEL-SANDKÖTTER 1977). It has been demonstrated that release of the nucleotides and the rise of ATP hydrolyzing activity are kinetically related (SCHUMANN and STROTMANN 1981). On the other hand, the activated enzyme is deactivated by tight re-binding of ADP to the previously depleted site (SCHUMANN and STROTMANN 1981, DUNHAM and SELMAN 1981, BAR-ZVI and SHAVIT 1982). It has also been shown that, in the presence of ADP and P_i, the exchange rate of 10–20% of the tightly bound nucleotides coincides with the initial rate of ATP synthesis when thylakoids are energized (SCHLODDER and WITT 1981, STROTMANN 1984). However, the majority of the tightly bound nucleotides is exchanged with a slower velocity. GRÄBER et al. (1977) concluded that only a certain fraction of the ATPase is involved in ATP synthesis at any one time. The active ATPase might be replaced continuously after they have carried out some 100 turnovers. Hence tightly bound nucleotides seem to monitor catalytically inactive enzyme molecules, and, on the other hand, tightly bound ATP and ADP are believed to be catalytic intermediates. The two apparently contradicting functions may be explained if we assume that they are brought about by two different conformational states of the same type of sites. Since the inactive enzyme exhibits the lowest energy state of all possible conformations, the whole population of enzyme molecules would fall into this "sink" upon de-energization. Catalytic sites would be converted to noncatalytic sites simply by the stop of the catalytic cycle while retaining the ligand as a tightly bound nucleotide. Such a mechanism would explain many of the reported results, e.g., the structural identity of the functionally different types of sites (see Sect. 4), the occurrence of newly formed ATP as a tightly bound ATP in the presence of an ADP-regenerating hexokinase trap (ROSEN et al. 1979), the efficient blockage of the ATP hydrolyzing activity after de-energization of unmodulated broken chloroplasts, and the necessity of release of tightly bound nucleotides preceding catalysis and following re-energization of the membrane. The hypothesis requires that none of the three functional units progresses in the catalytic cycle either in forward or backward direction as long as one of the sites is occupied by a tightly bound nucleotide. This might be fulfilled if the sites are working in a strict cooperative manner as proposed in the model of Fig. 1.

However, the situation is obviously different if we consider the isolated enzyme or the thiol-modulated membrane-bound enzyme. With isolated mitochondrial F_1, single-site catalysis of ATP hydrolysis has been observed (GRUBMEYER et al. 1982). Isolated heat-activated (BRUIST and HAMMES 1982) or octylglucoside-activated (PICK and BASSILIAN 1983) CF_1 is able to carry out ATP hydrolysis without losing a tightly bound ATP present on one of the three nucleotide binding sites. The thiol-activated membrane-bound enzyme, which has lost its tightly bound nucleotide during activation, incorporates and keeps one tightly bound ATP while catalyzing ATP hydrolysis (SCHUMANN 1981, BAR-ZVI and SHAVIT 1982, STROTMANN unpublished). Tightly bound ATP even seems to stabilize the ATP hydrolyzing activity under these conditions and to prevent the enzyme from becoming inactivated by tight binding of ADP. These results suggest that ATP hydrolysis may proceed on a different enzymatic pathway by employing only one or two of the three sites in catalysis. The participation of different enzyme conformations in ATP synthesis and ATP hydrolysis, respectively, has been postulated before, based on several independent lines of observation (FRANEK and STROTMANN 1981, SHERMAN and WIMMER 1982, SHAHAK and PICK 1983, SHAVIT et al. 1984). It should be emphasized that such an assumption does not affect the principle of microreversibility.

The three-site cooperative model shown in Fig. 1 also permits continuous ATP hydrolysis as a reversal of the synthesizing cycle, because the transfer of tightly bound to loosely bound ADP and phosphate is proposed to be exergonic. However, this may be a wrong presupposition. Nucleotide-binding studies (see above) have shown that enzyme molecules containing tightly bound ADP are in a low energy form, i.e., energy input is required to release the ADP. Uncoupler studies with chloroplasts indicate that a $\Delta\tilde{\mu}_{H^+}$ generated by ATP hydrolysis is needed to maintain the hydrolytic activity over a long period of time (BAKKER-GRUNDWALD 1974). However the energized state is required for reactivation of the slowly deactivating enzyme molecules rather than for the catalytic reaction. The obvious energetic problem might be overcome by assuming that ATP hydrolysis does not involve tightly bound ATP and ADP as catalytic intermediates, again suggesting different routes for ATP synthesis and hydrolysis.

7 Conclusion

The present state of research may be briefly summarized as follows:

1. There are probably three functional units with three nucleotide binding sites located in the β subunits in close vicinity to the α subunits. The three sites seem to be identical in structure.

2. The sites may exist in different conformations. Depending on the functional state they seem to carry out catalysis of ATP formation or ATP hydrolysis, or control of enzyme activity. The various conformations are determined by the energy state of the membrane, by the presence of ADP or ATP, and by thiol modulation.

3. The reversible formation of bound ATP on a catalytic site proceeds with an equilibrium constant around 1. Net synthesis of free ATP is driven by oscillating energy-linked changes in substrate affinities of the active sites. The three catalytic units seem to interact in a cooperative manner.

The inactive enzyme is characterized by the presence of one tightly bound nucleotide. However in the thiol-modulated membrane-bound ATPase and in isolated CF_1 the presence of one tightly bound ATP seems to stabilize the ATP hydrolyzing activity.

Although the basic concept of the energy-linked binding change mechanism is widely accepted, many experiments have to be done to prove it in detail. In addition there are completely unsolved problems, e.g., the nature and extent of the conformational changes, or the mechanism of the proposed site cooperativity. The most obscure point is the question how the protons of the electrochemical gradient are conducted to the catalytic subunits, and how they cause the conformational changes of the ATPase involved in ATP formation and in activation.

References

Abbott MS, Czarnecki JJ, Selman BR (1984) J Biol Chem 259:12271–12278
Aflalo C, Shavit N (1982) Eur J Biochem 126:61–68
Avron M, Sharon N (1960) Biochem Biophys Res Commun 2:336–339
Bakker-Grunwald T (1974) Biochim Biophys Acta 368:386–392
Bar-Zvi D, Shavit N (1982) Biochim Biophys Acta 681:451–458

Bar-Zvi D, Tiefert MA, Shavit N (1984) In: Sybesma C (ed) Advances in photosynthesis research, vol II. Nijhoff/Junk, The Hague, pp 539–542
Berzborn RJ, Müller D, Roos P, Andersson B (1981) In: Akoyunoglou G (ed) Photosynthesis III. Structure and molecular organisation of the photosynthetic membrane. Balaban, Philadelphia, pp 107–120
Bickel-Sandkötter S (1983) Biochim Biophys Acta 723:71–77
Bickel-Sandkötter S, Strotmann H (1981) FEBS Lett 125:188–192
Boyer PD, Kohlbrenner WE (1981) In: Selman BR, Selman-Reimer S (eds) Energy coupling in photosynthesis. Elsevier, Amsterdam New York Oxford, pp 231–240
Bruist MF, Hammes GG (1981) Biochemistry 20:6298–6305
Bruist MF, Hammes GG (1982) Biochemistry 21:3770–3377
Carlier MF, Hammes GG (1979) Biochemistry 18:3446–3451
Cerione RA, Hammes GG (1981) Biochemistry 20:3359–3365
Czarnecki JJ, Abbott MS, Selman BR (1982) Proc Natl Acad Sci USA 79:7744–7748
Dunham K, Selman BR (1981) J Biol Chem 256:212–218
Farron F (1970) Biochemistry 9:3823–3828
Feldman RI, Sigman DS (1982) J Biol Chem 257:1676–1683
Feldman RI, Sigman DS (1983) J Biol Chem 258:12178–12183
Franek U, Strotmann H (1981) FEBS Lett 126:5–8
Girault G, Galmiche JM, Lemaire C, Stulzaft O (1982) Eur J Biochem 128:405–411
Gräber P, Schlodder E, Witt HT (1977) Biochim Biophys Acta 461:426–440
Grubmeyer C, Cross RL, Penefsky HS (1982) J Biol Chem 257:12092–12100
Hackney DD, Rosen G, Boyer PD (1979) Proc Natl Acad Sci USA 3646–3650
Harris DA, Slater EC (1975) Biochim Biophys Acta 387:335–348
Kambouris NG, Hammes GG (1985) Proc Natl Acad Sci USA 82:1950–1955
Magnusson RP, McCarty RE (1976) J Biol Chem 251:7417–7422
McCarty RE, Racker E (1968) J Biol Chem 243:129–137
Merchant S, Shaner SL, Selman BR (1983) J Biol Chem 258:1026–1031
Mills JD, Mitchell P, Schürmann P (1980) FEBS Lett 112:173–177
Mitchell P (1974) FEBS Lett 43:189–194
Moroney JV, Lopresti L, McEwen BF, McCarty RE, Hammes GG (1983) FEBS Lett 158:58–62
Pick U, Bassilian S (1983) Eur J Biochem 133:289–297
Rosen G, Gresser M, Vinkler C, Boyer PD (1979) J Biol Chem 254:10654–10661
Rosing J, Kayalar C, Boyer PD (1977) J Biol Chem 252:2478–2485
Sakamoto J, Tonomura Y (1983) J Biochem (Tokyo) 93:1601–1614
Schlimme E, Groot EJ de, Schott E, Strotmann H, Edelmann K (1979) FEBS Lett 106:251–256
Schlodder E, Witt HT (1981) Biochim Biophys Acta 635:571–584
Schmid R, Jagendorf AT, Hulkower S (1977) Biochim Biophys Acta 462:177–186
Schumann J (1981) In: Selman BR, Selman-Reimer S (eds) Energy coupling in photosynthesis. Elsevier, Amsterdam New York Oxford, pp 223–230
Schumann J (1984) Biochim Biophys Acta 766:334–342
Schumann J, Strotmann H (1981) In: Akoyunoglou G (ed) Photosynthesis II. Photosynthetic electron transport and photophosphorylation. Balaban, Philadelphia, pp 881–892
Senter P, Eckstein F, Kagawa Y (1983) Biochemistry 22:5514–5518
Shahak Y, Pick U (1983) Arch Biochem Biophys 223:393–406
Shavit N, Aflalo C, Bar-Zvi D, Tiefert MA (1984) In: Sybesma C (ed) Advances in photosynthesis research, vol II. Nijhoff/Junk, The Hague, pp 493–500
Sherman PA, Wimmer MJ (1982) J Biol Chem 257:7012–7017
Shoshan V, Shavit N, Chipman DM (1978) Biochim Biophys Acta 504:108–122
Smith LT, Rosen G, Boyer PD (1983) J Biol Chem 258:10887–10894
Strotmann H (1984) In: Sybesma C (ed) Advances in photosynthesis research, vol II. Nijhoff/Junk, The Hague, pp 477–484
Strotmann H, Bickel-Sandkötter S (1977) Biochim Biophys Acta 460:126–135
Strotmann H, Hesse H, Edelmann K (1973) Biochim Biophys Acta 314:202–210

Strotmann H, Bickel-Sandkötter S, Edelmann K, Eckstein F, Schlimme E, Boos KS, Lüstorf J (1979a) Biochim Biophys Acta 545:122–130
Strotmann H, Bickel-Sandkötter S, Shoshan V (1979b) FEBS Lett 101:316–320
Süss KH, Schmidt O (1982) FEBS Lett 144:213–218
Sugiyama J, Mukohata J (1979) FEBS Lett 98:276–280
Viale A, Andreo CS, Vallejos HR (1982) Biochim Biophys Acta 682:135–144
Vinkler C, Rosen G, Boyer PD (1978) J Biol Chem 253:2507–2510
Wagenvoord RJ, Verschoor GJ, Kemp A (1981) Biochim Biophys Acta 634:229–236
Wagner R, Junge W (1980) FEBS Lett 114:327–333
Webb MR, Grubmeyer C, Penefsky HS, Trentham DR (1980) J Biol Chem 255:11637–11639
Yoshida M, Sone N, Hirata H, Kagawa Y, Ui N (1979) J Biol Chem 254:9525–9533

10.10 Sites of Herbicide Inhibition at the Photosynthetic Apparatus

G. SANDMANN and P. BÖGER

1 Introduction

Herbicides are phytotoxic to susceptible plants by interfering with metabolic processes. Many commercial herbicidal compounds are directed against chloroplast function and organization and against pigments. Their primary effects are on bioenergetic reactions or on the inventory of lipids and pigments. In this report, details are presented which outline the inhibitory mode of action of herbicidal compounds on photosynthetic electron transport, photophosphorylation, inhibition of carotenoids or chlorophyll biosynthesis, and the initiation of peroxidative destruction of chloroplast components, as indicated in Fig. 1.

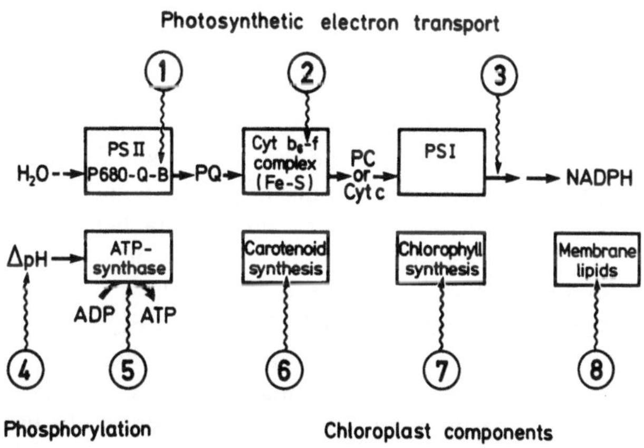

Fig. 1. Bioenergetic reactions in the chloroplast affected by phytotoxic compounds: *1* Inhibition of electron transport at the reducing side of photosystem II primarily at $B = Q_B$ (binding of, e.g., ureas, triazines); *2* inhibition of electron flow through the cytochrome b_6/f complex (diphenyl ethers, e.g., DNP-INT; DBMIB); *3* deviation of electrons from the reducing side of photosystem I (bipyridyliums, radical formation); *4* dissipation of the proton gradient (e.g., phenylhydrazones, carbanilates, diphenylamines); *5* inhibition of ATP formation at the ATPase (nitrofen); *6* inhibition of carotenoid biosynthesis (e.g., norflurazon, difunon, m-phenoxybenzamides); *7* inhibition of chlorophyll biosynthesis (oxadiazon, DTP, MK-616); *8* peroxidative destruction of membrane lipids (bipyridyliums, diphenyl ethers). *Cyt* cytochrome (*Cyt c* denotes cytochrome c-553 present in pro- and eukaryotic algae); *PC* plastocyanin; *Q* primary quencher; *P-680* reaction center chlorophyll of PS II

Several recent reviews and compilatory articles cover the relevant references on this subject (ARNTZEN et al. 1982, SANDMANN and BÖGER 1982a, ORR and HESS 1982a, BÖGER 1984, HIRSCHBERG et al. 1984, SANDMANN et al. 1984a). Since the number of references has to be limited, the authors refer to these reviews or to more recent publications only, and the reader should consult them for earlier literature. For chemical names of herbicides see list at the end.

2 Bioenergetic Reactions

2.1 Inhibition of Electron Transport Between the Two Photosystems

Various herbicides are known to inhibit photosynthetic electron transport (see ORT, Chapter 4 for an overview of the photosynthetic electron transport pathways). They interfere either at the so-called DCMU site or at the cytochrome b_6/f complex (see Fig. 1, sites 1 and 2). The relevance and phytotoxic consequences of bipyridylium compounds withdrawing electrons from photosystem I (PS I) will be discussed in Section 4.

The binding and inhibition site of DCMU or triazine-type herbicides is the B-protein of photosystem II (PS II). This peptide has been identified as a 32-kD thylakoid peptide (ARNTZEN et al. 1982). Its amino acid sequence has been deduced from the DNA sequence of the cloned chloroplast gene (HIRSCHBERG et al. 1984). In photoaffinity labeling experiments, azido-dinoseb was bound to a 47 kDa protein (OETTMEIER et al. 1982). This PS II reaction center polypeptide is thought to be the binding site for phenolic herbicides. An indication for functional interference of phenolic herbicides with PS II is their interaction with the Q_A-semi-quinone with its attached Fe(II) (RUTHERFORD et al. 1984). It has been hypothesized that DCMU and atrazine cause the replacement of a bound plastoquinone PQ at the binding protein B (Q_B). Recent reports on competitive replacement of DCMU by a PQ-analog in PQ-depleted thylakoids (OETTMEIER and SOLL 1983) and atrazine binding under different redox conditions of Q_B (JURISNIC and STEMLER 1983) have provided support for this concept. However, the data are still confusing, since PQ binding may involve several peptides.

Studies with mutants of higher plants (collected from fields controlled with triazines for several years) and also algal mutants selected against DCMU have led to a more detailed view of the receptor site(s). Resistance against herbicides other than atrazine or DCMU was also developed in all of these mutants. Nevertheless, resistance patterns in mutants selected against different herbicides were never identical, not even in the same species. Overlapping binding sites, i.e., several points of attachment (receptors) for one herbicide molecule instead of one binding site per molecule, have been proposed. According to this hypothesis, different herbicides share, in part, the same receptors. Consequently, herbicide resistance (of mutants) can be counteracted by properly substituted analogs matching the "intact" receptors (Böger 1983; Thiel and Böger 1984). A recent study has shown that a quantitative relationship exists between the rate of photosynthetic electron transport, amount of B-protein, and binding (sites) for electron-transport inhibitor herbicides (Herrmann et al. 1984).

In the case of atrazine-resistant *Solanum nigrum* and *Amaranthus hybridus* only a single amino acid of the 32 kDa peptide was found to be changed, i.e., a serine to glycine substitution (Hirschberg et al. 1984). More details of this work can be found in Chapter 9.4 of Arntzen and Pakrasi.

The oxidation of plastoquinone is prevented by the quinone analogs DBMIB and UHDBT, as well as by herbicidal inhibitors such as DNP-INT and other diphenyl ethers (cf. Böger 1984). Presumably, all of these herbicides interfere with the cytochrome b_6/f complex, most likely by altering the Rieske Fe–S center, as evidenced by changes in its EPR spectrum. In the presence of DBMIB, the Rieske Fe–S center is quantitatively converted to an altered form (Malkin 1982).

2.2 Inhibition of ATP Formation

Photophosphorylation can be inhibited either by uncouplers which dissipate the proton gradient (ΔpH) necessary to drive ATP synthesis, or by energy-transfer inhibitors which interfere with the ATP synthase (coupling factor CF_1). Uncoupling agents such as phenylhydrazones, carbanilates, diphenylamines, and ethane diamines, are also inhibitors of photosynthetic electron transport (see van den Berg and Brandse 1984, for further information).

Nitrofen and other chlorinated p-nitro diphenyl ethers inhibit by binding directly to the coupling factor (Lambert et al. 1979). It has been shown that nitrofen prevents exchange of ADP at CF_1; apparently the conformation state of the ATP synthase becomes affected, decreasing its ability to synthesize ATP (Huchzermeyer and Loehr 1983).

3 Pigment Formation

Bleaching herbicides decrease the carotenoid and chlorophyll contents of chloroplasts. This occurs either by inhibition of carotene or chlorophyll biosynthesis or by initiating pigment destruction (Sandmann et al. 1984a, b, c). The latter will be dealt with in Section 4.

3.1 Carotenoid Biosynthesis

Phytoene, formed by condensation of two molecules of geranylgeranyl pyrophosphate (GGPP), is the first carotene of the biosynthetic carotenoid pathway. This carotene is subjected to a series of dehydrogenation reactions followed by two cyclizations to give either α- or β-carotene. In addition, isomerization occurs somewhere between 15-cis phytoene, the product of phytoene synthase in photosynthetic tissues, and the colored carotenoids which have the all-trans configuration. Many herbicides, such as norflurazon, difunon, and fluridone, inhibit α- and β-carotene formation with concurrent accumulation of phytoene and some phytofluene (SANDMANN and BÖGER 1982a). Both of these colorless carotenes are substrates of a desaturase complex which converts phytoene to ζ-carotene in photoautotrophic plants.

Cell-free carotenogenic systems, developed from *Narcissus pseudonarcissus* and *Aphanocapsa,* convert prenylpyrosphosphates into β-carotene and have been used to demonstrate directly that the desaturase is involved in the inhibitory action of the herbicides mentioned above (SANDMANN et al. 1984a). In the case of S-3442, a m-phenoxybenzamide, noncompetitive inhibition of phytoene desaturase was found with the carotenogenic system from *Aphanocapsa* (CLARKE et al. 1985). Consequently, it was postulated that this herbicide could function through negative control of phytoene desaturase at an allosteric site similar to that found for the β-carotene precursors neurosporene, β-zeacarotene and γ-carotene in cell extracts of *Phycomyces*. Although many herbicides that inhibit phytoene desaturation show no chemical relationship, a feature common to all of them is high lipophilicity. In some of these compounds, electronic properties like σ_m or σ_p values of the substituents also influence inhibitory activities. Data on quantitative structure-activity relationships are available for the 2-phenylpyridazinones (SANDMANN and BÖGER 1983).

Recently, a newly developed dihydropyrone herbicide (LS 80.707) has been shown to be an inhibitor not only of phytoene desaturation, but also of the desaturation reactions which convert ζ-carotene to lycopene (SANDMANN et al. 1984b). This compound caused the accumulation of ζ-carotene, together with phytoene, both in vivo and in vitro. Amitrole, which has been reported to induce accumulation of ζ-carotene in plants, was not active in the *Aphanocapsa* cell-free system. FEIERABEND (1984) has reported that amitrole (and related compounds)-induced chlorophyll bleaching and inactivation of various enzymes differ from the action of inhibitor of phytoene desaturase.

The successive cyclization steps of lycopene into α- and β-carotene are other targets of inhibitors. In the cell-free *Aphanocapsa* system, lycopene conversion was decreased by CPTA and to a smaller extent by AMO-1618, the latter being a potent inhibitor of squalene cyclization and of GGPP cyclization to kaurene. In a similar manner, norflurazon is also nonspecific in that it can inhibit fatty-acid desaturation in addition to phytoene metabolism, however, only at a 100-fold higher concentration (ST. JOHN and CHRISTIANSEN 1976). Several CPTA analogs (substituted phenoxytriethylamines) are phytotoxic compounds and accumulate lycopene at the expense of β-carotene (WRIGHT et al. 1980).

3.2 Chlorophyll Biosynthesis

Oxadiazon, and MK-616 exert phytotoxic activities by interfering directly with chlorophyll biosynthesis (SANDMANN et al. 1984c). Other high affinity responses were excluded. However, the (enzyme) target is not yet known.

4 Peroxidation of Chloroplast Components

Bipyridylium salts and p-nitro- or p-chlorodiphenyl ethers (DPE's) initiate a general destruction of chloroplasts. Good markers of their effects are a decrease in chlorophylls and carotenoids, the formation of short-chain hydrocarbons from polyunsaturated fatty acids, especially ethane, the disappearance of ^{35}S radioactivity from prelabeled plastidic sulfolipid, and the occurrence of membrane leakage (SANDMANN and BÖGER 1982a). In addition, cellular and membrane damage can be seen on electron micrographs of herbicide-treated plants (ORR and HESS 1982a).

Peroxidative reactions are light-dependent. As shown in Fig. 2A, the activation processes of methylviologen and oxyfluorfen are different. Methylviologen is reduced by a one-electron step via PS I of the photosynthetic electron-transport chain, and can be reoxidized by molecular oxygen. The resulting superoxide-anion radical is then converted to H_2O_2 by superoxide dismutase and further to very active hydroxy radicals, presumably by a metal-catalyzed Fenton-type reaction. For oxyfluorfen and other DPE's the initial radicals are unknown. Two "activation" mechanisms of DPE's are evident: one is a photochemical activation observed with partially green cucumber cotyledons under comparably high light intensities. Under these conditions, carotenoids might be the photosensitizers for DPE's (ORR and HESS 1982b). Redox activation of DPE's by photosynthetic electron transport occurs in intact plants with an intact photosynthetic apparatus. It can also be brought about in isolated (spinach) chloroplasts using white or (even better) red light. This activation and subsequent phytotoxic DPE effects are inhibited by DCMU. Furthermore, lowering the cellular carotene level by difunon had no effect on peroxidation by oxyfluorfen as long as electron transport remained operative (cf. BÖGER 1984).

Little is known about the nature of the starter radicals. Radical quenchers like α-tocopherol alleviate peroxidation induced by DPE's, regardless of the activation reaction.

In recent spin-trap experiments, trace amounts of an endogenous radical signal exhibited by illuminated spinach (control) chloroplasts, were dramatically increased with peroxidative DPE's present. In the presence of DCMU, the radical signal disappeared (LAMBERT et al. 1984). Its shape may suggest that an alkyl or fatty-acid radical rather than a diphenyl-ether radical was observed. p-Nitro-DPE's are the best peroxidizing agents, provided that their neighboring (3)-substituents are -OR, -NHR, -COOR, or CONHR (R = alkyl). Radical signals were identical using either a peroxidizing 4-nitro derivative, or 4-chloro analog; they were similar to (the very small ones) of the untreated control with a nonsubstituted DPE. It should be emphasized that superoxide-anion radicals were essentially absent when DPE's were operative. EPR signals with methylviologen were different, requiring traps other than those necessary for DPE's.

The mechanism of peroxidative destruction of polyunsaturated fatty acids by starter radicals (R·) is outlined in Fig. 2B. After hydrogen abstraction from a fatty-acid molecule, the resulting radical reacts with oxygen in a chain reaction which yields alkylperoxide. This compound can undergo Fenton reactions simi-

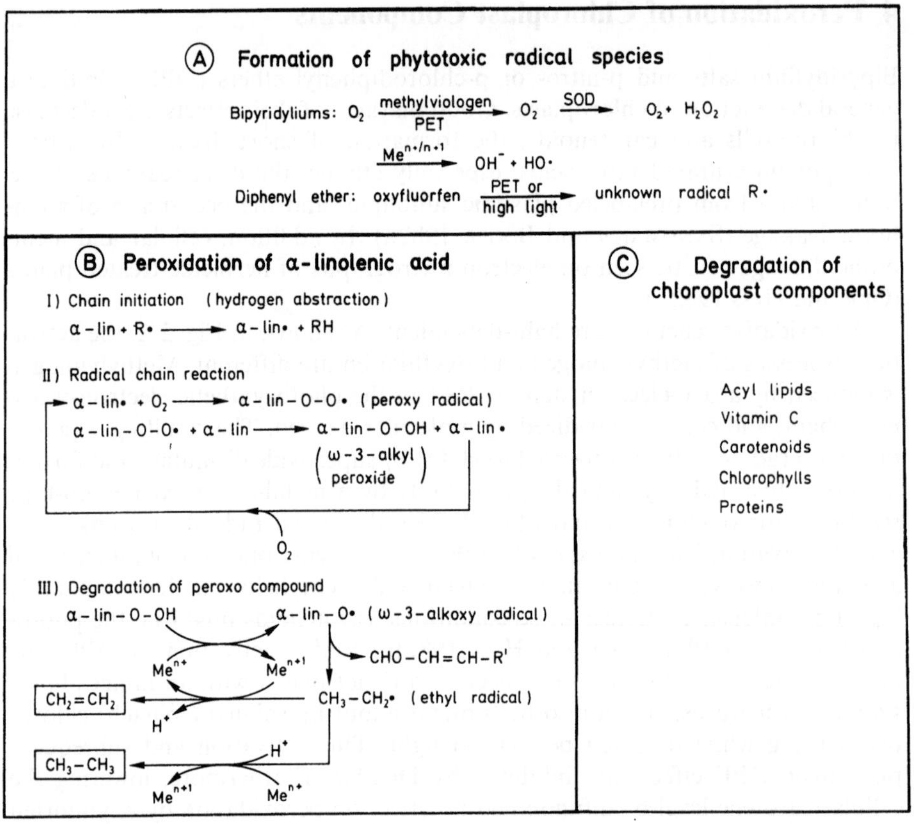

Fig. 2. Herbicide-induced radicalic peroxidation processes. **A** Radicals are originated by either bipyridylium salts or certain diphenyl ethers via photosynthetic electron transport (*PET*) or by photosensitizers. These radicals can abstract a hydrogen from α-linolenic acid (α-lin). After conjugation of two double bonds, this leads to the linolenic-acid radical (α-lin·) which reacts with molecular oxygen giving a peroxy radical. This, in turn, reacts with another linolenic-acid molecule thereby starting a chain reaction (**B** II). The resulting ω-3-alkyl peroxide is reduced to ω-3-alkoxy radical through a Fenton-type reaction (see **B** III) mediated by metal ions (Me) like Fe(II/III) or Cu(I/II). By β-scission this radical splits into the corresponding aldehyde and ethyl radical (R′ denotes the fatty-acid moiety carrying the carboxyl group). With copper ions present, the ethyl radical apparently reacts with Cu(II) to give ethylene and with Cu(I) to yield ethane. The chloroplast components which are predominantly peroxidized are shown in **C**

lar to hydrogen peroxide and form an alkoxy radical. About 0.3 to 0.9% of these radicals split into alkyl radicals leading to hydrocarbon gases (SANDMANN and BÖGER 1982b). The majority of the alkoxy radicals initiate radical destruction of other chloroplast constituents (Fig. 2C).

The chain length of the evolving hydrocarbons is determined by the position of the double bonds of the degraded dienoic or higher polyunsaturated fatty acids. The number of C-atoms is the same as that found beyond the double bond most distant from the carboxyl group (ω-1 rule; SANDMANN and BÖGER 1982b). The ratio of saturated versus unsaturated hydrocarbons formed during peroxidation is influenced by the nature (e.g., redox potential) of the redox-active metal ion present. In the case of Fe-ions, which are generally present in the cell, this ratio is strongly in favor of saturated short-chain hydrocarbons, however, on poisoning with Cu-ions, unsaturated hydrocarbons are dominant (SANDMANN and BÖGER 1982a).

An important aspect of herbicide-induced peroxidation is the activity of protective mechanisms in biological tissue. One of the endogenous antioxidants is vitamin C. Only after their exhaustion should peroxidative destruction proceed. Indeed, the level of vitamin C decreases inversely with increasing hydrocarbon-gas formation when mustard seedlings are treated with oxyfluorfen (KUNERT 1984). Furthermore, tolerance of certain higher-plant species toward oxyfluorfen has been correlated with their vitamin-C and α-tocopherol content (FINCKH and KUNERT 1985).

5 Conclusions

Herbicidal inhibitors are useful tools for the study of chloroplast function. DCMU and DBMIB, the classical inhibitors of photosynthetic electron transport, helped establish the arrangement of redox components in the "Z-scheme". Research on resistance to herbicidal photosynthesis inhibitors have substantially enlarged our knowledge of the reducing side of photosystem II. In carotene metabolism, specific inhibitors allow for the functional dissection of the complex biosynthetic pathway. Structural analyses of these carotenes should provide help for elucidating the stage at which cis-trans isomerization takes place. In addition, the effective structural elements of inhibitors may help researches determine the organization of the carotene-desaturase complex. The recent use of peroxidative diphenyl ethers has given further insight into peroxidative degradation processes and antioxidative protection mechanisms of the cell.

Basic research on the mode of action of herbicides should lead to a better understanding of herbicide-target interactions. Such knowledge is needed if reliable quantitative information on the relationship between structure and biochemical activity is to be obtained and utilized to improve the design of chemicals with predetermined properties.

6 List of Herbicides Mentioned in the Text

Amitrole, 3-amino-1,2,4-triazole; AMO-1618, (4-hydroxy-5-isopropyl-2-methylphenyl)-trimethylammonium chloride, 1-piperidinecarboxylate; atrazine, 2-chloro-4-ethylamino-6-isopropylamino-s-triazine; CPTA, 2-(4-chlorophenylthio)-triethylamine HCl; DBMIB, 2,5-dibromo-3-methyl-6-isopropyl-p-benzoquinone; DCMU (=diuron), 3-(3,4-dichlorophenyl)-1,1-dimethylurea; difunon (=EMD-IT 5914), 5-dimethylamino-methylene-2-oxo-4-phenyl-2,5-dihydrofurane-carbonitrile(3); LS 80.707, ethyl, cis(5-methyl-6-ethyl)-2-phenyl-5,6-dihydropyran-4-one-3 carboxylate; dinoseb, 2-sec-butyl-4,6-dinitrophenylacetate;

DNP-INT, 2,4-dinitrophenyl-2′-iodo-3′-methyl-4′-nitro-6′-isopropylphenyl ether; DTP, 1,3-dimethyl-4-(2,4-dichlorobenzoyl)-5-hydroxypyrazole; fluridone, 1-methyl-3-phenyl-5-(3-trifluoromethylphenyl)-4-(1H)-pyridinone; ioxynil, 4-hydroxy-3,5-diiodobenzonitrile; MK-616, N-(4-chlorophenyl)-3,4,5,6-tetrahydrophthalimide; methylviologen (=paraquat), 1,1′-dimethyl-4,4′-bipyridylium dichloride; nitrofen, 2,4-dichlorophenyl-4′-nitrophenyl ether; norflurazon (=SAN 9789), 4-chloro-5-methylamino-2-(3-trifluoromethylphenyl)-pyridazin-3(2H)one; oxadiazon, 3-(2,4-dichloro-5-isopropoxyphenyl)-5-*tert*-butyl-1,3,4-oxadiazole-2-(3H)one; oxyfluorfen, 2-chloro-4-trifluoromethylphenyl-3′-ethoxy-4′-nitrophenyl ether; S-3422, 3-(2,5-dimethylphenoxy)-N-ethylbenzamide; UHDBT, 5-*n*-undecyl-6-hydroxy-4,7-dioxobenzothiazole.

References

Arntzen CJ, Pfister K, Steinback KE (1982) In: Le Baron HM, Gressel J (eds) Herbicidal resistance in plants. Wiley Interscience, New York, pp 185–214

Berg G van den, Brandse M (1984) Z Naturforsch 39c:107–114

Böger P (1983) Biol Unserer Zeit 13:170–177

Böger P (1984) Z Naturforsch 39c:468–475

Clarke IE, Sandmann G, Bramley PM, Böger P (1985) Pest Biochem Physiol 23:335–340

Feierabend J (1984) Z Naturforsch 39c:450–454

Finckh B, Kunert KJ (1985) J Agric Food Chem 33:574–577

Herrmann G, Thiel A, Böger P (1984) Z Naturforsch 39c:430–433

Hirschberg J, Bleecker A, Kyle DJ, McIntosh L, Arntzen CJ (1984) Z Naturforsch 39c:412–420

Huchzermeyer B, Loehr A (1983) Biochim Biophys Acta 724:224–229

Jurisnic P, Stemler A (1983) Plant Physiol 73:703–708

Kunert KJ (1984) In: Bors W, Saran M, Tait D (eds) Oxygen radicals in chemistry and biology. De Gruyter, Berlin New York, pp 383–386

Lambert R, Kunert K-J, Böger P (1979) Pest Biochem Physiol 11:267–274

Lambert R, Kroneck P, Böger P (1984) Z Naturforsch 39c:486–491

Malkin R (1982) Biochemistry 21:2945–2950

Oettmeier W, Soll JH (1983) Biochim Biophys Acta 724:287–280

Oettmeier W, Masson K, Johanningmeier U (1982) Biochim Biophys Acta 769:383

Orr GL, Hess FD (1982a) In: Moreland DE, St John JB, Hess FD (eds) Biochemical responses induced by herbicides. ACS Symp Ser 181. Am Chem Soc, Washington DC, pp 131–152

Orr GL, Hess FD (1982b) Plant Physiol 69:502–507

Rutherford AW, Zimmermann JL, Mathis P (1984) FEBS Lett 165:156–162

Sandmann G, Böger P (1982a) In: Moreland DE, St John JB, Hess FD (eds) Biochemical responses induced by herbicides. ACS Symp Ser 181. Amer Chem Soc, Washington DC, pp 111–130

Sandmann G, Böger P (1982b) Lipids 17:35–41

Sandmann G, Böger P (1983) In: Miyamoto J, Kearney PC (eds) Pesticide chemistry: Human welfare and the environment, vol I. Pergamon, Oxford New York, pp 321–326

Sandmann G, Clarke IE, Bramley PM, Böger P (1984a) Z Naturforsch 39c:443–449

Sandmann G, Bramley PM, Böger P (1985) J Pestic Sci (Japan) 10:19–24

Sandmann G, Reck H, Böger P (1984b) J Agric Food Chem 32:868–872

St John JB, Christiansen MN (1976) Plant Physiol 57:257–259

Thiel A, Böger P (1984) Pest Biochem Physiol 22:232–242

Wright BJ, Baillie AC, Wright K, Dowsett JR, Sharpe TM (1980) Phytochemistry 19:61–65

11 Topography, Composition, and Assembly of Photosynthetic Membranes

11.1 Membrane Architecture of Anoxygenic Photosynthetic Bacteria

S.G. SPRAGUE and A.R. VARGA

1 Introduction

Green and purple bacteria utilize several distinct membrane morphologies to house the pigments and proteins necessary for anoxygenic photosynthesis. The different morphologies will be summarized here with an emphasis on recent advances in the understanding of membrane architecture. More extensive reviews and historical perspectives can be obtained from HOLT and MARR (1965), OELZE and DREWS (1972), REMSEN (1978), DREWS (1978), PIERSON and CASTENHOLZ (1978), OLSON (1980), KAPLAN (1981), and KAPLAN and ARNTZEN (1982). For more information on various aspects of the composition, synthesis, and regulation of anoxygenic photosynthetic membranes see DUTTON (Chap. 5), ZUBER (Chap. 6.1), COGDELL (Chap. 6.2), BLANKENSHIP and FULLER (Chap. 8.6), BACHOFEN and WIEMKEN (Chap. 11.2), DONOHUE and KAPLAN (Chap. 11.3) in this Volume.

The original designations of "green" and "purple" bacteria were made on the basis of the color of the growing cultures. Although examples exist in both groups of organisms that do not conform to this convention, these original group designations have retained their validity for most species with respect to the morphological organization of the photosynthetic apparatus. All green bacteria contain chlorosomes immediately adjacent and appressed to the inner surface of the cytoplasmic membrane (CM; Fig. 1). Although early researchers suspected that these ellipsoid inclusions contained the entire photosynthetic apparatus, it is now clear that the chlorosomes function as antenna structures similar to the phycobilisomes of red algae and cyanobacteria (see STAEHELIN, Chap. 1; GANTT, Chap. 6.3, this Vol.). The reaction centers (RC's) are contained entirely within the CM (see BLANKENSHIP and FULLER, Chap. 8.6, this Vol.). The situation in the purple bacteria is quite different: the RC's, as well as the light-harvesting (LH) complexes and, presumably, all the components involved in photosynthetic electron transport, are contained in a system of internal membranes termed intracytoplasmic membranes (ICM's). Figure 1 summarizes in diagramatic form the different kinds of ICM morphologies observed in purple bacteria. As seen in Fig. 1, these bacteria fall into several major groups in terms of the structural organization of their ICM's. Although electron micrographs of thin-sectioned bacteria frequently show cells containing many membrane "vesicles" in the cytoplasm, it is now clear that these are cross-sections through an extensive anastomosing network of vesicular-tubular or tubular membranes continuous with the CM. Examples of this kind of ICM

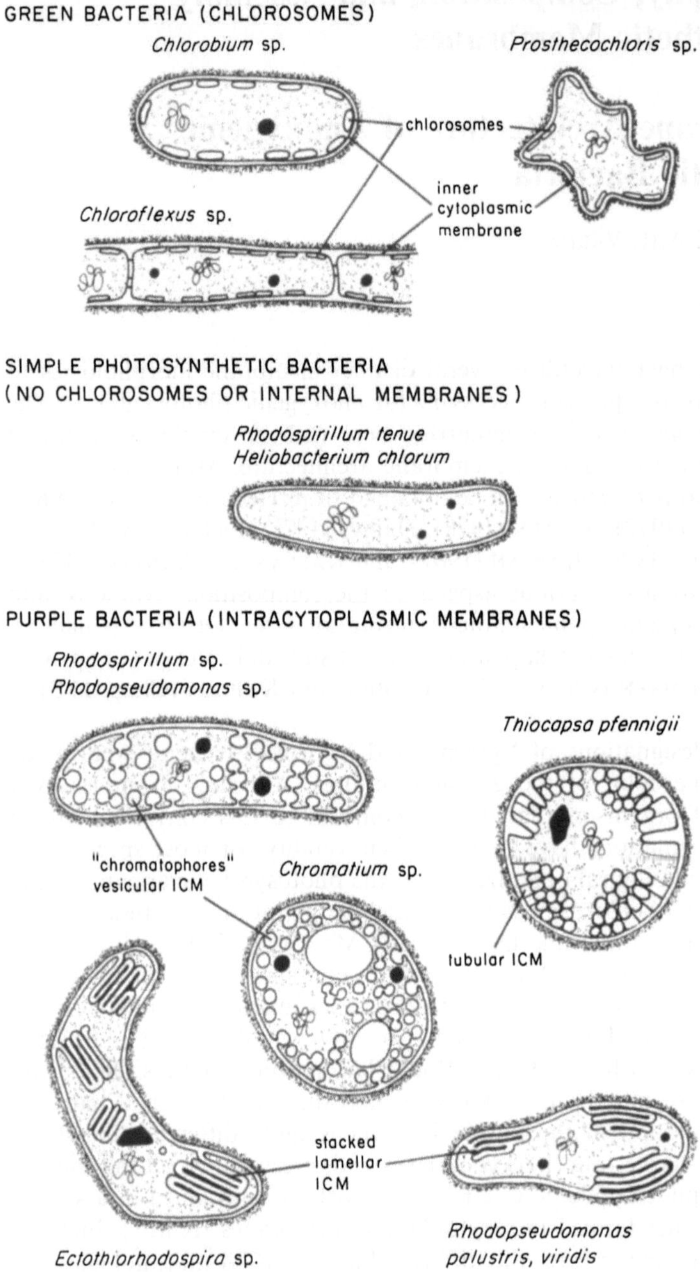

Fig. 1. Diagram depicting the morphologies observed in the anoxygenic photosynthetic bacteria for housing the photosynthetic apparatus. The green bacteria contain chlorosomes attached to the inner side of the CM. Most purple bacteria elaborate an extensive membrane system within the cytoplasm for the photosynthetic components. A few "simple" bacteria contain only a single CM which contains the pigments and proteins necessary for photosynthetic growth

organization can be found in *Rhodopseudomonas, Rhodospirillum, Chromatium* and *Thiocapsa* species. In contrast to these vesicular or tubular ICM's, lamellar membranes (thylakoids) are seen in several *Rhodopseudomonas* and *Ectothiorhodospira* species (Fig. 1). Where these lamellar membranes adhere to the CM, the latter membrane also becomes differentiated into a photosynthetic membrane (VARGA and STAEHELIN 1983).

There are a few exceptions to the morphologies described above. These photosynthetic bacteria contain neither chlorosomes nor ICM's. *Rhodospirillum (R.) tenue* and *Rhodopseudomonas (Rps.) gelatinosa* are purple bacteria with no ICM's. They house the photosynthetic apparatus in their CM. The newly described *Heliobacterium (H.) chlorum* (GEST and FAVINGER 1983) cannot be classified with either the green or purple bacteria despite the greenish color of healthy cultures. These cells also lack chlorosomes and ICM (FULLER et al. 1985), and contain a new bacteriochlorophyll (BChl; BROCKMANN and LIPINSKI 1983) that makes their reaction centers different from any of those previously characterized from green or purple bacteria.

Several of the recent advances in our understanding of membrane architecture in anoxygenic photosynthetic bacteria have come from the use of freeze-fracture electron microscopy. This technique allows visualization of integral membrane protein complexes as particles in replicas of membranes fractured along their central plane, and thus can provide detailed information about membrane organization. More information on this technique can be found in Chapter 1, this Volume, of STAEHELIN.

2 Green Bacteria

This group of organisms is composed of sulfur (Chlorobiaceae) and nonsulfur (Chloroflexaceae) families which are obligate and facultative anaerobes, respectively. Despite diverse cellular morphologies, the photosynthetic apparatus appears to be very similar in both families of bacteria (Fig. 1). All members of both families are characterized in electron micrographs of thin-sectioned cells by the presence of electron-transparent ellipsoid structures immediately underlying and appressed to the CM (Fig. 2). These structures were originally termed chlorobium vesicles (COHEN-BAZIRE et al. 1964), but have been renamed chlorosomes (STAEHELIN et al. 1980) because: (a) they are found in all green bacteria, not only in *Chlorobium (C.)* species; (b) the term "vesicle" has been confusing due to its conotation of a spherical structure bounded by a lipid bilayer (unit) membrane; and (c) they contain only antenna BChl and are functionally analagous to the phycobilisomes of the cyanobacteria and red algae.

2.1 Morphology

The photosynthetic apparatus of the green bacteria is composed of the chlorosomes and the regions of the CM immediately adjacent to each chlorosome.

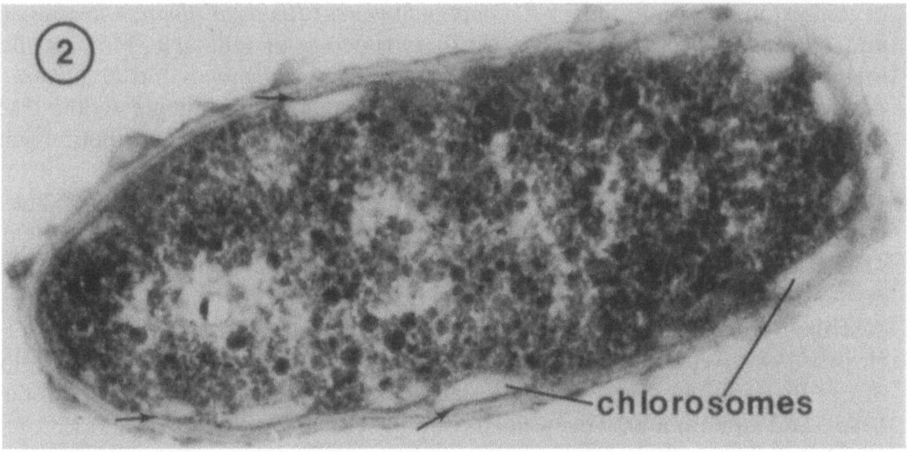

Fig. 2. Thin-section electron micrograph of *C. limicola* showing chlorosomes immediately underlying the cytoplasmic membrane. The baseplates (*arrows*) can be seen between the chlorosome and the membrane. ×90,000. (STAEHELIN et al. 1980)

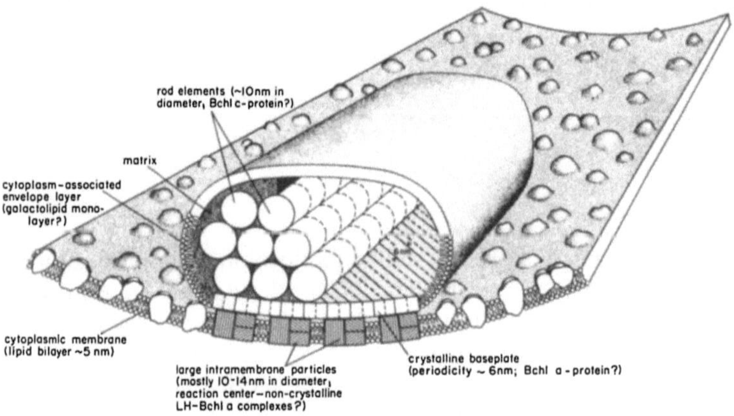

Fig. 3. Model of a chlorosome from *C. limicola* showing the fine structure obtained from freeze-fracture electron micrographs. (STAEHELIN et al. 1980)

The chlorosome contains only antenna BChl, while the CM contains the photosynthetically active RC as well as some antenna BChl (see BLANKENSHIP and FULLER, Chap. 8.6, this Vol.).

Our understanding of the fine structure of the chlorosomes and the photosynthetically active regions of the CM has been increased greatly in the past several years by the use of freeze-fracture electron microscopy. Although the size of the chlorosomes varies within a single cell, a single species and, more dramatically, between different genera and families (Table 1), we now have a good model of chlorosome structure (Fig. 3).

Table 1. Structural parameters of *Chlorobium* and *Chloroflexus (Cf.)* chlorosomes

Chlorosome	*Chlorobium*[a]	*Chloroflexus*	
Length	70–260 nm	90–150 nm[b];	106 ± 24 nm[c]
Width	40–100 nm	25–70 nm[b];	32 ± 10 nm[c]
Thickness	20–60 nm		12 ± 2 nm[c]
Envelope	Lipid-like layer with no substrucutre 3 nm thick	2 nm thick[c]	
Core	10–30 rod elements with ~10 nm diam	Rod elements ~5.2 nm diam with 6 nm globular subunits[c]	
Baseplate	5–6 nm thick 6 nm repeat that is oriented 40–60° from long axis of chlorosome	Crystalline with ~6 nm periodicity that is oriented 90° from long axis of chlorosomes[d]	
Membrane attachment site	20–30 very large (>12 nm) particles in field of 10–12 nm particles	~5 nm particles in crystalline lattice[d] 7 nm particles along the perimeter[c]	

[a] Staehelin et al. (1980) Measurements were from *C. limicola* strain 6230 by freeze-fracture electron microscopy.
[b] Madigan and Brock (1977) Measurements were made on negatively-stained cells of five strains of *Cf. aurantiacus*:J-10-fl, Ok-70-fl, Y-400-fl, 396-1, 254-2.
[c] Staehelin et al. (1978) Measurements were made on freeze-fractured cells of *Cf. aurantiacus* J-10-fl.
[d] Sprague et al. (1981a) Measurements were made on freeze-fractured cells of *Cf. aurantiacus* J-10-fl during development of the photosynthetic apparatus.

Chlorosomes are bounded on the cytoplasmic side by a 2–3 nm envelope which is presumed to be lipidic but is *not* a typical bilayer lipid membrane. Currently the envelope is thought to contain protein in a monolayer of lipid. It surrounds the hydrophobic core of the chlorosome, which contains many parallel rod elements extending the length of the chlorosome. The number and diameter of the rod elements varies between organisms (Table 1) and with growth conditions (see below). The rod elements are thought to contain the major antenna BChl (c, d, or e) and associated proteins (Betti et al. 1982).

The chlorosomes contain a baseplate composed of a crystalline lattice of particles (Fig. 4). The orientation of the lattice relative to the long axis of the chlorosome varies (Table 1), but the size of the particles in the lattice is similar in *Chlorobium* and *Chloroflexus* (Staehelin et al. 1980, Sprague et al. 1981a). The baseplate was suggested as the site of the antenna BChl a protein, the molecular structure of which has been characterized from *Prosthecochloris* (Matthews et al. 1979). The differences in baseplate structure between *Chlorobium* and *Chloroflexus* are consistent with this assignment, because *Chloroflexus* apparently has its own antenna BChl a protein that is structurally unrelated to the one found in *Prosthecochloris* (*P. aestuarii* is a green sulfur bacterium closely related to *Chlorobium*).

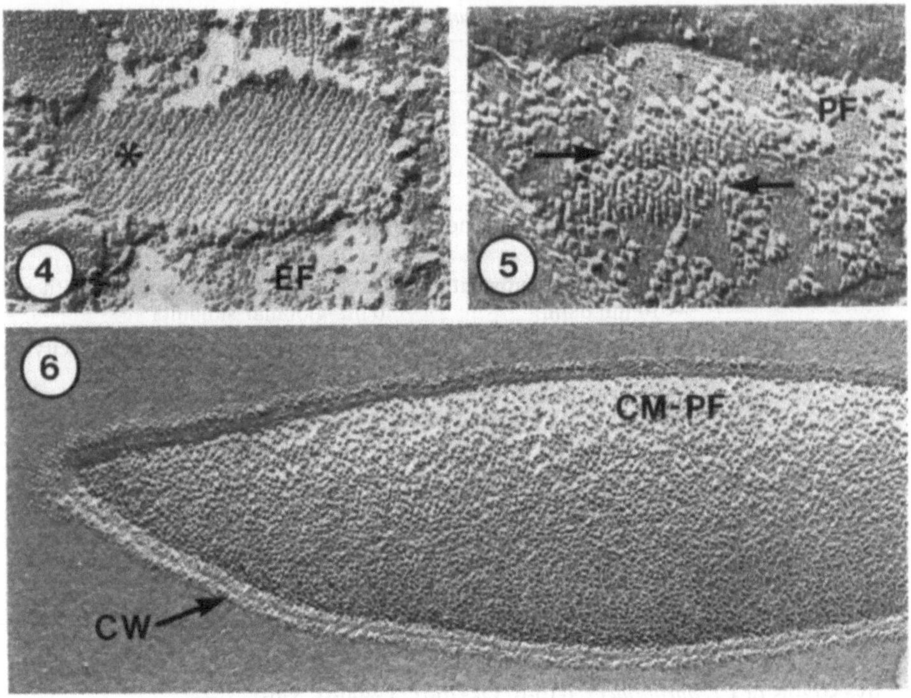

Fig. 4. Freeze-fracture electron micrograph of a chlorosome baseplate (*) in *C. limicola* showing the 6 nm periodicity. *EF* explasmic fracture face. ×225,000. (STAEHELIN et al. 1980)

Fig. 5. Freeze-fracture electron micrograph of the P fracture face (*PF*) of the CM showing the membrane attachment site (arrows) in *Chloroflexus*. ×180,000. (SPRAGUE et al. 1981a)

Fig. 6. Freeze-fracture electron micrograph of the *CM* (cytoplasmic membrane) of *H. chlorum* showing the extremely high particle density in the *P* (protoplasmic) fracture face (*PF*). No chlorosomes or invaginations for ICM's are seen. (*CW* cell wall) ×65,000

Underlying the chlorosome baseplate is the CM attachment site (Fig. 3). The differences in organization of the membrane attachment sites of *Chlorobium* and *Chloroflexus* are more marked than those noted for the baseplate (Table 1). In *Chlorobium* the membrane attachment site contains very large membrane particles randomly arrayed. In contrast, the same region in *Chloroflexus* contains a crystalline array of small membrane particles (Fig. 5). This difference is not unexpected when one considers that this region of the CM most likely contains the RC, and that the RC in *Chloroflexus* differs significantly from those of the Chlorobiaceae (see BLANKENSHIP and FULLER, Chap. 8.6, this Vol.).

As noted in the preceding paragraph, the differences in the organization of the chlorosome-CM junctions in *Chlorobium* and *Chloroflexus* cells are more marked than those in the body of the chlorosomes. The importance of these structural variations was supported by several biochemical studies where differences were reported in (a) the techniques necessary for isolating chlorosomes from these two organisms (SCHMIDT 1980) and (b) the fracturing behavior around the baseplate and membrane attachment site (STAEHELIN et al. 1978, 1980). The presence of 7 nm particles evenly spaced around the perimeter of the chlorosomes in *Chloroflexus* but not in *Chlorobium* may also have contributed to these experimental difficulties in separating *Chloroflexus* chlorosomes from the CM (SCHMIDT 1980, STAEHELIN et al. 1978). No role for these edge particles has been postulated.

Differences in the size and number of the rod elements and, therefore, the thickness of the chlorosome, and the density and number of particles in the membrane attachment site are consistent with the differences in habitat occupied by the two organisms. *Chlorobium* cells (and all green sulfur bacteria in general) grow in very low light conditions and require more antenna BChl per RC BChl than *Chloroflexus* cells (green nonsulfur bacteria), which grow under much higher light conditions. Hence we see larger chlorosomes containing more rod elements (antenna), and fewer particles in the membrane attachment site (RC's) in *Chlorobium* than in *Chloroflexus*.

2.2 Adaptation of the Photosynthetic Apparatus to Different Environmental Conditions

All photosynthetic bacteria respond to changes in light intensity by changing their BChl content. Regulation of the photosynthetic apparatus in green bacteria has been difficult to study because the green sulfur bacteria are very intolerant of high light intensities. Several researchers have shown that the antenna BChl (c, d, or e) content of cells increases when the light decreases (reviewed by PIERSON and CASTENHOLZ 1978). BROCH-DUE et al. (1978) used a combination of thin-section electron microscopy and sucrose density gradient centrifugation of isotopically labeled chlorosomes to show that chlorosomes in *C. limicola* forma sp. *thiosulfatophilum* strain 8327 (BChl d containing) increased in number and size after a transfer from high to low light conditions.

Investigation of the regulation of the photosynthetic apparatus in the green bacteria was greatly simplified with the discovery of a facultative anaerobe in this suborder, *Chloroflexus* (PIERSON and CASTENHOLZ 1974a). *Chloroflexus* cells grow aerobically in the dark, lacking both BChl (PIERSON and CASTENHOLZ 1974b) and chlorosomes (MADIGAN and BROCK 1977, SPRAGUE et al. 1981b). The cells also tolerate a wide range of light conditions during growth, and regulate the amounts of both BChl a and c and the size of chlorosomes in response to changes in light intensity (SCHMIDT et al. 1980).

In addition, *Chloroflexus* cells can be grown semi-aerobically to induce BChl synthesis (PIERSON and CASTENHOLZ 1974b) and assembly of the photosynthetic apparatus (SPRAGUE et al. 1981a, b). During semi-aerobic induction (see KAPLAN 1978, DONOHUE and KAPLAN, Chap. 11.3, this Vol.) of the photosynthetic apparatus in *Chloroflexus,* the small particles in the membrane attachment site, the adjacent crystalline baseplate material, and the envelope layer are assembled essentially simultaneously. Subsequently the longitudinal rod elements are added to the chlorosome cores (SPRAGUE et al. 1981 a). Addition of the rod elements probably corresponds to the filling of chlorosomes reported by SCHMIDT et al. (1980) after a shift from high to low light intensity growth conditions.

Unlike *R. tenue* (purple bacterium lacking ICM; WAKIM et al. 1978), *Chloroflexus* cells do not increase in size in conjunction with the shift from aerobic to semi-aerobic or photosynthetic growth (SPRAGUE et al. 1981 b). They appear rather to insert all the photosynthetic components into a pre-existing and, presumably, respiration-competent CM, instead of segregating respiratory and photosynthetic activities, as is seen in the purple bacteria. Localizing all the RC's immediately beneath the chlorosomes may be sufficient to maintain photosynthetic efficiency so that segregated membrane regions are superfluous in the green bacteria.

3 *Heliobacterium chlorum*

This recently described bacterium (GEST and FAVINGER 1983) cannot easily be classified with either the green or purple bacteria. It lacks chlorosomes and ICM's. On a morphological basis alone it could then be classified with *R. tenue* and *Rps. gelatinosa* as purple bacteria. However, its BChl content, and thus its RC, are unique, not resembling any of those characterized to date from either green or purple bacteria (FULLER et al. 1985).

The CM from *H. chlorum* has an extremely high particle density when viewed by freeze-fracture electron microscopy (Fig. 6). The CM contains ~70% protein on a dry weight basis in *H. chlorum* (FULLER et al. 1985), much more than the 27% protein reported for *Chlorobium* CM (CRUDEN and STANIER 1970). The composition and structure of the photosynthetic apparatus in *H. chlorum* remains to be determined. Maybe then it can be properly classified with the other photosynthetic bacteria.

4 Purple Bacteria

Like the green bacteria, the purple bacteria are divided into sulfur (Chromatiaceae) and nonsulfur (Rhodospirillaceae) families. Similarly we find the facultative anaerobes in the nonsulfur category. Although the morphology of the photosynthetic ICM's varies among the purple bacteria (Fig. 1), all of the photosynthetic membranes are derived from and are continuous with the CM. Most of the recent structural work on photosynthetic membranes from the purple bacteria has focused on the organization of the photosynthetic components within the membrane, and changes during induction of the photosynthetic apparatus.

4.1 Morphology

GOLECKI et al. have used freeze-fracture electron microscopy to visualize the integral membrane proteins in *Rps. capsulata* (1979), and *R. rubrum, Rps. sphaeroides,* and *R. tenue* (1980). Their results indicate that the number of particles in the membrane increases when the cells are switched from chemotrophic (high pO_2) to phototrophic (low pO_2) growth conditions, and that the number of protein particles per μm^2 is higher in the photosynthetic ICM than in the CM. These results correlate with the observations that the synthesis of several proteins is induced when cells are switched from chemotrophic to phototrophic growth, primarily the pigment proteins, and that insertion of new proteins is into the photosynthetic ICM (KAPLAN and ARNTZEN 1982, DONOHUE and KAPLAN, Chap. 11.3, this Vol.).

Rps. viridis is representative of a small group of purple photosynthetic bacteria that (a) utilize BChl b rather than BChl a for both RC and LH complexes; (b) have pigment proteins with absorption maxima much further into the near IR spectrum (960–1,020 nm) than the BChl a-utilizing species (800–880 nm); and (c) have lamellar ICM's that exhibit very distinct hexagonal lattices in electron micrographs of freeze-fractured and negatively-stained cells (MILLER 1979, ENGELHARDT et al. 1983a, b). This lattice structure has been examined by Fourier image analysis of electron micrographs to produce detailed models of membrane structure in these organisms. Two- and three-dimensional models at 3.3 nm (MILLER 1979, 1982) resolution show a large central structure believed to be the RC, surrounded by six foci of mass presumed to be LH units. These large RC-LH complexes would be arranged in the membrane to produce the hexagonal arrays seen in electron micrographs. More recent studies indicate that the lateral arrangement of the complexes in the membrane is intimately coupled to their composition. The RC has been isolated and reconstituted into large two-dimensional crystals which exhibit square arrays rather than the hexagonal ones seen in native membranes (MILLER and JACOB 1983). This transition indicates that the arrays seen in the membrane result from the interaction of the complete RC and LH units.

In addition to the two-dimensional crystals described above, isolated RC's from *Rps. viridis* readily form three-dimensional crystals. Such crystals have been used to determine an X-ray structure at 0.3 nm resolution (DEISENHOFER et al. 1984, MICHEL and DEISENHOFER, Chap. 8.4, this Vol.) and for kinetic analysis of early photochemical events (see NORRIS and VAN BRAKEL, Chap. 8.3, this Vol.).

Another organism with lamellar photosynthetic membranes is *Rps. palustris.* A recent study (VARGA and STAEHELIN 1983) showed that this organism, like *Rps. sphaeroides, R. rubrum,* and *Rps. capsulata,* has a higher integral membrane particle density in the photosynthetic ICM's than in the nonphotosynthetic CM. In addition, areas of the CM which adhere to the underlying photosynthetic ICM appear to be differentiated into photosynthetic membranes with a morphology that is distinct from that of nonphotosynthetic, nonadherent regions of the CM (Fig. 7). Examination of cells grown at different light intensities showed that protein particle diameters increased as light intensity decreased.

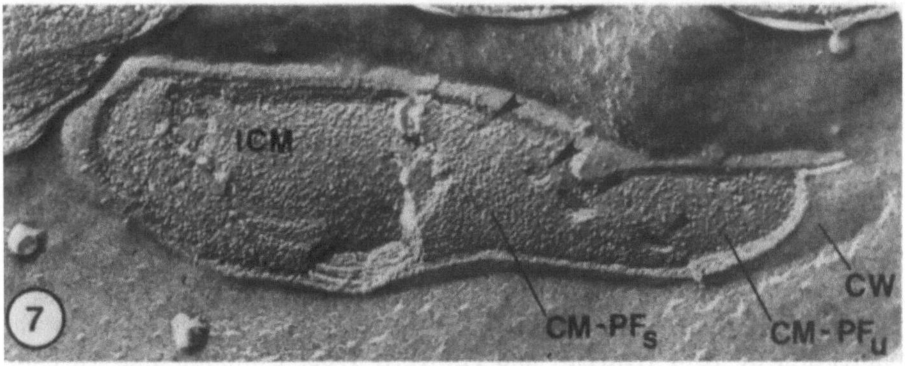

Fig. 7. P fracture face of a *Rps. palustris* cell showing differentiation of the CM. Invaginations of CM, connected to ICM's, are indicated by *arrowheads*. *S*tacked CM (*CM-PF$_s$*) is seen, with its characteristic high particle density. Nonadherent, *u*nstacked CM (CM-PF$_u$) areas are identified by their lower particle density. In this cell the fracture plane penetrates down into the cell to the bottom of an ICM stack (*ICM-PF$_u$*). ×85,000. (VARGA and STAEHELIN 1983)

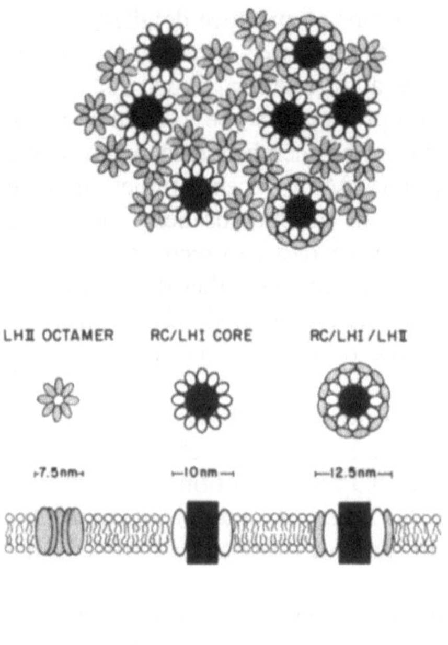

Fig. 8. Model showing possible arrangement of pigment–protein complexes in *Rps. palustris* membranes. The complexes shown are not representative of freeze-fracture images or true molecular structures, but rather illustrate possible membrane oligomers. Complexes are represented as close-packed, in agreement with freeze-fracture data. LH II (B800–850), 7.5 nm; RC-LH I (RC-B880), 10 nm; RC-LH I–LH II, 12.5 nm. (See VARGA and STAEHELIN 1985a for further discussion)

This inverse relationship is particularly evident in photosynthetic CM and ICM's, where the different particle size classes (7.5, 10.0, 1.25, 15 nm) vary in a light-dependent manner, with the larger particle sizes becoming predominant at low light levels. This light-dependent variation in particle size suggests that, as light decreases, specific subunits (probably LH units) are added to the membrane to yield incremental increases in both particle and photosynthetic unit size (VARGA and STAEHELIN 1983).

The correlation of particle sizes with in vivo structures seen in freeze-fracture replicas was pursued in a recent study involving the isolation of pigment–protein complexes from *Rps. palustris* membranes and reconstitution into liposomes (VARGA and STAEHELIN 1985a). Oligomeric complexes identified as RC-LH I (RC-B-880) core, LH II (B-800–850), and large complexes containing RC-LH I–LH II were isolated. Reconstitution of the complexes into liposomes produced the following size correlations: 7.5 nm particles for LH II oligomers, 10.0 nm for RC-LH I cores, and 12.5 nm for RC-LH I–LH II units. A model suggesting possible arrangments of the complexes in the membrane appears in Fig. 8.

4.2 Development of ICM Structure

The formation of photosynthetic pigment-proteins and membranes is induced by low oxygen tension in the cells' environment. The amounts of ICM and LH pigment-proteins are regulated at a secondary level by light intensity (DREWS 1978, DONOHUE and KAPLAN, Chap. 11.3, this Vol.). Since the ICM's and CM are physically connected (Figs. 1 and 9), major questions arise as to how the structural and functional domains of the two membrane systems are maintained. Several research groups have recently addressed this question of site specificity of membrane assembly and structure upon induction of membrane synthesis.

NIEDERMAN et al. (1979, 1981) examined the insertion of newly synthesized pigment proteins into the ICM by using pulse-chase techniques. In *Rps. sphaeroides* they found that newly synthesized pigment proteins are inserted into a membrane fraction that is distinct from the CM, but still less dense than mature ICM in a sucrose gradient. These proteins are later found in mature ICM, indicating a progressive development from new to mature photosynthetic membranes. The investigators propose that this new ICM fraction is derived from specific invagination sites in the CM. Labeling of cells with a tracer which does not permeate easily into the ICM lumen supports the notion that the developing ICM is closer to the CM than is the mature ICM (INAMINE et al. 1984). Similar experiments by DIERSTEIN et al. (1981), using *Rps. capsulata*, have shown that the newly synthesized pigment proteins are not chased into mature ICM but rather that the proteins seem to remain at their site(s) of insertion. The differences between the noninduced membrane structure in *Rps. sphaeroides* and *Rps. capsulata* may account for these seemingly contradictory results. Under aerobic conditions *Rps. sphaeroides* cells are completely devoid of ICM's, while under the same conditions *Rps. capsulata* cells retain a small amount of tubular ICM's. Upon induction of the photosynthetic apparatus, the CM must invaginate de novo in *Rps. sphaeroides* to initiate ICM synthesis, while in *Rps. capsulata* the tubular ICM's may differentiate into the vesicular type of ICM most frequently seen in these cells. Since de novo CM invagination is probably not required for ICM development in *Rps. capsulata*, developing and mature ICM fractions are not separable on sucrose gradients.

A recent combined biochemical and structural study of the development of new ICM's following anaerobic induction of the photosynthetic apparatus in *Rps. sphaeroides* cells has provided new insights into the mechanism of ICM

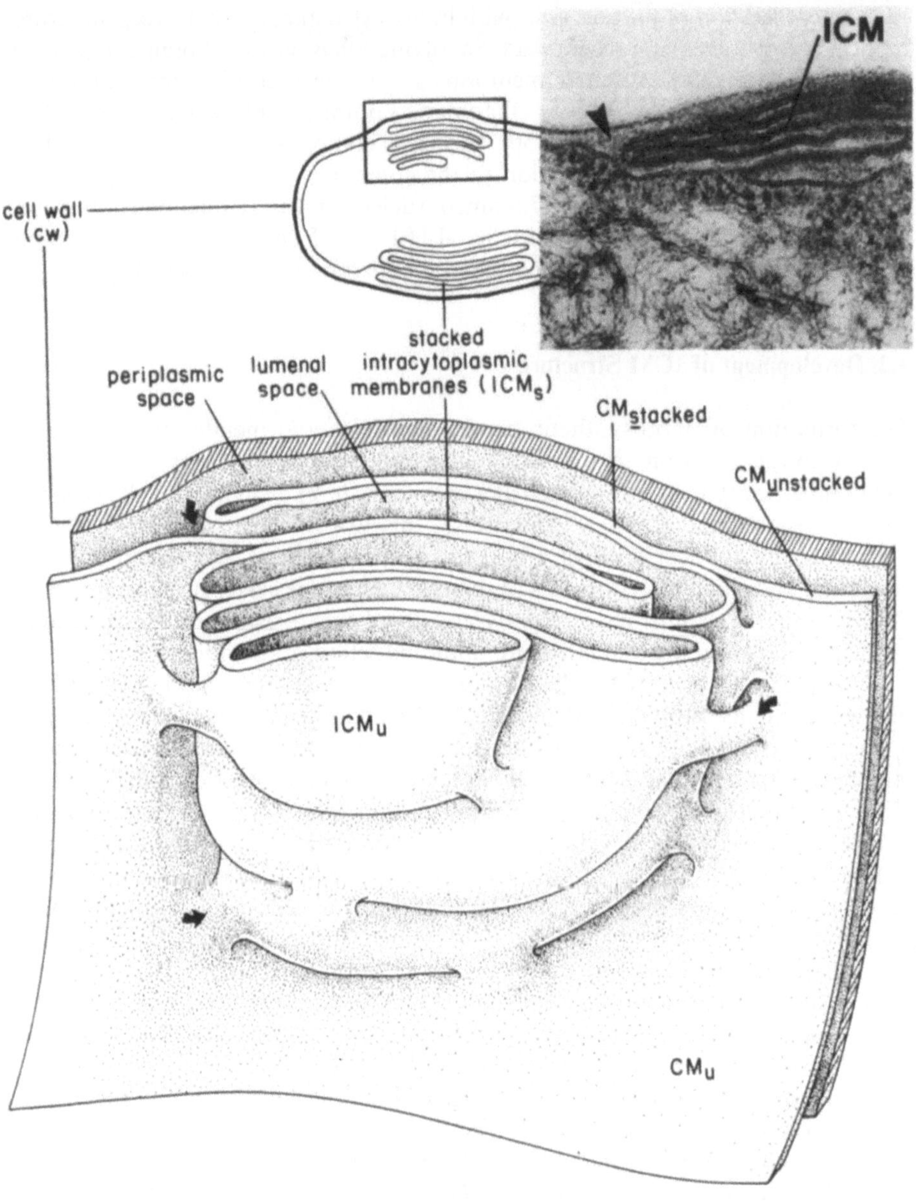

Fig. 9. Diagram representing the three-dimensional relationship of cell wall (*cw*), *CM*, and *ICM* of *Rps. palustris*. Some of the physical connections between the CM and the ICM's are indicated by *arrows*. *Inset:* thin-section micrograph of *Rps. palustris* cell illustrating an CM-ICM connection (*arrowhead*). ×42,000. (VARGA and STAEHELIN 1983)

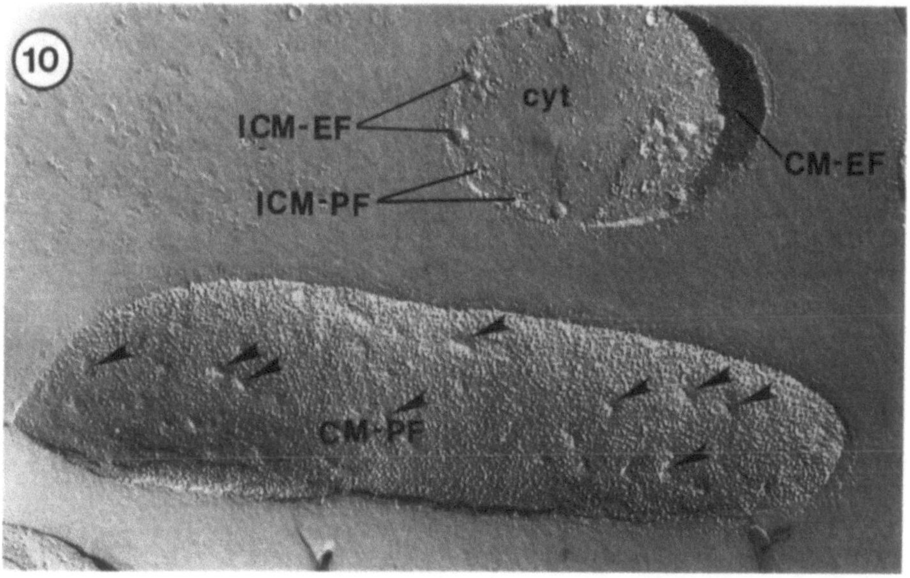

Fig. 10. Freeze-fracture micrograph of photosynthetic *Rps. sphaeroides* cells. In the lower cell some of the CM-ICM connections (invaginations in the *CM-PF*) are indicated by *arrowheads*. The fracture plane passes through the cytoplasm (*cyt*) and ICM's of the upper cell showing both *PF* and *EF* views of small vesicular ICM. ×55,000

development (CHORY et al. 1984). Prior to anaerobic induction, when neither ICM's nor pigment–protein complexes are found in the cells, indentations were present in the CM of aerobic cells. In addition to the indentations, low levels of the heavy subunit polypeptide (H) of the RC were found in the aerobic CM. It is intriguing to speculate that the two elements, H subunit and indentation, may serve as target sites for the insertion of ICM components when the cells are shifted to anaerobic conditions. A specific targeting of this sort would allow for rapid adaptation to the photosynthetic mode of growth. During intermediate stages of ICM induction, the insertion or transfer of membrane lipids from other cellular sources to the developing regions of the ICM's seems to temporarily outpace the synthesis of pigment proteins, leading to large, highly lipid-enriched ICM's connected to the CM. These high-lipid-content ICM's would have a lower buoyant density and may correspond in principle to the newly synthesized low-density ICM fraction found by NIEDERMAN et al. (1979, 1981). The synthesis and insertion of intact, functional pigment–protein complexes into the ICM's probably plays a role in determining the final structure of the ICM's. At late stages, the large, lipid-rich and tubular ICM's are converted into the more typical, small vesicular, protein-rich ICM's of *Rps. sphaeroides* (Fig. 10). Figure 11 illustrates, in diagramatic form, the stages of ICM development from putative assembly sites.

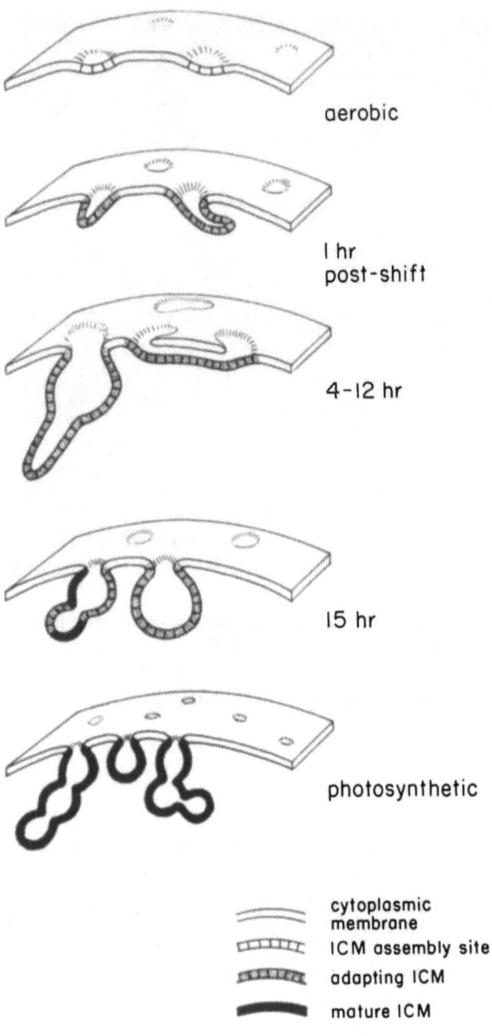

Fig. 11. Schematic representation of the sequence of morphological changes in the CM and ICM of *Rps. sphaeroides* when aerobic cells are shifted to strict anaerobic conditions. Proposed ICM assembly sites arise from indentations on the CM of aerobic cells. The process of ICM induction is shown as progressing from the assembly of small tubular membranes (1 h), first to the large, lipid-rich membranes seen at intermediate stages (4–12 h), then to membranes which appear as a mosaic of adapting and mature ICM (15 h), and finally to the fully mature vesicular ICM present in the photosynthetic cells (cf. Fig. 10). (CHORY et al. 1984)

4.3 Additional Factors Influencing Membrane Structure

The stacking and appression of photosynthetic membranes seen in *Rps. palustris* is apparently mediated by a pigment–protein complex (VARGA and STAEHELIN 1985b). Under appropriate ionic conditions, isolated photosynthetic membranes can be unstacked into individual thylakoid sacs, then restacked to resemble ICM's in vivo. Various isolated and reconstituted pigment–proteins were tested for their ability to promote membrane adhesion under the same conditions that promote unstacking and restacking of whole membranes. Samples containing LH I (B-880 in *Rps. palustris*) were found to restack in a manner resembling control membranes. Samples with LH II (B-800–850 in *Rps. palustris*) did not adhere under the same experimental conditions, implicating LH I as a main membrane adhesion factor (VARGA and STAEHELIN 1985b).

Irreversible destruction of the pigments, by photo-oxidation, in the ICM pigment–proteins of *R. rubrum* leads to a loss of the typical ordered membrane structure (GHOSH et al. 1984). The investigators proposed that the destruction of the pigments in the complexes led to dissociation of the proteins in the membrane which may be followed by random reaggregation.

Mutants deficient in carotenoids, such as *Rps. sphaeroides* R26, also exhibit altered membrane structure. Instead of the small vesicular membranes normally seen in *Rps. sphaeroides,* strain R26 has both vesicular and large lamellar ICM's (LOMMEN and TAKEMOTO 1978). Another *Rps. sphaeroides* mutant, 104, is lacking carotenoids and the B-800–850 complex (LH II). This mutant contains both vesicular and tubular ICM's. The cells also appear to be defective in their ability to complete cell division, since many cells remain connected by long membrane tubules (VARGA and KILEY unpublished results).

Another parameter involved in the maintenance of proper ICM structure is the lipid to protein ratio. Although the normal ratio in *Rps. sphaeroides* varies within a specific, limited range during the cell cycle (FRALEY et al. 1979), the structure of the ICM's appears to remain constant. However, if the amount of lipid is greatly increased, as occurs during the development of ICM's following induction, the ICM structure appears altered (CHORY et al. 1984). These high lipid ICM's are much larger and more tubular than mature vesicular ICM's. As the ICM's develop, with the insertion of pigment–proteins to the level found in mature ICM's, the tubular ICM's are converted to the small (40–70 nm diameter) round, protein-rich vesicles seen in photosynthetic cells (Fig. 10).

5 Summary

Despite gross morphological differences in cell and membrane structure in the anoxygenic photosynthetic bacteria, there are several features of the photosynthetic apparatus that are common to all, or major groups of, these species. The RC is localized within a membrane in all species. The green bacteria contain some antenna BChl in the CM, more in the baseplate, and the bulk of the antenna BChl in the chlorosome. The purple bacteria house LH pigment–proteins exclusively within a membrane. In most species the membrane is the large network of ICM's. The ICM's differ morphologically with the species, but all ICM's are derived from, and are continuous with the CM.

In the green bacteria the similarities between chlorosomes from different species are most noticeable at the antenna end of the light absorption and energy capture pathway. As one examines the ultrastructure of the photosynthetic apparatus in the intermediate antenna (baseplate) and primary RC (membrane attachment site) regions, the differences become more marked. These observations are in keeping with the biochemical data which suggests that the antenna BChl a proteins differ between the Chlorobiaceae and *Chloroflexus,* and that the RC's in *Chloroflexus* more closely resemble those from purple bacteria than any from the Chlorobiaceae. Despite the biochemical and morphological differences between the sulfur and nonsulfur green bacteria, the regulation and development of the photosynthetic apparatus appears to occur in the same manner: more rod elements (antenna BChl c, d, or e) are inserted into the chlorosome core in response to lowered light intensity (all species) or decreased oxygen tension (*Chloroflexus* sp.) and the number of chlorosomes per cell increases. Changes in antenna and RC BChl a concentrations also occur in response to changes in light intensity and oxygen tension, but no work has been done to correlate these with changes in the baseplates or membrane attachment sites.

For the purple bacteria one observes a diverse group of membrane morphologies. In all cases the ICM's are derived from and are continuous with the CM, and all the pigment-proteins are contained within a membrane. Recent work with the purple photosynthetic bacteria has shown that the protein content of the photosynthetic membranes plays a determining role in overall membrane morphology. The presence of intact, functional LH-RC complexes in *Rps. viridis* membranes yields the hexagonal array observed in the membranes in vivo. The presence of pigment–protein complexes in *Rps. palustris* membranes yields a membrane morphology distinct from nonphotosynthetic membranes, and a specific pigment–protein complex, LH I, appears to be the main membrane adhesion factor in this organism. In addition, specific pigment–protein complexes have been shown to correspond to certain particle size classes in freeze-fractured membranes.

Photosynthetic membrane induction studies on *Rps. sphaeroides* indicate that the final vesicular ICM morphology seen in these cells is only observed after all pigment proteins are inserted into the membranes in ratios appropriate both to each other and to membrane lipids. Cells defective in pigment synthesis, and therefore deficient in pigment–protein complexes, have grossly altered photosynthetic membrane morphology. Taken as a whole, these data support the notion that the ICM pigment proteins interact with each other and with membrane lipids to maintain structural integrity and long-range order in the photosynthetic membrane.

Acknowledgments. We thank J. LOGAN for her excellent art work. Part of the work described here was supported by grants from the National Institutes of Health (GM 22912) and the National Science Foundation (PCM 8118627) to L.A. STAEHELIN.

References

Betti JA, Blankenship RE, Natarajan LV, Dickinson LC, Fuller RC (1982) Biochim Biophys Acta 680:194–201
Broch-Due M, Ormerod JG, Fjerdingen BS (1978) Arch Microbiol 116:269–274
Brockmann H Jr, Lipinski A (1983) Arch Microbiol 136:17–19
Chory J, Donohue TJ, Varga AR, Staehelin LA, Kaplan S (1984) J Bacteriol 159:540–554
Cohen-Bazire G, Pfennig N, Kunisawa R (1964) J Cell Biol 22:207–255

Cruden DL, Stanier RY (1970) Arch Microbiol 72:115–134
Cruden DL, Cohen-Bazire G, Stanier RY (1970) Nature 228:1345–1347
Deisenhofer J, Epp O, Miki K, Huber R, Michel H (1984) J Mol Biol 180:385–398
Dierstein R, Schumacher A, Drews G (1981) Arch Microbiol 128:376–383
Drews G (1978) Curr Top Bioenerg 8:161–207
Engelhardt H, Baumeister W, Saxton WO (1983a) Arch Microbiol 135:169–175
Engelhardt H, Guckenberger R, Baumeister W (1983b) In: Bailey GW (ed) Proc 41 Ann Meet EM Soc Am. San Francisco Press, San Francisco, pp 740–741
Fraley RT, Yen GSL, Leuking DR, Kaplan S (1979) J Biol Chem 254:1987–1991
Fuller RC, Sprague SG, Gest H, Blankenship RE (1985) FEBS Lett 182:345–349
Gest H, Favinger JL (1983) Arch Microbiol 136:11–16
Ghosh R, Bachofen R, Hauser H (1984) Biochim Biophys Acta 765:97–105
Golecki J, Oelze J (1980) J Bacteriol 144:781–788
Golecki J, Drews G, Bühler R (1979) Cytobiologie 18:381–389
Holt SC, Marr AG (1965) J Bacteriol 89:1421–1429
Inamine GS, van Houten J, Niederman RA (1984) J Bacteriol 158:425–429
Kaplan S (1978) In: Clayton RK, Sistrom WR (eds) The photosynthetic bacteria, chap. 43. Plenum, New York London, pp. 809–840
Kaplan S (1981) Photochem Photobiol 34:769–774
Kaplan S, Arntzen CJ (1982) In: Govindjee (ed) Photosynthesis: Comparative aspects of bacteria and green plants. Academic Press, London New York, pp 65–153
Lommen MAJ, Takemoto J (1978) Arch Microbiol 118:305–308
Madigan MT, Brock TD (1977) J Gen Microbiol 102:279–285
Matthews BW, Fenna RE, Bolognesi MC, Schmid MF, Olson JM (1979) J Mol Biol 131:259–285
Miller KR (1979) Proc Natl Acad Sci USA 76:6415–6419
Miller KR (1982) Nature 300:53–55
Miller KR, Jacob JS (1983) J Cell Biol 97:1266–1270
Niederman RA, Mallon DE, Parks LC (1979) Biochim Biophys Acta 555:210–220
Niederman RA, Hunter CN, Inamine GS, Mallon DE (1981) In: Akoyunoglou G (ed) Photosynthesis V. Chloroplast development. Balaban Int Sci Serv, Philadelphia, pp 663–674
Oelze J, Drews G (1972) Biochim Biophys Acta 265:209–239
Olson JM (1980) Biochim Biophys Acta 594:33–51
Pierson BK, Castenholz RW (1974a) Arch Microbiol 100:5–24
Pierson BK, Castenholz RW (1974b) Arch Microbiol 100:283–305
Pierson BK, Castenholz RW (1978) In: Clayton RK, Sistrom WR (eds) The photosynthetic bacteria, chap 9. Plenum, New York London, pp 179–197
Pierson BK, Thornber JP (1983) Proc Natl Acad Sci USA 80:80–84
Remsen CC (1978) In: Clayton RK, Sistrom WR (eds) The photosynthetic bacteria, chap 3. Plenum, New York London, pp 31–60
Schmidt K (1980) Arch Microbiol 124:21–31
Schmidt K, Maarzahl M, Meyer F (1980) Arch Microbiol 127:87–97
Sprague SG, Staehelin LA, DiBartolomeis MJ, Fuller RC (1981a) J Bacteriol 147:1021–1031
Sprague SG, Staehelin LA, Fuller RC (1981b) J Bacteriol 147:1032–1039
Staehelin LA, Golecki JR, Fuller RC, Drews G (1978) Arch Microbiol 119:269–277
Staehelin LA, Golecki JR, Drews G (1980) Biochim Biophys Acta 589:30–45
Varga AR, Staehelin LA (1983) J Bacteriol 154:1414–1430
Varga AR, Staehelin LA (1985a) J Bacteriol 161:921–927
Varga AR, Staehelin LA (1985b) Arch Microbiol 141:290–296
Wakim B, Golecki JR, Oelze J (1978) FEMS Microbiol Lett 4:199–201

11.2 Topology of the Chromatophore Membranes of Purple Bacteria

R. BACHOFEN and V. WIEMKEN

1 Introduction

The concept of the biological membrane, based on the results of investigations with various physical, chemical, and biochemical methods has changed drastically in the course of the past 20 years. The fluid mosaic membrane model of SINGER and NICHOLSON (1972) represented an important step in this field of research. However, as more details about the secondary structure of integral and peripheral membrane proteins have become known, the original concept of "globular" protein has had to be revised. The emerging ideas point to structured peptide chains spanning the membrane once or several times. Thus segments of hydrophobic and hydrophilic character follow one another along the polypeptide chain. Membrane proteins are always asymmetric structures. In contrast, the lipids form a symmetric bilayer. However, the different lipid species present are not equally distributed on both shells of the bilayer, their transverse as well as lateral distribution may be governed by the molecular architecture of the membrane proteins.

The photosynthetic membranes of purple bacteria (Rhodospirillaceae and Chromatiaceae), called chromatophores (cph), are morphologically diverse and appear as vesicles, tubules, or thylakoids, depending on species and growth conditions. They consist of proteins and lipids in a weight ratio of about 2–3:1 (NIEDERMANN and GIBSON 1978). The membrane-associated proteins are those active in photosynthetic and respiratory electron transport, ATP synthesis and substrate translocation, functions which are described elsewhere in this volume (DUTTON, Chap. 5; PRINCE, Chap. 10.4; TIEDE, Chap. 8.2; MICHEL, Chap. 8.4; HEAST, Chap. 8.5; MELANDRI and VENTURULI, Chap. 10.6, this Vol.). For the proteins of photosynthetic membranes, not only is the topology of the polypeptide of interest, but the location and arrangement of bound cofactors, such as the pigments, the hemes, the quinones or the iron-sulfur centers are equally important.

A variety of physical, chemical, and biochemical methods have been employed to obtain information on the topology of the different components in the membrane. In most of the published studies membrane preparations from *Rhodopseudomonas sphaeroides*[1] and *Rhodospirillum rubrum* have been used.

The methodological approaches can be grouped as follows:

1. From a functional point of view the accessibility of external electron donors and acceptors, the direction of electron transport, or the site of binding

---

1 Recently renamed: *Rhodobacter sph.*

of small molecules (e.g., protons) can provide information on the orientation of the membrane. Furthermore, these parameters can also be used to evaluate the homogeneity of the membrane preparations, a critical problem in all topological investigations.

2. In biochemical studies, antibodies are used to probe the accessibility of specific groups, while the partial digestion of proteins or lipids provide insights into the topology of these membrane components.

3. A wide range of chemical compounds may be exploited as markers to investigate the localization of membrane components. Reactive groups on the membrane surface and in the hydrophilic phases on both sides of the membrane may be determined with nonpenetrating, hydrophilic substances. The microenvironment around the proteins and/or lipids within the membrane core is elucidated with hydrophobic markers. Cross-linkers of either type give indications on intermolecular distances and number and types of neighboring molecules. When the amino acid sequences of the polypeptides are known, the membrane-spanning segments may be determined according to one of the various prediction models (e.g., KYTE and DOOLITTLE 1982) and, with the use of appropriate markers, the localization of the C- and N-terminus is possible. The extractability with respect to detergents or organic solvents is a further indicator of the hydrophobicity of a given complex or polypeptide and of its integration within a membrane.

4. Finally physical methods provide insight into the precise arrangement of molecules. The distances of chlorophylls, quinones, or iron from the membrane surface or the angle of the chromophore plane to the membrane plane can be determined. Physical methods also provide information on the secondary structure of polypeptides and on the mobility of various compounds in the membrane.

2 Topology of Pigment–Protein Complexes

Two types of pigment–protein complex are found in the membranes of purple photosynthetic bacteria: reaction centers (RC) consisting in most cases of three subunits with an apparent molecular weight between 20 and 30 kDa (FEHER and OKAMURA 1978) and light-harvesting pigment protein complexes (LHC) which are composed of low molecular weight polypeptides (LHP, M_r less than 10 kDa; COGDELL and THORNBER 1980). All these polypeptides are highly hydrophobic integral membrane proteins, thus solubilization of these complexes can only be achieved with detergents. With the exception of subunit H of the RC, all polypeptides are soluble in organic solvents, e.g., chloroform-methanol.

The orientation of these proteins has been studied by labeling with differing hydrophilic and hydrophobic agents, as well as by digesting with various proteases. The methods of labeling and analysis have been greatly refined during the past 10 years, and, as mentioned above, the picture of the integral membrane proteins has evolved since the presentation of the fluid-mosaic model. Thus the results presented some years ago have to be discussed in a new light.

Subunit H of the RC was the first cph polypeptide of a pigment–protein complex to be labeled (REED et al. 1975, ZÜRRER et al. 1977) and also digested (OELZE 1978) from the cytoplasmic side. These and other studies led to the suggestion that a large portion of this rather hydrophilic polypeptide is exposed to the aqueous phase. By using a very unspecific protease, subunits M and L were also demonstrated to be exposed on the cytoplasmic membrane surface (WIEMKEN and BACHOFEN 1982). Exposure on the periplasmic side of the membrane is more difficult to prove experimentally by the use of labels or enzymes. Spheroplast-derived vesicles are less homogeneously oriented than cph, and the composition of the vesicles may differ due to contamination by the cytopasmic membrane. Problems are therefore encountered in the analysis of such samples by SDS-PAGE and peptide mapping. Subunit H of the RC was again labeled and digested from the periplasmic side of the membrane (BACHMANN et al. 1981). Since antibodies raised against the H, M, and L subunits are all able to bind to the periplasmic surface of cph (VALKIRS and FEHER 1982, JAY and LAMBILLOTTE 1985), one can conclude that all three subunits of the RC cross the membrane and are exposed on both membrane surfaces. Analysis of the structure of crystallized RC's of *Rps. viridis,* and of hydrophobicity plots of polypeptide sequences of *Rps. capsulata* confirmed the earlier findings (NORRIS and VAN BRAKEL, Chap. 8.3; MICHEL and DEISENHOFER, Chap. 8.4; YOUVAN et al. 1984).

The polypeptide part embedded in the hydrophobic core of the membrane has been investigated by the use of hydrophobic markers. Fluorescamine and dicyclo-hexylcarbodiimide reacted specifically with subunit M of the RC (BACHOFEN 1979, ZÜRRER et al. 1983). With iodo-naphthyl-azide all three subunits of the RC were labeled. In contrast, trifluoro-methyl-iodophenyl-diazirine and naphthyl-diazo-trifluoro-propionate, two less specific agents, only labeled the two more hydrophobic subunits of the RC, L and M, as well as the LHP (ODERMATT et al. 1980, MEISTER and BACHOFEN 1984).

Information on the lateral distribution of the membrane polypeptides has been obtained from studies with chemical cross-linkers (WIEMKEN et al. 1981, TAKEMOTO et al. 1982, BACHMANN and TAKEMOTO 1983). Such experiments indicate that the RC polypeptides are differently arranged or react differently in different species, e.g., in *R. rubrum, Rps. capsulata,* and *Rps. sphaeroides.*

The primary structure of the LHP's of several species of purple bacteria, e.g., *Rps. sphaeroides* R-26.1 and *R. rubrum* G-9 (ZUBER, Chap. 6.1, this Vol.) has been determined. Labeling and proteolytic degradation of the exposed polypeptide parts, and determination of the new N-terminal sequences, has led to the conclusion that the N-termini of the LHP's of *R. rubrum* are located on the cytoplasmic side of the cph membrane (BRUNISHOLZ et al. 1984). The distribution of label suggests that the C-terminus of the LHP's is exposed to the periplasmic space of the membrane or buried in the phospholipid head-group layer (WIEMKEN et al. 1983). The hydrophobic amino acids of the N-terminal part of the α-chain forming an amphipathic helix were labeled with the hydrophobic marker 3-trifluoromethyl-3-(m-iodophenyl)-diazirine (BACHOFEN et al.

1985). One half of the N-terminal α-helix is exposed to the aqueous phase, the other submerged into the phospholipid head group region. A model for the arrangement of the central hydrophobic segment of the B-850 complex is presented by THEILER and ZUBER (1984). The two chains form a tightly packed α-helical domain oriented perpendicular to the membrane plane. The interactions between the amino acids of the two chains and between the polypeptides and the chromophore are also shown. The N-termini of the α- and β-chain of the B-870 complex of *Rps. sphaeroides* wild type and the β-chain of the B-800–850 complex are cleaved by proteinase K from the cytosolic side of the membrane. Because the cph of *Rps. sphaeroides* R-26.2 do not form closed vesicles it is possible to identify polypeptide segments exposed on the periplasmic side of the membrane. In such specimens the C-terminal region of the α-chains of both light-harvesting complexes can be digested with proteinase K, while the β-chains with their very short hydrophilic C-terminal regions of only two and three amino acids respectively remain unchanged (WIEMKEN and BACHOFEN 1985, 1986). The primary structure of the LHP's reveals a conserved histidine within the hydrophobic domain of the membrane. It has been suggested that this histidine is involved in the binding of the pigment, resulting in the typical absorbance of the cph membrane in the near IR (850, 860, 870, and 880 nm respectively). The binding site for the BChl responsible for the absorbance at 800 nm is still a matter of speculation.

The N-terminal sequences of subunits L and M of the RC from *R. rubrum*, the whole sequence of the M and L subunits of *Rps. sphaeroides*, and the sequences of all three subunits the RC of *Rps. capsulata* have been determined (THEILER et al. 1983, SUTTON et al. 1982, WILLIAMS et al. 1983, 1984, YOUVAN et al. 1984, ZUBER, Chap. 6.1, this Vol.). The analysis of the sequences of the M and L subunits according to their hydropathy index suggests that these peptides span the membrane five times. Subunit H, however, has only one membrane-spanning segment. This segment seems to influence the electron transfer properties of Q_B (DEBUS et al. 1984). For subunit L of the RC, the N-terminus has been located on the cytoplasmic side of the membrane (BRUNISHOLZ et al. 1984). Subunit M of the RC of *Rps. sphaeroides* contains several histidines which have been suggested to be involved in the binding of the quinones (BUNKER et al. 1982, WILLIAMS et al. 1983). The herbicide inhibitor azidoatrazine, which has been shown to compete with Q_1 for binding to the RC, was found to bind to subunit L (DE VITRY and DINER 1984). At present, no definitive information is available on the location of the pigments on the L and M subunits. Proteolytic digestion of the RC of *R. rubrum* demonstrates that its spectral characteristics are retained as long as subunit L and a fragment of M remain intact (WIEMKEN and BACHOFEN 1984). The striking homology in the amino acid sequence for the known part of subunit L but not M of *R. rubrum* and *Rps. sphaeroides* is suggestive of a pre-eminent role of subunit L in the primary events of photosynthetic energy conversion in evolution (THEILER et al. 1983).

Spectroscopic studies of cph membranes have shown that around 60% of the membrane proteins have a α-helical structure. In RC's reconstituted into liposomes a 50% α-helical content has been observed. A small amount of β-structure is seen in the RC and in whole membranes as well. The orientation of these α-helices is close to perpendicular to the plane of the membrane (GANAGO and BROKLINE 1981, NABEDRYK et al. 1982). IR analyses of LHP and antenna complexes also suggest the presence of large α-helical regions (KOPP et al. 1979, THEILER and ZUBER 1984).

RC polypeptides and LHP's form complexes of defined stoichiometry in the photosynthetic membrane. These complexes can be seen in the electron microscope and have been isolated after treatment with mild detergent. In EM pictures of *Rps. palustris,* complexes of different sizes can be distinguished, depending on light intensity during growth (VARGA and STAEHELIN 1983, 1985, SPRAGUE and VARGA, Chap. 11.1, this Vol.). Distinct particle sizes between 7.5 and 15 nm suggest that the subunits consisting of LHP are attached to core particles containing the RC. Isolated complexes of RC and LHP alone and in combination (e.g., 1 RC + 12 B-880, 1 RC + 12 B-880 + 12 B-800–850) were inserted into liposomes. The particle sizes measured, corresponded to those seen in the native membrane. In *R. rubrum* a uniform population of particles of 13 nm diameter (MEYER et al. 1981) and in *Rps. sphaeroides* of 12 nm (REISS-HUSSON et al. 1977) is observed. In the latter organism there is a striking dependency of particle size and number on the light intensity during growth (YEN et al. 1982). The photosynthetic membranes of BChl b-containing bacteria have hexagonally arranged complexes with lattice spacings close to 13 nm (ENGELHARDT et al. 1983).

The size of solubilized RC and RC-LHP complexes in whole membranes has been studied with X-ray and neutron diffraction techniques (HODAPP and KREUTZ 1980, SADLER and WORCESTER 1982a, b). It was found that an RC-LHP complex has a diameter of 11 to 13 nm, while the RC in solution has a maximal size of 9 nm (KATAOKA and UEKI 1981, LEMAIRE and RIVAS 1983). From the determination of the size of the particles in the membrane, it was concluded that very little protein protruded from the lipid bilayer.

Using singlet-singlet quenching of the BChl fluorescence, the functional size of the antenna complexes (that is the number of light-harvesting BChl molecules that are connected through energy transfer reactions) has been determined (MONGER and PARSON 1977, VAN GRONDELLE et al. 1983). A large complex of B-800–850 (600 BChl a molecules) prepared by LDAO solubilization composed of smaller units of 25–30 BChl a molecules represents the smallest obtainable homogeneous form. A "lake" of B-800–850 complexes was postulated to surround the much smaller B-875 LH complex (12–18 units with 100–140 BChl a each). The distance between two BChl 800 was calculated to be 18–19 Å (KRAMER et al. 1984).

The location and orientation of the chromophores in RC's and LH complexes has been studied with various optical methods such as linear and circular dichroism and fluorescence polarization. A more detailed discussion of these techniques can be found in Chapter 7.5, this Volume, of BRETON and in the review of BRETON and VERMEGLIO (1982). X-ray and neutron diffraction have also provided information on the internal location of the pigments and the quinones in the RC, the binding site of the electron donor, and of cytochrome c_2. LH complexes of *R. rubrum* and of different strains of *Rps. sphaeroides* vary considerably with respect to their absorbance maxima, suggesting that these complexes provide different environments for the chromophores (BOLT et al.

1981, DAVIS et al. 1981). One of the interpretations used to explain the large spectral shift from the free bacteriochlorophyll to the bound form suggests that different proteins may provide different environments because of charges on nearby amino acids (ECCLES and HONIG 1983).

Several studies have shown that some of the photosynthetic pigments can be removed from the membrane and subsequently reconstituted. The B-800–850 antenna complex of *Rps. sphaeroides*, for instance, consists of two polypeptides carrying three BChl and one carotenoid molecule. The single BChl absorbing at 800 nm is more labile and is lost following physical, chemical, or enzymatic treatments, suggesting a possible binding site located near the cytoplasmic surface of the membrane (WEBSTER et al. 1980, CLAYTON and CLAYTON 1981). Free BChl added with detergents is able to bind back at the binding site for B-800. A possible candidate for this binding site is the second histidine in the β-chain (BRUNISHOLZ et al. 1984). The carotenoid is located close to the surface of the membrane in the B-800–850 complex (WEBSTER et al. 1980, SYMONS and SWYSEN 1983). This carotenoid has been identified as the one that displays spectral shifts in response to electrical field changes. Interestingly, several different carotenoids have been successfully and functionally reincorporated into the B-800–850 complex of the carotenoidless mutant R-26 of *Rps. sphaeroides* (DAVIDSON and COGDELL 1981).

RC's of wild-type *R. rubrum* and *Rps. sphaeroides* contain one carotenoid. In carotenoid-less mutants of both species the carotenoid binding site seems to be "free" and the RC can be reconstituted with respect to energy transfer by addition of different carotenoids (BOUCHER et al. 1977, AGALIDIS et al. 1980). However, no experimental information is available so far on the molecular location of the carotenoid-binding site.

Detailed results on the molecular structure of the RC have recently been obtained using X-ray diffraction with single crystals of the photosynthetic RC of *Rps. viridis* (MICHEL and DEISENHOFER, Chap. 8.4, this Vol.).

The quinone binding site in the RC was localized on the cytoplasmic side of the chromatophore membrane. From EPR measurements, a distance in the range of 10 Å was established between the bacteriochlorophyl dimer $(BChl)_2$ and the Q_I, the primary acceptor (TIEDE, Chap. 8.2, this Vol.; TIEDE et al. 1978). Furthermore, the plane of the aromatic ring of the quinone was found to be normal to the plane of the membrane (TIEDE and DUTTON 1981). The two quinones can be removed from the RC using mild treatment with detergent in the presence of o-phenanthroline and thereafter can be reconstituted. Reconstitution of a functionally intact RC is also possible with artificial quinones. Of special interest is the binding of azidoantraquinone, which is linked covalently to subunit M upon activation of the azidogroup to a nitrene (MARINETTI et al. 1979). From these experiments and others which compared electron transfer from Q_A to Q_B of antibody labeled and unlabeled subunits, it was concluded that both Q_A and Q_B are located on the M subunit of the RC (DEBUS et al. 1982). The nonheme iron found in the neighborhood of the quinones is probably complexed with N from histidine (BUNKER et al. 1982). A distance of 6.2 to 7.8 Å was calculated between Fe^{2+} and Q_I^- (DISMUKES et al. 1984). The reduction of the quinones not only needs electrons from the P-870 of the RC but also protons from the environment. Light-induced H^+ uptake has been demonstrated with isolated RC as well as with cph, supporting a location of the quinones on the cytoplasmic side of the cph membrane. Uptake occurs in the microsecond range, thus long diffusion distances are unlikely (DUTTON, Chap. 5, this Vol.; CHANCE et al. 1970).

3 Topology of Electron Carriers

Soluble cytochrome c_2 has been localized immunologically and by means of reconstitution experiments on the periplasmic side of the membrane as well as within whole cph's of purple nonsulfur and sulfur bacteria (PRINCE et al. 1975, KNAFF et al. 1980). In *Rps. sphaeroides* resonance X-ray and neutron diffraction studies place the cytochrome close to the RC in the membrane profile and give a distance of 44 Å between the iron of the cytochrome c_2 and the one in the RC. The charge separation between oxidized cytochrome and reduced FeQ_I^- thus spans around 2/3 of the membrane profile (BLASIE et al. 1983). Cytochrome c_2 is oxidized upon illumination by the RC in situ (RICKLE and CUSANOVICH 1979, BOWYER et al. 1979), by the RC in solution and also by the RC incorporated into liposomes (OVERFIELD and WRAIGHT 1980a, b). The apparent dissociation constant calculated was in the order of 10 μM. The binding of cytochrome c_2 or cytochrome c (horse heart) to the RC has been studied by equilibrium dialysis (ROSEN et al. 1980). Only one binding site was detected with a dissociation constant of 0.4 and 1 μM for cytochrome c and c_2 respectively. No difference in K_D was observed between the oxidized and reduced form of the cytochrome. In contrast to detergent-solubilized RC's, those incorporated into phosphatidylcholine vesicles bind three cytochromes and the dissociation constants vary depending on the oxidation state of them (PACHENCE et al. 1983). The lysines of the cytochrome c_2, which have been postulated to be involved in the binding to the RC, have been investigated by studying their accessibility to modification agents. Surprisingly, mostly amino acids located in the back of the heme crevice showed high shielding due to the binding to the RC or to the mitochondrial b/c_1 complex (RIEDER et al. 1985). Cytochrome c binding to the RC is inhibited by antibodies against subunits L and M, suggesting that one or both of these subunits are involved in the formation of an RC-cytochrome c complex (ROSEN et al. 1980). In *Chromatium vinosum* and *Rps. viridis* another firmly bound c-type cytochrome located on the periplasmic side mediates electron transfer from the soluble c_2 to the RC (VAN GRONDELLE et al. 1977, SHILL and WOOD 1984).

The topology of the cytochrome bc_1 complex of the Rhodospirillaceae has not yet been studied in detail. The Rieske-FeS cluster has been investigated in *Rps. sphaeroides* by EPR spectroscopy using dysprosium and lanthanum complexes as perturbants for spin relaxation in differently oriented membranes. The FeS complex was located toward the periplasmic side of the membrane within about 14 Å from the aqueous membrane interface (PRINCE 1983 and Chap. 10.4, this Vol.). It was concluded from the slow reactions of hydrophilic oxidants and reductants with cytochrome b that the hemes of this cytochrome are buried in a highly hydrophobic environment (HAUSKA et al. 1983, HAUSKA, Chap. 10.1, this Vol.). Recent analysis of the amino acid sequence data of cytochrome b from mitochondria and chloroplasts suggests that this protein spans the membrane five to nine times (WIDGER et al. 1984). The cytochrome f of chloroplasts, which is homologous to the cyt c_1 of the bacterial bc complex, spans the membrane once and the C-terminus is exposed on the stroma side. Approximately 85% of the amino acids are in the hydrophilic region of the inner thylakoid space (WILLEY et al. 1984).

3.1 Topology of the ATP Synthase, some Dehydrogenases and the Hydrogenase

The coupling factor complex F_0F_1 is a high molecular weight, multi-subunit complex catalysing ATP synthesis and hydrolysis that is coupled to proton flow across the membrane. The F_1 portion from various species of purple bacteria can be removed from cph's using extractions with EDTA and sonication, indicating a cytoplasmic orientation. The F_1 part is also accessible to hydrophilic modification reagents as well as to proteolytic enzymes (OELZE 1978, FILLINGAME 1981, KHANANSHVILI and GROMET-ELHANAN 1983). Knob-like particles attributed to the F_1 part of the ATPase are observed on the cytoplasmic side of the cph's in electron micrographs (LOW and AFZELIUS 1964). It is now firmly established that the F_1 portion of all energy-transducing membranes is on the negatively charged side of the energized membrane.

The F_0 part consists of three polypeptides, a, b, and c, which have been well characterized in *E. coli* (HOPPE and SEBALD 1984). The DCCD-binding protein (subunit c) from *R. rubrum* has been isolated and its sequence determined. In *E. coli* the N-terminus of subunit c is located on the periplasmic side and of subunit a on the cytoplasmic side of the membrane.

The location of several other enzymes related to energy metabolism has been studied by proteolytic digestion, chemical labeling and accessibility to antibodies. Parts of succinic dehydrogenase, L-lactate dehydrogenase, NADH dehydrogenase, and L-malate dehydrogenase have been located on the cytoplasmic side of cph's. The active site of some of these enzymes have also been reported to be exposed in spheroplast preparations, but this finding can probably be attributed to an inverse orientation of approximately 15% of spheroplast-derived vesicles (COLLINS et al. 1979, ELFERINK et al. 1979, OELZE 1978, TAKEMOTO and BACHMANN 1979). Furthermore, a hydrogenase has been located on the cytoplasmic side of the cph membrane of *Rps. capsulata* using antibodies (COLBEAU et al. 1983). This hydrogenase seems to be an integral membrane protein since detergents are necessary for its solubilization.

4 Topology of Membrane Lipids

The analysis of the lipids from members of the Rhodospirillaceae reveals a rather simple composition of only a few different phospholipids (SNOZZI and BACHOFEN 1982). Only little is known about their topographical arrangement.

An asymmetric distribution of one or two major phospholipids (phosphatidylethanolamine, PE, and phosphatidylglycerol, PG) has been demonstrated for cph membranes of *Rps. sphaeroides* and *R. rubrum* on the basis of trinitrobenzenesulfonic acid labeling and the action of phospholipase A_2 (AL-BAYATTI and TAKEMOTO 1981, DAPRA et al. 1982). In *Rps. spheroides* 33% of PE and 73% of PG is located in the outer shell of the bilayer. A similar asymmetric distribution of PE is also seen in *R. rubrum* and *Chromatium vinosum* (SHIMADA and MURATA 1976). Inhomogeneous lateral distribution of the membrane lipids is suggested from lipid analysis of whole membranes, isolated RC's and isolated antenna complexes of *R. rubrum* (SNOZZI and BACHOFEN 1979, PICOREL et al. 1983). During purification of the RC, the proportion of PE drops from 65 to 45%, while PG increases from 22 to 27 and DPG from 13 to 28% consistent with an inhomogeneous distribution of lipids between the bulk phase and the interphase of protein complexes. Linear dichroism spectra in the IR region as well as EPR spectra give an orientation for the fatty acids of the membrane lipids perpendicular to the membrane plane (NABEDRYK et al. 1982, GHOSH et al. 1984). This is also the axis of fast rotation of the whole P-lipid molecule or the phosphodiester group alone, as measured with ^{31}P-NMR. Comparative measurements of cph's and cph lipid dispersions show furthermore that orientation and motion of the P-lipid head group is not altered by the presence of membrane proteins (GHOSH et al. 1985).

5 Conclusions (Summarized in Fig. 1 A, B)

a) Polypeptide Location (Fig. 1A). All membrane proteins investigated so far are exposed on both membrane surfaces. The LHP's may have a nearly equal amount of their structures exposed on both sides of the membrane. Because of the highly asymmetric H subunit, the RC complex has a higher amount of protein exposed on the cytoplasmic surface. Furthermore, the F_1 part of the coupling factor is located on this side of the membrane. The active sites of various other enzymes and probably a considerable part of their structure appear to be oriented toward the cytosol. Thus, for the whole cph membrane it is concluded that a higher amount of peptide material is found on the cytosolic side. This is also the side from which the proteins are inserted and where regulation of their synthesis takes place. All polypeptides seem to cross the membrane at least once, but many several times. The picture of the three-dimensional organization of the different membrane-associated polypeptides will become more complete as additional information is gathered on their relationship to the hydrophobic, central, the hydrophilic, surface, and the interface regions of the bilayer membranes.

b) The Distribution of the Cofactors is Presented in Fig. 1 B. Their exact location will only be known with certainty when the protein structures and their binding sites have been determined within the sequences.

Models have been proposed for the lateral distribution of the RC-LHP polypeptides in the cph, but the remaining polypeptides, e.g., the cytochrome

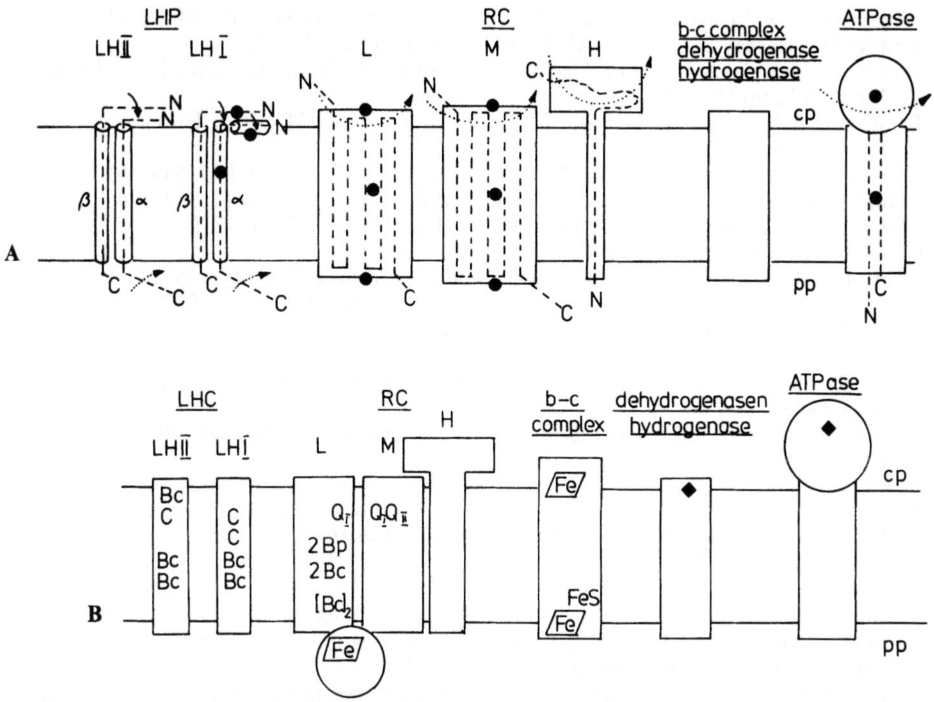

Fig. 1. A Topology of membrane proteins. Known sequence (-----); labeling site (•); proteolytic digestion observed, but cleavage site unknown (⤻); proteolytic digestion observed and cleavage site known (⤻); α-helical structure ⊂⊃; cytoplasmic side (*cp*); periplasmic side (*pp*); **B** Topology of the cofactors and active sites. Active site of enzyme ♦; bacteriochlorophyll (*Bc*), bacteriopheophytin (*Bp*), carotenoid (*C*), quinone (*Q*), non-heme-iron protein (*FeS*), heme (*Fe*), cytoplasmic side (*cp*), periplasmic side (*pp*)

bc complex or the coupling factor have not yet been included. This picture of the chromatophore membrane will have to be greatly refined in the future, before a complete topological map of the cph membrane can be formulated.

Acknowledgment. We gratefully acknowledge the financial support of our own work during the past decade by the Swiss National Science Foundation for Scientific Research.

References

Agalidis I, Lutz M, Reiss-Husson F (1980) Biochim Biophys Acta 589:264–274
Al-Bayatti KK, Takemoto JY (1981) Biochemistry 20:5489–5495
Bachmann RC, Takemoto JY (1983) Biochem Int 7:137–145
Bachmann RC, Gillies K, Takemoto JY (1981) Biochemistry 20:4590–4596
Bachofen R (1979) FEBS Lett 107:409–412
Bachofen R, Brunner J, Meister HP (1985) 3rd EBEC Conf, Hannover, 58–59
Blasie JK, Pachence JM, Tavormina A, Dutton PL, Stamatoff J, Eisenberger G, Brown G (1983) Biochim Biophys Acta 723:350–357
Bolt JD, Hunter CN, Niederman RA, Sauer K (1981) Photochem Photobiol 14:653–656

Boucher F, Rest M van der, Gingras G (1977) Biochim Biophys Acta 461:339–357
Bowyer JR, Tierney GV, Crofts AR (1979) FEBS Lett 101:207–212
Breton J, Vermeglio A (1982) In: Govindjee (ed) Photosynthesis. Academic Press, London New York, pp 153–194
Brunisholz RA, Wiemken V, Suter F, Bachofen R, Zuber H (1984) Hoppe-Seyler's Z Physiol Chem 365:689–701
Bunker G, Stern EA, Blankenship RE, Parson WA (1982) Biophys J 37:539–551
Chance B, Crofts AR, Nishimura M, Price B (1970) Eur J Biochem 13:364–374
Clayton RK, Clayton BJ (1981) Proc Natl Acad Sci USA 78:5583–5587
Cogdell RJ, Thornber JP (1980) FEBS Lett 122:1–8
Colbeau A, Chabert J, Vignais PM (1983) Biochim Biophys Acta 748:116–127
Collins ML, Mallon DE, Niederman RA (1979) J Bacteriol 139:1089–1092
DaPra E, Snozzi M, Bachofen R (1982) Arch Microbiol 133:23–27
Davidson E, Cogdell RJ (1981) Biochim Biophys Acta 635:295–303
Davis RC, Ditson SL, Fentiman AF, Pearlstein RM (1981) J Am Chem Soc 103: 6823–6826
Debus RJ, Valkirs GE, Okamura MY, Feher G (1982) Biochim Biophys Acta 682:500–503
Debus RJ, Feher G, Okamura MY (1984) Biochemistry 24:2488–2500
Dismukes G, Frank HA, Friesner R, Sauer K (1984) Biochim Biophys Acta 764:153–271
Eccles J, Honig B (1983) Proc Natl Acad Sci USA 80:4959–4962
Elferink MG, Hellingwerf KJ, Michels PAM, Seyen HG, Konings WN (1979) FEBS Lett 107:300–307
Engelhardt H, Baumeister W, Saxton WO (1983) Arch Microbiol 135:169–175
Feher G, Okamura MY (1978) In: Clayton RK, Sistrom WR (eds) The photosynthetic bacteria. Plenum, New York London, pp 349–386
Fillingame RH (1981) In: Sanadi DR (ed) Current topics in bioenergetics, vol 11. Academic Press, London New York, pp 35–106
Ganago AO, Erokhin YE (1981) Stud Biophys 82:9–16
Ghosh R, Bachofen R, Hauser H (1984) Biochim Biophys Acta 765:97–105
Ghosh R, Bachofen R, Hauser H (1985) Biochemistry 24:983–989
Grondelle R van, Duysens LNM, Wel JA van der, Wal HN van der (1977) Biochim Biophys Acta 461:188–201
Grondelle R van, Hunter CN, Bakker JGC, Kramer HJM (1983) Biochim Biophys Acta 723:30–36
Hauska G, Hurt E, Gabellini N, Lockau W (1983) Biochim Biophys Acta 97–133
Hodapp N, Kreutz W (1980) Biophys Struct Mech 7:65–95
Hoppe J, Sebald W (1984) Biochim Biophys Acta 768:1–27
Jay F, Lambillotte M (1985) Eur J Cell Biol 37:7–13
Kataoka M, Ueki T (1981) J Biochem (Tokyo) 89:71–78
Khananshvili D, Gromet-Elhanan Z (1983) FEBS Lett 159:271–274
Knaff DB, Whetstone R, Carr JW (1980) Biochim Biophys Acta 590:50–58
Kopp F, Cuendet PA, Mühlethaler K, Zuber H (1979) Biochim Biophys Acta 553:438–449
Kramer HJM, Grondelle R van, Hunter CN, Westerhuis WHJ, Amesz J (1984) Biochim Biophys Acta 765:156–165
Kyte J, Doolittle RF (1982) J Mol Biol 157:105–132
LeMaire M, Rivas E (1983) Biochim Biophys Acta 722:150–157
Low H, Afzelius BA (1964) Exp Cell Res 35:431–434
Marinetti TD, Okamura MY, Feher G (1979) Biochemistry 18:3126–3133
Meister HP, Bachofen R (1984) Biochim Biophys Acta 771:103–106
Meister HP, Bachofen R, Semenza G, Brunner J (1985) J Biol Chem 260:16326–16331
Meyer R, Snozzi M, Bachofen R (1981) Arch Microbiol 130:125–128
Monger TG, Parson WW (1977) Biochim Biophys Acta 460:393–407
Nabedryk E, Tiede DM, Dutton PL, Breton J (1982) Biochim Biophys Acta 682:273–280
Niedermann RA, Gibson KD (1978) In: Clayton RK, Sistrom WR (eds) The photosynthetic bacteria. Plenum, New York London, pp 78–118
Odermatt E, Snozzi M, Bachofen R (1980) Biochim Biophys Acta 591:372–380

Oelze J (1978) Biochim Biophys Acta 450–561
Overfield RE, Wraight CA (1980a) Biochemistry 19:3322–3327
Overfield RE, Wraight CA (1980b) Biochemistry 19:3328–3334
Pachence JM, Dutton PL, Blasie JK (1983) Biochim Biophys Acta 724:6–19
Picorel R, Belanger G, Gingras G (1983) Biochemistry 22:2491–2497
Prince R (1983) Biochim Biophys Acta 723:133–138
Prince RC, Baccarini-Melandri A, Hauska GA, Melandri BA, Crofts AR (1975) Biochim Biophys Acta 387:212–227
Reed DW, Raveed D, Reporter M (1975) Biochim Biophys Acta 187:368–378
Reiss-Husson F, Rivas E, Gulik-Krzywicki T (1977) Rev Biol Cell 30:27–32
Rickle GK, Cusanovich MA (1979) Arch Biochem Biophys 197:589–598
Rieder R, Wiemken V, Bachofen R, Bosshard HR (1985) Biochem Biophys Res Commun 128:120–126
Rosen D, Okamura MY, Feher G (1980) Biochemistry 19:5687–5692
Sadler DM, Worcester DL (1982a) J Mol Biol 159:467–484
Sadler DM, Worcester DL (1982b) J Mol Biol 159:485–499
Shill DA, Wood PM (1984) Biochim Biophys Acta 764:1–7
Shimada K, Murata N (1976) Biochim Biophys Acta 455:605–620
Singer SJ, Nicholson GL (1972) Science 175:720–731
Snozzi M, Bachofen R (1979) Biochim Biophys Acta 546:236–247
Snozzi M, Bachofen R (1982) In: Mitsui A, Black CC (eds) Handbook of biosolar resources, vol I. CRC Press, Boca Raton, pp 379–382
Sutton MR, Rosen D, Feher G, Steiner LA (1982) Biochemistry 21:3842–3849
Symons M, Swysen C (1983) Biochim Biophys Acta 723:454–457
Takemoto JY, Bachmann RC (1979) Arch Biochem Biophys 195:526–534
Takemoto JY, Peters J, Drews G (1982) FEBS Lett 142:227–230
Theiler R, Zuber H (1984) Hoppe-Seyler's Z Physiol Chem 365:720–729
Theiler R, Suter F, Zuber H (1983) Hoppe-Seyler's Z Physiol Chem 364:1765–1776
Theiler R, Suter F, Wiemken V, Zuber H (1984) Hoppe-Seyler's Z Physiol Chem 365: 703–719
Tiede DM, Dutton PL (1981) Biochim Biophys Acta 637:278–290
Tiede DM, Leigh JS, Dutton PL (1978) Biochim Biophys Acta 503:524–544
Valkirs GE, Feher G (1982) J Cell Biol 95:179–188
Varga AR, Staehelin LA (1983) J Bacteriol 154:1414–1430
Varga AR, Staehelin LA (1985) Pigment–protein complexes from *Rhosopseudomonas palustris:* Isolation, characterisation and reconstitution into liposomes. J Bacteriol 161:921–927
Vitry C de, Diner BA (1984) FEBS Lett 167:327–331
Webster GD, Cogdell RJ, Lindsay GJ (1980) FEBS Lett 111:391–394
Widger WR, Cramer WA, Herrmann RG, Trebst A (1984) Proc Natl Acad Sci USA 81:674–678
Wiemken V, Bachofen R (1982) Biochim Biophys Acta 681:72–76
Wiemken V, Bachofen R (1984) FEBS Lett 166:155–159
Wiemken V, Theiler R, Bachofen R (1981) J Bioenerg Biomembr 13:181–194
Wiemken V, Brunisholz R, Zuber H, Bachofen R (1983) FEMS Microbiol Lett 16:297–301
Wiemken V, Bachofen R (1985) Physiol Vég 23:789–800
Wiemken V, Bachofen R (1986) in preparation
Willey DL, Auffret AD, Gray JC (1984) Cell 36:555–562
Williams JC, Steiner LA, Ogden RC, Simon MI, Feher G (1983) Proc Natl Acad Sci USA 80:6505–6509
Williams JC, Steiner LA, Feher G, Simon MI (1984) Proc Natl Acad Sci USA 81:7303–7307
Yen GSL, Wraight CA, Kaplan S (1982) Biochim Biophys Acta 688:605–621
Youvan DC, Bylina MA, Begusch H, Hearst JE (1984) Cell 37:949–957
Zürrer H, Snozzi M, Hanselmann KW, Bachofen R (1977) Biochim Biophys Acta 460:273–279
Zürrer H, Snozzi M, Bachofen R (1983) FEBS Lett 153:151–155

11.3 Synthesis and Assembly of Bacterial Photosynthetic Membranes

T.J. DONOHUE and S. KAPLAN

1 Introduction

The anoxygenic photosynthetic bacteria (Rhodospirillaceae, Chromatiaceae and Chlorobiaceae), as well as the oxygenic blue-green algae, have provided a wealth of information relative to the role of specific pigment-protein complexes in converting incident light energy into chemical energy (see in this Vol. DUTTON, Chap. 5 for an overview, and PARSON and HOLTEN, Chap. 8.1, NORRIS and VAN BRAKEL, Chap. 8.3, MICHEL and DEISENHOFER, Chap. 8.4, BLANKENSHIP and FULLER, Chap. 8.6 for forther details). In particular, several of the purple nonsulfur bacteria are unique because of their ability to grow under both photoheterotrophic and chemoheterotrophic conditions. Since the photosynthetic or intracytoplasmic membrane (ICM) of these bacteria is physically continuous but functionally distinct from the inner or cytoplasmic membrane (CM) they are amenable to investigations of the synthesis, development, and regulation of this highly specialized membrane system.

In a previous review (KAPLAN 1978) we have defined the terms ICM *induction* and *inhibition* to refer to the de novo assembly or abrupt cessation of ICM synthesis which results from lowering or raising, respectively, the available oxygen tension (pO_2) relative to the species-specific threshold level at which ICM synthesis is controlled. Similarly, the concepts of ICM *repression* and *derepression* were introduced to denote changes in both the differential rate of ICM synthesis as well as the photosynthetic unit size which occur when light intensity is increased or decreased, respectively.[1]

In those species of Rhodospirillaceae which contain two light-harvesting (LH) antenna pigment-protein complexes the fixed photosynthetic unit (FPU) consists of aggregates of reaction center (RC) and LH I (B-875 or B-870) complexes in a ratio of approximately 10–15 LH I complexes per RC complex (KAPLAN and ARNTZEN 1982). The relative amount of variable antenna LH II (B-800–850) bacteriochlorophyll (BChl) present within a photosynthetic unit is inversely related to the incident light intensity and it is the size of the variable photosynthetic unit (VPU), as well as the amount of ICM per cell which changes as a result of ICM *repression* or *derepression*. Although the oxygen and light-dependent control of BChl synthesis seem to share a common site of action in vivo, there is evidence that these two stimuli also have unique sites of regulatory action (ARNHEIM and OELZE 1983).

1 Light intensities reported in the text are measured through a 620–1100 nm filter.

2 Assembly of the ICM in Steady-State Cells

Cultures of *Rhodopseudomonas sphaeroides* growing asynchronously in the presence of growth rate-saturating illumination (10 W m^{-2}) contain approximately 500 chromatophore equivalents per cell, with a packing density of approximately one VPU per 200 nm^2 of membrane surface and a protein to phospholipid ratio of approximately 3.2:1.0 (w/w) (KAPLAN and ARNTZEN 1982). Recent studies in *Rps. sphaeroides* (KAPLAN et al. 1983) and *Rps. palustris* (which contains stacked photosynthetic membranes, VARGA and STAEHELIN 1983, SPRAGUE and VARGA, Chap. 11.1, this Vol.) demonstrate an inverse relationship between light intensity and the diameter of the VPU, reflecting changes in the ratio of LH II BChl to RC BChl. In addition, they suggest that the increase in VPU unit size is the result of the random addition of discrete sized LH II complexes to a core RC·LH I FPU. Therefore in both of these species, as well as in *Rps. capsulata*, cells grown at low light intensity have both increased quantities of ICM per cell and a larger VPU. In *Rps. sphaeroides*, the VPU occupies a larger fraction of the membrane surface area in cells grown at low light intensity, but in *Rps. palustris*, they occupy less of the membrane surface in low light cells than in high light cells.

Studies employing synchronous photoheterotropic (10 W m^{-2}) cultures of *Rps. sphaeroides* have shown that only pre-existing ICM is used as the structural matrix for the insertion of ICM proteins and photopigments throughout the cell cycle (see KAPLAN 1981, and KAPLAN and ARNTZEN 1982 for a more thorough discussion). DNA sequence analysis of the genes for RC and both LH complex polypeptides from *Rps. capsulata* (YOUVAN et al. 1984) and the RC-M and RC-L proteins from *Rps. sphaeroides* (WILLIAMS et al. 1983, 1985, HEARST, Chap. 8.5, this Vol.) have failed to identify signal peptide sequences. Employing toluene-treated cells of *Rps. capsulata* it has been found that synthesis of the BChl-binding proteins of the LH complexes is directed by stable mRNA on nonmembrane-bound polysomes and is dependent on the synthesis of tetrapyrroles (DIERSTEIN 1984). The molecular mechanism by which these polypeptides correctly associate with photopigments during the formation of spectrally active complexes and their insertion into the membrane remains to be determined.

In vitro synthesis of the LH I and RC-M and L polypeptides using cloned genes from *Rps. sphaeroides* has shown that (1) the LH I and RC-M and L structural genes of *Rps. sphaeroides* are closely linked in a manner analogous to that found in *Rps. capsulata* (YOUVAN et al. 1984); (2) the LH I polypeptides are synthesized in vitro in substantially higher amounts relative to the RC-M and L polypeptides, perhaps mimicking the level of expression of these polypeptides seen in vivo; and (3) the in vitro synthesized LH I and RC-M and L polypeptides exhibit distinct changes in their electrophoretic mobility when solubilized for SDS-PAGE in the presence of purified ICM, suggesting that these polypeptides are capable of spontaneously associating with specific components of the ICM in vitro (KILEY et al. manuscript in preparation).

The insertion of phospholipid into preexisting ICM in *Rps. sphaeroides* occurs either just preceding or concomitant with cell division. This uncoupling between protein and phospholipid insertion into the ICM has been confirmed by quantitating cell cycle-specific changes in the ICM protein to phospholipid ratio, intrinsic specific density, packing density of VPU, and the measured microviscosity of the ICM bilayer (KAPLAN et al. 1983). Kinetic (CAIN et al. 1981, 1982) and enzymological analyses (CAIN et al. 1984; COOPER and LUEKING 1984) of phospholipid biosynthesis in *Rps. sphaeroides* have identified the CM as the subcellular locale for the majority of the phospholipid biosynthetic enzyme activities. Thus the in vivo and in vitro data confirm that there is a cell cycle-specific transfer of previously synthesized phospholipids from outside the ICM into the ICM at the time of cell division (CAIN et al. 1981). Other data, employing the inhibitor of fatty acid biosynthesis, cerulenin, independently demonstrate that protein and phospholipid accumulation into the ICM can be uncoupled (SHEPHERD and KAPLAN 1983).

These findings have several important implications for the assembly, partitioning, structure, and function of the ICM in steady-state photoheterotrophic cells of *Rps. sphaeroides*. First, considering the fact that the number of ICM invaginations remains constant during the cell cycle, ICM invaginations must be replicated by some unknown mechanism probably concomitant with the insertion of phospholipid into the ICM at cell division. Second, because the ICM is physically continuous with the cellular location of the majority of the phospholipid biosynthetic machinery (CM), the restriction to lateral diffusion of phospholipids from the CM to the ICM leads to the hypothesis that a barrier exists which prevents the free equilibration of phospholipids between these two membranes. The phospholipid transfer mechanism itself is relatively nonspecific, since all phospholipid species are mobilized into the ICM in amounts representative of their bulk cellular levels. Finally, at the high protein to phospholipid ratio found within the ICM prior to the influx of phospholipid, the membrane is relatively rigid with the vast majority of phospholipid immobilized, due to its association with protein.

Although net phospholipid accumulation into the ICM is restricted to the time of cell division, the initial observation (CAIN et al. 1981) that PE methylation to PC occurs throughout the cell cycle at the ICM suggested that the PE methyltransferase system might be associated with the ICM. Subsequent experiments have shown that this enzyme activity (activities) is not membrane-associated in cell lysates (CAIN et al. 1984). If the PE methyltransferase system of *Rps. sphaeroides* is only weakly membrane-bound in vivo, then this activity could methylate PE, presumably at the cytoplasmic leaflet of the bilayer, in both the ICM and CM throughout the cell cycle. The continued synthesis of PC at the ICM surface or within localized regions of the ICM during the cell cycle may be required for some unknown aspect of ICM structure, function, or the ability of the cell to mobilize phospholipids into the ICM during cell division or ICM replication.

3 Induction of ICM Synthesis

Several experimental protocols have been employed to lower the pO_2 in order to allow ICM induction. Early studies with *Rps. sphaeroides* demonstrated that

BChl and carotenoid synthesis began immediately upon transferring an exponential phase culture of aerobic cells to totally anaerobic conditions in the presence of light (*nongratuitous induction*) and that the length of the subsequent lag prior to the resumption of cell growth was inversely related to the light intensity.

Subsequent studies on ICM induction have been performed either by directly lowering the pO_2 below threshold levels or by reducing the aeration rate of a culture on a standard laboratory shaker (*gratuitous induction*). Cell growth lags only briefly under these conditions and in either case in both *Rps. sphaeroides* and *Rps. capsulata*, the coordinate insertion of RC and LH I complexes into the particulate fraction occurs before the accumulation of LH II complexes (HUNTER et al. 1982).

Experiments using the photosynthetic gene cluster of *Rps. capsulata* have shown that transcription from the genetic loci coding for the RC and LH I complex polypeptides is activated by lowering the pO_2 (CLARK et al. 1984). Expression of the genes for BChl biosynthesis has also been shown to be regulated by oxygen using individual β-galactosidase gene fusions within these genes (BIEL and MARRS 1983). Recent experiments using a DNA probe specific for the form II ribulose 1,5-bisphosphate carboxylase of *Rps. sphaeroides* (MULLER et al. 1985) have identified that specific mRNA species are isolable from intact cells and, using a Northern blot analysis, the relative levels of these mRNA's correlate with the levels of enzyme activity measured under different physiological conditions (ZHU and KAPLAN unpublished observations). The isolation and identification of unique mRNA species opens the way for detailed regulatory analyses of photosynthetic gene expression in these bacteria.

Several membrane fractions have been isolated from lysates of *Rps. capsulata* and *Rps. sphaeroides* during *gratuitous* induction experiments. An "upper pigmented band" described by NIEDERMAN and coworkers (INAMINE et al. 1984) has been isolated from both *Rps. sphaeroides* and *R. rubrum* (INAMINE et al. 1982), and this fraction is believed to represent peripheral regions of immature ICM invaginations into which the RC and LH I complexes are preferentially inserted during ICM induction. A "slowly sedimenting band" isolated from lysates of *Rps. capsulata* by DREWS and colleagues is also enriched in RC and LH I complexes relative to mature ICM, but a precursor-product relationship between this fraction and mature ICM has not been supported experimentally (GARCIA et al. 1981). A similar membrane fraction enriched in RC and LH I relative to LH II is isolable from aerobically grown *Rps. capsulata* (GARCIA et al. 1981). This final fact points to the lack of total inhibition of ICM synthesis by oxygen in *Rps. capsulata* and suggests the existence of subtle physiological differences in the oxygen-regulated control of ICM assembly between this species and *Rps. sphaeroides*.

Biochemical and morphological studies of *nongratuituous* ICM induction in *Rps. sphaeroides* (CHORY et al. 1984; SPRAGUE and VARGA, Chap. 11.1, this Vol.) have shown that whole cell protein and phospholipid accumulation parallel the approximately 12-h lag in cell growth, but that ICM invaginations are formed within 1 h after the shift, suggesting that phospholipid present in the aerobic cells is used as the structural matrix for the formation of new ICM. The aerobic cells used in this study were devoid of detectable BChl and ICM invaginations; however, electron microscopy demonstrated small indentations in the CM of the aerobic cells which were hypothesized as being converted to ICM invaginations when the cells were exposed to anaerobiosis. Microscopic analysis also revealed a series of morphological changes in ICM structure, organization and apparent ICM lipid enrichment during the lag period prior to the initiation of photosynthetic growth. The synthesis of individual phospholipid species during ICM induction suggested a role for PC metabolism in the observed changes in ICM structure.

The biochemical characteristics of the indentations within the CM of aerobic cells of *Rps. sphaeroides*, the nascent ICM, and the possible relationship of these transient ICM structures to the previously discussed membrane fractions isolated from cells shifted to semi-aerobic conditions remain to be determined. Although it was possible to document differential oxygen-dependent regulation of the synthesis of specific proteins associated with photoheterotrophic metabolism (i.e., cytochrome c_2, ribulose 1,5-bisphosphate carboxylase, BChl-binding proteins), no preferential synthesis of RC and LH I complexes prior to the insertion of LH II complexes into the ICM was found under these *nongratuituous* induction conditions. Finally, using antibody to purified RC complexes an immunochemically cross-reacting polypeptide (M_r 28,000) believed to be the RC–H was detected in the CM of aerobic cells. Further immunochemical studies are required to determine if this putative RC–H polypeptide is localized within the indentations present in the CM of aerobic *Rps. sphaeroides* and if the CM indentations serve as points of ICM assembly when aerobic cells are shifted to photoheterotrophic conditions. The recent cloning of the RC–H gene from *Rps. sphaeroides* (DONOHUE et al. unpublished observations) also permits studies aimed at identifing specific mRNA for this polypeptide in aerobic cells.

The observed differences in the kinetics of synthesis and insertion of LH II complexes during *gratuitous* versus *nongratuitous* conditions of ICM induction can be reconciled if the synthesis of the LH II complex polypeptides is regulated differently under semi-aerobic (*gratuitous*) versus anerobic (*nongratuitous*) ICM induction conditions. Additionally, since most of the *gratuitous* ICM induction experiments have used cells initially shifted to semi-aerobic conditions in the dark it is also possible that light is a critical variable. These possibilities are directly testable with antibody against purified LH polypeptides or individual LH antenna complexes, and specific DNA probes for the proteins within the pigment-protein complexes of *Rps. capsulata* (YOUVAN et al. 1984) and *Rps. sphaeroides* (WILLIAMS et al. 1983, 1985, KILEY, DONOHUE and KAPLAN unpublished experiments.

4 Inhibition of ICM Synthesis

Exposure of photoheterotrophically growing cells to oxygen has no dramatic effect on cell growth, but results in the virtually immediate cessation of BChl and ICM synthesis for as long as the oxygen level remains above the species-specific threshold (KAPLAN 1978, 1981). The discovery that penicillin binding protein (PBP) activity of *Rps. sphaeroides* is uniquely associated with the CM of both chemoheterotrophic and photohetrophically grown cells (SHEPHERD et al. 1981) allowed the intracellular distribution of these as well as other membrane markers to be examined in cells shifted from photoheterotrophic to chemoheterotrophic conditions. No significant amounts of the CM-specific PBP's were inserted into the residual ICM by two culture doublings after a shift to chemoheterotrophic growth, therefore it was concluded that the ICM is not converted to CM, but rather is remodeled into a membrane system enriched in aerobic electron transport activity after a shift to chemoheterotrophic growth (SHEPHERD et al. 1981). Subsequent experiments utilizing the differential sensitivity of chemoheterotrophic and photoheterotrophic cells to the antibiotic cerulenin also support this hypothesis (SHEPHERD and KAPLAN 1983).

5 Repression/Derepression of ICM Synthesis

In the only recent study of cells shifted from low to high light intensity (SCHUMACHER and DREWS 1979), it was shown in *Rps. capsulata* that BChl synthesis ceased for two culture doublings, the size of the VPU decreased, and LH I became the predominant spectral complex present.

Following a tenfold decrease in light intensity (ca. 300–30 W m^{-2}) in *Rps. sphaeroides*, cell growth, BChl, and bulk phospholipid synthesis ceased; bulk protein synthesis was reduced to approximately 15% of its pre-shift rate, with BChl and phospholipid synthesis resuming approximately midway through the lag (CHORY and KAPLAN 1983; CAMPBELL and LUEKING 1983). The derepression in BChl synthesis is due to the increased differential rate of synthesis of LH II polypeptides, thereby reflecting the increase in VPU size (CHORY and KAPLAN 1983). Similiar experiments confirm that *Rps. capsulata* responds to a decrease in light intensity by increasing the size of the VPU, the number of VPU per unit area of ICM as well as the number of ICM invaginations per cell. Unpublished experiments of KOSAWKOWSKI and KAPLAN employing *Rps. sphaeroides* equilibrated with D_2O and radiolabeled acetate suggest that the increase in the amount of ICM per cell which accompanies such a transition from high to low light is the result of the synthesis of totally new ICM, implying that the components of one ICM invagination can remain sequestered from those of another. This conclusion is compatible with the lack of free equilibration of components between the ICM and CM found in steady-state photoheterotrophic cells.

Finally, given the kinetics of macromolecular synthesis which accompany a reduction in light intensity, it is not surprising to find that a well-synchronized population of cells are obtained by returning a low light culture to high light conditions just prior to the end of the lag in cell growth (LUEKING et al. 1981). These light-synchronized cultures of *Rps. sphaeroides* show many of the same cell cycle-specific changes in ICM properties described in the earlier section on steady-state ICM assembly.

6 Future Prospects

With the diversity of information summarized in this volume, the photosynthetic membranes of the purple nonsulfur bacteria represent one of the most thoroughly studied subcellular organelles from a functional, biochemical, and structural standpoint. The body of knowledge accumulated to date allows us to pose more defined questions not only on the assembly and functional organization of the ICM itself but also as to how this assembly process is regulated by and related to the synthesis of the other cellular membrane systems.

For example, there is no data on the mechanism of protein targeting to the physically continuous yet functionally distinct CM and ICM. The PBP's and many of the enzymes involved in the early steps in phospholipid biosynthesis reside uniquely within the CM, the BChl-binding proteins of the LH II complexes are inserted solely into the ICM, while other proteins associated with the electron transport chain (cytochromes, ATPase etc.) are directed to both membrane systems or the periplasmic space (cytochrome c_2). The isolation of genes for some of these proteins (TYBULEWICZ et al. 1985) coupled with the recent observation that the lactose permease of the enteric bacterium *Escherichia coli* appears to be functionally incorporated into membranes of *Rps. sphaeroides* in vivo (ELERINK et al. 1983) provide tools to genetically and physiologically dissect the question of protein direction to and localization within these membrane systems.

The isolation of genes for various BChl-binding proteins of *Rps. capsulata* (YOUVAN et al. 1984) and *Rps. sphaeroides* (WILLIAMS et al. 1983, 1985, DONOHUE, KILEY and KAPLAN unpublished observations) provides investigators with the ability to probe numerous aspects of the regulatory pattern associated with ICM assembly under various physiological states. The unexpected initial observation that genes from these high mole percent guanine plus cytosine organisms were not expressed in *E. coli* (MARRS 1981; CHORY and KAPLAN 1982) will make studies on the factors controlling the oxygen and light-dependent regulation of gene expression in the photosynthetic bacteria very interesting not only in and of themselves, but also as a model for other organisms with similar DNA base composition.

Finally, understanding the means by which the movement of previously inserted proteins and phospholipids between the CM and ICM are controlled has profound implications for the fields of membrane assembly, structure, and function. The concepts of barriers to free molecular diffusion, transfer of components between membrane systems, existence of specific lipid or protein-containing domains within membranes and the possible role of such regions or processes in the assembly of functional membranes are all active fields of research in more complex biological systems. Thus the wealth of information available on the organization and regulation of the ICM in these bacteria will serve to provide a solid framework for exciting future research on membrane assembly in general and photosynthetic membrane biogenesis in particular.

Acknowledgments. Recent work in our laboratory has been supported by grants from the National Institutes of Health (GM 15590, GM 31667) and the National Science Foundation (PCM 80-20799).

References

Arnheim K, Oelze J (1983) Arch Microbiol 135:299–304
Biel AJ, Marrs BL (1983) J Bacteriol 156:686–694
Cain BD, Deal CD, Fraley RT, Kaplan S (1981) J Bacteriol 145:1154–1166
Cain BD, Donohue TJ, Kaplan S (1982) J Bacteriol 152:607–615
Cain BD, Donohue TJ, Shepherd WD, Kaplan S (1984) J Biol Chem 259:942–948
Campbell TB, Lueking DR (1983) J Bacteriol 155:806–816
Chory J, Kaplan S (1982) J Biol Chem 257:15110–15121
Chory J, Kaplan S (1983) J Bacteriol 153:465–474
Chory J, Donohue TJ, Varga AR, Staehelin LA, Kaplan S (1984) J Bacteriol 159:540–554
Clark WG, Davidson E, Marrs BL (1984) J Bacteriol 157:945–948
Cooper C, Lueking DR (1984) J Lipid Res 25:1222–1232
Dierstein R (1984) Eur J Biochem 138:509–518
Elerink MGL, Hellingwerf KJ, Nano FE, Kaplan S, Konings WN (1983) FEBS Lett 164:185–190
Garcia AF, Drews G, Reidl HH (1981) J Bacteriol 145:1121–1128
Hunter CN, Pennoyer JD, Niederman RA (1982) In: Cell function and differentiation. Liss, New York, pp 257–265
Inamine GS, Niederman RA (1982) J Bacteriol 150:1145–1153
Inamine GS, Houten J van, Niederman RA (1984) J Bacteriol 158:425–529
Kaplan S (1978) In: Clayton RK, Sistrom WR (eds) The photosynthetic bacteria. Plenum, New York London, pp 809–835
Kaplan S (1981) Photochem Photobiol 34:769–774
Kaplan S, Arntzen CJ (1982) In: Govindjee (ed) Photosynthesis: energy conversion by plants and bacteria, vol I. Academic Press, London New York, pp 67–151
Kaplan S, Cain BD, Donohue TJ, Shepherd WD, Yen GSL (1983) J Cell Biochem 22:15–29
Lueking DR, Campbell TB, Burghardt RC (1981) J Bacteriol 146:790–797
Marrs B (1981) J Bacteriol 146:1003–1012
Muller ED, Chory J, Kaplan S (1985) J Bacteriol 161:469–472
Schumacher A, Drews G (1979) Biochim Biophys Acta 547:417–428
Shepherd WD, Kaplan S (1983) J Bacteriol 156:1322–1331
Shepherd WD, Kaplan S, Park JT (1981) J Bacteriol 147:354–362
Tybulewicz VLJ, Falk G, Walker JE (1985) J Mol Biol 179:185–214
Varga AR, Staehelin LA (1983) J Bacteriol 154:1414–1430
Williams JC, Steiner LA, Ogden RC, Simon MI, Feher G (1983) Proc Natl Acad Sci USA 80:6505–6509
Williams JC, Steiner LA, Feher G, Simon MI (1984) Proc Natl Acad Sci USA 81:7303–7308
Youvan DC, Bylina EJ, Alberti M, Begusch H, Hearst JE (1984) Cell 37:949–957

11.4 Proteins of Cyanobacterial Thylakoids

T.M. BRICKER, J.A. GUIKEMA, H.B. PAKRASI, and L.A. SHERMAN

1 Introduction

The cyanobacteria are predominantly photoautotrophic prokaryotes capable of oxygenic photosynthesis. While the striking similarity between cyanobacterial and chloroplast photosynthesis has been recognized for some time, the full extent of this relationship is only now being appreciated (see ORT, Chap. 4, this Vol. for an overview of energy transduction in oxygenic photosynthesis). The accumulated cytological, physiological, and biochemical evidence over the past century gave rise to the endosymbiont hypothesis; this hypothesis states that chloroplasts are descended from cyanobacteria that became endosymbionts in an originally nonphotosynthetic host. This relationship has been refined recently by comparison of ribosomal and transfer RNA sequences, transcription and translation mechanisms, and through the comparison of protein and nucleic acid sequences (GRAY and DOOLITTLE 1982). The cloning of chloroplast genes and the ability to isolate related cyanobacterial genes by DNA-DNA hybridization has now emphasized the profound depth of these similarities. Thus, CURTIS and HASELKORN (1984) cloned the gene from *Anabaena* 7120 homologous to the chloroplast gene that codes for the 32 kDa, rapidly labeled membrane protein. The nucleotide sequence of these two genes was 75% conserved, and gave rise to derived amino acid sequences that were 87% conserved.

In this chapter, we will review the protein structure of the cyanobacterial thylakoids, with emphasis on the relationship of structure to function. Therefore, we will analyze the protein composition of the major thylakoid complexes (PS II, the cyt b_6/f complex, PS I, and the CF_1-CF_0 ATPase complex). For conciseness, we will concentrate on the similarities between cyanobacterial and chloroplast membrane structure. However, there are three features of the cyanobacterial photosynthetic mechanism that distinguish it from the mechanism of eukaryotic chloroplasts. First, cyanobacteria contain phycobiliproteins as a major light-harvesting pigment. The pigments are organized in phycobilisomes which are located on the external surface of the membrane (GANTT, Chap. 6.3, this Vol.). By contrast, the major light-harvesting apparatus of chloroplasts, the Chl b-containing light-harvesting protein complex (LHCP), is an integral membrane complex (THORNBER, Chap. 3, this Vol.). Secondly, the intracellular organization of the cyanobacterial thylakoids is markedly different from that of chloroplast thylakoids. In most cyanobacteria, the thylakoids are arranged peripherally in three to six concentric layers (GOLECKI and DREWS 1982, STAEHELIN, Chap. 1, this Vol.), and appear to act as separate units. Chloroplast thylakoids can be differentiated into stacked (granal) membranes and unstacked (stromal) membranes; furthermore, there is a good deal of lateral heterogeneity between these structural regions (STAEHELIN, Chap. 1, this Vol.). Finally, the cyanobacterial thylakoids are also thought to be the location of the respiratory apparatus. These features are important for they greatly affect the three-dimensional relationship of the complexes, and make topological comparisons between the two types of thylakoids quite difficult.

2 Protein Composition of Cyanobacterial Thylakoids

The thylakoids of *Anacystis nidulans* R2 and *Synechococcus cedrorum*, labeled with either ^{35}S or ^{14}C, revealed approximately 100 bands when analyzed by one-dimensional lithium dodecyl sulfate polyacrylamide gel electrophoresis (LDS-PAGE) (GUIKEMA and SHERMAN 1982). Similar gel patterns were obtained from a variety of cell breakage and membrane isolation procedures. These preparations also contain the plasmalemma which has been isolated by MURATA et al. (1981). However, the purified plasmalemma is generally only a minor contaminant of the membrane preparations. The topological organization of the membrane proteins has been analyzed by lactoperoxidase-catalyzed iodination (GUIKEMA and SHERMAN 1982), and by Triton X-114 phase fractionation (BRICKER and SHERMAN unpublished observations). At least 40 proteins could be iodinated, thus indicating the presence of surface-exposed domains. Triton X-114 partitions membrane proteins into intrinsic and peripheral fractions (BRICKER and SHERMAN 1982), since only proteins with extensive hydrophobic domains can form mixed micelles with the detergent. These studies indicated that approximately 50–60% of the proteins were intrinsic, and demonstrated that many of the photosynthetically important proteins were intrinsic with surface-exposed domains.

Labelling cells with ^{59}Fe yielded nine iron-containing bands on LDS-PAGE (GUIKEMA and SHERMAN 1982). Four of these bands, at 33, 19, 14, and 10 kDa, were also heme proteins as determined by tetramethylbenzidine staining, and represent cytochromes f, b_6, c-550, and c-554, respectively. The remaining iron proteins were highly sensitive to solubilization conditions, especially the presence of 2-mercaptoethanol, and it was suggested that these bands may be FeS proteins. The functional identification of other photosynthetic membrane proteins has been accomplished by detergent solubilization of the membrane and subsequent isolation of functional complexes. The properties and protein composition of these complexes will be described in the following sections.

3 Chlorophyll–Protein Complexes

Isolated thylakoid membranes, when monitored at liquid nitrogen temperatures, show three major Chl fluorescence emission peaks at 686, 696, and 716 nm (NEWMAN and SHERMAN 1978). These peaks are indicative of the microenviron-

Table 1. Composition of cyanobacterial chlorophyll-protein complexes resolved by SDS (LDS)-PAGE

Reference	Species	PS I-Associated			PS II-Associated		
		Complex	Composition[a] (kDa)	F1 Max[b] (nm)	Complex	Composition[a] (kDa)	F1 Max[b] (nm)
RUSCKOWSKI and ZILINSKAS (1980)	*Nostoc* sp.	CP I	58 (74,85)	735	CP III a	44,48	685
					CP III b	43,48	–
STEWART (1980)	*Phormidium luridum*	A 1	60 (12–17)	–	A 2	43,48	–
HLADIK et al. (1982)	*Plectonema boryanum*	Frac. II	60 (14,18)	680,726	Frac. I[c]	41,46 (52,28,15)	680
		Frac. III	60 (14,16,18)				
GUIKEMA and SHERMAN (1981, 1983)	*Anacystis nidulans*	CP I and CP II	64 (10,14,16,16.5)	676,716	CP VI	45,48	–
PAKRASI and SHERMAN (1984)		CP III	64 (<10,14,16, 16.5,35,36)	716	–	–	686,696
					CP VI a	48 (42,40)	696
		CP IV	64 (36)		CP VI b	45 (36)	686
		CP V	64	716			
TAKAHASHI and KATOH (1982)	*Synechococcus sp.*	CP 1 a	60,62 (10,13,14)	695,728	CP 2 a,b,c	47	686,696
		CP 1 b	60,62 (10,13,14)	723	CP 2 c	40	686
YAMAGISHI and KATOH (1983)		CP 1 c–d	60,62 (10,14)	721	–	–	
		CP 1 e	60,62	721	–	–	

[a] The presumed Chl-binding protein contained in each Chl-protein complex is shown. Additional polypeptides which are contained in the complex are shown in parentheses.

[b] Fluorescent emission peak(s) of chlorophyll monitored at liquid nitrogen temperatures (77 K).

[c] Fraction I was not recovered from SDS-PAGE. Fractions I–III represent DEAE column chromatography elution peaks of Triton X-100-treated membranes.

ment of cyanobacterial light harvesting chlorophyll, and have been ascribed to PS I (F-716) or to PS II (F-686 and F-696). A major effort in examining thylakoid substructure has been to identify the apoproteins involved in Chl binding, and to describe their roles in the function of PS I and PS II. Analysis of fluorescence spectra from specific detergent fractions or from Chl-protein complexes extracted from gels has played a key role in this identification.

Table 1 summarizes recent data using SDS or LDS-PAGE to visualize Chl association with protein. There is general agreement that two classes of Chl-binding proteins are present in cyanobacteria: one or two peptides between 58 and 64 kDa as an apoprotein for the PS I Chl, and two peptides between 40 and 48 kDa as Chl-binding proteins for PS II.

In contrast to work with higher plants (THORNBER, Chap. 3, this Vol.), Chl-binding proteins are not easily detected using SDS-PAGE. Instead, multiple forms of Chl-protein complexes are often observed as large Chl-protein aggregates. GUIKEMA and SHERMAN (1981, 1983) identified six Chl-containing bands. The five largest complexes (CP I–V) were associated with PS I, contained most of the chlorophyll, and had molecular masses of approximately 100–350 kDa. The peptide compositions of the PS I aggregates were complex, containing the PS I Chl-binding protein of 64 kDa, five to six peptides between 10 and 36 kDa, and a substantial amount of iron. The four largest aggregates were observed in membrane fractions enriched in PS I, which exhibited a Chl fluorescence peak at 714 nm. These findings suggested that the organization of PS I within the thylakoid was responsible for the larger multiple Chl-protein forms. Aberrant SDS: protein ratios preclude the use of SDS-PAGE to determine the molecular weights of the multiple Chl-protein forms; however, Ferguson plots suggest that they have similar surface charge densities in the presence of SDS, and that migration differences reflect size differences between them (TAKAHASHI and KATOH 1982). In addition, TAKAHASHI and KATOH (1982) observed P-700 activity, PS I acceptors A1 and A2, and cytochrome c-553 photooxidation associated with the large Chl-protein complexes. Since A1 and A2 are thought to be iron-sulfur proteins, this finding explains the observation of iron associated with these complexes, and implicates peptides at 10 to 14 kDa as these acceptors. Experiments of GUIKEMA and SHERMAN (1982, 1983) to monitor the topology of these peptides suggest that most of these small proteins are poorly surface-exposed.

The Chl proteins associated with PS II have been difficult to detect. GUIKEMA and SHERMAN (1983) used digitonin-treated, PS II-enriched thylakoid fractions to detect a comparatively lower molecular weight CP band (CP VI) that contained a 48 kDa and a 45 kDa polypeptide. RUSCKOWSKI and ZILINSKAS (1980) showed that similar bands possessed F-686. YAMAGISHI and KATOH (1983) observed four CP bands from PS II-enriched fractions after digitonin treatment of thylakoids from a thermophilic strain. According to these authors, the apoproteins in these CP bands have molecular weights of 47 and 40 kDa. Moreover, they showed that F-696 originates from CP bands associated with the 47 kDa protein, and the F-686 emanates from the CP band associated with the 40 kDa polypeptide. PAKRASI and SHERMAN (unpublished observations) have confirmed these observations by using dodecyl maltoside-treated PS II-enriched fractions from a nonthermophilic strain. They detected two CP bands in the CP VI region originally observed by GUIKEMA and SHERMAN (1983), and which they termed CP VIa and VIb. CP VIa gave rise to F-696 and consisted of polypeptides at 48, 42 and 40 kDa, whereas CP VIb gave rise to F-686 and had polypeptides at 45 and 36 kDa. Additional CP bands have been observed in the CP II and the CP IV regions (PAKRASI and SHERMAN unpublished observation). Some of these bands have F-686 and F-696 and also contain polypeptides in the 45–50 kDa region. Thus, they may represent higher order aggregates of PS II-related Chl-proteins.

4 Photosystem II

Isolation of completely functional PS II complexes from cyanobacteria has been difficult. Preparations enriched in PS II reaction centers were isolated from two different strains of *Synechococcus* (Table 2; for a comparison of PS II complexes of higher plants see DINER, Chap. 9.1, this Vol.). These membrane fractions had seven to eight polypeptides each, and were enriched in PS II-mediated DPC to DCPIP activity, but lacked oxygen-evolving activity. Recently, a number of laboratories have successfully isolated PS II preparations that evolve oxygen, all of which have a large number of polypeptides. PAKRASI and SHERMAN (1984a) isolated oxygen-evolving PS II particles from *Anacystis* that have functionally attached phycobilisomes. SCHATZ and WITT (1984) have suggested that a 110 kDa polypeptide in their PS II preparation may be a stable trimeric form of allophycocyanin. Both of these preparations probably contain a linker polypeptide that connects the PS II reaction center to the phycobilisome.

The composition of PS II from cyanobacterial membranes is very similar to that of PS II from higher plant chloroplasts. In addition to the two reaction center polypeptides (45–50 kDa) and cytochrome b-559, one or more polypeptides in the 30–35 kDa range are present. Experiments with *Aphanocapsa* and *A. nidulans* have shown that azidoatrazine binds to a protein in the 32 kDa region (ALLEN et al. 1983; PAKRASI and SHERMAN unpublished observation, ARNTZEN and PAKRASI, Chap. 9.4, this Vol.). Furthermore, this atrazine-binding protein is in the PS II oxygen-evolving preparation. Similar to the findings in higher plants, this polypeptide is not stained well with Coomassie blue and is readily detected only as an ^{35}S-labeled polypeptide or after azidoatrazine binding. Manganese is thought to have a role in the oxidizing side of PS II and oxygen evolution. Recently, a 13 kDa manganese binding polypeptide has been isolated from *Plectonema boryanum* (OKADA and ASADA 1983). Interestingly, antibodies raised against this polypeptide inhibited oxygen evolution from intact membranes. All of the recent PS II O_2-evolving preparations contain a 34 kDa protein (Table 2.I.A.) which may be an integral membrane protein required for O_2-evolution. PAKRASI and SHERMAN (unpublished observations) have demonstrated that this protein partitions as an integral membrane protein by Triton X-114 fractionation. Calcium has also been shown to have an effect

Table 2. Complexes obtained by detergent treatment of cyanobacterial thylakoids

I. Photosystem II

References	Protein molecular mass[a], kDa	Comments
A. Photosystem II – oxygen-evolving preparations		
1. Stewart and Bendall (1981)	(a)87,(b)83,(c)68,(d)57,(e)48, (f)46,(g)42,(h)40,(i)38,(j)34, (k)32,(l)30,(m)28,(n)18, (o)17,(p)16	*P. laminosum;* LDAO. 1300 µmol O_2 mg^{-1} Chl h^{-1}. (a) and (b) may be heat-aggregated complexes from (f) to (h). (k) cyt f; (p) cyt c-550
2. Bowes et al. (1983)	(a)59,(b)52,(c)43,(d)34,(e)30, (f)16,(g)15,(h)120,(i)58,(j)54, (k)51,(l)33	*P. laminosum;* LDAO + dodecyl maltoside. 2000 µmol O_2 mg^{-1} Chl h^{-1}. (f) and (g) may contain allophycocyanin. Proteins (a) to (g) were enriched in preparation. No 48 kDa protein. Suggested (b) through (e) required for O_2-evolution
3. Evans et al. (1983)	(a)77,(b)60,(c)40,(d)36,(e)33, (f)30,(g)27,(h)15,(j)14,(k)10	*A. nidulans* 625; LDAO. 700 µmol O_2 mg^{-1} Chl h^{-1}. (d) to (f) may be involved in Ca^{+2}-binding. Preparation contains at least 7 more proteins
4. Pakrasi and Sherman (1984a, b)	(a)97,(b)73,(c)68,(d)48,(e)45, (f)42,(g)36,(h)34,(k)32,(j)26, (k)25,(l)8	*A. nidulans* R2; 0.5 M Sucrose, 0.5 M phosphate, 0.3 M citrate buffer, pH 6.8; Triton X-100. 2500 µmol mg^{-1} Chl h^{-1}. Silver-stained gel. Particles contained attached phycobilisomes. Activity sensitive to calmodulin antagonists; enriched in cyt b-559 (l). Fl. Max. at 686 and 696 nm. Contains CP IV, VIa, and VIb. (h) is an intrinsic protein
5. Schatz and Witt (1984)	(a)95,(b)80,(c)67,(d)60,(e)52, (f)48,(g)42,(h)34,(i)28	*Synechococcus* sp. Zwittergents (Sulfobetaines). 2500 µmol mg^{-1} Chl h^{-1}. Contains allophycocyanin

Table 2 (continued)

References	Protein molecular mass[a], kDa	Comments
B. Photosystem II – non-oxygen-evolving preparations		
1. Newman and Sherman (1978) Sadewasser and Sherman (1981)	(a)71,(b)63,(c)52,(d)47,(e)36, (f)18,(g)14	*Synechococcus cedrorum*; digitonin. Semicarbazide to FeCN activity = 2000 µmol mg^{-1} Chl h^{-1}. 75% DCMU-sensitive. (b) is slight PS I contamination. (f) and (g) constitute about 65% total protein. Fl. max = 686 nm. 50 chl/b-559. (a) heavily labeled by lactoperoxidase, linker polypeptide to phycobilisomes
2. Guikema and Sherman (1982, 1983)	(a)70,(b)48,(c)45,(d)42,(e)35, (f)18.5,(g)18	*A. nidulans* R2; digitonin. Essentially same preparation as above. Particle enriched in CP VI which contains 48 kDa and 45 kDa proteins. These are main PS II chlorophyll-binding proteins equivalent to (c) and (d) above
3. Koenig and Vernon (1981)	(a)50,(b)48,(c)38,(d)31,(e)27, (f)14,(g)54,(h)35	*A. nidulans* 6301 Miranol. DPC to DPIP activity = 100 µmol mg^{-1} Chl h^{-1}. (a) and (b) are chlorophyll-binding proteins. Contaminated with CF_1, α, β, and γ
C. Photosystem II – consensus composition		
	(a) 70–80	– Phycobilisome linker
	(b) 48	– PS II reaction center Chl-binding protein (CP VIa or CP 2a); Fl. max = 696 nm
	(c) 45	– PS II antennae Chl-binding protein (CP VIb or CP 2c)-Fl. max = 686 nm
	(d) 40–42	–
	(e) 36	–
	(f) 34	– intrinsic membrane protein – O_2 evolution (?)
	(g) 32	– azidoatrazine-binding protein
	(h) 25	– O_2 evolution (?)
	(i) 14	–
	(j) 8	– cyt b-559

II. Photosystem I

References	Protein molecular mass[a], kDa	Comments
A. PS I preparations		
1. Newman and Sherman (1978)	(a)61,(b)18,(c)17.5,(d)16, (e)10	*Synechococcus cedrorum*; digitonin. 40 chlorophylls/P-700; DCPIP to MV = 350 μmol O_2 mg^{-1} Chl h^{-1}; Fl. max = 715 nm
2. Guikema and Sherman (1982)	(a)64,(b)20,(c)15.5,(d)11.5, (e)10	*Anacystis nidulans* R2; digitonin. DCPIP to MV = 380 μmol O_2 mg^{-1} Chl/h^{-1}; (d) and (e) are labeled with ^{59}Fe and may be FeS proteins. (a) is the Chl-binding protein
3. Williams et al. (1983)	(a)70,(b)18,(c)17.5,(d)16, (e)10	*Synechococcus* 6301; Triton X-100; 130 Chl/P-700; FeS centers A, B, and X are present
4. Nechushtai et al. (1983) Muster et al. (1984)	(a)70,(b)50,(c)16,(d)11,(e)10	*Mastigocladus laminosus*; Triton X-100; 50–100 Chl/P-700; active in cytochrome photo-oxidation. Unusually low molecular mass chlorophyll-protein I (a) and associated apoprotein (b) may be due to incomplete denaturation during electrophoresis. (a) and (c) cross-reacted to antibodies prepared against eukaryotic proteins. (Subunits I and II, respectively)
B. PS I Consensus composition		
	(a) 64–70	– PS I Chl-binding protein (present in all high molecular weight aggregates of PS I).
	(b) 18	– FeS center (?)
	(c) 17.5	– FeS center (?)
	(d) 16	– Equivalent to Subunit II in chloroplasts
	(e) 11.5	– FeS center (?)
	(f) 10	– FeS center (?)

Table 2 (continued)

III. F_1 preparations

References	Protein molecular mass[a] kDa	Comments
1. Binder and Bachofen (1979) Bohler et al. (1982)	$\alpha, \beta, \gamma, \varepsilon$; δ isolated separately	*Mastigocladus laminosus*; EDTA and chloroform treatment. 350 kDa molecular mass; reconstitutes photophosphorylation in presence of separately isolated delta subunit. Ca^{+2}-dependent ATPase activity
2. Lubberding et al. (1981)	No subunit data	*Synechococcus lividus*; EDTA and chloroform treatment. Ca^{+2}-dependent, trypsin-activated ATPase activity
3. Hicks and Yocum (1984)	(α)56,(β)54,(γ)36, (δ)21,(ε)15	*Spirulina plantensis*; 320 kDa molecular mass; reconstitutes photophosphorylation

IV. F_1-F_0 preparations

1. Binder et al. (1981)	No subunit data	*Mastigocladus laminosus*; active in ^{32}P-ATP exchange; DCCD and oligomycin sensitive; contaminated with phycocyanin
2. Lubberding et al. (1983)	α,β,γ	*Synechococcus* 6716; Active in ^{32}P-ATP exchange; DCCD and oligomycin sensitive; no phycocyanin. No recognizable subunits from F_0 portion of complex

[a] Proteins detected by Coomassie Blue staining unless otherwise stated

on PS II activity from cyanobacteria. EVANS et al. (1983) have isolated a PS II fraction from *Anacystis* that has a calcium-dependent, oxygen-evolving activity. Absence of calcium from the isolation buffer results in the decreased amounts of three polypeptides in the 30–36 kDa region, thus implying the role of these proteins in calcium binding. In addition, calmodulin antagonists such as chlorpromazine can greatly inhibit O_2 evolution from a PS II-O_2-evolving preparation (PAKRASI and SHERMAN, 1984b).

The only topological studies on the polypeptides of PS II have been done by GUIKEMA and SHERMAN (1982). Radioiodination and trypsin cleavage studies indicated that the 31.5 kDa atrazine binding protein, the 45 kDa reaction center protein and a 42 kDa protein have surface exposed domains.

5 Cytochrome b_6/f Complex

The only functional cytochrome b_6/f complex isolated from a cyanobacterium was obtained from *Anabaena variabilis* by KRINNER et al. (1982). The polypeptide composition of this complex is very similar to that found in higher plants. It contains cytochrome f (31 kDa), cytochrome b_6 (22.5 kDa), the Rieske FeS protein (22 kDa), and a 16.5 kDa protein of unknown function as the major proteins; two minor proteins at 9 kDa are also associated with the complex (HAUSKA et al. 1983; HAUSKA, Chap. 10.1). The stoichiometry of cyt f/cyt b_6/16.5 kDa polypeptides was estimated as 1:2:1, whereas the Rieske FeS protein was present in substoichiometric amounts. As a result, the activity of the complex was reduced compared to intact thylakoids. The Rieske FeS protein could not be detected separately in a SDS-PAGE system; however, it was purified to homogeneity and identified as a 22 kDa polypeptide. Interestingly, a 20.5 kDa polypeptide from *Anacystis nidulans* was labeled with ^{59}Fe; the label was lost in the presence of thiol reagents, indicating that this protein might be an FeS protein and possibly the Rieske center (GUIKEMA and SHERMAN 1982). The mobility of cyt f from both *A. variabilis* and *A. nidulans* decreases on heating the sample prior to electrophoresis (GUIKEMA and SHERMAN 1982; HAUSKA et al. 1983). This phenomenon could also be demonstrated by chloroform/methanol treatment, implying a close association of lipid to cytochrome f (GUIKEMA and SHERMAN 1982). Iodination and trypsin treatment studies indicated that cyt f, cyt b_6, as well as the presumed Rieske FeS protein have residues that are inaccessible to modification (GUIKEMA and SHERMAN 1982).

6 Photosystem I

Preparations competent in PS I-dependent electron transport have been isolated from a number of cyanobacteria (Table 2; for a comparison with PS I complexes of higher plants see WOLLMAN, Chap. 9.7, this Vol.). Most of these PS I preparations have polypeptide compositions which are quite similar (for exceptions, see KLEIN and VERNON 1977, and EVANS and PULLIN 1981). These preparations contain a Chl-binding apoprotein(s) (this component is often observed as a doublet) of about 60–70 kDa (50 kDa in *Mastigocladus*) and four or five low molecular mass subunits (20 to <10 kDa). While the binding of antennae Chl, P-700 (40–100 Chl/P-700), and the primary acceptor A_1 to the high molecular mass subunit has been well established, the function of the small polypeptides is less certain. It has been suggested that the secondary acceptors A_2 and P_{430} reside on two of the smaller components (14 and 10 kDa in *Synechococcus* sp.; TAKAHASHI and KATOH (1982). These results were based on photochemical activities of isolated Chl-protein complexes which were derived from a digitonin PS I preparation. In *A. nidulans* R2, GUIKEMA and SHERMAN (1982) observed ^{59}Fe bound to several nonheme proteins. Two of these components comigrated with subunits of a digitonin PS I preparation and possibly represent bound FeS centers. The topological orientation of the PS I subunits within the membrane has not been well studied; however, radioiodination and trypsin cleavage experiments by GUIKEMA and SHERMAN (1982) indicate that most of the PS I subunits possess surface-exposed regions.

The primary electron donor to PS I in the majority of cyanobacteria studied is the soluble cytochrome c-552 (c-554) (molecular mass 10–12 kDa). Both acidic and basic forms have been isolated from different cyanobacterial species. Plastocyanin (molecular mass 10–12 kDa), which is the primary electron donor to PS I in higher plants and most algae, has been isolated or detected in a few cyanobacteria and has also been isolated in acidic (STEWART and KAETHNER 1983) and basic forms. Some cyanobacteria apparently possess the ability to synthesize both plastocyanin and cytochrome c-552, and regulate the amount of these components based on the nutritional availability of iron and copper. In higher plants, it has been suggested that plastocyanin must bind to a 20 kDa component in the PS I complex to efficiently donate electrons to PS I. All of the cyanobacterial PS I complexes contain a 16–20 kDa component which may perform this function.

Soluble ferredoxin (2Fe−2S, molecular mass 8–12 kDa) is the principle acceptor of electrons from PS I in the cyanobacteria. Some species, however, can replace some of their ferredoxin complement with the FMN-containing electron carrier flavodoxin (molecular mass 19–22 kDa), when placed under iron-limiting conditions. Ferredoxin-NADP reductase (molecular mass about 33 kDa), the terminal electron transfer protein of PS I, has been isolated from a number of cyanobacteria (SERRANO and RIVAS 1983). Most preparations of this enzyme are heterogeneous, containing a number of active components of similar molecular weight, but which differ slightly in amino acid sequence. This heterogeneity has been observed in higher plants, but its functional basis has not been established.

7 F_1 and F_1-F_0

ATPase (F_1) and ATP synthetase (F_1-F_0) preparations have only been obtained from a few cyanobacteria (Table 2, see also MCCARTY and NALIN, Chap.

10.8, this Vol.). These complexes have been much more difficult to isolate than the analogous complexes from higher plants due to the presence of phycobiloproteins on the membrane exterior. Isolated F_1 from cyanobacteria contains five subunits (α, β, γ, δ, ε) of unknown stoichiometry. The apparent molecular mass of the holoenzyme is about 320–350 kDa. The molecular mass of the individual subunits has been determined for *Spirulina* (55.5, 54, 36, 21, and 15 kDa, respectively; HICKS and YOCUM personal communication). These preparations exhibit calcium-dependent, trypsin-activated, ATPase activity. F_1 extracted by chloroform treatment from *Mastigocladus* lacks the δ subunit and does not reconstitute photophosphorylation with F_1-depleted membranes (BINDER and BACHOFEN 1979). However, the δ subunit can be isolated separately, and combined with δ-deficient F_1 to yield reconstitution of photophosphorylation. F_1 isolated by EDTA extraction contains a full complement of polypeptides and reconstitutes photophosphorylation with depleted membranes.

Only two F_1-F_0 preparations have been reported from cyanobacteria (BINDER et al. 1981; LUBBERDING et al. 1983). Both preparations are active in $^{32}P-ATP$ exchange and are sensitive to DCCD and oligomycin. The polypeptide composition of the F_0 portion of the complex was not obtained from either preparation.

8 Discussion

Our suggestions as to the potential consensus protein compositions of PS I and PS II are listed in Table 2. Comparison of this table with the composition of chloroplast photosystems highlights the similarities of cyanobacterial and chloroplast photosynthesis that we discussed earlier. However, these identifications are tentative, and a great deal of work remains to be done in determining the structure and composition of cyanobacterial thylakoids. In particular, much more needs to be done in regard to the functional assignment of many of these proteins. However, recognition that the two membrane systems are so similar should serve to stimulate research in cyanobacteria. As photosynthetic prokaroytes, these organisms have much to offer in genetic and molecular biological studies that will be of great value to elucidation of oxygenic photosynthesis in general.

References

Allen MM, Turnburke AC, Lagace EA, Steinback KE (1983) Plant Physiol 71:1–14
Binder A, Bachofen R (1979) FEBS Lett 104:66–70
Binder A, Bohler MC, Wolf M, Bachofen R (1981) In: Akoygunoglou G (ed) Photosynthesis II. Electron transport and phosphorylation. Balaban Int Sci Serv, Philadelphia, pp 771–778
Bohler MC, Binder A, Bachofen R (1982) FEMS Microbiol Lett 15:115–118

Bowes JM, Stewart AC, Bendall DS (1983) Biochim Biophys Acta 725:210–219
Bricker TM, Sherman LA (1982) FEBS Lett 149:197–202
Curtis SE, Haselkorn R (1984) Plant Mol Biol 3:249–258
Evans EH, Pullin CA (1981) Biochem J 196:489–493
Evans EH, England R, Manwaring J (1983) In: Papageorgiou GC, Packer L (eds) Photosynthetic prokaryotes. Elsevier/North Holland Biomed Press, Amsterdam New York, pp 175–183
Golecki JR, Drews G (1982) In: Carr NG, Whitton BA (eds) The biology of cyanobacteria. Blackwell, Oxford London Edinburgh Melbourne, pp 125–141
Gray MW, Dootlittle WF (1982) Microbiol Rev 46:1–42
Guikema JA, Sherman LA (1981) Biochim Biophys Acta 637:189–201
Guikema JA, Sherman LA (1982) Biochem Biophys Acta 681:440–450
Guikema JA, Sherman LA (1983) Arch Biochem Biophys 220:155–166
Hauska G, Hurt E, Gabellini N, Lockau W (1983) Biochim Biophys Acta 726:97–133
Hicks DB, Yocum CF (1984) In: Sybesma C (ed) Proc 6th Int Congr of Photosynth. Nijhoff/Junk, The Hague, II.6.599–602
Hladik J, Pancoska P, Sofrova D (1982) Biochim Biophys Acta 681:263–272
Klein SM, Vernon LP (1977) Biochim Biophys Acta 459:364–375
Koenig F, Vernon LP (1981) Z Naturforsch 36C:295–304
Krinner M, Hauska H, Hurt E, Lockau W (1982) Biochim Biophys Acta 681:110–117
Lubberding HJ, Offerijns F, Vel WAC, Vries JR de (1981) In: Akoyunoglou G (ed) Photosynthesis II. Electron transport and phosphorylation. Balaban Int Sci Serv, Philadelphia, pp 779–788
Lubberding HJ, Zimmer G, Walraven HS van, Schrickx J, Kraayenhof R (1983) Eur J Biochem 137:95–99
Murata N, Sato N, Omata T, Kuwabara T (1981) Plant Cell Physiol 22:855–866
Muster P, Binder A, Bachofen R (1984) FEBS Lett 166:160–164
Nechushtai R, Muster P, Binder A, Liveanu V, Nelson N (1983) Proc Natl Acad Sci USA 80:1179–1183
Newman PJ, Sherman LA (1978) Biochim Biophys Acta 503:343–361
Okada S, Asada K (1983) Plant Cell Physiol 24:163–172
Pakrasi HB, Sherman LA (1984a) Plant Physiol 74:742–745
Pakrasi HB, Sherman LA (1984b) In: Sybesma C (ed) Proc 6th Int Congr of Photosynth. Nijhoff/Junk, The Hague, I.3.395–398
Rusckowski M, Zilinskas BA (1980) Plant Physiol 65:392–396
Sadewasser D, Sherman LA (1981) Biochem Biophys Acta 640:326–340
Schatz GH, Witt HT (1984) Photobiochem Photobiophys 7:1–14
Serrano A, Rivas J (1983) Anal Biochem 126:109–115
Stewart AC (1980) FEBS Lett 144:67–72
Stewart AC, Bendall DS (1981) Biochem J 194:877–887
Stewart AC, Kaethner TM (1983) Photobiochem Photobiophys 6:67–73
Takahashi Y, Katoh S (1982) Arch Biochem Biophys 219:219–227
Williams RC, Glazer AN, Lundell DJ (1983) Proc Natl Acad Sci USA 80:5923–5926
Yamagishi A, Katoh S (1983) Arch Biochem Biophys 225:836–846

11.5 Surface Electrical Charges and Protein Phosphorylation

J. BARBER

1 Introduction

All biological membranes are composed of positively and negatively charged components. Some of these charges will be exposed at the membrane surface but they generally do not balance each other so that the surface, taken as a whole, will possess a net electrical charge. At physiological pH, this net charge is, as far as I know, always negative for a wide range of biological membranes. Of course, this overall charged state takes no account of the localization and concentration of charges at various places on the surface which will be peculiar to a particular membrane type. It is also important to note that the density and polarity of the surface charge will be sensitive to the pH of the surrounding medium. Therefore, depending on the concentration and pK_a values of the various groups involved, there will be a pH value when the surface as a whole is electroneutral (isoelectronic) and a range of values when it has a positive polarity. All these various characteristics are observed with membranes of higher plant chloroplasts. It is a long time ago since the first study of the surface electrical properties of chloroplasts was conducted by MERCER et al. (1955), but it is only recently that the relevance of this type of information was recognized for understanding the processes of photosynthesis (BARBER et al. 1977; BARBER 1980a, 1982a). Before proceeding with detailed discussions of the chloroplast system, it is important to remember that when a membrane possessing fixed charges on its surface is immersed in an aqueous environment there will be, for the sake of electroneutrality, the formation of a diffuse layer of counterions immediately adjacent to the surface. The thickness of the diffuse layer can be estimated from the Gouy-Chapman theory (BARBER 1980a), which states that the electrical potential (ψ_x) governing the distribution of the counterions falls off exponentially as a function of distance x from its maximum value at the surface (ψ_o) according to the approximate expression (see BARBER, 1980a):

$$\psi_x = \psi_o \exp.(-\kappa x). \tag{1}$$

The surface potential ψ_o is related to the surface charge density (σ) and concentration of salt $\sum_i C_{ib}$, where C_{ib} is the concentration of ion i in the bulk medium. The relationship is expressed by combining the Poisson, Boltzmann and Gauss equations to yield (BARBER 1980a):

$$\psi_o = -\frac{RT}{Z_i F} \log_e \left[\frac{\sigma^2}{2\varepsilon_r \varepsilon_o RT \sum_i C_{ib}} + 1 \right], \tag{2}$$

where ε_r is the relative permittivity of the solution, ε_o is the permittivity of a vacuum, Z_i is charge carried by ion i, F is the Faraday constant and T is the absolute temperature. The thickness of the diffuse layer is often defined as the distance when ψ_x drops to 1/e of ψ_o and is therefore equal to $1/\kappa$.

It is very important to note that the value of ψ_o will vary with changes in the electrolyte concentration and be highly dependent on valency. On the other hand, σ is not sensitive to changes in salt levels unless specific ion binding or charge neutralization occurs. Another important parameter is the space charge density (ρ_x), since it is this quantity which relates directly to electrostatic screening at any point x in the diffuse layer and therefore to the control of coulombic repulsion between adjacent surfaces (BARBER 1980a, 1982a; RUBIN et al. 1981). Although changes in ψ_x and ρ_x often correlate, there are situations with mixed valency electrolytes when they do not, which has important implications in the control of thylakoid membrane organisation (BARBER et al. 1977; RUBIN and BARBER 1980; DUNIEC et al. 1979).

2 Surface Charge Densities

Although it is relatively easy to establish the electrical polarity of the surface of biological membranes, it is far more difficult to precisely determine the charge density. At best only estimates can be obtained which give a base for comparison. There are several reasons for this. Biological membranes are not homogeneous rigid systems with their surface electrical charges "smeared out", as assumed in the Gouy-Chapman theory. Rather, they are highly heterogeneous, being dynamic in structure and having their electrical charges localized on specific protein, carbohydrate and lipid components. Thus, at the microscopic level, the biological membrane surface could not be more different than the ideal surface assumed in the classical electrostatic theories used to calculate surface charge densities. No technique can account for the modification of the "observed" surface charge density by the existence of structural water and fixed ionic layers immediately adjacent to the surface and often termed the Stern or Helmholtz layer (BARBER 1980a). Moreover, it is impossible to allow for the topography of the surface which will not be smooth due to exposed segments of integral complexes and the attachment of bulky extrinsic proteins. Because of these problems no one technique is ideal and the best that can be accomplished is to compare the various approaches.

2.1 Particle Electrophoresis

This is a classical approach which involves using a microscope to measure the velocity of particles suspended in an electrically conducting solution moving under the influence of an external electric field (BIER 1967). The technique measures the electrical potential at the plane of shear, known as the zeta-potential. From this value an estimate of the surface charge density is possible by using Eq. (2) and assuming that the zeta-potential equals ψ_0. Because of this assumption σ will almost certainly be underestimated. The technique has been applied to: (a) isolated thylakoids (MERCER et al. 1955; NOBEL and MEL 1966; NAKATANI et al. 1978; NAKATANI and BARBER 1980; SCHAPENDONK et al. 1980), (b) the envelope surface of intact chloroplasts (NAKATANI et al. 1978), (c) inside-out partition membrane vesicles and right-way-round vesicles (MANSFIELD et al. 1982).

2.2 Differential Doppler Micro Electrophoresis

This technique is similar to that of classical particle electrophoresis but uses the Doppler shift of scattered laser light to measure mobility (PREECE and LUCKMAN 1981). It can, therefore, be used to obtain many measurements on any particular sample in very short times, as well as being able to detect the movement of small particles not normally observable in the classical optical systems. It has recently been used to investigate the surface charge properties of isolated thylakoids (VERNOTTE et al. 1983; TELFER and BARBER 1984), intact chloroplasts (TELFER and BARBER 1984), and various thylakoid membrane fragments (SUNDBY et al. 1984). Being based on electrophoresis this technique will, as before, underestimate the true overall surface charge density because it gives only an indication as to the electrical potential at the hydrodynamic plane of the shear.

2.3 Fluorescence Probe Techniques

The basis of this approach is that a probe molecule changes its fluorescence properties when it is located in the diffuse electrical double layer adjacent to the membrane surface. The probe molecule must also be electrically charged so that its concentration within the diffuse layer is governed by the magnitude of ψ_0. An ideal candidate is 9-aminoacridine, which is, at neutral pH, positively charged and has a high fluorescence yield in free solution, but is nonfluorescent within the diffuse layer (SEARLE and BARBER 1978; CHOW and BARBER 1980a). Other probe molecules used for estimating surface charge density are the sulphonic acid derivatives, 1-anilinonaphthalene-8-sulphonate (GIBRAT et al. 1983) and 2-p-toluidinonaphthalene-6-sulphonate (SEARLE and BARBER 1979). For 9-aminoacridine, quantitative analyses of the fluorescence signals have been carried out with chloroplast membranes using the Gouy-Chapman theory, and suggest that this technique can probe closely to the membrane surface and detect lateral reorganization of charges associated with cation-induced conformational changes (CHOW and BARBER 1980b). The technique has also been used to study surface charge characteristics of inside-out partition membrane vesicles (MANSFIELD et al. 1982) and other thylakoid membrane fragments (YA-

MAMOTO and KE 1981). In all cases the values of σ estimated are greater than those calculated from electrophoretic measurements.

2.4 Indirect Methods

Studies of electron transfer reactions involving charged exogenous donors and acceptors (ITOH 1978; ITOH et al. 1983; YERKES and BABCOCK 1980) and of cation-induced chlorophyll fluorescence changes (BARBER et al. 1977) give indirect estimates of surface charge densities. The former approach, unlike the latter, may give information about charge densities at specific sites on the inner and outer thylakoid membrane surface.

2.5 Summary of Surface Charge Properties

- The net negative electrical charge on the outer surface of the thylakoid membrane seems to be due to the carboxyl groups of glutamic and aspartic acid residues of exposed portions of integral proteins (MERCER et al. 1955; BERG et al. 1984; NAKATANI et al. 1978; NAKATANI and BARBER 1980).
- At pH 4.3 the thylakoid outer surface is electroneutral and below this pH becomes positively charged, probably due partly to the guanidine group of arginine (NAKATANI and BARBER 1980).
- The isoelectric point of the inner thylakoid surface, as determined with inside-out partition vesicles, is at pH 4.1 (ÅKERLÜND et al. 1979)
- There is no evidence that the charges of the acidic acyl lipids of the thylakoids, phosphatidylglycerol, and sulphoquinovosyldiacylglycerol are exposed at the outer membrane surface (GROSS and HESS 1974).
- The net charge on the thylakoid surfaces is not evenly distributed but is probably localized in different regions. The external surface of the nonappressed membranes (those exposed to the stroma) is thought to carry a higher charge density than the surfaces involved in membrane appression within the grana (BARBER 1980a; CHOW and BARBER 1980c; YAMAMOTO and KE 1981).
- The appressed membranes, and thus the vicinity of photosystem II, have an inner surface which is more charged than the outer (YERKES and BABCOCK 1980; MANSFIELD et al. 1982; SUNDBY et al. 1984).
- The outer surface of the chloroplast envelope is more negatively charged than the thylakoid surface (NAKATANI et al. 1978; TELFER and BARBER 1984).
- There are some reports that the charge density on the outer thylakoid surface increases on illumination (NOBEL and MEL 1966; SCHAPENDONK et al. 1980), while others have not detected this effect (NAKATANI and BARBER 1980).
- The average surface charge density may vary depending on growth conditions which alter the relative levels of appressed and nonappressed membranes (DAVIES et al. 1984).

- The values of the charge density for the outer surface of the thylakoid membrane, measured at neutral pH, range from -0.0046 to $-0.034\ C\ m^{-2}$ depending on the technique adopted, on the membrane fragements used and whether the measurement was for a "localised" or "average" value. For the sake of simplicity, an average value for the external surface can be assumed to be $-0.025\ C\ m^{-2}$ (BARBER 1980a).

3 Control of Photosynthetic Processes by Surface Charge

Surface charges on the thylakoid membrane can control a number of different processes, some of which relate to the surface potential (ψ_o) and some of which are controlled by the space charge density adjacent to the membrane surface (ρ_x). The surface potential effects are relatively easy to understand and include:

- Distribution of inorganic ions within the intact chloroplast (BARBER 1980c).
- Concentration of charged reactants at the membrane surface (BARBER 1980a; Itoh et al. 1983).
- The conductance and action of ionophores (BARBER 1980a).
- Transmembrane potentials (RUMBERG 1977; BARBER 1982a).

The ρ_x parameter becomes important when considering the interactions and spatial distances between electrically charged macroscopic surfaces. In this context it is possible to understand how surface electrical charges control the formation of grana stacks and perhaps how higher plants and green algae undergo State 1 – State 2 transitions. When thylakoids are suspended in low salt-containing media they unstack but can be made to restack to form their normal granal and stromal lamellae by the addition of cations. Associated with the restacking is an increase in room temperature chlorophyll fluorescence, a decrease in PS II to PS I energy transfer, and a lateral movement of intramolecular particles as observed by freeze-fracture. I have previously given the general properties of these cation-induced changes and argued how they can be understood in terms of electrostatic screening of surface charges due to changes in ρ_x (BARBER et al. 1977; BARBER 1980a, b, 1982a). The concept presented is that as ρ_x increases, the electrostatic repulsive forces between the adjacent membranes and protein surfaces is decreased so that long-range van der Waals attractive forces are able to dominate. However, although the van der Waals electrodynamic forces between macroscopic systems can be relatively large (PARSEGIAN 1973), their ability to draw two surfaces very close together will be limited by the existence of the fixed surface electrical charges even when there is good electrostatic screening (RUBIN et al. 1981; SCULLEY et al. 1980). To account for the close membrane distance which constitutes the grana partitions (about 40 Å), it is necessary to have a very low surface charge density. This could be achieved in two ways, either the added cations significantly bind to the surface and neutralize the surface charge, or there is a lateral movement of electrical charge away from the region of close appression. The neutralization model has been favoured by DUNIEC et al. (1981) while I have argued the case for the "lateral charge diffusion" model (BARBER 1980a, b, 1982a, b). My arguments have been based on a number of key experimental observations:

- The cation-induced stacking and associated chlorophyll fluorescence changes require a fluid membrane (BARBER et al. 1980).
- According to freeze-fracture studies, after restacking has occurred there is a lateral separation of intrinsic protein complexes within the plane of the membrane which were previously randomized in the unstacked membranes (STAEHELIN 1976).
- The neutralization of the surface charges by protonation or by adding certain polyvalent cations which bind to the surface (e.g., La^{3+}, polylysine) leads to extensive membrane stacking which does not resemble "normal" grana in that the membrane components maintain lateral randomization (SUNDBY et al. 1981; SCOUFFLAIRE et al. 1982) and the associated chlorophyll fluorescence changes differ from those induced by adding "nonbinding" cations (CHOW and BARBER 1980c).

The lateral charge displacement model had two important implications. It led to the idea that PS II and PS I complexes were partitioned into two separate regions; PS II in the appressed membranes of the grana where the net surface charge is low, while PS I was postulated to be localized in the highly charged nonappressed membranes which have their outer surface exposed to the aqueous stromal phase of the chloroplast (BARBER 1980b). Direct evidence for this extreme separation has come from the membrane fractionation studies of ANDERSSON and ANDERSON (1980) and from freeze-fracture investigations (reviewed by STAEHELIN, Chap. 1, this Vol.). The other important implication is that the analyses of the cation-induced effects has given a working hypothesis for describing the mechanism by which the phosphorylation of the light-harvesting chlorophyll a/chlorophyll b protein complex (LHC) can control energy distribution between PS II and PS I as manifested by the State 1–State 2 transitions.

The discovery of the State 1–State 2 phenomenon stems back to the early studies of MURATA (1969), DUYSENS and TALENS (1969) and BONAVENTURA and MYERS (1969). These workers demonstrated that within intact algae a mechanism exists to control the distribution of light energy between PS II and PS I in order to optimize photosynthesis for a particular illumination condition. The process is important when light is limiting and presumably has evolved to cope with changes in light quality and intensity which occur with multilayers of photosynthetic cells whether they be in an algal bloom or in the complex structure of a higher plant canopy (BARBER 1976). It was not until recently, however, that the State 1–State 2 adaptation mechanism was measured in the leaves of higher plants (CHOW et al. 1981; CANAANI et al. 1984; WEBBER et al. 1984). Although for a while it was thought that state transitions may be controlled by cation movements across the thylakoid membrane (MURATA 1971; BARBER 1976) it was the important studies of BENNETT et al. (1980) and HORTON and BLACK (1980) which implicated the phosphorylation of LHC with the regulatory mechanism. Since 1980 there have been many efforts to characterize this link to provide hints as to the molecular processes involved; the main findings are listed below (also see recent reviews by ALLEN 1983; BARBER 1983; BENNETT 1983; HORTON 1983; STAEHELIN and ARNTZEN 1983; STAEHELIN, Chap. 1, this Vol.).

- The phosphorylation of LHC is brought about by a membrane kinase which is activated when the plastoquinone pool is over-reduced (E_h ~0 V), either by excess PS II light or by added reductant (HORTON and BLACK 1980; ALLEN et al. 1981; TELFER and BARBER 1981).
- The phosphorylation occurs at one or two threonyl residues of a 2000 MW segment of the 25 to 30 kDa LHC polypeptides (BENNETT 1979; STEINBACK et al. 1979; Mullet 1983).
- When the kinase is not activated, the dephosphorylation of LHC is brought about by a membrane phosphatase which can be inhibited by fluoride or molybdate (BENNETT 1980).
- Associated with the phosphorylation of LHC is a decline in the maximum level of room temperature chlorophyll fluorescence (e.g., TELFER et al. 1983a) some unstacking of granal membranes (KYLE et al. 1983; TELFER et al. 1984a; BIGGINS 1982) and lateral movement of intramembrane particles as detected by freeze-fracture (KYLE et al. 1983; SIMPSON 1983).
- The phosphorylation-induced chlorophyll fluorescence decrease could be accounted for by an increase in the quantum efficiency of PS I photochemistry at the expense of PS II as measured by low temperature chlorophyll fluorescence yield changes (BENNETT et al. 1980; HORTON and BLACK 1980; KRAUSE and BEHREND 1983; SAITO et al. 1983), flash-induced absorption changes (TELFER et al. 1984b) and steady-state electron transfer (FARCHAUS et al. 1982).
- For the change in quantum efficiencies to occur the membrane must be fluid (HAWORTH 1983; SCOUFFLAIRE et al. 1983).
- Several other PS II polypeptides are phosphorylated (e.g., 9, 33, 35 and 45 kDa), but the kinetics of their dephosphorylation do not match the kinetics of the dephosphorylation of LHC and the associated changes in energy distribution (STEINBACK et al. 1982).
- Kinase activity is also controlled by the ATP/ADP ratio with ADP acting as an inhibitor of the enzyme even when the plastoquinone pool is reduced (BAKER et al. 1982; HORTON and FOYER 1983).
- Recently it has been shown that the kinase activity is affected by ΔpH across the thylakoid membrane being enhanced when this gradient is lowered (FERNYHOUGH et al. 1984).
- The nature of the phosphorylation-induced energy distribution change is dependent on the background level of screening cations; at levels when ρ_x is high (e.g., 5 to 10 mM Mg^{2+}), the main effect seems to be via changes in the absorption cross-section of PS II and PS I while at low cation levels, when ρ_x is reduced (e.g., 1 to 2 mM Mg^{2+}), both intersystem energy transfer (spillover) and absorption cross-section changes occur (HORTON and BLACK 1983; TELFER et al. 1983b, 1984a).

4 An Explanation of the Phosphorylation-Induced Control of Energy Transfer in Terms of Electrostatics

By analogy with the explanations for the cation-induced changes in energy transfer and membrane stacking, it has been relatively straight-forward to postulate a physical mechanism for the control of the State 1–State 2 transitions by LHC phosphorylation (Barber 1980a, 1982a, 1983). In the case of the cation-induced phenomena it was argued that changes in ρ_x bring about changes in coulombic repulsion between membrane surfaces which give rise to reorganisation of protein complexes in the plane of the membrane as depicted in Fig. 1. For the LHC phosphorylation-induced effects the same physical principle holds, except that in this case the cation level is constant but the surface charge density on the LHC is increased (see Fig. 1). As a consequence of this, coulombic repulsion will disturb the attractive interactions between adjacent surfaces and bring about membrane reorganization. It was suggested that phosphorylated LHC or PS II-LHC complexes would migrate to unappressed membranes, where they would be in more intimate contact with PS I (Barber 1982a, b, 1983). This prediction has received convincing experimental support (Andersson et al. 1982; Kyle et al. 1983; Staehelin, Chap. 1, this Vol.). According to the electrostatic theory, the degree and nature of the lateral migration and associated membrane unstacking would be dependent on the background level of screening cations (Barber 1983). This is indeed the case, for example, with 1 or 2 mM Mg^{2+} the phosphorylation of LHC can induce almost total unstacking and intermixing of LHC, PS II and PS I similar to that seen with nonphosphorylated membranes suspended in 0 mM Mg^{2+} (Telfer et al. 1983b, 1984a). At 5 to 10 mM Mg^{2+} the degree of unstacking is only about 20%, based on reduction of area, and under these conditions it seems that the main component which migrates into the appressed region is phosphorylated LHC. This sensitivity of the reorganisation of the complexes after LHC phosphorylation accounts for the dependence of the relative changes in absorption cross-section and spillover on the background cation levels as described above.

How much does phosphorylation change the charge density of the exposed surface of the LHC protein? It is believed that in peas, the phosphorylation occurs at the two threonine residues very close to the N-terminus of the LHC polypeptide (Mullet 1983). Detection and analyses of the nuclear gene which codes for one of the LHC polypeptides has been accomplished and its primary amino acid sequence published (Coruzzi et al. 1983). Recently Tobin et al. (1984) reported the amino acid sequence for LHC of *Lemna*. Both sequences are remarkably similar except for the first 15 amino acids at the N-terminal end. Using hydropathy indexes and the Eisenberg et al. (1983) method, a tertiary folding model has been constructed from the *Lemna* sequence (Tobin et al. 1984). Because of the very high degree of homology between the two sequences, the same folding pattern is applicable for peas. The model predicts three transmembrane segments with helical structures. According to the model, the portions of the LHC polypeptide exposed at the outer surface contain both acidic and basic amino acid residues with an overall balance of eight and six net negative

Poor electrostatic screening leads to unstacking and randomisation of pigment-proteins due to increased coulombic repulsion.

Increased electrostatic screening lowers coulombic repulsion and leads to appressed regions having low surface charge density and non-appressed regions with high surface charge. PS II and LHC are located in the former and PS I in the latter.

Introduction of negative charge on to the LHC surface due to phosphorylation induces partial unstacking and lateral movement of phosphorylated LHC to unappressed regions. The degree of unstacking and randomisation is dependent on the background level of screening cations.

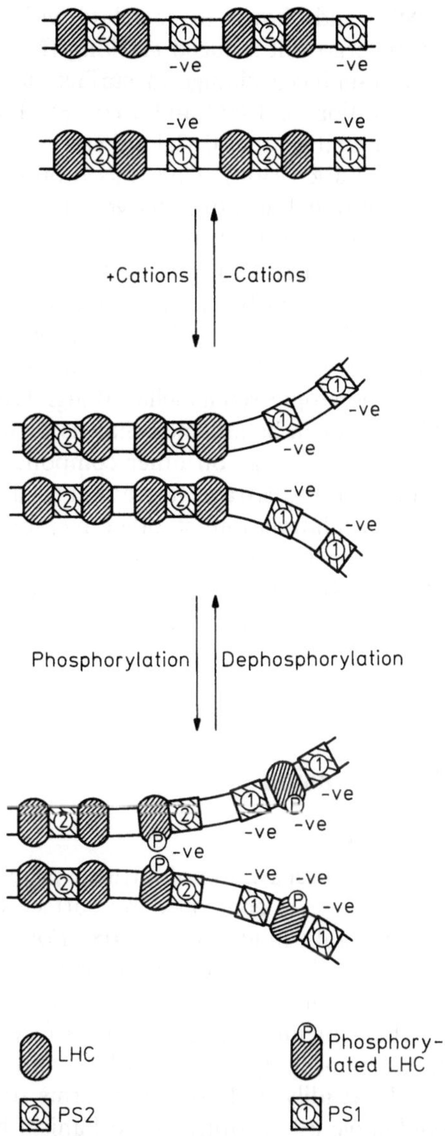

Fig. 1. Scheme showing how surface charge controls spatial relationships between intrinsic complexes of the thylakoid membrane and extent of membrane appression

charges on pea and *Lemna* LHC surfaces respectively. In both cases the trypsin cleavable N-terminal fragment contains a high level of basic amino acids, so that after proteolysis the LHC surface becomes even more negatively charged: eleven net negative charges for pea and ten for *Lemna*. Such an increase in the surface charge density has been detected experimentally (CHOW and BARBER

1980c) and shown to correlate with a loss of ability of isolated thylakoids to undergo salt-induced stacking (MULLET 1983). Therefore it seems that this trypsin-induced change in surface density is sufficient to inhibit salt-induced aggregation of LHC and LHC-PS II within the plane of the membrane and thus inhibit normal thylakoid stacking (acid stacking is still possible; see CHOW and BARBER 1980c). Thus, based on these considerations, it seems feasible that the additional negative charges incorporated onto the LHC surface after phosphorylation of the N-terminal threonines (note that *Lemna* has only one threonine) is sufficient to increase the charge density to a level which would favour its displacement from the stacked to the unstacked region. Increases in surface charge density due to LHC phosphorylation have been reported (KYLE and ARNTZEN 1983; SCOUFFLAIRE et al. 1983). Because phosphorylated LHC remain in the nonappressed lamellae at high levels of screening cations (e.g., 5 to 10 mM Mg^{2+}), it can be assumed that its surface density is as large as, or larger than, the surface charge on other components in this membrane region (e.g., PS I). Moreover, the extra negative charges due to LHC phosphorylation, and perhaps the phosphorylation of other membrane proteins, are sufficient to shift the cation concentration necessary for grana formation to a higher value (TELFER et al. 1984a), reminiscent of the same effect observed when artificially incorporating linolenic acid (VERNOTTE et al. 1983).

5 Conclusion

The lateral separation of intrinsic protein complexes within the thylakoid membrane plane is controlled by surface charge distribution. As this membrane has a highly fluid lipid matrix (FORD and BARBER 1983) any disturbance of the coulombic interactions between the charged surfaces of the complexes can lead to their lateral reorganisation. Such disturbances can be induced either by changing ionic conditions or surface charge characteristics. The phosphorylation/dephosphorylation of LHC controls the State 1–State 2 transitions and can be readily understood in terms of changes in surface charge density and coulombic interactions. A mechanism by which surface phosphorylation alters the spatial relationship between protein complexes by changing coulombic interactions could have implications for other membrane systems (BARBER 1982b). Also, the purpose of phosphorylating other thylakoid proteins, notably those of PS II, could be to regulate the conformation, and therefore the functional activity, of particular protein complexes (e.g., HORTON and LEE 1983).

Acknowledgements. I wish to thank the Science and Engineering Research Council and Agricultural and Food Research Council for financial support and also acknowledge my various personal colleagues who, over the past years, have been involved in the evolution of the concepts presented in this article, especially JOHN MILLS, ALISON TELFER, FRED CHOW, BARRY RUBIN, MICHAEL HODGES, DAVID CHAPMAN and PAUL MILLNER.

References

Åkerlünd H-E, Andersson B, Persson A, Albertsson PA (1979) Biochim Biophys Acta 552:238–346
Allen JF (1983) Trends Biochem Sci 8:369–373
Allen JF, Bennett J, Steinback KE, Arntzen CJ (1981) Nature 291:21–25
Andersson B, Anderson JM (1980) Biochim Biophys Acta 593:427–440
Andersson B, Åkerlünd H-E, Jergil B, Larsson C (1982) FEBS Lett 149:181–185
Baker NR, Markwell JP, Thornber JP (1982) Photobiochem Photobiophys 4:211–217
Barber J (1976) In: Barber J (ed) Intact chloroplast, vol II. Topics in photosynthesis. Elsevier, Amsterdam New York Oxford, pp 283–304
Barber J (1980a) Biochim Biophys Acta 594:253–308
Barber J (1980b) FEBS Lett 118:1–10
Barber J (1980c) In: Spanswick RM, Lucas WJ, Dainty J (eds) Plant membrane transport: Current conceptual issues. Elsevier, Amsterdam New York Oxford, pp 83–94
Barber J (1982a) Annu Rev Plant Physiol 33:261–295
Barber J (1982b) Bio Science Rep 2:1–13
Barber J (1983) Photobiochem Photobiophys 5:181–190
Barber J, Mills J, Love A (1977) FEBS Lett 74:174–181
Barber J, Chow WS, Scoufflaire C, Lannoye R (1980) Biochim Biophys Acta 591:92–103
Bennett J (1979) Chloroplast phosphoproteins. Eur J Biochem 99:133–137
Bennett J (1980) Chloroplast phosphoproteins. Eur J Biochem 104:85–89
Bennett J (1983) Biochem J 212:1–13
Bennett J, Steinback KE, Arntzen CJ (1980) Proc Natl Acad Sci USA 77:5253–5257
Berg S, Dodge S, Krogmann DW, Dilley RA (1974) Plant Physiol 53:619–627
Bier M (1967) Electrophoresis: theory, method and application, 2nd edn. Academic Press, London New York
Biggins J (1982) Biochim Biophys Acta 679:479–482
Bonaventura C, Myers J (1969) Biochim Biophys Acta 189:366 383
Canaani O, Barber J, Malkin S (1984) Proc Natl Acad Sci USA 81:1614–1618
Chow WS, Barber J (1980a) J Biochem Biophys Methods 3:173–185
Chow WS, Barber J (1980b) Biochim Biophys Acta 509:346–352
Chow WS, Barber J (1980c) Biochim Biophys 593:149–157
Chow WS, Telfer A, Chapman D, Barber J (1981) Biochim Biophys Acta 638:60–68
Coruzzi G, Broglie R, Cashmore A, Chua N-H (1983) J Biol Chem 258:1399–1402
Davies EC, Chow WS, Jordon BR (1986) Photosyn. Res. in press
Duniec JT, Sculley MJ, Thorne SW (1979) J Theor Biol 79:473–484
Duniec JT, Israelachvili JN, Ninham BW, Pashley RM, Thorn SW (1981) FEBS Lett 129:193–196
Duysens LMN, Talens A (1969) In: Metzner H (ed) Progress in photosynthesis research. Univ Tübingen Press, Tübingen, pp 1073–1081
Eisenberg D, Weiss RM, Terwilliger TC (1983) Nature 299:371–374
Farchaus JW, Widger WR, Cramer WA, Dilley RA (1982) Arch Biochem Biophys 217:362–367
Fernyhough P, Foyer CH, Horton P (1984) FEBS Lett 116:133–138
Ford RC, Barber J (1983) Biochim Biophys Acta 722:341–348
Gibrat R, Romieu C, Grignon C (1983) Biochim Biophys Acta 736:196–202
Gross EL, Hess SC (1974) Biochim Biophys Acta 339:334–346
Haworth P (1983) Arch Biochem Biophys 226:145–154
Horton P (1983) FEBS Lett 152:47–52
Horton P, Black MT (1980) FEBS Lett 119:141–144
Horton P, Black MT (1983) Biochim Biophys Acta 772:214–218
Horton P, Foyer C (1983) Biochem J 210:517–521
Horton P, Lee P (1983) FEBS Lett 162:81–84
Itoh S (1978) Plant Cell Physiol 19:149–166

Itoh S, Tamura N, Hashimoto K, Nishimura M (1983) In: Inoue Y, Crofts AR, Govindjee, Murata N, Renger G, Satoh K (eds) The oxygen evolving system of photosynthesis. Academic Press, London New York, pp 421–430
Krause GH, Behrend U (1983) Biochim Biophys Acta 723:176–181
Kyle DJ, Arntzen CJ (1983) Photobiochem Photobiophys 5:11–16
Kyle DJ, Staehelin LA, Arntzen CJ (1983) Arch Biochem Biophys 222:527–541
Mansfield RW, Nakatani HY, Barber J, Mauro S, Lannoye R (1982) FEBS Lett 137:133–136
Mercer FV, Hodge AJ, Hope AB, McLean JD (1955) Aust J Biol Sci 8:1–18
Mullet JE (1983) J Biol Chem 258:9941–9948
Murata N (1969) Biochim Biophys Acta 172:242–251
Murata N (1971) Biochim Biophys Acta 226:422–432
Nakatani HY, Barber J (1980) Biochim Biophys Acta 591:82–91
Nakatani HY, Barber J, Forrester JA (1978) Biochim Biophys Acta 504:215–225
Nobel PS, Mel HC (1966) Arch Biochem Biophys 113:695–702
Parsegian VA (1973) Annu Rev Biophys Bioeng 2:221–255
Preece AW, Luckman NP (1981) In: Preece AW, Light PA (eds) Cell electrophoresis in cancer and other clinical research. Elsevier/North-Holland Biomed Press, Amsterdam New York, pp 295–298
Rubin BT, Barber J (1980) Biochim Biophys Acta 592:87–102
Rubin BT, Chow WS, Barber J (1981) Biochim Biophys Acta 634:174–190
Rumberg B (1977) In: Trebst A, Avron M (eds) Photosynthesis I. Encyclopedia of plant physiology, New Ser, vol V. Springer, Berlin Heidelberg New York, pp 405–415
Saito K, Williams WP, Allen JF, Bennett J (1983) Biochim Biophys Acta 724:94–103
Schapendonk HCM, Hemricka-Wagner AM, Theuvenet APR, Wong Fong Sang HW, Vredenberg WJ, Kraayenhof R (1980) Biochemistry 19:1922–1927
Scoufflaire C, Lannoye R, Barber J (1982) Photobiochem Photobiophys 4:249–256
Scoufflaire C, Martens E, Lannoye R, Barber J (1983) Photosynth Res 4:191–202
Sculley MJ, Duneic JT, Thorne SW, Chow WS, Boardman NK (1980) Arch Biochem Biophys 201:339–346
Searle GFW, Barber J (1978) Biochim Biophys Acta 502:309–320
Searle GFW, Barber J (1979) Biochim Biophys Acta 545:508–518
Simpson DJ (1983) Biochim Biophys Acta 725:113–120
Staehelin LA (1976) J Cell Biol 71:136–158
Staehelin LA, Arntzen CJ (1983) J Cell Biol 97:1327–1337
Steinback KE, Burke JJ, Arntzen CJ (1979) Arch Biochem Biophys 195:546–557
Steinback KE, Bose S, Kyle DJ (1982) Arch Biochem Biophys 216:356–361
Sundby C, Andersson B, Albertsson P-A (1981) In: Akoyunoglou G (ed) Proc 5th Int Congr Photosynth, vol III. Balaban Int Sci Serv, Philadelphia, pp 67–76
Sundby C, Mansfield RW, Andersson B, Barber J (unpublished)
Telfer A, Barber J (1981) FEBS Lett 129:161–165
Telfer A, Barber J (unpublished)
Telfer A, Allen JF, Barber J, Bennett J (1983a) Biochim Biophys Acta 722:176–181
Telfer A, Hodges M, Barber J (1983b) Biochim Biophys Acta 724:167–175
Telfer A, Hodges M, Millner PA, Barber J (1984a) Biochim Biophys Acta 766:554–562
Telfer A, Bottin H, Barber J, Mathis P (1984b) Biochim Biophys Acta 764:324–330
Tobin EM, Wimpee CF, Silverthorne J, Stiekema WJ, Nenmann GA, Thornber JP (1984) In: Thornber JP, Staehelin LA, Hallick R (eds) Biosynthesis of the photosynthetic apparatus. Liss, New York, pp 325–334
Vernotte C, Solis C, Moya I, Maison B, Briantais JM, Arrio B, Johannin G (1983) Biochim Biophys Acta 725:376–384
Webber AN, Baker NR, Platt-Aloia K, Thomson WW (1984) Physiol Plant 60:171–179
Yamamoto Y, Ke B (1981) Biochim Biophys Acta 636:175–184
Yerkes CT, Babcock GT (1980) Biochim Biophys Acta 590:360–372

11.6 Freeze-Fracture Studies of Mutant Barley Chloroplast Membranes

D.J. SIMPSON

1 Introduction

Detailed knowledge of the supramolecular organization of chloroplast photosynthetic membranes (thylakoids) is essential for an understanding of the mechanisms by which light energy is trapped and converted into chemical energy (see STAEHELIN, Chap. 1; ORT, Chap. 4, this Vol.). One means of gaining such information is to analyze the ultrastructure of thylakoids from chloroplasts of mutants lacking one or more polypeptide components, using the technique of freeze-fracture electron microscopy. A comparison of the structure, composition, and physical properties of such mutant thylakoids with those from wild type, can be used to deduce the location and function of individual or a group of polypeptides in the thylakoid membrane.

When membranes are cooled to low temperatures (about −110 °C), there is a decrease in the free energy that stabilizes the hydrophobic interactions between the two halves of the lipid bilayer. Cleavage of a suspension of such frozen membranes (i.e., freeze-fracturing) tends to split the membrane along the internal plane between the lipid bilayer (see Fig. 22 of STAEHELIN, Chap. 1, this Vol.). This produces two semi-complementary fracture faces (designated EF for endoplasmic and PF for protoplasmic) which are visualized in the electron microscope using a thin platinum replica of the frozen surface. The thylakoid is usually differentiated into stacked (s) and unstacked (u) regions, known as grana and stroma lamellae. The differences between these regions can be recognized on both EF and PF fracture faces, from morphological details, and also differences in freeze-fracture particle density and size distribution. There are therefore four fracture faces, designated EFs, EFu, PFs, and PFu (see Fig. 22 of STAEHELIN, Chap. 1, this Vol.). The smooth fracture plane of the lipid is interrupted by pits and protruberances, the latter being the freeze-fracture particles. These particles are the morphological equivalents of the functional subunits of the membranes, and it is their composition which is of special interest.

Barley (*Hordeum vulgare* L.) is one of the few higher plant species from which mutants affecting the plastids have been selected, maintained, and genetically analyzed. Seeds of these single nuclear gene mutants have been made available by Professor von Wettstein of the Carlsberg Laboratory. Most of the mutants are seedling lethals, and are maintained as heterozygotic stocks. Thylakoids are isolated from 7-day-old homozygous seedlings which can be distinguished from wild type on the basis of their paler pigmentation, or higher fluorescence under UV light.

Although this article focuses on barley plastid mutants, it should be stressed that there have been investigations into the freeze-fracture ultrastructure of mutant thylakoids from other species, particularly maize (MILLER and CUSHMAN 1979; LETO et al. 1982), tobacco (MILLER 1980), pea (POPOV et al. 1981), *Chlamydomonas* (GOODENOUGH and STAEHELIN 1971; OLIVE et al. 1983; WOLLMAN et al. 1980; LADYGIN et al. 1980) and *Chlorella* (LACAMBRA et al. 1984). In all studies where thylakoid ultrastructure has been well characterized with respect to particle size and numbers per square micron, the conclusions have been consistent with respect to the location of the light-harvesting complex, and the reaction centres of photosystem I and II in PFs, large PFu and EFs particles respectively.

2 Location of Chlorophyll a/b Proteins

Each of the four thylakoid freeze-fracture faces can be characterized on the basis of the number of particles per square micron, and the size distribution of such particles. Objective measurements of these particles, pioneered by GOODENOUGH and STAEHELIN (1971), are valuable in characterizing mutant freeze-fracture faces and comparing them with those of wild type. Until relatively recently, unidirectional shadowing was used to produce replicas, which enables one to measure particle width. Unless each particle has a circular shape, the width measured from such replicas is a function of the orientation of the particle with respect to the direction of shadowing. Rotation of the specimen during platinum shadowing provides additional information about particle shape and uniformity, as well as improving the accuracy and reliability of particle density measurements (SIMPSON 1979, 1980). It has, for example, become clear that many particles have elliptical cross-sections at the fracture plane (Table 2). Values for particle size are a reliable basis for comparison between mutants,

Fig. 1. Rotary shadowed freeze-fracture electron micrograph of the *PFs* and *EFs* face of grana from wild type thylakoids

Fig. 2. The same fracture faces are shown for the mutant chlorina-f2 which lacks chlorophyll a/b-containing proteins. Most of the particles on the *PFs* face have disappeared and the *EFs* particles are slightly smaller than wild type

Figs. 3 and 4. The EFs particles protrude into the thylakoid lumen and are seen on the ESs surface as tetramers, sometimes in regular arrays. In spite of the absence of the LHC II, the chlorina-f2 ESs particles (Fig. 4) still have a tetrameric structure, but are slightly smaller than wild type (Fig. 3)

Fig. 5. The photosystem II mutant viridis-m^{29} has a normal *PFs* face, but lacks many of the *EFs* particles, which are the site of the photosystem II reaction centers

Fig. 6. The mutant viridis-zd^{69} totally lacks photosystem II activity and has only 5% of the EFs particles found in wild type. Figs. 1–6 magnification ×150,000

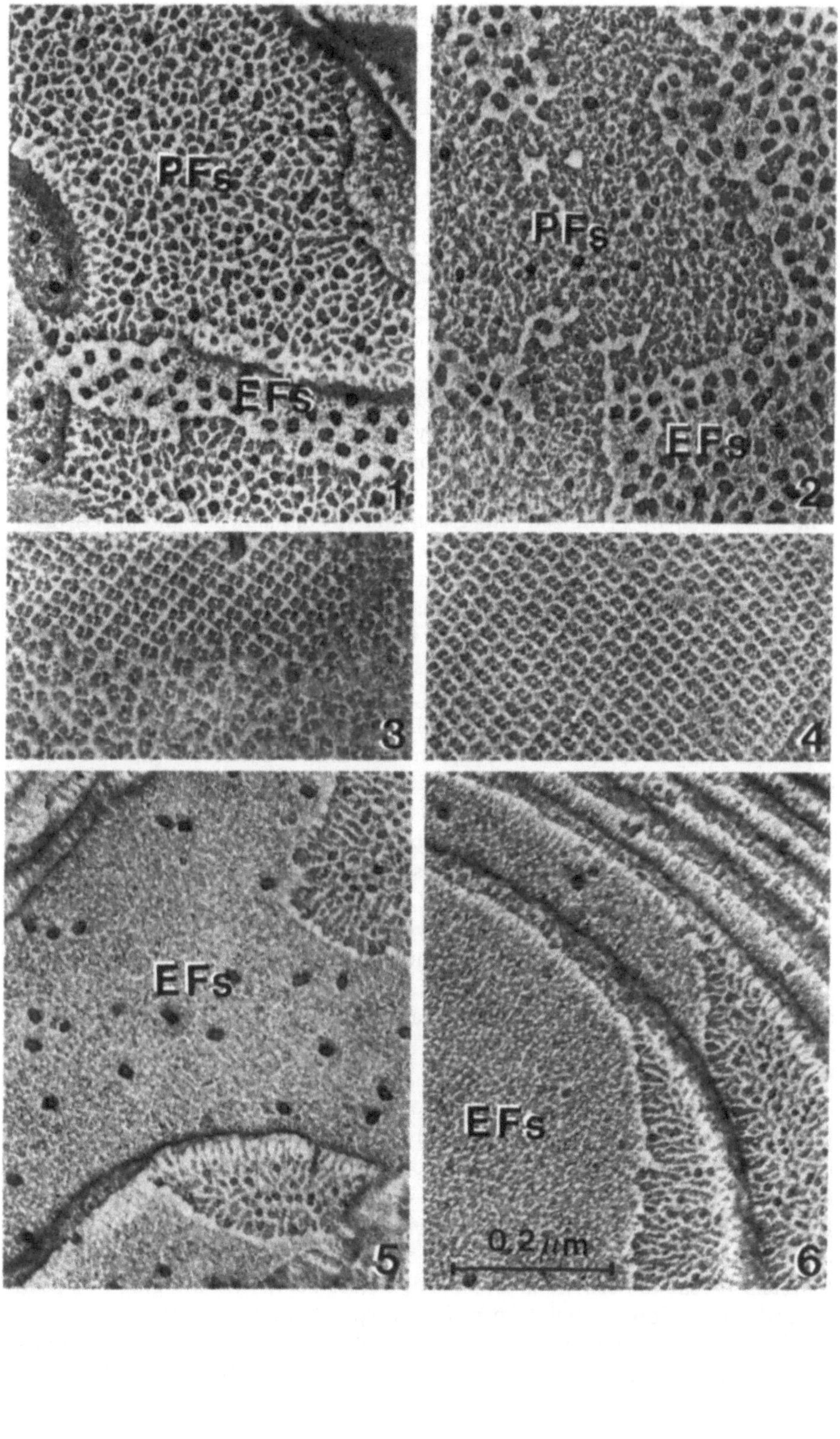
PFs
EFs
1
PFs
EFs
2
3
4
EFs
5
EFs
0.2 μm
6

Table 1. Numbers of freeze-fracture particles per square micron for some single nuclear gene mutants of barley

Mutant	EFu	PFu	EFs	PFs
Wild type	347	4553	1624	6257
chlorina-f2	491	4092	2004	1124
viridis-n^{34}	200	4331	1455	6219
viridis-zb^{63}	596	4591	1518	6276
viridis-m^{29}	282	4366	443	6523
viridis-zd^{69}	380	4201	66	6047

but since they are affected to unknown extents by plastic deformation and the thickness of the platinum layer, they should not be taken as absolute.

The single nuclear gene mutant chlorina-f2 lacks chlorophyll b and all chlorophyll a/b-proteins, but is otherwise competent in photosynthetic function. The thylakoids are organized into grana and stroma lamellae and appear near-normal by thin-section electron microscopy. Comparison of the freeze-fracture ultrastructure with wild-type thylakoids revealed major differences in the grana regions (Figs. 1, 2). The number of particles per square micron on the PFs face was reduced from 6257 in the wild type to 1124 in the f2 mutant (Table 1). It was concluded that most of the PFs particles contain the chlorophyll a/b-protein 2, also called the light-harvesting complex II (LHC II) or the chlorophyll a/b LHC (STAEHELIN, Chap. 1, this Vol.), since it is the major protein of wild-type barley thylakoids, and is absent in chlorina-f2 thylakoids (SIMPSON 1979, 1980).

While it is generally accepted that much of the LHC is located in PFs particles, there is some disagreement about its location elsewhere. Its postulated presence in EFs particles is partially based on the large increase in PFs particle density in a photosystem II mutant of *Chlamydomonas*, said to be due to LHC formerly located in EFs particles cleaving with the PFs face as the result of the loss of photosystem II reaction centers (WOLLMAN et al. 1980; LETO et al. 1982). But only small increases in PFs particle density have been observed in photosystem II mutants of barley (Table 1), tobacco (MILLER and CUSHMAN 1979), *Chlorella* (LACAMBRA et al. 1984) and manganese-deficient spinach (SIMPSON and ROBINSON 1984), all of which lack EFs particles to varying degrees. The most important point is that in vivo there is a close functional relationship between the LHC and photosystem II reaction centers, suggesting an inherent close physical association of the two complexes. What is in doubt is the path of the fracture plane between these two complexes, which determines on which face the LHC will be found. Localization of the LHC II in PFs particles instead of EFs particles offers a mechanism for excitation energy transfer between photosystem II reaction centers in EFs particles (see below) via bridging PFs particles, which are close enough to permit this type of energy transfer.

The relatively small decrease in EFs particle size (Table 2) and ESs tetramers (Figs. 3, 4) in the chlorina-f2 mutant, may more likely be attributable to the

Table 2. Freeze-fracture particle sizes for some single nuclear gene mutants of barley (in Å)

Mutant	EFu	PFu	EFs	PFs
Wild type	105 × 126	93 × 112	117 × 155	74 × 93
chlorina-f2	104 × 121	91 × 109	104 × 138	90 × 103
viridis-n^{34}	83 × 96	74 × 92	94 × 127	74 × 93
viridis-zb^{63}	101 × 121	72 × 87	112 × 153	81 × 94
viridis-m^{29}	76 × 88	74 × 92	90 × 120	66 × 78

absence of the less abundant chlorophyll a/b-protein 1 (Machold et al. 1979), rather than the light-harvesting complex II. No significant changes in EFu or PFu particle sizes were found (Table 2), perhaps indicating that chlorophyll a/b proteins are not major components of these particles.

3 Location of the Photosystem I Reaction Center

Viridis-n^{34} contains only 40% of the P-700 chlorophyll a protein 1 (or CP I), which contains the reaction center of photosystem I, and viridis-zb^{63} has no detectable levels of this chlorophyll protein (Møller et al. 1980; Hiller et al. 1980; Machold et al. 1979). A comparison of these two photosystem I mutants with wild-type barley revealed no consistent changes in thin-section ultrastructure, or the numbers of particles per square micron on the four thylakoid fracture faces (Simpson 1982, 1983). The PFu face of wild-type thylakoids contains two distinct populations of particles, two-thirds of which fall into the large (103 × 125 Å) category. Analysis of the mutant PFu faces revealed a loss of the large particles directly proportional to the loss of CP I (Figs. 7, 8). There was a corresponding decrease in the number of pits on the EFu face (Figs. 9, 10), which are therefore likely to be complementary to the large PFu particles. In combination with changes on the ESu (the true inner surface of unstacked thylakoid lamellae) this is putative evidence for a transmembrane structure of this photosystem I-containing particle. There were other, less dramatic changes in particle size on the other fracture faces of viridis-n^{34}, but these may be due to pleiotropic effects in this mutant, since particle sizes on the other faces of viridis-zb^{63} were not significantly different from wild type (Simpson 1983).

The location of the reaction center of photosystem I is thus assigned to the large particles of the PFu face. Because there was no change in PFu particle density resulting from the loss of CPI, it is concluded that the other components of this particle are able to assemble in the absence of CPI and form a somewhat smaller freeze-fracture particle. Grana ultrastructure was not measurably affected by the total loss of photosystem I, and in the light of biochemical studies showing no photosystem I components in grana (Møller and Høj 1983), photosystem I may be exclusively located in the large PFu particles of stroma lamellae.

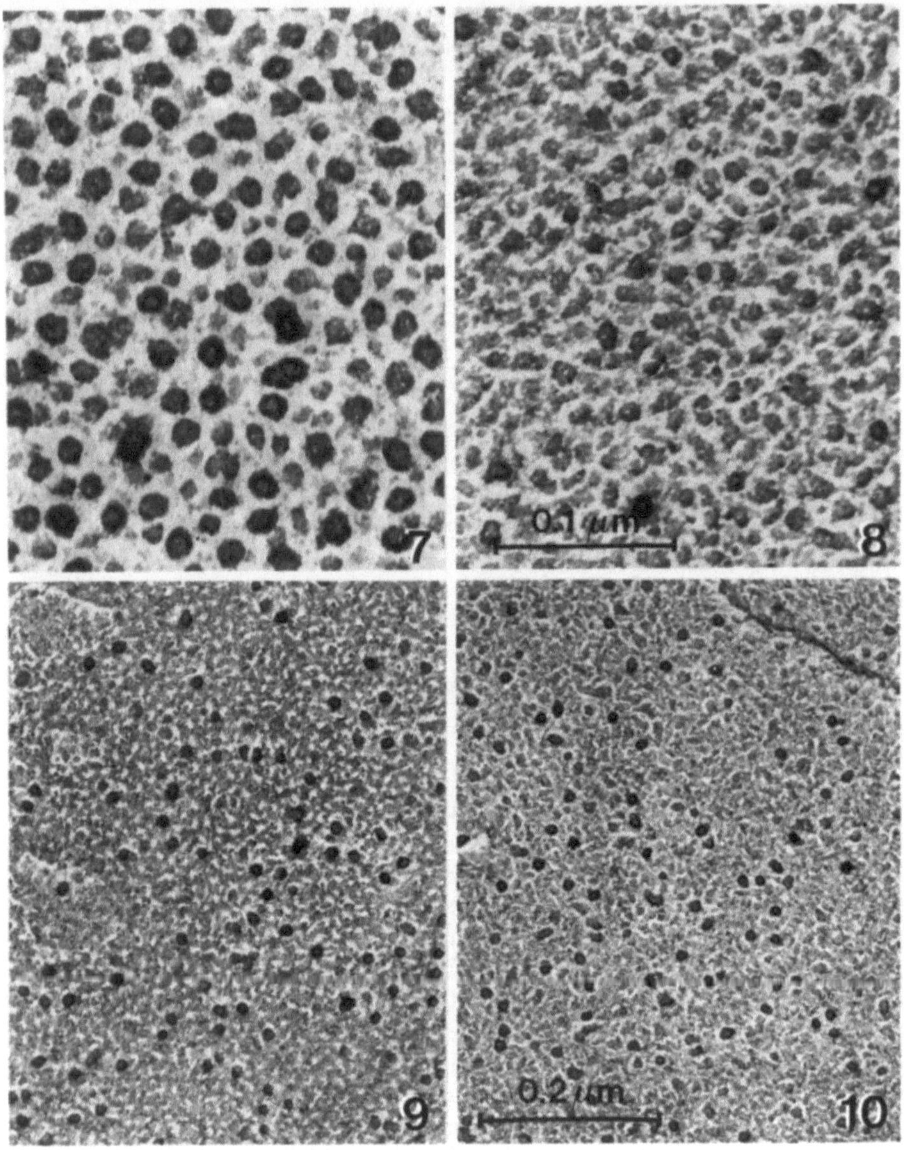

Fig. 7. The PFu face of wild-type thylakoids contains large particles which protrude above the lipid plane and appear dark due to a heavier coating of platinum. Also present are smaller particles which are not as heavily coated

Fig. 8. The photosystem I mutant viridis-zb^{63}, lacks the large particles on the PFu face, but contains many small particles. This indicates that the missing large PFu particles contain the photosystem I reaction center

Fig. 9. The EFu face of wild-type thylakoids is characterized by a low density of particles and a higher density of pits (*white dots*). No mutant has yet been described which lacks EFu particles

Fig. 10. The EFu face of the photosystem I mutant viridis-zb^{63} has EFu particles, but the pits are missing. They probably correspond to the large PFu particles and indicate that they span the membrane. Figs. **7, 8** magnification $\times 300{,}000$. Figs. **9, 10** magnification $\times 150{,}000$

Preliminary observations have been made on the double mutant viridis-zb^{63} × chlorina 2800, the thylakoids of which lack both photosystem I and chlorophyll b. The PFs face appears to be similar to that of chlorina-f2, indicating that the non-LHC particles on the PFs face are not photosystem I.

A *Chlamydomonas* mutant lacking the antenna chlorophyll-protein of photosystem I (CPO), but retaining photosystem I activity, is also deficient in the large PFu particles (OLIVE et al. 1983). This can be interpreted as indicating that the large PFu particles contain both the photosystem I reaction center and antenna chlorophyll protein. In the absence of the latter, the particles are smaller and cannot be distinguished from the other PFu particles. Since CPO is a chlorophyll b-containing protein, it is curious that there is no change in PFu particle size in the chlorophyll b-less mutant of barley (SIMPSON 1979, 1980), possibly indicating that it is stable in the thylakoid even in the absence of chlorophyll b.

4 Location of the Photosystem II Reaction Center

Two photosystem II mutants, viridis-m^{29} and viridis-zd^{69}, were examined. These were shown by partial electron transport activities and fluorescence induction kinetics, to contain 25% and 0% of wild-type activity, and to lack the chlorophyll a-protein 2 (MACHOLD et al. 1979; SIMPSON et al. 1977; SIMPSON and VON WETTSTEIN 1980). Viridis-m^{29} does contain diminished levels of chlorophyll a-protein 3, which may therefore contain the P-680 reaction center of photosystem II. The reduced numbers of particles on the EFs face of these mutants (Figs. 5, 6) reflects their reduced photosystem II activity (Table 1). No other fracture faces show significant changes in particle density compared with wild type, and it is reasonable to conclude that the EFs particles contain the reaction center of photosystem II. Since almost all EFs particles are missing in viridis-zd^{69}, and considering the uniform size of the ESs particles, which are the surface projections of EFs particles (MILLER and CUSHMAN 1979; SIMPSON 1978), it is conceivable that all EFs particles are identical in size, composition, and function, containing a single photosystem II reaction center. The slightly smaller EFs particle size in viridis-m^{29} thylakoids (Table 2) could be related to the absence of the chlorophyll a-protein 2, which may be located in these particles in wild-type grana and function as a photosystem II antenna (SIMPSON 1981).

The unexpected observation that these photosystem II mutants contain near-normal levels of EFu particles raises the question of their function and composition. They have been proposed as the site of photosystem II in stroma lamellae, but no photosystem II barley mutant has yet been found in which EFu particle numbers are reduced. The same anomaly has been observed in manganese-deficient thylakoids of spinach and barley. Limiting levels of this element during growth reduce the amount of functional water-splitting enzyme made. This results in a loss of EFs particles, without affecting the size or density of particles on other fracture faces. Thus the EFu particles are not affected, while EFs particle density is 1/4 of control thylakoids (SIMPSON and ROBINSON 1984). This implies a fundamental difference between EFu and EFs particles, not only in size and location, but in composition and function.

The EFs particle, based on the above results, probably consists of all electron transport components between the Mn-containing polypeptide and the plastoquinone-binding polypeptide, and contains besides the putative P-680-chlorophyll a-protein 3, two antenna chlorophyll proteins, namely chlorophyll a/b-protein 1 and chlorophyll a-protein 2. The tetrameric substructure of this particle, as is found on the ESs surface, is not dependent on the presence of the LHC, since chlorina-f2 has tetrameric ESs particles, but may house components of the water-splitting system.

5 Reconstitution Studies

The mutant work has been complemented by freeze-fracture analysis of functional thylakoid polypeptide complexes isolated by detergents and inserted into artificial liposomes. These investigations have been important, firstly in establishing that these polypeptide complexes form freeze-fracture particles, secondly in providing an independent estimate of the size of these particles, and thirdly in determining the polypeptide composition of such particles. This has been done so far for the light-harvesting complex (SIMPSON 1979), photosystem I (106 Å) (MULLET et al. 1980), coupling factor (CF_0CF_1) (95 Å) and cytochrome f/b_6 (85 Å) (MÖRSCHEL and STAEHELIN 1983). While this technique does not indicate with which face these particles cleave in thylakoids, or whether they are restricted to grana or stroma lamellae, the sizes found for reconstituted polypeptide complexes are consistent with those deduced from mutant analyses.

6 Effects of Single Gene Mutations on Functional Complexes

Although all the barley mutants described above are single gene mutants, the thylakoids lack several components. The polypeptides found in isolated complexes characterized by the function which is deficient in the mutant, are usually missing from the polypeptide pattern of the respective mutants. It is thus postulated that if one polypeptide is not synthesized due to genetic manipulation, its absence prevents the others in the complex from assembling into a structural unit, and the corresponding population of freeze-fracture particles is also missing from the thylakoid. An alternative or additional mechanism causing the loss of particles is suggested from studies of manganese-deficient thylakoids. In this instance, excitons arriving at EFs particles lacking a functional water-splitting system may cause photodestruction and subsequent turnover of the components of the particle, which thus disappears.

7 Concluding Remarks

The analysis of mutant chloroplast membranes has helped make the thylakoid one of the most thoroughly characterized membranes in terms of structure, composition, and function. The apparent structural and compositional complexity may not be as great as indicated by the range of particle sizes and number of different polypeptides. The membrane can be seen as being composed of discrete subunits, each subunit composed of a definite number of polypeptides in stoichiometric quantities, forming a structural and functional complex. Those so far isolated are: photosystem I, photosystem II, light-harvesting chlorophyll, cytochrome f/b_6 and coupling factor. Since each isolated functional complex has a fixed, unique composition, and their size in liposomes corresponds closely with that found in thylakoids, it is suggested that each of the different corresponding populations of freeze-fracture particles, large PFu, EFs, PFs, and small PFu, consists of particles of uniform size and composition. (The experimentally measured range of sizes may be an artifact of the freeze-fracture process. However, it could also reflect a variability in particle composition not hitherto considered, see STAEHELIN, Chap. 1, this Volume, for a somewhat different point of view.) This arrangement has obvious advantages in permitting the plant to alter the composition of its thylakoids in response to differing growth conditions by synthesizing varying amounts of the different complexes.

A higher degree of order is imposed upon the thylakoid by the lateral separation of these complexes into grana and stroma lamellae. Different types of particle cleave with the EF or PF face, depending on their mass distribution with respect to the lipid bilayer, the chemical nature of their components, and possibly their interaction with other thylakoid components. Lateral location may be determined by the strong affinity of one of the components of the LHC to itself (but not the chlorophyll-binding polypeptides, since the chlorophyll b-less mutant of barley has grana). Other PF particles tend to be excluded from regions where the LHC accumulates, due to entropic forces, while the EFs particles are retained in these regions. Mutual attraction of the LHC exposed at the protoplasmic surface (PS), stabilizes membrane appression and the formation of grana. A more detailed discussion of the forces responsible for segregating individual complexes into stacked and unstacked membrane regions can be found in Chap. 11.5, this Volume, of BARBER.

The questions that remain to be answered are:

1. what is the function of EFu particles?
2. what are the PFs particles that are not LHC II?
3. where are the cytochrome f/b_6 particles located?

The precise composition and arrangement of these components with respect to one another within each complex will be the goal of further research on this topic.

References

Goodenough UW, Staehelin LA (1971) J Cell Biol 48:594–619
Hiller RG, Møller BL, Høyer-Hansen G (1980) Carlsberg Res Commun 45:315–328
Lacambra M, Larsen U, Olive J, Bennoun P, Wollman FA (1984) Photobiochem Photobiophys 8:191–205
Ladygin VG, Semenova GA, Tageeva SV (1980) Fisiol Rast 27:91–98 (in Russian)
Leto KL, Keresztes A, Arntzen CJ (1982) Plant Physiol 69:1450–1458
Machold O, Simpson DJ, Møller BL (1979) Carlsberg Res Commun 44:235–254
Miller KR (1980) Biochim Biophys Acta 592:143–152
Miller KR, Cushman RA (1979) Biochim Biophys Acta 546:481–497
Mörschel E, Staehelin LA (1983) J Cell Biol 97:301–310
Møller BL, Høj P (1983) Carlsberg Res Commun 48:161–185
Møller BL, Smillie RM, Høyer-Hansen GM (1980) Carlsberg Res Commun 45:87–99
Mullet JE, Burke JJ, Arntzen CJ (1980) Plant Physiol 65:814–822
Olive J, Wollman FA, Bennoun P, Recouvreur M (1983) Biol Cell 48:81–84
Popov VI, Matorin DN, Gostimsky SV, Tageeva SV, Allakhverdov BL (1981) Planta 151:512–524
Simpson DJ (1978) Carlsberg Res Commun 43:365–389
Simpson DJ (1979) Carlsberg Res Commun 44:305–336
Simpson DJ (1980) Carlsberg Res Commun 45:201–210
Simpson DJ (1981) In: Akoyunoglou G (ed) Photosynthesis. Structure and molecular organisation of the photosynthetic apparatus, vol III. Balaban Int Sci Serv, Philadelphia, pp 107–116
Simpson DJ (1982) Carlsberg Res Commun 47:215–225
Simpson DJ (1983) Eur J Cell Biol 31:305–314
Simpson DJ, Robinson SP (1984) Plant Physiol 74:735–741
Simpson DJ, Wettstein D von (1980) Carlsberg Res Commun 45:283–314
Simpson DJ, Høyer-Hansen G, Chua NH, Wettstein D von (1977) In: Hall DO, Coombs J, Goodwin TW (eds) Photosynthesis 77. Proc 4th Congr Photosynth. Biochem Soc, London, pp 537–548
Wollman FA, Olive JA, Bennoun P, Recouvreur M (1980) J Cell Biol 87:728–735

11.7 Reconstitution of Thylakoid Components into Artificial Membranes

I.J. RYRIE

1 Introduction

Thylakoid membranes are extremely complex and heterogeneous structures. It is therefore not surprising that membrane reconstitution is finding increasing application in this area since the approach allows for the testing of structural and functional parameters of defined lipid and protein components without interference from other ones. Although contemporary work on model membranes has been comprehensively reviewed elsewhere (MONTAL 1976; RACKER 1979; HOKIN 1981; EYTAN 1982) no overview of the thylakoid work has appeared. Accordingly, this article will attempt to review the studies to date and to highlight advantages and problems of the technique as they apply to chloroplast membrane research.

2 Membrane Reconstitution – Overview, Techniques and Some Drawbacks

Our view of biomembranes is now firmly based on the Singer-Nicholson model of extrinsic (hydrophilic) and intrinsic (hydrophobic) proteins intimately associated with a bilayer of diacyl lipids. The reconstitution of such membranes from hydrophobic components frequently involves the mixing of lipid vesicles (liposomes) with the dispersed protein; some perturbation of the bilayer is then induced in order to facilitate protein insertion and the formation of stable hydrophobic interactions within the nonpolar interior of the bilayer. Unless systematic variation of the lipid components is under investigation, it is advantageous to use the natural composition of lipids present in the parent membranes, since lipid type and fatty acyl chain composition can dramatically influence function in some cases. A range of methods are available for the preparation of liposomes (SZOKA and PAPAHADJOPOULOS 1980) most of which yield small unilamellar vesicles (<0.1 µm diameter) provided the lipids are capable of forming bilayers. Large unilamellar liposomes (>1 µm) can be formed either by the ether injection method or by reversed phase evaporation. These approaches are attractive from an experimental point of view, provided incorporation of protein can be achieved without decreasing the vesicle size.

The variety of reconstitution procedures now used have been summarized elsewhere (RACKER 1979; HOKIN 1981; EYTAN 1982) and include detergent dialysis or dilution, sonication, direct incorporation, reversed-phase evaporation and freeze-thawing. In some cases a combination of these may be appropriate. In final analysis, the optimum procedure may depend critically on a range of parameters, including protein stability and the avidity of protein interaction with the bilayer, and may need to be selected empirically.

A poorly understood feature of many reconstituted systems is that high lipid:protein ratios, often greater than 10:1 (w/w), are required for functional activity. Furthermore, even though all of the vesicles may be identical in lipid composition, only a small proportion are able to interact productively with protein. Using cytochrome c oxidase, EYTAN and BROZA (1978) found that only liposomes with a minimal diameter of 22 nm accepted protein, suggesting that one factor which modulates incorporation may be the relief of the high surface energy associated with the acute curvature of small vesicles. Incomplete incorporation of proteins may also occur, but this problem can be overcome by subsequent column or density gradient separation of the components (e.g., RYRIE et al. 1980).

The reconstitution of a binary protein system may require special conditions, especially since vesicles containing a certain protein type may discriminate against a second protein species (EYTAN and RACKER 1977). Freeze-thawing can be a useful method of co-incorporation (RYRIE 1983). Alternatively, a sequential, two-step procedure can be employed (RYRIE et al. 1979; HAUSKA et al. 1980) or liposomes containing the individual proteins can be induced to fuse with each other (MILLER and RACKER 1976).

A fundamental problem in reconstitution work is the frequent inability to achieve an asymmetric, right-side-out orientation of the protein. For activity measurements, however, the unidirectional approach of the substrate can itself impose a vectorial character on the reaction of nonuniformly oriented protein. The problem of orientation is therefore more germaine where physical or structural aspects are under investigation.

3 Reconstitutions with Thylakoid Proteins

3.1 Dicyclohexylcarbodiimide-Sensitive CF_1–F_0 ATP Synthetase Complex

The stage was set for studies of CF_0–F_1 by the isolation of a highly purified complex, containing seven to nine subunits, by PICK and RACKER (1979). A useful preparative innovation was the inclusion of phospholipids throughout the sucrose density gradient, which preserved the activity considerably. The complex displayed uncoupler-sensitive ATP-P_i exchange and acid-base driven ATP formation following reconstitution with lipids, thereby confirming a chemiosmotic prediction that CF_1–F_0 contains the necessary enzymic machinery to transduce the protonmotive force into anhydro-bond energy. HAUSKA et al. (1980) subsequently reconstituted a phenazine methosulfate-dependent cyclic photophosphorylation system by co-incorporating photosystem 1. The rates were extremely low however, principally because many of the vesicles catalysed a useless proton efflux upon illumination due to reversed orientation of the photosystem.

Although soybean phospholipids were employed in the early studies, subsequent investigations of the lipid requirements clearly revealed some beneficial effects of the thylakoid diacyl lipids. Using the purified CF_0 proteolipid subunit SIGRIST-NELSON and AZZI (1980) found that galactolipids were required for formation of a proton channel sensitive to dicyclohexylcarbodiimide. PICK et al. (1984) observed a greater stimulation of both ATPase and ATP-P_i exchange, and changes in V_{max} and K_m values, when the complex was reconstituted with thylakoid lipids instead of soybean phospholipids. Further comparative lipid studies of this kind will undoubtedly prove informative, although we should recognize that a combination of factors may influence the measured activities. For instance, different lipids could variously affect the incorporation or orientation of the protein in the bilayer. Also, activities like ATPase and ATP-P_i exchange are complex and are directly modulated by the transmembrane electrochemical gradient of protons, the build-ups of which may vary depending on the proton permeability of the lipid. Clearly, careful studies are required to allow for a precise interpretation of the results in molecular terms.

Freeze-fracture analysis of reconstituted vesicles revealed that CF_0 gives rise to 96 Å particles in both phospholipid (MULLET et al. 1981) and digalactolipid (MÖRSCHEL and STAEHELIN 1983) bilayers. The dicyclohexylcarbodiimide-binding proteolipid alone forms particles of only slightly less diameter (83 Å), suggesting that 7500 kDa monomer can self-assemble into an oligomeric complex which may be a prerequisite for proton channelling through CF_0. Since CF_1–F_0 is excluded from the stacked regions of thylakoids, it will be interesting to determine whether a cation-induced adhesion between vesicles containing both CF_1–F_0 and LHC II (see below) causes exclusion of the former from the regions of LHC II appression.

3.2 The Light-Harvesting Complex of Photosystem II, LHC II

The now considerable studies with LHC II are probably the best illustration to date of the usefulness of the reconstitution approach for studying structural membrane parameters of a protein complex, including lateral heterogeneity and protein segregation. Following early indications of the essential role of LHC II in cation-induced thylakoid stacking, model LHC II membranes have been studied in several laboratories and found to accurately mimic the appression seen in thylakoids (Fig. 1, A versus B): divalent cations were more effective than monovalent ones, and trypsin treatment, which sharply attenuates adhesion in thylakoids by removal of a 2000-Da fragment from LHC II, was shown to abolish the cation effect (MCDONNEL and STAEHELIN 1980; MULLET and ARNTZEN 1980; RYRIE et al. 1980). These findings, and the fact that the cation-

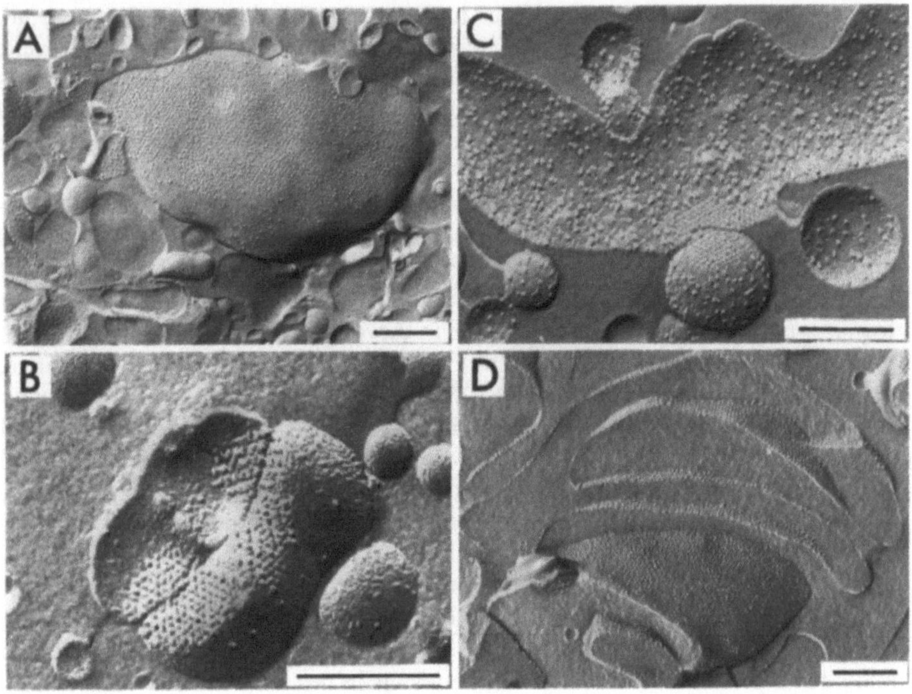

Fig. 1A–D. Model membranes containing chlorophyll-protein complexes from thylakoids. Vesicles were reconstituted using thylakoid diacyl lipids and the light-harvesting complex of photosystem II, LHC II. The membranes in **C** also contain the photosystem I complex. **A** LHC II vesicles at low salt, showing two-dimensional ordering of the particles. **B** Membrane adhesion in LHC II induced by 80 mM KCl. Note the co-alignment of particles on the opposing faces. **C** KCl-induced adhesion in vesicles containing both LHC II and photosystem I. **D** Destacking of appressed LHC II membranes at low salt concentrations. Details are given in RYRIE (1983). The calibration line corresponds to 200 nm

appressed membranes sometimes display precise co-alignment of particles on the two appressed faces (Fig. 1B and RYRIE 1983) is the most convincing evidence that thylakoid stacking is governed by a direct transmembrane interaction between opposing LHC II particles.

Unlike other thylakoid proteins so far examined, LHC II is not dispersed randomly in reconstituted membranes, but occurs as two-dimensional "patches" with quasi-crystalline ordering among the particles (Fig. 1A). These are not normally observed in thylakoids, presumably because much of the LHC II is attached to the core photosystem II complex, though SIEGEL et al. (1981) have observed such arrays following fusion-induced incorporation of exogenous phosphatidylcholine. Interestingly SPRAGUE et al. (1985) found that LHC II, when isolated by a different procedure, remained completely dispersed in digalactosyldiglyceride bilayers, mostly dispersed in phosphatidyl choline bilayers, and was reaggregated into patches by the addition of Mg^{2+}. Clearly, this propensity for self-association of LHC II can be influenced by various parameters including the method of preparation, the lipid environment, and cations. These aspects bear further investigation, especially since some of the LHC II is known to be "free" in the thylakoid membrane and to somehow migrate over large distances following surface phosphorylation during State 1–State 2 transitions (STAEHELIN, Chap. 1, this Vol., STAEHELIN and ARNTZEN 1983).

3.3 Photosystem I

This complex, in varying states of purity, has been reconstituted with lipids and shown to exhibit an electrogenic proton translocation in light (JAYNES et al. 1975; HAUSKA 1980). In a detailed study (ORLICH and HAUSKA 1980) showed that photosystem I proteoliposomes formed by sonication released protons in the presence of reduced phenazine methosulfate, in contrast to thylakoids, which take up protons. However, illumination simultaneously triggered an uncoupler-sensitive quenching of 9-aminoacridine fluorescence, indicating that internal acidification does occur in some vesicles. The reconstituted system is thus a heterogeneous mixture where only a small proportion of the vesicles contain the photosystem in its natural orientation. This illustrates a difficulty of reconstitution work mentioned previously, namely the inability to control the orientation during incorporation, and highlights the need for more detailed work on this problem.

Because of the dramatic effects of cations on the ultrastructure and inter-membrane adhesion in thylakoids and LHC II vesicles (see e.g. BARBER, Chap. 11.5, this Vol.), the effects of Mg^{2+} on photosystem I membranes were also examined. No adhesion occurred between membranes, but some particle clustering and the formation of lipid-rich domains was observed, probably due in part to a Mg^{2+}-induced phase separation of the negatively charged lipids. To establish unambiguously whether a liquid-crystalline to gel-state transition of lipids could cause exclusion of photosystem I particles, a phase separation was induced thermotropically in vesicles containing 40 mol% of dipalmitoylphosphatidylcholine: complete exclusion of particles from the gel-state domains was indeed observed, confirming that the distribution of the protein within the bilayer can be influenced by the fluidity and acyl chain configuration of the lipids (RYRIE 1983).

3.4 Cytochrome b_6/f

The recent advent of a more purified cytochrome b_6/f preparation from thylakoids (HAUSKA, Chap. 10.1, this Vol., HURT and HAUSKA 1981) has encouraged attempts to define its energy transducing functions in model membranes. The plastoquinol-plastocyanin oxidoreductase activity showed a clear stimulation by uncouplers and other ionophores with an additional proton ejection, and membrane potential generation, during electron flow (HAUSKA et al. 1983). The complex thus operates as an electrogenic proton translocator and therefore functionally resembles the cytochrome bc_1 preparations from other sources.

The detailed molecular mechanism of this transport function is clearly of fundamental importance and more information should be gained by further resolution of the complex. At present, the findings are insufficient to distinguish between Q- or b-cycle mechanisms, or other alternatives (see WHITMARSH, Chap. 10.2, JOLIOT and JOLIOT, Chap. 10.3, this Vol.).

As a first step in determining which freeze-fracture particles in thylakoids correspond to cytochrome b_6/f, MÖRSCHEL and STAEHELIN (1983) have used reconstituted systems to examine the ultrastructural effects of antibodies on lateral segregation of the complex. This approach is appealing, since it allows accurate definition of the particle size, and the effects and specificity of the antibodies, before applying the techniques to thylakoids. With an antiserum specific for cytochrome f, the particles became clustered into square or hexagonal arrays, which sometimes dislodged from the bilayer and became free in the medium. This inability of rigid planar assemblies to fit confortably into highly curved lipid vesicles is perhaps not surprising and has also been seen with bacteriorhodopsin (RYRIE et al. 1979).

3.5 Binary Protein Systems

Using low temperature fluorescence techniques LARKUM and ANDERSON (1982) examined the spectral features of reconstituted vesicles containing LHC II and either photosystem I or the core complex of photosystem II. That the spectral characteristics were preserved was shown by the coincidence of the absorption/emission peaks with those assigned to the complexes in situ. Excitation of chlorophyll b, which was assumed to be present only in LHC II, enhanced emission from both the core photosystem II complex (F-685 and F-695) and photosystem I (F-735) but no energy transfer occurred from photosystem II to photosystem I. Although the interpretation of these findings must now take into account the previously unrecognized presence of chlorophyll b in photosystem I, the results nevertheless provide evidence for potential excitation transfer from "free" LHC II units to either of the photosystems. In line with this observation, MURPHY et al. (1984) subsequently demonstrated an LHC II-dependent enhancement of photosystem II activity at subsaturating light intensities in vesicles containing LHC II and the oxygen-evolving complex. This shows that "free" LHC II can donate excitation energy to photosystem II already containing the LHC II assembly, and suggests that transfer can occur between free and bound units, a concept which has important implications for understanding thylakoid function.

In a freeze-fracture study, RYRIE (1983) examined whether the lateral segregation of the LHC II and photosystem I complexes seen in thylakoids in the presence of cations could be reproduced in a model system. Indeed either Mg^{2+} or a higher concentration of monovalent cations caused intervesicle adhesion between regions containing LHC II; photosystem I, with its presumed high surface charge density, was again confined almost exclusively to unappressed regions (Fig. 1C). As in thylakoids, a destacking of confluent LHC II regions occurred upon lowering the salt concentration (Fig. 1D) irrespective of whether photosystem I was present, showing that the inward migration of highly charged complexes from the unappressed regions is not required for this process.

4 Conclusions

Whilst artificial membrane reconstitution is still a relatively new intrusion into photosynthesis research, it seems clear from the foregoing that its further application affords considerable scope for definitive investigation of both the structure and function of thylakoid membrane complexes. An obvious conclusion that can be drawn from previous work is that this approach is uniquely suited for assigning unambiguously a function to a particular membrane protein. Several thylakoid complexes have already been functionally defined in this manner, and an attempt has been made to reconstruct the complete electron transport chain from individual components (LAM and MALKIN 1982). Membrane reconstitution studies also offer the opportunity for detailed investigations of lipid-protein interactions, and of the effects of different lipids on protein function. As discussed above, this latter approach has been already employed in studies of the CF_1–F_0 ATP synthetase complex.

In contrast to other membrane systems, chloroplasts manifest a striking, membrane adhesion-based lateral differentiation both in terms of structure and function. As illustrated here, this marvellous, natural example of membrane heterogeneity can usefully be examined in artificial systems, especially since the cation-dependent stacking phenomena can be reproduced in vesicles containing LHC II. From an ultrastructural standpoint, the molecular size and segregation behavior of individual proteins can be studied by freeze-fracture, possibly in combination with antibody techniques, and will most likely prove crucial in the eventual identification of individual particle size classes seen in the complicated fracture patterns of thylakoids. There thus remains an exciting potential for the future application of membrane reconstitution in photosynthesis research, and providing reasonable care is exercised in the interpretation of results, this approach will undoubtedly assist in our further understanding of basic physicochemical parameters of the photosynthetic apparatus.

References

Eytan GD (1982) Biochim Biophys Acta 694:185–202
Eytan GD, Broza R (1978) FEBS Lett 85:175–178
Eytan GD, Racker E (1977) J Biol Chem 252:3208–3213
Hauska G (1980) FEBS Lett 119:232–234
Hauska G, Samoray D, Orlich G, Nelson N (1980) Eur J Biochem 111:535–543
Hauska G, Hurt E, Gabellini N, Lockau W (1983) Biochim Biophys Acta 726:97–133

Hokin LE (1981) J Membr Biol 60:77–93
Hurt E, Hauska G (1981) Eur J Biochem 117:591–599
Jaynes JM, Vernon LP, Klein SM (1975) Biochim Biophys Acta 408:240–251
Lam E, Malkin R (1982) Proc Natl Acad Sci USA 79:5494–5498
Larkum AWD, Anderson JM (1982) Biochim Biophys Acta 679:410–421
McDonnel A, Staehelin LA (1980) J Cell Biol 84:40–56
Miller C, Racker E (1976) J Membr Biol 26:319–333
Mörschel E, Staehelin LA (1983) J Cell Biol 97:301–310
Montal M (1976) Annu Rev Biophys Bioeng 5:119–170
Mullet JE, Arntzen CJ (1980) Biochim Biophys Acta 589:100–117
Mullet JE, Pick U, Arntzen CJ (1981) Biochim Biophys Acta 642:149–157
Murphy DJ, Crowther D, Woodrow IE (1984) FEBS Lett 165:151–155
Orlich G, Hauska G (1980) Eur J Biochem 111:525–533
Pick U, Racker E (1979) J Biol Chem 254:2793–2799
Pick U, Gounaris K, Admon A, Barber J (1984) Biochim Biophys Acta 765:12–20
Racker E (1979) Methods Enzymol 55:699–711
Ryrie IJ (1983) Eur J Biochem 137:205–213
Ryrie IJ, Critchley C, Tillberg J-E (1979) Arch Biochem Biophys 198:182–194
Ryrie IJ, Anderson JM, Goodchild DJ (1980) Eur J Biochem 107:345–354
Siegel CO, Jordan AE, Miller KR (1981) J Cell Biol 91:113–125
Sigrist-Nelson K, Azzi A (1980) J Biol Chem 255:10638–10643
Sprague SG, Camm EL, Green BR, Staehelin LA (1985) J Cell Biol 100:552–557
Staehelin LA, Arntzen CJ (1983) J Cell Biol 97:1327–1337
Szoka F, Papahadjopoulos (1980) Annu Rev Biophys Bioeng 9:467–508

11.8 Prolamellar Bodies

C. LÜTZ

1 Introduction

Soon after the introduction of electron microscopy to the study of plant cells, botanists became aware of, and interested in, paracrystalline structures that were occasionally observed in plastids. In 1954 HEITZ, as well as LEYON, described crystal-like bodies in certain types of plastid which became known as Heitz-Leyon-crystals. These structures were found mainly in etiolated plants. Their involvement in the formation of thylakoid lamellae was soon recognized. This led to the introduction of the term prolamellar body (PLB). During the following years the three-dimensional arrangement of the tubular elements in PLB's and their conformational changes associated with greening were elucidated. Elaborate studies on PLB architecture were made by W. WEHRMEYER and by B.E.S. GUNNING; these have been reviewed in detail in the books of GUNNING and STEER (1975) and KIRK and TILNEY-BASSETT (1978).

The role of lipids in the formation of PLB's and of thylakoid membranes will be covered in this article. Particular emphasis is placed on work carried out during the last 6–10 years. An investigation of the PLB's proper became possible following the development of techniques for separating their components (LÜTZ 1978; A.R. WELLBURN and HAMPP 1979).

This review can only touch on some of the important and interesting features of PLB's in growing plants. PLB's cannot be understood without considering the effects of environmental conditions on plastids and on plant cells in general. A comprehensive view of etioplast development, diversity and composition is found in KIRK and TILNEY-BASSETT (1978). Formation of PLB's in the dark is associated with the accumulation of protochlorophyll(ide) (PChl[ide]). PLB-dispersal in the light is viewed as a direct consequence of pigment transformation. These relationships have been reviewed by SUNDQVIST et al. (1980) and by VIRGIN (1981). Development, inheritance, and evolution of plastids has been discussed by SCHIFF (1980). Similarly, A.R. WELLBURN (1982) has compared plastid development and PLB diversity in algae and in higher plants. Development of photosynthetic capacity in leaves of angiosperms is the main topic of the reviews of BRADBEER (1981) and of LEECH and BAKER (1983). They include also some aspects of membrane formation as well as lipid and protein biosynthesis.

Numerous articles on the role of lipids in plastid membranes exist (see MURPHY, Chap. 11.11, SIEGENTHALER and RAWYLER, Chap. 11.9, this Vol.), but none concerning the PLB's proper. Only TEVINI (1977) and KIRK and TILNEY-

BASSETT (1978) have reviewed data on etioplast lipids. Recent studies have shown that considerations of the structural role of lipids in the "green" thylakoid membrane are also of great value to understanding of the mechanisms of PLB formation and of the dispersal of PLB tubules during greening. These topics will also be discussed in this article.

2 Formation of Prolamellar Bodies

PLB's and PLB-like membraneous structures can be formed at specific stages of plastid development in all photosynthetic plants. They have been observed throughout the plant kingdom (SCHIFF 1980; A.R. WELLBURN 1982; MARIANI-COLOMBO et al. 1983). PLB's develop in young etioplasts in association with perforated prothylakoid (PT) membranes (Fig. 1, LÜTZ 1981 a). Optimal formation of PLB's with a high degree of order in the tubular lattice occurs under conditions of prolonged growth in the dark. The fact that this usually does not occur in vivo led to the statement of LEECH and BAKER (1983), that "etioplasts are unknown in normally grown plants." However, in plants still covered by earth after germination, or in coiled leaves or buds, well-developed PLB's are formed in plastids shielded from light. During greening, PLB's are transformed into the lamellae of thylakoids (cf. Fig. 2). This process depends on light intensity and light quality: red light in particular can induce thylakoid and grana formation, but simultaneously may allow recrystallization of PLB's. The same response is obtained with weak white light, while in strong light a rapid and complete transformation into membranes occurs. WHATLEY (1977) observed that PLB's are typical of *Phaseolus* leaves growing in the light; therefore concluding that darkness may not be necessary for the development of PLB's. Photosynthetically active chloroplasts are also capable of reforming PLB's, if the plant is transferred to darkness for several hours (ADAMSON et al. 1980). RYBERG and MINKOV (1984) used primary leaves of wheat to isolate reformed PLB's from greening tissues. The three-dimensional structure of these PLB's is very similar to normally formed ones (Figs. 4, 5). Thus a comparison of original and reformed PLB's with respect to lipids, proteins, and pigments is now possible.

The development of PLB's seems to be restricted to cells undergoing differentiation. Thus PLB's can develop as long as the cell is actively synthesizing and forming new thylakoid membranes, i.e., before reaching a steady-state situation. Consequently it is not possible to induce PLB's to reform by redarkening fully developed chloroplasts of older leaves (as in *Spirodela*, PORATH 1979).

The use of mutants in plastid biogenesis, or the application of growth regulators to follow membrane development will not be discussed in this review. However, nearly all such studies have focused on alterations of ultrastructure and changes in pigment and protein composition. Only little information is available on changes in lipid composition. More research on membrane lipids during plastid development is urgently needed.

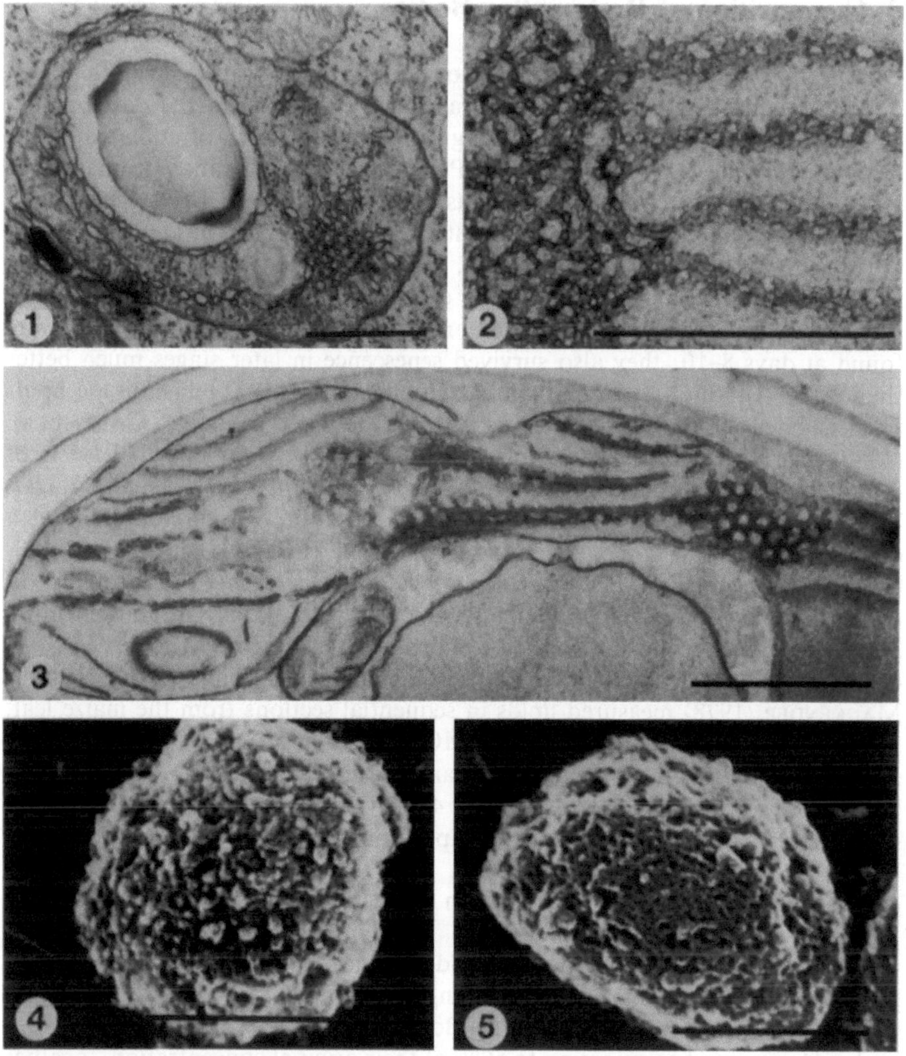

Fig. 1. 3-day-old oat etioplast. The assembly of a PLB in association with perforated sheets has just started in this stage

Fig. 2. 6-day-old oat etioplast after treatment with white light for 1 h (14,000 lx). The PLB is in the process of being transformed, and perforated thylakoids are seen protruding into the stroma. Fixation: glutaraldehyde/OsO_4 (both figures)

Fig. 3. *Scenedesmus* mutant C2A' after growing in 100 dark/light cycles (60 min dark, 2 min light). The perforated membranes resemble those seen in Figs. **1** and **2**. Fixation: 1% $KMnO_4$. (LÜTZ unpublished)

Figs. 4 and 5. Scanning electron micrographs of isolated original (Fig. **4**) and reformed (Fig. **5**) PLB's from wheat plants. (Figs. 4 and 5 kindly provided by Dr. M. RYBERG, Göteborg). Bar = 0.5 μm

3 Composition of Prolamellar Bodies

Data from experiments with etioplasts or etioplast membranes have often been directly related to PLB's proper. Ultrastructural studies on the development and aging of PLB's and of PT's during growth in the dark of etiolated plants have shown that the composition of PLB's differs significantly from that of PT's. In the case of etiolated oat plants, LÜTZ (1981a) and BERGWEILER et al. (1984) have followed the changes in plastid ultrastructure, PChl(ide) and protein during the lifespan of primary leaves. PT development reaches a maximum around 6–7 days after sowing, which coincides with peaks for pigment and membrane protein accumulation. In contrast, the best crystallized PLB's were found at days 8–10; they also survived senescence in later stages much better than PT's. Although a comparable developmental study for membrane lipids from isolated etioplasts is still lacking, some data are available. GALEY et al. (1980) investigated lipids of etioplasts prepared from three different levels of 8-day-old wheat plants which correspond to different developmental stages. With increasing age of the tissue (cells closer to the leaf tip), they found a constant increase in monogalactosyldiacylglycerol (MGDG) and digalactosyldiacylglycerol (DGDG) per protein, with DGDG increasing more rapidly. In contrast, sulfo- and phospholipids did not change markedly in concentration along the leaf. Fatty acid analysis of phosphatidylglycerol (PG) revealed a decrease in 16:0 and an increase in 18:3 moieties with aging of the plastids. MACKENDER (1978) measured lipids in sequential sections from the maize leaf. However, in this system the ratio of MGDG to DGDG rose with increasing age of the etioplasts from 0.7 to 1.2 in parallel with the galactolipid to phospholipid ratio, which changed from 0.5 to 4.5. PG increased from 20 to 41% of total phospholipids. As in the first example, an increase in PLB size and crystallinity accompanied the increase in PG. Lipid composition of fully developed etioplasts is similar to the composition of PLB preparations from 7-day-old oat or wheat plants (see Section 4).

STOBARD et al. (1982) followed lipid changes during senescence in dark-grown barley leaves. The observed changes in MGDG and in DGDG can be regarded as mostly etioplast-specific. Both types of lipid decline from day 7 to day 11 after seed germination to 50% of their original concentration, together with a shift in fatty acids from 18:3 to 18:2. This decline is not as strong as the decomposition during aging observed for PChl(ide) (to 20%) or membrane protein (to 30%) (derived from BERGWEILER et al. 1984). These studies could explain the appearance of "old" etioplasts, in which the stroma is dominated by large PLB's, while most of the PT membranes have already been decomposed (LÜTZ 1981a).

Prothylakoids contain most of the carotenoids found in etioplasts (A.R. WELLBURN and HAMPP 1979; LÜTZ 1981c). This is supported by studies with SAN-9789, which affects carotenoid synthesis (DAHLIN et al. 1983). The herbicide caused structural changes of PT's and plastoglobuli, but not of the PLB's proper. In another study with the herbicide difunon, which blocks up to 99% of carotenoid synthesis during plant growth in darkness, both normal-looking

PLB's and PT's were found (REISS et al. 1983). These results indicate a more functional than structural role for carotenoids in tubular as well as in lamellar systems of plastids.

Several methods have been developed in different laboratories since 1978 to separate PLB's from PT's (LÜTZ 1978; WELLBURN and HAMPP 1979; RYBERG and SUNDQVIST 1982; IKEUCHI and MURAKAMI 1983; LÜTZ et al. 1984). Chemical analysis of highly enriched PLB and PT fractions has revealed that the main components of PLB's are the glycolipids MGDG and DGDG. Smaller amounts of sulfolipid, PG and phosphatidylcholine (PC) are also present (SELSTAM et al. 1982; RYBERG et al. 1983; LÜTZ et al. 1984; LÜTZ and TÖNISSEN 1984; GIROUD and SIEGENTHALER 1984). The ratios of MGDG to DGDG and of PG to PC are higher in PLB's than in PT's. These experiments, performed with preparations from wheat and oat leaves, also demonstrated that 15–25% of the protein is in the PLB's, the rest in PT's; for PChl(ide) the values are 15–30% in PLB's, and the rest in PT's. These results have been incorrectly interpreted by A.R. WELLBURN et al. (1983) and by SURIN and SIRONVAL (1984). The low amount of protein in PLB's of older etiolated plants can be traced mainly to a reduction in the polypeptide of the protochlorophyllide reductase (PCR) (LÜTZ et al. 1981, 1984; RYBERG and SUNDQVIST 1982; IKEUCHI and MURAKAMI 1983). However, most of this protein and of its activity is located in the PT membranes, where the substrate PChlide is also primarily located. PT's rather than PLB's are the site of synthesis for chlorophyll precursors (RYBERG 1983).

The first studies performed on isolated PLB's and PT's used oat plants as source for etioplasts. These plants contain saponins in vacuoles. During their homogenization, the saponins are enzymatically converted to their des-gluco-forms (KESSELMEIER 1982; KESSELMEIER et al. 1984), which have a much higher affinity for PLB's than for PT's or green thylakoids, thereby leading to errors in the glycolipid assays (LÜTZ 1981c). Before this phenomenon was recognized, these saponins were believed to be involved in PLB formation (KESSELMEIER and BUDZIKIEWICZ 1979; LÜTZ 1981b). This problem has recently been overcome by modifications of isolation techniques (LÜTZ and NORDMANN 1983), leading to revised data (LÜTZ et al. 1984). The action of saponins on PLB lipids led to the formation of long, sometimes branched tubules in isolated PLB samples (TÖNISSEN and LÜTZ 1984). Such tubules had previously been observed in broken etioplast preparations, and their formation was used to explain the development of thylakoids from PLB's (A.R. WELLBURN and F.A.M. WELLBURN 1972), and the localization of PCR (MÜLLER and NISIUS 1982).

It should finally be stated that most analyses of PLB's have been made on material from one developmental stage, i.e., of PLB's that display a large size and a high degree of crystallinity. Studies on the dark development of etioplasts have revealed rapid changes in composition of such plastids. Therefore it is possible that PLB's found in younger plants have a composition more comparable to PT's, e.g., that they contain more pigment and protein. In older stages, senescence may also change PLB composition because a break-down of proteins and pigments takes place. This degradation process, which leads to a proportional increase in lipid content, could cause the higher crystallinity of PLB's in later stages. A further assumption also seems reasonable (A.R. Wellburn et al. 1983): some lamellar components, particularly lipids, may become reconverted into PLB material during aging.

4 The Role of Lipids in Prolamellar Body Formation

The synthesis of components necessary for PLB assembly seems to be coded by nuclear genes. Ribosome-free plastids obtained from plants grown at 32 °C can develop normal PLB's, but only rudimentary PT's (Feierabend et al. 1969). At this nonpermissive temperature the synthesis of glycolipids is not blocked (Kosmac and Feierabend 1984).

The structural role of glycolipids in PLB formation has been investigated by Lütz and Tönissen (1984). Isolated PLB-PT preparations were treated with different enzymes and the structural changes studied in the electron microscope. The cleavage of all structural proteins to small peptides with pronase P destroyed PT's, while only inducing a loss in crystallinity of the PLB's. Break-down of phospholipids disorganized the PLB lattice, but phospholipase A_2 was unable to decompose the tubules. A complete disappearance of PLB's and a partial destruction of PT's was obtained with a lipase from *Rhizopus arrhizus*, which primarily affects glycolipids.

The high concentration of MGDG in PLB's suggests a major role for this lipid in the formation of the PLB tubules (Selstam et al. 1982; Ryberg et al. 1983; Tönissen and Lütz 1984). The high ratio of MGDG and DGDG in PLB's (1.8 for wheat, 2.0 for oat), compared to the ratios determined for PT's (1.2 for wheat, 1.7 for oat) favors rounded or tubular lipid assemblies. MGDG has become well known as the best candidate for the formation of nonlamellar complexes (cf. Quinn and Williams 1983; Sprague and Staehelin 1983) due to its cone-shaped molecular configuration. This lipid is thought to be responsible for the curvatures at the margins of the thylakoids and should stabilize all nonplanar regions in membranes (Murphy 1982, Chap. 11.1, this Vol., Murphy and Woodrow 1983). This property of MGDG is in all probability the driving force for the assembly of the three-dimensional, tubular lattice of PLB's. MGDG may also be responsible for the numerous fenestrae in the thylakoids of young etioplasts, since the membranes also serve as assembly sites for PLB's (cf. Figs. 1, 3). Enrichment of PLB's with this glycolipid during dark development is paralleled by an increase in their size and crystallinity (Mackender 1978).

At present there is no evidence to support an essential role for phospholipids in the formation of the PLB structure. Phospholipids comprise around 20% of the total PLB lipids, primarily PG and PC. Phosphatidylethanolamine is absent. It is not clear why PG seems to be enriched in PLB preparations (ratio PG/PC: 2.5 in PLB's, 1.5 in PT's, Giroud and Siegenthaler 1984). Phospholipids may merely support the formation of PLB tubules.

It should be mentioned here that MGDG as well as certain phospholipids, can easily form inverted micells with structural similarities to PLB's (Murphy,

Chap. 11.11, this Vol.; SIEGEL 1984; VERKLEIJ 1984). Mixtures of PC and of cardiolipin (molar ratio 1:3) plus Ca^{2+} form bilayers, which are surprisingly similar in structure to the perforated sheets observed in greening plastids (BOROVJAGIN et al. 1982). The role of fatty acid composition of individual lipids in the formation of PLB's has not yet been investigated. No significant differences in fatty acid composition of PLB and PT glycolipids have been detected (RYBERG et al. 1983).

The native PLB is a complex macromolecular system composed of lipids and several proteins. A possible involvement of the predominant PLB protein PCR in the PLB structure has been evaluated by TÖNISSEN and LÜTZ (1984). The positive charges of this basic (pI 8.1–8.5) protein could neutralize negative charges of the lipids, thereby promoting the formation of inverted micelles. A reversible neutralization of positive charges could be responsible for changes in the degree of order of PLB's such as has been reported for isolated bean PLB's (SURIN and SIRONVAL 1984). These authors have found that incubation of PLB's with protein cross-linking agents for several hours leads to their disorganization and vesiculation. The most likely explanation is that the cross-linking blocks free amino groups of the basic PCR protein, thus removing the positive charges. A similar neutralization of positive charges could form the basis for converting the three-dimensionally organized PLB lipids into lamellar structures in vivo.

5 Prolamellar Bodies and Membrane Formation–An Outlook

The formation of PLB's in differentiating tissues raises many questions pertaining to pigment synthesis, the effects of environmental parameters, and the relationship to plastid development and membrane synthesis. Especially the latter aspect, i.e., the relationship between the synthesis of lipids and the assembly of membranes and PLB's is of particular interest. (In this context the reader may also want to consult Chap. 11.10, this Vol., of KEEGSTRA, in which both mechanisms and sites of thylakoid lipid synthesis are discussed.) Whenever a growing membrane in a plastid forms perforated lamellae, PLB-like bodies or tubular aggregates, it can be interpreted as resulting from a local enrichment in membrane lipids. This might be due to a reduction in the synthesis of membrane proteins, or due to a reduction of compounds that are normally complexed with lipids, like the pigments found in thylakoids. It further indicates the presence of a substantial amount of hexagonal II phase-forming molecules. F.A.M. WELLBURN et a. (1980) have also described an algal chloroplast mutant capable of forming an etioplast-like organelle in the dark: In contrast to wild-type cells of the green alga *Scenedesmus*, that develop normal chloroplasts when grown in the dark, the yellow mutant C2A' produces etioplast-like organelles that exhibit the same greening pattern as higher plants. Furthermore, when grown under 60 min, dark – 2 min light cycles the mutant develops PLB-like perforated sheets (Fig. 3). A chemical analysis of the membrane components of this mutant has not yet been reported.

Many problems remain to be investigated. Major progress in the field of membrane biogenesis will require a multidisciplinary approach to integrate information on lipid synthesis, the biophysics of macromolecules, cytological studies of mutants, and the effects of different hormones on membrane structure.

One of the central questions is: what causes the dispersal of PLB's in the light? An understanding of this problem will require knowledge of the steps required to transform tubules into lamellae in intact plants. It has already been shown that during the first hours of greening MGDG always decreases before new synthesis can be measured (TEVINI 1977; DAVIES and HARWOOD 1984). Does the light-induced formation of lamellae result from the degradation of lipid, similar to what has been reported for membrane proteins (HAMPP and DE FILIPPIS 1980)? ISRAELACHVILI and WOLFE (1980) have suggested an alternative mechanism, involving only limited amounts of energy for structural rearrangement of PLB components. Other questions related to the phenomenon of PLB transformation are as follows: Does the enhanced protein synthesis during greening cause a rearrangement of PLB lipids into lamellae? Does phytylation of the newly formed chlorophyllide change the amphipathic balance in the PLB/PT system (TREFFRY 1974) and induce a transfer of glycolipids to the esterified pigments in the PT's?

Our knowledge of how light and chloroplast development are tied into the synthesis of membrane lipids is still very limited. Studies are needed to evaluate the observed control of PLB transformation by phytochrome (GIRNTH et al. 1979). Furthermore, the negative effect of blue light on the dispersal of PLB tubes, as has been described for bean plants (BRADBEER et al. 1974) and for the fern *Dryopteris* (CRAN and DYER 1975), is not yet understood.

It is generally accepted that PLB's do not arise de novo, but rather develop from pre-existing membranes. PLB's have never been observed in plastids without any physical links to thylakoid membranes. Etioplasts may not represent a functional end-product in plastid differentiation. Nevertheless, this type of plastid can be used effectively to study in vivo the role of lipids in membrane formation; to date such studies have been largely restricted to in vitro systems.

References

Adamson HY, Hiller RG, Vesk M (1980) Planta 150:269–274
Bergweiler P, Röper U, Lütz C (1984) Physiol Plant 60:395–400
Borovjagin VL, Vergara JA, McIntosh TJ (1982) J Membr Biol 69:199–212
Bradbeer JW (1981) In: Stumpf PK, Conn EE (eds) The biochemistry of plants. Academic Press, London New York, pp 423–472
Bradbeer JW, Gyldenholm AO, Smith JW, Rest J, Edge HJW (1974) New Phytol 73:281–290
Cran D, Dyer AF (1975) Plant Sci Lett 5:57–65
Dahlin C, Ryberg H, Axelsson L (1983) Physiol Plant 59:562–566
Davies AO, Harwood JL (1984) Ann Bot 53:141–151
Feierabend J, Berger C, Meyer A (1969) Z Naturforsch 24b: 1641–1647

Galey J, Francke B, Bahl J (1980) Planta 149:433–439
Girnth C, Bergfeld R, Kasemir H (1979) Ber Dtsch Bot Ges 92:585–594
Giroud G, Siegenthaler PA (1984) Experientia 40 (6): p 607
Gunning BES, Steer MW (1975) Ultrastructure and the biology of plant cells. Arnold, London
Hampp R, Filippis LF de (1980) Plant Physiol 65:663–668
Heitz E (1954) Exp Cell Res 7:606–608
Ikeuchi M, Murakami S (1983) Plant Cell Physiol 24:71–80
Israelachvili JN, Wolfe J (1980) Protoplasma 100:315–321
Kesselmeier J (1982) Protoplasma 112:127–132
Kesselmeier J, Budzikiewicz H (1979) Z Pflanzenphysiol 91:333–344
Kesselmeier J, Laudenbach U, Urban B (1984) In: Sybesma C (ed) Proc 6th Int Congr Photosynth, Brussels, vol IV. Nijhoff/Junk, The Hague, pp 623–626
Kirk JTO, Tilney-Bassett RAE (1978) The plastids. Elsevier/North-Holland Biomed Press, Amsterdam New York
Kosmac U, Feierabend J (1984) Z Pflanzenphysiol 114:377–392
Leech RM, Baker NR (1983) In: Dale JE, Milthorpe FL (eds) The growth and functioning of leaves. Cambridge Univ Press, Cambridge, pp 272–307
Leyon H (1954) Exp Cell Res 7:609–611
Lütz C (1978) In: Akoyunoglou G (ed) Chloroplast development. Elsevier/North-Holland Biomed Press, Amsterdam New York, pp 481–488
Lütz C (1981 a) Protoplasma 108:83–98
Lütz C (1981 b) Protoplasma 108:99–115
Lütz C (1981 c) Z Pflanzenphysiol 104:43–52
Lütz C, Nordmann U (1983) Z Pflanzenphysiol 110:201–210
Lütz C, Tönissen H (1984) Isr J Bot 33:195–209
Lütz C, Röper U, Beer NS, Griffiths T (1981) Eur J Biochem 118:347–353
Lütz C, Nordmann U, Tönissen H, Bergweiler P, Röper U (1984) In: Sybesma C (ed) Proc 6th Int Congr Photosynth, Brussels, vol IV. Nijhoff/Junk, The Hague, pp 627–631
Mackender RO (1978) Plant Physiol 62:499–505
Mariani-Colombo P, Rascio N, Casadoro G (1983) New Phytol 93:457–465
Müller M, Nisius A (1982) Z Naturforsch 37c:476–480
Murphy DJ (1982) FEBS Lett 150:19–26
Murphy DJ, Woodrow IE (1983) In: Thomson WW, Mudd JB, Gibbs M (eds) Biosynthesis and function of plant lipids. Waverly Press, Baltimore, pp 104–125
Porath P (1979) New Phytol 82:733–737
Quinn PJ, Williams WP (1983) Biochim Biophys Acta 737:223–266
Reiss T, Bergfeld R, Link G, Thien W, Mohr H (1983) Planta 159:518–528
Ryberg M (1983) Physiol Plant 59:617–622
Ryberg M, Minkov I (1984) In: Sybesma C (ed) Proc 6th Int Congr Photosynth, Brussels, vol IV. Nijhoff/Junk, The Hague, pp 633–636
Ryberg M, Sundqvist C (1982) Physiol Plant 56:125–132
Ryberg M, Sandelius AS, Selstam E (1983) Physiol Plant 57:555–560
Schiff JA (1980) In: Stumpf PK, Conn EE (eds) The biochemistry of plants, vol I. Academic Press, London New York, pp 209–272
Selstam E, Lindblom G, Brentel I, Ryberg M (1982) In: Wintermans JFGM, Kuiper PJC (eds) Biochemistry and metabolism of plant lipids. Elsevier/North-Holland Biomed Press, Amsterdam New York, pp 389–392
Siegel DP (1984) Biophy J 45:399–420
Sprague SG, Staehelin A (1983) In: Thomson WW, Mudd JB, Gibbs M (eds) Biosynthesis and function of plant lipids. Waverly Press, Baltimore, pp 144–159
Stobart AK, El-Hussein SA, Hammond J (1982) Phytochemistry 21:2633–2640
Sundqvist C, Björn LO, Virgin H (1980) In: Reinert J (ed) Chloroplasts. Springer, Berlin Heidelberg New York, pp 201–224
Surin C, Sironval C (1984) In: Sironval C, Brouers M (eds) Protochlorophyllide reduction and greening. Nijhoff/Junk, The Hague, pp 275–294

Tevini M (1977) In: Tevini M, Lichtenthaler HK (eds) Lipids and lipid polymers in higher plants. Springer, Berlin Heidelberg New York, pp 121–145
Tönissen H, Lütz C (1984) Plant Cell Rep 3:116–120
Treffry T (1974) Eur J Biochem 43:349–351
Verkleij AJ (1984) Biochim Biophys Acta 779:43–63
Virgin H (1981) Annu Rev Plant Physiol 32:451–463
Wellburn AR (1982) Int Rev Cytol 80:133–191
Wellburn AR, Hampp R (1979) Biochim Biophys Acta 547:380–397
Wellburn AR, Wellburn FAM (1972) In: Forti G, Avron M, Melandri A (eds) Proc 2nd Int Congr Photosynth, Stresa. Junk, The Hague, pp 2349–2365
Wellburn AR, Gounaris I, Fäßler L, Lichtenthaler HK (1983) Physiol Plant 59:347–354
Wellburn FAM, Wellburn AR, Senger H (1980) Protoplasma 103:35–54
Whatley JM (1977) New Phytol 78:407–423

11.9 Acyl Lipids in Thylakoid Membranes: Distribution and Involvement in Photosynthetic Functions

P.A. SIEGENTHALER and A. RAWYLER

1 Introduction

Electron transport in thylakoid membranes is achieved through a spatial organization of chlorophyll-protein complexes linked together by protein and nonprotein redox components (ORT, Chap. 4, this Vol.). All these constituents are distributed within a lipid environment which is generally envisaged as a simple, homogenous bilayer.

The functional role of lipids in a membrane is an ambiguous concept because lipids, in contrast to proteins, have by themselves no recognized catalytic properties. Rather, they allow the maintenance of an appropriate conformation and orientation of proteins which may express their function only in the presence of specific (?) lipids. Thus, it is extremely difficult to dissociate the functional from the structural role of lipids. Several reviews deal with structural and biophysical roles of lipids in plant membranes (QUINN and WILLIAMS 1978, 1983; RAISON 1980; MURPHY and WOODROW 1983a; BISHOP 1983; GOUNARIS and BARBER 1983) but only few have focused on functional aspects of lipids (DUBACQ and TRÉMOLIÈRES 1983; JORDAN 1984). The purpose of this chapter is to review the currently available evidence for the distribution of acyl lipids in thylakoid membranes and for the possible involvement of these molecules in photosynthetic functions. The reader may also want to consult this Volume, Chapter 11.11 of MURPHY for a more detailed treatment of structural properties of thylakoid lipids, and Chapter 11.10 of KEEGSTRA, in which the biosynthesis of thylakoid lipids is discussed.

2 Aging of Thylakoids in Vitro

Although the aging approach is not easily controlable, it has brought about interesting information on the relationship between lipid composition and the structural and functional state of thylakoids. Dark aging in vitro of thylakoids is accompanied by an increase in both free fatty acids and interconversion products between galactolipids, together with a decrease in photochemical activities (SIEGENTHALER 1974; SIEGENTHALER and RAWYLER 1977 and refs. therein, SIEGENTHALER and DEPÉRY 1977; HENRY et al. 1982). Comparative studies on thylakoids from different species (spinach and lettuce) have ruled out the involvement of galactolipid interconversion products in the alteration of electron flow activities. Moreover, the use of purified and nonpurified thylakoids indicates that the identity and localization of the lipid(s) hydrolyzed during aging contribute to the impairment of photochemical activities as much as free fatty acids (SIEGENTHALER et al. 1981).

Another aspect of aging in vitro concerns the physical state of the membrane lipids as reflected by phase transition and fluorescence polarization studies. Dark aging of spinach thylakoids results in a gradual shift of the phase transition to lower temperatures when chlorophyll a is used as an intrinsic probe (VAN GINKEL and FORK 1981). However, when the extrinsic probe DPH is employed as an extrinsic probe in fluorescence polarization studies, an apparent decrease in fluidity of the lipid matrix is observed (FORD and BARBER 1980). This discrepancy may be explained in part by the different environment in which these two probes are embedded.

Thus, aging studies emphasize the importance of the compositional and dynamic aspects of the lipid matrix and of the localization of certain lipids within the membrane to sustain a particular function.

3 Extraction of Membrane Lipids by Organic Solvents

Among the methods used to determine the molecular associations in photosynthetic membranes, mild extraction of lipids from freeze-dried membranes has received special attention. COSTES et al. (1972) have made a careful study of the extractibility of the different lipids of wheat chloroplast membranes using various solvents. Each lipid shows a double heterogeneity in (a) the distribution between the fractions and (b) the acyl chain composition. This compartmentation has been explained on the basis of interactions (hydrophobic forces, ionic and hydrogen bonds) between lipids themselves and lipids and proteins. Applying this technique to *Rhodospirillum rubrum* chromatophores, COSTES et al. (1978) have found that phosphatidylglycerol (PG) is the only lipid to be isosoluble with bacteriochlorophyll. They suggest that PG might stabilize the pigments associated with the single photosystem of chromatophores. Similar experiments show that partial lipid depletion is accompanied by a loss of photochemical activities. All attempts to show a specific lipid requirement for the restoration of PS I activity have failed, since not only thylakoid lipids but also a wide range of neutral lipids are able to restore most of the activity (KRUPA and BASZYNSKI 1978 and ref. therein).

4 Modulation of Fatty Acid Composition

Changes in fatty acid composition of thylakoid lipids have been obtained by catalytic hydrogenation of fatty acyl chains in situ and by treatment of plants by substituted pyridazinones. Up to 40% of the double bonds can be hydrogenated without significant inhibition of PS II and PS I electron flow activities.

Moreover, the basic structure of chloroplasts remains unchanged and grana stacking is preserved (RESTALL et al. 1979). These results suggest that the number of residual double bonds in the bulk lipids is still sufficient to provide the required fluidity for maximal electron flow activities. An additional interpretation may be that the probability of the formation of nonbilayer structures, which are known to be associated with the inhibition of PS II-mediated reactions (GOUNARIS et al. 1983a), is greatly reduced by the decrease of the double bond index (GOUNARIS et al. 1983b). However, a minimal degree of unsaturation in thylakoid lipids seems to be required as suggested by (1) the failure of fully hydrogenated MGDG (monogalactosyldiacylglycerol) to restore the spectral properties of Triton X-100-treated thylakoids (SIEFERMANN-HARMS et al. 1982) and (2) the failure of the hydrogenated total lipid extract to stimulate the activity of O_2-evolving particles (GOUNARIS et al. 1983d). On the other hand, seasonal variations in the $H_2O/NADP^+$ electron flow activity can be nicely correlated with the unsaturation level of MGDG in thylakoids from *Pinus sylvestris* (ÖQUIST 1982). This author suggests that the decreased unsaturation degree of MGDG during winter months reduces the mobility of plastoquinones in thylakoid membranes, thus contributing to the inhibition of the electron transport activity.

The main effect of a treatment of seeds by substituted pyridazinones is to cause a reduction of $C_{18:3}$ (linolenic acid) to $C_{18:2}$ (linoleic acid) ratios and a small increase in the $C_{16:1-(3t)}$ (3-trans-hexadecenoic acid)-containing PG molecules, without interfering substantially with the development of chloroplasts (LEECH and WALTON 1983). After treatment of the plants with San 9785, there are fewer stroma thylakoids, leading to an increase in the ratio of appressed to exposed membranes, an increased proportion of the chlorophyll protein LHCP and a decreased proportion of CP I (chlorophyll protein I). The correlation between the increase in PG content and LHCP is consistent with the proposal of DUBACQ and TRÉMOLIÈRES (1983). All these changes do not affect the primary photochemistry of PS II, but do increase the spillover of excitation energy from PS II to PS I in treated plants. The latter result may be expected from the reduced number of PS I units due to the disappearance of part of the stroma lamellae.

Recently, LEHOCZKI and FARKAS (1984) have found that in cerulenin-treated *Chlorella* the appearance of herbicide resistance is associated with an alteration in acyl lipid composition, together with an increase in the degree of unsaturation and in membrane fluidity.

In conclusion, these experiments show that alterations in the lipid matrix can occur without significantly changing the structure and function of thylakoids, provided that the total lipid to chlorophyll ratio remains constant.

5 Lipid Composition of Thylakoids and Subchloroplast Fractions

Recent studies suggest that certain types of lipid may be preferentially associated with specific chlorophyll-protein complexes (see THORNBER, Chap. 3, this Vol.

for a more detailed discussion of these complexes). For instance, RAWYLER et al. (1980) have isolated LHCP and CP I complexes from spinach by nonionic detergents and found that their acyl lipid content different from that of intact thylakoid membranes. LHCP is enriched in anionic lipids, phosphatidylglycerol (PG) and sulfoquinovosyldiacylglycerol (SQDG), and markedly depleted in monogalactosyldiacylglycerol (MGDG). In contrast, CP I contains only SQDG and phosphatidylinositol (PI), characterized by a lower double bond index, i.e., essentially an increase in the $C_{18:0}/C_{18:3}$ ratio. The CP I particles photoreduce $NADP^+$ and their activity is further increased by the addition of ferredoxin-$NADP^+$ reductase. However, using another detergent technique to prepare LHCP from spinach, RYRIE et al. (1980) have observed that the lipid composition of the particles and of thylakoid membranes is similar. Another type of LHCP particle from tobacco isolated by electrophoresis contains all classes of lipids (TRÉMOLIÈRES et al. 1981; RÉMY et al. 1982). PG is the major lipid in the oligomeric form ($LHCP^1$), whereas MGDG is predominant in the monomeric form ($LHCP^3$). The oligomeric form of LHCP can be reconstituted from the monomeric one to the same extent by various acyl lipids but the reconstitution rate is maximal with PG containing $C_{16:1\text{-}(3t)}$ (RÉMY et al. 1983). These authors have suggested that this PG species is directly implicated in the maintenance of the oligomeric structure of LHCP and in the efficiency of light collection (DUBACQ and TRÉMOLIÈRES 1983). In contrast, both PS I and PS II reaction centers show a very low content of galactolipids and PG and a considerable enrichment in SQDG (RÉMY et al. 1982).

The common feature of the LHCP particles examined so far is that they contain all lipid classes encountered in the thylakoid membrane and that the lipid/chlorophyll molar ratio ranges from 0.2 to 0.8 as compared to about 2.5 in the original membrane. The common feature of CP I particles is a very low lipid/chlorophyll molar ratio, ranging from 0.05 to 0.1 and a high proportion of SQDG. This is in line with the observation that Triton X-100 subchloroplast particles enriched in photosystem I (TSF-1), which are completely depleted in galactolipids and phosphatidylcholine (PC) after lipolytic treatment but contain 50% of their anionic lipids (SQDG + PG) still display 50% of their initial electron flow activity (RAWYLER and SIEGENTHALER 1980). The variation observed in the lipid/chlorophyll ratios is mainly due to changes in the galactolipid content. This may be explained in part by nonspecific associations between some of these lipids and the chlorophyll-protein complexes (HEINZ and SIEFERMANN-HARMS 1981).

The lipid content of an oxygen-evolving PS II preparation from spinach thylakoid membranes obtained by Triton X-100 fractionation (BERTHOLD et al. 1981) was also examined. Detergent treatment results in a preferential loss of digalactosyldiacylglycerol (DGDG) and an enrichment of the particles in PG and PC. Addition of DGDG, PC or a total lipid extract with or without MGDG stimulates O_2 evolution. The hydrogenated total lipid extract has no effect on O_2 evolution and a mixture of PG + SQDG inhibits this activity completely (GOUNARIS et al. 1983d). Using the same particles, MATSUDA and BUTLER (1983) have reported that addition of DGDG or PC increases O_2 evolution and variable fluorescence yield, and restores cytochrome b-559 HP form. In contrast, SQDG has a deleterious effect on these parameters. A similar stimulation of O_2 evolution is observed when these particles are reconstituted with soybean PC liposomes. In addition, reconstitution in the presence of both PS II particles and LHCP enhances the quantum efficiency at subsaturating light intensities (MURPHY et al. 1984). Altogether, these results suggest that PC may be important for sustaining PS II activities, as also evidenced by phospholipase A_2 treatment studies (RAWYLER and SIEGENTHALER 1981a).

New methods based on the disruption of membranes by Yeda or French press followed by aqueous polymer two-phase partition (without detergents) have shown that PS I and PS II complexes are laterally segregated within the thylakoid membrane. This heterogeneity is also valid for acyl lipids, although to a lesser extent (GOUNARIS et al. 1983c; HENRY et al. 1983; MURPHY and WOODROW 1983a, b). For instance, stroma lamellae contain relatively low levels of acidic lipids (PG, SQDG) and MGDG/DGDG molar ratio close to 1, whereas inside-out vesicles derived from appressed thylakoids are enriched in these acidic lipids and show a ratio close to 3. In addition, on a chlorophyll basis, stroma lamellae are enriched, whilst stacked grana membranes are depleted in lipids compared to intact thylakoids (GOUNARIS et al. 1983c). MURPHY and WOODROW (1983a, b) obtained essentially the same results, but they point out that the apparent enrichment of appressed regions in PG, whilst true in terms of percent of total acyl lipids, is not found when lipid:protein molar ratios are considered. A direct consequence of this type of expression is that the lateral distribution of MGDG, and even more that of SQDG, differs significantly from the distribution proposed by GOUNARIS et al. (1983c). However, both types of calculations suggest the preferential localization of SQDG in nonappressed membranes which contain most PS I complexes (MURPHY and WOODROW 1983a, b). A similar enrichment in SQDG (and also in PC) has been found by TUQUET et al. (1977) in stromal and agranal thylakoids from maize compared to granal thylakoids. Moreover, these authors have observed that $C_{16:1-(3t)}$ is concentrated in PG of granal thylakoids, although the absolute amount of this lipid is similar in all types of thylakoids. Recently, CHAPMAN et al. (1984) have questioned the degree of the lateral heterogeneity of acyl lipids, by showing that in thylakoids in vivo no lipid type other than MGDG is selectively concentrated in PS II-LHCP membrane regions.

6 Modification of the Lipid: Protein Ratio in Thylakoid Membranes

The concept of lateral separation of PS II and PS I emphasizes the role of plastoquinones as important long-range diffusing molecules which connect intrinsic protein complexes to allow maximum electron flow from H_2O to $NADP^+$ (see ORT, Chap. 4; WHITMARSH, Chap. 10.2, both this Vol.). This implies that the fluidity of the thylakoid membrane must be optimal, which is true at room temperature (RAISON et al. 1982). The fluidity of a membrane is determined by the length and degree of unsaturation of its fatty acyl chains, by its composition of lipids and proteins, and by the temperature. Thus, stroma thylakoids, which are more fluid than grana thylakoids, have also a higher lipid:protein ratio (FORD et al. 1982). This has led to the hypothesis that thylakoid fluidity is determined by the lipid:protein ratio (HILLER and RAISON 1980; FORD and BARBER 1983a).

A comparison of thylakoid membranes isolated from *Pisum sativum* grown at 4°–7 °C and 14°–17 °C, shows that the lipid:protein ratio is higher in cold-grown plants, whereas no change in the relative amounts of lipid classes and degree of unsaturation occurs between the two growth conditions (CHAPMAN et al. 1983). The lipid:protein ratio has been manipulated artificially by incorporating lipids into thylakoid membranes. Incorporation of cholesteryl hemisuccinate inhibits progressively the dark reoxidation rate of Q, the dark steps of photoreduction of ferricyanide, the initial rate of proton uptake, and the cation-induced chlorophyll fluorescence changes. Simultaneously there is a decrease in membrane fluidity as probed by DPH fluorescence polarization (YAMAMOTO et al. 1981). Accordingly, the reductibility of flash-oxidized cytochrome f, but not its photooxidation, diminishes with increasing sterol incorporation (FORD and BARBER 1983b). Thus, the membrane fluidity appears to affect both the relative mobility of chlorophyll-protein complexes and plastoquinone-mediated electron flow between the two photosystems.

Lipid enrichment of thylakoid membranes has been achieved by addition of liposomes of mixed soybean and pure phospholipids without detectable loss of membrane components. This results in a decrease in the density of intramembrane particles, and in the appearance, in some regions of the membrane, of large hexagonal lattices identified as LHCP particles. Dissociation of LHCP from PS II reaction centers is observed only with pure PG or nonpurified PC, but not with pure PC (SIEGEL et al. 1981). Consistent with this observation, the activity of PS II electron flow at low light intensities is greatly reduced and the low 77 K fluorescence of LHCP is enhanced at the expense of PS I and PS II. Furthermore, the degree of electron flow inhibition increases and both flash-induced oxidation and dark reduction rates of cytochrome f are diminished with higher lipid enrichment levels (MILLNER et al. 1983). These results suggest that lipid incorporation alters the spatial relationship between membrane components. However, changes in surface charge density are also relevant. For example, an addition of $C_{18:3}$ obviously increases the lipid:protein ratio but does not change the membrane fluidity. This treatment increases the surface charge density and decreases the cation-induced chlorophyll fluorescence (BARBER et al. 1980). Moreover, the changes in 77 K fluorescence spectra (SHI-QING et al. 1983) are similar to those described by SIEGEL et al. (1981) upon addition of negatively charged phospholipids. The role of the surface charge density is further substantiated by the fact that only the ionized form of $C_{18:3}$ alters the absorption spectra of thylakoids (KONISHI et al. 1972). Thus it is likely that the $C_{18:3}$ inhibition of electron flow (SIEGENTHALER and RAWYLER 1977) can be explained not only in terms of interaction of $C_{18:3}$ with specific sites in the electron transport pathway (GOLBECK et al. 1980; SIEGENTHALER 1974), but also by a change in the spatial arrangement of the chlorophyll-protein complexes induced by the increased surface charge density.

Another way to manipulate the lipid:protein ratio is to add polypeptides. Melittin penetrates into the lipid bilayer of the thylakoid membrane, causing a restriction in motion of lipid molecules. At low melittin concentrations, the thylakoid membrane cannot build an electrochemical and proton gradient, whilst at high concentrations, the increased restriction of lipid motion interferes with plastocyanin-binding sites, thus inhibiting electron flow (BISHOP 1983).

7 Immunological Studies

Antibodies against lipids have been used to determine the transverse distribution and function of lipids in the thylakoid membrane. The following conclusions have been drawn (RADUNZ 1981 and references therein): (a) the lipids of the thylakoid membrane are more abundant at the internal than at the outer surface; (b) SQDG and phospholipids are mainly located at the outer surface, whilst galactolipids are preferentially found in the inner leaflet of the membrane, although the major part of SQDG and MGDG was not accessible to antibodies from either side of the membrane; (c) a large part of MGDG molecules is located near the coupling factor, as also suggested by reconstitution experiments (PICK et al. 1984); (d) PS II and PS I reactions are inhibited, on their oxidizing sides, with decreasing efficiency by antibodies against PC, MGDG and DGDG, and against DGDG, PG, MGDG, SQDG, PI, and PC, respectively (RADUNZ 1981; RADUNZ et al. 1984).

8 Lipolytic Treatments

The rationale of the enzymatic approach has been discussed recently (SIEGENTHALER 1982). To assess the transmembrane distribution of lipids, it is essential to obtain detailed measurements of the kinetics of lipid hydrolysis in order to obtain reliable data from which one can compute the most probable distribution of lipids.

When applied in limiting amounts, lipolytic acid hydrolase (LAH) degrades thylakoid lipids in a differential manner, phospholipids being hydrolyzed only when part of the galactolipids are destroyed. This may be due to a microheterogeneity of lipids at the membrane surface or to surface pressure requirements differing for each substrate (BISHOP 1983). Moreover, the kinetic studies show that each lipid class undergoes a rapid hydrolysis followed by a much slower one, suggesting a differential accessibility of the lipid molecules to the enzyme (RAWYLER and SIEGENTAHLER 1980). An obvious reason for this difference in accessibility may be lipid asymmetry. This hypothesis has been tested with phospholipases which specifically destroy PG and PC. Hydrolysis kinetics at various temperatures allow one to conclude that in spinach and oat thylakoids, as well as in oat prothylakoids, 70% of total PG is located in the outer leaflet of the membrane (RAWYLER and SIEGENTHALER 1981a; GIROUD and SIEGENTHALER 1984). The similarity of PG distribution in prothylakoids and mature thylakoids suggests that the localization of this lipid may not depend on the incorporation of chlorophyll-protein complexes during greening (GIROUD and SIEGENTHALER 1984). A preferential localization of PG in the outer monolayer may also be inferred from other reports (HIRAYAMA and NOMOTOBORI 1978; DUVAL et al. 1979; UNITT and HARWOOD 1982; JORDAN et al. 1983). A similar distribution (60–70% in the outer leaflet) is also found for PC in spinach and oat thylakoids (GIROUD and SIEGENTHALER 1984). On the other hand, preliminary results concerning the distribution of galactolipids are contradictory (UNITT and HARWOOD 1982; SUNDBY and LARSSON 1983). Recently, this distribution has been investigated by measuring the hydrolysis kinetics of galactolipids catalyzed by the lipase from *Rhizopus arrhizus* under a variety of ionic conditions, which allows for the control of surface electrical properties and of the degree of stacking of the membrane. The outside/inside distribution is 65/35 for MGDG and 15/85 for DGDG. Since these lipids represent 55 and 27 mol% of total acyl lipids in spinach thylakoids, it follows that total galactolipids are almost equally balanced within the two monolayers of the membrane (RAWYLER and SIEGENTHALER 1984). The above results do not exclude the possibility that lipid asymmetry might be different in stromal and granal thylakoid membranes.

Although lipolytic treatments have marked effects on photosynthetic electron transport, simple correlations, either temporal or quantitative between lipid hydrolysis and loss of activity, cannot be strictly established. Various LAH's, which generally hydrolyze both fatty acids of all classes of lipids (except SQDG), have been shown to readily degrade half of the lipid content of stacked thylakoids (MICHALSKI and KANIUGA 1980; KRUPA 1982) and cause a more rapid inhibition of noncyclic electron flow than of PS I activity, while leaving PS II activity largely unaffected (HIRAYAMA and MATSUI 1975; RAWYLER and SIEGENTHALER 1980). It is noteworthy that a treatment of thylakoids by bean LAH loosens the lipid-protein interactions in such a way that plastocyanin, LHCP[3] and polypeptides originating from the peripheral antennae proteins of PS I complex are released into the medium (KRUPA 1983). Unfortunately, these results do not allow a functional role to be attributed to a specific lipid, because of the nonspecificity of the LAH. Moreover, the actual extent of lipid hydrolysis often exceeds that effectively required to obtain comparable functional alterations. Indeed, the action of LAH at the membrane surface results in a peeling-off of most lipid polar groups (some of which may not be involved in the measured function) and in the formation of a large amount of free fatty acids which may increase the surface charge density (BARBER et al. 1980) and decrease the packing pressure (BISHOP 1983). These parameters may, in turn, change the relative spatial arrangement of the membrane complexes, disconnect PS II from PS I, and alter the peripheral antennae in the PS I complex (as suggested by KRUPA 1983). This may explain the inhibition of both noncyclic and cyclic electron flows. The relative insensitivity of PS II activity could be due to a limited access of the enzyme to the stacked membrane regions where most PS II is located.

In contrast to LAH's, the use of phospholipases offers the advantage of a selective degradation of two lipid classes only (PG and PC) and of lowering the extent of hydrolysis of membrane lipids to a maximum of about 15%, thereby limiting the potential damage to the bilayer structure. Several studies show that phospholipase A_2 (PLA_2) inhibits noncyclic and PS II electron flow and destroys 60–80% of PG and 40–60% of PC (HIRAYAMA and MATSUI 1976; RAWYLER and SIEGENTHALER 1981a; JORDAN et al. 1983). In the presence of BSA, the hydrolysis of lipids by PLA_2 is similar, but the activities are not significantly altered (HIRAYAMA and NOMOTOBORI 1978; DUVAL et al. 1979). This suggests that the hydrolysis products exert an inhibitory effect which can be relieved through BSA binding. Moreover, the extent of lipid degradation roughly accounts for the phospholipids which are located in the outer leaflet of the thylakoid membrane. Thus the presence of residual phospholipids may be the explanation for the persistance of PS II activity. Indeed, a detailed study of the kinetics of phospholipid degradation and of PS II activity shows that the inhibition of PS II occurs only when the internal pool of PC begins to be hydrolyzed. This emphasizes the role of this PC pool in sustaining PS II electron flow (RAWYLER and SIEGENTHALER 1981a). Further insight into the role of phospholipid is provided by the use of phospholipases of various specificities (phospholipase A_2, C, and D). Choline, phosphocholine, and glycerol, which represent hydrophilic portions of PC and PG, may not be directly involved in photosynthetic electron flow. On the other hand, the phosphoglycerol (hydrophilic portion) and the acyl chain at the sn-2 position (hydrophobic portion) appear to be required for maximal noncyclic electron flow activity. The role of the phosphoglycerol portion of PG may involve the connection between the two photosystems, but not the electron flow activities within the individual PS I and PS II complexes (SIEGENTHALER et al. 1984).

Although there is general agreement concerning the effect of PLA_2 on the extent of phospholipid hydrolysis, the mechanism by which photochemical reactions are affected is controversial. DUVAL et al. (1979) conclude that the efficiency of light collection by PS II antennae is lowered by PG depletion without affecting the activity of the photochemical centers. In support of this interpretation they have shown that PLA_2 treatment dissociates the oligomeric into the monomeric form of LHCP and that the reconstitution kinetics are more rapid with PG containing $C_{16:1-(3t)}$ than with any other type of lipid (RÉMY et al. 1982). On the other hand, we have suggested that PLA_2 treatment blocks the electron flow somewhere between the entry point of 1,5-diphenyl-carbazide and plastoquinone (RAWYLER and SIEGENTHALER 1981a). Moreover, KRUPA (1983) has concluded that PC plays an important role in PS II reaction centers (CP_a). JORDAN et al. (1983) have proposed a more precise site of PLA_2-induced inhibition, which may be at the electron transfer step from pheophytin to Q, and which may involve alteration of the relative distance and/or orientation between these two components. This interpretation is reminiscent of that proposed by GOLBECK et al. (1980) concerning the site of action of externally added $C_{18:3}$. A striking effect of PLA_2 is to increase PS I electron transport activity at high irradiances only. This stimulation is not due to increased spillover (JORDAN et al. 1983), but rather to an increased rate of diffusion or accessibility of the donors which feed electrons into PS I. In addition to this first target on the oxidizing side of PS I, a second one on the reducing side seems to modulate PLA_2-stimulated PS I activity. This stimulation has been ascribed to the hydrolysis of PG (RAWYLER and SIEGENTHALER 1981 b).

9 Biogenesis Studies

The biogenesis of functional photosynthetic membranes has been studied either during the greening of etiolated tissues or during the normal development of leaves. In both cases, chlorophyll synthesis occurs, accompanied by qualitative and quantitative changes in acyl lipid composition. Most authors find that during greening, galactolipids and $C_{18:3}$ accumulate, and that the ratio of galactolipid:chlorophyll remains more or less constant. PG increases in the same proportion, and becomes strongly enriched in $C_{16:1-(3t)}$ (KIRK and TILNEY-BASSETT 1978). A linear correlation between the biosynthesis of this acid and the development of appressed thylakoid membranes is observed during greening of primary leaves of etiolated maize seedlings (DUBACQ and TRÉMOLIÈRES 1983 and references therein).

A better experimental approach is to study the ultrastructural, biochemical, and functional properties of plastids in normally developing leaves. For example, along the developmental gradient of the maize leaf a differential accumulation of most lipid classes (MGDG, DGDG, SQDG, and PG) except PC is observed on a per-plastid basis. In the case of fatty acids, there is a large increase in the production of $C_{18:3}$ in all lipids at the expense of $C_{18:2}$, and the appearance of a new molecular species of PG containing $C_{16:1-(3t)}$ (LEECH and WALTON 1983). These changes parallel chlorophyll biosynthesis and a huge increase in the number of grana per plastid, whilst the average number of thylakoids per granum is much less affected. Moreover, experiments with isolated developing plastids show that the development of PS I electron flow activity is completed before that of PS II. For both PS I and PS II activities, the light saturation point decreases with increasing chlorophyll content, suggesting that the average photosynthetic unit size of both photosystems increases during development (BAKER and LEECH 1977). The newly formed chlorophyll-protein complexes become functionally integrated into the developing membranes, providing new lipids are available to pack them into an optimal environment. The appearance of highly unsaturated molecular species of MGDG and DGDG (MACKENDER and GILLANDERS 1984) and of PG containing $C_{16:1-(3t)}$ (LEECH and WALTON 1983) seems to satisfy these requirements.

10 Conclusions

In spite of the large number of studies concerning the distribution and function of lipids in thylakoid membranes, it is premature to propose a definite picture of the lipid topology. However, various approaches (aging, mild extraction, lipid composition and immunological studies, as well as lipolytic treatments) provide strong evidence for both lateral and transverse heterogeneity of acyl lipids in thylakoid membranes. Whether the transmembrane distribution is different between appressed and nonappressed thylakoids remains to be established.

Similarly, it is too early to relate in an unambiguous manner the presence of a specific lipid class with the expression of a specific function. For instance, we have emphasized that there is no strict correlation, either temporal or quantitative, between lipid hydrolysis and subsequent functional alterations. We propose that each lipid class consists of several topologically distinct pools which are separated laterally as well as transversally in the membrane. We can therefore visualize that each of these pools plays a specific role in the membrane. This is illustrated, for instance, by our finding that only the internal pool of PC sustains PS II electron flow. Some of these pools may be specifically associated with a function, whilst others may be involved in the maintenance of the biophysical properties of the membrane such as fluidity. Future work should attempt to identify the pools of all lipid classes and to attribute a role to each of them.

References

Baker NR, Leech RM (1977) Plant Physiol 60:640–644
Barber J, Chow WS, Scoufflaire C, Lannoye R (1980) Biochim Biophys Acta 591:92–103
Berthold DA, Babcock GT, Yocum CF (1981) FEBS Lett 134:231–234
Bishop DG (1983) In: Thomson WW, Mudd JB, Gibbs M (eds) Biosynthesis and function of plant lipids. Am Soc Plant Physiol, Rockville, pp 81–103
Chapman DJ, Felice J de, Barber J (1983) Plant Physiol 72:225–228
Chapman DJ, Felice J de, Barber J (1984) In: Siegenthaler PA, Eichenberger W (eds) Structure, function and metabolism of plant lipids. Elsevier/North-Holland Biomed Press, Amsterdam New York, pp 457–464
Costes C, Bazier R, Lechevallier D (1972) Physiol Vég 10:291–317
Costes C, Bazier R, Baltscheffsky H, Hallberg C (1978) Plant Sci Lett 12:241–249
Dubacq JP, Trémolières A (1983) Physiol Vég 21:293–312
Duval JC, Trémolières A, Dubacq JP (1979) FEBS Lett 106:414–418
Ford RC, Barber J (1980) Photobiochem Photobiophys 1:263–270
Ford RC, Barber J (1983a) Biochim Biophys Acta 722:341–348
Ford RC, Barber J (1983b) Planta 158:35–41
Ford RC, Chapman DJ, Barber J, Pedersen JZ, Cox RP (1982) Biochim Biophys Acta 681:145–151
Ginkel G van, Fork DC (1981) Photobiochem Photobiophys 2:239–243
Giroud C, Siegenthaler PA (1984) In: Siegenthaler PA, Eichenberger W (eds) Structure, function and metabolism of plant lipids, Elsevier/North-Holland Biomed Press, Amsterdam New York, pp 413–416
Golbeck JH, Martin IF, Fowler CF (1980) Plant Physiol 65:707–713
Gounaris K, Barber J (1983) Trends Biochem Sci 8:378–381
Gounaris K, Brain APR, Quinn PJ, Williams WP (1983a) FEBS Lett 153:47–52
Gounaris K, Mannock DA, Sen A, Brain APR, Williams WP, Quinn PJ (1983b) Biochim Biophys Acta 732:229–242

Gounaris K, Sundby C, Andersson B, Barber J (1983c) FEBS Lett 156:170–174
Gounaris K, Whitford D, Barber J (1983d) FEBS Lett 163:230–234
Heinz E, Siefermann-Harms D (1981) FEBS Lett 124:105–111
Henry LEA, Strasser RJ, Siegenthaler PA (1982) Plant Physiol 69:531–536
Henry LEA, Mikkelsen JD, Møller BL (1983) Carlsberg Res Commun 48:131–148
Hiller RG, Raison JK (1980) Biochim Biophys Acta 599:63–72
Hirayama O, Matsui T (1976) Biochim Biophys Acta 423:540–547
Hirayama O, Nomotobori T (1978) Biochim Biophys Acta 502:11–16
Jordan BR (1984) In: Siegenthaler PA, Eichenberger W (eds) Structure, function and metabolism of plant lipids. Elsevier/North-Holland Biomed Press, Amsterdam New York, pp 443–450
Jordan BR, Chow WS, Baker AJ (1983) Biochim Biophys Acta 725:77–86
Kirk JTO, Tilney-Bassett RAE (1978) The plastids: their chemistry, structure growth and inheritance. Elsevier/North-Holland Biomed Press, Amsterdam New York, pp 720–787
Konishi K, Ogawa T, Inoue Y, Shibata K (1972) Plant Cell Physiol 13:651–661
Krupa Z (1982) Photosynth Res 3:95–104
Krupa Z (1983) Photosynth Res 4:229–239
Krupa Z, Baszynski T (1978) Bull Acad Pol Sci Cl II 26:885–889
Leech RM, Walton CA (1983) In: Thomson WW, Mudd JB, Gibbs M (eds) Biosynthesis and function of plant lipids. Am Soc Plant Physiol, Rockville, pp 104–125
Lehoczki E, Farkas T (1984) Plant Sci Lett 36:125–130
Mackender RO, Gillanders BW (1984) In: Siegenthaler PA, Eichenberger W (eds) Structure, function and metabolism of plant lipids. Elsevier/North-Holland Biomed Press, Amsterdam New York, pp 401–404
Matsuda H, Butler WL (1983) In: Sybesma C (ed) Abstr 6th Int Congr Photosynth, Bruxelles, vol II. Nijhoff/Junk, The Hague, p 42
Michalski WP, Kaniuga Z (1980) Biochim Biophys Acta 589:84–99
Millner PA, Grouzis JP, Chapman DJ, Barber J (1983) Biochim Biophys Acta 722:331–340
Murphy DJ, Woodrow IE (1983a) In: Thomson WW, Mudd JB, Gibbs M (eds) Biosynthesis and function of plant lipids. Am Soc Plant Physiol, Rockville, pp 104–125
Murphy DJ, Woodrow IE (1983b) Biochim Biophys Acta 725:104–112
Murphy DJ, Crowther D, Woodrow IE (1984) FEBS Lett 165:151–155
Öquist G (1982) Plant Physiol 69:869–875
Pick U, Gounaris K, Admon A, Barber J (1984) Biochim Biophys Acta 765:12–20
Quinn PJ, Williams WP (1978) Prog Biophys Mol Biol 34:109–173
Quinn PJ, Williams WP (1983) Biochim Biophys Acta 737:223–266
Radunz A (1981) Ber Dtsch Bot Ges 94:477–489
Radunz A, Bader KP, Schmid GH (1984) Z Pflanzenphysiol 114:227–231
Raison JK (1980) In: Stumpf PK, Conn EE (eds) The biochemistry of plants, vol IV. Academic Press, London New York, pp 57–83
Raison JK, Roberts JKM, Berry JA (1982) Biochim Biophys Acta 688:218–228
Rawyler A, Siegenthaler PA (1980) Eur J Biochem 110:179–187
Rawyler A, Siegenthaler PA (1981a) Biochim Biophys Acta 635:348–358
Rawyler A, Siegenthaler PA (1981b) Biochim Biophys Acta 638:30–39
Rawyler A, Siegenthaler PA (1984) In: Siegenthaler PA, Eichenberger W (eds) Structure, function and metabolism of plant lipids. Elsevier/North-Holland Biomed Press, Amsterdam New York, pp 409–412
Rawyler A, Henry LEA, Siegenthaler PA (1980) Carlsberg Res Commun 45:443–451
Rémy R, Trémolières A, Duval JC, Ambard-Bretteville F, Dubacq JP (1982) FEBS Lett 137:271–275
Rémy R, Ambard-Bretteville F, Dubacq JP, Trémolières A (1983) In: Sybesma C (ed) Abstr 6th Int Congr Photosynth, Bruxelles, vol I. Nijhoff/Junk, The Hague, p 144
Restall CJ, Williams P, Percival MP, Quinn PJ, Chapman D (1979) Biochim Biophys Acta 555:119–130
Ryrie IJ, Anderson JM, Goodchild DJ (1980) Eur J Biochem 107:345–354

Shi-qing L, Shi-qing L, Qi-de Z, Chong-gin T, Ting-yun K, Pei-sung T (1983) In: Sybesma C (ed) Abstr 6th Int Congr Photosynth, Bruxelles, vol I. Nijhoff/Junk, The Hague, p 120
Siefermann-Harms D, Ross JW, Kaneshiro KH, Yamamoto HY (1982) FEBS Lett 149:191–196
Siegel CO, Jordan AE, Miller KR (1981) J Cell Biol 91:113–125
Siegenthaler PA (1974) FEBS Lett 39:337–340
Siegenthaler PA (1982) In: Wintermans JFGM, Kuiper PJC (eds) Biochemistry and metabolism of plant lipids. Elsevier/North-Holland Biomed Press, Amsterdam New York, pp 351–358
Siegenthaler PA, Depéry F (1977) Plant Cell Physiol 18:1047–1055
Siegenthaler PA, Rawyler A (1977) Plant Sci Lett 9:265–273
Siegenthaler PA, Rawyler A, Henry LEA (1981) In: Akoyounoglou G (ed) Proc 5th Int Congr Photosynth, vol II. Balaban Int Sci Serv, Philadelphia, pp 167–174
Siegenthaler PA, Smutny J, Rawyler A (1984) In: Siegenthaler PA, Eichenberger W (eds) Structure, function and metabolism of plant lipids. Elsevier/North-Holland Biomed Press, Amsterdam New York, pp 475–478
Sundby C, Larsson C (1983) In: Sybesma C (ed) Abstr 6th Int Congr Photosynth, Bruxelles, Vol I. Nijhoff/Junk, The Hague, p 312
Trémolières A, Dubacq JP, Ambard-Bretteville F, Rémy R (1981) FEBS Lett 130:27–31
Tuquet C, Guillot-Salomon T, Lubac M de, Signol M (1977) Plant Sci Lett 8:59–64
Unitt MD, Harwood JL (1982) In: Wintermans JFGM, Kuiper PJC (eds) Biochemistry and metabolism of plant lipids. Elsevier/North-Holland Biomed Press, Amsterdam New York, pp 359–362
Yamamoto Y, Ford RC, Barber J (1981) Plant Physiol 67:1069–1072

11.10 Biosynthesis of Chloroplast Membrane Lipids

K. KEEGSTRA

1 Introduction

Chloroplast membranes are the most abundant membranes in green tissues of plants, and their lipids have been estimated to be the most abundant in the biosphere (GOUNARIS and BARBER 1983). Given this situation, it is rather surprising that chloroplast lipid synthesis has received as little attention as it has. However, during the last 10–15 years, several research groups have made significant contributions to our understanding of chloroplast membrane lipid biosynthesis. As a result, a general understanding of this important area has emerged, although many question still need to be answered.

Many aspects of chloroplast lipid synthesis have been covered in several extensive reviews and the reader should consult these for details on this topic (DOUCE and JOYARD 1979, 1980; ROUGHAN and SLACK 1982). This very brief review can only attempt to give a short description of the current view and to summarize some recent developments. As a result, many important contributions had to be eliminated from consideration and the author apologizes to those whose contributions are not discussed.

2 Structure of Chloroplast Membrane Lipids

Chloroplast membranes have a glycerolipid composition very different from other plant cell membranes. In addition to the phospholipids found in most biological membranes, all three chloroplast membranes contain large amounts of galactolipids and smaller amounts of sulfolipids. Galactolipids are the major species in thylakoids and the inner envelope membrane constituting 80–90% of the total membrane lipids (CLINE et al. 1981; DOUCE and JOYARD 1980). The outer envelope membrane contains nearly equal amounts of galactolipids and phospholipids (CLINE et al. 1981). Thus, the outer envelope is intermediate in composition between the galactolipid-rich thylakoid and inner envelope membranes and the other membranes of the cell which contain only phospholipids.

The glycolipid and phospholipid categories of chloroplast glycerolipids can each be further subdivided according to the type of diacylglycerol (DG) residues that form the hydrophobic regions of the glycerolipids. Glycerolipids containing DG residues with a 16 carbon fatty acid at position 2 of glycerol are known as prokaryotic lipids, while those with DG containing an 18 carbon fatty acid at position 2 are called eukaryotic lipids (ROUGHAN and SLACK 1982).

The designation of a lipid as prokaryotic or eukaryotic can be made independently of its designation as phospholipid or glycolipid. For example, phosphatidylglycerol (PG) of chloroplast membranes is a prokaryotic lipid, whereas phosphatidylcholine (PC) is a eukaryotic lipid (ROUGHAN and SLACK 1982). The galactolipids contain both types of DG residue, with the quantity of each dependent upon the species of plant under consideration. The galactolipids of some species, e.g., pea, are exclusively eukaryotic, whereas the galactolipids of other species, e.g., spinach, are split almost equally between prokaryotic and eukaryotic types (DOUCE and JOYARD 1980). Both prokaryotic and eukaryotic galactolipids contain a high proportion of polyunsaturated fatty acids, i.e., 16:3 and 18:3 respectively. Thus species with prokaryotic galactolipids are known as 16:3 plants (JAMIESON and REID 1971), and species which contain predominantly eukaryotic galactolipids are known as 18:3 plants (DOUCE and JOYARD 1980).

Recognition of the prokaryotic and eukaryotic categories of lipid may appear to add unneccessary complexity to the consideration of chloroplast lipid composition. However, recognition of these two categories is very helpful in attempting to understand the biosynthesis of chloroplast lipids.

Because the galactolipids are the most abundant chloroplast lipids, their synthesis has received the greatest amount of study and will receive the greatest amount of attention in this review. Very little is known about the synthesis of sulfolipids or the chloroplast phospholipids, and their synthesis will be considered only briefly. A discussion of the physicochemical properties of these lipids can be found in Chapter 11.11, this Volume of MURPHY.

3 Biosynthesis of Galactolipids

Three different but complementary approaches have been applied to the study of galactolipid synthesis. First, galactolipid synthesis has been studied in whole tissues or whole cells, using radioactive precursors to label the galactolipids (ROUGHAN and SLACK 1982; WILLIAMS et al. 1983). These studies have been very useful in developing hypotheses regarding the relationship between galactolipid metabolism and the overall pattern of lipid metabolism in plant cells. The second general strategy is to examine galactolipid synthesis in isolated, intact chloroplasts (BERTRAMS et al. 1981; HEINZ and ROUGHAN 1983; ROUGHAN and SLACK 1982). Finally galactolipid synthesis has been examined using purified proteins or membrane fractions to examine the details of the individual enzymatic reactions (HEINZ 1977; DOUCE and JOYARD 1979). This latter approach is useful for determining where in the cell each of the steps occurs. No single approach has provided a satisfactory overall picture of galactolipid synthesis in green tissues. However, by combining results from all three approaches, a general picture of galactolipid synthesis has emerged.

A detailed consideration of the methods and results of the first two approaches is beyound the scope of this limited review. The reader should consult the excellent review of Roughan and Slack (1982) for more details on these two approaches. Very briefly summarized, these approaches have led to the hypothesis that the DG portion of prokaryotic lipids is synthesized in the chloroplast, whereas the DG portion of eukaryotic lipids is synthesized in the endoplasmic reticulum (ER). The remainder of this brief review focuses primarily on the enzymology of galactolipid synthesis, and in particular on some recent results which support the hypothesis stated above.

Galactolipid synthesis can be divided into three distinct stages: (1) synthesis of the fatty acids from acetyl CoA, (2) synthesis of DG from fatty acids and glycerol-3-P and (3) addition of the galactosyl head group. Each of these stages is briefly described below before considering the specifics of the prokaroytic and eukaryotic pathways.

Fatty acid synthesis has been reported to occur exclusively in the stromal space of plastids (Ohlrogge et al. 1979). The enzymes involved in this process have been purified and characterized in considerable detail (Stumpf 1980). The final product of fatty acid synthesis is a long chain fatty acid attached to acyl carrier protein (ACP). The major product synthesized by the multienzyme complex involved in fatty acid synthesis is palmitoyl ACP (C16:0). Chloroplasts also contain the enzymes necessary to elongate palmitoyl ACP to stearoyl ACP (C18:0) and desaturate it to the oleoyl ACP (C18:1). The fatty acyl ACP's can either be hydrolyzed to release free fatty acids for export to the ER (Roughan and Slack 1982) or employed as the acyl donor for DG synthesis (Frentzen et al. 1983).

Diacylglycerol synthesis involves two sequential additions of fatty acids from activated thioesters to glycerol-3-P, forming phosphatidic acid (PA). Polar head groups may then be added to PA in the synthesis of some glycerolipids. For the synthesis of other glycerolipids the phosphate group is removed from PA to produce DG, which is the substrate for the addition of the polar head group. The biosynthesis of PA and DG occurs in two locations in plant cells, the chloroplast envelope (Douce and Joyard 1979) and the ER (Roughan and Slack 1982).

For galactolipid synthesis, a galactosyl residue is transferred from UDP-galactose to DG to form monogalactosyldiacylglycerol (MGDG).[1] The mechanism for the synthesis of digalactosyldiacylglycerol (DGDG) is somewhat controversial. Most probably it is formed from MGDG by an intergalactolipid galactosyl transferase which transfers a galactosyl residue from one molecule of MGDG to another to produce DGDG and DG (van Besouw and Wintermans 1978). These galactosylation reactions occur in the chloroplast envelope membranes (Douce and Joyard 1979).

In addition to the reactions involved in these three stages of galactolipid synthesis, the fatty acids need to be desaturated to produce the polyunsaturated fatty acids found in galactolipids. The conversion of stearic acid (C18:0) to oleic acid (C18:1) occurs before the fatty acid is incorporated into the galactolipids. The remainder of the double bonds are formed on fatty acids that are

1 The reader should note that in Chapter 11.11, this Volume of Murphy MGDG is referred to as diacylgalactosylglycerol (DGG) and DGDG as diacyldigalactosylglycerol (DDG). Both nomenclatures are used in the literature.

part of galactolipids, although the details vary according to the type of lipid under consideration.

3.1 Synthesis of Prokaryotic Galactolipids

In vitro studies using isolated intact chloroplasts from spinach (and other 16:3 plants) have demonstrated that these chloroplasts are able to conduct de novo synthesis of galactolipids when provided with the appropriate precursors, i.e., acetate, glycerol-3-P and UDP-galactose (HEINZ and ROUGHAN 1983). Thus, in spinach chloroplasts not only are all the enzymes necessary for galactolipid synthesis present, but they are organized in such a way that they function as a biosynthetic pathway. Several recent studies have examined various parts of this pathway and concluded that it is located primarily in the inner envelope membrane and produces prokaryotic MGDG (FRENTZEN et al. 1983; BLOCK et al. 1983).

In a very elegant study, FRENTZEN et al. (1983) have examined the specificity of the two acytransferases involved in DG synthesis in both spinach and pea chloroplasts. The first acyltransferase is a soluble stromal enzyme that adds a fatty acid residue to position 1 of glycerol-3-phosphate, forming lysophosphatidic acid. The second acyltransferase, which is located in the chloroplast envelope, adds a fatty acid to position 2 of glycerol to give PA. Both enzymes prefer acyl ACP as the fatty acid donor. More importantly, the first enzyme prefers oleoyl (18:1) ACP, while the second enzyme has a strong preference (requirement) for palmitoyl (16:0) ACP. From this evidence FRENTZEN et al. (1983) conclude that this pathway in the chloroplast envelope produces prokaryotic PA residues. The precise location of the second acyl transferase is not known for spinach chloroplasts, but the specific phosphatase which converts PA to DG has been shown to be located in the inner envelope membrane (BLOCK et al. 1983).

The addition of the polar head group to this prokaryotic DG residue is catalyzed by a galactosyl transferase that uses cytoplasmic UDP-galactose as the galactosyl donor. This enzyme is located in the inner envelope membrane in spinach chloroplasts (BLOCK et al. 1983). The UDP-galactose presumably gains access to the inner membrane via pores in the outer envelope membrane. The lipid product of the galactosyl transferase reaction is MGDG.

The final stage in prokaryotic galactolipid synthesis is desaturation of the fatty acids. The enzymes involved in the desaturation process have not been well characterized, but a considerable amount of evidence suggests that desaturation of the oleoyl and palmitoyl residues occurs after galactose addition (DUBACQ et al. 1983; HEINZ and ROUGHAN 1983; ROUGHAN and SLACK 1982).

The net result is that the assembly of prokaryotic galactolipids from fatty acids, glycerol-3-P and UDP-galactose occurs primarily in the chloroplast inner envelope membrane. Some precursors originate inside the chloroplast, whereas others originate in the cytoplasm. Some steps, such as the acyltransferase reactions, have been well characterized, whereas other steps, such as the desaturation of the fatty acids, are poorly understood.

3.2 Synthesis of Eukaryotic Galactolipids

Synthesis of eukaryotic galactolipids follows a very different route. When isolated intact chloroplasts from peas, a species with exclusively eukaryotic galactolipids, are incubated under conditions where spinach chloroplasts make MGDG, very little galactolipid synthesis occurs (HEINZ and ROUGHAN 1983). These chloroplasts have the ability to synthesize prokaryotic PA residues, but a very limited ability to convert PA to DG. Moreover, virtually none of the DG is used for galactolipid synthesis. GARDINER and ROUGHAN (1983) have provided evidence that the activity of the phosphatase that produces DG may be important in regulating the rate of lipid flow into prokaryotic galactolipids, and that this activity is very low in pea chloroplasts. Of greater importance in understanding the lack of galactolipid synthesis in intact pea chloroplasts may be the fact that the galactosylation enzymes are in a different membrane than the enzymes involved in DG synthesis. Recent evidence from our laboratory indicates that pea chloroplasts are similar to spinach chloroplasts in that the second acyltransferase which produces PA and the phosphatase which produces DG are both in the inner envelope membrane (ANDREWS et al. 1985). However, the galactosyl transferase involved in MGDG synthesis is located in the outer envelope membrane of pea chloroplasts (CLINE and KEEGSTRA 1983). Thus it appears as if the enzymes which function as a biosynthetic pathway producing prokaryotic galactolipids in spinach chloroplasts do not carry out a similar function in pea chloroplasts.

This raises two questions about lipid metabolism in pea chloroplasts. First, what is the function of the pathway which synthesizes prokaryotic DG in the inner envelope membrane? Second, what is the source of the eukaryotic DG that serves as the acceptor for the galactosyl transferase in the outer envelope membrane?

The first question is discussed in Section 4 below. With respect to the second question, the hypothesis of ROUGHAN and SLACK (1982) described earlier predicts that eukaryotic DG residues are synthesized in the ER. This biosynthetic process involves several reactions, many of which are poorly characterized. Fatty acids synthesized in chloroplasts are exported to the ER, presumably as fatty acyl CoA's which are synthesized in the chloroplast outer envelope membrane (ANDREWS and KEEGSTRA 1983). Long chain fatty acyl CoA's are the substrates for acyltransferases present in the ER (MURPHY et al. 1984). Pulse chase experiments indicate that the initial product is PC. The PC residues are rapidly labeled, but quickly chase into other lipids, e.g., the galactolipids of chloroplast membranes (ROUGHAN and SLACK 1982). Thus, PC synthesized in the ER may serve as the precursor for the DG residues which are acceptors for the galactosyl transferase in the pea chloroplast outer envelope membrane.

The DG residues derived via the eukaryotic pathway are thought to contain two linoleic acid (C18:2) residues (MURPHY et al. 1984). The conversion of these to the linolenic acid (C18:3) found in eukaryotic galactolipids occurs after galactosylation to MGDG (ROUGHAN and SLACK 1982; HEINZ and ROUGHAN 1983; LEM and WILLIAMS 1983). The enzymes involved in these desaturation steps are not well characterized. The eukaryotic MGDG synthesized in the outer envelope membrane serves as the substrate for the interlipid galactosyl transferase that is also located there (DORNE et al. 1982; CLINE and KEEGSTRA 1983) producing the eukaryotic DGDG found in chloroplast membranes.

This complex pathway for the biosynthesis of eukaryotic galactolipids not only functions in 18:3 plants such as peas, but also functions in 16:3 plants such as spinach (ROUGHAN and SLACK 1982; DUBACQ et al. 1983; HEINZ and ROUGHAN 1983). However, many questions remain unanswered concerning the details of this hypothetical pathway. For example, the chemical form and mechanism of transfer for the transport of eukaryotic glycerolipids from the ER to the chloroplast envelope have not been identified. It has been suggested that PC is transported via phospholipid exchange proteins (OHNISHI and YAMADA 1982; DUBACQ et al. 1983). However, it is not clear whether these proteins can carry out the unidirectional transport required for the operation of this pathway. Furthermore, an enzyme capable of converting PC to DG in the chloroplast envelope has not been identified. Despite the many unresolved problems, the hypothesis of ROUGHAN and SLACK (1982) for the involvement of the ER in the synthesis of eukaryotic galactolipids provides the best explanation for many different observations and a theoretical basis for future experiments.

4 Biosynthesis of Other Chloroplast Lipids

The biosynthesis of phosphatidylglycerol (PG) and sulfolipid has received much less study than the synthesis of the galactolipids. Several hypotheses have been proposed to explain the synthesis of sulfolipids, although most lack experimental support (HARWOOD 1980). HAAS et al. (1980) demonstrated that isolated intact spinach chloroplasts can incorporate radioactive sulfate into sulfolipids. The enzymes responsible for this and other steps in sulfolipid synthesis still need to be identified.

Progress in understanding the biosynthesis of PG has been achieved recently. PG is unique in that it contains prokaryotic DG, regardless of whether it derives from 16:3 or 18:3 plants (ROUGHAN and SLACK 1982). In addition, it has a unique fatty acid, trans-Δ^3-hexadecanoic acid at the 2 position (SPARACE and MUDD 1982). PG can be synthesized by isolated intact spinach and pea chloroplasts (SPARACE and MUDD 1982; ANDREWS and MUDD 1985). Moreover, ANDREWS and MUDD have demonstrated that pea chloroplast envelope membranes contain the enzymes to synthesize CDP-DG from CTP and PA and to synthesize PG from CDP-DG and glycerol-3-P (J. ANDREWS personal communication). Thus, the question raised earlier concerning the reason for synthesizing prokaryotic PA in pea chloroplasts which do not make prokaryotic galactolipids can now be answered. This PA is used, at least in part, for the synthesis of PG.

5 Summary

Following the discovery in the mid to late 1970's that the chloroplast envelope contained many of the enzymes involved in galactolipid biosynthesis, the hypothesis was put forward that the chloroplast envelope was the only cellular location for the assembly of galactolipids (DOUCE and JOYARD 1979). The hypothesis stated that fatty acids synthesized in the chloroplast were incorporated into DG in the envelope membranes and that the DG was converted into galactolipids in the envelope membranes. It is now clear that this hypothesis is only partly correct in that is applies only to prokaryotic galactolipids. Present evidence indicates that eukaryotic galactolipids are synthesized by a very different pathway. Eukaryotic DG is assembled in the ER and then transported to the chloroplast envelope where galactosylation occurs. The reasons for having two pathways leading to the biosynthesis of galactolipids are unknown.

References

Andrews J, Keegstra K (1983) Plant Physiol 72:735–740
Andrews J, Mudd B (1985) Plant Physiol 79:259–265
Andrews J, Ohlrogge J, Keegstra K (1985) Plant Physiol 78:459–465
Bertrams M, Wrage K, Heinz E (1981) Z Naturforsch 36c:62–70
Besouw A van, Wintermans JFGM (1978) Biochim Biophys Acta 529:44–53
Block MA, Dorne AJ, Joyard J, Douce R (1983) J Biol Chem 258:13281–13286
Cline K, Keegstra K (1983) Plant Physiol 71:366–372
Cline K, Andrews J, Mersey B, Newcomb EH, Keegstra K (1981) Proc Natl Acad Sci USA 78:3595–3599
Dorne AJ, Block MA, Joyard J, Douce R (1982) FEBS Lett 145:30–34
Douce R, Joyard J (1979) Adv Bot Res 7:1–116
Douce R, Joyard J (1980) In: Stumpf PK (ed) The biochemistry of plants, vol IV. Lipids: Structure and function. Academic Press, London New York, pp 321–362
Dubacq JP, Drapier D, Tremolieres A (1983) Plant Cell Physiol 24:1–9
Frentzen M, Heinz E, McKeon TA, Stumpf PK (1983) Eur J Biochem 129:629–636
Gardiner SE, Roughan PG (1983) Biochem J 210:949–952
Gounaris K, Barber J (1983) Trends Biochem Sci 8:378–381
Haas R, Siebertz HP, Wrage K, Heinz E (1980) Planta 148:238–244
Harwood JL (1980) In: Stumpf PK (ed) The biochemistry of plants, vol IV. Lipids: Structure and function. Academic Press, London New York, pp 301–320
Heinz E (1977) In: Tevini M, Lichtenthaler HK (eds) Lipids and lipid polymers in higher plants. Springer, Berlin Heidelberg New York, pp 102–120
Heinz E, Roughan PG (1983) Plant Physiol 72:273–279
Jamieson GR, Reid EH (1971) Phytochemistry 10:1837–1843
Lem NW, Williams JP (1983) Biochem J 209:513–518
Murphy DJ, Mukherjee KD, Latzko E (1984) Planta 161:249–254
Ohlrogge JB, Kuhn DN, Stumpf PK (1979) Proc Natl Acad Sci USA 76:1194–1198
Ohnishi J, Yamada M (1982) Plant Cell Physiol 23:767–773
Roughan PG, Slack CR (1982) Annu Rev Plant Physiol 33:97–132
Sparace SA, Mudd JB (1982) Plant Physiol 70:1260–1264
Stumpf PK (1980) In: Stumpf PK (ed) The biochemistry of plants, vol IV. Lipids: Structure and function. Academic Press, London New York, pp 302–319
Williams JP, Kuhn MV, Mitchell K (1983) In: Thomson WW, Mudd JB, Gibbs M (eds) Biosynthesis and function of plant lipids. Am Soc Plant Physiol, Rockville, pp 28–39

11.11 Structural Properties and Molecular Organization of the Acyl Lipids of Photosynthetic Membranes

D.J. MURPHY

1 Introduction

The photosynthetic membrane system of oxygenic organisms is characterised by an extremely unusual acyl lipid composition compared to other plant or animal membranes (see ORT, Chap. 4, this Vol.). Photosynthetic membranes contain 35% $^{w}/_{w}$ acyl lipid, of which 70% is contributed by the uncharged polyunsaturated glycolipids, diacylgalactosylglycerol (DGG) and diacyldigalactosylglycerol (DDG). The reader should note that in Chapter 11.10 of KEEGSTRA, this Volume DGG is referred to as monogalactosyldiacylglycerol (MGDG) and DDG as digalactosyldiacylglycerol (DGDG). Both nomenclatures are used in the literature. DGG alone accounts for 40–50% of the total thylakoid lipid – a fact that has led to its being termed "the most abundant polar lipid in Nature" (GOUNARIS and BARBER 1983). DGG is of great interest, not only due to its great abundance, but also to its lack of a bilayer-forming capacity. Both DGG and DDG are further characterised by their extremely high polyunsaturated fatty acid content. The dominant galactolipid molecular species is dilinolenin (C18:3), which may constitute up to 95% of the total acyl species of this lipid. The remaining 30% of the thylakoid acyl lipid is made up of the anionic lipids diacylsulfoquinovosylglycerol (DSQ) and phosphatidylglycerol (PG), with the zwitterionic lipid phosphatylcholine (PC) as a minor component. A discussion of where and how these lipid are synthesized in different plant cells can be found in Chapter 11.10 of KEEGSTRA.

The role of acyl lipids in the function and molecular organisation of thylakoid membranes is currently the subject of intense research. Although many of these studies are still at a relatively early stage, it is already clear that thylakoid acyl lipids are not merely an inert matrix into which the various membrane proteins are inserted. Indeed, the pronounced lateral and transverse heterogeneities of acyl lipids in the thylakoid membrane and their strong interactions with many membrane proteins (see SIEGENTHALER and RAWYLER, Chap. 11.9, this Vol.) suggest that these lipids play a dynamic and essential role in both the structure and function of the photosynthetic apparatus.

2 Composition

The acyl lipid and fatty acid composition of a representative thylakoid membrane, as shown in Table 1, demonstrates the overwhelming preponderance of

Table 1. Acyl lipid and fatty acid molecular species composition of typical higher plant photosynthetic membranes

Lipid molecular species	DGG	DDG	DSQ	PG	PC
18:3/18:3	92	70			
18:3/16:0	4	24	62	2	51
18:2/16:0			33	5	34
18:1/16:0				7	8
any C_{18}/16:1^3t				45	
16:0/16:1^3t				37	
Total	38	29	12	14	6

18:3, linolenoyl; 18:2, linoleoyl; 18:1, oleoyl; 16:0, palmitoyl; 16:1^3t, Δ_3-trans-hexadecanoyl. These values are averages of data from a range of higher plant species, including maize, spinach, oleander, and soybean

polyunsaturated galactolipids which is characteristic of all such photosynthetic membranes. In contrast, nongreen plant membrane systems, such as those of the mitochondria and endoplasmic reticulum, are mainly made up of relatively saturated phospholipids whose compositions broadly resemble those of the equivalent organelles in animal membranes.

3 Structures

The chemical formulae of the thylakoid acyl lipids have long been known. However, the chemical and physical structures of these lipids have only very recently been investigated, and as yet this has mainly involved studies of model systems.

3.1 Chemical Structures

The complete description of the conformation and molecular packing of an acyl lipid molecule in a lamellar bilayer has so far only been attempted in the case of crystalline model systems consisting of purified preparations of the isoelectric lipids phosphatidylcholine (PC) and phosphatidylethanolanine (PE), using samples of defined and homogeneous acyl composition (HAUSER et al. 1981). Even here, only a partial description of their molecular conformations, mainly involving head-group interactions, was forthcoming. The nature of acyl chain dynamics above gel to liquid-crystalline phase transitions, even in purified model systems, remains essentially unresolved. The extension of the techniques for the elucidation of lipid molecular conformation, e.g., ^{2}H–NMR, ^{31}P–NMR, X-ray crystallography and X-ray and neutron diffraction, to thylakoid acyl lipids is still awaited. It is possible, however, to deduce information about

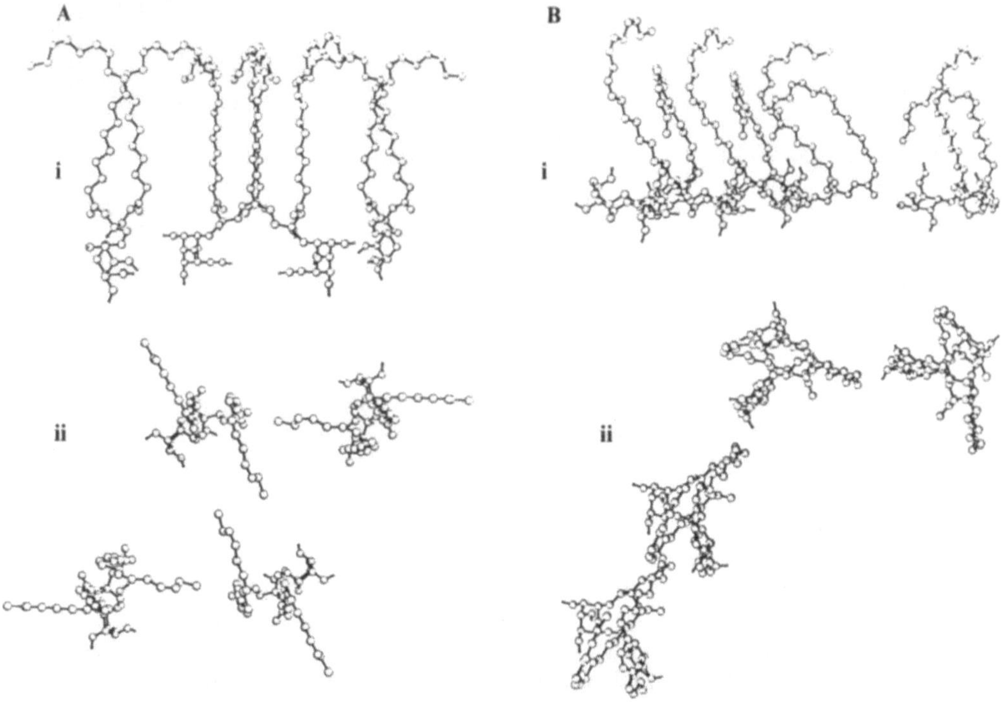

Fig. 1. Conformations of **A** diacylgalactosylglycerol and **B** diacyldigalactosylglycerol. In each case four molecules of the most probable comformer are shown from a side view (**i**) and from an above view (**ii**). (BRASSEUR et al. 1983)

the overall shape of thylakoid acyl lipids from model-building and physical studies, as well as more detailed information from computer simulations.

DGG has a single, neutral galactose moiety on its polar head-group, plus two relatively bulky polyunsaturated acyl residues in its hydrocarbon phase. The other thylakoid acyl lipids have larger head-group moieties and/or more saturated and thus less bulky acyl residues. Therefore the general prediction from model-building studies is that DGG should be more or less cone-shaped, with the galactose group at the apex, while the other lipids should be more cylindrical. Measurements of ^{13}C-longitudinal relaxation times indicate that the head-group motions of DDG, SQD, and PG are similar, but that of DGG is considerably faster (CODDINGTON et al. 1981 a). This implies that the DGG head-group is significantly smaller than those of the other lipids. Surface pressure vs. area isotherms of galactolipid monolayers show the DDG to be more expanded than the DGG film, and that the area of the latter film appears to be governed by acyl chain interactions above the gel to liquid crystalline transition (CODDINGTON et al. 1981 b). Again, this implies that DGG has a relatively less bulky head-group and more bulky acyl residues. Together these data imply that, like the nonchloroplast lipid PE, DGG is roughly cone-shaped.

These conclusions have been strengthened by the results of computer simulation studies by BRASSEUR et al. (1983) designed to discover the most probable

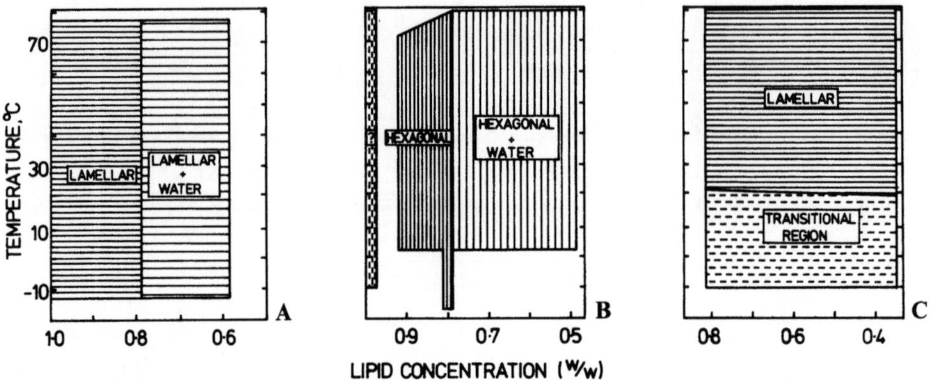

Fig. 2. Phase diagrams of acyl lipid-water systems: **A** diacyldigalactosylglycerol, **B** diacylgalactosylglycerol, **C** diacylsulfoquinovosylglycerol. (After SHIPLEY et al. 1973)

conformers of DGG and DDG (Fig. 1). The conformers were generated by calculating the minimal conformation energy arising from the sum of van der Waal's interactions, tortional potentials, electrostatic interactions, and transfer energy at a simulated bilipid-water interface. The structures are of great interest since they independently predict a cone shape for DGG and a more cylindrical shape for DDG.

3.2 Physical Structures

Purified acyl lipids exhibit both lyotropic and thermotropic mesomorphism (Fig. 2). This has important consequences for the structures adopted by the mixtures of these lipids present in thylakoid membranes.

3.2.1 Diacylgalactosylglycerol

Purified DGG exhibits complex physical properties in aqueous systems depending on the length and degree of unsaturation of its acyl groups (QUINN and WILLIAMS 1983). However, the physiological mixture of molecular species, over 90% of which is normally dilinolenate, exhibits more straightforward behaviour. Low-angle X-ray diffraction analysis has shown that, over a wide range of temperatures and aqueous concentrations, an inverted hexagonal phase consisting of H_{II} structures predominates (SHIPLEY et al. 1973), see also Fig. 3A. This lipid is apparently unable to form lamellar bilayer structures and is therefore unable to form either vesicles or liposomes unless other lipids are present (see Sect. 3.2.5). Differential scanning calorimetry studies have demonstrated that purified DGG exhibits a single thermal transition, almost certainly from a liquid-crystalline to a gel phase, at −30 °C. There seems to be a relatively strong attraction between DGG head-groups since this lipid forms more condensed films than phospholipids with corresponding degrees of unsaturation

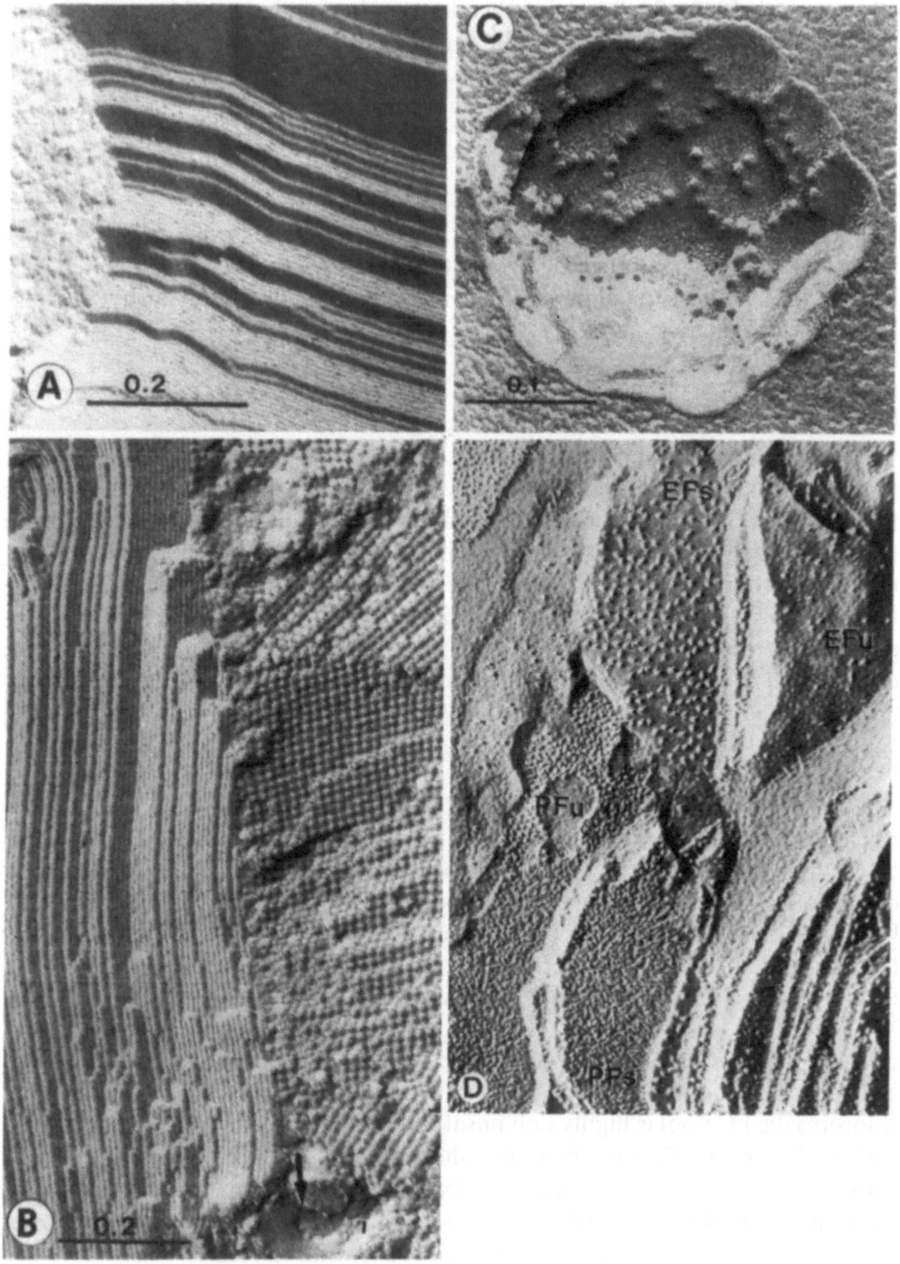

Fig. 3. Electron micrographs of freeze-fracture replicas prepared from aqueous dispersions of **A** diacylgalactosylglycerol, **B** equimolar mixture of diacylgalactosylglycerol and diacyldigalactosylglycerol, **C** total thylakoid acyl lipid extract and of **D** whole thylakoid membranes. *PFu* protoplasmic face, unstacked; *EFu* exoplasmic face, unstacked; *PFs* protoplasmic face, stacked, *EFs* exoplasmic face, stacked (Sprague and Staehelin 1984)

(BISHOP et al. 1980). Such a strong head-group interaction would tend to favour the formation of H_{II} phases in a similar manner to PE, where there are strong hydrogen-bond and electrostatic interactions between N and P atoms on adjacent head-groups.

3.2.2 Diacyldigalactosylglycerol

DDG mixtures purified from thylakoid total lipid extracts exhibit only lamellar structures over a wide range of temperatures and concentrations in water. The lipid is in a liquid-crystalline state until its thermal transition to a gel phase at −50 °C (SHIPLEY et al. 1973).

3.2.3 Diacylsulfoquinovosylglycerol

DSQ containing about 50% palmitate, which is the principal form found in thylakoids, is in a lamellar liquid-crystalline phase above 20 °C. Below this temperature it is gradually replaced by the gel phase until at −10 °C the system consists entirely of gel phase lipid (SHIPLEY et al. 1973). The pK_a value of the sulfonate group of DSQ is about 2.0 in open solution, suggesting that the head-group is ionised under physiological conditions. However, the surface-active properties of this lipid, as judged from surface pressure vs. area isotherms, do not seem to be greatly affected by electrostatic repulsion (BISHOP et al. 1980).

3.2.4 Phospholipids

The thermotropic phase behaviour of PG over the physiological temperature range appears to be highly dependent on the molecular species composition of the lipid, which in turn depends upon the type of plant from which it was obtained. In PG purified from a chilling-tolerant species such as spinach, which has under 50% saturated or trans-monoenoic acyl residues, the onset of the liquid-crystalline to gel phase transition, as deduced from measurements of the fluorescence polarisation of trans-parinaric acid, was below 5 °C (MURATA and YAMADA 1984). In the case of PG from chilling-sensitive species such as sweet potato or sponge cucumber, which have over 80% saturated/trans-monoenoic acyl residues, the onset of this phase transition occurred above 30 °C. The chloroplastic PC pool is highly polyunsaturated (MURPHY and WOODROW 1983), and total soybean PC extracts do not show a liquid-crystalline to gel transition above −30° to −20 °C. hence, PC forms a fluid lamellar liquid-crystalline phase in vivo. Both PC and PG form only lamellar bilayer structures over their entire physiological temperature and concentration ranges.

3.2.5 Thylakoid Lipid Mixtures

Extensive nonlamellar-bilayer structures are readily seen in binary mixtures of DGG and DDG (Fig. 3B) at ratios as low as 1:1.5, the physiological ratio being about 2:1. These structures, as analyzed by freeze-fracture electron microscopy and optical diffraction, have been interpreted as consisting of hexagonally packed tubes, presumably consisting of lipids in an H_{II} phase, and also of rows of inverted micelles sandwiched inside lipid monolayers. The H_{II} phases, inverted micelles, and lamellar phases apparently coexist in these mixtures in a dynamic equilibrium. Similar inverted micelle and H_{II} structures are seen in lipid extracts of whole thylakoid membranes (Fig. 3C), where DGG only constitutes 50% of the total lipid. Neither inverted micelles or H_{II} structures are normally seen in freeze-etch replicas of native thylakoid membranes (Fig. 3D). They can, however, be induced by short exposures to high temperatures (QUINN and WILLIAMS 1983, see also STAEHELIN, Chap. 1, this Vol. for a general discussion of temperature-induced phase changes in thylakoid membranes). This suggests that there is a subtle balance between bilayer-forming and nonbilayer-forming lipids in the membranes.

3.3 Physiological Significance of Lipid Structures

Information about the molecular shape of individual lipid molecules has been used in attempts to explain their physical behaviour in bulk solutions. This has been done for model systems (ISRAELACHAVILI et al. 1977), thylakoid membranes (MURPHY 1982; SPRAGUE and STAEHELIN 1984), and other biological membranes (WIESLANDER et al. 1981). The tendency of lipids to form bilayer or nonbilayer phases has also been interpreted in terms of the relative hydrophobicity and hydrophilicity of their headgroup and acyl chain regions (QUINN and WILLIAMS 1983). It is probable that both steric factors and the relative hydrophobic/hydrophilic balance play a role in determining the eventual structures adopted by such lipid assemblies.

4 Asymmetric Distribution

Asymmetry is now recognized to be a widespread phenomenon in biological membranes from both plants and animals. It has been proposed that such factors as differential lipid-packing properties, entropic effects, and relative hydrophobicity/hydrophilicity can lead to spontaneous thermodynamically stable transverse and lateral asymmetries in membrane lipid distribution (ISRAELACHAVILI et al. 1977).

4.1 Transverse Asymmetry

Studies of the transverse distribution of lipid across membranes are fraught with problems since, in order to assay individual lipids, the organization of the membrane must of necessity be disrupted (see also SIEGENTHALER and RAWYLER, Chap. 11.9, this Vol.). This is particularly true in the case of inner-monolayer lipids. Such disruptions may well induce changes in the trans-membrane lipid distribution itself. The results of such studies should therefore be treated with some caution.

4.1.1 Antibody Studies

These studies have shown a preponderance of PG and PC on the outer thylakoid monolayer and of DGG and DSQ on the inner monolayer (RADUNZ 1980). It has also been deduced that much of the galactolipid pool is inaccessible to antibodies. When the surface-exposed part of the coupling factor was removed by EDTA washing, twice as many galactolipid molecules were bound to antibodies. This indicates that much galactolipid is located in the vicinity of the coupling factor and/or on the inner thylakoid surface. In general it was found that more lipid was located on the inner surface, while more protein was located on the outer surface.

4.1.2 Lipolytic Studies

This technique suffers from problems due to the incomplete specificity of some lipases and membrane disruption by lipid breakdown products. Following studies in which a lipolytic acyl hydrolase and a phospholipase A_2 were employed, SIEGENTHALER (1982) has reported that in spinach thylakoids (a) galactolipids were more accessible to the lipase than ionic lipids and (b) phospholipids were asymmetrically distributed, with 70% of PG but only 35% of PC on the outer monolayer. In a separate study, about 50% DDG but only 20% DGG and 10% DSQ were on the outer thylakoid leaflet (UNITT and HARWOOD 1982). It was found that most of the PG was localized on the outer monolayer and, in particular, essentially all of the Δ_3-trans-hexadecenoic was accessible to the lipase. In contrast, the results of a recent study have been interpreted as suggesting that most (65%) of the DGG was on the outer monolayer, while most (85%) of the DDG was on the inner monolayer (RAWYLER and SIEGENTHALER 1984; SIEGENTHALER and RAWYLER, Chap. 11.9, this Vol.).

4.2 Lateral Heterogeneity

Relatively small-scale lateral heterogeneities in lipid distribution have been inferred in several biological membranes from indirect techniques such as ferritin/lectin/antibody capping, fluorescent probes, or non-penetrating cross-linking probes. The recent development of techniques for the mechanical separation of appressed from nonappressed thylakoid regions has made possible the direct measurement of lipid composition along a contiguous but structurally differentiated membrane. The appressed regions are very depleted in acyl lipids and enriched in chlorophyll relative to the nonappressed regions. The acyl lipid class distributions also show a marked lateral heterogeneity (MURPHY and WOODROW 1983; GOUNARIS et al. 1983b). When calculated on a protein basis, the appressed regions are found to be very depleted in both DGG, DDG, and DSQ but have similar amounts of PG. Appressed thylakoid regions prepared by mild detergent treatment also showed a severe depletion of DGG and DDG and an enrichment in PG, PC, and DSQ relative to whole thylakoids (GOUNARIS et al. 1983c).

4.3 Extent and Functions of Lateral and Transverse Asymmetries

The general conclusion from the above studies is that the acyl lipids of thylakoid membranes are very asymmetrically distributed, in terms of the total amount of lipid and also of the individual lipid classes, both in the transverse and lateral planes of the membrane. The bulk of the available evidence on transverse asymmetry supports the following general qualitative conclusions: (although the location of some lipids, particularly the galactolipids, is still in dispute)

a) more acyl lipid is located on the inner than on the outer monolayer,
b) most of the DGG and DSQ is on the inner monolayer,
c) DDG may be more evenly distributed across the membrane,
d) most of the PG is on the outer monolayer.

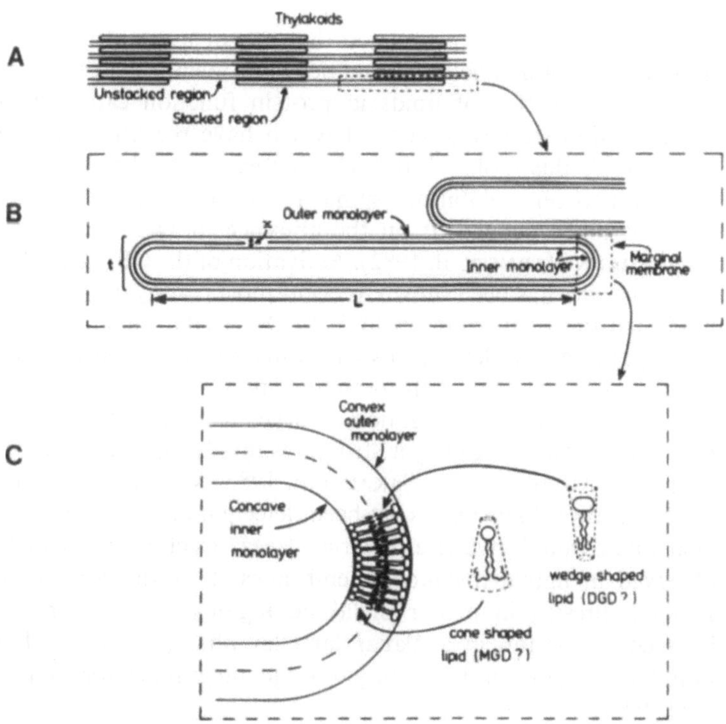

Fig. 4. A and B Schematic representation of thylakoid structure in cross-section. Membranes are represented as elongated flattened vesicles with variable degrees of overlap depending on short- and long-term environmental changes. **C** packing constraints in highly curved margins favour the presence of cone shaped lipids (e.g., DGG) in the inner, concave monolayer and more wedge-shaped lipids on the outer convex monolayer. (MURPHY 1982)

These findings, together with those on lateral heterogeneity, are consistent with the hypothesis (MURPHY 1982) that up to one third of the total thylakoid acyl lipid is required for stabilization of the tightly curved marginal regions consisting of an inner DGG monolayer and an outer DDG monolayer (Fig. 4). It should be noted that a substantial amount of DGG will still be present in planar thylakoid regions, where it may interact with membrane proteins. The above results are also consistent with recent reports on the importance of ionic lipids, and especially PC and PG, in the function of some of the proteins present in appressed regions of thylakoids. The extent and significance of these putative lipid-protein interactions in thylakoid membranes will now be considered.

5 Lipid-Protein Interactions

5.1 Galactolipids

It has frequently been postulated that one or more of the thylakoid acyl lipids may interact in a functional manner with membrane proteins. Definitive data concerning the role of lipids in protein function can only be gleaned from reconstitution studies, several of which have recently been reported (see also Chap. 11.7, this Vol., of RYRIE). For example, the reconstitution of energy transfer between solubilized spinach chlorophyll-protein complexes has been reported to be dependent on the presence of DGG, but not on other lipids (SIEFERMANN-HARMS et al. 1982). Activation of the CF_0–CF_1 ATP synthase from spinach leaves has been shown to occur most readily in the presence of thylakoid lipid mixtures containing up to 50% DGG (PICK et al. 1984). In this case the ATP synthase complex was also activated in the presence of soybean PC, albeit to a lesser extent than with the chloroplast lipid mixture. Several models (see Fig. 5) depicting the stabilization of large intrinsic protein complexes by inverted micellar or nonplanar bilayer structures involving DGG and/or DSQ have been proposed (MURPHY 1982; SAKAI et al. 1983; WILLIAMS et al. 1984). Nonbilayer lipidic particles have not as yet been detected, and inverted-micelle-plus-protein assemblies would give rise to larger freeze-fracture particles than are actually observed in native thylakoid membranes. It is therefore more likely that, if DGG is present in planar thylakoid regions at concentrations too high for its accommodation in a planar lamellar phase, it will tend to partition into nonplanar regions in the vicinity of large integral membrane proteins (WILLIAMS et al. 1984).

5.2 Phospholipids

In contrast to the above reconstitution studies, experiments with solubilized pea thylakoids have shown relatively poor reconstitution of energy transfer

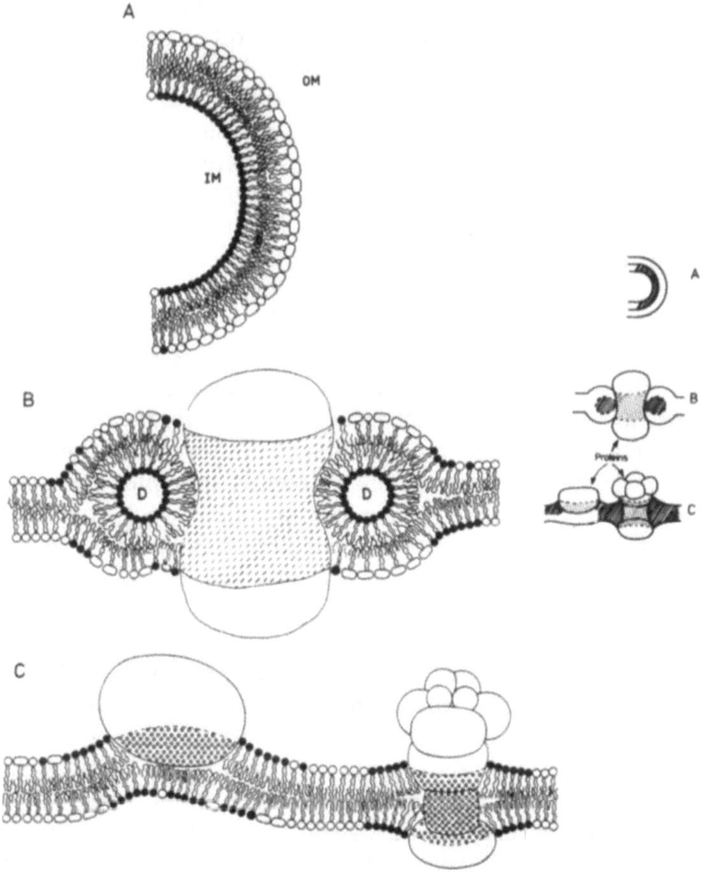

Fig. 5 A–C. Diagrammatic models showing possible lipid-protein associations in thylakoid membranes. **A** Highly curved membrane regions, e.g., thylakoid margins. In such a region, packing constraints will tend to favour the partitioning of wedge-shaped lipids, such as DDG, into the outer monolayer (*OM*) of the membrane and of cone-shaped lipids, such as DGG, into the inner monolayer (*IM*). This will give rise to a spontaneous transverse asymmetry of lipid distribution. **B** Possible stabilisation of a large membrane-spanning protein complex by localized nonbilayer structures of DGG sandwiched inside a lamellar bilayer. The nonbilayer structures (*D*) could be either individual spherical inverted micelles or a continuous cylinder around the protein. **C** Deformations of the thylakoid membrane from a planar bilayer configuration imposed by the presence of large extrinsic or intrinsic membrane proteins would result in the formation of localised domains of both concave and convex membrane curvature. Lipid molecules with differential packing properties would tend to partition into those membrane regions in which their presence would be most thermodynamically favoured, e.g., DGG would partition into highly concave domains. In this way, the protein components of thylakoid membranes may make possible the accommodation of large amounts of nonbilayer-forming lipids, such as DGG, in a lamellar bilayer phase without the formation of nonbilayer structures

in the presence of DGG, but very good results with PG or PC (MURPHY 1984). The latter findings are consistent with other reports on the efficacy of PC as a reconstituting agent for photosystem II preparations, in particular for energy transfer between purified light-harvesting chlorophyll-protein complexes and photosystem II (MURPHY et al. 1984), for electron transfer in both photosystem I and photosystem II preparations (MURPHY and PRINSLEY unpublished results) for oxygen evolution (GOUNARIS et al. 1983c), and for the restoration of high-potential cytochrome b-559 (MATSUDA and BUTLER 1983). PG has been found to be the best acyl lipid for reconstitution of the oligomeric form of the light-harvesting chlorophyll-protein complex from the monomeric form (REMY et al. 1984).

Studies involving lipase treatment of thylakoids have demonstrated inhibitory effects on various photosynthetic activities. However, some of these may arise as secondary consequences of the membrane disruption caused by lipase action and the results must be treated with some caution. Lipolytic acyl hydrolase, which mainly attacks phospholipids, disrupted electron transport between plastoquinone and P-700 but did not affect PS I or PS II electron transport (RAWYLER and SIEGENTHALER 1981). Phospholipase A_2 treatment, which results in the hydrolysis of most of the PG and PC stimulated PS I but inhibited PS II electron transport (JORDAN et al. 1983). These studies appear to imply a relatively nonspecific role for galactolipids in stabilization of protein function and a more direct role for phospholipids, especially in PS II function.

Results obtained from spin-probes, such as TEMPO[1], 7N14[2], or 12-doxyl-stearate, indicate a greatly increased microviscosity in the hydrocarbon phase of appressed membrane regions compared to whole thylakoids or nonappressed regions (ARONSON et al. 1983; FORD et al. 1982). These differences in lipid fluidity have been ascribed to the differences in the lipid/protein in the various fractions. However, spin-probes such as these, which are not derived from endogenous membrane components, may well be reporting on an atypical membrane environment, e.g. many of these probes tend to partition into more fluid domains and may not be representatively sampling regions of less mobile lipid. Better estimates of the degree of lipid-protein interaction may be obtained from spin-probes attached to endogenous lipids such as PC and PG. The results show that much of the phospholipid in whole thylakoids is motion-restricted, and that in appressed regions even more of the PG and most of the PC is immobilized (on the ESR time scale) over the temperature range of 0°–30 °C (MURPHY and KNOWLES 1984). The conclusion is that the PG and PC present in appressed regions, which are relatively enriched in these lipids, is nearly all directly associated with protein to form a relatively immobile pool of bound (but not neccessarily tightly bound) lipid. The greatly increased viscosity of the hydrocarbon phase of appressed membranes may have important consequences for the lateral mobility of plastoquinone between PS II and the cytochrome b/f complex. This is of particular interest since the reoxidation of plastohydroquinone is believed to be the rate-limiting step of photosynthetic electron transport.

1 2,2,6,6-tetramethylpipiridine-N-oxyl.
2 2-heptyl-2-hexyl-5,5-dimethyloxazolidine-N-oxyl.

References

Aronson H, Waggoner C, Moore J, Berg SP (1983) Biochim Biophys Acta 725:519–528

Bishop DG, Kenrick SR, Bayson JH, Macpherson AS, Johns SR (1980) Biochim Biophys Acta 602:248–259

Brasseur R, Meutter J de, Goormaghtigh E, Ruysschaert JM (1983) Biochem Biophys Res Commun 115:666–672

Coddington JM, Johns SR, Leslie DR, Willing RI, Bishop DG (1981a) Aust J Biochem 34:357–363

Coddington JM, Johns SR, Leslie DR, Willing RI, Bishop DG (1981b) Biochim Biophys Acta 663:653–660

Ford RC, Chapman DJ, Barber J, Pederson JC, Cox RP (1982) Biochim Biophys Acta 681:145–151

Gounaris K, Barber J (1983) Trends Biochem Sci 8:378–381

Gounaris K, Sundby C, Andersson B, Barber J (1983) FEBS Lett 156:170–174

Gounaris K, Whitford D, Barber J (1983) FEBS Lett 163:230–234

Hauser H, Pascher I, Pearson RH, Sundell S (1981) Biochim Biophys Acta 650:21–51

Israelachavili JN, Mitchell DJ, Ninham BW (1977) Biochim Biophys Acta 470:185–201

Jordan B, Chow WS, Baker AJ (1983) Biochim Biophys Acta 725:77–86

Matsuda H, Butler WL (1983) Biochim Biophys Acta 725:320–324

Murata N, Yamada J (1984) Plant Physiol 74:1016–1024

Murphy DJ (1982) FEBS Lett 150:19–26

Murphy DJ (1984) In: Siegenthaler PA and Eichenberger W (eds) Structure, function and metabolism of plant lipids, pp 465–470, Elsevier-North Holland, Amsterdam

Murphy DJ, Knowles PF (1984) In: Siegenthaler PA and Eichenberger W (eds) Structure, function and metabolism of plant lipids. pp 425–428, Elsevier-North Holland, Amsterdam

Murphy DJ, Woodrow IE (1983) Biochim Biophys Acta 725:104–112

Murphy DJ, Crowther D, Woodrow IE (1984) FEBS Lett 165:151–155

Pick U, Gounaris K, Admon A, Barber J (1984) Biochim Biophys Acta 765:12–20

Quinn PJ, Williams WP (1983) Biochim Biophys Acta 737:223–266

Radunz A (1980) Z Naturforsch 35c:1024–1031

Rawyler A, Siegenthaler PA (1981) Biochim Biophys Acta 638:30–39

Rawyler A, Siegenthaler PA (1984) In: Siegenthaler PA and Eichenberger W (eds) Structure, Function and Metabolism of Plant Lipids, pp 409–412, Elsevier-North Holland, Amsterdam

Sakai WS, Yamamoto HY, Miyazaki T, Ross JW (1983) FEBS Lett 158:203–207

Shipley GG, Green JP, Nichols BW (1973) Biochim Biophys Acta 311:531–544

Siefermann-Harms D, Ross JW, Kaneshiro KH, Yamamoto HY (1982) FEBS Lett 149:191–196

Siegenthaler PA (1982) In: Wintermans JFGM, Kuiper PJC (eds) Biochemistry and metabolism of plant lipids. Elsevier, Amsterdam New York Oxford, pp 351–358

Sprague SG, Staehelin LA (1983) In: Biosynthesis and function of plant lipids (Thomson WW, Mudd JR and Gibbs M, eds) pp 144–159. Am Soc Plant Physiol, Rockville MD, USA. Biochim Biophys Acta

Trémolières A, Dubacq JP, Duval JC. Lemoine Y, Remy R (1982) In: Biochemistry and Metabolism of Plant Lipids (Wintermans JFGM and Kuiper PJC eds) pp 369–372. Elsevier-North Holland, Amsterdam

Unitt MD, Harwood JL (1982) In: Wintermans JFGM, Kuiper PJC (eds) Biochemistry and metabolism of plant lipids. Elsevier, Amsterdam New York Oxford, pp 331–334

Wieslander A, Christianson A, Rilfors L, Lindblom G (1981) Biochemistry 19:3650–3655

Williams WP, Gounaris K, Quinn PJ (1984) In: Sybesma E (ed) Proc 6th Int Congr Photosynth, vol III. Nijhoff/Junk, The Hague, pp 123–130

Author Index

Page numbers in *italics* refer to the references

Subject Index

A must for all plant physiologists – by internationally recognized experts

Encyclopedia of Plant Physiology

New Series · Editors: A. Pirson, M. H. Zimmermann

Volume 9

Hormonal Regulation of Development I

Molecular Aspects of Plant Hormones

Editor: **J. MacMillan**

1980. 126 figures. XVII, 681 pages. ISBN 3-540-10161-6

"... this volume will be of value to anyone requiring an up-to-date overview of plant hormones. ... It is impeccably edited, extensively indexed and beautifully produced and is a worthy addition to the New Series." *Phytochemistry*

Volume 10

Hormonal Regulation of Development II

The Functions of Hormones from the Level of the Cell to the Whole Plant

Editor: **T. K. Scott**

1984. 42 figures. XV, 309 pages

ISBN 3-540-10196-9

Volume 11

Hormonal Regulation of Development III

Role of Environmental Factors

Editors: **R. P. Pharis, D. M. Reid**

1985. 117 figures. Approx. 870 pages

ISBN 3-540-10197-7

A discount of 20% is available when volumes 9, 10, 11 are purchased together

Volume 12 A

Physiological Plant Ecology I

Responses to the Physical Environment

Editors of Volumes 12A–D: O. L. Lange, P. S. Nobel, C. B. Osmond, H. Ziegler

1981. 110 figures. XV, 625 pages. ISBN 3-540-10763-0

Volume 12 B

Physiological Plant Ecology II

Water Relations and Carbon Assimilation

1982. 153 figures. XI, 747 pages. ISBN 3-540-10906-4

Volume 12 C

Physiological Plant Ecology III

Responses to the Chemical and Biological Environment

1983. 104 figures. XI, 799 pages. ISBN 3-540-10907-2

Volume 12 D

Physiological Plant Ecology IV

Ecosystem Processes: Mineral Cycling, Productivity and Man's Influence

1983. 61 figures. XI, 644 pages. 340,–. ISBN 3-540-10908-0

"... one must be impressed by the amount of information which has been pressed into four volumes. The provision of a subject index at the end of the quartet will enable ecologists of all shades to gain rapid access to a wealth of up-to-date, well-reviewed material with extensive bibliographies. ... these four books will render redundant about two shelves of older ecology texts, so it is well worth making room for them." *Nature*

Volume 13 A

Plant Carbohydrates I

Intracellular Carbohydrates

Editors: **F. A. Loewus, W. Tanner**

1982. 103 figures. XXII, 918 pages

ISBN 3-540-11060-7

Volume 13 B

Plant Carbohydrates II

Extracellular Carbohydrates

Editors: **W. Tanner, F. A. Loewus**

1981. 124 figures. XXII, 769 pages

ISBN 3-540-11007-0

"These two, large, multi-author volumes (1,700 pages in all)... will be generally invaluable sources of information and inspiration for research work and advanced teaching." *Nature*

"... I can vouch for their accuracy, thoroughness, timeliness and general excellence. ... magnificent efforts [of the authors] have led to this outstanding production. ... undoubtedly the standard work of reference for the 1980's... should be widely available in all biochemistry and plant science libraries." *Phytochemistry*

Springer-Verlag Berlin Heidelberg New York Tokyo

Springer

MIX
Papier aus verantwortungsvollen Quellen
Paper from responsible sources
FSC® C105338

If you have any concerns about our products,
you can contact us on
ProductSafety@springernature.com

In case Publisher is established outside the EU,
the EU authorized representative is:
Springer Nature Customer Service Center GmbH
Europaplatz 3, 69115 Heidelberg, Germany

Printed by Libri Plureos GmbH
in Hamburg, Germany